C0-AZW-473

HEAT TRANSFER IN ELECTRONIC AND MICROELECTRONIC EQUIPMENT

PROCEEDINGS OF THE INTERNATIONAL CENTRE FOR HEAT AND MASS TRANSFER
NAIM AFGAN, EDITOR
J. T. ROGERS, SENIOR CONSULTING EDITOR

Belgrade

1. Afgan and Schlünder **Heat Exchangers: Design and Theory Sourcebook**
2. Afgan and Beer **Heat Transfer in Flames**
3. Blackshear **Heat Transfer in Fires**
4. deVries and Afgan **Heat and Mass Transfer in the Biosphere**
5. Denton and Afgan **Future Energy Production Systems**
6. Hartnett **Alternative Energy Sources**
7. Spalding and Afgan **Heat Transfer and Turbulent Buoyant Convection**
8. Zarić **Thermal Effluent Disposal From Power Generation**
9. Hoogendoorn and Afgan **Energy Conservation in Heating, Cooling, and Ventilating Buildings**
10. Durst, Tsklauri, and Afgan **Two-Phase Momentum, Heat and Mass Transfer in Chemical, Process, and Energy Engineering Systems**
11. Spalding and Afgan **Heat and Mass Transfer in Metallurgical Systems**
12. Jones **Nuclear Reactor Safety Heat Transfer**
13. Bankoff and Afgan **Heat Transfer in Nuclear Reactor Safety**
14. Zarić **Structure of Turbulence in Heat and Mass Transfer**
15. Taborek, Hewitt, and Afgan **Heat Exchangers: Theory and Practice**
16. Metzger and Afgan **Heat and Mass Transfer in Rotating Machinery**
17. Afgan **Measurement Techniques in Power Engineering**
18. Soloukhin and Afgan **Measurement Techniques in Heat and Mass Transfer**
19. Abramson **Guidebook to Light Water Reactor Safety Analysis**
20. van Swaaij and Afgan **Heat and Mass Transfer in Fixed and Fluidized Beds**
21. Radovanovic **Fluidized Bed Combustion**
22. Mori, Sheindlin, and Afgan **High Temperature Heat Exchangers**
23. Sheindlin **High Temperature Equipment**
24. Bougard and Afgan **Heat and Mass Transfer in Cryoengineering and Refrigeration**
25. Afgan **Transient Phenomena in Multiphase Flow**
26. Spalding and Afgan **Heat and Mass Transfer in Gasoline and Diesel Engines**
27. Markatos **Computer Simulation for Fluid Flow, Heat and Mass Transfer, and Combustion in Reciprocating Engines**
28. Kline and Afgan **Near-Well Turbulence: 1988 Zoran Zarić Memorial Conference**
29. Bergles **Heat Transfer in Electronic and Microelectronic Equipment**

HEAT TRANSFER IN ELECTRONIC AND MICROELECTRONIC EQUIPMENT

Edited by

A. E. Bergles
Rensselaer Polytechnic Institute
Troy, New York, USA

HEMISPHERE PUBLISHING CORPORATION
A member of the Taylor & Francis Group

New York Washington Philadelphia London

HEAT TRANSFER IN ELECTRONIC AND MICROELECTRONIC EQUIPMENT

1 2 3 4 5 6 7 8 9 BCBC 9 8 7 6 5 4 3 2 1 0

Cover design by Debra Eubanks Riffe.

A CIP catalog record for this book is available from the British Library.

Library of Congress Cataloging-in-Publication Data

Heat transfer in electronic and microelectronic equipment / edited by A. E. Bergles.
p. cm.—(Proceedings of the International Centre for Heat and Mass Transfer : 29)
Proceedings of a symposium held Aug. 29-Sept. 2, 1988 in Dubrovnik, Yugoslavia.

1. Electronic apparatus and appliances—Cooling—Congresses.
2. Heat—Transmission—Congresses. I. Bergles, A. E., date.
II. Series.
TK7870.25.H39 1990
621.381'54—dc20
ISBN 0-89116-277-1

89-49572
CIP

Contents

LIQUID COOLING

CONDUCTION ASPECTS

THERMAL ANALYSIS

Preface

The International Centre for Heat and Mass Transfer chose Heat Transfer in Electronic and Microelectronic Equipment as the subject of its 1988 Symposium, August 29-September 2, in Dubrovnik, Yugoslavia. It is fitting that the Symposium was immediately preceded by a Scientific Council Meeting commemorating the 20th Anniversary of the Centre. This volume constitutes the archival proceedings of the Symposium.

The successful operation of modern electronic devices, ranging from microelectronic devices to large power tubes, is critically dependent on efficient and reliable heat removal. Thermal control has always been necessary, of course, and publications on electrical equipment cooling date back over 60 years. But now the thermal problems are so pervasive that they are spotlighted, and the thermal engineers are recognized for their contributions. The need for improved electronic cooling technologies has attracted the attention of a large number of industrial and university researchers. In many respects, the thermal control of electronic and microelectronic equipment emerged as the foremost heat transfer problem of the 1980's. The 1988 Symposium was intended to bring together experts from around the world to share their knowledge on this increasingly important subject. Indeed, this was the most international of the conferences on the subject held up to that time.

The symposium was attended by 120 experts in the field of thermal control representing 20 countries. Sixty papers were scheduled to be presented. The review and contributed papers showed both the diversity and complexity of the subject as well as the numerous experimental and analytical/numerical approaches. It was evident from the presentations that the full force of heat transfer research and development capability is being directed toward the cooling problem.

Prof. Naim Afgan and The Executive Committee of the Centre are acknowledged for making the Symposium possible. Very special thanks are due to Prof. Eyup Ganic for his contributions as Scientific Secretary, to Dr. Jovica Riznic for his efforts as Symposium Secretary, and to the staff of the Centre. Contributing to development of the program were the other members of the Symposium Organizing Committee: C.-K. Chen, G. N. Dulnev, F. P. Incropera, J.-P. Le Jannou, C.-F. Ma, L. F. Milanez, W. Nakayama, and M. M. Yovanovich. Dr. A. N. Belova also provided valuable assistance with program development. Keynote lectures were delivered by W. Nakayama, R. E. Simons, R. J. Hannemann, A. D. Kraus, C.-F. Ma, F. P. Incropera, M. M. Yovanovich, and G. N. Dulnev. Session chairmen were A. D. Kraus, L. F. Milanez, M. Misale, M. M. Yovanovich, C.-F. Ma, F. P. Incropera, W. Nakayama, G. N. Dulnev, and E. N. Ganic. Above all, the authors are acknowledged for their presentations and subsequent transmittal of high-quality manuscripts.

Support for the Symposium was provided by the United Nations Educational, Scientific, and Cultural Organization, Paris, France; The Boris Kidric Institute of Nuclear Sciences, Belgrade, Yugoslavia; Hitachi, Ltd., Tsuchiura, Japan; IBM Corp., Poughkeepsie, New York; 3M, St. Paul, Minnesota; EI, Belgrade, Yugoslavia; Iskra-Delta, Ljubjna, Yugoslavia; Bell-Northern Research, Ltd., Ottawa, Canada; Hemisphere Publishing Corp., and Rensselaer Polytechnic Institute, Troy, New York. Elizabeth Schoonmaker of Rensselaer deserves special thanks for her conscientious attention to countless details of the symposium and editing of the manuscripts. The staff of Hemisphere Publishing Corporation are acknowledged for their excellent cooperation beginning with Symposium publicity and ending with this volume.

A. E. Bergles

OVERVIEW

Cooling Electronic Equipment: Past, Present and Future

WATARU NAKAYAMA
Mechanical Engineering Research Laboratory
Hitachi, Ltd.
502 Kandatsu, Tsuchiura
Ibaraki 300, Japan

ARTHUR E. BERGLES
Department of Mechanical Engineering
Aeronautical Engineering, and Mechanics
Rensselaer Polytechnic Institute
Troy, New York 12180-3590, USA

ABSTRACT

The evolution of cooling technology for electronic equipment is reviewed, with special emphasis on the thermal control of solid-state devices. The rising levels of overall power dissipation and chip-level heat fluxes in computers are discussed. Techniques for calculating and improving conduction heat flow are outlined. Natural and forced convection air cooling are still very popular; however, increasingly complex flow paths are being encountered. Indirect liquid cooling, used traditionally for cold plates, is being extended to novel configurations. Much research is currently directed at use of inert, dielectric liquids in intimate contact with the devices and wiring; the modes of primary interest are natural convection, forced convection, and subcooled pool boiling. It is concluded that research and development will be increasingly important to provide adequate thermal control for future microelectronic devices of extremely small scale.

INTRODUCTION

Heat transfer is an ever challenging subject for designers of electronic equipment. Concerns over the short life of vacuum tubes due to overheating seemed to end with the invention of the transistor in December 1947 [1]. Since that time, solid-state technology has undergone a dramatic growth, not only making such space-charge devices as vacuum tubes obsolete but also creating a multitude of new areas of application. However, this growth of solid-state technology has not relieved the equipment designer from the task of thermal management. Instead, it is demanding the development of more advanced heat transfer engineering to advance further the frontiers of the microelectronic revolution.

To understand this, one needs only to recall the proliferation of silicon chips in almost all aspects of our lives today. Silicon chips are operating in computers of various classes, control and diagnostic devices, medical equipment, communications equipment, and a variety of instruments. They are operating not only in air-conditioned environments, but in diverse thermal environments of factories, power plants, offices, homes, automobiles, trains, ships, aeroplanes, outdoor communication stations, satellites, and space ships. The thermal design of equipment and devices must insure that chip temperatures are maintained below 80 - 100 °C. It is also important to reduce thermal stresses and strains on the encapsulants of chips and other supporting components such as ceramic substrates and printed wiring boards. Further, in large systems like mainframe computers, there are mechanical devices, such as magnetic disk files, tape data files, and printers, that require a great deal of attention to the maintenance of their thermal environments.

Heat transfer and electronics have a long history of mutual interaction that began in the days of vacuum tubes. Particularly since the advent of microelectronics, research and development work on new electronic equipment has led to the milestone works in heat transfer, and the results of this heat transfer research have benefited subsequent generations of electronic equipment by providing designers with the data and correlations for thermal design. Today we are at the threshold of a new era, where we anticipate the development of novel methodologies of thermal design to solve complex thermal problems that are now controlling the pace of the advancement of electronics technology. As a prelude to this Symposium we will look at the history of this technology, review the present, and look ahead to the future.

HEAT IN ELECTRONIC EQUIPMENT - WHAT WE ARE DEALING WITH

Heat transfer considerations were, of course, important to the successful design of electric equipment such as transformers; however, it was many years before thermal management was recognized as a distinct discipline [2]. Advanced air cooling or water cooling schemes were required for the high power vacuum tubes of the 1930s and 1940s but the heat produced by the many vacuum tubes in the first digital computers was dissipated simply by an array of industrial cooling fans. As noted above, it was initially thought that solid state devices would pose no cooling problems. To the contrary, the cooling of microelectronic devices has provided enormous challenges to heat transfer specialists. Cooling for the entire range of equipment, from microelectronic chips to high power devices, is considered a major specialty area of heat transfer. Worldwide, cooling of electronic equipment has become the heat transfer problem of the 1980s.

An introductory review is presented here on the rising levels of heat load in computers that have been observed since around 1980 [3]. Computers, mainframes, and supercomputers, in particular, embody the coordinated work of electronic and heat transfer engineering, and serve as good illustrations of thermal management required for modern electronic equipment. The focus on the recent development of heat loads is intended to call attention to the urgency of industrial needs to find better solutions of thermal management. The typical figures given below illustrate the levels of heat flux and volumetric

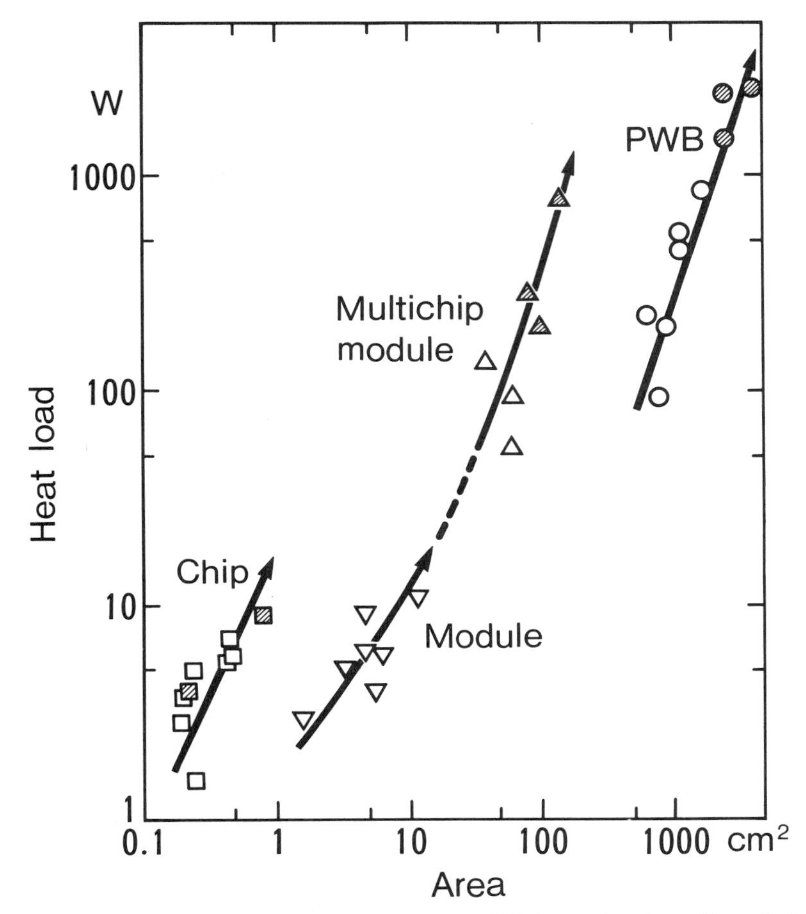

FIGURE 2. Increase in heat load at different structural levels of mainframe computers since around 1980 (shaded symbol = indirect water-cooled machine open symbol = air-cooled machine).

the system due to the high temperature of the coolant surrounding them. When air is used as the coolant, the flow rate needs be sufficient to maintain a reasonable air temperature at the exit of the cooling channel. There exists a limit to the air velocity, in most instances imposed by the necessity to maintain low acoustic noise levels. Where liquid is employed as coolant, the temperature rise of the coolant is usually minimal because of the large heat capacity. Water is the best coolant in this respect as it has large specific heat and high thermal conductivity. However, because of the need to guarantee electrical insulation, water cannot be brought into direct contact with the chips. In what are called indirect water-cooled modules, somewhat elaborate designs are adopted to conduct heat from the chips to water-cooled cold plates.

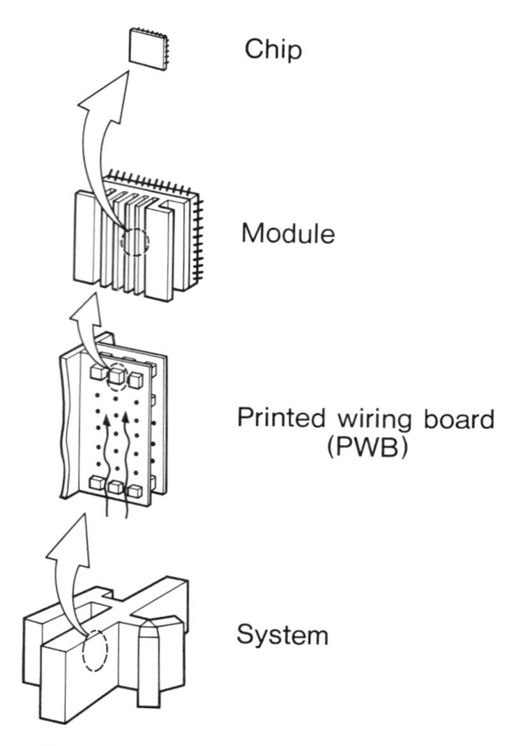

FIGURE 1. Structural levels of an electronic computer.

density of heat dissipation which heat transfer researchers have to deal with.

The thermal management of a complex system like a computer needs be planned by dividing the system into different structural levels. Figure 1 illustrates the chip, which is the minimum component in the system; the module, whose function is to isolate the chip(s) from the atmosphere and at the same time provide the leads for transmission of signals to and from the chip(s) and the supply of power; and the printed wiring board (PWB), which carries modules. Heat produced by the chip has to travel along multiple conduction paths inside the module overcoming what is called internal thermal resistance. The internal thermal resistance is a complex function of the module structure and materials constituting the module.
On the surface of the module, heat is transferred to the coolant, thereby overcoming the external thermal resistance. Techniques of heat transfer enhancement are beginning to play an important role in coping with increasing rates of heat dissipation. At the level of the PWB, the temperature of the coolant increases as the coolant absorbs heat from the modules along its flow path. This is referred to as the system level thermal resistance. Modules near the exit of the coolant flow are subject to the most severe thermal environment within

Figure 2 shows the data for heat loads on chips, modules, and PWBs. The horizontal axis shows the area of the components: for chips and modules it shows the projected area on the board. The data belong to the machines already in commercial operation in the field; those carrying open symbols represent air-cooled machines, and the shaded symbols represent machines where indirect water cooling is employed. The arrows indicate the trends of increasing heat load at all structural levels of the computer. In terms of the heat flux, it may not be long before we have to deal with several tens of watts per square centimeter of heat dissipation on the chip. One study [4] projects a chip level heat flux of 100 W/cm^2 for the early 1990s. At the module level, the use of indirect water cooling now allows heat fluxes in the range 2 - 6 W/cm^2 [5,6]. At the board level, the heat flux is somewhat low, at most 0.5 W/cm^2 on air-cooled boards [7], and 1 W/cm^2 in machines with indirect water-cooling [8]. At the board level, however, the critical

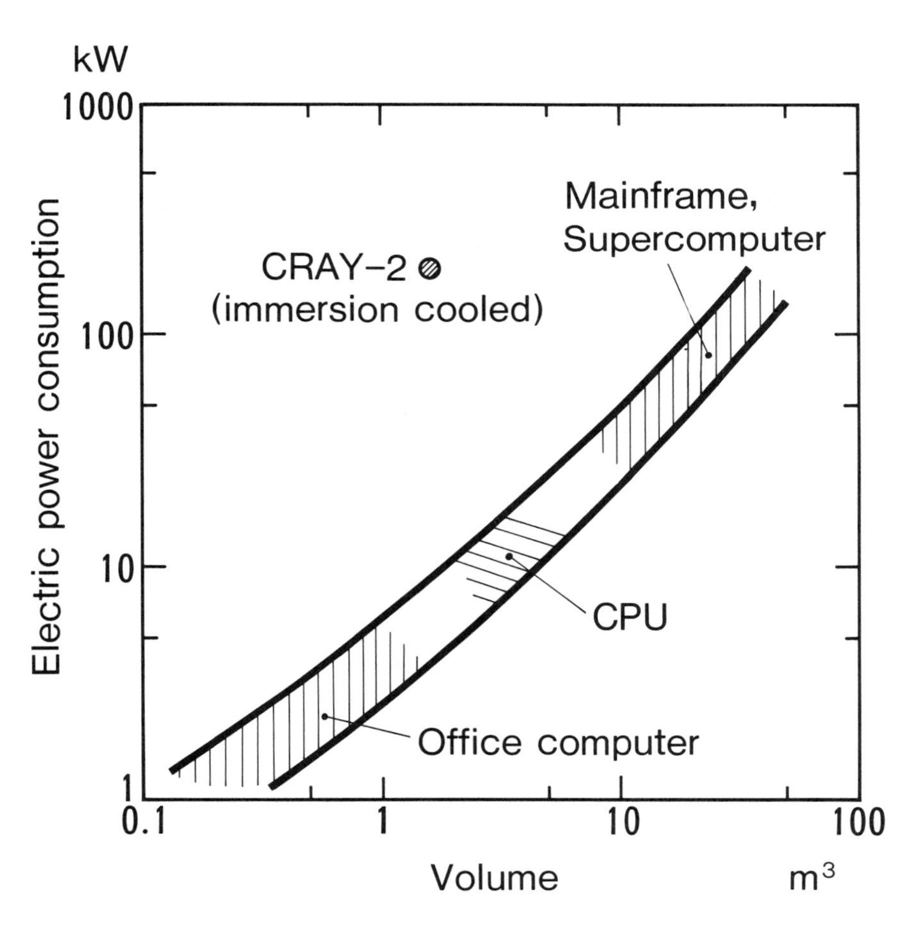

FIGURE 3. Electric power consumption and volume of computers.

factor is the absolute rate of heat dissipation from the board rather than the level of heat flux. When the heat load per board exceeds 1 kW, one must use liquid as the heat carrier in order to suppress the temperature rise of the coolant.

Figure 3 shows another view of the current state of the hardware design of computers. The vertical axis shows the rate of electrical power consumption by the computer system, which is approximately equal to the rate of heat generation by the system. This figure refers to not only the data of mainframes and supercomputers but also to medium-sized computers for office use. It should be noted that there are some uncertainties in data supplied by the manufacturers. In most cases the volume includes the space where heat loading is very low, such as cable housing and service space. Despite the crude nature of the data, Fig. 3 reflects certain facts observed in modern computers. The shaded band covers the data for office computers, central processor volume of some typical computers, and system volume of mainframes and supercomputers. Most of these data belong to the air-cooled machines, with only a few characterizing the indirect water-cooled machines. The system volume of the latter often does not significantly differ from the air-cooled machine because, except for the thermally critical parts, the system is cooled by air. An examination of this data collection indicates that the volumetric power density of medium and small scale computers is comparable to that of large-scale computers. Hence, the design of air cooling for small computers is just an equally demanding task as for large systems.

A data point located far above the shaded band belongs to the supercomputer CRAY-2 which employs immersion cooling for its central processor. Four processing units are housed in a highly compact frame, 135 cm diameter and 113 cm high [9], and directly cooled by liquid Fluorinert FC-77. Only immersion cooling allows such a high volumetric power density.

Figure 4 presents the situation strictly in terms of heat flux. Air cooling is limited to chip level heat fluxes of about 10^4 W/m^2. The inert, dielectric liquids, such as R-113, will accommodate heat fluxes in excess of 10^5 W/m^2 with saturated pool boiling from plain surfaces. With surface enhancements or convective assists (to be discussed later) allowable heat fluxes are above 10^6 W/m^2. Water-cooled chips with microgrooves (also to be mentioned later) raise the limits to about 10^7 W/m^2. As an absolute reference, the highest steady heat flux that has been accommodated at a reasonable surface temperature is close to 4×10^8 W/m^2.

In some other electronic equipment, the heat dissipation density is equally high. For instance, on the face of thyristor power inverters, the heat flux runs as high as 20 W/cm^2, and the volumetric heat density in the inverter housing is up to 600 kW/m^3 [10]. In this instance, the thyristor elements have to be cooled by boiling refrigerant. The heat dissipation density in those power devices continues to increase with the mounting demand on handling more electric power in confined spaces [11]. There are still other examples of electronic equipment where we see higher heat dissipation densities;

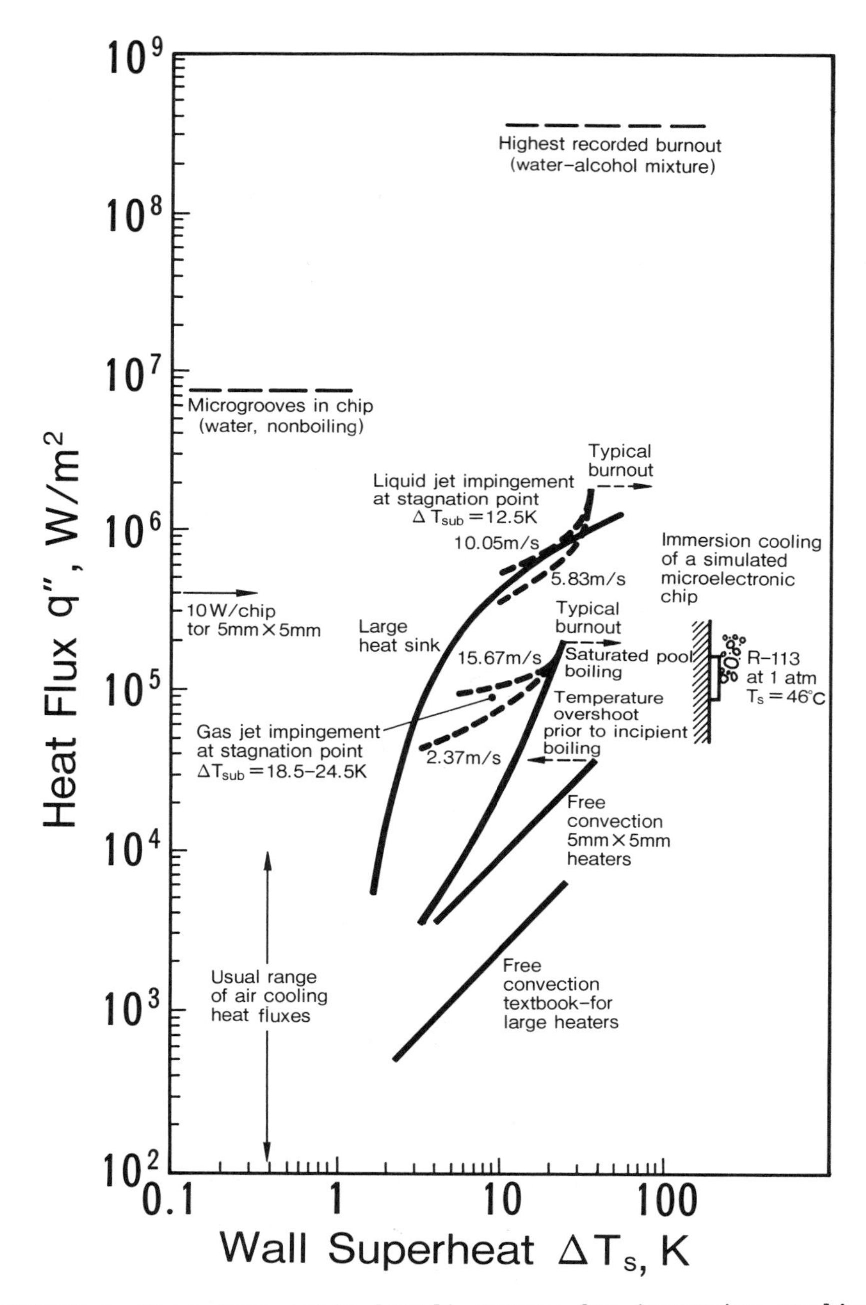

FIGURE 4. Heat flux (chip level) ranges for immersion cooling of microelectronic chips; air and water capabilities are provided for reference.

however, due to the complexity of its structure and the stringency of temperature control, the computer is presenting the most challenging problems to the designer.

As already stated, the knowledge base established with the past generations of electronic equipment serves as the vehicle for the development of future generations of equipment. Moreover, we need to expend a great deal of effort to further enrich our knowledge base as well as to devise new engineering methodologies to solve a host of complex thermal problems anticipated in future generations of electronic equipment. In the following sections, perspectives of the technology and heat transfer research will be presented according to the modes of heat transfer, namely, conduction, natural convection and radiation, forced convection, and phase-change heat transfer.

HEAT CONDUCTION - MULTIPLYING HEAT SOURCES AND THERMAL INTERFACES

The problems of heat conduction in electronic equipment quite closely reflect the development of electronics technology. About ten years after the emergence of transistors came the invention of the integrated circuit [12]. The production of integrated circuits in mass quantity began in the early 1960s. The number of components, such as transistors and resistors on a chip was then around 10, in what is called small-scale integration (SSI). Medium-scale integration (MSI), where components per chip were of the order of 100, appeared in the mid-1960s. In the early 1970s, the technology was advanced to produce large-scale integration (LSI) chips that bear more than 1000 components per chip. The 1980s is the era of very-large-scale integration (VLSI) and ultra-large-scale integration (ULSI), where the number of components per chip now exceeds 500,000 on memory chips. On logic chips, the number of elementary circuits, each circuit (referred to as gate) composed of several transistors and other components, is on the order of several thousand to ten thousand today. The integration has been most advanced on certain memory chips, where the degree of integration is most appropriately expressed in terms of the capacity for storing information in a specific unit called a "bit". Today, one-million-bit chips are being introduced in commercial equipment. The integration is also proceeding at the module and board levels. The progress in circuit integration has been accompanied by the increased complexity of package structure; heat conduction problems have intensified accordingly. This trend will continue into the future.

Figures 5 and 6 contrast the structural simplicity of a canned-transistor package and the complex composition of a silicon-on-silicon package. The former illustration is from the book by Kraus [13], the first comprehensive treatise ever written on heat transfer in electronic equipment. The package of the type illustrated in Figure 5 is still in use today, but with elevated heat dissipation from the transistor and the use of advanced packaging materials. The main feature of this packaging was, however, established in the early days of transistor development. Figure 6 illustrates the concept which semiconductor manufacturers are currently striving to materialize into commercial products [14]. When viewed as a thermal system, chips are multiple heat sources on a few layers of substrate. Similar thermal systems are already present in the current generations of multichip modules and printed wiring boards.

The lumped parameter approach like that depicted in Fig. 5 is a standard tool to estimate the thermal resistance from the transistor junction to the surface of the package. Due to the increase in the number of heat sources placed in close proximity on a substrate as well as the increase in the number of thermal resistance components in the package, the thermal resistance network tends to become complex. Fortunately, however, the high-density packaging is often accomplished by placing chips and modules in regular geometrical patterns so that the temperature field can be analyzed utilizing the geometrical regularity. For instance, the temperature field produced by rectangular heat sources on a multilayer substrate is expressed as a Fourier series expansion in a computer program developed by Ellison [15].

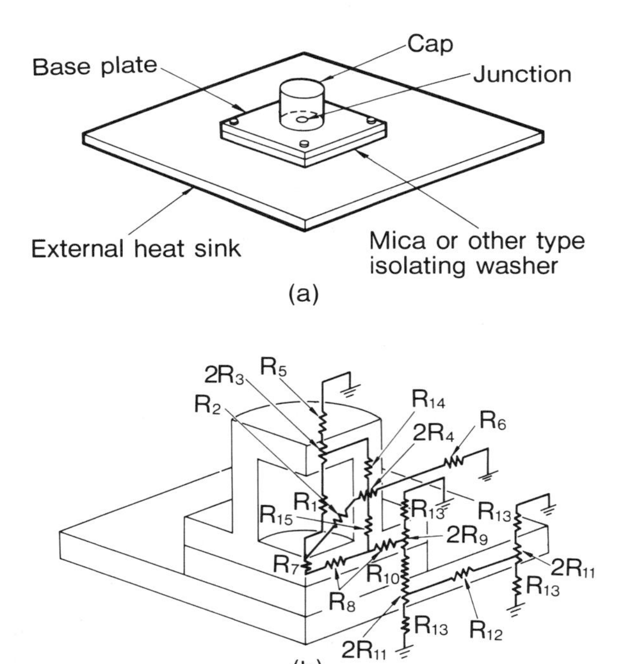

FIGURE 5. Transistor mounting arrangement and heat flow paths [13].

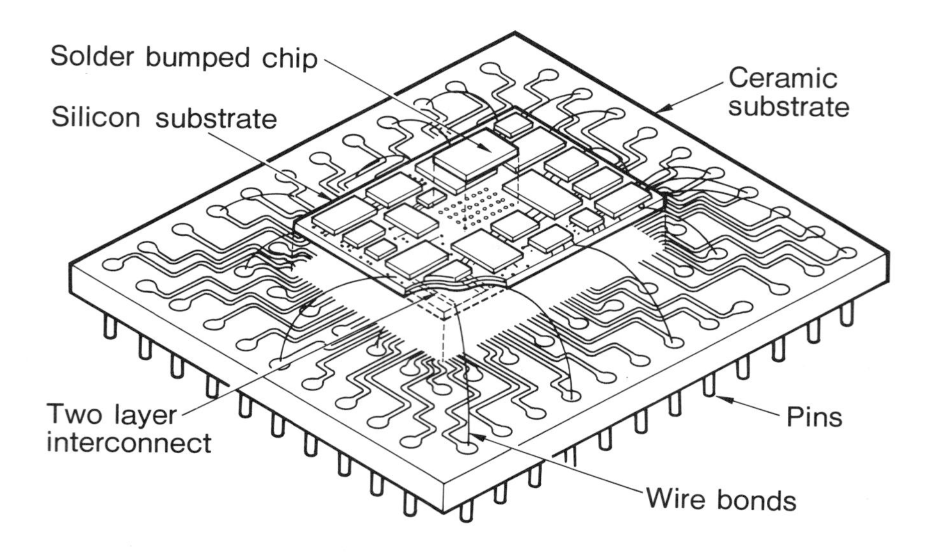

FIGURE 6. Silicon-on-silicon packaging [14].

The temperature field has to be analyzed not only for the estimation of internal thermal resistance of the package but also for the thermal stress within the package. Accuracy in the prediction of thermal stress in the package needs to be increased, because electronic circuits and structural components of the package are being miniaturized. An illustration is given by Figure 7, which shows the areas occupied by a DRAM (dynamic random access memory) chip within a plastic encapsulant of a limited size [16]. As the memory capacity has been increased from 16 kilobits in 1977, to 256 kilobits in 1984, and to 1 megabits today, the size of the chip has also been increased, leaving less area of encapsulant around the chip. With the overall size constant, packaging of larger chips can only be accomplished by sophisticated management of thermal stress in the encapsulant. Moreover, the thermal stress on the chip also needs to be considered for purposes ranging from the prevention of cracking of the chip to the protection of miniaturized circuits from harmful effects of excessive stress. For the task of analyzing temperature and stress fields in the module, general purpose computer programs, such as ADINA [17] and ANSYS [18], are being extensively used today. In the future, more attention will be directed to temperature and thermal stress fields of microscopic scales, such as those in the complex microstructure of Figure 8, which depicts a cross section of the integrated circuit whose feature length is close to 1 micron. If one attempts an analysis dividing the chip into cells of microscopic dimension in a straightforward manner, the number of cells soars up to an astronomical order which even the today's largest supercomputer cannot handle. A new analytical method needs to be developed to deal with what are called large matrix problems.

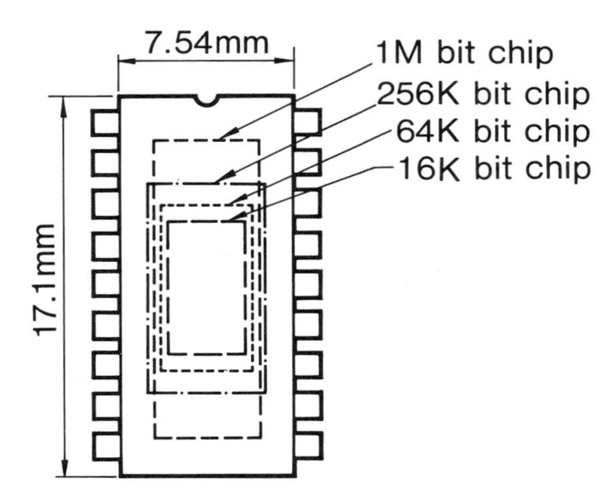

FIGURE 7. Growing size of a DRAM chip molded in a dual-in-line (DIP) package [16].

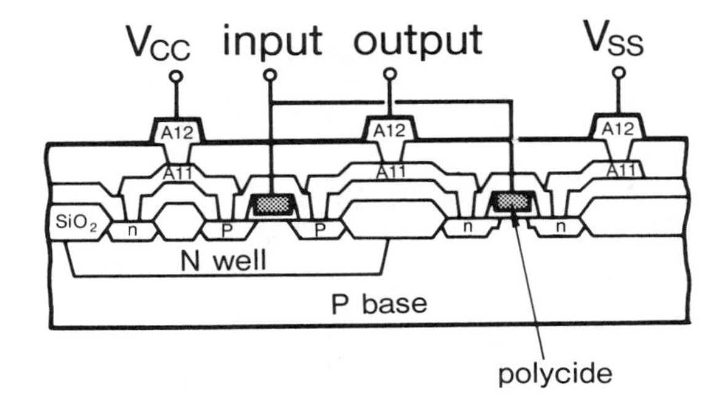

FIGURE 8. Cross section of a CMOS inverter used on the periphery of a memory cell array.

Another significant development in the packaging of electronic equipment is the growing number of thermal interfaces due to the increasing structural complexity. There are a variety of techniques of thermal interfacing, some of which date back to the days of vacuum tubes. Figure 9 shows an example of the cold plate scheme applied to cooling vacuum tubes in the 1950s [19], where the shields served to minimize hot spots in the glass and provide a conduction path to the heat sink. In the early 1970s, Control Data Corporation introduced conduction cooling for single-chip modules in computers. As shown in Figure 10, modules are pressed onto the water-cooled cold channel by a special socket and hold-down spring [20].

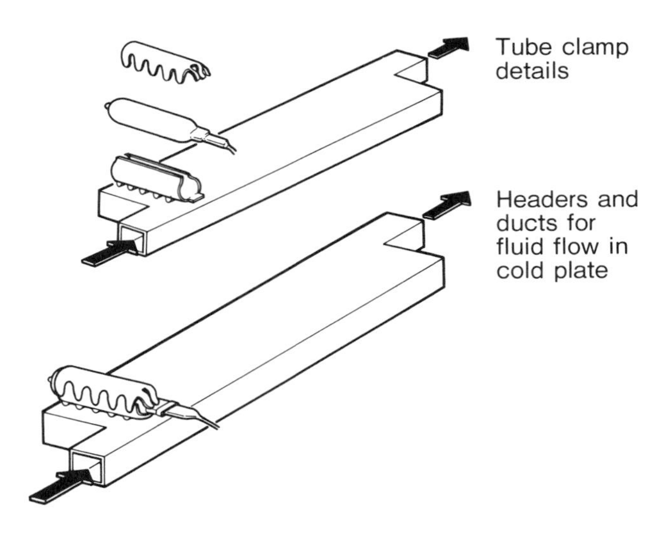

FIGURE 9. Air-cooled cold plate for vacuum tubes [19].

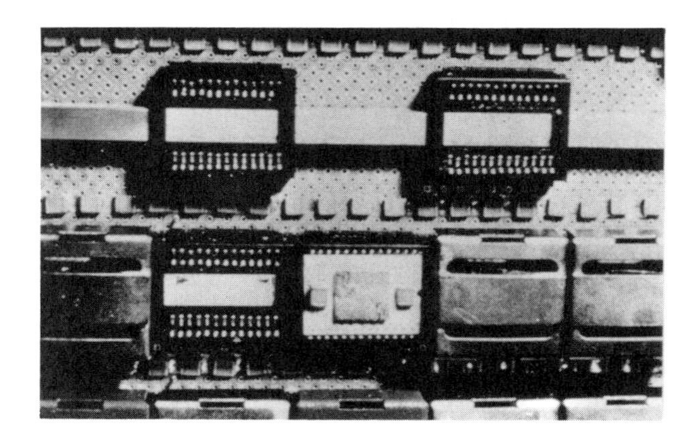

FIGURE 10. Rectangular module carrier and cold channel in the Cyber 205 [20].

By the end of the 1970s, the demand on high-density packing of chips led to the development of multichip modules. The thermal management of multichip modules requires unique cooling designs under the constraints imposed by the geometrical arrangement of associated components and for ease of maintenance work. The thermal conduction module developed by IBM, shown in Fig.11 [21], houses as many as 118 chips on a 90 mm square ceramic substrate. Each chip dissipates up to 4 W, and is contacted by a spring-loaded piston which provides an important part of the conduction path from the chip to the water-cooled cold plate. In the 1980s we have seen some other designs of thermal interfacing employed in water-cooled as well as in air-cooled modules of large-scale computers [16].

The fundamental understanding of heat transfer across the interface of component surfaces has been advanced by a number of investigations. Yovanovich [22] provided a summary of the physical understanding gained by the late 1970s. Very recently Eid and Antonetti [23] presented a review of the measurement methods for the estimation of contact resistance in electronic equipment. The problems of thermal interfaces in electronic equipment are characterized by relatively low contact pressure, which is required not to subject chips and other components to excessive stresses. In such circumstances, the kind of fluid filling the interstice has a significant effect on the contact resistance. Where thermal compound or solder is used to fill the interstice, the formation of voids should be avoided. Table 1 summarizes the factors involved in the management of contact resistance in actual electronic equipment [16].

At this point a note is made on the use of heat pipes in cooling electronic equipment. Since the invention of heat pipe in the mid-1940s [24], heat pipe technology has been developed for diverse industrial applications. Heat pipes are effective tools, particularly for cooling aerospace electronic equipment, where the demands on saving of space and weight are great.

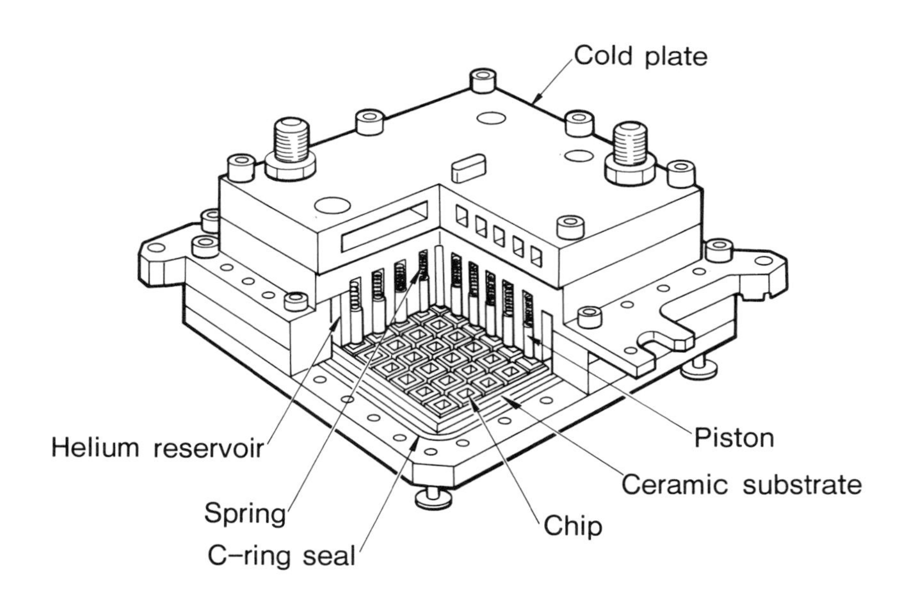

FIGURE 11. Exploded view of thermal conduction module [21].

TABLE 1 Factors affecting thermal resistance across interface

Phase	Parts	Factors
pre-assembling	parent	bow, wave, microroughness
	interstice	selection of material (gas, liquid, adhesive, solder)
assembling	parent	non-alignment, tilting, elastic-plastic deformation
	interstice	void formation
service	parent	variation of interstitial gap, cleavage formation, due to TEC mismatching of components
	interstice	deterioration of interstitial material

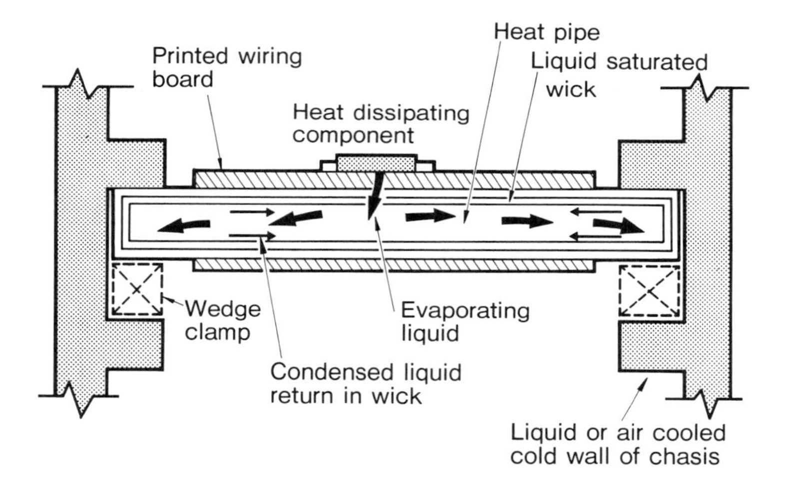

FIGURE 12. Edge cooled heat pipe heat exchanger [25].

Figure 12 shows a cross section of a flat plate heat pipe which provides thermal paths from the heat dissipating component to the cold walls of the chassis. According to Scott and Tanzer [25] the replacement of a copper conduction plate to a heat pipe allows the maximum power load on the PWB to increase by a factor of 1.3 to 2. In order to utilize the great advantages of high thermal conductance in the heat pipe, it is necessary to make good thermal contact at the heat source and heat sink ends of the heat pipe.

NATURAL CONVECTION COOLING

Deviations From Elenbaas' Basic Models

With the spread of electronic equipment in many corners of our lives, the demands are growing for the elimination of acoustic noise inherent in forced air-cooling of equipment. Natural convection cooling is also increasing in importance for equipment that operates in hostile thermal environments and at remote locations, where the service life of air-moving devices is a matter of concern.

In 1942 Elenbaas published a paper on heat transfer from vertical and inclined parallel plates to natural convection of air [26]. As reviewed recently by Landis [27], Elenbaas' semiempirical formula withstood the test of the time, and has served as a benchmark reference for many subsequent works on natural convection heat transfer from fins. Elenbaas' experiment was conducted using pairs of isothermal plates with the intention to model the thermal situation produced by a row of naturally cooled extended surfaces. Finned heat sinks, like the ones attached to a transistor as shown in Figure 13, will continue to be an important item of hardware for electronic equipment.

Where the finned heat sink is designed to dissipate heat to open air, for instance, by its attachment to the exterior of the equipment housing, the design is now a relatively easy task owing to confidence in the Elenbaas correlation and additional information developed by other investigators. In actual electronic equipment, however, the presence of components and other obstacles imposing constraints on the convection of air is almost the rule, particularly in equipment where increased packing density of components is inevitable. Figure 14 shows a situation commonly found in electronic equipment: a row of vertical card boards, each board carrying different types of packages. The convection pattern of air is complex, due to the difference of heat loads among the boards, the entrainment of air from the sides of the board

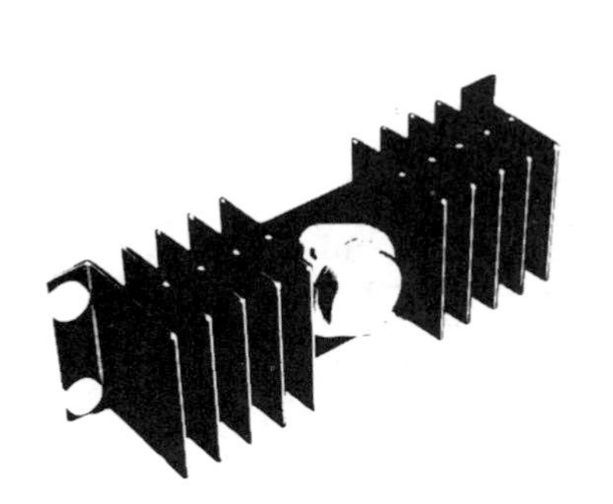

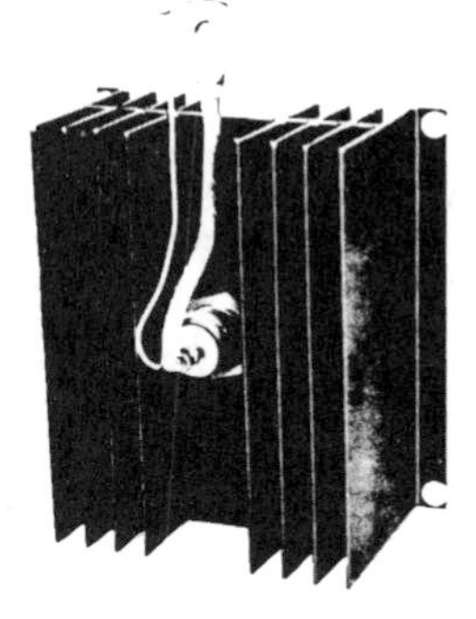

FIGURE 13. Typical transistor mountings on air-cooled heat sinks [13].

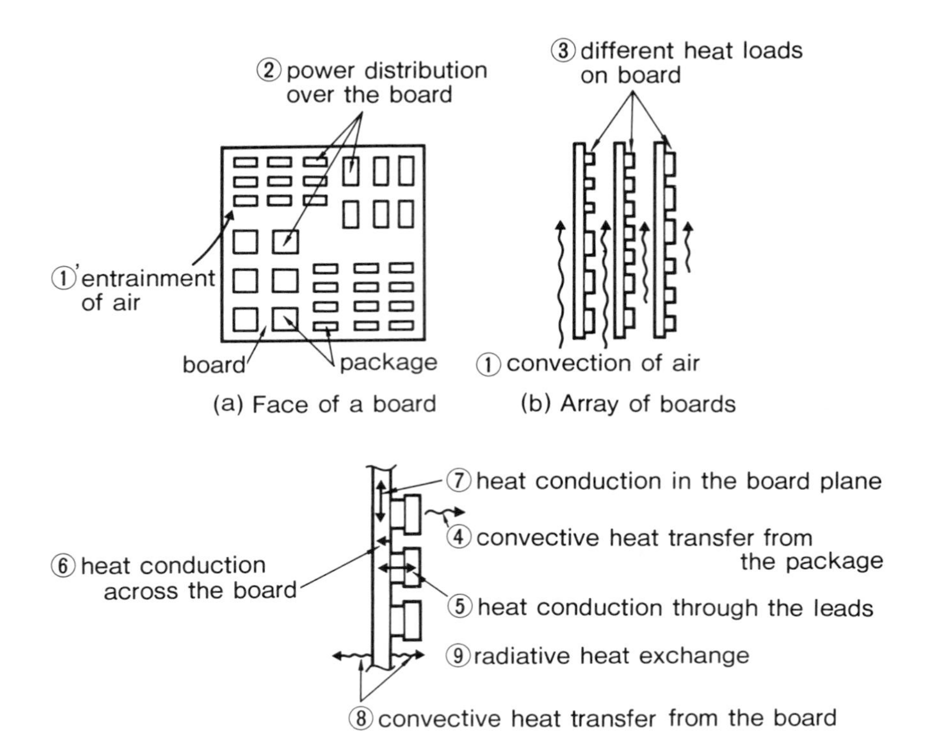

FIGURE 14. Factors involved in natural convection cooling of modules on printed wiring cards.

array, and the packages causing local disturbance in the flow. Radiation also plays a role in heat transfer. Moreover, conduction from the package to the board and that within the board is coupled with convective heat transfer on the surfaces of the packages and the board. Any attempt to estimate temperatures of the components in such a complex thermal system requires modeling of the system. In a simple model, the boards are approximated by a row of isothermal smooth plates. Where the boards are surrounded by an enclosure, a vertical rectangular duct could be a first-order approximation to the convection path between the boards. We find in another of Elenbaas' papers [28], on the natural convection heat transfer in vertical channels of different cross sections, a useful guide to estimate the approximate temperatures of the boards. Recent studies have been directed to more accurate models of this physical system, such as a pair of vertical plates with non-symmetric thermal boundary conditions, a vertical wall having discrete heat sources embedded in it, or a model allowing air entrainment from the opening on the side of a parallel plate channel. The actual boards may be used for a heat transfer test. Bar-Cohen and Rohsenow [29] review studies of thermally non-symmetric vertical channels. Some of the additional works on sophisticated models are mentioned by Nakayama [16].

One of the most difficult tasks for the equipment designer is to predict the distribution and the intensity of convection driven by different heat sources in equipment housing. An enclosure containing discrete heat sources serves as a physical model of naturally cooled equipment. The tools to solve such problems are numerical integration of the Navier-Stokes and energy equations, flow visualization to be conducted with models, and diagnosis from a thermography picture of the actual equipment. Figure 15 shows a sketch of the thermography setup. Many aspects of this problem will be treated in papers presented at this Symposium.

It should be noted at this point that small heat sources in large enclosures display natural convection heat transfer characteristics different from those predicted by conventional correlations. However, this is not an important concern for

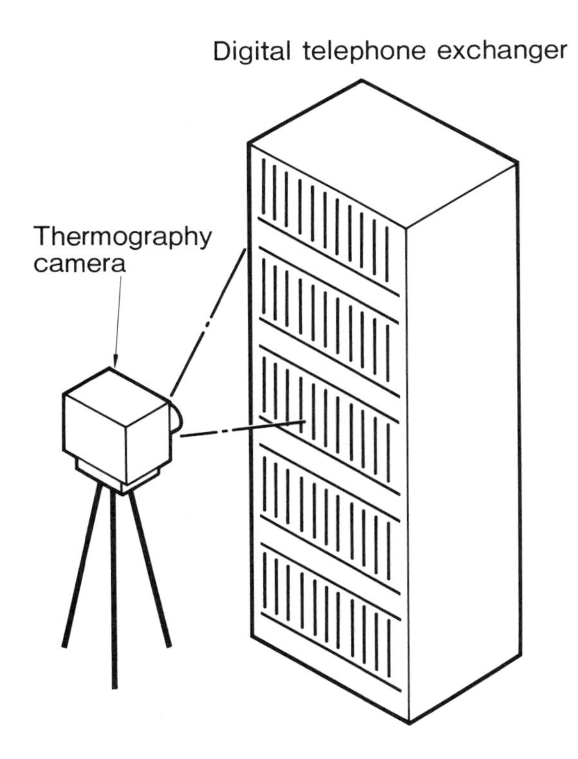

FIGURE 15. Thermograhic diagnosis of naturally cooled digital telephone exchanger.

cases where the natural draft is driven by the chimney effect in an array of parallel boards, and its velocity is close to a level of forced convection. The aforementioned deviations in heat transfer behavior will be discussed further in connection with liquid cooling with phase change.

Radiation

The analysis of radiative heat exchange among the components and the environment is another challenging task for the equipment designer. Radiation can account for a substantial portion of the heat transfer if natural convection of air is utilized. For airborne equipment, in particular, the contribution of radiation to the total heat transfer can be more than twice the convective component [30]. The determination of view factors and the solution of heat balance equations are the necessary steps to determine the component temperature. Along with the tasks of convective heat transfer analysis, those tasks demand the development of user-friendly computer programs for designers who are not necessarily familiar with heat transfer analysis.

FORCED CONVECTION COOLING

Air Cooling - New Dimensions In the Classical Cooling Method

Blowing air past heat generating components has been the most popular method of cooling. In 1942, Mouromtseff [31] documented the design analysis for cooling a high-power vacuum tube, shown in Figure 16, which was performed applying the correlations presented by McAdams [32]. In this illustration, the extended surfaces surround the heat generating portion of the device. In most similar cases of cooling devices by forced convection of air, the thermal design is synonymous with the design of extended surfaces. Kraus [33] recently appraised Gardner's classic work [34] on the graphical presentations of the fin efficiency which was published back in 1945. The technique of heat transfer enhancement on extended surfaces has also been needed for devices of high-power dissipation since the early years of electrical and electronics technology. An example is found in the report by London [35] on the use of louvered fins to improve the thermal design of a new type of vacuum tube (microwave) dissipating up to 25 kW. The now popular reference book by Kays and London [36] originated from the need to provide a database of heat transfer and flow friction for such design work.

With the advent of integrated circuit technology, new dimensions were added to the problem of air cooling. This is explained by drawing an example from present day air-cooled mainframe computers. Figure 17 shows a printed wiring board, 28 cm wide and 42 cm long, carrying 72 modules, and dissipating 512 W [37]. Grouped in the middle of the board are the high-speed memory modules, and the rest are the logic modules. Heat sinks are designed to accommodate the heat dissipation, 6 W from the logic module, and 11 W from the memory module. A row of such boards is set in the housing where air is

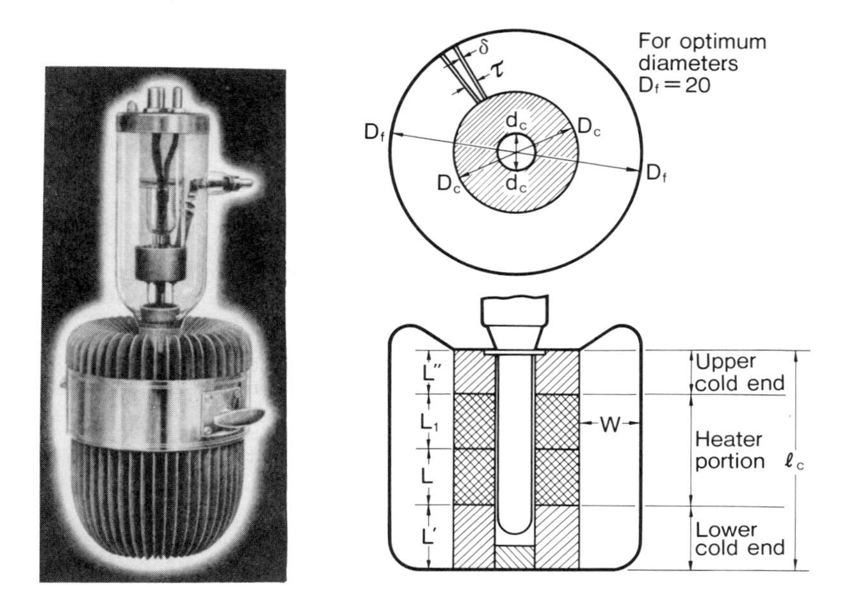

FIGURE 16. High-power vacuum tube with air-cooled anode [45].

FIGURE 17. Array of logic modules and memory modules on printed wiring board [37].

forced through by the blowers. Hence, the physical situation is described by a model where the modules rest on one side of the parallel plate channel leaving free-flow space between the fin tips and the other side of the channel, and also in between the columns of the modules. Heat transfer problems involved in this example embody the following generic issues for the present and future generations of electronic equipment.

(1) Optimum design of finned heat sinks

An individual heat sink has to be provided with an optimum number of fins for a given fin height and a given air velocity. Figure 18 shows an example of a design study, where the base of the module is assumed to have an area L^2, the relative fin height is 0.5 or 1, the relative thickness of the fin is 0.025, and the air velocity is expressed by the Reynolds number. The Nusselt number, defined in terms of the heat flux on the module base, is shown to have a peak

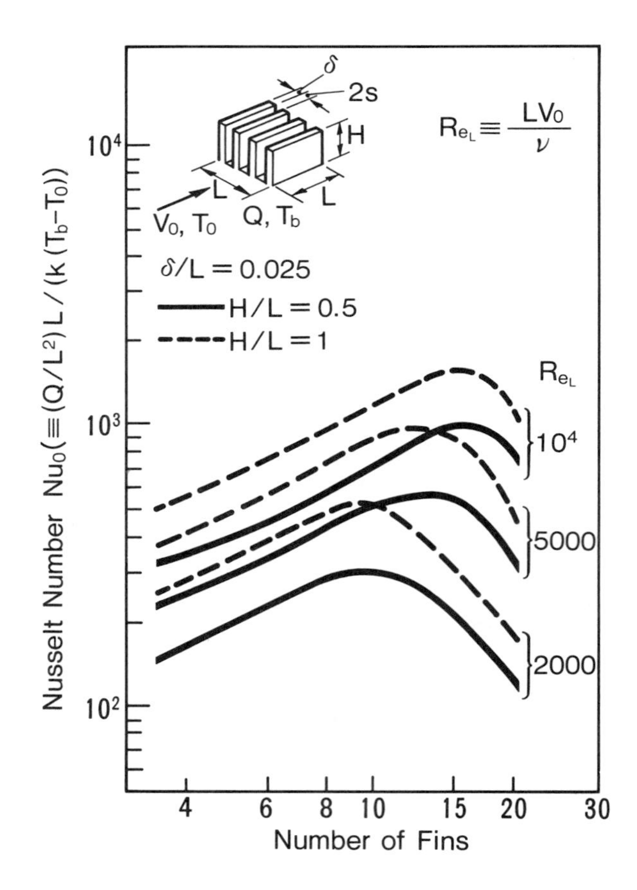

FIGURE 18. Heat transfer performance of finned heat sinks convectively cooled in an open environment [16].

at a certain number of fins. A peak is produced by the contention between the increase in the heat transfer area and the decrease of the rate of air flow in the inter-fin passages, both increasing with fin number. In actual operative situations, the presence of a neighbor board tends to affect the diversion of air flow from the module. Sparrow and his co-workers [38, 39] have considered the effect of the spacing between the fin tips and the channel wall on heat transfer from the fins.

(2) Heat transfer from multiple heat sources in a channel

The air in the channel has a complex temperature distribution resulting from the fact that the air flow over the columns of the modules has only a finite rate of mass exchange with the air through the free flow areas between the module columns. The air temperature over the modules is higher than that in the adjacent free flow areas. The variation of air temperature in the channel is governed by the geometry of the channel, the size and structure of the module, the placement pattern of the modules, the rate of heat dissipation from the modules, and the air velocity. It is important for the equipment designer to predict the distribution of air temperature in order to hold the temperature of modules below a tolerable level near the channel exit, as well as to constrain the temperature variation among the modules within a permissible range. Only a limited body of experimental data is available for such design work: Arvizu and Moffat [40] and Wirtz and Dykshoorn [41], for arrays of rectangular solid blocks; Nakayama et al. [42], for arrays of finned modules.

(3) Management of air flow in the system

Present day electronic equipment involves complex systems with many components and multiple paths for coolant flow. Figure 19 shows examples of air flow schemes employed in mainframe computers. In Figure 19 (a), the air is forced from the bottom to the top of the vertical board arrays, in (b) the parallel paths are formed by horizontal boards for once-through flows of air [3], and in (c) air jets impinge onto the multichip module [43]. The determining factors in choosing air flow schemes are the heat load of boards, the permissible air velocity, and geometrical constraints imposed by the architectural design to achieve required electronic functions.

The developments in packaging technology will promote a proliferation of multichip modules and surface-mount modules. Successful thermal management of those devices can only be realized by designs which are based on detailed knowledge of heat transfer on and near the modules. Presently, the level of such knowledge is very low; for instance, data that can be used to estimate the distribution of local heat transfer coefficient over the surface of the module are scarce. Computer programs capable of dealing with complex flows, for example, the one reported by Karniadakis et al. [44], will prove effective in understanding the process of momentum and heat transfer in flows over modules.

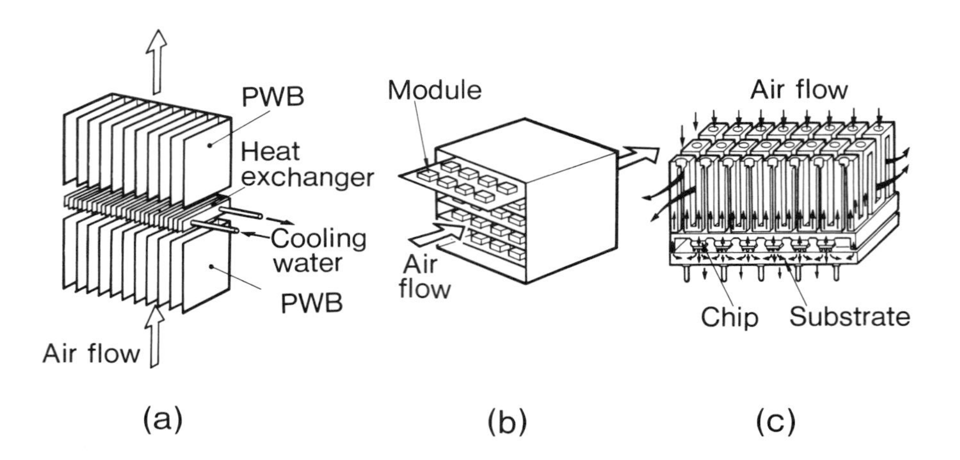

FIGURE 19. Management of airflow; (a) vertical throughflow in stack of board arrays, (b) once-through flow in horizontal stack of boards, (c) jet impingement onto multichip module.

Liquid Cooling - Recurring Schemes In New Applications

Liquid cooling of electronic equipment dates back to the era of vacuum tubes. An example is found in the 1935 paper by Mouromtseff and Kozanowski [45] on the design of high-power vacuum tubes (transmitters). The reference book by Kraus [13] contains another example, shown in Figure 20, a liquid-cooled traveling-wave tube. Concurrent with those instances of liquid cooling, the consideration of air versus liquid cooling has been a task for the designer of new equipment. Mouromtseff [31] wrote in his paper of 1942,

> "The decision whether one or the other type of cooling is preferable in any particular tube application depends on the general design of the transmitter, relative cost of installation, and on certain specific conditions of operation."

This observation still holds today.

Where liquid coolant is brought into direct contact with electronic components, the coolant must have high enough dielectric strength. Despite its highest heat transfer capability, water is disqualified as the coolant in most instances due to its low dielectric strength. Major efforts have been waged since the 1950s to develop coolants of high dielectric strength and good chemical stability. Those developed coolants are given brand names by the manufacturers: "FCs" (3M), "Coolanols" (Monsanto), "DCs" (Dow Corning), "Freons" (Du Pont). A measure of the heat transfer capability of the coolant (for single-phase forced convection) is given by the dimensional factor, $\rho^{0.8}$ $k^{0.6}$ $c^{0.4}$ / $\mu^{0.4}$, now referred to as the Mouromtseff number.

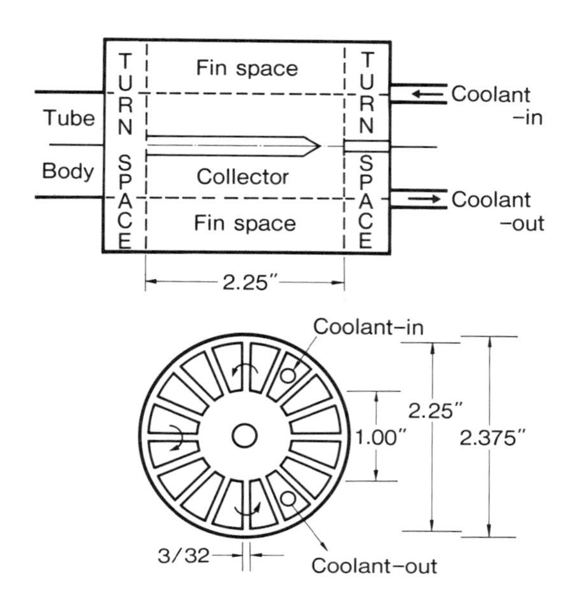

FIGURE 20. High-power traveling-wave tube with liquid cooling [13].

Figure 21 shows this parameter and its dependence on temperature for different fluids [46].

In the design of cooling systems for computers, easy access to components for replacement or repair work is often a dominant requirement. This requirement has long made air cooling the most popular mode of cooling computers. In cases where the use of liquid was unavoidable, the cold plate scheme has been developed to meet this requirement. The water-cooled cold plate, of course, is still widely used for power supplies.

The application of what is termed "immersion cooling" to cooling computers is a very recent event. The CRAY-2 supercomputer, announced in 1985, is the first machine that has the central processing unit immersed in liquid. The liquid is FC-77 supplied by 3M. The schematic of the cooling system is shown in Figure 22. The cooling liquid is pumped through the sealed computer chassis at a rate of 260 litres per minute, keeping a bulk fluid temperature in a range of 21 - 27 °C [47].

The advantages offered by liquid coolant are twofold: first, a small variation of module temperature within the system, effected by the relatively large heat capacity of liquid, and second, high heat transfer coefficient on the module surface. The first of those advantages manifests itself in the indirect-cooling or cold-plate schemes already employed in some of the presently marketed computers. The cold plate scheme, however, inherently involves a relatively large thermal resistance between the chip and the coolant. In order to fully utilize

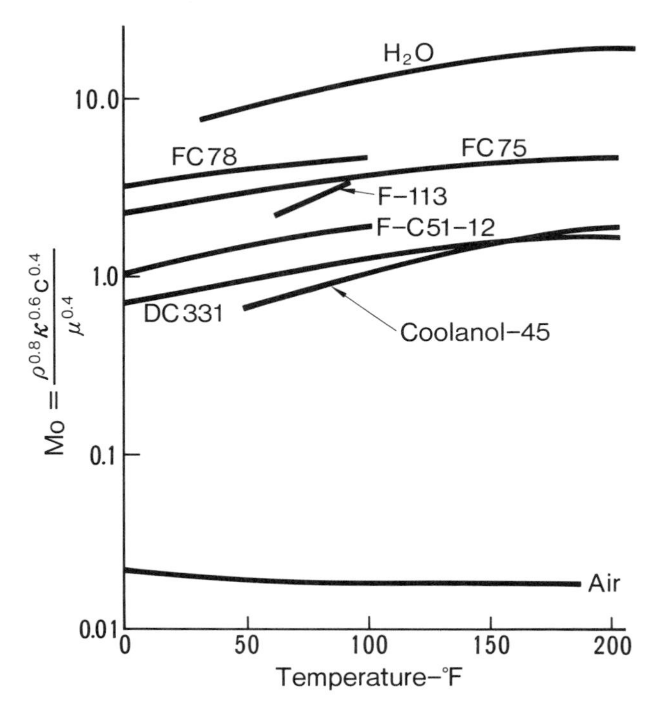

FIGURE 21. Mouromtseff number for fully developed turbulent flow in smooth tubes [46].

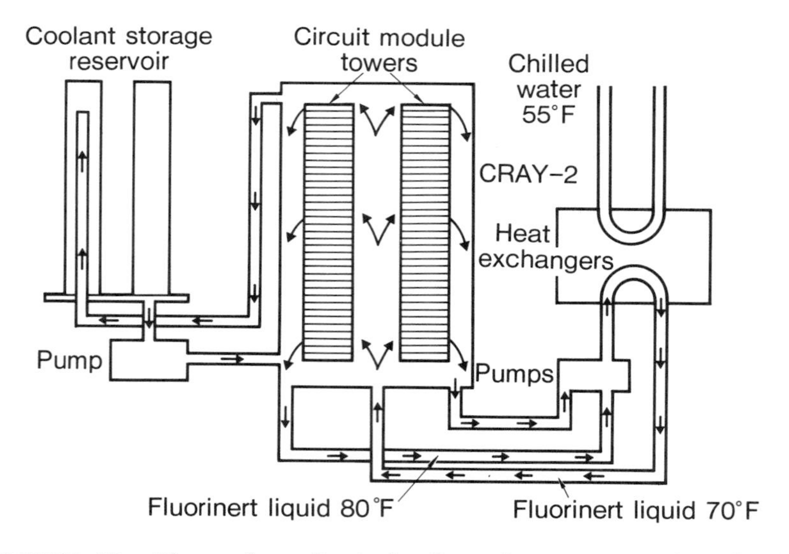

FIGURE 22 Flow of coolant in Cray 2 supercomputer [47]

the potential of liquid cooling, the intermediate thermal resistance has to be minimized. This is accomplished in direct immersion cooling, where the convective heat transfer on the module surface or on the bare chip becomes a dominant factor in the chip-to-fluid thermal resistance. Although liquid coolant has a heat transfer capability two orders of magnitude higher than air (Figs. 4 and 21), anticipation of rapid increases of heat fluxes at the module and chip levels has already motivated the search for measures of heat transfer enhancement on liquid-cooled surface. Figure 23 shows the scheme studied by Ramadhyani and Incropera [48]. The model module, having a base area of 12.7 mm × 12.7 mm, is equipped with copper pins. Subfins are provided on each pin, resulting in the real heat transfer area on the module being about 12.8 times the base area. With FC-77 flowing through the module array at a velocity of 13 - 60 cm/s, the heat transfer coefficient (defined in terms of the base area) is in the range of 8500 - 16000 $W/m^2 K$.

The ultimate form of cooling chips using forced convection of a liquid was proposed by Tuckerman and Pease [49]. Figure 24 is the schematic of their concept which reduces the thermal resistance between transistor junctions and the coolant to a bare minimum. On a 1 cm × 1 cm silicon substrate, microscopic cooling channels are formed by a directional etching process. The dimensions of the microchannels are optimized to provide the highest heat transfer performance for a given pumping power. With 300 μm deep and 50 μm wide channels at a pitch of 100 μm, a thermal resistance of 0.09 °C/W is shown to be

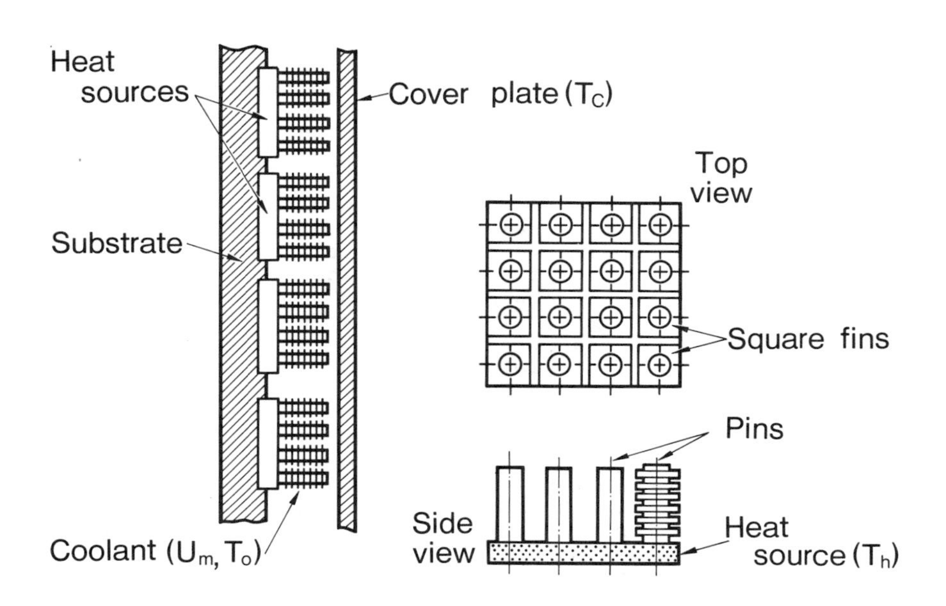

FIGURE 23. Liquid cooled finned-pin array [48].

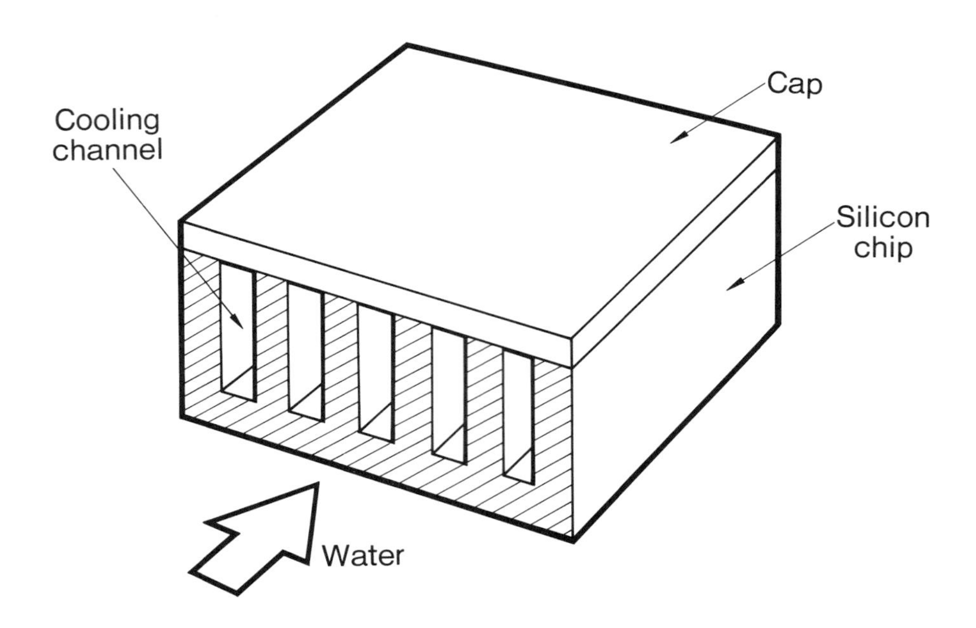

FIGURE 24. Microchannels on silicon chip [49].

possible with water as the coolant. Their work has aroused considerable interest in microchannel cooling because of its great potential in cooling future electronic equipment. There is, of course, difficulty in using water because of dielectric requirements; with dielectric coolants, the thermal resistance is increased considerably. The other problem not yet satisfactorily addressed is how to provide the coolant hookup and seal against high pressures resulting from large pressure drops.

Natural convection cooling with liquids is also of interest. The chips are mounted in a sealed chamber filled with dielectric liquid that has an immersed cooling coil or a cooled wall. Baker [50] was the first to indicate that heat transfer coefficients for single simulated chips in R-113 increase as chip size decreases, specifically, from 20 mm wide × 10 mm high to 4.6 mm × 2.3 mm. Park and Bergles [51] systematically varied the height and width of simulated chips in R-113 and concluded that short, but wide, heaters have coefficients typically 15 % higher than predicted because of preheating from substrate conduction and fluid conduction. As width decreases, the heat transfer coefficients increase very substantially, with the coefficients for 2 mm wide heaters being 150 % above those for 70 mm wide heaters. Heat transfer coefficients for protruding heaters are greater by about 15 % than those for smooth heaters. The coefficients for in line or staggered arrays may increase or decrease depending on vertical and horizontal chip spacing. Referring to Fig. 4, these generally higher single-phase coefficients are important in determining the transition to boiling conditions, as discussed in the next section.

PHASE-CHANGE COOLING - RENEWED INTEREST IN COPING WITH HEAT IN THE FUTURE

For many years evaporative cooling has been considered for severe service applications by equipment designers and heat transfer researchers. Interest in its application to cooling electronic components dates back to the late 1940s and the early 1950s [52]. In those early years, however, the investigations were centered around equipment for military and aerospace equipment, where the demand on high reliability of components in a relatively short period of service preceded any other concerns. Figure 25 shows examples of cooling airborne equipment described by Kaye and Choi [53]; Figure 25(a) shows a simple expendable spray cooling system and Figure 25(b) depicts a combined forced air and spray system.

Laboratory work on cooling microelectronic chips and modules by direct immersion in a liquid became active in the 1960s [5]. In due course, boiling heat transfer to dielectric organic liquids became a subject of interest to engineers in the electronics industry. Nucleate boiling has several major advantages in that the coolant effectively reaches all areas of the package and wide variations in heat flux are accompanied by modest changes in surface temperature. Reflecting this heightened interest in phase-change cooling was the report by Armstrong [54], where an attempt is made to obtain a single pool nucleate boiling curve for organic liquids at 1 atm. This, however, served mainly to underscore the need for the variable surface-fluid coefficient of Rohsenow [55].

Throughout the 1970s, interest in this mode of cooling was sustained in anticipation of rapid increases of heat loads on

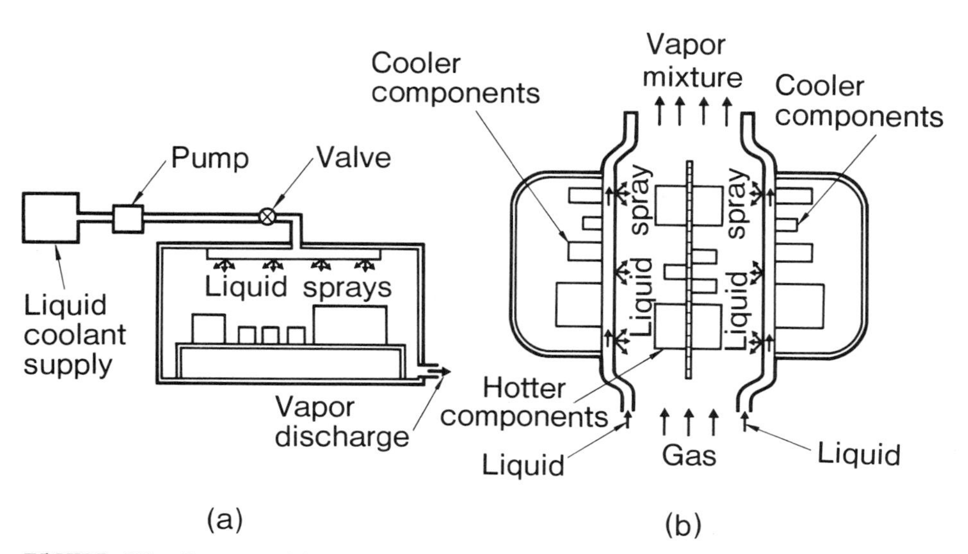

FIGURE 25. Evaporative spray cooling techniques for electronics [53].

electronic components. Actual applications to cooling thyristor devices were started in this period, and have continued to this day [10, 56]. Figure 26 shows an example, where a thyristor controller for the traction motors of a train is cooled by boiling R-113. In the area of microelectronics, work was undertaken at IBM to develop what was termed the Liquid Encapsulated Module [57], shown in Figure 27. A substrate carrying an array of chips is mounted within the liquid holding module. Heat from the chips is transferred to fluorocarbon coolant, then to the internal fins in the module, and ultimately to the water-cooled cold plate. Boiling heat transfer on the chips produces high heat transfer coefficients, and the cooling system is designed to achieve a heat removal capability of 300 W from the module. This endeavor, however, did not materialize in commercial equipment, for reasons that are still shaping the focus of today's research. One of the early concerns was the maintenance of purity of the coolant during the service life of equipment. Long experience with the "electronic liquids", such as Fluorinerts, has demonstrated the compatibility and stability of these coolants [58].

Another continuing concern with liquids is the uncertainty of bubble nucleation; in case of nucleation failure on a chip, an unacceptably high chip temperature could result. This is due to the high wettability of the typical coolants; all but the smallest nucleation sites are "snuffed out" with the result that high superheats are required for incipient boiling. (See the "temperature overshoot" in Fig. 4.)

As the above examples illustrate, immersion cooling of commercial electronic components requires a closed environment where condensation of generated vapor is an equally important process. Figure 28 is an excerpt from the thesis by Bravo [59], where two principal condensing schemes are shown, vapor space

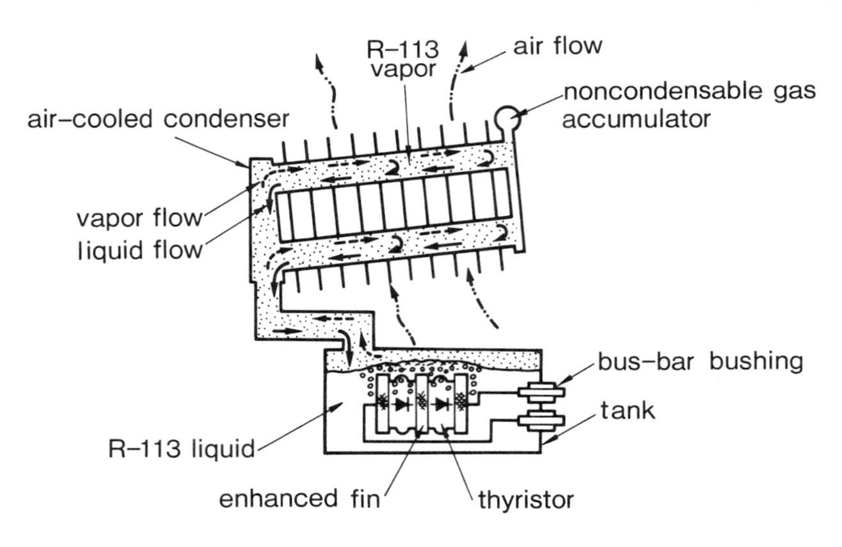

FIGURE 26. Thermosiphon cooling of chopper thyristors [10].

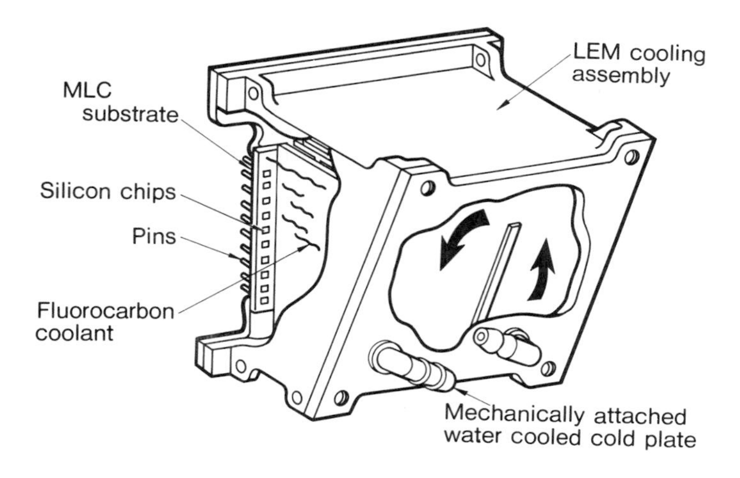

FIGURE 27. Liquid encapsulated module with integral water-cooled cold plate [57].

condensation and submerged condensation. The design of a condenser has to meet the requirements to maintain the internal pressure of the enclosure near atmospheric, keeping it almost free from variations of external thermal environment. The presence of noncondensable gas, resulting from inadequate degassing or air that seeps in the enclosure during inactive periods, cannot be ruled out. Bravo and Bergles [60] documented the effect of a noncondensable gas on the system pressure and temperature. The designer has to carefully identify the parametric bounds where a satisfactory operation of submerged condensation is guaranteed. Bar-Cohen [61] gives an overview on this topic in his article on immersion cooling.

For the microelectronics technology in the 1980s and beyond, phase-change cooling has renewed its appeal by its potential to provide high heat transfer coefficients on the component surface. The availability of chemically stable dielectric coolants has supported this cooling technology. The list of technical problems to be solved before the adoption of phase-change cooling in commercial equipment is, however, not limited by those mentioned above. Rapid development of circuit integration currently underway is requiring a fresh look at the need to enhance heat transfer from the chip to boiling coolant. Such attempts were made by Oktay [62] by providing roughness, microchannels, and microporous dendrites on the chip surface, and some of the measures are reported to be effective in reducing the temperature overshoot of the chip as well as raising the critical heat flux. Figure 29 shows a scheme studied by Nakayama et al. [63], where a porous cylindrical stud is bonded to the heat dissipating component. Microtunnels crisscrossing in the stud provide numerous bubble nucleation sites, and once the vaporization of the coolant is started, become the avenues of dynamic two-phase flows, thereby producing a very high heat transfer coefficient (refer-

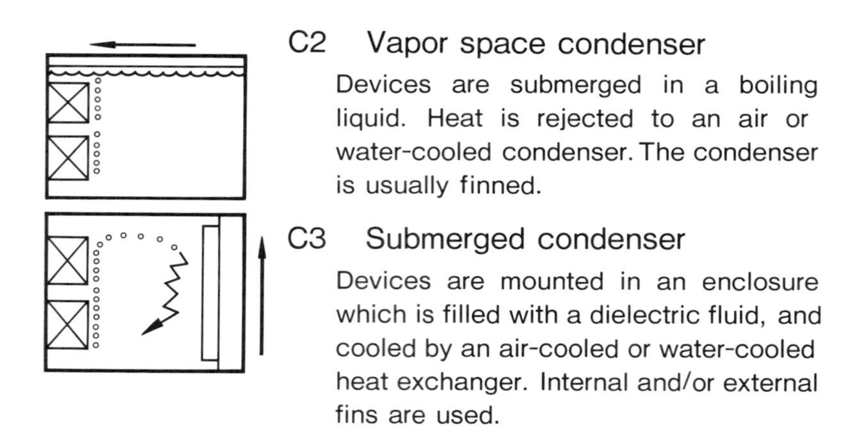

C2 Vapor space condenser
Devices are submerged in a boiling liquid. Heat is rejected to an air or water-cooled condenser. The condenser is usually finned.
C3 Submerged condenser
Devices are mounted in an enclosure which is filled with a dielectric fluid, and cooled by an air-cooled or water-cooled heat exchanger. Internal and/or external fins are used.

FIGURE 28. Condensation of vapor in enclosures [59].

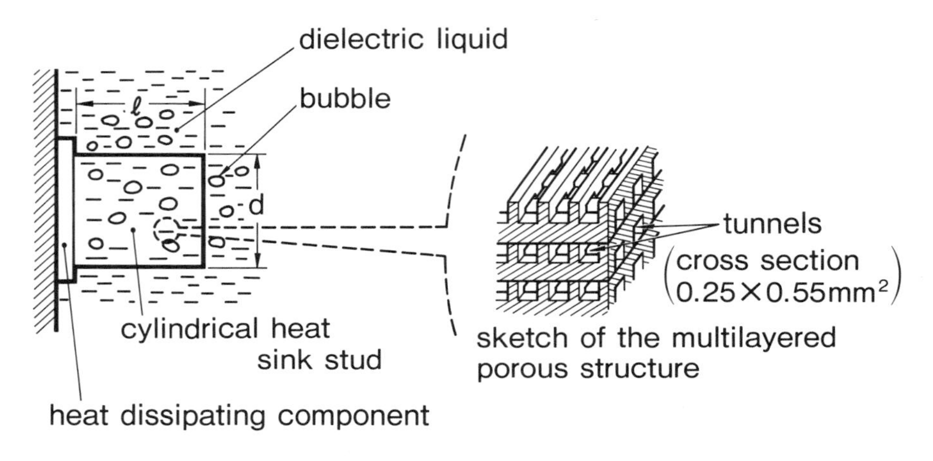

FIGURE 29. Porous stud attached to the face of a heat dissipating component [63].

red to the base surface). As shown in Fig. 4, with a stud of 1 cm in diameter and 5 mm in length, a heat load of several tens of watts is permissible if the component temperature is to be below 70 °C, and the burnout heat load is shown to be at the level of 100 W.

For future generations of microelectronic equipment, the predictability of temperature overshoot on evaporation-cooled components shall become an increasingly important issue. Success in the industrial use of evaporative cooling hinges on the development of novel approaches toward this problem. Simon [64] proposes an interesting concept, in which a mass of data is used to locate the temperature overshoot of least probability, a "safe" threshold to be observed in equipment design. Modes of evaporation other than pool boiling are also considered with expectations to eliminate temperature overshoot. Submerged jet impingement cooling studied by Ma and Bergles [65], and evaporation of falling liquid film on the board studied by Mudawwar and Incropera [66], are the examples that proved effective in reducing temperature overshoots to negligible level.

Another important problem is the prediction of the transient thermal field to be experienced by miniaturized circuits on a chip throughout the period from bubble nucleation to established boiling. Tracing the transient development of the temperature field on a chip requires a solution of conjugate problems of heat conduction and surface heat transfer, where numerical analysis is difficult because of the large number of nodal points.

Finally, a note is made on one of the important developments in the hardware technology of supercomputers. The supercomputer ETA 10 developed by ETA Inc. [67] has its processors submerged in the liquid nitrogen bath (77 K) to utilize the increase in the data processing speed in such cryogenic environments. Heat dissipation from chips is held low in this circumstance. However, before the machine starts operation, the processors are cooled down from the room temperature to a cryogenic temperature, and the management of thermal stress on structural components of the processors during the cool-down period is an important aspect of the design of this type of machine.

CONCLUDING REMARKS

None of the thermal design parameters, such as the heat flux on the component surface and the rate of heat dissipation from the equipment, can be a yardstick to describe the advances of cooling technology for electronic equipment. Heat fluxes of the order of 100 W/cm^2 were already dealt with in the design of a high-power vacuum tube in the early 1950s [35]. One of the earliest electronic computers in the 1940s, the Electrical Numerical Integrator and Computer (ENIAC), required an array of fans to remove 140 kW dissipated from the 18,000 vacuum tubes. This rate of heat dissipation is comparable to the value associated with the largest present day computer systems that execute data processing at incomparably greater speeds. Neither is the type of coolant an indicator of the evolution of cooling technology. Conversion of air-cooling to liquid-cooling with the progress of product development has not always been the rule. For instance, in Mouromtseff's account of high-power vacuum tubes [31] the coolant first employed was water, then air-cooling was examined to solve the problems of water freezing in cold climates and also to eliminate the complexity of water piping.

Glancing at the list of research topics of heat transfer, it seems that problems of similar nature have been resurfacing from time to time in phase with the development of electronics technology. In reality, however, developing configurations of electronic components and systems, and increasing stringency in thermal design criteria have always been introducing new dimensions to heat transfer problems. Those changes have been accelerating in recent years, requiring certain breakthroughs in the method of thermal management for future electronic circuits of extremely small scale. A description of these trend have been attempted in this paper, with a view on the extension of the semiconductor technologies currently prevalent in the industry. In the future, we will see the advent of new technologies such as integrated circuits composed of high temperature superconducting elements, which holds a promise of finally relieving the designer from cooling problems. However, the laboratory work on these embryonic technologies is showing the requirements for very stringent control of thermal environments for the manufacturing and operation of electronic circuits.

It should be noted that electronic equipment today is no longer confined to performing tasks for specialists, but is also making inroads into the vital control centers of society. The reliability of equipment operation is gaining unprecedented seriousness in setting the design goals of new equipment, for which heat transfer research is expected to play an ever greater role.

It is our expectation that this Symposium will significantly advance the international understanding of important issues related to cooling of electronic equipment and promote the resolution of these issues.

Nomenclature

c = specific heat capacity J/(K kg)
H = fin height m
k = thermal conductivity W/(m K)
L = side length of heat sink m
Mo = Mouromtseff number
Nu_0 = Nusselt number $=(Q/L^2)L/(k(T_b - T_0))$
Q = heat load W
q'' = heat flux W/m^2
R = thermal resistance K/W
Re_L = Reynolds number $=LV_0/\nu$
s = half of inter-fin gap m
T_b = temperature of heat sink base °C
T_0 = temperature of cooling air °C
ΔT_S = wall superheat K
ΔT_{sub} = subcooling of coolant K
V_0 = approach velocity of cooling air m/s

Greek symbols
δ = fin thickness m
ρ = mass density kg/m^3
μ = dynamic viscosity kg/(s m)
ν = kinematic viscosity m^2/s

References

1. Jones, M. R., Holton, W. C., and Stratton, R., Semiconductors: The Key to Computational Plenty, Proc. IEEE, Vol.70, No.12, pp.1380-1409, 1982.

2. Bergles, A. E., Evolution of Cooling Technology for Electrical, Electronic, and Microelectronic Equipment, Heat Transfer Engineering, Vol.7, No.3-4. pp.97-106, 1986.

3. Nakayama, W., A Survey of Design Approaches in Japanese Computers (keynote lecture), International Symposium on Cooling Technology for Electronic Equipment, Honolulu, 1987.

4. Bar-Cohen, A., Mudawwar, I., and Whalen, B., Future Challenges (Session 6), Research Needs in Electronic Cooling, in Proc. Workshop Sponsored by NSF and Purdue Univ., ed. F. P. Incropera, pp.70-77, Andover, MA, 1986.

5. Simons, R. E., Direct Liquid Immersion Cooling: Past, Present, and Future, Proc. 1987 International Symposium on Microelectronics (ISHM), Minneapolis, pp.186-197, 1987.

6. Hwang, U. P. and Moran, K. P., Cold Plates for IBM Thermal Conduction Module Electronic Modules, International Symposium on "Heat Transfer in Electronic and Microelectronic Equipment", August 29 - September 2, 1988, Dubrovnik, Yugoslavia.

7. Izutani, Y., Kuwata, M., Bando, K., Narita, Y., Hashimoto, K., Yoshimoto, R., Akagi, M., Nakamura, T., Inoue, M., Ishibashi, M., Watari, T., Matsuo, H., and Ishikawa, K., Hardware of ACOS System 1500 Series, NEC-Giho, Vol.38, No. 11, pp.9-21, 1985.

8. Watari, T. and Murano, H., Packaging Technology for the NEC SX Supercomputer, IEEE Trans., Vol.CHMT-8, No.4, pp.462-467, 1985.

9. "A Sleek, Superpowered Machine", TIME, p.45, June 17, 1985.

10. Nakayama, W., Okada, S., and Kuwahara, H., Experience in Cooling Power Electronic Devices by Phase-Change Refrigerants and the Prospect of Its Application to Microelectronic Devices, Proc. 1987 International Symposium on Microelectronics (ISHM), Minneapolis, pp.165-174, 1987.

11. Chen, D. Y., Power Semiconductors: Fast, Tough, and Compact, IEEE Spectrum, Vol. 24, No.9, pp.30-35, 1987.

12. Kilby, J. S., Miniaturized Electronic Circuits, U.S. Patent 3 138 743, June, 1964.

13. Kraus, A. D., Cooling Electronic Equipment, Englewood Cliffs, N.J., 1965.

14. Spielberger, R. K., Huang, C. D., Nunne, W. H., Mones, A. H., Fett, D. L., Hampton, F. L., Silicon-on-silicon Packaging, IEEE Trans., Vol.CHMT-7, No.2, pp.193-196, 1984.

15. Ellison, G. N., Thermal Computations for Electronic Equipment, VanNostrand Reinhold, New York, 1984.

16. Nakayama, W., Thermal Management of Electronic Equipment: A Review of Technology and Research Topics, Applied Mechanics Reviews, Vol.39, No.12, pp.1847-1868, 1986.

17. Bathe, K. J., ADINA - A Finite Element Program for Automatic Dynamic Incremental Nonlinear Analysis, Acoustic and Vibration Laboratory, Dept. Mechanical Engineering, Massachusetts Institute of Technology, Report 82448-1, 1975, 1978.

18. Desalvo, D. J. and Swanson, J. A., ANSYS Engineering Analysis System Theoretical Manual, Swanson Analysis Systems, Inc., Houston, Pennsylvania, 1983.

19. Kaye, J., Review of Industrial Applications of Heat Transfer in Electronics, Research Laboratory of Heat Transfer in Electronics Rept. No.RLHTE-12, Massachusetts Institute of Technology, Cambridge; Proc. IRE, Vol.44, pp.977-991, 1956.

20. Lyman, J., Supercomputers Demand Innovation in Packaging and Cooling, Electronics, pp.136-143, Sept. 22, 1982.

21. Chu, R. C., Hwang, U. P., and Simons, R. E., Conduction Cooling for an LSI Package: A One-Dimensional Approach, IBM J. Res. Dev., Vol.26, No.1, pp.45-54, 1982.

22. Yovanovich, M. M., Thermal Contact Resistance in Microelectronics, NEPCON Proceedings, pp.177-188, 1978.

23. Eid, J. C. and Antonetti, V. W., Thermal Contact Resistance Measurement Techniques for Microelectronic Packages, in Temperature/Fluid Measurements in Electronic Equipment, ed. J. Bartoszek, S. Furkay, S. Oktay, and R. Simons, ASME HTD-Vol.89, pp.25-30, 1987.

24. Gaugler, R. S., Heat Transfer Device, U.S. Patent 2350348, June, 1944.

25. Scott, G. W. and Tanzer, H. J., Evaluation of Heat Pipes for Conduction Cooled Level II Avionic Packages, in Heat Transfer in Electronic Equipment - 1986, ed. A. Bar-Cohen, ASME HTD-Vol.57, pp.67-75, 1986.

26. Elenbaas, W., Heat Dissipation of Parallel Plates by Free Convection, Physica, Vol.9, No.1, pp.1-28, 1942.

27. Landis, F., W. Elenbaas' Paper on "Heat Dissipation of Parallel Plates by Free Convection", in Heat Transfer in Electronic Equipment - 1986, ed. A. Bar-Cohen, ASME HTD-Vol.57, pp.11-21, 1986.

28. Elenbaas, W., The Dissipation of Heat by Free Convection-The Inner Surfaces of Tubes of Different Shapes of Cross Section, Physica, Vol.9, No.8, pp.865-874, 1942.

29. Bar-Cohen, A. and Rohsenow, W. H., Thermally Optimum Spacing of Vertical, Natural Convection Cooled, Parallel Plates, ASME J. Heat Transfer, Vol.106, pp.116-123, 1984.

30. Scott, A. W., Cooling of Electronic Equipment, pp.44-63, John Wiley & Sons, Inc., 1974.

31. Mouromtseff, I. E., Water and Forced-Air Cooling of Vacuum Tubes, Proc. IRE, Vol. 30, pp.190-205, 1942.

32. McAdams, W. H., Heat Transmission, McGraw-Hill, New York, 1933.

33. Kraus, A. D., An Appraisal of Gardner's Pioneering Extended Surface Effort, in Heat Transfer in Electronic Equipment, ed. A. Bar-Cohen, ASME HTD-Vol.57, pp.35-39, 1986.

34. Gardner, K. A., Efficiency of Extended Surfaces, Trans. ASME, Vol.67, pp.621-631, 1945.

35. London, A. L., Air-Coolers for High Power Vacuum Tubes, Trans. IRE, Professional Group on Electron Devices, Vol.ED-1, pp.9-26, 1954.

36. Kays, W. M. and London, A. L., Compact Heat Exchangers, National Press, Palo Alto, Calif., 1955.

37. Kobayashi, F., Murata, S., Watanabe, H., Kawashima, S., Anzai, A., Murakami, K., and Ikuzaki, K., Hardware Technology for M-680/682H, Nikkei Electronics, pp.268-288, 1985.

38. Sparrow, E. M., Baliga, B. R., and Patankar, S. V., Forced Convection Heat Transfer From a Shrouded Fin Array With and Without Tip Clearance, ASME J. Heat Transfer, Vol.100, pp.572-579, 1978.

39. Sparrow, E. M. and Kadle, D. S., Effect of Tip-to Shroud Clearance on Turbulent Heat Transfer from a Shrouded, Longitudinal Fin Array, ASME J. Heat Transfer, Vol.108, pp.519-524, 1986.

40. Arvizu, D. C. and Moffat, R. J., The Use of Superposition in Calculating Cooling Requirements for Circuit Board Mounted Electronic Components, Proc. Electron Components Conf., Vol.32, pp.133-144, 1982.

41. Wirtz, R. A. and Dykshoorn, P., Heat Transfer from Arrays of Flat Packs in a Channel Flow, Proc. 4th Int. Electronic Packaging Conf., pp.318-326, 1984.

42. Nakayama, W., Matsushima, H., and Goel, P., Forced Convective Heat Transfer from Arrays of Finned Packages, Cooling Technology for Electronic Equipment, ed. W. Aung, pp.195-210, Hemisphere Publishing Corporation, New York, N.Y., 1988.

43. Biskeborn, R. G., Horvath, J. L., and Hultmark, E. B., Integral Cap Heat Sink Assembly for the IBM 4381 Processor, Proc. 1984 Int. Electronic Packaging Soc. Conf., pp.468-474, 1984.

44. Karniadakis, G. E., Mikic, B. B., and Patera, A. T., Heat Transfer Enhancement by Flow Destabilization; Application To The Cooling of Chips, Proc. Int. Symposium on Cooling Technology for Electronic Equipment, Honolulu, pp.498-521, 1987.

45. Mouromtseff, I. E. and Kozanowski, H. N., Comparative Analysis of Water-Cooled Tubes as Class B Audio Amplifiers, Proc. IRE, Vol.23, pp.1224-1251, 1935.

46. Chu, R. C., Seely, J. H., Antonetti, V. W., and Pascuzzo, A. L., Thermal Design Optimization in Large Digital Computers, TR 00.2039, IBM Corporation Poughkeepse, N. Y., Presented at the 1970 IEEE Computer Group Conf., 1970.

47. Danielson, R. D., Krajewski, N., and Brost, J., Cooling a Superfast Computer, Electronic Packaging & Production, pp.44-45, July, 1986.

48. Ramadhyani, S. and Incropera, F. P., Forced Convection Cooling of Discreet Heat Sources With and Without Surface Enhancement, Proc. Int. Symposium on Cooling Technology for Electronic Equipment, Honolulu, pp.249-264, 1987.

49. Tuckerman, D. B. and Pease, R. F., Ultrahigh Thermal Conductance Microstructures for Cooling Integrated Circuits, IEEE 32nd Electronics Components Conf. Proc., pp.145-149, 1982.

50. Baker, E., Liquid Immersion Cooling of Small Electronic Devices, Microelectronics and Reliability, Vol.12, pp.163-173, 1973.

51. Park, K.-A. and Bergles, A. E., Natural Convection Heat Transfer Characteristics of Simulated Microelectronic Chips, ASME J. Heat Transfer, Vol.109, pp.90-96, 1987.

52. Kraus, A. D. and Bar-Cohen, A., Thermal Analysis and Control of Electronic Equipment, p.377, Hemisphere Publishing Corporation, New York, N.Y., 1983.

53. Kaye, J. and Choi, H. Y., General Aspects of Cooling Airborne Electronic Equipment, IRE Trans. Aernaut. Navig. Electron., Vol.ANE-5, No.1, pp.1-9, 1958.

54. Armstrong, R. J., The Temperature Difference in Nucleate Boiling, Int. J. Heat Mass Transfer, Vol.9, pp.1148-1149, 1966.

55. Rohsenow, W. M., A Method of Correlating Heat Transfer Data for Surface Boiling of Liquids, Trans. ASME, pp.969-976, (pp.51-58 of same volume), 1953.

56. Yamada, Y., Itahana, H., and Okada, S., Evaporation Cooling System for Chopper Controller, Hitachi Review, Vol.29, No.1, pp.25-30, 1980.

57. Aakalu, N. G., Chu, R. C., and Simons, R. E., Liquid Encapsulated Air-Cooled Module, U.S. Patent 3741292, 1973.

58. Danielson, R. D., Tousignant, L., and Bar-Cohen A., Saturated Pool Boiling Characteristics of Commercially Available Perfluorinated Inert Liquids, Proc. 1987 ASME/JSME Thermal Engng. Joint Conf., Vol.3, pp.419-430, 1987.

59. Bravo, H. V., Limits of Heat Transfer in Liquid-Filled Enclosures, M. S. Thesis, Iowa State University, Ames, 1975.

60. Bravo, H. V. and Bergles, A. E., Limits of Boiling Heat Transfer in a Liquid-Filled Enclosure, Proc. 1976 Heat Transfer Fluid Mechanics Institute, pp.114-127, Stanford University Press, 1976.

61. Bar-Cohen, A., Thermal Design of Immersion Cooling Modules for Electronic Components, Heat Transfer Engineering, Vol.4, pp.35-50, 1983.

62. Oktay, S., Departure from Natural Convection (DNC) in Low-Temperature Boiling Heat Transfer Encountered in Cooling Microelectronic LSI Devices, Heat Transfer - 1982, Hemisphere Publishing Corporation, New York, Vol.4, pp.113-118, 1982.

63. Nakayama, W., Nakajima, T., and Hirasawa, S., Heat Sink Studs Having Enhanced Boiling Surfaces for Cooling of Microelectronic Components, ASME Paper No.84-WA/HT-89, 1984.

64. Simon, T., University of Minnesota, private communication, October 1, 1987.

65. Ma, C.-F. and Bergles, A. E., Boiling Jet Impingement Cooling of Simulated Microelectronic Chips, in Heat Transfer in Electronic Equipment, ASME HTD-Vol.28, pp.5-12, 1983.

66. Mudawwar, I. A., Incropera, T. A., and Incropera, F. P., Microelectronic Cooling by Fluorocarbon Films, Proc. Int. Symp. Cooling Technology for Electronic Equipment, Honolulu, pp.340-357, 1987.

67. Purcell, C. J., An Introduction to the ETA10, in Supercomputers and Fluid Dynamics, ed. K. Kuwahara, R. Mendez, and S. A. Orszag, pp.184-200, (Lecture Notes in Engineering 24), Springer-Verlag, 1986.

Evolution of Cooling Technology in Medium and Large Scale Computers—An IBM Perspective

R. C. CHU and R. E. SIMONS
Poughkeepsie Laboratory
Data Systems Division
IBM Corporation
Poughkeepsie, New York, USA

ABSTRACT

This paper provides a chronological review of the evolution of cooling technology used in IBM medium and large scale mainframe computers. Package cooling technology and its evolution leading to the thermal conduction module (TCM) is described. Air cooling system technology is discussed along with its enhancements in the form of impingement and air-liquid hybrid cooling. Water-cooled systems are discussed in terms of cold plate and cooling distribution unit development. The paper concludes with a brief description of an application of direct liquid immersion cooling within IBM today.

INTRODUCTION

Electronic computers, from desktop personal computers to large scale mainframes, are rapidly permeating virtually every aspect of modern life. The applications of these machines vary from games for the entertainment of children (and sometimes adults) to highly sophisticated programs supporting vital health, economic, scientific, and defense activities. In some of these applications failure of a computer results only in minor inconvenience. In a growing number of applications, however, a computer failure can result in a major disruption of vital services, and could even have life threatening consequences. As a result, efforts to improve the reliability of electronic computers have become as important as efforts to improve their speed and storage capacity.

In a paper describing the evolution of cooling technology for electrical, electronic and microelectronic equipment, Bergles [1] termed the development of the modern digital computer as "the major event of the 1940's." Since the development of these first electronic digital computers, the effective removal of heat has played a key role in insuring the reliable operation of successive generations of computers. The Electrical Numerical Integrator and Computer (ENIAC) which was dedicated in 1946, has been described as a "30 ton, boxcar-sized machine requiring an array of industrial cooling fans to remove the 140 KW dissipated from its 18,000 vacuum tubes" [1].

As with ENIAC, all early IBM computers up to 1957 used vacuum-tube electronics [2] and were cooled with forced air. On January 27, 1948, T.J. Watson, Sr. dedicated the Selective Sequence Electronic Calculator (SSEC) "to the use of science throughout the world" [3]. Designed, built, and placed in operation in only two years, the SSEC contained 21,400 relays and 12,500 vacuum tubes. Also introduced in 1948, the IBM 604 contained more than 1400 vacuum tubes, and was built using "pluggable units" consisting of a vacuum tube and its closely associated resistors and capacitors packaged together in a removable assembly. This "unitized" packaging concept greatly increased the packing density of electronic equipment, and with appropriate evolutionary changes, was adopted for all subsequent electronic calculators and computers built by IBM. The first "true" large-scale electronic computer that IBM put into production was the 701. Its success, along with that of the less powerful 650 and the large-scale 702, led to the development of the 704, 705, and 709.

The invention of the transistor, by Bardeen, Brattain, and Shockley at Bell Laboratories in 1947 [4], foreshadowed the development of generations of computers yet to come. As a replacement for vacuum tubes, the miniature transistor generated less heat, was much more reliable, and promised lower production costs. The first IBM product to use transistor circuits without any vacuum tubes was the 608 calculator, first shipped in 1957. The 608 contained more than 2100 transistors on 600 five inch printed circuit cards. For a while it was thought that the use of transistors would greatly reduce if not totally eliminate cooling concerns. Such thoughts were short-lived as engineers sought to improve computer speed and storage capacity by packaging more and more transistors on printed circuit boards, and then ceramic substrates.

PACKAGE COOLING TECHNOLOGY

The hierarchy of electronic packages within IBM mainframe computers has been typified by three levels [5]. The first level package (or module) houses the integrated circuit chip(s) providing the proper mechanical, thermal, and electrical environment while interconnecting chip terminals and providing pins to interface to the next level package [6]. The second level package has typically been a printed circuit card providing the means to mount, power, and interconnect the chip carriers. The next level of packaging has usually been printed circuit boards on which many cards may be mounted. IBM mainframes in the past have typically contained many stacks of card-carrying printed circuit boards. These three packaging levels are closely paralleled by three levels of thermal resistance termed component, package, and system [7]. The methods which IBM engineers have used, to control or minimize these resistances and enhance heat transfer, will be subsequently discussed.

The first IBM chip carriers were small single-device carriers, mounted to the next level card package, as were the resistors, capacitors, and other discrete components. The card was 2.5 inches by 4.5 inches with printed lines on the back of the card. This packaging technology developed in the 1950s was called the Standard Modular System (SMS) [8]. In the early 1960s, IBM engineers developed the Solid Logic Technology (SLT) module which contained all the components and interconnections for a circuit on a one-half inch ceramic module [9]. By the time Monolithic System Technology (MST) was introduced in 1969, many circuits could be placed on a single integrated circuit chip.

During the 1960s - 1970s IBM mainframes were built using the single chip module (SCM) package geometry with its thermal paths shown in Figure 1. Heat generated within the chip flowed by conduction through the interconnecting

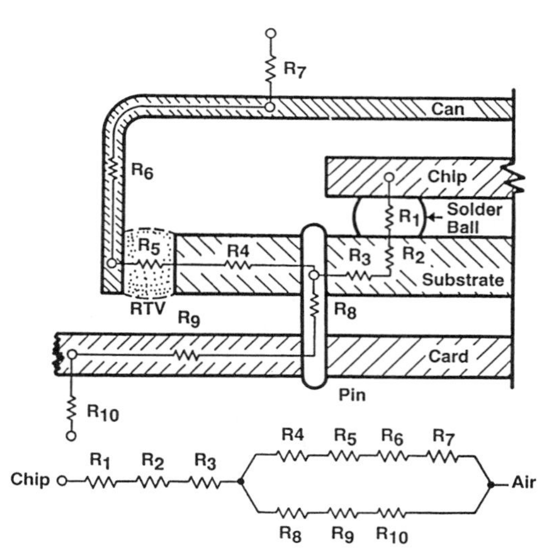

FIGURE 1. Cutaway view and thermal paths for single chip module on printed circuit card.

solder pads to the ceramic substrate, and then to the can, and finally to the printed circuit card. It may be noted that although the rest of the industry used back-down mounted chips, IBM developed and implemented the chip-substrate solder joining technology called C-4, for Controlled Collapse Chip Connections [10]. Although, this technology resulted in a higher chip substrate thermal resistance, this thermal disadvantage was more than offset by the manufacturing advantages it offered. Heat was removed from the SCM(s) and card by forced convection with air. Thermal engineers often had little influence on the early design of these single chip modules, and thermal control was provided after the fact at the card/board and system level through design of the air-cooling system.

By the early 1970s it was already apparent that to achieve the circuit/chip packaging densities for machines yet to come, multi-chip modules (MCM) with many chips on a single substrate would have to be developed. It was clear that thermal design considerations would play a pivotal role in the development of these new modules. Early thermal studies indicated that the chip to substrate to can/card thermal paths would be adequate for small MCMs with 4-9 chips. As the proposed substrate size and chip count increased, however, the C-4 thermal resistance (chip-substrate) and the thermal conduction spreading resistance (in the substrate) could no longer be tolerated. Some means would have to be developed to remove heat directly from the back of the chip. Based upon cooling studies conducted during the late 1960s direct liquid immersion was considered as a possible means of providing cooling off the back of the chip [11].

Liquid Encapsulated Module

Although earlier direct liquid immersion cooling efforts concentrated on circulating-liquid schemes, with a fluorocarbon coolant being pumped through modules [12,13], a module cooling design evolved using a self-contained liquid-encapsulated module (LEM) [14]. As shown in Figure 2, the substrate carrying the integrated circuit chips was mounted within a sealed module-cooling assembly containing a fluorocarbon coolant. Boiling at the exposed chip surfaces resulted

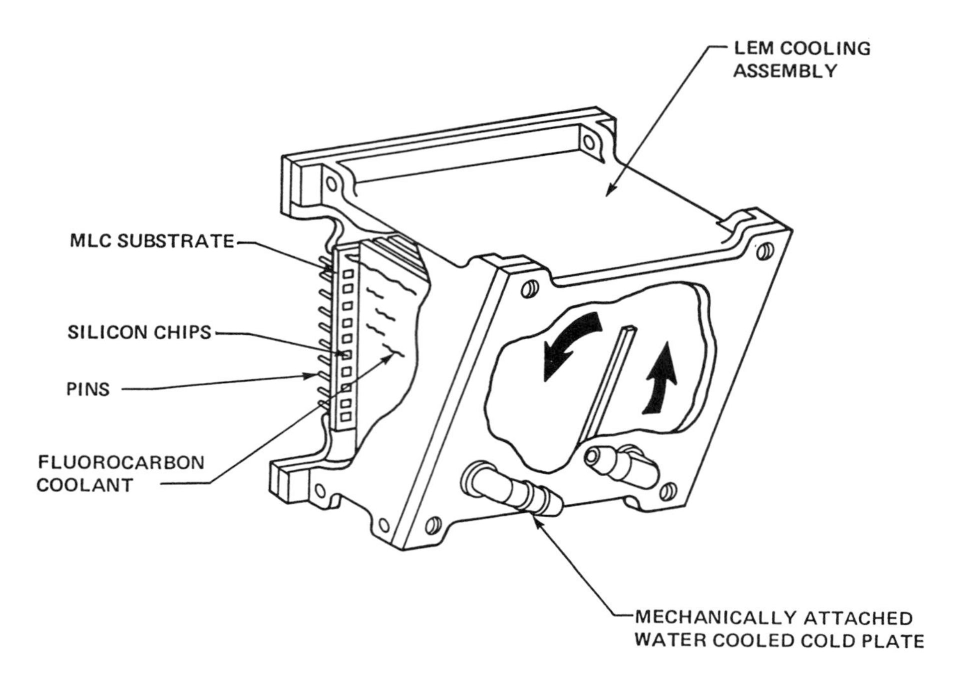

FIGURE 2. Liquid encapsulated module (LEM) with water-cooled cold plate.

in high heat transfer coefficients ($0.17 - 0.57\ W/cm^2K$) with which to meet the chip cooling requirements. Heat removed from the chips was transferred from the fluorocarbon coolant to internal fins, and then to water flowing through an externally attached cold plate.

Although this technique appeared to be capable of meeting the requirement to cool a 4 W chip within a 300 W module, cooling related concerns had arisen by the mid-1970s. It was essential that the liquid be pure and that all residues from chip and module joining processes be removed, since these contaminants could be dissolved and redeposited at the C-4 pads as part of the boiling process. It was feared that the end result could be corrosion and failure of the interconnecting pads. In addition, single chip boiling experiments showed significant variability in the initiation of chip boiling. In some instances, significant superheating or temperature overshoot of the chip occurred before boiling began. In other instances, no boiling occurred and the chip was cooled by natural convection, but at unacceptably high chip temperatures. It was in the light of these concerns, that efforts were put in place to develop a viable cooling alternative.

Thermal Conduction Module

Conceptually, it was desired to bring the water-carrying cold plate surface as close "thermally" to the chip heat sources as possible. At the same time, it was necessary to allow for variations in chip heights and locations, resulting from the manufacturing process. Allowances also had to be made for nonuniform thermal expansion or contraction across whatever thermal path was provided. The concept was thus conceived of a spring-loaded mechanical piston touching the chip to provide a thermal path from chip to case, with point contact and minute air gaps between the chip and piston and between the piston and module housing. To minimize thermal resistance across the gaps and achieve the desired internal thermal resistance, air within the module was replaced with helium gas. The total module-cooling assembly, patented as the gas-encapsulated module [15] and later named the thermal conduction module (TCM), provided cooling for 100 or more chips mounted on 90 mm substrates [16]. The successful implementation of this package cooling technology made it possible and practical to support chip level heat fluxes of 20 W/cm^2 and greater.

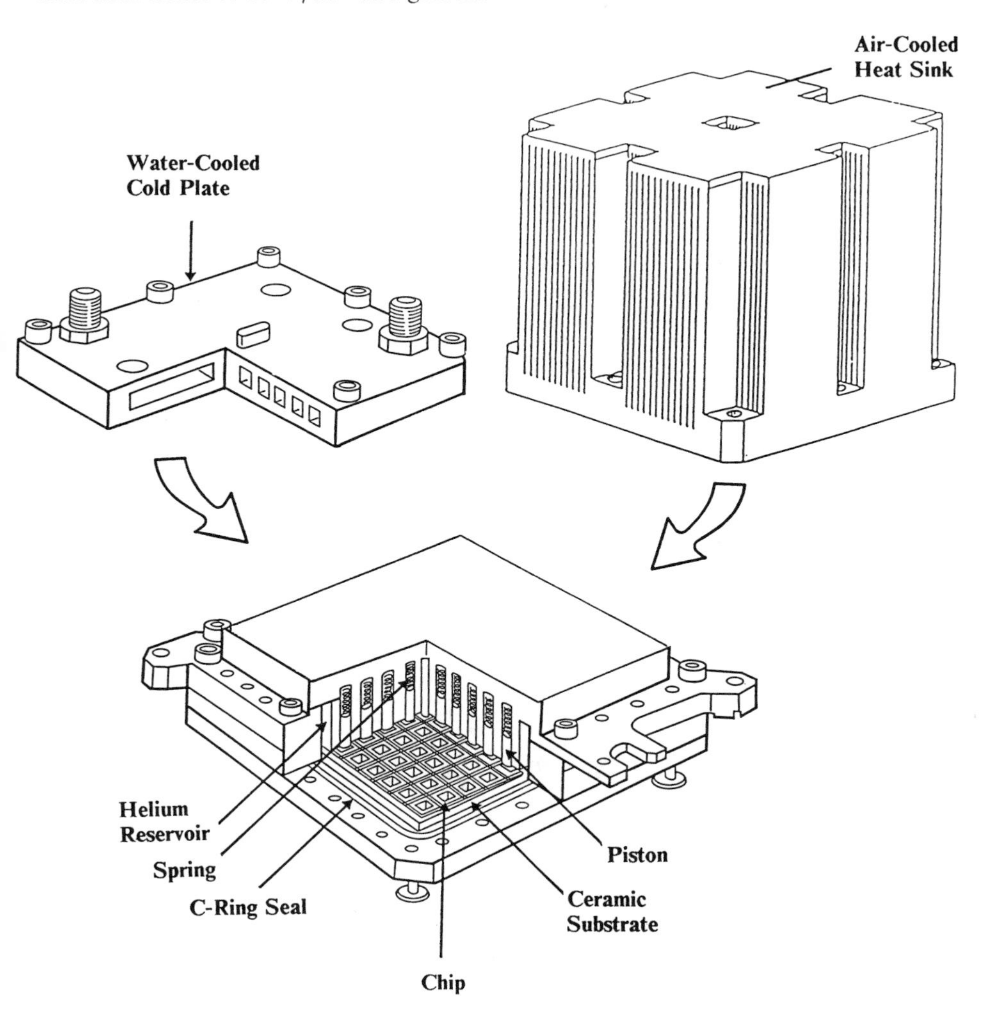

FIGURE 3. Cutaway view of thermal conduction module (TCM) with water-cooled cold plate or air-cooled heat sink.

Although originally designed to use water-cooled cold plates, as shown in Figure 3, depending upon module power the TCM could also be adapted to use air-cooled heat sinks. All the processor logic, cache, memory, and control storage for the recent IBM 9370 Model 90 is contained in a single air-cooled TCM [17].

4381 Module

In a parallel effort a somewhat smaller multi-chip module was developed to meet the needs of the mid-range IBM 4381 processor then under development [18]. This module was to utilize a 64 mm square substrate with as many 36 chips mounted on it. A direct thermal path from the back of the chip to the ceramic cap was provided using a unique thermal paste developed at IBM [19]. This paste has a thermal conductivity of 1.25 W/m-K, is electrically nonconductive, and is easily removable for rework if necessary. In order to remove the module heat load from the cap a novel heat sink was designed to be cooled by an impinging air flow [20]. As shown in Figure 4, this heat sink provides multiple segmented fin structures attached to a ceramic cap. These structures allow air to readily flow through the heat sink, and provide sufficient compliance to take up the thermal mismatch between the metal heat sink and the ceramic cap.

As shown in Figure 5, the development of the package technologies described here led to significant increases in circuit packaging densities at all levels [21]. The most dramatic increase in circuits per module was achieved with the TCM. For example, with chips up to about 700 circuits each, each TCM used in the IBM 3081 contained the equivalent of all the logic gates of the earlier IBM 370/145 Processor [22]. As shown in Figure 6, the combined result of increasing chip circuit density and substrate chip density, has been significant increases in module heat fluxes [23]. In addition to the internal package thermal improvements already discussed, it was also necessary to provide external cooling improvements in order to adequately accommodate these heat fluxes. The remainder of this paper will discuss the improvements that were made over the years in IBM air and water-cooled systems.

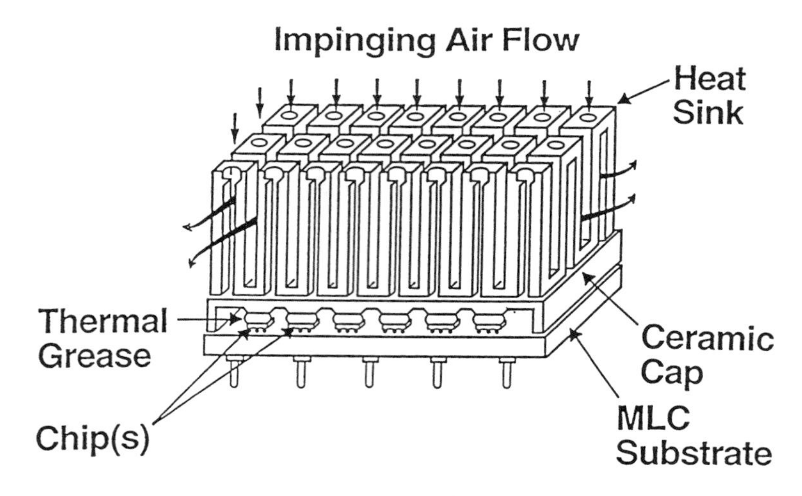

FIGURE 4. Impingement air-cooled multi-chip module used in IBM 4381 Processor.

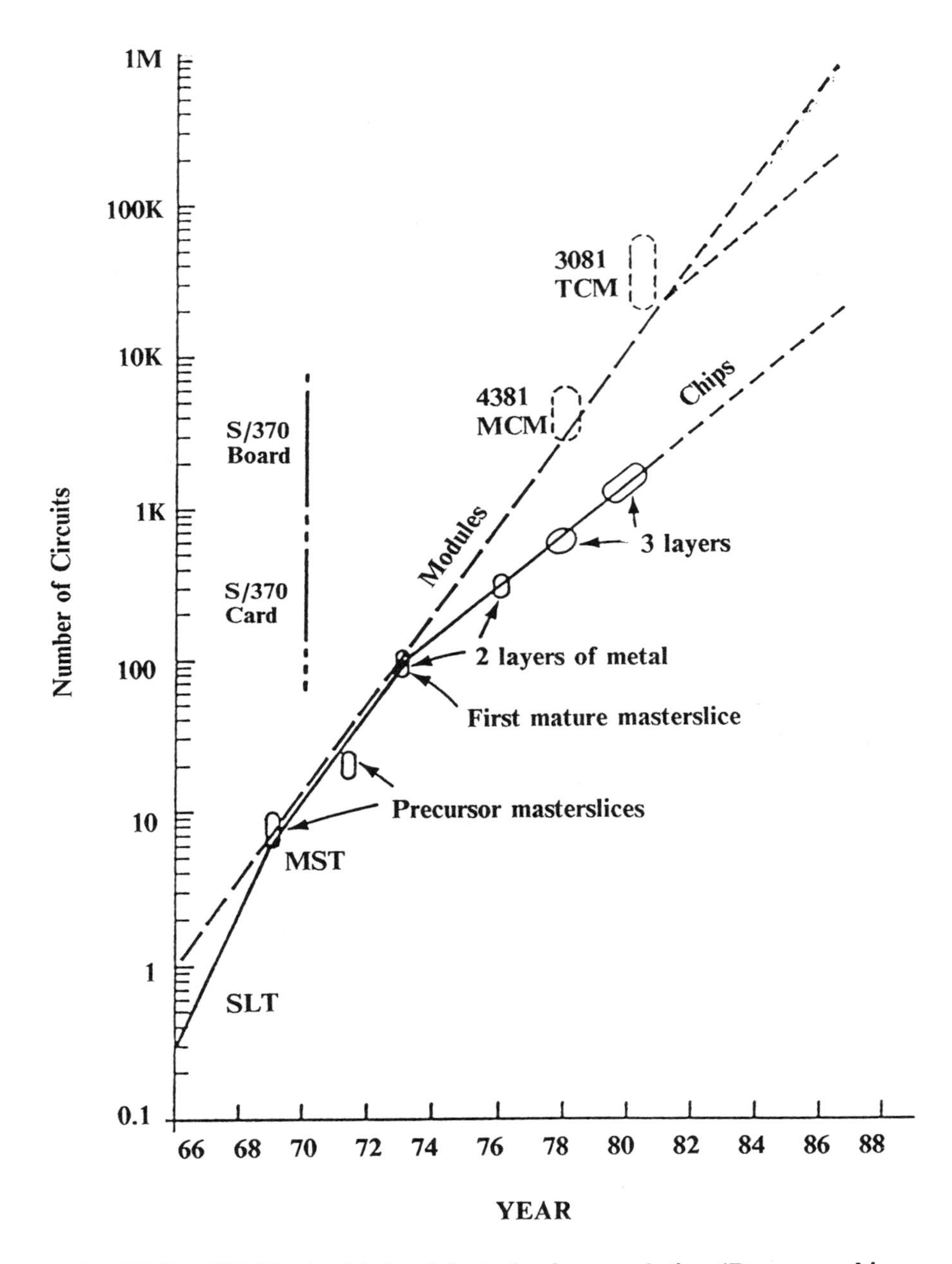

FIGURE 5. IBM logic chip/module technology evolution (Rymaszewski et al., 1981).

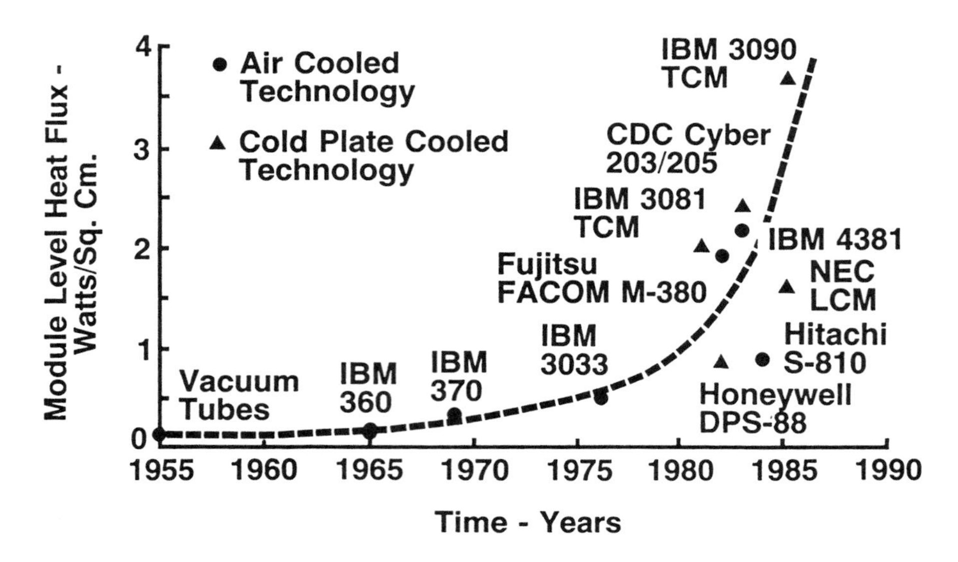

FIGURE 6. The trend in heat flux at the module level.

AIR-COOLED SYSTEMS

As discussed earlier, forced convection air cooling has been used in IBM since the earliest vacuum tube machines. The principal advantages of this approach have been, of course, the ready availability of air and its ease of application. Even today, a number of air-cooled boards are used in the 308X/3090 families of machines [24].

Typically, IBM air-cooling systems have been serial flow systems with the same cooling airstream passing over a number of printed circuit cards or boards stacked one above the other. Although in many cases blowers at either the bottom or top of a column of boards provided adequate cooling air flow, in some cases more flow was required. In these cases, a push-pull air cooling configuration with blowers at both the top and bottom was used. The push-pull configuration not only provided more air flow, it also allowed open-door time for servicing, without a discontinuity of air flow [25].

In some instances, module external thermal resistance was reduced through the use of boundary layer turbulators [26]. As shown in Figure 7, turbulator strips have been placed at appropriate locations on printed circuit cards to trip the flow and break up the boundary layer. For a fixed air flow rate, the heat transfer coefficient on a module behind a turbulator can be as much as 42% higher than when no turbulator is present [27]. Of course, the added pressure drop due to the turbulators will cause the blower to shift its operating point, and deliver less air flow. Taking this into account the actual improvement realized in heat transfer coefficient is around 15% [28]. Nonetheless, turbulator strips have been used in practice to achieve modest reductions in chip temperature.

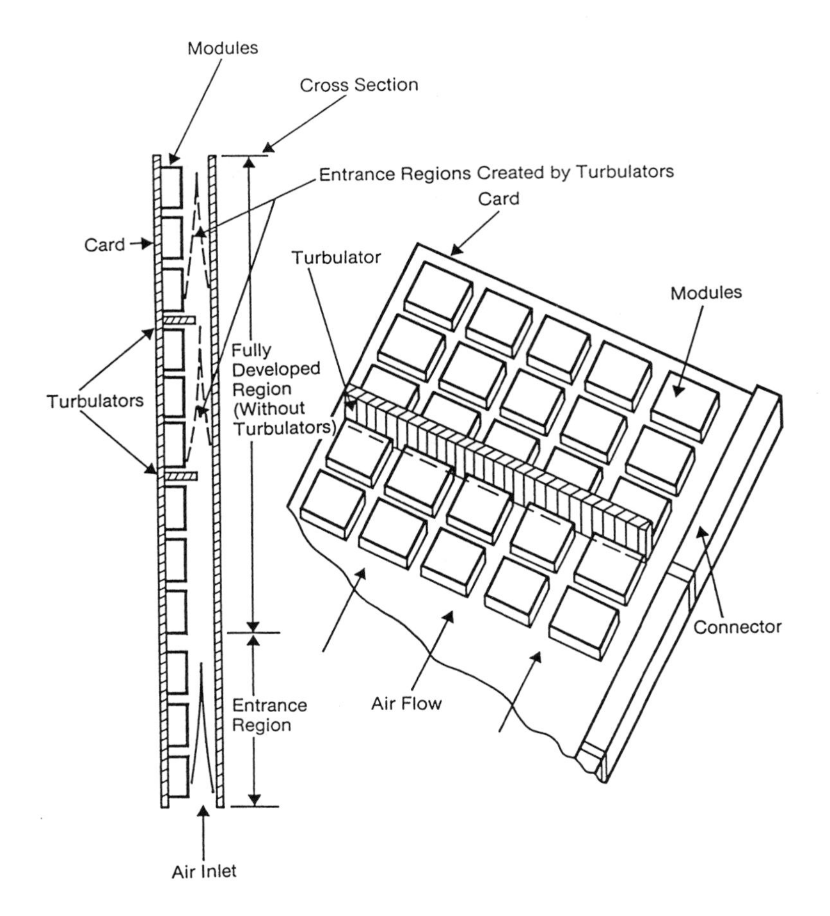

FIGURE 7. Air-cooled modules on printed circuit card with turbulators for boundary-layer control.

Impingement Cooling

As noted earlier, the multi-chip modules used in the IBM 4381 (announced in 1983) were designed for air-impingement cooling [29]. As shown in Figure 8, air is delivered to a plenum containing flow nozzles to supply a cooling air jet to each module in a 5 x 5 array on the 4381 logic board. In addition to the high heat transfer coefficients attained with the impinging flow, each module receives its own supply of "fresh" cooling air. The blower used for this application is an IBM developed blower providing a volumetric flow rate of 490 CFM [30], and air velocities of 2000 fpm impinging on the heat sink. In addition to meeting the air flow requirement, the blower underwent an extensive acoustic design to meet stringent acoustic specifications. As a result, the 4381 system was 3 db quieter than the earlier 370/145 system (announced in 1970), even though the 4381 volumetric air flow rate is over 60% higher [31].

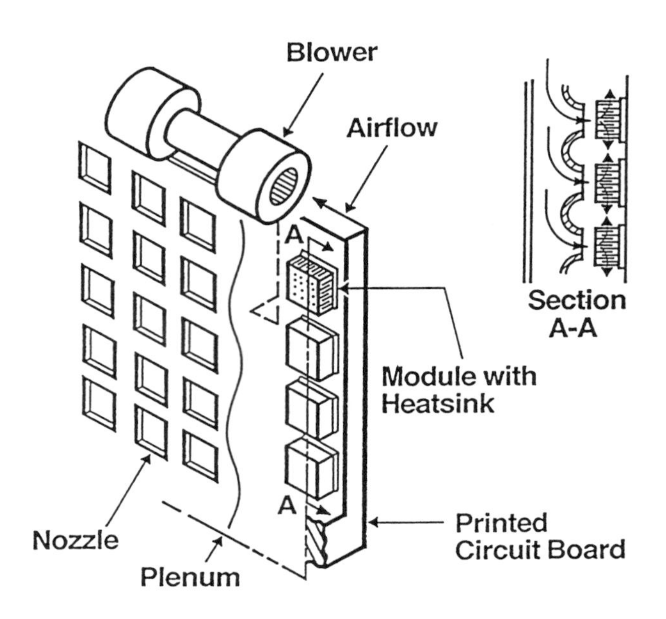

FIGURE 8. Impingement air flow cooling scheme used in the IBM 4381 Processor.

Hybrid Cooling

Large mainframe computers preceding the System/360 Model 91 (announced in 1964) were cooled by air moving devices that took in room air and provided a forced serial flow of air over each column of boards. In the case of the System/360 Model 91 CPU, the power density was increased by a factor of three [32]. Under these conditions, a substantial temperature rise would occur in the cooling air from inlet to outlet. Increases in cooling air temperature would result in corresponding increases in chip operating temperatures. Since it was not practical to provide each board its own supply of "fresh" cooling air, some other means of controlling air temperature was required. The problem was solved through the development of a hybrid cooling scheme. As shown in Figure 9 this scheme incorporated a water-cooled finned tube heat exchanger between each successive row of boards [33]. With this scheme, heat continues to be removed from the modules on cards by the flowing air. The hot air exiting a board, however, passes through the air-liquid heat exchanger before arriving at the next board. As a result, much of the heat put into the air passing over a board is taken back out by the heat exchanger. As shown in Figure 9, the temperature rise in the cooling air is reduced throughout the board column. Analysis of the circuit device temperatures in an electronics frame utilizing air-liquid hybrid cooling, has shown reductions in the maximum device temperature, mean device temperature, and the variation from minimum to maximum device temperature [34].

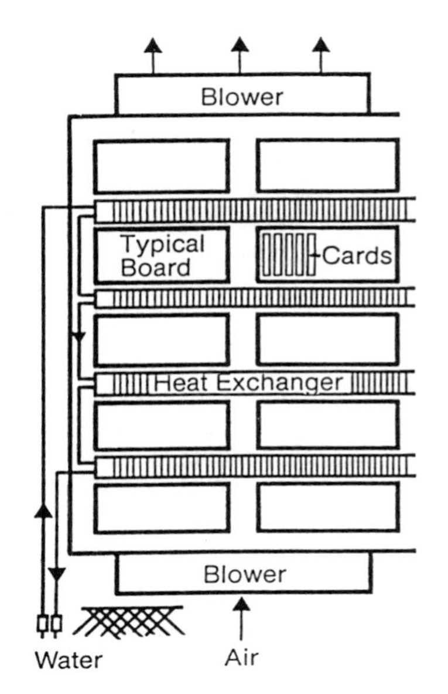

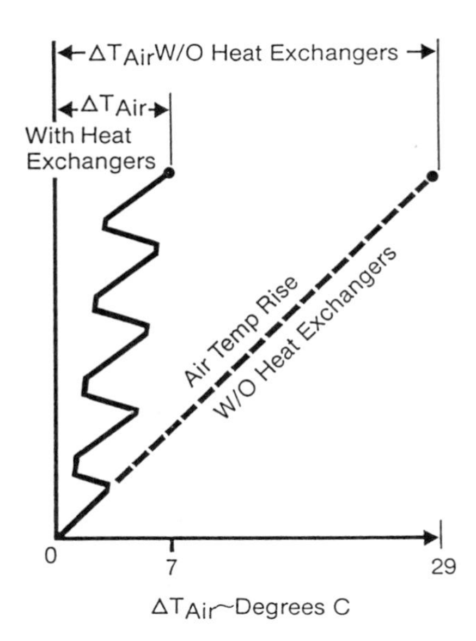

FIGURE 9. Air-liquid hybrid cooling scheme for control of air temperature rise across an electronics frame.

The use of air-liquid hybrid cooling offered another benefit as well. Approximately 50% of the heat generated within the frame is transferred to the cooling water flowing through the heat exchangers, thereby reducing the heat load to customer room air [35].

In addition to the interboard heat exchangers, the introduction of water as part of the hybrid cooling system required additional cooling hardware in the form of hoses and piping required to distribute the water. Some means also had to be provided to circulate and control the temperature of the cooling water. As described in the next section, the coolant distribution unit (CDU) was designed to perform this function, and act as a buffer between the system water and customer water. Although it was not realized at the time, the introduction of water into IBM mainframes in the 1960s, provided the system cooling technology base upon which the water-cooled systems of the 1980s would be built.

WATER-COOLED SYSTEMS

As dramatic as the improvements in air cooling have been over the years, air cooling could not support the packaging densities and associated module heat flux levels attained in the 308X and 3090 series of processors. The concept of "a liquid cooling system of modular construction for cooling modular electronic components . . . . wherein the modular electronic components may be serviced or removed without affecting the cooling system" [36], was implemented to accomplish this.

Cold Plate

As shown in Figure 2, the cold plate developed to provide external cooling for TCMs is bolted to the module top surface. This allows TCMs to be removed without disrupting water distribution lines. The thermal interface resistance between the cold plate and module surface is controlled by a combination of plating and surface finish. Passages are provided in the cold plate to guide and control the water flow through the cold plate. In addition, the side walls of these passages serve as "fins" to increase the convective surface area within the cold plate. As designed for use in the 3081 Processors, the thermal resistance of the cold plate from the surface of the TCM housing to the water within the cold plate was 0.020 °C/W, providing an equivalent overall surface heat transfer coefficient of 4470 W/cm^2K [37]. The IBM TCM cold plate accommodates the highest reported module heat fluxes in the industry at 2.2 and 3.8 W/cm^2 in the IBM 3081 and 3090 Processors [38]. Although attaining the desired cooling performance was a key objective in the design of the cold plate, equal emphasis was placed on providing a design that would be rugged, highly reliable, and manufacturable in the required quantities.

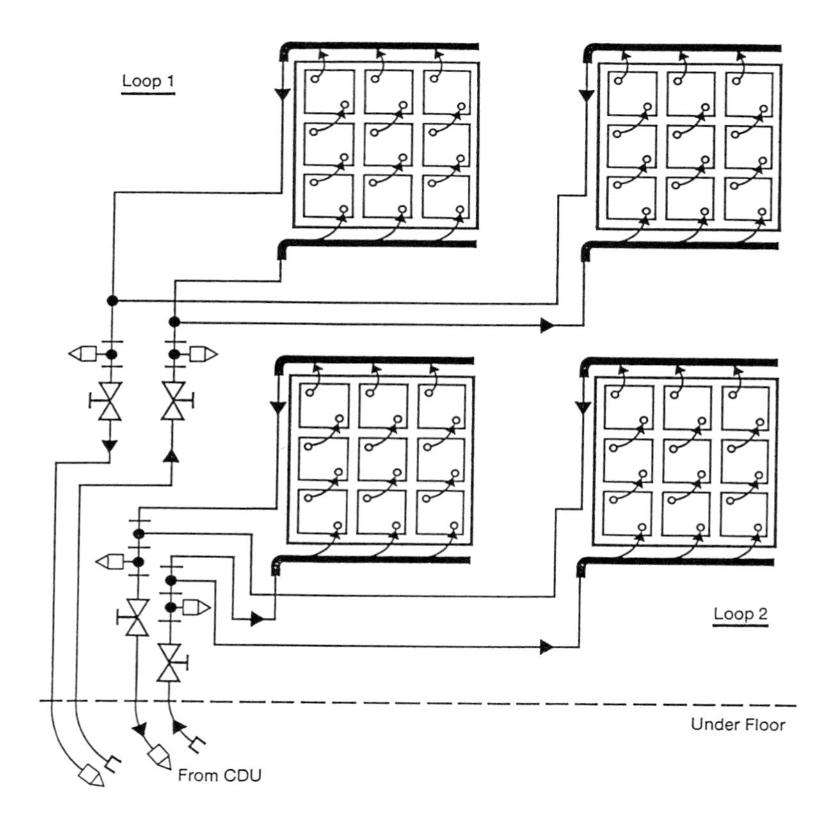

FIGURE 10. TCM water cooling distribution loops within IBM 3081 processor/storage frame.

It is interesting to note that the development of the cold plate and the associated water distribution system was actually begun to support cooling of the LEMs. When the module technology changed from LEMs to TCMs the board level packaging and system cooling remained the same. TCMs within the 3081 processor/storage frame, were mounted on boards capable of holding 6 or 9 TCMs. As shown in Figure 10, a combination of parallel-series water loops was used to minimize the pressure drop and water temperature rise across the cold plates, while minimizing the total flow required [39].

Coolant Distribution Unit

As noted previously, water cooling in IBM mainframes was first introduced in air-liquid hybrid systems. It was realized at the time that water quality, temperature, and flow rates available at the customer's facility could be subject to considerable variation from customer to customer. In order to insure the integrity, performance, and reliability of the cooling system it was decided not to run customer water directly through the water-cooling components in the electronic frames. Instead, it was decided to develop a coolant distribution unit (CDU) to provide cooling water to the mainframes as shown in Figure 11 [40].

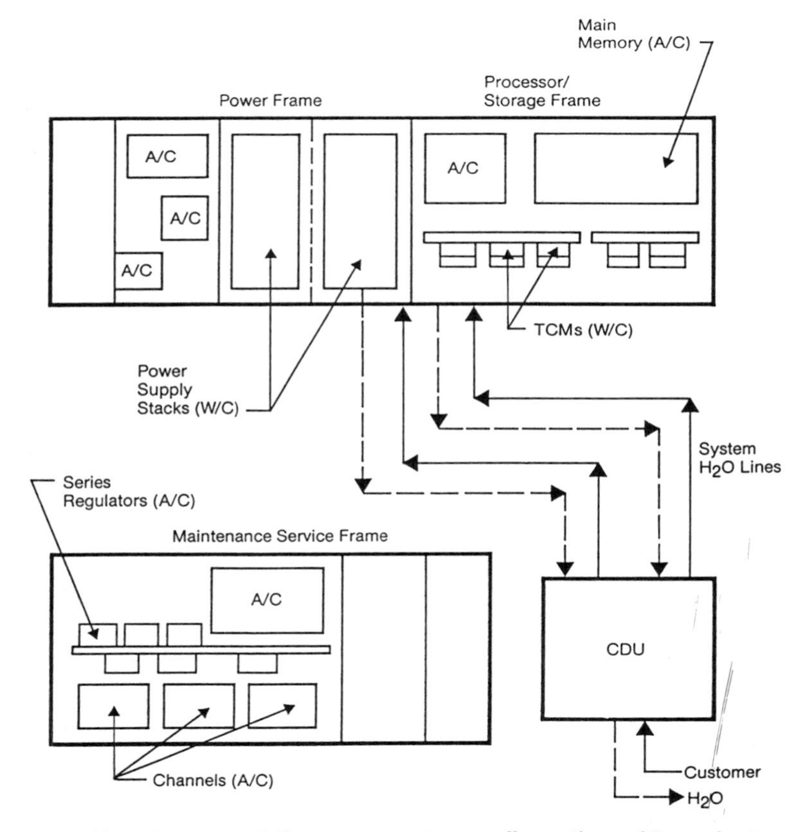

FIGURE 11. Large mainframe computer configuration with coolant distribution unit (CDU).

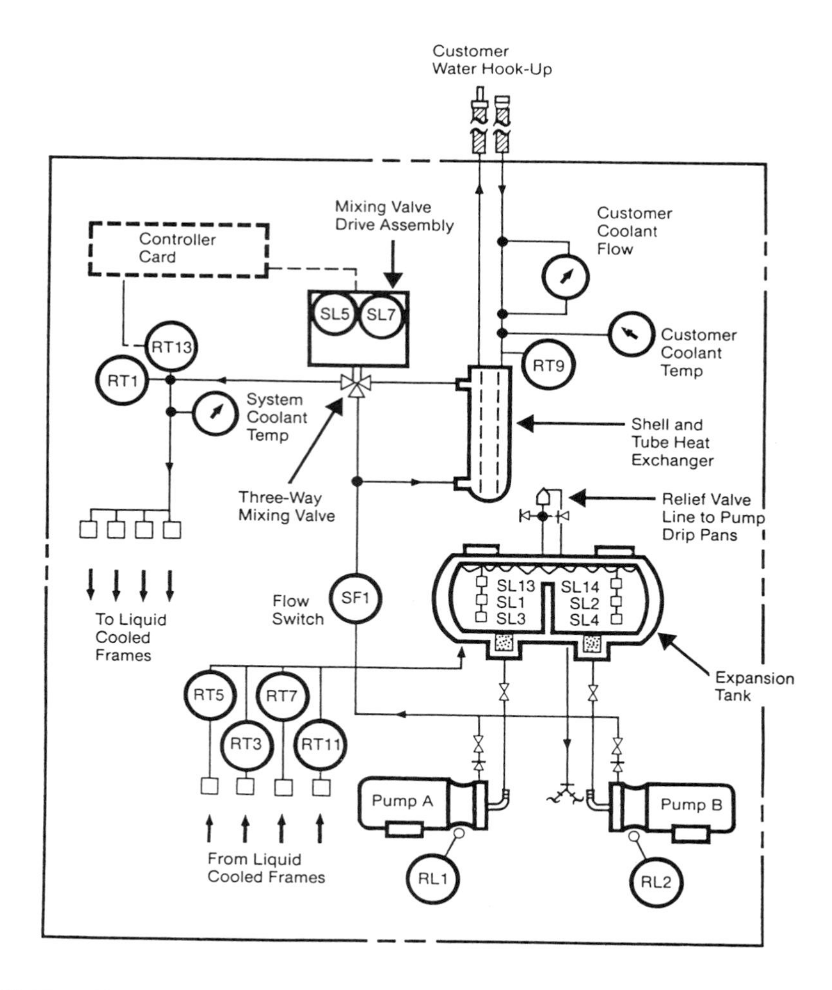

FIGURE 12. Cooling schematic of a Coolant Distribution Unit (CDU).

Although the detailed design of the CDU has changed to meet the specific requirements of each new generation of IBM mainframe, from the 360 Model 91 to the 3090 Model 600E, the basic cooling architecture has changed little. As shown in Figure 12, the basic flow and heat transfer components within the CDU consist of a heat exchanger, temperature controller, pumps, expansion tank, and water supply/return manifolds. Water flow in the primary loop is provided at a fixed flow rate by a single operating pump, with a stand-by pump to provide uninterrupted operation should the operating pump fail. Temperature control of the primary loop is obtained using a blend valve to regulate the fraction of the flow which is allowed to pass through the heat exchanger, forcing the remainder to bypass the heat exchanger.

The fact that water is a highly active chemical has been recognized and taken into account since the design of the first IBM water-cooling system and CDU [41]. The heat exchanger which transfers the heat load from the system water to customer water, is sized with an appropriate degree of conservatism, to ensure long life and adequate cooling capability for all conditions of customer water quality, temperature, and flow. On the system side, every step is taken to make certain that all of the materials which come into contact with water are compatible with the water and each other. All water-carrying components and the assembled frames undergo a cleaning procedure and hydrostatic test before entering service. In addition, appropriate corrosion inhibitors are added to the water to provide further protection against any possibility of corrosion. IBM experience has demonstrated, that with appropriate attention to materials selection and manufacturing quality control, water cooled-systems can be designed to provide both long life and highly effective cooling for mainframe computers.

DIRECT LIQUID IMMERSION

Direct immersion liquid cooling has long been recognized as a potential method to satisfy integrated circuit chip cooling requirements [42]. As noted earlier, IBM interest in direct immersion cooling techniques can be traced back to the 1960s, and development of the LEM preceded the TCM.

Although, the TCM replaced the LEM in production machines, some aspects of the direct immersion cooling technology that was developed are still being utilized within IBM today [43]. Prior to final module assembly, each TCM substrate with its chips undergoes an electrical test. To obtain the required electrical measurements the surface of the substrate and the area around each chip site must be ac-

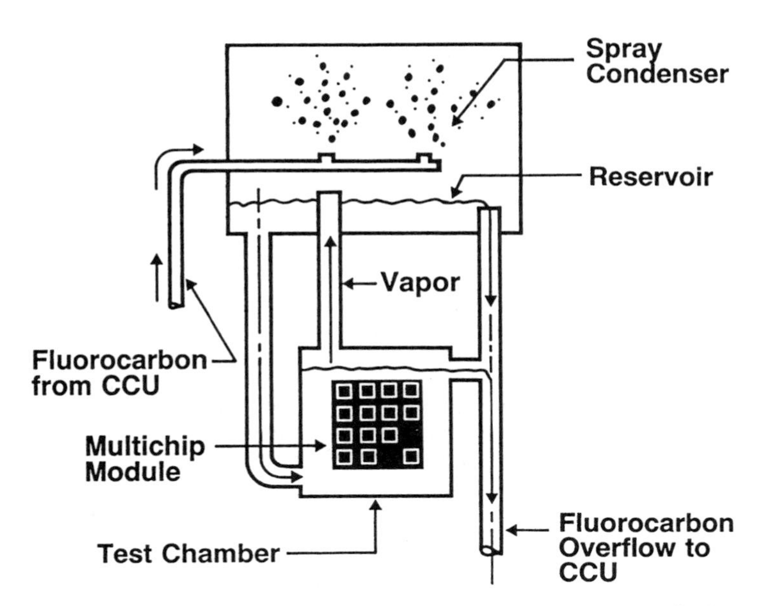

FIGURE 13. Direct immersion liquid cooled test system for TCM substrates with chips.

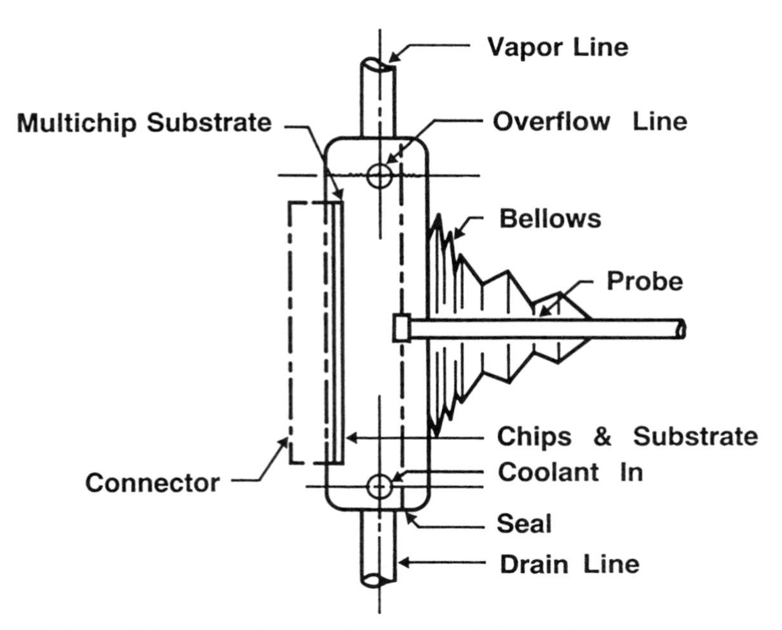

FIGURE 14. Side view of view of test chamber showing movable electrical probe assembly.

cessible for electrical probing. To fulfill this requirement and provide satisfactory cooling the substrate assembly to be tested is placed in a test chamber filled with fluorocarbon coolant as shown in Figure 13. As shown in Figure 14, a flexible bellows is attached to one wall of the test chamber so that the test probe can be moved horizontally and vertically to test each chip on the substrate. A remote cooling and circulating unit supplies subcooled fluorocarbon through spray nozzles in a chamber above the test chamber. The spray provides both direct contact condensation of vapor from the test chamber, and a continuous supply of liquid to the test chamber [44].

SUMMARY

In this paper, the authors have sought to trace the evolution of cooling technology for IBM mainframe computers from the era of the vacuum tube to that of very large scale integration (VLSI) on silicon chips. Covering a period of more than 30 years in the brief space of this paper, it has been impossible to do more than outline the developments that took place, and highlight some of the more significant aspects of these developments. The authors have had the good fortune of being part of the IBM team for a good part of that period and participated in much of the developments discussed here. It was an exciting and challenging period, and we share a sense of pride, along with the many other IBM engineers who shared in the effort, in the technology that evolved. If the past is any indicator of the future, the density of integrated circuit packaging will continue to increase, and with it so will the heat flux. It is the authors' belief that the development of new and more effective cooling technologies will continue to play a pivotal role in the development of generations of computers yet to come.

REFERENCES

1. Bergles, A.E., "The Evolution of Cooling Technology For Electrical, Electronic, and Microelectronic Equipment," ASME HTD-Vol. 57, pp. 1-9, June 1986.
2. Bashe, C.J., Bucholz, W., Hawkins, G.V., Ingram, J.J., and Rochester, N., "The Architecture of IBM's Early Computers," IBM Jour. Res. and Dev., Vol. 25, No. 5, pp. 363-375, September 1981.
3. Bashe, C.J., Johnson, L.R., Palmer, J.H., and Pugh, E.W., IBM's Early Computers , pp. 54-55, MIT Press, Cambridge, MA, 1986.
4. Hanson, D., The New Alchemists, pp. 76-78, Avon Books, New York, NY, 1982.
5. Blodgett, A.J., "Microelectronic Packaging," Scientific American, pp. 86-89, July 1983.
6. Seraphim, D.P., and Feinberg, I., "Electronic Packaging Evolution in IBM," IBM Jour. of Res. and Dev., Vol. 25, No. 5, pp. 617-629, September 1981.
7. Seely, J.H., and Chu, R.C., Heat Transfer in Microelectronic Equipment, pp. 7-10, Marcel Dekker, Inc., New York, NY, 1972.
8. Johnson, A.H., "Electrical Interconnecting and Mounting Device for Printed Circuit Boards," U.S. Patent 3,008,112, 1961.
9. Davis, E.M., Harding, W.E., Schwartz, R.S., and Corning, J.J., "Solid Logic Technology: Versatile, High-Performance Microelectronics," IBM Jour. of Res. and Dev., Vol. 8, pp. 102-114, 1964.
10. Miller, L.F., "Controlled Collapse Reflow Chip Joining," IBM Jour. Res. and Dev., Vol. 13, pp. 239-250, 1969.
11. Simons, R.E., "Direct Liquid Immersion Cooling: Past, Present, and Future," Proceedings of the 1987 International Symposium on Microelectronics, pp. 186-197, Minneapolis, MN, October 1987.
12. Chu, R.C., Gupta, O.R., Hwang, U.P., Moran, K.P., and Simons, R.E., "Cooling System for Data Processing Equipment," U.S. Patent 3,586,101, June 22, 1971.
13. Simons, R.E., and Moran, K.P., "Immersion Cooling Systems for High Density Electronic Packages," National Electronic and Packaging Production Conference (NEPCON) Proceedings, pp. 396-409, February 1977.
14. Aakalu, N.G., Chu, R.C., and Simons, R.E., "Liquid Encapsulated Air Cooled Module," U.S. 3,741,292, June 26, 1973.
15. Chu, R.C., Gupta, O.R., Hwang, U.P., and Simons, R.E., "Gas Encapsulated Cooling Module," U.S. Patent 3,741,292, November 23, 1976.

16. Chu, R.C., Hwang, U.P., and Simons, R.E., "Conduction Cooling for an LSI Package: A One-Dimensional Approach," IBM Jour. of Res. and Dev., Vol. 26, No. 1, pp. 45-54, January 1982.

17. Dvorak, P.J., "Packaging Computers to Survive in the Real World," Machine Design, Vol. 60, No. 12, pp. 70-76, May 1988.

18. Oktay, S., Dessauer, B., and Horvath, J.L., "New Internal And External Cooling Enhancements For The Air-Cooled IBM 4381 Module," presented at IEEE International Conference on Computer Design: VLSI in Computers, Port Chester, NY, November 1983.

19. Mondou, E., Young, S., "Thermal Compound for Semiconductor Packages," IBM Technical Disclosure Bulletin, Vol. 25, No. 76, December 1982.

20. Bisekeborn, R.G., Horvath, J.L., and Hultmark, E.B., Integral Cap Heat Sink Assembly for the IBM 4381 Processor," Fourth Annual International Electronic Packaging Society (IEPS) Conference Proceedings, pp. 299-304, October 1984.

21. Rymazewski, E.J., Walsh, J.L., and Leehan, G.W., Semiconductor Logic Technology in IBM," IBM Jour. of Res. and Dev., Vol. 25, No. 5, pp. 603-616, September 1981.

22. Clark, B.T., "Design of the IBM Thermal Conduction Module," Proceedings of the 31st Electronics Component Conference, Atlanta, GA, May 1981.

23. Chu, R.C., "Heat Transfer in Electronic Systems," Proceedings of the Eighth International Heat Transfer Conference, Vol. 1, pp. 293-305, August 1986.

24. Simons, R.E., Moran, K.P., Antonetti, V.W., and Chu, R.C., "Thermal Design of the IBM 3081 Computer," National Electronic Packaging and Production Conference Proceedings, 1982.

25. Antonetti, V.W., and Pascuzzo, A.L., "Cooling Large Scale Computer Systems," ASHRAE Journal, pp. 25-30, August 1971.

26. Chu, R.C., Cohen, M.G., and Seely, J.H., "Thermal Considerations for Electronic Circuit Packages in Modern Digital Computers," Proceedings of the 9th International Electronic Circuit Packaging Symposium, Los Angeles, CA, 1968.

27. Sparrow, E.M., Niethammer, J.E., and Chaboki, A., "Heat Transfer and Pressure Drop Characteristics of Arrays of Rectangular Modules Encountered in Electronic Equipment," Int. J. Heat and Mass Transfer, Vol. 25, No. 7, July 1982.

28. Hwang, U.P., "Thermal Design Using Turbulators for Air-Cooled Electronic Modules on a Card Package," National Electronic and Packaging Production Conference (NEPCON) Conference Proceedings, pp. 441-449, March 1984.

29. Timko, N., and Brzyski, R., "Air-Impingement Method on IBM 4381 Keeps Dense Logic Cool," Computer Systems Equipment Design, pp. 32-34, March 1985.

30. Timko, N., "Blower Performance Enhancements in the 4381 Computer," Fourth Annual International Electronics Packaging Society (IEPS) Conference Proceedings, pp. 475-482, October 1984.

31. Timko, N., and Plucinski, M.D., "A History of Thermal Control in IBM Intermediate Mainframes," Sixth Annual International Electronic Packaging Society (IEPS) Conference Proceedings, pp. 489-500, November 1986.

32. Antonetti, V.W., Chu, R.C., and Seely, J.H., "Thermal Design for IBM System/360 Model 91," IBM TR 00.1617, presented at the Eighth International Electronics Circuit Packaging Symposium, San Francisco, CA, August 1967.

33. Chu, R.C., Cunavelis, P.J., and Gerstenhaber, J., "Cooling Electrical Apparatus," U.S. Patent 3,317,798, May 2, 1967.

34. Antonetti, V.W., Simons, R.E., and Arent, G.R., "Hybrid Cooling Systems Design for Computer Electronics," IBM TR 00.2392, December 1972.

35. Antonetti, V.W., "Cooling High-End Computers," IBM TR 00.3067, October 1980.

36. Chu, R.C., "Modular Cooling System," U.S. Patent 3,481,393, December 2, 1969.

37. Chu, R.C., and Simons, R.E., "Heat Transfer in Electronic Equipment," Proceedings of the U.S.-Japan Joint Heat Transfer Seminar, San Diego, CA, September 1985.

38. Simons, R.E., "A Perspective on Single Phase Liquid Cooling," IBM TR 00.3388, presented at NSF/Purdue University Workshop on Research Needs in Electronic Cooling, Andover, MA, June 1986.

39. Simons, R.E., "Thermal Sensing and Control for a Large-Scale Digital Computer," SEMI-THERM1 Proceedings, Southwest Seminars, Phoenix, AZ, 1984.

40. Chu, R.C., and Simons, R.E., "Thermal Management of Large Scale Digital Computers," Int. J. for Hybrid Microelectronics, Vol. 7, No. 3, pp. 35-43, September 1984.

41. Chu, R.C., Simons, R.E., and Moran K.P., "System Cooling Design Considerations For Large Mainframe Computers," presented at the International Symposium on Cooling Technology for Electronic Equipment, Honolulu, HI, March 1987.

42. Simons, R.E., and Chu, R.C., "Direct Immersion Cooling Techniques for High Density Electronic Packages and Systems," Proceedings of the 1985 International Symposium on Microelectronics, pp. 314-321, Anaheim, CA, November 1985.

43. Hwang, U.P., and Moran, K.P., "Boiling Heat Transfer of Silicon Integrated Circuits Chip Mounted on a Substrate," ASME HTD-Vol. 20, pp. 53-59, 1981.

44. Antonetti, V.W., Gupta, O.R., and Moran, K.P., "Cooling System Providing Spray Condensation," U.S. Patent 3,774,677, November 27, 1987.

Thermal Control for Mini- and Microcomputers: The Limits of Air Cooling

R. HANNEMANN
Digital Equipment Corporation
Andover, Massachusetts, USA

Abstract

Mini- and microcomputers have traditionally used ambient air cooling with either free or forced convection as the primary mode of heat transfer. The advantages of air cooling in the cost bands and user environments for which these computers are targeted makes it attractive to develop packaging and thermal control techniques that allow air convection cooling to continue to be used. This paper examines the fundamental limitations of forced air cooling in a pragmatic way and compares these limits to the other dominant limiting physical factor in the design of electronic assemblies, the problem of interconnecting (wiring) arrays of increasingly complex semiconductor packages. The conclusion is that for the forseeable future (through the mid-1990s), there are no fundamental barriers that will force abandonment of forced air cooling for mainstream computing equipment.

1. INTRODUCTION

Progress in the development and manufacture of ever more powerful computing machines has been remarkable over the past three decades. This progress has mainly been slowed by the engineering or technical limits of the day, which govern the size, cost, or energy consumption of a particular system design. It has long been recognized that two of the fundamental limiting physical technologies for computing equipment (whether at the semiconductor, electronic assembly, or system levels) are the ability to interconnect and communicate between various components and manage the resulting wiring complexity and the ability to manage the flows of energy in the system.

A number of excellent examinations of the fundamental limitations in the design and realization of information processing equipment have been published [c.f. 1,2,3,4,5]. This paper is more narrowly focused than these works and is intended to be more pragmatically oriented. It seeks to provide an answer to an important question for the practitioners of electronics thermal control technology: what are the practical limits to the use of forced air cooling, and will a departure from this method of thermal control be required in the forseeable future for mini- and microcomputers? (In this rapidly advancing arena, the forseeable future is taken to be the middle of the next decade.)

An answer to this question cannot be formulated without simultaneous consideration of the development of integrated circuit (IC) technology and circuit packaging and interconnection technology. In fact, the interplay between circuit packaging (which attempts to provide ever more densely packed assemblies of ICs, thus creating increasing thermal densities) and cooling is the major focus of this paper.

The primary motive force for the electronics revolution has come from semiconductor technology. From a starting point of a single circuit element (transistor) per device, by the mid-1990s ICs with well over 1 million transistors will be in widespread commercial use. These devices will be operating at speeds such that interchip communications will limit overall system performance unless significant progress can be made in increasing chip packing density. Thus, the increasing scale of circuit integration has dramatically impacted the complexity of the required electronics packaging technology, the number of chips needed per system, the wiring needed for interchip communications, and the cooling approaches that are necessary.

For purposes of this paper, it is assumed that increased system complexity will require multiple-chip (but single board assembly) solutions in spite of increased on-chip integration. The complexity of the electronic modules that result is directly related to the parameters of the IC technology used to implement the system. In order to inject practical system requirements into the problem of assessing requirements and limitations of thermal control technology, the integrated circuit paradigms shown in Table 1 will be referenced several times. Note that these paradigms apply to mini- and microcomputers, not mainframes or supercomputers whose requirements will differ. Also note that they represent reasonable extrapolations of current technology with the 1996 paradigm biased conservatively and guided by CMOS (Complementary Metal Oxide Semiconductor) technology scaling rules [6]. In the table, the "TTL" acronym refers to Transistor-Transistor Logic, the predominant bipolar semiconductor logic technology of the 1970s.

The remainder of this paper focuses on electronics physical architecture and the role of thermal engineering, an examination of the topological limitations that impact the requirements for cooling technology, and an examination of the basic limitations posed by forced air cooling technology.

To restate the basic question to be explored, will forced air cooling limit progress for mini- and microcomputer design, or will wiring prove to be the limiting factor for the forseeable future?

1945-1955	First generation: Vacuum tube electronics
1955-1965	Second generation: Transistors
1965-1975	Third generation: Integrated circuits (SSI)
1975-1985	Fourth generation: Large-scale integration (LSI)
1985-	Fifth generation: LSI→VLSI→ULSI

Table 2. Computer Technology Generations.

sitive to temperature in that overtemperature operation led to rapid wearout and failure. In some cases, liquid cooling was required. Thermal engineering was very important in the first generation.

The transistor era offered relief from many thermal problems. The power dissipation of discrete logic transistors was orders of magnitude below that of an equivalent vacuum tube. Typical packing densities allowed the use of natural convection air cooling in many instances.

The invention of the integrated circuit and the application of ICs in the small scale integration (SSI) era allowed for a continuation of the decreased emphasis on cooling technology. Typical feature sizes, packing densities, and power consumptions of SSI integrated circuits allowed for natural and forced air cooling of most computer equipment. However, vastly more complex systems were being built, using hundreds or thousands of ICs. While the individual reliability of the devices was very high, the aggregate reliability of the system required that careful cooling analysis and design be accomplished toward the end of the third generation.

In the large scale integration (LSI), very large scale integration (VLSI), and ultra- large scale integration (ULSI) generations, cooling has once again become a very important issue. Although per-transistor power dissipation has declined substantially as devices have been scaled, the increased packing density on chips has caused a significant increase in per-chip power. High performance computers are once again employing liquid cooling, either conductively coupled (e.g., IBM 30xx series) or direct (Cray 2). New high performance logic technologies requiring cryogenic cooling are coming into use.

However, a dramatic change has occurred in computing: the development of the

	1980	1988	1996
Semiconductor technology	TTL	CMOS	CMOS
Relative density	1	200	1250
Chip power (W)	2	10	40
Chip power density (W/cm^2)	8	17	25
Chip area (cm^2)	0.25	1.0	1.6
Pins/chip (max)	64	200	400
System clock (MHz)	5	25	125

Table 1. IC Technology Trends: Mini- and Microcomputers.

2. THE ELECTRONICS COOLING PROBLEM

In order to provide a context for understanding the computer cooling problem and its strong relationship to other limiting factors, it is appropriate to briefly examine electronics physical technology, the historical background of electronics thermal engineering, and future technology trends.

Electronics physical technology is the application of basic physical principles to the tasks of electrically connecting active and passive circuit elements, physically protecting these elements, managing the distribution of power to and from the circuits, and controlling the electrical and thermal environment in which they operate. These tasks must be accomplished with designs that are within cost constraints and are manufacturable and serviceable. It is easy to understand that any thermal control system approach is constrained by and must be consistent with a large number of other considerations. An overview of the larger physical technology problem is provided in reference [7].

Electrical and electronics engineering and heat transfer engineering have historically been intertwined. Thermal control has always been an important consideration in the design of electronic equipment; early designs spurred the development of some fundamental analytical and experimental techniques in heat transfer. Computer technology eras are shown in Table 1.

Vacuum tubes were inefficient and of fragile construction. They were very sen-

Primary Cooling Mechanism	Typical HTC (W/m^2K)	Relative Effectiveness	Achievable Density	Complexity
Natural convection (air)	10	0.1	Low	Very low
Forced convection (air)	100	1.0	Medium	Low
Natural convection (liquid)	100	1.0	Medium	Medium
Forced convection (liquid)	1000	10.0	High	High
Phase change (liquid)	5000	50.0	High	High

Table 3. Thermal Control Methods.

microprocessor has allowed the migration of powerful computers from the protected environment of the computer room to office and factory floors, and the cost of these machines has dropped by two orders of magnitude. Thus, while per-chip power continues to increase, mainframe computers are migrating toward the use of liquid cooling, and supercomputers are moving to immersion and even cryogenic cooling, an increasing fraction of computing equipment remains air cooled and for cost and environmental reasons this situation is unlikely to change in the near term. For a more detailed discussion of the historical background of electronic equipment cooling, see reference [8].

The possible methods of thermal control vary widely. Broad classifications include conduction cooled systems, direct and indirect air cooled systems (natural vs. forced convection being a sub-classification), direct and indirect single phase liquid cooling, and phase-change liquid cooling. The relative performance and features of these cooling modes are shown in Table 3.

It is difficult to assess cooling techniques and capabilities without reference to the typical materials and constructions actually used in producing computer equipment. Using typical circuit packaging techniques and constraints, the thermal map of Figure 1 [7] was constructed. This thermal performance map assumes planar arrays of components, with the primary thermal variables of importance being maximum chip power and overall array thermal flux. Note that high values of either of these parameters pose difficulties for the thermal engineer.

Current trends are causing the operating point for some computers and other

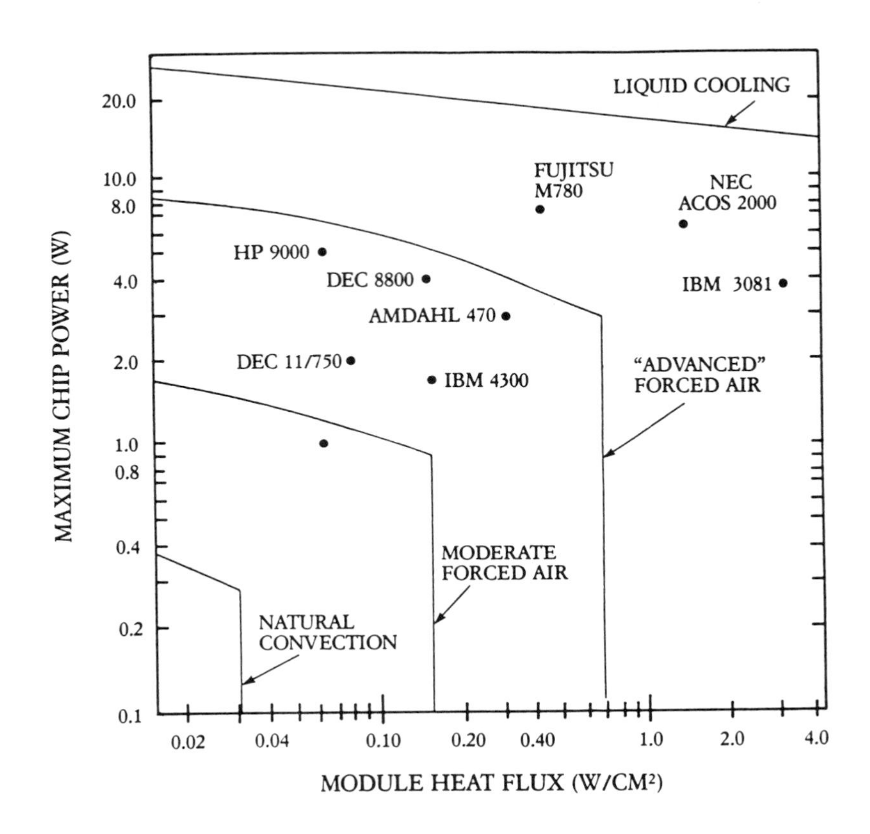

FIGURE 1. Cooling Technology Map.

electronics to move upward and to the right on this map. Current generation high end mainframe computers have chip powers approaching 20 W and array heat fluxes of perhaps 3.0 W/cm^2. Clearly, advanced cooling techniques including liquid cooling and phase change cooling will become more important for a specialized segment of the computer market.

However, as has been previously pointed out, it is safe to say that air cooling will still be dominant for the ubiquitous mini- and microcomputer segment, at least in the near term. For this class of computers, the maximum chip power has risen at a moderate pace while the array-level flux remains an order of magnitude lower than that for the most aggressive mainframes. Historical trend data for a series of Digital Equipment Corporation minicomputers is shown in Figure 2.

The basic reason that mini- and microcomputers have remained substantially more thermally benign than their mainframe counterparts has been the use of very high density custom MOS semiconductors, which have typically been an order of magnitude more power efficient and more dense than the corresponding highest-performance semiconductors required for mainframe machines. Recent trend data, however, as typified in the paradigms of Table 1, indicate that increasing microprocessor performance can only be gained at the cost of significantly higher pincount and significantly increased power.

Essentially, mini- and microcomputers lag the technology performance demands of mainframes and supercomputers by one generation. Therefore, given strong pressures to retain the use of air cooling it is appropriate to examine the limits of air cooling given the dramatic changes in chip package and interconnect technology caused by rapidly rising chip pincount and interconnect requirements. The hypothesis of this paper is that chip interconnect is at least as limiting as cooling technology and that there are no fundamental reasons demanding a departure from air cooling for mini- and microcomputers.

3. TOPOLOGICAL LIMITS

For purposes of this limit analysis, three major assumptions have been made:

- The basic electronics module configuration that is ubiquitous today will continue to be used: an array of ICs on a planar substrate with embedded wires to accomplish the interconnection of the terminals on the chips.
- The primary measure of goodness of a packaging scheme is the areal density of chips on the plane.
- The limiting factors for increasing the chip density are topological (primarily, whether the requisite wiring density is supportable with the available wiring channels) and thermal (whether or not the overall cooling task can be accomplished at a given density).

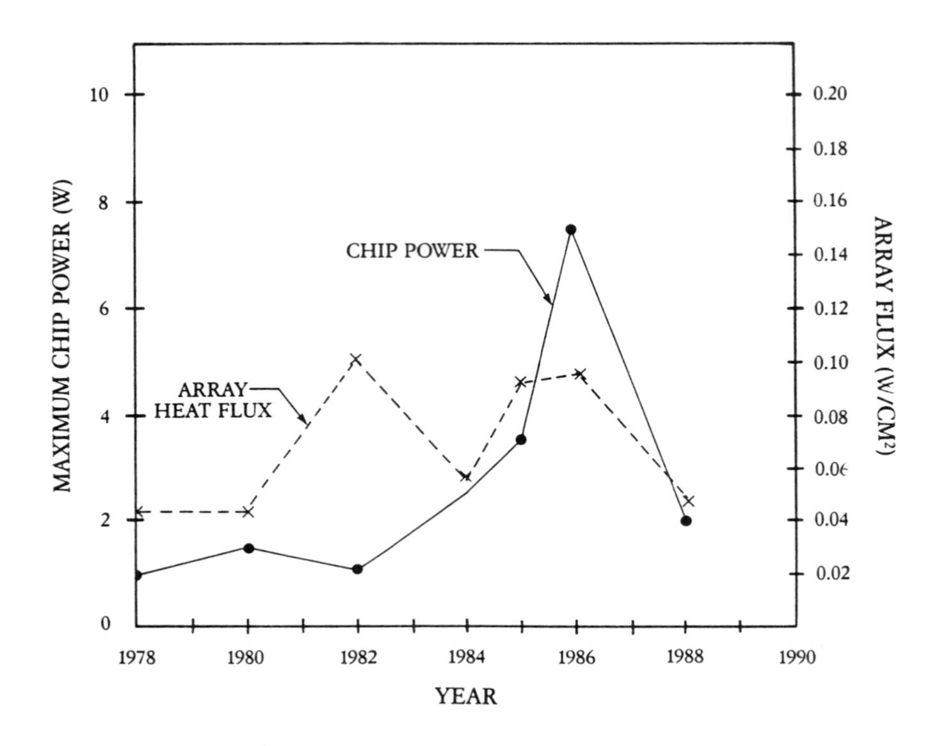

FIGURE 2. Thermal Trends: DEC Minicomputers.

Note that other limiting factors can be examined as a subset of either the topological limits or the thermal limit. For example, electrical signal integrity is strongly influenced by the wiring density and wire geometries. Signal integrity may limit the achievable substrate wiring density, which then limits the chip packing density. This in turn sets the average wire length and ultimately the maximum achievable interchip interconnect performance.

In examining the limits resulting from wiring and cooling capabilities, it is convenient to consider chip terminal count (pincount) as a primary independent variable. It is well known (c.f. [7]) that the number of required pins per chip is related to the number of circuits per chip by an approximate relationship known as Rent's Rule. Thus, progress at the IC level results in higher levels of on-chip integration and more pins per chip. The future requirements for wiring density can be expressed as a function of chip pincount. Similarly, in the cooling limit analysis the chip pincount is taken as the independent variable with chip heat flux density as a parameter. All limits are compared in the pincount-chip density space.

This section examines the topological limits to increasing chip packing density taking as a given the trends in IC technology. Given the planar array assumption, two limits can be studied. First, the *tiling limit* that results when the entire plane is tiled with chips (thus giving the maximum achievable chip density) is analyzed. Second, the *wiring limit* that occurs due to limitations in available wiring is examined.

3.1 Tiling Limit

Determining the plane tiling limit in the chip packing density - device pincount space is straightforward.

The achievable pincount for a given chip package area A_p is dependent on the package pin pitch p_p and the style of pin attachment (either peripheral, in which the pins extend in a single row around the periphery of the package, or area array, in which the pins are attached in a matrix over the entire area of the package).

For peripheral leads, the package area as a function of pin pitch and number of leads is given by

$$A_p{}^{-1} = \frac{16}{p_p{}^2(n_t + 4)^2}. \tag{1}$$

In the chip tiling limit, the number of chips per unit area is simply given by the inverse of the package area.

Similarly, for area array interconnections the formula for maximum chip packing density is

$$A_p{}^{-1} = \frac{1}{p_p{}^2(n_t{}^{0.5} + 1)^2}. \tag{2}$$

These relationships are plotted in Figure 3 for a variety of realizable package lead pitches. For a given lead pitch, the admissable design space is to the left and below the appropriate envelope boundary.

Current electronics packages in common use have lead pitches as low as $0.64mm$ for peripheral leaded devices and $1.27mm$ for area array (pin grid array) packages. In the extreme, the package body itself can be eliminated; typical resulting minimum lead pitches are $0.25mm$ for peripheral leads (tape automated bonding, or TAB) and $0.25mm$ for area array pins (flip chip mounting).

3.2 Wiring Limit

Obtaining the limits placed on chip packing density by the availability of wires in the interconnect substrate depends on the development of a relationship between block (chip) pitch on the substrate, the size of the substrate, the number of terminals on each block, and the geometric parameters that describe the wiring capabilities of the substrate.

As the number of circuits per chip increase, the required number of package pins also increases. An approximate relationship between these two quantities that is valid for a number of applications (and which has a topological justification for relatively random circuitry) is Rent's Rule, which can be expressed in the form

$$n_t = aB^b, \tag{3}$$

where B is the number of circuits per chip and a and b are constants.

This relationship is very valuable as an estimating tool and can be used to derive wiring models (c.f. [9]), but in general is inapplicable for microprocessors and other highly integrated systems. Thus, wiring models which rely on the recursive application of Rent's Rule at the chip and chip array levels are flawed in many cases of interest. Fortunately, a simple wiring model that does not depend on invocation of Rent's Rule can be developed.

Consider an interconnect substrate with side L. N components are to be placed and wired on this substrate using m layers of wiring with wires that have a characteristic feature size of λ.

Figure 4 describes a typical local region of a given wiring plane. "Vias" are structures used to interconnect between wiring layers. In general, k wires may pass between vias placed on a grid of dimension g. As measured along a line orthogonal to the wiring direction, an average of $k + 1$ wires are available over a length g for a single wiring layer. The characteristic feature size of the wiring plane is then

$$\lambda = \frac{g}{1 + k}. \tag{4}$$

Defining α as the average number of connections per pin or fanout factor (which depends on design style but is typically about 0.67), the total number of connections

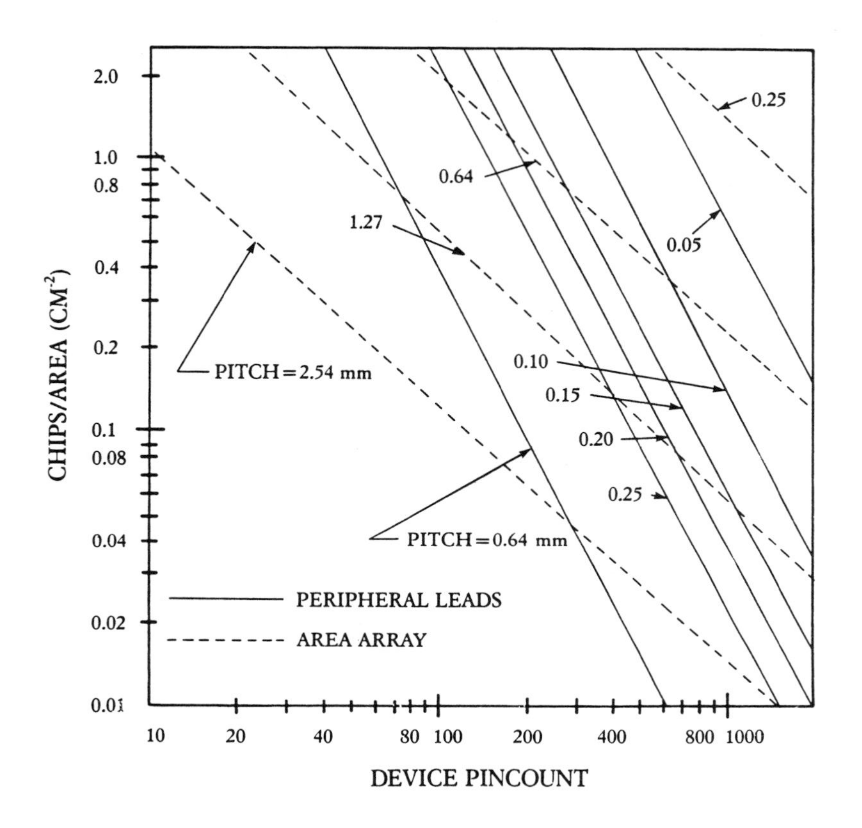

FIGURE 3. Chip Tiling Limit.

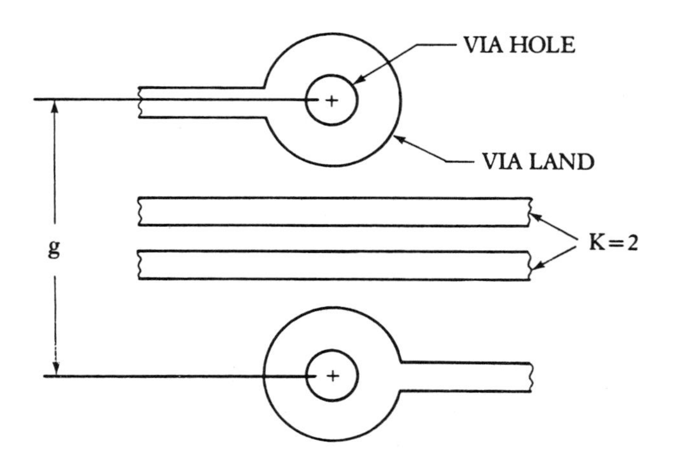

FIGURE 4. Wiring Geometry.

to be made is just $n_t\alpha$. Then, with the average device pitch p and the average wire length in units of device pitches β, the total length of wire needed to interconnect the array of N devices is just

$$l_w = n_t\alpha\beta p. \tag{5}$$

But $p \simeq \frac{L}{N^{0.5}}$ for a relatively homogeneous placement of components, so

$$l_w = \alpha\beta n_t \frac{L}{N^{0.5}}. \tag{6}$$

Now, the total available wiring over all m layers is given by

$$l_a = m\frac{L}{\lambda}L = m\frac{L^2}{\lambda}; \tag{7}$$

with a wiring usage efficiency (which depends on the automatic routing system used and various design rules) of η, this must just equal the total wiring needed.

So we have the wiring equation

$$\eta m\frac{L^2}{\lambda} = \alpha\beta n_t \frac{L}{N^{0.5}}. \tag{8}$$

Expressed in terms of the number of wiring layers needed for a particular array wiring problem, the relationship is

$$m = cH, \tag{9}$$

where $c = \frac{\alpha\beta}{\eta}$ is a constant and H is a dimensionless number given by

$$H = \frac{n_t\lambda}{LN^{0.5}}. \tag{10}$$

In most cases, α is about 0.67, β is about 1.5, and η can vary between 0.2 and 0.6. Thus c should be in the range 1.5-5.0. If the value $c = 3.9$ is chosen, a variety of experimental data points can be fit using this wiring equation, usually to within a single wiring plane pair or better. The correlation is shown in Figure 5.

Equation 10 may be manipulated to give the achievable chip packing density on a given substrate as a function of the overall wiring channel capacity m/λ and the number of pins per chip n_t:

$$\frac{N}{L^2} = \frac{0.0657(m/\lambda)^2}{n_t^2}. \tag{11}$$

This limit relationship has been plotted in the chip packing density - device pincount space in Figure 6 with m/λ as a parameter. Currently achievable values for m/λ can be as high as $630cm^{-1}$ but are more typically on the order of $100cm^{-1}$ for cost-sensitive mini- and microcomputer applications [7].

It is perhaps of some interest to compare the limit envelopes from the tiling limit and the wiring limit. Depending on the micro-joining technique being used,

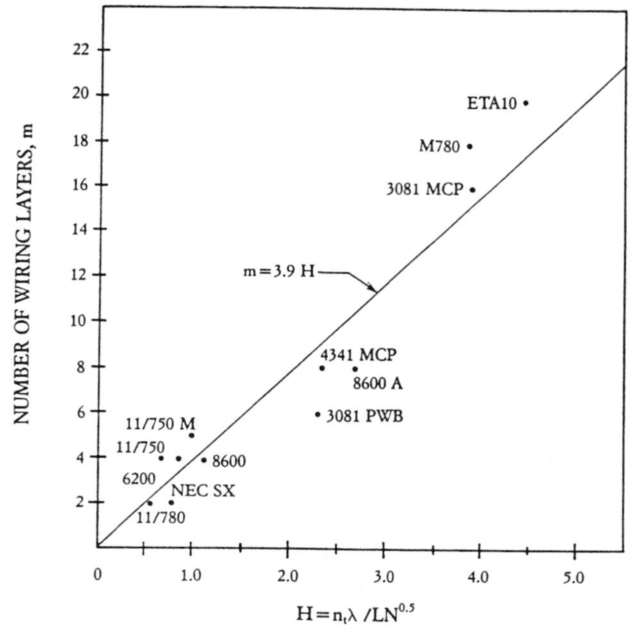

FIGURE 5. Wiring Correlation.

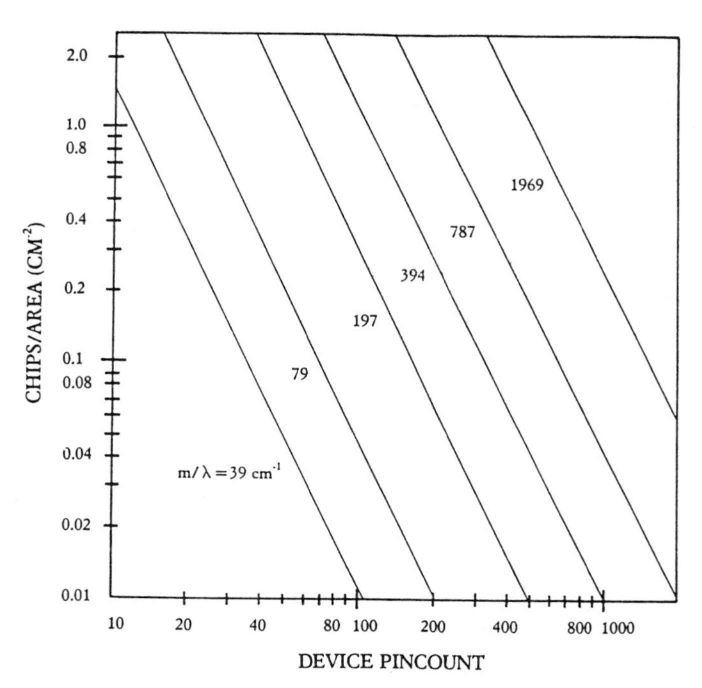

FIGURE 6. Substrate Wiring Limit.

the target for m/λ should be set accordingly. Increasing wiring channels beyond this point is not productive.

4. COOLING LIMITS

Given the topological limitations described and analyzed in the previous section, we wish to analyze the practical limits of forced convection cooling using ambient air and express the results in terms of N/L^2 vs. n_t for this cooling limit.

We begin with the basic equation describing forced convection cooling of an array of electronic components [10]:

$$\Delta T = \frac{\Psi_1 Q}{\dot{m} c_p} + \Psi_2 R_{da} P. \tag{12}$$

The variables in this equation are described in the nomenclature. The first term represents the air temperature rise experienced by the cooling stream upstream of a given component; the second gives the temperature rise of the component due to the heat transfer resistances between the chip surface (the heat source) and the cooling airstream, including conduction resistances in the package and the convective resistance between the package and the airstream. ΔT is the overall temperature rise above ambient conditions for any particular component in the array.

Both serial and parallel cooling schemes can be devised that have acceptable pressure drops and mechanical complexity; since we are interested in practically achievable limits, we assume a parallel air supply system and $\Psi_1 = 0$. $\Psi_2 R_{da}$ is then the effective chip package thermal resistance (die surface to airstream). The limiting case will have $\Psi_2 = 1$, with no position-dependent degradation in the standard thermal resistance R_{da}. This is achievable with parallel cooling and a careful coolant supply system design.

In the limit, the elimination of internal package resistance will result in the maximum capabilities for forced air cooling; while the total elimination of internal resistance is not achievable, we assume that the convective resistance dominates and write the governing thermal performance equation as

$$\Delta T = \frac{1}{hA} P = \frac{q A_c}{hA}. \tag{13}$$

Here q is the chip heat flux, A_c is the chip area, h is the forced convection heat transfer coefficient, and A is the *effective* area for heat transfer, inclusive of the effects of extended surfaces. While this equation ignores the package thermal resistance, which can be minimized, the internal thermal resistance will be accounted for in a gross sense when practical thermal limiting performance is discussed in the Summary.

The use of extended surfaces is very common in the cooling of microelectronic components [10]. Of interest for present purposes is the situation sketched in Figure

7. Here an extended surface of base area A_b, fin height y, and fin pitch s is in thermal contact with a chip of area A_c; the chip package area A_p is shown but does not enter directly into the analysis. The chip is part of an overall array.

The number of chips per unit area in the array is given by

$$\frac{N}{L^2} = \frac{1}{A_b}. \tag{14}$$

But, from equation (13),

$$\Delta T = \frac{qA_c}{h\gamma A_b}, \tag{15}$$

where γ represents the effective area ratio A/A_b.

So the chip area that may be cooled with an extended surface of base area A_b and chip heat flux q is given by

$$A_c = \frac{h\gamma\Delta T}{q}A_b. \tag{16}$$

The area enhancement ratio γ depends on fin spacing s, fin height y, and fin length l_f. The fin length is related to the base area A_b as $l_f = A_b^{0.5}$. The fin height y is of course variable, but in practical cases in which electronic modules are connected through backplanes in a stacked configuration, y is limited by the card-to-card spacing. For this analysis a value $y = 2cm$ is assumed. With a reasonable choice of fin material, degradation of fin effectiveness due to conduction losses will be minimal with this choice (but depends on the area ratio A_b/A_c as well).

The total heat transfer area of the extended surface is approximately

$$A = A_b + 2yl_f\left(\frac{l_f}{s} + 1\right), \tag{17}$$

which results in the simple equation

$$\gamma = 1 + \frac{2y}{s} \tag{18}$$

when l_f is significantly greater than s.

The optimum choice of the parameter s depends on a number of factors, including the acceptable pressure drop and the interaction of boundary layers in the fin channels. Keeping in mind the orientation of this limit analysis toward currently practical technology, for "reasonable" air velocities (say, $\simeq 5m/s$) the thermally-optimal choice of fin spacing has been experimentally determined in [10] as

$$s = 3.41\left(\frac{\nu L}{v}\right)^{0.5}, \tag{19}$$

where v is the air velocity and ν is the air kinematic viscosity.

Assuming room temperature atmospheric conditions, and assuming a practically achievable velocity of $5m/s$, equations (16), (18) and (19) yield the relationship

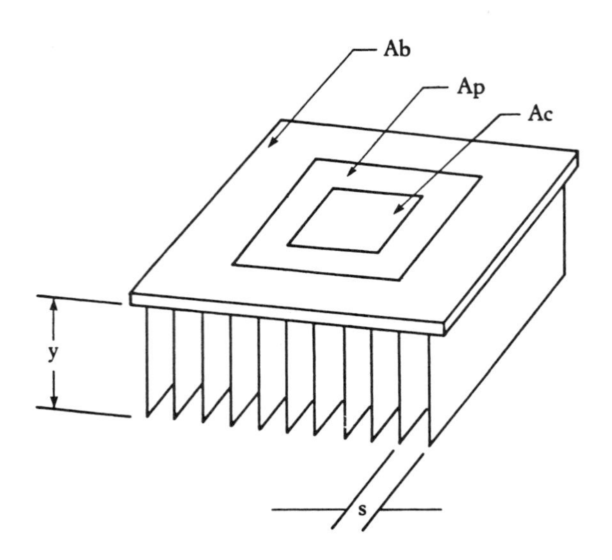

FIGURE 7. Extended Surface Chip Cooling.

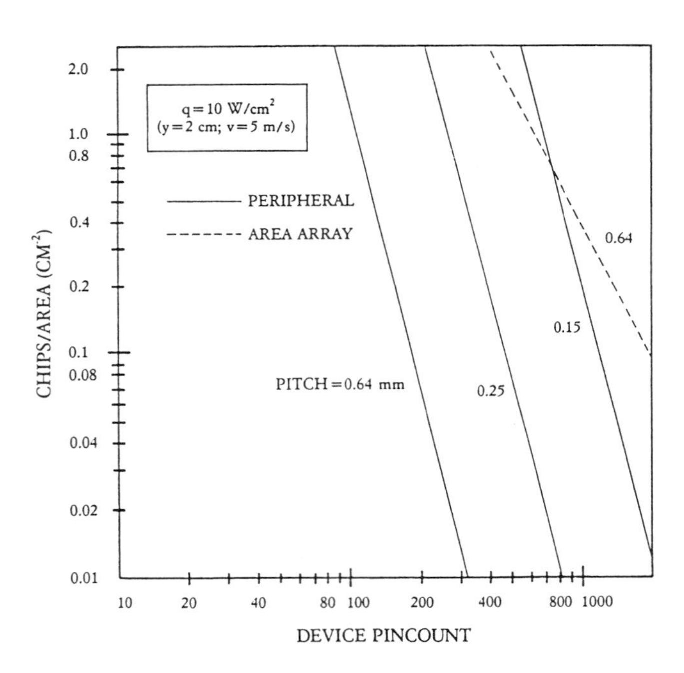

FIGURE 8. Thermal Limits - $q = 10W/cm^2$.

$$A_c = \frac{h \Delta T A_b}{q} \left(1 + \frac{6.39}{l_f^{0.5}}\right) \tag{20}$$

in S.I. units.

Again based on experimental data generated for micro-extended surfaces [10], the heat transfer coefficient can be expressed (with S.I. units understood) at room temperature as

$$h = 3.78 \left(\frac{v}{l_f^{0.5}}\right) = \frac{8.45}{l_f^{0.5}} \tag{21}$$

for $v = 5m/s$.

Recognizing that $l_f^2 = A_b$, and combining equation (20) with equation (21), we are left with the relationship

$$A_c = \frac{\Delta T}{q} \left(8.45 A_b^{0.75} + 54 A_b^{0.5}\right), \tag{22}$$

which again is valid only for S.I. units.

The envelope equation for thermally-limited chip density can be derived from equation (22) by recognizing that $N/L^2 = A_b^{-1}$ and that the chip area A_c is a function of the chip pincount that depends on the feature sizes and pinning style (peripheral or area array) that are used.

With a maximum allowable ΔT of $50K$, which is an eminently practical value, the envelope curves shown in Figures 8, 9, and 10 can be drawn. These figures show the maximum thermally-allowed chip density at a given chip pincount and pinning style with the chip heat flux P/A_c as a parameter.

These envelope curves are valid only if the internal thermal resistance is neglected. Typically, in large VLSI packages the external convective resistance is at least a factor of 2 higher than the internal resistance; new package designs can be envisioned in which direct contact between the heat sink and the chip can be achieved with minimal interface resistance. In any case, the effect of internal thermal resistance has been taken into account in the limit curves of Figure 11, in which the effective heat transfer coefficient has been discounted by 50% (thus doubling the no-internal-resistance R_{da}). These curves will be used in the summary discussion below.

5. SUMMARY AND CONCLUSION

In order to understand the limits placed on microsystem design by chip tiling, wiring, and the thermal limits resulting from the use of forced air cooling, the composite limit chart of Figure 12 was constructed.

Pragmatic assumptions were used for the limits plotted in Figure 12. First, the tiling limit is shown for a $0.25mm$ peripheral leaded packaging approach. This lead

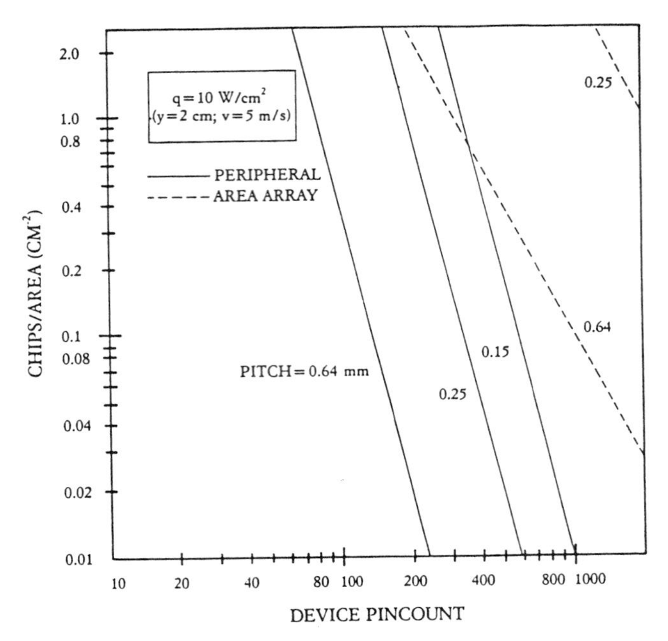

FIGURE 9. Thermal Limits - $q = 20W/cm^2$.

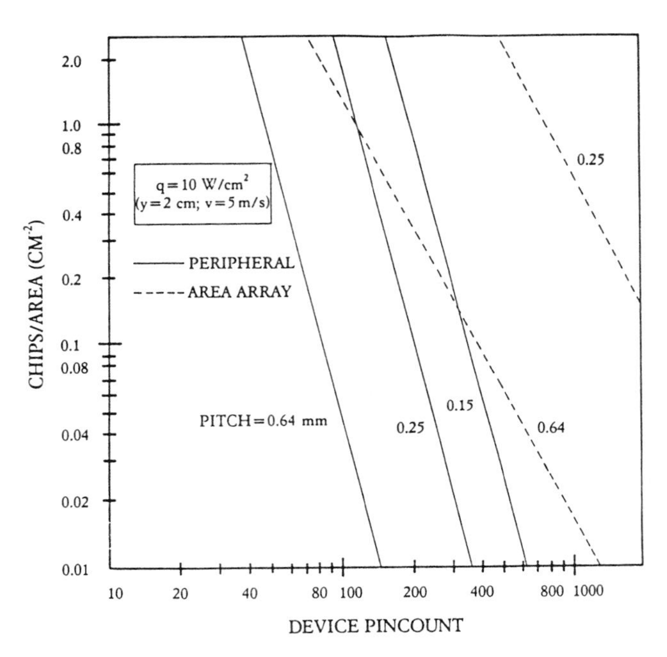

FIGURE 10. Thermal Limits - $q = 50W/cm^2$.

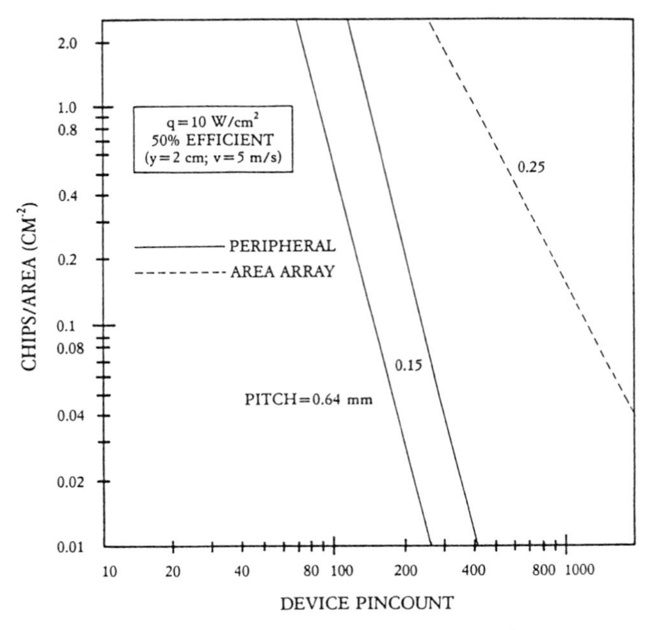

FIGURE 11. Thermal Limits - $q = 50W/cm^2$, 50% efficiency.

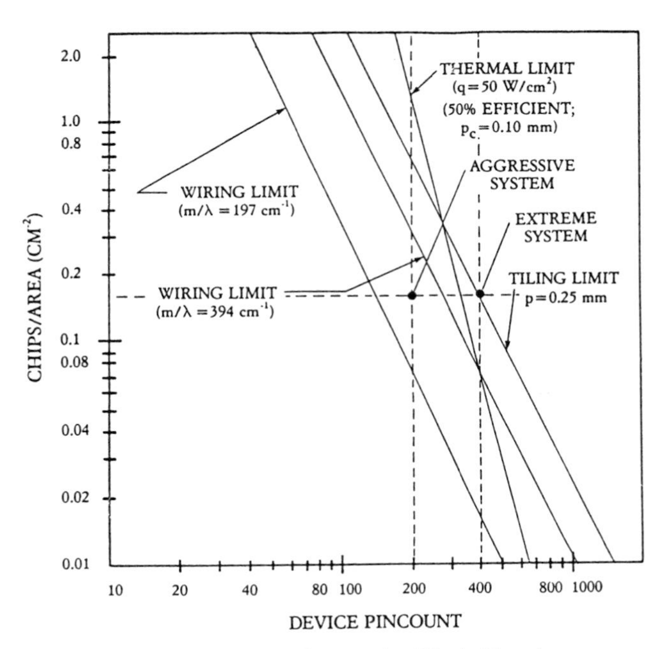

FIGURE 12. Composite Limit Envelopes.

pitch is readily achievable using micro-joining techniques such as Tape Automated Bonding (TAB).

Next, two wiring limit curves are plotted for $m/\lambda = 197cm^{-1}$ and $m/\lambda = 394cm^{-1}$. It is very unlikely that these values will be exceeded for commercial microsystems within the time horizon assumed in this paper.

Finally, a thermal limit curve is plotted assuming a chip heat flux $q = 50W/cm^2$, an internal package resistance equal to the external convective resistance (i.e., 50% efficiency), and a $0.10mm$ peripheral pad pitch on the chip. The heat flux is aggressive (twice as large as the posited chip heat flux for the 1996 paradigm system); the chip pad pitch is achievable today using TAB bonding at the chip leads. Therefore, the thermal limit curve is biased towards being more rather than less limiting than can be expected to be achievable.

The limit curves of Figure 12 may be examined in various lights. Assume, for example, that a chip packing density of $0.16cm^{-2}$ is required (this is a reasonable value given historical packaging results; a typical wire length of less than $1cm$ will result). Also assume a maximum chip pincount of 400 leads but an average chip pincount of 200 (termed "aggressive") or simply an average pincount of 400 (termed "extreme"). Then the system operating points shown in Figure 12 result. The "aggressive" system is more likely to be realized in a normal microsystem in which low leadcount memory and peripheral chips are mixed with the high leadcount packages. In this case, it can be seen that if an interconnect substrate with $m/\lambda \geq 394cm^{-1}$ is used the system can be realized and cooled using forced air convection. The "extreme" system will be limited by tiling, wiring, and thermal limits, but more by wiring density than by thermal dissipation limits.

Another way of examining the limits is to assume an average chip pincount and increase chip packing density. Then the first limit line to be encountered will be the factor that limits system performance. At 200 leads, wiring channel density clearly is limiting. This corresponds to a realistic microsystem in the 1996 timeframe. At this level, tiling becomes a limit before heat transfer is an issue. At an average leadcount of 400 (which is an extreme assumption for 1996), both wiring and thermal performance will be limiting at a density of $0.07cm^{-2}$.

While there may be reasons to depart from air cooling in any particular case, the conclusions to be drawn from this admittedly assumption-filled analysis are that there is no fundamental reason for departing from forced air cooling for mini- and microcomputers in the forseeable future, and that research in creative packaging and cooling approaches using forced air will be beneficial in designing high performance microsystems well into the 1990s.

6. NOMENCLATURE

Symbol	*Quantity*	SI Units
A	Effective heat transfer area	m^2
A_b	Extended surface base area	m^2
A_c	Chip area	m^2
A_p	Package area	m^2
a	Coefficient in Rent's Rule	*Dimensionless*
B	Number of blocks in Rent's Rule	*Dimensionless*
b	Exponent in Rent's Rule	*Dimensionless*
c_p	Fluid specific heat	$J/kg - K$
g	Via grid dimension	m
H	Wiring parameter, equation (10)	*Dimensionless*
h	Heat transfer coefficient	$W/m^2 - K$
k	Number of wires between vias	*Dimensionless*
L	Length of array of chips	m^2
l_a	Available length of wire	m
l_f	Fin length	m
l_w	Needed length of wire	m
m	Number of wiring planes	*Dimensionless*
$\dot{m}$	Coolant mass flow rate	kg/s
N	Number of chips in an array	*Dimensionless*
n_t	Number of pins per chip	*Dimensionless*
P	Chip power dissipation	W
p	Chip pitch	m
p_c	Pin pitch on chip	m
p_p	Pin pitch on package	m
Q	Array heat dissipation	W
q	Chip heat flux	W/m^2
R_{da}	Device thermal resistance	K/W
s	Fin pitch	m
v	Air velocity	m/s
y	Fin height	m
α	Connections per pin	*Dimensionless*
β	Average connection length in chip pitches	*Dimensionless*
γ	Area ratio	*Dimensionless*
ΔT	Chip temperature rise	K
η	Wiring usage efficiency	*Dimensionless*
λ	Wiring pitch	m
ν	Kinematic viscosity	m^2/s
Ψ_1, Ψ_2	Correlation coefficients	*Dimensionless*

7. REFERENCES

[1] Keyes, R., Fundamental Limits in Digital Signal Processing, *Proc. IEEE*, vol.69, no.4, pp.267-278, 1981.

[2] Hoeneisen, B. and Mead, C., Fundamental Limitations in Microelectronics Technology-I. MOS Technology, *Solid-State Electronics*, vol.15, pp.819-829, 1972.

[3] Seitz, C. and Matisoo, J., Engineering Limits on Computer Performance, *Physics Today*, pp.38-45, May 1984.

[4] Pence, W. and Krusius, J., The Fundamental Limits for Electronic Packaging and Systems, *IEEE Trans. CHMT*, vol.CHMT-10, no.2, pp.176-183, 1987.

[5] Folberth, O., The Interdependence of Geometrical, Thermal, and Electrical Limitations for VLSI Logic, *IEEE J. Solid-State Circuits*, vol.SC-16, no.1, pp.51-53, 1981.

[6] Mead, C. and Conway, L., *Introduction to VLSI Systems*, p.33, Addison-Wesley, Reading, MA, 1980.

[7] Hannemann, R., Physical Technology for VLSI Systems, Proceedings IEEE Conference on Computer Design, pp.48-53, 1986.

[8] Bergles, A., The Evolution of Cooling Technology for Electrical, Electronic, and Microelectronic Equipment, *Heat Transfer in Electronic Equipment*, HTD-Vol.57, pp.1-11, American Society of Mechanical Engineers, New York, 1986.

[9] Schmidt, D., Circuit Pack Parameter Estimation Using Rent's Rule, *IEEE Trans. on CAD*, vol.1, no.4, 1982.

[10] Hannemann,R., Fox, L., and Mahalingham, M., Thermal Design for Microelectronic Components, Proceedings 1987 International Symposium on Cooling Technology for Electronic Equipment, to be published.

AIR COOLING

Analysis of Extended Surface Arrays for Air-Cooled Electronic Equipment

ALLAN D. KRAUS
Department of Electrical and Computer Engineering
Naval Postgraduate School
Monterey, California, USA

ABSTRACT

Previous work, by the author and others, pertaining to parameterizations for individual fins is reviewed. These are the thermal transmission matrices and ratios which were devised to facilitate the analysis of an assembly of individual fins in an array of extended surface. An elaboration of the validity of these parameters, particularly with regard to their superiority over the notion of fin efficiency or fin effectiveness is made. The concept of reciprocity is developed and the representation of an individual fin as a connection of just three simple resistances is developed. A procedure for the nodal analysis of finned arrays is developed via a matrix-oriented approach.

INTRODUCTION

As indicated in Fig. 1, finned arrays composed of a conglomeration of individual fins are used for the thermal management of electronic equipment. The analysis of such arrays is often complicated and a realistic assessment of the array performance is often impossible without an actual test.

Fins of various geometries and thermal conductivities respond differently to identical and uniform heat sources and sinks. Similarly, there are many ways in which the temperature and heat transfer coefficient relating the fin to the sources and sinks may vary. Important to the analysis of any fin geometry are the constraints or assumptions which are employed to define and limit the problem and, of course, to simplify its solution. Simplified constraint analysis of an extended surface employs the limiting assumptions that are attributed to Murray [1] and Gardner [2] which are:

1. The heat flow and temperature distribution throughout the fin are independent of time, i.e., the heat flow is steady.
2. The fin material is homogeneous and isotropic.
3. There are no heat sources in the fin itself.
4. The heat flow to or from the fin surface at any point is directly proportional to the temperature difference between the surface at that point and the surrounding fluid.
5. The thermal conductivity of the fin is constant.
6. The heat transfer coefficient is the same over all the fin surface.
7. The temperature of the surrounding fluid is uniform.
8. The temperature of the base of the fin is uniform and the joint between the fin and the prime surface is assumed to offer no bond resistance.
9. The thickness is so small compared to its height that temperature gradients normal to the surface may be neglected.

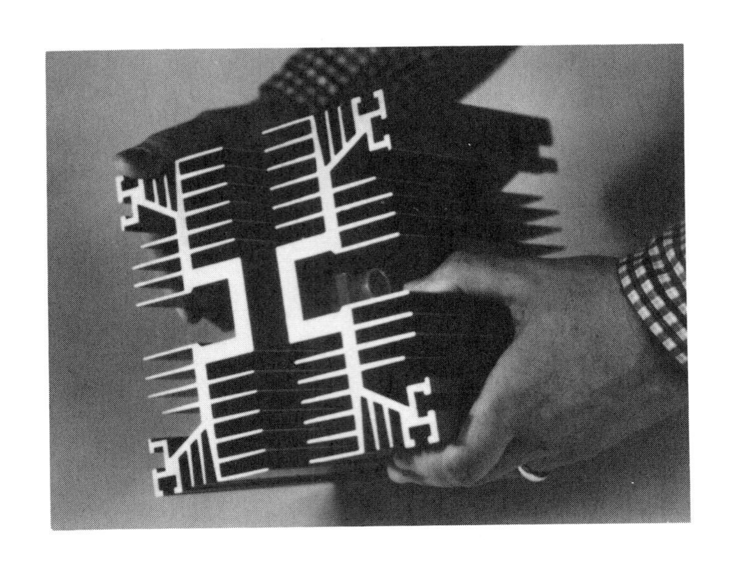

FIGURE 1. Two examples of finned arrays used in the thermal management of electronic equipment (*Photographs courtesy of TRANTEC, Columbus, Nebraska*).

10. The heat transferred through the outermost edge of the fin is negligible compared to that passing through the faces.

To be sure, analyses based on the foregoing assumptions are not *real world* analyses. Indeed, there have been literally thousands of investigators who have revealed, in the historical advance of the technology, the inadequacies of the Murray - Gardner assumptions. Yet, while this attempt to adopt more realistic constraints has put the mathematical analysis models in closer agreement with the actual physical situation, the design of heat transfer equipment utilizing extended surfaces is still based on the simplified constraints that employ the limiting assumptions.

The next section presents a discussion of the design parameters of fin efficiency, fin effectiveness and the recently proposed fin or array input admittance. This is followed by a section that deals with the mathematical representation of individual fins in terms of three matrices; the thermal transmission, the inverse thermal transmission and the thermal admittance matrices. After a discussion of what are believed to be the deficiencies and limitations of the fin efficiency as a design parameter, it is shown that the use of the input admittance overcomes these deficiencies.

A detailed procedure for the determination of the thermal transmission matrix for any fin shape is then provided and it is shown how the input admittance can be obtained from the thermal transmission matrix. Algorithms for the assembly and combination of individual fins into an array are then presented and the paper concludes with a general method of finned array analysis.

PARAMETERIZATIONS

The convective fin efficiency, apparently first proposed by Harper and Brown [3] and Parsons and Harper [4] and then discussed in detail by Gardner [2], is defined as the ratio of the heat dissipated by the fin to the amount of heat that would be dissipated if the fin were to operate throughout at the base temperature. An alternate definition is the ratio of the heat dissipated by the fin to the heat that would be dissipated by a fin of the same dimensions but with infinite thermal conductivity. Efficiencies were given by Gardner [2], in a comprehensive and pioneering paper, for eleven common profiles of longitudinal fins, radial fins and spines.

The fin efficiency is a design parameter that has been in use for half a century. It merely modifies (as a multiplier) the total fin surface to account for the fact that every point on the fin operates at a different temperature. Thus, the total surface to be employed in

$$q = hS\theta_b \tag{1}$$

is

$$S = S_b + \eta S_f \tag{2}$$

where the temperature excess, $\theta = T - T_e$, and, where η is the fin efficiency.

Another design parameter is the fin effectiveness defined by Gardner [2] as the ratio of the heat transferred through the base of a fin to that which would be transferred, at the same temperature, through the same base area (prime surface area) if the fin were not present. Gardner showed that the fin effectiveness is related to the fin efficiency by

$$\phi = \frac{S_f}{A_b}\eta \tag{3}$$

Gardner [2] was quick to point out that, in most practical cases, the addition of extended surface to a metal prime surface changes the base temperature. Two studies confirming this fact were conducted by Sparrow and Hennicke [5] and Sparrow and Lee [6]. Thus, the use of the fin effectiveness as a design parameter is somewhat limited and the employment of the fin efficiency as the design parameter has prevailed. Trumpler [7] pointed out that the simplistic approach utilizing the limiting assumptions should cause no concern because in most commercial installations, the fin efficiency is equal to or greater than 90 percent.

Manzoor, Ingham and Heggs [8], in a study that considered radiation as well as convection from the fin faces, contended that the heat flow through a finned assembly could be expressed conveniently in the form of an augmentation factor defined as the ratio of the heat dissipated by the fin assembly to that of the unfinned surface operating under the same conditions. The augmentation factors proposed for both one-dimensional and two-dimensional heat transfer in the fins differ from the conventional fin effectiveness in that the latter fails to account for either the conductive heat flow within the supporting or prime surface or the convective heat dissipation from the unfinned side of the prime surface.

A completely new parameterization was proposed by Kraus, Snider and Doty [9]. This parameterization, also based on the limiting assumptions, is a thermal transmission ratio or "input admittance" for an individual fin or a finned array. This input admittance is defined as the ratio of the fin or array base heat flow to the fin or array base temperature excess

$$Y_i = \frac{q_b}{\theta_b} \tag{4}$$

and is related to the fin efficiency by

$$Y_i = \eta h S_f \tag{5}$$

Its use in conjunction with the base or prime surface is additive; the total input admittance, Y_{iT}, is merely the sum of Y_{iP} and Y_i and the heat dissipation is

$$q_T = Y_{iT}\theta_b = (Y_{iP} + Y_i)\theta_b \tag{6}$$

This new parameter is also particularly useful in the analysis and evaluation of finned arrays and the next section will be devoted to a short discussion of its origin.

THE FIN OR ARRAY INPUT ADMITTANCE

Intuition dictates that conditions of temperature excess and heat flow at the tip of a fin (θ_a and q_a) are induced by similar conditions (θ_b and q_b) at the fin base. One may therefore assume that θ_b and q_b are independent analysis variables and θ_a and q_a are dependent variables related to the independent variables by the linear transformation

$$\begin{bmatrix} \theta_a \\ q_a \end{bmatrix} = \mathbf{\Gamma} \begin{bmatrix} \theta_b \\ q_b \end{bmatrix} = \begin{bmatrix} \gamma_{11} & \gamma_{12} \\ \gamma_{21} & \gamma_{22} \end{bmatrix} \begin{bmatrix} \theta_b \\ q_b \end{bmatrix} \tag{7}$$

where the matrix $\mathbf{\Gamma}$ is called the thermal transmission matrix with elements that are called the thermal transmission parameters.

If one wants to represent conditions at the fin base in terms of conditions at the fin tip, it is easy to see that

$$\begin{bmatrix} \theta_b \\ q_b \end{bmatrix} = \mathbf{\Gamma}^{-1} \begin{bmatrix} \theta_a \\ q_a \end{bmatrix} = \mathbf{T} \begin{bmatrix} \theta_a \\ q_a \end{bmatrix} = \begin{bmatrix} \tau_{11} & \tau_{12} \\ \tau_{21} & \tau_{22} \end{bmatrix} \begin{bmatrix} \theta_a \\ q_a \end{bmatrix} \tag{8}$$

where the matrix $\mathbf{T}$ is the inverse of the matrix $\mathbf{\Gamma}$ and is called the inverse thermal transmission matrix having elements designated as the inverse thermal transmission parameters.

If the multiplication indicated by equation (8) is carried out, one may represent the heat flow and temperature excess at the base of the fin as a superposition of two effects, one due to the tip temperature excess and one due to the heat leaving the tip:

$$\theta_b = \tau_{11}\theta_a + \tau_{12} q_a \tag{9a}$$

$$q_b = \tau_{21}\theta_a + \tau_{22} q_a \tag{9b}$$

and from these, one may form the ratio called the input admittance:

$$Y_i = \frac{q_b}{\theta_b} = \frac{\tau_{21}\theta_a + \tau_{22} q_a}{\tau_{11}\theta_a + \tau_{12} q_a} = \frac{\tau_{21} + \tau_{22}(q_a/\theta_a)}{\tau_{11} + \tau_{12}(q_a/\theta_a)} \tag{10}$$

Observe that the tip heat flow may be negligible in which case, $q_a = 0$ and the input admittance simply becomes $Y_i = \tau_{21}/\tau_{11}$. It is also possible to consider fins or spines that taper to an edge or a point. In this case $q_a = 0$ because the metal cross-sectional area for the flow of heat is zero at the tip of the fin or spine.

Fins or spines are categorized by the presence or lack of fin tip cross-sectional area and different parameterizations are required for each category; regular fins and spines possess a finite metal cross-sectional area at their tip and singular fins and spines do not.

The input admittance has been proposed as the new paramaterization for the regular fins. As indicated by equation (10), it is easily obtained from the elements of the thermal transmission matrix which have been provided by Kraus and Snider [10] for the commonly used fin shapes. However, an attempt to follow the same reasoning for a singular fin quickly falls apart because the determinant of the thermal transmission parameter matrix is equal to zero. This means that this matrix does not possess an inverse; the matrix is singular and hence the name singular fin. It is sufficient to parameterize the singular fin with a single parameter

$$\mu = q_b/\theta_b \tag{11}$$

called the thermal transmission ratio which is seen to be, like the input admittance, the ratio of the fin base heat flow to the fin base temperature excess. For the longitudinal fin of triangular profile, the thermal transmission ratio is given by

$$\mu = \frac{k\delta_b}{b} nL \frac{I_1(2nb)}{I_0(2nb)} \tag{12}$$

where $n = (h/k \sin\kappa)$ and κ is the taper ratio ($\kappa = \arctan \delta_b/2b$).

For regular fins, it is convenient to relate both heat flow flows (at the base and tip of the fin) to both temperature excesses. This linear transformation is defined by the thermal admittance matrix

$$\begin{bmatrix} q_b \\ q_a \end{bmatrix} = \mathrm{Y} \begin{bmatrix} \theta_b \\ \theta_a \end{bmatrix} = \begin{bmatrix} y_{11} & y_{12} \\ y_{21} & y_{22} \end{bmatrix} \begin{bmatrix} \theta_b \\ \theta_a \end{bmatrix} \tag{13}$$

whose elements are easily obtained, either from a conversion chart (Table 1) or from some simple manipulations involving matrix algebra.

It is contended that the fin or array input admittance should be considered as a viable alternative to the fin efficiency for design purposes. Most of this paper will be devoted to a review of the derivation and uses of this idea. However, the limitations and inadequacies of the fin efficiency should be further delineated and this is done in the next section.

TABLE 1. Conversions between Parameters

	Γ		$\mathbf{T}$		$\mathbf{Y}$	
Γ	γ_{11}	γ_{12}	τ_{22}	$-\tau_{12}$	$-y_{11}/y_{12}$	$1/y_{12}$
	γ_{21}	γ_{22}	$-\tau_{21}$	τ_{11}	$-\det \mathbf{Y}/y_{12}$	$y_{22}/{}_{12}$
$\mathbf{T}$	γ_{22}	$-\gamma_{12}$	τ_{11}	τ_{12}	$-y_{22}/y_{21}$	$1/y_{21}$
	$-\gamma_{21}$	γ_{11}	τ_{21}	τ_{22}	$-\det \mathbf{Y}/y_{21}$	y_{11}/y_{21}
$\mathbf{Y}$	$-\gamma_{11}/\gamma_{12}$	$1/\gamma_{12}$	τ_{22}/τ_{12}	$-1/\tau_{12}$	y_{11}	y_{12}
	$-1/\gamma_{12}$	γ_{22}/γ_{12}	$1/\tau_{12}$	$-\tau_{11}/\tau_{12}$	y_{21}	y_{22}

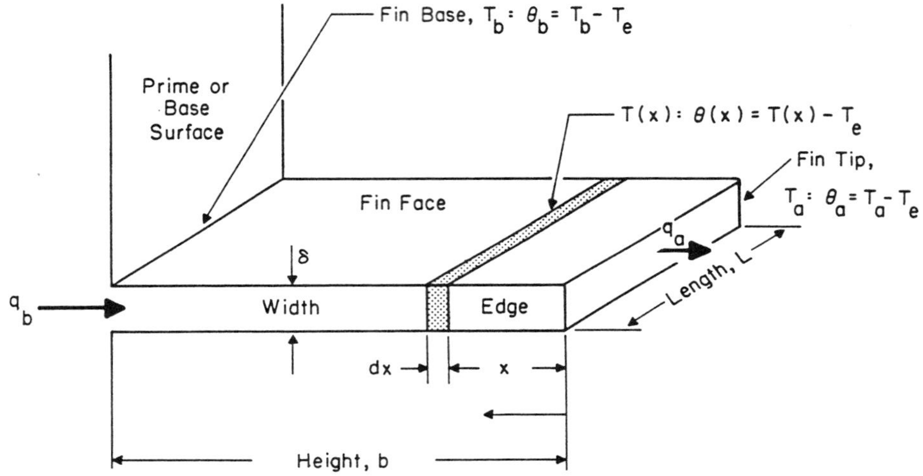

FIGURE 2. Terminology for longitudinal fin of rectangular profile.

THE LIMITATIONS OF THE FIN EFFICIENCY

The general idea of an efficiency as a performance parameter is sound; it is a dimensionless ratio comparing performance with a certain standard. However, the way that the efficiency has been defined for fins compares every fin with a different standard; what it does as compared with what it could do if conditions were perfect. Two fins in the same environment may have the same efficiency but they may transmit different quantities of heat. A simple example, based on a real-world optimization, can show that one fin can transmit more heat than another under identical environmental conditions and operate at a lower value of fin efficiency.

Using the terminology in Fig. 2, one may consider an aluminum longitudinal fin of rectangular profile ($k = 202\ W/m -^{o} K$) with a base temperature of 200° C dissipating to an environment at 100° C under natural convection conditions were h is taken as $h = 10\ W/m^2 -^{o} K$. The fin dimensions are: height, $b = 10$ cm; width, $\delta = 0.2286$ cm and length, $L = 25$ cm. Using

$$q_b = (2hk\delta)^{1/2} \, L\theta_b \tanh mb \tag{14}$$

where $\theta_b = 200 - 100 = 100°$ C is the base temperature excess and where $m = (2h/k\delta)^{1/2}$ m^{-1} is the fin performance parameter, one may compute the heat dissipation as 43.85 W. Then, using

$$\eta = \frac{\tanh mb}{mb} \tag{15}$$

the efficiency may be computed as $\eta = 0.877$.

In a desire to save mass, it is proposed that a magnesium fin ($k = 148 \; W/m -^o K$) with the same dimensions operating at identical conditions be employed. Using the dissipation of 43.85 W, one may solve equation (14) to obtain $b = 10.64 \; cm$ which shows that, because of its poorer conductivity, the height of the magnesium fin must be increased in order to accommodate an identical dissipating requirement. For the magnesium fin, the efficiency, computed from equation (15) is $\eta = 0.824$. Observe that the magnesium fin which has a huge advantage in a weight optimization dissipates the same quantity of heat but at a lower fin efficiency. Yet, for both fins, the input admittance, $Y_i = 43.85/100 = 0.439 \; W/^oC$, is identical.

If the heat transfer coefficient is reduced to say $h = 8 \; W/m^2 -^o K$, equations (14) and (15) show that the magnesium fin (with the greater fin height) will dissipate 36.30 W at an efficiency of 0.853 while the aluminum fin will dissipate less heat, 35.94 W, at the higher efficiency of 0.899. In this case, the input admittances are $0.363 W/^oC$ for the magnesium fin and 0.359 $W/^oC$ for the aluminum fin. The fins are no longer behaving identically and the magnesium fin is clearly outperforming its aluminum *design-point twin* at a lower efficiency.

At this point, it is noted that the fault is probably not in the efficiency concept but in the efficiency definition. If the efficiency had originally been called "the surface utilization factor" (which is exactly what it is) instead of the fin efficiency, it might never have entered either the preceding discussion or the calculations.

The usefulness of the fin efficiency is debatable when a finned array is considered. With the fin efficiency defined as the ratio of the heat dissipated, Q_1, to the heat dissipated if the fin possessed infinite thermal conductivity, $Q_0 = hS\theta_b$

$$\eta = Q_1/Q_0 \tag{16}$$

For a single fin without tip heat dissipation, $Q_1 = q_b$ and equation (16) prevails. For a single fin with heat dissipation from the tip governed by $q_a = h_a A_a$, where, as Sparrow, Baliga and Patankar [11] and others have shown, h_a, the coefficient of heat transfer at the fin tip, does not necessarily equal the coefficient of heat transfer on the fin faces. In this case, A_a, the fin cross-sectional area at the fin tip is, of course, equal to the fin surface area at the fin tip and the entire convective dissipation passes through the base of the fin. Thus, with $Q_1 = q_b$, the efficiency is also given by equation (16).

But, in a finned array, if the fin tip heat flow, q_a, is injected into the base of another fin or a cluster of fins, then the heat dissipated is $Q_1 = q_b - q_a$ and the fin efficiency becomes

$$\eta = \frac{Q_1}{Q_0} = \frac{q_b - q_a}{Q_0} \tag{17}$$

and it is observed that the efficiency of a fin in a finned array depends on where the fin is mounted in the array. This, it is felt, is another inadequacy of the concept of the fin efficiency.

An example will help to illustrate the differences between η, Γ, and μ. Consider the finned array shown in Fig. 3 where the base temperature excess is 50° C. Fins 1, 2, 4 and 5 are identical longitudinal fins of rectangular profile and fins 3 and 6 are identical triangular profile fins. The specifications are as follows:

$$\text{Fins 1,2,4 and 5; } \delta = 0.635 \text{ cm and b} = 5.08 \text{ cm}$$

$$\text{Fins 3 and 6; } \delta_b = 0.635 \text{ cm and b} = 4.00 \text{ cm}$$

For all fins, $L = 30.48$ cm, $k = 180\ W/m - ^\circ K$ and $h = 100\ W/m^2 - ^\circ K$. All of the heat flows and temperature excesses displayed in Fig. 3 were determined using the techniques developed by Kraus, Snider and Doty [9] and which will be discussed here in a later section.

One may compute the six fin efficiencies using equation (17) with $hS = 3.0968\ W/^\circ C$ for fins 1, 2 4 and 5 and $hS = 2.4384\ W/^\circ C$ for fins 3 and 6.

$$\eta_1 = \frac{135.09}{50(3.0968)} = 0.872$$

$$\eta_2 = \frac{190.62 - 68.55}{50(3.0968)} = 0.788$$

$$\eta_3 = \frac{68.55}{31.78(2.4384)} = 0.885$$

$$\eta_4 = \frac{219.34 - 103.99}{50(3.0968)} = 0.745$$

$$\eta_5 = \frac{103.99 - 37.40}{27.28(3.0968)} = 0.788$$

$$\eta_6 = \frac{37.40}{17.33(2.4384)} = 0.885$$

Thus, the identical rectangular fins, all operating in the same environment, all have different efficiencies. Yet all have the same thermal transmission matrix which can be evaluated from equations presented in the next section

$$\mathbf{\Gamma} = \begin{bmatrix} 1.2344 & -0.1570 \\ -3.3351 & 1.2344 \end{bmatrix}$$

On the other hand, the singular triangular fin, which can only be used as the most remote fin in any leg within the array, possesses a thermal transmission ratio computed from equation (12), $\mu = 2.1567\ W/^\circ C$ and it must have the same efficiency wherever it is employed.

FINDING THE THERMAL TRANSMISSION MATRIX

The temperature excess in any longitudinal or radial fin or spine is governed, in the steady state, by a differential equation of the form

$$\frac{d}{dx}\left[k(x)A(x)\frac{d\theta(x)}{dx}\right] - h(x)\frac{dS(x)}{dx}\theta(x) = 0 \tag{18}$$

In most cases, x is a height coordinate measured from fin tip to base, but for radial fins, x is the radial coordinate measured from base to tip. Here $h(x)$ is the heat transfer coefficient and $k(x)$ is the thermal conductivity which may vary with x. The cross-sectional area for the flow of heat by conduction is $A(x)$ and $dS(x)$ is the infinitesimal surface area of the fin faces over which a steady state energy balance may be taken:

$$q(x) = k(x)A(x)[\theta_b \lambda_1'(x) + q_b \lambda_2'(x)] \tag{21b}$$

In matrix form, equations (21) become

$$\begin{bmatrix} \theta(x) \\ q(x) \end{bmatrix} = \begin{bmatrix} 1 & 0 \\ 0 & k(x)A(x) \end{bmatrix} \begin{bmatrix} \lambda_1(x) & \lambda_2(x) \\ \lambda_1'(x) & \lambda_2'(x) \end{bmatrix} \begin{bmatrix} \theta_b \\ q_b \end{bmatrix} \tag{22}$$

The second matrix to the right of the equal sign can be seen to resemble the familiar Wronskian. The thermal transmission matrix is generated when x is set equal to a (a may equal zero) in equation (22)

$$\begin{bmatrix} \theta_a \\ q_a \end{bmatrix} = \mathbf{\Gamma} \begin{bmatrix} \theta_b \\ q_b \end{bmatrix} = \begin{bmatrix} \gamma_{11} & \gamma_{12} \\ \gamma_{21} & \gamma_{22} \end{bmatrix} \begin{bmatrix} \theta_b \\ q_b \end{bmatrix} \tag{7}$$

where

$$\gamma_{11} = \lambda_1(a) \tag{23a}$$

$$\gamma_{12} = \lambda_2(a) \tag{23b}$$

$$\gamma_{21} = k(a)A(a)\lambda_1'(a) \tag{23c}$$

and

$$\gamma_{22} = k(a)A(a)\lambda_2'(a) \tag{23d}$$

Observe that if $A(a) = 0$, $\gamma_{21} = \gamma_{22} = 0$ and

$$\mathbf{\Gamma} = \begin{bmatrix} \lambda_1(0) & \lambda_2(0) \\ 0 & 0 \end{bmatrix}$$

This matrix has a determinant equal to zero and is termed singular because it has no inverse. This is why fins and spines that taper to a zero cross section are called singular and why the thermal transmission ratio μ was proposed to parameterize them. An example of the foregoing procedure used to determine $\mathbf{\Gamma}$ for the longitudinal fin of rectangular profile now follows.

EXAMPLE: FINDING THE PARAMETERS

In Fig. 2, consider that the origin of the coordinate is at the fin tip with positive orientation toward the fin base. Here, the fin width is constant, $\delta(x) = \delta$ so that $A(x) = \delta L$ and $dS(x) = L dx$. With constant thermal conductivity, $k(x) = k$, equation (18) reduces to

$$\frac{d^2\theta}{dx^2} - m^2\theta = 0$$

where $m = (2h/k\delta)^{1/2}$.

This differential equation has a general solution

$$\theta(x) = C_1 e^{mx} + C_2 e^{-mx}$$

where the arbitrary constants C_1 and C_2 are to be evaluated from the *initial value data*

$$\theta(x = b) = \theta_b$$

and

$$q(x = b) = q_b$$

This makes

$$\theta_b = C_1 e^{mb} + C_2 e^{-mb}$$

and

$$q_b = k\delta mL[C_1 e^{mb} - C_2 e^{-mb}]$$

or

$$q_b = Y_o[C_1 e^{mb} - C_2 e^{-mb}]$$

where $Y_o = k\delta mL = (2hk\delta)^{1/2} L$ is designated as the characteristic admittance of the fin.

It is then a matter of algebra to evaluate C_1 and C_2 such that

$$\theta(x) = \cosh\,(b - x)\,\theta_b - \frac{1}{Y_o}\sinh m(b - x)\,q_b$$

and

$$q(x) = Y_o \sinh m(b - x)\,\theta_b + \cosh m(b - x)\,q_b$$

Here reference to equations (19) shows that

$$\lambda_1(x) = \cosh m(b - x) \qquad ;\lambda_1(b) = 1$$

$$\lambda_2(x) = -\frac{1}{Y_o}\sinh m(b - x) \qquad ;\lambda_2(b) = 0$$

$$\lambda_1'(x) = -m \sinh m(b - x) \qquad ;\lambda_1'(b) = 0$$

$$\lambda_2'(x) = \frac{m}{Y_o}\cosh m(b - x) \qquad ;\lambda_2'(b) = \frac{m}{Y_o} = \frac{1}{k\delta L}$$

and at $x = a = 0$ equations (23) provide

$$\gamma_{11} = \lambda_1(0) = \cosh mb \tag{24a}$$

$$\gamma_{12} = \lambda_2(0) = -\frac{1}{Y_o}\sinh mb \tag{24b}$$

$$\gamma_{21} = k\delta L\lambda_1'(0) = -k\delta Lm \sinh mb = -Y_o \sinh mb \tag{24c}$$

and

$$\gamma_{22} = h\delta L\lambda_2'(0) = \frac{k\delta Lm}{Y_o}\cosh mb = \cosh mb \tag{24d}$$

Thus, for the longitudinal fin of rectangular profile

$$\mathbf{\Gamma} = \begin{bmatrix} \cosh mb & -(1/Y_o)\sinh mb \\ -Y_o \sinh mb & \cosh mb \end{bmatrix}$$

The inverse thermal transmission matrix is, as the name suggests, the inverse of the thermal transmission matrix. With the determinant of $\mathbf{\Gamma}$ equal to

$$det\,\mathbf{\Gamma} = \cosh^2 mb - \sinh^2 mb = 1$$

$\mathbf{T}$ becomes

$$\mathbf{T} = \mathbf{\Gamma}^{-1} = \begin{bmatrix} \cosh mb & (1/Y_o)\sinh mb \\ Y_o \sinh mb & \cosh mb \end{bmatrix}$$

Here

$$\tau_{11} = \cosh mb \tag{25a}$$

$$\tau_{12} = \frac{1}{Y_o} \sinh mb \tag{25b}$$

$$\tau_{21} = Y_o \sinh mb \tag{25c}$$

and

$$\tau_{22} = \cosh mb \tag{25d}$$

The thermal admittance matrix can then be obtained from an exercise in matrix algebra or from Table 1

$$\mathbf{Y} = \begin{bmatrix} Y_o \coth mb & -Y_o \operatorname{csch} mb \\ Y_o \operatorname{csch} mb & -Y_o \coth mb \end{bmatrix}$$

with

$$y_{11} = Y_o \coth mb \tag{26a}$$

$$y_{12} = -Y_o \operatorname{csch} mb \tag{26b}$$

$$y_{21} = Y_o \operatorname{csch} mb \tag{26c}$$

and

$$y_{22} = -Y_o \coth mb \tag{26d}$$

For radial fins, a minus sign must be introduced into equation (20) because x, now the radial coordinate, is measured in the same direction as q (base to tip). The net result of this modification is to change the sign of γ_{12} and γ_{21} in equations (24b) and (24c).

THE ARRAY INPUT ADMITTANCE

The input admittance, which is the proposed parameterization

$$Y_i = \frac{q_b}{\theta_b} \tag{4}$$

is obtained from the elements of either the thermal transmission matrix

$$Y_i = \frac{\gamma_{21} - \gamma_{11}(q_a/\theta_a)}{-\gamma_{22} + \gamma_{12}(q_a/\theta_a)}$$

or the inverse thermal transmission matrix

$$Y_i = \frac{\tau_{21} + \tau_{22}(q_a/\theta_a)}{\tau_{11} + \tau_{12}(q_a/\theta_a)} \tag{27}$$

In either event, the ratio q_a/θ_a is known. For single fin analysis, $q_a = 0$ or, if there is fin tip dissipation, $q/\theta_a = h_a A_a$. In this case, the input admittance, Y_i, is equal to the individual fin thermal transmission ratio, μ. In array analysis, it must be recognized that the most remote connection in any array leg must either be an individual fin or a cluster of fins. If the connection is an individual fin, considerations of tip heat loss or no tip heat loss pertaining to individual fins prevail. If the connection is a cluster of fins, q_a/θ_a can be evaluated from the cluster algorithm which is developed in a subsequent section. In this case, Y_i is called the array input admittance.

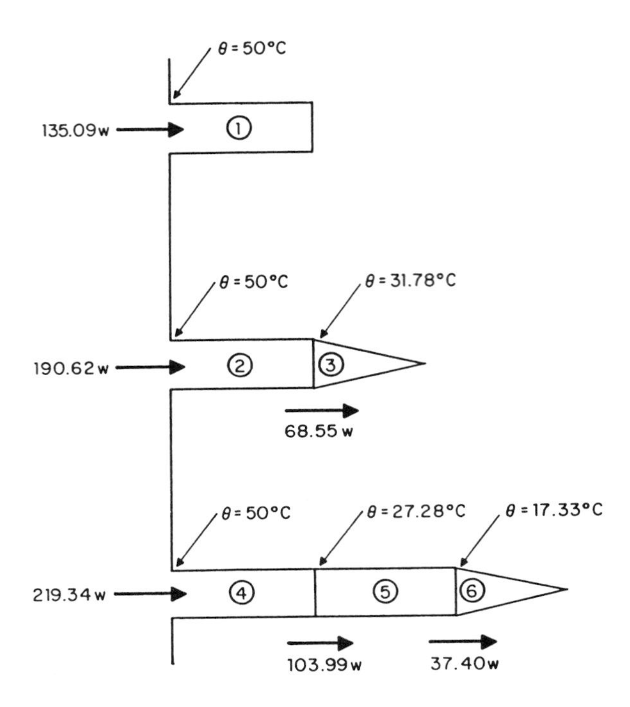

FIGURE 3. A finned array containing six fins.

The theory of linear, homogeneous, second-order differential equations (with or without variable coefficients) shows that equation (18) is singular at all points where $k(x)A(x) = 0$ and is regular otherwise. Physically, because the thermal conductivity is not identically zero, singular points may only occur where the fin width is zero. In turn, this can take place at fin tips where $x = 0$ or $x = a$ depending on the origin of the height coordinate. The theory further dictates that equation (18) possesses two independent solutions $\lambda_1(x)$ and $\lambda_2(x)$ which satisfy the initial conditions at the base of the fin where $x = b$

$$\lambda_1(b) = 1 : \lambda_1'(b) = 0 \tag{19a}$$

where the prime indicates a first derivative and

$$\lambda_2(b) = 0 : \lambda_2'(b) = \frac{1}{k(b)A(b)} \tag{19b}$$

The heat flow $q(x)$ in the fin is always taken as positive from base to tip. Thus for non-radial fins, $q(x)$ is given by

$$q(x) = k(x)A(x)\frac{d\theta(x)}{dx} \tag{20}$$

Therefore, one can use the solutions λ_1 and λ_2 to assemble the expressions for the temperature excess $\theta(x)$ and heat flow $q(x)$ at any point in the fin in terms of their values θ_b and q_b at the fin base

$$\theta(x) = \theta_b\lambda_1(x) + q_b\lambda_2(x) \tag{21a}$$

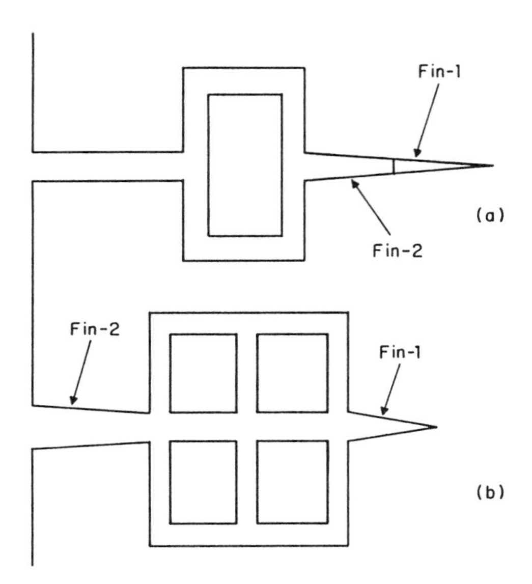

FIGURE 4. Two arrays of extended surface, each with three longitudinal fin profiles. The array in (a) contains no loops in a graph theoretical sense and in both arrays, the fin of triangular profile is fin−1, the fin of trapezoidel profile is fin−2 and all other fins are fins of rectangular profile.

ALGORITHMS FOR THE COMBINATION OF FINS

Figure 4 displays two finned arrays. Each of them contains three longitudinal fin profiles and in each, the fin of triangular profile is the most remote fin in the array. The array shown in Fig. 4b contains four loops in the graph theoretical sense; four loops that cannot be removed by a simple parallel combination. The array in Fig. 4a contains 9 fins. Algorithms for the combination of fins in an array that does not contain loops in the graph theoretical sense will be developed in this section.

The Cascade Algorithm.

Consider the simplest case of two fins in cascade shown in Fig. 5. Here, continuity dictates that the heat leaving fin-2 must enter fin-1 and that at the point of intersection, the temperature excesses must match. Thus, one may set down a matrix equality embracing continuity and compatibility using a for tip conditions and b for base conditions

$$\begin{bmatrix} \theta_{a2} \\ q_{a2} \end{bmatrix} = \begin{bmatrix} \theta_{b1} \\ q_{b1} \end{bmatrix} \tag{28}$$

In addition, each fin has a $\mathbf{T}$ parameterization which is a linear transformation from tip conditions to base conditions. For fin-1

$$\begin{bmatrix} \theta_{b1} \\ q_{b1} \end{bmatrix} = \mathbf{T}_1 \begin{bmatrix} \theta_{a1} \\ q_{a1} \end{bmatrix} \tag{29}$$

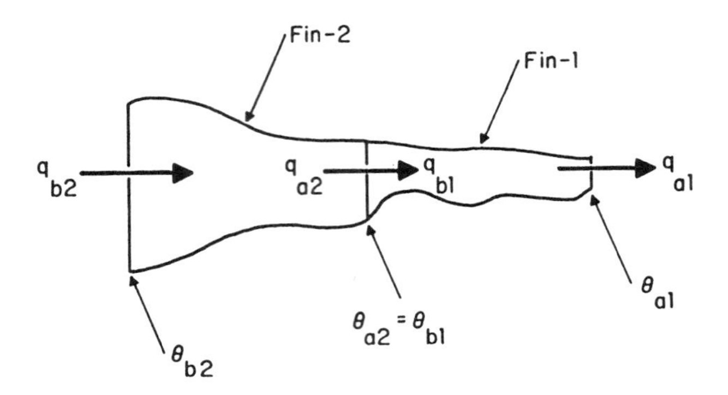

FIGURE 5. Two fins in cascade.

and for fin-2

$$\begin{bmatrix} \theta_{b2} \\ q_{b2} \end{bmatrix} = \mathbf{T}_2 \begin{bmatrix} \theta_{a2} \\ q_{a2} \end{bmatrix} \tag{30}$$

A simple exercise in matrix algebra shows that there is an equivalent **T** representation that maps conditions from the tip of fin-1 to the base of fin-2 which is at the base of the array. First take equation (29) and use equation (28)

$$\begin{bmatrix} \theta_{b1} \\ q_{b1} \end{bmatrix} = \begin{bmatrix} \theta_{a2} \\ q_{a2} \end{bmatrix} = \mathbf{T}_1 \begin{bmatrix} \theta_{a1} \\ q_{a1} \end{bmatrix}$$

and then use equation (30)

$$\begin{bmatrix} \theta_{b2} \\ q_{b2} \end{bmatrix} = \mathbf{T}_2 \begin{bmatrix} \theta_{a2} \\ q_{a2} \end{bmatrix} = \mathbf{T}_2 \mathbf{T}_1 \begin{bmatrix} \theta_{a1} \\ q_{a1} \end{bmatrix} = \mathbf{T}_e \begin{bmatrix} \theta_{a1} \\ q_{a1} \end{bmatrix} \tag{31}$$

This shows the existence of an equivalent thermal transmission matrix that is a simple matrix product of the of the two individual inverse thermal transmission matrices (in the proper order because matrix multiplication, in general, is not commutative).

This may be extended to n fins in cascade

$$\mathbf{T}_e = \mathbf{T}_n \mathbf{T}_{n-1} \mathbf{T}_{n-2} \cdots \mathbf{T}_2 \mathbf{T}_1 \tag{32}$$

and equation (32) is the cascade algorithm.

The array input admittance can be obtained using equation (27) with the elements of $\mathbf{T}_e$ with $q_a/\theta_a = 0$ or $q_a = h_{a1} A_{a1}$. Alternatively, the input admittance may be determined from two applications of equation (27), first using the elements of $\mathbf{T}_1$ for fin-1 with $q_a/\theta_a =$ 0 or $h_{a1} A_{a1}$ to find the input admittance at the base of fin-1. This input admittance, by the continuity and compatibility conditions of equation (28) is the value of q_a/θ_a for fin-2. Another application of equation (27), this time using the elements of $\mathbf{T}_2$ for fin-2 will provide the sought after result.

The foregoing seems to indicate an apparent *toss-up* between the use of equation (27) which is frequently called the "reflection relationship" and the use of the equivalent $\mathbf{T}_e$. Indeed, the usefulness of the cascade algorithm may be concealed by the simplicity of equation (27).

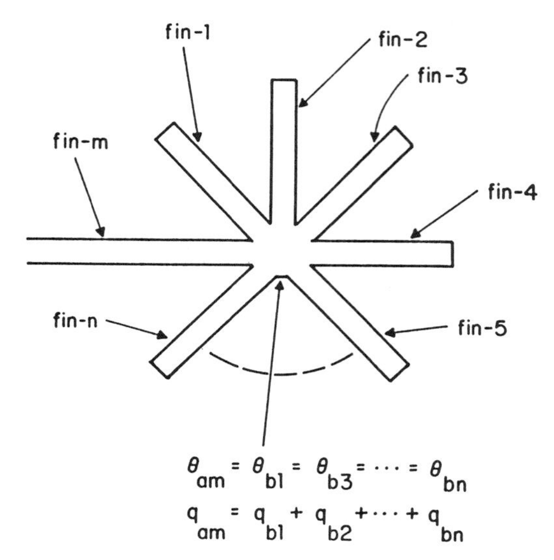

FIGURE 6. Fins in cluster. There are n fins attached to the tip of fin-m and the tip of fin-m is designated as the cluster point.

Consider, however, that the determination of an array input admittance for any convective heat transfer coefficient variation over the faces of any or all of the fins in the array is easily accomplished by an application of the cascade algorithm. One merely makes a piecewise continuous approximation (to the accuracy that is required) of the heat transfer coefficient. This dictates the number of sub-fins that must be employed to represent the original fin. The elements of $\mathbf{T}$ (via $\mathbf{\Gamma}$) can then be obtained for each sub-fin using the average value of h in the individual segments of the h-curve. A repeated matrix multiplication then gives the elements of $\mathbf{T}_e$ and then, equation (27) with $q_a = 0$ gives the required Y_i.

To be sure, efficiencies have been developed for various heat transfer coefficient variations on longitudinal fins of rectangular profile. Gardner [12] proposed a power series variation, Han and Lefkowitz [13] used a parabolic variation and Chen and Zyskowski [14] employed an exponential variation. However, the cascade algorithm will efficiently handle convective coefficients of any distribution on the faces of any fin or spine.

Fins in Cluster.

Figure 6 shows n-fins that are appended to fin-m. These fins are said to be in a cluster arrangement and the tip of fin-m is called a cluster point. Here, continuity dictates that the heat that leaves fin-m must divide among fins-1 through n. Moreover, compatibility dictates that the temperature excesses of all fins involved in the cluster must match. Thus, with the usage of subscripts a and b, which has now become customary, it is observed that

$$q_{am} = q_{b1} + q_{b2} + q_{b3} + \cdots + q_{bn} \tag{33a}$$

and

$$\theta_{am} = \theta_{b1} = \theta_{b2} = \theta_{b3} = \cdots = \theta_{bn} \tag{33b}$$

Consider the admittance at the tip of fin-m which is termed the cluster admittance, Y_c,

$$Y_{am} = Y_c = q_{am}/\theta_{am}$$

and employ equations (33). This yields

$$Y_{am} = Y_c = \frac{q_{b1} + q_{b2} + q_{b3} + \cdots + q_{bn}}{\theta_{am}}$$

or

$$Y_{am} = Y_c = \frac{q_{b1}}{\theta_{b1}} + \frac{q_{b2}}{\theta_{b2}} + \frac{q_{b3}}{\theta_{b3}} + \cdots + \frac{q_{bn}}{\theta_{bn}} \tag{34}$$

Each term on the right side of equation (34) represents an individual fin input admittance or, a thermal transmission ratio. Thus a cluster admittance may be defined as

$$Y_c = \sum_{i=1}^{n} Y_i = \sum_{i=1}^{n} \mu_i \tag{35}$$

This addition of input admittances to form the cluster input admittance is the cluster algorithm.

Fins in Parallel.

A parallel arrangement of n fins is shown in Fig. 7. Notice that a single fin, designated as fin-m is delivering heat through its tip to the parallel combination and that a single fin, designated as fin-p, is picking up, at its base, all of the heat leaving the parallel combination. In this case, continuity demands that

$$q_{am} = q_{b1} + q_{b2} + q_{b3} + \cdots + q_{bn} \tag{36a}$$

and

$$q_{bp} = q_{a1} + q_{a2} + q_{a3} + \cdots + q_{an} \tag{36b}$$

In addition, compatibility requires that at point-m

$$\theta_{am} = \theta_{b1} = \theta_{b2} = \theta_{b3} = \cdots = \theta_{bn} \tag{37a}$$

and at point-p

$$\theta_{bp} = \theta_{a1} = \theta_{a2} = \theta_{a3} = \cdots = \theta_{an} \tag{37b}$$

Use of the admittance parameter matrix defined by equation (13) permits a linear transformation of temperature excess to heat flow for any individual fin. Thus, from equations (36)

$$\begin{bmatrix} q_{am} \\ q_{bp} \end{bmatrix} = \begin{bmatrix} q_{b1} \\ q_{a1} \end{bmatrix} + \begin{bmatrix} q_{b2} \\ q_{a2} \end{bmatrix} + \begin{bmatrix} q_{b3} \\ q_{a3} \end{bmatrix} + \cdots + \begin{bmatrix} q_{bn} \\ q_{an} \end{bmatrix}$$

and the use of equation (13) provides

$$\begin{bmatrix} q_{am} \\ q_{bp} \end{bmatrix} = \mathbf{Y}_1 \begin{bmatrix} \theta_{b1} \\ \theta_{a1} \end{bmatrix} + \mathbf{Y}_2 \begin{bmatrix} \theta_{b2} \\ \theta_{a2} \end{bmatrix} + \mathbf{Y}_3 \begin{bmatrix} \theta_{b3} \\ \theta_{a3} \end{bmatrix} + \cdots + \mathbf{Y}_n \begin{bmatrix} \theta_{bn} \\ \theta_{an} \end{bmatrix} \tag{38}$$

But, by equations (37)

$$\begin{bmatrix} \theta_{am} \\ \theta_{bp} \end{bmatrix} = \begin{bmatrix} \theta_{b1} \\ \theta_{a1} \end{bmatrix} = \begin{bmatrix} \theta_{b2} \\ \theta_{a2} \end{bmatrix} = \begin{bmatrix} \theta_{b3} \\ \theta_{a3} \end{bmatrix} = \cdots = \begin{bmatrix} \theta_{bn} \\ \theta_{an} \end{bmatrix}$$

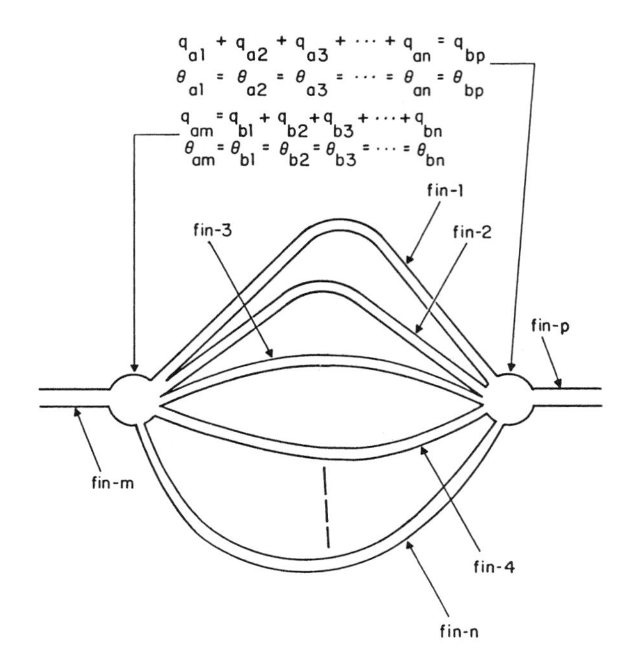

FIGURE 7. Parallel combination of n-fins.

and when this is incorporated into equation (38)

$$\begin{bmatrix} q_{am} \\ q_{bp} \end{bmatrix} = [\mathbf{Y}_1 + \mathbf{Y}_2 + \mathbf{Y}_3 + \cdots + \mathbf{Y}_n] \begin{bmatrix} \theta_{am} \\ \theta_{bp} \end{bmatrix} \tag{39}$$

This shows that there is an equivalent thermal admittance matrix, $\mathbf{Y}_e$, that relates the total heat flows to the temperature excesses at the base and tip of a parallel combination of fins

$$\begin{bmatrix} q_b \\ q_a \end{bmatrix} = \mathbf{Y}_e \begin{bmatrix} \theta_b \\ \theta_a \end{bmatrix}$$

where

$$\mathbf{Y}_e = \mathbf{Y}_1 + \mathbf{Y}_2 + \mathbf{Y}_3 + \cdots + \mathbf{Y}_n \tag{40}$$

Equation (40) is the parallel algorithm.

EXAMPLE

The versatility of the algorithms developed in the preceding section can be demonstrated through an example. The array shown in Fig. 8 containing 15 individual longitudinal fins will be analyzed. Assume that the following groups of fins of rectangular profile possess identical dimensions:

fins-1, 2 and 3 (with no tip heat loss),

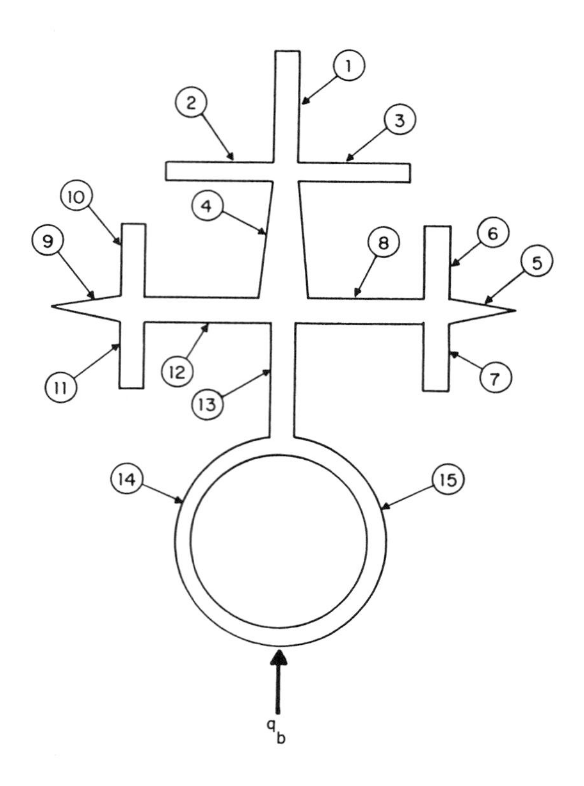

FIGURE 8. Finned array composed of 15 individual fins.

fins-6, 7, 10 and 11 (also with no tip heat loss),

fins-8, 12 and 13 and

fins-14 and 15

Observe that fin-4 is a fin of trapezoidal profile and that fins-5 and 9 are identical fins of triangular profile.

The key equation is equation (27) which will be referred to as the reflection relationship. A computationally efficient step-by-step procedure for the determination of the array input admittance now follows:

1. Determine the inverse thermal transmission parameters for fins-1 through 15 (except for fins-5 and 9) using the catalog contained in Kraus, Snider and Doty [9] or Kraus and Snider [10]. For all of the rectangular profile fins, equations (25) pertain.
2. Determine the thermal transmission ratio for fins-5 and 9. For the triangular profile fin, equation (12) should be employed.
3. For fins-14 and 15, the thermal admittance parameters will be needed and these can be

determined from the inverse thermal transmission parameters by using Table 1.

4. Use the reflection relationship with $q_a = 0$ to obtain the input admittance of fins-1, 2 and 3. Then use the cluster algorithm of equation (35) to find q_a/θ_a at the tip of fin-4. Another application of the reflection relationship will provide the input admittance at the base of fin-4.
5. Use the reflection relationship with $q_a = 0$ to determine the input admittance of fins-6 and 7. By the cluster algorithm, the value of q_a/θ_a at the tip of fin-8 will be equal to the sum of the input admittances of fins-6 and 7 plus the thermal transmission ratio of fin-5. Using this value of q_a/θ_a in the reflection relationship provides the input admittance at the base of fin-8.
6. An identical procedure to that given in step-5 will yield the input admittance at the base of fin-12. The input admittances at the bases of fin-8 and 12 will be equal because the applicable fin dimensions are all equal.
7. The cluster algorithm then yields the value of q_a/θ_a at the tip of fin-13. It says that the value of this *"tip admittance"* must be equal to the sum of the input admittances at the bases of fins-4, 8 and 12. The reflection relationship can then be used to determine the input admittance at the base of fin-13.
8. Fins-14 and 15 are identical and are in parallel. Table 1 was employed in step-3 to obtain the thermal admittance parameters for each of these fins. The equivalent Thermal Admittance Matrix is then obtained from the sum of the two individual Thermal Admittance Matrices. An equivalent Thermal Transmission Matrix can then be obtained from an application of Table 1. These are then used in the reflection relationship with the q_a/θ_a value determined at the end of step-7 to determine the sought after total input admittance for the array.

There is no need to obtain the fin efficiency because the input admittance may be multiplied by the base temperature excess to obtain the heat dissipated by all fifteen fins in the array.

ANALYSIS OF FINNED ARRAYS CONTAINING LOOPS

Finned arrays containing loops in the graph theoretical sense such as the array shown in Fig. 4b may be analyzed by the general array algorithm proposed by Snider and Kraus [15] or by a method of node analysis proposed by Kraus, Snider and Landis [16]. The node analysis method is described here in a matrix oriented method believed to be more general than the more specific procedure given in the 1982 work.

All regular fins and spines possess an important property known as reciprocity. The regular fins possess a thermal transmission matrix whose determinant is equal to unity (-1 in the case of the radial fins) and, because of this, the off-diagonal elements of the thermal admittance matrix are related by

$$y_{12} = -y_{21} \tag{41}$$

This permits any reciprocal fin to be represented as the equivalent pi-network consisting of three thermal admittances shown in Fig. 9.

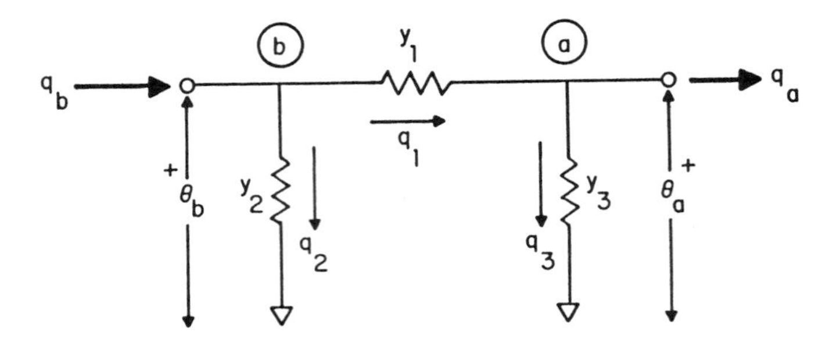

FIGURE 9. Equivalent pi-network.

The Equivalent Pi-network.

Application of continuity at points a and b, called node points or merely nodes, gives

$$q_b = q_1 + q_2$$
$$q_a = q_1 - q_3$$

However, each heat flow, q_1, q_2 and q_3 in Fig. 9 can be represented as a product of an admittance and a temperature excess:

$$q_1 = Y_1(\theta_b - \theta_a)$$
$$q_2 = Y_2\theta_b$$
$$q_3 = Y_3\theta_a$$

so that

$$q_b = Y_1(\theta_b - \theta_a) + Y_2\theta_b$$
$$q_a = Y_1(\theta_b - \theta_a) - Y_3\theta_a$$

or

$$q_b = (Y_1 + Y_2)\theta_b - Y_1\theta_a \tag{42a}$$
$$q_a = Y_1\theta_b - (Y_1 + Y_3)\theta_a \tag{42b}$$

Equations (42) may be put into matrix form:

$$\begin{bmatrix} q_b \\ q_a \end{bmatrix} = \begin{bmatrix} (Y_1 + Y_2) & -Y_1 \\ Y_1 & -(Y_1 + Y_3) \end{bmatrix} \begin{bmatrix} \theta_b \\ \theta_a \end{bmatrix}$$

and then compared to equation (13), which defines the admittance parameter matrix,

$$\begin{bmatrix} q_b \\ q_a \end{bmatrix} = \begin{bmatrix} y_{11} & y_{12} \\ y_{21} & y_{22} \end{bmatrix} \begin{bmatrix} \theta_b \\ \theta_a \end{bmatrix} \tag{13}$$

The comparison shows that

$$Y_1 + Y_2 = y_{11}$$
$$Y_1 = -y_{12} = y_{21}$$

and

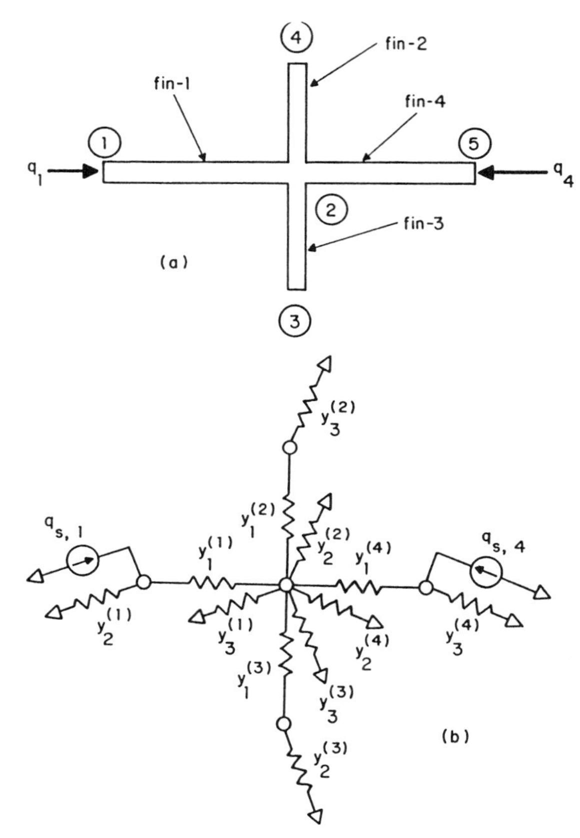

FIGURE 10. (a) An array of four fins with heat input at opposite ends and, (b) an equivalent network composed of four pi-networks to represent the array.

$$Y_1 + Y_3 = -y_{22}$$

Then, it is only a matter of algebra to show that

$$Y_1 = -y_{12} = y_{21} \tag{43a}$$
$$Y_2 = y_{11} + y_{12} = y_{11} - y_{21} \tag{43b}$$
$$Y_3 = y_{12} - y_{22} = -(y_{21} + y_{22}) \tag{43c}$$

It is noted that as long as $y_{12} = -y_{21}$, any fin may be represented as an equivalent pi. All regular fins are reciprocal and possess this important property. Hence, arrays composed of regular fins may be treated as a combination of equivalent pi-networks representing the individual fins in the array and connected as the array is connected. An example of a four fin array with unequal heat inputs provided at the opposite ends is displayed, along with its network representation, in Fig. 10.

A simplification is shown in Fig. 10a where

$$Y_a = Y_3^{(1)} + Y_2^{(2)} + Y_3^{(3)} + Y_2^{(4)}$$

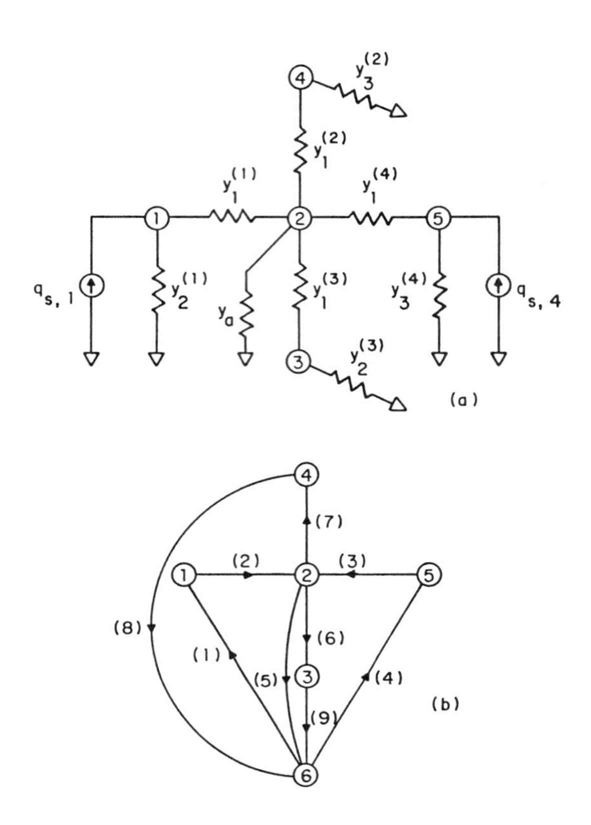

FIGURE 11. (a) A simplification of the network in Fig. 10b and, (b) an oriented graph for the network of (a). The oriented graph contains six nodes and nine branches.

and Fig. 11b shows an oriented graph of the network in Fig. 11a. The use of circles to designate nodes (points where two or more branches intersect) and numbers in parentheses to designate branches may be noted. The heat sources are included in branches-1 and 4 and the branch orientations are discretionary except for the branches containing the heat sources where the positive orientation must correspond to the direction of the heat input. The objective is to obtain all node temperature excesses in a computationally efficient manner and then, knowing the temperature excesses and heat flows at nodes-1 and 5, to determine the input admittances at nodes-1 and 5.

The General Branch.

The most general case of the "jth" branch is the case, shown in Fig. 12, where a heat source, q_{sj}, and a temperature source, ΔT_{sj}, is present. The branch must also contain one of the thermal admittances.

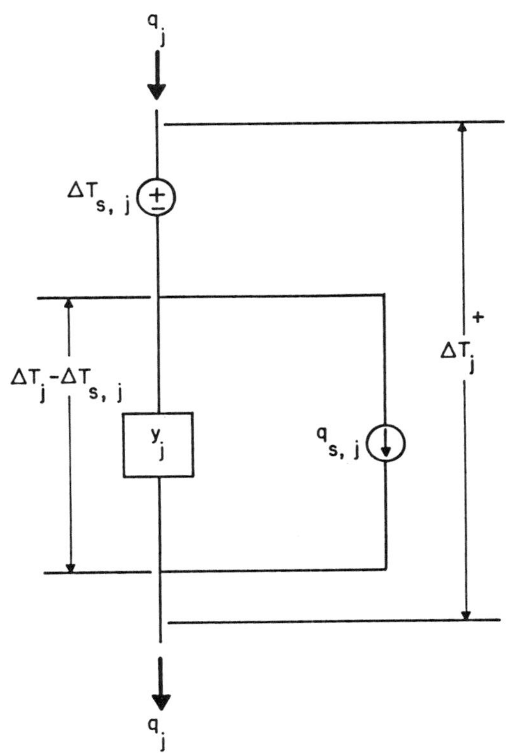

FIGURE 12. General branch containing an admittance, a heat source and a temperature source.

Because the total temperature drop across each branch must be the sum of component temperature drops, it is easy to see that the temperature drop across the parallel combination of the heat source, q_{sj}, and the admittance, Y_j, must be $\Delta T_j - \Delta T_{sj}$ where ΔT_{sj} is the temperature source in branch j. Continuity then dictates that the branch heat flow, q_j, is represented, in matrix form, for all j branches, by

$$\mathbf{Q} = \mathbf{Q}_s + \mathbf{Y}(\mathbf{\Delta T} - \mathbf{\Delta T}_s) \tag{44}$$

Here, with b branches

$\mathbf{Q}$ is a $b \times 1$ heat flow vector
$\mathbf{Q}_s$ is a $b \times 1$ heat source vector (all nodes may possess heat sources)
$\mathbf{Y}$ is a $b \times b$ branch admittance matrix which is diagonal
$\mathbf{\Delta T}_s$ is a $b \times 1$ branch temperature source vector which may be (and most often is) null
$\mathbf{\Delta T}$ is a $b \times 1$ branch temperature vector.

The Node Branch Incidence Matrix.

For a graph containing n_t nodes and b branches, a matrix that exactly represents the graph may be formulated. This matrix which is called the node-branch incidence matrix will have n_t rows and b columns and will contain elements $a_{jk} = +1$, -1 or 0 in accordance with the following scheme.

$$a_{jk} = \begin{cases} +1 & \text{if branch } j \text{ leaves node } k \\ -1 & \text{if branch } j \text{ enters node } k \\ 0 & \text{if branch } j \text{ does not touch node } k \end{cases} \tag{45}$$

For example, the node-branch incidence matrix for the oriented graph in Fig. 11b will be $n = 6 \times b = 9(6 \times 9)$:

$$\mathbf{A_a} = \begin{bmatrix} -1 & 1 & 0 & 0 & 0 & 0 & 0 & 0 & 0 \\ 0 & -1 & -1 & 0 & 1 & 1 & 1 & 0 & 0 \\ 0 & 0 & 0 & 0 & 0 & -1 & 0 & 0 & 1 \\ 0 & 0 & 0 & 0 & 0 & 0 & -1 & 1 & 0 \\ 0 & 0 & 1 & -1 & 0 & 0 & 0 & 0 & 0 \\ 1 & 0 & 0 & 1 & -1 & 0 & 0 & -1 & -1 \end{bmatrix} \tag{46}$$

It can be noted that every column contains a single +1 and a single −1 and that a summation of all elements in each column yields a zero.

The reduced incidence matrix contains $n - 1$ rows and b branches. It is obtained from $\mathbf{A}_a$ by merely deleting the row that represents the surrounding environment. In Fig. 11, node-6 represents the environment and when it is deleted

$$\mathbf{A} = \begin{bmatrix} -1 & 1 & 0 & 0 & 0 & 0 & 0 & 0 & 0 \\ 0 & -1 & -1 & 0 & 1 & 1 & 1 & 0 & 0 \\ 0 & 0 & 0 & 0 & 0 & -1 & 0 & 0 & 1 \\ 0 & 0 & 0 & 0 & 0 & 0 & -1 & 1 & 0 \\ 0 & 0 & 1 & -1 & 0 & 0 & 0 & 0 & 0 \end{bmatrix} \tag{47}$$

Because the row representing the environment has been deleted, the analysis becomes a node-to-datum analysis with the environment as the datum and all node temperatures to be considered as temperature excesses.

Continuity.

The product **AQ** which postmultiplies an $n \times b$ matrix by a $b \times 1$ column vector presents a statement of continuity at each node

$$\mathbf{AQ} = \begin{bmatrix} -1 & 1 & 0 & 0 & 0 & 0 & 0 & 0 & 0 \\ 0 & -1 & -1 & 0 & 1 & 1 & 1 & 0 & 0 \\ 0 & 0 & 0 & 0 & 0 & -1 & 0 & 0 & 1 \\ 0 & 0 & 0 & 0 & 0 & 0 & -1 & 1 & 0 \\ 0 & 0 & 1 & -1 & 0 & 0 & 0 & 0 & 0 \end{bmatrix} \begin{bmatrix} q_1 \\ q_2 \\ q_3 \\ q_4 \\ q_5 \\ q_6 \\ q_7 \\ q_8 \\ q_9 \end{bmatrix} = \begin{bmatrix} q_2 - q_1 \\ q_5 + q_6 + q_7 - q_2 - q_3 \\ q_9 - q_6 \\ q_8 - q_7 \\ q_3 - q_4 \end{bmatrix} = \mathbf{0}$$

and this can be confirmed by an inspection of the oriented graph in Fig. 11b. Thus the equation

$$\mathbf{AQ} = \mathbf{0} \tag{48}$$

is a matrix statement of the continuity at every node in the array.

A matrix that links the branch temperature drops can also be found. Define the elements c_{jk} of the matrix **C** in

$$\mathbf{\Delta T} = \mathbf{C\Theta} \tag{49}$$

by

$$c_{jk} = \begin{cases} +1 \text{ if branch } j \text{ leaves node } k \\ -1 \text{ if branch } j \text{ leaves node } k \\ \ \ 0 \text{ if branch } j \text{ does not touch node } k \end{cases} \tag{50}$$

The matrix C will contain b rows and n columns and the $b \times n$ matrix for the oriented graph of Fig. 11b (without node 6) will be

$$\mathbf{C} = \begin{bmatrix} -1 & 0 & 0 & 0 & 0 \\ 1 & -1 & 0 & 0 & 0 \\ 0 & -1 & 0 & 0 & 1 \\ 0 & 0 & 0 & 0 & -1 \\ 0 & 1 & 0 & 0 & 0 \\ 0 & 1 & -1 & 0 & 0 \\ 0 & 1 & 0 & -1 & 0 \\ 0 & 0 & 0 & 1 & 0 \\ 0 & 0 & 1 & 0 & 0 \end{bmatrix} \tag{51}$$

There are two kinds of branches. Those that touch two nodes, say nodes r and s, will posses a temperature drop, $\Delta T = \theta_r - \theta_s$. Those that touch a node and the datum node will have a temperature drop equal to the temperature excess for that node. In both cases, the branch temperature drops depend on the node temperature excesses. If equation (51) is put into equation (49), the result is

$$\mathbf{C\Theta} = \begin{bmatrix} -1 & 0 & 0 & 0 & 0 \\ 1 & -1 & 0 & 0 & 0 \\ 0 & -1 & 0 & 0 & 1 \\ 0 & 0 & 0 & 0 & -1 \\ 0 & 1 & 0 & 0 & 0 \\ 0 & 1 & -1 & 0 & 0 \\ 0 & 1 & 0 & -1 & 0 \\ 0 & 0 & 0 & 1 & 0 \\ 0 & 0 & 1 & 0 & 0 \end{bmatrix} \begin{bmatrix} \theta_1 \\ \theta_2 \\ \theta_3 \\ \theta_4 \\ \theta_5 \end{bmatrix} = \begin{bmatrix} -\theta_1 \\ \theta_1 - \theta_2 \\ \theta_5 - \theta_2 \\ -\theta_5 \\ \theta_2 \\ \theta_2 - \theta_3 \\ \theta_2 - \theta_4 \\ \theta_4 \\ \theta_3 \end{bmatrix} = \mathbf{\Delta T}$$

which is entirely correct. Moreover, a comparison of equations (47) and (51) shows that $a_{jk} = c_{kj}$ which indicates that

$$\mathbf{C} = \mathbf{A^T}$$

and that

$$\mathbf{\Delta T} = \mathbf{A^T \Theta} \tag{52}$$

Node-to-Datum Analysis of Finned Arrays.

The method of node-to-datum analysis is based on the branch equation, which is a form of equation (44),

$$\mathbf{Q} = \mathbf{Q}_s - \mathbf{Y\Delta T}_s + \mathbf{Y\Delta T} \tag{53}$$

the expression of continuity

$$\mathbf{AQ} = \mathbf{0} \tag{48}$$

and the relationship between the branch temperature drops and the node temperature excesses

$$\mathbf{\Delta T} = \mathbf{A^T \Theta} \tag{52}$$

If equation (53) is premultiplied by **A** and set equal to zero in accordance with equation (48)

$$\mathbf{AQ} = \mathbf{0} = \mathbf{AQ}_s - \mathbf{AY}\Delta\mathbf{T}_s + \mathbf{AY}\Delta\mathbf{T}$$

or

$$\mathbf{AY}\Delta\mathbf{T} = \mathbf{AY}\Delta\mathbf{T}_s - \mathbf{AQ}_s \tag{54}$$

Then define a heat source vector

$$\tilde{\mathbf{Q}} = \mathbf{AY}\Delta\mathbf{T}_s - \mathbf{AQ}_s \tag{55}$$

so that

$$\mathbf{AY}\Delta\mathbf{T} = \tilde{\mathbf{Q}}$$

If equation (52) is inserted here

$$\mathbf{AYA}^T\boldsymbol{\Theta} = \tilde{\mathbf{Q}}$$

the node equations result

$$\mathbf{Y}_n\boldsymbol{\Theta} = \tilde{\mathbf{Q}} \tag{56}$$

where the node admittance matrix $\mathbf{Y}_n$ is defined by

$$\mathbf{Y}_n = \mathbf{AYA}^T \tag{57}$$

The solution of equation (56)

$$\boldsymbol{\Theta} = \mathbf{Y}_n^{-1}\tilde{\mathbf{Q}} \tag{58}$$

yields the temperature excess at each node and, if desired, input admittances can then be evaluated at each point where heat is injected into the array.

CONCLUDING REMARKS

As heat rejection systems become more and more complicated, computationally efficient methods must be developed for their analysis and evaluation., This paper has attempted to summarize two methods for the analysis of complex finned arrays. One of them pertains to the array with no loops in the graph theoretical sense in which the analysis is carried out using three combinational algorithms based on a mathematical representation of a fin as either an inverse thermal transmission matrix or a thermal transmission ratio. The other is a general and matrix oriented procedure that is used for arrays containing loops in the graph theoretical sense and is based on the principle of reciprocity and the representation of each individual fin as an equivalent pi-network.

ACKNOWLEDGEMENT

Some of the techniques presented in this article were developed under National Science Foundation Grant ENG–77–01297. The author also wishes to achnowledge the assistance of Mrs. Robert Limes who generated the text and Mr. Alvin W. Lau who prepared the illustrations.

NOMENCLATURE

A	cross-sectional area	m^2
$\mathbf{A}$	node-branch incidence matrix	
b	fin height	m
C	arbitrary constant	
$\mathbf{C}$	a matrix	

h	heat transfer coefficient	W/m^2K
I	designates modified Bessel Function	
k	thermal conductivity	W/mK
L	fin length	m
m	fin performance factor	m^{-1}
n	fin performance factor for longitudinal fin of triangular profile	m^{-1}
Q	heat flow	W
$\mathbf{Q}$	heat flow vector	W
q	heat flow	W
S	surface area	m^2
T	temperature	K
$\mathbf{\Delta T}$	branch temperature vector	K
x	height coordinate	m
Y	thermal admittance	W/K
$\mathbf{Y}$	thermal transmission matrix, branch admittance matrix, or node admittance matrix	
y	elements of the Thermal Admittance matrix	W/K
$\mathbf{\Gamma}$	thermal transmission matrix	
γ	elements of thermal transmission matrix	
δ	fin width or thickness	m
η	fin efficiency	
θ	temperature excess	K
$\mathbf{\Theta}$	temperature excess vector	
κ	taper angle	rad
λ	a solution of a differential equation	
μ	thermal Transmission Ratio	W/K
$\mathbf{T}$	inverse Thermal transmission matrix	
τ	elements of inverse thermal transmission matrix	
ϕ	fin Effectiveness	

Subscripts

a	designates tip of fin or augmented matrix
b	designated base of fin
c	designates a cluster
e	designates an equivalent or environment
f	designates fin
i	designates input condition
j	designates a matrix element
k	designates a matrix element
n	designates a total number of fins or a node or the node admittance matrix.
o	designates characteristic value
P	designates prime surface
s	designates a source
T	designates a total

REFERENCES

1. Murray, W M (1938). *Heat Transfer Through an Annular Disk or Fin of Uniform Thickness*, Trans. ASME, J. Applied Mech., 60, A78.

2. Gardner, K A (1945). *Efficiency of Extended Surface*, Trans. ASME, 67, 621.

3. Harper, D R, and Brown, W B (1922). *Mathematical Equations for Heat Conduction in*

the Fins of Air Cooled Engines, NACA Report No. 158.

4. Parsons, S R and Harper, D R (1922). *Radiators for Aircraft Engines*, U S Bureau of Standards, Technical Paper No. 211, 327.

5. Sparrow, E M, and Hennecke, D K (1970). *Temperature Depression at the Base of a Fin*, J. Heat Transfer, 92, 204.

6. Sparrow, E M, and Lee, L (1975). *Effects of Fin Base Depression in a Multifin Array*, J. Heat Transfer, 97, 63.

7. Trumpler, P R (1945). *Discussion of Gardner, K A (1945). Efficiency of Extended Surface*, Trans. ASME, 67, 630.

8. Manzoor, M, Ingham, D B, and Heggs, P J (1983). *Improved Formulations for the Analysis of Convecting and Radiating Finned Surfaces*, AIAA J., 21, 120.

9. Kraus, A D, Snider, A D, and Doty, L F (1978). *An Efficient Algorithm for Evaluating Arrays of Extended Surface*, J. Heat Transfer, 100, 288.

10. Kraus, A D, and Snider, A D (1980). *New Parameterizations for Heat Transfer in Fins and Spines*, J. Heat Transfer, 102, 415.

11. Sparrow, E M, and Baliga, B R, and Patankar, S V (1978). *Forced Convection Heat Transfer from a Shrouded Fin Array with and without Tip Clearance*, J. Heat Transfer, 100, 572.

12. Gardner, K A (1951), "Discussion on paper of M L Ghai", *General Discussion on Heat Transfer*, I Mech., London UK, 1951.

13. Han, L S, and Lefkowitz, S G (1960). *Constant Cross Section Fin Efficiencies for Non-uniform Surface Heat Transfer Coefficients*, ASME Paper 60-WA-41.

14. Chen, S Y, and Zyskowski, G L (1963). *Steady State Heat Conduction in a Straight Fin with Variable Heat Transfer Coefficient*, ASME Paper 63-HT-1, Sixth ASME-AIChE Heat Transfer Conference, Boston, MA.

15. Snider, A D and Kraus, A D (1981), *A General Extended Surface Analysis Method*, J. Heat Transfer, 103, 699.

16. Kraus, A D, Snider, A D, and Landis, F (1982). *The Reciprocity of Extended Surface and the Node Analysis of Finned Arrays*, Proc. 7th Int. Heat Transfer Conf., Munich, FRG., 6, 223.

Channel Natural Convection Air Cooling—The Variable Property Effect and Thermal Drag*

ZENGYUAN GUO, ZHIXIN LI, and YEWEI GUI
Department of Engineering Mechanics
Tsinghua University
Beijing, PRC

ABSTRACT

In this paper, the variable property effects on channel air natural convection are analyzed based on solving the variable property governing equations, and the thermal drag phenomena related to the variable property effect are discussed. A dimensionless criterion, Hv, which reflects the variable property effect, is derived from the governing equations. The numerical results show that the property variation due to heating will lead to a reduction both in mass flow rate and heat transfer. A critical heat flux occurs also with single phase gas flow.

INTRODUCTION

Channel natural convection has been widely adopted in electronic equipment, such as fins of large power electronic components, air cooling on modules and boards of computer systems and much other engineering equipment. There have been extensive analytical and experimental studies of this problem [1-5]. The constant property approximation (Boussinesq approximation) is commonly introduced in the analysis and the variable property effect has been given little attention before. In many other types of natural convection problems (e.g., natural convection on single plates, single cylinders, enclosures and so on), the variable property effect is commonly considered by means of the reference temperature method [6-8]. In this paper, the emphasis is on the mechanism of the variable property effects on heat transfer and mass flow rate of channel air natural convection. Also, the phenomena of thermal drag related to the variable property in channel natural convection are discussed.

GOVERNING EQUATIONS

The vertical flat plate channel is shown in Fig.1. The channel height is l and width is 2b. The entrance velocity of gas with the ambient temperature is regarded as uniform. By using the following transformations:

----

* Project Supported by National Natural Science Foundation of China.

$$U = \frac{b^2 u}{l \nu_\infty Gr} \qquad V = \frac{bv}{\nu_\infty}$$

$$X = \frac{x}{lGr} \qquad Y = \frac{y}{b} \qquad (1)$$

$$P = \frac{b^4 p'}{\rho_\infty l^2 \nu_\infty^2 Gr^2}$$

$$\theta = \begin{cases} \dfrac{T - T_\infty}{T_1 - T_\infty} & \text{for isothermal plates} \\ \dfrac{(T - T_\infty) k_\infty}{\dot{q} b} & \text{for isoflux plates} \end{cases}$$

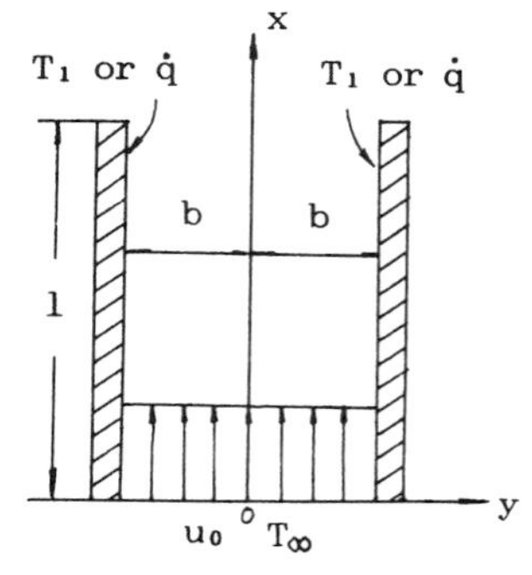

FIGURE 1. Channel.

where

$$p' = p - p_\infty \qquad (2)$$

$$Gr = \begin{cases} \dfrac{g\beta (T_1 - T_\infty) b^4}{l \, \nu_\infty^2} & \text{for isothermal plate channel} \\ \dfrac{g\beta \dot{q} b^5}{l \, k_\infty \nu_\infty^2} & \text{for isoflux plate channel} \end{cases} \qquad (3)$$

Assuming that $\rho \mu = \rho_\infty \mu_\infty$ and $\rho k = \rho_\infty k_\infty$ [9], the governing equations for two dimensional, steady, laminar natural convection are transformed to

$$\frac{\partial(\rho U/\rho_\infty)}{\partial X} + \frac{\partial(\rho V/\rho_\infty)}{\partial Y} = 0 \qquad (4)$$

$$U\frac{\partial U}{\partial X} + V\frac{\partial U}{\partial Y} = \frac{\rho_\infty}{\rho} \frac{\partial}{\partial Y} \left(\frac{\rho_\infty}{\rho} \frac{\partial U}{\partial Y} \right) + \theta - \frac{\rho_\infty}{\rho} \frac{dP}{dX} \qquad (5)$$

$$U\frac{\partial \theta}{\partial X} + V\frac{\partial \theta}{\partial Y} = \frac{\rho_\infty}{Pr \rho} \frac{\partial}{\partial Y} \left(\frac{\rho_\infty}{\rho} \frac{\partial \theta}{\partial Y} \right) \qquad (6)$$

where

$$Pr = \frac{\rho_\infty Cp \, \nu_\infty}{k_\infty} \qquad (7)$$

It can be seen that with variable properties an additional factor ρ_∞ / ρ appears in the governing equations compared with the constant property equations in [1]. This factor can be expressed in dimensionless form as

$$\frac{\rho_\infty}{\rho} = \theta Hv + 1 \qquad (8)$$

where

$$Hv = \begin{cases} \dfrac{T_1 - T_\infty}{T_\infty} & \text{for isothermal plate channel} \\ \dfrac{\dot{q} b}{k_\infty T_\infty} & \text{for isoflux plate channel} \end{cases} \qquad (9)$$

Hence, Hv is the dimensionless parameter which reflects the variable property effects in air natural convection. Consequently, Hv, as well as Gr and Pr, is a characteristic parameter in air natural convection.

NUMERICAL SOLUTION AND RESULTS

After adopting Illingworth-Stewartson transformation [10]

$$
\begin{aligned}
u &= u \\
v &= \frac{\rho v}{\rho_\infty} + u\int_0^y \frac{\partial}{\partial x}\left(\frac{\rho}{\rho_\infty}\right)dy \\
x &= x \\
y &= \int_0^y \frac{\rho}{\rho_\infty}\, dy
\end{aligned}
\qquad (10)
$$

and the non-dimensional expression (1), equations (4)-(6) are simplified as follow:

$$\frac{\partial U}{\partial X} + \frac{\partial V}{\partial Y} = 0 \qquad (11)$$

$$U\frac{\partial U}{\partial X} + V\frac{\partial U}{\partial Y} = \frac{\partial^2 U}{\partial Y^2} + \theta - \frac{\rho_\infty}{\rho}\frac{dP}{dX} \qquad (12)$$

$$U\frac{\partial \theta}{\partial X} + V\frac{\partial \theta}{\partial Y} = \frac{1}{Pr}\frac{\partial^2 \theta}{\partial Y^2} \qquad (13)$$

with boundary conditions

$$
\begin{aligned}
&X = 0,\ 0 < Y < B : U = U_0;\ V = 0;\ \theta = 0 \\
&Y = 0,\ X > 0 \quad : \frac{\partial U}{\partial Y} = 0;\ V = 0;\ \frac{\partial \theta}{\partial Y} = 0 \\
&Y = B,\ X > 0 \quad : U = 0;\ V = 0;\ \begin{cases} \theta = 1 & \text{for isothermal plates} \\ \dfrac{\partial \theta}{\partial Y} = 0 & \text{for isoflux plates} \end{cases} \\
&X = 0 \text{ or } X = L : P = 0
\end{aligned}
\qquad (14)
$$

where $B=\int_0^b (\rho/\rho_\infty)dy/b$; and $L=1/Gr$ is the dimensionless channel height.

By using the finite difference method [1], equations (10)-(14) can be numerically solved. Figure 2 shows the illustrative curves of the dimensionless mass flow rate, M, versus dimensionless height, L. The dimensionless mass flow rate is

$$M = \frac{bm}{\rho_\infty l\, \nu_\infty Gr} = \int_0^B U\, dY \qquad (15)$$

where the mass flow rate per unit depth of the channel, m, is

$$m = \int_0^b \rho u dy = \rho_\infty u_0 b \qquad (16)$$

The dimensionless relation between heat transfer and channel height is plotted in Fig.3, The dimensionless heat transfer Q is

$$Q = \frac{bq}{\rho_\infty C p \nu_\infty l Gr(T_1 - T_\infty)} = \int_0^B U\theta \, dY \tag{17}$$

where the heat transfer per unit depth of the channel, q, is

$$q = \int_0^b \rho \, Cp \, u(T-T_\infty) dy \tag{18}$$

ANALYSIS AND DISCUSSION

Variable Property Effects

Dimensionless mass flow rate and heat transfer. As shown in Figs.2 and 3, the property variation with temperature will lead to a reduction in mass flow rate and heat transfer. For fully-developed channel natural convection with isothermal boundary conditions, the following correlations of mass flow rate and heat transfer between constant property and variable property cases can be derived from equations (5), (15)-(18):

$$M_{var} = \frac{M_{Bou}}{(1+Hv)^3} \tag{19}$$

$$Q_{var} = \frac{Q_{Bou}}{(1+Hv)^3} \tag{20}$$

where subscript "var" represents variable properties and "Bou" represents Boussinesq approximation (constant properties). It can be seen that the variable property effect are quite large even though the dimensionless criterion, Hv, is small, because of the factor $(1+Hv)^3$.

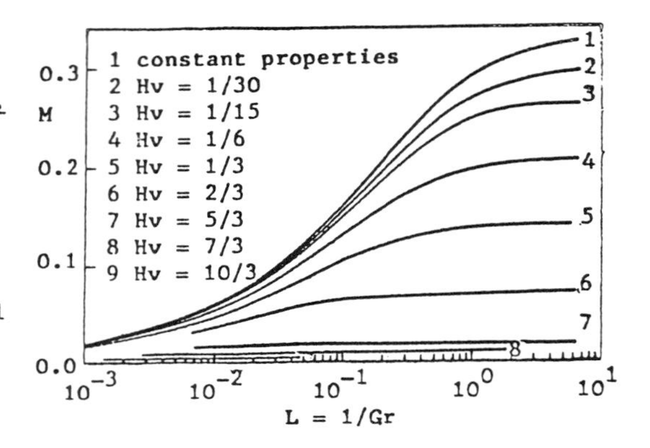

a. Isothermal plates

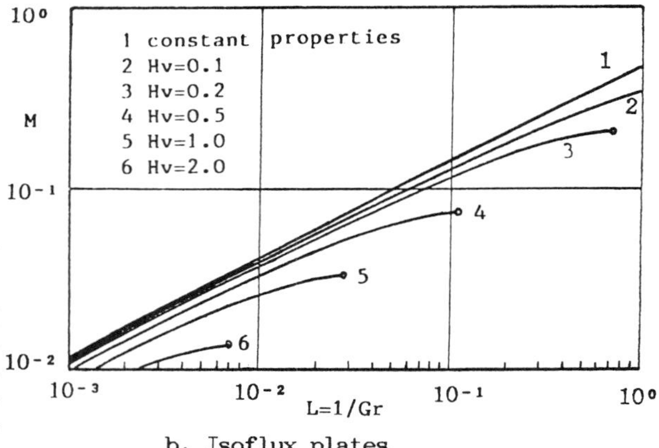

b. Isoflux plates

FIGURE 2. Dimensionless mass flow rate.

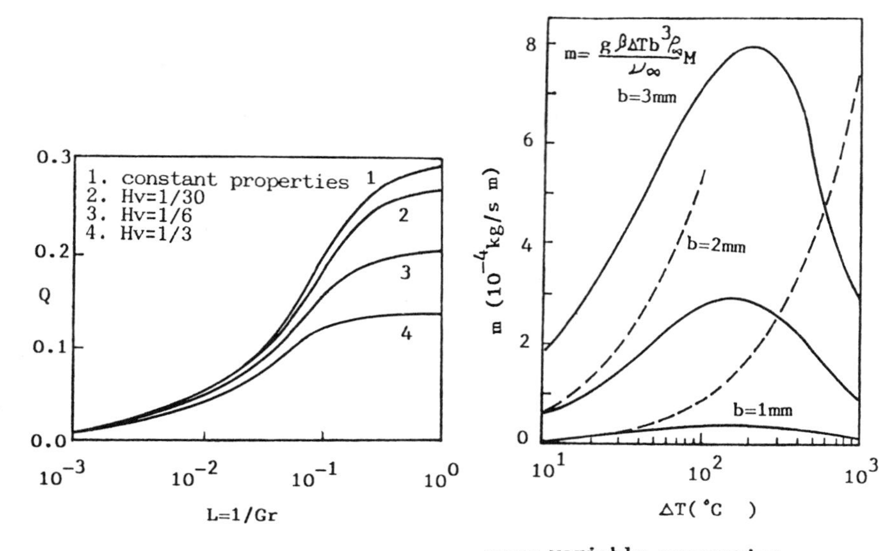

—— variable properties
---- constant properties

FIGURE 3. Dimensionless heat transfer.

FIGURE 4. Dimensional mass flow rate.

Maximum in mass flow rate and heat transfer. According to the usual solution, the larger the temperature difference between channel surface and ambient, the larger the mass flow rate and heat transfer since the driving force increases. On the contrary, when the property variations with temperature are taken into consideration, a maximum for both mass flow rate and heat transfer does exist as the temperature difference increases(see Fig.4). For fully developed flow, the maximum values can be analytically obtained from equations (3), (9), (11), (15) and (17) :

$$m_{max} = \frac{4\rho_\infty g\, b^3}{84\nu_\infty} \tag{21}$$

for mass flow rate when Hv=1/2; and

$$q_{max} = \frac{4\rho_\infty g\, Cp\, b^3 T_\infty}{84\nu_\infty} \tag{22}$$

for heat transfer when Hv=2 .

Thermal drag. The channel natural convection originates from the temperature difference (density difference) --- induced buoyancy. However, it can be seen from the results in the previous section that taking the variation of density and viscosity due to heating into consideration will lead to a reduction in mass flow rate and heat transfer. This implies that the thermal drag does exist in channel natural convection as well as in channel forced convection. The physical mechanism of thermal drag as described by present authors [11] is that the heat addition to the flowing gas causes a pressure drop along the channel, restricting the channel flow. Consequently, heating gas flow produces both thermal driving force (buoyancy force) and

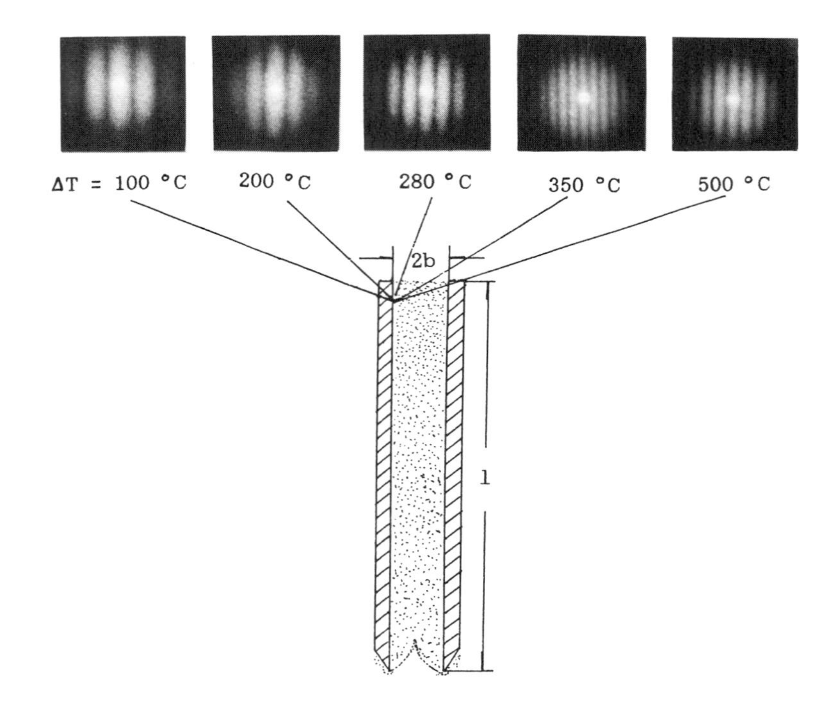

FIGURE 5. Laser speckle photography.

thermal drag. It can be shown that, as heating intensity (Hv) increases, the buoyancy force will approach a limit, while the drag will rise monotonically without limit. This results in a maximum in mass flow and heat transfer for fixed channel natural convection. This also means that the thermal drag plays a dominant role in case of heavy heating.

Figure 5 shows the laser speckle photography of channel natural convection [12]. It is clear that the density of the laser speckle fringes in Fig.5, which reflect the temperature gradient, comes at first from rare to dense and then from dense to rare as the temperature difference continuously increases. This implies that increasing temperature difference will at first promote and then prevent the channel flow to become fully developed. On the other words, it means that the mass flow rate and heat transfer will at first increase and then decrease as the temperature difference increases. This is also an evidence for the existence of the heat transfer maximum resulting from thermal drag.

Mean Nusselt number. The mean Nusselt number obtained from solving the governing equations are shown in Fig.6. For comparison, the analytical result of constant properties is plotted in Fig.6. It is found that the Nusselt number decreases with the increase of the temperature difference between plate and ambient. For example, the Nusselt number decreases from 6.2 to 3.8 (see Fig.6) when the temperature difference is 100°C and Rayleigh number is 4.

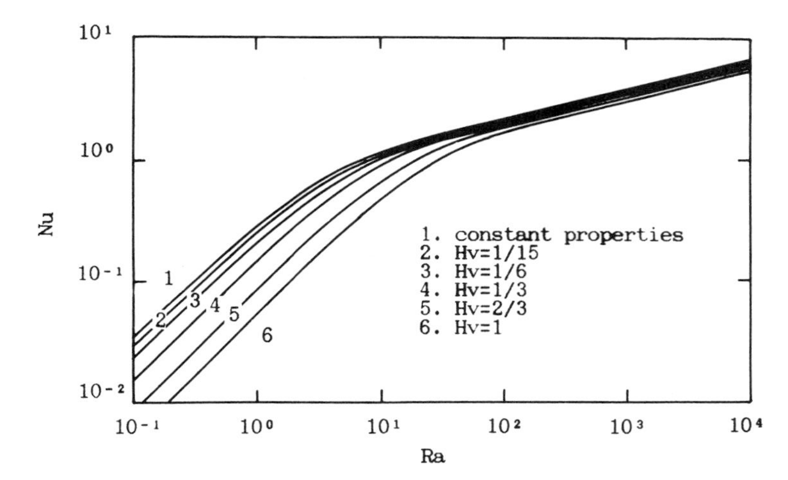

FIGURE 6. Nusselt number of isothermal plate channel.

Ability of Channel to Transfer Heat in Natural Convection

According to constant property approximation, the channel natural convection heat transfer might always increase without limit as the temperature difference goes up. When the property variations are taken into consideration, no matter what boundary conditions of the channel are, the heat transfer ability of channel natural convection is limited.

Now integrate the variable property momentum equation (5) from y to b at the exit of a channel with symmetric heating and the buoyancy force

$\int_y^b (\rho_\infty - \rho) g dy < \rho_\infty g(b-y)$ implies

$$\frac{\partial u_e}{\partial y} < \rho_\infty g \ (b-y) \tag{23}$$

Because $\mu > \mu_\infty$ for air, so

$$\int_0^b u_e dy < \int_0^b \frac{g(by-y^2/2)}{\nu_\infty} dy = \frac{gb^3}{3\nu_\infty} \tag{24}$$

The channel heat transfer is given by

$$q = \int_0^b \rho_e Cp \ u_e T_e dy - \rho_\infty Cp \ u_0 T_\infty b \tag{25}$$

According to the equation of state and $p_e = p_\infty$ at the channel exit (see (14)) $\rho_e T_e = p_\infty / R = \rho_\infty T_\infty$, and from equations (3), (9), (15), (16), (24)

$$q < \rho_\infty Cp T_\infty b(\frac{gb^2}{3\nu_\infty} - u_0) = \rho_\infty g Cp T_\infty b^3(\frac{1}{3} - HvM) \tag{26}$$

Because HvM is greater than zero and very small, that means, the heat transfer value of the channel natural convection can not be beyond $\rho_\infty g \ CpT_\infty b^3/(3\nu_\infty)$. Consequently, the heat transfer limit can be approximated as follow:

$$q_{lim} = \rho_\infty g \ CpT_\infty b^3/(3\nu_\infty) \tag{27}$$

It can be seen that the ability of a vertical channel to transfer energy by natural convection is sensitive to channel width and is independent of channel height. As an example, for a channel 0.2 m in depth and 5 mm in spacing (i.e., b= 2.5 mm), then the natural convection heat transfer cannot be larger than 240 J/s per plate no matter how high the channel is and what the boundary conditions of the channel are.

Consequently, if the energy to be dissipated from the isoflux walls of the channel is beyond the limiting ability of heat transfer, either the heat must be partly brought away by an other heat transfer mode or the system will be damaged due to the endless increase of plate temperature. This phenomenon is called the critical phenomenon of an isoflux plate channel in this paper.

From equations (22) and (27), the critical heat transfer may be assumed to be proportional to the following parameter group.

$$q_{cr} \propto \frac{\rho_\infty g \, Cp \, T_\infty b^3}{\nu_\infty} \tag{28}$$

The critical correlation is obtained by fitting numerical results as follows :

$$(Ra/Hv^2)_{cr} < 25 \tag{29}$$

The above correlation for the critical condition coincides with that of a numerical calculation (seen in Fig.2 b, the mass flow rate does not increase continuously with the increasing of dimensionless channel height). For example, an isoflux plate channel of 10 cm in height and 5 mm in spacing (i.e., b = 2.5 mm) will reach its critical condition when the surface heat flux is equal to 1390 J/m²s.

CONCLUDING REMARKS

1. Property variations due to heating will lead to a reduction of both dimensionless mass flux and heat transfer. For fully-developed channel natural convection with isothermal boundary conditions, the correlations of mass flow rate and heat transfer between constant property and variable property cases are $M_{var}=M_{Bou}/(Hv+1)^3$ and $Q_{var} = Q_{Bou}/(Hv+1)^3$, where the criterion, Hv, reflects the variable property effects.

2. For an isothermal plate channel, a maximum both in dimensional mass flow rate and heat transfer exists as the criterion Hv increases.

3. The heat transfer ability of channel natural convection is limited. For the isoflux plate channel, the critical heat flow rate occurs and the critical correlation is $(Ra/Hv^2)_{cr} = 25$.

4. All above mentioned phenomena, which cannot be found in channel natural convection with Boussinesq approximation, should be attributed to the variable property effects and, in particular, the existence of thermal drag.

5. The mean Nusselt number decreases with the increase of the temperature difference if the variable property effects are taken into consideration.

NOMENCLATURE

Symbol	Quantity	SI Unit
b	channel half width	m
Cp	specific heat capacity	J/(K kg)
g	gravitational acceleration	m/s^2
k	thermal conductivity	J/(m s K)
l	channel height	m
m	mass flow rate per unit channel depth	kg/(s m)
p	pressure	N/m^2
q	heat transfer rate per unit channel depth	J/(m s)
R	universal gas constant, R = 8.3144	J/(mol K)
T	temperature	K
u, v	velocity	m/s

Greek Letters

β	volumetric expansion coefficient	K^{-1}
μ	dynamic viscosity	kg/(s m)
ν	kinetic viscosity	m^2/s
ρ	air density	kg/m^3

Coordinates

x,y	cartesian coordinates

Symbol and Definition	Name

Dimensionless parameters

Symbol and Definition	Name
$Gr = \begin{cases} \dfrac{g\beta(T_1-T_\infty)b^4}{l\nu_\infty^2} \\ \dfrac{g\beta\dot{q}b^5}{lk_\infty\nu_\infty^2} \end{cases}$	Grashof number
$Hv = \begin{cases} (T_1-T_\infty)/T_\infty & \text{for isothermal plate} \\ \dot{q}b/(k_\infty T_\infty) & \text{for isoflux plate} \end{cases}$	dimensionless heating number
$L = l/Gr$	dimensionless height
$M = b\,m/(\rho_\infty l\ \nu_\infty Gr)$	dimensionless mass flux
$P = b^4 p'/(\rho_\infty l^2 \nu_\infty^2 Gr^2)$	dimensionless pressure
$Pr = \rho_\infty Cp \nu_\infty/k_\infty$	Prandtl number
$Q = b\,q/[\rho_\infty Cp \nu_\infty l Gr(T_1-T_\infty)]$	dimensionless heat transfer
$Ra = Gr\ Pr$	Rayleigh number
$U = b^2 u/(l\nu_\infty Gr)$	dimensionless velocity
$V = b\,v/\nu_\infty$	

$X = x/(l\ Gr)$	dimensionless coordinates
$Y = y/b$	
$\theta = \begin{cases} (T - T_\infty)/(T_1 - T_\infty) & \text{for isothermal plate} \\ k_\infty (T - T_\infty)/(\dot{q}\ b) & \text{for isoflux plate} \end{cases}$	dimensionless temperature

Subscripts

0	channel entrance
1	channel surface
∞	ambient
Bou	Boussinesq approximation
cr	critical
e	channel exit
lim	limiting value
max	maximum value
var	variable properties

REFERENCE

1. Bodoia, J. R. & Osterle, J. F., The Development of Free Convection between Heated Vertical Plates, *ASME Trans., J. Heat Transfer*, vol.84, no.1, pp. 40-44, 1962.

2. Aung, W., Fletcher, L. S. & Sernas, V., Developing Laminar Free Convection between Vertical Flat Plates with Asymmetric Heating, *Int. J. Heat Mass Transfer*, vol.15, no.11, pp.2293-2307, 1972.

3. Carpenter, J. R., Briggs, D. G. & Sernas, V., Combined Radiation and Developing Laminar Free Convection between Vertical Flat Plates with Asymmetric Heating, *ASME Trans., J. Heat Transfer*, vol.98, no.1, pp. 95-100, 1976.

4. Wirtz, R. A. & Stutzman, R. J., Experiments on Free Convection between Vertical Plates with Symmetric Heating, *ASME Trans., J. Heat Transfer*, vol.104, no.3, pp. 501-507, 1982.

5. Guo, Z. Y., Li, Z. X. & Gui,Y. W., Analytical and Experimental Study on Natural Convection between Isothermal Vertical Plates,*Proceedings of the International Symposium on Cooling Technology for Electronic Equipment*, Honolulu, Hawaii, U.S.A., pp. 23-31, 1987.

6. Clausing, A. M. & Kempka,S. N., The Influence of Property Variations on Natural Convection from Vertical Surfaces, *ASME Trans., J. Heat Transfer*. vol.103, no.4, pp. 609-612, 1981.

7. Morgan, V. T., Overall Convection Heat Transfer from Circular Cylinders, *Advances in Heat Transfer*, vol.11, Academic Press, New York, pp. 199-264, 1975.

8. Zong, Z. Y., Yang, K. T. & Lloyd, J.R., Variable Property Effects in Laminar Natural Convection in a Square Enclosure, *ASME Trans., J. Heat Transfer*, vol.107, no.1, pp. 133-138, 1985.

9. Sparrow, E. M. & Gregg, J. L., The Variable Fluid Property Problem in Free Convection, *Asme Trans., J. Heat Transfer*, vol.80, no.2, pp. 879-886, 1958.

10. Cebeci, T. & Bracshow, P., *Physical and Computational Aspects of Convective Heat Transfer*, Springer-Verlag Inc. New York , pp. 302-312, 1984.

11. Guo, Z. Y., Thermal Drag and Thermal Roundabout Flow in Convective Problems, *Proc. 8th Int. Heat Transfer Conf., San Francisco, CA U.S.A.*, vol.1, pp. 59-68, 1986.

Studies on Natural Convection Heat Transfer from Arrays of Block-Like Heat-Generating Modules on a Vertical Plate

Y. M. CHEN and Y. KUO
Department of Mechanical Engineering
National Taiwan University
Taipei, Taiwan, ROC

ABSTRACT

Natural convection from block-like heated elements located on a vertical plate has been investigated both numerically and experimentally. The problem considered in the numerical study is laminar and two-dimensional. Assumptions are made of constant properties and the Boussinesq approximation. Solutions are obtained by solving the Navier-Stokes and energy equations using a finite-difference numerical scheme. Results are demonstrated for Ra numbers ranging from 10^2 to 10^5 at Pr=0.7. Streamline and isotherm plots illustrate the effect of the block-like heaters on the flow patterns. Experimental corroboration of the numerical results is provided by measuring temperatures with a laser holographic interferometer. Comparison of the local heat transfer coefficient distribution between the numerical and experimental results shows that good agreement is achieved by setting the interface boundary condition properly in the numerical procedure. The results indicate also the need for considering the conjugate transport arising from longitudinal conduction in the plate.

1. INTRODUCTION

Natural convection heat transfer induced by heated electronic modules mounted on a printed circuit board has received growing attention because of the important role it plays in the cooling of electronic equipments. The energy dissipated in the heated modules must be effectively carried away in order to maintain the electronic components operating under the adequate limiting temperature, otherwise the electronic system may not function properly.

In applications involving low power and packaging densities, the most convienient and reliable way to control the component temperature is to use natural convection cooling. Accordingly the understanding of the natural convective processes in situations encountered in the cooling of electronic equipments is in a great urgent need. The processes are complicated by the presence of the components on the plate, which is similar to the flow across blocks; therefore, flow separations may exist.

A problem somewhat similar to the flow considered in the present study has been investigated by Sparrow and Faghri [1]. They considered two heated isothermal plates seperated by a specified distance, so that a free plume arises between the two plates and numerically determined the heat transfer from the two heated surfaces. In their study, the boundary layer equations for the problem of two vertical flush mounted heaters were solved. Jaluria [2] also numerically solved this problem for multiple heaters, dissipating energy with a uniform surface heat flux. In his study, the induced flow is also treated as a boundary layer problem. Jaluria [3] also solved the more general equations for two vertical flush mounted heaters. It was found that the differences between the solutions of the boundary layer theory and the more general equations increase as Gr number decreases.

The two-dimensional natural convection flow arising from a single flush mounted heater located on a vertical or inclined plate was studied experimentally by Jaluria [4]. The importance of the conjugate transport effect was emphasized. Zinnes [5] also considered the effect of heat conduction into the plate on the thermal field generated by flush mounted heaters. In his study, the boundary layer assumptions were used and a supporting experimental investigation was also carried out.

In the above mentioned work, investigations were all concentrated on flush mounted heaters. Recently, an experimental study on arrays of small heaters was carried out in natural convection by Park and Bergles [6]. In this study, it was found that the protruding heaters have higher heat transfer coefficients then the flush mounted ones. Data were also obtained with varied distance between heaters. These results show the difference between two types of heaters, indicating the complicated development and interaction of the plumes of the heat sources.

In reviewing the existing literature, there seems still a great lack of work on protruded types of heaters. In the present study, therefore, natural convection from block-like heated elements located on a vertical surface is considered. The problem is laminar and two-dimensional and assumptions are made of constant properties and the Boussinesq approximation. Solutions are obtained by solving the Navier-Stokes and energy equations using a finite-difference numerical procedure. Of particular interest in the problem under consideration are the local heat transfer coefficients arround the block-like heaters located in the wake of others.

Experimental corroboration of the numerical results is also provided by measuring the temperatures with a laser holographic interferometer.

2. NUMERICAL ANALYSIS

The problem consists of flow induced by block-like heated elements located on an adiabatic vertical surface (Fig. 1). The elements are heated by some arrangment of surface heaters, each having a uniform heat flux q or the elements are heated and maintained at a uniform wall temperature T . The ambient fluid is at a characteristic temperature T . Assumptions of constant fluid properties and the Boussinesq approximation are made. The nondimensional stream function, vorticity and energy equations are as follows:

$$\frac{\partial^2 \bar{\Psi}}{\partial \bar{x}^2} + \frac{\partial^2 \bar{\Psi}}{\partial \bar{y}^2} = \bar{\omega} \tag{1}$$

$$\frac{\partial \bar{\omega}}{\partial t} + \bar{u}\frac{\partial \bar{\omega}}{\partial \bar{x}} + \bar{v}\frac{\partial \bar{\omega}}{\partial \bar{y}} = Pr\left(\frac{\partial^2 \bar{\omega}}{\partial \bar{x}^2} + \frac{\partial^2 \bar{\omega}}{\partial \bar{y}^2}\right) + Pr\ Ra^* \frac{\partial \bar{T}}{\partial \bar{y}} \tag{2}$$

$$\frac{\partial \bar{T}}{\partial t} + \bar{u}\frac{\partial \bar{T}}{\partial \bar{x}} + \bar{v}\frac{\partial \bar{T}}{\partial \bar{y}} = \frac{\partial^2 \bar{T}}{\partial \bar{x}^2} + \frac{\partial^2 \bar{T}}{\partial \bar{y}^2} \tag{3}$$

the boundary conditions can be expressed as

$$\frac{\partial \bar{\omega}}{\partial \bar{x}} = \frac{\partial \bar{\Psi}}{\partial \bar{y}} = \frac{\partial \bar{T}}{\partial \bar{x}} = 0 \qquad \text{on AB, CD surfaces} \tag{4}$$

$$\frac{\partial \bar{\omega}}{\partial \bar{y}} = \frac{\partial \bar{\Psi}}{\partial \bar{y}} = \bar{T} = 0 \qquad \text{on BC surface} \tag{5}$$

$$\bar{\omega} = \frac{\partial^2 \bar{\Psi}}{\partial \bar{y}^2},\ \bar{\Psi}=0,\ \frac{\partial \bar{T}}{\partial \bar{y}} = 0 \qquad \text{on AD surface} \tag{6}$$

$$\bar{\omega} = \frac{\partial^2 \Psi}{\partial \bar{n}^2},\ \bar{\Psi}=0,\ \frac{\partial \bar{T}}{\partial \bar{n}} = -1 \qquad \text{along the surfaces of the heated elements} \tag{7}$$

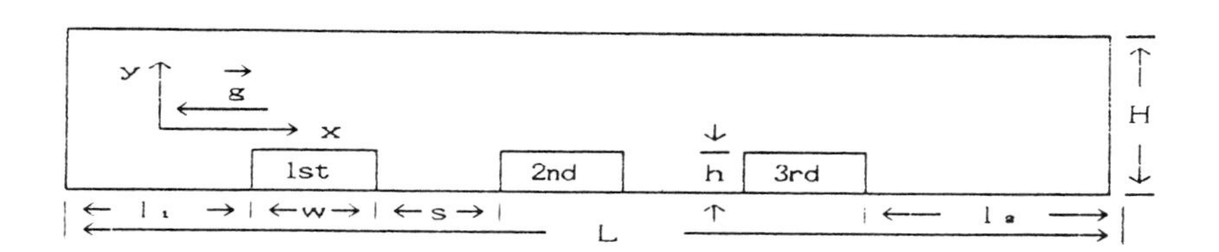

FIGURE 1. Geometry and configuration.

The equations were solved by a finite-difference method. The temperature and vorticity equations are parabolic, while the stream function equation is elliptic. The stream function equation is discretized by the central difference scheme and the resulting Poisson equation can be solved by the successive over-relaxation method. The energy and vorticity equations are solved by the 2nd upwind method. The convective terms in both equations are discretized by a hybrid scheme. Details of the discretization equations can be found in [7].

Solution procedures will be shown in the following:

(1). Specify the initial values for $\bar{\Psi}$, $\bar{\omega}$, $\bar{T}$.

(2). Solve the energy equation for $\bar{T}$ at each interior grid point at new time step.

(3). Substitute the new temperature values calculated from (2) into the buoyancy term of the vorticity transport equation and use the vorticities at previous time step to solve the vorticity transport equation.

(4). The new vorticity values obtained in (3) are then introduced into the stream function equation which is solved for the new stream function by the method of S.O.R.

(5). Calculate the boundary values for temperature and vorticity.

(6). Calculate the velocity by the definition of the stream function.

(7). All the values calculated at this time step are up-dated to be the initial values for the next time step. The procedure returns to (1).

This builds up an iterative process, which stops when some preestablished convergence criterion is reached. The convergence criterion used in this study is

$$\frac{\max|\varphi_{ij}^{(n)} - \varphi_{ij}^{(n-1)}|}{\max|\varphi_{ij}^{(n)}|} < 10^{-4} \tag{8}$$

where φ stands for $\bar{\Psi}$, $\bar{\omega}$, and $\bar{T}$.

The computational domain (see Fig.1) is extended four block widths in the Y direction, and six block widths downstream in the X direction and four block widths upstream in the X direction. A nonuniform mesh with a large concentration of nodes in the region of steep gradients, such as close to the walls of the blocks, is employed. The computational region consists of 86 grid lines in the X direction and 26 grid lines in the Y direction. To ensure the grid independence, solution was performed with different grid density of 112x36. These two results are in agreement as the maximum difference of the Nu number is within 1%.

3. EXPERIMENTAL METHOD

In order to corroborate the above numerical procedure, a laser holographic interferometer was used to measure the temperatures on the heater surfaces and in the flow.

The test modules were first milled from aluminum blocks to the dimensions of 1cm x 2cm x 15cm. Each module was then hollowed out to accomodate Nichrome electrical heaters, which were attached to the inside surfaces using thermally conductive expoxy. Each such module was also instrumented with copper-constantan thermocouples. These were soldered to its inner surface near the modules´ centroid. The vertical plate to which the heated modules were mounted consists of 8 mm thick bakelite. The modules were held to the plate with threaded fasteners and electrical leads were routed out via small holes drilled through the plate behind each element. The test modules were heated by direct current from a power source. The voltage across the heaters and the current through them were measured. During the experiments, the test assembly was fixed vertically on a support in draft free air.

The holographic interferometer has been described extensively in the literature. The principle of this measuring technique shall not be explained here. The arrangement of the required optical set-up is shown in Fig.2. A 25 mW He-Ne laser was used as a light source. To perform the interferometric measurements, the real-time method was used. The interferogram obtained was analyzed with the aid of an optical profile projector (PJ300, Mitutoyo, resolution 5μ). Further details of the experimental procedure and the interferogram evaluation technique can be found in [7]. It was found that the edge effect and the refraction error in this experiment can be neglected. Verification of the measurements with thermocouples shows that the present interferometric measurement is accurate to ±7% for the temperature. The local heat transfer coefficient can be deduced from the measured temperature field as follows:

$$h = \frac{-k\left(\frac{\partial T}{\partial n}\right)_w}{(T_w - T_\infty)} \qquad (9)$$

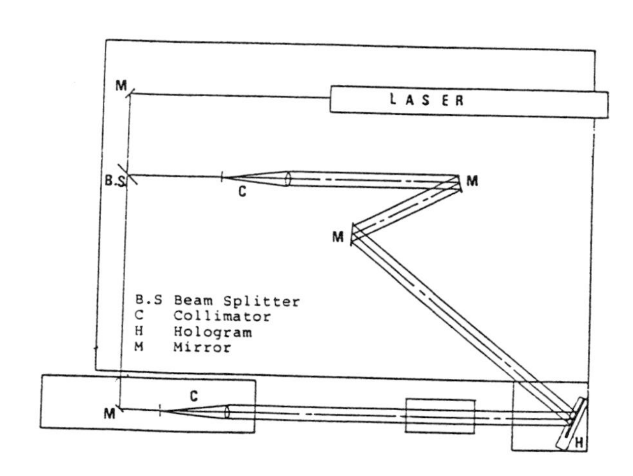

FIGURE 2. Optical set-up of the holographic interferometer.

4. RESULTS AND DISCUSSION

Numerical results

From the governing equations, it can be shown that the buoyancy induced flow from block-like heat-generating modules on a vertical plate is governed by nondimensional groups, the Prandtl number, the Rayleigh number and the nondimensional geometrical parameters about the modules, namely the height of the modules and the distance between the two adjacent modules. Since four basic dimensionless parameters are required to characterize the system, a comprehensive analysis of all combinations of problems is complicated. The results presented here will at first limited to illustrate the effect of Rayleigh number on the flow. The fixed input parameters that were used in all cases were: $s/w=1$, $h/w=0.5$, $l_1/w=4$, $l_2/w=6$. This is representative of the most common size of an IC component on a PC board. The Prandtl number is assigned the value of 0.7, corresponding to air. The Rayleigh numbers are varied from 10^2 to 10^5.

To verify the computer program for steady-state solution of the Navier-Stokes and energy equations, laminar free convection on a vertical plate was selected as a test problem. The solutions obtained at various Ra numbers are compared with the correlation developed in [8]. Fig.3 shows excellent agreement between them.

Fig. 4 and Fig. 5 show streamline and isotherm patterns for various Ra numbers. Several interesting features are observed from these plots. In general the air in the ambient

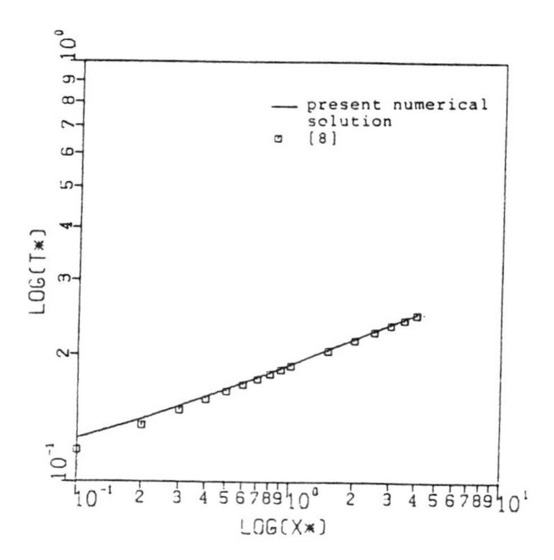

FIGURE 3. Comparison of the numerical results with that developed in [8] for laminar free convection on a vertical plate.

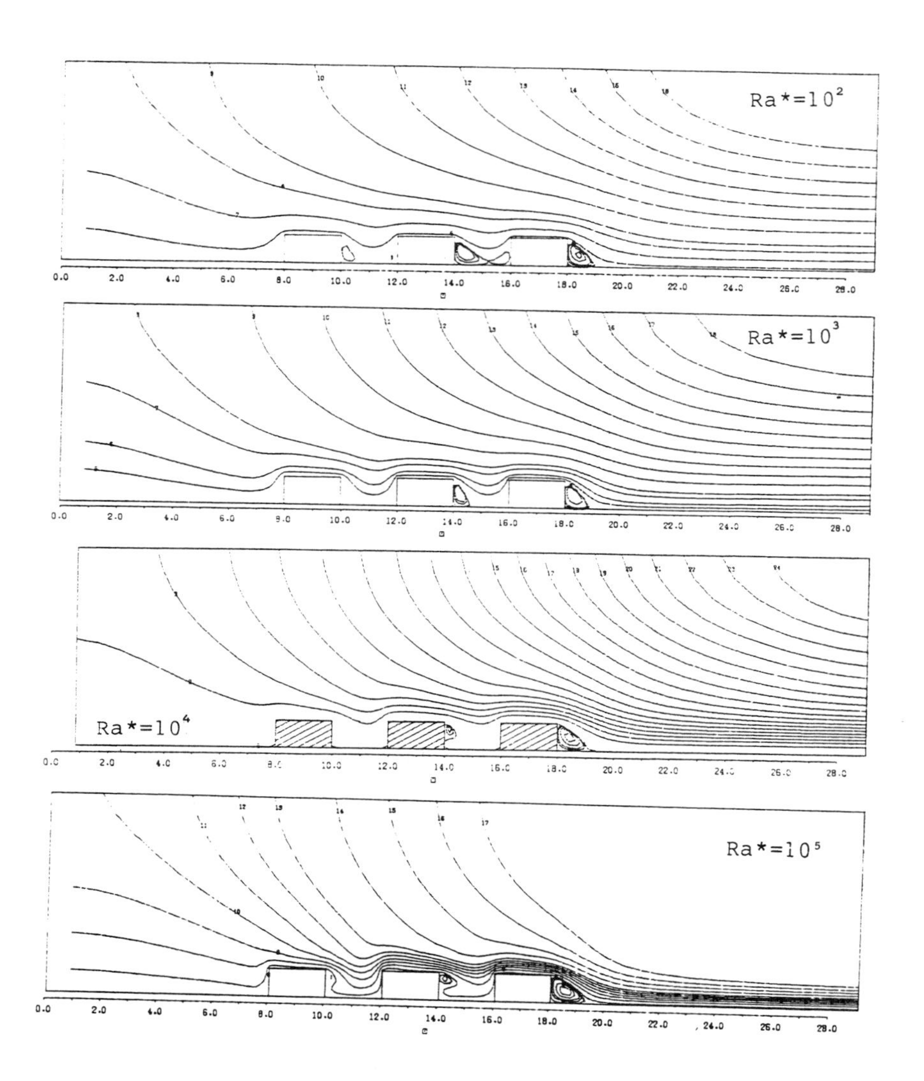

FIGURE 4. Streamlines for flow over three blocks.

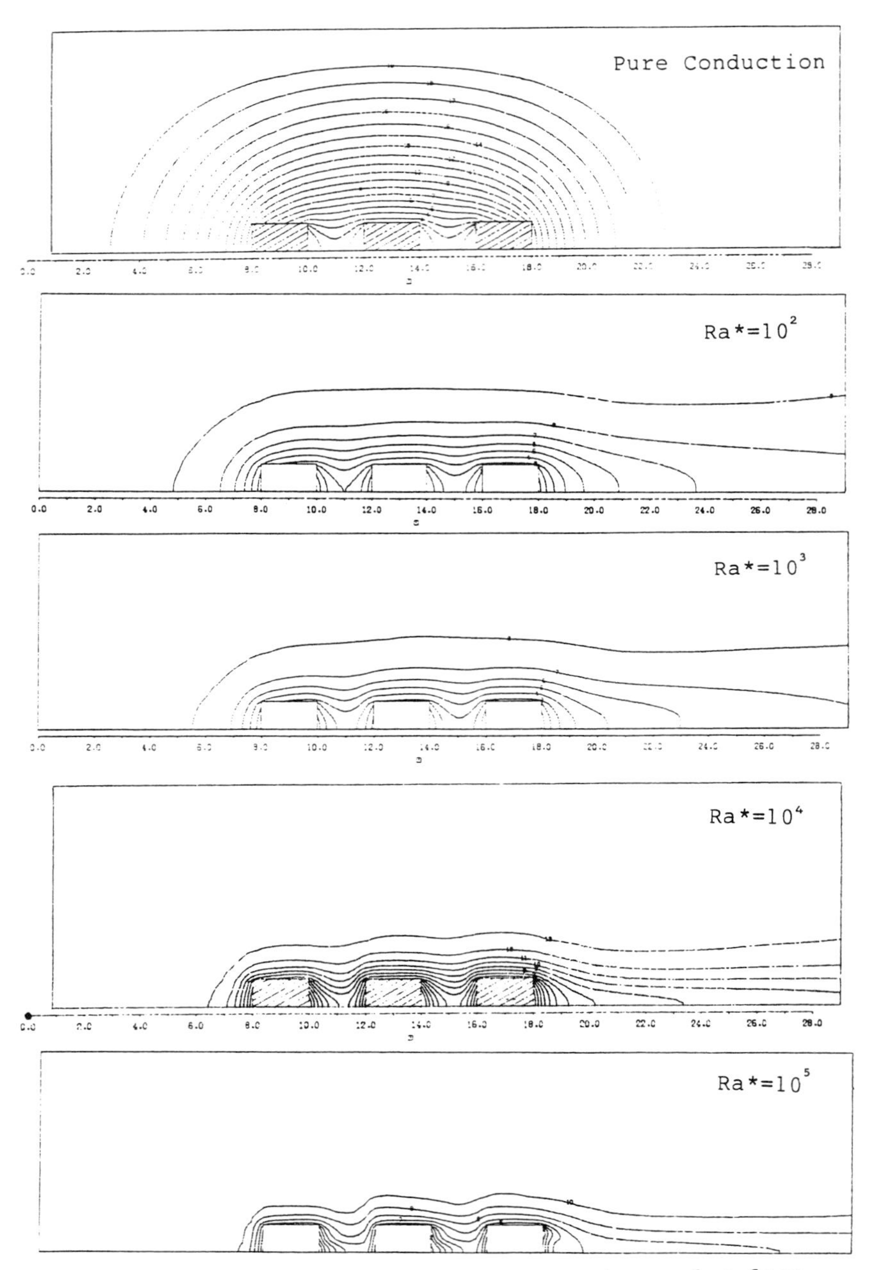

FIGURE 5. Dimensionless temperature contours for free convection from three heated blocks.

is drawn from left and bottom toward the heated elements as can be seen. The streamlines are considerably distorted near the heaters due to their presence. At a low Ra number of 10^2, recirculation regions can be observed at all the corners near the vertical plate. The flow field between the heated blocks is still quite symmetric. The heat tranfer mechanism is conduction dominated. The temperature distribution is only slightly distorted by the flow due to convection. For comparison, the temperature field by pure conduction is also shown in Fig. 5. As Ra number increases, the recirculation cells near the upstream wall disappear and the flow penetrates deeper into the spacings between the blocks. Isotherms are spaced closer together near the lower part of the heated wall, indicating better heat transfer in these regions. The flow field between the heaters is no longer symmetric and small separation bubbles may exist near the edge at the downstream surfaces. These recirculating regions are initiated at separation which occurs at or near the edge of the downstream surface and reattaches again on the same surface. One such separation bubble can be observed between the middle and top heater at Ra = 10^4. A similar separation bubble exists even at the downstream surface of the first block below Ra = 10^5. The recirculation inside the region generally influences the temperature field greatly as can be seen from the corresponding temperature contours. The heat transfer in these regions is poor. The size of the recirculation cell downstream from the upper heater and also the reattachment length seems not to be affected by the Ra number in the range discussed here. Figure 6 presents X direction velocity profiles at Ra = 10^5. It is observed that the velocity level increases downstream due to buoyancy, and the boundary layer thickness in the outer region increases also. The velocity profile in the cavity is distorted. Negative velocities adjacent to the downstream surfaces of the lower and middle heaters, implying a flow separation in these regions, are also shown in the same figure. The separated flow at the downstream surface of the upper heater reattaches as can be inferred from the velocity profiles near X=9.67. The thermal boundary thickness decreases as expected as the Ra number increases. An interesting feature is that at high Ra number the distribution of temperature contours penetrate deep into the space between the heated blocks, creating a large potential difference between the block and the fluid temperature. The wavy-like thermal boundary at Ra= 10^5 can also be observed from the experimentally obtained interferometric fringe pattern in Fig. 8.

The local Nusselt number distribution along the surface of the heated blocks can be deduced from the calculated temperature profiles as

$$Nu = \frac{1}{\bar{T}_w} \tag{10}$$

The local Nu number distribution for each of the three blocks is presented in Fig.7 as a function of Ra number. In this plot, the X coordinate system used includes the

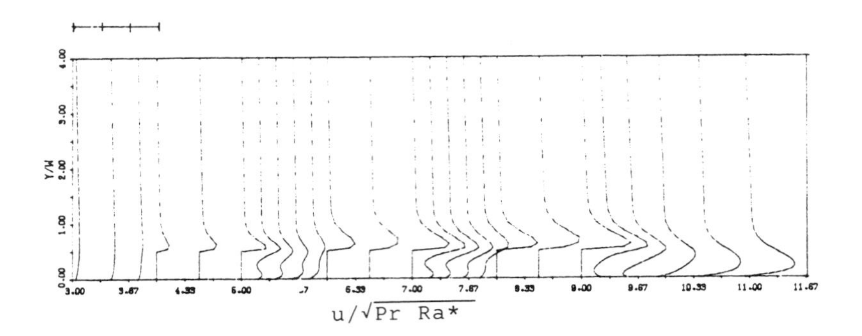

FIGURE 6. Velocity plofiles at $Ra^*=10^5$.

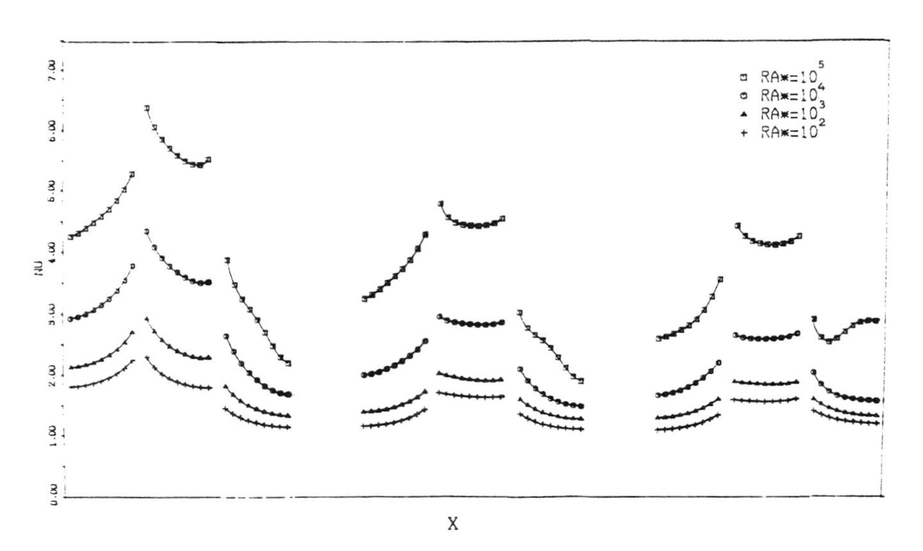

FIGURE 7. Local Nusselt number distribution as a function of Ra^*

horizontal surfaces of the blocks. As can be seen, the maximum heat flux occurs at the vertical surface of each block. As the fluid turns around the corner, the Nu number reaches a peak. This effect is more pronounced at high Ra number. In general the Nu number increases as Ra number increases. And for all Ra numbers in the range studied here, there is a common pattern wherein the highest Nu number is attained in the first block below, and with increasing distance downstream the Nu number decreases. The quite different Nu number distribution at the lee side of the blocks at Ra = 10^5 reflects the complicated flow pattern and complicated relation between the heat input and the geometrical parameters.

Experimental verification

In order to corroborate the numerical results, the laser holographic interferometer was used to measure the temperature field. The interferogram in Fig. 8 shows the temperature distribution corresponding to the case of Ra = 7.0×10^4. Comparison can be first made between this experimentally obtained fringe pattern and the corresponding theoretical isotherms from the numerical solutions in Fig.5. The agreement seems good, especially in the outer boundary layer region. The temperature distributions between the blocks compare favorably. It is seen in Fig. 8 that the temperature starts rising upstream of the first heater below. This is a consequence of the axial conduction transport in the plate. The incoming ambient fluid thus gets heated by the plate upstream of the first heated block. The surface temperature rises sharply in the neighborhood of the heated elements. Isotherms that surround the heaters can be observed in the close vicinity of the wall, indicating a uniform surface condition at the heated wall. This is due to the high thermal conductivity of the aluminum casings used in the test modules. Reasonable comparison of the results between the numerical model and the experiment can only be made if the proper assumption of the constant wall temperature boundary condition is used in the numerical procedure. For this reason the numerical calculation has been carried out for constant wall temperature boundary condition.
Numerical results of the streamline distribution and temperature profile at Ra =7.0×10^4 are shown in Fig. 9. The flow pattern is different from that of the case with the constant heat flux condition. There exist two similar recirculation cells and the size and form of them differ from that observed in Fig. 4. The temperature profile compares better with the experimental obtained interferometric pattern. Slight differences may still exist near the vertical plate because of the idealization of the adiabatic boundary condition on the wall. The v-shaped isotherms observed in the interferogram near the vertical plate demonstrate that in these regions the heat is not only transported to the ambient air but also to the cold plate.

Comparison of the calculated local Nu number distribution with that obtained from the experiments is depicted in Fig. 10. The agreement is quite good, especially

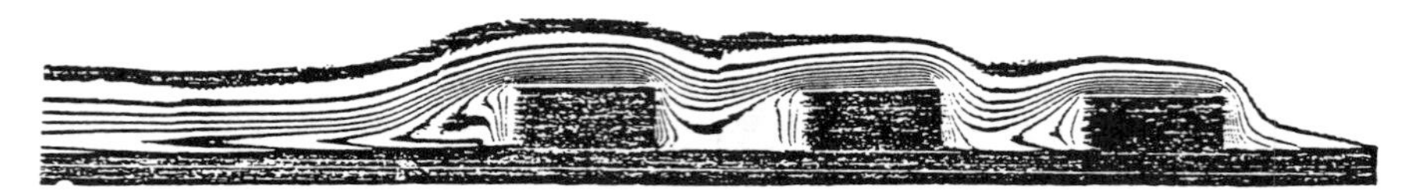

FIGURE 8. Interferogram for natural convection from three block-like heaters on a vertical plate.

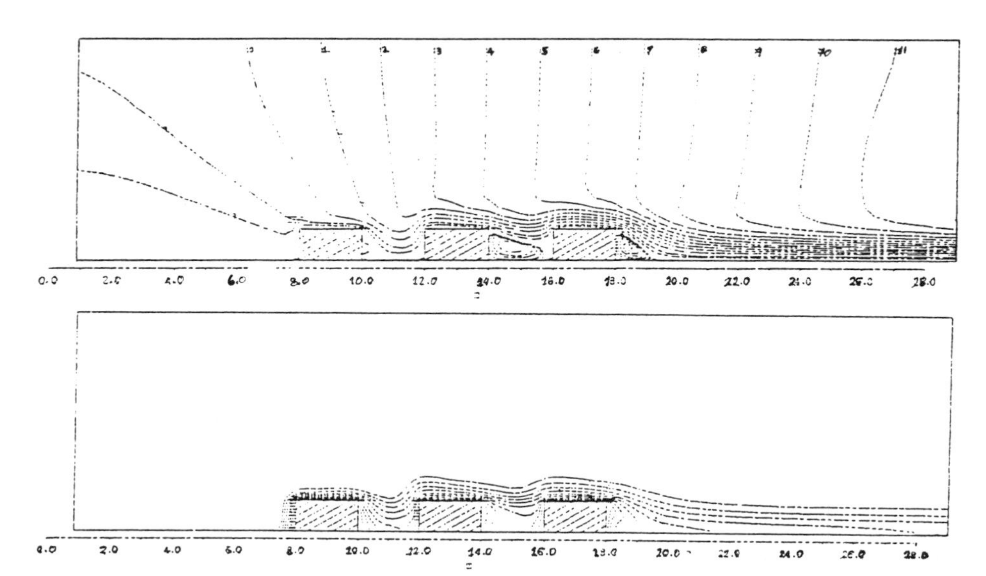

FIGURE 9. Streamline distribution and isotherms at RA= $7x10^4$ for constant wall temperature boundary condition.

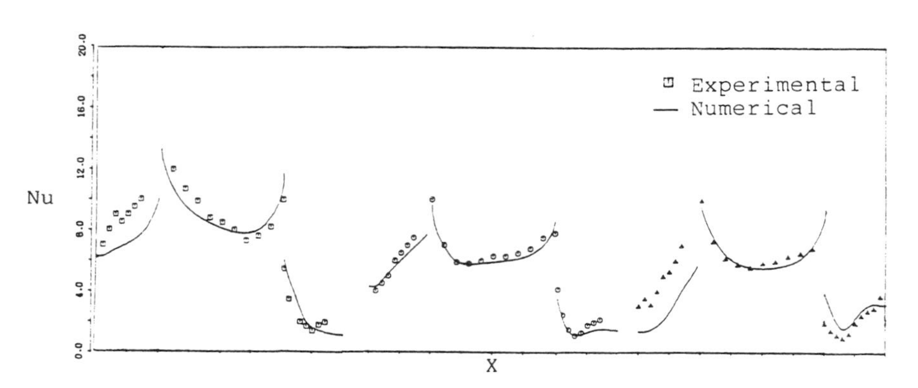

FIGURE 10. Comparison of the local Nusselt number distribution between the numerical and experimental results.

in the boundary layer region. The small discrepancy results from the idealized adiabatic assumption along the vertical plate. This suggests the need for considering the conjugate transport in the plate in the numerical calculation.

In the above analysis, emphasis is on the effect of Ra number on the flow field and heat transfer charateristics. The influence of the geometrical parameters is not discussed. The computer program developed in this study can be used without difficulty to study the geometrical effect. Further studies on the geometrical effects and also for higher Ra numbers are suggested.

5. CONCLUSIONS

Solutions to the Navier-Stokes and energy equations have been obtained for natural convection heat transfer from arrays of block-like heat-generating modules on a vertical plate. The effect of the block-like heaters on the flow pattern is quite complex. The influnce of the Ra number on the flow and heat tranfer has been examined. The local Nu number distributions show that the heat tranfer on the upstream side of the heated surface is in general higher than that on the downstream side. The maximum heat flux occurs at the vertical surfaces of the heaters. The heat transfer coefficient for the upper heater is affected by the flow generated by the lower heater and this effect is more pronounced as Ra number increases.

Experimental corroboration with a laser holographic interferometer shows good agreement betwen the numerical and experimental results. The small discrapancy suggests the need for considering the conjugate transport, arising from longitudinal conduction in the plate, in the numerical study.

6. ACKNOWLEDGEMENT

The authors wish to acknowledge the National Science Council for the financial support under grant NSC77-0401-E002-11.

7. NOMENCLATURE

g	acceleration of gravity
h	height of the block, heat transfer coefficient
H	width of the calculation domain
k	thermal of conductivity
l_1	length of the plate upstream from the heated block
l_2	length of the plate downstream from the heated block
L	length of the calculation domain
n	normal direction
Nu	Nusselt number, Nu = hw/k
q	heat flux
Ra	Rayleigh number, $Ra = g\beta(T_W - T_\infty)w^4/\alpha\nu$
Ra*	modified Rayleigh number, $Ra^* = g\beta\, qw^4/k\alpha\nu$

s	space between block
t	time
$\bar{t}$	non-dimensional time, $\bar{t} = t\alpha/W^2$
$\bar{T}$	non-dimensional temperature, $\bar{T} = (T - T_\infty)/(qw/k)$
T_W	wall temperature
T_∞	ambient fluid temperature
u	axial velocity component
$u, \bar{v}$	non-dimensional velocity, $\bar{u} = uw/\alpha$, $\bar{v} = vw/\alpha$
v	transverse velocity component
w	block length
x	axial position
$\bar{x}, \bar{y}$	non-dimensional coordinates, $\bar{x} = x/w$, $\bar{y} = y/w$
y	transverse position
$\bar{\Psi}$	non-dimensional stream function, $\bar{\Psi} = \Psi/\alpha$
$\bar{\omega}$	non-dimensional vorticity, $\bar{\omega} = \omega w^2/\alpha$
α	themal diffusivity
β	coefficient of volumetric thermal expansion
ν	kinematic viscosity

8. REFERENCES

(1) Sparrow E.M. and Faghri M., Natural Convection Heat Transfer From the Upper Plate of a Colinear Separated Pair of Vertical Plates. ASME Journal of Heat Transfer, Vol. 102, pp. 623-629, 1980.

(2) Jaluria Y., Buoyancy-induced Flow Due to Isolated Thermal Sources on a Vertical Surface. ASME Journal of Heat Transfer, Vol. 104, pp. 223-227, 1982.

(3) Jaluria Y., Interaction of Natural Convection Wakes Arising from Thermal Sources on a Vertical Surface. Fundamentals of Natural Convection/ Electronic Equipment, HTD-Vol.32, ASME. pp. 67-76, 1984

(4) Jaluria Y., Thermal Transport from an Isolated Heat Source on a Vertical or Inclined Surface. Proceedings of the Eighth International Heat Transfer Conference. Vol. 3, pp. 1341-1346, 1986.

(5) Zinnes A.E., The Coupling of Conduction with Laminar Natural Convection from a Vertical Flat Plate with Arbitrary Surface Heating. ASME Journal of Heat Transfer, Vol.92, pp.528-535, 1970.

(6) Park K.A. and Bergles A.E., Natural Convection Heat Characteristics of Simulated Microelectronic Chips. ASME Journal of Heat Transfer, Vol.109, pp.90-96, 1987.

(7) Chen Y.M., NSC-report (NSC77-0401-E002-11).

(8) Sparrow E. M., Laminar Free Convection on a Vertical Plate with Prescribed Nonuniform Wall Heat Flux or Prescribed Nonuniform Wall Temperature, NACA TN 3508, July 1955.

A Fundamental Investigation of Laminar Natural Convection Heat Transfer from a Vertical Plate with Discontinuous Surface Heating

KOKI KISHINAMI, HAKARU SAITO, and IKUO TOKURA
Department of Mechanical Engineering
Muroran Institute of Technology
Muroran, 050, Hokkaido, Japan

ABSTRACT

Laminar natural convective heat transfer from a vertical plate with heated and unheated elements has been studied by numerical analysis and experiments, as a coupled system of convective heat transfer with heat conduction, by considering the effect of heat flux direction in unheated elements. Clarification of the interaction between the fields of fluid convection and heat conduction in unheated elements is the most important part of explaining the heat transfer behavior of this problem. New parameters $S_p(= R_\lambda \cdot B/Gr_L^{1/4})$ for parallel and $S_n(= R_\lambda /B/Gr_L^{1/4})$ for normal directions of conduction heat flux to the surface were introduced through a vectorial dimensional analysis. Heat transfer behavior is discussed, based on numerical calculation and experiment, by using these new parameters. Formulae for both cases have been derived to correlate the heat transfer characteristic, $\overline{Nu}/Gr_L^{1/4}$, and the parameters S_p and S_n.

1. INTRODUCTION

Convective heat transfer from a vertical plate having discontinuous heat generation has become a subject of considerable interest in several technological applications. It is particularly important in the fields of modern electronic technology [1-5], which have had rapid growth over the last 2 decades. The reasons usually cited for this concern about excessively high temperature in semiconductor devices are degradation of reliability due to both performance variations and component failures [2,3]. In such fundamental and basic thermal cases as mentioned above, convective heat transfer from a composite plate with heated and unheated elements should analytically be treated as a coupled convection system with heat conduction where the conduction occurs in various directions in the unheated elements [1].

Problems with flow situations similar to those considered here have been studied by many investigators [6-8]. Wake flow from an isothermal vertical plane by Yang [6], buoyancy induced flow due to isolated heat sources on a vertical adiabatic surface by Jaluria [7] and Sparrow et al. [8], and coupled conduction with natural convection from a vertical plate with arbitrary surface heating by Zinnes [9]. Almost all of the works mentioned above have treated heat transfer by considering the heated element as being isolated thermally, and have paid little attention to the effect of heat conduction between multiple isolated heat sources.

In this study, laminar natural convection heat transfer from a vertical plate with the combination of isothermally heated and unheated elements has been studied, numerically and analytically, considering the direction of the heat conduction flux in the unheated elements. Two cases are studied: case (A) with parallel, and case (B) with normal flux directions to the surface of main heat conduction. Both cases have the same temperatures for heated surfaces and the same thermal properties for the unheated elements. The effects of heat conduction as differentiated by heat flux direction in unheated elements are considered through numerical analysis and a vectorial dimensional analysis [1,5,10].

Heat transfer behavior predicted by the numerical calculations and observed in experiments are discussed based on the new parameters $S_p(= R_\lambda \cdot B/Gr_L^{1/4})$ and $S_n(= R_\lambda /B/Gr_L^{1/4})$ for cases (A) and (B). These S parameters are derived through a vectorial dimensional analysis [1,5]. Finally, proposed equations for $\overline{Nu}/Gr_L^{1/4}$ in both cases (A) and (B) are given for predicting the convective heat transfer by using the parameters S_p and S_n. Then, some discussion is presented on the effect of thermal radiation from the surfaces in this study.

2. ANALYSIS AND EXPERIMENT

2.1 Analysis

The problem of coupled natural convection heat transfer with heat conduction considered here is a case which actually corresponds to the heat delivery from electronic boards or IC chips. Heat conduction in boards or chips by an embedded heat generation circuit must be fundamentally divided into two cases by their heat flux components.

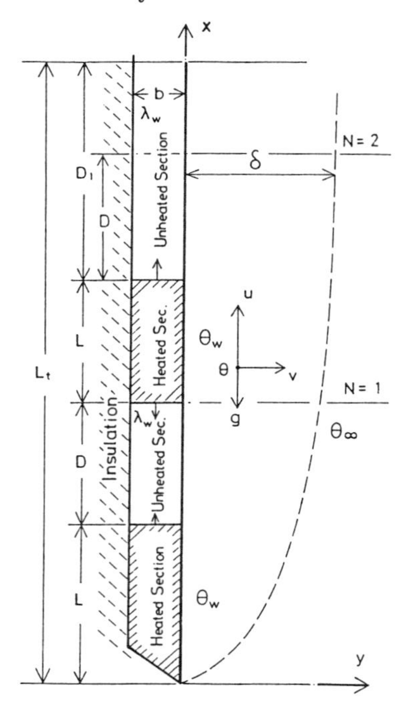

(A) X-directional conduction heat flux

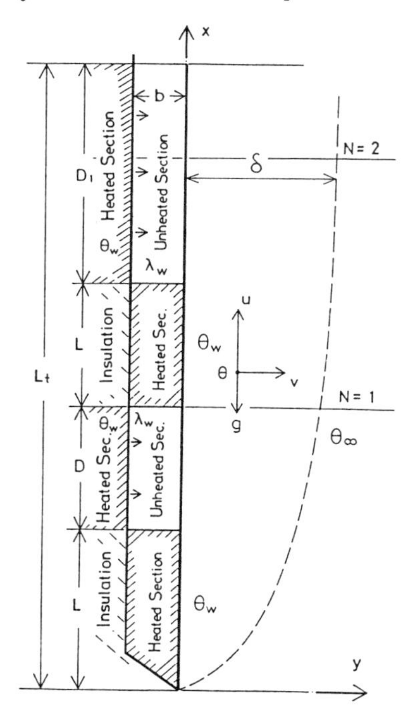

(B) Y-directional conduction heat flux

FIGURE 1. Physical Model.

The schematic diagram for the situation of natural convection under consideration is shown in Fig. 1, subdivided into the two cases (A) and (B) considering the heat flux direction in unheated elements. Case (A) has an adiabatic condition for the bases of the unheated elements, and the predominant heat conduction in the unheated section is in the x-direction (parallel to the surface). Case B has isothermal bases for the unheated elements. Heat conduction is in the y-direction, opposite to case (A). The coupled system has steady, laminar flow over a vertical surface with total length L_t. Unheated elements are made of a material with thickness b and thermal conductivity λ_w. Those elements are heated by conduction through their contact with adjoining heated elements for case (A) or with from their bases for case (B). Both cases (A) and (B) then, set the elements in approximate thermal equilibrium with the buoyancy induced flow over the composite surface.

In treating the multi-stage heated elements considered here, the following conditions were set: Both the heated and unheated elements are all the same length, the same isothermal heating temperature is put on the heated elements and bases of the unheated elements, and identical thermal properties are used for all unheated elements.

Governing equations

The following dimensionless variables commonly used for the isothermal heating condition are made by taking the element length L as the characteristic length [1,8].

$$X = x/L, \quad Y_f = (y/L)\cdot Gr_L^{1/4}, \quad T = (\theta - \theta_\infty)/(\theta_w - \theta_\infty), \quad Pr = \nu/\alpha$$

$$Gr_L = g\cdot\beta\cdot L^3\cdot(\theta_w - \theta_\infty)/\nu^2, \quad U = u\cdot L/\nu\cdot Gr_L^{-1/2}, \quad V = v\cdot L/\nu\cdot Gr_L^{-1/4} \tag{1}$$

$$B = b/L, \quad Y = y/L, \quad R = \lambda_w/\lambda_f \tag{2}$$

The governing equations for the laminar boundary layer and the conduction fields are given in the following dimensionless form by applying the dimensionless variables in Eqs. (1) and (2) [1].

Continuity Equation

$$\frac{\partial U}{\partial X}+\frac{\partial V}{\partial Y_f}=0 \tag{3}$$

Momentum and Energy Equations

$$U\frac{\partial U}{\partial X}+V\frac{\partial U}{\partial Y_f}=T+\frac{\partial^2 U}{\partial Y_f^2} \tag{4}$$

$$U\frac{\partial T}{\partial X}+V\frac{\partial T}{\partial Y_f}=\frac{1}{P_r}\left(\frac{\partial^2 T}{\partial Y_f^2} + G_{rL}^{-1/4}\cdot\frac{\partial^2 T}{\partial X^2}\right) \tag{5}$$

Heat Conduction Equation

$$\frac{\partial^2 T}{\partial X^2}+\frac{\partial^2 T}{\partial Y^2}=0 \tag{6}$$

where the standard boundary layer assumptions have been employed. except that x-direction conduction in the fluid is included in order to accommodate possible large temperature gradients at localized regions near the contact point between the heated and unheated elements. For the present conditions, this temperature gradient in the x-direction has little effect on the velocity field and mean or total Nusselt number over the plate. Therefore, the Prandtl number Pr is the only parameter appearing in the governing equations, if the unheated elements are simply regarded as an adiabatic plate (neglecting conduction Eq.(6)). However, a heat balance equation between the solid and fluid regions of the unheated elements has to be considered to confirm the validity of the coupled model considered here.

In the unheated elements, the heat flux due to convection on the surface at the interface between the plate and fluid is equal to that transferred by conduction. Accordingly, the dimensionless heat balance equation (interface coupling condition) can be expressed by adopting the same variables as in Eqs. (1) and (2) as follows:

$$R_\lambda \frac{\partial T}{\partial Y}\bigg|_{-0} = G_{rL}^{1/4} \frac{\partial T}{\partial Y_f}\bigg|_{+0} \tag{7}$$

A dimensionless convective heat transfer coefficient (local Nusselt number) based on the difference in temperature between the heated wall and ambient fluid ($\theta_w - \theta_\infty$) may be defined by adapting the dimensionless variables listed above as follows:

$$N_u = \frac{hL}{\lambda_f} = -G_{rL}^{1/4} \frac{\partial T}{\partial Y_f}\bigg|_0 \tag{8}$$

In the present problem, it is most important to estimate the total heat delivery from the entire system and the behavior of local heat transfer on the heated and unheated elements. Therefore, the average and total Nusselt numbers, $\overline{Nu}$ and Nu_t, over the plate and the average Nusselt number Nu_i over each of the elements are defined as follows:

$$\overline{Nu} = 1/L_t \cdot \int_0^{L_t} Nu \cdot dx = 1/(2N) \cdot \int_0^{L_t/L} Nu \cdot dX \tag{9}$$

$$Nu_t = 1/L \cdot \int_0^{L_t} Nu \cdot dx = \int_0^{L_t/L} Nu \cdot dX = 2 \cdot N \cdot \overline{Nu}$$

$$= \sum_{i=1}^{N} \left(Nu_i \big|_{heated} + Nu_i \big|_{unheated} \right) \tag{10}$$

where $L_t = 2 \cdot L \cdot N$. It is clear that the total Nusselt number, i.e., integration of Nu over the plate with respect to X, is important to discuss the behavior of Nu_i for each element of stages, and is equal to the normal average Nusselt number $\overline{Nu}$ multiplied by $2 \cdot N$ (N: stage number).

The boundary conditions of the two heated elements corresponding to Fig. 1 express only the limits of the difference between cases (A) and (B).

For case (A): $0 < X < L_t$, $Y = -B$; $\partial T/\partial Y = 0$ (11)

For case (B): $0 < X < L_t$, $Y = -B$; $T = 1$ (12)

Outline of the numerical solution technique

The governing equations for conduction in the plate and convection in the boundary layer can be written in finite-difference form based on the integration over the finite cell. Here, non-uniform node-spacing is used both in x- and y-directions, coupled through the common heat flux at the plate-fluid interface (equation (7)), and solved numerically by an iterative technique [1,9,11] for both cases. In this numerical calculation, a fine longitudinal node spacing near the leading edge and a fine transverse node spacing in the vicinity of the wall were employed to obtain precise results [11].

2.2 Experimental Apparatus and Procedure

Outlines of the experimental apparatus with N= 3 and L= 150 mm for case (B), in which the bases of the unheated elements have an isothermally heated condition, is shown in Fig. 2. The apparatus consists of N= 3 sets of isothermal elements (aluminum plates of 3 mm thickness) and unheated elements (phenol resin plates of 10 and 5 mm thicknesses). The back sides of unheated elements were insulated with a foam urethane material of 100 mm thickness for case (A) [1] and were separated from the isothermally heated hot water box by a thin (1.5 mm) plastic spacer in case (B) of Fig. 2. The apparatus is located in a large thermostatic chamber with a constant room temperature. The surfaces of the heated elements are warmed by hot water from an electric boiler 1 and controlled at a temperature between 35°C and 75°C in both cases (A) and(B) and also 99.4 °C by steam for case (A) [1,11]. The convection surface is 400 mm wide. A 100 mm thick sheet of foam urethane insulating material is installed on the back surface. Side walls of transparent acrylate plates (800 mm in height and 400 mm in depth) prevent external disturbances. A thin aluminum foil with an emissivity of 0.09 and with 15 μm thickness is bonded tightly to the surface of the plate in order to minimize the effect of thermal radiation. The foil is notched every 4-5 mm to retard conduction across the plate's surface. The heated and unheated elements can be replaced and the number of stages changed.

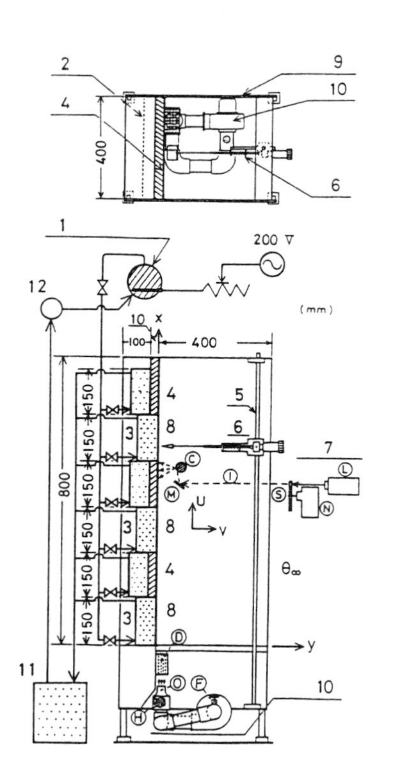

1 : Electrical Boiler
2 : Styrofoam Insulation
3 : Hot Water Box
4 : Unheated Convective Wall made of Phenol Resin Board
5 : Scale Guide
6 : C-C Thermocouple Probe
7 : Optical System
- L : Laser or Projector
- S : Rotary Shutter
- I : Intermittent Beam
- M : Flat Mirror
- C : Cylindrical Mirror

8 : Isothermal Convective Surface made of Aluminum
9 : Side Baffle made of Acryl Resin Board
10 : Micro Particle Ejector
- F : Fan
- H : Honeycomb
- O : Outlet Nozzle
- D : Screen Damper

11 : Reservoir
12 : Pump

For case (B) with N= 3

FIGURE 2. Experimental Apparatus (L=D= 150 mm).

The temperature profile in the boundary layer and surface temperature distribution on the plate for cases (A) and (B) are measured by a traversing thermocouple 6, and the velocity distribution and flow pattern are measured by the particle trajectory method [5].

In addition to the above large scale apparatus, a small scale apparatus was also made for case (A). It had less than 60 mm length for L, and was heated by an electric heater. The unheated elements were made of solid balsa wood with thin aluminum foil bonded to the surface.
A Mach-Zehnder interferometer was used to measure the behavior of the temperature boundary layer [1,5].

3. RESULTS AND DISCUSSION

The characteristic behavior of natural convection heat transfer of this system studied is considerably different from that of the ordinary isothermal plate due to the interaction between the fields of convection and thermal conduction in the unheated elements.

Vectorial Dimensional Analysis

Carrying out a vectorial dimensional analysis [1,5,10] for the coupled problem of natural convection considered here, the following expressions for cases (A) and (B) are obtained for the wall temperature and the heat transfer coefficient when the predominant heat flux direction by conduction in unheated elements is assumed to be parallel or normal to the surface. Here, the characteristic length L, the wall thickness b and thermal conductivity λ_w of the unheated elements are considered in addition to the fundamental physical quantities used for ordinary natural convection.

For case(A): Heat flux by conduction is parallel to the surface

$$T_w = (\theta_w(X) - \theta_\infty)/(\theta_w - \theta_\infty) = F_1(X, Pr, (R_\lambda \cdot B)^4/Gr_L) \quad (13)$$

$$Nu/Gr_L^{1/4} = (R_\lambda \cdot B)/Gr_L^{1/4} \cdot F_2(X, Pr, (R_\lambda \cdot B)^4/Gr_L) \quad (14)$$

For case (B): Heat flux by conduction is normal to the surface

$$T_w = (\theta_w(X) - \theta_\infty)/(\theta_w - \theta_\infty) = F_1(X, Pr, (R_\lambda/B)^4/Gr_L) \quad (15)$$

$$Nu/Gr_L^{1/4} = (R_\lambda/B)/Gr_L^{1/4} \cdot F_2(X, Pr, (R_\lambda/B)^4/Gr_L) \quad (16)$$

In this dimensional analysis, the length was distinguished by the fluid region L_f and solid region L_w [1, 5, 10] in addition to x, y and z vectorial directions. In these expressions the parameters $(R_\lambda \cdot B/Gr_L^{1/4})$ and $((R_\lambda/B)/Gr_L^{1/4})$ must be considered as the most important factor in the respective cases. They are defined as abbreviations S_p and S_n.

$$S_p = R_\lambda \cdot B/Gr_L^{1/4} \quad \text{for case (A)} \quad (17)$$

$$S_n = (R_\lambda/B)/Gr_L^{1/4} \quad \text{for case (B)} \quad (18)$$

The results of the numerical computation of the governing equations, as transformed by the Zinnes variables [9], reveal that the surface temperature distribution $T_w(x)$, and the local heat transfer characteristic $Nu/Gr_L^{1/4}$ for cases (A) and (B) remain constant as long as the parameters S_p or S_n is constant in spite of the values of $R_\lambda \cdot B$ or R_λ/B and Gr_L, within the calculation range of $0.012 < B < 0.07$, $4 < R_\lambda < 200$, and $10^5 < Gr_L < 10^7$ for case (A) and $0.002 < B < 0.067$, $3.1 < R_\lambda < 97$, and $4.4x10^6 < Gr_L < 3.6x10^7$ for case (B). The temperature variation in the solid region is found to be distributed similarly as long as S_p or S_n is constant when B<0.067, and has a one-dimensional distribution in the x-direction for case (A) and in the y-direction for case (B).

Therefore, the heat transfer characteristic $Nu/Gr_L^{1/4}$ and dimensionless surface temperature $T_w(x)$ obtained by numerical calculations and experiments are discussed based on the new parameters S_p and S_n for cases (A) and (B) with the only limitation of geometrical factors being L/D =1.0.

3.1 The Surface Temperature Distribution

Figures. 3 and 4 show the surface temperature distributions on the plate for cases (A) and (B) from the results of numerical calculation for two heated elements corresponding to Fig. 1 for various values of the parameters S_p and S_n. These results expressed in solid line are interpreted as generalized or similar distributions in relation to the parameters S_p and S_n.

The distribution of temperature on the each elements expressed by the solid lines (no radiation) is as follows: The surface temperature on the first unheated element for case (A) in Fig. 3 is found to have a parabolic curved distribution as the value of S_p increases due to conduction heat flux from both ends in contact with adjoining heated elements. The surface temperature on the second unheated element for case (A) are found to have sharp drops near the lower contact point (x=3.0) and to follow a gradual decay downstream. That happens even though the temperature for larger $S_{p,n}$ is higher than that for smaller $S_{p,n}$ for both cases. The distribution of temperature for case (B) in Fig. 4 is found to flatten as the value of S_n increases because of the conduction heat flux from the isothermally heated bases of the element. The temperature on the 2nd unheated element for case (B) in Fig.4 is found to not drop so sharply with X as compared to that for case (A), as shown in the dotted chain line. On the contrary the distributions flatten at high values of S_n due to directly normal heat flux by conduction through the bases. The surface temperatures for the low ranges of S_p and S_n on the first and second unheated elements for cases (A) and (B) become similar to each other in their distributions, and then are completely the same at $S_p = S_n = 0$ as shown in Figs. 3 and 4.

In these figures it is worth noting that the sharp drops in the temperature on the first unheated element for low values of S_p or S_n cause the so-called edge effect on the next heated element, which enhances convective heat transfer.

To examine the effect of thermal radiation from the surface [11] of this model in addition to heat conduction, the numerical results of the surface temperature for cases (A) and (B) under the same conditions are shown as dotted lines in Figs. 3 and 4. Here, F is the radiation shape factor and ε is the emissivity of the plate. So $F\cdot\varepsilon = 0$ shown as the solid lines correspond to the result with no radiation, which is the focus of this discussion, and $F\cdot\varepsilon = 1.0$ shown as the dotted lines is the result considering radiation heat flux. The temperatures of unheated elements with thermal radiation (dotted lines) in Figs. 3 and 4 are found to drop quite sharply in comparison to those with no thermal radiation. That would indicate more effective heat transfer and represents a

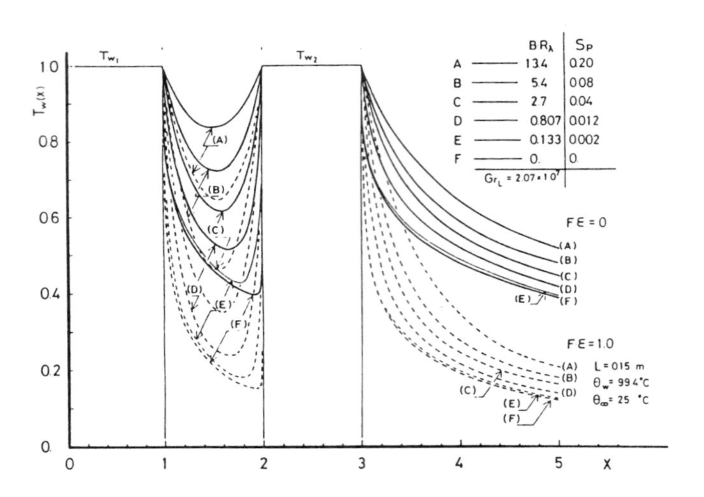

FIGURE 3. Surface Temperature Distribution (N= 2) on the Plate for Case (A).

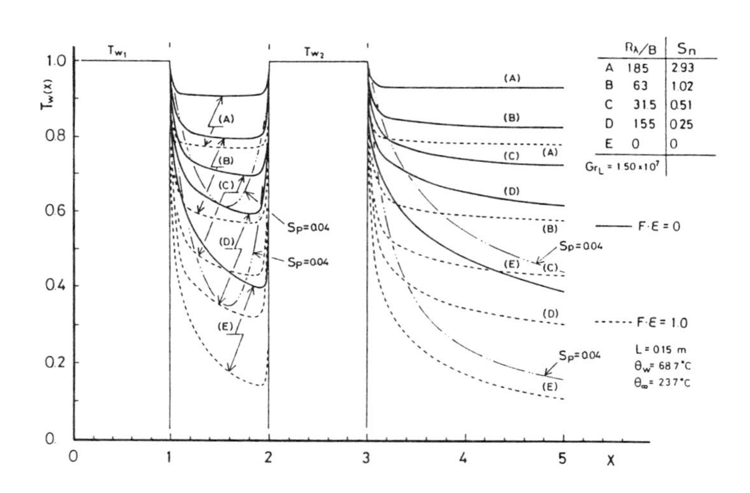

FIGURE 4. Surface Temperature Distribution (N= 2) on the Plate for Case (B).

distinguishable difference from the heat transfer characteristics with no radiation, for both cases.

3.2 Local Heat Transfer Characteristic

Local Nusselt numbers for cases (A) and (B) are given in Figs. 5 and 6, which are obtained by numerical analysis for the same conditions as in the preceding figures. These are generalized or similar solutions in the relation between the $Nu/Gr_L^{1/4}$ in transferred coordinates and the parameter S_p or S_n. It is seen that $Nu/Gr_L^{1/4}$ on the first heated element for both cases changes in the same manner as an ordinary isothermal vertical plane, but $Nu/Gr_L^{1/4}$ on the following elements is very dependent on the value of S_p or S_n. That is, its variation is similar to the surface temperature distribution on the unheated elements in Figs. 3 and 4.

The heat transfer of case (A) decreases considerably with decreasing S_p, having a parabolic curved distribution on the first unheated element.

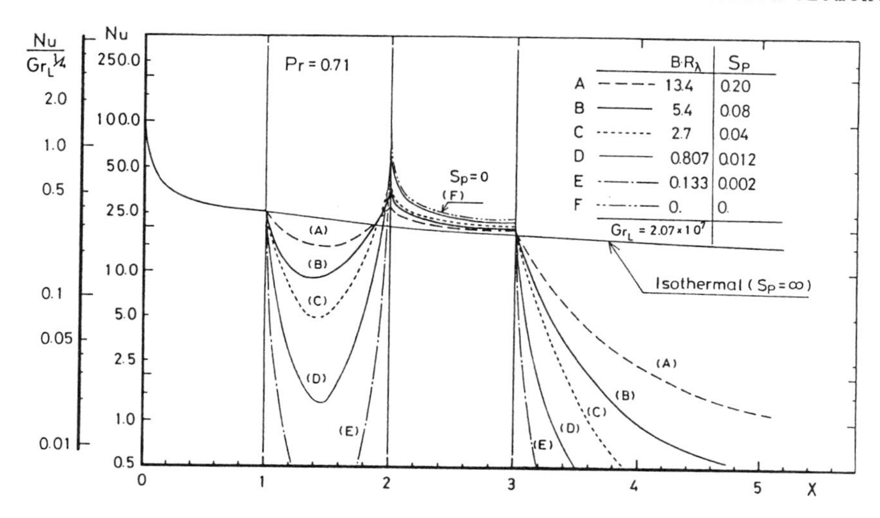

FIGURE 5. Local Nusselt Number (N= 2) for Case (a).

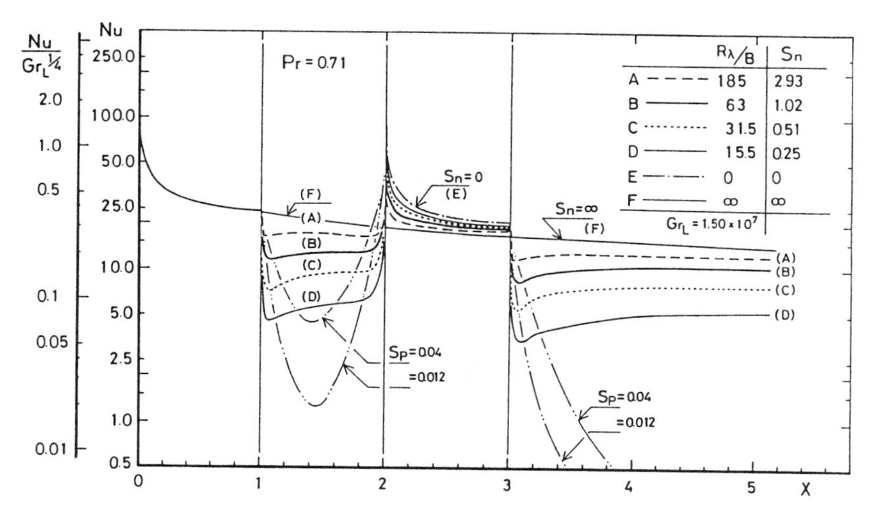

FIGURE 6. Local Nusselt Number (N= 2) for Case (B).

$Nu/Gr_L^{1/4}$ or Nu on the following heated element increases abruptly at the lower contact point (X=2.0) due to the edge effect. The edge effect has considerable effectiveness in the low range of S_p; it is superior to the isothermal situation as shown by comparison with the thin solid line of the figure. On the last unheated element Nu or $Nu/Gr_L^{1/4}$ is found to decrease gradually downstream, similar to the temperature distribution of Fig. 3.

The convective heat transfer behaviors for case (B) (Fig. 6) on the first and last unheated elements show very different distributions from those of the previous case (A). Nu or $Nu/Gr_L^{1/4}$ on the unheated elements drops sharply near the edge of the contact point (X=1.0, 3.0), and subsequently tends to gradually increase with X. This is much different from case (A) as shown in the two dotted chain lines in the figure. The distributions are, however, similar to (A) in the first and last heated elements. The edge-effect heat transfer enhancement takes place at the intermediate (N =2) heated elements in the same manner as above.

The numerical results for adiabatic conditions of the unheated elements ($S_p = S_n = 0$) for both cases (A) and (B) are seen to be completely coincident with each other, that is, Nu or $Nu/Gr_L^{1/4}$ can be neglected on the unheated elements but show effective enhancement of heat transfer on the second heated elements due to the edge effect.

It can be shown that the total heat transfer from the plate for case (B) is superior to that for case (A) due to the greater contribution of the so-called fin effect on the unheated elements.

Verifiing the numerical results by experiments

Figure 7 shows typical surface temperature distributions and local Nusselt numbers for N= 4 stages obtained by experiment for case (B) (R_λ/B= 134, Gr_L= 6×10^6 1.53×10^7). In this figure the analytical result (S_n= 2.3) for N= 2 is plotted as a solid line. This is in good agreement with experimental data up to N= 2 for surface temperature and local Nusselt

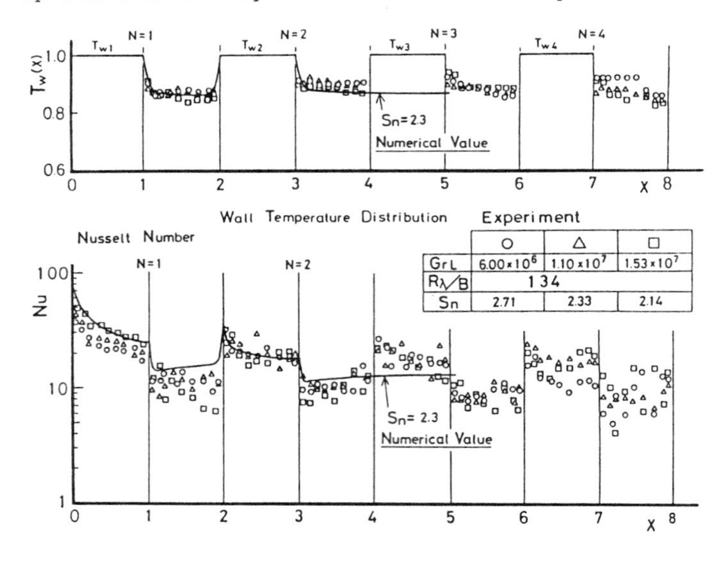

FIGURE 7. Verifying the Numerical Results by Experiment for Case (B).

numbers. In the figure, it is found experimentally that the distribution of Nu above N= 3 stages is similar to that for the numerical result up to the 2nd stage. This similarity seems to be due to the edge effect and fin effect on the heated and unheated elements.

An observation of the flow reveals that the stable laminar flows begin to fluctuate at the middle point of the 4th stage (x = 970 mm, Gr_L= 1.10 x10^7) for case (B), and at the 3d stage (x = 720 mm, Gr_L= 1.47x10^7) for case (A), which produce irregular results after the points, respectively.

3.3 Total and Average Nusselt Numbers

Figure 8 shows the relation between the total Nusselt number over the plate and Gr_L with $R_\lambda B$ for case (A) and R_λ/B for case (B) as parameters. The results of the relation for case (A) (N = 2) were obtained by the numerical calculation of the governing equations due to the Zinnes transformation of variables [9]. The range of the numerical analysis are L = 0.1, 0.15, 0.20 m, θ_w= 70, 99.4, 200, 300 °C for $R_\lambda B$ = 0, 0.807, 2.7, 5.4, and 13.4. From the figure it is seen that Nu_t is proportional to

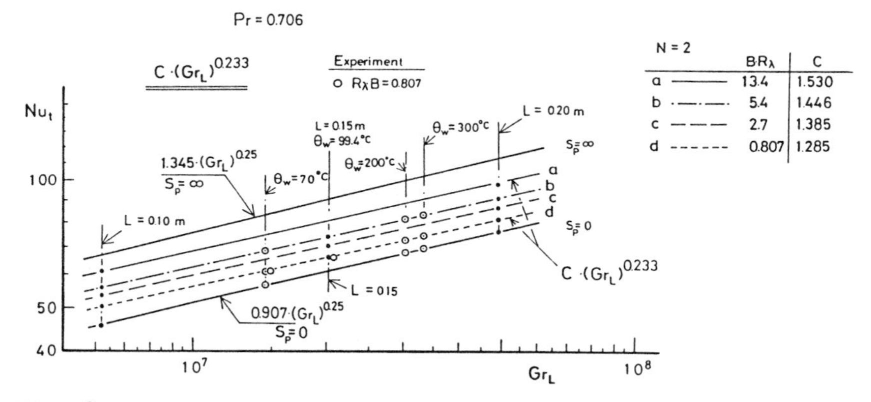

FIGURE 8. Relation between Total Nusselt Number Nu_t (N= 2) and Gr_L for Case (A).

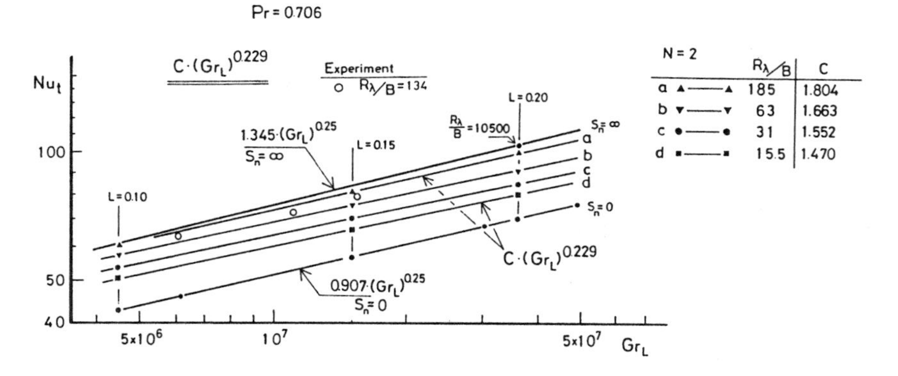

FIGURE 9. Relation between Total Nusselt Number Nu_t (N= 2) and Gr_L for Case (B).

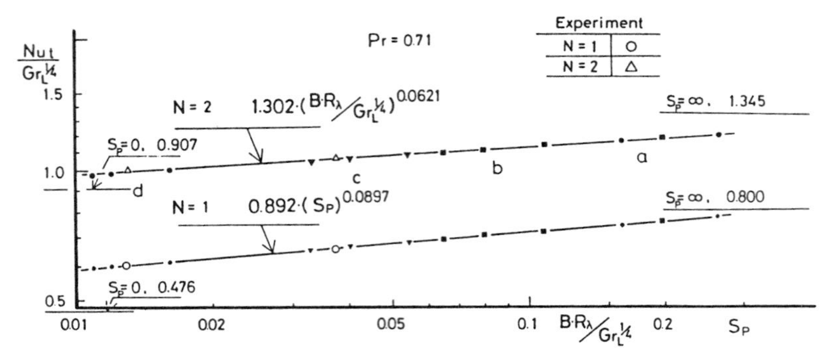

FIGURE 10. Relation of Total Heat Transfer Characteristic $Nu_tGr_L{}^{1/4}$ and Parameter S_p for Case (A).

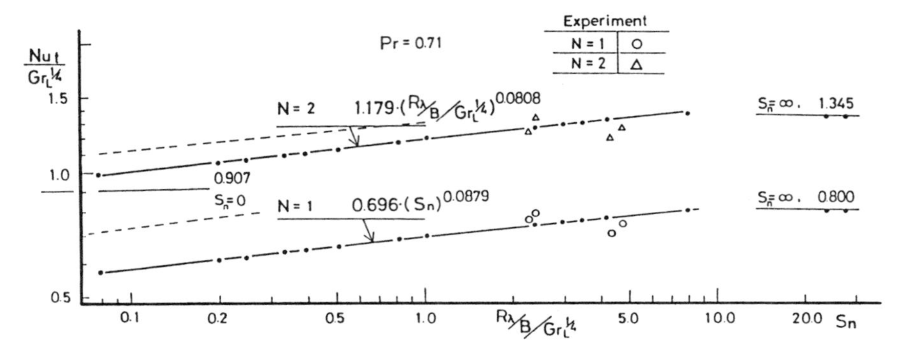

FIGURE 11. Relation of Total Heat Transfer Characteristic $Nu_t/Gr_L{}^{1/4}$ and Parameter S_n for Case (B).

$Gr_L{}^{0.233}$ except for $R_\lambda B = 0$ ($S_p = 0$) and ∞ ($S_p = \infty$), and increases as the parameter $R_\lambda B$ increases, in which the experimental results for $R_\lambda \cdot B = 0.807$ are in good agreement with the corresponding numerical line. The values of proportionality constant C to $Gr_L{}^{0.233}$ for each $R_\lambda \cdot B$ are given in the attached table. It is very interesting to note that the results for $R_\lambda B = 0$ and ∞ depend completely on $Gr_L{}^{0.25}$ and each proportionality C are given by 0.907 and 1.345 ($0.475 \times 4^{0.75}$). Accordingly, Nu_t for $R_\lambda \cdot B = \infty$ (isothermal plate) is superior by 48 % to that for $R_\lambda \cdot B = 0$ corresponding to the adiabatic condition.

In the same manner as above, Fig. 9 shows the numerical results of the relation between Nu_t (N= 2) and Gr_L with R_λ/B as parameter for case (B). The ranges for the numerical analysis are L= 0.1, 0.15, 0.20 m, θ_w= 68, 100, 200 ℃, for R_λ/B= 0, 15.5, 31, 63, 185, and 10500. It is also seen that Nu_t for case (B) is proportional to $Gr_L{}^{0.229}$ except for R_λ= 0 where the power of Gr_L is slightly smaller than that for case (A), and increases as the parameter R_λ/B increases. In the figure the experimental results for R_λ/B= 134 are found to be within the range of numerical corresponding lines. Nu_t for R_λ/B= 0 is exactly the same as in the previous case of Fig. 8.

Furthermore, the results in Figs. 8 and 9 are applied to the relations between $Nu_t/Gr_L^{1/4}$ and $S_p(= R_\lambda \cdot B/Gr_L^{1/4})$ for case (A) or $S_n(= R_\lambda/B/Gr_L^{1/4})$ for case (B), as expressed by Eqs.(14) or (16) in the dimensional analysis. The modified results for N= 2 and 1 are obtained as shown in Fig. 10 for case (A) and Fig. 11 for case (B). In these figures the experimental results S_p= 0.013 and 0.037 for case (A) and S_n= 2.3 and 4.6 for case (B) are plotted along with the limited conditions for $S_p=S_n$= 0 and $S_p=S_n=\infty$ for reference.

It can be concluded that the average heat transfer characteristic $\overline{Nu}/Gr_L^{1/4}$ for cases (A) and (B) are completely dependent on the parameters S_p and S_n as expressed in the following relations.

For case (A): adiabatic condition for the bases $(0.01 < S_p < 0.28)$

$$\overline{Nu}/Gr_L^{1/4} = 0.326\ S_p^{0.0621} \quad (N= 2) \tag{19}$$

$$= 0.446\ S_p^{0.0897} \quad (N= 1) \tag{20}$$

For case (B): isothermal condition for the bases $(0.08 < S_n < 8.0)$

$$\overline{Nu}/Gr_L^{1/4} = 0.295\ S_n^{0.0808} \quad (N= 2) \tag{21}$$

$$= 0.348\ S_n^{0.0879} \quad (N= 1) \tag{22}$$

From Fig. 11, it is predicted that $\overline{Nu}$ over $S_n > 8.0$ for case (B) and over $S_p > 1.5$ for case (A) expressed by dotted line (N = 2) must be considered as an isothermal condition.

Practical Significance/Usefulness

From the view point of heat transfer promotion technology for the electronic board or IC chips [12], it is very significant that in coupled systems similar to this convection model it is important to consider the effect of heat conduction in the unheated elements, particularly with regard to the direction of the conduction heat flux. So, it is generally recommended that a high value of λ_w and large value of b in unheated element material must be selected for the case (A), and a high value of λ_w and small value of b for the case (B). In the small range of Gr_L, however, a considerable convective heat transfer promotion can be expected regardless of the thermal properties in unheated elements.

4. CONCLUSIONS

In this work, natural convective heat transfer from a vertical plate with the heated and unheated elements, which is considered as the most fundamental model for the coupled convection system with heat conduction, was studied numerically and analytically by considering the effect of heat conduction in the unheated elements. The effect of heat conduction on this natural convection was investigated in detail by considering the heat flux direction in unheated elements.

The heat transfer characteristics $Nu/Gr_L^{1/4}$ and Tw(X) were found to depend completely on the dimensionless parameters $S_p(=R_\lambda \cdot B/Gr_L^{1/4})$ for case (A) and $S_n(= (R_\lambda/B)/Gr_L^{1/4})$ for case (B) derived through a vectorial dimensional analysis. Prediction formulas for $\overline{Nu}/Gr_L^{1/4}$ could be given in terms of S_p and S_n for both cases (A) and (B).

Acknowledgment

The authors extend their appreciation to Dr. N. Seki, President at Hokkaido Polytechnic College and the Honorary Professor at Hokkaido University.

NOMENCLATURE

- b: thickness of unheated element
- B: dimensionless thickness of unheated element, $= b/L$
- D,L: length of heated and unheated elements
- F: radiation shape factor from area element of surface to environment
- Gr_L: Grashof number based on the temperature difference between the heated element and ambient, $(\theta_w - \theta_\infty)$, and characteristic length L, $= g \cdot \beta \cdot L^3 \cdot (\theta_w - \theta_\infty)/\nu^2$
- h: heat transfer coefficient based on $(\theta_w - \theta_\infty)$
- L_t: total length of a plate with heated and unheated elements
- N: stage number from leading edge, counting a pair with one heated and one unheated element as one stage
- Nu: local Nusselt number by convection
- $\overline{Nu}$: average Nusselt number over the plate by convection
- Nu_i: average Nusselt number over either the heated or the unheated element at i-th stage by convection
- Nu_t: total of Nu_i or integration of Nu over the plate, defined by Eq. (10)
- Pr: Prandtl number
- R_λ: ratio of thermal conductivities of wall and fluid, $= \lambda_f/\lambda_w$
- S_p: dimensionless parameter combining heat conduction and natural convection fields for case (A), $= R_\lambda \cdot B/Gr_L^{1/4}$
- S_n: dimensionless parameter combining heat conduction and natural convection fields for case (B), $= (R_\lambda/B)/Gr_L^{1/4}$
- T: dimensionless temperature, $= (\theta - \theta_\infty)/(\theta_w - \theta_\infty)$
- U,V: dimensionless vertical and transverse velocities defined by Eq. (1)
- X: dimensionless vertical coordinate, $= x/L$
- Y: dimensionless transverse coordinate in solid region, $= y/L$
- Y_f: dimensionless transverse coordinate in fluid region, $= (y/L) \cdot Gr_L^{1/4}$

- α : thermal diffusivity
- ε : emissivity of radiation
- θ : temperature
- λ : thermal conductivity
- ν : kinematic viscosity

Subscript

- w: wall or heated element
- f: fluid
- ∞ : ambient

REFERENCES

1. Kishinami, K.,Saito, H., and Tokura, I., Natural Convective Heat Transfer on a Vertical Plate with Discontinuous Surface-Heating (Effect of Heat Conduction in Unheated Elements), Proc. 2nd ASME-JSME Thermal Eng. Joint Conf., Vol. 4, 1987, pp.61- 68.

2. Negus,K.J., Yovanovich, M,M., and Roulston, D.T., An Introduction to Thermal-Electrical Coupling in Bipolar Transistors, Proc. 2nd ASME-JSME Thermal Eng. Joint Conf., Vol. 3, 1987, pp.359-401.

3. Bergles, A.E.,The Evolution of Cooling Technology for Electrical, Electronic and Microelectronic Equipment, AIAA/ASME 4th Thermophysics and Heat Transf. Conf., HTD-Vol. 57, 1986, pp.1-9.

4. Cengel, Y.A.,and Zing, P.T.L., Enhancement of Natural Convective Heat Transfer From Heat Sinks by Shrouding, Proc. 2nd ASME-JSME Thermal Eng. Joint Conf., Vol. 3, 1987, pp.451-457.

5. Kishinami,K., Saito, H., and Tokura, I., An Experimental Study on Natural Convective Heat Transfer from a Vertical Wavy Surface Heated at Convex/Concave Elements, Proc. 1st World Conf. on Experimental Heat Transf., Fluid Dynamics, and Thermodynamics, 1988, pp.177-184.

6. Yang, K-T., Laminar Free-Convection Wake Above a Heated Vertical Plate, ASME J. of Heat Transf., Vol. 86, 1964, pp.131-138.

7. Jaluria, Y., Buoyancy-induced Flow Due to Isolated Thermal Sources on a Vertical Surface,ASME J. of Heat Transf., Vol. 104, 1982, pp.223-227.

8. Sparrow, E.M., Patanker, S.V., and Abdel-Washed, R.M., Development of Wall and Free Plumes Above a Heated Vertical Plate, ASME J. of Heat Transf., Vol. 100, 1978, pp.184-190.

9. Zinnes, A.E., The Coupling of Conduction with Laminar Natural Convection From a Vertical Plate with Arbitrary Surface Heating, ASME J. of Heat Transf., Vol, 92, 1970, pp.528-535.

10. Chida, K., and Katto, Y., Study on Conjugate Heat Transfer By Vectorial Dimensional Analysis,International J. of Heat and mass Transf., Vol. 19, 1976, pp.453-459.

11. Kishinami,K., and Seki, N., Natural Convective Heat Transfer on an Unheated Vertical Plate Attached to an Upstream Isothermal Plate, ASME J. of Heat Transf., Vol. 105, 1983, pp.759-766.

12. Nakayama, W., Thermal Management of Electrical Equipment: A Review of Technology and Research Topics, Applied Mechanics Reviews, Vol. 39, No.12, 1986, pp.1847-1868.

Laminar Mixed Convection between Vertical Plates with Isolated Thermal Sources

PAULO I. F. DE ALMEIDA
Departamento de Tecnologia Mecânica
UFPb
58000, João Pessoa, Paraida, Brazil

LUIZ F. MILANEZ
Departamento de Energia
FEC/UNICAMP
13081, Campinas, São Paulo, Brazil

ABSTRACT

This work presents numerical results of laminar mixed convection flow due to two finite thermal sources located in one of the adiabatic walls of a vertical channel. The sources are supplied with a uniform heat flux. Two-dimensional elliptic equations governing the flow are solved numerically by a line-by-line implicit finite difference method using a stream function-vorticity formulation. With the hypothesis of insulated walls, fully developed flow at the channel exit is assumed, the channel height being arbitrarily chosen. Results for wall temperature and local values of the Nusselt number are presented for Gr/Re^2 equal to 1 and 15. The power dissipated in the sources and their relative position were varied and the effects analyzed. The space between the channel walls and the plate dimensions are chosen so as to produce a developing flow near the walls.

INTRODUCTION

Prediction of generated heat in electrical equipment and electronic devices is important in the identification of safe levels for the electronic junction temperature of the components. For effective removal of the energy being dissipated, the location of the electronic components should be based on their heat transfer characteristics, particularly when the components are positioned in the wake of other sources.

Natural convection flow due to multiple thermal sources located on a vertical adiabatic surface has been studied numerically by Jaluria using boundary layer approximations [1] and also by solving the more general equations considering a horizontal surface at the temperature of the extensive ambient at the leading edge [2]. Sparrow and Faghri [3] numerically solved the boundary layer equations for the natural convection problem in air originated by two-in-line, wide, vertical, flush-mounted, isothermal heaters. Mixed convection flow iver localized multiple thermal sources on a vertical surface has been investigated by Jaluria [4].

Some situations of practical interest have been dealt with experimentally.Milanez and Bergles [5] obtained experimental results for natural convection flow originated by two horizontal wide strips and two horizontal cylinders (adjacent to or displaced from the surface) on a vertical adiabatic surface in both air and water. Tewari et al. [6] carried out an experimental study of the interactions between two finite-sized heat sources located on a flat adiabatic plate; both natural and mixed convective flows were investigated for the horizontal orientation of the surface and only natural convection was studied for the vertical surface.

Laminar natural convection heat transfer in vertical open channels is also receiving considerable attention. Oosthuizen [7] carried out a numerical study of laminar natural convection flow between partially heated vertical plates using a forward marching implicit finite procedure. Yan and Lin [8] solved the boundary layer equations for the problem of natural convection in a channel with discrete heating.

In many situations of electronic packaging a small power blower may be required to improve the heat removal capability. In this case, as a result of an aiding flow configuration, temperature levels within the equipment are lower when compared to the levels attained with natural convection only, and the study of mixed convection is necessary to model the phenomenon. A comprehensive analysis of mixed convection in channels with discrete heat sources was presented by Chow et al. [9]. An elliptic formulation was used with special attention to the effects of axial heat conduction in the fluid. This effect was found to be significant only for small values of the Peclet number, mainly for liquid metals. Tomimura and Fujii [10] analyzed laminar mixed convection between parallel plates where one of the plates is insulated and the other has seven equally spaced and equally powered heat sources and obtained a correlation to predict the maximum temperature on each heat source. Nickell et al. [11] considered combined natural convection and radiation between parallel plates with discrete heat sources.

The present study is concerned with mixed convection heat transfer between vertical plates with isolated thermal sources. Two thermal sources, taken as long planar sources of same finite height, are located in one of the adiabatic walls of the channel. A two-dimensional flow is considered and the full elliptic equations governing the flow are solved numerically for air (Pr=0.72) flowing through the channel.

GOVERNING EQUATIONS

The flow under consideration is shown in Fig. 1 along with the coordinate system employed. The flow is assumed to be two-dimensional and steady, and the fluid properties are constant except for the variation of density in the buoyancy term of the momentum equation. The dimensionless equations of motion and energy with the Boussinesq approximation in rectangular Cartesian coordinates are employed together with the stream function-vorticity formulation and can be written as

$$\frac{\partial^2 \psi}{\partial X^2} + \frac{\partial^2 \psi}{\partial Y^2} = - \xi \tag{1}$$

$$U \frac{\partial \xi}{\partial X} + V \frac{\partial \xi}{\partial Y} = \frac{\partial^2 \xi}{\partial X^2} + \frac{\partial^2 \xi}{\partial Y^2} - Gr \frac{\partial \theta}{\partial Y} \tag{2}$$

$$U \frac{\partial \theta}{\partial X} + V \frac{\partial \theta}{\partial Y} = \frac{1}{Pr} \left[\frac{\partial^2 \theta}{\partial X^2} + \frac{\partial^2 \theta}{\partial Y^2}\right] \tag{3}$$

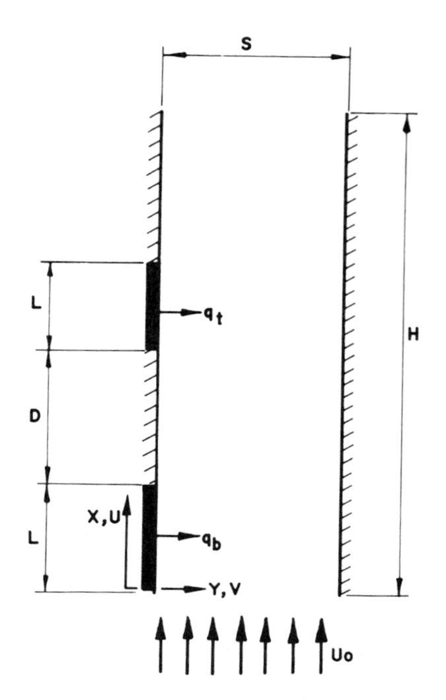

FIGURE 1. Flow configuration and coordinate system.

where $U=\partial\psi/\partial Y$ and $V=-\partial\psi/\partial X$. The formulation being elliptic, the values of ψ, ξ and θ must be specified along all boundaries. For the channel entry, uniform velocity and temperature profiles are assumed and at the channel exit velocity and temperature profiles ate taken as fully developed. The boundary conditions are

at the inlet ($X=0$, $0<Y<1$)

$$\psi = ReY \qquad \xi = 0 \qquad \theta = 0 \tag{4}$$

at the exit ($X=H/S$, $0<Y<1$)

$$\frac{\partial\psi}{\partial X} = 0 \qquad \frac{\partial\psi}{\partial X} = \frac{\partial^2 V}{\partial X^2} \qquad \frac{\partial\theta}{\partial X} = 0 \tag{5}$$

at the left wall ($Y=0$, $X\geqslant 0$)

$$\psi = 0 \qquad \xi = -\frac{\partial^2\psi}{\partial Y^2}$$

$$\frac{\partial\theta}{\partial Y} = -1 \qquad \text{for} \qquad 0\leqslant X\leqslant 1$$

$$\frac{\partial \theta}{\partial Y} = - \frac{qt}{qb} \qquad \text{for} \qquad 1 + \frac{D}{L} \leqslant X \leqslant 2 + \frac{D}{L} \tag{6}$$

$$\frac{\partial \theta}{\partial Y} = 0 \qquad \text{for} \qquad 1 < X < 1 + \frac{D}{L} \quad \text{and} \quad X > 2 + \frac{D}{L}$$

at the right wall ($Y=1$, $X \geqslant 0$)

$$\psi = Re \qquad \xi = - \frac{\partial^2 \psi}{\partial Y^2} \qquad \frac{\partial \theta}{\partial Y} = 0 \tag{7}$$

Boundary condition (5) is a consequence of $V=0$ and $\partial U/\partial X=0$ for fully developed flow. As the dimensionless transverse velocity V is not known outside the channel it is necessary to evaluate $\partial^2 V/\partial X^2$ by means of a four point, one sided, second-order accurate difference formula.

SOLUTION

The governing system of equations is characterized by a strong interdependence of the variables ψ, ξ and θ. The convective terms in the momentum and energy equations depend on ψ, the source term of the Poisson's equation defining the stream function is the negative of ξ, the source term in the vorticity transport equation depends on the derivative of θ, and the vorticity values at the channel walls are obtained from a second order approximation in ψ. A classical solution for this set of equations is obtained by converging it into a transient one so that it may be solved to obtain the steady state, but there are several restrictions to choose the time interval necessary to assure the stability and convergence of this false transient method. The present work develops a method of solution for steady-state equations with the utilization of an underrelaxation ccefficient for the vorticity in order to obtain a faster convergence. The method, based on a line-by-line implicit finite - difference procedure, is described below.

From an arbitrary vorticity field the stream function equation (1) is solved and the velocity field (U and V) is determined at the channel walls and its normal derivative at the exit is then found. The energy equation is solved and the coefficients of the discretized equation calculated with the residual being determined. The residual together with the ratio

$$\frac{\sum_{i}^{ni} \sum_{j}^{ni} \left| \phi_{i,j}^{k-1} - \phi_{i,j}^{k} \right|}{\sum_{i}^{ni} \sum_{j}^{ni} \left| \phi_{i,j}^{k-1} \right|}$$

where ϕ represents variables ξ or θ and k is the number of iterations,are used as a double check for the convergence of the iterative process.If at least one of these values decreases, the new vorticity field is found by solving the vorticity transport equation with an underrelaxation factor, initially taken as unity, and the preceding steps are repeated. If both values increase, the underrelaxation factor is reduced by multiplying it by a fraction, the vorticity field is recalculated for this new factor and the preceding procedure repeated. Finally, every time the convergence tests are positive the underrelaxation factor is increased by dividing it by the fraction. This way there is an automatic adjustment of the underrelaxation coefficient that tends to be the largest possible to obtain a faster convergence.

The control volume formulation described by Patankar [12] was used in the discretization of the governing differential equations. For the vorticity and energy transport equations an exponential discretization [13] was implemented by a truncated Taylor series and the central difference scheme was used to obtain the finite difference equation from the Poisson's equation of the stream function.

The solution of the two-dimensional Poisson's equation for ψ with a source term $F(X, Y)$, where F represents negative values of the vorticity in grid points, and with boundary conditions of normal derivatives (Neumann's boundary conditions), requires the following integral equation to be satisfied, as a consequence of Green's first theorem:

$$\int_A F(X,Y)dA = \int_c \frac{\partial \psi}{\partial n} d\ell \qquad (8)$$

where A is the area bounded by c, the contour of the solution domain, n is the normal to c and ℓ the distance along c.

The implementation of Neumann's boundary conditions, due to truncation errors, usually does not satisfy the integral restriction. The Poisson's equation of the stream function is then solved with the source term modified by a small corrective term. The quantity E, defined as

$$E = \int_A F(X,Y)dA - \int_c \frac{\partial \psi}{\partial n} d \qquad (9)$$

was computed numerically and Poisson's equation solved with the insertion of the corrective term E/A:

$$\nabla^2 \psi = F(X,Y) - E/A \qquad (10)$$

This procedure has been previously used by Briley [14] in solving the governing equations for viscous flow in ducts, with primitive variables formulation. The line and surface integrals in this work were evaluated by Simpson's rule.

After the temperature field $\theta(X,Y)$ is obtained, the bulk temperature can be determined by

$$\theta_m(X) = \frac{\int_0^1 U.\theta.dY}{\int_0^1 U.dY} \qquad (11)$$

The local Nusselt number can be expressed as

$$Nu(X) = 1/[\theta_w(X) - \theta_m(X)] \tag{12}$$

NUMERICAL RESULTS AND DISCUSSION

For a flat plate with localized heat sources Jaluria [4] concluded that boundary layer assumptions are valid for Reynolds number Re larged than around 100 and the utilization of the full elliptic equations is necessary to obtain results for Re<100. In the present work Re=50 and the solutions are obtained for Prandtl number Pr equal to 0.72 which applies for air. The height L of the heated elements was taken equal to the plate spacing S. A uniform mesh size $\Delta X = \Delta Y = 0.05$ was employed throughout the channel and results were obtained for $Gr/Re^2 = 1$ where forced convection is dominant and also $Gr/Re^2 = 15$ where natural convection dominates. In the results the ratio H/S=12 was used.

Figures 2 and 3 show the axial velocity distribution at two different positions corresponding to the trailing edge of the sources for ratio of heat flux input qt/qb=1 and for D/L=4 and Gr/Re^2 equal to 1 and 15, respectively. For $Gr/Re^2 = 1$ where the flow is dominated by forced convection, the profiles are approximately parabolic as shown in Fig. 2. Figure 3 shows that for $Gr/Re^2 = 15$, with significant buoyancy forces, the profiles exhibit a different behavior with the velocity maximum migrating towards the plate where the sources are located.

Downstream variation of the temperature of the surface containing the

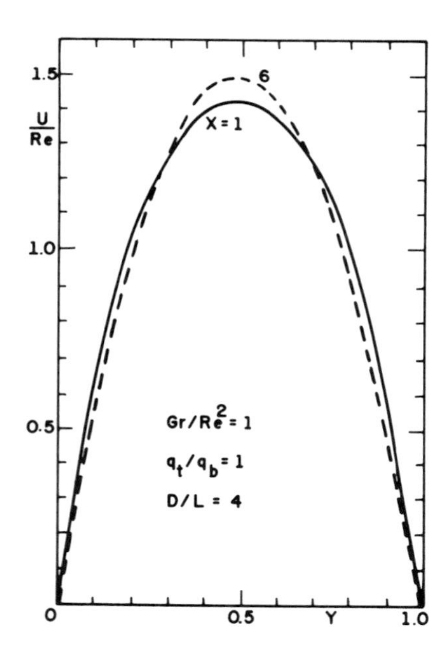

FIGURE 2. Velocity distribution across the channel.

sources are shown in Fig. 4 and 5 for Gr/Re^2 equal to 1 and 15, respectively. It can be seen that the temperature reach a maximum value in a position near the trailing edge of the heated elements. This result is a little bit different from that obtained by assuming boundary layer approximations for in this case the velocity maximum is always located at the trailing edge of the sources. This is an indication of the limitations of the parabolic formulation to accurately represent the phenomenon. An inspection in these figures reveals that for $Gr/Re^2=1$ highter surface temperatures are observed relative to $Gr/Re^2=15$, situation where buoyancy forces are relevant. Figure 4 shows higher surface temperatures when the distance between the heated elements is reduced for $Gr/Re^2=1$; this effect is less felt for $Gr/Re^2=15$ as shown in Fig. 5.

In Fig. 6 the distributions of the local values of Nusselt number defined by Eq. (12) for $Gr/Re^2=1$ are shown for different distances between heated elements and compared with those for forced convection when one of the plates is uniformly heated along its entire surface with the same heat flux imput of the sources. The character of thermal boundary layer increases the local values of Nusselt number at the leading edge of each heat source. The figure shows that if the results of a channel with one wall uniformly heated and the other wall insulated were used to evaluate the local Nusselt number of discrete sources the results would be overestimated. Temperature levels obtained this way would be lower than those actually attained.

The heat transfer coefficient for the source located at the top is affected by the presence of the source at the bottom. Velocities induced by the lower heater enhance the heat transfer from the upper heater; however, this effect is offset by preheating of the fluid. The resulting effect on the heat transfer coefficient is seen Fig. 7 and 8, where the ratio of

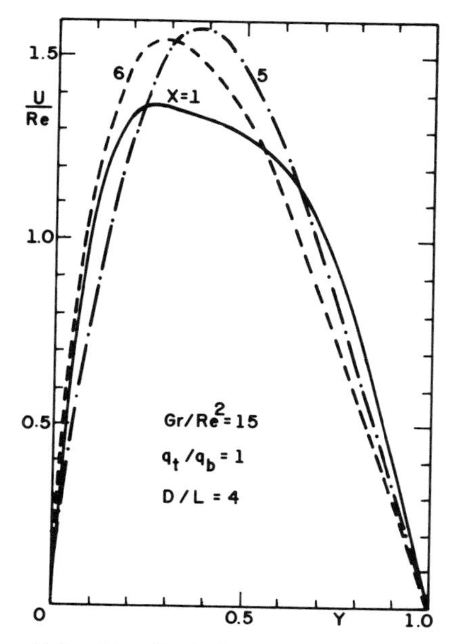

FIGURE 3. Velocity distribution across the channel.

the average Nusselt number for the top heater $(Nu)_t$ to that for the bottom heater $(Nu)_b$ is plotted against the distance between the heaters as a function of the ratio of the heat flux inputs to the top and bottom heaters qt/qb. From these figures it is clear that sources dissipating more energy should be positioned downstream of sources with less energy. For $Gr/Re^2=15$ the presence of the bottom heater is more efficient in enhancing the heat transfer from the source downstream than the situation where forced convection dominates, $Gr/Re^2=1$. The appropriate utilization of the buoyancy force in the aiding flow configuration in mixed convection may enhance the heat transfer relative to the situation of forced

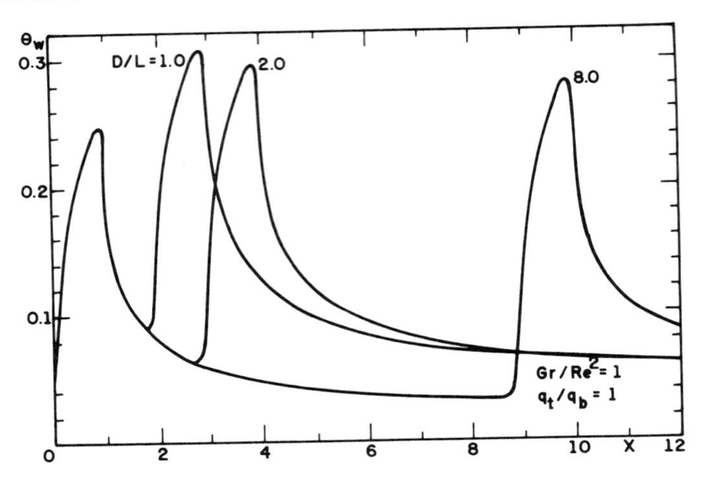

FIGURE 4. Downstream variation of the temperature θ_W of the surface containing the source.

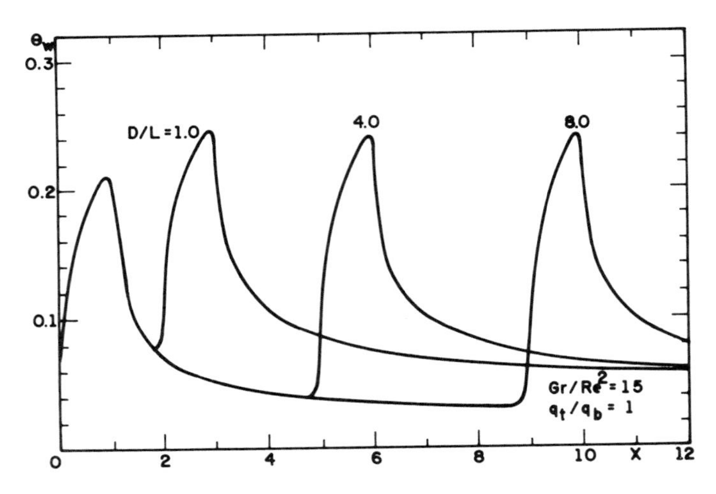

FIGURE 5. Downstream variation of the temperature θ_W of the surface containing the source.

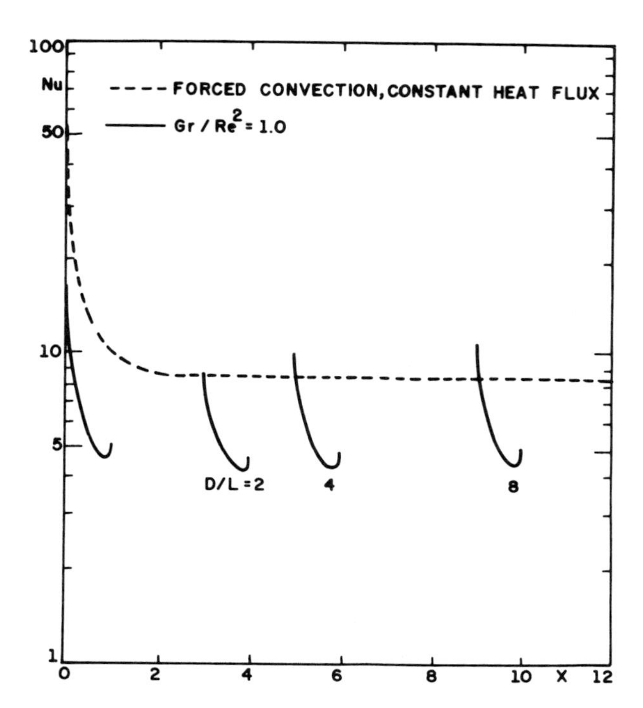

FIGURE 6. Distributions of local Nusselt number.

convection.

Figures 9 and 10 give qualitative presentation of the isotherms obtained for some of the configurations analyzed in the present study. To facilitate the presentation, dimension Y was multiplied by a factor 1.6.

NOMENCLATURE

a = thermal diffusivity
A = surface area
D = distance between heated elements
g = acceleration due to gravity
Gr = Grashof number = $g\gamma qL^4/k\nu^2$
H = channel height
k = thermal conductivity
L = height of heated element
Pr = Prandtl number = ν/a
q = heat flux input
Re = Reynolds number = $u_o S/\nu$
S = interplate spacing
T = temperature
u = axial velocity
U = dimensionless axial velocity = uS/ν

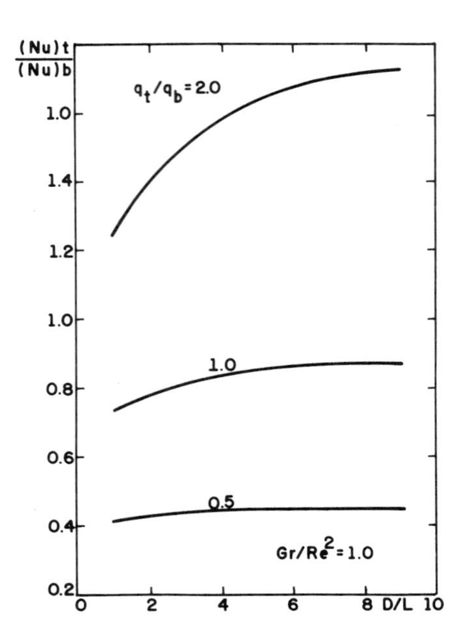

FIGURE 7. Variation of $(Nu)_t/(Nu)_b$ with D/L for different values of q_t/q_b.

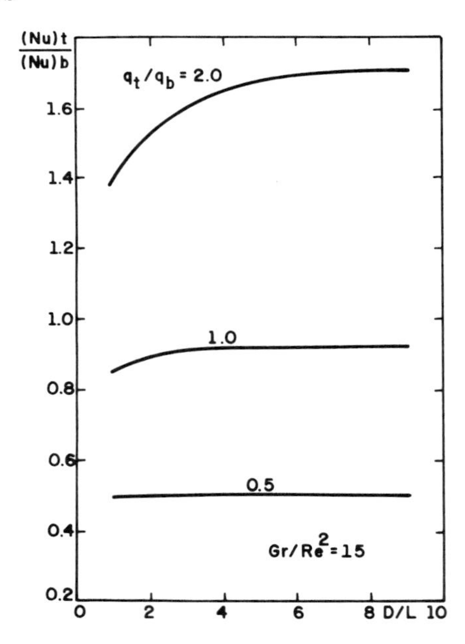

FIGURE 8. Variation of $(Nu)_t/(Nu)_b$ with D/L for different values of q_t/q_b.

FIGURE 9. Isotherms for $Gr/Re^2=15$ and $q_t/q_b=1$.

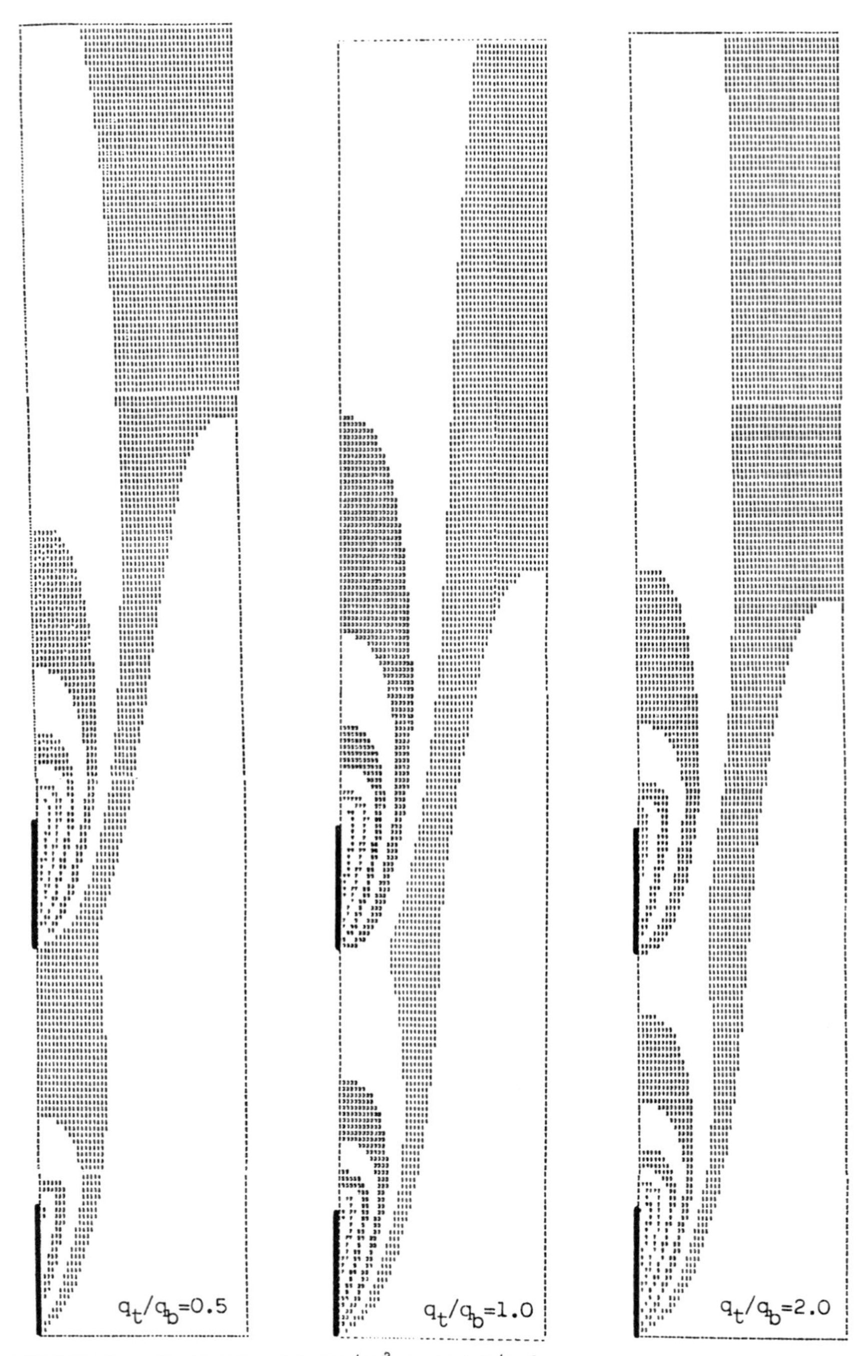

FIGURE 10. Isotherms for $Gr/Re^2=1$ and $D/L=2$.

v = transverse velocity
V = dimensionless transverse velocity = vS/ν
x = axial coordinate
X = dimensionaless axial coordinate = x/S
y = transverse coordinate
Y = dimensionless transverse coordinate = y/S
γ = coefficient of thermal expansion
ν = kinematic viscosity
ρ = density
θ = dimensionless temperature = $(T-T_\infty)k/q_b S$
ψ^* = stream function
ψ = dimensionless stream function = ψ^*/ν
ξ^* = vorticity
ξ = dimensionless vorticity = $\xi^* S^2/\nu$

Subscript

b = relative to the bottom source
m = bulk value
o = value at channel entrance
t = relative to the top source
w = value at the wall

REFERENCES

1. Jaluria, Y., Buoyancy - Induced Flow Due to Isolated Thermal Sources on Vertical Surface, J. Heat Transfer, vol. 104, pp. 223-227, 1982.

2. Jaluria, Y., Interaction of Natural Convection Wakes Arising from Thermal Sources on a Vertical Surface, Fundamentals of Natural Convection/Eletronic Cooling, ASME HTD - vol. 32, pp. 67-76, 1984.

3. Sparrow, E.M. and Faghri, M., Natural Convection Heat Transfer from the Upper Plate of a Colinear, Separated Pair of vertical Plates, J. Heat Transfer, vol. 102, pp. 623-629, 1980.

4. Jaluria, Y., Mixed Convection Flow Over Localized Multiple Thermal Sources on a Surface, Phys. Fluids, vol. 29, pp. 934-940, 1986.

5. Milanez, L.F. and Bergles, A.E., Studies on Natural Convection Heat Transfer from Thermal Sources on a Vertical Surface, Proc. 8th Int. Heat Transfer Conf., San Francisco, vol. 3, pp. 1347-1352, 1986.

6. Tewari, S.S., Jaluria, Y. and Goel, S., Natural and Mixed Convection Transport from Finite Size Heat Sources on a Flat Plate in Cooling of Electronic Equipment, Temperature/Fluid Measurements in Electronic Equipment, ASME HTD - vol. 89, p. 1-9, 1987.

7. Oosthuizen, P.H., A Numerical Study of Laminar Free Convective Flow Through a Vertical Open Partially Heated Plane Duct, Fundamentals of Natural Convection/Electronic Equipment Cooling, ASME HTD - vol. 32, pp. 41-47, 1984.

8. Yan, W.M. and Lin, T.F., Natural Convection Heat Transfer in Vertical Open Channel Flows with Discrete Heating, Int. Comm. Heat Mass

Transfer, vol. 14, pp. 187-200, 1987.

9. Chow, L.C., Husain, S.R. and Campo, A., Effects of Free Convection and Axial Conduction on Forced Convection Heat Transfer Inside a Vertical Channel at Low Peclet Numbers, J. Heat Transfer, vol. 106, pp. 297-303, 1984.

10. Tomimura, T. and Fujii, M., Laminar Mixed Convection Heat Transfer Between Parallel Plates with Localized Heat Sources, Proc. Int. Symp. on Cooling Technology for Electronic Equipment, Honolulu, pp. 701-715, 1987.

11. Nickell, T.W., Ulrich, R.D. and Webb, B.W., Combined Natural Convection and Radiation Heat Transfer from Parallel Plates with Discrete Heat Sources, ASME Winter Annual Meeting, Boston, paper 87-WA/EEP-1, 1987.

12. Patankar, S.V., Numerical Heat Transfer and Fluid Flow, Hemisphere Publ. Co., 1980.

13. Allen, D.N. de G and Southwell, R.V., Relaxation Methods Applied to Determine the Motion in 2 Dimensions of a Viscous Fluid Past a Fixed Cylinder, Quart. J. Mech. Appl. Math., vol. 8, pp. 129-145, 1955.

14. Briley, W.R., Numerical Method in Predicting Three-Dimensional Steady Viscous Flow in Ducts, Journal of Computational Physics, vol. 14, pp. 8-28, 1974.

Mixed Convective Heat Transfer in Vertical Cavities with Heated Bottom Surface

HEIU-JOU SHAW
Department of Naval Architecture and Marine Engineering
National Cheng-Kung University
Tainan, Tawain 70101

CHA'O-KUANG CHEN
Department of Mechanical Engineering
National Cheng-Kung University
Tainan, Taiwan 70101

J. W. CLEAVER
Department of Mechanical Engineering
University of Liverpool
UK

ABSTRACT

In the present work, numerical studies of the mixed convective heat transfer on the heated bottom surface of a cavity in a vertical channel are reported. Flow and temperature patterns in the channel as well as local heat transfer rates on the heated bottom surfaces are evaluated by a cubic spline collocation numerical method for various values of Gr/Re^2.

Introduction

Convective heat transfer in a vertical channel flow over a cavity has received much of attention[1] because of many industrial applications, such as the enhancement of heat transfer on rough surfaces, ventilation of an underground structure, heat dispersion from a building on the ground and cooling of cavities of an electronic system between discrete circuit devices. Many theoretical heat transfer investigations[2-3] and experimental studies[4-5] have been presented for a channel flow over a rectangular cavity with a heated bottom surface. However, relatively few studies have been reported for mixed convection from a heated cavity in a vertical channel. Humphrey and Jacobs[6] presented numerical solutions for laminar mixed convection from a square cavity in a channel with various inclinations. The effect of the Reynolds number was not considered in their investigation.

In the present study, laminar mixed convective heat transfer from a heated cavity in a vertical channel is studied. Numerical results are obtained using a cubic spline collocation method over wide ranges of the mixed convection parameter, Gr/Re^2. The Reynolds number, Re, is based on the inlet mean velocity, U_m. The streamlines and isotherms for the channel flow as well as the variation of the local heat transfer rates on the heated bottom surfaces are presented.

Governing Equations

The system under consideration consists of a vertical channel flow over a rectangular cross section cavity with a heated bottom surface as shown in Fig. 1. In the present study, the bottom surface of the cavity is maintained at a

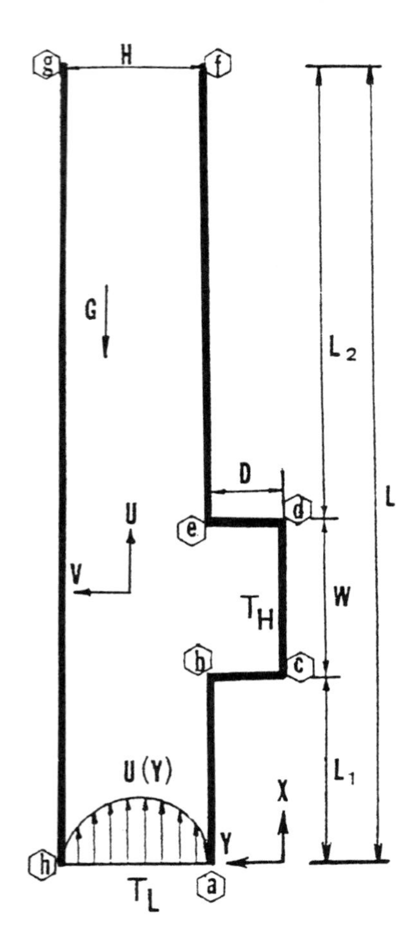

FIGURE 1. Schematic of the physical situation.

constant temperature, T_H, while the other two sides of the cavity and both sides of the channel are adiabatic.

The governing equations for the laminar two-dimensional vertical channel flow are the Navier-Stokes and energy equations. Owing to the presence of the cavity, the flow is elliptic in nature. With the usual Boussinesq approximation, the governing equations are expressed in dimensionless form as follows:

$$- \Omega = \frac{\partial^2 \Psi}{\partial X^2} + \frac{\partial^2 \Psi}{\partial Y^2} \tag{1}$$

$$\frac{\partial \Omega}{\partial \tau} + \frac{\partial (U\Omega)}{\partial X} + \frac{\partial (V\Omega)}{\partial Y} = 4 \frac{Gr}{Re^2} \left(- \frac{\partial \theta}{\partial Y}\right) + \frac{2}{Re} \left(\frac{\partial^2 \Omega}{\partial X^2} + \frac{\partial^2 \Omega}{\partial Y^2}\right) \tag{2}$$

$$\frac{\partial \theta}{\partial \tau} + \frac{\partial (U\theta)}{\partial X} + \frac{\partial (V\theta)}{\partial Y} = \frac{2}{Re \cdot Pr} \left(\frac{\partial^2 \theta}{\partial X^2} + \frac{\partial^2 \theta}{\partial Y^2}\right) \tag{3}$$

where the stream funtion, Ψ, and vorticity, Ω, are defined by

$$U = \frac{\partial \Psi}{\partial Y}, \qquad V = \frac{- \partial \Psi}{\partial X} \tag{4}$$

$$\Omega = \frac{\partial V}{\partial X} - \frac{\partial U}{\partial Y} \tag{5}$$

The boundary conditions can be expressed as

$$X=0, \qquad U=U(Y), V=0, \qquad \theta=1 \tag{6}$$

$$X=L_1/L, (L_1+W)/L, \ Y=0 \text{ to } -D/L, \qquad U=V=0, \qquad \frac{\partial \theta}{\partial X} = 0 \tag{7}$$

$$X=1 \qquad \frac{\partial U}{\partial X} = \frac{\partial V}{\partial X} = 0, \quad \frac{\partial \theta}{\partial X} = 0 \tag{8}$$

$$Y=0, \ H/L, \qquad U=V=0, \qquad \frac{\partial \theta}{\partial Y} = 0 \tag{9}$$

$$Y=-D/L, X=L_1/L \sim (L_1+W)/L, \qquad U=V=0, \qquad \theta=1 \tag{10}$$

The local Nusselt number is defined as

$$Nu_x = \frac{hx}{K} \tag{11}$$

where
$$h = \frac{- K \left.\frac{\partial T}{\partial Y}\right|_{Y = 0}}{T_H - T_L} \tag{12}$$

Numerical Procedures

The coupled vorticity transport, energy and stream function equations (1)-(3) may be solved by different numerical methods. In an attempt to reduce the computing time and storage requirements in a personal computer, the cubic spline collocation method[7-8] is developed. Comparing the cubic spline collocation method and the upwind finite difference method[9], it is found that the cubic spline collocation method can achieve the same accuracy with fewer grid points. Therefore, the governing equations in the present work are solved with the aid of the cubic spline collocation method.

The main adavantages of using the cubic spline procedure are:

1) The governing matrix system is always tridiagonal so that well-developed and highly efficient inversion algorithms are applicable, such as the Thomas algorithm.

2) The spatial accuracy of the spline approximation has shown that the first derivative has fourth-order accuracy for a uniform mesh and third-order accuracy for a nonuniform mesh. The second derivative has second-order accuracy for uniform as well as nonuniform grids.

3) Derivative boundary conditions can be directly incorporated into the inversion procedure.

The solutions of the governing equations are obtained by an iterative SADI (Spline Alternating Direction Implicit) method. The following procedure is used by Shaw[9].

Stream function equation

Step 1: $$\Psi_{ij}^{n+1,s+1/2} = \Psi_{ij}^{n+1,s} + \frac{\Delta\sigma}{2} \{ (\Psi_{YY})_{ij}^{n+1,s+1/2} + (\Psi_{XX})_{ij}^{n+1,s} + \Omega_{ij}^{n+1} \} \tag{13}$$

Step 2: $$\Psi_{ij}^{n+1,s+1} = \Psi_{ij}^{n+1,s+1/2} + \frac{\Delta\sigma}{2} \{ (\Psi_{YY})_{ij}^{n+1,s+1/2} + (\Psi_{XX})_{ij}^{n+1,s+1} + \Omega_{ij}^{n+1} \} \tag{14}$$

where $\Delta\sigma$ is a fictitious time step and $\sigma=s\Delta\sigma$. Solutions for equations (13) and (14) are the steady-state limit, $(\tau \rightarrow \infty)$, of equation (1).

Vorticity equation

Step 1: $$\Omega_{ij}^{n+1/2} = \Omega_{ij}^{n} + \frac{\Delta\tau}{2}\{-(\Psi_Y)_{ij}^{n}(\Omega_X)_{ij}^{n} + (\Psi_X)_{ij}^{n}(\Omega_Y)_{ij}^{n+1/2} + Pr(\Omega_{XX})_{ij}^{n} + Pr(\Omega_{YY})_{ij}^{n+1/2} + (Gr/Re^2)\cdot(\theta_X)_{ij}^{n}\} \quad (15)$$

Step 2: $$\Omega_{ij}^{n+1} = \Omega_{ij}^{n+1/2} + \frac{\Delta\tau}{2}\{-(\Psi_Y)_{ij}^{n}(\Omega_X)_{ij}^{n+1} + (\Psi_X)_{ij}^{n}(\Omega_Y)_{ij}^{n+1/2} + Pr(\Omega_{XX})_{ij}^{n+1} + Pr(\Omega_{YY})_{ij}^{n+1/2} + (Gr/Re^2)\cdot(\theta_X)_{ij}^{n}\} \quad (16)$$

Energy equation

Step 1: $$\theta_{ij}^{n+1/2} = \theta_{ij}^{n} + \frac{\Delta\tau}{2}\{-(\Psi_Y)_{ij}^{n}(\theta_X)_{ij}^{n} + (\Psi_X)_{ij}^{n}(\theta_Y)_{ij}^{n+1/2} + 1/(Re\cdot Pr)[(\theta_{XX})_{ij}^{n} + (\theta_{YY})_{ij}^{n+1/2}]\} \quad (17)$$

Step 2: $$\theta_{ij}^{n+1/2} = \theta_{ij}^{n+1/2} + \frac{\Delta\tau}{2}\{-(\Psi_Y)_{ij}^{n}(\theta_X)_{ij}^{n+1} + (\Psi_X)_{ij}^{n}(\theta_Y)_{ij}^{n+1/2} + 1/(Re\cdot Pr)(\theta_{XX})_{ij}^{n+1} + (\theta_{YY})_{ij}^{n+1/2}\} \quad (18)$$

After some rearrangement, equations (13) - (18) may be written in the following form:

$$\phi_{ij}^{p+1/2} = F_{ij}^{p} + G_{ij}^{p}\, l_{\phi ij}^{p+1/2} + S_{ij}^{p}\, L_{\phi ij}^{p+1/2} \quad (19)$$

$$\phi_{ij}^{p+1} = F_{ij}^{p+1/2} + G_{ij}^{p+1/2}\, m_{\phi ij}^{p+1} + S_{ij}^{p+1/2}\, M_{\phi ij}^{p+1} \quad (20)$$

where ϕ's represent the functions Ψ, Ω and θ.

In this relation between the function and its first two derivatives, the quantities F_{ij}, G_{ij} and S_{ij} are known coefficients evaluated at previous time steps (Table 1 and Table 2). It should be noted that equations (19) and (20) are of a very general nature and do not depend on the method for the spatial integration.

Equations (19) and (20) may be written in tridiagonal form as

$$a_{i,j-1}\,\Phi_{ij}^{n+1} + b_{ij}\,\Phi_{ij}^{n+1} + c_{i,j+1}\,\Phi_{ij}^{n+1} = d_{ij} \quad (21)$$

where Φ represents the function (Ψ,Ω,θ) or its derivatives. Equation (21) is easily solved by use of the Thomas algorithm. A double iterative procedure is described as follows:

(1) Initially the stream function is assumed as zero everywhere.

(2) The temperature distribution is obtained with the SADI technique as presented in equation (17) and (18). At the initial state the solution describes the temperature distribution for the pure conduction case.

(3) The temperature distribution and the associated stream function field are then substituted into the vorticity equation, from which a distribution for vorticity is obtained. The boundary conditions are determined by

$$\Omega_{i,0} = \left.\frac{\partial^2\Psi}{\partial x^2}\right|_x + \left.\frac{\partial^2\Psi}{\partial y^2}\right|_{y=0} \tag{22}$$

$$\Omega_{0,j} = \left.\frac{\partial^2\Psi}{\partial x^2}\right|_{x=0} + \left.\frac{\partial^2\Psi}{\partial y^2}\right|_y \tag{23}$$

(4) The stream function field is calculated from the obtained vorticity by using a fictitious unsteady-state term. The solution for the stream function is ascertained by ensuring the following residual-error criterion is satisfied:

$$\left| \frac{\Psi_{ij}^{s} - \Psi_{ij}^{s-1}}{\Psi_{max}^{s}} \right| < 10^{-4} \tag{24}$$

where superscripts s and (s-1) denote the current and the previous iterations, respectively.

(5)(i) If only the steady-state solution is required, the calculation proceeds to the next time step by returning to step (2) with $n \rightarrow n+1$. The iteration process is employed until the maximum relative change in temperature and flow fields satisfies the following criterion:

$$\left| \frac{\phi_{ij}^{z} - \phi_{ij}^{z-1}}{\phi_{max}^{z}} \right| < 10^{-4} \tag{25}$$

In the above criterion, ϕ refers to Ψ,θ and ω, and z denotes the number of false time steps.

(ii)If an accurate transient solution is required, the iterative calculations proceed from step (2) to step (4). This process continues until convergence. The steady-state condition is also reached by using the criterion defined above. Although accurate transient computations have been performed in a number of cases, only the steady-state results are presented in this paper.

The convergence behavior of the numerical method is checked by computing with different mesh sizes. The overall Nusselt number was used to develop an understanding of what grid fineness is necessary for accurate numerical simulations. It was found that 60 x 25 grid points were sufficient to provide accurate results.

In addition, the use of the Eq.(11) as a boundary condition is proved to be valid and effective by comparing temperature and velocity profiles obtained by two different calculations with the same Re and Gr but for different channel length L. The profiles at X=8(the exit of the channel) obtained for L=8H agree well with those at the same X(the middle of the channel) obtained for L=15H.

Results and Discussions

The plots of the streamlines and isotherms as well as local Nusselt number distributions along the isothermal heated bottom surface of the cavity are presented and the results are discussed.

Effect of Gr/Re^2 for Re=100

Streamlines and isotherms for different Gr/Re^2 are given in Figs. 2 and 3. It is shown in Fig. 2 that the increase of Gr/Re^2 will increase the buoyancy effect. The direction of the buoyancy force is the same for the flow in the cavity of the vertical channel. As for the effect of varying Gr/Re^2 on mixed convection is greater than that in a horizontal channel with heated cavity[9]. It can be seen from Fig. 2 that as Gr/Re^2 increases the velocity of the fluid near the cavity is accelerated. As a result, a streamline is more concentrated near the cavity and deflected to the right side of the vertical channel. In Fig.3, it can be observed from the isotherms that the thermal boundary layer becomes thinner as the buoyancy flow increases.

The variations of the local Nusselt number Nu_X are shown in Fig. 4 for Gr/Re^2=0, 0.1, 1 and 10, respectively. It is shown in Fig. 4 that the local Nusselt number decreases as Gr/Re^2 increases.

Effect of the Reynolds number for Gr/Re^2=1

This effect can be seen in Figs. 5 and 6 which presents the streamlines and isotherms for different Reynolds numbers at Gr/Re^2=1. As the Reynolds number increases, the buoyancy flow at the cavity increases significantly, and the thickness

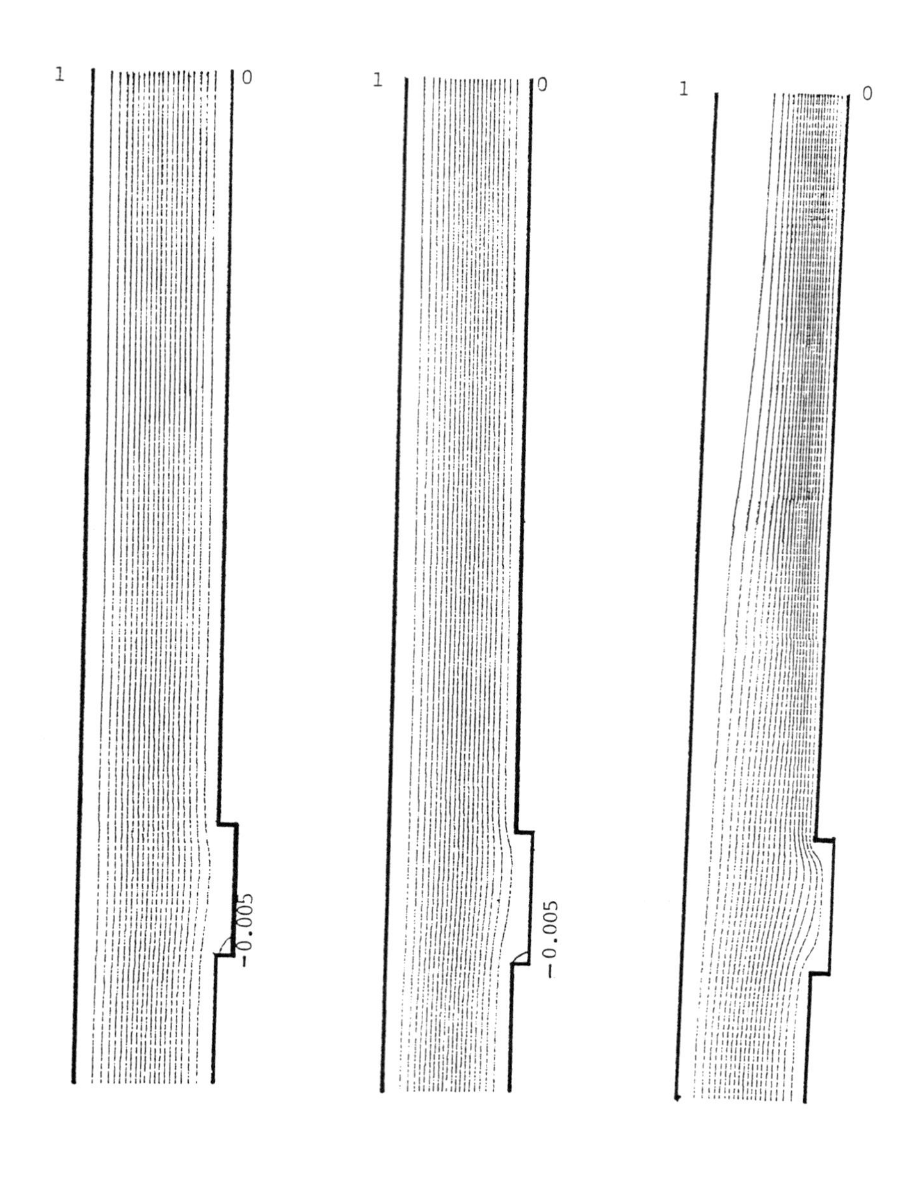

(a) $Gr/Re^2=0$ (b) $Gr/Re^2=1$ (c) $Gr/Re^2=10$

$\Delta\Psi=0.05$

FIGURE 2. Streamlines for Re=100 (a) $Gr/Re^2=0$, (b) $Gr/Re^2=1$, (c) $Gr/Re^2=10$.

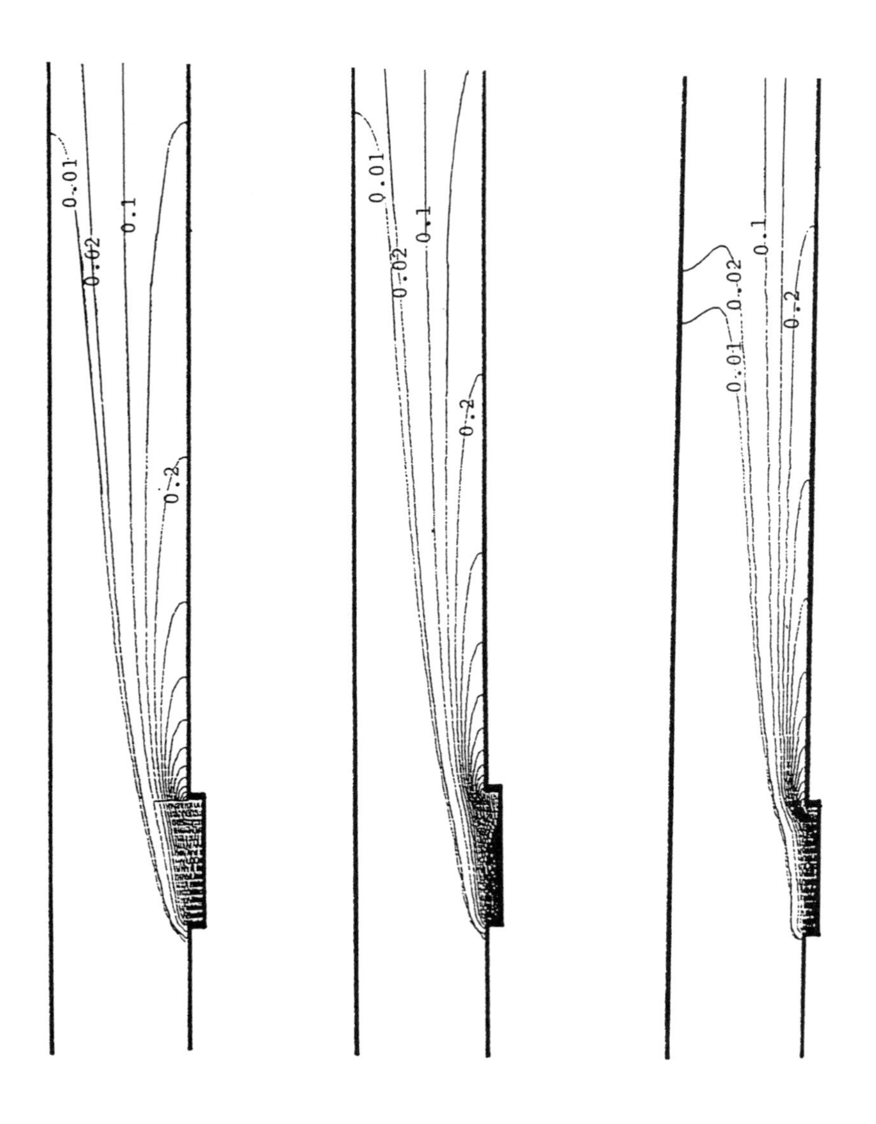

(a) $Gr/Re^2=0$ (b) $Gr/Re^2=1$ (c) $Gr/Re^2=10$

$\Delta\theta=0.05$

FIGURE 3. Isotherms for Re=100 (a) $Gr/Re^2=0$, (b) $Gr/Re^2=1$, (c) $Gr/Re^2=10$.

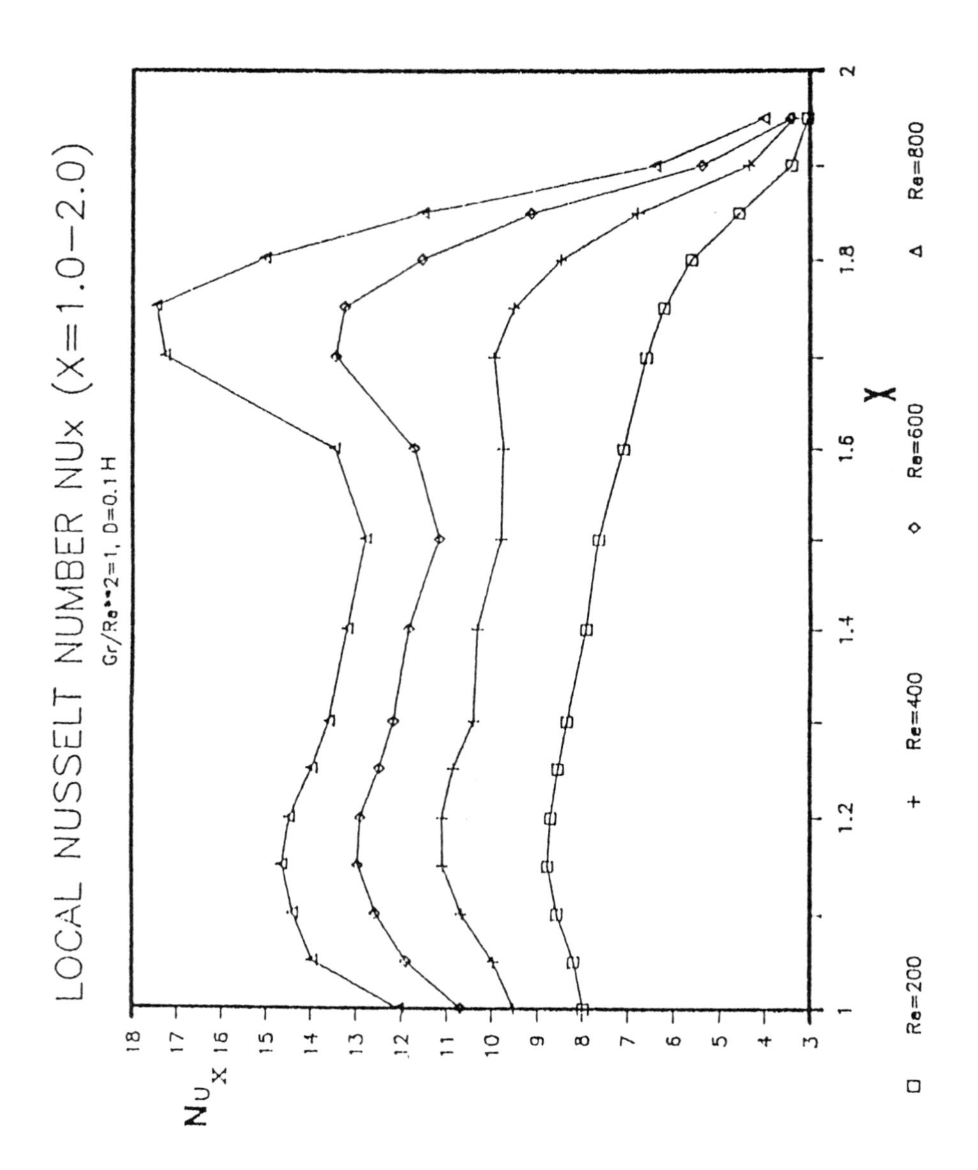

FIGURE 4. Local Nusselt number distributions along the heated bottom surface for Re=100.

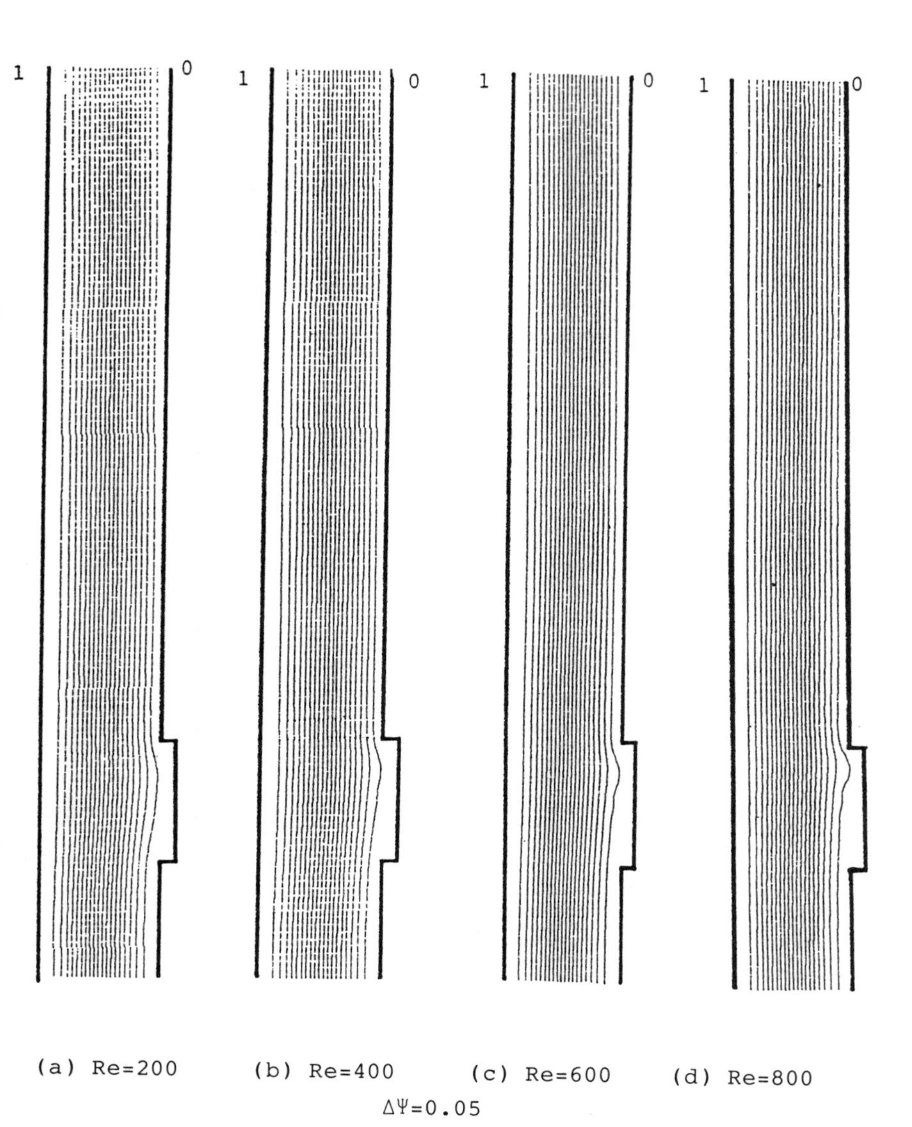

(a) Re=200 (b) Re=400 (c) Re=600 (d) Re=800

$\Delta\Psi=0.05$

FIGURE 5. Streamlines for $Gr/Re^2=1$ (a) Re=200, (b) Re=400, (c) Re=600, (d) Re=800.

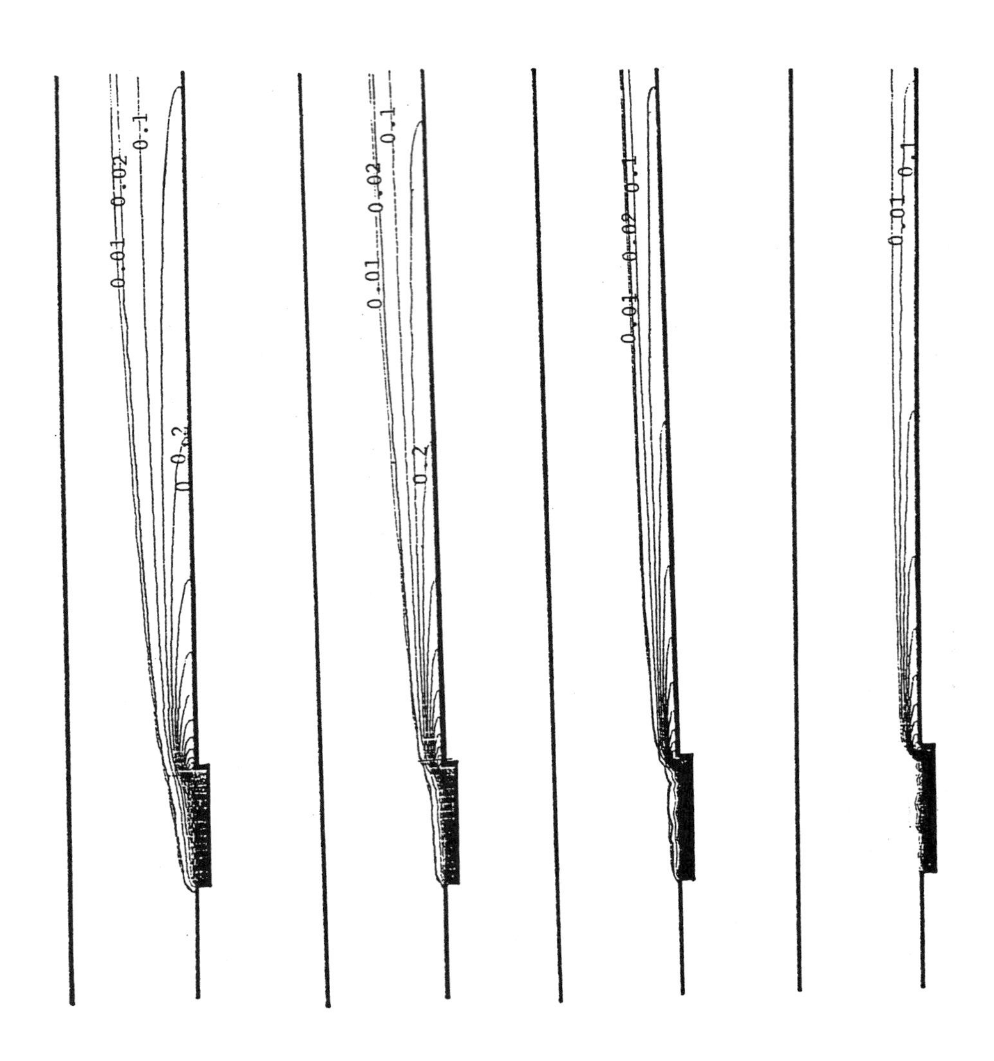

(a) Re=200 (b) Re=400 (c) Re=600 (d) Re=800

$\Delta\theta=0.05$

FIGURE 6. Isotherms for $Gr/Re^2=1$ (a) Re=200, (b) Re=400, (c) Re=600, (d) Re=800.

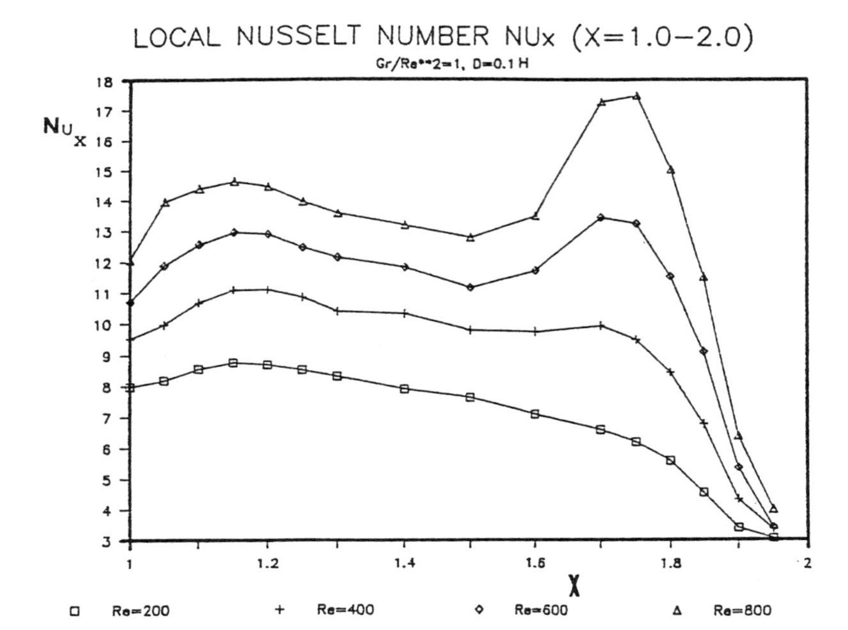

FIGURE 7. Local Nusselt number distributions along the bottom surface for $Gr/Re^2=1$.

of the thermal boundary becomes thinner. Fig. 7 indicates the distributions of local Nusselt number Nu_x for Re=200, 400, 600 and 800, respectively. It can be seen in Fig. 7 that the local Nusselt number increases significantly with an increase in the Reynolds number.

Conclusions

The mixed convective heat transfer in a vertical channel flow over a rectangular cavity with a heated bottom surface is studied by a numerical method. The solutions are obtained for the range of Re from 100 to 800 and Gr from 0 to 10^6. The main conclusions are as follows

(1)For the same value of Re, the effect of the buoyant force, resulting from increasing Gr, is significant.

(2)For the same value of Gr/Re^2, Nu_x of the heated surface increases significantly with an increase in the Reynolds number.

Nomenclature

D	depth of the cavity
g	magnitude of gravitational acceleration
H	width of the channel
Gr	Grashof number = $\frac{g\beta(T_H - T_o)H^3}{\nu^2}$
K	thermal conductivity of fluid
L_1	distance between entrance of the channel and the leading edge of the cavity
L_2	distance between the trailing edge of the cavity and the outlet of the channel
L	total channel length
m	number of vertical grid lines
n	number of horizontal grid lines
Nu_x	local Nusselt number
Pr	Prandtl number = $\frac{\nu}{\alpha}$
Q	dimensionless heat transfer rate $= q/K(T_H - T_o)$
q	heat transfer rate to fluid per unit depth of channel from bottom surface of the cavity
Ra	Rayleigh number = Pr•Gr
t	time
T	local temperature
T_H	temperature of the bottom surface of the cavity
T_L	inlet-flow temperature
$U_o(Y)$	inlet-flow velocity
U_m	inlet mean velocity
u	dimensional velocity
U	dimensionless velocity = u/U_m
v	dimensional velocity
V	dimensionless velocity = v/U_m
W	width of the cavity
x	dimensional coordinate
X	dimensionless coordinate = x/H
y	dimensional coordinate
Y	dimensionless coordinate = y/H
α	fluid thermal diffusivity
β	coefficient of thermal expansion
μ	viscosity
ν	kinematic viscosity

ψ dimensional stream funtion
Ψ dimensionless stream function = $\psi/(U_m H)$

θ dimensionless temperature = $(T-T_o)/(T_H-T_o)$
ω dimensional vorticity
Ω dimensionless vorticity = $\omega/(U_m H)$
τ dimensionless time = $t/(H/U_m)$

References

1. Chilcott, R. E.,"A Review of Separated and Reattaching Flows with Heat Transfer,"Int. J. Heat Mass Transfer, Vol. 10,
2. Inaba, H. and Wada, Y.,"Convective Heat Transfer in an Inclined Cavity with Heated from Bottom Surface in a Forced Laminar Flow,"Trans. Japan Soc. Mech. Engrs.(in Japanese), Vol. 51,
3. Bhatti, A. and Aung, W.,"Finite Difference Analysis of Laminar Separated Forced Convection in Cavities,"Trans. ASME J. Heat Transfer, Vol. 106,
4. Yamamoto, H., Seki, N. and Fukusako, S.,"Forced Convection Heat Transfer on Heated Bottom Surface of a Cavity,"Trans. ASME J. Heat Transfer, Vol. 101, PP. 475-479, 1979.
5. Aung, W.,"An Interferometric Investigation of Separated Forced Convection in Laminar Flow Past Cavities,"Trans. ASME J. Heat Transfer, Vol. 105, PP.505-512, 1983.
6. Humphrey, J. A. C. and Jacobs, E. W.,"Free-Forced Laminar Flow Convective Heat Transfer from a Square Cavity in a Channel with Variable Inclination,"Int. J. Heat Mass Transfer, Vol. 24,
7. Rubin, S. G. and Graves, F. P., "Viscous Flow Solution with a Cubic Spline Approximation," Computers and Fluids, Vol. 3, pp.1-36, 1975.
8. Wang, P. and Kahawita, R. "Numerical Integration of Partial Differential Equations Using Cubic Splines," Int. J. Comput. Math., Vol.13, pp. 271-286, 1983.
9. Shaw, H. J."Study on the Natural and Mixed Convectie Heat Transfer in Confined Region," Doctor's Dissertation of National Cheng Kung University, Tainan, Taiwan,

Experimental and Numerical Investigation of Natural Convection in Square Enclosures with a Nonuniformly Heated Vertical Surface

G. CESINI, M. PARONCINI, and R. RICCI
Dipartimento di Energetica
Università di Ancona, Italy

ABSTRACT

This paper describes a numerical and experimental study of free convective heat transfer in a square enclosure which has one vertical side partially heated by an isothermal horizontal strip and the opposite vertical side kept at a lower uniform temperature.
The effect of heated strip location is investigated. The temperature distribution in the air layer and average heat transfer coefficients have been measured for two configurations by using real-time holographic interferometry for Rayleigh numbers between $1 \cdot 10^3$ and $3 \cdot 10^5$.
Experimental results are compared with theoretical predictions obtained by solving the two-dimensional time-dependent Boussinesq equations by a finite-difference method.

INTRODUCTION

Natural convection is a convenient and inexpensive mode of heat transfer. It is commonly employed in cooling of electronic equipment.
In this application the free convective heat transfer is very important in confined spaces, such as ducts, enclosures and cells, with distributed or concentrated heat sources.
Natural convection in enclosures with two opposite vertical walls differentially heated at uniform temperatures has been extensively studied. Comprehensive rewiews of these studies have been presented by Ostrach [1,3] and Catton [2].
The investigation of the effect of nonuniformities on the surface temperature of enclosures has received only limited attention.
A number of works are concerned with the case where the bottom wall of the cavity is non-uniformly heated [4-9].
Chu and Churchill [10] studied theoretically and experimentally the effect of the size and location of an isothermal horizontal strip in an otherwise insulated vertical surface of a rectangular channel.

---

* Work supported by Ministero Pubblica Istruzione (Italy).

Poulikakos [11] reported a numerical simulation of free convective flow and heat transfer in an enclosure heated and cooled along a single vertical wall.
Kuhn and Oosthuizen [12] presented a study concerned with three-dimensional unsteady natural convective flow and heat transfer within a rectangular cavity or box with a localized heat source on a vertical wall.
In the present work is described an experimental and numerical study of natural convective heat transfer in a square enclosure which has one vertical wall partially heated and the opposite wall kept at a lower uniform temperature.
On the non-isothermal wall is located a horizontal heating strip. The remaining portion of the wall is insulated.
Fig. 1 shows a schematic drawing of the enclosure.
The temperature distribution in the air layer is measured by holographic interferometry techniques.
Heat transfer coefficients are obtained for two different locations of the heating strip.
Theoretical results are obtained by solving the unsteady equations. An Implicit Alternating Direction finite-difference method is used to obtain the distributions of vorticity, stream function, velocity and temperature in the fluid for a Rayleigh number range between $1\ 10^3$ and $3\ 10^5$.
Numerical correlations of heat transfer coefficients are evaluated.
Numerical and experimental results are compared and the effect of heat source location is investigated.

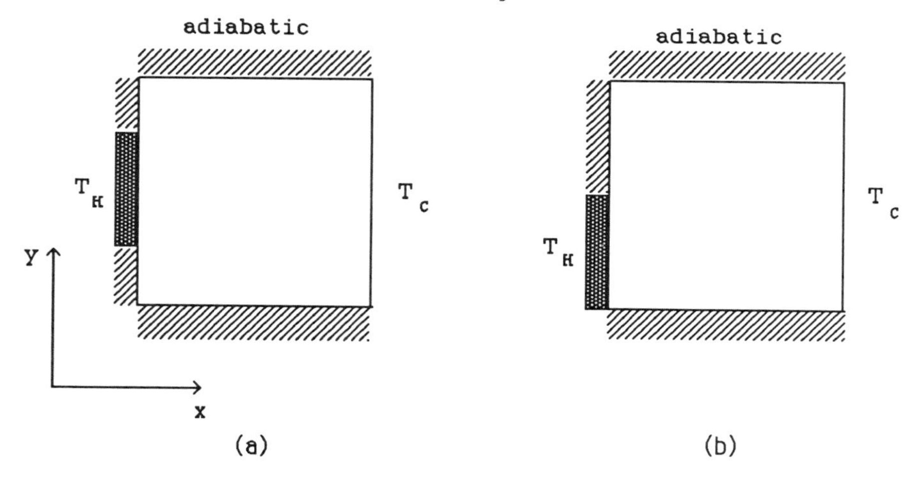

FIGURE 1. Schematic of the enclosure with thermal boundary conditions.
a) The heating strip is located at the center of the wall
b) The heating strip is located at the bottom of the wall

GOVERNING EQUATIONS AND NUMERICAL SOLUTION

Employing the Boussinesq approximation the dimensionless equations describing the laminar flow of an incompressible fluid with the strem-function and vorticity formulation are:

$$\frac{\partial\theta}{\partial t} + \nabla\cdot(\vec{V}\theta) = \frac{1}{Pr}\nabla\cdot(\nabla\theta) \tag{1}$$

$$\frac{\partial\omega}{\partial t} + \nabla\cdot(\vec{V}\omega) = \nabla\cdot(\nabla\omega) + Gr\frac{\partial\theta}{\partial x} \tag{2}$$

$$\nabla\cdot(\nabla\psi) = -\omega \tag{3}$$

where equations (1) and (2) are in "Conservative" form. For the energy and vorticity equations we have used the integral approach over a control volume.
In this formulation we used the Second Upwind Differencing Scheme [13,14] or "Donor Cell" to obtain the convection term and a central finite-difference approximation to obtain the diffusion and source terms. The A.D.I. method [15] has been adopted to solve the transient problem.
The final form of these equations is

$$-B_1 F_{i-1,j}^{n+1/2} + B_2 F_{i,j}^{n+1/2} - B_3 F_{i+1,j}^{n+1/2} = B_4 (F_{i,j+1}^{n} - 2 F_{i,j}^{n} - B_3 F_{i,j-1}^{n}) + \\ + B_5 F_{i,j}^{n} + B_6 F_{i,j-1}^{n} - B_7 F_{i,j+1}^{n} + B_8 (\theta_{i+1,j}^{n} - \theta_{i-1,j}^{n}) \tag{4}$$

$$-C_1 F_{i,j-1}^{n+1} + C_2 F_{i,j}^{n+1} - C_3 F_{i,j+1}^{n+1} = C_4 (F_{i+1,j}^{n+1/2} - 2 F_{i,j}^{n+1/2} - B_3 F_{i-1,j}^{n+1/2}) + \\ + C_5 F_{i,j}^{n+1/2} + C_6 F_{i-1,j}^{n+1/2} - C_7 F_{i+1,j}^{n+1/2} + B_8 (\theta_{i+1,j}^{n+1/2} - \theta_{i-1,j}^{n+1/2}) \tag{5}$$

where

$$B_1 = (\Delta t/4\,\Delta x)(u_L + |u_L|) + C_4 \tag{6a}$$

$$B_2 = 1 + (\Delta t/4\,\Delta x)(u_R + |u_R| - u_L + |u_L|) + (a\,\Delta t/\Delta x^2) \tag{6b}$$

$$B_3 = C_4 - (\Delta t/4\,\Delta x)(u_R - |u_R|) \tag{6c}$$

$$B_4 = a\,\Delta t/2\,\Delta y^2 \tag{6d}$$

$$B_5 = 1 + (\Delta t/4\,\Delta y)(v_B - |v_B| - v_T - |v_T|) \tag{6e}$$

$$B_6 = (\Delta t/4\ \Delta y)(v_B+|v_B|) \quad (6f)$$

$$B_7 = (\Delta t/4\ \Delta y)(v_T-|v_T|) \quad (6g)$$

$$B_8 = b\ Gr\ \Delta t/4\ \Delta x \quad (6h)$$

$$C_1 = (\Delta t/4\ \Delta y)(v_B + |v_B|) + B_4 \quad (7a)$$

$$C_2 = 1 + (\Delta t/4\ \Delta y)(v_T+|v_T|-v_B+|v_B|) + (a\ \Delta t/\Delta y^2) \quad (7b)$$

$$C_3 = B_4 - (\Delta t/4\ \Delta y)(v_T-|v_T|) \quad (7c)$$

$$C_4 = a\ \Delta t/\ 2\ \Delta x^2 \quad (7d)$$

$$C_5 = 1 + (\Delta t/4\ \Delta x)(u_L-|u_L|-u_R-|u_R|) \quad (7e)$$

$$C_6 = (\Delta t/4\ \Delta x)(u_L+|u_L|) \quad (7f)$$

$$C_7 = (\Delta t/4\ \Delta x)(u_R-|u_R|) \quad (7g)$$

$$u_L = (u_{i-1,j}+u_{i,j})/2\ ; \qquad u_R = (u_{i+1,j}+u_{i,j})/2$$

$$v_B = (v_{i,j-1}+v_{i,j})/2\ ; \qquad v_T = (v_{i,j+1}+v_{i,j})/2 \quad (8)$$

and

$$a = 1/Pr \text{ and } b=0 \quad \text{if } F_{i,j} = \theta_{i,j}$$

$$a=1 \text{ and } b=1 \quad \text{if } F_{i,j} = \omega_{i,j}. \quad (9)$$

The above finite-difference scheme is also "Transportative", [16].

The stream-function equation (3) has been discretized by using the Liebmann's iterative formula [17] and S.O.R. method with underrelaxation of convergence parameter.

This solution scheme shows non-physical spatial oscillations near the steady-state solution for Grashof number greater than $4 \cdot 10^5$. This behaviour strongly depends on the hot surface boundary condition and on the number of grid points, which can result in a great vorticity gradient at the heated wall.

When the heater is positioned on the bottom of the vertical surface, the spatial oscillations occur for higher Grashof numbers.

The boundary conditions for the horizontal walls and the heated and cooled vertical walls are schematically described in Fig.1. The remaining verical walls are supposed adiabatic.

A uniform initial condition is assumed.

The time evolution of vorticity, stream-function and temperature is obtained by solving the above equations.

Local and average heat tranfer coefficients at the hot strip are directly obtained from temperature distribution at steady-state conditions.

EXPERIMENTAL APPARATUS

The test cell is an enclosure filled of air.A schematic drawing of test cell is shown in Fig.2.
One vertical wall includes an isothermal strip, electrically heated, which can be located on the central region (Fig.1a) or at the bottom (Fig.1b) of the wall. The remaining part of the wall is made of plexiglass.
The opposite vertical wall is made of aluminium and is cooled by water circulation through a jacket attached to the back surface.
The top and bottom surfaces of the enclosure are made of plexiglass while the remaining vertical walls are constituted of glass to allow optical access to the cavity.
The dimension of the test cell in the direction parallel to the laser beam is large enough that end effects can be neglected. Then the air layer may be considered as a two-dimensional object.
The temperature distribution in the air layer is measured by a holographic interferometer.
The light source is an argon-ion laser with a nominal power rating of 5 W with etalon for the 514.5 nm wavelength. Both object and reference beams have a maximum diameter of 0.15 m.
The optical set-up allows use of both double-exposure and real-time holographic interferometry techniques. The first method is utilized for steady-state measurements, the second one to study the temporal evolution of transfer processes.

By using a micropositioning plate holder it is possible to position the interferometer in finite fringes or infinite fringes field.
A copper-constantan thermocouple is located in the core region of the air layer to provide a reference temperature for the evaluation of the interferograms. The temperature uniformity of the heated and cooled surfaces is controlled by copper-constantan thermocouples.
The interferograms are evaluated by using a travelling microscope to obtain the intensity distribution. Density and temperature distributions are obtained by the usual methods of inversion [18].

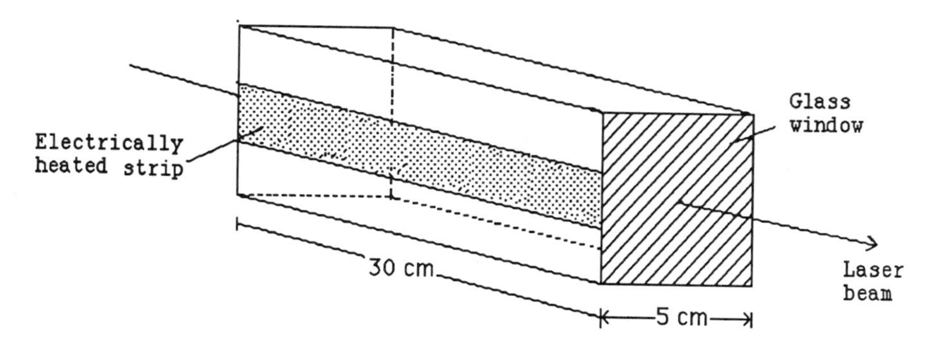

FIGURE 2. Schematic of the test-cell.

RESULTS

A number of computations have been carried out to verify the ability of the numerical code to describe the convective heat process inside the cavity. The numerical results have been then compared with the experimental data.
Figs.3a and 5a show typical interferograms obtained for $Ra=1\cdot10^5$ with the heating strip located, respectively, at the center and at the bottom edge of the wall. The holograms were obtained with the interferometer positioned in infinite fringe field, so they display contours of constant value of temperature.
The experimental temperature distribution can be directly compared with that obtained numerically for the same Rayleigh number (Figs.3b and 5b).
Employing the least-squares technique, numerical data, obtained with the heating strip located at the wall center, are correlated by the following equation:

$$Nu_d = 0.254\ Ra_d^{0.283} \tag{10}$$

with a standard deviation of ± 2% for $1\ 10^3 < Ra < 3\ 10^5$.
A comparison of numerical and experimental results is reported in Fig.4; the agreement is very good. The maximum standard deviation between experimental results and numerical correlation is 7%.
Fig.6 shows numerical and experimental results for case of the heat source positioned at the lower edge of the vertical wall. Numerical data are correlated by the equation

$$Nu_d = 0.235\ Ra_d^{0.298} \tag{11}$$

with a standard deviation of ± 1%.

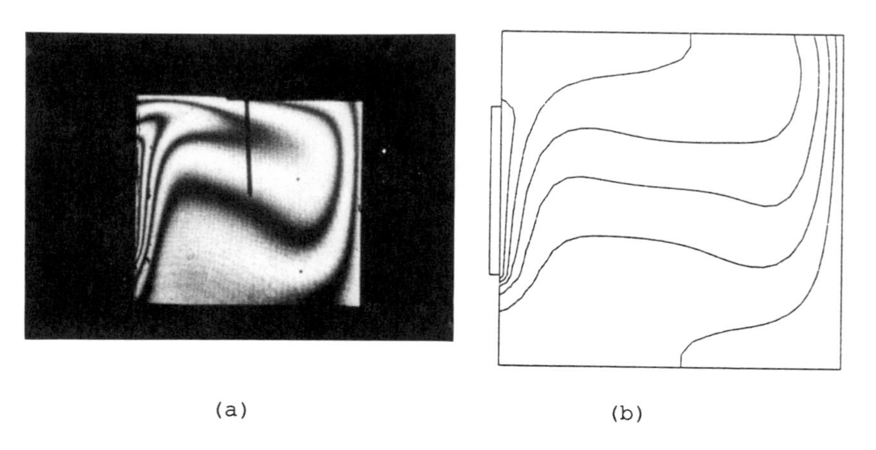

(a) (b)

FIGURE 3. Steady state temperature contours for $Ra_d = 1\cdot10^5$ with the heating strip at the center of the wall.
(a) Experimental interferogram
(b) Numerical distribution

Comparison with the experimental data shows a maximum standard deviation of 17%. The experimental results are systematically lower than the numerical correlation. This is probably due to the difficulties in interpreting the interferograms near to the bottom corner of the enclosure, due to the high concentration of fringes. Furthermore the effect of thermal conduction of the insulating walls should be taken into consideration.

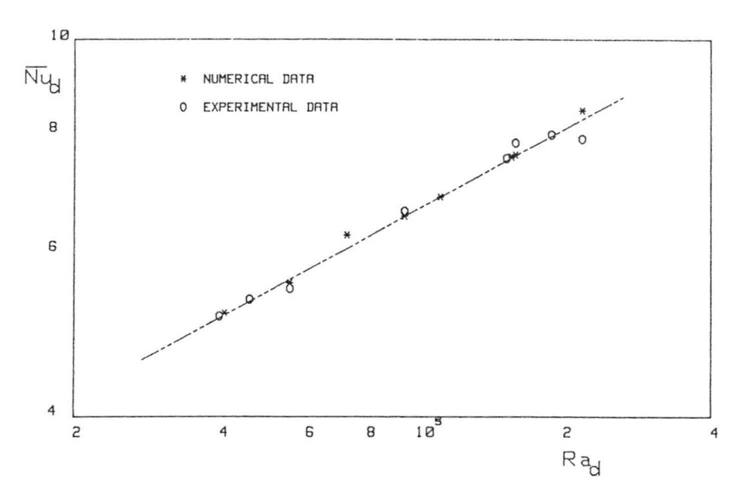

FIGURE 4. Comparison of numerical and experimental results obtained with the heating strip located at the center of the wall.

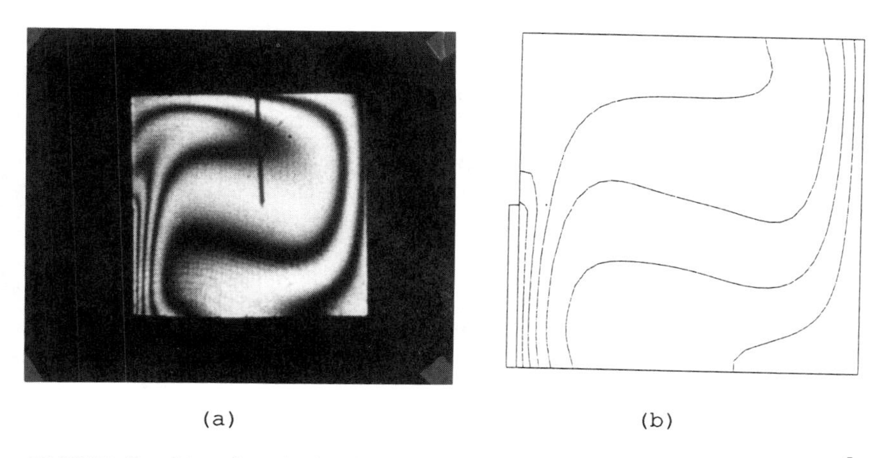

FIGURE 5. Steady-state temperature contours for $Ra_d = 1\ 10^5$ with the heating strip in the lower half of the wall
(a) Experimental interferogram
(b) Numerical distribution

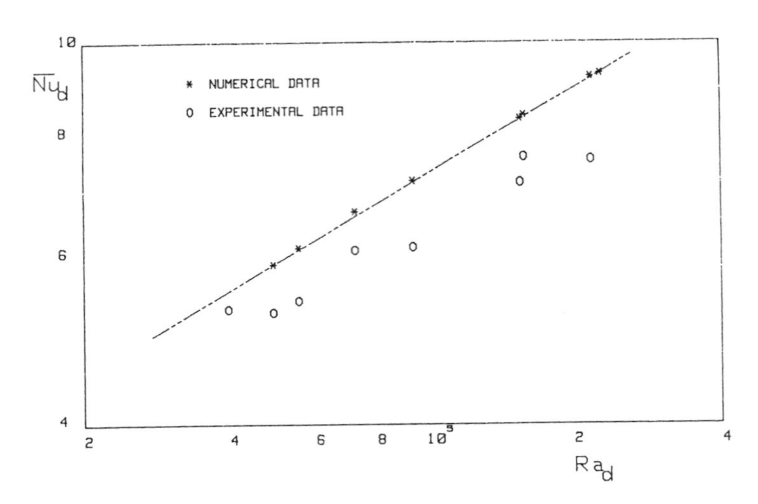

FIGURE 6. Comparison of numerical and experimental results obtained with the heating strip located in the lower half of the wall.

CONCLUSIONS

The problem of natural convection in an enclosure with one vertical wall non-uniformly heated has been numerically and experimentally studied.
Accurate results have been obtained solving the governing differential equation by a finite-difference scheme.
Holographic interferometry has been used to determine the temperature distribution in the air layer and to obtain the convective heat transfer coefficients.
Numerical and experimental results are in good agreement.
Correlations for the average Nusselt number have been obtained with the heating source located at the center and at the bottom of the vertical wall.
Comparison between the two correlations shows that the average Nusselt number is about 10% higher when the heating strip is in the central position.

ACKNOWLEDGEMENT

The authors wish to thank Renzo Paoloni for his contribution to the experimental work.

NOMENCLATURE

a,b	defined in equations (9)
B_1, ,B_8	defined in equations (6)
C_1, ,C_8	defined in equations (7)
d	width of the cavity
g	gravitational acceleration
h	heat transfer coefficient
H	height of cavity
Gr	Grashof number = $gd^3\gamma\Delta T/\upsilon^2$
Pr	Prandtl number = υ/α
Nu_d	average Nusselt number = hd/λ
Ra_d	Rayleigh number
t	dimensionless time $\upsilon\tau/d^2$
Δt	dimensionless time step
T	temperature
ΔT	temperature difference = $T_H - T_C$
u	dimensionless velocity in x direction = Ud/υ
U	velocity in x direction
v	dimensionless velocity in y direction = Vd/υ
V	velocity in y direction
x,y,z	dimensionless cartesian coordinates = X/d,Y/d,Z/d
X,Y,Z	cartesian coordinates
Δx	dimensionless grid spacing in x direction
Δy	dimensionless grid spacing in y direction
α	thermal diffusivity of fluid
γ	volumetric thermal expansion coefficient of fluid
λ	thermal conductivity of fluid
υ	kinematic viscosity of fluid
τ	time
θ	dimensionless temperature = $(T-T_C)/(T_H-T_C)$
Ψ	stream function
ψ	dimensionless stream function = Ψ/υ
Ω	vorticity
ω	dimensionless vorticity = $\Omega d^2/\upsilon$

Superscripts

n	time iteration number

Subscripts

C	cooled isothermal surface
H	heated isothermal strip
i,j	subscripts denoting x and y grid node

REFERENCES

1. Ostrach, S., Natural Convection in Enclosures, *Advances in Heat Tranfer*, Vol.8, pp.161-227, 1972.

2. Catton, I., Natural Convection in Enclosures, *Proc. 6th Int. Heat Transfer Conf.*, Toronto, Vol.6, No.2, pp.13-31, 1978.

3. Ostrach, S., Natural Convection Heat Transfer in Cavities and Cells, *Proc. 7th Int. Heat Transfer Conf.*, Vol. , pp.365-379, 1982.

4. Torrance, K.E., Orloff, L., and Rockett, J. A., Experiments on Natural Convection in Enclosures with Local Heating from Below, *J. of Fluid Mech.*, Vol.36, pp.21-31, 1969.

5. Torrance, K. E., Rockett, J. A., Numerical Study of Natural Convection in an Enclosure with Localized Heat From Below - Creeping Flow to the Onset of Laminar Instability, *J. of Fluid Mech.*, Vol.36, pp.33-54, 1969.

6. Greenspan, D., Schultz, D., Natural Convection in an Enclosure with Localized Heating from Below, *Computer Meth. Appl. Mechanics and Engineering,* Vol.3, pp.1-10, 1974.

7. Chao, P. K., Ozoc, H., and Churchill, S. W., The effect of a Non Uniform Surface Temperature on Laminar Natural Convection in a Rectangular Enclosure, *Chem. Eng. Commun.*, Vol.9, pp.245-254, 1981.

8. Chao, P. K., Ozoe, H., Churchill, S. W., and Lior, N., Laminar Natural Convection in an Inclined Rectangular Box with the Lower Surface Half-Heated and Half-Insulated, *ASME J. of Heat Transfer*, Vol.105, pp. 425-432, 1983.

9. Han, S. M., Chen, H., Heat Transfer in Enclosures, *ASME HTD* Vol.39, pp. 21-27,

10. Chu, H. H. S., Churchill, S. W., The Effect of Heater Size, Location, Aspect-Ratio and Boundary Conditions on Two Dimensional Laminar Natural Convection in Rectangular Channel, *ASME J. of Heat Transfer*, Vol.98,pp.194-201, 1976.

11. Poulikakos, D., Natural Convection in a Confined Fluid Filled Space Driven by a Single Vertical Wall with Warm and Cold Regions, *ASME J. of Heat Transfer,* Vol.107, pp.867-876 1985.

12. Kuhn, D., Oosthuizen, P. H., Three-Dimensional Transient Natural Convective Flow in a Rectangular Enclosure with Localized Heating, *ASME HTD*-Vol.63, pp.55-62, 1986.

13. Lilly, D. K., On the Computational Stability of Numeric Solutions of Time Dependent Non-Linear Geophysical Dynamics Problems, *Monthly Weather Rev.*, p.93, 1986.

14. Kubleck, K., Merker, G. P., and Straub, J., Advanced Numerical Computation of Two-Dimensional Time-Dependent Free Convection in Cavities, *Int. J. of Heat and Mass Transfer*, Vol.23, pp.203-217, 1980.

15. Peaceman, D. W. and Rachford, H. H. Jr., The Numerical Solution of Parabolic and Elliptic Differential Equations, *J. Soc. Industrial Appl. Math.*, Vol.3, 1955.

16. Roache, P. J., *Computational Fluid Dynamics*, Hermosa Publishers, Albuquerque, New Mexico, 1972.

17. Bejan, A., *Convection Heat Transfer*, John Wiley & Sons, 1984.

18. Hauf,W., and Grigull,U., Optical methods in Heat Transfer, in *Advances in Heat Transfer* (ed. by J. P. Hartnett and T. F. Irvine Jr.), Vol.6, pp. 133-366, Academic Press, New York, 1970.

Numerical Simulation of Natural Convection in Rectangular Enclosures with Discrete Heated Elements

GUO-XIANG WANG, HAI-LIN ZHANG, and WEN-QUAN TAO
Department of Power Machinery Engineering
Xi'an Jiaotong University
Xi'an, Shaanxi 710049, PRC

ABSTRACT

Natural convection in rectangular enclosures with finite-sized heated elements on a vertical adiabatic wall is studied numerically. The problem considered is relevant to the cooling of electronic equipment. Two kinds of enclosures are studied, one with two elements and the other with three elements. The effects of separating distance between elements and the aspect ratio of enclosures are investigated. The results show that the per-element average Nusselt number of the lower element is always higher than that of upper one. The separating distance between elements has a rather significant effect on the Nusselt number of the upper element. For a two-element enclosure, the average Nusselt number of upper element can be enhanced by as much as 17-24% when it is at the optimum location. Detailed velocity and temperature fields are obtained. The isotherms and the streamlines for some typical cases are presented. They are in good agreement with the results of flow visualization

INTRODUCTION

Natural convection heat transfer from isolated finite-sized heated elements located on a vertical adiabatic surface is of interest in several practical problems, such as the cooling of electronic equipment by natural convection. It is often of importance to determine the energy input and the location of the electronic elements to avoid overheating the elements and the surface on which they are located. The energy dissipated results in a bouyancy driven flow that rises above the elements as a wake or plume. The interaction of such a wake with the flow arising from other thermal elements is an important consideration in the positioning of elements, since the heat transfer from an element is strongly affected by the flow generated in its neighborhood.

The natural convection flow due to isolated finite-sized heated elements on a vertical adiabatic surface in an infinite medium or in an enclosure has received some attention in the literature. The wall plumes and the interaction of flows generated by finite-sized heated elements in an infinite medium have been studied numerically by employing boundary layer approximations or by solving the full governing equations by the finite difference method [1,2] . The numerical results show that the heat transfer coefficient of thé upper element is affected by the separating distance between the two elements. Chu and co-workers [3] have numerically and experimentally

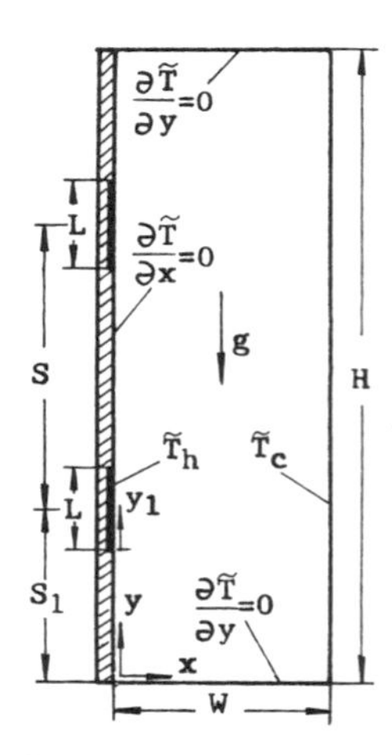

FIGURE 1. Coordinate system and thermal boundary conditions.

investigated the natural convection of one element on an adiabatic vertical surface of rectangular enclosures. They especially studied the effects of heater size, location, aspect ratio and boundary conditions. Later, Turner and Flack [4] experimentally studied the same problem but at higher Rayleigh numbers. Recently, the authors of this paper experimentally studied the natural convection of two and three heated elements in a rectangular cavity and visualized the flow pattern by smoke technique [5] . The results show that the flow is very complicated.

This paper numerically studies the natural convection due to multiple discrete heated elements on a vertical adiabatic surface of a rectangular enclosure by the finite difference method. The geometry and thermal boundary conditions of the enclosure are shown in Figure 1. It is a two-dimensional rectangular cavity with height H and width W. On one vertical adiabatic wall there are two or three heated isothermal elements of width L with temperature T_h. The separating distance between the top and bottom elements is S. The ratio S/H is one of the parameters in the study. The opposing vertical wall is maintained at a uniform lower temperature T_c. The two horizontal surfaces are insulated. Results have been obtained under two-dimensional steady laminar flow conditions for S/H=0.125-0.8, aspect ratio H/W from 3.0 to 10.0, a range of Rayleigh numbers from 3×10^3 to 10^6 and for Pr=0.7. In setting up the governing equations, the Boussinesq approximation regarding the density and the assumption that other properties remain constant have been used.

MATHEMATICAL MODEL AND NUMERICAL PROCEDURE

The nondimensional governing equations of the steady laminar flow considered can be written as follows:

$$\partial U/\partial X + \partial V/\partial Y = 0 \quad (1)$$

$$U(\partial U/\partial X) + V(\partial U/\partial Y) = -\partial P/\partial X + \partial^2 U/\partial X^2 + \partial^2 U/\partial Y^2 \quad (2)$$

$$U(\partial V/\partial X) + V(\partial V/\partial Y) = -\partial P/\partial Y + \partial^2 V/\partial X^2 + \partial^2 V/\partial Y^2 + (Ra/Pr)\cdot T \quad (3)$$

$$U(\partial T/\partial X) + V(\partial T/\partial Y) = (1/Pr)\cdot(\partial^2 T/\partial X^2 + \partial^2 T/\partial Y^2) \quad (4)$$

The boundary conditions are
at X=0,

$$\partial T/\partial X = 0, \text{ for } 0 \leqslant Y < (S_1/W-0.5L/W) \quad (5a)$$

$$T = 1, \text{ for } (S_1/W-0.5L/W) \leqslant Y \leqslant (S_1/W+0.5L/W) \quad (5b)$$

$$\partial T/\partial X = 0, \text{ for } (S_1/W+0.5L/W) < Y < (S/W+S_1/W-0.5L/W) \quad (5c)$$

$$T = 1, \text{ for } (S/W+S_1/W-0.5L/W) \leqslant Y \leqslant (S/W+S_1/W+0.5L/W) \quad (5d)$$

$$\partial T/\partial X = 0, \text{ for } (S/W+S_1/W+0.5L/W) < Y \leqslant H/W \quad (5e)$$

at Y=0 or 1,

$$\partial T/\partial Y = 0, \text{ for } 0 \leqslant X \leqslant 1 \quad (5f)$$

at X = 1,

$$T = 0, \text{ for } 0 \leqslant Y \leqslant H/W \quad (5g)$$

on the all surfaces,

$$U = V = 0 \quad (5h)$$

The nondimensional variables are defined as follows:

$$U = uW/\nu, \quad V = vW/\nu, \quad X = x/W, \quad Y = y/W, \quad P = (p+\rho_0 gy)/(\rho_0(\nu/W)^2) \quad (6a)$$

$$T = (\tilde{T} - \tilde{T}_c)/(\tilde{T}_h - \tilde{T}_c), \quad Pr = \nu/a, \quad Ra = g\beta(\tilde{T}_h - \tilde{T}_c)W^3/(\nu a) \quad (6b)$$

The local Nusselt number is defined by

$$Nu = \alpha L/\lambda = -(L/W)\cdot(\partial T/\partial X)\Big|_{X=0} \quad (7)$$

The per-element average Nusselt number is defined as

$$\overline{Nu} = \bar{\alpha} L/\lambda = \int^{L/W} Nu\,dY/(L/W) \quad (8)$$

The conservation equations were solved numerically using the finite-volume discretization method described in [6] . The SIMPLEC method of [7] was used. The resulting algebraic equations were solved by the SLUR procedure. The results presented here were calculated on a grid of 22x32 nodes. The maximum difference of the average Nusselt numbers between the solutions of 22x32 and 32x42 was less than 4%. As this difference was small, a 22x32 grid was used. The code developed was first used to compute a limiting case, natural convection in a square enclosure with two vertical isothermal walls. The Nusselt numbers of the present study agreed well with those given by Davis and Jones [8] (the maximum difference was less than 2%).

After each iteration, the following relative change of the dependent variables was calculated:

$$R_\phi = \left|\phi(i,j) - \phi^*(i,j)\right|_{max} / \left|\phi(i,j)\right|_{max} \quad (9)$$

where the subscript max means the maximum value over all interior nodes, and the star stands for the solution of previous iteration. Iteration is continued until the following criterion is satisfied:

$$R_{\phi} \leqslant \mathcal{E} \tag{10}$$

The value of $\mathcal{E}$ is taken as 10^{-3} in this study.

RESULTS AND DISCUSSIONS

First, the flow pattern and the isotherms at different Rayleigh numbers will be examined. Figures 2 and 3 show two typical flow fields and isotherms for the case with two heated elements. As compared with a simple single isothermal vertical plate, the streamlines and isotherms are all distorted, which shows the significant effects of the thermal boundary conditions. At lower Rayleigh number($Ra \leqslant 10^5$), only one vortex appears. Along the lower element, heated fluid flows up and develops as a wall plume along the otherwise adiabatic surface. This wall plume then flows over the place where the upper heated element is located. Due to the buoyancy force generated by the upper element, the streamlines are forced close to the element. Because of the restriction of the rigid boundaries, the fluids turn to the cold plate at the left top corner and then flow down along the cold wall. Finally, one vortex is formed. But at larger Rayleigh number, for example, at $Ra=10^6$, three vortices are formed in the recirculating flow, as shown in Figure 3(b). At this Rayleigh number, fluid temperature stratification

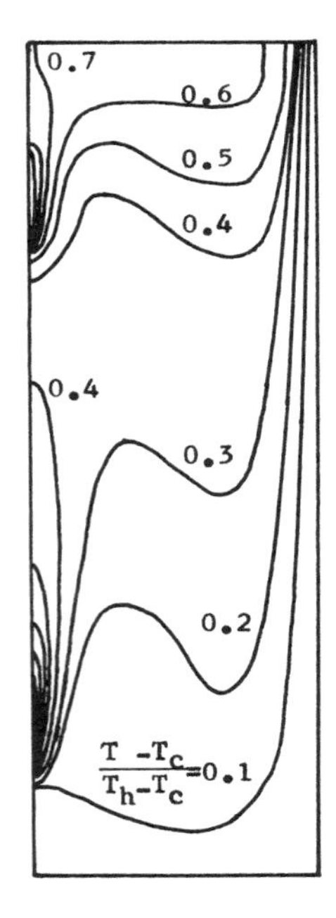

(a) Isotherms

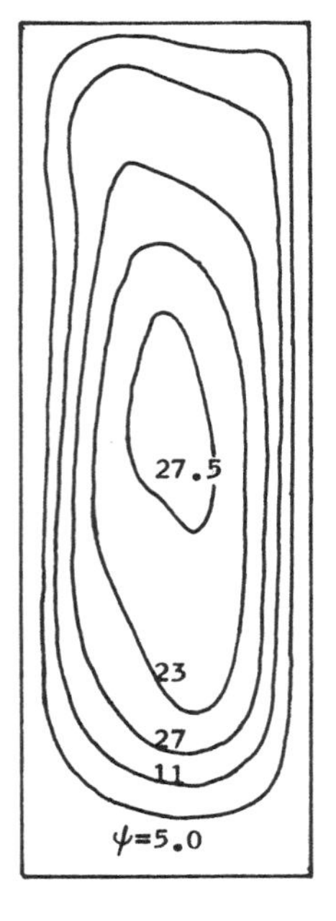

(b) Streamlines

FIGURE 2. Isotherms and streamlines($H/W=0.3$, $S/H=0.6$, $Ra=10^5$).

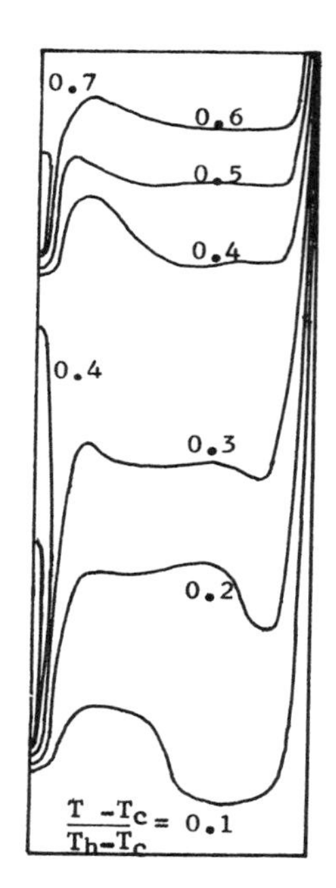

(a) Isotherms

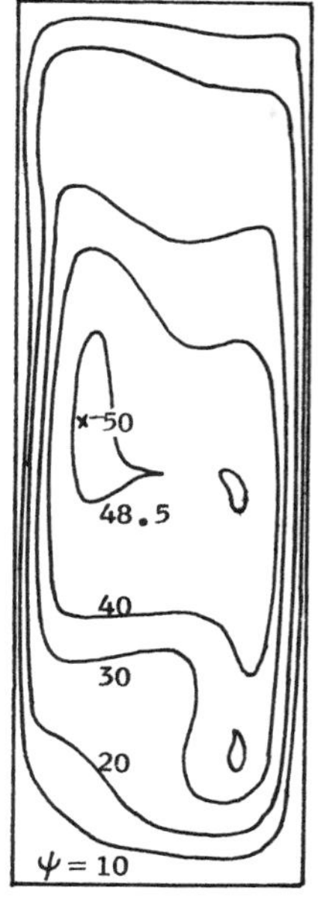

(b) Streamlines

FIGURE 3. Isotherms and streamlines(H/W=3, S/H=0.6, L/H=0.125, Ra=10^6).

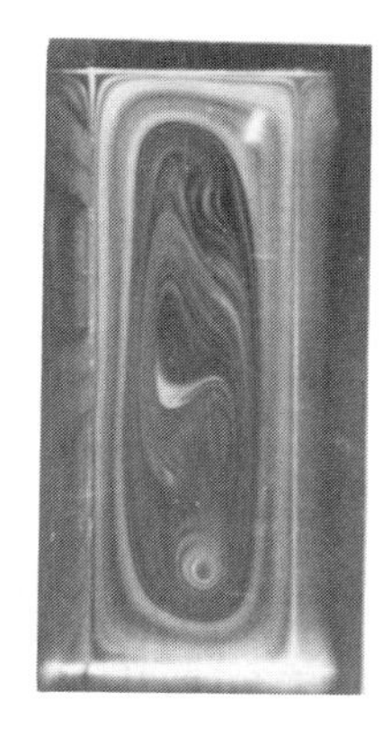

FIGURE 4. Experimental streamlines(obtained by smoke technique, H/W=3, L/H=0.125, S/H=0.67, Ra=$3x10^5$) [5].

occurs. This flow pattern is in good agreement with the flow visualization results obtained by the present authors [5] , as shown in Figure 4. This indicates that the correct model was used.

Attention will now be turn to the per-element average Nusselt number. Because the lower element is surrounded by cold fluid and the upper one by hot fluid, the heat transfer rate of the lower element is higher than that of the upper one. The heat transfer coefficient is based on the temperature defference of ($\tilde{T}_h-\tilde{T}_c$) which is the same for both elements. Thus, the per-element average Nusselt number of the lower element is always higher than that of the upper one within the range of the Rayleigh number investigated($3x10^3$-10^6)(Figure 5). The same behavior exists for all other geometric parameters studied.

Figures 6, 7 and 8 are the streamlines and isotherms at various Rayleigh numbers for the case with the third heated element added at the middle of the two mentioned above. At low Rayleigh number(Ra=10^4), there is a single vortex flow. Increasing Rayleigh number to 10^5 leads to the appearance of secondary vortex(Figure 7). At this Rayleigh number, the inversion of isotherms occurs, indicating the separation of the hot and cold thermal boundary layers. With further increases in Rayleigh number the temperature stratification core will appear as in the case of two heated elements, which indicates that the convection heat transfer dominates.

The effects of geometric parameters will now be examined. According to

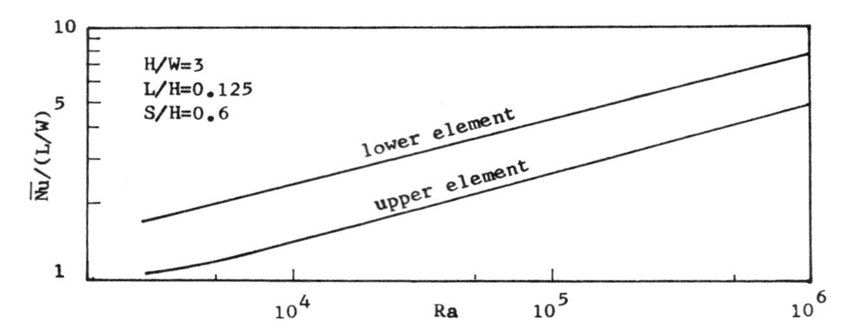

FIGURE 5. Variation of the average Nusselt numbers of the upper and lower elements with Rayleigh number.

Jaluria [1] , in an infinite medium increasing the separation distance between the two heated elements will increase the heat transfer of the upper element but have no effect on the lower one. This is because the upper element is located at the wake region generated by the lower one, while the existence of upper element can not be felt by the lower one. As for the case of an enclosure, the situation is more or less the same. The

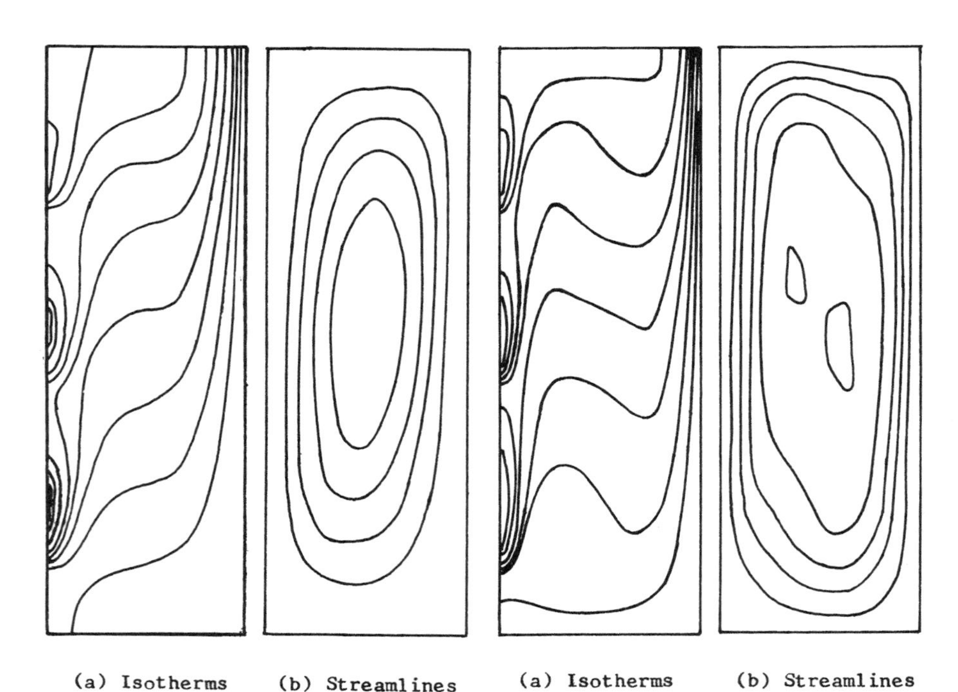

(a) Isotherms (b) Streamlines

FIGURE 6. Isotherms and streamlines ($H/W=3$, $L/H=0.125$, $S/H=0.6$, $Ra=10^4$).

(a) Isotherms (b) Streamlines

FIGURE 7. Isotherms and streamlines ($H/W=3$, $L/H=0.125$, $S/H=0.6$, $Ra=10^5$).

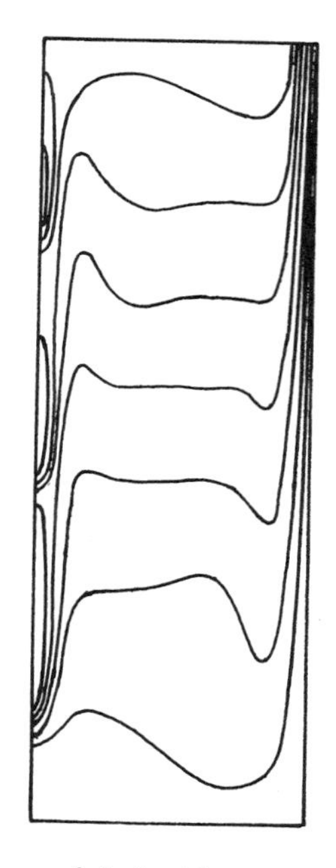

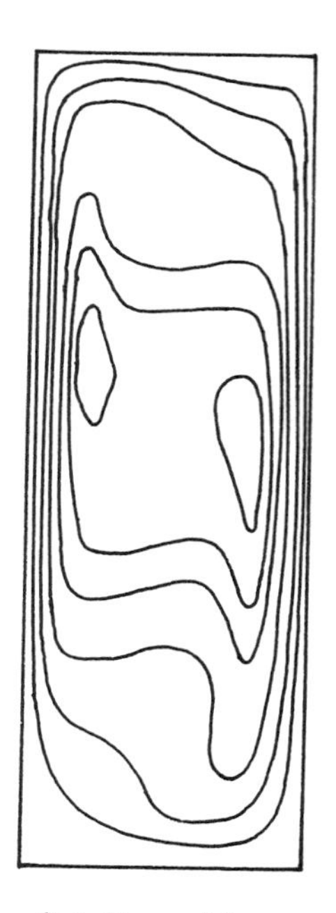

(a) Isotherms (b) Streamlines

FIGURE 8. Isotherms and streamlines
($H/W=3$, $L/H=0.125$, $S/H=0.6$, $Ra=10^6$).

fluid coming from the lower element has been heated and the upper element is exposed to a moving fluid at a higher temperature. For this moving fluid, the buoyancy force will make its velocity level increase downstream and the entrainment of cold air outside the stream will decrease its temperature level. Therefore, increasing the separating distance S/H, the upper element will exposed to fluid with higher velocity and lower temperature, resulting in an inceased heat transfer coefficient.

Figures 9 and 10 show the effect of the separating distance S/H on the per-element average Nusselt number of the upper and lower elements, respectively. For the upper element shown in Figure 9, when the two elements bound with each other($S/H=0.125$), the Nusselt number is low. Separating the two elements will increase the Nusselt number. With the increase of S/H, the heat transfer intensity of the upper element increases first, reaches its maximum and then gradually decreases. The location for upper element to obtain the highest Nusselt number is at about $S/H=0.4$, where the heat transfer rate can be enhanced by 17-24%, compared with the case of $S/H=0.125$. As shown in Figure 10, the lower element has different characteristics. At low Rayleigh numbers($Ra \leqslant 10^5$), the separating distance S/H has little effect on its heat transfer. At large Rayleigh number($Ra > 10^5$), the Nusselt number variation has the same nature as that of the upper one. But the location for lower element to obtain the highest heat transfer is at about $S/H=0.6$ and the heat transfer rate is only enhanced by about 6%.

However, as for the case of enclosures, besides the interaction between the

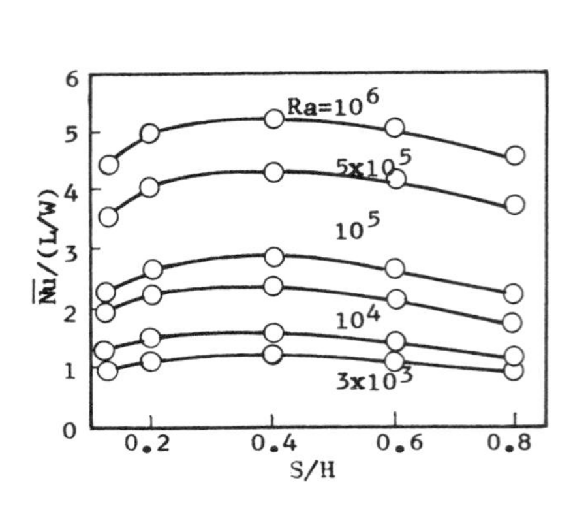

FIGURE 9. Effect of S/H on average Nusselt number of upper element ($H/W=3$, $L/H=0.125$).

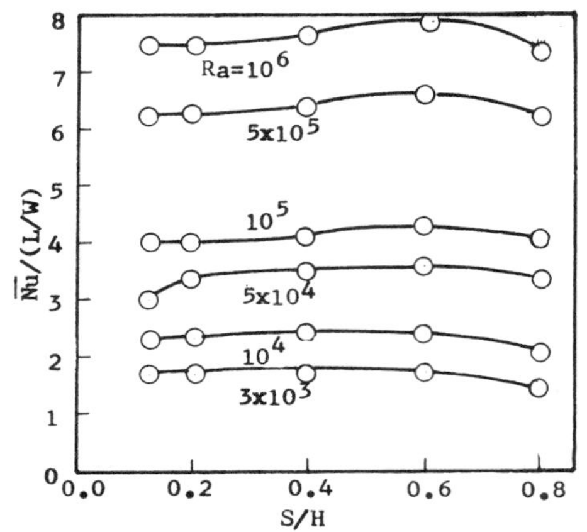

FIGURE 10. Effect of S/H on average Nusselt number of lower element($H/W=3$, $L/H=0.125$).

two heated elements, the rigid boundaries will also affect the heat transfer of both elements. Therefore, apart from the separating distance S/H, the aspect ratio, H/W, is another important parameter.

In an infinite medium, increasing S/H will increase the heat transfer of the upper element. When S/H is not too large, this is also true for the enclosure. But when S/H is large, the effect of the two horizontal end will be felt by the two elements. In this case, the fluid flow is restricted and forced to turn at the corner, which will decrease the heat transfer. This behavior is clearly seen in the Figure 11, which shows the profiles of the local heat transfer coefficients of the two elements. When two ele-

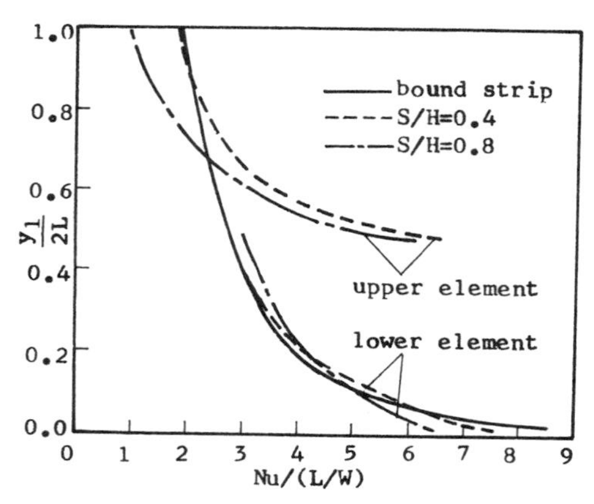

FIGURE 11. Distributions of local Nusselt numbers along the heated elements($H/W=3$, $L/H=0.125$, $Ra=10^5$).

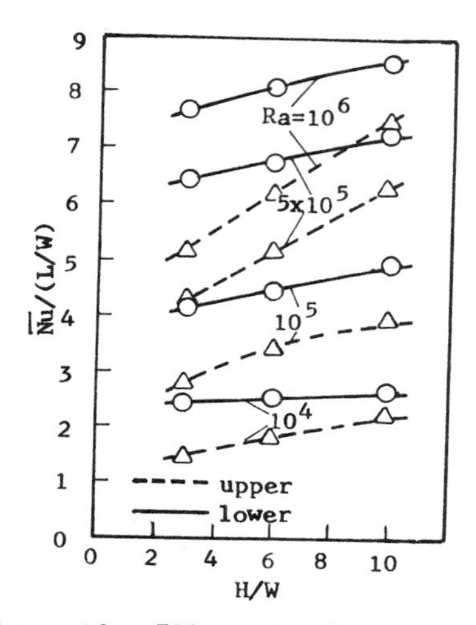

Figure 12. Effect of H/W on the Nusselt numbers of both upper and lower elementa(L/W=0.375, S/H=0.4).

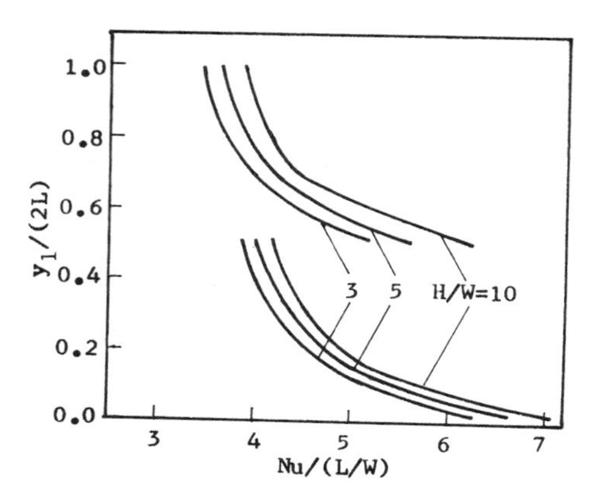

FIGURE 13. Distributions of local Nusselt number at different H/W (L/W=0.375, S/H=0.4,Ra=10^5).

ments are bound with each other(S/H=0.125), the bound strip behaves as a single one and its local heat transfer coefficients decrease monotonically from the bottom to the top. Once the two elements are separated, because the temperature level of the fluid flowing over the upper element will decrease, its local Nusselt number is much higher than that of its counterpart on the bound strip. However, if the upper and lower elements are quite close to the top and bottom, respectively, or if the value of S/H is large enough, e.g., o.8, the local Nusselt numbers and the per-element average Nusselt numbers decrease, which can be attributed to the restricted flows around the two corners, where the fluid is forced to turn and the slow velocity decreases the convection heat transfer.

Figure 12 shows the effect of the aspect ratio H/W. The average Nusselt numbers of both upper and lower elements increase as H/W increases, with the Nusselt number of the upper one showing a much stronger dependence. This behavior can be seen more clearly if the distributions of the local Nusselt numbers on the different places of the two elements at various H/W are inspected. As shown in Figure 13, when H/W increases, the local Nusselt numbers at all places of both upper and lower elements increase, but the increase of the upper one is larger than that of the lower one. At the same size of heated elements and S/H, increasing H/W means increasing the separating distance between two elements. As the forgoing discussion indicated, this will increase the heat transfer of the upper and lower elements.

Finally, the effect of S/W on the per-element average Nusselt numbers for a three-element enclosure will be briefly discussed. Adding a third heated element into the middle of the two elements will make the heat transfer characteristics of the elements different, as can be seen in Figure 14. Here S represents the distance between the upper most element and the lowest one. S/H=0.175 means these three are bound to form a single strip. Similar to the case of two elements bound together, the local Nusselt numbers of the bound strip monotonically decrease from bottom to top(which is not

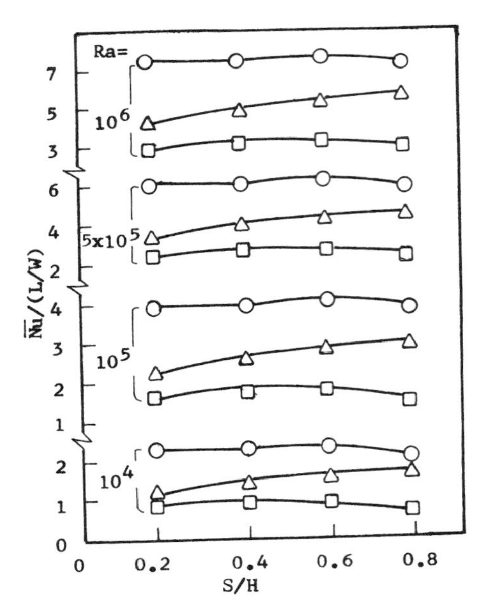

FIGURE 14. Effect of S/H on three-element enclosure natural convection heat transfer(H/W=3, L/H=0.125).
○ bottom element; △ middle element; □ top element

shown in the figure), and the average Nusselt numbers of the upper two elements are much lower than those of the lowest one. Separating the three elements will cut off the thermal boundary layer, which will be reestablished when flowing over the upper two elements, resulting in heat transfer enhancement of the upper two elements. The larger S/H is, the more the enhancement will be. As for the case of two elements, when S/H is too large, the top and bottom elements will be too close to the end wall and, therefore, the heat transfer rates of these elements will decrease. Thus, there also exists an optimum S/H at which the Nusselt numbers of the elements reach their maximum. For this case, the optimum value of S/H is about 0.6. Compared with the case of S/H=0.175, the heat transfer intensity of the upper element will be enhanced by about 6-12% and the lowest one only 4%. For the middle one, the Nusselt numbers will always increase as S/H increases because the two horizontal end plates have little effect on them. The largest enhancement of heat transfer is about 30-38% at S/H=0.8.

CONCLUSIONS

The natural convection heat transfer due to discrete heated elements on a vertical adiabatic surface of a rectangular enclosure has been numerically investigated. The overall Rayleigh number was varied from $3x10^3$ to 10^6. The effects of the separating distance and the aspect ratio on the element heat transfer characteristics were parametrically investigated. The streamlines and isotherms of several typical cases were presented. The variation of the local and per-element average Nusselt numbers with the two geometric parameters were also given. Some interesting and important results are obtained :

(1) For the enclosure with two heated elements, the per-element average Nusselt number of the lower element is always higher than that of the upper one.

(2) With the increase in the distance between the two elements, the heat transfer intensity of the upper element increases first, reaches its maximum and then gradually decreases. The optimum value of S/H is 0.4, for which the heat transfer rate can be enhanced by 17-24%.

(3) Increasing the aspect ratio H/W will increase the average Nusselt numbers of both upper and lower elements, with the Nusselt number of the upper one showing much stronger dependence.

(4) For the enclosure with three elements, the trends of the average Nusselt numbers of upper and lower elements with the parameter S/H are qualitatively the same as those of the two-element enclosure, while the average Nusselt number of the middle element increases monotonically with increasing S/H.

(5) At lower Rayleigh number, there was one vortex. But at higher Rayleigh number, three vortices existed in the enclosure, which was in good agreement with the flow visualization results obtained by the present authors.

These results should be useful for the design of electronic pakages cooled by natural convection.

NOMENCLATURE

a	thermal diffusivity, m^2/S
g	gravitational acceleration, m/S^2
H	height of enclosure, m
L	width of heating element, m
Nu	local Nusselt number, Eq. (7)
$\overline{Nu}$	per-element average Nusselt number, Eq. (8)
p	pressure, N/m^2
P	dimensionless pressure, Eq.(6)
Pr	Prandtl number
R_ϕ	relative change between consecutive iterations, Eq.(9)
Ra	Rayleigh number, Eq.(6)
S	separating distance between two elements, m
S_1	height of lower element from bottom horizontal surface, m
$\tilde{T}$	temperature, K

T	dimensionless temperature
u	velocity in x direction, m/S
U	dimensionless velocity in x direction
v	velocity in y direction, m/S
V	dimensionless velocity in y direction
W	width of enclosure, m
x,y	Cartisian coordinates, m
X,Y	dimensionless Cartisian coordinates
y_1	local vertical coordinate along the heated element, m
α	local heat transfer coefficient, $W/(m^2\ K)$
$\bar{\alpha}$	average heat transfer coefficient, $W/(m^2\ K)$
β	coefficient of thermal expansion of fluid, K^{-1}
ε	parameter governing convergence of iteration
λ	thermal conductivity of fluid, W/(m K)
ν	kinematic viscosity of fluid, m^2/S
ρ_0	fluid density, kg/m^3
ϕ	general dependent variable
ψ	dimensionless stream function

Subscripts

h	hot
c	cold
max	maximum

Superscript

*	previous iteration

REFERENCES

1. Jaluria, Y., Buoyancy-Induced Flow due to Isolated Thermal Sources on a Vertical Surface, J. Heat Transfer, vol. 104, 223-227, 1982

2. Jaluria, Y., Interaction of Natural Convection Wakes Arising from Thermal Sources on a Vertical Surface, J. Heat Transfer, vol. 107,

883-892, 1985

3. Chu, H. H.-S., Churchill, S. W., and Patterson, C. V. S., The Effect of Heater Size, Location, Aspect Ratio, and Boundary Conditions on Two-Dimensional, Laminar, Natural Convection in Rectangular Channels, J. Heat Transfer, vol. 98, 194-201, 1976

4. Turner, B. L., and Flack, R. D., The Experimental Measurement of Natural Convection Heat Transfer in Rectangular Enclosures with Concentrated Energy Sources, J. Heat Transfer, vol. 102, 236-241, 1980

5. Zhang, H. L., Wang, G. X., and Wu, Q. J., Natural Convection in Enclosure with Discrete Heat Sources on a Vertical Wall, 5th Heat and Mass Transfer Conference of the Chinese Society of Engineering Thermophysics, paper no. 873142, 1987

6. Patankar, S. V., Numerical Heat Transfer and Fluid Flow, Hemisphere, Washington, D. C., 1980

7. Van Doormaal, J. P., and Rathby, G. D., Enhancement of the SIMPLE Method for Predicting Incompressible Fluid Flows, Numerical Heat Transfer, vol. 7, 147-163, 1984

8. De Vahl Davis, G., and Jones, I. P., in Numerical Methods in Thermal Problems, ed. R. W. Lewis, K. Morgan, and B. A. Schrefler, pp. 552-572, Pineridge Press, Swansea, U. K., 1981

Cooling Performance of Fins under an Obstacle

YASUYUKI YOKONO, TOMIYA SASAKI, and MASARU ISHIZUKA
Mechanical Engineering Laboratory
Toshiba R&D Center
4-1, Ukishima-cho, Kawasaki-ku
Kawasaki, 210 Japan

ABSTRACT

In order to examine the heat dissipation capability of fins in practical electronic equipment, the effect of an obstacle, located above the fins, on the cooling performance was experimentally investigated. The reduction of the clearance between the fin tips and the obstacle affected the natural air convection from the fin and reduced the heat dissipation capability of the fin. The critical value of clearance at which the cooling performance started to decrease was not significantly affected by the supply power, the fin height or the inter-fin spacing. Through flow visualization, it was found that the natural air convection from the fin was retarded by the obstacle while the flow pattern remained unchanged regardless of the variation in the clearance, the inter-fin spacing or the fin height. The correlation between the fin height and the variation in the fin cooling performance resulting from changes in clearance did not reverse in any case.

INTRODUCTION

Recent progress in micro-electronics technology has brought a significant reduction in electronic equipment size; as a result, there occurred an increase in heat generation per unit volume. Therefore, the capability of handling high heat dissipation per unit volume has become an important issue. It is shown by Kraus and Bar-Cohen [1] that fins are widely adopted for heat sinks on electronic equipment. Starner and McManus [2] and Welling and Wooldridge [3] have studied heat transfer from fins and derived several equations for the cooling performance with natural air convection. Harahap and McManus [4] have studied the flow from fins and presented average heat transfer coefficients, and Jones and Smith [5] have suggested the optimum arrangement of rectangular fins. Yokono et al. [6] have derived equations for small fins used for LSI packages under both natural and forced air convection cases. However, fins were set in open air in such studies. In actual electronic equipment, heat sources are often set in a confined region, and the natural air heat transfer is considerably altered in comparison with that in open air. In connection with the thermal design for electronic equipment, Sparrow and Kadle [7] have made a study on the forced air convection heat transfer for shrouded fins. However, there have been few investigations reported on natural air convection heat transfer from fins interrupted by the surroundings, and the cooling performance for fins in practical electronic equipment is still largely unexplored.

Because of the above uncertainties, an experimental study was performed on the cooling performance of fins under an obstacle. The effect of the clearance between the fin tips and the obstacle on the temperature rise for several kinds of fins with different heights and spacings was examined. Furthermore, the influence of the supply power, the fin height and the inter-fin spacings on the correlation between the clearance and the thermal conductance of fins was explored.

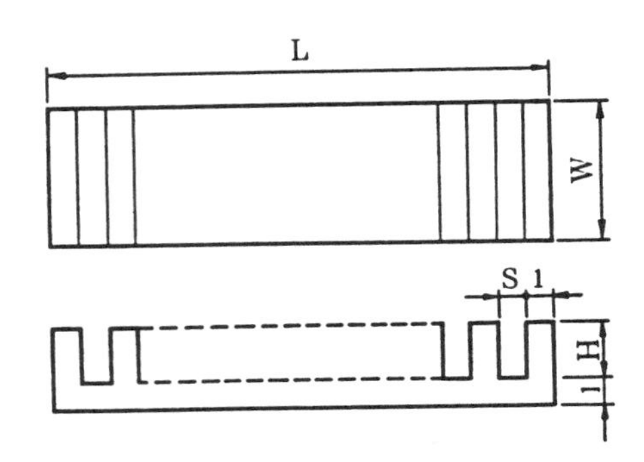

No.	1	2	3	4	5	6	7	8	9	10	11	12
LXW	60.0×20.0											
S	2	2	2	2	5	5	5	5	9	9	9	9
H	2	5	9	14	2	5	9	14	2	5	9	14
S/H	1	0.4	0.22	0.14	2.5	1	0.56	0.36	4.5	1.8	1	0.64
W/H	10	4	2.22	1.43	10	4	2.22	1.43	10	4	2.22	1.43

FIGURE 1. Fin geometry.

EXPERIMENTAL APPARATUS AND PROCEDURE

The geometry of the fins used in the present study is shown in Fig.1. The fins were made of aluminum and the fin base area was 20.0mm X 60.0mm. Both the thickness of the fin and the base had constant values of 1mm and inter-fin spacings S and fin height H were changed from 2mm to 9mm and from 2mm to 14mm, respectively.

The experimental set-up is shown schematically in Fig.2. The fins were set such that the base remain horizontal on the fin table. The table surface in contact with the fins was made of teflon (i.e., polytetrafluoroethylene) to prevent melting and was covered by a foamed styrene for thermal insulation. A 100mm square box made of acrylic resin plate was set above the fin as an obstacle. The variable clearance C between the fin tips and the obstacle was measured by a micrometer. Sasaki et al. [8] have made a study varying obstacle dimension and reported that the obstacle size did not impose any prominent effect on the correlation between heat dissipation capability of the fins and the clearance when the dimension of the cube obstacle was chosen longer than the width of the fin base. Based on those results, the obstacle having a dimension beyond the width of the fin base was chosen.

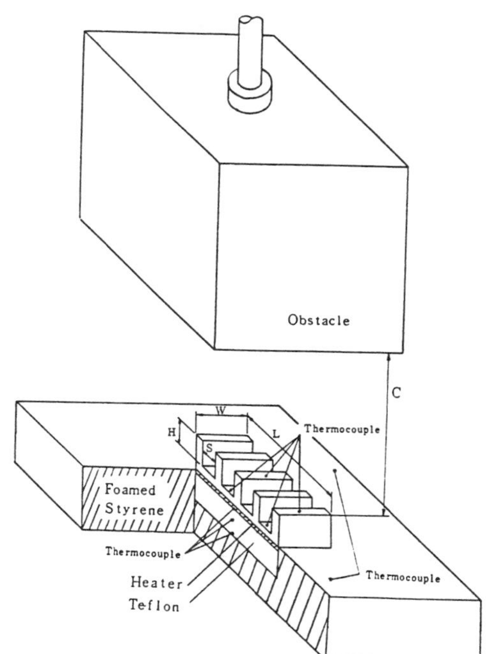

FIGURE 2. Schematic diagram of experimental apparatus.

The fin was heated up by 0.25mm thickness sheet heater, whos area was kept same as the fin base area. A specified heat flux could be supplied to the fin. Fin temperature was measured b 0.1mm diameter T type thermocouples attached to a total of fou points on the fin tip and the fin base as shown in Fig.2. Th measurements were carried out in a 500mm x 500mm x 500mm tes case, which had some holes preventing extreme temperature rise to prevent extraneous air currents. The difference between th mean temperature of the fin and ambient air temperature, whic was measured at two points of the fin front in the test case, wa defined as the fin temperature rise Δ T. Experimental results in dicated that there were little differences in temperature of th four measured points of fin and the measured ambient temperature in which the perturbation was not observed, was almost identical with the room temperature outside of the box. Accordingly the fi was considered to be isothermal and the used test case was bi enough to carry out a natural convection experiment. In order t assess the heat leak from the bottom of the heater, the tempera tures of both sides of the teflon plate were measured by ther mocouples. The energy dissipated through the teflon fin table calculated on the basis of the teflon thermal conductivity an temperature differences between the surface and the bottom of th teflon plate, was found to be less than 2% of the total supplie power. The heat leak from the heater sides was considered to b very small, because the heater side area was extremely small com pared with the bottom area, and was not taken into consideration

Furthermore, through the flow visualization using cigarette smok as a tracer, the natural air convection from the fins was ex amined. A 2KW xenon slit light was adopted as a light source i photographing the flow utilizing a 35mm camera equipped with 50m focal distance f1.2 lens and ASA400 film.

EXPERIMENTAL RESULTS

Flow Visualization

Figure 3 shows the flow visualization photographs in the case o inter-fin spacing S=9mm, fin height H=9mm and the supplied powe Q=4W, observed from the front and the side of the fin. The air contact with the fin surface flowed vertically upward smooth through natural convection when the obstacle was not present. Th side view indicated that the ambient air flowed horizontal along the fin table and was heated up by the fin and convert into a vertical flow. These flows from right and left joined a the fin center and flowed vertically upward stably. This flow known as single chimney flow and it results in the best hea transfer coefficient in laminar flow. All fins used in this stud showed the single chimney flow when the obstacle was not presen These results are consistent with those reported by Sane and Suk hatme [9]. When the clearance between the fin tips and th obstacle C was 20mm, the upward vertical flow from the fins wa

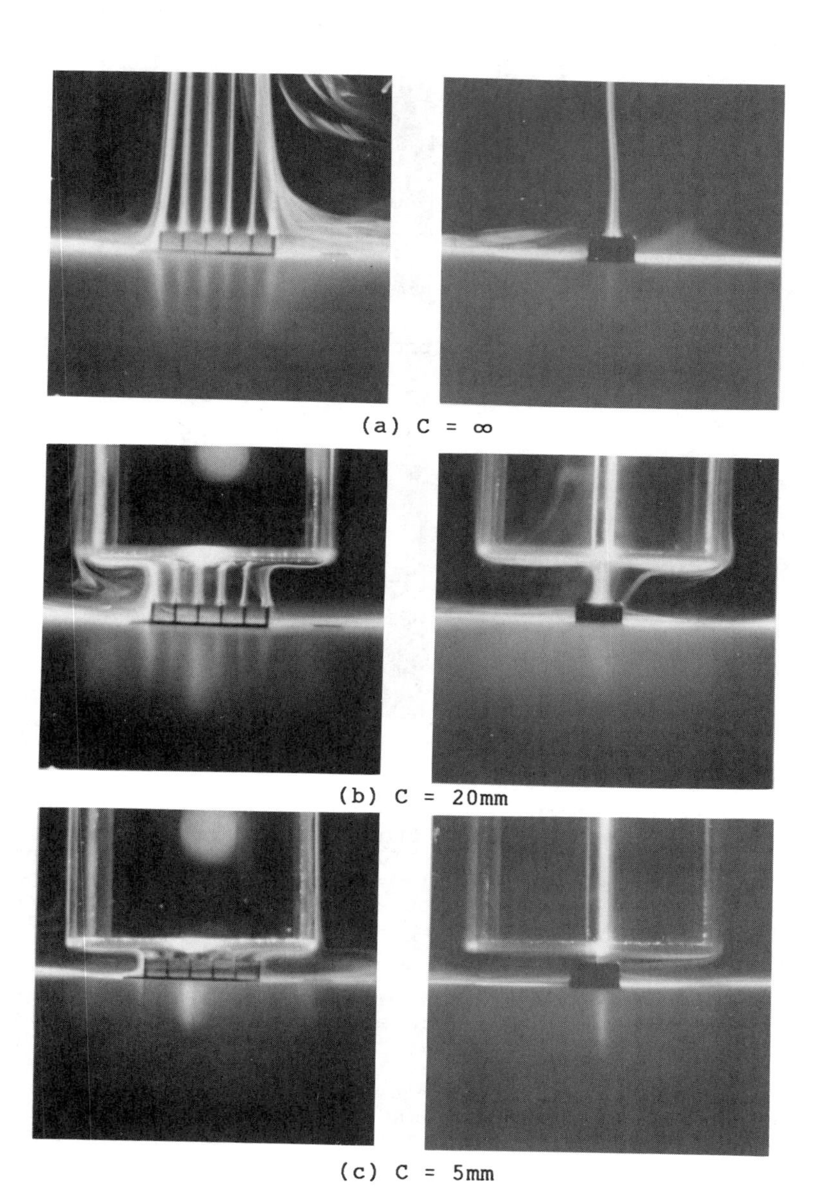

FIGURE 3. Flow visualization photographs observed from the front and the side of the fin for S=9mm and H=9mm.

(a) C = 3mm (b) C = 1mm

FIGURE 4. Flow visualization photographs observed from the fi side for S=9mm and H=9mm.

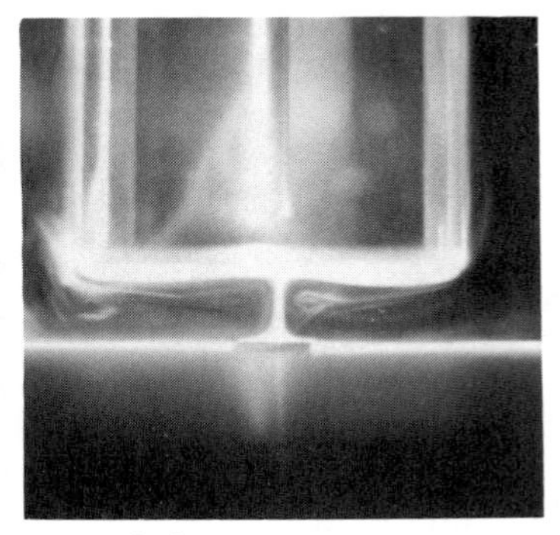

(a) S=2mm, H=9mm (b) S=2mm, H=2mm

FIGURE 5. Flow visualization photographs for C=20mm.

still observed even though the natural air convection encountere the obstacle. The observation from the side indicated that th flow was still a single chimney flow as was observed in the cas without an obstacle, and the flow pattern remained unchanged. I the case of C=5mm, although natural convection from fin occurre as a single chimney flow, the flow velocity decreased and stag nated flow was observed.

Figure 4 shows the flow visualization photographs taken from th side in the case when the clearance is as small as C=3mm ar C=1mm. Vortices were observed between the fin tips and th obstacle, and hot air heated by the fin was cooled down at th obstacle and then re-entered into the fin array. This led to th conclusion that the thermal boundary condition of the obstacl could change the cooling performance of fins. Figure 5 shows th flow visualization photographs of different inter-fin spacing and fin height for clearance C=20mm. The single chimney flow was observed in all the cases, indicating that the inter-fin spacing and the fin height did not alter the flow pattern.

Fin Temperature Rise

The effect of the clearance C between the fin tips and the obstacle on the fin temperature increase ΔT was examined. Typical results in the case of inter-fin spacing S=9mm and fin height H=9mm are shown in Fig.6. For each case of supplied power, when the clearance C was larger than 30mm, fin temperature remained stable at essentially the same value as with the case without the obstacle. Fin temperature started to increase at a clearance smaller than a critical value. Under a constant supplied power, the reduction of fin temperature rise means a reduction of the fin heat dissipation capability. Accordingly, it was found that the cooling performance of fins decreased when the clearance between the fin tips and the obstacle became less than a critical value. This critical value of clearance was not significantly affected by the supplied wattage.

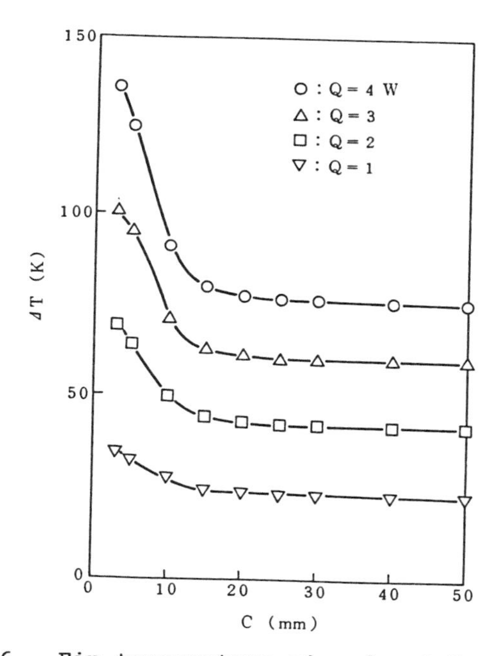

FIGURE 6. Fin temperature rise for S=5mm and H=9mm.

Thermal Conductance

The fin thermal conductance, $Q/\Delta T$, which indicates the heat dissipation capability per unit temperature difference, was derived to examine the cooling performance. Figure 7 shows the relation between $Q/\Delta T$ and ΔT for S=9mm and H=9mm. For each clearance, the thermal conductance of the fin increased with its temperature rise.

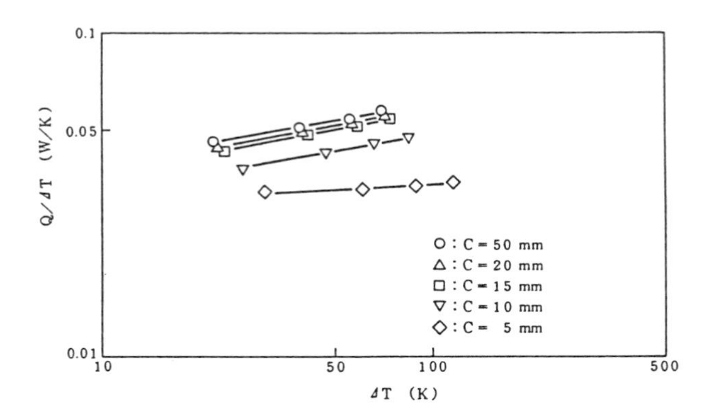

FIGURE 7. Correlation of fin thermal conductance with the fin temperature for S=9mm and H=9mm.

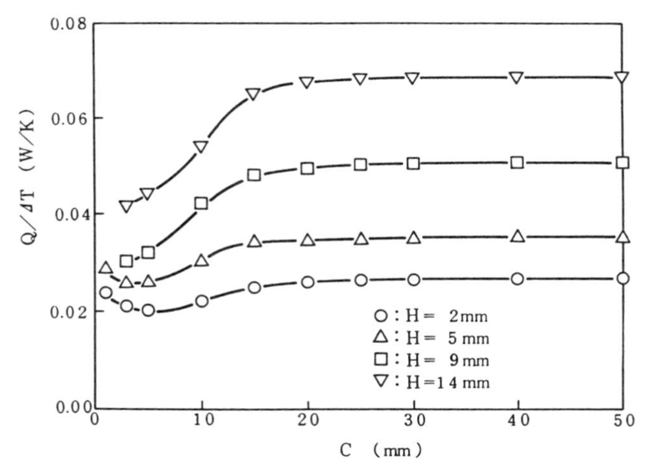

FIGURE 8. Fin height effect on thermal conductance variation according to the clearance between the fin tips and the obstacle for Δ T=50K and S=5mm.

The effect of the temperature rise on the thermal conductance was consistent when the clearance between the tips and the obstacle was over 10mm; however, there was a slight decrease in the slope when the clearance was as narrow as 5mm. Accordingly, it is concluded that the effect of temperature rise on the cooling performance gets altered when the clearance is small and a further extensive analysis including the obstacle thermal boundary condition would be necessary in this region.

The fin thermal conductance for $\Delta T=50K$ was calculated and the representative results are shown in Fig.8 to examine the fin height effect on the relation between the thermal conductance and the clearance. For each fin height, when the clearance was larger than 30mm, the thermal conductance did not change and corresponded to the case when the obstacle was not present. The thermal conductance began to decrease when the value of the clearance was below a critical value, which was not significantly affected by the change in the fin height. When the obstacle was very close to the fin tip, the thermal conductance was less depended on the clearance, and the thermal conductance actually increased in the case of H=2mm and 5mm due to the cooling effect of the obstacle wall. As observed by the flow visualization, the obstacle was not adiabatic. This explains the variation in the fin thermal conductance as clearance is decreased. In the present experiments, the thermal conductance increased with increasing fin height for any given clearance. The various curves of the thermal conductance caused by the clearance do not contact or cross with each other by changing the fin height.

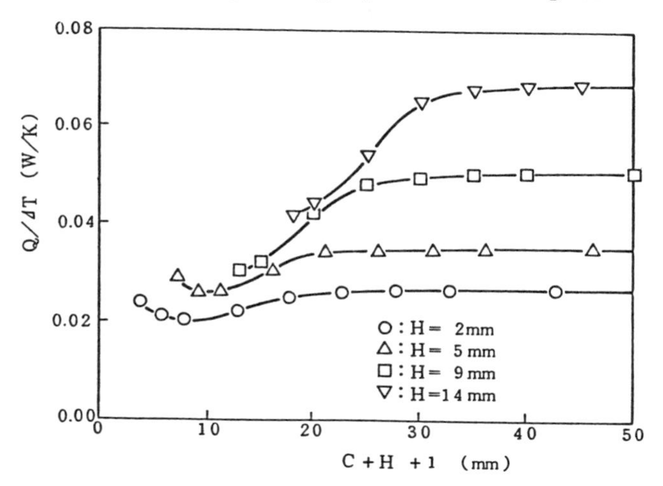

FIGURE 9. Fin height effect on thermal conductance variation according to the distance from the fin table to the obstacle for $\Delta T=50K$ and S=5mm.

Figure 9 indicates the variation of the thermal conductance with the distance from the fin base to the obstacle, which was equal to the sum of the clearance C, the fin height H and the fin base thickness of 1mm. The figure leads to the conclusion that for a similar thermal boundary condition of an obstacle, as is considered in the present study, the relation between the thermal conductance and this distance is consistent as the fin height is changed. Therefore, it can be predicted that, for example, for a specified spacing of electronic circuit boards, the tallest fin shows the best cooling performance.

Figure 10 shows the inter-fin spacing effect on the correlation between thermal conductance and the clearance for the fin height H=5mm at the fin temperature rise Δ T=50K. The figure indicates that the inter-fin spacing has no major influence on the thermal conductance. As the clearance becomes narrow, the thermal conductance for S=5mm and S=9mm decreased in the same manner while that for S=2mm showed less change. The thermal conductance of S=2mm was lager than that of S=5mm when the clearance C was greater than 5mm but showed the smaller value than S=5mm in the case when the clearance was below 5mm. Figure 11 shows the inter-fin spacing effect for the fin height H=14mm. In the case of H=5mm, a discrepancy of the thermal conductance caused by altering the inter-fin spacing was scarcely observed at the large clearance, while in case of H=14mm, the thermal conductance was affected by the inter-fin spacing. As the clearance became less than the critical value which was not significantly affected by the inter-fin spacing, the thermal conductance began to decrease and the thermal conductance variation with inter-fin spacing reversed.

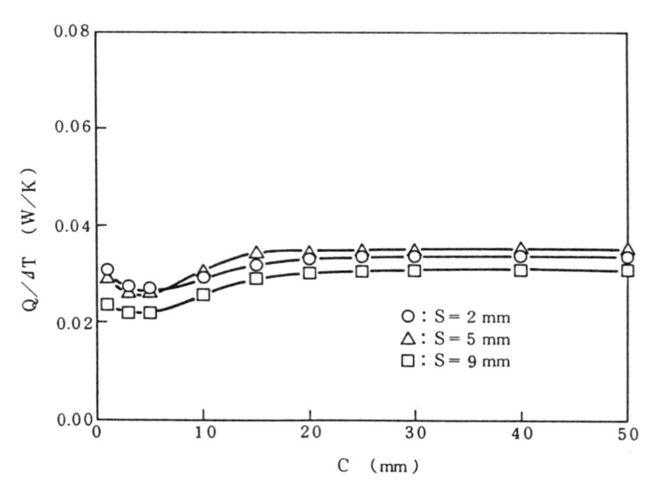

FIGURE 10. Inter-fin spacing effect on correlation between thermal conductance and the clearance for Δ T=50K and H=5mm.

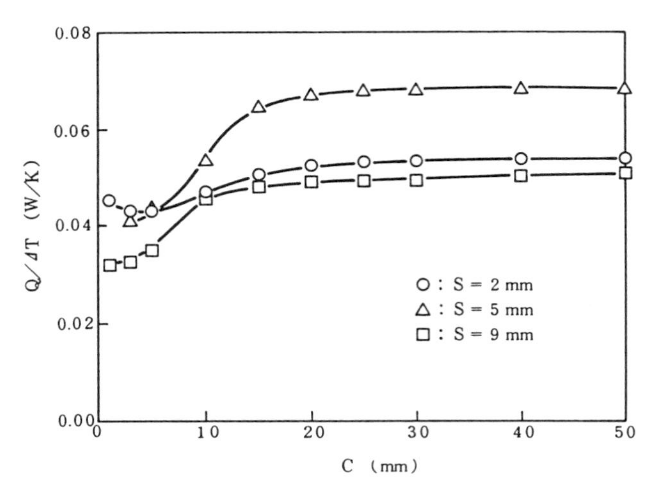

FIGURE 11. Inter-fin spacing effect on correlation between thermal conductance and the clearance for Δ T=50K and H=14mm.

It was found that the inter-fin spacing has a significant influence on the correlation between the cooling performance and the clearance. Therefore, it is expected that there has to be an optimized inter-fin spacing for the cooling performance for each fin height and clearance.

CONCLUSIONS

The studies of the cooling performance of fins under an obstacle led to the following results. The cooling performance of the fin decreased as the clearance between the fin tips and the obstacle became smaller than a critical value which was not significantly affected by the supply power, the fin height and the inter-fin spacing. Also, through flow visualization, it was found that the natural air convection from the fin was disrupted by the obstacle while the flow pattern remained unchanged regardless of the variation in the clearance, the inter-fin spacing and the fin height. The correlation between the fin height and the variation in the fin cooling performance due to the clearance did not reverse in any case.

NOMENCLATURE

C = clearance between fin tips and obstacle
H = fin height
L = width of fin base
Q = supply power
S = inter-fin spacing
W = fin length
Δ T = fin temperature rise

REFERENCES

1 Kraus, A.D. and Bar-Cohen, A., Thermal Analysis and Control of Electronic Equipment, Hemisphere Pub. Corp., Washington, D.C., 1983

2 Starner, K.E. and McManus, H.N. Jr., An Experimental Investigation of Free-Convection Heat Transfer From Rectangular-Fin Arrays, Journal of Heat Transfer, Trans. ASME, Series C, Vol.85, 1963, 273

3 Welling, J.R. and Wooldridge, C.B., Free Convection Heat Transfer Coefficients from Rectangular Vertical Fins, Journal of Heat Transfer, Trans. ASME, Series C, Vol.87, 1965, 439

4 Harahap, F. and McManus, H.N. Jr., Natural Convection Heat Transfer From Horizontal Rectangular Fin Arrays, Journal of Heat Transfer, Trans. ASME, Series C, Vol.89, 1967, 32

5 Jones, C.D. and Smith, L.F., Optimum Arrangement of Rectangular Fins on Horizontal Surface for Free Convection Heat Transfer, Journal of Heat Transfer, Trans. ASME, Series C, Vol.92, 1970, 6

6 Yokono, Y., Sasaki, T. and Ishizuka, M., Small Cooling Fin Performances for LSI Packages, Proceedings of the International Symposium on Cooling Technology for Electronic Equipment, 1987, 679

7 Sparrow, E.M. and Kadle, D.S., Effect of Tip-to-Shroud Clearance on Turbulent Heat Transfer From a Shroud, Longitudinal Fin Array, Journal of Heat Transfer, Trans. ASME, Series C, Vol.108, 1986, 519

8 Sane, N.K. and Sukhatme, S.P., Proceedings 5th International Heat Transfer Conference Tokyo, 3, NC 3.7, 1974

9 Sasaki, T., Yokono, Y. and Ishizuka, M., Cooling Performances for Fin under an Obstacle Plate, National Convention Record, 1987, The Institute of Electronics, Information and Communication Engineers, 1-128 (in Japanese)

Enhancement of Forced Convection Air Cooling of Block-Like Electronic Components in In-Line Arrays

Y. P. GAN, S. WANG, D. H. LEI, and C. F. MA
Department of Thermal Engineering
Beijing Polytechnic University
Beijing 100022, PRC

ABSTRACT

The objective of this work is to investigate the enhanced heat transfer from rectangular electronic components in in-line arrangements in airflow. Two types of enhancement devices, displaced enhancement promoters of two configurations and finned heat sinks, were employed for improving the air cooling of electronic components. Enhancement in the forced convection heat transfer coefficient up to 51% and 153% were obtained for displaced promoters and finned heat sinks, respectively. A set of empirical formulas is presented to correlate the experimental data for each specific configuration.

INTRODUCTION

Forced convection air cooling has been most widely employed in electronic and microelectronic equipment. Heat transfer from arrays of components mounted on the wall of a parallel plate channel is an important problem commonly encountered in the design of cooling systems for these devices. The heat transfer characteristics of an array of rectangular electronic components in air flow have been studied experimentally by Sparrow et al.[1,2], Moffat et al.[3], Lehmann and Wirtz[4], and Chen and Wang[5]. Hwang[6] reported an experimental investigation of enhancement techniques using airflow turbulators, which are placed at key locations on a card populated with electronic modules. When forced air flows over the card, the turbulators are used to break up the boundary layer formed along the card. Convective heat transfer from finned components in airflow has been investigated by Nakayama[7,8]. It was reported that for the first row the finned package yields heat transfer coefficients about 1.6 times higher than those on the plain block. More recently, a new enhancement technique was presented by Ratts et al.[9]. Augmentation of the forced convection heat transfer coefficient of component arrays up to 83% was found when cylinders were placed periodically above the back edge of each row of the components.

The objective of the present work is to investigate the enhanced heat transfer from rectangular electronic components in in-line arrangements in airflow. Two types of enhancement devices, displaced enhancement promoters of two configurations and finned heat sinks, were employed for improving the air cooling of the electronic components. The effect of the number and position of these two types of promoters on the heat transfer from the components was experimentally investigated. Moderate improvements about 51% were obtained with the displaced promoters.

Enhancement of heat transfer as high as 153% was recorded with finned heat sinks. Pressure losses were also measured with or without heat transfer enhancement devices. A set of empirical correlations was presented to predict the heat transfer coefficients as well as friction factors. The experiments were conducted in the range of Re=2800-22000.

EXPERIMENTAL APPARATUS AND PROCEDURE

A schematic diagram of the experimental apparatus used in the investigation is shown in Fig. 1. Laboratory air is drawn through the flow channel by a blower. A valve is used to regulate the flow rate of air. The test section is shown in Fig. 2. It consists of three parallel plate channels of the same size: 500mm wide, 600mm long and 125mm height. On each of the lower walls (floors) of the channels 45 block-like components are mounted in the same in-line arrangement as shown in the figure. The array has 5 rows and 9 columns, the rows being normal to the air flow. Both the spanwise and longitudinal spacing between the components is 28mm. All the components have a base area of 28mmx28mm and a height of 10mm. Being used as hydrodynamic guard elements, most components are not heated. They are made of plexiglas. Only the 5 components of the central column in the low channel, which was used for measurement, are active and made of bakelite. A strip of $8\mu m$ stainless steel is cemented to the bakelite block and used as a heating element as well as a heat transfer surface. Regulated DC power is supplied to the heaters. The temperature of each active component is measured by 4 copper-constantan thermocouples of 0.1mm wire diameter, which were electrically insulated from the heating foil, yet in close thermal contact. Two thermocouples were located along the centerline of the top surface 7mm from the each side. Two other thermocouples were located at the center of the front side and rear side, as shown in Fig. 3. The wall temperature of the heater was measured by a precision digital MV-meter. The bakelite blocks were hollowed out from the back side to minimize conductive heat loss from the heaters into the supporting channel wall. The five active test components were simultaneously heated in all the experiments.

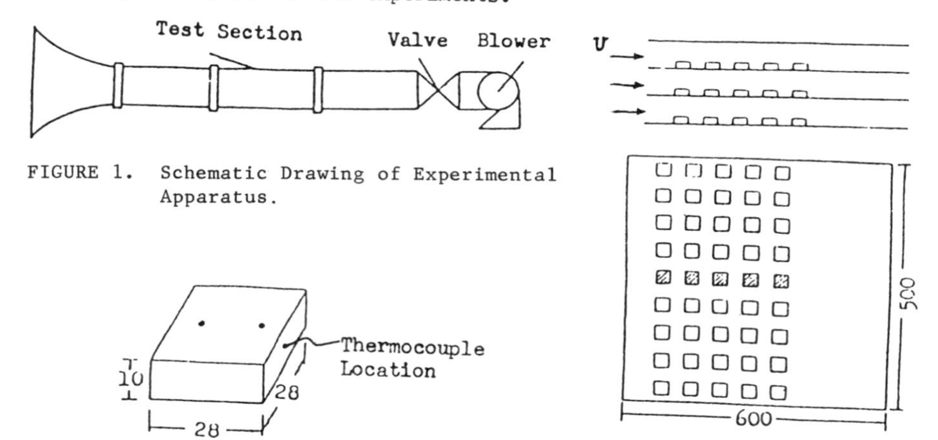

FIGURE 1. Schematic Drawing of Experimental Apparatus.

FIGURE 3. Block-Like Component Specification.

FIGURE 2. Block-Like Arrangement in the Test Section.

The two types of displaced promoters used in the experiments are illustrated in Fig. 4. Promoter 1 is a half cylinder while promoter 2 is of streamline body shape. Their sizes are given in the same figure.

The finned heat sink used in present work is shown in Fig. 5. It is made of brass and includes seven fins of 0.5mm thickness and 10mm height. The spacing between fins is 4.1mm. The finned heat sinks were attached very closely on the test components. A special thermal grease, which is dielectric but of high thermal conductivity, was filled between the heater surfaces and the heat sinks to minimize the thermal contact resistance. Two 40-gage iron-constantan thermocouples were fixed at the base of the fins to measure the base temperature of the fins.

The temperature of incoming air was measured by a thermometer with an estimated accuracy of 0.1°C. The air velocity was measured by a calibrated pitot tube. The air pressure was measured by a precision inclined manometer.

DATA REDUCTION

The average heat transfer coefficient is defined as

$$h=q/(T_w-T_c) \tag{1}$$

where $T_w=T_{w1}+T_{w2}$ for heat transfer from components without finned heat sinks with T_{w1} and T_{w2} being the measured temperatures on the top surface of the components. For heat transfer from finned heat sinks they are the temperatures of the base of the fins. T_c is the inlet temperature of the air. The heat flux q is calculated from the electrical power supplied to the heater and the heat transfer area of the heater.

The Nusselt number is defined as

$$Nu=hL/K \tag{2}$$

where k is the thermal conductivity of air evaluated at inlet temperature and L is the length of the components.

The Reynolds number is also based on the component length:

$$Re=\frac{UL}{\nu} \tag{3}$$

where ν is the kinematic viscosity of air evaluated at the temperature of the incoming air. U is the mean flow velocity in the unobstructed channel upstream of the array.

The friction factor is defined as

$$C_f=\frac{\Delta p}{\frac{1}{2}\rho U^2} \tag{4}$$

where Δp is the pressure drop along the test section and ρ is the air density evaluated at inlet temperature.

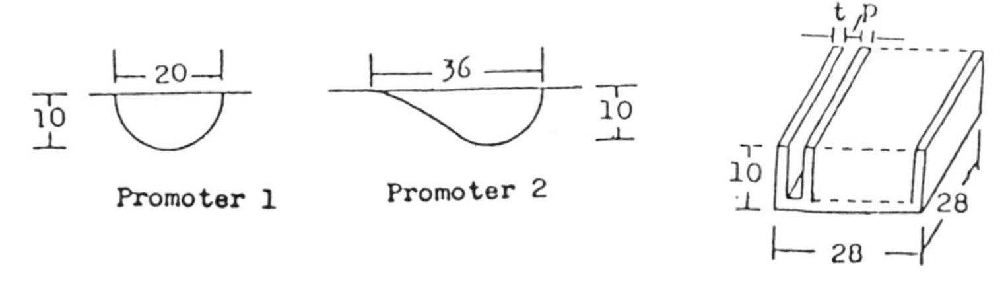

FIGURE 4. Promoter Specification.

FIGURE 5. Finned Heat Sink Specification.

Conduction heat losses from the heated components to the channel wall were estimated to be less than 5%. Radiation heat loss was found to be very small and could be neglected.

RESULTS AND DISCUSSION

Convective Heat Transfer without Enhancement Devices

Experiments were conducted with four flow velocities U=1.5, 4.5, 7.5, and 10.0m/s. It was found in the experiments that the heat transfer coefficient varies from component to component inside the array, because of the variation of the local air velocity, the turbulence intensity and the pressure gradient. The experimental results are plotted in Fig. 7. The heat transfer coefficient is largest in the first row and decreases smoothly with the row number to a so called "fully developed" value, which appears after the third row in present study. This general trend has been found by previous investigators[3,4]. Correlations have been obtained from the present data by the least squares method both for "developing flow" and "fully developed flow":

For the first row

$$Nu=0.848Re^{0.558} \text{(with 1.8\% scatter)} \tag{5}$$

For fully developed flow (the fourth and fifth rows)

$$Nu=0.899Re^{0.53} \text{(with 3.2\% scatter)} \tag{6}$$

The data for fully developed flow are plotted in Fig. 6. together with the experimental results of Wirtz[4]. It is seen in the figure that the present data agree in general with Wirtz's result. The maximum discrepancy is only 13.7%. This fact somewhat shows the reliability of the present experimental apparatus and instrumentation, as the geometry of the two test sections is similar.

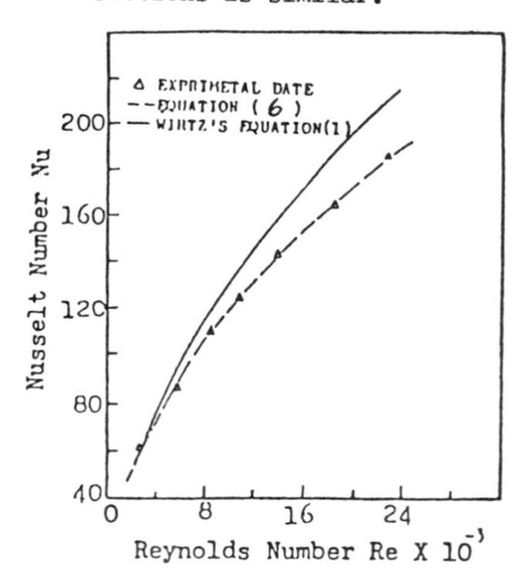

FIGURE 6. Fully Developed Nusselt Number (Row 4 and 5) vs. Reynolds Number.

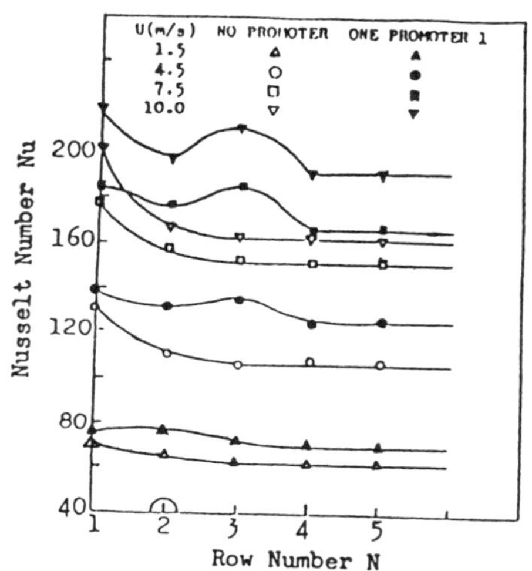

FIGURE 7. Nusselt Number vs. Row Number; the Effect of One Promoter 1.

Convective Heat Transfer with Displaced Promoters

Displaced promoters of two types were employed in the present study. They were placed on the upper wall (ceiling) of the test channel. The effect of the configuration, number and location of the promoters on the enhancement was studied running experiments with

(1) a promoter of type 1 placed above the second row (Fig. 7).
(2) a promoter of type 2 placed above the second row (Fig. 8).
(3) two promoters of type 1 placed above the second row and third row (Fig. 9).
(4) two promoters of type 1 placed above the second row and the center-line between the third row and fourth row (Fig. 10).

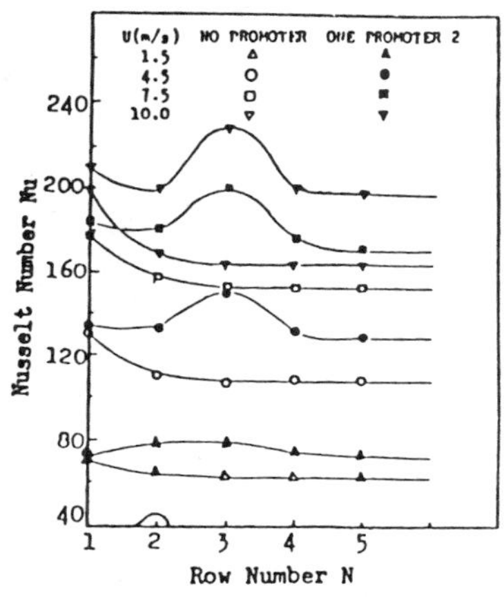

FIGURE 8. Nusselt Number vs. Row Number; the Effect of One Promoter 2.

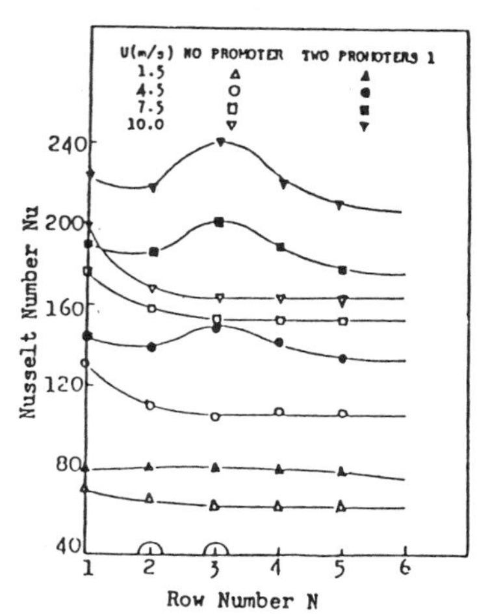

FIGURE 9. Nusselt Number vs. Row Number; the Effect of Two Promoter 1 (A).

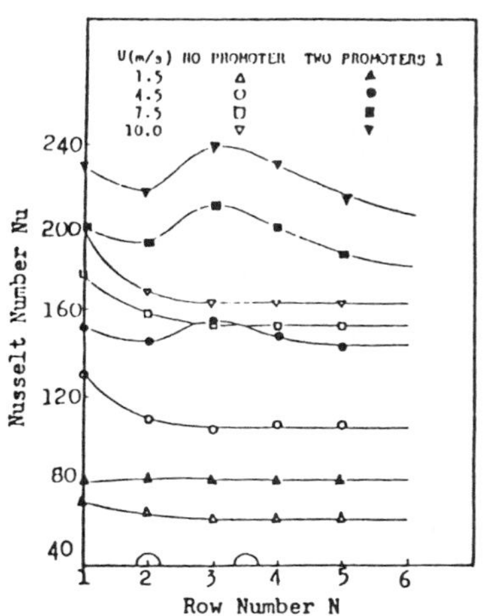

FIGURE 10. Nusselt Number vs. Row Number; the Effect of Two Promoter 1 (B).

All the improvements in the heat transfer coefficient for the four cases are illustrated in Fig. 7-10 in comparison with the results of heat transfer without enhancement.

It is noted that the enhancement varies from row to row inside the array as shown in the figures. The variation of enhanced heat transfer with the row number is not monotonic for most cases. For the case of promoter 1, the maximum enhancement always appears on the third row, with the exception of the low velocity (1.5m/s) condition, as shown in Figs.7 and 8. The enhancement is also influenced by the configuration of the promoters. Comparing the results plotted in Fig. 7 and 8, we find that the promoter 2 may enhance the convective heat transfer more highly than promoter 1. The effect of two promoters (type 1) placed in different locations was also examined experimentally. As illustrated in Fig. 9 and 10, the enhancement created by two promoters is always higher than that by a single promoter. The maximum heat transfer rate also appears on the third row. Considerable improvement is also obtained with the fourth row. The influence of the locations of the two promoters on heat transfer was investigated. The experimental results showed that the heat transfer enhancement with two promoters placed above the second row and the centerline between the third row and the fourth row respectively (case B) is higher than that with the two promoters above the second row and the third row respectively (case A). As shown in Fig. 10 improvement in heat transfer coefficient as high as 51% may be obtained with two promoters. However, in order to determine the optimal promoter locations more research work is still required.

A set of empirical correlations has been obtained as follows:

For single promoter

For the first row

Type 1 $\quad Nu=1.116Re^{0.535}$ (with 8.1% scatter) (7)

Type 2 $\quad Nu=0.735Re^{0.575}$ (with 1.3% scatter) (8)

For the third row

Type 1 $\quad Nu=0.575Re^{0.574}$ (with 4.6% scatter) (9)

Type 2 $\quad Nu=0.455Re^{0.633}$ (with 23% scatter) (10)

For the fourth or fifth row

Type 1 $\quad Nu=1.044Re^{0.530}$ (with 4.3% scatter) (11)

Type 2 $\quad Nu=0.938Re^{0.544}$ (with 3.0% scatter) (12)

For two promoters (Type 1)

For the first row

Case A $\quad Nu=0.952Re^{0.555}$ (with 2.2% scatter) (13)

Case B $\quad Nu=0.922Re^{0.564}$ (with 3.6% scatter) (14)

For the second row

Case A $\quad Nu=0.861Re^{0.562}$ (with 3.3% scatter) (15)

Case B $Nu=0.986Re^{0.552}$ (with 1.1% scatter) (16)

For the third row

Case A $Nu=0.738Re^{0.586}$ (with 1.5% scatter) (17)

Case B $Nu=0.764Re^{0.588}$ (with 3.2% scatter) (18)

For the fourth row

Case A $Nu=0.953Re^{0.553}$ (with 1.8% scatter) (19)

Case B $Nu=0.755Re^{0.585}$ (with 1.9% scatter) (20)

For the fifth row

Case A $Nu=1.04Re^{0.539}$ (with 1.4% scatter) (21)

Case B $Nu=1.15Re^{0.533}$ (with 1.5% scatter) (22)

The variations of Nusselt number with Reynolds number in the four cases are presented in Figs. 11, 12 and 13 for the first row, third row and fifth row, respectively. In comparison with the data of heat transfer without enhancement, it is obvious that the enhancement is increased with increasing Reynolds number. All the experimental data are very well described by the empirical formulas presented above as shown in the figures.

The enhancement may be attributed to two factors: acceleration of local flow velocity and increase of turbulence intensity. The enhancement probably is also caused by the flow modulation induced by vortex shedding from the half-cylinder-like promoters in air crossflow as described by Ratts et al.[9]. There has been very little work done on the mechanism of the enhanced heat transfer. More effort should be devoted to the study of the enhancement mechanism, both experimentally and theoretically.

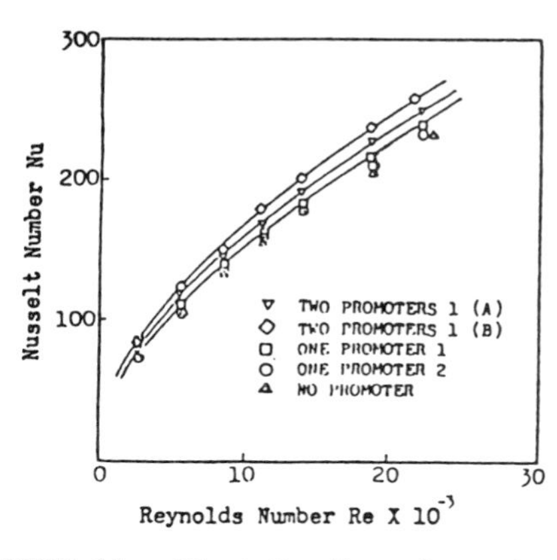

FIGURE 11. First Row Nusselt Number vs. Reynolds Number; all Cases with Promotors.

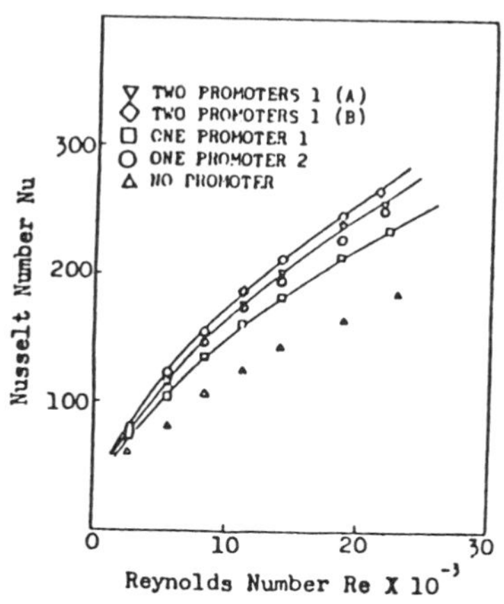

FIGURE 12. Third Row Nusselt Number vs. Reynolds Number; all Cases with Promoters.

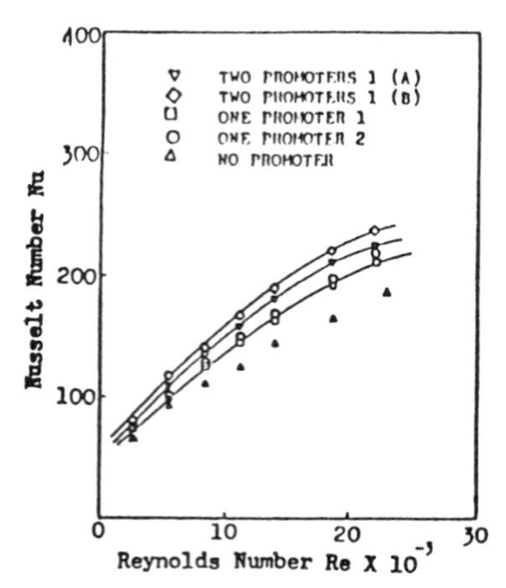

FIGURE 13. Fifth Row Nusselt Number vs. Reynolds Number; all Cases with Promotors.

Convective Heat Transfer with Finned Heat Sinks

Experiments were performed to investigate the characteristics of heat transfer from block-like components with finned heat sinks. The results are presented in Fig. 14. It is found that the variation of heat transfer coefficient with row number for the components with finned heat sinks is similar with that for the components without heat sinks. The variation is monotonic as shown in the figure. Also the heat transfer pattern may be divided into a "developing region" and a "fully developed region". The third row may be still considered as the line of demarcation between the two flow patterns. However, very significant enhancement can be obtained with the finned heat sinks. Figs. 15 and 16 show the variations of Nusselt number with Reynolds number for the first row and the fifth row, respectively, with or without finned heat sinks. As shown in Fig. 15, the finned components yield heat transfer coefficients about 2.53 times larger; this value is close to 2.6 reported by Nakayama et al.[8] higher than that on the components without heat sinks.

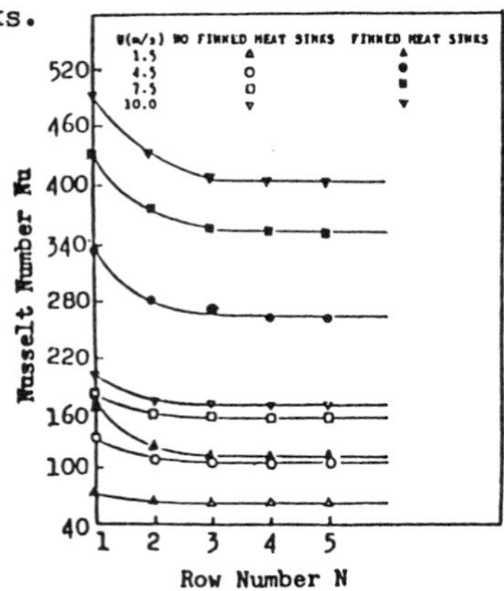

FIGURE 14. Nusselt Number vs. Row Number; the Effect of Finned Heat Sinks Attached to the Experimental Components.

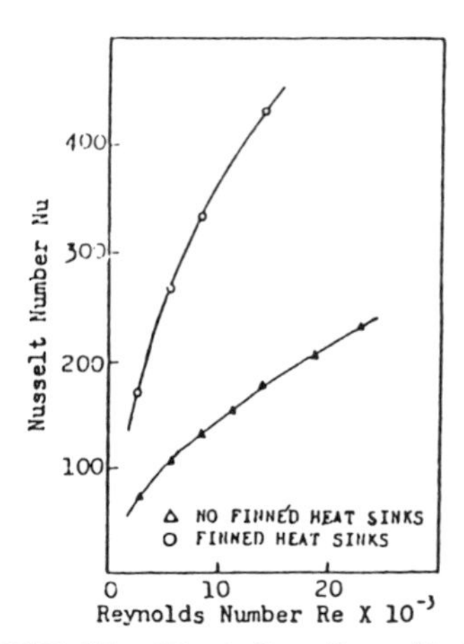

FIGURE 15. First Row Nusselt Number vs. Reynolds Number; the Effect of Finned Heat Sinks Attached to the Experimental Components.

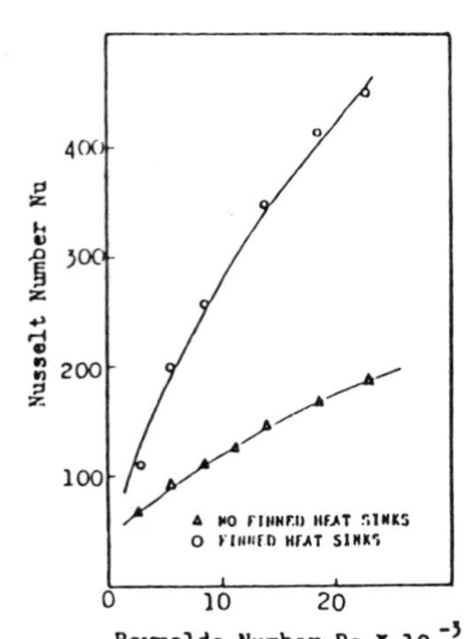

FIGURE 16. Fifth Row Nusselt Number vs. Reynolds Number; the Effect of Finned Heat Sinks Attached to the Experimental Components.

The experimental data for finned heat sinks may be well described by the following empirical formulas as shown in Figs. 15 and 16.

For the first row

$$Nu=2.578Re^{0.535} \text{(with 4.7\% scatter)} \tag{23}$$

For the fourth and fifth rows

$$Nu=0.995Re^{0.611} \text{(with 15.6\% scatter)} \tag{24}$$

Pressure Drop Performance

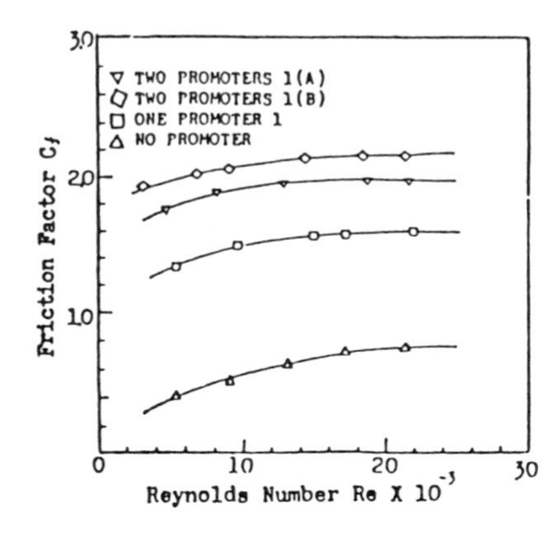

FIGURE 17. Friction Factor vs. Reynolds Number; Some Cases with Promoters.

Measurements without or with displaced promoters (type 1) were made to determine the friction factors of an array of block-like components. The experimental results are presented in Fig. 17. The friction factor for the component array with one promoter is much higher than that for the array without a promoter, but lower than that for the array with two promoters. It is also noted that the friction factor for two promoters placed above the second row and the centerline between the third and fourth rows respectively (case B)is higher than that for two promoters placed above the second row and the third row respectively (case A). However, the heat transfer coefficient for case B is also higher than that for case A as discussed in previous part of this report.

CONCLUSIONS

(1) Measurements were made to determine the heat transfer coefficient from block-like electronic components in air flow with or without enhancement devices, within the range of $Re=2.8x10^3-2.2x10^4$. Friction factors were also measured with or without the enhancement devices.
(2) Displaced promoters of two configurations were employed in the experiments. The effect of configuration, number and location of the promoters on the enhanced heat transfer was studied. Improvement in heat transfer coefficient as high as 51% were obtained.
(3) Finned heat sinks were tested for improving the convective heat transfer. Enhancement of heat transfer as high as 153% was recorded.
(4) A set of empirical formulas was presented, with which the experimental data both with and without enhancement are well described.

NOMENCLATURE

L	Characteristic length, component length
T	temperature
q	heat flux
k	thermal conductivity of air
h	heat transfer coefficient
u	velocity in the unobstructed channel upstream of the arrays
g	gravity vector
Δp	pressure drop
Re	Reynolds number, UL/ν
Nu	Nusselt number, hL/k
C_f	friction factor, $\Delta p/\frac{1}{2}\rho U^2$

Greek symbols

ν	kinematic viscosity
ρ	density

Subscripts

w	wall
c	air

ACKNOWLEDGEMENTS

This study was supported by the National Natural Science Foundation of China. The assistance of Mrs. S.F. Lee of the Foundation is appreciated.

REFERENCES

1 Sparrow, E.M., Neithammer, J.E. and Chabocki, A. "Heat Transfer and Pressure Drop Characteristics of Arrays of Rectangular Modules

Encountered in Electronic Equipment", Int. J. Heat Mass Transfer 25, 1983, pp. 961-973.

2 Sparrow, E.M., Yanezmoreno, A.A. and Otis, D.R. "Convection Heat Transfer Response to Height Differences in An Array of Block-Like Electronic Components", Int. J. Heat Mass Transfer 27, 1984, pp. 469-473.

3 Moffat, R.J., Arvizu, D.E. and Ortega, A. "Cooling Electronic Components: Forced Convection Experiments with Air-Cooled Array", ASME HTD-Vol. 48, 1985, pp. 17-27.

4 Lehmann, G.L. and Wirtz, R.A., "The Effect of Variations in Stream-Wise Spacing and Length on Convection from Surface Mounted Rectangular Components", ASME HTD-Vol. 48, 1985, pp. 39-47.

5 Chen, Z.Q. and Wang, Y.H., "Heat Transfer Performance of Rectangular Components in Electronic Devices" (in Chinese), Journal of Xian Jiaotong University, Vol. 21, No. 5, pp. 23-31, 1987.

6 Hwang, V.P., "Thermal Design Using Turbulator for Air-cooled Electronic Modules on A Card Package", Proc. NEPCON west, March, 1984, pp. 441-449.

7 Ashiware, N., Nakayama, W. and Dairoku, T., "Forced Convection Heat Transfer from LSI Packages in an Air-Cooled Wiring Card Array", ASME HTD-Vol. 28, 1983, pp. 35-42.

8 Nakayama, W., Matsushima, H. and Goel, P., "Forced Convection Heat Transfer from Arrays of Finned Packages", Proceedings of the International Symposium on Cooling Technology for Electronic Equipment, March, 1987, Honolulu.

9 Ratts, E., Amon, C.H., Mikic, B.B. and Patera, A.T., "Cooling Enhancement of Forced Convection Air Cooled Chip Array Through Flow Modulation Induced Vortex-Shedding Cylinders in Cross-Flow", Proceedings of the International Symposium on Cooling Technology for Electronic Equipment, March, 1983, Honolulu.

Heat Exchange of Rectangular and Cylindrical Elements of Radioelectronic Equipment under Forced and Natural Air Motion

A. A. KHALATOV, V. V. ORLYANSKY, and A. F. VASILYEV
Ukrainian Academy of Sciences
Institute of Engineering Thermophysics, Kiev, USSR
Polytechnical Institute, Kiev, USSR

ABSTRACT

Results of an experimental heat exchange research of rectangular and cylindrical elements located on a flat surface are presented in the report. The experiments have been fulfilled under natural and forced convection. Similarity equations both for individual element heat exchange and for numerous elements have been obtained.

1. INTRODUCTION

The problem of radioelectronic (microelectronic) equipment cooling appeared in connection with its intensive development which resulted in complication of the construction, dissipated power increase and raising of the requirements concerning reliability of the cooling system. There are many factors which influence component temperature, the main ones being equipment design features. Nevertheless, well grounded heat transfer calculation methods have been devised [1 - 4] which can be applied in equipment of various classes.

The present calculation methods are based on typical radioelectronic applications. The heat exchange equations employed do not completely reflect the radioelectronic equipment heat transfer process peculiarities, even in such relatively simple cases as heat exchange of rectangular and cylindrical bodies located on flat surfaces. These elements are of small height and usually are cooled by constrained, as well as natural, air flow. The distinguishing feature of the hydrodynamics and heat exchange of such bodies is the presence of flow separation spaces on the sides and ends which make a considerable contribution to the heat exchange.

The present literature data on the local heat exchange of wall-located bodies concerns mainly cylinders [5 - 7] and long rectangular projections [8 - 9] with, respectively, diameter and height of the projection taken as a determining dimension. In the mentioned papers one should note the obtained results for irregularities of local heat irradiation from different parts of finite length cylinders, caused by the mentioned particularities of the flow.

Reference [5] ascertains that heat or mass exchange from a transverse cylinder in the near-wall layer considerably exceeds that in the channel's center and the heat exchange within the distance of only 3.5 diameters from the channel wall is 38 percent of the total. In paper [6] was investigated the tip and adjacent surface heat exchange in the range of Reynolds number variation from $2 \cdot 10^3$ to $2 \cdot 10^4$. The results showed that, first, the tip heat exchange depends on Reynolds number to a greater degree than that of the cylindrical surface adjacent to the tip and, second, the heat exchange of the said surfaces is higher than that of the surfaces lying at some distance from the tip. References [7,8] also note a heat exchange improvement for similar surfaces of the rectangular recess [7] at the laminar regime flow, and of the rectangular prism [8] in the turbulent flow regime by 10-30 percent if compared with a flat plate heat exchange. The authors in references [10 - 12] deal with the heat exchange of different shape bodies located on a flat surface with forced convection. They analyze heat transfer of cylindrical fins in an in-line array and in a staggered array. The microchip temperature dependence on Reynolds numbers distances between plates and chip spacing was investigated in [13] .

A limited number of papers [13 - 15, etc.] concerns natural convection heat exchange of individual flat surfaces. As a rule, only the separate aspects of this topic are investigated. There are no generalized dependencies for a calculation of heat exchange of various shaped individual bodies and their systems located on the channel walls, and that can be explained mainly by the fact that it is very difficult to choose a determining dimension of such bodies.

The results of an experimental investigation of individual rectangular and cylindrical elements, located on the channel wall, under their natural and forced air circulation, and rectangular elements under forced air circulation are presented in this report. The investigation conforms to the thermal mode expected for elements located in radioelectronic equipment subblocks. An attempt is made to generalize the experimental data on the basis of equivalent dimensions which make allowances for end tip surface heat exchange.

2. THE EXPERIMENTAL PLANT

The forced convection heat transfer characteristics of

individual rectangular and cylindrical elements and an array of rectangular elements was studied in an experimental facility, the scheme of which is presented in Fig.1. The operating part of the apparatus is a horizontal channel of rectangular section of 168 mm height, 120 mm width and 300 mm length. In the top and bottom of the channel walls were milled mortises, in which the glass-textolite plates of 240 mm length, forming the parallel channels, were installed. When the heat exchange of individual elements was investigated, two similar elements were installed on both sides of one of the plates. These elements were at the same temperature. This arrangement decreased the heat losses from the elements to the wall of the channel. For the same purpose elements were fixed to the wall by a cotton thread. Nichrome electric heaters were placed inside the steel elements of both shapes, and on their surfaces a uniform heat flux was created. The main geometrical dimensions of investigated bodies are given in Table 1.

The walls were made of 2 mm thick glass-textolite plates with sharp edges at the air inlet side. The distance from the inlet was: for rectangular elements - $l_0 = 100$ mm, and for cylindrical elements - $l_0 = 60$ mm, and the width of the channel where they were installed was, respectively, 29 and 37.5 mm.

Room air was circulated. Its flow rate was measured by a graduated flowmeter. The element surface temperature was measured by thermocouples, the places of location and quantity of which were determined according to element size. Six thermocouples were mounted on the cylinder to allow for its comparatively large size.

Differential thermocouples were fixed along the investigated element's contour to determine the heat losses to the wall. The thermoelectric power measurement and primary processing of all information was done by a mini-computer.

The same experimental facility was used to study the in-line arrays of rectangular elements. For this purpose the elements given in Table 1, Number 3, were used. From 9 to 30 elements were mounted on one side of the glass-textolite plate of 240 mm length. All thermocouple connections and current-conducting wires of these elements led through the plate to its opposite side. Then 2 of such plates were connected by these sides with one another. This double channel wall construction greatly decreased, as in case of individual elements, heat losses from an element to the wall in the perpendicular direction. Depending on the experiment conditions one or two double walls with elements were installed in the operating part of the apparatus.

For investigation of the heat exchange of individual elements under free convection the operating part was disconnected from the described plant and fixed in the vertical position. Then it was placed into a chamber of large size to reduce the influence of environmental convection movement on

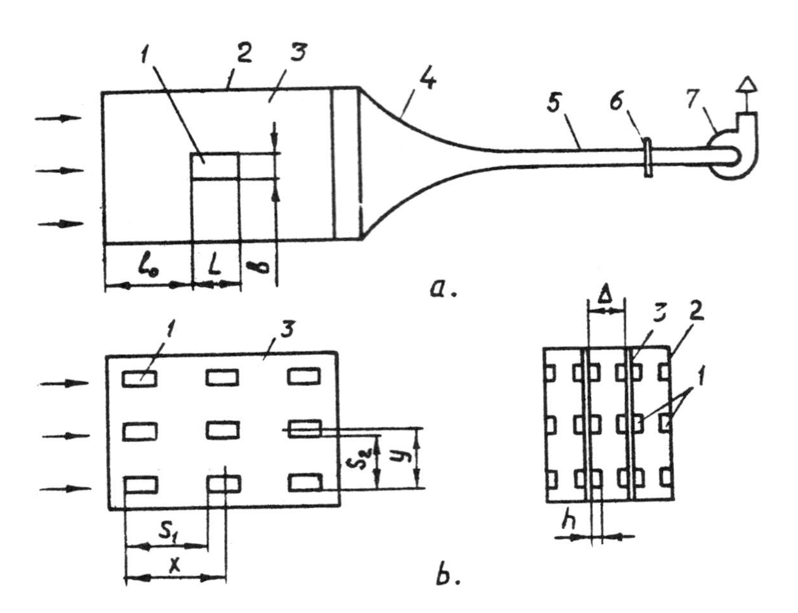

FIGURE 1. (a) Experimental facility and (b) schematic of the in-line array of element:
1 - element; 2 - operating part; 3 - channel wall; 4 - diffusar; 5 - tube; 6 - flowmeter; 7 - fan.

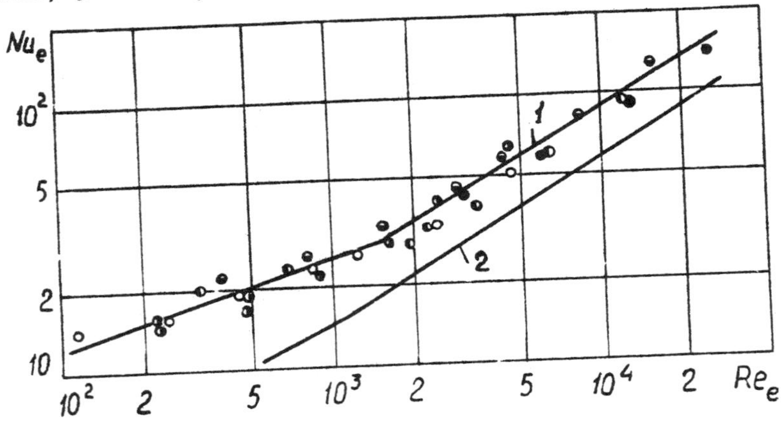

FIGURE 2. Correlation of heat transfer data for individual rectangular and cylindrical elements under forced convection: O - element No.1 (Table 1); ◑ - No.2; ◐ - No.3; ◒ - No.4; ● - No.5. Curves: 1 - equations (2) and (3); 2 - horizontal cylinders in a large enclosure [18].

TABLE 1. Dimensions of the Elements Investigated (Channel Flow)*

Element shape	No. of element	L	b	h	d	Δ	d_e
Rectangular	1	10	7	3.7	-	29	10.04
	2	16.1	16.7	6	-	29	19.43
	3	22.5	18.5	6	-	29	26.71
	4	46.4	16.9	6	-	29	33.9
Cylindrical	5	-	-	28	58	37.5	58

* in millimeters (mm)

on the heat exchange under investigation. To reduce heat losses from the rectangular elements a square orifice of 65 x 65 mm size, covered at both sides by 0.05 mm thick lavsan film, was made in the plate. The elements under investigation were fixed to the both sides of this film, being thus separated by an air clearance of 2 mm. The distance of the elements from the lower edge of the test section, as well as the channel width during investigations with cylindrical elements, had the same values as in the case of forced convection. The channel width with rectangular elements was varied from 18 to 63 mm. The dimensions of the investigated bodies, made of aluminium (except for the cylinder), are presented in Table 2. The main share of experiments with element number 3 was made with the long fin located vertically. In some experiments the element was also fixed with the fin located horizontally.

In all experiments carried out, the convective heat exchange constituent was determined. All heat losses (radiation, transfer to the channel wall through both the electrodes and the current-conducting wires) were determined by calculation using experimental results.

The heat exchange by radiation was done under certain allowances: elements and channel wall surfaces represent grey bodies and are diffuse irradiators, each of the surfaces participating in the heat exchange is isothermal and channel walls dimensions are not limited. With such approximations the individual elements radiant heat exchange determination was not of great difficulty.

TABLE 2. Geometrical Dimensions of the Elements Investigated (Natural Convection)*

Element shape	No. of element	L	b	h	d	Δ	d_e
Rectangular	1	9.8	6.5	2.9	-	18-63	9.5
	2	15.1	15.3	5.2	-	-"-	18
	3	45.2	15.2	5.4	-	-"-	31.9
Cylinder	4	-	-	28	58	37.5	58

* in millimeters (mm)

In the calculations for the radiation exchange, we used a computer to calculate angular coefficients of facet radiation of a concrete element in relation to other facets of all adjacent elements located both on one channel wall and, if necessary, on the opposite channel wall.

The heat losses through the current-conducting wires and the thermocouple electrodes were determined by standard heat conduction calculations for a constant section rod with heat exchange from the rod surface under forced [16] and free [17] convection. The heat loss from the elements to the channel wall was determined by the temperature gradient in the plate, measured by the differential thermocouples, and its thermal conductivity. The experiments show that the total heat losses for individual elements under forced convection constitute 25-40 percent and under natural convection the losses were 35-60 percent. For arrays in forced convection the losses were 15-30 percent of the total heat emission.

While investigating heat exchange in forced convection the inlet temperature of the air was the reference temperature. For individual elements it was the environment temperature, and for the system of elements it was the local mixed-mean temperature.

3. HEAT TRANSFER FROM INDIVIDUAL ELEMENTS

While processing the results for individual elements under forced convection, the following values were taken successively as determining dimensions:linear size of an element along the air flow direction - L , height of an element - h , and a flow path length, determined as a ratio of body surface to the longest perimeter of the body cross section in the direction perpendicular to the flow [16]. In all cases $Nu = f(Re)$. It was possible to generalize the data by a common function.

Considering the peculiarities of heat exchange with elements of small height (existence of separation zones on the element surface, tip surface heat emission, etc.), it is suggested to use as a determining dimension the cylinder diameter whose height h and heat exchange surface F being equal to those of the rectangular element. The equivalent diameter for rectangular elements is determined by

$$d_e = 2\sqrt{h^2 + F/\Pi} - 2h \qquad (1)$$

Numerical values of this diameter are cited in Table 1. As indicated in Figure 2 by using d_e as a measure of the linear dimension the data are correlated quite well. Laminar and transition flows, separated by a critical value of Reynolds number Re_e = 1550, are quite evident. The least square method was applied for determination of common equations appropriate to above mentioned modes:

for $120 \leq Re_e \leq 1550$ $\quad Nu_e = 2.57 \cdot Re_e^{0.33}$ (2)

for $1550 \leq Re_e \leq 2550$ $\quad Nu_e = 0.36\, Re_e^{0.6}$ (3)

Generalization of the obtained experimental data (Fig.3) led to the following equation:

$$Nu_e = 0.8\, Gr_e^{0.22} \tag{4}$$

which may be applied in the range $Gr_e = 1.5\ 10^3 \ldots 1.1\ 10^6$, relative height $h/d_e = 0.17 \ldots 0.48$ and relative body length (along the air flow direction) $l/d_e = 0.48 \ldots 1.42$, where $Pr = 0.7$. The change of orientation of the elements on the wall and the relative width of the channel $\Delta/2h$ practically does not affect the equation. The maximum deviation of the calculated results according to the cited curve in comparison with experimental data does not exceed 8.4 percent.

Figure 3 represents data for long [17] and short [15] horizontal cylinders. It is obvious that exchange of the rectangular elements is 15 - 24 percent higher than for a long cylinder. This may be explained, we suppose, by the deceleration of the rising boundary layer with simultaneous cooling by an unheated channel wall. The thermal effect is more intensive than the hydrodynamic one. In other words, the presence of the unheated wall next to the element might affect to some extent the rising boundary layer; it decelerates and at the same time cools it. In view of the obtained results this leads to the decreased influence of the Grashof number on heat exchange, and simultaneously increases the Nusselt number as compared with large volume heat exchange.

Considering the obtained results, we undertook a recalculation of the experimental data [19] which deals with an investigation of heat transfer under free convection of isothermal spherical segments with different central angles using as the characteristic dimension d_e (Fig.4). The correlation of those results corresponds closely to the correlation of heat transfer for a horizontal cylinder in a large enclosure. This is due to the weak effect of the horizontal non-heated surfaces located below the spherical elements.

4. HEAT TRANSFER FROM RECTANGULAR ELEMENTS IN A FORCED AIR FLOW

The elements were positioned in an in-line array and all elements were heated. An equivalent diameter was used as the characteristic dimension when reducing the experimental data for a constant Prandtl number ($Pr = 0.7$). The Nusselt number is given by the following function:

$$Nu_e = f(Re_e, Gr_e, x/L, y/B, \Delta/2h, S_1/L, S_2/B) \tag{5}$$

Function (5) was obtained with application of supplements given in [20] which are related to introduced three length dimensions for the x, y and z axes, respectively. One of the dimensionless ratios including d_e value and element dimensions did not vary and that is why it was not introduced into function (5).

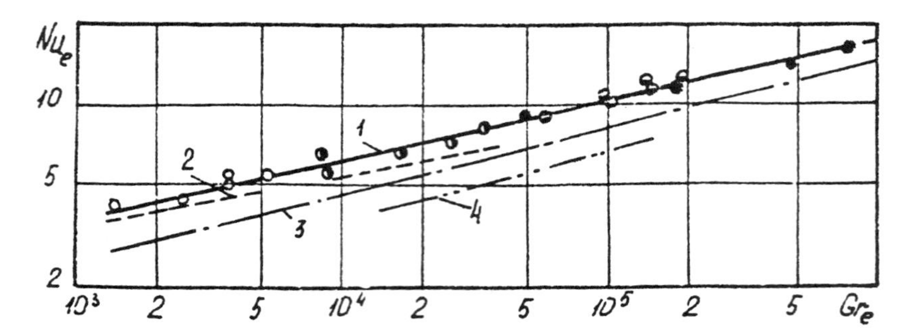

FIGURE 3. Free convection heat transfer for individual rectangular and cylindrical elements:
0 - element No.1(Table 2); ◑ - No.2; ⊖ - No.3; ● - No.4; Nu - calculated values for horizontal cylinders according to 1 - equation (4), 2 - correlation for a channel with non-heated walls [17] , 3 - correlation for a cylinder located in a large enclosure [17] , 4 - correlation for a cylinder attached to a vertical plate at the same temperature [15].

The maximum discrepancy of the data does not exceed [17] percent. Compared with data for the transverse flow past long cylinders [13] the end effects (presence of tip end, vicinity of channel wall) considerably intensifies the heat exchange process, as noted by other investigators.

For generalization of experimental data on individual elements under natural convection conditions, an equivalent diameter was chosen as a determining dimension for all elements, according to equation (1).

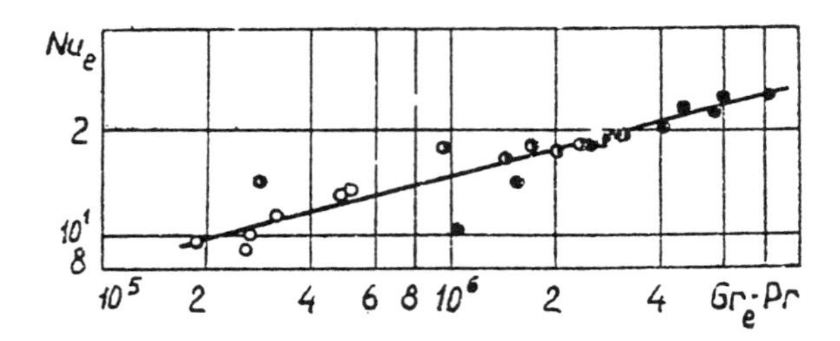

FIGURE 4. Free convection heat transfer from isothermal spherical segments [19] (the determining dimension is d_e). Segment's central angle: 0 - φ = 60° ; ◑ - 90°; ● - 120°.

The following exponential functions were obtained:
- for a channel with elements located on one wall when $Re_e < 1550$

$$Nu_e = 0{,}84 \cdot Re_e^{0{,}38} \cdot (S_1/L)^{0{,}27} \cdot (S_2/b)^{0{,}24} \qquad (6)$$

and when $Re_e > 1550$

$$Nu_e = 0{,}22 \cdot Re_e^{0{,}6} (S_1/L)^{0{,}17} \cdot (S_2/b)^{0{,}15} (x/L)^{-0{,}19 + 0{,}11 \ln (S_1/L)} \qquad (7)$$

- for a channel with elements located on two walls when $Re_e < 1550$

$$Nu_e = 0{,}79 \cdot Re_e^{0{,}38} \cdot (S_1/L)^{0{,}27} \cdot (S_2/L)^{0{,}24} (\Delta/2h)^{0{,}1} \cdot (x/L)^{-0{,}19 + 0{,}11 \ln (S_1/L)} \qquad (8)$$

and when $Re_e > 1550$

$$Nu_e = 0{,}18 \cdot Re_e^{0{,}6} (S_1/L)^{0{,}17} (S_2/b)^{0{,}15} (\Delta/2h)^{0{,}1} \cdot (x/L)^{-0{,}19 + 0{,}11 \ln (S_1/L)} \qquad (9)$$

Equations (6 - 9) may be applied for the following ranges of variable range: S_1/L = 1.7...4.2, S_2/b = 1.6...3.2, $\Delta/2h$ =1.3...3.8, x/L = 0.5...8.8, y/b = 0.5...3.7, Re_e =550...1550 and Re_e = 1550...7000. For elements located on one wall Gr_e = 1.5 10^4...1 10^5 and for the elements located on two walls Gr_e = 1 10^4...7 10^4. The experimental data are generalized by the cited characteristic curves to within [15] percent.

5. CONCLUSIONS

An equivalent determining dimension is suggested to generalize the results of the fulfilled experiments for individual rectangular elements heat exchange under forced and free convection. Generalized heat exchange dependencies for forced air motion are obtained for the element system with an in-line array when the elements are located both on one wall and on two main walls of a flat channel. In the dependencies the element location is taken into consideration. These dependencies can be recommended for calculation of a convective constituent for electronic equipment parts heat exchange.

SYMBOLS

F_ℓ - surface area
d_e - equivalent diameter of the elements
ℓ - element length along the air flow direction
b - element width
h - element height
Δ - distance between channel main walls
S_1, S_2 - element arrangement spacings, longitudinal and transversel to the air flow direction, respectively
ℓ_0 - distance between the channel input and the element
x - distance between the middle of an element being examined and a front facet of the first one along the air flow direction
y - distance between the middle of an element being examined and a lower facet of a lower element
q - a convective constituent of a heat flow density
ν - kinematic viscosity coefficient
u - air flow velocity
β - volumetric expansion coefficient
g - acceleration of gravity
λ - coefficient of thermal conductivity
t_e - element temperature
t - mixed mean temperature in front of the element
$Nu_e = q d_e / \ell \lambda (t_e - t)$ - Nusselt number
$Re_e = u d_e / \nu$ - Reynolds number
$Gr = g \cdot \beta \Delta t d_e^3 / \nu^2$ - Grashof number

REFERENCES

1. Dulnev, G.N., Teplo- i masssbmen v radioelectronnoy apparature, pp. 1-247, Vissh.shck., M, 1984.
2. Dulnev, G.N., Tarnovsky, N.N., Teplovye regimy electronnoy apparatury, pp.1-248, Energiya, L., 1971.
3. Rotcop, L.L., Spokoiny, U.E., Obespechenie teplovykh regimov pri construirovanii REA, pp.1-229, Sov.radio, M., 1976.

4. Glushitskiy, I.V., Okhlagdenie bortovoi apparatury aviatsionnoi techniki, pp. 1-184, Mashinostroyenie, M.,1987.

5. Sparrow, E.M., Samie F., Measured Heat Transfer Coefficients at and Adjacent to the Tip of a Wall-Attached Cylinder in Crossflow-Application to Fins, Journal of Heat Transfer, vol.103, pp.193-201, 1981.

6. Goldstein, R.J., Karni J., The Effect of a Wall Boundary Layer on Local Mass Transfer from a Cylinder in Crossflow, Journal of Heat Transfer, vol.106, pp.1-9, 1984.

7. Young, M.F., Ozel T., Mixed Free and Forced Convection from a Short Vertical Cylinder Placed in a Laminar Horizontal Flow, Journal of Heat Transfer, vol.107, pp.213-217, 1985.

8. Aung W., An Experimental Study of Laminar Heat Transfer Downstream of Backsteps, Journal of Heat Transfer, vol.1o5, pp. 143-150, 1983.

9. McCormick, D.C., Test, F.L., Lessman, R.C., The Effect of Free-Stream Turbulence on Heat Transfer from a Rectangular Prism, Journal of Heat Transfer, vol.106, pp.9-17, 1984.

10. Spokoinyi, U.E., Kaydash, E.V., Lerner, V.M. et al., Convectivnyi teploobmen v plostkostnych blokakh na mikroskhemakh, Voprosy radioelektroniky, ser. TRTO, 3 ed., pp.17-26, 1971.

11. Kawamura T., Hiwanda M., Mabushi J., Humada M. Augmentation of Turbulent Heat Transfer on a Flat Plate with Three-Dimensional Protuberence, 2nd report, Shape Effect and Row of Protuberances, Bull. ISME, 28, No.236, pp.283-291, 1985.

12. Sparrow, E.M., Ramsey, J.W., Altemani, C.A.C., Experiments on In-Line Pin Fin Arrays and Performance Comparisons with Staggered Arrays, Journal of Heat Transfer, vol.102,pp.48-56, 1980.

13. Polyacheck, G.P., Tareev, A.E., Terpigorova, V.M. Teploobmen v nekotorykh konstruktsiyakh blokov na mikroskhemakh, Voprosy radioelektroniky, ser. TRTO, 3 ed., pp.3-18,1971.

14. Gidalevich, V.B., Davydov, V.F., Meshkov, V.N. et al. Teplootdacha corpusov mikroskhem pri yestestvennom okhlazhdenii, Voprosy radioelektroniki, ser. TRTO, 3 ed., pp.20-23, 1979.

15. Sparrow, E.M., Chrysler, G.M., Natural Convection Heat Transfer Coefficients for a Short Horizontal Cylinder Attached to a Vertical Plate, Journal of Heat Transfer, vol.103, pp.26-35, 1981.

16. Kast W., Krischer O., Reinicke H., Wintermantel K., Konvektive Wärme- und Stoffübertragung, Springer-Verlag, Berlin, Heidelberg, New York, pp.1-49, 1974.

17. Martinenko, O.G., Sokovishin, U.A., Svobodno-konvektivnyi teploobmen. Spravochnik, Nauka i technika, Mn., pp.1-400, 1982.

18. Sukauskas A., Jiugde I. Teploperedacha tsylindra v poprechnom potoke jidkosti, Mocslas, Vilnjus, pp.1-240, 1979.

19. Stewart,Jr., W.E., Johnson, J.C. Experimental Natural Convection Heat Transfer from Isothermal Spherical Zones, Journal of Heat Transfer, vol.107, pp.200-202, 1985.

20. Huntley, H.E., Dimensional Analysis, Dover Publications, Inc., New York, pp.1-176, 1967.

Experimental Study: Effect of Surface Orientation on Mixed Convection Heat Transfer from Discrete Sources on Conducting Plates

J. R. CULHAM and M. M. YOVANOVICH
Microelectronics Heat Transfer Laboratory
University of Waterloo
Waterloo, Ontario, Canada N2L 3G1

ABSTRACT

An experimental study of mixed convection from a conductive flat plate is carried out to determine the effect of plate orientation, and in turn the alignment of the forced air stream with the buoyancy driven air stream, on wall temperature. Tests are performed for a range of flow velocities between 1.75 and 5.0 m/s and heat fluxes of 1350 and 2700 W/m^2 with four plate orientations typically found in microelectronic applications.
Results indicate that the orientation of the plate has a significant effect on the wall temperature over the full range of velocities examined.

INTRODUCTION

The thermal design of microelectronic circuit boards and system enclosures has received much attention in recent years as higher power densities impede system performance and long term reliability. System designers are faced with maintaining IC junction temperatures at acceptable operating levels without adversely affecting electrical performance. The orientation of the circuit board and the heat dissipating components with respect to the gravity vector is not always left to the discretion of the designer. Very often the orientation of the circuit board is controlled by extraneous factors such as the profile of the overall system unit, the number of cards or the type of air flow being used to cool the system unit. A buoyancy driven air flow over the vertical plane of the circuit board may assist or retard the forced air stream, causing the surface temperature profile predicted by conventional forced or free convection models to be in error by some unknown amount. Jaluria [1] used a numerical model to examine the significance of mixed convection over a vertical plate with stripwise, uniform flux heat sources. However, little information, either experimental or numerical, is available for mixed convection from plates of different orientations.

The following experimental study examines the effect of changing the orientation of the circuit board and in turn the approach angle of the forced air stream with respect to the gravity vector. A flat plate with stripwise, surface mounted heaters is used to simulate a circuit board with low profile components. Four circuit board orientations commonly observed in microelectronic applications are examined.

APPARATUS AND TEST PROCEDURES

Wind Tunnel

The wind tunnel used to control the direction and velocity of the forced air stream is an open loop design, where the test section is in the open jet produced by the 610 mm × 610 mm discharge section of the wind tunnel. Air velocities between 1.75 and 20.0 m/s are attainable using a single inlet centrifugal fan, driven by a 20 hp, 550 volt AC motor. The air speed is regulated by varying the pitch of damper blades with the aid of a mechanical linkage that can be maneuvered while the fan is operational. The fan feeds air into a 2000 mm × 2000 mm smoothing chamber containing perforated screens, to provide an even flow distribution into a final contraction section with a contraction ratio of 9:1. The velocity profile across any traverse is sensibly flat with a sharp cut off at the boundary, providing a free air stream of 600 mm × 600 mm.
Additional discharge sections can be added to the wind tunnel to obtain various jet cross sectional dimensions and discharge directions such as an upward vertical jet.

Test Plate

Commercially available foil heat strips can be used for simulating stripwise sources with a uniform heat flux condition; however, the uncertainty of the bond established between the foil heater and the underlying substrate leads to problems in obtaining desired levels of heating. The problem of maintaining uniform adhesion between the heat sources and the underlying substrate can be alleviated by using a printed circuit board (PCB) with sources etched into a thin layer of copper bonded to a layer of fiberglass reinforced epoxy (FR4). Normally a DC current passing through the layer of copper would be sufficient to provide Joulean heating with a heat flux proportional to the product of the current and the applied voltage. However, the total resistance of the copper is not sufficient to provide a means of adequately controlling the voltage drop across the heater, making precise control of the imposed heat flux nearly impossible. One method of reducing the resistance of the surface heaters is to reduce the thickness of the copper layer. The typical thickness of copper applied to most PCB's is 1 ounce copper (approximately 0.0356 mm in thickness). The minimum thickness for which the copper layer can be uniformly applied is 1/2 ounce copper, but even this does not provide sufficient resistance to allow for precise control of the surface heat flux.
The method chosen to provide a voltage drop of at least 1 volt across a 58 cm^2 strip heater is to use a 1/2 ounce copper layer with a serpentine path etched within each strip to increase the total path length and in turn increase the total resistance of the heat strip. As shown in Fig. 1, a conventional 305 mm × 229 mm PCB was etched to produce 16 stripwise heaters of alternating 25.4 mm and 12.7 mm widths. Each strip can be individually heated or connected in a series arrangement to maintain a constant I^2R loss, thereby imposing a uniform heat flux over the PCB.
Although the etched PCB eliminates the bonding problems encountered in gluing heat sources to the surface of a flat plate, the etch lines used to control the flow path of the applied current create a problem in determining the effective conductivity of the laminated board because of the discontinuous flow path in the copper layer. Normally, the effective

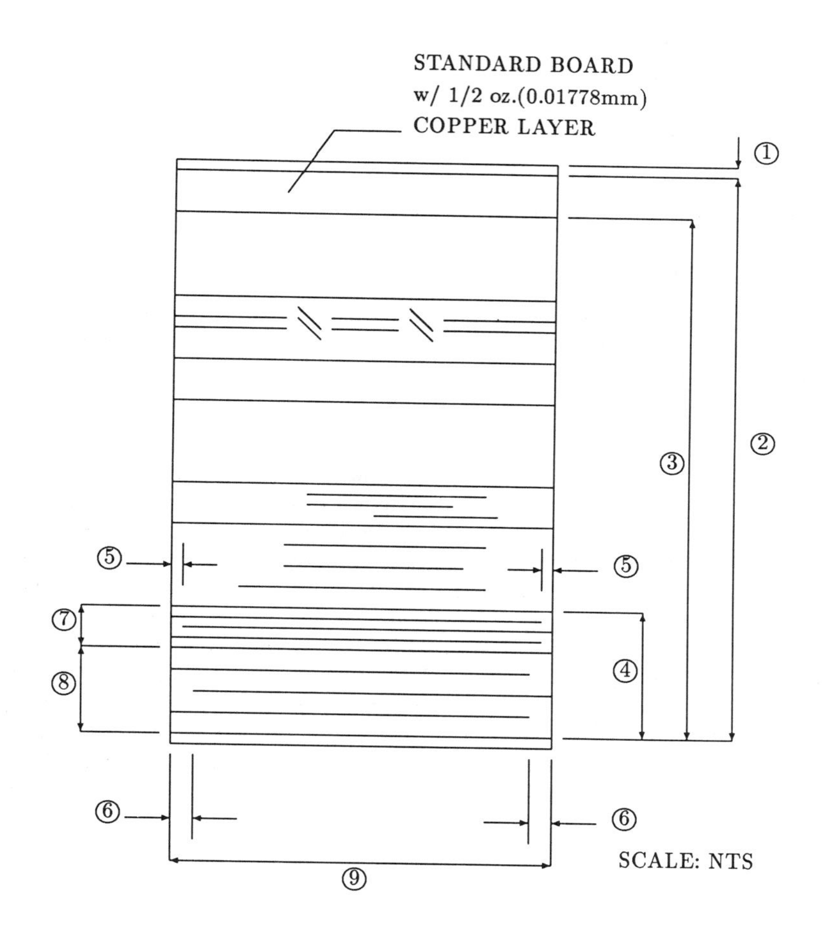

Dimensions – mm			
1	2.54	5	3.18
2	304.80	6	6.35
3	292.10	7	12.70
4	38.1 repeated 8 times	8	25.40
		9	228.60

FIGURE 1. Test Plate.

thermal conductivity of two homogeneous layers can be readily calculated by taking the geometric mean (Gautesen [2]) of the thermal conductivities obtained from the parallel resistive network, given as

$$(k_s)_{pl} = (k_s)_1 \frac{t_1}{t_{tot}} + (k_s)_2 \frac{t_2}{t_{tot}} \tag{1}$$

and the series resistance network, given as

$$(k_s)_{sr} = \frac{t_{tot}}{\dfrac{t_1}{(k_s)_1} + \dfrac{t_2}{(k_s)_2}} \tag{2}$$

where the geometric mean is given as

$$(k_s)_e = \sqrt{(k_s)_{pl} \cdot (k_s)_{sr}} \tag{3}$$

Through a comparison of a uniformly heated plate with analytical models, the effective thermal conductivity of the PCB used in all tests presented herein was determined to be 1.4 W/(m K).

Data Acquisition

Temperature measurements were obtained using 30 gauge (0.254 mm), type 'T', copper-constantan thermocouples with a guaranteed accuracy of 0.75% of the full scale reading. The voltage signal produced by the thermocouples was monitored using a Fluke 2204B 40 channel, programmable data logger. The Fluke data logger provides automatic reference junction compensation with an error of less than $\pm$ 0.05 °C for temperature readings between 15 and 35 °C. The overall system accuracy using type 'T' thermocouples is 0.5 °C for temperature readings between 15 and 35 °C, with a worst case resolution of 0.1 °C. Voltage readings are accurate to within $\pm$(0.02% reading + 0.015% range + 5 μV) with a resolution of 100 μV on the $\pm$4 V range and a resolution of 1 mV on the $\pm$40 V range.
The air velocity readings are obtained using a standard alcohol based inclined manometer connected to a 4 mm ellipsoidal nose pitot tube which is situated 50 mm below and approximately at the center of the heated surface of the test plate. The inclined manometer was calibrated using a Taylor 3132 Jewelled Anemometer.

Test Procedures

The test plate was suspended in the open jet of the wind tunnel, where the applied flow was introduced at various angles to the gravity vector as shown in Fig. 2. In each case the mean cross sectional temperature of the plate was measured at 16 equally spaced points along the centerline of the board by imbedding thermocouples in holes drilled on the back side of the board. The free lengths of thermocouple wire were bound together using tygon tubing to minimize the disruption of the boundary layer as it formed over the back surface of the plate.
For each configuration a single 25.4 mm stripwise source, located 40.6 mm from the leading edge of the plate, was heated using a constant current power supply. The heat flux was determined by measuring the voltage drop across the source and the voltage drop across a current shunt which was calibrated to allow the current across the source to be monitored.

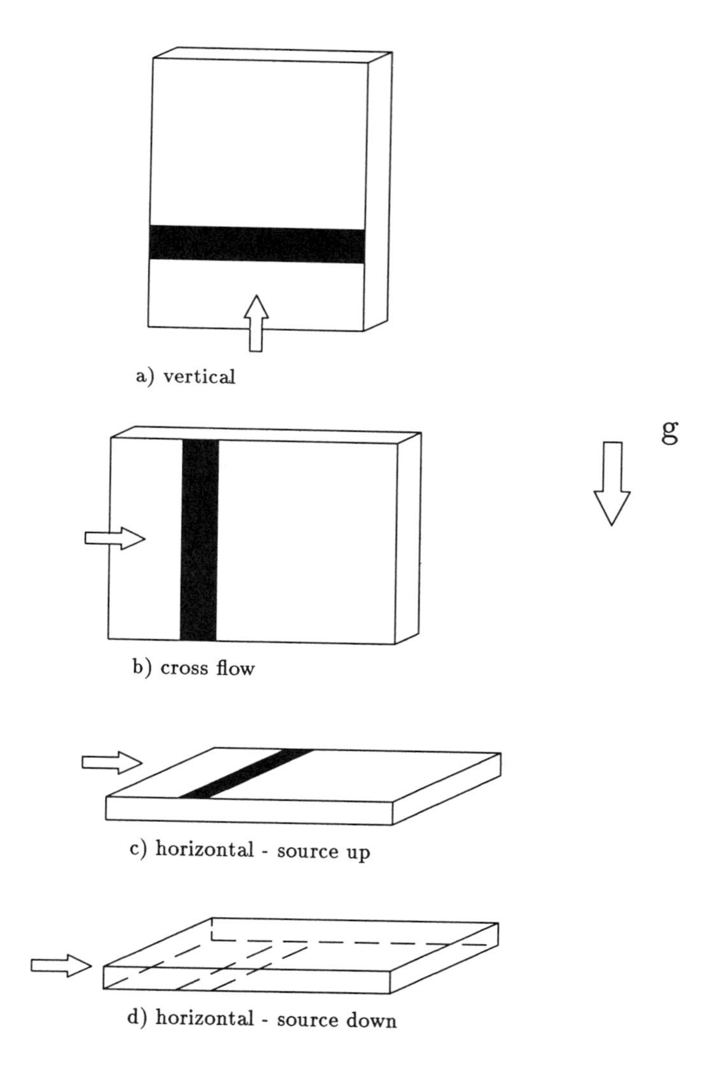

FIGURE 2. Plate Orientations.

Since the thermal resistance of the copper heater fluctuates with changes in temperature, the applied current had to be closely monitored to insure that the heat flow rate to the strip was maintained at the appropriate level.

TEST RESULTS AND DISCUSSION

Sixteen tests were conducted for each orientation, where the applied flow velocity was varied between 1.75 and 5.0 m/s for a heat flux of approximately 1350 and 2700 W/m^2. Figures 3a - 3d show the dimensionless temperature profiles for a vertical plate, a plate in cross flow, a horizontal plate with the source facing up and a horizontal plate with the source facing down, respectively. The dimensionless temperature is defined as the measured local temperature excess divided by the theoretical temperature excess for the case of total heat flow uniformly distributed over the total plate area,

$$\theta^* = \frac{0.454 \cdot \theta \cdot \mathrm{Re_L}^{1/2} \mathrm{Pr}^{1/3} \cdot k_f}{q \cdot \ell} \tag{4}$$

The normalizing function used in Eq. (4) allows the temperature excess data to collapse onto a single curve for a range of flow velocities and heat fluxes only if conjugate, buoyancy and radiation effects are negligible. Therefore any differences between the curves in Figs. 3a - 3d, which are outside the range of experimental uncertainty, can be attributed to these three effects. Although it is difficult to quantify the significance of each of these effects, several conclusions can be drawn from the trends observed in the plots of dimensionless temperature versus dimensionless position.

As shown in Figs. 3a - 3d, the dimensionless temperature is consistently lower over the non-heat source regions of the plate as the flow velocity becomes higher. At a velocity of 1.75 m/s the film resistance over the heat source is high, and a large fraction of the heat dissipated at the source is conducted through the plate, which has a relatively lower thermal resistance than the boundary layer above the source. This results in higher temperatures off the source. But as the velocity is increased a larger fraction of the dissipated heat enters the boundary layer directly above the heat source, resulting in lower temperatures over the non-source locations.

The shape of the temperature profile curves and the relationship between the curves for various flow velocities remains consistent for the four orientations examined. But the magnitude, especially over the source regions, changes significantly as a function of the plate orientation. Figure 4 depicts the average dimensionless temperature for the range of flow velocities and heat fluxes examined, as a function of plate position, for the four orientations discussed herein. The curves presented in Fig. 4 are obtained by using a cubic spline curve fitting routine to obtain a best fit of experimentally obtained data points. It is clearly seen that the vertical orientation, where the buoyancy driven flow assists the forced flow, has the lowest temperatures over the full length of the plate. The horizontal plate with the source facing down has a lower temperature profile than a similar plate with the source facing up. This is contrary to Gryzagoridis' [3,4] results for natural convection, where the lower surface of a flat plate had lower heat transfer rates and in turn higher temperatures than the upper surface of the plate. Gryzagoridis monitored the heat transfer rate above and below an isothermal plate inclined at various angles between 0 and 90 degrees. Under purely natural convection conditions the hot fluid on the underside of

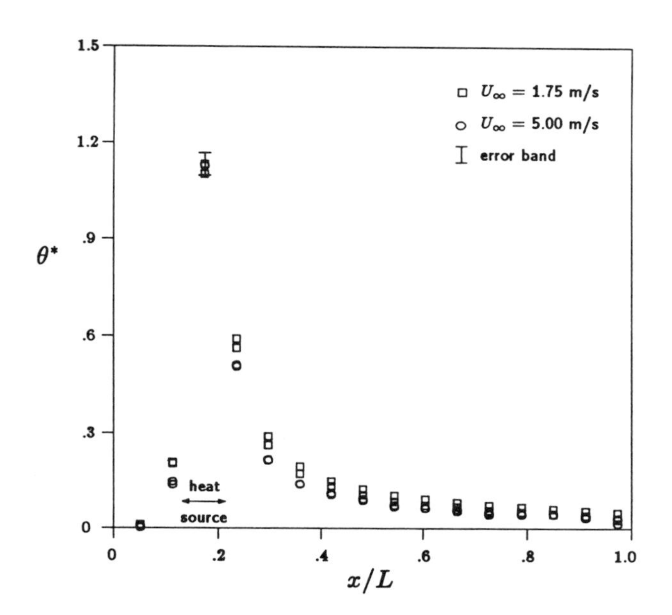

FIGURE 3a. Dimensionless Temperature Profile for a Vertical Plate.

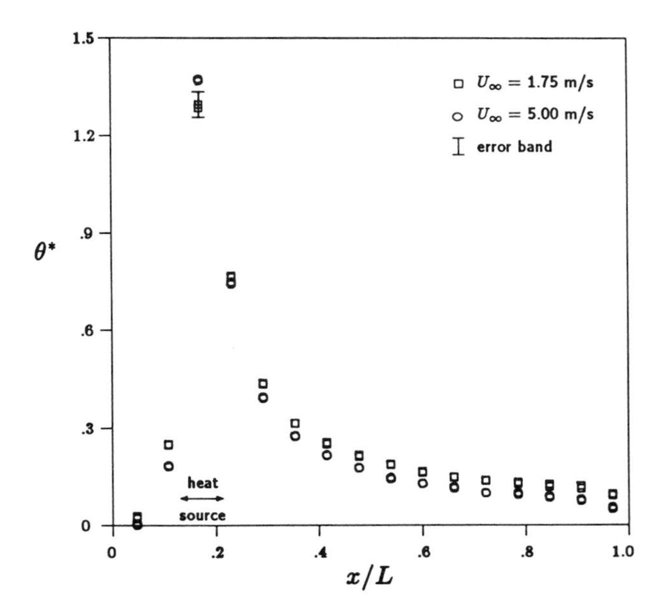

FIGURE 3b. Dimensionless Temperature Profile for a Plate in Cross Flow.

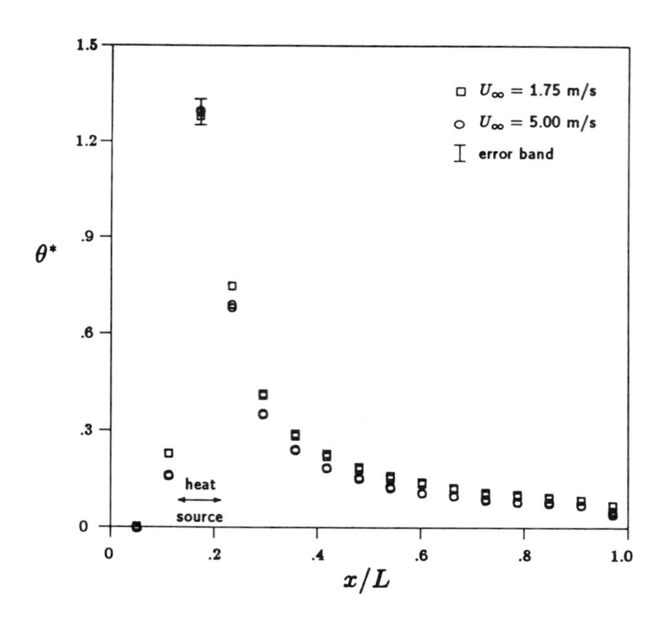

FIGURE 3c. Dimensionless Temperature Profile for a Horizontal Plate with the Source Facing Up.

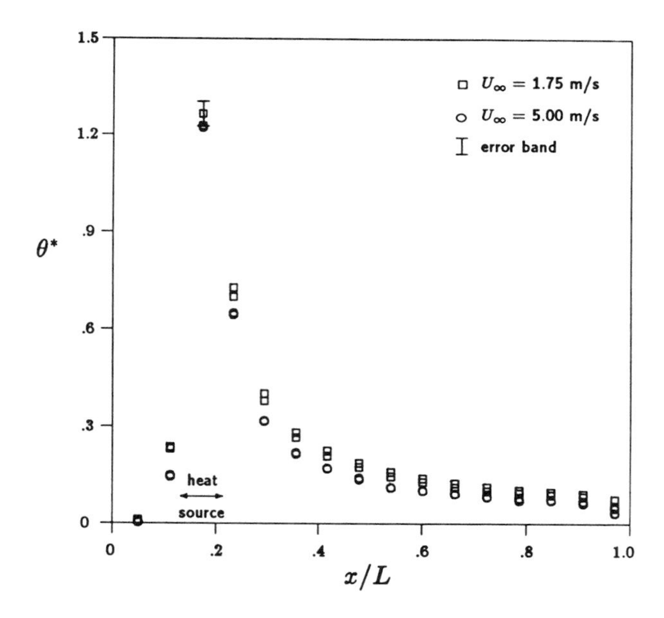

FIGURE 3d. Dimensionless Temperature Profile for a Horizontal Plate with the Source Facing Down.

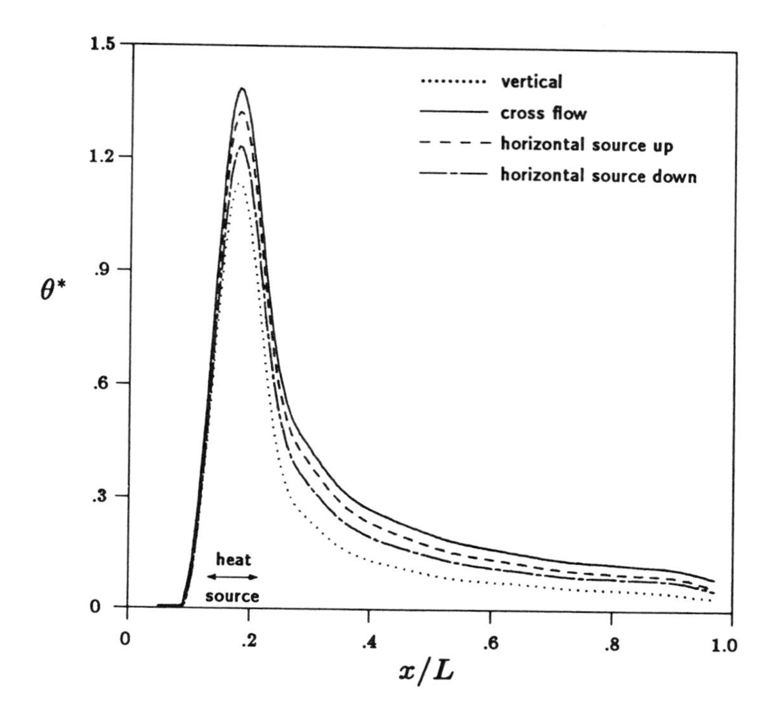

FIGURE 4. Average Dimensionless Temperature Profile for Various Plate Orientations.

the plate remains in the vicinity of the wall causing a general increase in the mean plate temperature. However, in the present study the forced air stream purges the underside of the plate of the heat dissipated on the bottom surface. In fact, as the boundary layer forms on the under side of the plate the air movement due to buoyancy disrupts the normal formation of the boundary layer inducing turbulence, thus lowering the overall wall temperature. The plate with forced air flowing perpendicular to the buoyancy driven flow showed the highest temperature of the four plate configurations examined. The air flow due to natural convection appears to impede the forced air stream, resulting in a higher overall temperature.

Table 1 shows the difference in the peak temperature compared to the horizontal plate with the source facing up. The peak temperature varies by approximately 26 percent about the temperature observed for the horizontal plate with the source facing up. As expected the vertical plate has a lower temperature, by approximately 14 percent on average. The horizontal plate with the source facing up and the plate in the cross flow position are in good agreement at low flow velocities, as shown in Table 1, but show some disagreement for velocities over 3.0 m/s. At higher velocities the wall temperatures of plate in the cross flow position are higher by as much as 10 percent. The horizontal plate with the heat source facing down shows an overall lower wall temperature than a similar configuration

TABLE 1. Difference Between the Peak Temperature of a Horizontal Plate Compared with Other Orientations.

Percent Difference in Peak Temperature	$\frac{\theta_i - \theta_{hu}}{\theta_{hu}} \times 100$		
Velocity	cross flow	horizontal source down	vertical
		$q \approx 2700$ W/m^2	
1.84	0.1	-2.1	-12.3
2.09	0.7	-6.3	-12.3
2.65	0.8	-12.1	-14.2
3.15	4.6	-9.6	-13.9
3.55	9.5	-6.4	-15.5
3.99	8.5	-5.7	-12.7
4.51	7.9	-4.7	-11.9
4.97	10.0	-5.7	-12.9
		$q \approx 1350$ W/m^2	
1.77	0.4	-4.0	-13.8
2.14	1.5	-9.7	-17.0
2.69	1.6	-11.2	-16.6
3.07	3.4	-9.2	-16.1
3.55	8.3	-5.8	-12.8
3.96	10.1	-5.4	-11.5
4.34	9.2	-4.3	-12.1
4.92	6.0	-5.3	-12.8

with the source facing up, over the full range of flow velocities examined.
Heat losses due to radiative heat transfer can be significant, even for low temperature applications such as those described here. A conjugate heat transfer model (Culham, [5]) is used to predict the surface temperature of the test plate for the case of a horizontal plate with the source facing up. The results as shown in Fig. 5 indicate a substantial difference in the plate temperature for a change in the surface emissivity from ϵ=0 to ϵ=0.5. Figure 5 also shows good agreement between the predicted temperature and the experimental data when the measured emissivity of 0.5 is used in the conjugate model.

CONCLUSIONS

An experimental examination of mixed convection heat transfer in a conductive flat plate with a single stripwise heat source is carried out. The orientation of the plate and, in turn, the alignment of the forced air stream with the buoyancy driven air stream has a significant effect on the wall temperature of a flat plate with an isolated heat source. A plate with the forced and buoyant air stream in perfect alignment shows the lowest temperature while a

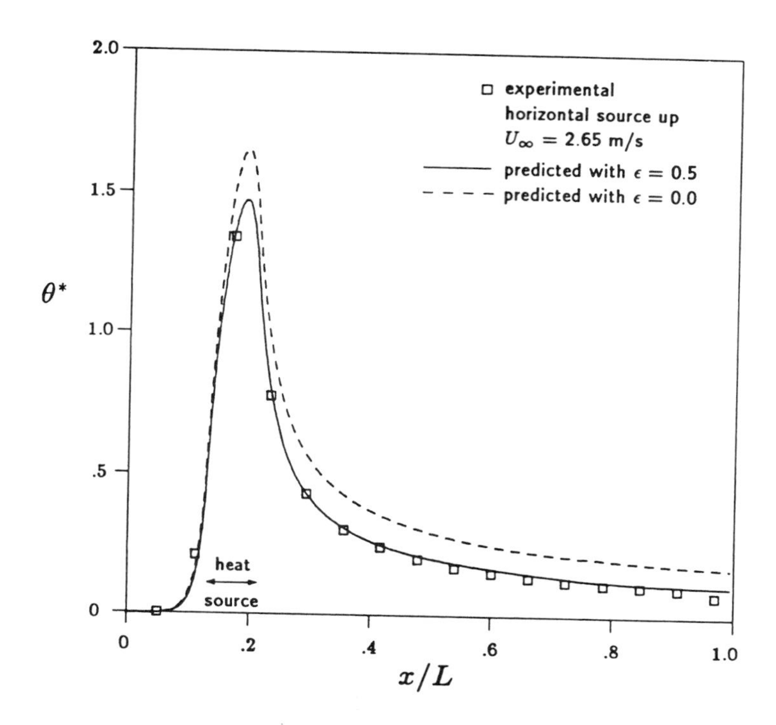

FIGURE 5. Effect of Radiation in Conjugate Heat Transfer.

plate with the forced air stream perpendicular to the buoyant air stream has the highest temperature. A temperature difference of between 2 and 12 percent exists between a horizontal plate with the source facing up and an identical plate with the source facing down.

ACKNOWLEDGEMENTS

The authors wish to thank the Natural Sciences and Engineering Research Council of Canada for financial support under CRD contract P-8322. The authors also wish to thank Kevin Graham of Northern Telecom Limited, Ottawa, for providing the test plates used in this study.

NOMENCLATURE

g	gravitational acceleration, m/s^2
I	current flow, ampere
k	thermal conductivity, $W/(m \cdot K)$
ℓ	source length, m
L	total plate length, m
Pr	Prandtl number
q	heat flux, W/m^2
R	electrical resistance, ohms
Re_L	Reynolds number $\equiv UL/\mu$
t	thickness, m
T	plate temperature, °C
U	velocity, m/s
x	plate position parallel to air flow, m

Greek Symbols

ϵ	emissivity
θ	temperature excess, $\equiv (T - T_\infty)$
θ^*	dimensionless temperature excess $\equiv 0.454 \cdot \theta \cdot Re_L^{1/2} Pr^{1/3} k_f / (q \cdot \ell)$
μ	dynamic viscosity, $kg/(s \cdot m)$

Subscripts

e	effective
f	fluid
hu	horizontal source facing up
pl	parallel
s	solid
sr	series
tot	total
1,2	plate layers
∞	free stream

REFERENCES

1. Jaluria, Y., Mixed Convection Flow Over Localized Multiple Thermal Sources on a Vertical Surface, *Phys. Fluids*, vol. 29, no. 4, pp. 934-940, April 1986.

2. Gautesen, A.K., The Effective Conductivity of a Composite Material With a Periodic Rectangular Geometry, *SIAM J. Appl. Math.*, vol. 48, no. 2, pp. April 1988.

3. Gryzagoridis, J., Natural Convection From an Isothermal Downward Facing Horizontal Plate, *Int. Comm. Heat Mass Transfer*, vol. 11, pp. 183-190, 1984.

4. Gryzagoridis, J. and Klingenberg, B.E., Natural Convection From Upper and Lower Surfaces of an Inclined Isothermal Plate, *Int. Comm. Heat Mass Transfer*, vol. 13, pp. 163-169, 1986.

5. Culham, J.R., Ph.D. Thesis in progress, University of Waterloo, Waterloo, Ontario, Canada, 1989.

A Detailed Analysis of the Conductive/ Convective Heat Transfer in a Single Electronic Board

Y. BERTIN and J. B. SAULNIER
Laboratoire de Thermique, ENSMA
UA au CNRS n° 1098
University of Poitiers
Poitiers, France

ABSTRACT

The understanding of thermal exchanges in a printed circuit board (PCB) and components requires in particular a detailed knowledge of both conduction and convection.
The paper is divided into three parts for which the subject is sufficiently simplified to isolate particular aspects of the aforementioned problem:

- the first study utilized a PCB fabricated with epoxy resin. The PCB had parallel resistive heating lines that could be powered separately. This configuration shows the complex interaction between conduction and convection,
- the second constitues a step to a more realistic situation where the dissipating areas are considered as small squares on the PCB surface. In this case, the conduction in the board is tridimensional,
- the third situation is a logical progression from the previous and aims at identifying the convective heat transfer coefficient field on both faces of the PCB.

1. INTRODUCTION TO THE CONDUCTION/CONVECTION INTERACTION

Six different copper networks powered independently are etched onto a smooth epoxy PCB to simulate the electronic components (cf. Fig. 1).

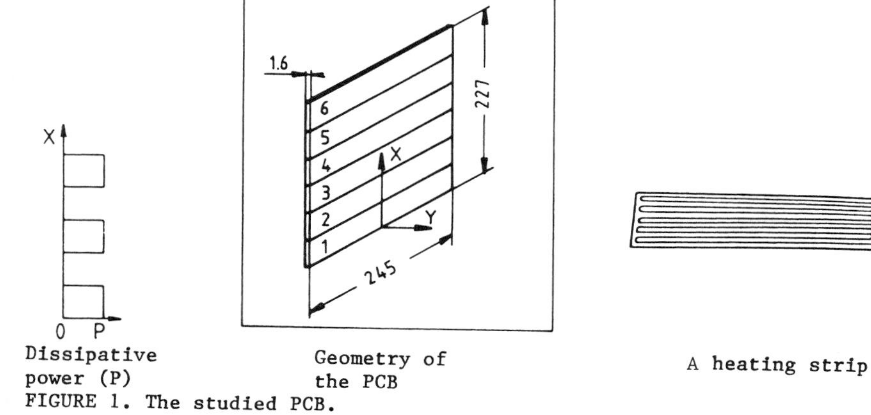

Dissipative power (P)

Geometry of the PCB

A heating strip

FIGURE 1. The studied PCB.

In fact, a symmetrical board was considered with the same network on both sides. The power dissipated on each side is identical.
Initially, the card is supposed isothermal across its thickness. A detailed study was developed numerically and experimentally concurrently:

- numerically, by using a nodal model [1] for the board together with a finite difference method to solve the parabolic equations of the fluid flow,
- experimentally (this part of the work was performed by the French CNET), an Infra-Red (I.R.) Camera evaluated the wall temperature field. Measurements made with LDA and thermocouples permitted mapping of the dynamical and thermal characteristics in the fluid (velocity and temperature).

One of the examples presenting interesting results [2] will be treated here.
Three strips on the board were heated and a non heated strip separated each of them (cf. Fig. 1). The heat transfer to the environment was purely by natural convection. Figure 2 shows the evolution of the surface temperature on the board. A satisfactory correlation between measurements and numerical simulations was observed.
The heat transfer coefficient field - obtained by the computation - (cf. Fig. 3) varies considerably: large values for the heated zones and small values for the unheated zones. In particular, negative values were obtained for the regions where the air (near the wall) is locally hotter than the wall : the heat is transferred from the air to the wall (even when "macroscopically" it's rather the reverse !). This local aspect is also confirmed by the inversion of temperature gradient in the fluid (cf. Fig. 5).
As a conclusion of this first study, it must be kept in mind that this distribution (certainly coarse) of the components leads to a discontinuous distribution of the dissipated power. In these conditions a classical convection model (with a uniform heat flux or wall temperature imposed) is unable to represent the heat exchanges. The full knowledge depends upon both the conductive (small in this PCB) and the complex convective heat transfer studies. Rightly, the convective transfers are highly dependent on the conduction level in the board. Figure 4 shows the dependence of heat transfer coefficient on the variation of the thermal conductivity of the PCB material.

▾ : experimental
\- : numerical

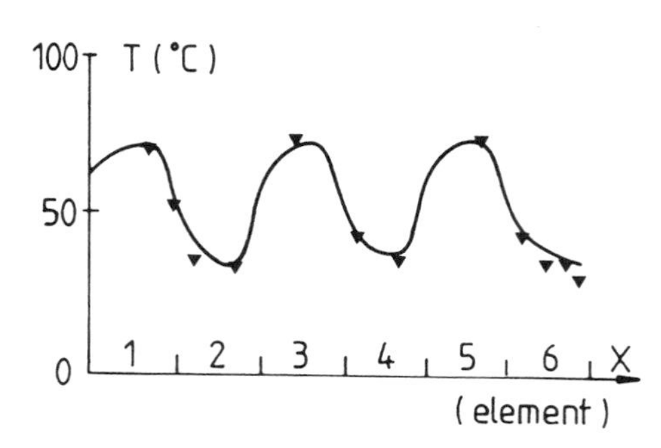

FIGURE 2. The wall temperature field.
(▾ : experimental results)
(- : numerical results)

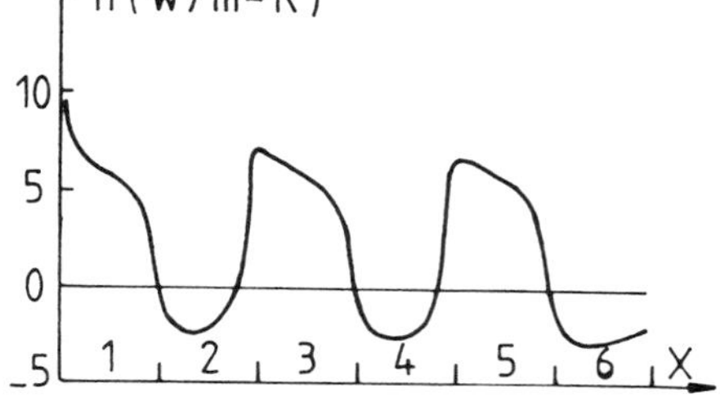

FIGURE 3. The convection coefficient field.

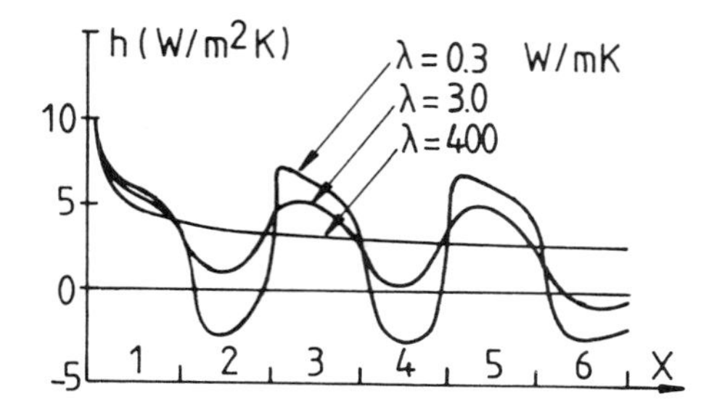

FIGURE 4. Influence of the thermal conductivity of the PCB material on the heat transfer coefficient field.

2. TRIDIMENSIONAL BEHAVIOR OF THE CONDUCTION IN THE BOARD (3)

This second part of the study will consider a board with small heated zones of 1 cm square (each chip is electrically powered thanks to two very thin (37 μm thickness) copper strips). All these elements are located on the same side (cf. Fig. 6). The heat flux is mostly dissipated by forced convection. Owing to the heat source distribution, the problem is obviously at least bidimensional, but the conductive transfer across the thickness of the same board creates tridimensionality as can be seen from the experimental measurements of temperature profiles shown by Figures 8.a and 8.b where a large temperature difference appears between the two sides of the board. For instance, in unit 6, we observe a difference of 27°C for a thickness of 1.6 mm with 1.2 W dissipated on 1 cm^2 (cf. Figs 7, 8.a and 8.b).
At first a bidimensional nodal model (i.e., a first dimension in the plane of the board, a second one in the direction of its thickness, and with an isotropic conductivity $\lambda = 0.45$ W/mK) provided an insufficient temperature difference between the two faces (heated and non-heated) near a heat source (cf. the list below).

TABLE 1. Sensitivity of the "isotropic" model.

h	λ_{epoxy}	thermal dissipation on one node	thermal dissipation by a chip	T^{max}	T^{min}	ΔT
30	0.46	0.024	0.96	152	136	16
30	1	0.024	0.96	125	117	8
30	2	0.024	0.96	102	98	4
40	0.46	0.024	0.96	126	111	15
70	0.46	0.03	1.20	101	83	18

Clearly, this isotropic model with $\lambda = 0.45$ W/(m.K) is not able to simulate the experimentally observed gradient (about 27°C). Then an anisotropic distribution of thermal conductivity was adopted: $\lambda_x = \lambda_y = 0.45$ W/(m.K) in the plane of the board and $\lambda_z = 0.29$ W/(m.K) across its thickness. This last value (50% less than λ_x) resulted from a consequent adjustment of our anisotropic model, in such a way that the gradient, across the thickness of the board was 26.6°C for the concerned temperatures. This value $\lambda_z = 0.29$ was also confirmed by its direct measurement.

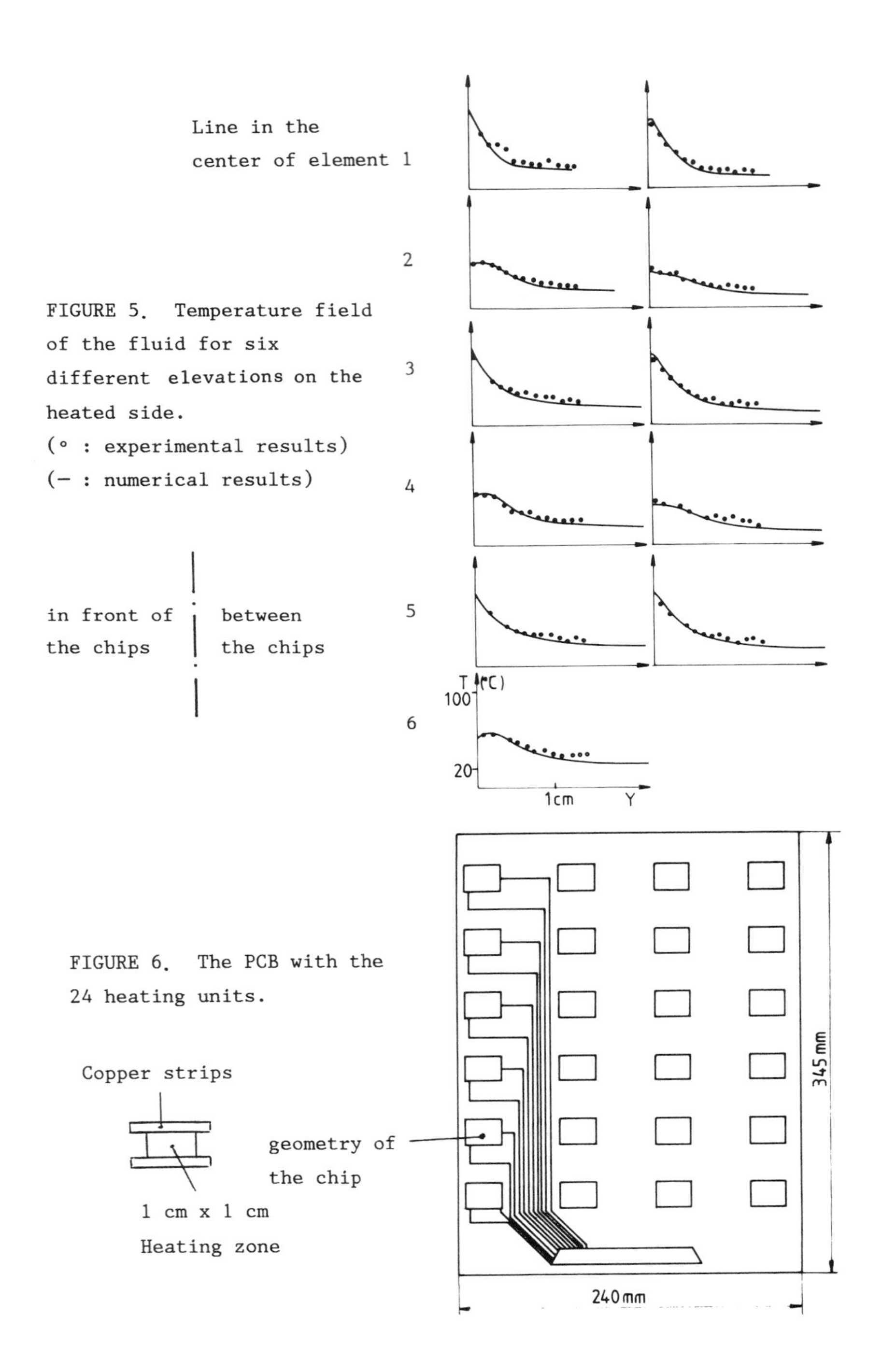

FIGURE 5. Temperature field of the fluid for six different elevations on the heated side.
(° : experimental results)
(— : numerical results)

FIGURE 6. The PCB with the 24 heating units.

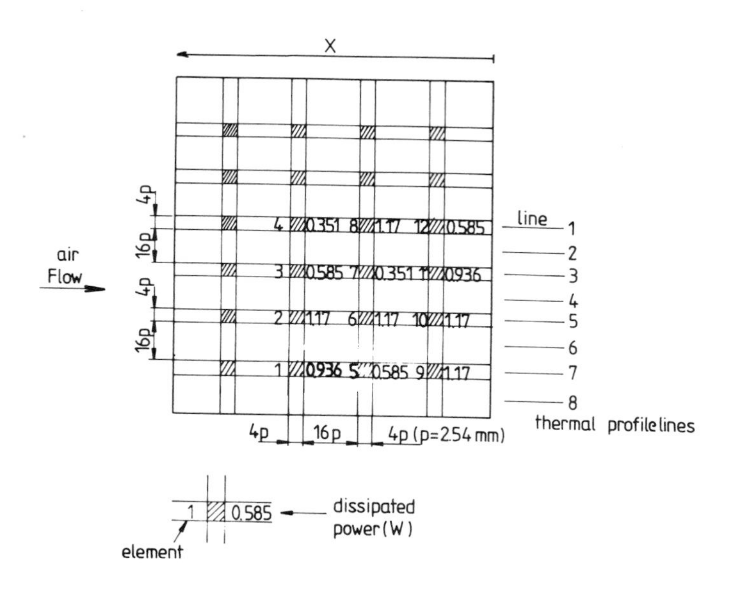

FIGURE 7. Heating unit distribution.

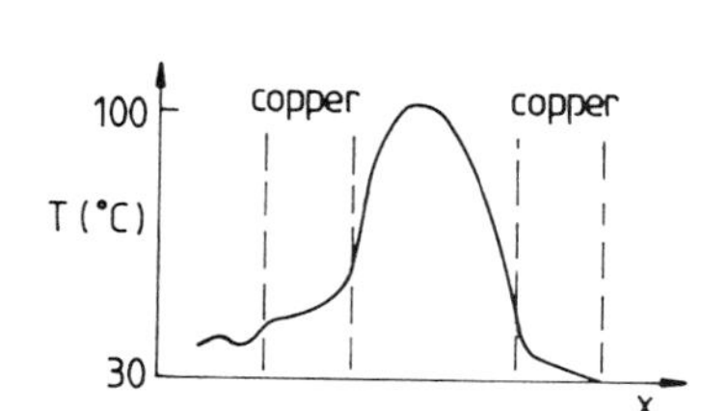

FIGURE 8a. Measured temperatures in front of the unit 6 on the heated side.

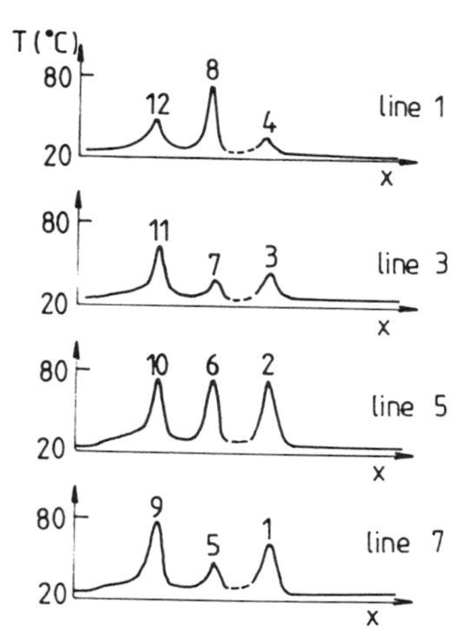

FIGURE 8b. Measured temperatures in front of the units on the non-heated side.

Now, the nature of the problem is reasonably well understood. The question is to restore :

- Firstly, the tridimensional aspect of the conduction, by taking into consideration the non-isotropic behavior of the PCB.
- Secondly, the complex nature of local convective exchanges analogous to the previous mentioned (cf. § 1).

3. TRIAL OF SIMULATION USING A GLOBAL REPRESENTATION: THE LINKING OF TRIDIMENSIONAL CONDUCTION/LOCAL CONVECTIVE EXCHANGE

In this example, four electronic components of the previous board are considered (cf. Fig. 9). A preliminary investigation showed a negligible interaction between the components for the spatial pitch and for the dissipated power adopted.
The final objective is here to obtain a complete 3D model which will involve the knowledge that was gained from the two preceeding studies:

- the convection coefficients are not uniform (study 1)
- the conduction should be treated as anisotropic and was identified as such in study 2.

But instead of solving completly the velocity and temperature fields in the fluid, as in the first study, the numerical method used will be different here: given values will be locally assumed for the heat exchange coefficient, and they will be optimized in such a way to retrieve the surface temperatures: this was done by using a network analyzer.

3.1. Model with a Uniform Value of the Heat Exchange Coefficient

An initial trial was performed with the anisotropic model, identified in study 2, but with a uniform heat transfer coefficient. The best estimation was then:

$\lambda_x = \lambda_y = 0.45$ W/(m.K)

$\lambda_z = 0.29$ W/(m.K)

$h = 61$ W/(m^2.K).

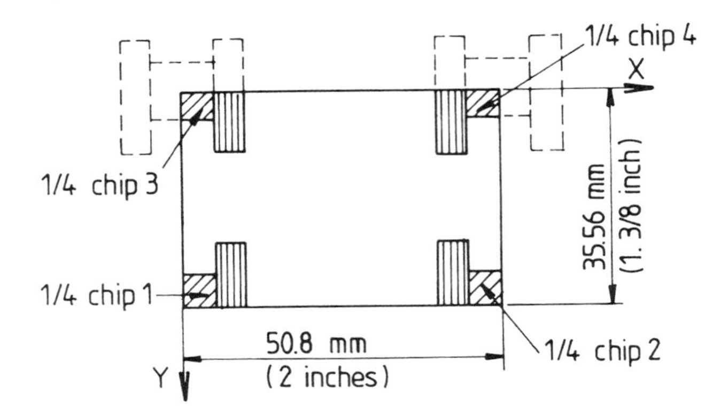

FIGURE 9. Geometry and dimensions of the "four components" element of the board.

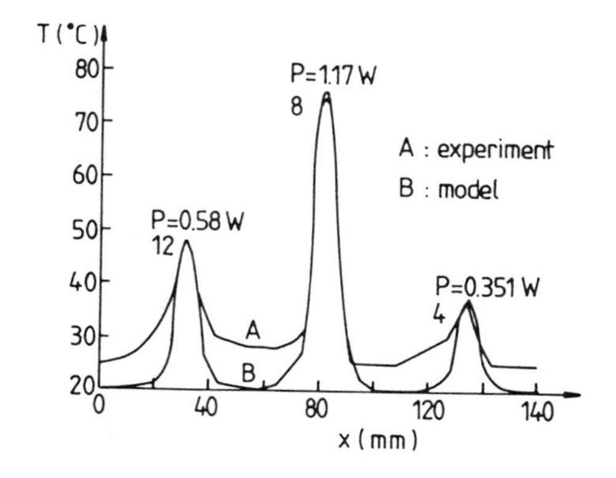

FIGURE 10. Temperature comparison between experimental and numerical results with a uniform heat transfer coefficient.

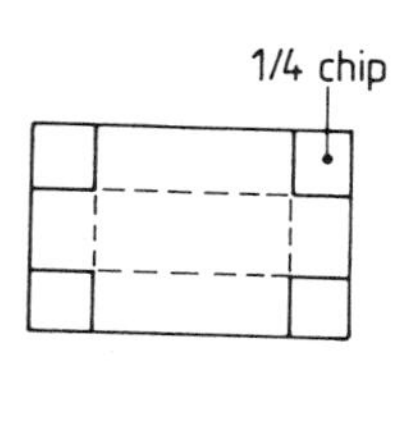

FIGURE 11. Zones of the heat transfer coefficients application.

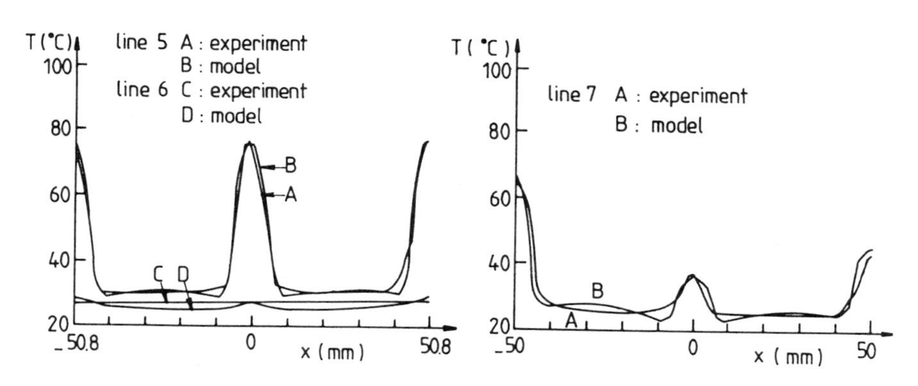

FIGURE 12. Line 5 and 6

FIGURE 13. Line 7

Temperature comparison between experimental and numerical results with a non-uniform heat transfer coefficient.

"over" (flow ←)

68	-0.3	68
5	5	5
68	-0.3	68

"under" (flow ←)

68	5	68
5	5	5
68	5	68

FIGURE 14. Identified values of the heat transfer coefficient.

Unfortunately, it led in the space between the chips to a great temperature discrepancy with the experimental results (as high as 10°C) (cf. fig. 10). By refering to the law of variation of this coefficient shown in the first study, this difference was thought be connected to the fact that the heat transfer coefficient was choosen uniform. Its relatively high value permitted to retrieve the temperature immediately in front of the heated components but it remained excessive at the unheated zones.

3.2. Model with Different Values of the Heat transfer Coefficient

Starting from an already tested optimization strategy and by assuming the heat transfer coefficient uniform in each new elementary zone of the card unit (cf. Fig. 11), several sets of values have been identified. The validated results expressed a sensible improvment of the spatial evolution temperatures shown on Figures 12 and 13.
The identified values of the heat transfer coefficient are set on Figure 14. In particular, we observe:

- a uniform value between the heated zones and in the air flow direction ($h \cong 5\ W/(m^2.K)$),
- an important variation in the flow direction relatively to a thin streak of air going over the component zones.

More, a lightly negative value of the interested coefficient is locally found ($\cong -0.3\ W/m^2.K$). This may attest of a local phenomenon - like in the study of § 1 -, where air gets warm in front of the components and injects then some heat flux in the non-heated zone of the board.

4. CONCLUSIONS

The different studies presented here try to bring a better knowledge of the heat exchange mechanisms between component, board, and surrounding in the electronic devices context. We have certainly simplified the study by considering some smooth dissipative units, though it's a good approach to the chip carrier board technology representation. But even with this simplified sketch, we identified some particularly interesting results, a priori not so foregone.
For instance :

- the acting part of the anisotropy of the PCB,
- the full dependence of the heat transfer coefficient on the conductive transfer.

The interaction between conduction and convection involved some complex phenomena of which the effects cannot be evaluated a priori.
In fact, this detailed numerical analysis needs a lot of nodes : we must think now to perform the simulation, with this rather fair quality of representation but with, of course, less nodes and CPU time involved. A solution could be, for instance, in the use of what is known as model reduction techniques (4).

NOMENCLATURE

h	heat transfer coefficient	$W/(m^2.K)$
P	thermal dissipation of the chips	W
p	pitch between components	m
T	wall temperature	°C
$\lambda_x, \lambda_y, \lambda_z$	thermal conductivity of the PCB in the x, y, and z directions.	W/(m.K)

REFERENCES

1. SAULNIER J.B., La modélisation thermique et ses applications aux transferts couplés et au contrôle actif. Thèse de Doctorat d'Etat, E.N.S.M.A., Poitiers, France, 1980.
2. HUCLIN J.C., Contribution à la modélisation des transferts combinés conduction, rayonnement, convection. Application au couplage des échanges thermiques solide - fluide le long d'une plaque. Thèse de Docteur-Ingénieur, E.N.S.M.A., Poitiers, France, Juin 1984.
3. PIPITONE M., Contribution à l'analyse de l'interaction entre conduction et convection sur les cartes de composants électroniques. D.E.A. d'Aérodynamique, Combustion et Thermique. E.N.S.M.A., Poitiers, Avril 1987.
4. BERTIN Y., GORIN B., SAULNIER J.B., Application de la réduction de modèles à la thermique des systèmes électroniques et électrotechniques. Journée d'Etudes GUT/SFT, Paris, France, Avril 1987.

A Methodology for Optimization of Convective Cooling Systems for Electronic Devices

B. B MIKIĆ, H. KOZLU, and A. T. PATERA
Department of Mechanical Engineering
Massachusetts Institute of Technology
Cambridge, Massachusetts 02139, USA

Abstract

We report here on a new procedure for selection of flow and geometric parameters (Reynolds number and hydraulic diameter) that yield minimum pumping power for a chosen convective cooling system with specified thermal requirements. In general, different cooling systems, satisfying the same functional constraints and each individually designed for the optimal performance, would yield different pumping power at the design point. Comparison of these minimum dissipation values for all cooling choices considered (e.g. laminar or turbulent flow in smooth channels, or various augmentation schemes) would lead to a selection of an optimal heat removal method and identification of geometric and flow parameters around which the chosen system should be designed and operated.

The procedure requires specification of the thermal load (the total heat removal rate), the maximum allowable surface temperature, the choice of coolant fluid, the length and the shape of the cooling channel(s) and ($f = f(Re)$ and $Nu = Nu(Re, Pr)$). The results cover both constant heat flux and constant wall temperature conditions. The analysis currently assumes fully developed flow although it can be extended to developing flows with entrance and exit losses. The procedure can also be modified to select the best cooling system that would yield minimum pressure drop (rather than the minimum pumping power).

The study provides a strong insight into the relationship between the physics of transport enhancement and heat transfer system design, allowing *a priori* evaluation of various enhancement schemes. In particular it can be shown that the dissipation-minimizing enhancement scheme proceeds from macro-scale eddy-promoters to micro-groove roughness elements with increasing thermal load. Significant dissipation savings result from properly designed cooling systems.

1 Introduction

Thermal performance of cooling systems could limit the power density and directly affect the cost and size of electronic equipment; for these reasons heat removal is one of the major considerations in the development of high-density electronic equipment [1,2].

Previous studies on forced convection cooling systems for electronic devices were approached with consideration of material and manufacturing costs, pressure drop and pumping power [3-5]. Much of this work has focused on the use of micro-channel cooling geometries for laminar flows without addressing the general aspects of thermal-hydraulic design. In this work we introduce a methodology for design of convective cooling systems and show the relationship between the optimization and heat transfer augmentation. Potential for significant dissipation (or pressure drop) savings resulting from properly designed cooling systems, with or without augmentation, is demonstrated in a sample optimization study for heat transfer in a channel with air as the working fluid ($Pr = 0.71$). In the case of systems with heat transfer augmentation, it was found that the best results are achieved by flow destabilization at appropriate spatial scales of motion. The results of this work can be applied to other heat removal systems that have similar constraints.

We organize the presentation as follows: in Section 2 we consider the general framework for thermal-hydraulic design. In Section 3 we solve the nondimensional optimization problem. Results of optimization for smooth-channel flows (laminar and turbulent) are presented in Section 4. In Section 5 we present the effects of heat transfer augmentation for macro-scale (eddy-promoter) and micro-scale (micro-grooved channels) augmentation schemes. Lastly, in Section 6, we present the conclusions that can be drawn from the results of the present study.

2 Thermal-Hydraulic Design

2.1 General Considerations

We consider the problem, Figure 1a, of incompressible flow in a channel of length L and hydraulic diameter D_H, with uniform heat flux q'' imposed around the perimeter with a constraint of specified maximum wall temperature $T_{w_{max}} (= T_{w_{exit}})$. The latter is imposed to assure functionality and/or extended life of the device. The coolant flow having average velocity V is forced by a pressure gradient $\frac{\Delta P}{L}$. The fluid is assumed to have constant (temperature-independent) properties (thermal conductivity k, specific heat c_p, kinematic viscosity ν, and density ρ).

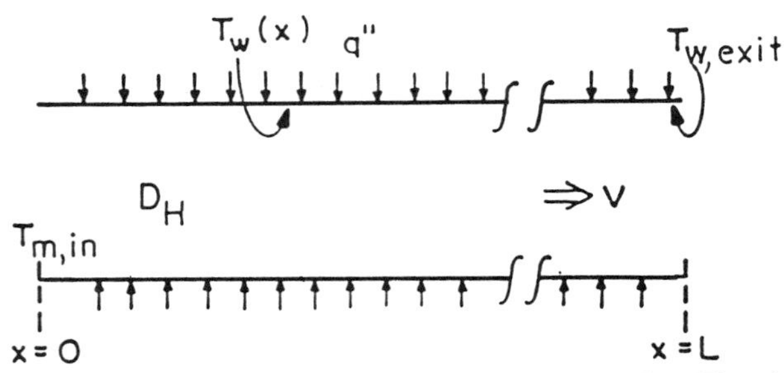

FIGURE 1a. Basic channel geometry for the optimization study. The channel is of length L, and hydraulic diameter D_H, with uniform heat flux q'' imposed around the perimeter.

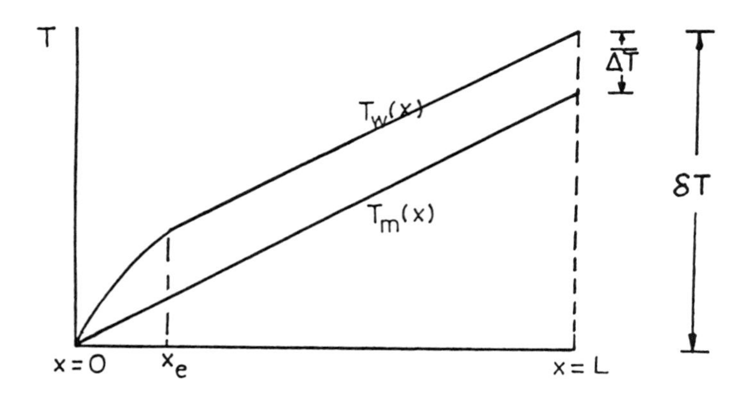

FIGURE 1b. Thermal behavior of flows in channels with uniform heat flux q''.

The energy balance gives

$$q''PL = VA\,\rho c_p(T_{m_{out}} - T_{m_{in}}) \quad , \tag{1}$$

P is the perimeter, A is the channel cross section, $T_{m_{in}}$, $T_{m_{out}}$ are the mixed mean temperature of the fluid at the inlet ($x = 0$), and at the exit ($x = L$), respectively. Since $T_{m_{out}} - T_{m_{in}} = (T_{w_{max}} - T_{m_{in}}) - (T_{w_{max}} - T_{m_{out}})$ (cf. Figure 1b) we can write equation (1) as

$$\frac{q''PL}{VA\,\rho c_p} = \delta T - \overline{\Delta T} \quad , \tag{2}$$

where $\delta T = T_{w_{max}} - T_{m_{in}}$, and $\overline{\Delta T} = T_w(L) - T_m(L)$. δT is then specified and $\overline{\Delta T}$ is a function of heat transfer coefficient at a given heat flux. Two hydraulic-performance variables, pressure drop and pumping power, can be written as

$$\Delta P = 4\ f_{average}(Re)\ \frac{L}{D_H}\ \frac{\rho V^2}{2} \quad , \tag{3}$$

where $f_{average}$ is the average friction factor, and

$$\Psi = \Delta P\ V\ A \quad . \tag{4}$$

2.2 Nondimensionalization

In terms of Prandtl number, Reynolds number, and Nusselt number

$$Pr = \frac{\nu}{\alpha} \tag{5}$$

$$Re = \frac{VD_H}{\nu} \tag{6}$$

$$Nu(Re, Pr) = \frac{q''D_H}{k\,\overline{\Delta T}} \quad , \tag{7}$$

together with normalized hydraulic diameter

$$\overline{D}_H = \frac{D_H}{L} \quad , \tag{8}$$

and thermal load parameter, defined as

$$\Lambda = \frac{q'' L}{k \, \delta T} \quad , \tag{9}$$

the energy balance equation becomes

$$\Lambda = \frac{1}{\dfrac{\overline{D}_H}{Nu(Re, Pr)} + \dfrac{4}{Re\,Pr}} \quad . \tag{10}$$

Equation (10) must be satisfied for any thermal-hydraulic system; it provides a fundamental relationship between thermal load, system geometry, coolant properties, heat transfer performance (Nusselt number) and flow rate (Reynolds number).

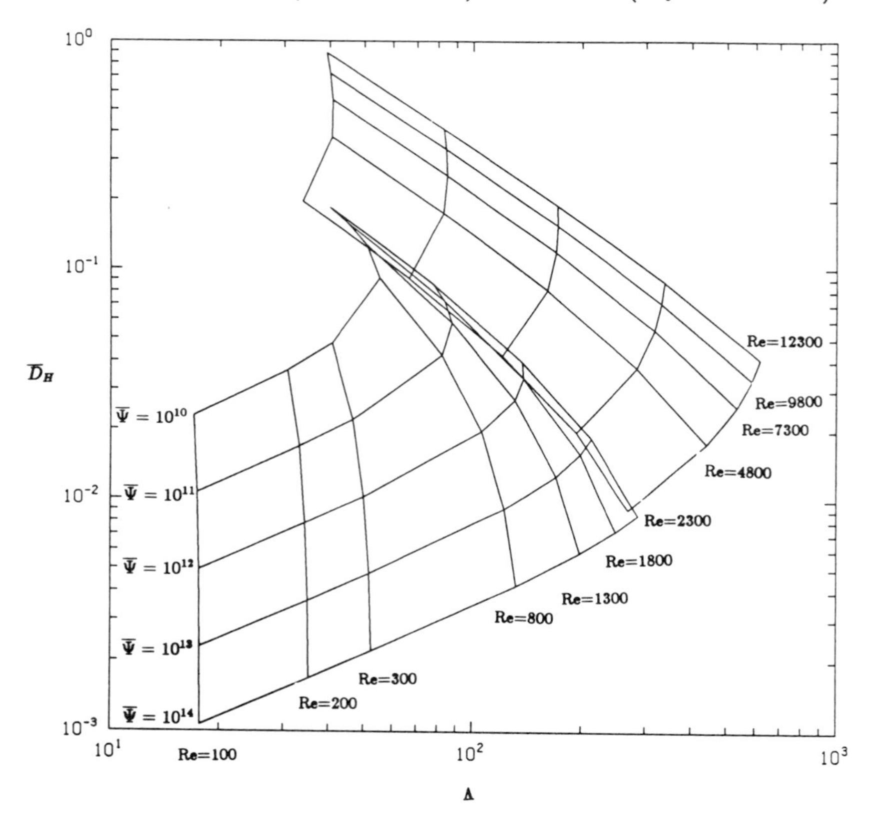

FIGURE 2a. A plot of dimensionless hydraulic diameter versus thermal load for high-aspect ratio channels with uniform wall heat flux at fixed Reynolds numbers and pumping powers. Air is the working fluid ($Pr = 0.71$).

We complete the nondimensionalization by introducing dimensionless pumping power and pressure drop as

$$\overline{\Psi} = \frac{\Psi}{\frac{\rho \nu^3 P}{L^2}} = \frac{f_{average}}{2} \frac{Re^3}{\overline{D}_H^3} \quad , \tag{11}$$

and

$$\overline{\Delta P} = \frac{\Delta P}{\frac{\rho \nu^2}{L^2}} = 2 f_{average} \frac{Re^2}{\overline{D}_H^3} \quad . \tag{12}$$

In Figure 2a we plot dimensionless hydraulic diameter versus thermal load for fixed Reynolds numbers and pumping powers using equations (10) and (11) for high-aspect ratio channels which are typically required in electronic systems [4]. We also present the same plot for circular channels in Figure 2b. Air is the working fluid ($Pr = 0.71$) and the flow is assumed to be hydrodynamically and thermally

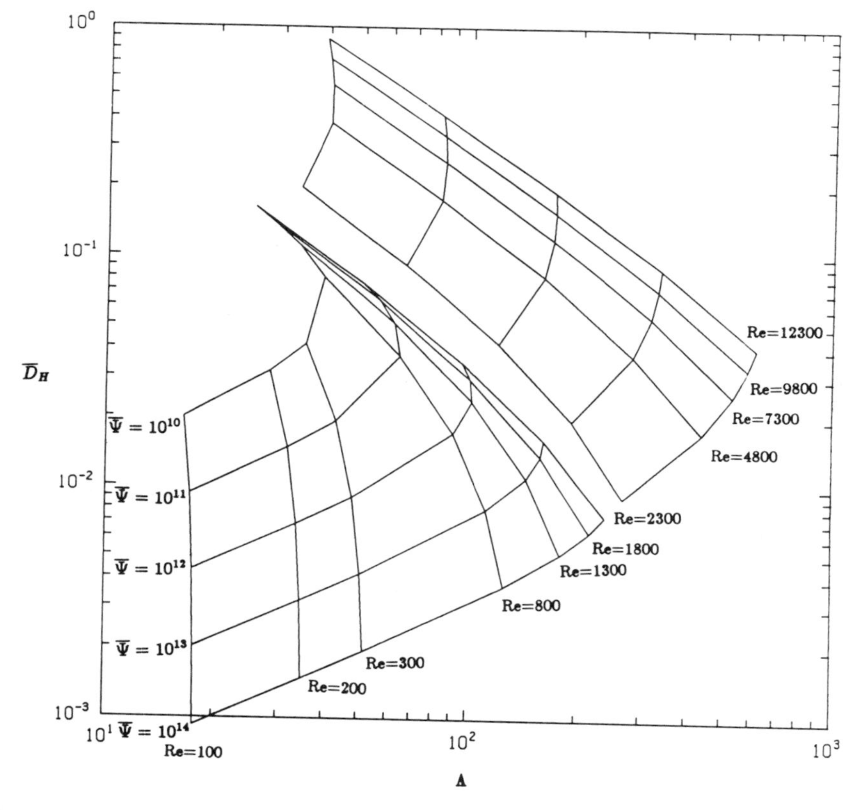

FIGURE 2b. A plot of dimensionless hydraulic diameter versus thermal load for circular channels with uniform wall heat flux at fixed Reynolds numbers and pumping powers. Air is the working fluid ($Pr = 0.71$).

fully developed. For laminar flow in high-aspect ratio smooth channels we use the exact solutions [6]

$$f = \frac{24}{Re} \quad , \tag{13}$$

$$Nu = 8.235 \quad . \tag{14}$$

For laminar flow in circular channels we use $f = \frac{16}{Re}$ and $Nu = 4.364$. For turbulent smooth-channel flows (high-aspect ratio and circular) we use correlations for the friction factor and Nusselt number (for $Re \leq 10000$) as given in [7],

$$f = 0.25\,(1.82\,logRe - 1.64)^{-2} \quad , \tag{15}$$

$$Nu = \frac{\frac{f}{2}\,(Re - 1000)\,Pr}{1.0 + 12.7\,(\frac{f}{2})^{0.5}(Pr^{\frac{2}{3}} - 1)} \quad . \tag{16}$$

For Reynolds number higher than 10000 we use the correlation developed by Petukhov (changing $(Re - 1000)$ with Re and 1.0 with 1.07 in equation (16)) [7]. It can be seen from Figures 2a and 2b that for a fixed Λ, depending on the choice of different hydraulic diameter, dissipation may vary roughly by a factor of 10^3. This illustrates the importance of optimization even for the case of *unenhanced* heat transfer. The savings in pumping power can be further increased by employing enhancement techniques, as will be discussed in Section 5.

3 The Optimization Procedure

Power dissipation and pressure drop affect the costs of operation, material, and manufacturing. Frequently in practical situations an important task would be to minimize the pumping power or pressure drop, while maintaining a fixed thermal load for a given system. In the following we present the procedure for minimization of pumping power (minimization of pressure drop follows similar lines). We seek

$$\min_{Re,\overline{D}_H} \overline{\Psi} \quad \text{for given} \quad (q'', \delta T, L, k, \rho, c_p, \nu) \quad . \tag{17}$$

An equivalent procedure would be to find maximum thermal load (Λ) for a fixed pumping power ($\overline{\Psi}$). As can be seen from Figures 2a and 2b, there is only one maximum thermal load for each constant value of pumping power. For each optimum operating condition ($\overline{\Psi}_{min}$ for fixed Λ, or Λ_{max} for fixed $\overline{\Psi}$) there is a set of corresponding values of Reynolds number (Re) and hydraulic diameter ($\overline{D}_H$).

We proceed by assuming power-law relationships for friction factor and Nusselt number

$$f = \frac{a}{Re^n} \qquad a = \text{constant} \tag{18}$$

$$Nu = b\,Re^m \qquad b = b(Pr) \quad . \tag{19}$$

Equation (19) together with equation (10) yields

$$\overline{D}_H = \frac{bRe^m}{\Lambda}\,(1 - \frac{4\Lambda}{PrRe}) \quad . \tag{20}$$

Substituting friction factor (f) and hydraulic diameter ($\overline{D}_H$) from above in equation (11), we obtain for the dimensionless pumping power

$$\overline{\Psi} = \frac{a\Lambda^3}{2b^3} \frac{Re^{3-n-3m}}{(1 - \frac{4\Lambda}{PrRe})^3} \quad . \tag{21}$$

Since we are looking for $\overline{\Psi}_{min}$ at constant Λ, we continue by differentiating $\overline{\Psi}$ in equation (21) with respect to the control variable, Reynolds number, and equate the result to zero. We obtain

$$Re^{OPT} = \frac{4\Lambda}{Pr} \frac{6-n-3m}{3-n-3m} \quad , \tag{22}$$

and after substituting the above into equation (21)

$$\overline{\Psi}^{OPT} = \frac{a\Lambda^3}{2b^3} \left(\frac{4\Lambda}{Pr} \frac{6-n-3m}{3-n-3m}\right)^{3-n-3m} \left(1 - \frac{3-n-3m}{6-n-3m}\right)^{-3} \quad . \tag{23}$$

The required value for $\overline{D}_H$ follows now from equations (20) and (22)

$$\overline{D}_H^{OPT} = \frac{b}{\Lambda} \left(\frac{4\Lambda}{Pr} \frac{6-n-3m}{3-n-3m}\right)^m \left(1 - \frac{3-n-3m}{6-n-3m}\right) \quad . \tag{24}$$

It can be seen from equations (22-24) that the optimal operating conditions depend on the Prandtl number, thermal load Λ, and the thermal ($Nu(Re, Pr)$) and hydraulic ($f(Re)$) behavior of the system. Thus, for a given thermal load (Λ), coolant-fluid properties (Pr), and known thermal-hydraulic characteristics of the system ($Nu(Re, Pr)$ and $f(Re)$), we can evaluate the required system parameters (Reynolds number, hydraulic diameter, and minimum pumping power) directly from equations (22)-(24). Optimization with a constant wall temperature condition (rather than the constant heat flux) is presented in the Appendix.

4 Results

We present here results of the optimization procedure developed in Section 3. Results are presented for high-aspect-ratio smooth-channel flows with $a = 24, n = 1, b = 8.235$, and $m = 0$ for laminar flows (cf. Equations (18)-(19)) , and $a = 0.046, n = 0.2, b = 0.0174$ and $m = 0.8$ for turbulent flows [6-7]. Air is the coolant fluid ($Pr = 0.71$). The value of b is obtained using the Petukhov correlation for air and Reynolds numbers of interest.

In Figure 3 we plot optimum pumping power ($\overline{\Psi}^{OPT}$) versus thermal load (Λ) for laminar and turbulent smooth-channel flows using equation (23). There are two important regions in Figure 3. For very low Λ, laminar flow performs better than the turbulent flows. Optimum Reynolds number, pumping power and hydraulic diameter for laminar channel flows can be expressed as (using equations (22), (23), and (24))

$$Re^{OPT} = 14.09 \ \Lambda \tag{25}$$

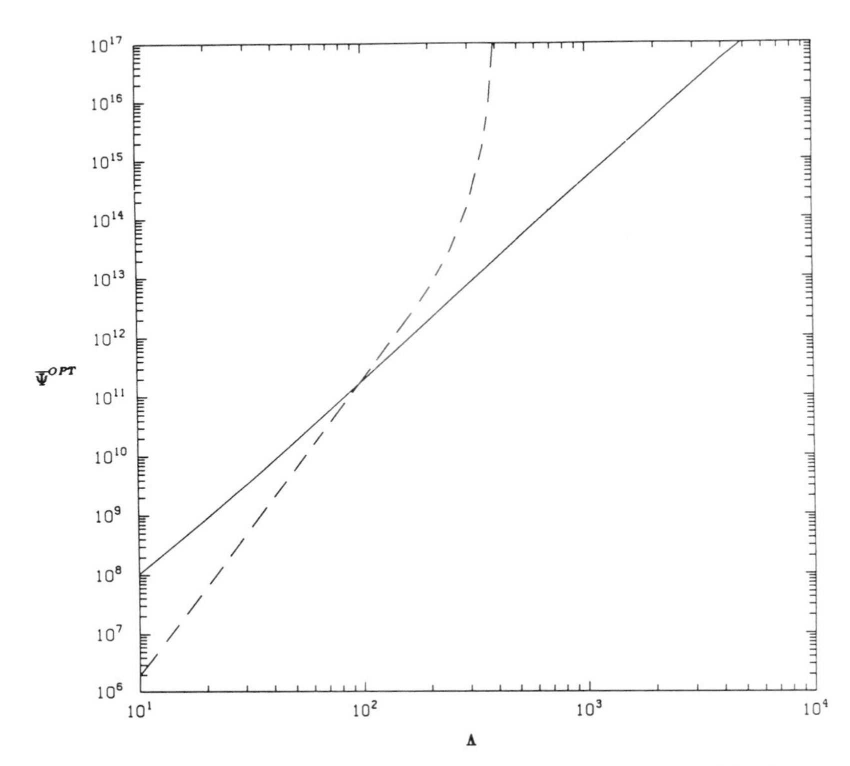

FIGURE 3. A plot of optimum pumping power versus thermal load for laminar and turbulent smooth-channel flows.— — laminar flow. —— turbulent flow. Air is the working fluid ($Pr = 0.71$)

$$\overline{\Psi}^{OPT} = 19.74 \;\; \Lambda^5 \tag{26}$$

$$\overline{D}_H^{OPT} = \frac{4.94}{\Lambda} \quad , \tag{27}$$

and they are valid for a limited range of Λ, constrained with $Re \leq Re_{tr}$ (We take here $Re_{tr} = 2300$). This implies (cf. equation (22)) that we could minimize $\overline{\Psi}(Re)$ only for $\frac{\Lambda}{Re} \leq \;\; 0.071$. However, if this is not the case, minimum $\overline{\Psi}$ occurs at the highest Reynolds number for laminar flow (Re_{tr}). Hence, our final result for pumping power for the laminar regime can be expressed as

$$\overline{\Psi}^{OPT} = 19.74\,\Lambda^5 \quad for \quad \Lambda \leq \; 0.071 \; Re_{tr} \tag{28}$$

$$\overline{\Psi}^{OPT} = 113670\,\Lambda^3\,(1 - \frac{\Lambda}{408.25})^{-3} \quad for \quad 0.071\;Re_{tr} < \;\Lambda < \; 0.177\;Re_{tr}. \tag{29}$$

Turbulent flows can only exist for $Re \geq Re_{tr}$. Thus, there is a cut-off for optimum power dissipation (cf. equation (22)), and it can be expressed as $\Lambda_{turbulent}^{cut-off} = 0.021\;Re_{tr}$. For Λ smaller than $\Lambda_{turbulent}^{cut-off}$ optimum pumping power should be chosen at Re_{tr}. Optimum Reynolds number, pumping power, and hydraulic diameter for urbulent channel flows ($\Lambda \geq 0.021\ Re_{tr}$) can be expressed as

$$Re^{OPT} = 47.89\ \Lambda \tag{30}$$

$$\overline{\Psi}^{OPT} = 29870 \quad \Lambda^{3.4} \tag{31}$$

$$\overline{D}_H^{OPT} = \frac{0.339}{\Lambda^{0.2}} \tag{32}$$

ls Λ increases turbulent flow becomes more efficient than laminar flow since tran-ition to the turbulence causes a relatively larger increase in Nusselt number than he corresponding increase in pumping power, resulting in reduced dissipation at he same thermal load.

5 Optimization and Heat Transfer Augmentation

Previous studies on heat transfer augmentation typically presented the per-ormance as a ratio between enhanced and base case heat transfer coefficients, to-ether with corresponding changes in friction factor, all at the same Reynolds num-er [9-10]. This information does not directly provide sufficient guidance for the election of an appropriate enhancement system for a specified task (thermal load nd geometry). Our aim here is to present the relationship between optimum heat ransfer design and enhancement which will determine the proper augmentation cheme at a given thermal load. We consider two different augmentation schemes orresponding to the use of a) periodic eddy promoters and b) micro-grooves, shown chematically in Figures 4a and 4b, respectively. These two enhancement schemes re typical examples of flow destabilization at macro and micro scales of the mo-on, respectively. Heat transfer augmentation by flow destabilization are given in 8,11]. The details of the experiments, which were carried out in air ($Pr = 0.71$), re presented in [8].

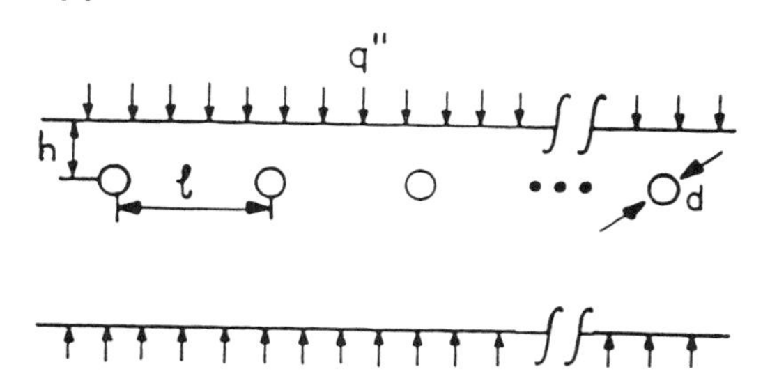

IGURE 4a. The geometry of the periodic eddy-promoter channel is described y the distance between successive eddy-promoter cylinders, l, the diameter of he eddy-promoters, d, and the distance of the eddy-promoters from the wall, h l/D_H=1.85, d/D_H=0.111, h/D_H=0.139).

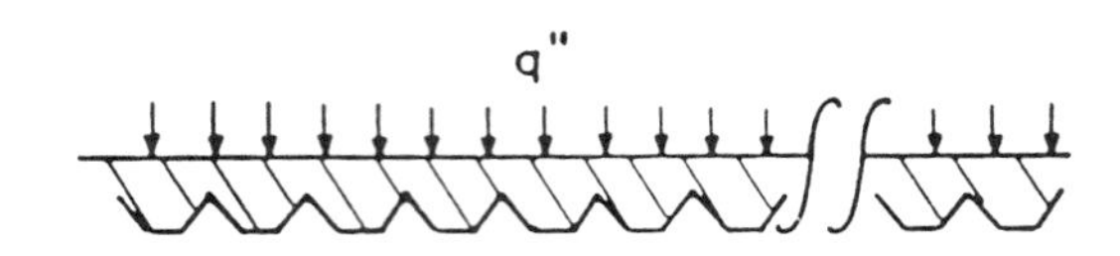

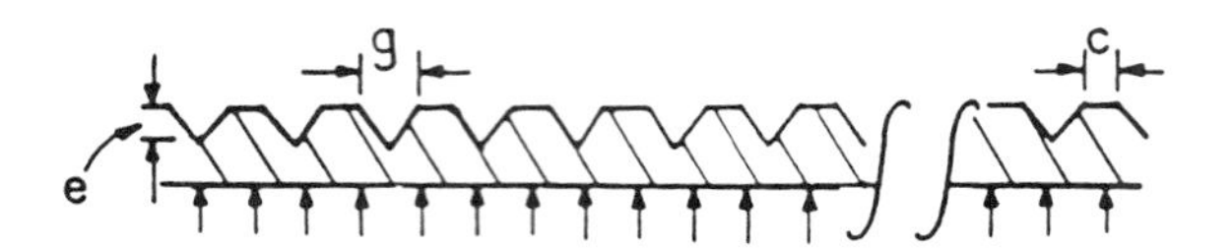

FIGURE 4b. The geometry of the periodic microgrooved channel is described by the groove dwell, c, the depth of the groove, e, and the groove pitch, g (c/D_H=0.0083, e/D_H=0.0139, g/D_H=0.019).

Thermal-hydraulic data ($Nu(Re, Pr = 0.71)$, $f(Re)$) for these augmentation schemes are given in Figures 5a and 5b. From the thermal-hydraulic data we obtained the minimum dissipation for a fixed thermal load as follows: for a given thermal load (Λ) using equations (10) and (11) we calculated $\overline{D}_H$ and $\overline{\Psi}$ for all data points for the augmentation scheme of interest. We take the Reynolds number and the hydraulic diameter resulting in the minimum pumping power ($\overline{\Psi}$) over all data points as the optimum operating conditions.

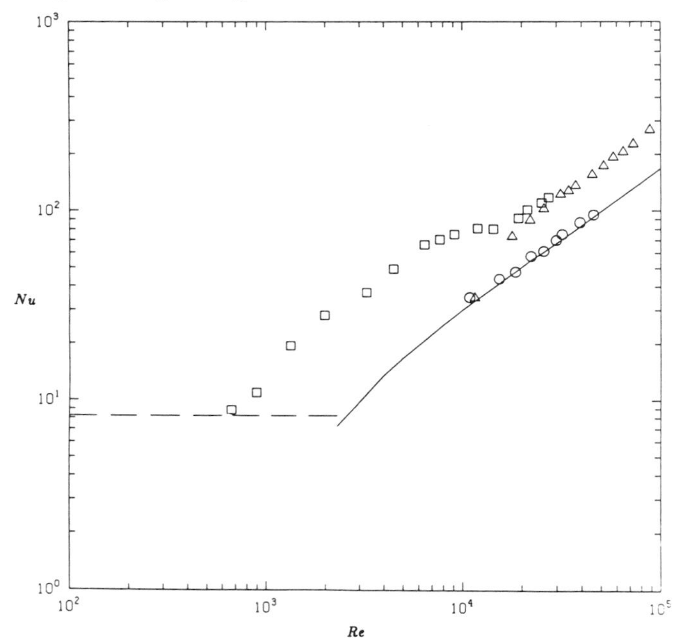

FIGURE 5a. Heat transfer data for Pr = 0.71. Smooth channel: ○ experiment; — — laminar analytical solution, and —— turbulent correlation. □ Eddy-promoters. △ Microgrooves.

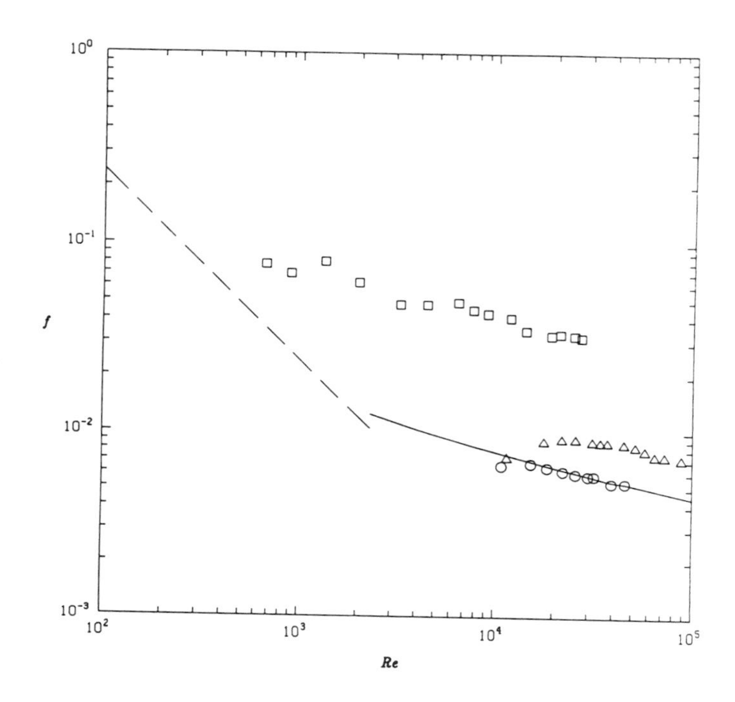

FIGURE 5b. Friction coefficient data. Smooth channel: ○ experiment; — —laminar analytical solution, and —— turbulent correlation. □ Eddy-promoters. △ Microgrooves.

We present the results as minimum dissipation versus Λ for the augmentation schemes of interest and smooth-channel flows in Figure 6. There are three important regions in Figure 6. First for very low Λ, laminar flow performs the best, indicating that in this range enhancement has little effect. Second, as Λ increases, the macro-scale eddy promoters become relatively more efficient than smooth channels; this is due the fact that the eddy promoters destabilize the flow at naturally stable spatial scales of motion which causes an increase in Nusselt number, thereby a decrease in dissipation. As Λ increases further, use of eddy promoters is no longer efficient for enhancement since for these turbulent Reynolds numbers, the flow is already naturally unstable at the spatial scales of eddy promoters. At this point the micro-grooves become important as they match the stable part of the flow, the viscous sublayer. Thus, Figure 6 illustrates the practical value of our methodology in selecting an appropriate augmentation scheme for a cooling system with specified heat removal requirements.

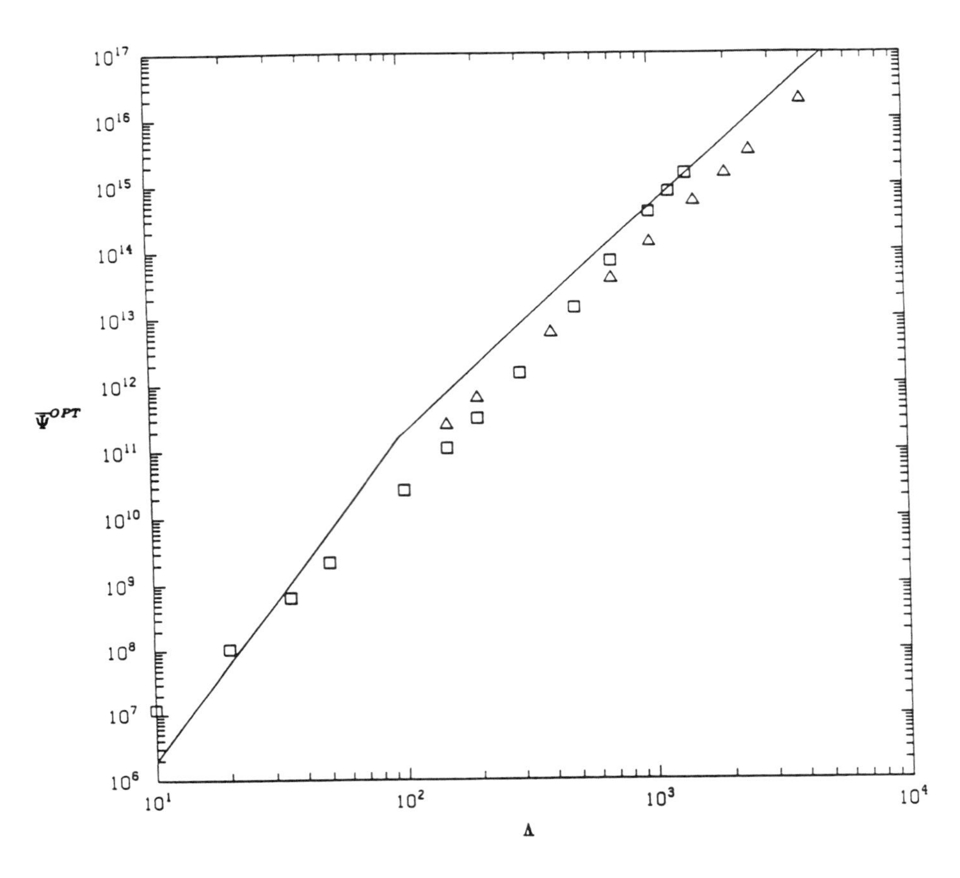

FIGURE 6. Optimum dissipation versus thermal load for smooth channel and augmentation schemes. —— Smooth-channel analytical solution and correlations. □ Eddy-promoters. △ Microgrooves. Air is the working fluid ($Pr = 0.71$).

6 Conclusions

A methodology for optimization of convective cooling systems is presented. The optimal heat transfer design is accomplished by addressing a general theory for thermal-hydraulic design which also relates transport enhancement and power dissipation. The study enables us to select flow and geometric parameters (Reynolds number and hydraulic diameter) that yield minimum pumping power (or pressure drop) for a chosen cooling system with specified thermal requirements. The significant dissipation savings possible through optimization, with or without augmentation, are demonstrated in a sample study for heat transfer in a channel with air as the working fluid ($Pr = 0.71$). The study also allows for *a priori* evaluation of various enhancement schemes for a chosen convective cooling system.

NOMENCLATURE

a	constant in equation (18)
A	area of the channel cross section
b	constant in equation (19)
c	groove dwell (Fig. 4b)
c_p	specific heat at constant pressure
d	diameter of the eddy promoters (Fig. 4a)
D_H	hydraulic diameter
$\overline{D}_H$	non-dimensional hydraulic diameter (cf. equation (8))
$\overline{D}_H^{OPT}$	optimum non-dimensional hydraulic diameter
e	groove depth (Fig. 4b)
$f_{average}$	average Fanning friction factor defined by equation (3)
g	groove pitch (Fig. 4b)
h	distance of the eddy promoters from the wall (Fig. 4a)
h	heat transfer coefficient
k	thermal conductivity
l	distance between successive eddy promoters (Fig. 4a)
L	channel length
m	constant in equation (19)
n	constant in equation (18)
Nu	Nusselt number defined by equation (7)
P	heated perimeter
Pr	Prandtl number
q''	heat flux per unit area
$q_{wall}^{average}$	average heat flux (cf. Appendix, equation (3))
Q	total heat transfer rate
Re	Reynolds number
Re_{tr}	transitional Reynolds number
Re^{OPT}	optimum Reynolds number
T_m	fluid mean temperature
T_w	wall temperature
V	channel average velocity
x	streamwise coordinate
x_e	entrance length for thermally fully developed flow

Greek

α	thermal diffusivity
ΔP	pressure drop
$\overline{\Delta P}$	non-dimensional pressure drop (cf. equation (12))
δT	total temperature difference defined by equation (2)
$\overline{\Delta T}$	time-averaged temperature difference between wall temperature and fluid mean temperature

ΔT_{out} temperature difference between the wall and exit-fluid-mean temperature (cf. Appendix, equation (2))

Λ thermal load (cf. equation (9))

ν kinematic viscosity

Ψ dissipation

$\overline{\Psi}$ non-dimensional dissipation (cf. equation (11))

$\overline{\Psi}^{OPT}$ optimum non-dimensional dissipation

ρ density

References

[1] W. Nakayama, Thermal Management of Electronic Equipment: a Review of Technology and Research Topics, *Applied Mechanics Reviews*, **39**, No.12, 1847-1868 (1986).

[2] R. W. Keyes, Physical Limits in Digital Electronics, *Proc. IEEE*, **63**, 740-767 (1975).

[3] D. B. Tuckerman and R. F. W. Pease, High Performance Heat Sinking for VLSI, *IEEE Electron Device Letters*, EDL-2, No.5, 126-129 (1981).

[4] D. B. Tuckerman, Heat Transfer Microstructures for Integrated Circuits, Ph.D. Thesis, Stanford University, Electrical Engineering Department (1984).

[5] N. Goldberg, Narrow Channel Forced Air Heat Sink, *IEEE Transactions on Components, Hybrids, and Manufacturing Technology*, **CHMT-7**, No., 154-159 (1984).

[6] W. M. Kays and M. E. Crawford, Convective Heat and Mass Transfer, McGraw-Hill, New York (1980).

[7] W. M. Kays and H. C. Perkins, Forced Convection, Internal Flow in Ducts, in Handbook of Heat Transfer Fundamentals, Eds. W. M. Rohsenow, J. P. Hartnett, and E. N. Ganić, Chap. 6, McGraw-Hill, New York (1985).

[8] H. Kozlu, Ph. D. Thesis, MIT, Department of Mechanical Engineering (1989).

[9] A. E. Bergles, Techniques to Augment Heat Transfer, in Handbook of Heat Transfer Applications, Eds. W. M. Rohsenow, J. P. Hartnett, and E. N. Ganić, Chap. 3, McGraw-Hill, New York (1986).

[10] R. L. Webb, and A. E. Bergles, Performance Evaluation Criteria for Selection of Heat Transfer Surface Geometries Used in Low Reynolds Number Heat Exchangers, in Low Reynolds Number Flow Heat Exchangers, Eds. S. Kakaç, R. K. Shah, and A. E. Bergles, Hemisphere Pub. Corp., Washington, D.C., 735-752 (1983).

[11] G. E. Karniadakis, B. B. Mikić and A. T. Patera, Minimum Dissipation Transport Enhancement by Flow Destabilization: Reynolds' Analogy Revisited, *J. of Fluid Mechanics*, **192**, 365-391 (1988).

APPENDIX

We consider here the problem stated in Section 2.1 with constant wall temperature (T_w) instead of uniform heat flux. The energy balance gives

$$V A \, \rho c_p \, (T_{m_{out}} - T_{m_{in}}) = Q \quad , \tag{1}$$

where Q is the total energy that has to be removed. Since $T_{m_{out}} - T_{m_{in}} = (T_w - T_{m_{in}}) - (T_w - T_{m_{out}})$ as shown in Figure 7 we can write equation (1) as

$$\frac{Q}{V A \, \rho c_p} = \delta T - \Delta T_{out} \quad , \tag{2}$$

where $\delta T = T_w - T_{m_{in}}$, and $\Delta T_{out} = T_w - T_{m_{out}}$. In terms of Prandtl number, and Reynolds number, defined as before, and thermal load parameter, redefined as

$$\Lambda = \frac{(Q/LP)\, L}{k\, \delta T} = \frac{q_{wall}^{average}\, L}{k\, \delta T} \quad , \tag{3}$$

where $q_{wall}^{average}$ is the average heat flux around perimeter, the global energy balance equation becomes

$$\Lambda = \frac{Re\, Pr}{4}(1 - \frac{\Delta T_{out}}{\delta T}) \quad . \tag{4}$$

Applying energy balance for an infinitesimal control volume and integrating over the channel length (assuming a constant heat transfer coefficient h) yields

$$\frac{\Delta T_{out}}{\delta T} = exp\,(-\frac{h\, P\, L}{V\, A\, \rho\, c_p}) \quad . \tag{5}$$

In terms of the non-dimensional hydraulic diameter and Nusselt number we can write equation (5) as

$$\frac{\Delta T_{out}}{\delta T} = exp\,(-\frac{4\, Nu}{Re\, Pr\, \overline{D}_H}) \quad . \tag{6}$$

Substituting the above into equation (4) we obtain

$$\Lambda = \frac{Re\, Pr}{4}\,(1 - exp(-\frac{4\, Nu}{Re\, Pr\, \overline{D}_H})) \quad , \tag{7}$$

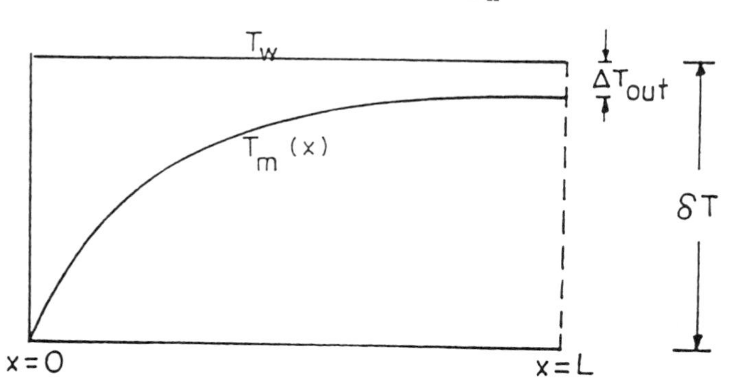

FIGURE 7. Thermal behavior of flows in channels with uniform wall temperature.

which is the equivalent of equation (10) that was obtained for the constant heat flux case. In Figures 8a and 8b we plot dimensionless hydraulic diameter versus thermal load for fixed Reynolds numbers and pumping powers for high-aspect ratio and circular channels, respectively. For laminar flow in high-aspect ratio channels we use $Nu = 7.54$ whereas for circular channels we use $Nu = 3.66$. For friction factor and turbulent flow correlations we use the correlations given in equations (13),(15) and (16). Assuming power-law correlations for friction factor and Nusselt number and using equation (7) we obtain for hydraulic diameter

$$\overline{D}_H = - \frac{4\, b\, Re^{m-1}}{Pr\, ln(1 - \frac{4\,\Lambda}{Re\, Pr})} \quad . \tag{8}$$

Substituting friction factor (f) and hydraulic diameter ($\overline{D}_H$) from above in equation (11) we obtain for the dimensionless pumping power

$$\overline{\Psi} = - \frac{a\, Pr^3}{2^7\, b^3}\, Re^{6-n-3m}\, (ln\,(1 - \frac{4\,\Lambda}{Re\, Pr}))^3 \quad . \tag{9}$$

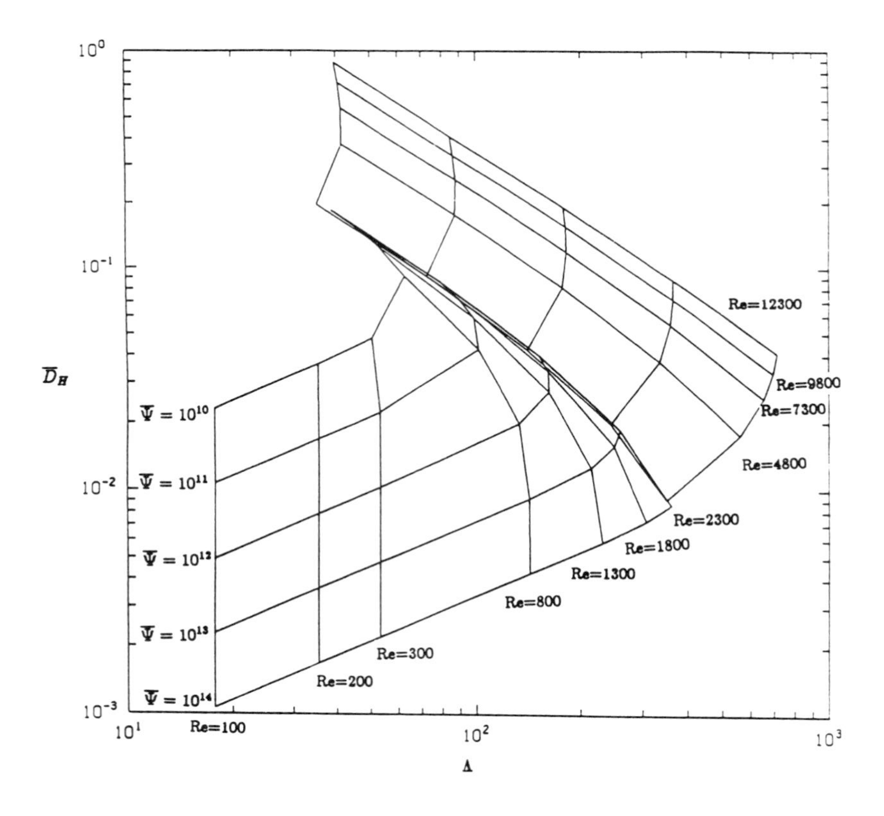

FIGURE 8a. A plot of dimensionless hydraulic diameter versus thermal load for high-aspect ratio channels with uniform wall temperature at fixed Reynolds numbers and pumping powers. Air is the working fluid ($Pr = 0.71$).

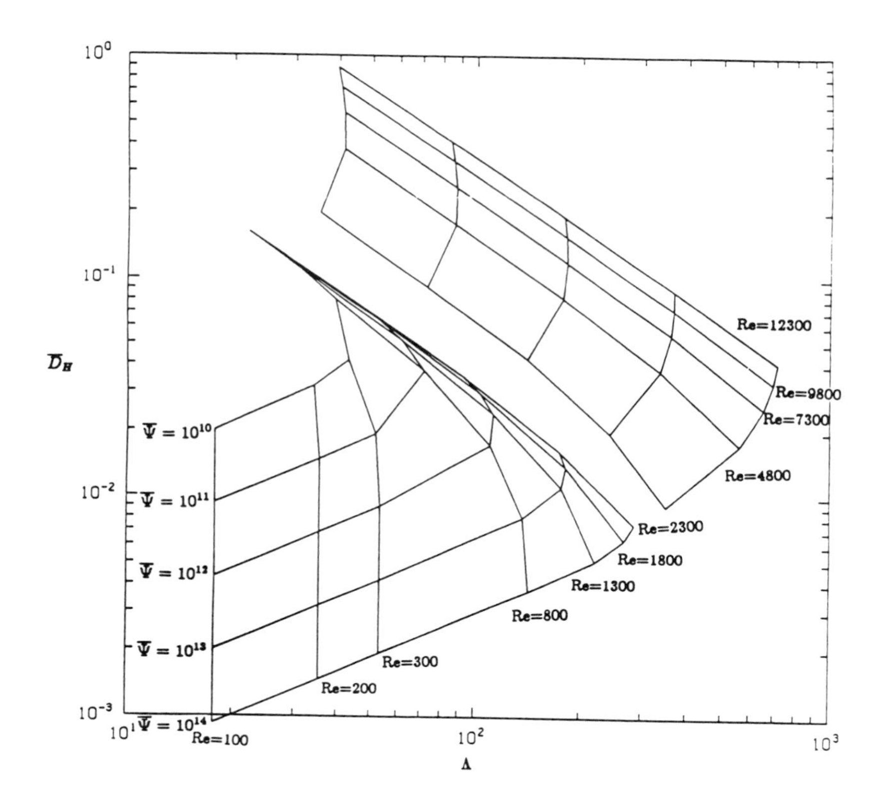

FIGURE 8b. A plot of dimensionless hydraulic diameter versus thermal load for circular channels with uniform wall temperature at fixed Reynolds numbers and pumping powers. Air is the working fluid ($Pr = 0.71$).

To proceed we differentiate $\overline{\Psi}$ in equation (9) with respect to control variable, Reynolds number, and equate the result to zero. We obtain

$$\left(\frac{Re\,Pr}{4\,\Lambda} - 1\right) ln\left(1 - \frac{4\,\Lambda}{Re\,Pr}\right) = \frac{3}{3m + n - 6} \quad . \tag{10}$$

To find the optimum Reynolds number for the given thermal load, Prandtl number, and thermal ($Nu(Re, Pr)$) and hydraulic ($f(Re)$) data, equation (10) has to be solved iteratively. Then, substituting this optimum Reynolds number in equation (9) optimum pumping power and using equation (8) the required hydraulic diameter can be obtained.

Laminar smooth-channel flow

We consider high-aspect ratio smooth-channel laminar flow with $a = 24, n = 1, b = 7.54$, and $m = 0$ [6]. Air is chosen as the coolant ($Pr = 0.71$) and the flow is assumed to be hydrodynamically and thermally fully developed. Equation (10) gives for the optimum Reynolds number

$$Re^{OPT} = 9.19\,\Lambda \quad . \tag{11}$$

To obtain pumping optimum pumping power we substitute the above into Equation (9)

$$\overline{\Psi}^{OPT} = 8.78\,\Lambda^5 \quad . \tag{12}$$

Required hydraulic diameter can be found using equation (8)

$$\overline{D}_H^{OPT} = \frac{4.87}{\Lambda} \quad . \tag{13}$$

Turbulent smooth-channel flow

We use $a = 0.046, n = 0.2, b = 0.0174$, and $m = 0.8$ [6-7] for thermal-hydraulic data and air as the coolant ($Pr = 0.71$). Optimum operating conditions are obtained as (using equations (10),(9), and (8))

$$Re^{OPT} = 25.89\,\Lambda \tag{14}$$

$$\overline{\Psi}^{OPT} = 23015\,\Lambda^{3.4} \tag{15}$$

$$\overline{D}_H^{OPT} = \frac{0.208}{\Lambda^{0.2}} \quad . \tag{16}$$

For turbulent and laminar flows there are cut-off points for the optimum dissipation as discussed in the case of constant heat flux. Thus, above results hold for $\Lambda \leq 0.1087\ Re_{tr}$ for laminar flows, and $\Lambda \geq 0.0386\ Re_{tr}$ for turbulent flows.

Cooling Techniques for Electronic and Microelectronic Equipment— Chinese Research

C. F. MA
Beijing Polytechnic University
Beijing 100022, PRC

D. S. ZHAO
Xidian University
Xian, PRC

ABSTRACT

The rapid Development of the Chinese Computer industry has established the basic demand for thermal control of associated electronic components. The Chinese research activity in this area has been active since early 1980's. A great deal of effort has been devoted to the study of natural and forced convection air cooling both in fundamental and applied aspects. The Chinese supercomputer YH-1 has been successfully operated with a forced air convection cooling system. Considerable attention has also been given to the study of liquid cooling. Some important achievements, including immersion cooling, liquid jet impingement cooling, spray cooling and heat pipe cooling, have been reported. This paper provides a review and summary of Chinese heat transfer research in this area.

1. Introduction

The reliability of electronic equipment is strongly affected by the working temperature, power dissipation, and environmental conditions of the electronic components. It is known that a near exponential dependence of failure rate on working temperature exists for most electronic components [1]. The major problem in thermal design of electronic equipment is to control the internal temperature of the components as well as the internal and external temperatures of the enclosure. The rapid development of the Chinese computer industry has established the basic demand for thermal control of associated electronic components. Supported by the National Natural Science Foundation of China, Chinese Government, and some computer firms, many Chinese thermal engineers and heat transfer scientists have been involved in research and development of cooling techniques for electronic and microelectronic equipment. Their technical papers are mainly presented in conferences sponsored by the Chinese Society of Engineering Thermophysics and the Chinese Society of Computer Engineering. Some papers in English are also presented in international symposia.

As direct air cooling is the most widely used cooling technology in China, Chinese research in this area has been focused on air cooling technology. A great deal of effort has been devoted to the study of natural convection cooling that may be effectively employed in electronic devices with low levels of heat dissipation. Recently, with the development of large scale computers in China, considerable attention has also been given to the study of forced convection cooling, especially investigations involving enhancement techniques. Since the early 1980's, increased attention has also been directed to the study of direct liquid cooling. Some novel cooling methods have been tested and reported. Significant achievement has been made not only in academic research but also in engineering practice. One important example is the Chinese supercomputer YH-1 (Fig. 1). Several sets of YH-1 computers have been successfully operated in China for five years. The success of the supercomputer may be partly attributed to the design of the forced convection air cooling system.

FIGURE 1. Chinese Super Computer YH-1.

The objective of the present report is to provide a review and summary of Chinese heat transfer research associated with electronic and microelectronic cooling. Covered are natural and forced convection air cooling, immersion and impingement liquid cooling, cooling enhancement, and heat pipe applications. Because of space limitations, we can not mention all the Chinese research work in this field. Only part of the open literature will be involved in the present review.

2. Natural Convection Air Cooling and Radiation

Natural convection air cooling is still the most important cooling method for microelectronic devices in China. Considerable effort has been devoted to the study of this cooling technique both in academic and practical aspects. Guo et al. [1] studied combined natural convection and radiation heat transfer between isothermal plates both analytically and experimentally. Their experimental and numerical results showed that radiation may make an important contribution and can not be ignored in some cases with small Rayleigh number. In order to estimate the relative importance of radiation, they presented a general criterion Crn:

$$Crn=\begin{cases} \dfrac{45b\sigma\bar{T}^3}{k(\frac{1}{b}-\frac{3}{\varepsilon}-3)}\cdot\dfrac{1}{Ra_l} & Ra_l<1 \\ \dfrac{45b_d^2\sigma\bar{T}^3}{k\,f_a}Ra_l^{\,lg(0.5333fa/fb)} & 1\leqslant Ra_l\leqslant 10 \\ \dfrac{24b\sigma\bar{T}^3}{k(\frac{1}{b}-\frac{3}{\varepsilon}-3)}\cdot Ra_l^{-0.25} & Ra_l>10 \end{cases} \tag{1}$$

where $\bar{T}=(T_l+T_\infty)/2$

$b_d=4\sqrt{(l\nu^2/g\beta\Delta T)\,Pr}$

$$fa = \frac{1}{b_d} + \frac{3}{\varepsilon} - 3$$

$$fb = \frac{1}{1.78b_d} + \frac{3}{\varepsilon} - 3$$

The definition of all symbols in above formulae can be found in Ref. [1]. The relation between the ratio of radiation and natural convection heat transfer Q_r/Q_n and Crn is illustrated in Fig. 2. It can be seen that Q_r/Q_n is approximately a linear function of Crn with a constant of proportionality not far from unity.

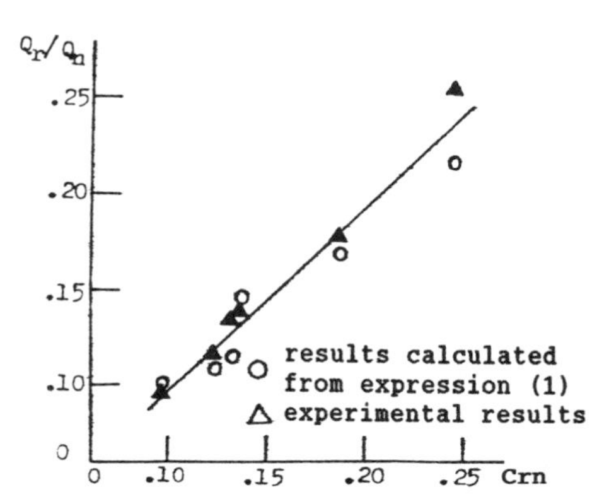

FIGURE 2. The Ratio of Radiation Heat Transfer to Natural Convection Heat Transfer.

In practice, the cooling of microelectronic equipment may be considerably improved by increasing the radiation heat transfer. Zhao [2] studied the effect of the surface emissivity of the cabinet on the cooling of microelectronic components. The experimental results showed that the surface temperature of the components can be decreased as much as 20°C by increasing the emissivities of both the internal and external cabinet walls.

Guo and his colleagues studied the side effect of natural convection channel flow between two isothermal vertical plates [3]. Their experiments showed that the side effect depends strongly on channel geometry. A reduction in the ratio of channel length to height will always strengthen the side effect, while the influence of the channel width on the side effect is non-monotonic. It was also found that the side effect is not very sensitive to the variation of the temperature difference between the wall and the ambient.

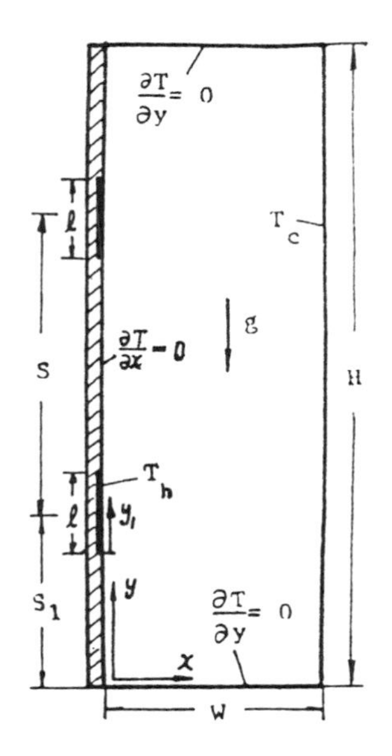

FIGURE 3. Natural Convection of Rectangular Enclosures with Discrete Heating Elements.

Natural convection in a rectangular enclosure with finite-size heated elements on a vertical adiabatic wall is of interest in cooling problems related to electronic equipment. Wang et al. [4] studied the effects of geometric parameters on the heat transfer characteristics of the discrete heating elements as shown in Fig. 3. Their numerical study showed that the Nusselt number of the lower element is always much higher than that of the upper one. They found that Nu of both the upper and lower elements increases with increasing H/W, with Nu of the upper element showing a much stronger dependence on this parameter. It is also noted that the maximum Nu of the upper element is obtained at about S/H=0.4, where the heat transfer rate can be enhanced by 8-10% compared with the case of S/H=0.125.

Along with the development of microelectronic technology, the density of components on PCB's becomes higher and higher. In order to improve the component cooling and reduce the equipment size, it is necessary to control the spacing of PCB's. The study of optimal spacings of PCB's is of great significance. Zhao [5] studied this problem experimentally. His study showed that the optimal spacing of natural air-cooled PCB's (vertically installed) is about 19-21mm. Ye [6] studied in detail the effect of cabinet structure on component cooling. The vent area and location were the main structural factors in the experimental study. With combinations of these structural factors, 150 structure types were prepared and tested. It was found that the average temperature of the components is inversely proportional to the total area of the vents located on the top or side faces. However, the component temperature is only slightly influenced by the area of the ventilation holes when the area is more than 20-30% of the total area of the internal surface of the cabinet. It was also found that the cooling may be improved by increasing the distance between the input and output vents. An empirical formula was presented to predict the area of input vents, which are slightly smaller than output vents.

3. Forced Convection Air Cooling

Lo and Ren [7] reported research on forced convection air cooling of high-output silicon controlled rectifiers. Based on their experimental study, a physical model with its corresponding mathematical representation was proposed. The design procedure was given in detail to predict the enhanced heat transfer of the air cooler. Several new types of air coolers used for different purposes have been designed with this method and applied in industry.

Forced convection heat transfer from a rectangular block array was studied by Chen and Wang [8]. The convective heat transfer coefficients of the blocks were obtained by applying the analogy between heat and mass transfer. The mass transfer coefficient was measured via the naphthalene sublimation technique. The effect of channel height H on the heat transfer was studied experimentally and the following non-dimensional expression was obtained:

$$Nu=0.0118\ Re^{0.877}\ (H/L)^{0.26} \qquad (2)$$

Where both Nu and Re are based on the length of the block L. The deviation between the calculated values and the experimental data is within 2.3-4.0% as shown in Fig. 4. A non-standard array with one or two rows missing was also studied. It was found that significant heat transfer enhancement may be obtained in the downstream rows with a negligible increase of pressure drop.

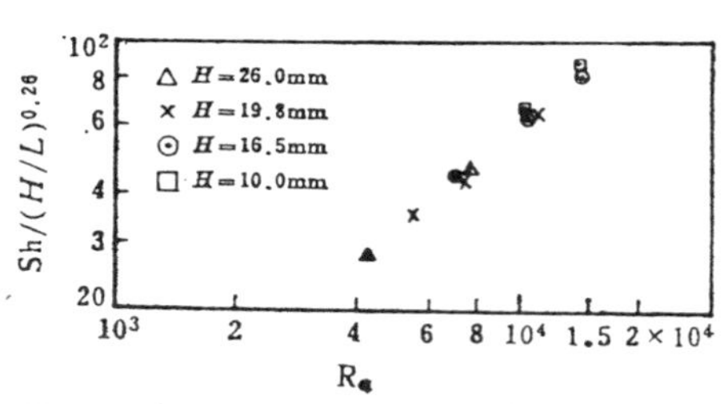

FIGURE 4. Mass Transfer from Array of Rectangular Naphthalene Blocks in Air Flow - Effect of the Channel Height.

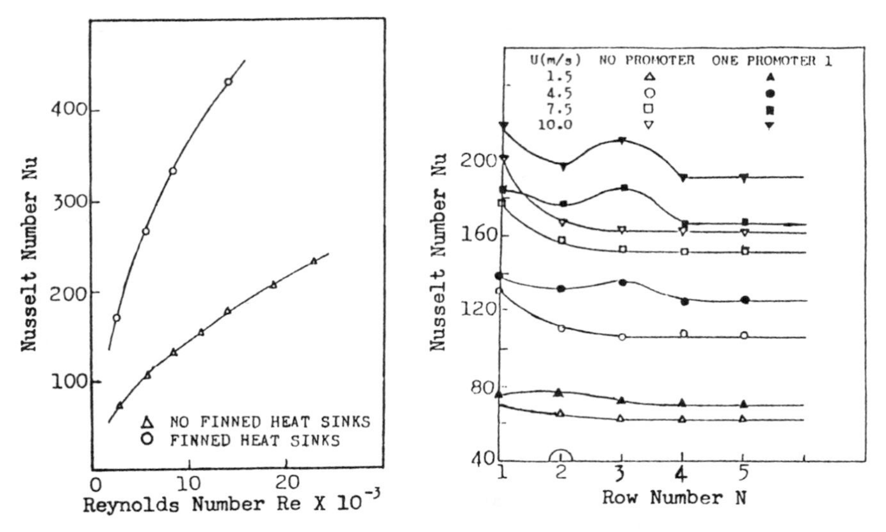

(a) With Finned Heat Sinks (b) With Promoter of Half Cylinder Type

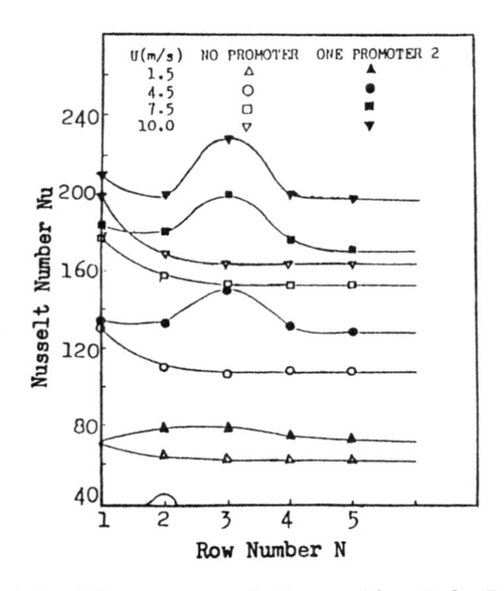

(c) With Promoter of Stream-line Body Type

FIGURE 5. Enhancement of Forced Convective Heat Transfer from Array of Rectangular Electronic Components in Air Flow.

components in in-line arrays was studied in Ref.[9]. Two types of enhancement devices, displaced enhancement promoters of two configurations and a finned heat sink, have been tested for improving the air cooling of eletronic components. It was found that heat transfer may be significantly enhanced as much as 153% by using the finned heat sink shown in Fig. 5(a). With the other two promoters, moderate improvements in heat transfer performance were obtained as illustrated in Fig. 5(b,c).

The thermal characteristics of PCB assemblies was studied by Wei[10]. The effects of structural factors on the thermal performance were studied both experimentally and numerically. A general code for the calculation of fluid flow and heat transfer was presented and a group of subroutines for plotting curves was developed. The numerical results are in good agreement with the experimental data.

As mentioned before, the Chinese supercomputer YH-1 is successfully cooled by forced air convection. The details of the design procedure and the heat transfer characteristics of the cooling system are presented in Ref. [11]. In order to improve the design of the exhaust air system for large computers, a properly designed exit plenum should be provided. A mathematical model of ductwork that allows for static pressure recovery is proposed in Ref. [12].

4. Liquid Cooling

Water cooling has been successfully used in a vehicle-carried radar [13]. Many techniques related to the design and operation, including sealing, filtering, water-gas separation, nitrogen filling, ion exchange, and noise control, are discussed in detail in this Reference.

Working with Prof. Bergles, one of the present authors finished an investigation on boiling jet impingement cooling of microchips [14, 15]. Heat flux as high as $10^6 w/m^2$ was recorded from a simulated microelectronic chip to a submerged subcooled R-113 jet [14]. The characteristics of nucleate jet impingement boiling on simulated microchips were

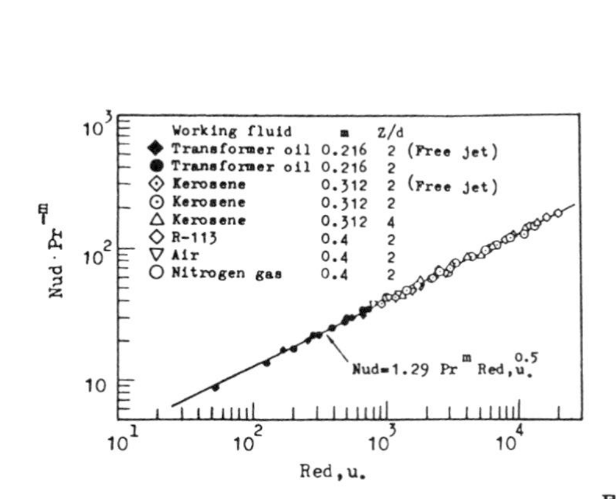

FIGURE 6. Impingement Heat Transfer at Stagnation Point of Circular Jet.

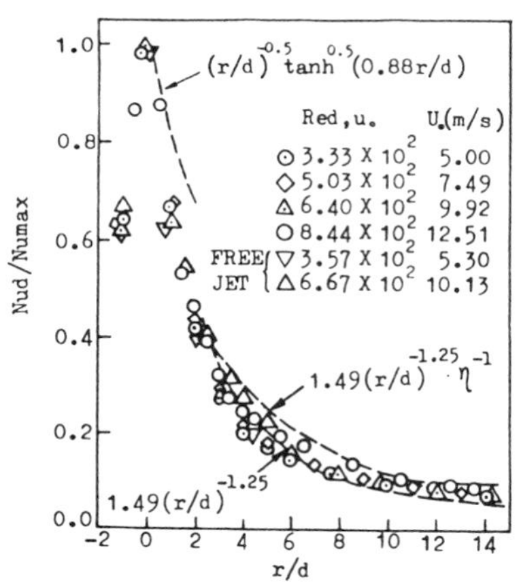

FIGURE 7. Radial Distribution of Impingement Heat Transfer Coefficients with The Plate within The Potential Core.

investigated in Ref. [15]. Included are the effects of velocity, subcooling, flow direction and surface condition on fully developed boiling and incipient boiling. More recently, single phase jet impingement cooling with transformer oil is proposed and tested [16]. Fig. 6 shows the heat transfer at the stagnation point for a transformer oil jet together with the data of R-113, kerosene, air and nitrogen gas [17, 18]. It is noted that all the data may be well correlated with the correlation in the figure. The exponents on Pr number decrease with increasing of Pr number itself as shown in the figure. The radial distribution of heat transfer coefficients with the plate within the potential core is presented in Fig. 7. All the transformer oil data are in good agreement with the predicted curve of semi-empirical formulas which are given in the same figure.

Immersion cooling of microelectronic devices may be considerably enhanced by foreign gas jet impingement [19, 20]. Fig. 8 shows the heat transfer from a simulated microchip submerged in an R-113 bath with air jet impingement. It is found that heat flux from the heater to the coolant as high as 3.3×10^4 w/m^2 was recorded while the wall temperature was lower than the temperature of the coolant. Based on a photographic study [21], a physical model of simultaneous heat and mass transfer was presented and a semi-empirical method was developed, with which all non-boiling data may be well correlated as shown in Fig. 8. This enhancement method is also found to be very effective to reduce the temperature over-shoot (Thermal hysteresis) as shown in Fig.9.

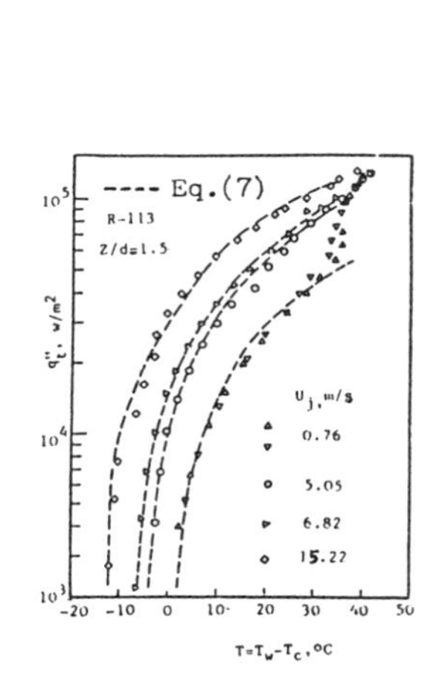

FIGURE 8. Impingement for R-113 at Z/d=1.5.Enhanced Heat Transfer with Air Jet.

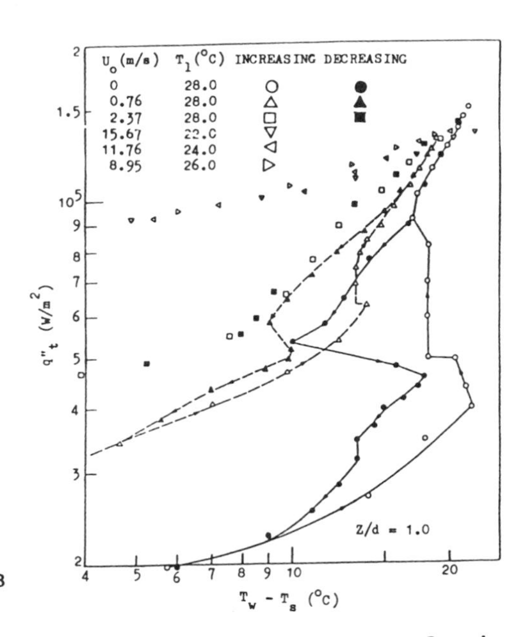

FIGURE 9. Pool Boiling Heat Transfer in R-113 Bath with Foreign Gas Impingement.

Thermal hysteresis is a critical problem for boiling immersion cooling of microelectronic components submerged in fluorocarbon liquid.

To develop anti-hysteresis enhanced boiling surfaces is a important target for heat transfer engineers working in this field. Recently it was found in the Heat Transfer Laboratory of Beijing Polytechnic University that temperature overshoot may be totally eliminated with wire-wrapped plain tube if the geometric parameters can be carefully controlled [22]. Fig. 10 shows the boiling curves of this new enhanced boiling surface in R-113. It is seen in the figure that nucleate boiling takes place at very low superheat less than 0.8°C. To the best of our knowledge, no other anti-hysteresis surface has been reported. More attention should be directed to the study of this topic.

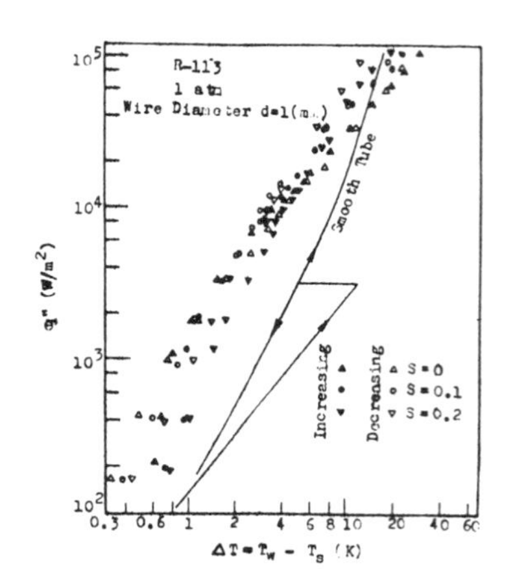

FIGURE 10. Nucleate Boiling from a Wire-Wrapped Plain Tube.

A new cooling technique for electronic devices is being tested in Beijing Polytechnic University. Spray cooling of mixture of water and nitrogen gas has been testified to be a very attractive enhancement method [23]. This cooling method is even more effective than pool boiling or jet impingement boiling as shown in Fig. 11. Heat flux as high as $3x10^6$ w/m²was recorded, with the wall temperature of the simulated microchip being only 63 C. Fig. 12 shows the heat transfer coefficient variation with the liquid

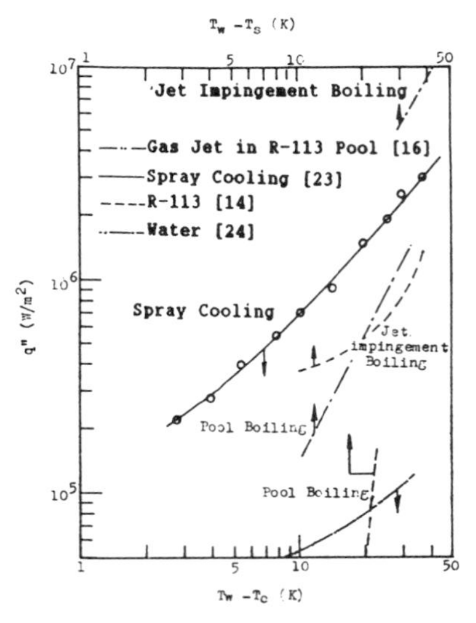

FIGURE 11. Spray Cooling of a Simulated Microchip.

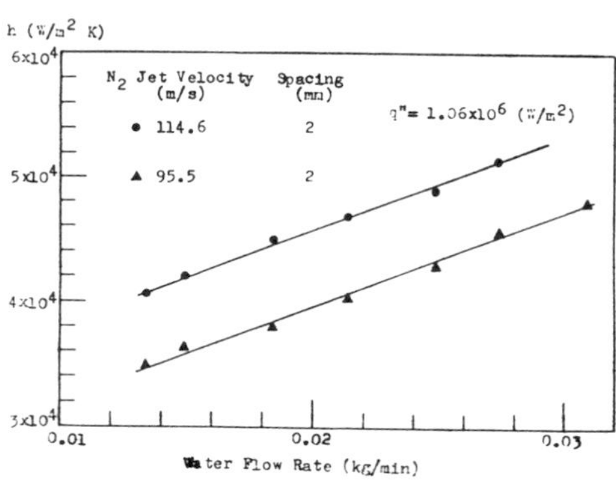

FIGURE 12. Spray Cooling of A Simulated Microchip.

flow rate. Very high heat transfer coefficients can be seen in the figure. An experimental and theoretical study is still going on in Beijing Polytechnic University.

5. Heat Pipe Cooling and Other Cooling Techniques

Heat pipe techniques have been used in electronic cooling in China. One example is reported by Xin and his colleagues [25]. Their report describes a water heat pipe with an optimal wick structure with low radial thermal resistance and high transport capability, which has been successfully applied to an air-cooled heat-pipe radiator for a 1000 ampere thyristor.

Gu and Wu [26] reported a experimental method for measuring packaging material characteristic parameters and the package thermal resistance (including contact resistance). A physical model based on one-dimensional analysis and quasi-steady conduction process was developed. A continuous laser beam with uniform intensity was used as a heat source. The thermal diffusivity of beryllia ceramic and the package thermal resistance have been measured. It is expected that their method may directly be used to check the efficiency of the process of package fabrication.

6. Concluding Remarks

Before we conclude our keynote lecture, we should emphasize that only the research work finished by the investigators in the mainland is reported in this lecture. We recognize that important achievements have also been made in Taiwan in this field. Without including the contributions of our colleagues in Taiwan, this lecture can not be considered as a complete review paper for Chinese research.

ACKNOWLEDGEMENT

This study was partially supported by the National Natural Science Foundation of China. The assistance of Mrs. S.F. Lee of the Foundation is appreciated.

REFERENCES

1 Guo, Z.Y. et al. ," Analytical and Experimental Study on Natural Convection Between Isothermal Vertical Plates", Proceedings of the International Symposium on Cooling Technology for Electronic Equipment, March 1987, Honolulu.

2 Zhao, D.S., "Natural Cooling Methods for Electronic Equipment" (in Chinese), Proceedings of Second CIE Structure and Technology in Radar, Shanghai, 1980.

3 Li, Z.X., Gui, Y.W. and Guo, Z.Y., "Experimental Investigation on the Side Effect of Natural Convection between Two Vertical Plates" , ICHMT 20th International Symposium, Dubrovnik, Yugoslavia, 1988.

4 Wang, G.X., Zhang, H.L. and Tao, W.Q., " Numerical Simulation of Natural Convection in Rectangular Enclosures with Discrete Heated Elements",ICHMT 20th International Symposium, Dubrovnik, Yugoslavia, 1988.

5 Zhao, D.S., " The Optimal Spacing of PCBs ", Proceedings of 7th International Electronic Packaging Society Conference, Boston, pp.473-479, 1987.

6 Ye,J.R., "The Effect of Structural Factors on the Results of Natural Cooling for Electronic Equipment" (in Chinese), Electronic

Mechanism Engineering No. 4, 1987.

7 Lo, D.A. and Ren, Z.P., " A Design Method for the Intensification of Air Cooling of the High-output Silicon Controlled Rectifier" (in Chinese), Journal of Engineering Thermophysics, Vol. 1, No. 2, pp. 176-184, 1980.

8 Chen, Z.Q. and Wang, Y.H., "Heat Transfer Performance of Rectangular Components in Electronic Devices" (in Chinese), Journal of Xian Jiaotong University, Vol. 21, No. 5, pp. 23-31, 1987.

9 Gan, Y.P., Wang, S., Lei, D.H. and Ma, C.F., "Enhancement of Forced Convection Air Cooling of Block-like Electronic Components in in-line Arrays", ICHMT 20th International Symposium, Dubrovnik, Yugoslavia, 1988.

10 Wei, K., " Numerical and Experimental Study of the Thermal Characteristics of Forced Convection Air-Cooling of PCBs" (in Chinese), Research Report of Northwest Telecommunication Engineering Institute, Xian, China, 1988.

11 Deng,S.G., "Design of Forced Convection Cooling System of Electronic Computer" (in Chinese), Electronic Mechanism Engineering, No. 3,1987.

12 Teng, M.S., "The Static Pressure Regain Blast Pipe Equalizing Static Pressure is Applied in Blowout Wind System of Electronic Computer (in Chinese), Journal of National Defense University of Science and Technology, pp. 123-136, 1984.

13 Fan, G.C., " A Design and Study of Pure Water Cooling System for Vehicle-carrried Radar", Proceedings of 7th International Electronic Packaging Society, Boston, 1987.

14 Ma, C.F. and Bergles, A.E., " Boiling Jet Impingement Cooling of Simulated Microelectronic Chips", ASME Publication HTD-Vol. 28, New York, 1983, pp. 5-12.

15 Ma, C.F. and Bergles, A.E., "Jet Impingement Nucleate Boiling" , International Journal of Heat and Mass Transfer, Vol. 29, No. 8, pp. 1095-1101, 1986.

16 Lei, D.H., Sun, H., Tian, Y.Q. and Ma, C.F., " Impingement Heat Transfer of a Single Axisymmetric Oil Jet", to be Presented in the 4th Asia Conference on Fluid Mechanics, Hong Kong, 1989.

17 Ma, C.F., Tian, Y.Q., Sun, H., Lei, D.H., "Local Characteristic Of Heat Transfer from a Small Heater to an Impinging Round Jet of Liuid of Large Pr Number", International Symposium on Heat Transfer Enhancement and Energy Conservation, Guang-Zhao, China, 1988.

18 Ma, C.F. and Bergles, A.E., " Convective Heat Transfer on a Small Vertical Heated Surface in an Impinging Circular Liquid Jet", Presented in the 2th Beijing International Symposium on Heat Transfer, Beijing, 1988.

19 Ma, C.F., Gan, Y.P., Tang, F.J., and Bergles, A.E., "A New Method of Heat Transfer Augmentation by Means of Foreign Gas Jet Impingement in Liquid Bath", Heat Transfer Science and Technology, Edited by B.X. Wang, pp. 789-797, Hemisphere Publishing Corporation, 1987.

20 Ma, C.F. and Bergles, A.E. , " Enhancement of Immersion Cooling of Microelectronic Devices by Foreign Gas Jet Impingement", ICHMT 20th International Symposium, Dubrovnik, Yugoslavia, 1988.

21 Gan, Y.P. and Ma, C.F., " Photographical Study of a New Method of Heat Transfer Augmentation with Foreign Gas Jet Impingement in Liquid Pool" (in Chinese), Journal of Beijing Polytechnic University, Vol. 12, No. 2., 1986.

22 Tang, F.J. , "Nucleate Boiling Heat Transfer from Wire-Wrapped Plain Tubes", M.S. Thesis, Beijing Polytechnic University, 1986.

23 Tian, Y.Q., Lei, D.H. and Ma, C.F., " Experimental Investigation on Spray Cooling of Simulated Microelectronic Chips", to be Published.

24 Katto, Y. and Monde, M., "Study of Mechanism of Burnout in A High Heat Flux Boiling System with an Impinging Jet", Heat Transfer 1974, Proceedings of the 4th International Heat Transfer Conference, Vol.2, Paper FC. 5.2, 1974.

25 Xin, M.D., Chen, Y.G., Lei, H.S. and Zhang, H.J., " Heat Transfer Performance of a Heat Pipe Radiator for a 1,000 Ampere thyristor with Air Cooling", ASME Publication HTD-Vol. 20, pp. 61-64, 1981.

26 Gu, Y.Q. and Wu, Y.J., "An Experimental Study of Thermal Resistance of A Power Semiconductor Package", Proceedings of Intersociety Conference on Thermal Phenomena in the Fabrication and Operation of Electronic Components, Los Angeles, 1988, pp. 37-40.

Effects of Rear and Lateral Ducts on the Thermal Behavior of Digital Transmission Equipment

NILO R. KIM and LUIZ F. MILANEZ
Departamento de Energia
FEC/UNICAMP
13081, Campinas, São Paulo, Brazil

ABSTRACT

The slim rack is a digital transmission equipment where printed circuit boards (PCBs) are arranged horizontally. Such equipment has a high height/width ratio and is assembled side by side and back to back thus forming lateral and rear ducts, respectively. The present work deals with the steady-state thermal dissipation in the slim rack considering the effects of these ducts. The Nusselt number for each duct is evaluated and the results are used to analyze the thermal behavior of the slim rack.

INTRODUCTION

Natural convection is a very interesting heat transfer mode due to its relative structural simplicity and near zero operational cost. Despite the low heat transfer coefficients associated with this case, most devices rely on this mechanism for their cooling. Particularly in digital transmission equipment, the performance and reliability of the electronic components are heavily dependent on natural convection heat transfer. In most applications the electronic components are mounted on printed circuit boards that, in turn, are placed inside modules or assembled one on top of the other, side by side and back to back thus forming lateral and rear ducts. A schematic view of such equipment is shown in Fig. 1.

In this work the heat dissipation in the slim rack is analyzed. A model originally developed by Carvalho et al.[1] to predict the temperature distribution in critical regions of the system for a given power supply and ambient air temperature is employed. This model is concerned with the steady-state thermal dissipation in the so-called thermal unit of the slim rack made of the PCB, the magnetic shield above it and the bounding walls constituting an enclosure. The heat transfer in each thermal unit is investigated by assuming isothermal enclosure surfaces with radiosities. Effects of rear and lateral ducts are included in the boundary conditions by considering a numerical analysis for natural convection in vertical ducts.

APPARATUS

Six columns of the slim rack were used in the experiments with one as

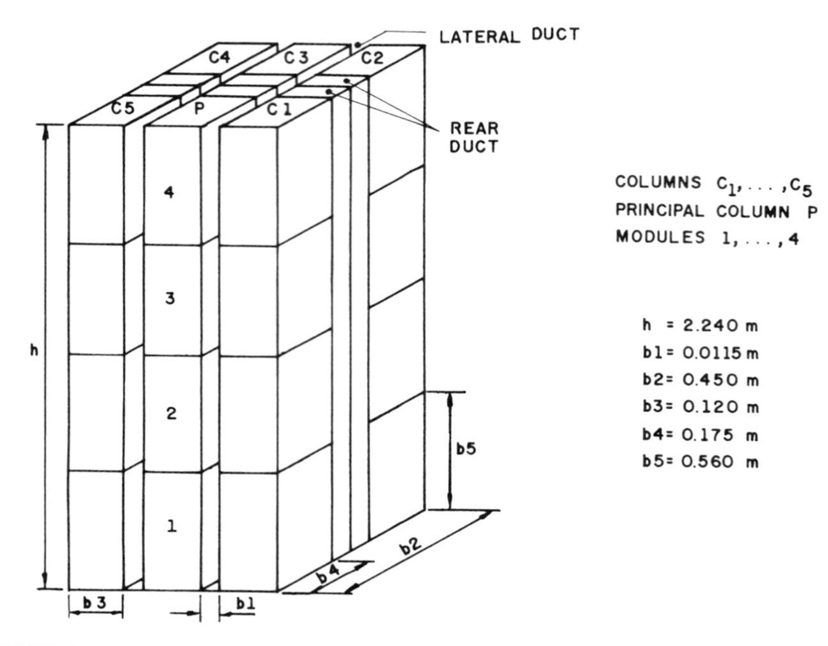

FIGURE 1. Schematic representation of test equipment.

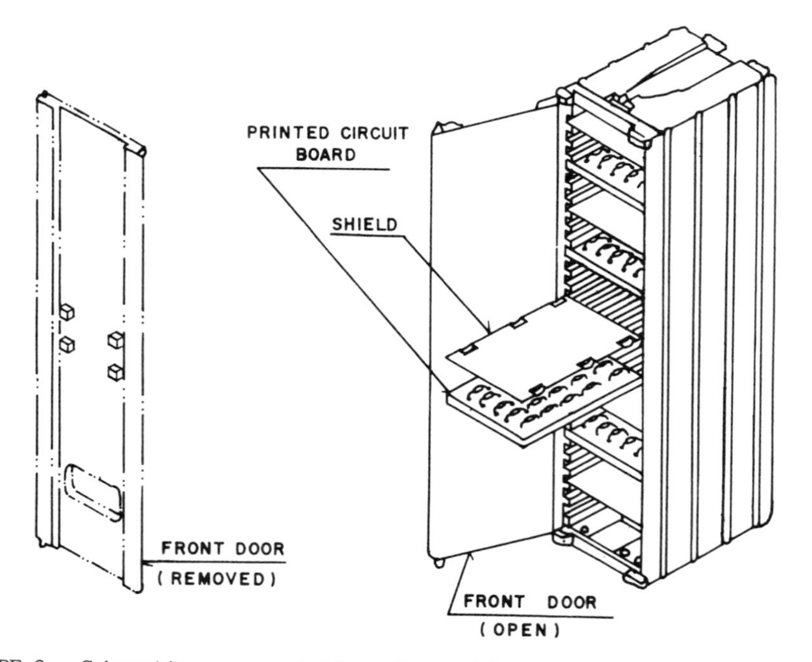

FIGURE 2. Schematic representation of a module.

the main column where the measurements were taken and the other five acting as sort of guard heaters to provide the necessary boundary conditions. The slim rack column arrangement is such that a space is allowed between them, forming channels. Dimensions of the ducts can be seen in Fig. 1. Each column of the slim rack used in this study is made of four modules. A module is illustrated in Fig. 2. The back side of the module is held tightly against a U-beam made of aluminum. The beam surface together with the module back side form a duct through which the electrical wiring is directed to the electrical connectors. The internal surfaces of the module side walls have extruded grooves to allow for the insertion of a pair of PCB and a magnetic shield. Each module accommodates 27 PCB and an equal number of magnetic shields.

In the actual digital transmission equipment the power is dissipated by several different components, but in this study only resistors were used to simulate these components. The PCBs designed for this experiment had 25 resistors uniformly distributed, as shown in Fig. 3. The power supply to each PCB may vary from 0.5 to 2.5W. Copper - constantan thermocouples were used to measure the temperatures of side and back walls, front door and air in the ducts. Approximately four hours were necessary to attain steady-state conditions in each test.

MODELING PROCEDURE

Analysis

Modeling was undertaken by identifying and equating the heat fluxes in a thermal unit, from the resistors where heat is generated to the ambient, along many paths. Energy is irradiated and convected from the resistors to the walls of the thermal unit cavity, and conducted through the resistor leads to the printed circuit board. The copper circuits printed in the epoxy boards conduct heat to the cavity side walls through contact resistances. A fraction of the energy arriving at the circuit board is conducted and irradiated to the lower magnetic shield through a thin air layer of 4mm thickness. This shield is associated with the thermal unit immediately below the unit under consideration. Assuming this is a repetitive pattern - one thermal unit stacked over another similar unit - this heat flux was converted and included in the unit itself where it was generated. From the shields, the energy is conducted to the walls. Finally the heated walls and front door dissipate the heat generated internally to the ambient air by natural convection and radiation. The heat irradiated by the structure is analyzed by considering a very large room, the walls of which are maintained at the ambient air temperature. As for natural convection in flat vertical surfaces two different conditions are frequently considered: uniform wall temperature or uniform heat flux.

The thermal resistance network corresponding to the thermal paths through which the heat flows from the electronic component where it is generated, to the ambient, is shown in Fig. 4. The PCB was divided into six subsections (see Fig. 3), with each subsection assumed isothermal. The resistors were assumed to be flush with the epoxy circuit board, whose temperature was taken as the area weighed average of the resistors' temperature and of the board itself, thus forming a virtual plane. Radiant heat was equated by assuming energy exchange among gray isothermal surfaces with uniform radiosities. Thermal resistances associated with heat

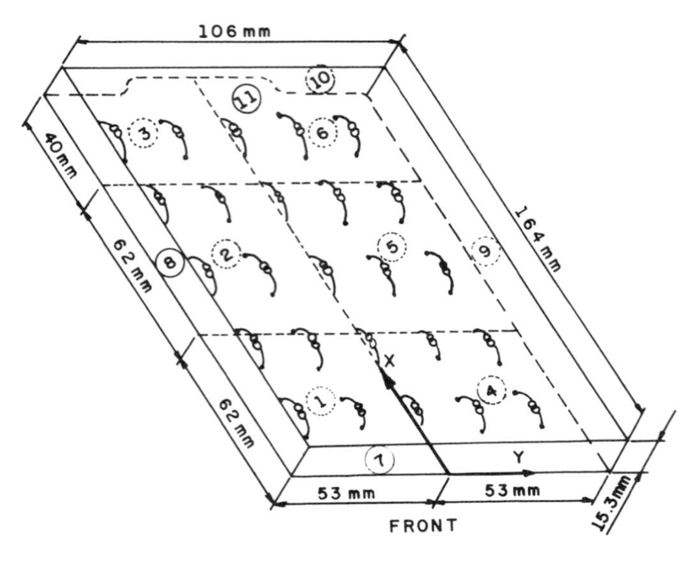

FIGURE 3. PCB and module thermal unit.

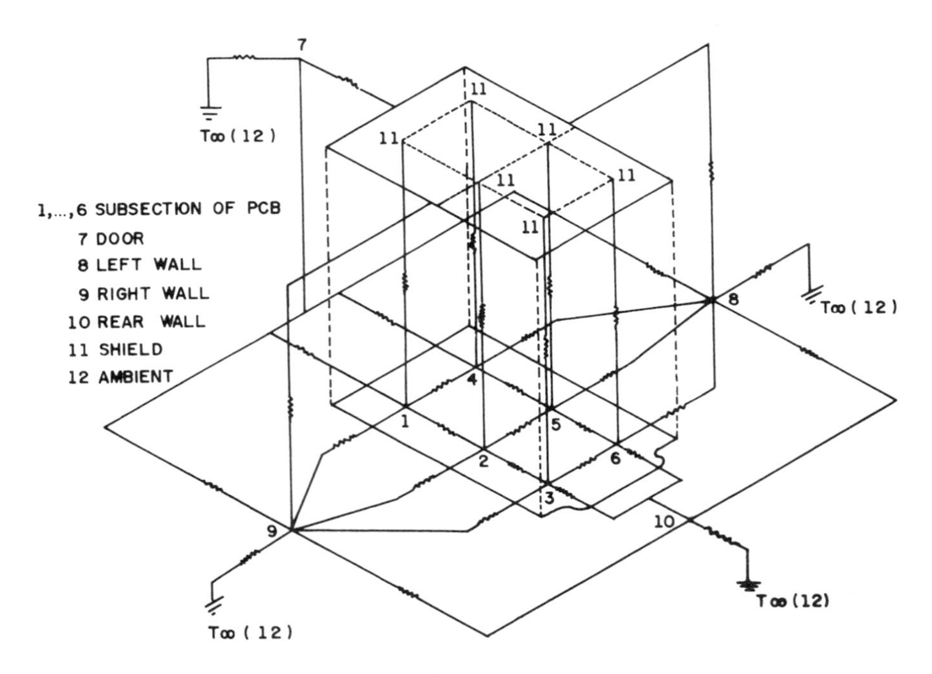

FIGURE 4. Thermal resistance network.

conduction were considered constant for the temperature levels studied; but for convection and radiation heat transfer with the environment they are temperature dependent.

External thermal resistances R_t related to the thermal unity cavity (7-12, 8-12, 9-12 and 10-12) are described below. All others are calculated according to the procedure developed by Carvalho et al. [1] considering a separate test module, out of the column of the slim rack. They considered as boundary conditions an insulated back wall and side walls open to the ambient air having a thermal behavior similar to the front door, that is, natural convection from an isothermal vertical flat surface. Their analysis is extended in the present work to include the effects of rear and lateral ducts, as well as the significant height of the columns. In the numerical procedure an initial temperature distribution is assumed for each node. After the temperature dependent thermal resistances are calculated, heat fluxes due to conduction and convection are determined. From the radiosities of each surface the radiant heat fluxes are obtained. Kirchhoff's current law is applied to the circuit, which states that the sum of the currents entering a node is zero. If the sum is not zero, the temperature field is conveniently adjusted and the process repeated. The set of equations, linear for the radiosisties and nonlinear for the temperatures, is solved by the Newton-Raphson method in a Fortran program, according to Stoecker [2].

Lateral Duct

In the literature the majority of the works is concerned with the flow between vertical flat plates considering uniform velocity and ambient temperature at the inlet. Actually, as the columns are positioned directly on the floor, the lower end of the duct is not open. Therefore , the air enters the duct perpendicular to gravity. Due to the complexity of this configuration, the situation is simulated as an axial flow along a vertical channel. Therefore, the analytical expression to be used to calculate the heat transfer to the lateral ducts does not take into consideration the fact that the lower end is blocked. This can be true if the columns are lifted a little to allow for air entrainment at the lower end, thus improving heat transfer coefficients. Another assumption to be made concerns the boundary condition at the walls: uniform wall temperature or uniform heat flux. Carvalho et al. [1] used the first alternative because the test module was relatively short, namely 0.28m. It was observed experimentally that the door temperature and the wall temperatures of a module were almost uniform (the difference between the extremes and the center were always below 4%) due to the highly conductive material they are made of (aluminum), but varying from module in the column. The geometry considered is shown in Fig. 5. Despite the actual physical situation in which uniform heat flux is present, the external surface of each module is very close to isothermal. Along the door height the heat transfer coefficient is evaluated according to Carvalho Fº and Goldstein Jr. [3]. The lateral duct was divided into 4 subsections originating 4 consecutive ducts, with the temperature of each duct assumed as constant.

The value of the Rayleigh number varied from 5 to 25 in the tests. According to Aung et al. [4] for this range of Rayleigh number the flow can be assumed as fully developed. For this situation Aung [5] obtained a relation between the average Nusselt number $\overline{Nu}$ and the average Rayleigh number $\overline{Ra}$ for the heat exchanged between the walls and the air:

$$\overline{Nu} = \overline{Ra}/24 \qquad (1)$$

After the heat transfer coefficients at the door and lateral ducts are calculated,the thermal resistances due to natural convection are given by

$$R_t(i-12) = 1(\overline{h}_i\ A_i) \qquad i=7,\ldots,10 \qquad (2)$$

In order to evaluate the influence of the lateral ducts in extracting the heat generated in the slim rack the two adjacent columns C1 and C5 were removed. By doing so, the boundary condition at the side walls became the same as at the front door; that is, natural convection from an isothermal vertical surface to the ambient air.

The radiation thermal resistances are given by

$$R_t(i-12) = 1(A_i F_{ij}) \qquad i=8,9 \qquad (3)$$

Assuming uniform radiosities, the high duct height/width ratio (about 200) combined with symmetry of the wall temperatures result in heat transfer by radiation about three times less than natural convection.

Less than 15% of the total heat generated within the column is dissipated through the lateral ducts.

Rear Duct

The rear duct is made of four walls and its top view can be seen in Fig. 6. Wall 1 is in direct contact with the PCB and its temperature is slightly below that of the module lateral surfaces. Walls 2 and 3 receive a heat flux by conduction and wall 4 is assumed as insulated due to symmetry. The four walls were considered to be at the same temperature. The duct hydraulic diameter is 0.051m. By observing that the electric wiring occupies the central portion of the rear duct corresponding to 75% of the cross sectional area, two "channels" are available for the air flow, one adjacent to wall 2 and other adjacent to wall 3, with hydraulic diameter approximately 0.02m and height of 2.366m.

The actual arrangement of the structure allows an air flow through the rear duct, although the heat removed in this process is not significant, accounting for less than five percent according to the relation developed by Davis and Perona [6]

$$Nu_r = \frac{1}{16} Gr_r . Pr \qquad (4)$$

where r, the characteristic dimension, is the hydraulic radius. In the rear duct the heat transfer by radiation is negligible.

RESULTS AND DISCUSSION

Two series of experiments were conducted, one with the complete structure as shown in Fig. 1, the other with columns C1 and C5 removed in order to evaluate the influence of the lateral ducts. It was impossible to ignore the rear duct in this simulation because it conveys the electrical wiring to the connectors;but, as it discussed earlier, it is not effective in removing the heat.

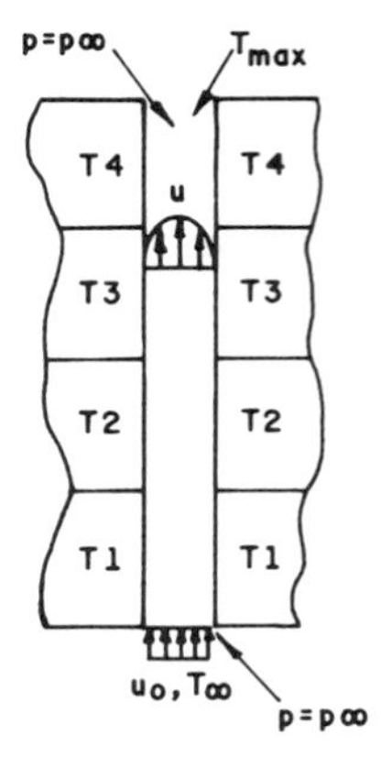

FIGURE 5. Geometry considered for the lateral duct analysis.

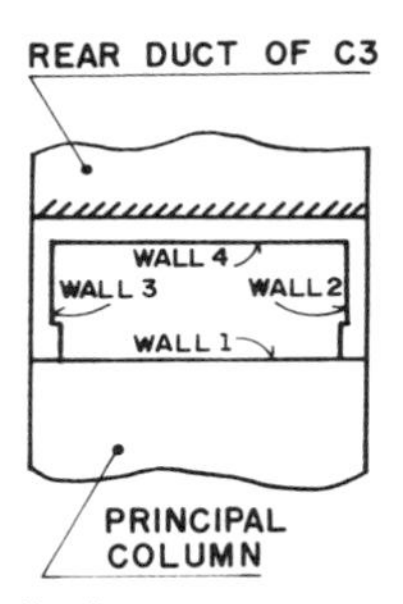

FIGURE 6. Top view of rear duct.

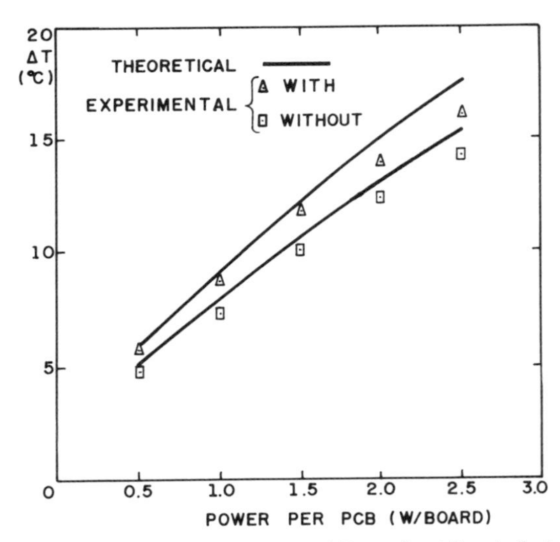

FIGURE 7. Door excess temperature with and without lateral ducts.

Figures 7 to 12 compare numerical and experimental data. Measurements were taken in modules 1 and 2. In module 1 the sixth thermal unit from the bottom was instrumented (Figs. 7, 8 and 9) and in module 2 the nineteenth from the base of this module, that is, at approximately 2/3 of its height, was considered (Figs. 10, 11 and 12). In all these figures we may observe the variation of the temperature excess above room temperature, ΔT, against the power dissipated per board, from 0.5W to 2.5W in intervals of 0.5W. As it would be expected, the temperature excess always increases with an increase of the power dissipated, note also the differences between numerical and experimental results because in the modeling the con duction thermal resistances were assumed invariant for the power range considered. Besides, for high powers, turbulence may occur in the air flow due to surface irregularities, increasing the heat transfer. Actually, resistances vary slightly because some of them are functions of the air transport properties. These figures depict temperature excess of the door, side walls and back wall.

In Fig. 7 for module 1 and Fig. 10 for module 2 it can be seen that the presence of lateral ducts does not influence the door temperature signifi cantly. There is a reduction of only $3^{\circ}C$ in the temperature excess when the adjacent columns are removed, for maximum power dissipation. However, for the back and lateral walls the reduction of the temperature excess is more pronounced, about $10^{\circ}C$, for the same situation of maximum power dissipation (Figs. 8 and 9 for module 1 and Figs. 11 and 12 for module 2). It may be noticed in Fig. 11 that module 2 which has walls at higher tem perature than module 1 is more affected by the presence of lateral ducts. In this case the increase of its temperature excess due to the presence of lateral ducts is about $13^{\circ}C$.

By examining Figs. 8 and 9 it is possible to observe that the temperature excess distribution for the side walls as a function of the power dissipated per board almost coincides with the temperature excess distribution for the back wall for module 1. If the adjacent columns are removed with the boundary condition being without lateral ducts, the back wall tends to exhibit a higher temperature excess distribution because the side walls are now exchanging heat with the environment with a higher heat transfer coefficient. For module 2 (Figs. 11 and 12) the back wall temperature excess is higher than that of the side walls for any circumstances, because the cooling air has already been preheated after interacting with module 1.

Also shown in Figs. 8 and 11 are the temperature excess for the very central resistor used to simulate the electronic component. It is important to observe how the boundary conditions at the side walls affect the temperature of the components inside the cabinet.

CONCLUSIONS

The present work shows that the thermal behavior of the slim rack, a typical digital transmission equipment, is significantly affected by the presence of lateral ducts. According to Figs. 8 and 11 from a situation where the ducts are present to the limit situation where the adjacent col mns are removed, the temperature excess of the electronic component (simulated by a resistor in this study) located at the center of the PCB is reduced by 17%. The rear duct, however, is not effective in removing the heat from the thermal unit.

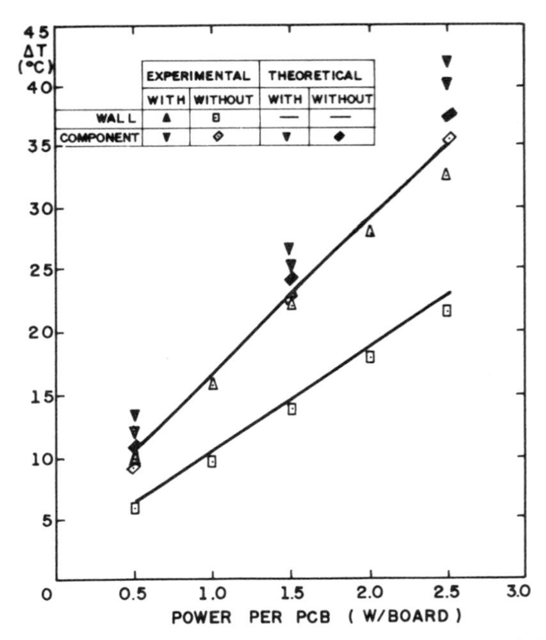

FIGURE 8. Side wall and component excess temperature, with and without lateral ducts (Module 1).

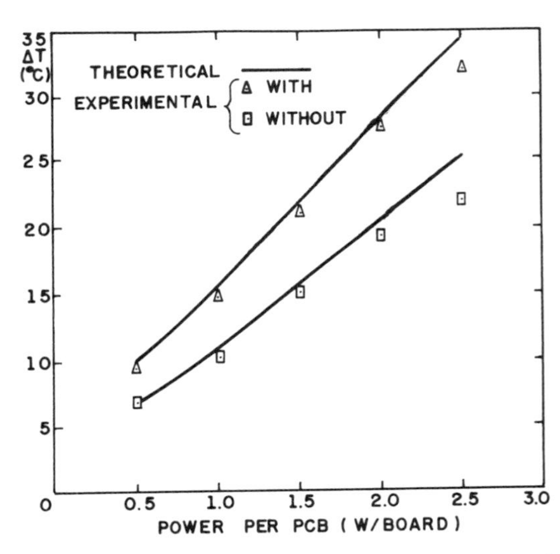

FIGURE 9. Rear wall excess temperature, with and without lateral ducts. (Module 1).

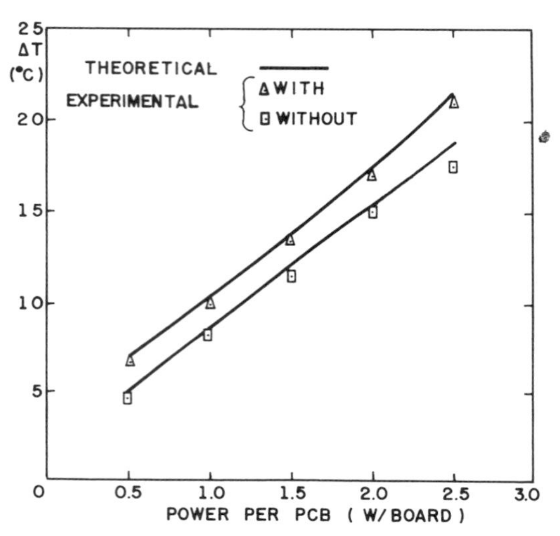

FIGURE 10. Door excess temperature, with and without lateral ducts (Module 2).

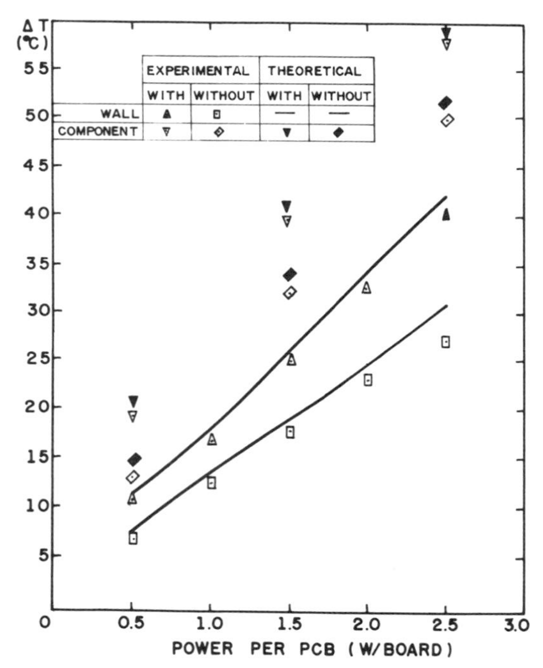

FIGURE 11. Side wall and component excess temperature, with and without lateral ducts (Module 2).

ACKNOWLEDGEMENTS

This research was performed under the auspices of TELEBRÁS - Telecomunicações Brasileiras S.A. and with support to N.R. Kim from CNPq - Conselho Nacional de Pesquisa e Desenvolvimento.

REFERENCES

1. Carvalho, R.D.M., Goldstein Jr., L. and Milanez, L.F., Heat Transfer Analysis of Digital Transmission Equipment With Horizontally Arranged Printed Circuit Boards, Heat Transfer in Electronic Equipment - 1986, ASME HTD - vol. 57, pp. 145-152, 1986.

2. Stoecker, W.F., A Generalized Program for Steady-State System Simulation, ASHRAE Semiannual Meeting, Philadelphia, 1971.

3. Carvalho Fº, P. and Goldstein Jr., L., Free Convection from a Vertical Flat Plate With Non Uniform Temperature Distribution, ASME HTD - vol. 96, pp. 213-218, 1988.

4. Aung, W., Fletcher, L.S. and Sernas, V., Developing Laminar Free Convection Between Vertical Flat Plates With Asymmetric Heating, Int. J. Heat Mass Transfer, vol. 15, pp. 2293-2308, 1972.

5. Aung, W., Fully developed Laminar Free Convection Between Vertical Plates Heated Asymmetrically, Int. J. Heat Mass Transfer, vol. 15, pp. 1577-1580, 1972.

6. Davis, L.P. and Perona, J.J., Development of Free Convection Flow of a Gas in a Heated Vertical Open Tube, Int. J. Heat Mass Transfer, vol. 14, pp. 889-903, 1971.

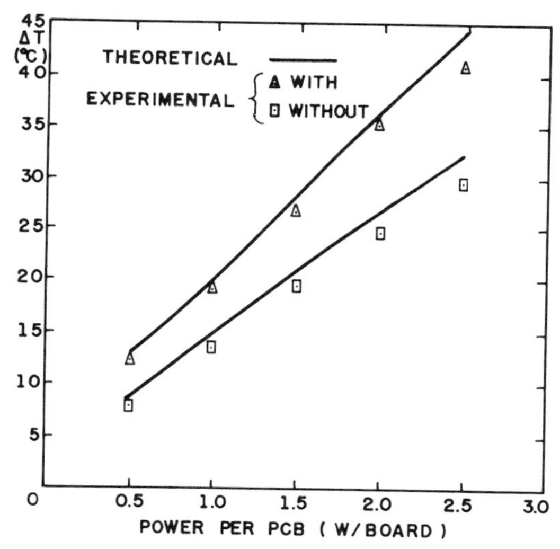

FIGURE 12. Rear wall excess temperature, with and without lateral ducts (Module 2).

Some Factors Influencing the Optimum Free Air Cooling of Electronic Cabinets

G. GUGLIELMINI, G. MILANO, and M. MISALE
Energy Engineering Department
University of Genoa
Via all'Opera Pia 15/a
16145 (I) Genova, Italy

ABSTRACT

The free air cooling of parallel vertical cards packaged within a box provided with vents has been experimentally investigated. In particular, the influence of the following factors has been examined: card spacing, shape and size of vents, and the insertion of some unheated metallic slabs between the heated cards. For a given volumetric power density there is an optimum spacing to which corresponds a minimum temperature rise inside the box. The opening area has a great influence on the air temperature rise while the opening geometry has a minor effect. Finally, a simple relationship has been proposed able to correlate all experimental data to within ±10%.

1. INTRODUCTION

Natural convection of air is often applied as a cooling technique for thermal control of electronic equipment with low power density. The main advantage of natural convection is its inherent reliability, because air movement is simply generated by density gradient in a body-force field. Consequently no fans or blowers are required and thus the reliability of the electronic device is improved and the cost reduced.

The optimum thermal design of electronic devices, cooled by natural convection, consists in an accurate choice of the geometrical configuration and of the heat source distribution able to generate the air flow rate that minimizes the temperature rise inside the cabinet. The solution of this problem becomes more difficult with increasing volumetric power density, and requires a detailed knowledge of the heat transfer and fluid flow in confined or partially confined spaces.

Although many correlations exist in the literature for external free convection heat transfer on simple geometries, limited experimental data are available on natural convection in enclosures such those encountered in electronic equipment. Moreover, for electronic devices containing printed circuit cards, the heat transfer and fluid flow are generally complex and three dimensional, so that experimental modelling studies are necessary.

Some simple thermal design criteria for electronic cabinets cooled by natural convection were presented in [1]. More recently, Ellison [2], Ngai [3], and Ishizuka et al. [4], have developed procedures able to analyze the main characteristics of natural ventilated boxes. In [5] and [6] some effects on the thermal field inside the box have been considered, as, for example, the position and size of the openings, the internal heat source locations and the dissipated power removed by the thermocirculation flow rate. A simple correlation for volumetric air flow rate has been also suggested in [6] able to predict the inner air temperature rise as a function of the main thermal and geometrical characteristics of the box.

In this work, the natural air cooling of parallel vertical cards packaged within a confined volume provided with vents has been further investigated. In particular, the influence of the card spacing, the shape and size of the vents located on the frontal wall of the box have been considered. Moreover, the effect of vertical unheated metallic slabs inserted between the heated cards on the inner temperature rise has been also examined. All experimental results obtained in this work have been represented by a correlating equation proposed in [6] to within ±10%.

2. EXPERIMENT

2.1 Experimental apparatus

The experimental apparatus is shown in Fig. 1 (a,b,c) and consists of a box having external dimensions (length x width x height) 100 mm x 152 mm x 254 mm.

Some equispaced vertical cards are packaged within the box whose electronic circuitry is simulated by 12 rows of electrical resistors arranged in three equal sectors as indicated in Fig. 1 b,c. The number of cards was varied during experiments from 3 to 9 and a series of tests was also performed inserting between the heated cards some additional metallic plates without heat sources.

On the frontal wall of the box two openings are made at the top and near the lower region of the wall itself. The position of the barycentric lines of the openings with respect to the first and last row of resistors has been fixed in all experiments as follows: h_1=0, h_2=25, 35, and 40 mm. These values give rise to a more efficient air flow circulation as found in previous investigations [5,6].

Different shapes of the openings have been investigated with a total area varying in the range 760-2000 mm^2, respectively. For the most part the tests have been performed using two single rectangular vents both having the area 10x100 mm^2 or 20x100 mm^2; in some other tests the two openings have been realized by several rows of circular holes with different diameters (D=3.25, 5.5, 7.0, and 9.75 mm) or by a series of vertical slots 3x26 mm^2 and 5.2x26 mm^2.

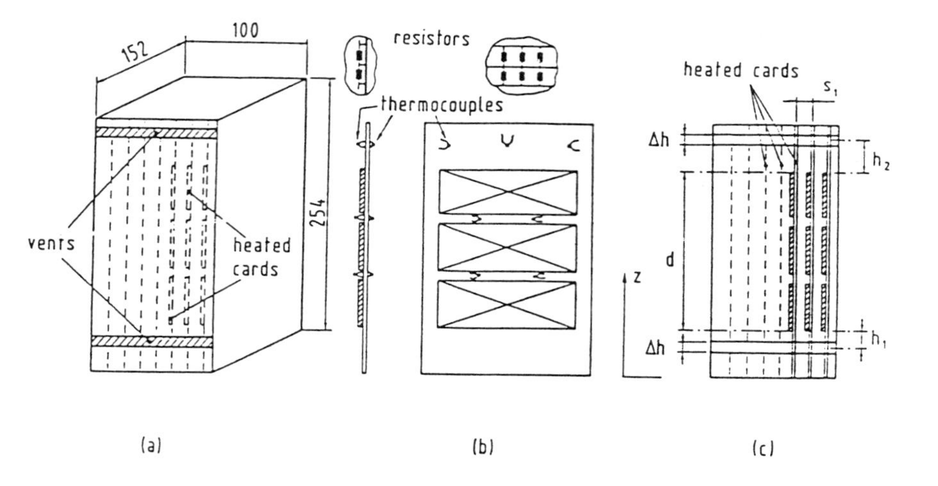

FIGURE 1. Geometrical configuration of the box (a); instrumentation of the cards (b);heat source arrangement and vent coordinates (c).

The cooling air gets into the box through the lower opening, removes a fraction of heat dissipated inside and comes out through the upper opening. Both the vents are located on the frontal wall because in many equipments for telecommunication, several boxes are placed adjacent to each other in closed racks, the only free surface being the frontal one. Moreover, in order to simulate the thermal behavior under the effective working conditions all the box surfaces, except the frontal one, have been carefully insulated simulating the presence of adjacent boxes. In spite of the thermal insulation a not negligible percentage of the total heat dissipated inside the box is transferred outside through the walls. The heat losses have been evaluated by means of a preliminary series of tests performed with the box fully closed and varying the difference between the mean inner air temperature and the external one. In this way the thermocirculation flow rate is absent and all the heat flux dissipated inside is transferred outside due to the heat leakage of the walls. Hence, the heat flux removed by thermocirculation during the test with open vents has been obtained by subtracting from the total dissipated power the contribution due to the heat loss evaluated with closed vents at the same mean temperature difference between the inner air and the ambient air.

Some Schlieren visualizations of the convective flow patterns inside the box have also been performed. To this aim the frontal and the back walls of the box have been made of glass; in this way a series of photographs has been taken from which qualitative information on the thermal boundary layers arising near the hot cards have been drawn.

2.2 Temperature measurements

Air temperature distributions inside the box have been measured by means of thermocouples arranged on each card at three different elevations and partly located near the thermal boundary layer generated by the resistors and partly on the air surrounding the back of the card itself (Fig. 1b). Measured temperatures have been averaged to obtain the integrated value $\bar{T}_z$ over the section at the elevation z and the integrated value $\bar{T}$ over the whole volume of the box. Moreover, for a given test the maximum value of air temperature T_M occurring at a particular location inside the box has been also detected. The corresponding temperature excess values in respect to the ambient temperature T_o are defined as follows: $\overline{\Delta T}_z=\bar{T}_z-T_o$, $\overline{\Delta T}=\bar{T}-T_o$ and $\Delta T_M=T_M-T_o$.

Finally, a series of thermocouples has been also placed near the two openings for the measurement of inlet and outlet air temperatures. The difference $\Delta T_{1,2}$ between the outlet and inlet air temperature has been used to evaluate the experimental thermocirculation air flow rate $\dot{V}$ using an enthalpy balance.

In all experiments the outlet air temperature was very close to the mean temperature $\bar{T}_z$ at the highest elevation while the inlet temperature was practically equal to the ambient temperature.

3. FACTORS INFLUENCING THE AIR COOLING EFFICIENCY

3.1 Effect of vent position

The influence of vent position on the natural cooling efficiency has been previously investigated in [5,6] for rectangular vents having different area. In particular, the inner air temperature profile was measured at different values of the opening elevations h_1 and h_2 with respect to the first and the last row of the heated resistors.

The main results found in [5,6] were that the vent position greatly affects the inner air temperature profile and the best cooling conditions can be obtained by placing the upper vent at the top of the box and the lower one just under the first row of heat sources. Taking into account that result,all the experiments in this work were performed with the above-mentioned position of the openings.

3.2 Spacing between the cards

For vertical two-dimensional channels formed by parallel dissipating plates in natural convection cooling, it was observed that the rate of heat transfer from each plate decreases as the plate spacing is reduced. However, for a given volume, reducing the spacing between the cards, the total surface area of the plates increases and an optimum spacing may be expected which maximizes the total power transferred to the fluid. This problem was studied by many authors [7,8], and recently Bar-Cohen and Rohsenow [9] derived analytic relations for the natural convection heat transfer coefficient along the channel with symmetric and asymmetric as well as isothermal or isoflux surface conditions. Those relations were also applied to find the plate spacing at which the product of total surface area of the plates and the local heat transfer coefficient is a maximum [10].

In this work the thermal conditions along the channels inside the box are quite different, the flow is three-dimensional and partially confined by the box, with relevant lateral edge effects. However, an attempt has been made to verify if an optimum spacing exists also in this situation.

The number of heated cards was varied from 3 to 9, with a corresponding spacing s_1 ranging within 31.5 mm to 9 mm.
Figures 2 and 3 show the results of a series of tests performed varying the card spacing for different values of heat flux transferred to the fluid P' and of opening area A. Examination of Figs. 2 and 3 shows that an optimum spacing may exist also in this case. In fact, for a given volumetric power density P'/V (where V is the box volume) there is an

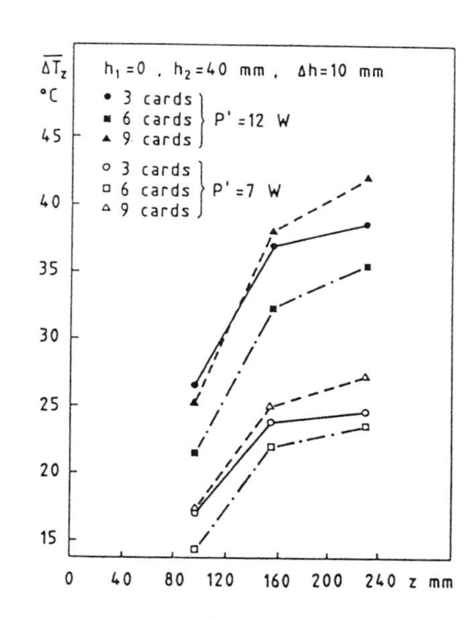

FIGURE 2. Mean temperature rise $\overline{\Delta T}_z$ inside the box at various elevations z: A=1000 mm².

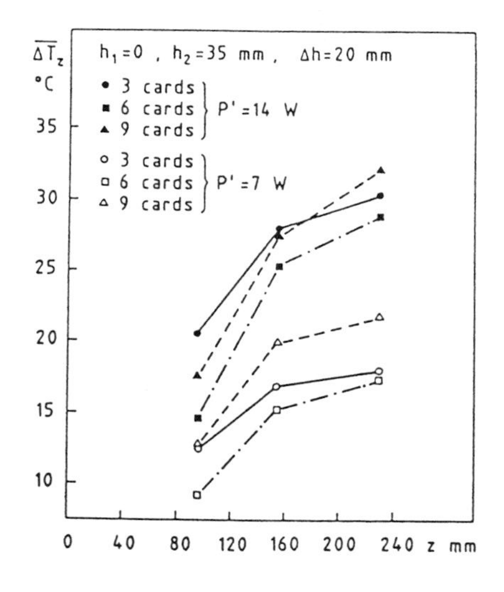

FIGURE 3. Mean temperature rise $\overline{\Delta T}_z$ inside the box at various elevations z: A=2000 mm².

optimum number of cards to which corresponds a minimum inner temperature rise and, therefore, a minimum value $\Delta T_{1,2}$ (maximum thermocirculation flow rate $\dot{V}$). With respect to the results obtained with 3 or 9 cards using 6 cards there is a sensible reduction of the inner air temperature and this favorable effect occurs also when the power P' and the opening area are varied.

3.3 Influence of additional metallic plates inserted between the heated cards

A method commonly used to improve the cooling of cards dissipating high power density consists of a copper frame-work placed inside the card itself which distributes and conveys heat to the cooling air and to the metallic supports of the box [11].
If the number of cards inside the box is not too large the cooling efficiency can be improved in a more economical way by placing some thin metallic plates, made of aluminium or copper, between the heated cards as shown in the sketches of Figs. 4 and 5.
These additional plates remove heat by radiation and convection in the regions at higher temperature, generally the upper part of the box, and

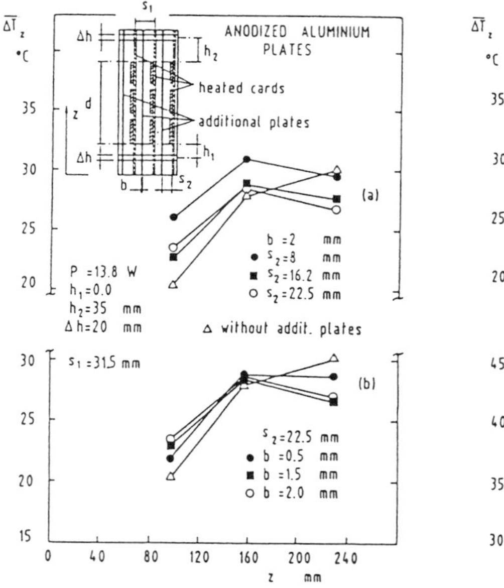

FIGURE 4. Influence of location (a) and thickness (b) of additional metallic plates, inserted between the heated cards, on the local mean temperature difference $\overline{\Delta T}_z$.

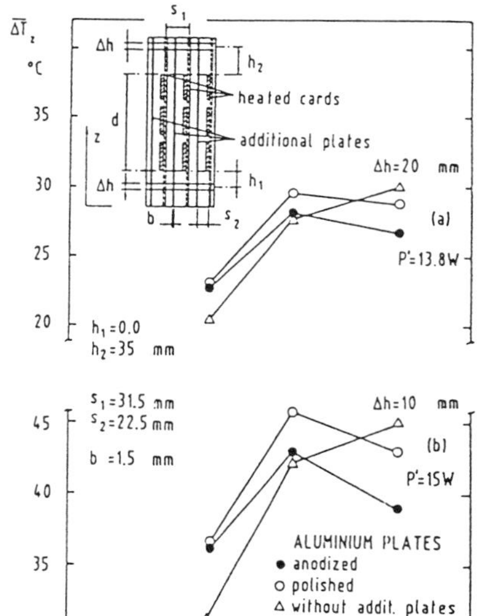

FIGURE 5. Influence of surface finish of the additional metallic plates, inserted between the heated cards, on the local mean temperature difference $\overline{\Delta T}_z$ for two different opening areas.

convey it in the regions at lower temperature. In this way the peak temperature is reduced and the air temperature is more uniformly distributed inside the box.

The results of a series of tests performed varying position, thickness, material and surface finish of the additional plates show that the more favourable effects can be obtained when
- the plates are located far enough from the heated side of electronic cards to avoid interferences with the thermal boundary layer developing near the heat sources (Fig. 4a);
- the parameters $\lambda \cdot b$ (where λ and b, are respectively, thermal conductivity and thickness of the metallic plates) which is proportional to the conductive conductance in the z-direction is not less than about 0.3 W/K (Fig. 4b);
- the emissivity of the surfaces is high in order to increase the radiant heat exchange (Fig. 5b). Radiation plays a relevant role in the heat transfer process inside the box.

The results of Fig. 5 show a possible percentage reduction of the local mean temperature difference in the top region of the box of about 15% with respect to the same conditions but without additional plates.

The fact that the local air mean temperature in the upper region of the box may be lower than that of the central region can be explained bearing in mind that in the upper region the heat flux is removed by metallic slabs and by the heat losses through the box walls. Moreover, the measured inner air temperature $\overline{T}_z$ represents a conventional mean value (a static spatial weighted value) rather than an effective mixing value.

Further tests, performed with additional plates having a certain number of holes, showed the tendency to decrease the favourable effect of the temperature reduction obtainable using plates without holes. This behavior can be related both to the reduction of total surface participating in radiative exchanges and to a reduction of the global conductive conductance of the plate in the z-direction. On the contrary the air cross flow through the holes seems to produce negligible effects.

3.4 Shape and size of the opening

In Figs. 6 and 7 the heat transfer results for a box having different kinds of openings are summarized.

The box contains 9 heated cards and the total surface area of each opening is A=1450 mm^2 for the tests reported in Fig. 6 and A=760 mm^2 for those of Fig. 7.
The openings have been realized by several rows of circular holes with different diameters: D=3.25, 5.5, 7.0, and 9.75 mm or by a row of vertical rectangular slots 3x26 mm^2 and 5.2x26 mm^2.
The results are compared with the reference case of openings simply formed by two rectangular vents having the area 7.6x100 mm^2 and 14.5x100 mm^2, respectively. The results of Figs. 6 and 7 show that the mean $\overline{\Delta T}$

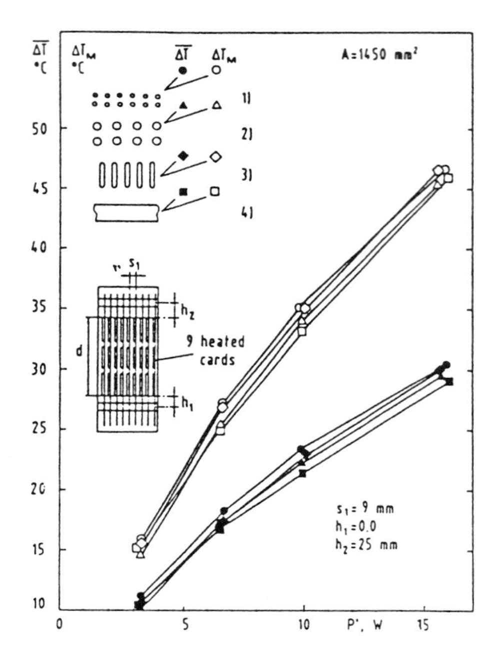

FIGURE 6. Mean temperature rise $\overline{\Delta T}$ and maximum temperature rise ΔT_M as a function of power transferred to the air P' for four different shapes of the vents:

1) 4 rows of circular holes D=5.5 mm,
2) 2 rows of circular holes D=9.75 mm,
3) 1 row of rectangular slots 5.2x26 mm^2,
4) 1 rectangular opening 14.5x100 mm^2.

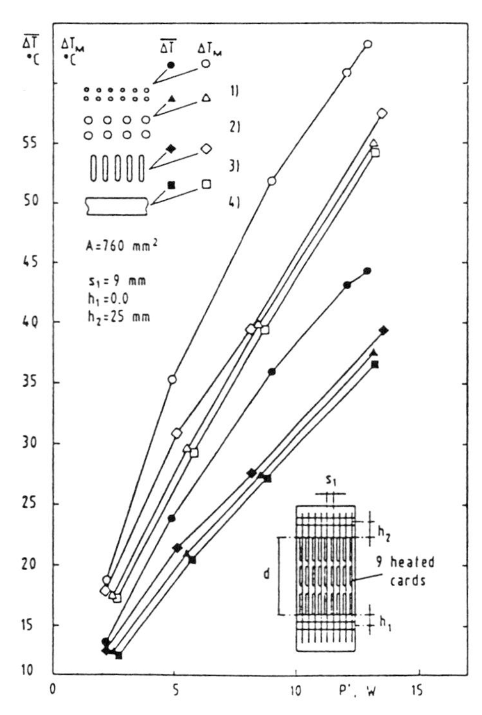

FIGURE 7. Mean temperature rise $\overline{\Delta T}$ and maximum temperature rise ΔT_M as a function of power transferred to the air P' for four different shapes of the vents:

1) 6 rows of circular holes D=3.25 mm,
2) 2 rows of circular holes D=7.0 mm,
3) 1 row of rectangular slots 3x26 mm^2,
4) 1 rectangular opening 7.6x100 mm^2.

and the ΔT_M air temperature difference are greatly influenced by the value of the total surface area of the opening as already observed in [6] whereas the opening geometry has a minor effect. In fact, only in the case of circular holes having the smallest diameter D=3.25 mm is there a sensible increase of $\overline{\Delta T}$ and ΔT_M at the higher P' value.
Therefore, the holes diameter D=4-5 mm represents the limiting dimension below which the concentrated pressure losses at the openings tend to increase rapidly in respect to the reference configuration of fully open rectangular vent at the same total area.

4. A CORRELATING EQUATION FOR THERMOCIRCULATION FLOW RATE

A preliminary thermal design of the box, cooled by natural air convection, involves the evaluation of the mean inner air temperature which chiefly depends on the number of heated cards and on the total power transferred to the air as well as on the position, shape and size of the openings. A simple correlating equation can be obtained relating the main parameters to the volumetric air flow rate through the box. In the steady-state condition the heat flux P' transferred to the air may be written by the inlet-outlet enthalpy variation $\Delta H_{1,2}$

$$P'=\bar{\rho}\dot{V}\Delta H_{1,2}=\bar{\rho}\dot{V}C_p\Delta T_{1,2} \tag{1}$$

and the buoyant potential provides the fluid kinetic energy

$$(\rho_o-\bar{\rho})gh=K_e\bar{\rho}\frac{u^2}{2} \tag{2}$$

where ρ_o and $\bar{\rho}$ are, respectively, the external air density and the internal density at the mean temperature value $\bar{T}$; h is the chimney height, i.e., the distance between the barycentric lines of the openings; K_e is an equivalent fluid resistance coefficient which takes into account the friction losses along the whole air pattern; $\dot{V}$ is the volumetric flow rate evaluated at the mean temperature $\bar{T}$; and u is a reference value for air mean velocity.

Denoting by $C=\Delta T_{1,2}/\overline{\Delta T}$ the ratio between the inlet-outlet air temperature difference and the mean inner temperature rise, the relation (1) can be re-written

$$\overline{\Delta T}=\frac{P'}{\bar{\rho}C_p\dot{V}C} \tag{3}$$

Assuming the air density inside the box is a function of temperature only we have

$$\rho=\rho_o\frac{T_o}{\bar{T}} \tag{4}$$

and from a combination of equations (2), (3), and (4), one can obtain

$$\dot{V}^3 = \frac{2gA^2hP'\overline{T}}{\rho_o C_p K_e C T_o^2} \tag{5}$$

where A is the vent area and for simplicity we have assumed $u=\dot{V}/A$. Finally, taking for air the values C_p=1007 J/kgK and $\rho_o=p_a/RT_o$ (p_a=101.3 kPa atmospheric pressure; R=286.7 J/kgK air gas constant), equation (5) can be expressed as follows

$$\dot{V}^3 = 5.51\ 10^{-5}\ \frac{A^2hP'}{K_e C}\ \frac{\overline{T}}{T_o} \tag{6}$$

Equation (6) allows the evaluation of the volumetric flow rate $\dot{V}$ (m^3/s) knowing the total power removed by the fluid P' (W), the opening area A (m^2), and the chimney height h (m) for a fixed value of average air temperature inside the box $\overline{T}$ (K). Moreover, the foregoing equation (6) includes the coefficient C whose value depends upon the actual air temperature profile which is established inside the box.
As already noted in [6] C is chiefly influenced by the position of the upper and lower vents with respect to the heated regions of the cards. The coefficient C also appears to increase weakly with the power transferred to the fluid P' and to a lesser extent to the opening area A. However, in the case here considered in which the lower opening is placed just below the first row of heat sources and the upper one is located a little above the dissipation height, the coefficient C can be considered only a function of the power removed P', as shown in Fig. 8. As a design criteria, for want of more accurate data, the coefficient C can be taken equal to

$$C=1.18+0.19\log P' \tag{7}$$

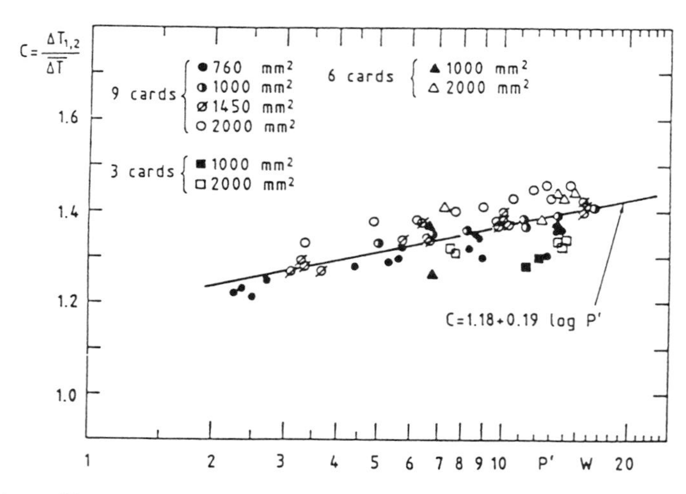

FIGURE 8. Influence of card number and vent area on the coefficient C for different values of power P' transferred to the air.

Equation (7) is valid in the range of $760 \leq A \leq 2000$ mm², for spacing between cards s_1 ranging within 9 to 31.5 mm and for vents positioned as indicated above.

The equivalent fluid resistance coefficient K_e can be evaluated taking into account the friction losses Δp along the whole air path

$$\Delta p = 2\Delta p_{op} + \Delta p_{ch} \tag{8}$$

where

$$\Delta p_{op} = K' \frac{1}{2} \bar{\rho} u^2 \tag{9}$$

and

$$\Delta p_{ch} = K'' \frac{h}{D_h} \frac{1}{2} \bar{\rho} w^2 \tag{10}$$

In relations (9) and (10) K' is a proper fluid resistance coefficient characteristic of the opening and of the change air path, K" represents the friction factor which takes into account the distributed pressure losses along the channels, w is the air mean velocity inside the heated channels and D_h is a proper hydraulic diameter.

Combining equation (8) with (9) and (10), and referring also the distributed pressure losses to the mean velocity u, the total pressure Δp can be re-written

$$\Delta p = \left[2K' + K'' \frac{h}{D_h} \left(\frac{A}{NA_c} \right)^2 \right] \frac{1}{2} \bar{\rho} u^2 = K_e \frac{1}{2} \bar{\rho} u^2 \tag{11}$$

where A_c is the channel cross section area to which corresponds the mean velocity w as shown in the sketch of Fig. 9; N represents the number of

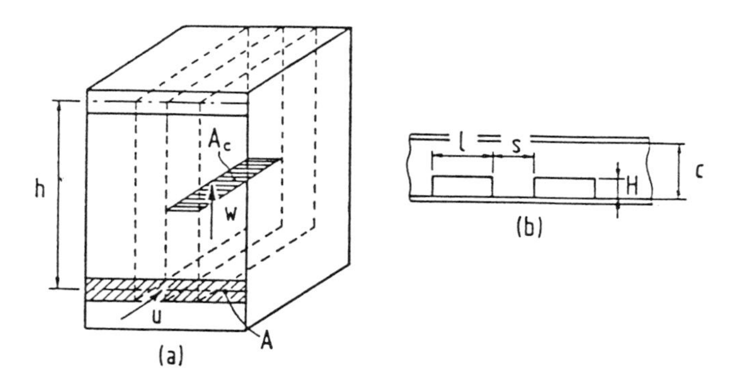

FIGURE 9. Sketch of the air flow pattern inside the box (a) and geometrical parameters of the channels with two-dimensional obstructions (b).

equal channels formed by the heated cards, assuming an equally distributed air flow rate in each channel.
From equation (11) the equivalent fluid coefficient K_e can be defined

$$K_e = 2K' + K'' \frac{h}{D_h} \left(\frac{A}{NA_c} \right)^2 \quad (12)$$

Concerning the values of the coefficients K' and K" the bibliography is generally scarce and only a limited number of geometrical configuration has been considered.

In the case of rectangular openings in sheet metal walls, and for Reynolds number varying in the typical range $200 \leq Re_{op} \leq 400$ the coefficient K' varies between 2.1 and 2.8 [1].
For perforated sheet metal plates the value of K' depends upon the hole size and the Reynolds number and for opening made of a grid of small circular holes, in line or staggered, some correlations have been developed in [4,12].

Our results indicate that the better agreement between experimental and theoretical values of the volumetric air flow rate $\dot{V}$ can be obtained assuming for all the geometric configurations tested a value of K'=2.6, except in the case of vents made of a grid of the smallest holes (D=3.25 mm) for which a value K'=4.0 is more appropriate.
In general, the value of K' found in this work agree with those recommended in [1] for rectangular openings and in [12] for perforated plates except the configuration having the holes diameter D=3.25 mm for which the correlation proposed in [12] gives a value of K' more than three times greater.

Some reference values for the friction factor K" can be obtained assuming perfectly smooth channels [13] or parallel cards with some particular kinds of obstructions [2]. Furthermore, in case of channels containing two-dimensional obstructions theoretical studies of mixed and forced convection, including an analysis of the distributed friction losses, have been developed in [14,15]. In this work the coefficient K" has been evaluated, in approximate way, using the correlation proposed in [15] and originally valid for forced laminar flow inside channels with two-dimensional obstructions

$$K'' = \frac{77.4}{Re_{ch}} \quad (13)$$

where the hydraulic diameter in the Reynolds number ($Re_{ch} = wD_h/\nu$) is defined as follows:

$$D_h = 2 \left[\frac{c(l+s) - Hl}{l+s+H} \right] \quad (14)$$

(For the meaning of the symbols used in (13) and (14) see the sketch of Fig. 9).

Finally, equation (12) shows that the overall pressure loss Δp depends not only upon the value of the coefficients K' and K" but also upon the ratio $(A/NA_c)^2$ which influences the relative magnitude of the two friction terms. If the opening area A is relatively small compared to the sum of cross section area of the channels NA_c, the distributed pressure losses are generally negligible with respect to those concentrated near the openings. On the contrary in a box with relatively large value of A the second term of equation (12) may give a contribution comparable to the first one, depending upon the value of K" and a correct evaluation of the distributed friction losses is necessary to give a satisfactory agreement with the experimental results.

The tests performed in this investigation have shown that for a box containing 9 cards and with opening area of about $A \geq 1000\ mm^2$ the pressure loss Δp_{ch} has the same order of magnitude of Δp_{op}. In all the other cases Δp_{ch} was negligible with respect to Δp_{op}.

In Fig. 10 all the experimental results obtained for boxes with 3, 6, and 9 cards and for openings with different shape and area are correlated using equation (6). The theoretical predictions are in satisfactory agreement with experimental data and the scattering of the correlated results is within ±10%.

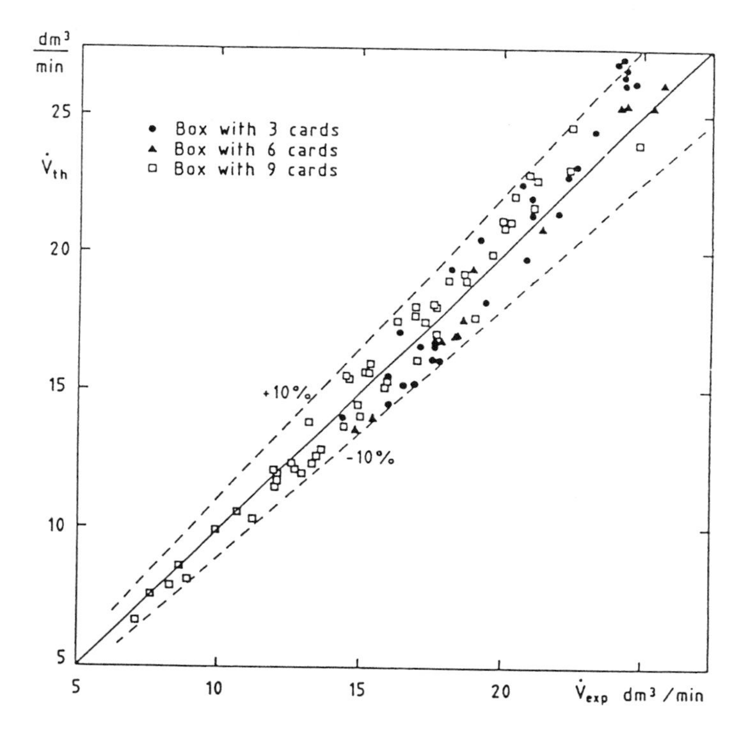

FIGURE 10. Comparison between theoretical and experimental air flow rate $\dot{V}$ for boxes with 3, 6, and 9 cards.

Equation (6) can be used as a preliminary tool for the thermal design of a natural ventilated box. For a given dissipated power P' and at a fixed average inner temperature $\bar{T}$, knowing the value of coefficient C (equation (7)) the inlet-outlet temperature rise $\Delta T_{1,2}$ can be obtained and from the balance equation (1) the volumetric air flow rate $\dot{V}$ can be evaluated. Finally, using equation (6) the opening area A can be established to which corresponds the temperature rise $\Delta T_{1,2}$.

5. CONCLUSIONS

The effects of some parameters on the cooling efficiency of a natural ventilated box containing electronic cards have been experimentally investigated. The main results obtained in this work are

- the inner air temperature rise and the thermocirculation flow rate are considerably affected by the spacing between the cards; for a given volumetric power density there is an optimum spacing to which corresponds a minimum temperature rise inside the box. The value of the optimum spacing is s_1=15 mm;
- the insertion of some unheated vertical metallic slabs between the dissipating cards may improve the efficiency of the natural cooling, reducing the air temperature rise in the upper region of the box;
- the opening surface area has a great influence on the temperature rises $\overline{\Delta T}$ and ΔT_M while the opening geometry has a minor effect. In the case of openings made of circular holes the diameter D=4-5 mm represents a critical value below which the concentrated pressure losses at the openings tend to increase rapidly;
- a simple relationship is proposed able to correlate all experimental data with a scattering contained within ±10%. This correlation can be used as a preliminary tool for the thermal design of a box.

REFERENCES

1. Heat Transfer and Fluid Flow-Data Books, General Electric, Section 504.4, April 1974, pp. 1-4.

2. Ellison, G.N., *Thermal Computations for Electronic Equipment*, Van Nostrand Reinhold Company, 1984.

3. Ngai, P., Mathematical Modelling in Microcomputers and Electronic Boxes by Natural Ventilation, *Proc. 3d Int. Elec. Pack. Soc. (IEPS)*, pp. 91-101, October 1983.

4. Ishizuka, M., Miyazaki, Y. and Sasaki, T., On the Cooling of Natural-air-cooled Electronic Equipment Casing (Proposal of a Practical Formula of Thermal Design), *Bulletin of JSME*, vol. 29, no. 247, January 1986.

5. Guglielmini, G., Milano, G. and Misale, M., Electronic Cooling by Natural Convection in Partially Confined Enclosures, *Heat and Technology*, vol.3, no. 3/4, pp. 43-57, 1985.

6. Guglielmini, G., Milano, G. and Misale, M., Free Convection Air Cooling of Ventilated Electronic Enclosures, *Proc. 2d U.K. Nat. Heat Transfer Conference*, Glasgow, 14-16th September 1988.

7. Elenbaas, W., Heat Dissipation of Parallel Plates by Natural Convection, *Physica*, vol. 9, no. 1, Holland, 1942.

8. Levy, E. K., Optimum Plate Spacing for Laminar Natural Convection Between Vertical Plates with Symmetric Heating, *J. of Heat Transfer*, vol.104, pp. 501-507, 1982.

9. Bar-Cohen, A. and Rohsenow, W.M., Thermally Optimum Spacing of Natural Convection Cooled Parallel Plates", *J. of Heat Transfer*,vol. 106, pp. 116-123, 1984.

10. Bar-Cohen, A. and Rohsenow, W.M., Thermally Optimum Arrays of Cards and Fin in Natural Convection, *IEEE Trans.*, vol. CHMT-6, no. 2, June 1983.

11. Steinberg,D.S., *Cooling Technique for Electronic Equipment*, J. Wiley and Sons. Inc., 1980.

12. Ishizuka, M., Miyazaki, Y. and Sasaki, T., Air Flow Coefficients for Perforated Plates in Free Convection, *J. of Heat Transfer*, vol. 109, pp. 540-543, 1987.

13. Rohsenow, W.M., Hartnett, J.P. and Ganic', E.N., *Handbook of Heat Transfer Fundamentals*, 2d Edition, McGraw-Hill, Chapter 7, 1985.

14. Sparrow, E.M. and Chukaev A., Forced-convection Heat Transfer in a Duct Having Span-Wise-Periodic Rectangular Protuberances, *Numerical Heat Transfer*, vol. 3, pp. 149-167, 1980.

15. Braaten, M.E. and Patankar, S.V., Analysis of Laminar Mixed Convection in Shrouded Arrays of Heated Rectangular Blocks, *Int. J. Heat Mass Transfer*, vol. 28, no. 9, pp. 1699-1709, 1985.

Modeling of PCB's in Enclosures

M. CADRE
Centre National d'Etudes des Télécommunications
Route de Trégastel
22301, Lannion Cedex, France

A. VIAULT, V. PIMONT, and A. BOURG
CISI Ingénierie
57, rue Pierre-Semard
38000, Grenoble Cedex, France

ABSTRACT

The designers of electronic equipment are more and more often obliged to enclose the heat dissipating printed circuit boards in impervious compartments. This constraint is imposed upon them mainly because of use of the equipment in all kinds of environments (polluted conditions electromagnetic problems, etc.). In such a configuration the cooling of the components is limited, as it is necessary to optimize the geometric configuration of the system in order to be able to obtain maximum power while at the same time keeping the components under reasonable thermal conditions.

At the present time, the design of electronic enclosures relies essentially on experiments in special conditions, generally ideal, and only a good deal of common sense finally leads the designer towards an entirely satisfactory arrangement of the enclosure. In fact, it is only possible to detect potential local difficulties (hot spot on a component) by numerical simulation of the cooling of the enclosure.

The aim of the present paper is to describe the method used in the computer code ECORCE to model the combined convection, radiation and conduction in walls and PCB's taking place in such enclosures.

The results obtained on enclosures consisting of 2 to 3 PCB's are presented, showing the usefulness of the model for simulating thermal exchanges inside as well as outside of the enclosure.

INTRODUCTION

Due to technological requirements, such as electromagnetic protection and environmental protection, electronic devices are becoming more and more confined in enclosures which are often impervious. For such configurations, the cooling of the printed circuit boards is limited and thus it is necessary to optimize the geometrical configuration of the system so as to extract a maximum amount of heat and to keep the integrated circuits under reasonable thermal conditions.

The power dissipated to the outside air by a component passes through the walls of the enclosure, by conduction within the walls, and then by radiation and convection.

Heat is exchanged between the component and the enclosure by one of the following 3 ways :

- directly by radiation

- through the printed circuit boards and then the contact surfaces boards- enclosure by conduction (especially when the printed circuit boards are of the metal core type).

- through the air which transfers the heat beween the components and the enclosure walls by convection.

Consequently, the amount of heat transfered to and from each surface of the device is highly dependent on the temperature of the surrounding surfaces (conduction); but due to the convection and radiation, it is also dependent upon the characteristics of other elements in the enclosure.

Numerical simulations of natural convection air flow in a channel [ref. 1, 2, 3] or in cavities free of obstacles [ref. 4, 5, 6] have shown that, a priori, none of the modes of thermal transfer can be neglected.

The computer software THEBES deals with air flows and temperature simulation in complex electronic devices, in natural as well as forced convection. The software ECORCE, presented in this paper, is a specialized version of THEBES, oriented towards analysis of the air flow and thermal design of enclosures containing vertical printed circuit boards.

1 - MAIN FEATURES OF THE NUMERICAL MODEL

The enclosures considered for the study are commonly used in the telecommunications industry ; they contain printed circuit boards which are set up vertically and parallel to one another.

Boards are modelled by smooth plates dissipating uniform or non-uniform heat.

Due to the different possible locations of the boards inside the enclosure relative to the position of the walls, the numerical simulation of the air flow must be three-dimensionnal.

This 3-dimensionnal aspect also allows for the heat exchange from the walls to the ambient air. These transfers are modelled by means of a heat transfer coefficient relevant to the type of air flow surroundings the enclosure (natural or forced convection) and to the external structure of the walls (with or without fins).

The natural convection air flow inside the device is simulated by using discrete values for the Navier-Stokes equations and the energy equation according to the "finite volumes" method. The energy equation meets the Boussinesq requirement [ref. 7]. The boundary conditions assume no slip condition for all solid parts inside the enclosure, such as that required in the computer software THEBES for air flow calculation inside cabinets [ref. 8, 9, 10]. The thermal coupling between fluid and wall, or board, is assured by a thermal balance, which is applied for each element of the network of the solid surfaces. This network is derived from the network for the air flow.

The walls are assumed to be thin enough to be isothermal within their thickness.

The heat is generated on one side of the printed circuit board and then conducted to the other side. The description of such phenomena would have required a specific software and, therefore, the board is only modelled by two distinct parallel networks, one for each side of the board. The power dissipated on the board is distributed to each face.

2 - THERMAL BALANCE FOR A SOLID ELEMENT

The thermal balance for every node I of the walls or of the boards must take into consideration the convective flux φ_{FI} to the volume of air adjacent to the considered element, the conductive flux φ_{CI} with the surface element J in contact with node I and, finally, the radiative flux φ_{RI} from all the surface elements in view of the lateral surface of element I.

- the convective flux φ_{FI} :

On the surface, the air has no velocity and thus the exchange between the surface element at temperature T_{PI} and the fluid element J at temperature T_{FJ} is considered to be conductive (figure 1) :

$$\varphi_{FI} = \frac{\lambda_F}{D_{IJ}} (T_{FJ} - T_{PI}) \qquad (1)$$

where λ_F is the thermal conductivity of air and D_{IJ} is the distance between the two nodes I and J. Such an approximation is correct as long as the mesh is fine enough.

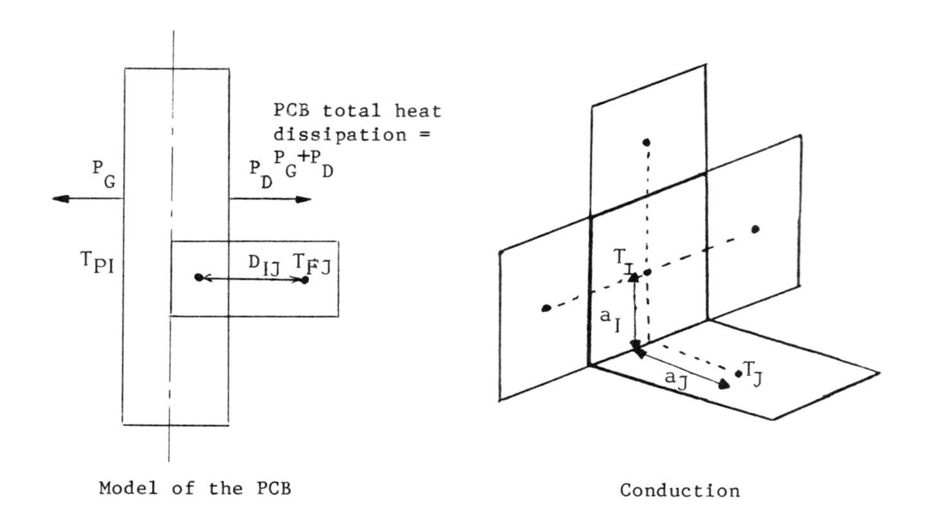

FIGURE 1. Schematization of conduction.

- the conductive flux φ_{CI} :

This flux, related in reality to the section S_I of the element I, is expressed as a function of the lateral face A_I :

$$\varphi_{CI} = \frac{S_I}{A_I} \sum_{j=\text{contact element}} \frac{\lambda_{PIJ}}{a_{IJ}} (T_{PJ} - T_{PI}) \tag{2}$$

a_{IJ} is the thermal resistance between the nodes I and J (figure 1) :

$$\frac{a_{IJ}}{\lambda_{PIJ}} = \frac{a_I}{\lambda_{PI}} + \frac{a_J}{\lambda_{PJ}} \tag{3}$$

- the radiative flux φ_{RI} :

The different walls and boards are assumed non-transparent and gray, with an isotropic emissivity. Under these conditions, the total emissivity of the element I is equal to its absorptivity.

As the temperature expected inside the enclosure should not exceed 100°C, the radiation is infra-red, and it can thus be assumed that the air is transparent.

The radiative fluxes are modelled according to well-known concepts [ref. 11 and 12] such as the irradiance E of the face (flux effectively received from other solid elements of the enclosure) and the face emissivity ϵ_I where σ is the Planck Constant :

$$\varphi_{RI} = \epsilon_I (E_I - \sigma T_{PI}^4) \tag{4}$$

To make allowances for possible multiple reflections, E_I is expressed as a function of the radiosity J_L of all elements L of the walls and boards. The radiosity J_L symbolizes the total flux (emitted and reflected) from the face L :

$$J_L = (1 - \epsilon_L) E_L + \epsilon_L \sigma T_{PL}^4 \tag{5}$$

The first term in equation (5) expresses the reflection and the second term is related to the emission.

Finally, only a certain percentage of the energy leaving L will reach the face of the element I. The view factor F_{LI}, which relates this percentage of energy reaching I, assumes that the faces I and L are isothermal and the irradiance is uniform :

$$F_{LI} = \frac{\text{radiosity of L reaching I}}{\text{radiosity of L}} \tag{6}$$

Thus, F_{LI} is only dependant on the relative geometry between L and I. If A_L is the lateral surface of the element L, F_{LI} can be written as follows :

$$F_{LI} = \int_{A_L} \int_{A_I} \frac{\cos \beta_I \cos \beta_L \, dA_I \, dA_L}{\pi R^2} \tag{7}$$

Notations are shown on figure 2.

Futhermore, F_{LI} must satisfy the following equations :

$$A_I F_{IL} = A_L F_{LI}$$
$$\sum_L F_{IL} = 1 \tag{8}$$

Using these relations, E_I can be written as follows :

$$E_I = \sum_L F_{IL} J_L \tag{9}$$

Thus, the radiative flux φ_{RI} is determined entirely from the resolution of the set of equations (9) and (5) where the radiosities J_L are finally the unknows.

The thermal balance of a wall or a board element I is established from the fluxes φ_{RI}, φ_{FI}, and φ_{CI} written for the lateral face of I and from a last flux φ_I, which expresses either :

- the power dissipated by the element if I is a board element

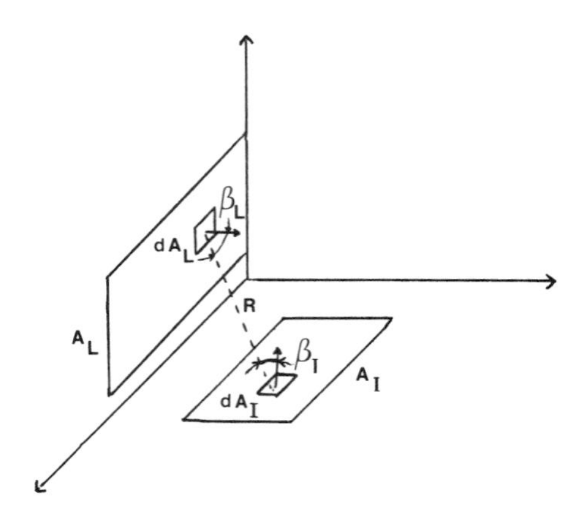

FIGURE 2. View factor.

- or the heat exchange with the outside air, modelled by a heat transfer coefficient, h_{CI}, if I is a wall element

$$\varphi_I = h_{CI} (T_\infty - T_{PI}) \tag{10}$$

If the thermal inertia of the wall and the board is neglected, the temperature distribution on the surfaces is obtained by solving the non-linear system which is based upon the radiative equations (5) and (9) and the energy balance of the solid elements of the enclosure (equation (11)) as follows :

$$\varphi_{FI} + \varphi_{CI} + \varphi_{RI} + \varphi_I = 0 \tag{11}$$

3 - RESOLUTION OF THE SET OF EQUATIONS MODELLING THE FLUID-SOLID SURFACES SYSTEM

In order to reach the steady state and the thermal equilibrum of the enclosure, a time iterative calculation is performed for the thermo-convective state fo the total fluid-solid system.

From one iteration n to the next one at time t_{n+1}, the calculations are followed as indicated below :

a) evaluation of the air velocity field $\vec{V}^{n+1}$ using an explicit method, such as the SOLA method [ref. 7 and 13].

b)evaluation of the irradiance E^n by solving the set of equations (5) and (9), the temperatures $T_{PL}{}^n$ being fixed.

c) evaluation of the solid temperatures $T_{PI}{}^{n+1}$ from equation (11) in which the irradiances are given by the evaluation at stage b. (Here, radiation is expressed in an explicit way, so as to obtain a linear system $T_p{}^{n+1}$. Convection is also expressed in an explicit way at this stage).

d) evaluation of the fluid temperatures T_F^{n+1} from T_p^{n+1} and $\vec{V}^{n+1}$.

4 - SUITABLE CONFIGURATIONS FOR USE OF ECORCE SOFTWARE

Much detailled work must be carried out during the development of the software in order to obtain a precise model of the thermal board and thermal walls and to be able to describe the geometry (i.e., the shape factors, contact between elements, etc.). Consequently, not only the computational time, but also the memory size needed, particularly to calculate the radiation, rapidly become extremely large. Sometimes it is even impossible to accommodate if the mesh is highly sophisticated in order to outline the boundary layers.

Under these conditions, when the mesh calculations are close to the solid surfaces, equation (1) is used. This type of model is restricted to enclosures including no more than 2 or 3 printed circuit boards (cf fig.3 and 4 for a Rayleigh-Benard air flow type), however in industry, such enclosures can have 10 boards or more.

Furthermore, if the components on the PCB's have a non negligible thickness it may be necessary to carry out a specific study on each one.

Thus the first technical problem encountered when trying to evaluate industrial devices is to **reduce the size of the physical model** ; this implies that a formulation of the physical phenomena must be proposed for use with a coarse mesh ; that is to say, an integration in space of the local model must be performed. Obviously, such correlations should not alter the preditive character of the model (i.e., its capability to take into account the interactions between the different simulated phenomena). Under such conditions, the software can be useful **to compare two different possible configurations** for a specific device. It can only guarantee indications for the dimensions of the apparatus.

For such models, the reduction of the size of the mesh will be associated with

- a simplification of the calculations of the shape factors

- the use of correlations to model the heat exchange and the friction on the boards and the walls.

The second problem set up by the use of such reduced models is that the software user must be capable of choosing or verifying the appropriateness of the models for the studied system (according to the mesh, the area of air flow, etc.)

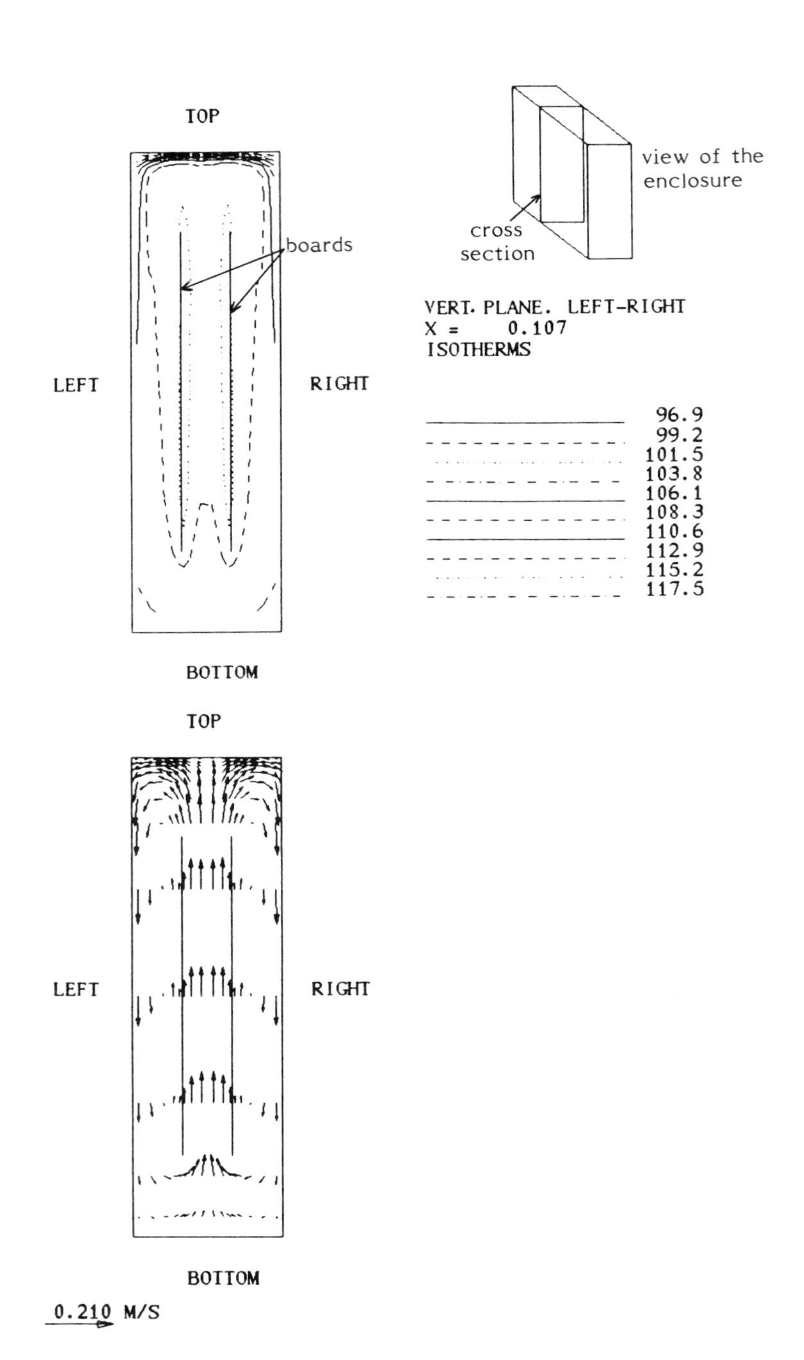

FIGURE 3. Rayleigh - Benard air flow in an enclosure containing 2 PCB's. Isotherms and velocity vectors in a vertical plane, left-right section.

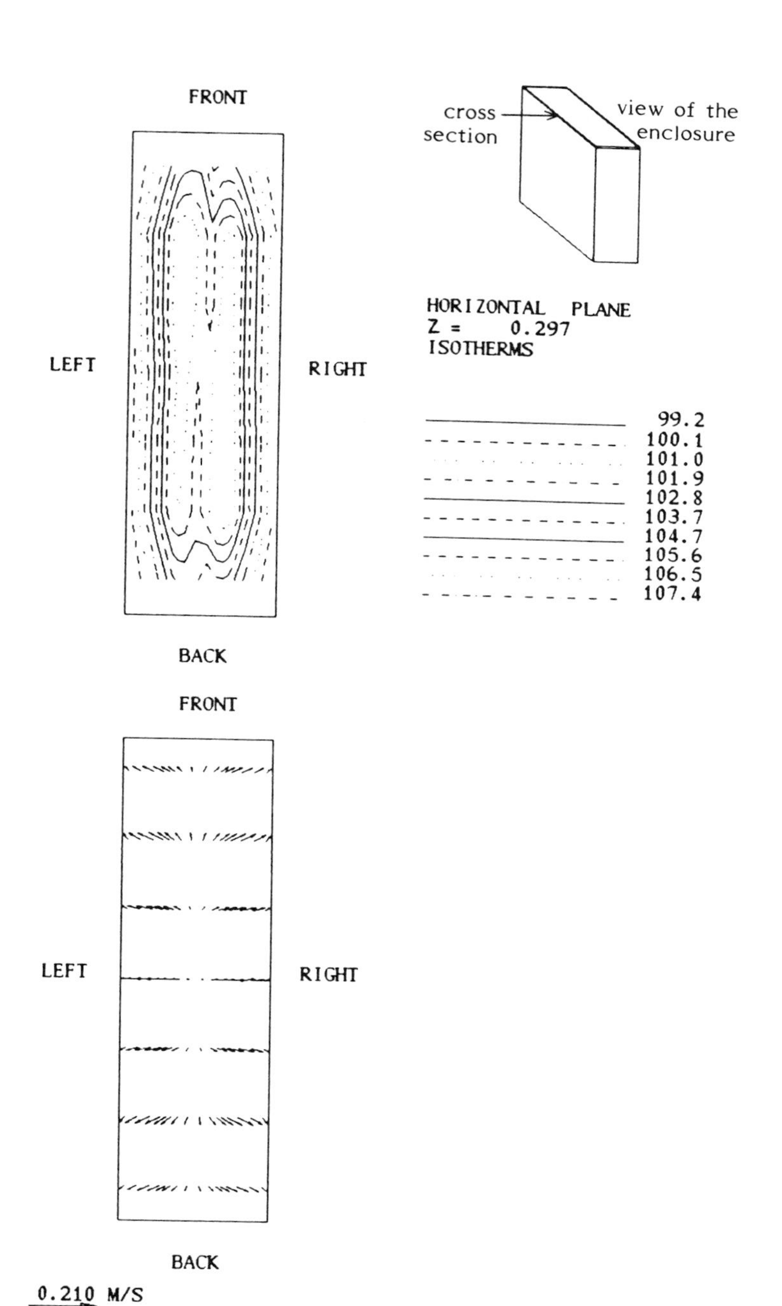

FIGURE 4. Rayleigh - Benard air flow in an enclosure containing 2 vertical PCB'S. Isotherms and velocity vectors in an horizontal plane-cross section located close to the cover of the enclosure.

5 - CONCLUSIONS

The computer software ECORCE, developed by a joint programme between "CISI Ingenierie" and the French National Center for Telecommunication (C.N.E.T.), is a tool for analyzing the cooling of enclosures containing vertical printed circuit boards.

Sophisticated models can be used on computers with a large amount of CPU time and a large virtual storage for analyzing complex enclosures or to evaluate correlations on a simple model with a thin mesh.

A rough model with the following characteristics has also been developed :

- there are not many elements between two boards (sometimes only one elementary volume)

- the boundary layers on solid surfaces are simulated by thermal and hydraulic correlations adjusted by experiments or tests made locally with a thin mesh.

- an average radiative model for each zone is used to reduce the virtual storage of the view factors.

This second release of the software requires at least scientific workstations and is useful for rapidly eliminating configurations which are not acceptable as far as the thermal aspects are concerned.

NOMENCLATURE

A_I = lateral surface of the solid element I (m^2)

E = irradiance (W/m^2)

F_{LI} = view factor between I and L

J = radiosity (W/m^2)

S_I = section of surface I (m^2)

T_F = temperature of a fluid element (K)

T_P = temperature of a solid element (K)

T_∞ = temperature of the air surrounding the enclosure (K)

a = thermal resistance (K m/W)

h_c = convective heat transfer coefficient (W/m^2 K)

ϵ = emissivity

σ = Stefan-Boltzmann Constant = $5.67\ 10^{-8}\ W/m^2\ K^4$

λ = thermal conductivity (W/m K)

φ_{CI} = conductive flux leaving I (W/m^2)

φ_{FI} = convective flux leaving I (W/m^2)

φ_{RI} = radiative flux leaving I (W/m^2)

Subcripts

F = fluid

P = solid surface (walls or boards)

REFERENCES

1. Capenter, J.R., Briggs, D.G., and Sernas, V., Combined radiation and developing laminar free convection between vertical flat plates with asymmetric heating, **Journal of Heat Transfer**, pp. 95-100, 1976.

2. Sparrow, E.M., Shah, S., and Prakash, C., Natural convection in a vertical channel : 1 Interacting convection and radiation. 2 The vertical plate with and without shrouding, **Numerical Heat Transfer**, vol.3, pp. 297-314, 1980.

3. El Yahyioui, M., Influence des effets radiatifs et conductifs sur la convection naturelle entre deux plaques verticales parallèles chauffées à flux constant - thèse 3ème cycle, Université de Poitiers, France, 1983.

4. Lauriat, G., A numerical study of a thermal insulation enclosure : influence of the radiative heat transfer, Natural convection in Enclosures, 19 th National Heat Transfer Conference, ASME, Ed. TORRANCE and CATTON, 1980.

5. Larson, D.W., Viskanta, R., Transient combined laminar free convection and radiation in a rectangular enclosure, **Journal Fluid Mech.**, vol. 78, part 1, pp. 65-85, 1984.

6. Kim, D.M., Viskanta, R.N., Effect of wall conduction and radiation on natural convection in a rectangular cavity, **Numerical Heat Transfer**, vol. 7, pp. 449-470, 1984.

7. Villand, M., Discrétisation par volumes finis, Thermohydraulique monophasique, cours INSTN, novembre 1987.

8. Latrobe, A., Viault, A., et Chabanne, J., Calculs Thermiques dans les baies électroniques en convection naturelle ou forcée. 2e colloque national sur la thermique, l'énergie et l'environnement des matériels de télécommunications, informatique, bureautique et autres systèmes électroniques, 1985.

9. Cadre, M., Latrobe, A., Le Jannou, J.P., Simulation and experimentation of air flow in electronic racks, Intelec'86, Toronto, 1986.

10. Le Jannou, J.P., Cadre, M., Latrobe, A., Viault A., Thermal field prediction in electronic equipment, Heat Transfer 1986, Proceedings of the International Heat Transfer Conference, Vol. 6, Hemisphere Publishing Corp., Washington, D.C. 1986, pp. 2983-2988.

11. Sparrow, E.M., Cess, R.D., **Radiation Heat Transfer**, Mc Graw Hill, pp. 366, 1978.

12. Siegel, R., Howell, J.R., **Thermal Radiation Heat Transfer**, Mc Graw-Hill, pp. 814, 1972.

13. Hirt, Nickols, Romero, SOLA A numerical Solution Algorithm for transient fluid flows. Los Alamos scientific laboratory Report LA 5852, 1975.

Convective Heat Transfer in Electronic Facilities

G. N. DULNEV, V. A. KORABLYOV, and A. V. SHARKOV
Institute of Precision Mechanics and Optics
Leningrad, USSR

ABSTRACT

Results of experimental investigations are presented for free convective heat transfer in enclosed cavities parallelepiped in shape and in horizontal channels bounded by isothermal plates. The influence of different conditions is investigated at inlet of channels of cooling systems under forced convection heat transfer. The geometric parameters of radiators with heat flow of 2×10^6 W/m^2 density for cooling power semiconductor facilities are determined.

The accuracy of computed temperature fields and the correctness of parameters of systems providing satisfactory thermal environment facilities depends on the reliability of available information, in particular, on data concerning local and average characteristics of the convective heat transfer.

The determination and refinement of the heat transfer relations leads not only to greater design accuracy but show possibilities of application of the least power intensive cooling methods that in the end decrease weight and power characteristics of electronic devices and increase their operating reliability and quality.

Considered here are the results of an investigation of heat transfer intensity under free and forced conditions in cavities and channels common to electronic facilities.

1. FREE-CONVECTIVE HEAT TRANSFER IN ENCLOSED CAVITIES

A cavity having two opposite surfaces with different temperatures and four other adiabatic surfaces is the most common investigated model [1]. The heat transfer in an enclosed cavity with all surfaces taking part in this process (Figure 1) is not nearly so often investigated. Reference [2] contains a relation for estimating heat transfer in a cavity where one of the surfaces (vertical or lower horizontal) has a higher temperature as compared with the other five isothermal surfaces.

Reference [5] describes the heat transfer in a cube-like cavity, four wall of which had different temperatures and the two horizontal surfaces were thermally isolated. We have not succeeded in computing the free convective heat transfer in enclosed cavities all surfaces of which take part in heat transfer and have different temperatures. The experimental facility shown in Figure 2 has been developed to solve this problem.

The facility comprises a chamber formed with five rigidly fixed plates, 1, 2, 3, 4 and 5, and mobile plate 6 (Figure 2). The plates are made of copper. Their temperatures are maintained by a temperature controlled liquid flowing through pipes 7. To reduce heat flow there are ebonite gaskets between plates. Heat insulation 9 is set outside preventing plate heat losses and decreasing the influence of the environment upon heat transfer in the chamber. The chamber dimensions can vary up to 0.2 x 0.2 x 0.2 m. The chamber is suspended, providing the possibility to turn it around the horizontal axis. For the experiment four surfaces were vertical and two were horizontal (Figure 1).

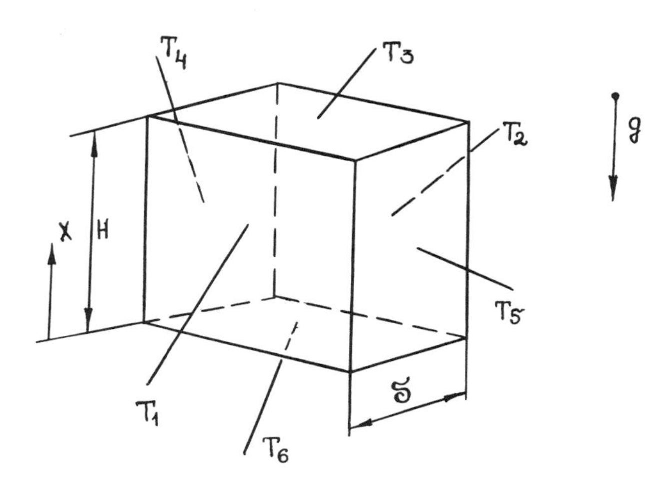

FIGURE 1. Definition sizes and temperatures of enclosed cavity.

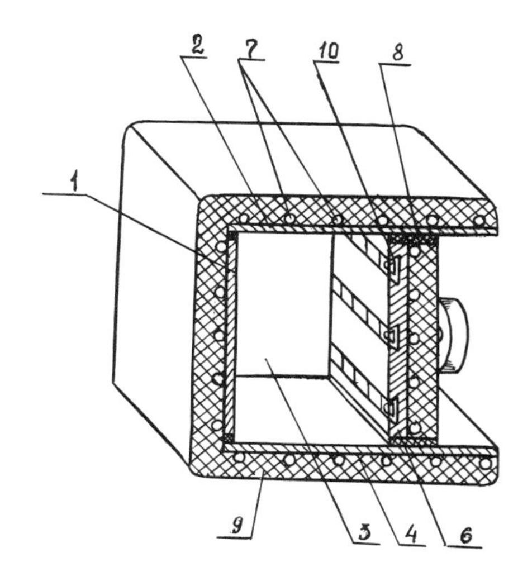

FIGURE 2. Experimental facility.

The temperature of the chamber surfaces is measured by means of 24 thermocouples. Heat flow meters of α -calorimeter type [6] are used to measure local heat flows. Six heat flow meters of disk shape 15 mm in diameter and 2 mm thick are installed in specially made depressions so that their outer surface is coplanar to the plane surface. The possibility is proposed to move these flow heat meter along the plate surfaces. The heat flow measurement error analysis showed that it does not exceed 3% at 0.95 confidence level.

The main dimensions and designations of the cavity are given in Figure 1. The distance between the horizontal plates was changed from 0.01 to 0.20 m. The investigation was carried out for the following thermal combinations:

1. $T_1 > (T_2 = T_3 = T_4 = T_5 = T_6)$;
2. $(T_1 = T_2) > (T_3 = T_4 = T_5 = T_6)$;
3. $(T_1 > T_2) > (T_3 = T_4 = T_5 = T_6)$;
4. $(T_1 < T_2) > (T_3 = T_4 = T_5 = T_6)$.

The convective component of heat flow q_i was determined by subtracting the radiant component q_Λ from heat flow q_Π determined by heat flow meter. The value q_Λ was calculated according to the radiant heat transfer law and usually didn't exceed 15% of q_Π.

It is proposed to estimate local and average coefficients of convective heat transfer $\alpha_i(x)$ and $\bar{\alpha}_i$ at the i-th surface by the formulas

$$\alpha_i(x) = \frac{q_i}{T_i - \bar{T}} \quad ; \quad \bar{\alpha}_i = \frac{\bar{q}_i}{T_i - \bar{T}} , \tag{1}$$

where q_i and $\bar{q}_i$ are local and average convective heat fluxes, T_i is the temperature of the i-th surface. T_i is weighted average temperature, which is found by the formula

$$\bar{T} = \sum_{i=1}^{6} T_i A_i \left(\sum_{i=1}^{6} A_i \right)^{-1} , \tag{2}$$

where A_i are areas of the cavity surfaces.

Generalization of results of experimental studies is made by means of local and average Nusselt and Rayleigh numbers defined as follows:

$$Nu_{xi} = \frac{\alpha_i(x) \cdot X}{\lambda} \; ; \quad \overline{Nu}_{Hi} = \frac{\bar{\alpha}_i \cdot H}{\lambda} , \qquad (3)$$

$$Ra_{xi} = \beta g \frac{X^3}{\nu^2} (T_i - \overline{T}) Pr , \quad Ra_{Hi} = \beta g \frac{H^3}{\nu^2} (T_i - \overline{T}) Pr , \qquad (4)$$

where thermophysical properties were determined at temperature $\bar{T}$.

The correlation between Nu_{xi} and Ra_{xi} was determined by the least squares method with the following dependence used as a model:

$$Nu_x = C_1 \, Ra_x^{C_2} . \qquad (5)$$

where C_1 and C_2 are constant coefficients calculated by the method of least squares, Nu_x and Ra_x are local values of Nusselt and Rayleigh numbers.

Within $0.17 < xH^{-1} < 0.83$, $0.1 < \delta H^{-1} < 1$ and $5 \times 10^4 < Ra_x < 2 \times 10^7$ the obtained experimental data are governed by the relation

$$Nu_x = 0,53 \, Ra_x^{0,25} . \qquad (6)$$

This equation was obtained on the basis of generalization of experimental results on the convective heat transfer intensity from the vertical surfaces of cavities which width is approximately equal to height. For the other vertical surfaces the accuracy of the determined relationship needs additional studies.

A comparison of this relationship (6) with experimental data is given in Figure 3. The experiments were carried out at the above combination of surfaces temperature and the results are labeled with marks.

Agreement of experimental results with the values estimated by equation (6) is ± 22% at 0.95 confidence level.

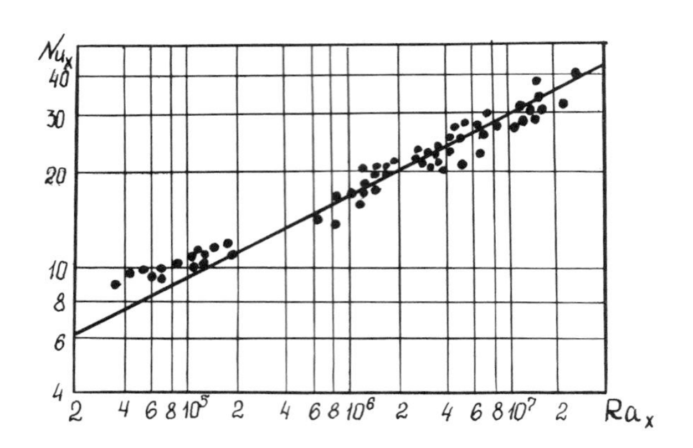

FIGURE 3. Results of free convective heat transfer investigation from the vertical surfaces of an enclosed cavity.

The average parameters of heat transfer in an enclosed cavity were also determined. For this purpose surface average heat flow densities $\bar{q}_i$ were calculated from

$$\bar{q}_i = \frac{1}{n} \sum^{n} q_i \,, \tag{7}$$

where n is the number of measurements of local heat flows.

For this case the correlation is

$$\overline{Nu}_H = 0{,}71 \; Ra_H^{0{,}25} . \tag{8}$$

This equation describes the obtained experimental data at $0.1 < \delta H^{-1} < 1$ and $10^7 < Ra_H < 10^8$ with 12% error at 0.95 confidence level.

Studies of heat transfer at the horizontal surfaces of enclosed cavities were conducted under following conditions (Figure 1):

1. $T_4 > (T_1 = T_2 = T_3 = T_5 = T_6)$;

2. $(T_4 = T_3) > (T_1 = T_2 = T_5 = T_6)$;

3. $T_3 > (T_2 = T_1 = T_4 = T_5 = T_6)$.

Studies and processing of measurement results were made according to the abovementioned procedure of measuring heat transfer from the vertical surfaces. The temperatures of all cavity surfaces and local heat flows were measured at three points on every one of the horizontal surfaces as well as weighted average temperature, and the associated coefficients of heat transfer from surfaces were determined by (1).

The experiments were carried out at 90°C on hot surfaces and at 17°C on cold ones. Distance H (Figure 1) between the horizontal surfaces discretely varies from 0.01 m to 0.20 m in this case two other dimensions of cavity the width and the height remained unchangeable and were equal to 0.20 m.

Analysis of the measurement results of heat transfer intensity from horizontal surfaces showed that heat transfer coefficient is distributed nonuniformly along the surface and also with time. Besides that, convective heat transfer intensity differs considerably for surfaces transferring heat upward or downward. On surfaces transferring heat upward the local heat transfer coefficient variation with time is of random character, the average period is approximately 100 ± 30 s and the amplitude of the fluctuations deviates not less than $\pm$ 50% from the average value (heavy line in Figure 4). The time averaged local Nusset numbers are given in Figure 4 with dotted line. The width of the heat transfer surface is used for δ in the dimensionless numbers. The following relationship to estimate surface averaged Nusselt number was obtained:

$$\overline{Nu}_\delta = 0{,}245\, Ra_\delta^{0{,}33}. \qquad (9)$$

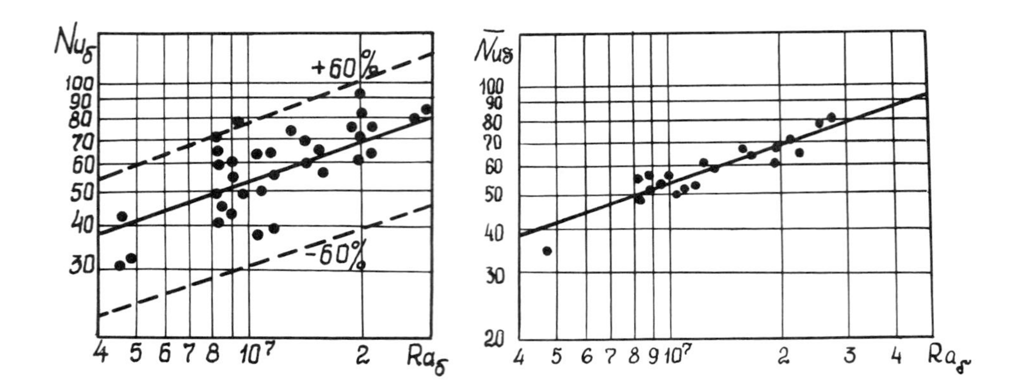

FIGURE 4. Local characteristics of heat transfer from the horizontal surface of enclosed cavity.

FIGURE 5. Average characteristics of heat transfer from horizontal surface of enclosed cavity.

A correlation is presented in Figure 5 between surface averaged Nusselt numbers and criterion relation (9). Disagreement between the experimental data and those obtained from (9) does not exceed 12% at 0.95 confidence level.

On the cavity surface transferring heat downward the heat flow fluctuations are practically absent. The heat exchange intensity maximum is usually at one of the edges and can differ from average value $\overline{Nu}_\delta$ by 30%. At $H > 0.3\ \delta$ the convective heat exchange contributed to heat transfer from upper hot surface to lower and vertical cold surfaces not more than 30% (convection coefficient is 1.3). At $H > 0.3\ \delta$ convective heat exchange intensity no longer varies with increasing distance between surfaces H and is defined by

$$\overline{Nu}_\delta = 0.1\ Ra_\delta^{0.33} \qquad (10$$

This relationship describes experimental data at $10^7 < Ra < 1$ and $0.3 < H\delta^{-1} < 1$ with 21% error at 0.95 confidence level.

A comparison of these relationships with data available in the literature [11, 12, 13, 14] showed reasonable coincidence.

2. FREE-CONVECTIVE HEAT TRANSFER IN HORIZONTAL OPEN CAVITIES

The plane horizontal cavity which has no vertical walls appears to be a typical element of instrumentation systems. For example, instrument boxes placed in racks are generally separated by air layers, the boxes contain PCBs that can be oriented horizontally, etc.

Experimental investigations were carried out on a facility (Figure 6) described in [7]. Plates of 0.25 m x 0.25 m and 0.11 m x 0.24 m were suspended and the distance between them was adjusted from 0.005 to 0.1 m. The temperature of plates was maintained equal and varied from +20°C to +90°C by temperature controlled water forced through pipes located within the plates. Heat flow going from the plate into the horizontal cavity was measured by a heat flow meter of the "additional wall" type. The temperature of the convective heat transfer surfaces was measured by thermocouples.

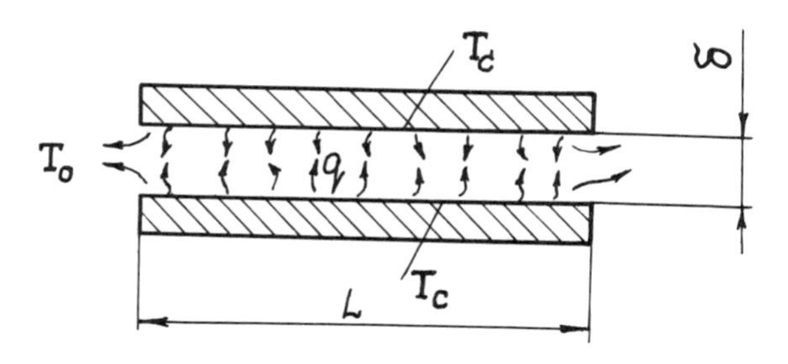

FIGURE 6. Experimental facility to investigate free convective heat transfer in a closed cavity.

The average heat transfer coefficient in a channel was calculated by the formula

$$\alpha = \frac{\Phi - \Phi'}{A\,(T_c - T_o)}\,, \tag{11}$$

where Φ is the heat flow measured by heat flow meter; A - area of heat transfer surface; T_c and T_o - temperature of heat transfer surface and ambient air; Φ' - heat loss from the plate to the surrounding space, which is determined by calculations.

During the test the distance δ between plates was set up at 5, 10, 15, 20, 30, 40, 50, 60 and 70 mm and the surface temperature changed from +30 °C to +90 °C.

The range of Rayleigh numbers was $5 \times 10^6 < Ra < 1 \times 10^8$. The air thermophysical properties were chosen at the temperature $T_s = 0{,}5\ (T_c + T_o)$. The width of the heat transfer surface L (Figure 7) was taken as the determining dimension. In case the heat transfer surface was not square the least dimension of it was taken as the determining one.

The results of the experiment are partially given in Figure 7. It is shown that the convective heat transfer intensity grows with the rise of temperature of heat transfer surfaces and increase in distance between them. During heat transfer in unbounded airspace ($\delta \rightarrow \infty$), the obtained Nusselt numbers correspond to the formulae recommended in [8, 9, 10]:

$$Nu_\infty = \frac{\alpha L}{\lambda} = 0{,}155\ Ra^{1/3}\,, \tag{12}$$

if a heat flow is upward from the heated surface

$$Nu_\infty = 0{,}095\ Ra^{1/3}. \tag{13}$$

if a heat flow is downward from the heated surface.

The convective heat transfer intensity between horizontal plates with isothermal walls can be described by the following relationship:

$$Nu = Nu_{\infty} \cdot \varepsilon_{\delta} , \tag{14}$$

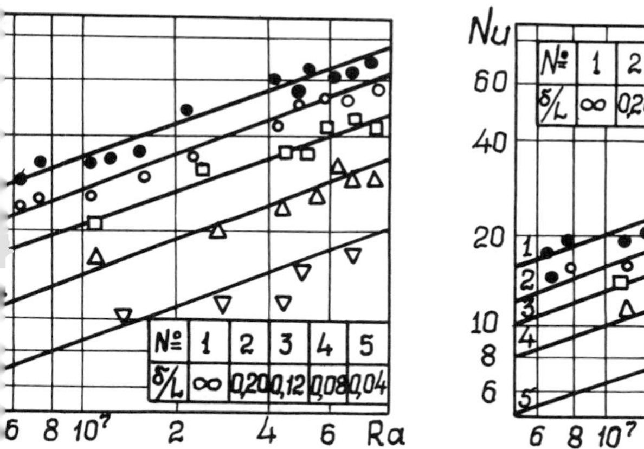

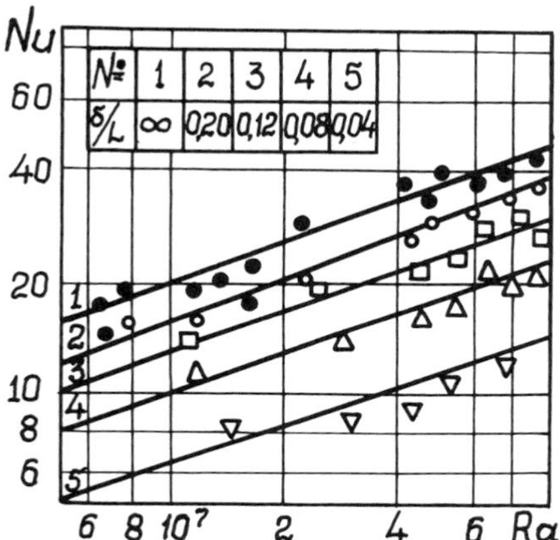

FIGURE 7. Results of free convective transfer in open cavity from its upper (a) and lower (b) surfaces.

where ε_{δ} is an empirical coefficient depending upon parameter δL^{-1} ; Nu is the Nusselt number for the parallel plate system; Nu_{∞} is the Nusselt number, estimated for the same surface transferring heat in the unbounded airspace.

On the basis of our studies for estimating heat transfer intensity in a horizontal channel with isothermal wall with natural convection of a gas the following relationships can be recommended:

for a surface transferring heat upward:

$$Nu = 0{,}155\, Ra^{1/3} \left[1 - \exp\left(0{,}23 - 10{,}3\ \delta L^{-1}\right)\right], \tag{15}$$

for a surface transferring heat downward:

$$Nu = 0{,}095\, Ra^{1/3} \left[1 - \exp\left(0{,}02 - 8{,}8\ \delta L^{-1}\right)\right]. \tag{16}$$

The equations (15) and (16) describe the derived experimental data with an error not exceeding $\pm$ 16% and at 0.95 confidence level. A comparison of the experimental data with the equations (15) and (16) is presented in Figure 7.

3. CONVECTIVE HEAT IN ANNULAR CHANNELS

The provision of adequate heat conditions for a variety of facilities such as lasers, electrovacuum instruments and others requires application of forced air and fluid cooling. The

available estimation procedures enable determination of the heat transfer intensity within the initial sections of cooling channels only at separate kinds of orifices. We have not succeded in finding information allowing estimation of local and mean intensity of convective heat transfer within the initial section of an annular channel when the incoming flow enters at various angles through various size orifices. Also we have not discovered works describing results of convective heat transfer investigations for different conditions of fluid exit from the channel. These effects are particularly important for short channels.

The test section shown in Figure 8 has been developed to carry out investigation of convective heat transfer in annular channels. The thin wall tube 1 of stainless steel has outer diameter of 10 mm. The temperature of the outer tube surface was obtained by measuring the temperature of inner surface with consideration of temperature difference across the tube wall. The inner temperature was measured by means of two thermocouple whose junctions were situated at the ends of springs pressing them to the inner surface of the pipe. By means of a special device thermocouples were moved along the tube 1.

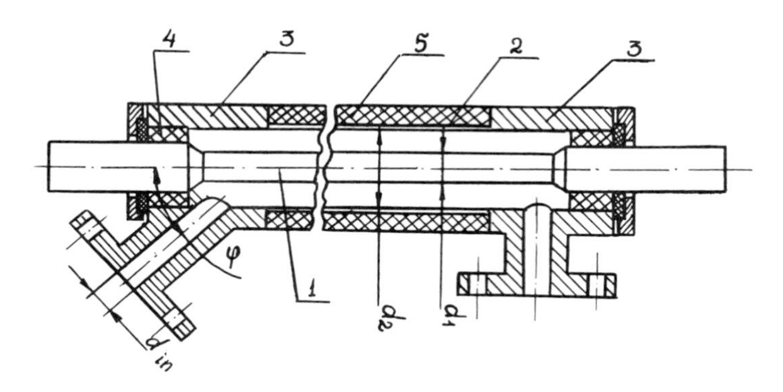

FIGURE 8. Test section to investigate heat transfer in an annulus

The inner tube 1 was inserted into the tube 2 which is connected with inlet and outlet branch pipes. Coincidence of the tube 1 and the tube 2 axes was provided with the gaskets 4. Moreover, these gaskets provided airtightness of the operating section. The outer surface of the tube 2 is covered with thermal insulation 5 to decrease heat loss from the operating section.

The branch pipes 3 differ from one another by diameter of inlet opening d_{in}, by diameter of channel d_2, by incoming liquid angle (Figure 8) and by displacement of inlet opening relative to symmetry axis e.

The characteristics of the branch pipes are given in Table 1.

The inner diameter of the tube 2 was 20, 15.6 or 13.6 mm and the length was 200 mm. There was a possibility of connecting two tubes of the same diameter in series. In this case the length of channel including inlet and outlet openings was 440 mm. The local coefficient of convective heat transfer $\alpha(x)$ was determined by

$$\alpha(x) = \frac{\Phi}{A[T_w(x) - T_f(x)]}, \tag{17}$$

TABLE 1.

Branch No.	ID d_{in}, mm	OD d_e, mm	Liquid flow turn angle, $\varphi,^o$	Inlet displacement, e, mm
1	12.5	20	90	0
2	12.5	20	90	4
3	12.5	20	45	0
4	12.5	20	45	4
5	8	20	90	0
6	8	20	90	4
7	6	15.6	90	0
8	6	15.6	90	4
9	6	15.6	45	0
10	6	15.6	45	2
11	6	13.6	90	0
12	6	13.6	45	0

where Φ - electric power supplied to the rod, A - area of tube heat transfer surface, $T_w(x)$ - temperature of the rod surface at x distance from inlet of the channel, $T_f(x)$ - mean discharge water temperature at distance x. $T_f(x)$ was estimated by

$$T_f(x) = T_{in} + \frac{\Phi \cdot x}{l\, c\, \rho\, G}, \quad (18)$$

where c, ρ, G are the heat capacity, density and water mass flux, respectively. T_{in} is the temperature of the liquid at the inlet of the channel. Electric power supplied to the tube was 1-2 kW. Water temperature at inlet T_{in} was kept at 30°C. The results of measurement with the branch pipe 1 are given in Figure 9. Rayleigh number variation was provided by changing heat flux rate. Local coefficients are plotted in Figure 9 as a function of non-dimensional position.

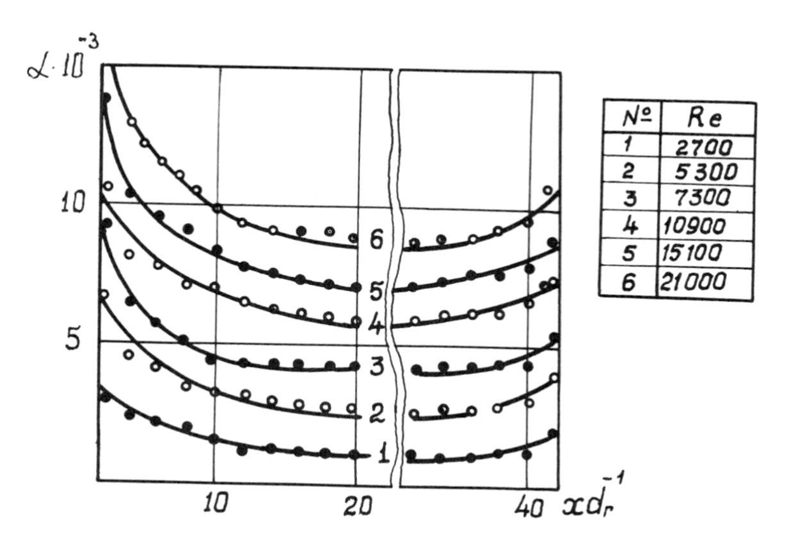

FIGURE 9. Distribution of local intensity of heat transfer coefficient along the channel.

Distinctive variations of the convective heat transfer intensity allow the overall length of channel to be divided into three sections: (a) inlet section, (b) section of stabilized heat transfer and (c) outlet section. Within the inlet section the convective heat transfer coefficient falls from the maximum at the channel inlet to nearly fully developed values. At $\frac{x}{d_r} = 20$ it differs from the fully developed values by 5-7%.

The outlet section shows the increase of heat transfer intensity from stabilized values to maximum ones at the channel outlet. At distance of four hydraulic diameters from the outlet heat transfer coefficient exceeds its stabilized value by 5%. The fully developed heat transfer section occupies channel space at $20 < x/d_r < (\ell/d - 4)$. The change in heat transfer coefficient within this region is no more than about 5%.

Within the stabilized heat transfer section the obtained data with an error not more than 12% for water and 8% for air at 0.95 confidence level, correlate well with the Mikheev and Ramm [3] relations:

$$Nu_{st} = \alpha(x)\, d_r/\lambda = 0{,}021 \cdot Re_f^{0,8}\, Pr_f^{0,43}\,(Pr_f/Pr_w)^{0,25}, \quad Re_f \geq 10^4;$$
$$Nu_{st} = 0{,}0225\, Re_f^{0,8}\, Pr_f^{0,43}\,(1-6\cdot 10^5\, Re_f^{-1,8})(Pr_f/Pr_w)^{0,25}, \quad 2\cdot 10^3 < Re_f < 10^4. \qquad (19)$$

Immediately at the channel inlet the heat transfer intensity is maximum (Figure 9). The flow laterally overflows the rod under application of branch pipes 1, 5, 7 and 11 (Table 1). The dependence of the convective heat transfer coefficient at the channel inlet with diameter d_{in} for water and air is presented in Figure 10. Marks 1 and 3 fit heat transfer at inlet for water and air, respectively.

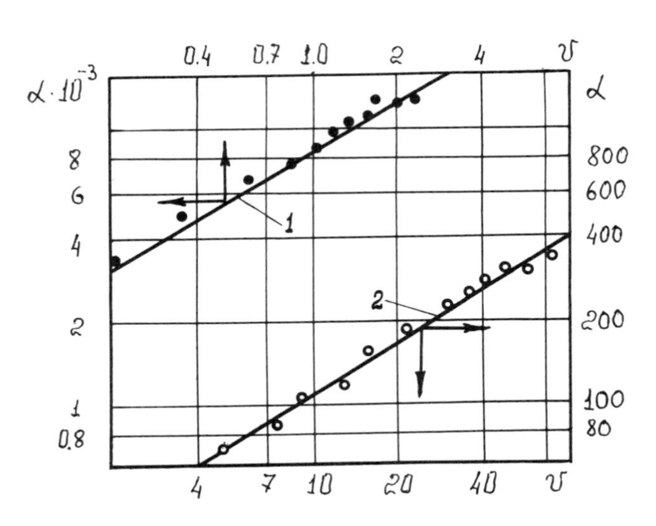

FIGURE 10. Heat transfer coefficient at inlet opening of channel.

The Zhukauskas formula [4] describing heat transfer at outer cross flow around cylinder in an unbounded air space can be recommended for generalizing obtained experimental data on heat transfer at the outlet of the channel:

$$\alpha(0) = 0{,}25\,\frac{\lambda}{d_1}\left(\frac{v_{in}\cdot d_1}{\nu}\right)^{0,6} Pr_f^{0,4}. \tag{20}$$

In this formula the velocity of flow v_{in} at the channel inlet opening is used.

Equation (20) turned out to be applicable to an annular channel for $0.5 < d_1\, d_2^{-1} < 0.74$, $0.6 < d_{in}\, d_1^{-1} < 1.25$ and for water and air velocities $0.2 < V_{in} < 2$ m/s and $2 < V_{in} < 80$ m/s. The discrepancy between experimental and estimated values from the condition does not exceed 17% for water and 11% for air at 0.95 confidence level.

Within the inlet section the heat transfer coefficient decreases monotonically from values associated with equation (20) to developed ones. The decreasing coefficient can be accounted for by introducing $\varepsilon_{in}(x)$:

$$Nu(x) = Nu_{st}\cdot \varepsilon_{in}(x), \tag{21}$$

where Nu (X) is the Nusselt number within the initial section.

The maximum value of $\varepsilon_{in}(x)$ will be at $x = 0$. Considering equations (20) and (21), $\varepsilon_{in}(x)$ at $x = 0$ can be estimated by following relation:

$$\varepsilon_{in}(0) = 11{,}9\left(\frac{d_2 - d_1}{d_1}\right)^{0,4}\left(\frac{A_n}{A_{in}}\right)^{0,6} Re_f^{-0,2}, \tag{22}$$

where A_n and A_{in} are flow cross sectional areas of the annular channel and the inlet opening respectively.

The monotonic decrease can be described by the exponential equation:

$$\varepsilon_{in}(x) = [\,\varepsilon_{in}(0) - 1\,]\exp\left(-0{,}5\sqrt{x(d_2 - d_1)^{-1}}\right) + 1, \tag{23}$$

where the exponent is determined from the data shown in Figure 9, as well as of other test data obtained in experiments with branch pipes 5, 7 and 11. Substituting equation (22) in (23) gives an equation to estimate heat transfer intensity within the initial section of the annular channel in the case when the flow enters channel at $\varphi = 90°$:

$$\varepsilon_{in}(x) = \left[11{,}9\left(\frac{d_2 - d_1}{d_1}\right)^{0{,}4}\left(\frac{A_n}{A_{in}}\right)^{0{,}6} Re_f^{-0{,}2} - 1\right]\exp\left[-0{,}5\sqrt{x(d_2 - d_1)^{-1}}\right] + 1. \quad (24)$$

Equation (24) represents the experimental investigation results with 18% error at 0.95 confidence level for the following ranges of the main parameters:

$$0{,}36 < \frac{d_2 - d_1}{d_1} < 1\ ; \quad 1{,}9 < A_n A_{in}^{-1} < 8{,}3\ ; \ 3 \cdot 10^3 < Re_f < 2 \cdot 10^4.$$

The influence of the entrance angle in channel was studied with branch pipes 3, 9 and 12 (Table 1). It was found that on decreasing entrance angle to 45° heat transfer was reduced by 20% on the average.

An investigation of the influence of tangential swirl of the inlet flow the heat transfer was also carried out by means of branch pipes 2 and 6. The heat transfer in the immediate entrance region was slightly below that without displacement of the inlet pipe, but at $x(d_2 - d_1)^{-1} > 2$ the heat transfer was improved compared to that without swirl by 25%. On the whole the effects of inlet pipe displacement and angle change can be expressed by:

$$\varepsilon_{in}(x) = \left[11{,}9\left(\frac{d_2 - d_1}{d_1}\right)^{0{,}4}(A_n/A_{in})^{0{,}6} Re_f^{-0{,}2}(1 + 2l/d_2)^{0{,}78}(1 - 0{,}25 \cdot \cos^{0{,}8}\varphi) - 1\right]\exp\left(-0{,}5\sqrt{x(d_2 - d_1)^{-1}}\right) + 1,$$

which is valid for $0.36 < \frac{d_2 - d_1}{d_1} < 1$; (25)

$1.9 < A_n A_{in}^{-1}$ 8.3; $2.7 \cdot 10^3 < Re_f < 2\ 10^4$;

$0.5 < d_1 d_2^{-1} < 0.73$; $0 < l d_2^{-1} < 0.25$ and

$0° < \varphi < 90°$; $x(d_2 - d_1)^{-1} > 2$.

Tests performed with branch pipes 4 and 10 through which the liquid was supplied with $\varphi = 45°$ and inlet opening displacements of 4 and 2 mm, showed that equation (25) is also valid when both factors occur. The difference between experimental and estimated data did not exceed 14% at 0.95 confidence level.

The convective heat transfer intensity within the outlet section was investigated in the same manner as for the inlet opening of the channel. It was discovered that the heat transfer within the liquid exit section is considerably intensified (Figure 9). This is associated with the flow velocity in the outlet channel, as shown in Figure 10. The angle of the axis pipe and the outlet opening displacement do not affect the heat transfer intensity significantly. The local Nusselt number is given by

$$Nu(x) = Nu_{st} \cdot \varepsilon_{out}(x),$$
$$\varepsilon_{out}(x) = \left[6,7\left(\frac{d_2 - d_1}{d_1}\right)^{0,4}\left(\frac{A_n}{A_{in}}\right)^{0,6} Re_f^{-0,2} - 1\right]\exp\left(-\sqrt{(l-x)(d_2-d_1)^{-1}}\right)+1, \qquad (26)$$

As a result of these studies a general formula for estimating the convective heat transfer coefficient in annular channels, considering the shape of inlet and outlet, channel length and the heat flux is proposed:

$$Nu(x) = Nu_{st} \cdot \varepsilon_{in}(x) \cdot \varepsilon_{out}(x), \qquad (27)$$

where Nu_{st} - Nusselt number given by equation (19), $\varepsilon_{in}(x)$ and $\varepsilon_{out}(x)$ - corrections for the inlet and outlet sections, estimated by equations (25) and (26). Equation (27) is applied with an error not less than 13% at 0.95 confidence level in channels of length more than 25 hydraulic diameters d_r in length.

4. INTENSIFICATION OF HEAT TRANSFER IN EVAPORATION-CONDENSATION COOLING SYSTEMS

Evaporation-condensation cooling systems have rather small weight, high reliability and good efficiency. Boiling and condensation processes provide high intensity of heat transfer without application of pumps, fans and other facilities. But these systems have some disadvantages. The heat flow from semiconductor elements is limited by the critical heat flux because at the high temperatures in film boiling the elements are usually put out of operation. Available techniques of overcoming this problem, for example, the vapotron cannot be applied to semiconductor facilities because this effect depends on operation in film or transition boiling. Along with the heat flux in semiconductor facilities can be 1-2 MW/m^2 which exceeds critical by several times for acceptable heat transfer liquids.

The heat transfer intensification in evaporation-condensation cooling systems can be achieved by selecting optimal values of the fluid pressure and temperature, by providing special microstructure at the heat transfer surfaces, by coating these surfaces with dispersive layers of particles and so on. Consider a simple and effective technique of heat transfer intensification due to ribbing the surfaces of electronic devices.

When choosing the ribbing type the main purpose is to provide maximum heat transfer under low wall temperatures. The maximum heat transfer intensity during boiling occurs with nuclear boiling where the heat transfer coefficient α depends on the difference between the surface temperature and the liquid saturation temperature ϑ according to expression [4] :

$$\alpha = K \cdot \vartheta^{m} \tag{28}$$

where K and m are coefficients depending on pressure, thermophysical properties of the liquid and surface roughness.

Calculating the heat transfer relation for a semi-infinite rib on whose face the wall superheat ϑ_0 is maintained and whose the side wall heat transfer is governed by equation (28) allows us to obtain the heat flow, dissipating by this rib,

$$\Phi = \sqrt{\frac{2PAK\lambda}{m+2}}\ \vartheta_0^{\frac{m}{2}+1}, \tag{29}$$

where P and A are the perimeter and area of the rib cross section, respectively, and λ is its heat conductivity.

An analysis of the heat transfer in gaps between the ribs as well as of the heat flow dissipated by them allows one (Figure 11) to choose the shape and dimensions of a heat transfer surface (a = 5.5 mm, δ = 1.6 mm, h = 20 mm, γ = 60°) which displays the maximum heat transfer capacity under conditions of nucleate boiling. The relation of the heat transfer flow has the form

$$q = 23\sqrt{\frac{K\lambda}{m+2}}\ \vartheta_0^{\frac{m}{2}+1}. \tag{30}$$

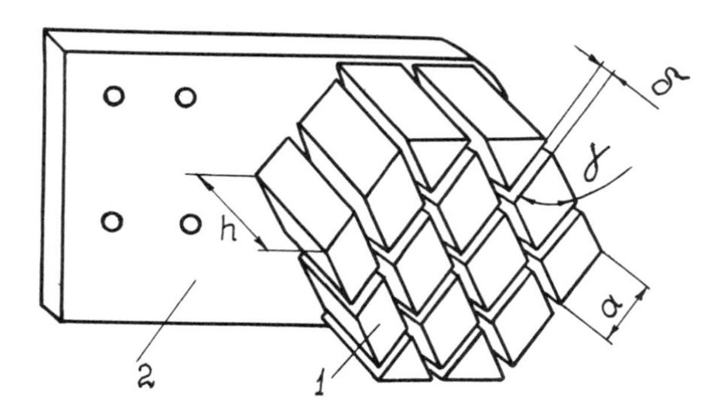

FIGURE 11. Outer view of heat transfer surface.

It was supposed that ribs (Figure 11) had the ideal contact with the base.

The results of testing the obtained relations showed good agreement of equations (29) and (30) with the experimental data. When R 113 was used as the working fluid maximum heat flux at the surface was 1.2 MW/m^2 with a wall superheat of only 23 K. In case of the ribbing surface maximum heat flux was 0.3 MW/m^2, under other similar conditions.

CONCLUSIONS

Key features are summarized as follows:

1. This paper presents relationships for estimating local and average coefficients of free convective heat transfer in enclosed cavities formed by plane walls of different temperatures and in horizontal channels limited by isotherma plates.
2. Equations for estimation of local heat transfer intensity under forced convection in channels with different inlet heat transfer were obtained.
3. The estimation procedure of geometrical parameters of the heat transfer surfaces providing good heat transfer intensity under evaporation cooling was developed and experimentally proved.

NOMENCLATURE

ymbol	Quantity	SI Unit
A	area	m^2
a	thermal diffusivity	m^2/s
C	heat capacity	J/(K kg)
d	diameter	m
g	local gravitational acceleration	m/s^2
H	height	m
P	perimeter	m
Q	heat flow	W
q	heat flux density	W/m^2
T	thermodynamic temperature	K
V	velocity	m/s

reek Letters

α	heat transfer coefficient	$W/(m^2\ K)$
β	volumetric expansion coefficient	K^{-1}
δ	thickness	m
λ	thermal conductivity	W/(m K)
ν	kinematic viscosity	m^2/s
ρ	mass density	kg/m^3

nbol and Definition	Name
nensionless parameters	
$Gr = \frac{g L^3 \beta \Delta T}{\nu^2}$	Grashof number
$Nu = \alpha L/\lambda$	Nusselt number
$Pr = \nu/a$	Prandtl number
$Ra = Gr \cdot Pr$	Rayleigh number

REFERENCES

1. Martynenko, O.G., Sokovishin, Yu.A., Free Convective Heat Transfer: Reference book, Nauka; Tekhnika Press, Minsk, 1982.
2. Dulnev, G.N., Semyashkin, E.M., Heat Transfer in Electronic Facilities, Energia Press, Leinigrad, 1968.
3. Andreev V.A., Heat Transfer Facilities for Viscous Liquids, Energia Press, Moscow, 1972.
4. Kutateladze, S.S. Fundamentals of Heat Transfer Theory, Atomizdat Press, Moscow, 1979.
5. Bohn, M.S., Kirkpatrick, A.T., Olson, D.A., Experimental Study of Three-Dimensional Natural Convection High-Rayleigh Number, Trans ASME, J. Heat Transfer, vol. 106, no. 2, pp. 339-345, 1984.
6. Zarichnyak, Yu. P., Platunov, E.S., Sharkov, A.V., Method of Measuring Local Heat Flows, J. of Engineering Physics, vol. 28, no. 2, pp. 249-256, 1975.
7. Begunkova A.F., Zarichnyak, Yu.P., Korablyov, V.A., Sharkov A.V., Plant for Measuring Effective Coefficient of Heat Conductivity of Heat Insulating Materials, Izv. Vuzov SSSR, Instrument Engineering, vol. 26, no. 4, pp. 84-88, 1983.
8. Leontyev, A.I., Kirdyashkin, A.G., Heat Transfer under Free Convection in Horizontal Slots and Significant Volume over Horizontal Surface, J. of Engineering Physics, vol. 9, no. pp. 9-14, 1965.
9. Brdlik, G.M., Turchin I.A. Heat Transfer under Natural Convection from Horizontal Surfaces Presented with Heat Radiating Surfaces Downward, J. of Engineering Physics, vol. 14, no. 3, pp. 470-475, 1968.

10. Yousef W.W., Tarasuk, I.D., McKeen W.I. Free Convection Heat Transfer from Upward-Facing Isothermal Horizontal Surfaces, Trans. ASME, J. Heat Transfer, vol. 104, no 3, pp. 85-92, 1982.

11. Leontyev, A.I., Heat Transfer Theory, Wisshaja Shkola Press, Moscow, 1979.

12. Schinkel, W.M.M., Linthorst, S.J.M., Hoogendoorn, C.J., The Stratification in Natural Convection in Vertical Enclosures, Trans. ASME, J. Heat Transfer, vol. 105, no. 2, pp. 57-64, 1983.

13. Elsherbiny, S.M., Rathby, G.D., Hollands, K.G.T., Heat Transfer by Natural Convection Across Vertical and Inclined Air Layers, Trans ASME, J. Heat Transfer, vol. 104, no. 1, pp. 96-102, 1982.

14. Emery, A., Chu, N.C., Heat transfer across vertical layers, Trans ASME, J. Heat Transfer, vol. 18, no. 1, pp. 14-20, 1965.

Intensification of Air Cooling by Inclined Inlet of Air

M. JÍCHA and J. HORSKÝ
Technical University
Faculty of Mechanical Engineering
Department of Thermomechanics
Technická 2, 61669 Brno, Czechoslovakia

In the paper the influence of the inclined inlet of air into the cooling channel on heat transfer performance is studied. Two main cases were investigated:
1. the air is pushed into the channel
2. the air is sucked out of the channel
The angle of attack varied from 0° (axial entrance) up to 70°. The experiments were carried out for one flow rate corresponding to a Reynolds number of 42520. For the whole channel length (which has been of 15 diameters) only a small difference exists in the mean heat transfer coefficient (about 15%) in favor of the case when the air is pushed into the channel. For a shorter channel length this difference considerably increases. As to the influence of the various angles of attack, on one hand, it is more substantial when the air is pushed into the channel but, on the other hand, it is more intensively damped towards the end of the channel.

1. INTRODUCTION

Cooling channels are often used in electric motors especially in stators into which the air enters obliquely, i.e., the direction of the entering air is not parallel to the axis of the channel but it is inclined at a certain angle. Accordingly, the influence of the oblique inlet of the air on the heat transfer was studied. Two main cases were realized. First of all, the air is pushed into the channel and second, the air is sucked out of the channel.

2. EXPERIMENTAL SET-UP AND MEASUREMENT

The schematic view is in fig.1. The experimental stand consists of the cooling channel, entrance chamber, inlet pipe and other parts necessary to complete the measurements. The cooling channel (fig.2) is a circular tube 1m long and 58mm inner diameter. There are 8 positions - radial holes - along the channel (Tab.1) which served to introduce a hot-wire probe for velocity and temperature profile measurements. The profiles were measured in planes both of symmetry and asymmetry, the plane of asymmetry being that defined by the inlet

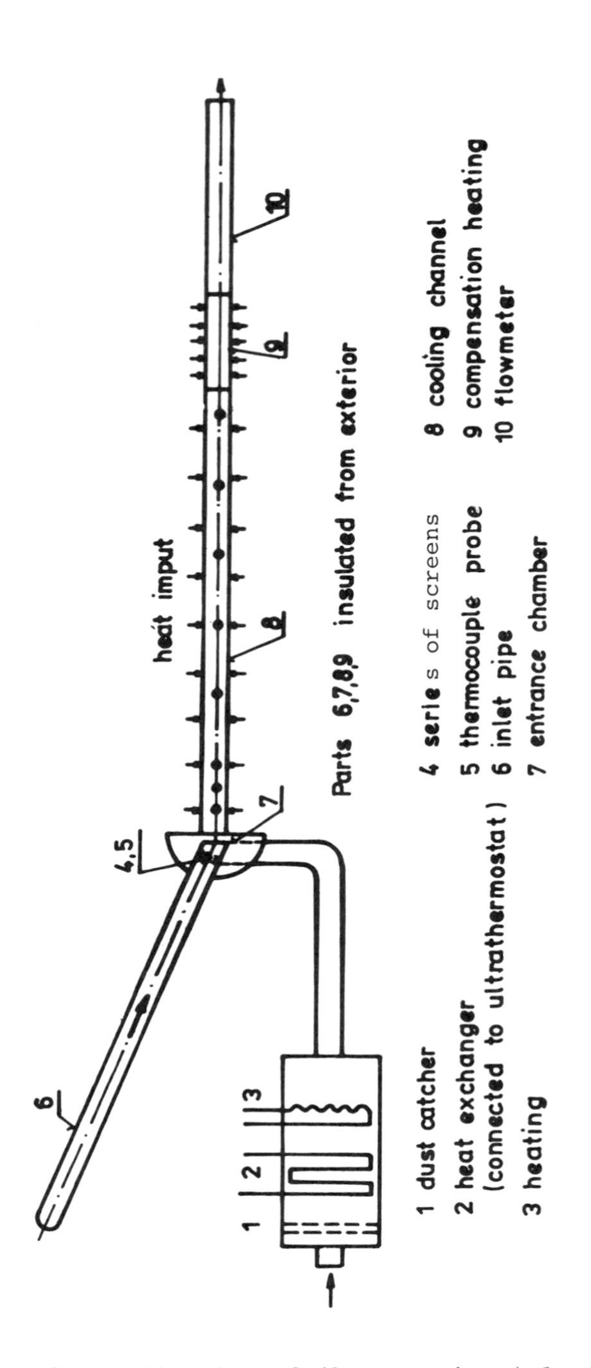

FIGURE 1. Schematic view of the experimental stand.

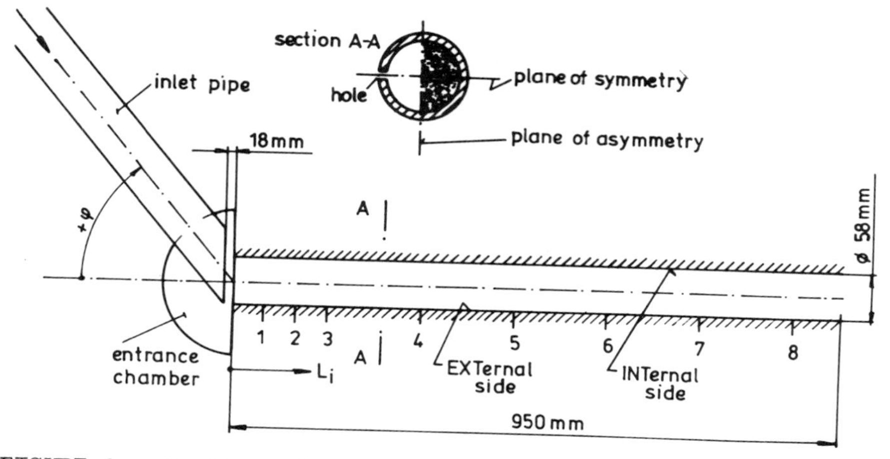

FIGURE 2. Cooling channel with traverse stations.

pipe and the cooling channel. The plane of symmetry is then that perpendicular to the plane of asymmetry (see onset of fig.2). In the plane of asymmetry we distinguish between the so-called External side of the channel or the windward side that is directly attacked by the entering air flow and the Internal side or the leeward side on which the separation of the air flow in a very short entrance region of the channel occurs.

TABLE 1. Positions along the channel

Position No.	1	2	3	4	5	6	7	8
L_i/d	0.86	1.72	2.58	5.17	7.75	10.34	12.93	15.51
x_i [mm]	50	100	150	300	450	600	750	900

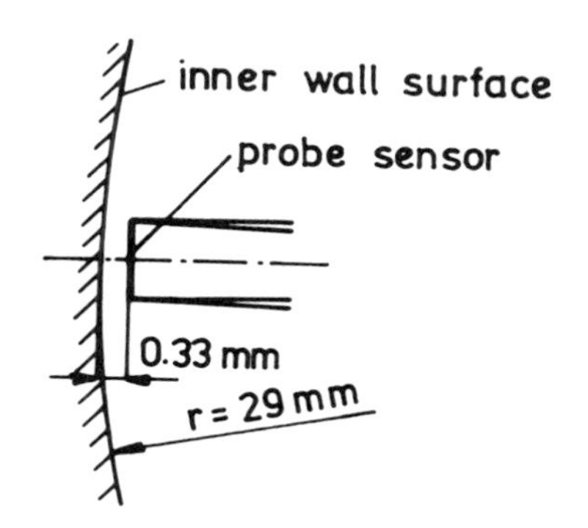

FIGURE 3. Location of the probe near the wall.

The air enters the cooling channel from the inlet pipe that is fixed in the entrance chamber so that it could be positioned with regard to the cooling channel at an angle up to 70°. A gap of 18 mm is fixed between the exit cross section of the inlet pipe and the entry cross section of the cooling channel, the planes of these two cross sections being parallel at any angle of attack. It means that the inlet pipe is at its exit end beveled under the corresponding angle.

The cooling channel is heated electrically from the outer surface using the resistance heating belt wound on.

The air velocity and temperature profiles in the 8 above mentioned positions were measured with the fiber-film probe DISA R01 with a thin quartz coating. The diameter of the probe sensor is 70 μm, the sensitive length 1.25 mm. The minimum distance from the wall was fixed at 0.33 mm (i.e., outer surface of the probe sensor - fig.3). The temperature of the air entering the cooling channel was maintained at $30^{o}C \mp 0.25^{o}C$. The inner duct wall temperature was always measured in the horizontal plane in the line opposite to the holes for introducing the hot-wire probe. There were fixed 27 thermocouples along the channel. The thermocouple was inserted in the wall in the manner that is depicted in fig.4. The thermocouple junction was coated with an insulating paint except the lower spherical cap. After we had pressed the copper plug into the hole we supposed a perfect contact between the thermocouple lower cap and the wall of the channel. The thickness of the wall material between the lower cap of the thermocouple junction and the inner surface of the channel is estimated at 0.02 ∓ 0.01 mm. To illustrate it better the thermocouple junction is enlarged in fig.4.

Generally, the measurements of the velocity and temperature profiles were always carried out - from the point of view of the holes - in the more distant half of the cross channel section (in the shaded area in the onset of fig.2) to eliminate the disturbances caused by the holes for introducing the hot-wire probe. This fact was checked by the measurements with the hot-wire probe introduced in the longitudinal direction (against the flow air) in the shorter channel of the same inner diameter and for the same flow rate. The flow rate had a constant value corresponding to the Reynolds number of 42520.

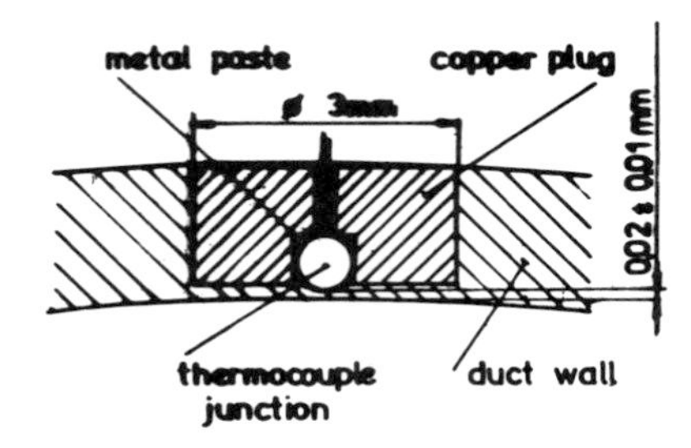

FIGURE 4. Location of the thermocouple in the duct wall.

The mean heat transfer coefficient $\bar{\alpha}$ was evaluated from the measured velocity and temperature profiles (measured at about 100 points in each of the 8 cross sections) and from the entrance air temperature and duct wall temperature by means of the well known calorimetric method using the following formula:

$$\bar{\alpha} = \frac{\pi(d^2/4)\bar{u}\ c_p(T_{bi} - T_{bO})}{\pi d L_i(\bar{T}_w - \bar{T}_b)} \qquad (1)$$

The mean wall temperature $\bar{T}_w$ and mean bulk temperature $\bar{T}_b$ of the air were evaluated while using the B-spline analysis for the wall temperatures T_w and bulk air temperatures T_{bi} (for each of the specified 8 positions along the channel).

The evaluation of the mean heat transfer coefficient in the first 3 regions of the channel, i.e., for channel lengths $L/d = 0.86$, 1.72 and 2.58 is charged with a certain error due to the fact that the difference between the mean bulk temperatures doesn't exceed 0.75 K for the channel length $L/d=2.58$; for the channel lengths $L/d = 0.86$ and 1.72 this difference is not greater than 0.3 - 0.5 K. The inlet air temperature is measured with the accuracy of 0.05°C so that the relative error is in the range from 5% to 30% at the channel position $L/d = 2.58$. From the mean heat transfer coefficient $\bar{\alpha}$ the mean Stanton number $\overline{St}$ was calculated by the formula

$$\overline{St} = \frac{\bar{\alpha}}{\rho c_p \bar{u}} \qquad (2)$$

Also the local Stanton numbers were calculated by the formula

$$St = \frac{\lambda(dT/dr)_w/(T_w - \bar{T}_b)}{\rho c_p \bar{u}} \qquad (3)$$

The accuracy of the local Stanton number in this manner evaluated is influenced by the fact that we suppose the velocity and temperature profiles linear between the wall and the first measured point from the wall (i.e., to the distance 0.33 mm from the wall). When we calculate the universal distance y^+ for the case of the axial entrance (angle of attack of 0°) e.g., for position 8 (at the end of the channel) then we obtain $y^+ = 11$. This is the distance in which the velocity and temperature profiles begin to deviate from the linear slope. At the positions closer to the entrance of the cooling channel, the y^+ distances are somewhat greater due to the fact that the boundary layer thickness is smaller and the friction coefficient is greater. Yet the uncertainty in the measured duct wall temperature affects the local heat transfer coefficient and/or the local Stanton number. Taking into account the possible nonperfect contact between the thermocouple and the wall and the fact that the thermocouples were calibrated with the accuracy of up to 0.1°C we estimate this uncertainty to be 5%. From this fact it follows that the temperature gradient used to calculate the local heat transfer coefficient is charged with an error about 10 to 15%.

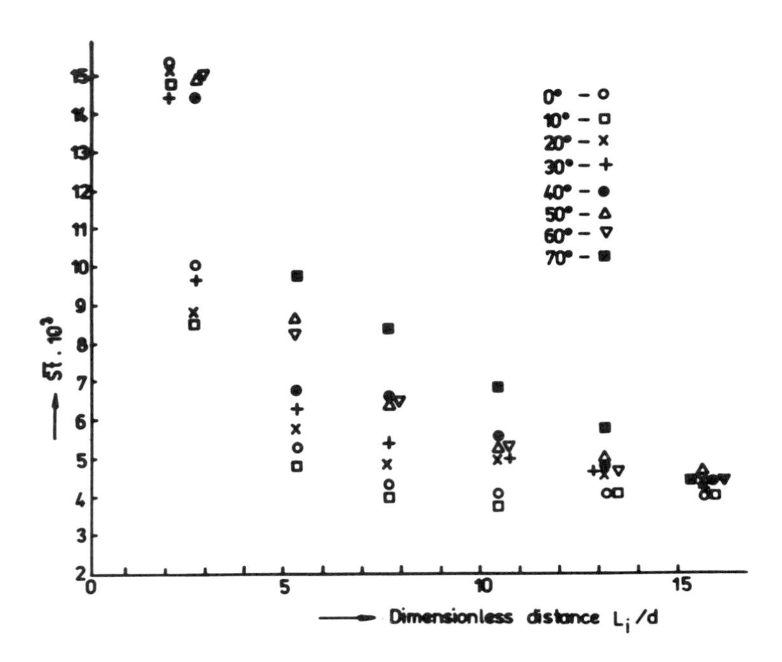

FIGURE 5. Mean Stanton number for the case when the air is pushed into the channel.

3. ANALYSIS OF THE RESULTS

In figs.5 and 12 the mean Stanton number is presented. At first glance a difference is evident between the case when the air is pushed into the channel and sucked out of the channel. From fig.5 we can see that the greater is the angle of attack the greater is the heat transfer. The most remarkable improvement of the heat transfer - as one could expect - is for the angle of attack of 70°. For the channel length of approx. 5d this improvement is 85%, for L ≐ 10d it is 72% and even for the whole channel (L ≐ 15d) the heat transfer is still about 10% higher than for the case of the axial entrance (angle of attack of 0°). A certain anomaly can be observed for the angle of attack of 10°. In this case the heat transfer is for any length of the channel lower than for the axial entrance. This result is supported by the measurement of the local heat transfer (see figs.6 and 7). When the flow attacks the channel under the angle of 10°, the heat transfer on the External side and in the plane of symmetry is almost identical with the results for the axial entrance; but on the Internal side of the channel the heat transfer is lower till the location L ≐ 10d. With this fact the mean heat transfer is consistent. The decrease of the heat transfer with respect to the axial entrance is very large up to the location L ≐ 10d. Later we return once more to the local heat transfer.

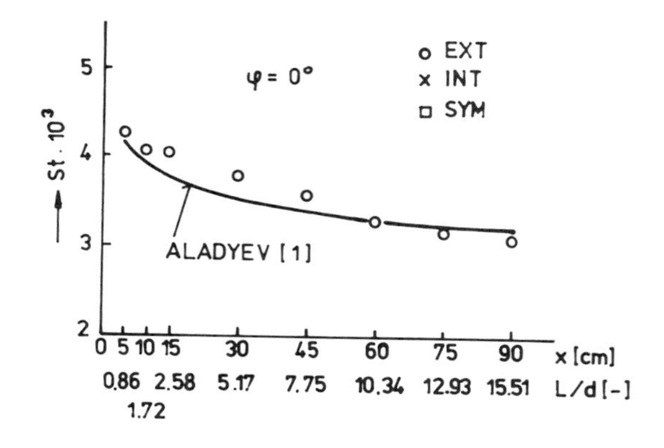

FIGURE 6. Local Stanton number-air is pushed into the channel.

A comparison with some other experimental results is very difficult because of the lack of these data. In fig.10 we can observe the comparison made for the axial entrance and for the angle of attack of 40°. The solid line represents the results of Aladyev [1] for the entrance region of a circular tube in which both the velocity and temperature profiles are being developed. The results of Aladyev correspond to those obtained by Boelter et al. (quoted from[2]) for the "bell-mouth" entrance with screen. Our type of axial entrance is similar to that of Boelter but our results are somewhat higher. Also the results are presented for the angle of attack of 40° compared with the 45°-angle-bend entrance after Boelter et al. Our results are again higher than those of Boelter. But the discrepancy is much greater than for the case of the axial entrance. Notice that the entrance configuration of our channel is considerably different in the respect that the air flow from the inlet pipe is directed under the corresponding angle into the channel while in Boelter's entrance con-

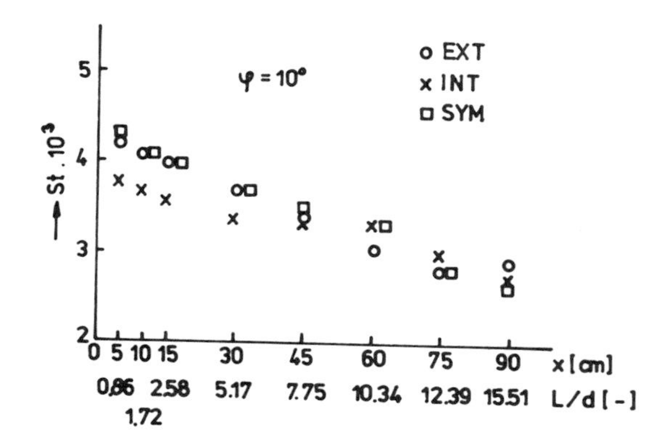

FIGURE 7. Local Stanton number-air is pushed into the channel.

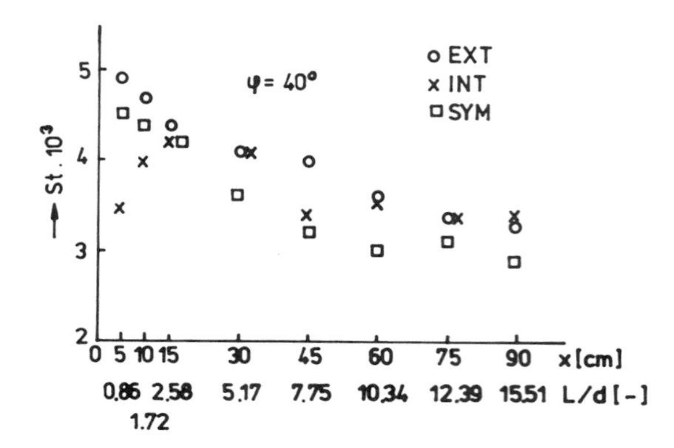

FIGURE 8. Local Stanton number-air is pushed into the channel.

figuration the inclined tube (under the angle of 45°) is followed by a portion of a straight tube.

In fig.11 we can observe the influence of the angle of attack for 3 various lengths of the channel: L/d = 5.17, 10.34 and 15.51. It is seen how the history of the entrance is successively damped for the longer channel.

Now return to figs.6 to 9 to examine the local heat transfer. For the reason of a certain inaccuracy in the measurements of the local heat transfer coefficient due to the fact described in section 2., the local heat transfer coefficient serves only for qualitative comparison of various angles of attack and to support qualitatively the mean heat transfer coefficient measurements. In fig.6 the local Stanton number for the axial entrance is compared with the results of Aladyev. The slope of our measurements is steeper and should be even steeper with respect to the inaccuracy mentioned above.

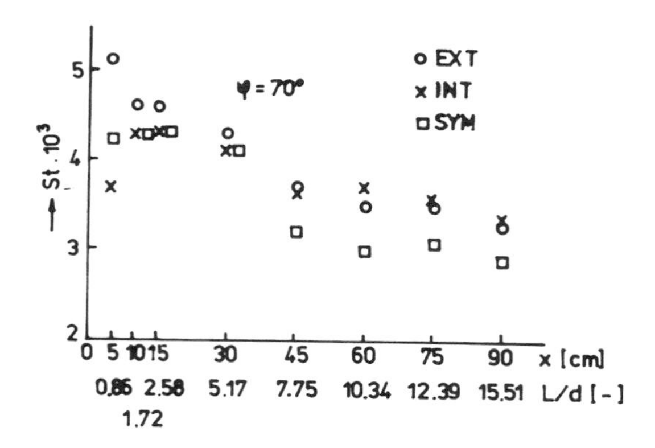

FIGURE 9. Local Stanton number-air is pushed into the channel.

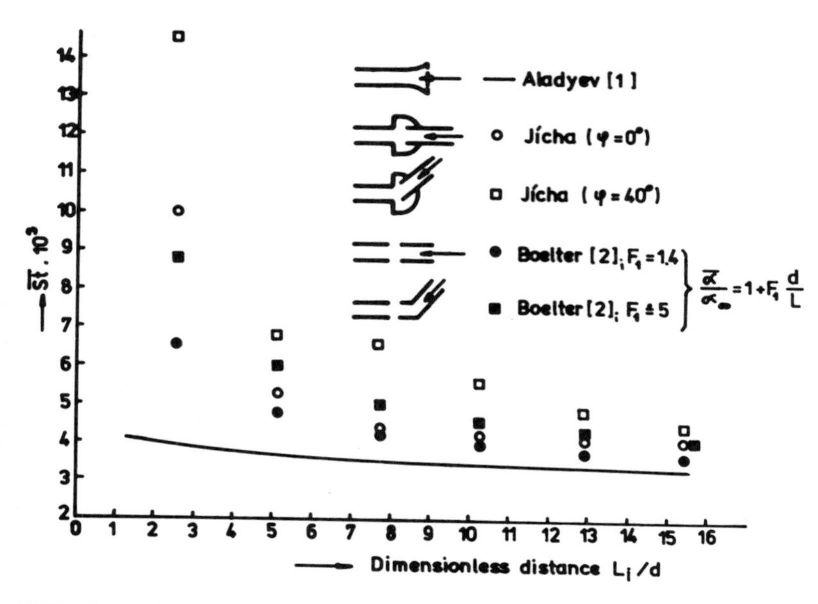

FIGURE 10. Comparison of mean Stanton number with published data - air is pushed into the channel.

In fig.7 for the angle of attack of 10° we can observe that on the External side and in the plane of symmetry the heat transfer is almost identical with the case of the axial entrance but on the Internal side the heat transfer is much lower. This is certainly the reason why the mean heat transfer coefficient (mean Stanton number) for the angle of attack of 10° is lower than for the axial entrance (see fig.5).

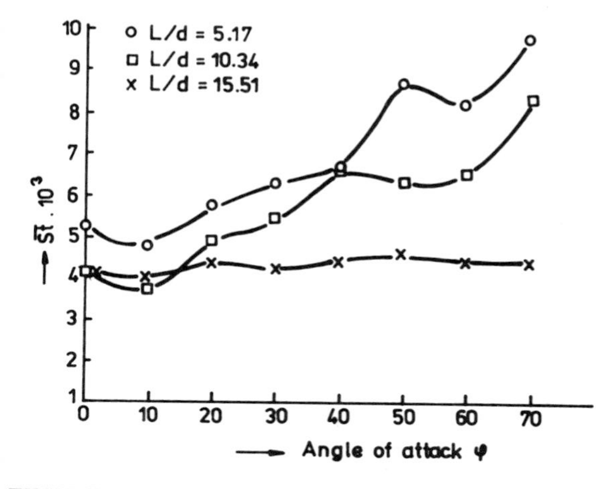

FIGURE 11. Mean Stanton number versus angle of attack.

In figs.8 and 9 we can see a successively changing character of the heat transfer on the Internal side (the local maximum occurs) and even in the plane of symmetry for an angle of attack of 70°.

In fig.12 the mean Stanton number for the case when the air is sucked out of the channel is presented. At first glance we can see the evident difference compared with the case when the air is pushed into the channel (see fig.5). The damping of the influence of the entrance configuration is much faster and stronger than when the air is pushed into the channel. And again-which is interesting-a certain anomaly is observed for the angle of attack of 10°. Towards the end of the channel the heat transfer increases. For the shorter channel lengths up to about 7d the high Stanton numbers result probably from the data scatter. Having in mind this fact the curve representing Stanton numbers from the length about 7d towards the end of the channel would lie at lower values but with an increasing tendency towards the end of the channel. This increasing tendency is proved also by the local Stanton number in fig.16 where we can observe the increase from the position approx. 10d towards the position 15d.

In fig.13 the comparison of the same kind as in fig.10 is shown. We see for the axial entrance a very good agreement with the results of Aladyev from the position L = 7.75d towards the end of the channel. Due to the entrance configuration the heat transfer in the first half of the channel length is much higher. In comparison with the published data of Boelter et al. the agreement is good enough for the whole channel length within the range of approx.10% to 15%. (Note that Boelter's formula is recommended for L/d>5.) Also the measurements for the angle of attack of 40° and 50° compared

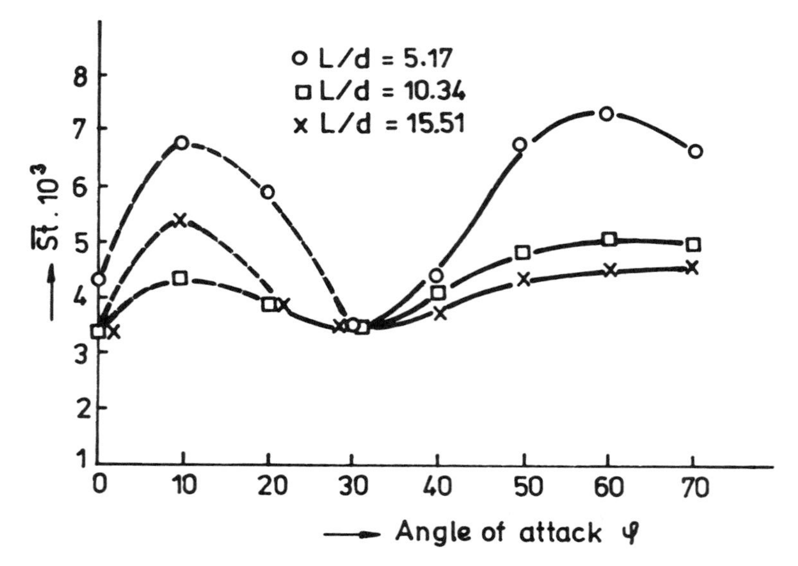

FIGURE 14. Mean Stanton number versus angle of attack.

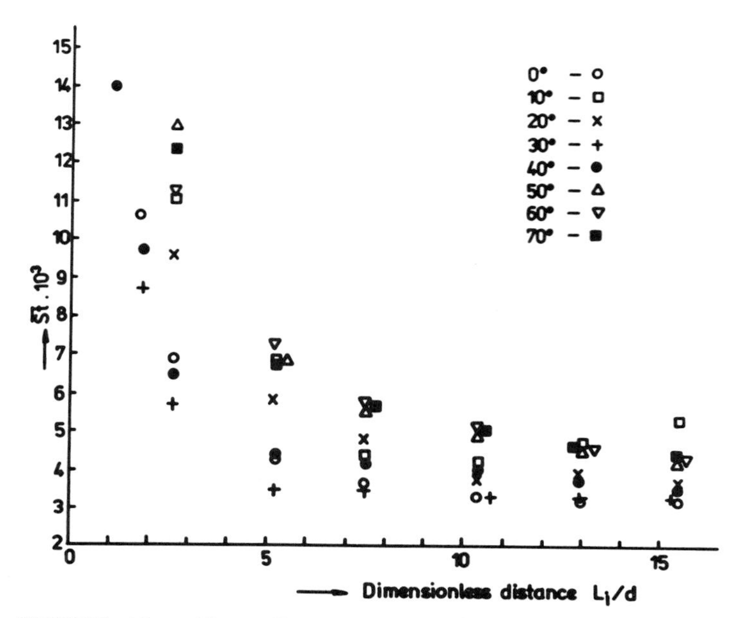

FIGURE 12. Mean Stanton number for the case when the air is sucked out of the channel.

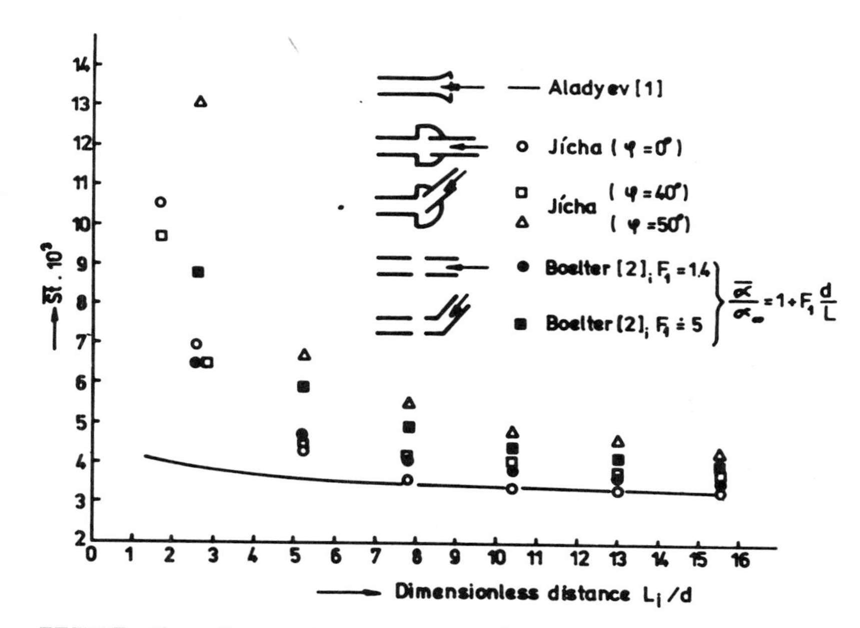

FIGURE 13. Comparison of mean Stanton number with published data - air is sucked out of the channel.

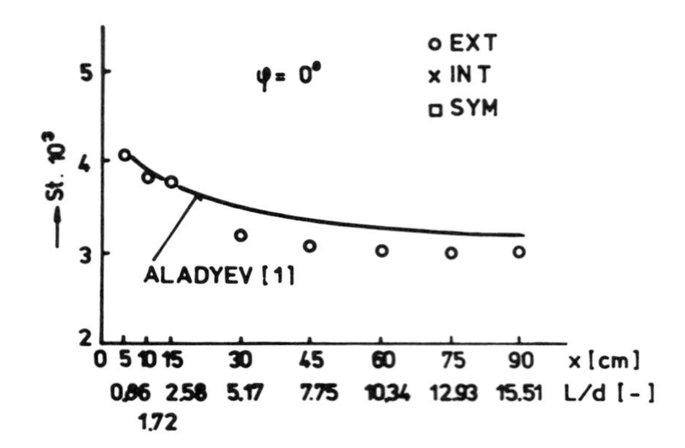

FIGURE 15. Local Stanton number - air is sucked out of the channel.

with Boelter's results for the 45°-angle-bend entrance are shown in this figure. Our results are spread very regularly on both sides of the curve representing Boelter's results.

In fig.14 we can observe explicitly the influence of the angle of attack on the heat transfer and on the Stanton number. With dotted line we mark the values loaded with a considerable error due to what was said about the angle of attack of 10° and taking into account the uncertainty in the Stanton numbers for the channel lengths below $L/d = 5$.

In figs.15 to 19 the local Stanton numbers are presented. While for the angle of attack of 10°(fig.16) the local Stanton number has a continuous decreasing character on the Internal side, for the angle of attack of 20° and particularly 30° the maximum in the position $L/d = 1.72$ occurs, giving evidence of the separation of the flow on the Internal side of the channel immediately after the channel entrance. To illustrate

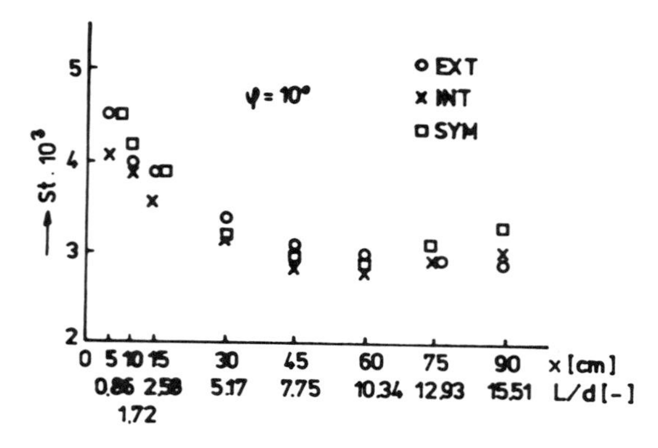

FIGURE 16. Local Stanton number - air is sucked out of the channel.

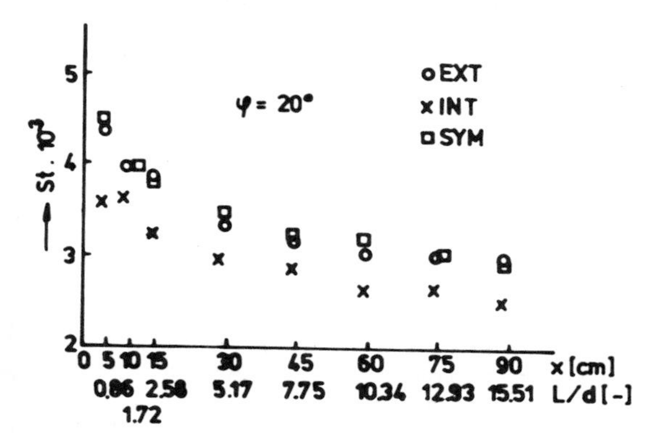

FIGURE 17. Local Stanton number - air is sucked out of the channel.

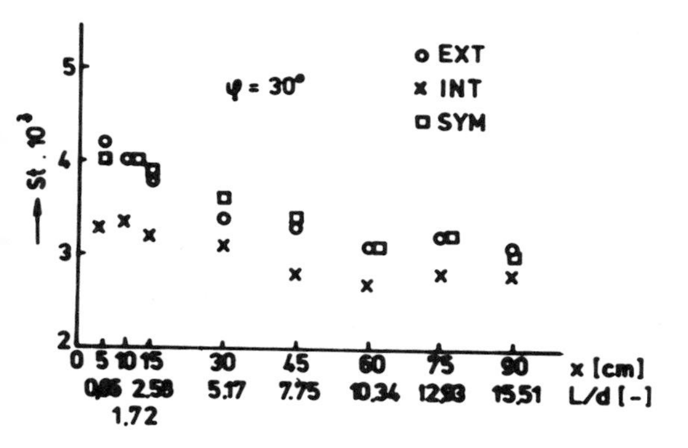

FIGURE 18. Local Stanton number - air is sucked out of the channel.

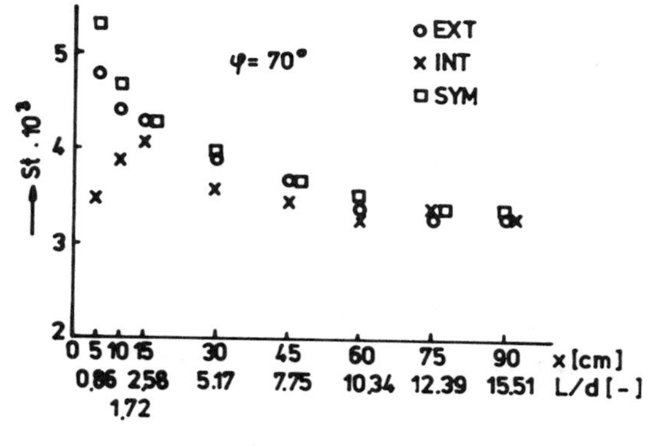

FIGURE 19. Local Stanton number - air is sucked out of the channel.

the behavior of the local heat transfer for steeper angles of attack we add fig.19 for the angle of attack of 70°.

Yet another presentation of the results is given. In fig.20 and 21 the longitudinal inner duct wall temperature for the angles of attack of 10° and 70° normalized with the longitudinal inner duct wall temperature for an axial entrance is shown. In fig.20 we can observe for the angle of attack of 10° that the wall temperature is higher than that for the axial entrance, so the cooling effect in the channel is worse This fact is in a very good agreement with the heat transfer coefficient (Stanton number) - fig.5 - which is for the whole channel length lower for the angle of attack of 10° than for the axial entrance. Particularly, it applies to the wall temperature on the Internal side on which the local Stanton number (see fig.7) is considerably lower than that on the External side and in the plane of symmetry. In fig.21 for the case when the air is sucked out of the channel we have a support (and the consequence as well) for the improving heat transfer towards the end of the channel for the angle of attack of 10°. The value 1.0 is attributed to the same cooling effect as has the axial entrance. So we can see a better cooling effect, particularly towards the end of the channel.

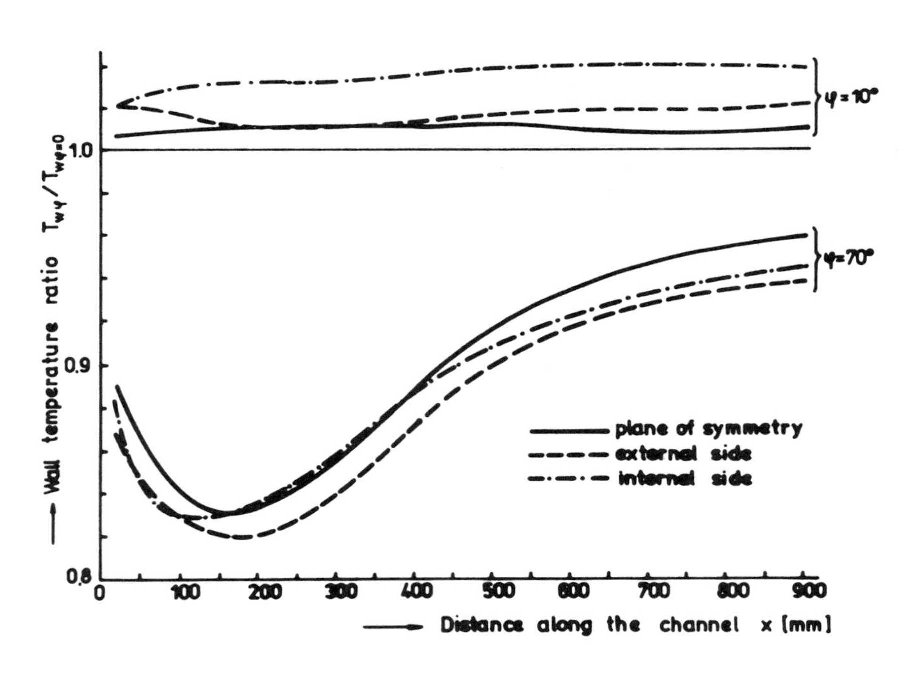

FIGURE 20. Inner duct wall temperature ratio for the case when the air is pushed into the channel.

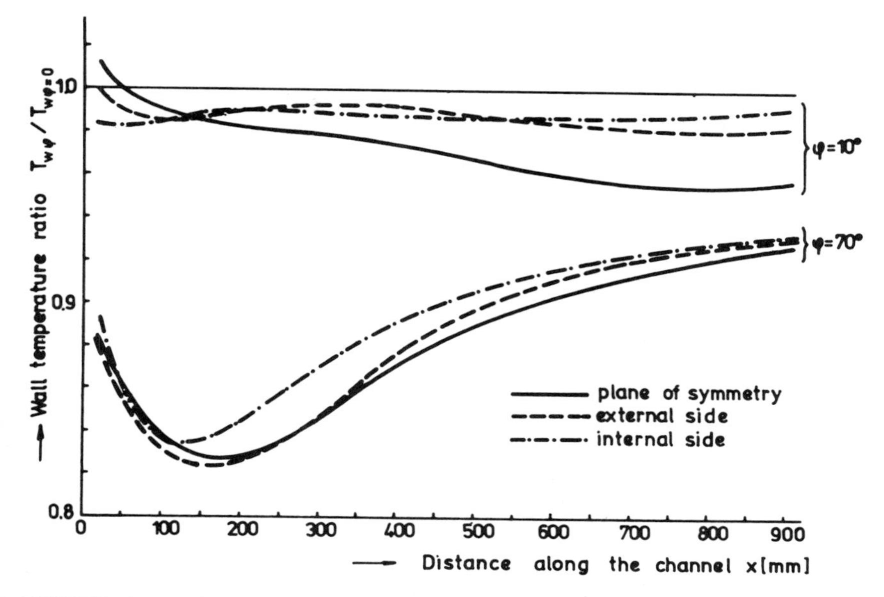

FIGURE 21. Inner duct wall temperature ratio for the case when the air is sucked out of the channel.

4. CONCLUSIONS

While comparing the two cases - when the air is pushed into the channel and sucked out of it we can conclude for the whole channel length $L/d \doteq 15$ there is not a very great difference in the heat transfer data (except for angle of attack of 10°). The difference is about 10% in favor of the case when the air is pushed into the channel. For the channels successively shorter the increase in the heat transfer for the case when the air is pushed into the channel is much greater and, for example, for the channel length $L/d \doteq 5$ for the angle of attack of 70° this increase is about 50%.

Even though the measurements were carried out only for one flow rate or one Reynolds number of 42520 we can estimate the behavior for the other Reynolds numbers. Probably for higher Reynolds numbers the influence of the angle of attack on the heat transfer will be more substantial so that the mean Stanton number will increase. But we can expect this effect to be very rapidly damped along the channel. For lower Reynolds numbers this influence will probably decrease with decreasing differences for various angles of attack.

NOMENCLATURE

c_p	specific heat capacity at constant pressure	J/(K kg)
d	inner duct diameter	m
L	channel length	m
T_b	bulk air temperature	K
T_w	inner duct wall temperature	K
u	axial velocity	m/s

Greek letters

α	heat transfer coefficient	$W/(m^2K)$
λ	thermal conductivity	W/(m K)
ϱ	mass density	kg/m^3
φ	angle of attack	deg

REFERENCES

1. Aladyev, I.T., Experimentalnoje opredelenije lokalnych i srednich koeficientov teplotdači pri turbulentnom tečeniji židkosti v trubach, Izv.AN SSSR, OTN, N.11, 1951.

2. Knudsen, J.G., Katz, D.L., Fluid Dynamics and Heat Transfer, pp.402,403, McGraw-Hill, New York, 1958.

An Approach to the Design of a Forced-Air Cooling System for I/O Cards in Standard 19-inch Card Frames

MIROSLAV ŽIVANOVIĆ
ILR-Lola Institute
Bulevar revolucije 84
11000 Belgrade, Yugoslavia

ABSTRACT

The peripheral equipment of industrial computers is commonly cooled by natural or forced-air convection. Fan trays are offered by standard 19" equipment manufacturers as a solution for the problem of power dissipation.

In this paper, a new solution for forced-air cooling system of I/O cards placed in standard 19" card frames is presented. This solution is compared with that of a 19" equipment manufacturer and the advantages and a short-coming of it are given.

INTRODUCTION

The trend in greater packaging density of electronic equipment today increases the importance of heat management. Most electronic components are sensitive to heat, and excessive heat or cold can impair the reliability of the equipment. It is very important to implement a good thermal design for electronic equipment in the earlier stages because there are cases in which the whole system can be damaged by a malfunction or fault in a single electronic component. It is necessary to keep electronic components within an acceptable temperature range (from +5° to +70°C) in order for the equipment to function properly.

Some electronic equipment may be designed to be cooled, by using a direct liquid cooling (e.g., 3M - Fluorinert^{TM}) or a liquid-cooled plate, which are more efficient means of heat transfer than air. Environmental air, however, remains the most popular cooling medium, especially for electronics packaged in standard 19" card frames according to DIN 41494.

As an air moving device, the standard 120 mm - square small axial fan (Fig.1) is most commonly used in electronic cooling. It can have an AC or brushless

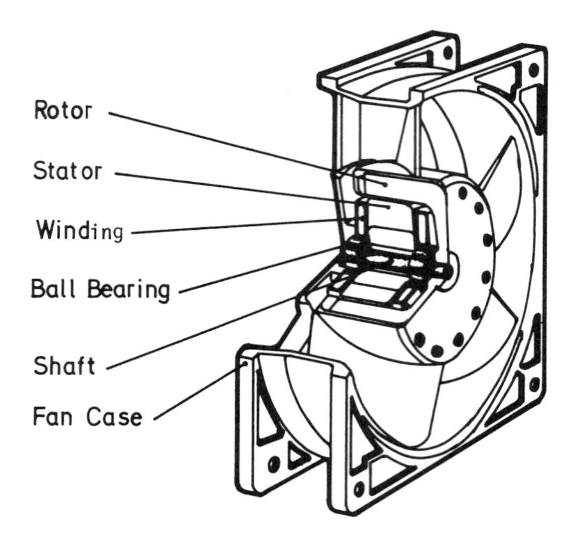

FIGURE 1. The standard 120 mm - square small axial fan.

DC motor, ball bearings or sleeve bearings. These small fans are often applied in 19" fan tray assemblies to keep cool printed circuit boards (Fig.2)

Designers find the small fans easy to use as they are compact, individually inexpensive and available off-the-shelf from various manufacturers. They are adequate for many electronic equipment cooling applications, but it is easy to apply them improperly.

If a designer has left no room for an inlet or outlet distribution plenum, then the actual installed fan tray assembly performance will be significantly lower than expected because of adverse interaction with printed circuit boards, and poor air distribution as well as "dead spots" will result.

Axial fans are most efficient when moving large volumes of air at low pressure, i.e., they should not be used at much more than half of the maximum pressure.

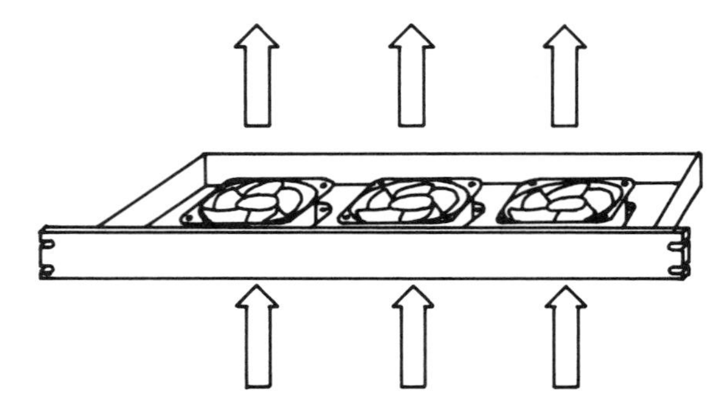

FIGURE 2. The 19" fan tray assembly.

DESIGN PROCEDURE

An innovative design approach has been used to develop the forced-air cooling system for I/O cards in standard 19" card frames of the new series Lola industrial computers, such as Lola 33 CNC and Lola LPA 512 L.

Design Demands

The basic approach of air flow system in previous use was to flow air through the system of 19" card frames with printed circuit boards from the bottom to the top of the computer, the so-called serial-flow system (Fig. 3). Through the computer application in industrial environment, the shortcomings of the serial-flow cooling system have been evident. Thus, the poor air distribution was a result of applying standard 19" fan tray assembly without a distribution plenum between two 19" card frames (Lola 30 CNC). Air "dead spots" resulted from the fan motor position. The fan reliability was poor since it was used in a horizontal position, and fan motor bearings could not work properly. The fan efficiency was decreased as fans were working with a large amount of pressure drop, and according to the particular position in the system, with unequal air temperature. Maintenance was difficult because of the need to withdraw the complete cooling unit out of the 19" rack.

Two main design demands were established: the first one was to eliminate the above-mentioned shortcomings of the air flow system and the second to determine the needed fan properties and fan number in the cooling unit, according to a power dissipation of I/O cards and accepted air temerature difference as requirements of good electronic cooling.

Air Flow System

The differences between the old and new forced-air cooling system are obvious from a glance at the schemes in Fig. 3 and Fig. 4. The parallel-flow system was chosen because of its advance in the application of small fans, rather than as they had been used in previous serial-flow systems. The scheme of a new cooling unit which enables the parallel flow of air through the system of 19" card frames is shown in Fig. 4. It enables a flow of large air volumes at low pressure and equal inlet air temperature to all 19" card frames in the system. The distribution plenum is an integral part of the cooling unit, which enables better air distribution than in the old one, and air "dead spots" are avoided. The shape of the inlet plenum is designed according to [1].

Cooling Unit Performances

The analysis and schematics of heat transfer in the industrial computer have been made with the aim of obtaining a thermal model. The structure of the computer has been combined from many components with unequally placed heat sources and heat sinks. The temperature field might be very complex and can depend on the disposition of heat sources and heat sinks, as well as structural element geometry and their thermophysical properties. The thermal model is presented with an idealized heat transfer and simplified structural elements.

Conduction heat transfer from the components to the printed circuit board, plastic guide bars, housing elements etc. is neglected because of the high thermal resistance of used materials. Radiation heat transfer is insignificant for the relatively small difference in surface temperature between the electronic and the structural components, and it can be neglected. It is

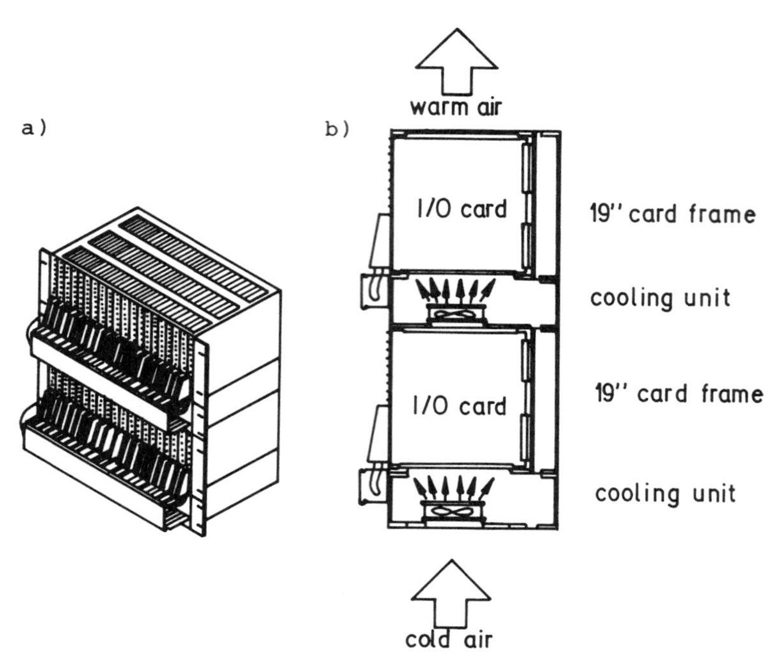

FIGURE 3. The standard forced-air cooling system for 19" equipment: a) 19" card frame set with cooling units, b) cross section.

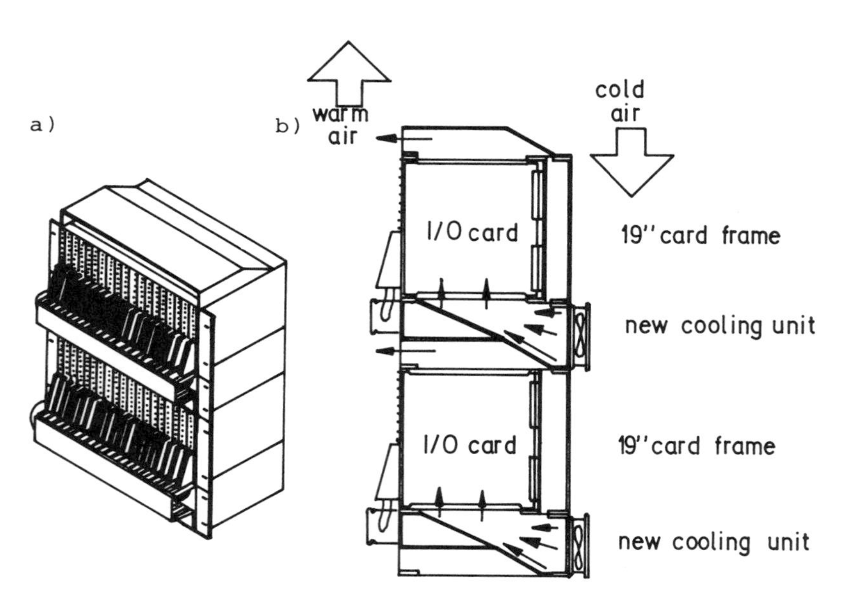

FIGURE 4. The new forced-air cooling system for 19" equipment: a) 19" card frame set with new cooling units, b) cross section.

considered that the heat dissipation of components from the printed circuit board is completely transferred to the air by forced convection.

One of the possibilities for simplification of the thermal model is to replace the real complex form of the electronic components, which have different heat dissipations, with a rectangular parallelepiped. The rectangular parallelepiped has a constant mean surface temperature and uniformly placed heat sources (Fig. 5). This replacement is based on an arithmetic average principle. The effective thickness of the parallelepiped is given by

$$\delta_{eff} = \delta + V_C / (L_Z L_Y) \qquad (1)$$

where δ is the printed circuit board thickness, V_C the total sum of pc board component volumes, L_Z the pc board height, and L_Y the pc board depth. The effective thickness of the pc board in this particular design case is

$$\delta_{eff} = 3\ 10^{-3}\text{m}\ (\delta = 1.6\ 10^{-3}\text{m},\ V_C = 4.674\ 10^{-5}\text{m}^3,\ L_Z = 0.233\text{m},\ L_Y = 0.22\text{m}).$$

The space formed by two pc boards (I/O cards) might be observed as a vertical asymmetrically heated channel. In this channel, the primary mechanism for heat transfer is forced-air convection. A housing of I/O cards is combined by these parallel channels. It is considered that heat dissipation in each channel is the same. The channels in standard 19" card frame (I/O card housing) have the same geometric characteristics, and it is possible to observe heat transfer in one channel whose height is the same as the pc board (Fig.6).

The air flow rate and pressure drop in the system are factors that need to be known, so that the cooling unit fans can be chosen correctly. These factors are determined for the cooling of I/O cards with a maximum power dissipation of 10W per channel.

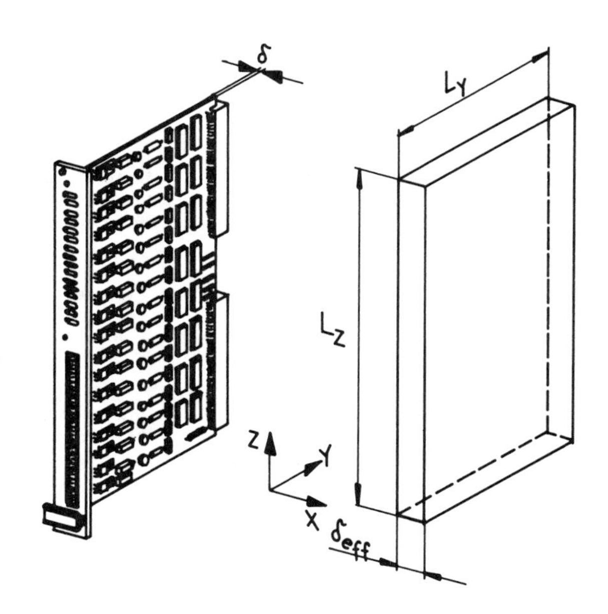

FIGURE 5. The I/O card and its simplified model.

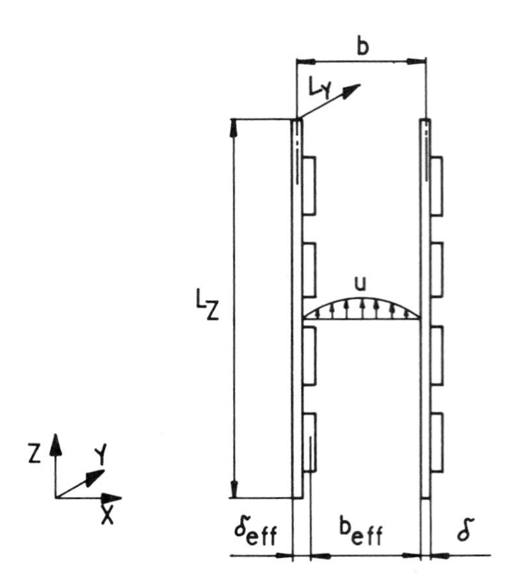

FIGURE 6. The vertical channel formed by two I/O cards.

Air flow rate. The necessary air flow rate is determined according to [2]. In this case, the air flow regime for forced convection is turbulent. The card temperature field is unequal in direction of all three X, Y, Z axes. For forced air cooling, the temperature inequality in the direction of X and Y axis is neglected [2, 4].

It is considered that the air temperature in the surroundings of the electronic components is equal to the one at the channel exit. The air temperature difference in the channel is given by

$$\Delta T = T_C - T_a \tag{2}$$

where T_C is the admissible case temperature of the electronic component and T_a the admissible air temperature in the surroundings of the component. According to Eq. (2), the value of the admissible temperature difference is $\Delta T = 15^oC$ ($T_C = 70^oC$, $T_a = 55^oC$).

The required mean value of the heat transfer coefficient in the channel is calculated according to

$$h = P_d/(\Delta T A) \tag{3}$$

where P_d is the power dissipation in the channel and A the total surface area of the components on the pc board, including the surface area of the pc board which is free of components. The heat transfer coefficient value is h = 12.4 $W/(m^2K)$ (P_d = 10W, A = 5.38 $10^{-2}m^2$).

The channel effective width is given by

$$b_{eff} = b - \delta_{eff} \tag{4}$$

where b is the channel width. The value of the effective channel width is $b_{eff} = 0.017$ m (b = 0.020 m).

The mean value of Nusselt number, according to definition, is given by

$$Nu = h\, D_e / k_f \qquad (5)$$

where D_e is the equivalent diameter of the narrow channel whose cross section shape is rectangular and k_f the thermal conductivity of the air at T_a temperature. According to [2, 3, 4], the equivalent diameter is $D_e = 2b_{eff}$ in the case of turbulent forced-air flow in narrow rectangular channels. Based on Eq. (5), the value of Nusselt Number is Nu = 14.7 ($D_e = 0.034$ m, $k_f = 2.87\ 10^{-2}$ W/(mK) at $T_a = 55°C$).

The ratio L_Z/D_e is important for the type determination of turbulent flow - a developing or fully developed flow. According to [2, 3, 4], in the case of $L_Z/D_e \le 20$, the turbulent flow is developing, and in case of $L_Z/D_e > 20$, the turbulent flow is fully developed. Based on the above-mentioned criteria, there are two eqs. for determining a Nusselt number mean value in channels according to [2], as follows:

$$Nu = 1.165\ Nu_\infty\ (20\ D_e/L_Z)^{0.167},\quad L_Z/D_e \le 20 \qquad (6)$$

and

$$Nu = Nu_\infty\ (1 + 3.3\ D_e/L_Z),\quad L_Z/D_e > 20 \qquad (7)$$

where Nu_∞ is the Nusselt number mean value at the fully developed turbulent flow in an infinite tube with D_e diameter.

According to [3], the Nusselt number value for turbulent flow in an infinitely long tube is determined as follows:

$$Nu_\infty = 0.023\ Re^{0.8}\ Pr^{0.43}\ (Pr_f/Pr_w)^{0.25} \qquad (8)$$

where Re is the Reynolds number, Pr the Prandtl number, Pr_f the Prandtl number at T_a air temperature, and Pr_w the Prandtl number at T_C air temperature. Since the L_Z/D_e ratio is $<$ 20, the turbulent flow is developing and from Eq. (6) it is possible to determine the Nusselt number - Nu_∞ and its value is $Nu_\infty = 10.6$.

The Prandtl number value at the temperature range which is observed here is Pr = 0.7. Since the difference between T_C and T_a temperature is small, the term $(Pr_f/Pr_w)^{0.25}$ in Eq. (8) is neglected. Based on the simplified formula (8), it is possible to determine the Reynolds number as follows:

$$Re = (Nu_\infty\ /0.0197)^{1.25} \qquad (9)$$

The Reynolds number value is Re = 2576 and in the case of channel abrupt entrance, according to [5], the air flow is turbulent.

According to definition, the Reynolds number is given by

$$Re = u\ D_e / \nu \qquad (10)$$

where u is the mean value of air velocity in the channel and ν the kinematic viscosity of the air at T_a temperature. Based on Eq. (10, it is possible to determine the mean value of air velocity in the channel, and it is u = 1.4 m/s ($\nu = 18.46\ 10^{-6}\ m^2/s$).

The air flow rate in one channel is given by

$$\dot{V}_{ch} = f_s \, u \, A_{ch} \tag{11}$$

where f_s is the safety factor for leakage in the housing of I/O cards and A_{ch} the physical free space for air flow through the channel. The necessary air flow rate for one channel is determined and its value is $\dot{V}_{ch} = 4.62 \; 10^{-3}$ m^3/s ($f_s = 1.25$, $A_{ch} = 2.64 \; 10^{-3} \; m^2$).

The air flow rate in the 19" card frame with I/O cards is given by

$$\dot{V} = N_{ch} \, \dot{V}_{ch} \tag{12}$$

where N_{ch} is the number of I/O card formed channels in the 19" card frame. The necessary air flow rate, which has to be delivered by the cooling unit for 19" card frame is $\dot{V} = 9.702 \; 10^{-2} \; m^3/s$ ($N_{ch} = 21$).

Pressure drop. The pressure drop in the system that is formed by the cooling units and I/O cards in the 19" card frames is determined according to [1]. Since the great part of the pressure drop is caused by the inlet and outlet distribution plena of the cooling unit, the part of the pressure drop which is made in the I/O card formed channels is neglected.

As the air flow through the system is parallel, the pressure drop in the inlet plenum is given by

$$\Delta P_i = \rho_i \, u_i^2/2 + 0.822 \, \rho_o u_o^2/2 \tag{13}$$

where ρ_i is the inlet air density, u_i the mean value of air velocity at the inlet plenum, ρ_o the outlet air density, and u_o the mean value of air velocity at the outlet plenum.

The pressure drop in the outlet plenum is given by

$$\Delta P_o = 0.645 \, \rho_o u_o^2/2 \tag{14}$$

The inlet and outlet air velocity is determined according to the following equations:

$$u_i = \dot{V}/A_i \tag{15}$$

and

$$u_o = \dot{V}/A_o \tag{16}$$

where A_i is the free inlet area and A_o the free outlet area of the cooling unit. The inlet and outlet air velocity values are calculated respectively and they are $u_i = 2.2$ m/s, $u_o = 3.9$ m/s ($A_i = 0.0441 \; m^2$, $A_o = 0.0252 \; m^2$).

According to Eqs. (13) and (14), the pressure drop is determined for the inlet and outlet plena of the cooling unit which are, respectively, $\Delta P_{io} = 9.5$ Pa, $\Delta P_o = 5.3$ Pa ($\rho_i = 1.128 \; Kg/m^3$ at the inlet air temperature of 40^oC, $\rho_o = 1.077 \; kg/m^3$ at the outlet air temperature of 55^oC).

Pressure drop in the system is determined according to the following equation:

$$\Delta P = f_c \; (\Delta P_i + \Delta P_o) \tag{17}$$

where f_c is the correction factor which considers the deviation of the inlet and outlet plenum shapes from the theoretical one and the neglected channel pressure drop. The pressure drop in the system is $\Delta P = 17$ Pa ($f_c = 1.15$).

Fan choice. According to the determined values of the necessary air flow rate and pressure drop in the system, the choice of standard 120 mm - square small fan manufactured by ISKRA, type MIV 0201 is made. The fan performance yields an air flow rate $\dot{V}_F = 3.5\ 10^{-2}\ m^3/s$ at a pressure drop of $\Delta P_{st} = 17.5$ Pa. The fan number in the 19" cooling unit is given by

$$N_F = \dot{V}/\dot{V}_F \tag{18}$$

where $\dot{V}_F$ is the air flow rate of one small fan. The necessary fan number in the 19" cooling unit is calculated and it is $N_F = 3$.

System operating point. The pressure - vs. - volume curves of the cooling unit performance and the cooled equipment characteristic are shown in Fig. 7. The system operating point is given by the intersection of the above-mentioned curves.

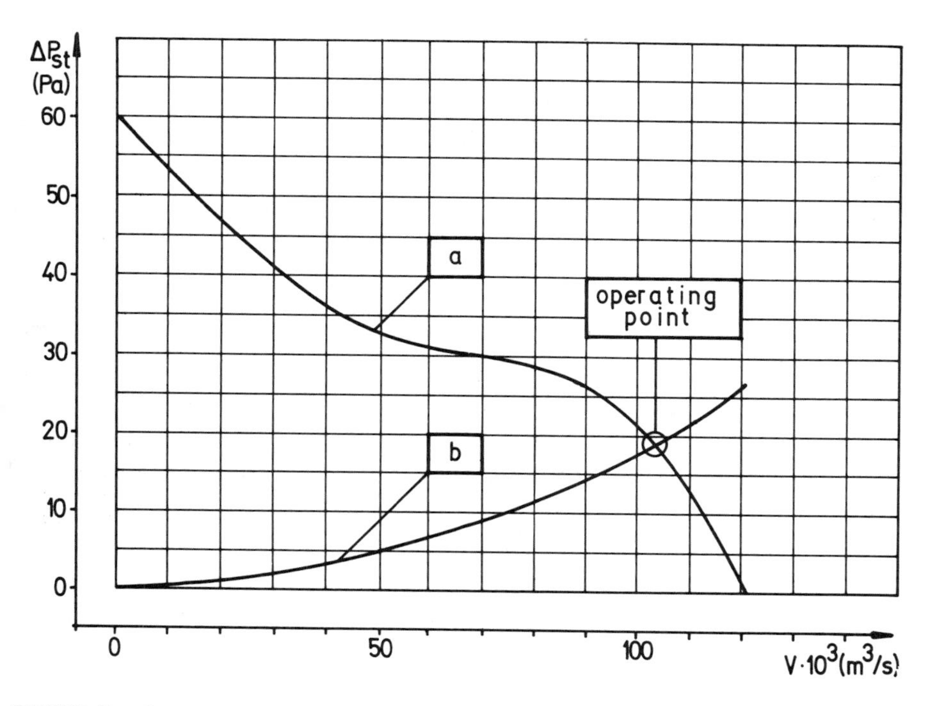

FIGURE 7. Pressure-vs.-volume curves: a) the cooling unit performance, b) the cooled equipment characteristic.

DISCUSSION

The real temperatures of the I/O card electronic components are measured in the new cooling system of the 19" card frame as well as in the old one. The mean temperature values of the components in the new cooling system are $5^{\circ}C$ or more below the corresponding components in the old one, and are more uniform, because each type of component is at the same air flow level and air temperature. At full load on the output cards and the ambient temperature of $40^{\circ}C$, none of the components exceeds $70^{\circ}C$.

Since the needs of different applications of industrial computers demand different combinations of I/O cards and packaging technologies, it is necessary to take this into account to achieve uniform air flow. In the case of free slots in the 19" card frame, it is necessary to insert so-called "dummy" cards in it, and in that way to keep the same geometric characteristics in the channels. These "dummy" cards have surfaces with elevated parts which simulate electronic components in order to keep a uniform pressure drop at air flow through the channels in the 19" card frame. They can be made from ABS plastic sheets by thermoforming technology.

The obvious shortcomings of the old serial air flow system are avoided by the introduction of the new parallel air flow system. The parallel-flow system advantages in relation to the serial-flow system are as follows:

1. Each 19" card frame is provided at its inlet with cold air of the same temperature.
2. The fan efficiency is increased.
3. The fan working time is increased as a result of a fan position in the system and less inlet air temperature.
4. Air "dead spots", the consequence of the fan motor position are avoided.
5. A better air stream pattern across the card components is gained.
6. The fan maintenance is simplified.

The shortcoming of the new cooling system is in the need to use more material for the auxiliary rotating rack in the case of wall installation.

CONCLUSIONS

The improvements made in the new cooling system described in this paper over the old one are in the fields both of better cooling of I/O cards and of fan reliability. In the case of new 19" cooling unit design, the design procedure described in this paper is useful for determining the necessary air flow rate and pressure drop in the system with a different number of I/O cards in the 19" card frame (i.e., channel number - N_{ch}), a different power dissipation of the I/O cards (P_d) and a different admissible temperature difference (ΔT) from the ones which are presented in this paper. According to this and the chosen fan performances, it is possible to find a new fan number in the 19" cooling unit (N_F). In this way, the electronics would be maintained in the admissible temperature range by the 19" cooling unit fans.

Since the design demands are not always achieved in practice with the application of fans which have brushless DC motors, and since the speed of the DC variety is easily controlled with a thermistor according to the ambient temperature, it is possible to improve their working time, and to reduce fan noise, by decreasing speed at a lower ambient temperature, as well as to protect the electronic modules from overheating at a short-time overload by increasing speed at higher ambient temperature.

NOMENCLATURE

Symbol	Quantity	SI Unit
A	area, cross section	m^2
a	thermal diffusivity	m^2/s
b	width	m
D_e	equivalent diameter	m
h	heat transfer coefficient	$W/(m^2K)$
k	thermal conductivity	W/(mK)
L	length	m
N	number	-
P_d	power dissipation	W
P	pressure	N/m^2
T	temperature	^oC
u	velocity	m/s
V	volume	m^3
$\dot{V}$	volume flow rate	m^3/s
Greek Letters		
δ	thickness	m
Δ	indicates a difference between variables	-
ν	kinematic viscosity	m^2/s
ρ	density	kg/m^3
Coordinates		
X, Y, Z	cartesian coordinates	
Subscripts		
a	indicates air condition	
C	indicates component condition	
ch	indicates channel condition	
eff	indicates effective condition	
F	indicates fan condition	
f	indicates fluid condition	
i	indicates inlet condition	
o	indicates outlet condition	
st	indicates static condition	
w	indicates wall condition	
∞	indicates infinite condition	

Symbol and Definition	Name
Dimensionless parameters	
$Nu = h\ L/k$	Nusselt number
$Pr = \nu/a$	Prandtl number
$Re = u\ L/\nu$	Reynolds number

REFERENCES

1. Shah, R. K., and Mueller, A. C., Heat Exchangers, in **Handbook of Heat Transfer Applications**, eds. Rohsenow, W. M., Hartnett, J. P., and Ganić, E. N., ch. 4, pp. 4-266 to 4-270, McGraw-Hill, New York, 1985.

2. Savelev, A. J., and Ovchinnikov, V. A., **Konstruirovanie EVM i sistem**, pp. 108-128, Vysšaja Škola, Moskva, 1984.

3. Dulnev, G. N., **Teplo- i massoobmen v radioelektronnoj apparature**, pp. 80-81, Vysšaja Škola, Moskva, 1984.

4. Kraus, A. D., and Bar-Cohen, A., **Thermal Analysis and Control of Electronic Equipment**, pp. 321-338, McGraw-Hill, New York, 1983.

5. Kays , W. M., and Perkins, H. C., Forced Convection, Internal Flow in Ducts, in **Handbook of Heat Transfer Fundamentals**, eds. Rohsenow, W. M., Hartnett, J. P., and Ganić, E. N., ch. 7, pp. 7-89 to 7-92, McGraw-Hill, New York, 1985.

Thermal Enhancement for Simulated LSI Packages by External Rods

J. H. CHOU and K. F. CHIANG
Department of Engineering Science
National Cheng Kung University
Tainan, Taiwan, PRC

ABSTRACT

An experimental investigation has been conducted to study the feasibility of using external rods to increase local heat transfer rates for simulated LSI packages. Parameters examined included the flow Reynolds number and the rod location. It is observed that, for high Reynolds number flows, a single rod at a proper location is an effective local heat transfer promoter. However, at low Reynolds numbers, the rod may have no effect at all.

INTRODUCTION

To remove dissipated heat effectively from an electronic device is very critical to the design of densely packaged electronic components. From the thermal design point of view, the goal is to achieve a proper junction temperature and a uniformly distributed temperature field. A lower junction temperature can increase not only the reliability but also the lifetime of an electronic component. A uniformly distributed temperature field can reduce thermal stresses related fatigue problems, and solder cracks can thus be avoided.

The best approach for a thermal designer to achieve the above goals is to get involved in the early design stage so that all thermal related problems can be carefully considered. However, this might not be possible all the time, and alternative approaches might be sought.

It has been observed that staggering the array of LSI packages can reduce the air temperature difference so that a more uniform temperature distribution can be obtained[1]. It would be very helpful if one can apply the staggered array technique to circuit design, but this might not be practical because of circuit routing requirements.

Sparrow et al.[2,3] used barriers to increase local heat transfer rates. Chou and Lee[4] and Chou et al.[5] used vortex generators to enhance local heat transfer from simulated LSI packages and thus reduce temperature non-uniformities for these packages. The feasibility of using cylindrical rods as turbulence promoters to enhance local heat transfer for chip arrays has been demonstrated recently by both theoretical and experimental means[6-9]. Barriers, vortex generators and cylinders are devices external to chip arrays that can be installed at desired locations without any need for concern about the routing rules. If the pros and cons of these techniques can be documented, then they might be very useful to packaging designs.

The purpose of this study is to further exploit the merits of using external rods (i.e., cylinders) as heat transfer promoters and to see whether they can be used to reduce thermal non-uniformities. The investigation was conducted by experimental methods using simulated LSI packages.

EXPERIMENTIONAL APPROACH

All the experiments presented in this study were conducted in the recently established Microelectronics Heat Transfer Laboratory in the Department of Engineering Science, National Cheng Kung University. A thermal mockup approach was used in this study so that unnecessary thermal damage to electronic components could be avoided. The simulated LSI packages were made of ceramic resistors. There are a total of 40 resistors. Each resistor represents an LSI package. They are arranged in a configuration, as shown in Fig. 1, of five columns and eight rows in the direction of forced air flow. These resistors are mounted on a flat insulation base plate and serially connected to each other so that the current running through each resistor is the same. The dimension for each resistor is 21.5 mm in length (L), 9.5 mm in width (W) and 10.5 mm in height (H). The typical power consumption is approximately 1.01 watts for each resistor.

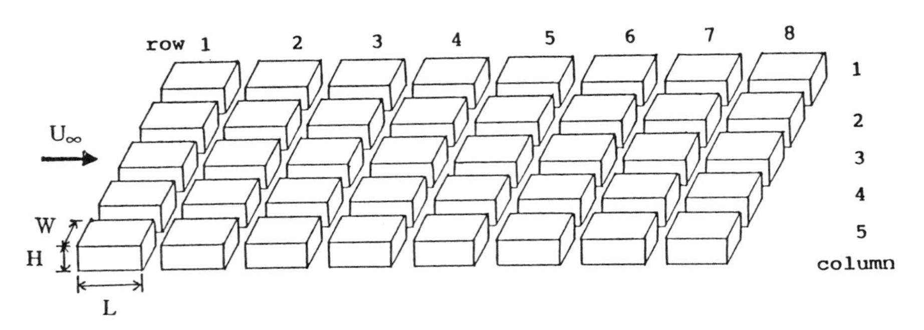

FIGURE 1. Thermal mockup module for simulated LSI packages.

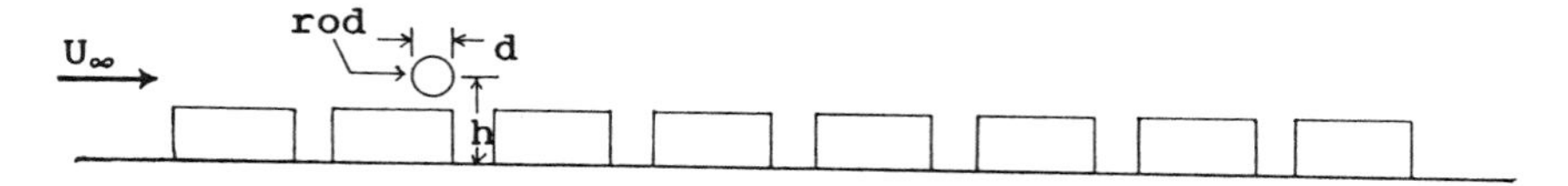

FIGURE 2. An external rod and the thermal mockup module (side view).

The insulation base plate of the tested thermal mockup module was mounted flush to a low speed wind tunnel wall at the exit plane of the tunnel. The tunnel is a continuous, open-loop type with a contraction ratio of 9:1. Laboratory air was the working fluid. The room temperature was approximately 22°C.

Cylindrical rods of various diameters were used as external devices for thermal enhancement. A typical configuration is shown in Fig. 2 where a glass rod is mounted near the aft location of the second row package. The rod axis is perpendicular to the incoming flow direction. The rod is long enough to cover the range of the second row so that rod two-dimensionality can be insured.

Flow velocities were measured by heat-sensitive flowmeters. The air velocities tested were between 1 m/sec and 6 m/sec. Temperature distributions for the simulated LSI packages were measured by thermocouples and an infrared imaging radiometer. Thermocouples are conventional T-type sensors with a range from -180°C to 350°C and an error of ±0.1 °C. The range of the infrared radiometer is from -20°C to 850°C with an error of ±0.1°C at a measuring temperature of 30 °C. In order to facilitate the thermal scan by the radiometer, the thermal mockup module is completely painted with a black paint. The emissivity of this non-reflective black paint is 0.95.

Two types of rod configuration were examined. One is with a single rod installed at the second row. The other is with 5 nearly identical rods installed simultaneously, one rod per row, at rows 2,3,4,5 and 6. The former type is to investigate the basic mechanisms. The latter type is to test the possibility of making a uniform temperature distribution for the whole mockup module. Flow fields with three rod heights and six Reynolds number conditions are studied.

RESULTS AND DISCUSSION

For comparison purposes, an average temperature T for each simulated LSI package was calculated from the infrared radiometer according to the following formula:

$$T = \sum T_i A_i / A$$

where T_i is a representative temperature for the gray scale i, A_i is the area of the gray scale i on the top surface of a simulated LSI package and $A = \sum A_i$ is the top surface total area which is 21.5 mm by 9.5 mm. A typical value for i is five in this study. This implies that there are usually five gray scales from the infrared scanner. Since there are 40 simulated LSI packages on the thermal mockup module, there are 40 average temperatures for each test condition. As described in the previous section, the forty simulated packages are arranged into an 8 row by 5 column configuration. But rather than show the temperature distributions for the whole mockup module, only the temperature distributions for the center column (i.e., column number 3) are presented in the following paragraphs. This is mainly to keep the description as simple as possible so that comparisons for different test conditions can be made conveniently.

A typical result for the Fig. 2 configuration is depicted in Fig. 3. It can be seen that by applying a rod at the second row, the rod influence can be felt at the second, the third and the fourth row, At these three rows, the average surface temperatures of the LSI packages are reduced significantly. The temperature reduction is as much as 10 degrees centigrade at the third row. The rod has no effect on rows 5 and 6. But the effect is felt again, though not as significant as the front rows, at rows 7 and 8. Over the

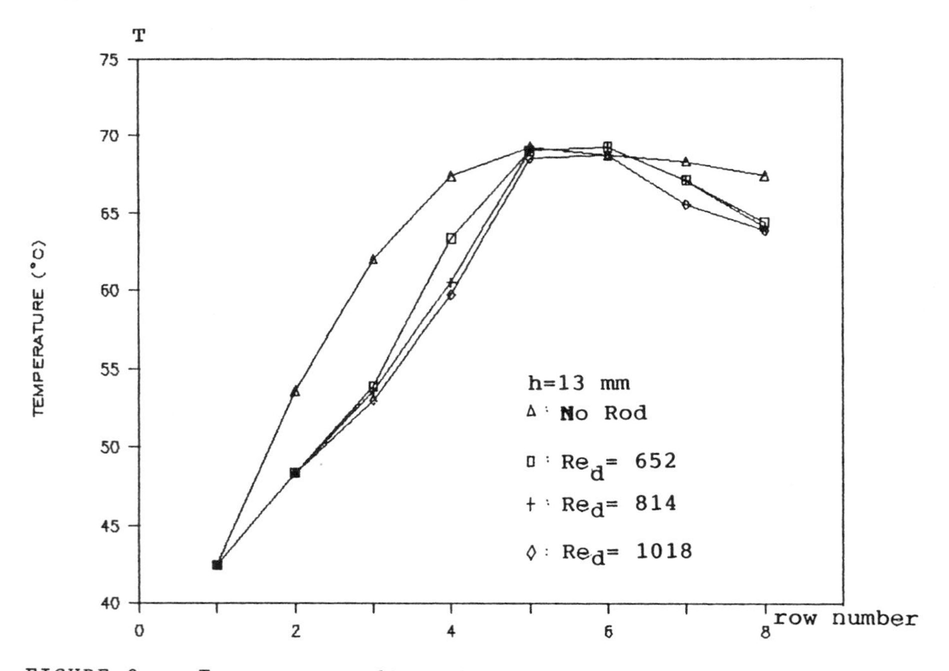

FIGURE 3. Temperature distributions for the LSI module.

range of Reynolds numbers given in Fig. 3, there seems to be no rod Reynolds number dependence on the temperature reduction mechanism. This trend is fairly typical for relatively large Reynolds numbers. It seems to indicate that the role of an external rod is to shed unsteady vortices to modulate the wake flow behind the rod. Therefore, as far as the diameter of the rod is large enough to trigger vortex pairing, which is of course not the Karman vortex, then its effect is minimum. The pairing vortices promote momentum exchange between fluid in the thermal region and the outer non-thermal region by its rolling mechanism. This rolling behavor is the key reason that there is no temperature reduction at rows 5 and 6.

The effect of rod height is shown in Fig. 3 and Fig. 4. The height is measured from the surface of the module base plate to the centerline of the cylindrical rod. The values are 13 mm, 13.4 mm and 13.9 mm for the results shown in Fig. 3, Fig. 4a and Fig. 4b, respectively. By comparing results in these figures, one can see that in this range of rod heights, the height has no significant influence on the temperatures for packages at row 1 through row 5. However, for rows 6,7 and 8, there is a temperature overshoot as the height increases. Here, overshoot means that the temperature is higher than that of the same module without installation of any external rod. Since a larger height tends to induce large unhindered pairing vortices, this overshoot phenomena is reasonable.

At lower Reynolds numbers, the results show some different features, as depicted in Fig. 5. The Reynolds numbers based on rod diameters are 326, 163 and 102 for Figs. 5a, 5b and 5c, respectively. At a Reynolds number of 326, there is a mixed convection region near rows 6,7 and 8. But this mixed mode is not significant; thus, the general trend of rod effect is the same as that described in Figs. 3 and 4. However, at Reynolds numbers of 163 and 102, the effect of mixed convection heat transfer is obvious. It can be observed that at these two Reynolds numbers, the temperature distributions for the module without any external rod increase rapidly from row 1 to row 4 (which is also true for higher Reynolds numbers), then levels off from row 4 to row 6, and finally decreases from row 6 to row 8. The rod effect shown in Fig. 5b is quite complicated. Since the rod height has a significant influence on the temperature distributions in the region between row 3 and row 7, the effect is probably a combination of flow instabilities and pairing momentum exchange, and needs to be clarified in future studies.

At a Reynolds number of 102, which corresponds to an air speed of 1 m/sec and a rod diameter of 2 mm, the rod has essentially no influence on the temperature field of the mockup module. For flows at this condition or at even lower air speed, other mechanisms might have to be considered to enhance the local heat transfer coefficient.

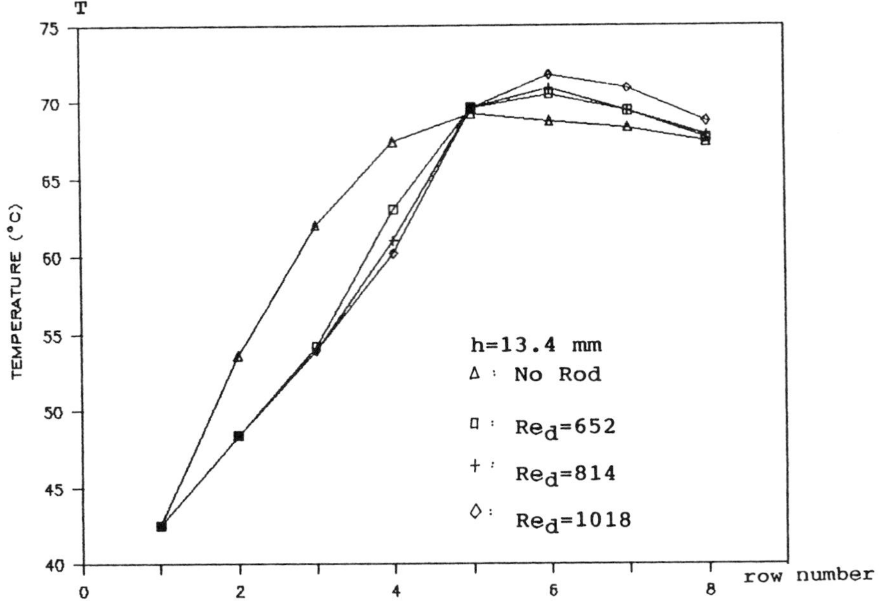

FIGURE 4a. Temperature distributions for the LSI module (height effect).

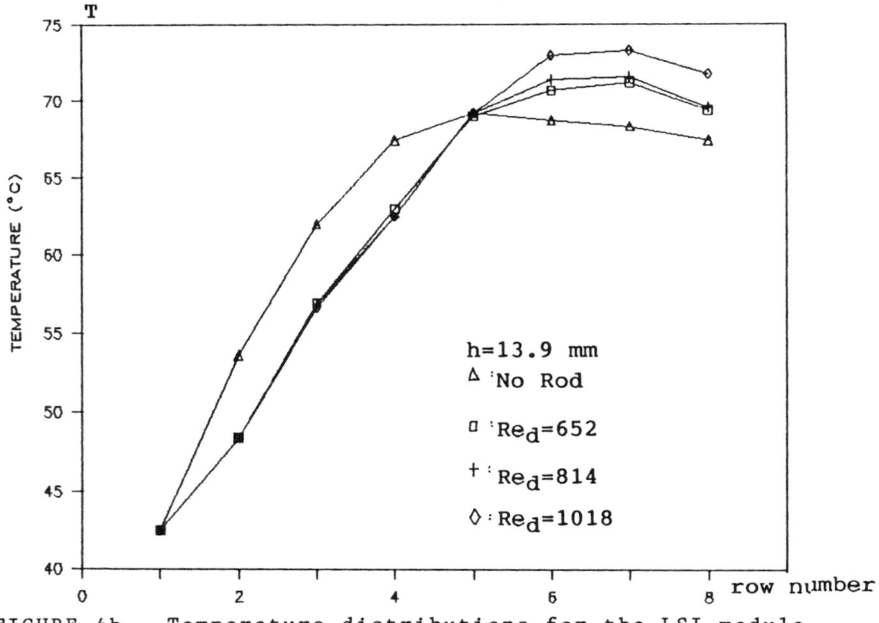

FIGURE 4b. Temperature distributions for the LSI module (height effect).

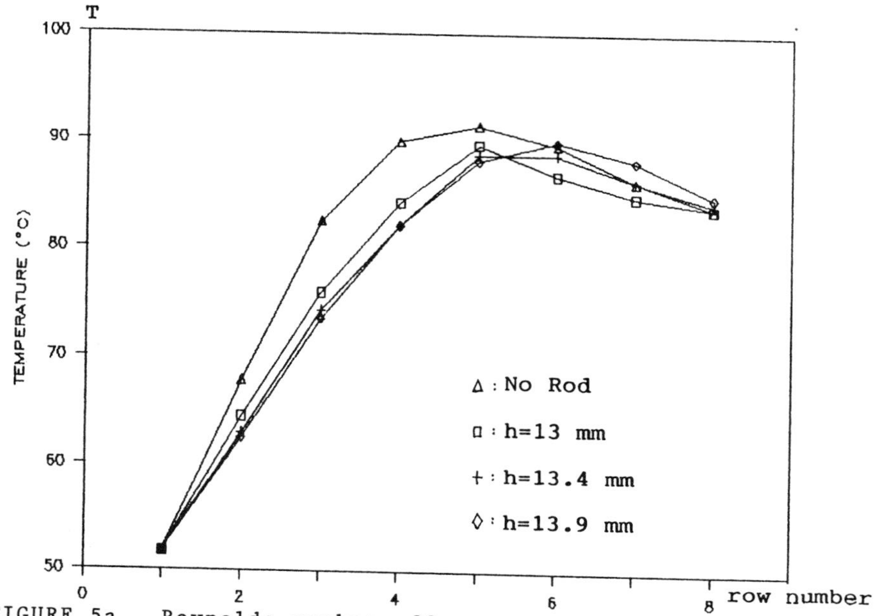

FIGURE 5a. Reynolds number effects on temperature distributions (Re_d = 326).

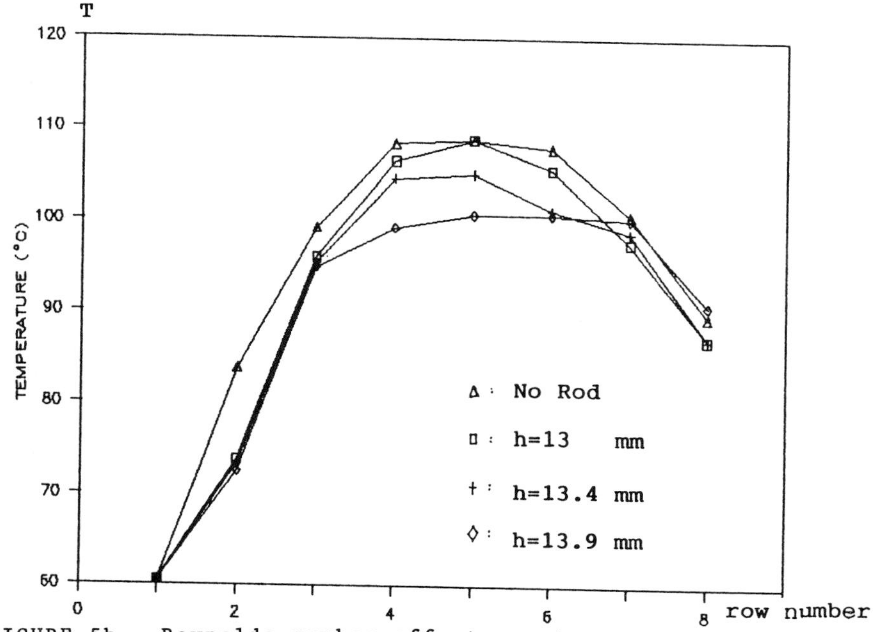

FIGURE 5b. Reynolds number effects on temperature distributions (Re_d = 163).

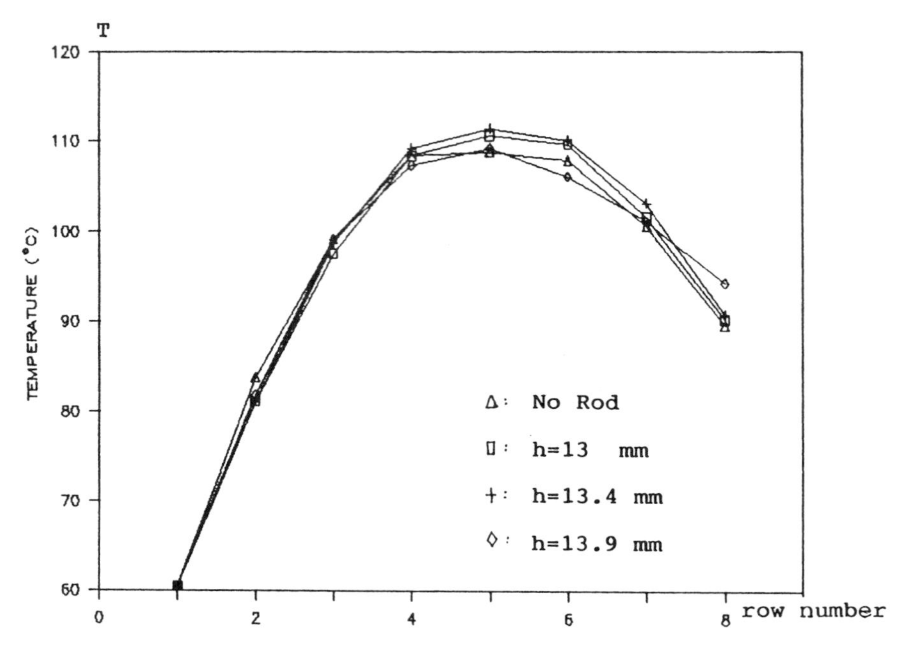

FIGURE 5c. Reynolds number effects on temperature distributions (Re_d = 102).

All the results described above are for flows either without any external rod or with an external rod installed at the aft location of the second row. The results clearly indicate that for higher Reynolds numbers, a rod can be used as a heat transfer promoter to enhance local heat transfer, thus reducing package temperatures. Another question that arises naturally is whether one can take this temperature reduction capability to make the temperature of the whole thermal mockup module become more uniform. In order to answer this question, nearly identical rods are installed at rows 2,3,4,5 and 6. The relative position of each rod in the corresponding simulated LSI package is the same as the rod at the second row package. A typical result is shown in Fig. 6. It can been that for the three Reynolds numbers shown, the temperature distributions are essentially independent of Reynolds number. This trend is the same as that for the single rod case with the same Reynolds numbers. By comparing the results in Fig. 3 and Fig. 6, it is clear that there is a temperature reduction for row number 5 and a temperature overshoot at rows 7 and 8 for the 5 rod installation case. There is no significant temperature difference for rows 2,3 and 4 between single rod cases and 5 rods cases. Since multiple rod application causes

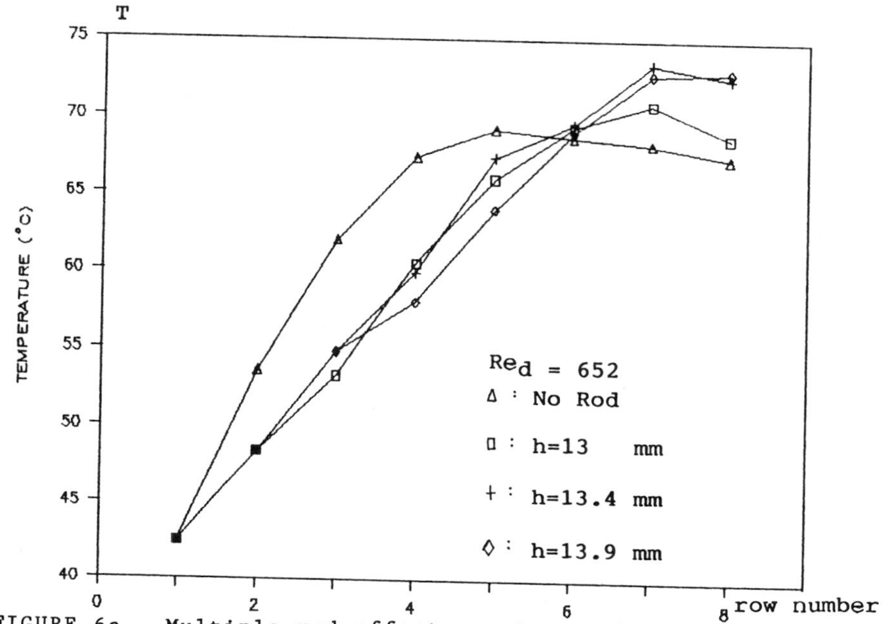

FIGURE 6a. Multiple rod effects on temperature distributions at high Reynolds number.

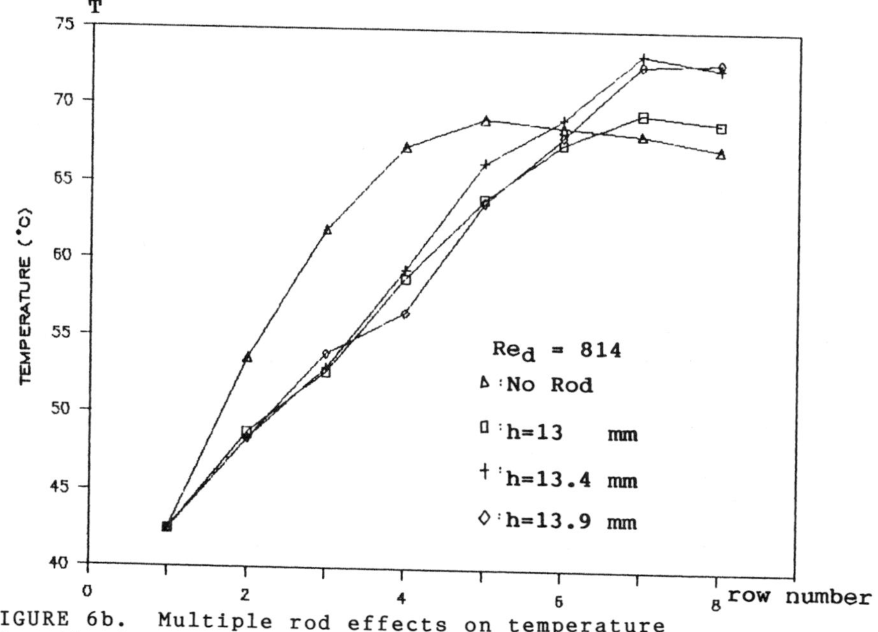

FIGURE 6b. Multiple rod effects on temperature distributions at high Reynolds number.

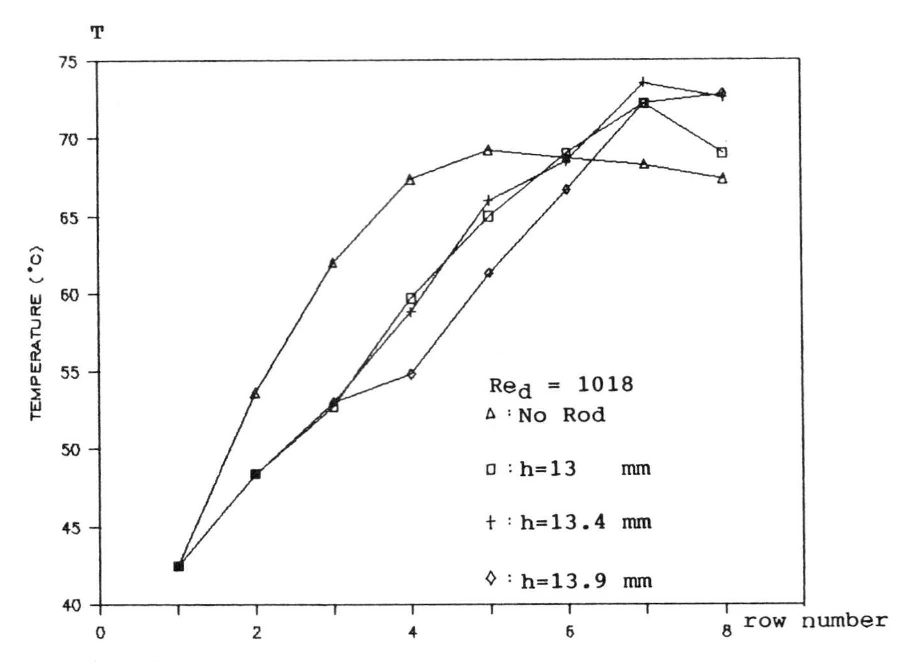

FIGURE 6c. Multiple rod effects on temperature distributions at high Reynolds number.

temperature overshoot and temperature reduction occurs only at row number 5, the goal of achieving a uniformly distributed temperature field is not met.

At lower Reynolds numbers, multiple rod application also causes temperature overshoot. Besides this phenomenon, the difference in effect between single rod and multiple rods configuration is very small.

CONCLUSIONS

An experimental investigation has been conducted to study the characteristics of flow fields with external rods as local heat transfer promoters. Key features are summarized as follows:

1. For high Reynolds number flows, a single rod at the proper location is an effective local heat transfer promoter. At low Reynolds numbers, there may be no effect at all.

2. The trend of rod effect on temperature distributions

is very similar for single rod and multiple rod installations.

3. Multiple rod application is not an effective way to reduce temperature non-uniformities.

NOMENCLATURE

Symbol	Quantity	SI Unit
A_i	top surface total area of an LSI package	mm^2
A	top surface area for gray scale i of an LSI package	mm
d	diameter of the external rod	mm
H	height of an LSI package	m
h	rod height	m
L	length of an LSI package	m
Re_d	Reynolds number based on U_∞, d and ν	
T	top surface average temperature of an LSI package	°C
T_i	top surface temperature for gray scale i of an LSI package	°C
U_∞	freestream velocity	m/s
W	width of an LSI package	cm
ν	kinematic viscosity of air	cm^2/s

REFERENCES

1. Ashiwake, N.,Nakayama,W., Daikoku, D.and Kobayashi, F., Forced Convective Heat Transfer from LSI Packages in an Air-Cooled Wireing Card Array, Heat Transfer in Electronic Equipment, ASME HTD-Vol. 28, PP. 35-42, 1983.

2. Sparrow, E.M. Neithammer, J.E. and Chaboki, A., Heat Transfer and Pressure Drop Characteristics of Arrays of Rectangular Modules Encountered in Electronic Equipment, Int. J. Heat Mass Transfer, Vol.25, No.7, PP. 961-973, 1982.

3. Sparrow, E.M., Vemuri, S.B. and Kadle, D.S.,Enhanced and Local Heat Transfer, Pressure Drop, and Flow Visualization for Arrays of Block-Like Electronic Components, Int. J. Heat Mass Transfer, Vol.26, No.5, PP.

689-699, 1983.

4. Chou, J.H. and Lee, J., Reducing Flow Non-Uniformities in LSI Packages by Vortex Generator, Proceedings of the International Symposium on Cooling Technology for Electronic Equipment, Honolulu, Hawaii, USA, PP.583-594, 1987.

5. Chou, J.H., Chiang, K.F. and Lee, J., The Effect of Vortex Generators on Temperature Distributions of LSI Packages, Proceedings of the 4th Nat. Conf. on Mech. Eng., CSME, Hsinchu, Taiwan, PP.463-470,1987.

6. Ghaddar, N.K., Korczak, K.Z., Mikic, B.B. and Patera, A.T., Numerical Investigation of Incompressible Flow in Grooved Channel:Stability and Self-Sustained Oscillations, J.of Fluid Mechanics, Vol.163, PP.99-127,1986.

7. Patera, A.T. and Mikic, B.B., Exploiting Hydrodynamic Instabilities, Resonant Heat Transfer Enhancement, Int. J. Heat Mass Transfer, Vol. 29, No.7, PP.1127-1138, 1986.

8. Karniadakis, G.Em., Mikic, B.B. and Patera, A.T., Heat Transfer Enhancement by Flow Destabilization; Application to the Cooling of Chips, Proceedings of the International Symposium on Cooling Technology for Electronic Equipment, Honolulu, Hawaii, USA, PP. 498-521, 1987.

9. Ratts, E., Amon, C.H., Mikic, B.B. and Patera, A.T.,Cooling Enhancement of Forced Convection Air Cooled Chip Array through Flow Modulation Induced by Vortex-Shedding Cylinders in Cross-Flow, Proceedings of the International Symposium on Cooling Technology for Electronic Equipment, Honolulu, Hawaii, USA, PP. 651-662, 1987.

LIQUID COOLING

Liquid Immersion Cooling of Electronic Components

F. P. INCROPERA
Heat Transfer Laboratory
School of Mechanical Engineering
Purdue University
West Lafayette, Indiana 47907, USA

ABSTRACT

As circuit densities on a single silicon chip continue to increase and as more chips are packed in closer proximity on multi-chip modules, power densities continue to increase at both the chip and module levels. To dissipate this power, while maintaining acceptable chip temperatures, it is becoming difficult to rely on air or indirect liquid cooling. Hence, increasingly more emphasis is being placed on the development of direct liquid (*liquid immersion*) cooling schemes, which may involve single-phase convection as well as pool or forced convection boiling. The chips may be mounted to a substrate which forms one wall of a liquid-filled enclosure and in which heat is dissipated by free convection or boiling. Alternatively, the chips may be cooled by liquid jet impingement, or the substrate to which they are mounted may form one wall of a channel through which the liquid is forced. In both cases heat transfer may occur with or without boiling. Moreover, the ability to dissipate large power densities may necessitate the use of extended heat transfer surfaces. In this paper, available heat transfer data, correlations and predictions are reviewed for single chips and for multi-chip arrays, with and without extended surfaces.

INTRODUCTION

Since development of the first electronic digital computers, heat removal has played an important role in ensuring reliable operation. An important trend, which initially alleviated and subsequently exacerbated the heat removal problem, involved the integration of monolithic circuits on a silicon chip and the development of ever larger scales of circuit integration. From the large scale integration (LSI) technologies of the 1970s, which involved up to 1000 gates per chip, to the very large scale integration (VLSI) of the 1980s, which involved up to 100,000 gates per chip, there has been a steady increase in heat dissipation at the chip, module and system levels. Such increases have made the role of heat transfer and thermal design more important than ever, and the development of future large scale, high speed circuits may well be limited by the inability to maintain effective cooling.

With recognition that the limits of air cooling were being exceeded by rapid developments in circuit integration, there came the impetus for developing liquid cooling technologies. A distinction is made between *direct* liquid cooling, for which there is intimate contact between the coolant and the electronics, and *indirect* cooling, for which the electronics are physically separated from the liquid. In the late 1970s and 1980s, indirect cooling systems were devised to exceed the heat transfer capabilities of air cooling, while avoiding difficulties which were perceived to be associated with direct liquid cooling. As applied to multi-chip modules, indirect liquid cooling involves attachment of a water-cooled cold plate to the module, and a well known example of this technology is the Thermal Conduction Module (TCM) used on the IBM 308X/3090 series of computers [1]. As shown in Fig.1, contact is made between each chip and the spherical head of a spring loaded, aluminum piston. The pistons are seated in helium filled cavities of an aluminum housing which is bolted to a cold plate. Heat transfer from a chip is primarily by conduction across the helium gap at the chip/pin interface and then by conduction along the pin and across the helium gap between the pin and housing. The total thermal resistance is determined primarily by internal components, of which the helium gap resistances are the most prominent.

To substantially reduce internal resistances associated with indirect liquid cooling schemes, the coolant flow must be brought in closer proximity to the electronics. One approach [2] involves mounting flat-leaded chip carriers to both sides of a printed wiring board and interfacing each side with a cold plate assembly (Fig.2). The interface is provided by a bellows within which water flow maintains large convection heat transfer coefficients characteristic of a submerged jet. The stack separating the bellows from a chip consists of a compliant substance sandwiched between conductive heat transfer materials. Alternatively, interface resistances between the cold plate and a multi-chip module may be eliminated by making the

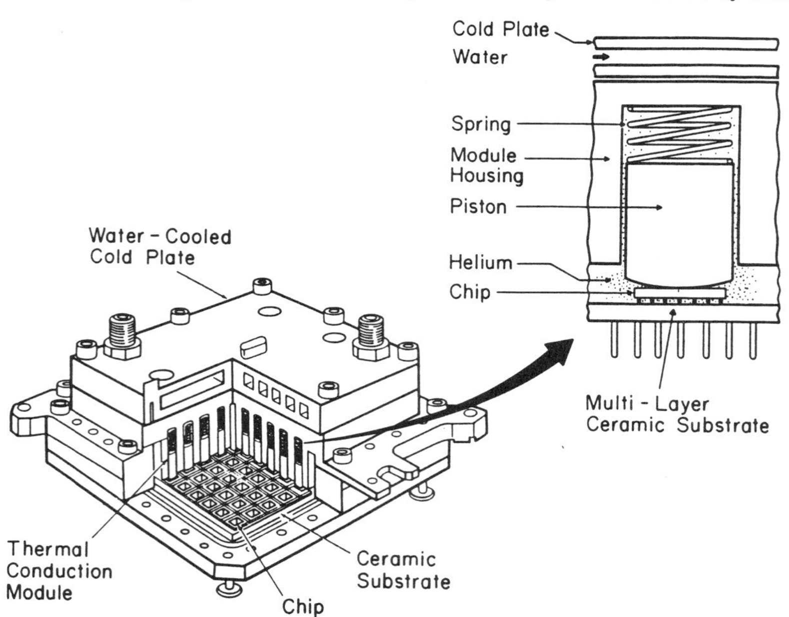

FIGURE 1. IBM 3080 X/3090 thermal conduction module (adapted from [1]).

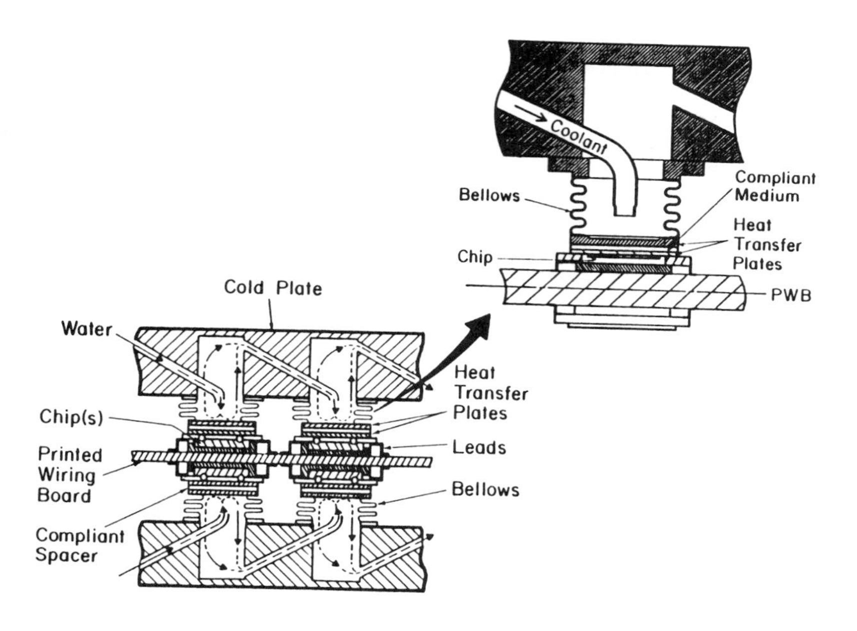

FIGURE 2. Fujitsu FACOM M-780 water-cooled module with bellows interface (adapted from [2]).

cold plate an integral part of the module [3]. As shown in Fig.3, water is pumped through small (sub-millimeter) coolant channels in a multi-layer alumina substrate to which an array of VLSI chips is mounted. In addition to greatly reducing internal resistances, the substrates may be stacked, one on top of another, to achieve a large volumetric packaging density.

The ultimate in indirect liquid cooling is achieved when the cold plate and chip are frabricated as an integral unit. This concept is illustrated in Fig.4. With heat dissipating IC elements attached to one side of a silicon substrate, microchannels may be chemically etched or precision machined (sawed or milled) on the opposite side [4,5]. The resulting structure resembles a miniature, extruded, longitudinally finned heat sink. Closure of the channels is achieved with a cover plate, and liquid is continuously pumped through the channels. In this way the conduction and interface resistances associated with conventional indirect liquid cooling schemes are virtually eliminated. Ultimately, however, application of microstructure cooling techniques will depend on the ability to establish flow, chip, and module interfaces which are both manufacturable and reliable. To minimize the crippling effects of fouling and/or blockage of flow passages, stringent requirements must be placed on coolant purity and hence on the coolant distribution unit. In this respect, direct cooling methods may be preferable.

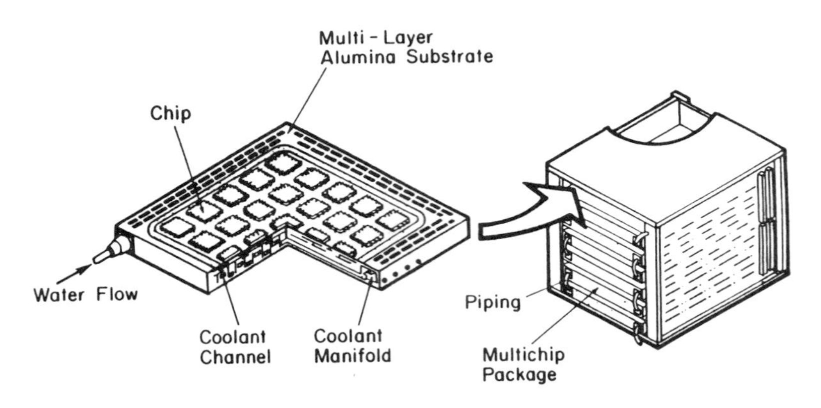

FIGURE 3. NTT multi-layer substrate with integral water cooling (adapted from [3]).

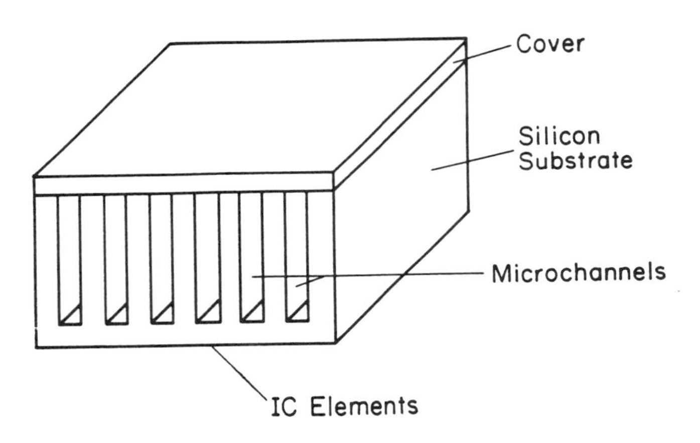

FIGURE 4. Microchannel cooling of silicon substrate (adapted from [4]).

Since direct liquid cooling maintains physical contact between the coolant and the electronic components, the coolant must have a very large dielectric strength and good chemical compatibility with the components. Typically, such coolants are characterized by low boiling points, and their use could involve cooling by pool or forced convection boiling, as well as by single phase convection (natural, forced, or mixed). Although the option of direct liquid cooling has been known for many years, actual application to commercial computers is limited to a single system [6]. As shown in Fig.5, vertical flow of the coolant (3M FC-77) between stacks of circuit modules precedes horizontal flow through the modules. Memory and logic chips attached to the modules are thereby immersed in the liquid, and heat transfer is likely to occur under single-phase, mixed convection conditions.

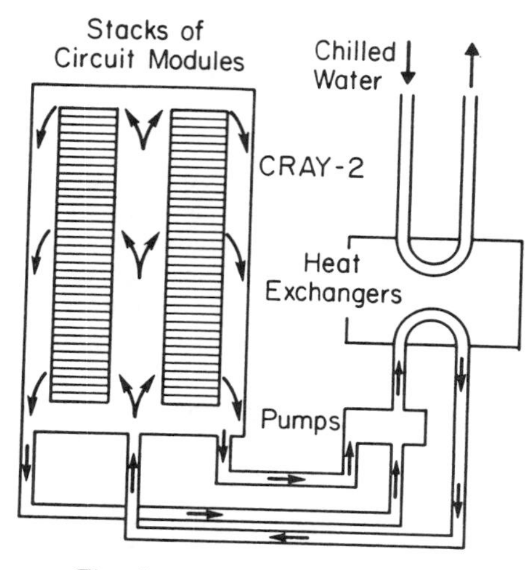

FIGURE 5. Cray-2 immersion cooling system (adapted from [6]).

Despite the paucity of existing applications of direct liquid cooling, there are clearly many options and prospects for the future are bright. Specific options involve single phase free, mixed and forced convection, as well pool and forced convection boiling. Since immersion cooling necessitates the use of inert dielectric liquids having comparatively poor heat transfer properties, options for dissipating large power densities may require the use of extended surfaces. In the following sections, the relationship of the heat transfer literature to immersion cooling is reviewed, and emphasis is placed on results for discrete, chip-like heat sources, with and without extended surfaces.

NATURAL CONVECTION

Although immersion cooling is usually associated with the need to dissipate large power densities, and hence with the use of single phase forced convection or boiling, applications involving single phase natural convection are not precluded. Natural convection is, for example, the requisite heat transfer mode in passive cooling systems and, relative to air, dielectric liquids offer significant thermal advantages. Three configurations are suitable for electronic cooling and they include: (i) parallel plate channels, (ii) discrete sources in a quiescent ambient, and (iii) closed rectangular cavities. It is noteworthy that, with respect to liquid immersion cooling, extended surfaces have not yet been implemented for any of these configurations.

Parallel Plate Channels

Traditionally, one of the most common electronic equipment cooling configurations has involved free convection in vertical (or inclined) parallel plate channels which are open to the ambient. As in an array of PCBs (Fig.6a), components may be mounted to the plates, and although the manner in which heat is dissipated varies with specific packaging and operating conditions, many applications are suitably approximated by assuming smooth plates with isothermal or isoflux surface thermal conditions (Fig.6b). Conditions may be symmetrical ($T_1 = T_2$; $q_1 = q_2$) or asymmetrical ($T_1 \neq T_2$;$q_1 \neq q_2$).

For vertical channels ($\theta = 0$) buoyancy acts exclusively to induce motion in the streamwise (x) direction and, beginning at $x = 0$, boundary layers develop on each surface. For short channels and/or large spacings, independent boundary layer development occurs at each surface and conditions approximate those for an isolated plate in an infinite quiescent medium. For large L/S, however, boundary layers developing on opposing surfaces eventually merge to yield a fully developed condition. If the channel is inclined, there is a component of the buoyancy force normal, as well as parallel, to the streamwise direction, and conditions may be strongly influenced by development of a three-dimensional, secondary flow.

Beginning with the benchmark paper by Elenbaas [7], the vertical channel orientation has been studied extensively for symmetrically and asymmetrically heated plates with isothermal or isoflux surface conditions. In a more recent treatment of the subject, Bar-Cohen and Rohsenow [8] derived limiting Nusselt number relations for fully-developed laminar flow in symmetrical isothermal and isoflux channels, as well as for channels with an insulated wall and an adjoining isothermal or isoflux wall. Using an accepted formalism for inferring general correlations from limiting results, the fully-developed limits were combined with those for the isolated plate to obtain the following expressions for isothermal and isoflux conditions, respectively:

$$\overline{Nu_S} = \left[\frac{C_1}{(Ra_S S/L)^2} + \frac{C_2}{(Ra_S S/L)^{1/2}}\right]^{-1/2} \quad (1)$$

$$Nu_{S,L} = \left[\frac{C_1}{Ra_S^* S/L} + \frac{C_2}{(Ra_S^* S/L)^{2/5}}\right]^{-1/2} \quad (2)$$

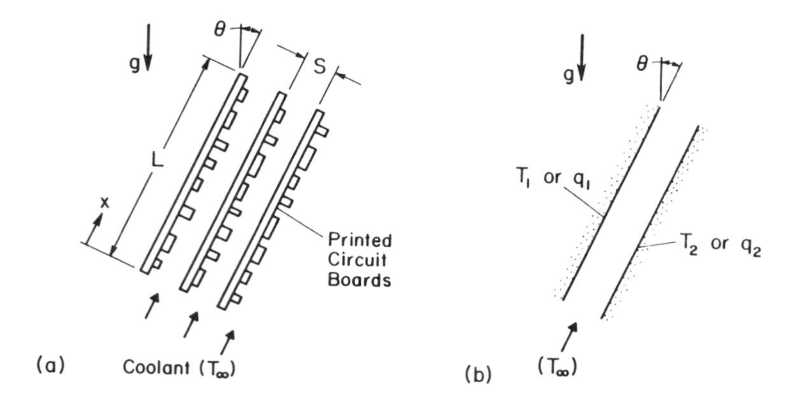

FIGURE 6. Natural convection in parallel plate channels: (a) PCB array, (b) smooth plate approximation.

Equation 1 provides the average Nusselt number for an isothermal plate, while equation (2) provides the local Nusselt number at the trailing edge of an isoflux plate. Values of the constants C_1 and C_2 differ according to the existence of isothermal or isoflux conditions, while C_2 depends weakly on Prandtl number. In each case the fully-developed and isolated plate limits correspond to $Ra_S(\text{or } Ra_S^*)S/L \lesssim 10$ and $Ra_S(\text{or } Ra_S^*)\, S/L \gtrsim 100$, respectively. Although data used to validate the correlations were originally associated with air, other comparisons have validated the correlations for water [9].

In water, additional experiments have been performed for the isolated plate limit ($Ra_S S/L > 200$), and the data were correlated by [10]

$$\overline{Nu_S} = C\,(Ra_S S/L)^{1/4} \tag{3}$$

where $C = 0.675$ for symmetric plates and $C = 0.642$ for asymmetric (heated/adiabatic) plates. The slightly larger (~5%) Nusselt numbers for the symmetric condition were attributed to an enhancement of the mass flow due to heating from both sides. No such distinction is associated with the isolated plate limit of the Bar-Cohen [8,9] correlations. For asymmetric heating, a recirculating flow, which had little effect on heat transfer, was observed along the upper portion of the unheated wall.

Azevedo and Sparrow [10] also performed experiments for inclined channels in water ($0 \le \theta \le 45$ deg), and three-dimensional longitudinal vortices driven by the normal component of the buoyancy force associated with bottom heating were observed. The vortices were periodically distributed in the spanwise direction and were more pronounced for an insulated, rather than isothermal, top surface. For heated and insulated top and bottom plates, respectively, a two-dimensional recirculating flow was observed to exist near the unheated wall and the channel outlet. Its depth of penetration increased with increasing $(S/L)Ra_S$. Data for all experimental conditions were correlated to within ± 10% by

$$\overline{Nu_S} = 0.645(Ra_S S/L)^{1/4} \tag{4}$$

Departures of the data from the correlation were most pronounced for inclined channels with the bottom surface heated and the top surface insulated and were attributed to heat transfer enhancement by the longitudinal vortices. For inclined channels with the top and bottom walls heated and insulated, respectively, an improved correlation of data was given by

$$\overline{Nu_S} = 0.644\,\left[(Ra_S S/L)\cos\theta\right]^{1/4} \tag{5}$$

However, for bottom heating, the $\cos\theta$ term failed to provide an improved correlation of data.

Factors which may influence applicability of the foregoing results to immersion cooling include Prandtl number effects, edge effects, board conduction, and the effect of protruding components. Although experiments have yet to be performed for dielectric liquids, the proposed Prandtl number dependence is weak [9] and, at least to a first approximation, the foregoing correlations are likely to be valid.

Moreover, although the correlations presume two-dimensional flow, and hence ignore the entrainment of ambient fluid at the lateral edges, such effects have been shown to be negligible for 152 mm $\times$ 152 mm plates when Ra_S S/L > 2 and for 76 mm $\times$ 76 mm plates when Ra_S S/L > 10. However, the influence of board conduction, which is related to combined conduction/convection (conjugate) effects, can be significant, particularly for the nonuniform heat dissipation associated with discrete components on the board. Studies concerning applicability of isothermal or isoflux correlations to PCBs, which have been performed for regular and irregular arrays of wall protrusions in air, have yielded mixed results [11,12].

Discrete Sources in a Quiescent Ambient

Energy dissipated by an electronic component in a quiescent ambient induces a buoyancy-driven flow that ascends from the component as a wake or plume. A common example concerns free convection heat transfer from a small rectangular component, flush-mounted to a vertical substrate. The problem was first considered by Baker [13,14], who performed experiments for surface areas ranging from 1.06 to 200 mm^2. From measurements performed in air, silicone oil, and R-113, the heat transfer coefficient was found to increase significantly with decreasing heater size and to be underpredicted by accepted free convection correlations. Inapplicability of the accepted correlations was attributed to three-dimensional boundary layer edge effects. Heat transfer coefficients in excess of standard two-dimensional predictions were also measured for a 4.7 mm $\times$ 4.7 mm heater in water [15] and for heaters of varying length and width ($L = 5$ mm, 10 mm; 2 mm $< W <$ 70 mm) in water and R-113 [16]. In the latter study, the coefficient increased with decreasing width (coefficients for the 2 mm heater were reported to be 150% larger than for the 70 mm heater), and the effect was attributed to an induced flow of ambient fluid at the sides of the heater. Such flow is driven by a spanwise pressure gradient resulting from acceleration of fluid in the free convection boundary layer, and its effect on heat transfer increases with decreasing heater width. Underprediction of the data was attributed to upstream conduction from the boundary layer, as well as to induced flow at the heater sides.

In the case of a vertical, in-line array of heat sources, the plume ascending from a lower source can strongly influence heat transfer from an upper source. From numerical simulations of the laminar, two-dimensional flow associated with plumes ascending from two isolated sources flush mounted on a vertical wall [17-19], the competing effects on heat transfer from the upper source were delineated. In particular, heat transfer from the upper source is enhanced and degraded, respectively, due to fluid acceleration and preheating by the lower source. Hence, depending upon the height, spacing, and relative heating of the two sources, convection coefficients at the upper source may be less than or greater than those at the lower source. The problem has been considered experimentally, and for two in-line heater strips in air or water [20], it was verified that, for large heater spacings, flow induced by the lower heater can enhance, rather than suppress, heat transfer from the upper heater. However, for vertical arrays of two or three in-line heaters [16], each 5 mm on a side, data revealed top surface heat transfer coefficients which were consistently smaller than those of the bottom surface(s). In experiments for an in-line array of two protruding heaters mounted on a vertical substrate in water and R-113 [16], convection coefficients for a 1 mm protrusion were found to exceed those for flush heaters by approximately 14% and the coefficient for the top heater exceeded that for the bottom heater.

Rectangular Cavities

Numerous applications exist for which electronic components are packaged within rectangular enclosures (Fig.7). The components may be mounted to one wall of the enclosure, while one or more of the other walls is cooled. Buoyancy forces induce a recirculating flow within the enclosure, and heat is transferred by natural convection from the component surfaces. Such a configuration is favored when coolant contamination must be minimized and is a viable candidate for immersion cooling of aircraft and missile systems, as well as commercial computers.

Although the literature on natural convection in enclosures is voluminous, comparatively little has been reported on the effects of discrete heat sources. The first such study [21] involved a two-dimensional numerical solution for a single, isothermal (T_h) heater strip, flush mounted ($t = 0$) to a vertical wall of the cavity. The opposite wall was cooled (T_c), while all other surfaces were adiabatic ($\partial T/\partial n = 0$). The heater size L and position S were varied, and the anticipated

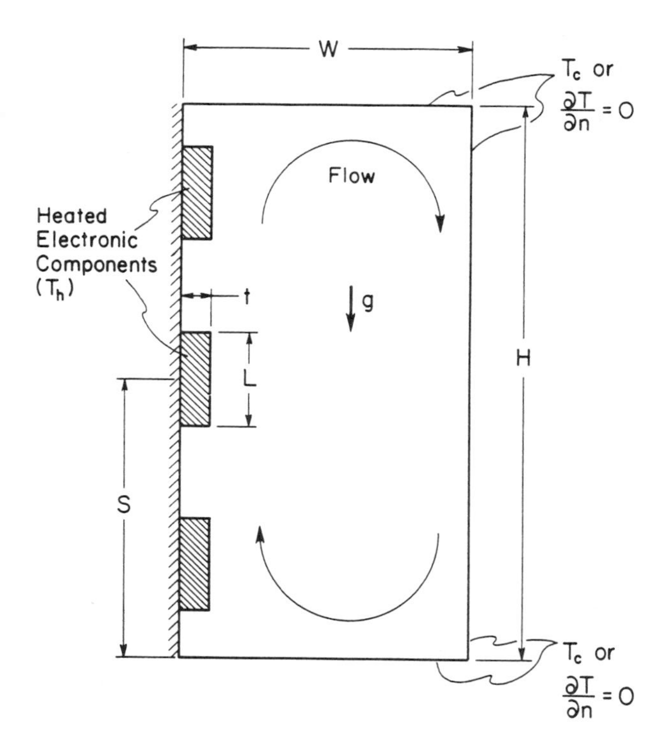

FIGURE 7. Buoyancy driven flow due to wall mounted heaters in an enclosure.

functional dependence of the average Nusselt number was of the form

$$\overline{Nu} = \overline{Nu}(Gr,Pr,H/W,L/H,S/H) \tag{6}$$

Over the range of conditions considered, $\overline{Nu}$ was found to increase as $Gr^{0.3}$. It also increased with increasing L/H, although the increase was small for $L/H > 0.25$, exhibited a maximum for an optimal heater location of $S/H \approx 0.4$, and was approximately independent of aspect ratio. Subsequent experiments [22,23] confirmed the predicted L/H, S/H and H/W trends.

Although the foregoing studies were performed for air, confirmation of the $Gr^{0.3}$ dependence has been obtained in a two-dimensional numerical simulation for water [24], which considered a single heater on a side wall and cooling from the top. Numerical simulations [25] and experiments [26] have also been performed for an isothermal protruding heater (T_h) mounted to one vertical wall of a water-filled rectangular cavity. The remainder of the heater wall and the opposing wall were insulated, while the top and bottom surfaces were cooled (T_c). The aspect ratio was fixed at $H/W = 4$, as were the heater length ($H/L = 6$) and protrusion ($t/L = 0.5$). Three heater locations corresponding to values of $S/H = 0.25$, 0.50 and 0.75 were considered for Rayleigh numbers in the range $1\times10^6 < Ra_L < 8\times10^6$. At large Rayleigh numbers, Nusselt number data were in good agreement with the predictions, but at small Rayleigh numbers the data, which were characterized by large uncertainty, were significantly overpredicted. The data and predictions yielded opposing trends for the effect of S/H on $\overline{Nu}_L$. Irrespective of heater location, both the numerical simulations and visualization revealed that the buoyancy driven flow was concentrated in the region above the heater. With stable stratification existing in the region below the heater, flow within the region is much weaker and is driven by shear forces induced by the overlying buoyancy driven flow.

Experiments have been performed for an array of eleven equally spaced heater strips flush mounted to one vertical wall of a tall ($H/W = 16.5$) cavity, with the opposing wall cooled [27]. In ethylene glycol ($Pr \sim 150$), a complex multicellular flow was observed and the convection coefficient decreased with increasing elevation of the heater strip. Heater Nusselt numbers were correlated by

$$Nu_y = 1.009(Ra_y^*)^{0.1805} \tag{7}$$

where y is measured from the bottom of the cavity to the heater midheight. The correlation is in good agreement with that obtained for a flush mounted heater on a vertical surface in water [16].

A three-dimensional simulation has been performed for flow and heat transfer from a 3×3 array of block-like electronic components mounted to one vertical wall of a rectangular cavity filled with a dielectric liquid [28]. Vertical surfaces unoccupied by the components were insulated, while the top and bottom surfaces were cooled. Except for the cavity width W, geometrical and thermal features of the simulation were fixed, with a uniform heat flux prescribed at the component surfaces. Except for very small cavity widths, $(W - t) \lesssim 5$ mm, maximum temperatures were predicted to occur on the upper horizontal faces of components in the top row, while minimum temperatures were associated with the lower horizontal faces of components in the bottom row. With increasing W, component temperatures decreased significantly for $(W - t) \lesssim 12$ mm and much more gradually for larger values of W.

FORCED CONVECTION

Immersion cooling by single phase forced convection may involve an impinging jet or flow through a rectangular duct. Results may be differentiated according to whether components are mounted flush with or protrude from a substrate and to whether measures for heat transfer enhancement are employed.

Rectangular Ducts

In this configuration, components are mounted to the walls of a rectangular channel and conditions are complicated by the existence of multiple length scales. Considering, for example, an array of discrete sources flush mounted to a substrate, relevant scales would be associated with heater lengths parallel and perpendicular to the direction of the mainflow, the spacings (longitudinal and spanwise pitches) between heaters, and the hydraulic diameter of the channel. Moreover, for an array, heat transfer from one source may be strongly influenced by its neighbors, as when a source is located in the thermal boundary layer generated by upstream sources. In general, conditions are three-dimensional, since intermittent heating is associated with the spanwise, as well as streamwise, directions. Conditions are further complicated in applications for which the resistance to heat transfer by convection from the chip to the coolant is not much less than that due to conduction from the chip to its substrate. In such cases coupled conduction and convection (conjugate) heat transfer processes strongly influence component thermal behavior, and to characterize convective aspects of the problem, it is necessary to simultaneously consider substrate conduction effects.

Early experimental studies of forced convection heat transfer from flush mounted sources considered silicone oil and R-113 in parallel flow over small, chip-like heaters [13,14]. Single-phase forced convection coefficients increased significantly with decreasing surface area from 200 to 1 mm^2 and exceeded results associated with nucleate pool boiling. For the smallest heaters, the data also exceeded predictions based on two-dimensional boundary layer theory, and differences were attributed to three-dimensionality of the flow associated with small heating elements.

Studies related to heat transfer from an array of discrete (12.7 mm $\times$ 12.7 mm) isothermal sources flush mounted to one wall of a rectangular channel have also been performed. The array (Fig.8) consisted of four equally spaced rows, with three heaters in each row. Initially, experiments were performed for water and FC-77 without the pin fins shown in the figure [29]. Reynolds numbers were in the range $1000 < Re_D < 14,000$, and results were compared with predictions based on two-dimensional, conjugate forced convection models for laminar and turbulent flow. The data were significantly underpredicted in laminar flow, and differences were attributed to the effects of buoyancy on the experimental results. In turbulent flow, however, agreement between the predicted and measured results was good, suggesting that three-dimensional boundary layer effects were negligible. Convection from a single source was quantified in terms of an average

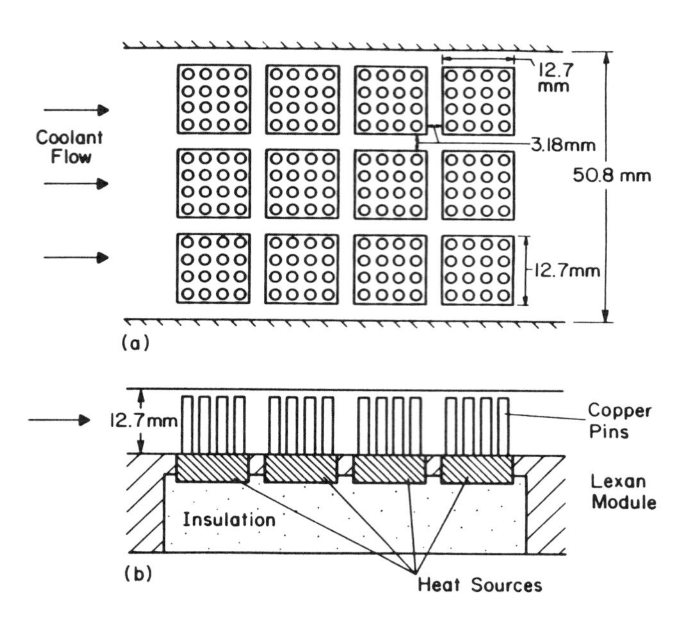

FIGURE 8. Array of discrete sources mounted to one wall of a rectangular duct (experiments performed with and without pin fins).

Nusselt number $\overline{Nu}_L$, and variations between sources of a row were negligible. As shown in Fig. 9, however, average Nusselt numbers in the second row were approximately 25% less than those of the first row, while Nusselt numbers in the third row were approximately 10% less than those of the second row. This reduction in $\overline{Nu}_L$ with increasing row number was attributed to the effects of upstream thermal boundary layer development. However, the decrease between the third and fourth rows was only 3%, suggesting that a fully developed condition is approached for which thermal boundary layer development on one row is balanced by boundary layer dissipation in the unheated region between rows. For each row the data were correlated by an expression of the form

$$\frac{\overline{Nu}_L}{Pr^{0.38}(\mu_o/\mu_h)^{0.11}} = C\, Re_D^m \tag{8}$$

where C and m decreased and increased, respectively, with increasing row number. Parametric calculations performed using the two-dimensional model for a *single* heat source mounted to one wall of a parallel plate channel in turbulent flow yielded a correlation of the form

$$\overline{Nu}_L = 0.062\, Re_L^{0.75}\, Pr^{0.35}\, (L/H)^{0.1}(k_f/k_s)^{0.02} \tag{9}$$

The weak dependence on H suggests that, although hydrodynamic conditions correspond to internal flow, thermal conditions are more representative of external flow.

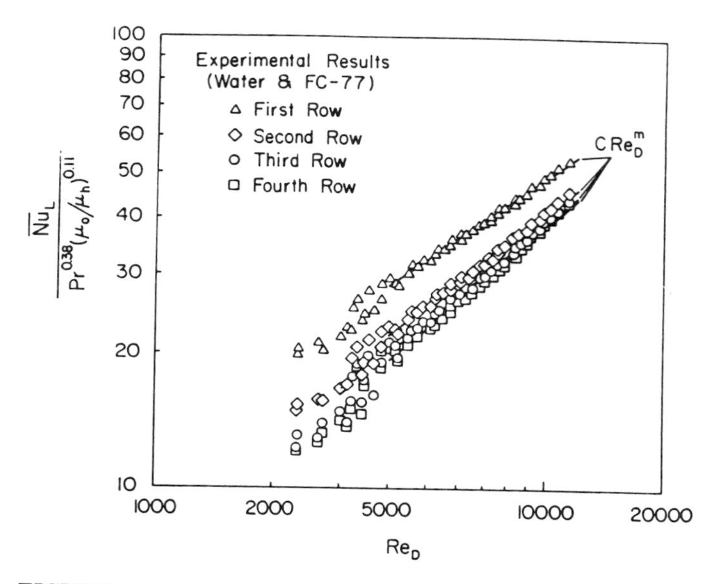

FIGURE 9. Average Nusselt numbers associated with forced convection heat transfer from each row of an in-line array of discrete heat sources.

In a related study of heat transfer from a single 0.25 mm long by 2.0 mm wide source mounted flush with one wall of a rectangular duct [30], results obtained for FC-72 and R-113 in the Reynolds number range $7000 \lesssim Re_H \lesssim 1.5 \times 10^5$ were correlated by the expression

$$\overline{Nu_H} = 0.47\, Re_H^{0.58}\, Pr^{1/2} \tag{10}$$

In R-113 a peak heat flux of 200 W/cm^2 was achieved without nucleation. More recently [31], experiments were performed in FC-72 for a single, flush mounted, 12.7 mm×12.7 mm heater over the extended Reynolds number range $2800 < Re_L < 1.5 \times 10^5$. The Reynolds number dependence is consistent with previous results [29], but in the overlapping Reynolds number range, Nusselt numbers are approximately 37% larger.

Although resistor network, finite-difference or finite-element methods have been used to predict heat transfer from a discrete source, it is common practice to ignore the coupling between heat transfer by conduction in substrate materials and convection to the coolant. However, the need to consider such coupling is dictated by the mutual dependence of conditions in the fluid, substrate and heated component. The extent to which the component is more strongly coupled to the fluid or the substrate depends on the nature of the flow, as well as on the fluid and substrate thermophysical properties. For heat sources mounted flush to one wall of a parallel plate channel, this dependence has been considered for both laminar [32] and turbulent [33] flows. For hydrodynamically developed laminar flow over two sources of length L and separation S (Fig.10), the ratio of the substrate to total heat loss Q_S/Q_T is approximately the same for the two sources

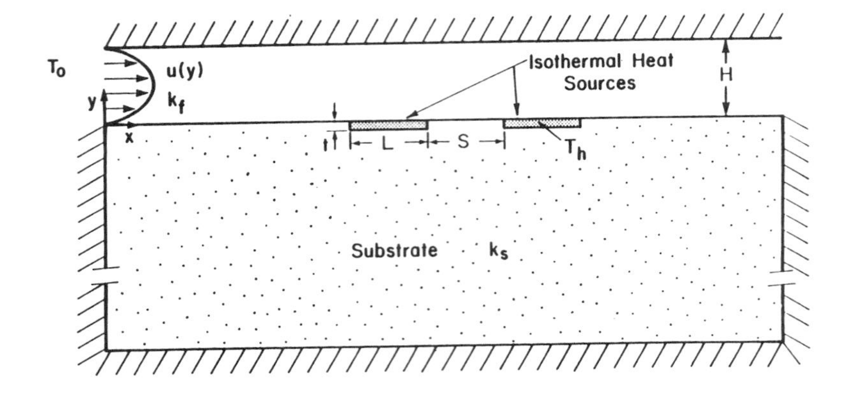

FIGURE 10. Configuration for two-dimensional simulation of conduction/convection from discrete sources in a parallel plate channel.

(Fig.11) but decreases appreciably with increasing Peclet number and decreasing conductivity ratio. For small values of the dimensionless pitch, S/L, Q_S/Q_T is slightly smaller for the downstream source. The effect of S/L on the local Nusselt number is shown in Fig.12, where the results are normalized with respect to the Nusselt number for hydrodynamically and thermally developed flow in a channel, with one wall isothermal and the other adiabatic ($Nu_{fd} = 4.86$). Although the effect becomes less pronounced with increasing S/L, thermal boundary layer development on the upstream source attenuates local Nusselt numbers on, and therefore decreases convection heat transfer from, the downstream source.

Several studies have considered means by which heat transfer from flush mounted, liquid cooled heat sources may be enhanced. Experiments have been performed for a 4×3 in-line array of 12.7 mm flush-mounted heat sources, with a 4×4 array of 2 mm diameter copper pins attached to each source (Fig.8) [34]. Use of the pins reduced the thermal resistance by factors up to 10 for water and 25 for FC-77. In a sequel study [35], square fins integrally machined from each pin were found to further reduce the thermal resistance by 20%. The effects of attaching strip fins [36] and low profile microstuds [31] to a single 12.7 mm × 12.7 mm heat source have also been considered. In the strip fin study, the smallest thermal resistance was provided by an offset arrangement, and for an equivalent duct Reynolds number in water or FC-77, the resistance was approximately 20% less than that for the heat source with finned pins [35]. In FC-72, 1.02 mm long microstuds reduced the thermal resistance by up to a factor of 6, but at comparable Reynolds numbers, resistances exceeded those associated with the finned pin or offset strip fin arrangements. Thermal resistances associated with the different enhancement schemes are compared in Fig.13. The thermal resistance of 0.09 K/W achieved by Tuckerman and Pease [4] for water flow through microchannels in a 10 mm×10 mm silicon substrate is also included.

Other surface arrangements may be used to enhance heat transfer from electronic components, and the success of such arrangements is determined as much by their ability to disrupt thermal boundary layer development and to deliver cold fluid directly to the heated region, as by the extent to which they increase the amount of surface area. Tailored surface arrangements which maximize the velocity and

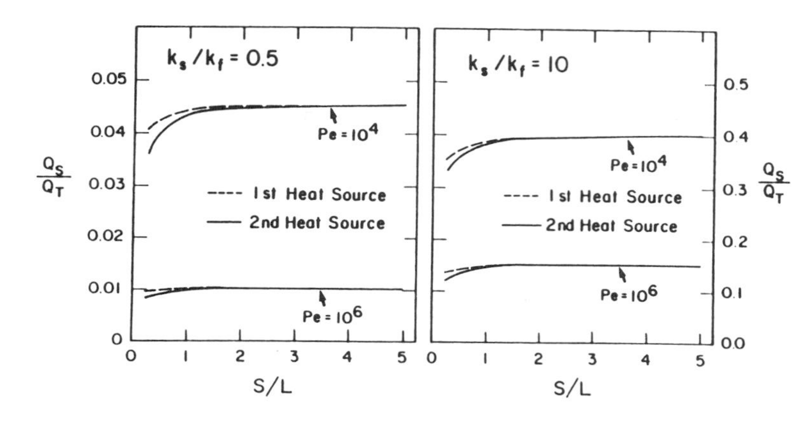

FIGURE 11. Ratio of substrate to total heat loss for each of two adjoining heat sources flush mounted to one wall of a parallel plate channel.

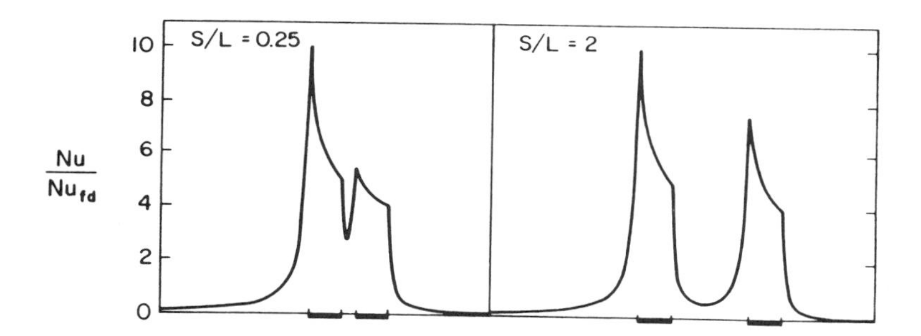

FIGURE 12. Variation of normalized local Nusselt number along the substrate and heater surfaces for adjoining heaters flush mounted to one wall of a parallel plate channel.

minimize the temperature of liquid flow over the heated surface are preferred. For example, deflectors or implanted ribs may be used to disrupt or turbulate a flow, thereby increasing convection heat transfer from downstream components. Heat transfer enhancement may also be realized by modulating low Reynolds number laminar flows in a channel. The process, known as resonant heat transfer enhancement, may be effected in a grooved channel by actively modulating the flowrate at the natural frequency of the system [37] or by achieving passive modulation through insertion of small vortex-shedding cylinders in cross-flow [38].

Jet Impingement

Due to its association with large convection coefficients at the stagnation point, jet impingement cooling has long been favored in large heat flux applications. Jets immersed in a miscible ambient fluid, such as a liquid-liquid combination, are said to be submerged, while jets immersed in an immiscible fluid, such as liquid-gas,

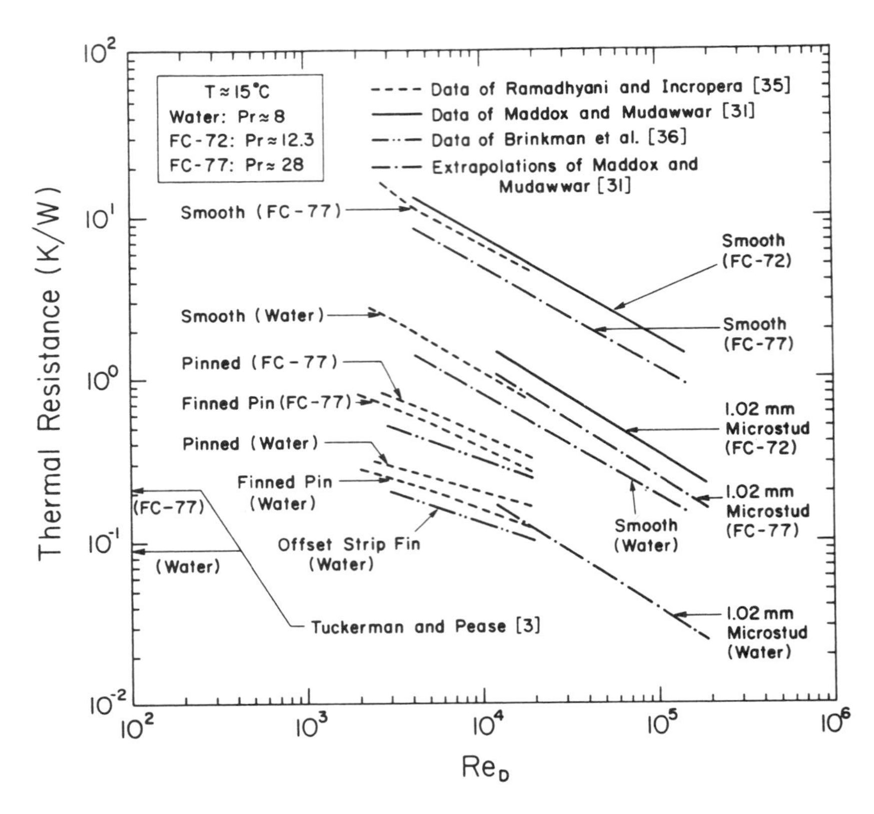

FIGURE 13. Comparison of thermal resistances for a 12.7 mm×12.7 mm heat source with heat transfer enhancement by pins, finned pins and microstuds.

are said to be *free*. Momentum exchange between a submerged jet and its ambient is often significant, causing expansion of the jet between the nozzle and the stagnation region. In contrast, momentum exchange with the ambient is less significant for the free jet, and momentum is more efficiently delivered to and redirected along a solid surface. A free jet can also be significantly accelerated by gravity, causing it to contract, rather than to expand. Whether submerged or free, applications typically involve round or planar (slot) jets arranged independently or in an array.

Heat transfer from a flat surface to a free, circular jet has been analyzed by Choudbury [39], and average heat transfer coefficient data have been obtained by Metzger et al. [40]. The data revealed little influence of the nozzle-to-surface spacing but did reveal a pronounced increase in the average convection coefficient with decreasing surface-to-jet diameter ratio. Local and average convection heat transfer coefficient data have also been obtained for free, planar, liquid jets. McMurray et al. [41] divided the flow field into stagnation and downstream regions for which the freestream velocity component parallel to the surface is, respectively, less than and approximately equal to the jet velocity. Local Nusselt number correlations of the form $Nu_x \sim Re_x^{1/2} Pr^{1/3}$ were obtained for both regions. The results were confirmed by Miyasaka and Inada [42] and by Inada et al. [43], who also considered the effect of the ratio of the nozzle height to jet width. The effect became pronounced for ratios less than approximately 0.8 and could be correlated by adjusting the Prandtl number exponent and the correlation coefficient.

Specific consideration of impinging liquid jets to electronic cooling is relatively recent. Ma and Bergles [44] made heat transfer measurements at the stagnation point of a 5 mm×5 mm source cooled by a submerged 1 mm diameter jet of R-113. The source was mounted vertically and positioned 2 mm from the jet orifice. The data were correlated by an expression of the form $Nu \sim Re^{1/2} Pr^{0.4}$.

The most extensive study of single-phase jet impingement cooling for microelectronic applications has been performed by Jiji and Dagan [45]. Heat transfer measurements were made for a single source, as well as for 2×2 and 3×3 heat source arrays. The sources were 12.7 mm on a side, and the dimensionless longitudinal and transverse pitches of the array were each 1.5. Using FC-77, each heat source was cooled by a single jet or a multi-jet array. Square arrays of four and nine jets were considered with jet diameters of 0.5 and 1.0 mm. For all of the cooling configurations, the average Nusselt number was correlated by a single expression, which included the effects of the number of jets and the heater length to nozzle diameter ratio. The results were independent of the nozzle to heat source spacing. The results also indicated that the thermal resistance associated with jet impingement cooling can be reduced by reducing the jet diameter and increasing the number of jets. Resistances as low as 0.15 $cm^2 \cdot C/W$ and 1 $cm^2 \cdot C/W$ were obtained with water and FC-77, respectively, and it was suggested that further improvements in thermal performance could be realized by seeking optimal arrays of small diameter jets.

Yamamoto et al. [2] reported local and average Nusselt numbers for the 3 mm diameter submerged water jet used to cool the heat transfer plate of the Fujitsu FACOM M-780 computer (Fig.2). The Nusselt number at the stagnation point exceeds the average value by approximately 75%, and both are correlated by an expression of the form $Nu_d \sim Re_d^{1/2} Pr^{0.4}$. In water, convection resistances as low as 0.7 $cm^2 \cdot C/W$ were obtained.

MIXED CONVECTION

Under mixed convection conditions, flow is driven by an externally imposed pressure gradient, as well as by buoyancy forces. If the channel is vertical, buoyancy acts to augment or retard the flow, according to whether the buoyancy force aids or opposes the imposed flow. If the channel is horizontal and heating occurs at the bottom surface, buoyancy may induce a secondary flow which, in combination with the main flow, produces a system of longitudinal vortices. If the channel is inclined, the buoyancy force has two components, one parallel to the surface, acting to accelerate or decelerate flow in the streamwise direction, and the other normal to the surface, acting to drive the secondary flow. The relative influence of these effects depends on the inclination angle.

Heat transfer enhancement is most pronounced when the channel is horizontal and heat is transferred from the bottom surface. In experiments performed for laminar water flow between asymmetrically heated parallel plates [46], convection coefficients at the bottom plate were found to exceed those corresponding to pure forced convection by up to a factor of 6. Enhancement was due to a buoyancy driven flow, which replaced warmer parcels of fluid ascending from the plate with cooler fluid descending from the main flow. In contrast, conditions near the top plate were thermally stratified and convection coefficients corresponded to pure forced convection.

Braaten and Patankar [47] considered laminar, fully developed flow of freon between horizontal parallel plates, with heated rectangular blocks attached to the top or bottom plate. In both cases, the buoyancy driven flow about each block consisted of single or multiple vortex pairs, although thermal stratification associated with blocks on the top plate resulted in much weaker flow conditions. In both cases heat transfer was enhanced above that for pure forced convection, although enhancement was much more pronounced when the heated blocks faced upward.

Although the benefits of mixed convection appear to have been exploited in the Cray-2 immersion cooling system (Fig.5), basic studies of related phenomena have yet to be performed. In view of the inherently low pressure losses, mixed convection may well be suited for a wide range of applications and careful studies are clearly warranted.

POOL BOILING

Liquid immersion cooling in pool boiling is by no means a new concept, with related studies and system designs traceable to the 1950's and 60's [48-50]. However, the need for chemically inert coolants of high dielectric strength and low boiling point limits consideration to fluids for which boiling incipience may be characterized by a *temperature overshoot.* That is, once boiling is initiated, the surface temperature drops sharply, following the typical boiling curve with increasing heat flux. If the heat flux is subsequently reduced, the surface temperature continues to follow the typical boiling curve, thereby exhibiting a *hysteresis* phenomenon. The phenomenon is attributed to the small surface tension of dielectric fluids, and is exacerbated by the small cavities (bubble nucleation sites) associated with smooth chip surfaces. Boiling in dielectric fluids is also characterized by small values of the critical heat flux (CHF), and additional problems could potentially include cyclical thermal stresses and electrical noise related to temperature fluctuations, device contamination due to impurities, and increased complexity associated with the need for sealed chambers [14].

Temperature overshoots up to 25 °C have been observed for smooth silicon chips in Flutec PP liquids [51] and perfluorinated liquids [52,53], as well as for metallic surfaces in R-113 [54-56] and FC-72 [55]. The nature of the phenomenon is revealed by the representative results of Fig.14, which also indicate the comparatively small CHF values (~ 20 W/cm^2) associated with dielectric fluids.

Bar-Cohen and Simon [57] have reviewed existing literature on wall temperature overshoot and have suggested possible mechanisms for delayed nucleation. Boiling incipience at a heated surface is due to the nucleation of vapor bubbles from embryonic vapor/gas pockets in the surface cavities. If a vapor bubble is released from a cavity, surrounding liquid enters the cavity and the shape of the advancing liquid front is determined by its wetting angle β (insert, Fig.14).

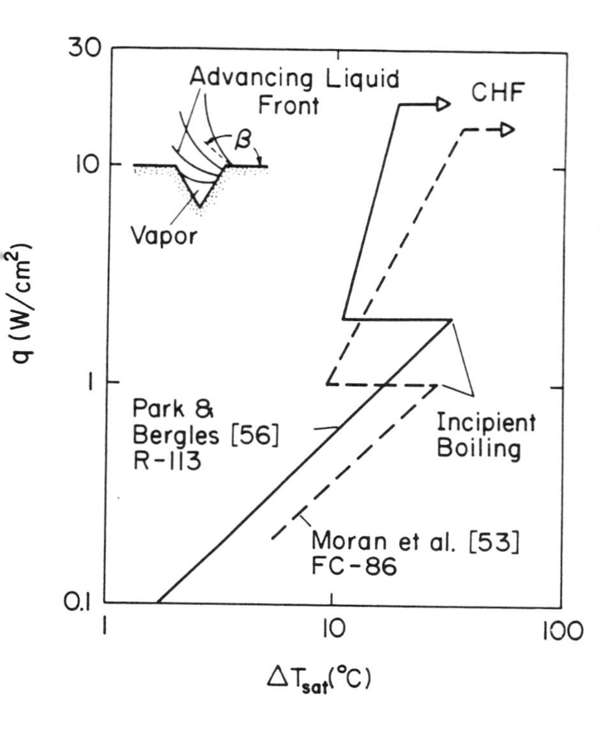

IGURE 14.Representative pool boiling curves with temperature overshoot.

;ommonly, β exceeds the effective cone angle of the cavity, and the liquid cannot ll the cavity, thereby trapping residual vapor and restoring the nucleation site. lowever, dielectric liquids have uncommonly small surface tensions, and hence ery small wetting angles. For all but the smallest cavities, an advancing liquid ront may therefore flood a cavity, depleting it of the vapor embryo needed for he next bubble. Extensive deactivation of nucleation sites by flooding would hen require elevated temperatures to initiate bubbles at the remaining, smaller avities or through homogeneous nucleation in the bulk liquid. It is noteworthy hat, for superheats up to $(T_s - T_{sat}) = 46\,^\circ C$, Reeber and Frieser [58] observed o nucleation from a polished silicon surface immersed in FC-72. However, they lso investigated the effect of different surface treatments on boiling incipience at ilicon surfaces and concluded that sandblasting, followed by a KOH—H_2O etch, ould be effective in promoting bubble nucleation. It is also noteworthy that 'okouchi et al. [59] report a significant reduction in temperature overshoot for an 0/20 volumetric ratio of 3M FX-3250 and FX-3300 fluorocarbon coolants. lowever, the authors admit to the need for further research to clarify the henomenon. In any case, once a steady source of bubbles is provided, the ctivation of surface cavities spreads rapidly and normal boiling characteristics re observed.

3ar-Cohen and Simon [57] obtained an upper bound for the incipience superheat xcursion from a model based on homogeneous nucleation in the bulk liquid. The nodel and limited data indicate that the temperature excursion decreases with increasing subcooling and pressure, as well as with increasing velocity (forced convection boiling).

Options for reducing the temperature overshoot and enhancing pool boiling heat transfer have been widely considered. Considerable attention has been given to either altering the surface or attaching it to a heat sink. For example, Chu and Moran [60] achieved enhancement by laser drilling holes in a chip surface, thereby forming re-entrant cavities, while Oktay [52] also reported a reduction in the temperature overshoot for a sandblasted/KOH treated chip.

Extensive consideration has been given to the use of heat sinks which can be attached to the chips. For example, Oktay and Schmeckenbecher [61] found that, in FC-86, dendritic surfaces prepared by electrolytic deposition eliminated the temperature overshoot and enhanced heat transfer in the nucleate boiling regime. Oktay [52] drilled 0.8 mm diameter vertical holes in a 1 mm thick copper block and, relative to a plain silicon surface, the tunnel heat sink yielded enhanced nucleate boiling and a reduced temperature overshoot (from approximately 20 °C to 10 °C) in FC-86.

Extensive consideration has also been given to the use of commercially available surfaces [54,55,62-64]. The primary function of such surfaces is to provide re-entrant cavities which enhance the availability of active nucleation sites. Bergles and Chyu [54] studied nucleate pool boiling of R-113 from porous metallic coatings formed by brazing copper particles to a base surface (Linde High Flux surface). Although nucleate boiling heat transfer was significantly enhanced above that of an uncoated surface, the coating failed to eliminate the temperature overshoot, which was as high as 9 °C. This result was confirmed by Marto and Lepere [55], who also studied the High Flux surface in R-113. The inability to eliminate the temperature overshoot by using porous boiling surfaces was further confirmed by Kim and Bergles [63], who measured superheats for incipient boiling ranging from 9 to 30 °C for etched Cu-Nb surfaces and for copper surfaces sintered with spherical copper powders.

Marto and Lepere [55] also studied boiling of R-113 and FC-72 from commerically available grooved fin surfaces (Hitachi Thermoexcel E and Wieland Gewa-T), as well as from the High Flux porous coating. The Thermoexcel-E surface consists of tunnels and pores formed by bending the ridges of microfin surfaces, while the Gewa-T surface is formed by compressing the fin tips. Pool boiling curves obtained for the three surfaces in FC-72 are shown in Fig.15, and each curve is characterized by an overshoot of approximately 5 °C. Although the High Flux surface performs best in the nucleate boiling region, its CHF, as well as that of the Thermoexcel-E surface, remains small ($< 20\ W/cm^2$), thereby differing little from that for a smooth surface. The Gewa-T surface performs best at high fluxes, and although not measured, CHF may exceed $20\ W/cm^2$. The improved performance of this surface was attributed to an increased surface area and to a larger spacing between reentrant grooves, which inhibited the coalescence of vapor columns. Collectively, the structured surfaces enhanced nucleate boiling heat transfer from two- to five-fold relative to a plain surface. In R-113, enhancement ranged from a factor of two to ten, and the temperature overshoot was approximately 10 °C. Superior performance of the High Flux surface in nucleate pool boiling has also been indicated by Nakayama [65].

In an effort to enhance the probability of bubble nucleation from small surfaces and hence boiling heat transfer, Nakayama et al. [62] considered the effect of

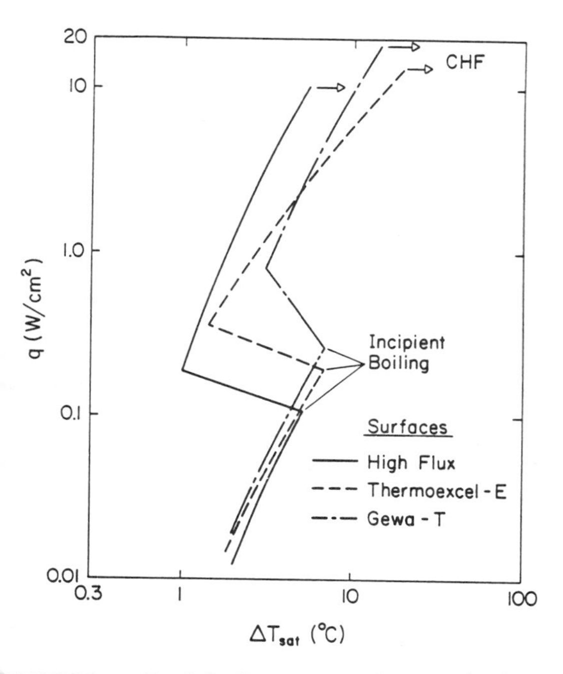

FIGURE 15. Pool boiling curves for standard porous and grooved fin surfaces (adapted from [55]).

attaching a horizontal, cylindrical copper stud with fine structured surfaces to the heat dissipating component. The structures consisted of microfins, microfins covered by a porous plate, and a multilayered porous surface. Superior performance was achieved for a stud length-to-diameter ratio of $L/D \sim 1$, with microfins near the base and multilayered porous plates near the tip. With negligible hysteresis in FC-72, up to 100 W could be dissipated for a 1 cm base diameter, providing base heat fluxes in excess of 100 W/cm^2 at base temperatures up to 80 °C. Park and Bergles [64] evaluated the performance of several heat sinks attached to a 5 mm×5 mm foil heater. The heat sinks included 1 mm thick copper blocks with microholes or microfins machined parallel to the surface, as well as the High Flux and Thermoexcel-E surfaces. In R-113 the best overall performance was provided by the High Flux surface, which was, in turn, followed by the Thermoexcel-E surface and the copper block with 0.71 mm diameter vertical holes.

In a more recent study [66], the effect of various enhancement surfaces was considered for boiling from a vertically oriented 12.7 mm × 12.7 mm heat source in FC-72. Four surface categories were tested: smooth, drilled (creating artificial cavities of 0.36 mm diameter), low-profile structures (microfin, microstud, and microgroove), and a short (4.27 mm) square stud with a structured (microfin) perimeter. The artificial cavities were ineffective in reducing the incipience temperature or in enhancing nucleate boiling and CHF. Of the low profile surfaces, the microstuds yielded the largest CHF (~ 50 W/cm^2), although the temperature overshoot was large (> 5 °C). Negligible temperature overshoot was associated with the 4.27 mm stud and CHF values up to 60 W/cm^2 were achieved.

Selected results are compared in Fig.16, where the temperature overshoot is excluded by plotting the data for decreasing heat flux. Superior performance is clearly associated with the microfin/porous surface stud of Nakayama et al. [62]. The good performance of the heat sink with vertical holes [52] is attributed to liquid pumping, and hence enhanced velocities, due to bubble formation in the holes.

The effect of heater size on pool boiling in R-113 has been considered by Baker [14] and Park and Bergles [56]. Although Baker reported a pronounced shift in the fully established nucleate boiling curve to lower superheats with decreasing heater size, Park and Bergles observed no such effect. However, Park and Bergles do report increasing CHF with decreasing heater height and width. They also report that, for two, in-line vertical heaters, boiling incipience occurs at smaller superheats for the top surface but that there is little difference in the established boiling curve for the two heaters.

FORCED CONVECTION BOILING

Forced convection boiling may occur when cooling is effected by an impinging jet, a channel flow, or a falling film. Impinging jets may be directed normal to the surface or at any angle off the normal. Katto and Ishii [67] considered free,

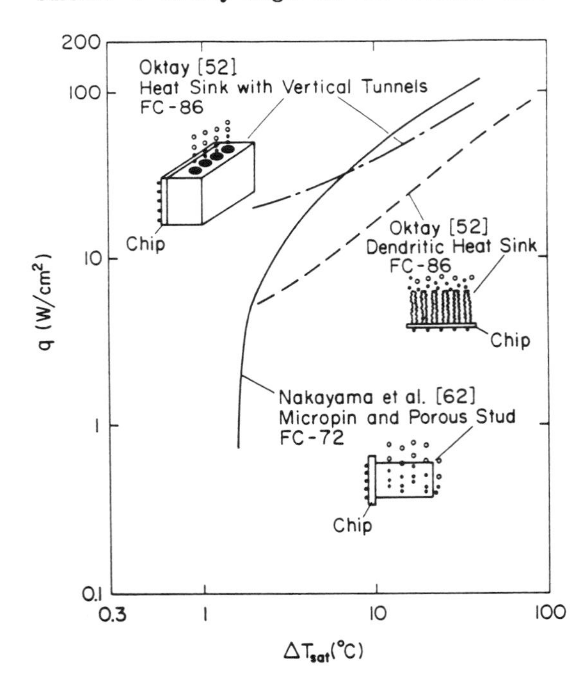

FIGURE 16. Pool boiling curves for various heat sink configurations.

planar liquid jets impinging at various angles on rectangular surfaces which were 10 to 20 mm long and 15 mm wide. They observed a thickening of the wall jet due to bubble formation, and as CHF was approached, most of the liquid was driven from the surface by intense vapor effusion, leaving a very thin film at the surface. CHF was associated with evaporation of the film and the appearance of dry patches on the surface. Monde and Katto [68] observed similar behavior for free jets impinging on circular heat sources with diameters ranging from 11 to 21 mm. They also proposed a critical heat flux correlation of the form

$$\frac{q_c}{\rho_g h_{fg} V} = C_1 \left(\frac{\rho_f}{\rho_g}\right)^{m_1} We^{m_2} \left[1 + C_2 \left(\frac{\rho_f}{\rho_g}\right)^{n_1} Ja^{n_2}\right] \tag{11}$$

where $We = \rho_f V^2 L/\sigma$ and $Ja = c_{p,f} \Delta T_{sub}/h_{fg}$ are the Weber and Jakob numbers, respectively, V is the characteristic velocity of the forced flow, and L is the heater length. The effects of heater length and subcooling are accommodated by the Weber and Jakob numbers, respectively. For subcooled, normally impinging circular jets, the correlating coefficients $C_1 = 0.0745$, $m_1 = 0.725$, $m_2 = -0.333$, $C_2 = 2.7$, $n_1 = 0.5$, and $n_2 = 2.0$ are recommended [68], while for saturated planar jets, Katto and Ishii [67] recommend the coefficients $C_1 = 0.0164$, $m_1 = 0.867$ and $m_2 = -0.333$. For such a jet and representative values of $V = 1$ m/s and $L = 10$ mm, equation (11) yields the somewhat low value of 9.4 W/cm² for FC-72.

Ma and Bergles [44] performed experiments for submerged jets of R-113 impinging normally on a 5 mm×5 mm heat source. Consistent with previous results [67,68], CHF was found to increase with the cube root of velocity. It was also found to increase with the degree of subcooling, although the effect was not as pronounced as that observed by Monde and Katto [68]. CHF values up to 100 W/cm² were obtained, and temperature overshoots were negligible.

Katto [69] considered forced convection boiling of water and R-113 at the heated lower and upper surfaces of a rectangular duct. For heater lengths as small as 10 mm, CHF was found to increase with decreasing length and with increasing velocity and subcooling. Samant and Simon [30] investigated forced convection boiling of FC-72 at a 0.25 mm long by 2.0 mm wide heater attached to the lower surface of a rectangular duct. Fully-developed, subcooled, turbulent flow was considered, with velocities and subcoolings extending to $V = 16.9$ m/s and $\Delta T_{sub} = 68.1\,^\circ C$, respectively. Although a large temperature overshoot (27 °C) was observed for small values of $V = 2.1$ m/s and $\Delta T_{sub} = 13.4\,^\circ C$, the overshoot decreased with increasing V and ΔT_{sub} and was essentially eliminated for $\Delta T_{sub} \gtrsim 50\,^\circ C$. The critical heat flux increased with increasing V and ΔT_{sub}, reaching a value of 426 W/cm² for the maximum values of 16.9 m/s and 68.1 °C.

Working with forced convection boiling data obtained for free and submerged jets [44,68] and channel flow [30], Lee et al. [70] were able to correlate the data by using equation (11) with $C_1 = 0.0742$, $m_1 = 0.761$, $m_2 = -0.365$, $C_2 = 0.952$, $n_1 = 0.118$, and $n_2 = 1.414$. The data, and hence the correlation, are restricted to heater lengths in the range $0.25 \leq L \leq 5.0$ mm.

More recently, experiments were performed for forced convection boiling of FC-72 from a single 12.7 mm heater flush mounted to one wall of a rectangular duct [31]. Although increasing fluid velocity resulted in significant heat transfer enhancement for the single-phase and nucleate boiling portions of the boiling curve, the increase in CHF was much smaller. However, significant CHF enhancement was associated with increased fluid subcooling and with use of a low profile (1.02 mm) microstud surface. CHF values as large as 93.5 W/cm^2 were achieved. The degree of hysteresis at boiling incipience decreased with increasing subcooling and velocity and with use of the microstuds.

Forced convection boiling may also occur in free-falling, liquid films used to cool vertically mounted electronic chips. Related studies have been performed by Ueda et al. [71] and Mudawwar et al. [72]. In the latter study liquid films of FC-72 were injected over vertical heaters ranging in length from 12.7 to 127 mm. Film thickness and velocity were varied over the ranges $0.25 \leq \delta \leq 1.50$ mm and $0.50 \leq V \leq 2.0$ m/s, while subcooling was maintained below $\Delta T_{sub} = 6\,^{\circ}C$. Boiling hysteresis was nonexistent, and conditions in the fully developed nucleate boiling regime were approximately independent of film thickness and velocity. The onset of CHF was characterized by separation of most of the film from the heater and subsequent dryout of a thin liquid subfilm which maintained contact with the heater. The critical heat flux increased with increasing velocity and decreasing heater length and, for zero subcooling, could be correlated by equation (11) with $C_1 = 0.121$, $m_1 = 0.667$ and $m_2 = -0.42$. The largest critical heat flux ($q_c = 24$ W/cm^2) was obtained for the 12.7 mm heater and a velocity of 2 m/s.

Attempts to enhance CHF for falling films have been made by several investigators. Nakayama et al. [73] considered a long (300 mm) heater in R-11 and studied the effects of structured surfaces, which included grooved (microfinned) surfaces with the grooves aligned (vertical) or perpendicular (horizontal) to the flow direction and a porous cover plate with horizontal subsurface tunnels. Although the horizontally grooved surface nucleated first, the vertically grooved surface provided superior performance at low heat fluxes and the two boiling curves converged in the fully developed nucleate boiling regime. The largest heat transfer enhancement (relative to pool boiling) was generally provided by the porous plate, whose performance was independent of flowrate, although comparable performance was exhibited by the vertically grooved surface, when operated at an optimal flowrate. CHF data were not reported.

For a falling film of FC-72 on a 63.5 mm long heater, Grimley et al. [74] studied enhancement effects associated with the use of structured surfaces (vertical microfins or microstuds), subcooling, and a louvered plate mounted parallel to and 2.54 mm from the heated surface. Although both the microfin and microstud surfaces enhanced nucleate boiling heat transfer relative to a smooth surface, only the microfin surface provided significant enhancement of CHF. CHF was again observed to be due to dryout of a thin subfilm which remained on the boiling surface after the bulk of the fluid in the falling film had separated due to intense vapor generation. It was argued that the microfins extended CHF by allowing surface tension forces to more effectively maintain the liquid film on the surface and by inhibiting the lateral spread of dry patches after film separation. In contrast, the microstud surface acted to break up the film, thereby hastening film separation and decreasing CHF. CHF was also enhanced by subcooling the liquid and by installing a louvered flow deflector. While subcooling decreased the intensity of vapor effusion, the deflector inhibited film separation.

Previous studies [72,74] indicated that CHF in a falling liquid film may be enhanced by reducing the length of the heated surface, machining longitudinal grooves in the surface, shrouding the surface with a louvered flow deflector, or subcooling the liquid. To establish upper limits for CHF enhancement, the collective effect of such measures was considered by Grimley et al. [75], who performed experiments for a subcooled film of FC-72 falling over a 12.7 mm microfinned surface with an attached flow deflector (Fig.17). It was found that, although CHF may be improved by the combination of reduced heater length, subcooling, and flow deflection, the effects are not independent and superposition does not apply. CHF could not be enhanced much above 30 W/cm^2, and as shown in Fig.18, the large heat fluxes projected for VLSI devices are much more likely to be achieved with jet impingement or channel flows than with a falling film.

TWO-PHASE THERMOSYPHONS

Of the many possibilities for two-phase, liquid immersion cooling of microelectronic assemblies, complete encapsulation is an attractive option. The option, was first used for high power components and airborne electronic packages and was subsequently advocated for densely populated PCBs and chip arrays [76-78]. Typically, it provides for saturated or subcooled boiling at the surfaces of the electronic components and condensation within the liquid and/or at an encapsulated cold plate. As shown in the representative systems of Fig.19, chips may be mounted to a ceramic substrate forming one wall of a module filled with a fluorocarbon. Heat is transferred from the chips to the coolant and from the coolant to the internal fins of a water-cooled cold plate or an air-cooled heat sink. Use of a submerged condenser facilitates maintenance of subcooled conditions in the coolant. Alternatively, the thermosyphon may use a vapor space condenser, in which case the liquid is saturated and condensate is returned from the upper portion of the chamber.

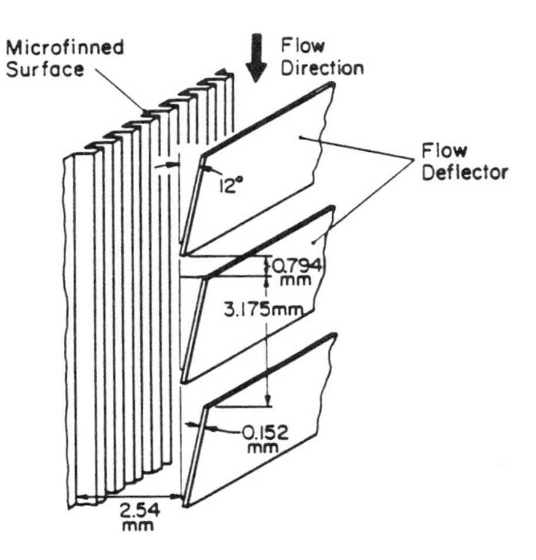

FIGURE 17. Flow deflector for CHF enhancement in a falling liquid film.

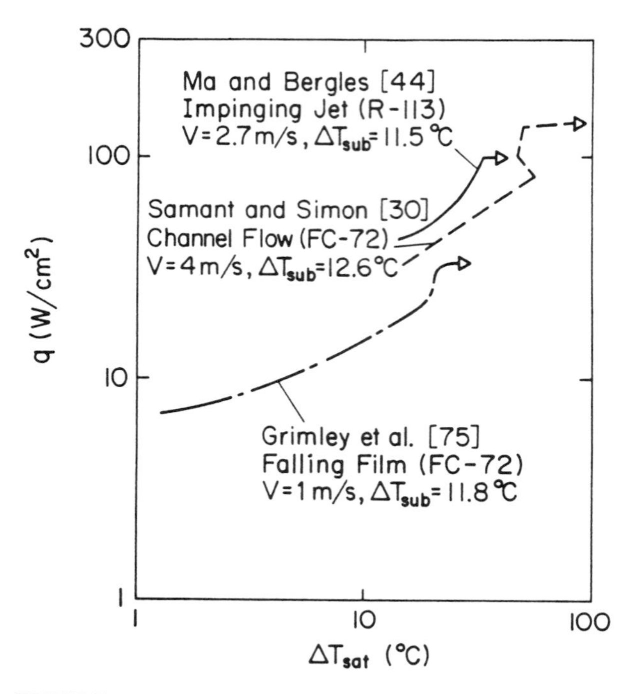

FIGURE 18. Comparison of selected forced convection boiling schemes.

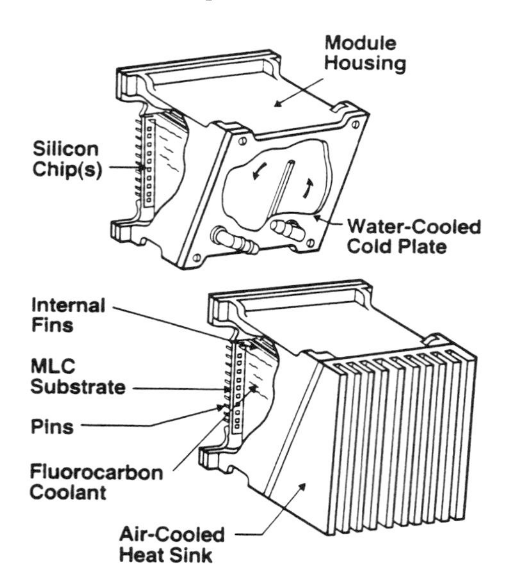

FIGURE 19. Encapsulated two-phase thermosyphons.

Megerlin and Vingerhoet [76] considered boiling and condensation of FC-86 in a narrow cavity, with one vertical wall consisting of nine chips on an 8.5 cm^2 substrate and the opposite wall cooled. Heat fluxes up to 15.5 W/cm^2 were reported for a chip temperature of 69 °C, and equivalent results were obtained for cavity widths of 3.4 mm and 6.8 mm. Mosinski et al. [79] considered a similar system (a 3×3 array of simulated chips on one wall and a submerged condenser at the opposite wall), but for narrow cavity widths ranging from 0.43 to 0.81 mm. The wall superheat for a prescribed heat flux and the critical heat flux were found to decrease with decreasing cavity width. In FC-72, CHF values up to 20 W/cm^2 were achieved and were associated with formation of a vapor space and attendant dryout of the top row of heaters. Although subcooling reduced the size of the vapor bubbles, it did not condense all of the bubbles, thereby limiting CHF due to formation of the vapor space.

In an alternative configuration intended to enhance condensation, Yokouchi et al. [59] installed condenser tubes in both the liquid and vapor spaces and used porous nickel plates to trap bubbles in proximity to the submerged tubes. With forty-nine simulated chips mounted on each of two opposing circuit boards in FC-72, 10W/cm^2 could be dissipated at the chip level and 1 kW could be dissipated in the 900 cm^3 module volume. It was further determined that, by using an 80:20 mixture by volume of FC-72 and FC-75, respectively, temperature overshoot could be eliminated. The problem of boiling between vertical arrays of printed circuit boards has also been considered by Bar-Cohen and Schweitzer [80]. Idealizing conditions with a pair of flat isoflux plates, they developed a model for liquid flow through the channel and experimentally determined wall temperature as a function of axial location, heat flux and plate spacing.

Matters which have inhibited application of the foregoing concepts relate to concern for the temperature overshoot at incipient boiling and for maintaining liquid purity. Unless totally removed, residues from chip and module joining processes could be dissolved and redeposited at connecting pads during boiling, with subsequent corrosion and failure of the pads.

An alternative to the foregoing *full-immersion* thermosyphons has been suggested by Mudawwar et al. [81]. As shown in Fig.20, cooling is maintained by flow of a falling fluorocarbon film over a vertical multichip module. Vapor generated by nucleate boiling in the film is condensed on the finned surfaces of a vapor space condenser, and the condensate is collected in a constant head reservoir above the modules. A static head of 10 cm in the reservoir would be sufficient to maintain a film velocity in excess of 1 m/s. Ideally, operation as a thermosyphon would require that all of the liquid film be evaporated at the lower edge of the module. However, since such a condition would be difficult to maintain, excess liquid flow would be needed to prevent film dryout on the lowermost heaters. To circulate the excess liquid, an auxiliary heater could be submerged in the bottom reservoir, or, as shown, a small pump could be used to return liquid to the constant head reservoir. Such a *semi-passive* thermosyphon would provide some of the advantages of forced convection boiling, such as negligible hysteresis and moderate heat fluxes ($q_c \sim 30$ W/cm^2), while having minimal pump power requirements and being adaptable to three-dimensional packaging schemes.

Although widely adopted for other applications, the use of heat pipes in electronic cooling has been limited. Nelson et al. [82] and Scott and Tanzer [83] have

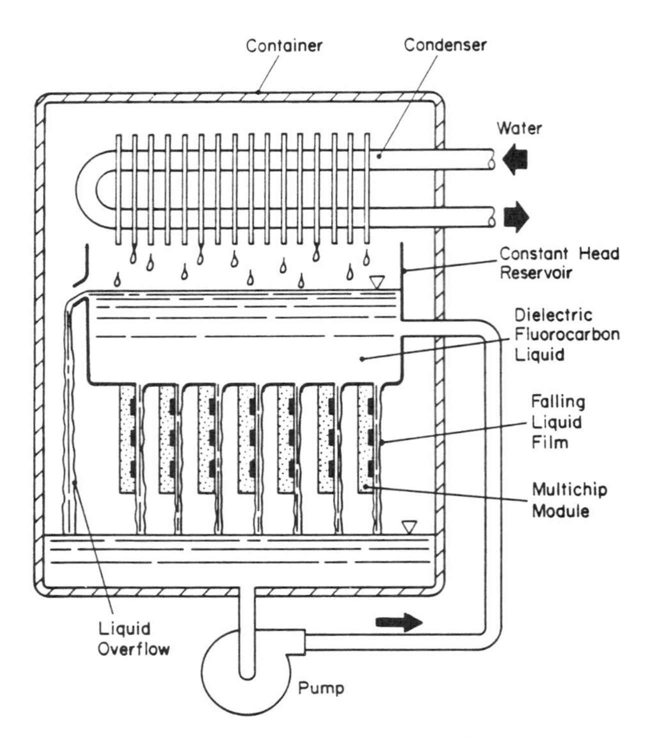

FIGURE 20. Two-phase thermosyphon with vapor space condenser.

mounted heat pipes on one side of a printed circuit board to improve heat transfer paths for components mounted on the opposite side. Heat pipes have also been coupled to the components in a hermetically sealed package to facilitate direct heat removal [84,85]. In this approach there is direct contact between the heat pipe fluid and the components. Kromann et al. [85] obtained a significant reduction in junction-to-case thermal resistance by placing a pentane saturated wick in contact with the non-circuit sides of 16 chips in a multi-chip assembly. Vapor generated at the wick/chip interface condenses on grooves machined in a cold plate, and liquid returns by gravity and capillary action to the chip surface. Module thermal resistances of less than 2 $cm^2 \cdot C/W$ were reported. Ultimate applicability of the integral chip/heat pipe concept will depend on the compatibility and stability of the working materials, as well as on thermal performance.

Kiewra and Wayner [86] performed experiments for which a disc heat source of 5 mm diameter was cooled by placing it at the base of a cylindrical cavity of height and diameter equal to 25 mm and 5 mm, respectively. The cavity was partially filled with hexane or decane which formed a thin circular film shaped as an extended meniscus. Heat transfer from the source was sustained by evaporation from the meniscus and condensation at the top surface of the cavity. Although the thermal resistance of this thermosyphon is large ($\sim$50 $cm^2 \cdot C/W$), optimization of design and operating parameters may yield some improvement.

SUBMERGED AND VAPOR SPACE CONDENSERS

As discussed in the previous section, cooling electronic assemblies by complete immersion in an encapsulated liquid of large dielectric strength has long been considered as a thermal control option. However, a major constraint on system volume is imposed by the ability to remove heat from the immersion fluid. When boiling occurs at the component surfaces, heat removal must be effected by a vapor-space or a submerged condenser. Vapor-space condensers typically consist of cooled vertical or horizontal (downward facing) surfaces mounted at the top of the enclosure (Fig.20).

Although boiling from electronic components has been and continues to be widely considered, comparatively little attention has been given to attendant condensation requirements. Dropwise and film condensation on horizontal surfaces have been studied [87-91]. Noteworthy results pertain to the deleterious effects of noncondensables and to the existence of an optimum surface width corresponding, approximately, to twice the wavelength of the Taylor instability. An accumulation of noncondensables in the vapor space will impede vapor transport to the condensing surface, thereby decreasing the surface convection coefficient and condenser performance. For condensation of FC-87 on closely spaced, vertical surfaces, Aakalu [92] observed a precipitous drop in the convection coefficient when the mass of air increased above a critical value. The critical condition was attributed to a *gas flooding* phenomenon corresponding to merger of hypothetical gas (air) films on adjoining surfaces. An accumulation of noncondensables will also increase the pressure and hence the saturation temperature of the vapor. In turn, junction temperatures will increase and the structural integrity of the container may be threatened.

To circumvent limits on the maximum convection coefficient imposed by an optimum width, Yanadori et al. [91] considered the effect of installing vertical fins on a downward facing horizontal surface. In R-113 significant heat transfer enhancement resulted from use of the fins, with maximum enhancement corresponding to a fin pitch of approximately 4 mm. At this optimum a thin condensate layer provides complete wetting of the surface and its curvature enhances the surface area available for condensation. Below the optimum pitch, space between the fins is filled with a thick condensate layer and heat transfer is reduced.

The adverse effects of noncondensables can be circumvented by using submerged condenser surfaces. Sidewall (vertical), as well as top (horizontal), surfaces may be used to subcool the liquid, causing ascending vapor bubbles to collapse or to impact the surfaces with diminished volume. For water and R-113 in a cubical enclosure, Markowitz and Bergles [93] determined the performance characteristics of a horizontal submerged condenser. For low levels of heating (no boiling), heat transfer at the condenser surface is governed by natural convection. With increasing heat flux, however, *bubble-pumped convection* due to boiling from submerged heaters can substantially enhance heat transfer at the condenser surface, despite the fact that only a small amount of vapor (less than 1% of its initial volume) may reach the surface. With increased heating and decreased subcooling, it is condensation, rather than single-phase convection, which dominates heat transfer at the condensing surface. The upper limit to heating corresponds to vapor blanketing of the condenser surface. Arguing that vapor generation induces an additional buoyancy force, which substantially enhances heat transfer relative to single-phase natural convection, Markowitz and Bergles developed semi-analytical expressions which successfully correlated heat transfer data for the condenser surface.

For boiling and condensation of R-113 in a sealed module with vertically finned and unfinned cold plates, Bravo and Bergles [90] found that condensation could be significantly enhanced through the use of fins. Bar-Cohen and Distel [94] were able to correlate vertical condenser plate data using the expression developed by Markowitz and Bergles [93] for a horizontal plate, and Bar-Cohen [95] provided a comprehensive review of limits and performance relations for submerged horizontal and vertical condenser surfaces. More recently, Bar-Cohen et al. [96] confirmed the significant influence of bubble-pumped convection on heat transfer at a vertical submerged condenser plate and extended the range of data by an order of magnitude.

SUMMARY

From the foregoing review it is evident that significant progress has been made in establishing a convection heat transfer knowledge base pertinent to the problems of immersion cooling. For a wide range of geometries, this knowledge base encompasses conditions corresponding to single-phase free, forced and mixed convection, as well as pool and forced convection boiling. It allows the packaging engineer to assess, at least in an approximate sense, the relative merits of various cooling options, several of which permit the dissipation of large heat fluxes ($> 100\ W/cm^2$) in a dielectric liquid. However, the knowledge base is by no means complete. For example, there is much that could still be done to establish generalized correlations or predictive methodologies, particularly for arrays of discrete sources, to develop improved enhancement schemes, and to clarify upper limits for each of the cooling options.

NOMENCLATURE

c_p	specific heat
D	pin diameter; duct hydraulic diameter
g	gravitational acceleration
Gr	Grashof number
H	plate, enclosure, or channel height
$\underline{h}$	local convection coefficient
h	average convection coefficient
h_{fg}	heat of vaporization
Ja	Jakob number, $c_p(T_{sat} - T_f)/h_{fg}$
k	thermal conductivity
L	plate or heater length
$\underline{Nu}$	local Nusselt number, $hx/\underline{k}$
Nu	average Nusselt number, $\bar{h}L/k$
Pe	Peclet number
Pr	Prandtl number
q	local heat flux
Q	heat rate
Ra_S	Rayleigh number, $g\beta(T_s - T_\infty)S^3/\alpha\nu$
Ra_S^*	modified Rayleigh number, $g\beta qS^4/k\alpha\nu$

Re_D	Reynolds number based on duct hydraulic diameter
S	spacing between parallel plates or heat sources
T	temperature
t	plate or heater thickness
V	velocity
W	enclosure or heater width
We	Weber number, $\rho_f V^2 L/\sigma$
x,y	streamwise and transverse coordinates
α	thermal diffusivity
β	thermal expansion coefficient
ν	kinematic viscosity
σ	surface tension
ρ	density
θ	plate inclination

Subscripts

c	chilled wall; critical heat flux
f	fluid
g	vapor
h	heater
o	inlet
s	surface
S	substrate
sat	saturated
sub	subcooled
T	total

REFERENCES

1. Chu, R.C., Hwang, U.P., and Simons, R.E., Conduction Cooling for an LSI Package: A One-Dimensional Approach, *IBM Journal of Research and Development*, vol.26, pp.45-54, 1982.
2. Yamamoto, H., Udagawa, Y., and Suzuki, M., Cooling System for FACOM M-780 Large Scale Computer, *Proc. Int. Symp. on Cooling Technology for Electronic Equipment*, Honolulu, March 17-24, pp.96-109, 1987.
3. Kishimoto, T., and Ohsaki, T., VLSI Packaging Technique Using Liquid-Cooled Channels, *IEEE Trans. on Comp., Hybrids, and Manuf. Tech.*, vol.CHMT-9, pp.328-335, 1986.
4. Tuckerman, D.B., and Pease, R.F., High-Performance Heat Sinking for VLSI, *IEEE Electronic Device Letters*, vol.EDL-2, pp.126-129, 1981.

5. Phillips, R.J., Glicksman, L.R., and Larson, R., Forced-Convection, Liquid Cooled Microchannel Heat Sinks for High-Power-Density Microelectronics, *Proc. Int. Symp. on Cooling Technology for Electronic Equipment*, Honolulu, March 17-21, pp.227-248, 1987.

6. Danielson, R.D., Krajewski, N., and Brost, J., Cooling of a Superfast Computer, *Electronic Packaging and Production*, July, pp.44-45, 1986.

7. Elenbaas, W., Heat Dissipation of Parallel Plates by Free Convection, *Physica*, vol.9, pp.1-28, 1942.

8. Bar-Cohen, A., and Rohsenow, W.M., Thermally Optimum Spacing of Vertical, Natural Convection Cooled, Parallel Plates, *J. Heat Transfer*, vol.106, pp.116-123, 1984.

9. Bar-Cohen, A., and Schweitzer, H., Convective Immersion Cooling of Parallel Vertical Plates, *Proc. Fourth Int. Electronics Packaging Conf.*, pp.596-615, 1984.

10. Azevedo, L.F.A., and Sparrow, E.M., Natural Convection in Open-Ended Inclined Channels, *J. Heat Transfer*, vol.107, pp.893-901, 1985.

11. Ortega, A., and Moffat, R.J., Heat Transfer from an Array of Simulated Electronic Components: Experimental Results for Free Convection with and without a Shrouding Wall, *Heat Transfer in Electronic Equipment-1985*, S. Oktay and R.J. Moffat, eds., ASME HTD-vol.48, pp.5-15, 1985.

12. Johnson, C.E., Evaluation of Correlations for Natural Convection Cooling of Electronic Equipment, *Heat Transfer Engineering*, vol.7, Nos.1-2, pp.36-45, 1986.

13. Baker, E., Liquid Cooling of Microelectronic Devices by Free and Forced Convection, *Microelectronics and Reliability*, vol.11, pp.213-222, 1972.

14. Baker, E., Liquid Immersion Cooling of Small Electronic Devices, *Microelectronics and Reliability*, vol.12, pp.163-173, 1973.

15. Carey, V.P., and Mollendorf, J.C., The Temperature Field Above a Concentrated Heat Source on a Vertical Adiabatic Surface, *Int. J. Heat Mass Transfer*, vol.20, pp.1059-1067, 1977.

16. Park, K.A., and Bergles, A.E., Natural Convection Heat Transfer Characteristics of Simulated Microelectronic Chips, *J. Heat Transfer*, vol.109, pp.90-96, 1987.

17. Sparrow, E.M., and Faghri, M., Natural Convection Heat Transfer from the Upper Plate of a Colinear, Separated Pair of Vertical Plates, *J. Heat Transfer*, vol.102, pp.623-629, 1980.

18. Jaluria, Y., Buoyancy-Induced Flow due to Isolated Thermal Sources on a Vertical Surface, *J. Heat Transfer*, vol.104, pp.223-227, 1982.

19. Jaluria, Y., Interaction of Natural Convection Wakes from Thermal Sources on a Vertical Surface, *J. Heat Transfer*, vol.107, pp.883-892, 1982.

20. Milanez, L.F., and Bergles, A.E., Studies in Natural Convection Heat Transfer from Thermal Sources on a Vertical Surface, *Heat Transfer-1986*, Proc.8th Int. Heat Transfer Conf., vol.3, pp.1347-1352, 1986.

21. Chu, H.H.-S., Churchill, S.W., and Patterson, C.V.S., The Effect of Heater Size, Location, Aspect Ratio, and Boundary Conditions on Two-Dimensional, Laminar Natural Convection in Rectangular Channels, *J. Heat Transfer*, vol.98, pp.194-201, 1976.

22. Flack, R.D., and Turner, B.L., Heat Transfer Correlations for Use in Naturally Cooled Enclosures with High-Power Integrated Circuits, *IEEE Trans. on Components, Hybrids and Manufacturing Technology*, vol.CHMT-3, pp.449-452, 1980.

23. Turner, B.L, and Flack, R.D., Experimental Measurement of Natural Convective Heat Transfer in Rectangular Enclosures with Concentrated Energy Sources, *J. Heat Transfer*, vol.102, pp.236-241, 1980.

24. Yaghoubi, M.A., and Incropera, F.P., Analysis of Natural Convection due to Localized Heating in a Shallow Water Layer, *Numerical Heat Transfer*, vol.3, pp.315-330, 1980.

25. Lee, J.J., Liu, K.V., Yang, K.T., and Kelleher, M.D., Laminar Natural Convection in a Rectangular Enclosure due to a Heated Protrusion of one Vertical Wall - Part II: Numerical Simulations, *Proc. Second ASME-JSME Thermal Engineering Joint Conference*, P.J. Marto and I. Tanasawa, eds., vol.2, pp.179-185, 1987.

26. Kelleher, M.D., Knock, R.H., and Yang, K.T., Laminar Natural Convection in a Rectangular Enclosure due to a Heated Protrusion on One Vertical Wall-Part 1: Experimental Investigation, *Proc. Second ASME-JSME Thermal Engineering Joint Conference*, P.J. Marto and I. Tanasawa, eds., vol.2, pp.169-177, 1987.

27. Keyhani, M., Prasad, V., and Cox, R., An Experimental Study of Natural Convection in a Vertical Cavity with Discrete Heat Sources, *J. Heat Transfer*, In Press, 1988.

28. Liu, K.V., Yang, K.T., and Kelleher, M.D., Three-Dimensional Natural Convection Cooling of an Array of Heated Protrusions in an Enclosure Filled with a Dielectric Fluid, *Proc. Int. Symp. Cooling Technology for Electronic Equipment*, Honolulu, March 17-21, pp.486-497.

29. Incropera, F.P., Kerby, J., Moffatt, D.F., and Ramadhyani, S., Convection Heat Transfer from Discrete Sources in a Rectangular Channel," *Int. J. Heat Mass Transfer*, vol.29, pp.1051-1058, 1986.

30. Samant, K.R., and Simon, T.W., Heat Transfer from a Small, High-Heat-Flux Patch to a Subcooled Turbulent Flow, ASME Paper 86-HT-22, 1986.

31. Maddox, D.E., and Mudawwar, I., Single- and Two-Phase Convective Heat Transfer from Smooth and Enhanced Microelectronic Heat Sources in a Rectangular Channel, National Heat Transfer Conference, Houston, Texas, 1988.

32. Ramadhyani, S., Moffatt, D.F., and Incropera, F.P., Conjugate Heat Transfer From Small Isothermal Heat Sources Embedded in a Large Substrate, *Int. J. Heat Mass Transfer*, Vol.28, pp.1945-1952, 1985.

33. Moffatt, D.F., Ramadhyani, S., and Incropera, F.P., Conjugate Heat Transfer from Wall Embedded Sources in Turbulent Channel Flow, *Heat Transfer in Electronic Equipment - 1986*, A. Bar-Cohen, ed., ASME HTD-Vol.57, pp.177-182, 1986.

34. Kelecy, F.J., Ramadhyani, S., and Incropera, F.P., Effect of Shrouded Pin Fins on Forced Convection Cooling of Discrete Heat Sources by Direct Liquid Immersion, *Proc. Second ASME-JSME Thermal Engineering Joint Conference*, P.J. Marto and I. Tanasawa, eds., vol.3, pp.387-394, 1987.

35. Ramadhyani, S., and Incropera, F.P., Forced Convection Cooling of Discrete Heat Sources with and without Surface Enhancement. *Proc. Int. Symp. on Cooling Technology for Electronic Equipment*, Honolulu, March 17-21, pp.249-264, 1987.

36. Brinkman, R., Ramadhyani, S., and Incropera, F.P., Enhancement of Convective Heat Transfer from Small Heat Sources to Liquid Coolants Using Strip Fins, *Experimental Heat Transfer*, In Review, 1988.

37. Greiner, M., Ghadder, N.K., Mikic, B.B., and Patera, A.T., Resonant Convective Heat Transfer in Grooved Channels, in *Heat Transfer 1986*, C.L. Tien, V.P. Carey, and J.K. Ferrell, eds., Hemisphere, New York, vol.6, pp.2867-2872, 1986.

38. Karniadakis, G.E., Mikic, B.B., and Patera, A.T., Heat Transfer Enhancement by Flow Destablization: Application to the Cooling of Chips, *Proc. Int. Symp. on Cooling Technology for Electronic Equipment*, Honolulu, March 17-21, pp.498-521, 1987.

39. Choudbury, Z.H., Heat Transfer in a Radial Liquid Jet, *J. Fluid Mechanics*, vol.20, pp.501-511, 1964.

40. Metzger, D.E., Cummings, K.N., and Ruby, W.A., Effects of Prandtl Number on Heat Transfer Characteristics of Impinging Liquid Jets, *Proc. 5th Int. Heat Transfer Conference*, vol.2, pp.20-24, 1974.

41. McMurray, D.C., Myers, P.S., and Uyehara, O.A., Influence of Impinging Jet Variables on Local Heat Transfer Coefficients along a Flat Surface with Constant Heat Flux, *Proc. 3rd Int. Heat Transfer Conference*, vol.2, pp.292-299, 1966.

42. Miyasaka, Y., and Inada, S., The Effect of Pure Forced Convection on Boiling Heat Transfer between Two-Dimensional Subcooled Water Jet and a Heated Surface, *Chemical Engineering*, Japan, vol.13, pp.22-28, 1980.

43. Inada, S., Miyasaka, Y., and Izumi, R., A Study on the Laminar Flow Heat Transfer Between a Two-Dimensional Water Jet and a Flat Surface with Constant Heat Flux, *Bulletin of the JSME*, Vol.24, pp.1803-1810, 1981.

44. Ma, C.F., and Bergles, A.E., Boiling Jet Impingement Cooling of Simulated Microelectronic Chips, *Heat Transfer in Electronic Equipment-1983*, S. Oktay and A. Bar-Cohen, ed., ASME HTD-Vol.28, pp.5-12, 1983.

45. Jiji, L.M., and Dagan, Z., Experimental Investigation of Single Phase Multi-Jet Impingement Cooling of an Array of Microelectronic Heat Sources, *Proc. Int. Symp. on Cooling Technology for Electronic Equipment*, Honolulu, March 17-21, pp.265-283, 1987.

46. Osborne, D.G., and Incropera, F.P., Laminar Mixed Convection Heat Transfer for Flow between Horizontal Parallel Plates with Asymmetric Heating, *Int. J. Heat and Mass Transfer*, vol.28, pp.207-217, 1985.

47. Braaten, M.E., and Patankar, S.V., Analysis of Laminar Mixed Convection in Shrouded Arrays of Heated Rectangular Blocks, *Int. J. Heat Mass Transfer*, vol.28, pp.1699-1709, 1985.

48. Mark, M.M., Stephenson, M., and Goltsos, C.E., An Evaporative-Gravity Technique for Airborne Equipment Cooling, *IRE Transactions*, vol.ANE-5, pp.47-52, 1958.

49. Goltsos, C.E., and Mark, M.M., Packaging with a Flexible Container for Oil-Filled or Evaporative Cooled Electronic Equipment, *IRE Transactions*, vol.PEP-6, pp.44-48, 1962.

50. Armstrong, R.J., Cooling Components with Boiling Halocarbons, *IEEE Transactions*, vol.PMP-3, No.4, pp.135-142, 1967.

51. Preston, S.B., and Shillabeer, R.N., Direct Liquid Cooling of Microelectronics, *Proc. Int. Electronic Packaging and Production Conf.* (INTER/NEPCON), pp.10-31, 1970.

52. Oktay, S., Departure from Natural Convection (DNC) in Low Temperature Boiling Heat Transfer Encountered in Cooling Micro-Electronic LSI devices, *Heat Transfer-1982*, Proc. 7th Int. Heat Transfer Conf., Hemisphere Publishing Corp., vol.4, pp.113-118, 1982.

53. Moran, K.P., Oktay, S., Buller, L., and Kerjilian, G., Cooling Concepts of IBM Electronic Packages, *Proc. Second Annual Conf. Int. Electronic Packaging Society (IEPS)*, pp.120-140, 1982.

54. Bergles, A.E., and Chyu, M.C., Characteristics of Nucleate Pool Boiling from Porous Metallic Coatings, *J. Heat Transfer*, vol.104, pp.279-285, 1982.

55. Marto, P.J., and Lepere, V.J., Pool Boiling Heat Transfer from Enhanced Surfaces to Dielectric Fluids, *J. Heat Transfer*, vol.104, pp.292-299, 1982.

56. Park, K.A., and Bergles, A.E., Effects of Size of Simulated Microelectronic Chips on Boiling and Critical Heat Flux, *Heat Transfer in Electronic Equipment-1986*, ASME HTD-vol.57, A. Bar-Cohen, ed., pp.95-102, 1986.

57. Bar-Cohen, A., and Simon, T.W., Wall Superheat Excursions in the Boiling Incipience of Dielectric Fluids," *Heat Transfer in Electronic Equipment - 1986*, A. Bar-Cohen, ed., ASME HTD-vol.57, pp.83-94, 1986.

58. Reeber, M.D., and Frieser, R.G., Heat Transfer of Modified Silicon Surfaces, *IEEE Trans. Components, Hybrids and Manufacturing Technology*, vol.CHMT-3, pp.387-391, 1980.

59. Yokouchi, K., Kamerhara, N., and Niwa, K., Immersion Cooling for High Density Packages, *Proc. IEEE Electronic Components Conference*, Boston, May 11-13, pp.545-549, 1987.

60. Chu, R.C., and Moran, K.P., Method for Customizing Nucleate Boiling Heat Transfer from Electronic Units Immersed in Dielectric Coolant, U.S. Patent No.4,050,507, 1977.

61. Oktay, S., and Schmeckenbecher, A.F., Preparation and Performance of Dendritic Heat Sinks, *J. Electrochemical Society*, vol.21, pp.912-918, 1974.

62. Nakayama, W., Nakayama, T., and Hirasaura, S., Heat Studs Having Enhanced Boiling for Cooling of Microelectronic Components, ASME Paper 84-WA/HT-89, 1984.

63. Kim, C.-J., and Bergles, A.E., Structured Surfaces for Enhanced Nucleate Boiling, Report HTL-36, ERI Project 1544, Iowa State University, 1985.

64. Park, K.A., and Bergles, A.E., Boiling Heat Transfer Characteristics of Simulated Microelectronic Chips with Detachable Heat Sinks, *Heat Transfer-1986*, C.L. Tien, V.P. Carey, and J.K. Ferrell, ed., Hemisphere, New York, vol.4, pp.2099-104, 1986.

65. Nakayama, W., Thermal Management of Electronic Equipment: A Review of Technology and Research Topics, *Applied Mechanics Reviews*, vol.39, pp.1847-1868, 1986.

66. Anderson, T.M., and Mudawwar, I., Microelectronic Cooling by Enhanced Pool Boiling of a Dielectric Fluorocarbon Liquid, National Heat Transfer Conference, Houston, Texas, HTD-96, vol.1, pp.551-560, 1988.

67. Katto, Y., and Ishii, K., Burnout in a High Heat Flux Boiling System with a Forced Supply of Liquid through a Plane Jet, *Proc. Sixth Int. Heat Transfer Conference*, Toronto, Canada, vol.1, pp.435-440, 1978.

68. Monde, M., and Katto, Y., Burnout in High Heat-Flux Boiling System with an Impinging Jet, *Int. J. Heat Mass Transfer, vol.21*, pp.295-305, 1978.

69. Katto, Y., General Features of CHF for Forced Convection Boiling in Uniformly Heated Rectangular Ducts, *Int. J. Heat Mass Transfer*, vol.24, pp.1413-1419, 1981.

70. Lee, T.Y., Simon, T.W., Bar-Cohen, A., An Investigation of Short-Heating-Length Effect on Flow Boiling Critical Heat Flux in a Subcooled Turbulent Flow, *Proc. Int. Symp. on Cooling Technology for Electronic Equipment*, Honolulu, March 17-21, pp.358-373, 1987.

71. Ueda, T., Inoue, M., and Nagatome, S., Critical Heat Flux and Droplet Entrainment Rate in Boiling of Falling Liquid Films, *Int. J. Heat and Mass Transfer*, vol.24, pp.1257-1266, 1981.

72. Mudawwar, I.A., Incropera, T.A., and Incropera, F.P., Boiling Heat Transfer and Critical Heat Flux in Liquid Films Falling on Vertically Mounted Surfaces, *Int. J. Heat Mass Transfer*, vol.30, pp.2083-2095, 1987.

73. Nakayama, W., Daikoku, T., and Nakajima, T., Enhancement of Boiling, and Evaporation on Structured Surfaces with Gravity Driven Film Flow of R-11, *Heat Transfer 1982*, Hemisphere Publishing Corporation, New York, pp.409-414, 1982.

74. Grimley, T.A., Mudawwar, I., and Incropera, F.P., CHF Enhancement in Flowing Fluorocarbon Liquid Films Using Structured Surfaces and Flow Deflectors, *Int. J. Heat Mass Transfer*, vol.31, pp.55-65, 1988.

75. Grimley, T.A., Mudawwar, I., and Incropera, F.P., Limits to Critical Heat Flux Enhancement in a Liquid Film Falling over a Structured Surface which Simulates a Microelectronic Chip, *J. Heat Transfer*, In Press, 1988.

76. Megerlin, F.E., and Vingerhoet, P., Thermal Control of Densely Packaged Microelectronics in Dielectric Fluids, *Proc. National Aeronautics Electronics Conference*, pp.254-259, 1971.

77. Aakalu, N.G., Chu, R.C., and Simons, R.E., Liquid Encapsulated Air Cooled Module, U.S. Patent 3,741,292, 1973.

78. Ciccio, J.A., and Thun, R.E., Ultra-High Density VLSI Modules, *IEEE Trans. Components, Hybrids and Manufacturing Technology*, vol.CHMT-1, pp.242-248, 1978.

79. Mosinski, T.A., Chen, S.J., and Chato, J.C., Liquid Enhanced Cooling of Microchips, *Proc. Int. Symp. on Cooling Technology for Electronic Equipment*, Honolulu, March 17-21, pp.321-339, 1987.

80. Bar-Cohen, A., and Schweitzer, H., Thermosyphon Boiling in Vertical Channels, *J. Heat Transfer*, vol.107, pp.772-778, 1985.

81. Mudawwar, I., Incropera, T.A., and Incropera, F.P., Microelectronic Cooling by Fluorocarbon Liquid Films, *Proc. Int. Symp. on Cooling Technology for Electronic Equipment*, Honolulu, March 17-21, pp.340-357, 1987.

82. Nelson, L.A., Sekhon, K.S., and Ruttner, L.E., Applications of Heat Pipes in Electronic Modules, AIAA Paper no.78-449, 1978.

83. Scott, G.W., and Tanzer, H.G., Evaluation of Heat Pipes for Conduction Cooled Level II Avionic Packages, *Heat Transfer in Electronic Equipment - 1986*, A. Bar-Cohen, ed., ASME HTD-Vol.57, pp.67-75, 1986.

84. Nelson, L.A., Sekhon, K.S., and Fritz, J.E., Direct Heat Pipe Cooling of Semiconductor Devices, AIAA Paper No.78-450, 1978.

85. Kromann, G.B., Hannemann, R.J., and Fox, L.R., Two-Phase Internal Cooling Technique for Electronic Packages, *Heat Transfer in Electronic Equipment - 1986*, A. Bar-Cohen, ed., ASME HTD-Vol.57, pp.61-65, 1986.

86. Kiewra, E.W., and Wayner, P.C., Jr., A Small Scale Thermosyphon for the Immersion Cooling of a Disc Heat Source, *Heat Transfer in Electronic Cooling - 1986*, A. Bar-Cohen, ed., ASME HTD-Vol.57, pp.77-82, 1986.

87. Gerstman, J., and Griffith, P., Laminar Film Condensation on the Under Side of Horizontal and Inclined Surfaces, *Int. J. Heat Mass Transfer*, vol.10, pp.561-580, 1967.

88. Kroger, D.G., and Rohsenow, W.M., Condensation Heat Transfer in the Presence of a Noncondensable Gas, *Int. J. Heat Mass Transfer*, vol.11, pp.15-26, 1968.

89. Markowitz, A., Mikic, B.B., and Bergles, A.E., Condensation on a Downward-Facing Horizontal Rippled Surface, *J. Heat Transfer*, vol.94, pp.315-320, 1972.

90. Bravo, H.V., and Bergles, A.E., Limits of Boiling Heat Transfer in a Liquid-Filled Enclosure, *Proc. 1976 Heat Transfer and Fluid Mechanics Institute*, Stanford University Press, pp.79-99, 1976.

91. Yanadori, M., Hijikta, K., Mori, Y., and Uchida, M., Enhancement of Filmwise Condensation Heat Transfer to the Downward Horizontal Cooled Surface of a Small Enclosure by Vertical Fins, *Proc. Int. Symp. Cooling Technology for Electronic Equipment*, Honolulu, March 17-24, pp.374-384, 1987.

92. Aakalu, N.G., Condensation Heat Transfer for Dielectric Vapors in Presence of Air, *Heat Transfer in Electronic Equipment - 1983*, S. Oktay and A. Bar-Cohen, eds., ASME HTD-Vol.28, pp.21-27, 1983.

93. Markowitz, A., and Bergles, A.E., Operational Limits of a Submerged Condenser, *Progress in Heat and Mass Transfer*, vol.6, Pergamon Press, Oxford, pp.701-716, 1972.

94. Bar-Cohen, A., and Distel, H., Bubble-Pumped Augmented Natural Convection in Submerged Condenser Systems, *Heat Transfer - 1978, Proc. Sixth Int. Heat Transfer Conference*, Hemisphere Publishing Corporation, vol.3, pp.197-202, 1978.

95. Bar-Cohen, A., Thermal Design of Immersion Cooling Module for Electronic Components, *Heat Transfer Engineering*, vol.4, Nos.3-4, pp.35-50, 1983.

96. Bar-Cohen, A., Perelman, G., and Sabag, A., Bubble-Pumped Convective Augmentation on Vertical-Condenser Surfaces, *Proc. Second ASME-JSME Thermal Engineering Joint Conference*, P.J. Marto and I. Tanasawa, eds., vol.3, pp.431-439, 1987.

Natural Convection Immersion Cooling of an Array of Simulated Electronic Components in an Enclosure Filled with Dielectric Fluid

YOGENDRA JOSHI and MATTHEW D. KELLEHER
Department of Mechanical Engineering
Naval Postgraduate School
Monterey, California 93943, USA

T. J. BENEDICT
Lieutenant, United States Navy

ABSTRACT:

An experimental investigation of natural convection cooling of a 3 by 3 array of heated protrusions in a rectangular enclosure filled with dielectric fluid FC75 is carried out. Each of the nine rectangular components geometrically simulated a 20 pin dual-inline package. The top and bottom surfaces of the chamber were maintained at uniform temperatures while all other boundaries were insulated. Detailed flow visualizations and component surface temperature measurements were made for a number of power levels in the range 0.1 - 3.1 W through each component. The flow patterns in six different vertical planes indicated three-dimensional transport with increasing complexity as the dissipation levels increased. For the lowest heating rate, the flow structure was largely determined by the thermal conditions at the enclosure surfaces. With increasing power levels, an upward flow near the protrusions was established. The flow away from the components was time dependent and highly three dimensional. These trends were also confirmed by thermocouples embedded within the elements. Surface temperatures were measured at the centers of various fluid exposed faces of each component. These were then used to calculate non-dimensional heat transfer relationships for the geometry investigated.

1. INTRODUCTION

As microelectronic components continue to decrease in size and increase in circuit density, the problem of heat dissipation becomes increasingly critical in the design of electronic equipment. A strong need to keep the devices operating generally below 85°C exists, since for every 20°C decrease in the junction temperature, the failure rates are reduced by half (Oktay [1]). An extensive discussion of the available cooling techniques for the thermal control of electronic equipment is provided by Kraus and Bar-Cohen [2] and Chu [3].

The direct immersion of components into inert dielectric liquids provides significantly higher heat transfer coefficients than air cooling. Both single phase and phase change cooling schemes have been investigated as described by Chu [3]. Natural convection in liquids offers high heat transfer coefficients coupled with added advantages such as low noise and high reliability. Despite these potential benefits, only a limited number of studies have addressed transport in geometries characteristic of electronic packaging. Typically these contain an array of discrete, flush or protruding heated elements mounted on a substrate.

A number of recent studies have examined natural convection air cooling of discrete heat sources. Jaluria [4,5] considered the transport resulting from flush sources on a vertical surface. Ortega and Moffat [6,7] and Moffat and Ortega [8] investigated the heat transfer from an array of cubical protrusions mounted on one wall of a vertical channel. Kuhn and Oosthuizen [9] computed the transient three-dimensional transport in a rectangular enclosure following a sudden temperature change on a rectangular portion of one vertical wall. Oosthuizen and Paul [10] numerically studied the heat transfer from a square element mounted on an adiabatic wall of a tilted square enclosure. The early work by Baker [11] examined liquid immersion cooling of a discrete heat source using both forced and natural convection. In a following study, Baker [12] investigated the effect of heat source size on the heat transfer coefficient with both natural and forced convection in two different liquids. An order of magnitude increase in the heat transfer coefficient was found as the heat source size was decreased from 2.0 to 0.01 square centimeters.

Park and Bergles [13] experimentally investigated natural convection from discrete flush mounted and protruding heaters of 5 and 10 mm height and varying widths in the range 2-70 mm, in water and Freon. For the flush heaters, the midpoint heat transfer coefficients were higher than the boundary layer estimates for all heater widths. This increase was attributed to three-dimensional effects. The protruding heaters were found to have about 15% higher heat transfer coefficients than the flush heaters.

Kelleher and Knock [14] presented flow visualizations and heat transfer measurements from a single long protrusion, mounted on a vertical insulated wall of a long enclosure filled with water. All other vertical surfaces were also insulated, while the top and bottom walls were maintained at prescribed temperatures. The observed flow patterns and heat transfer responses were confirmed by the accompanying numerical computations of Lee et al. [15].

A three dimensional computational study of heat transfer in a rectangular enclosure filled with a dielectric fluid was carried out by Liu et al. [16]. A 3 by 3 array of uniformly heated protrusions was mounted on an otherwise adiabatic vertical wall. The top and bottom were considered to be at a given temperature while all other boundaries were considered adiabatic. The resulting transport was found to be inherently unsteady. In the absence of experimental evidence, these predictions could not be confirmed.

The present investigation experimentally examines the natural convection transport resulting from a 3 by 3 array of heated protrusions mounted on one vertical wall of a rectangular enclosure filled with FC-75, a dielectric fluid. The top and bottom surfaces of the enclosure are maintained at prescribed temperatures and all other boundaries are insulated. Flow visualizations are presented at various vertical planes within the enclosure. Detailed heat transfer measurements have also been carried out using thermocouples embedded within the protrusions.

2. EXPERIMENTAL ARRANGEMENT

A schematic sketch of the arrangement is provided in Fig. 1. The experiments were carried out in a 13 mm thick plexiglass chamber with the inside dimensions of 120 mm length, 144 mm height and 30 mm width. The chamber was filled with FC-75, a commercially available dielectric fluid through a tubing at the bottom of the chamber. A pressure equalization port was provided near the top of one of the chamber sidewalls.

The top and bottom surfaces of the chamber were 3 mm thick aluminum plates. These could be maintained at prescribed temperatures by means of two heat exchangers and two separate water circulation baths. The aluminum plates were placed over the O-rings provided along the top and bottom perimeters of the plexiglass chamber. The heat exchangers were then attached to these plates using C-clamps. The O-rings ensured a proper seal for the chamber fluid. All chamber walls except the top and bottom were insulated using foam insulation.

A 3 by 3 array of discrete protrusions was mounted on a 12.7 mm thick plexiglass card. This card slid vertically into the enclosure along grooves on the chamber sidewalls. The geometrical arrangement of the elements is seen in Fig. 2(a,b). Each heated component was a block of aluminum, 8 mm by 24 mm and 6 mm high. These dimensions were chosen to approximately simulate a 20 pin dual-in-line-package (DIP). A nearly uniform heat flux condition was maintained at the base of each block by attaching a foil type heater.

The foil heaters contained a network of Inconel foil mounted on a Kapton backing. The resulting thickness of the heaters was 0.18 mm. The power leads of the heaters were gold plated to ensure a low electrical resistance. The heaters were 23.6 mm by 7.6 mm and were bonded to the base of each

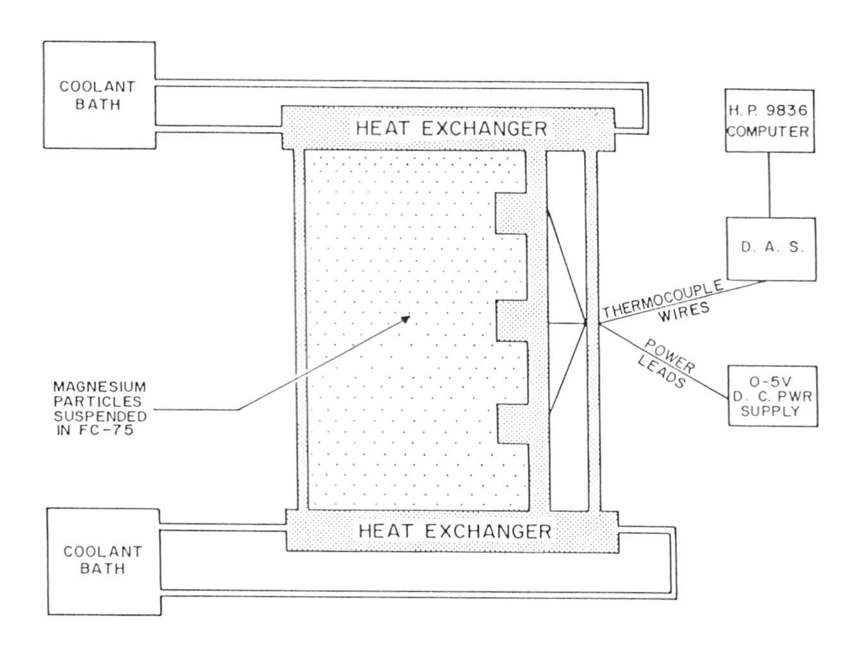

FIGURE 1. Schematic sketch of the experimental arrangement.

aluminum block using Omega Bond 101, a high thermal conductivity epoxy. Holes were provided within the heaters to pass various thermocouple wires.

Face temperatures of each block were determined using 0.254 mm diameter copper constantan thermocouples. These were placed within 0.5 mm deep grooves on each surface. Thermocouple locations on each component are seen in Fig. 3. All grooves were filled with high thermal conductivity epoxy prior to attachment to the plexiglass card. The thermocouple and power leads were routed out of the back of the card and eventually through a wiring passage hole in the chamber. These were then connected to an HP 3497 automatic data acquisition system controlled by an HP 9836 microcomputer. All thermocouples were referenced against an electronic ice point.

Each heater had a 2Ω precision resistor connected in series. The nine heater and precision resistor combinations were in turn connected in parallel to a 0-40 V, 0-1A D.C. power supply. Voltage drop across each precision resistor was measured. A simultaneous measurement of the overall voltage drop allowed the computation of the power dissipation through individual heaters.

Flow visualization was carried out with a 4 mW Helium-Neon laser for illumination. The laser beam was spread into a sheet of light using a cylindrical lens. Finely ground Magnesium particles were added to the FC-75 to visualize the flow. Magnesium has a specific gravity of 1.92 compared to 1.76 for FC-75 at 25°C. This makes the particles almost neutrally buoyant in the fluid. Time-exposure photographs

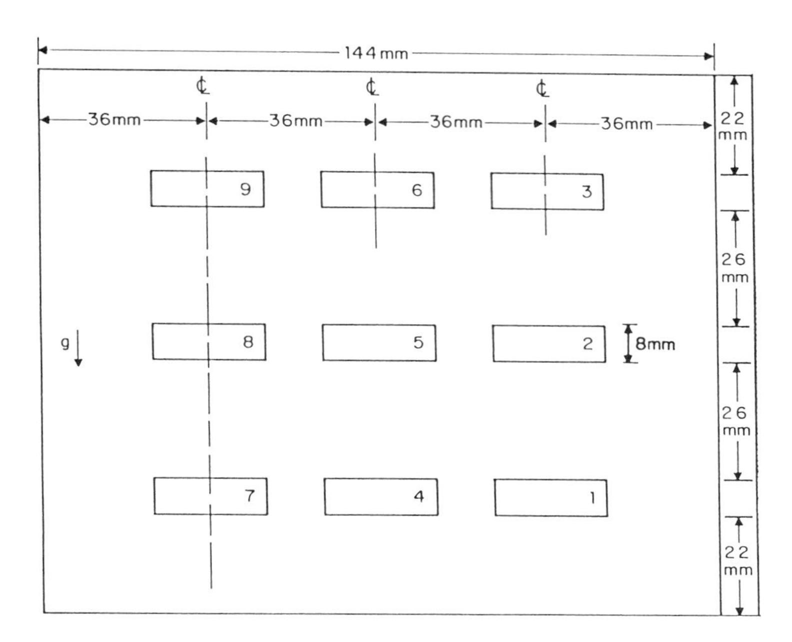

FIGURE 2(a). Front view of the simulated circuit card with the component locations.

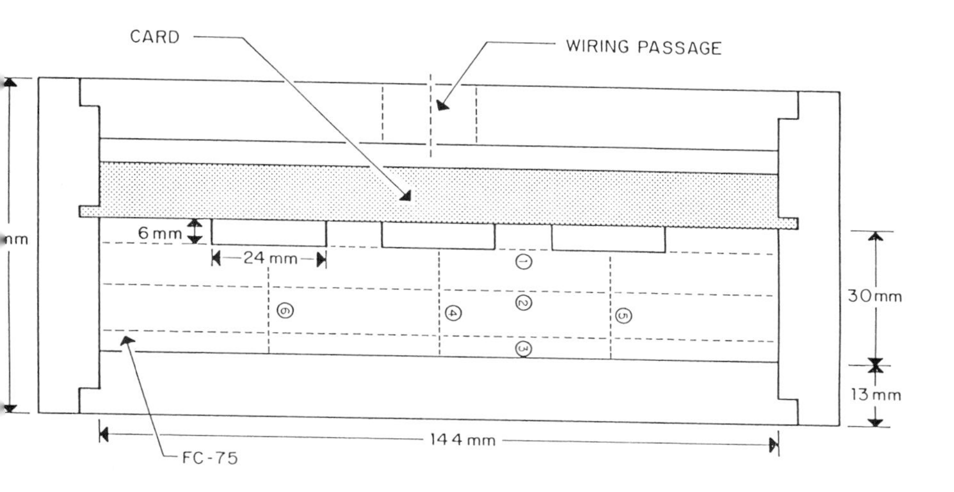

FIGURE 2(b). Top view of the enclosure with the card placed in position. The six vertical flow visualization planes are identified in the sketch.

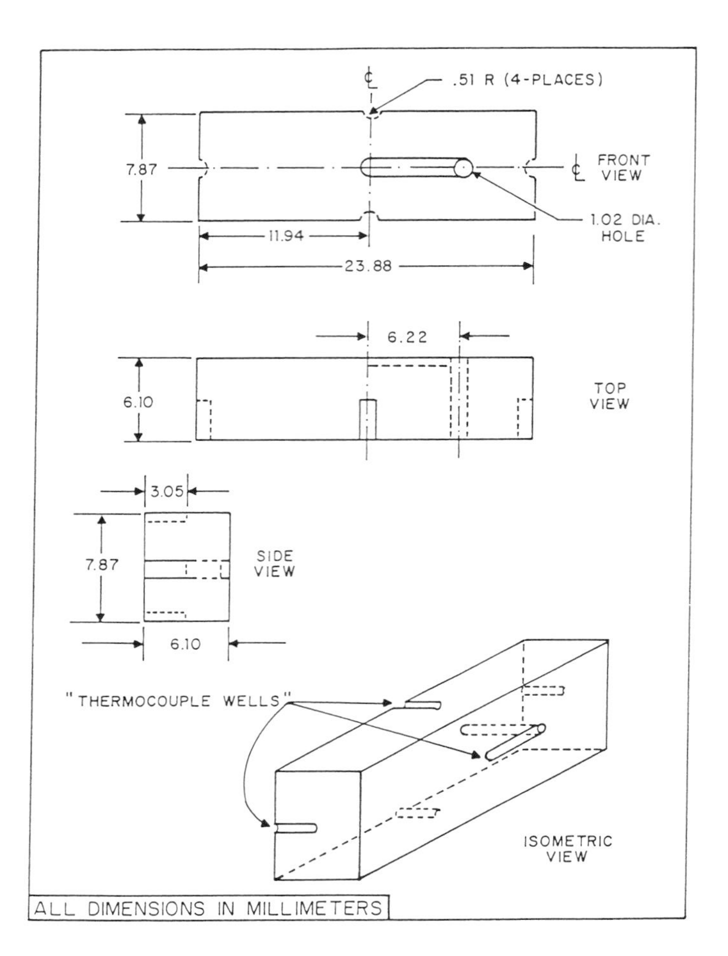

FIGURE 3. Details of the aluminum protrusions and the thermocouple locations. The foil heaters were bonded to the component back faces.

of the flow were obtained using a motor driven Nikon F-3 camera controlled by an intervalometer.

Prior to energizing the power supply or the constant temperature baths, all thermocouples were scanned to ensure uniformity. The baths were then started to maintain the heat exchangers at prescribed temperatures. The heat exchanger inlet and outlet temperatures were recorded using copper constantan thermocouples. Approximately four hours were allowed after powering the components before steady temperature measurements began. All surface temperatures were measured by the data acquisition system and stored on magnetic tapes for subsequent data analysis. Selected thermocouples were also monitored as a function of time to investigate timewise fluctuations as the component power levels increased.

Once the component surface temperature measurements were completed, the flow visualization phase of the investigation was initiated. Magnesium particles 60 μm in size were injected into the chamber by way of the bottom vent tube. After allowing them to disperse throughout the chamber, time exposure photographs were taken at the various power levels corresponding to the measurements. Flows in several different vertical planes were visualized by varying the locations of the laser-lens assembly and the camera.

In the following, a detailed account of the flow patterns and heat transfer characteristics is presented for component power levels in the range of 0.1-3.1 W. Flow visualizations discussed in the next section clearly indicate the complex three-dimensional transport mechanisms involved. In Section 4, component surface temperature variations and the resulting heat transfer relations are examined. Non-dimensional correlations for the heat transfer data are also developed.

3. FLOW VISUALIZATIONS:

Flow patterns in six vertical planes were visualized to study the three-dimensional transport responses in the range of power dissipation levels investigated. These planes are identified in Fig. 2(b). Collectively, these two-dimensional records of particle traces provided considerable insight into the actual three-dimensional nature of these flows. Detailed descriptions of the observed flows are provided next. In all the visualizations reported here, the top and bottom heat exchangers were maintained at uniform temperatures of 10°C ± 0.5°C.

3.1 Flows in Planes Parallel to Component Front Faces: Three such planes examined in this study are indicated in Fig. 2(b). These are at increasing distances out from the component front faces. The flow patterns observed at several power dissipation levels are collected in Figs. 4-6.

For the lowest power level of 0.1 W, Fig. 4(a), the flow consisted of two large clockwise cells, one on each side of the central component column. Along the central column, a buoyant upflow was observed with a meandering trend in time. This overall flow is established largely as a result of the thermal boundary conditions at the enclosure walls. The buoyancy forces generated due to power dissipation are relatively weak. They influence the flow primarily near the central column. A noticeable feature in Fig. 4(a) is the dark region below the bottom component row indicating almost quiescent conditions. This occurs as a result of the stable stratification maintained by the bottom heat exchanger. The presence of this relatively inactive zone was evident at all power levels studied.

Particle traces in plane 2, at the same power level are seen in Fig. 4(b). A strong descending flow is found in the central region of the photograph. This downflow is seen to persist up to the level of the bottom row of components. The cold fluid below causes a horizontal spreading of this sinking layer as it becomes neutrally buoyant. This layer moves

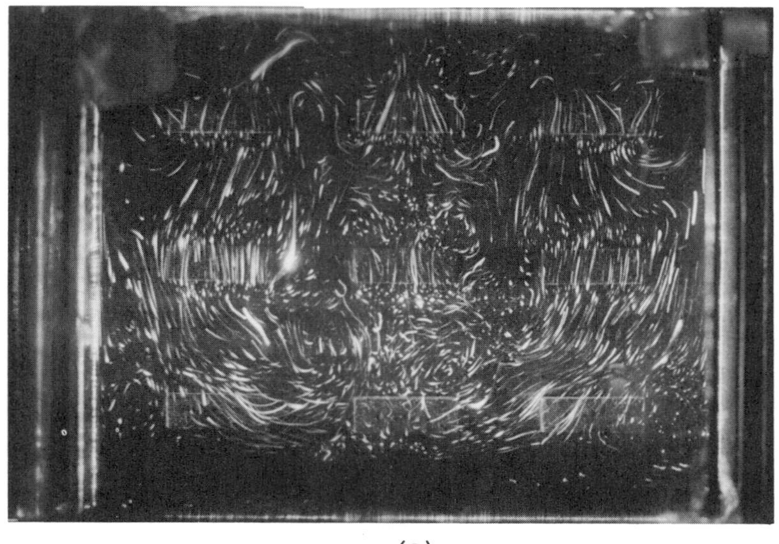

(a)

FIGURE 4. Flow visualization for component dissipation level of 0.1 W in planes parallel to component faces. The exposure times are 10 seconds; (a) Patterns in plane 1, (b) Patterns in plane 2 and (c), Patterns in plane 3.

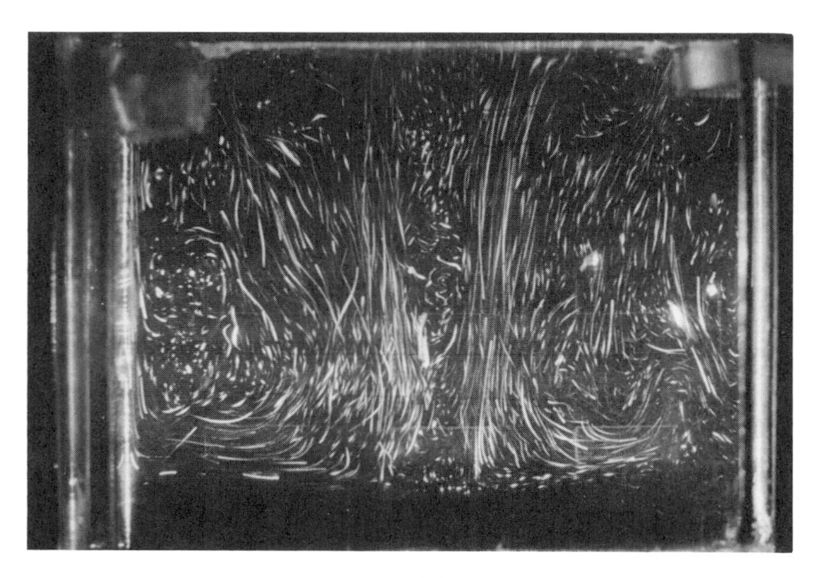

(b)

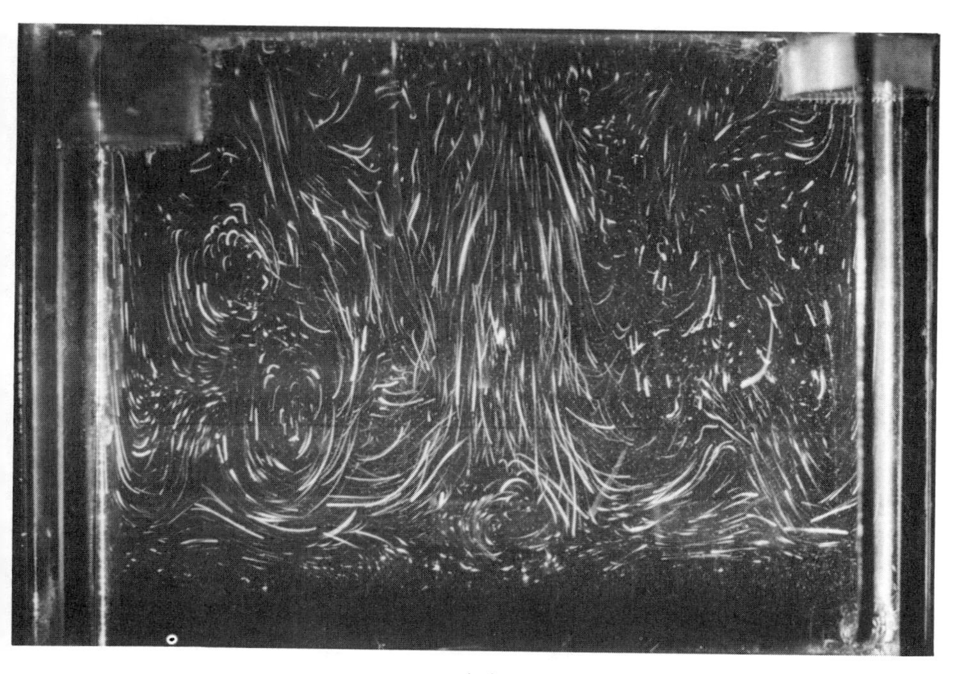

(c)

towards the lateral edges of the chamber and eventually moves out of the plane.

Further out from the component faces in Fig. 4(c), flow features quite different from Fig. 4(b) are found. A strong upflow was found along the central component column. This occurs due to the upward buoyancy force caused by the warmer walls of the chamber. It is noted that the component surface temperatures for this power input were significantly below the ambient temperature which was always at 23 ± 0.5 °C.

At a higher power level of 0.3 W, the flow patterns in planes 1 and 2 are seen in Fig. 5. Adjacent to the component front faces in (a), a strong upward flow is now seen. In plane 2, strongly time dependent patterns were observed. These varied between distinct downward layers adjacent to each column of components to highly three-dimensional patterns. One such flow visualization is seen in Fig. 5(b). The flow in plane 3, Fig. 5(c), is upward everywhere. This again is a result of the warmer enclosure surface.

With further increase in power dissipation, the flow in plane 1 continued to exhibit strong upflow near the components. Further away from the protrusion front faces in plane 2, increasing three dimensional nature of the flow was evident from the very short particle traces. In plane 3, a generally downward motion was found as the fluid temperatures increased sufficiently and the vertical walls were no longer warmer. Visualizations in the three planes at the highest

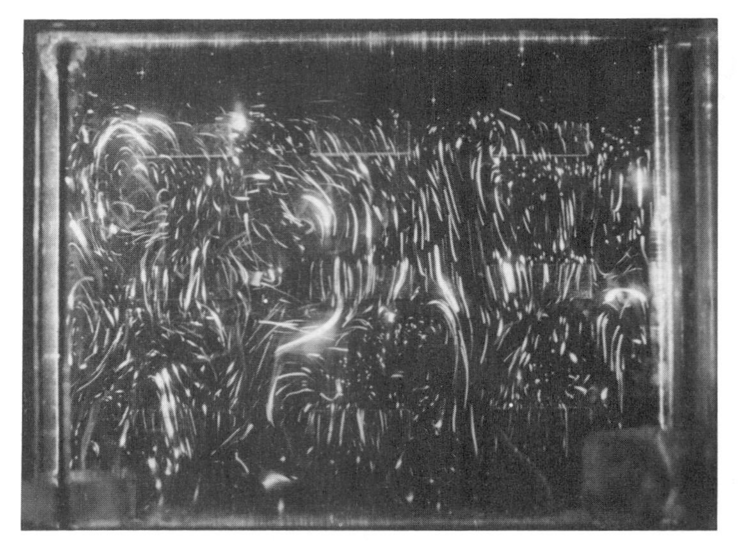

(a)

FIGURE 5. Flow visualization for component dissipation level of 0.3 W in planes parallel to component faces. The exposure times are 10 seconds; (a) Patterns in plane 1, (b) Patterns in plane 2 and (c), Patterns in plane 3.

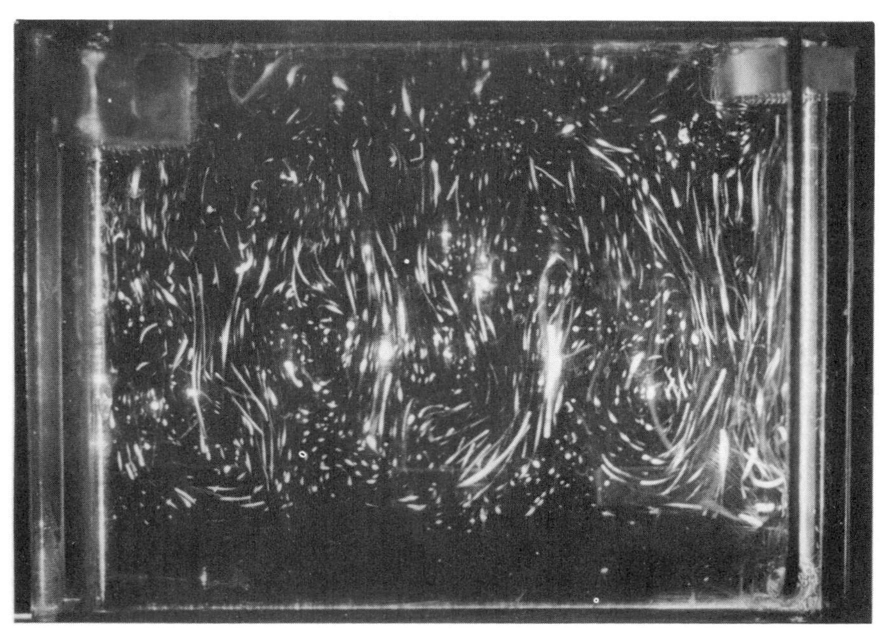

(b)

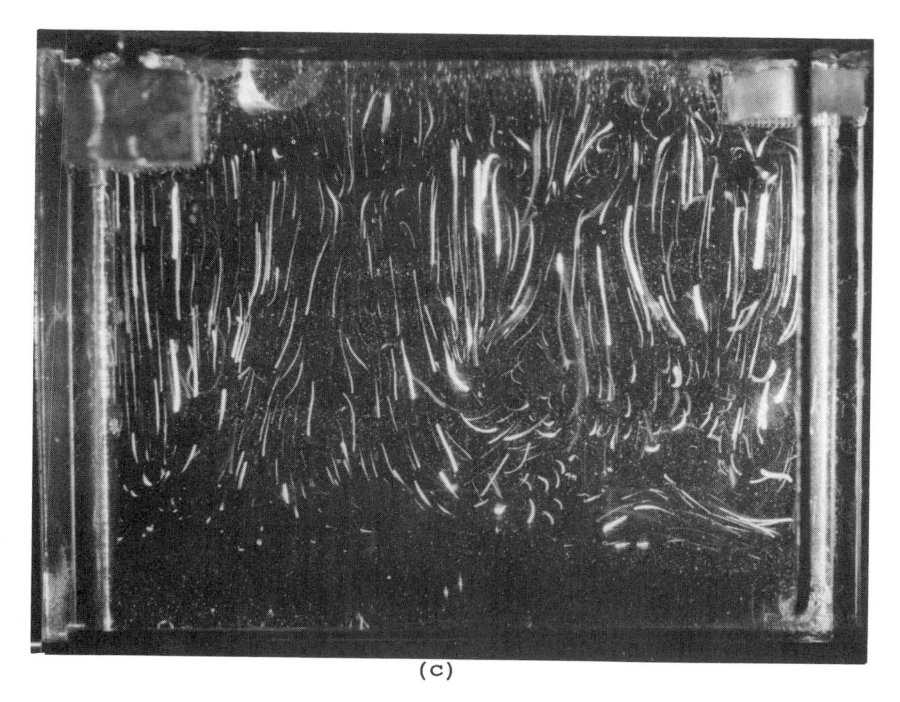

(c)

power level of 3.1 W, provided in Figs. 6 (a)-(c), confirm these trends.

3.2 Flows in Planes Containing Single Columns of Components: Each of the three selected vertical planes passes through the center of components along a single column. The resulting visualizations are seen in Figs. 7-9 for representative component power levels. Each figure documents the changes in flow structure as the power dissipation is increased. The planes are identified as in Fig. 2(b) in the following discussions.

Fig. 7 shows the flow patterns observed in plane 4. The heated protrusions are clearly visible on the left wall of the enclosure. In (a), at the lowest power level of 0.1 W per component, the fluid follows the contours of the protrusions. In the unheated regions between components, an upward flowing layer is maintained in this plane. The warm fluid transfers heat to the top surface and descends along approximately half the chamber width. This strong downflow along the central column was also seen in Fig. 4(b). The upflow found in Fig. 4(c) is confined very close to the right vertical wall. Two distinct recirculation zones of approximately same size are also seen to form in the region between the components in Fig 7(a).

Increase in the power level to 0.3 W in (b) results in a thinner buoyant layer adjacent to the components. The downward flow along the opposite sidewall is also confined to a narrower region than in (a). In most of the central region of the enclosure, the flow exhibits significant three-dimensional trends. This is also confirmed in Fig. 5(b) in

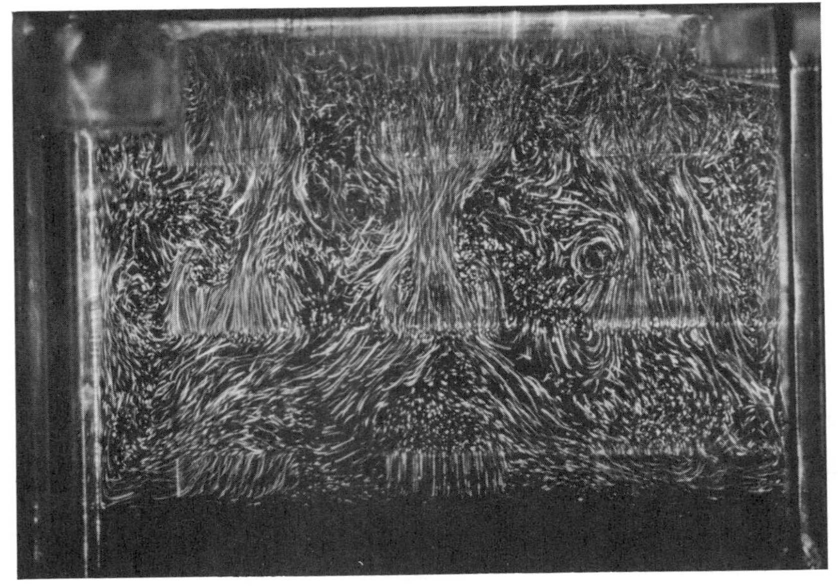

(a)

FIGURE 6. Flow visualization for component dissipation level of 3.1 W in planes parallel to component faces. The exposure times are 8 seconds; (a) Patterns in plane 1, (b) Patterns in plane 2 and (c), Patterns in plane 3.

(b)

(c)

plane 2. Two re-circulation regions are still observed, as for 0.1 W, with the upper one occupying almost half of the chamber. At a higher power level of 0.7 W in (c), these recirculation regions and the downward flow are not distinct. The flow is strongly three-dimensional and time-dependent. At the highest input power of 3.1 W, (d), the buoyant upflow adjacent to the components persists. The transport appeared to be turbulent at this dissipation level.

Flow visualizations in planes 5 and 6 are seen in Figs. 8 (a) and (b) for a power level of 0.3 W. In (a), the upflow adjacent to the components is also similar to that observed in plane 4 in Fig. 7(b). Also, a cell is seen to form between the bottom two components, approximately at the same height as in Fig. 7(b). The flow was time varying and three-dimensional away from the close proximity of the components. The flow patterns in Fig. 8(b) show variations from Fig. 8(a), despite their symmetrical location with respect to plane 4. This could be a result of the minor differences in the individual component power levels, which were inevitably present. With an increase in component power dissipation rates, the flow visualizations in these planes exhibited no regular trends aside from the strong up-flow near the components.

In summary, Figs. 4-8 show the complex flow patterns over a wide range of component heating. At the lowest level of 0.1 W, the flow pattern is largely determined by the thermal conditions at the enclosure surfaces, with only a weak influence of the element heating. At higher heating rates,

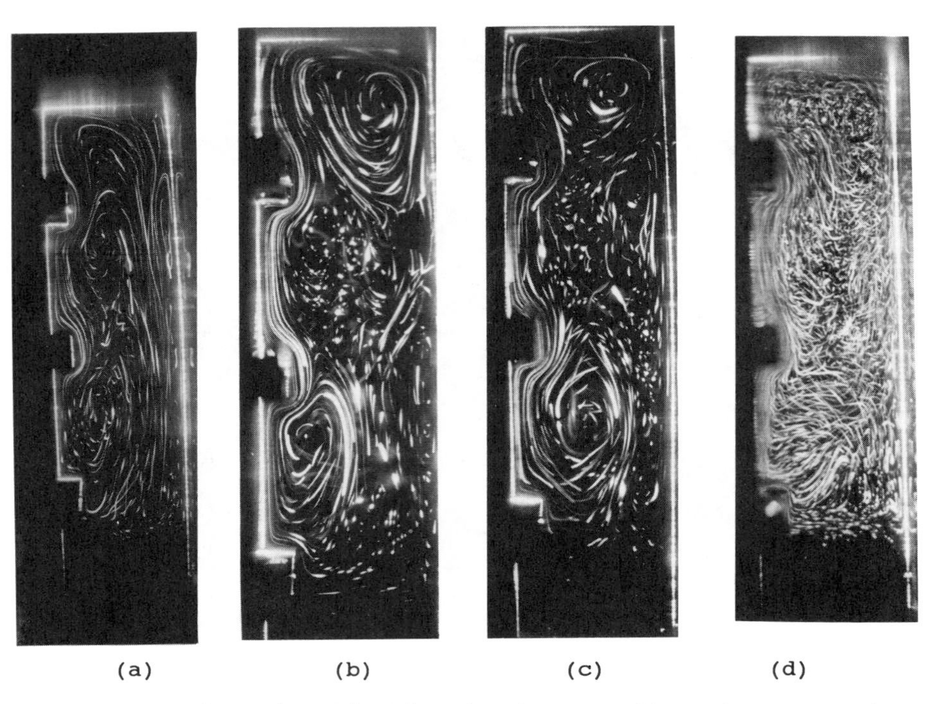

FIGURE 7. Flow visualization in plane 4, through components in the central column at various component power levels, (a) 0.1 W, (b) 0.3 W, (c) 0.7 W and (d) 3.1 W. The exposure times are 10 seconds for (a) - (c) and 8 seconds for (d).

distinct regions of upflow adjacent to the components develop. Flow away from these regions exhibited three-dimensional, time varying patterns. Detailed temperature measurements on component surfaces for these dissipation levels support the trends found in the flow visualizations. They also allow computations of the heat transfer performance of such systems. These are discussed next.

4. MEASUREMENTS OF COMPONENT SURFACE TEMPERATURES:

Surface temperatures at all component face centers were measured with embedded thermocouples. At selected locations, the temperatures were monitored over extended time periods, to obtain the timewise fluctuations from steady mean levels. As discussed in the following, these confirm the time dependent nature of the transport, also seen in the visualizations. Surface temperature data are also presented in a non-dimensional form as heat transfer correlations for this configuration.

4.1 Longtime Surface Temperature Fluctuations: At three locations, temperature variations with time were recorded

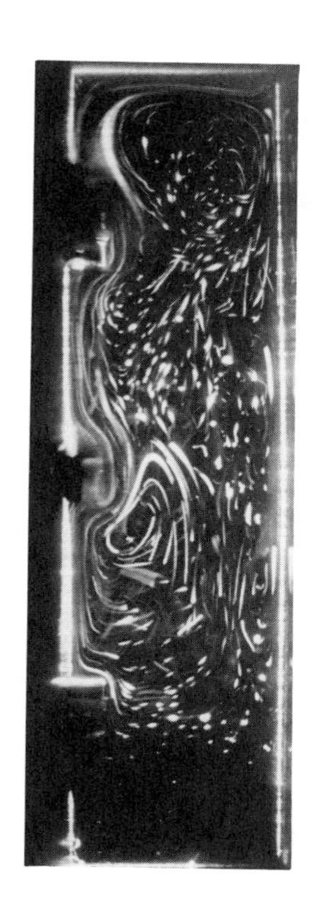

(a) (b)

FIGURE 8. Flow visualization in vertical planes through off center component columns, (a) plane 5 and (b) plane 6, at power level of 0.3 W. The exposure times are 10 seconds.

after nominally steady conditions had been established, for several heating levels. These are seen in Figs. 9-11. The front face temperatures on a component in the lowest row, Fig. 9, show almost no changes with time, except at 3.1 W. This is also supported by the visualizations in Section 3, where the fluid below the bottom row is almost stagnant. At the highest power, some intrusion of the fluid from the top results in the variations seen in Fig. 9.

Temperatures on the top surface of a component in the highest row, Fig. 10, show fluctuation amplitudes increasing with input power. At the highest heating condition of 3.1 W, these are approximately 0.4°C. Yet, larger temperature disturbances are measured on the front face of a component in the top row and off-center location in Fig. 11. As in Fig. 10, these increase with heating levels, attaining an amplitude of approximately 0.8°C at 3.1 W.

These timewise fluctuations provide qualitative details of transport changes at selected locations. Even for the

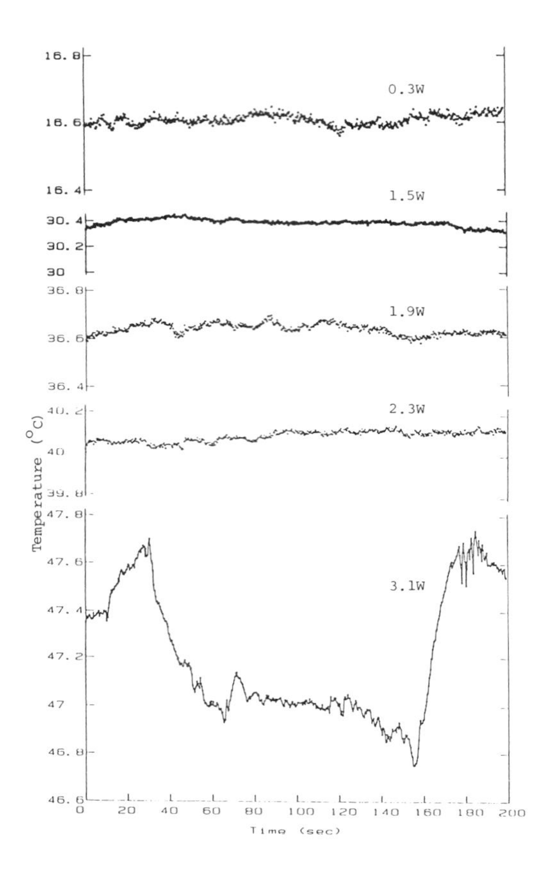

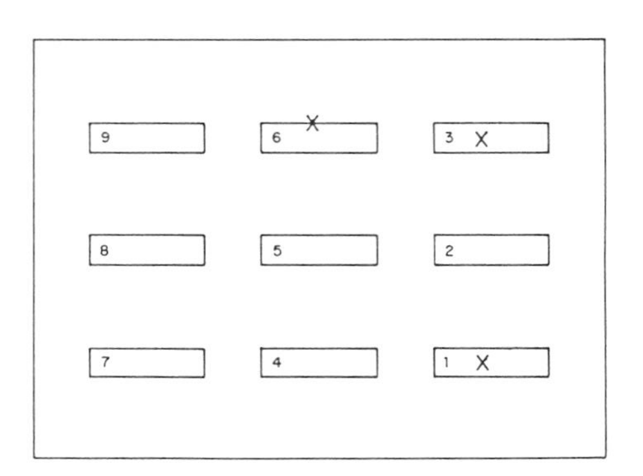

FIGURE 9. Timewise surface temperature variation measured on component 1 in the bottom row with increasing heating. The thermocouple was embedded within the center face as shown.

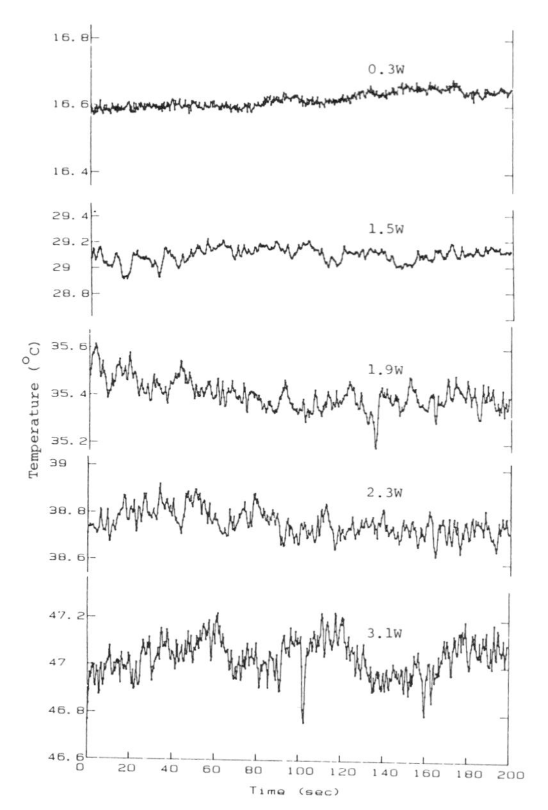

FIGURE 10. Timewise surface temperature variation measured on component 6, with increasing heating at the location indicated in Fig. 9.

largest power level, the amplitude of temperature variations in Fig. 11 is within 2% of the difference between the mean face temperature and temperature of the heat exchanger. Since these measurements are responses of embedded thermocouples, the disturbance levels in the adjacent fluid are expected to be larger. However, for the heat transfer calculations, a "snapshot" scan of the entire thermocouple array following the achievement of nominally steady conditions was considered appropriate.

4.2 **Heat Transfer Characteristics:** Surface temperatures at the various faces were found to be within 0.5°C of each other for the smallest component power input of 0.1 W. For larger dissipation levels, the differences were larger, being as much as 5°C for 3.1 W. The highest temperature was always on the front faces, except for the top component in the central column, for which the top surface was the warmest. The location of the coolest surface varied from component to component.

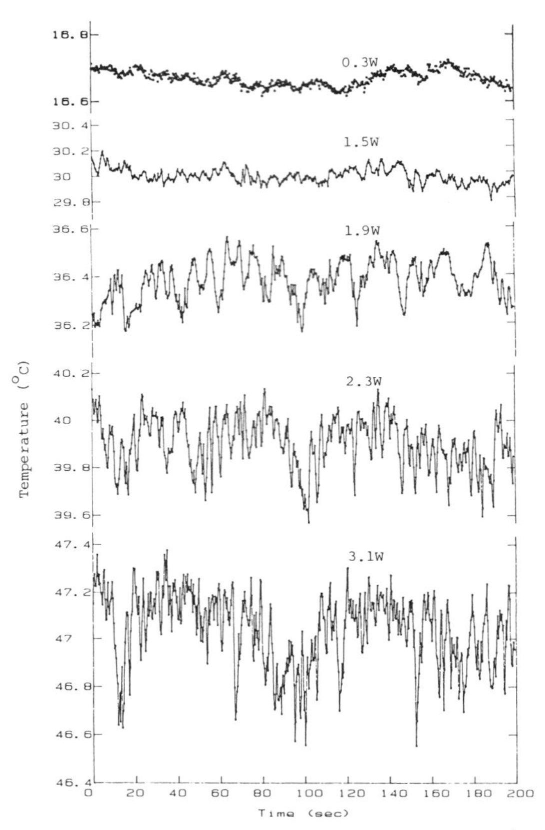

FIGURE 11. Timewise surface temperature variation measured on component 3, with increasing heating at the location indicated in Fig. 9.

The face temperatures for each heating level allowed the computation of non-dimensional thermal transport rates for the configuration studied. The energy lost by conduction through the back of the plexiglass card was estimated by measuring the temperature difference across the card thickness at several component locations. The loss was subtracted from the component input power to arrive at the net energy rate transferred to the fluid per element, Q.

The heat transfer coefficient was next defined as

$$h = Q \;/\; A \;(T_{avg} - T_c) \qquad (1)$$

where A is the total wetted surface area of the component and T_c is the temperature of the heat exchangers. Also, T_{avg} is the surface area averaged temperature for a given component

$$T_{avg} = \sum_i A_i \, T_i / \sum_i A_i \qquad (2)$$

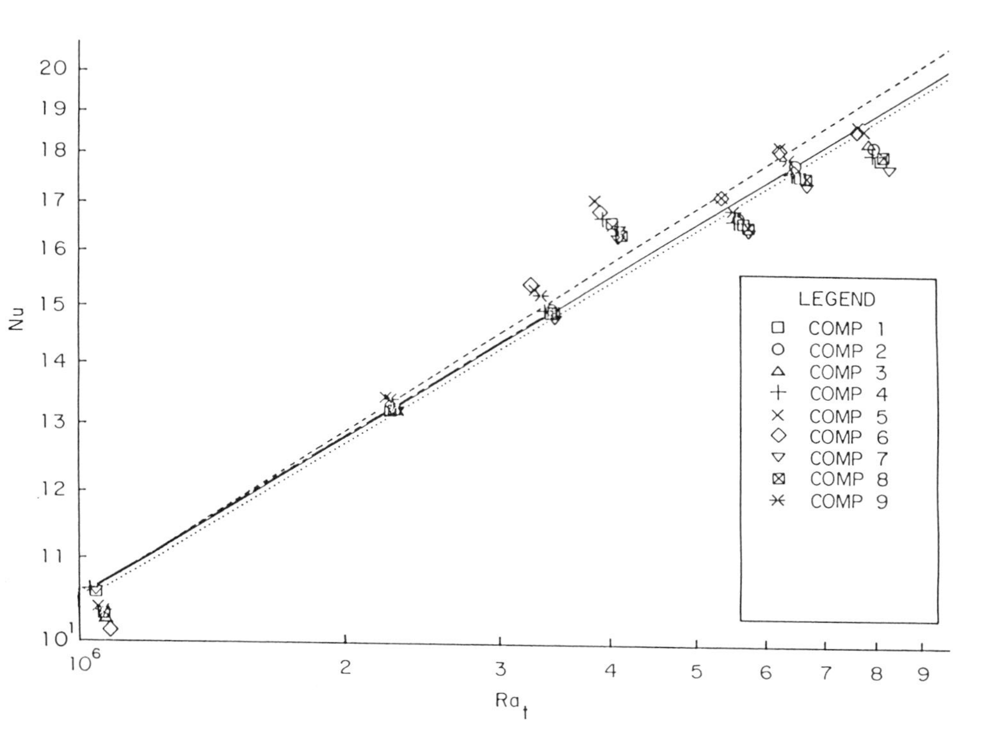

FIGURE 12. Non-dimensional surface temperature data for all components seen as a variation of Nusselt number with a Rayleigh number based on area averaged component temperature. The various best fit lines correspond to; ----data for central column only, — – — data for left column only, ... data for right column only and, ——— entire data set.

In eq. 2, the summation is over the five component faces exposed to the fluid.

In order to obtain a non-dimensional representation of the heat transfer performance, the Nusselt number was defined as

$$Nu = h\ L \ /\ k \tag{3}$$

where L was taken as the vertical dimension of each component. Rayleigh numbers based on component average surface temperatures were calculated as

$$Ra_t = g\ \beta\ L^3\ (T_{avg} - T_c)\ /\ (\nu\ \alpha) \tag{4}$$

A heat flux based Rayleigh number was also determined as

$$Ra_f = g\ \beta\ L^4\ Q\ /\ (\nu\ k\ \alpha\ A) \tag{5}$$

In eqs. (3)-(5) all fluid properties were evaluated at a film temperature defined as the average of T_{avg} and T_c.

Computed Nu levels are seen in Fig. 12 as a function of Ra_t for the nine components at the various dissipation levels. The Nusselt numbers increase with Ra_t in almost a linear trend. The scatter in data is partially accounted for by the columnwise variations in the heat transfer rates. The three dotted lines are the best fits for the data for each column. It is seen that the Nusselt numbers for the central column of components are slightly higher than the two other columns. This is expected in light of the more vigorous fluid motion observed in that region in the flow visualizations. The solid line in Fig. 12 is the best fit for the entire entire data set given as

$$Nu = 0.201\ Ra_t^{\ 0.287} \qquad (6)$$

In the range of $10^6 < Ra_t < 10^7$ and $14 < Pr < 31$, the deviation of any measurement from eq. 5 is within 10%. An alternative presentation of the above data is made in terms of the flux-based Rayleigh number, Ra_f, in Fig. 13. The data again show a linear trend as in Fig. 12, with a best fit described by

$$Nu = 0.279\ Ra_f^{\ 0.224} \qquad (7)$$

in the range $10^7 < Ra_f < 2*10^8$ and $14 < Pr < 31$. The correlation in eq. 7 allows a direct computation of the average component temperature from the power dissipation and may often be more convenient to use than eq. 6 which contains a temperature dependence on both sides.

5. CONCLUSIONS

The flow visualizations and component surface temperature measurements presented here provide considerable insight into the complex, three-dimensional, transient transport in this configuration. The smallest heating level of 0.1 W produced only a weak change in the flow pattern determined by the thermal conditions at the enclosure surfaces. At higher power levels, an upward fluid motion developed adjacent to each column of components. The flow away from the elements became strongly three-dimensional and time-dependent with increasing thermal inputs. These trends were also confirmed by time records of surface embedded thermocouples. Measured component surface temperatures were used to arrive at non-dimensional heat transfer correlations. These measurements provide both a predictive capability for heat transfer, as well as, a data-base for comparisons with future numerical computations.

Acknowledgements: The authors acknowledge support of this work by the Naval Postgraduate School Research Council (Y.J.) and the Naval Weapons Support Center (M.D.K.). The expert efforts of Ms. Pamela Ellis in preparing the final manuscript are appreciated.

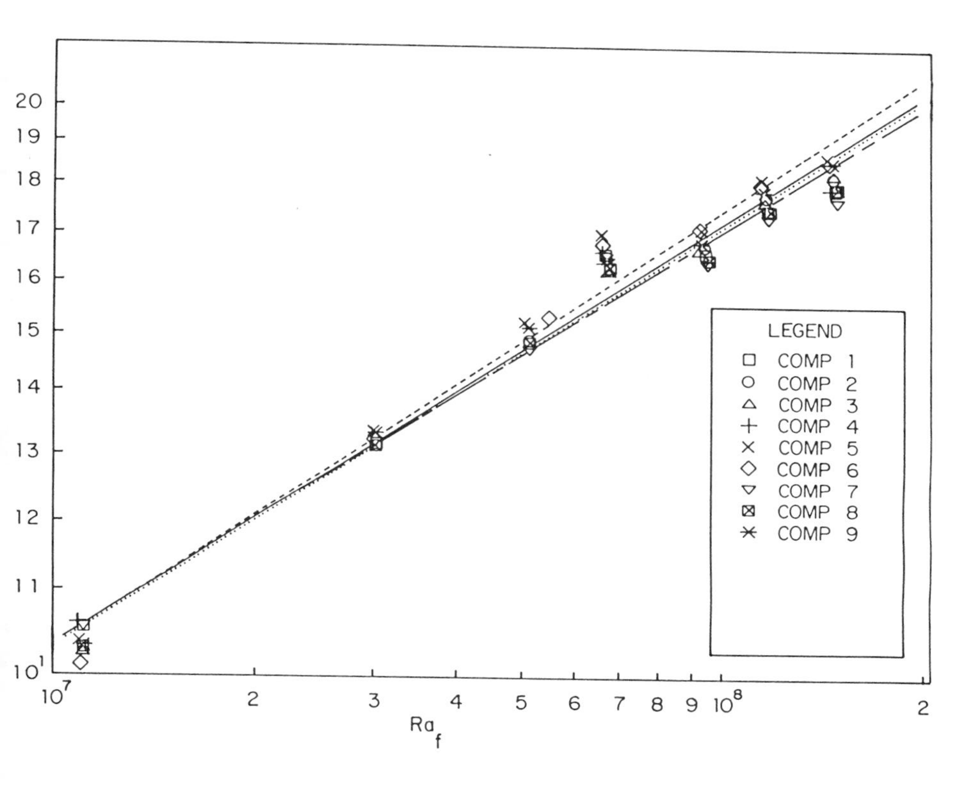

FIGURE 13. Non-dimensional surface temperature data for all components seen as a variation of Nusselt number with a Rayleigh number based on a heat flux. The various best fit lines correspond to; ----- data for central column only, — – — data for left column only, ... data for right column only and —— entire data set.

NOMENCLATURE

A	Total wetted surface area of each component	m^2
α	Thermal diffusivity	m^2/s
β	Thermal expansion coeficient	1/K
c_p	Specific heat	J/kg-K
g	Acceleration due to gravity	m/s^2
h	Heat transfer coefficient	W/m^2-K
k	Thermal conductivity	W/m-K
L	Component height	m
Nu	Nusselt Number	Dimensionless
ν	Kinematic viscosity	m^2/s
Pr	Prandtl number	Dimensionless
Q	Energy added to the fluid per element	W
Ra_t	Temperature based Rayleigh number	Dimensionless
Ra_f	Flux based Rayleigh number	Dimensionless
T_{avg}	Area averaged block temperature	°C
T_c	Average Heat exchanger temperature	°C

REFERENCES:

1. Oktay, S., 1986, "Electronic Package Cooling," Paper Presented at the Second Annual International Electronics Packaging Conference, San Diego, CA.

2. Kraus, A.D. and Bar-Cohen, A., 1983, "Thermal Analysis and Control of Electronic Equipment," Hemisphere Publishing Corporation.

3. Chu, R.C., 1986, "Heat Transfer in Electronic Systems," Proceedings of the Eighth International Heat Transfer Conference, San Francisco, CA, pp. 293-305.

4. Jaluria, Y., 1982, "Buoyancy Induced Flow Due to Isolated Thermal Sources on a Vertical Surface," J. Heat Transfer, vol. 104, pp. 223-227.

5. Jaluria, Y., 1984, "Interaction of Natural Convection Wakes Arising From Thermal Sources on a Vertical Surface," J. Heat Transfer,vol. 107, pp. 883-892.

6. Ortega, A. and Moffat, R.J., 1985, "Heat Transfer From an Array of Simulated Electronic Components: Experimental Results For Free Convection With and Without a Shrouding Wall," ASME HTD-Vol.48, pp. 5-15.

7. Ortega, A. and Moffat, R.J., 1986, "Buoyancy Induced Convection in a Non Uniformly Heated Array of Cubical Elements on a Vertical Channel Wall," ASME HTD-Vol.57, pp. 123-134.

8. Moffat, R.J. and Ortega, A., 1986, "Buoyancy Induced Forced Convection," ASME HTD-Vol.57, pp. 135-144.

9. Kuhn, O., and Oosthuizen, P.H., 1986, "Three-Dimensional Transient Natural Convective Flow in a Rectangular Enclosure with Localized Heating," ASME Publication HTD-Vol 63, pp 55-62.

10. Oosthuizen, P.H. and Paul, J.T., 1987, "Natural Convective Heat Transfer From a Square Element Mounted on the Wall of an Inclined Enclosure," Paper Presented at the AIAA 22nd Thermo-physics Conference, Honolulu, Hawaii.

11. Baker, E., 1972, "Liquid Cooling of Microelectronic Devices by Free and Forced Convection," Microelectronics and Reliability, vol. 11, pp. 213-222.

12. Baker, E., 1973, "Liquid Immersion Cooling of Small Electronic Devices," Microelectronics and Reliability, vol. 12, pp. 163-173.

13. Park, K.A. and Bergles, A.E., 1987, "Natural Convection Heat Transfer Characteristics of Simulated Microelectronic Chips," ASME J. Heat Transfer, vol. 109,

pp. 90-96.

14. Kelleher, M.D. and Knock, R., 1987, "Laminar Natural Convection in a Rectangular Enclosure Due to a Heated Protrusion on One Vertical Wall-Part I: Experimental Investigation," Proc. 2nd ASME/JSME Thermal Engineering Joint Conference, Honolulu, Hawaii, pp. 169-177.

15. Lee, J.J., Liu, K.V., Yang, K.-T. and Kelleher, M.D., 1987, "Laminar Natural Convection in a Rectangular Enclosure Due to a Heated Protrusion on One Vertical Wall-Part II:Numerical Simulations," Proc. 2nd ASME/JSME Thermal Engineering Joint Conference, Honolulu, Hawaii,pp. 179-185.

16. Liu, K.V., Yang, K.-T and Kelleher, M.D., 1987, "Three Dimensional Natural Convection Cooling of an Array of Heated Protrusions in an Enclosure Filled with a Dielectric Fluid," Proc. Int. Symposium on Cooling Technology for Electronic Equipment, Honolulu, Hawaii, pp. 486-497.

Mechanism of Boiling Hysteresis of Enhanced Surfaces

YI-DING CAO and MING-DAO XIN
Institute of Engineering Thermophysics
Chongqing University
Chongqing, PRC

ABSTRACT

Boiling hysteresis or temperature overshoot of enhanced surfaces was studied. Flooding criteria derived show that most commercial enhanced surfaces are subject to flood at moderate subcooling with fluorocarbon liquids. The predicted results agree with experimental data from the literature. Based on the nucleation criterion theory, it is concluded that boiling incipience within liquid-filled cavities is determined by the liquid temperature fields inside the cavities. With appropriate structures, enhanced surfaces can still eliminate significant temperature overshoot.

INTRODUCTION

Enhanced surfaces for use in boiling heat transfer have received an ever growing interest in the augmentation of heat transfer and immersion cooling of electronic elements, such as microelectronic chips and thyristors. Although enhanced surfaces offer high heat transfer rates, their application, especially to cooling of electronic systems with fluorocarbon liquid coolants, has not been without problems. One major problem which has been encountered is that of temperature overshoot or boiling hysteresis, which was also described as departure from nature convection (DNC) by Oktay [1], as shown in Fig. 1. Since semiconductor devices require operation over a fairly narrow temperature range, when nucleation of bubbles fails to commence within a sufficiently short time after the application of power, or nucleation does not commence at all, the temperature may increase to an intolerable level.
Bergles and Chyu [2] and Marto and Lepere [3] studied boiling hysteresis of three commercial enhanced surfaces (High Flux, Thermoexcel-E and Gewa-T) and they found that the temperature overshoot and the resulting nucleate boiling hysteresis pattern is severe with R-113 and FC-72. They concluded that this is due to flooding of the porous matrix or surface cavities with liquid so that only relatively small sites are available for nucleation.

On the other hand, Oktay [1] found that an irregular "dentritic" structure of nickel plated directly on the chips reduced the wall superheat in established boiling and almost eliminated the temperature overshoot with FC-88; Nakayama et al. [4] reported large enhancements with complex structured studs in FC-72. The temperature overshoot was reported to be small. But they did not give detailed explanation. It can be anticipated that this will be an area for continued investigation [5].

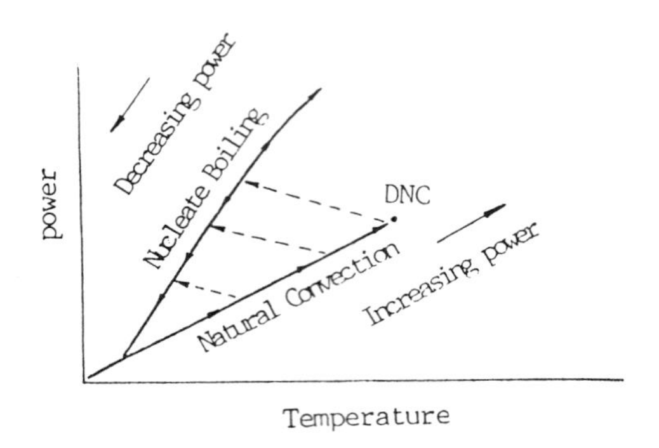

FIGURE 1. Departure from natural convection.

The purpose of this paper is to look deeper into the mechanism of boiling hysteresis, give some reasonable explanation of the phenomenon for various enhanced surfaces and shed some light on avoiding boiling hysteresis.

FLOODING CRITERIA OF RE-ENTRANT CAVITIES

Figure 2 shows different types of porous surface structures. They are characterized by numerous re-entrant cavities which have the ability to entrap vapor, thereby becoming active nucleation sites. The surfaces may trap air when they are immersed in liquids. However, after several hours of operation, they are gradually degassed. From the view of engineering, it is the retention of vapor, not gas, that is important to the incipience of boiling. The question is whether the cavities can still retain vapor after operation is over. This depends on cavity, geometry, liquid subcooling, properties of liquids used, etc. If cavities have already entrapped vapor before restarting of operation, the temperature overshoot or boiling hysteresis might be avoided. Unfortunately, most enhanced surfaces are subject to such boiling hysteresis.

In order to derive flooding criteria for re-entrant cavities,

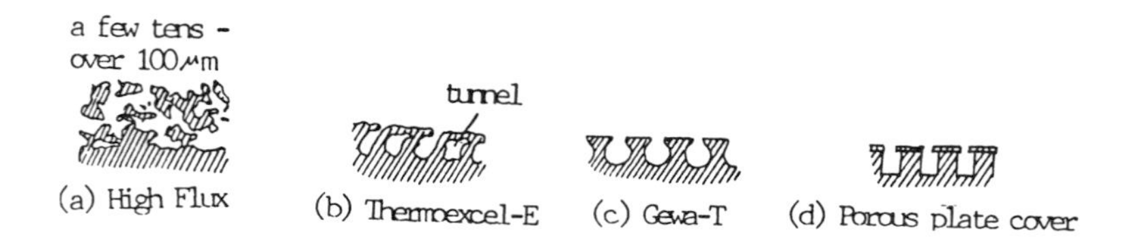

FIGURE 2. Examples of porous surface structures.

let's study a typical re-entrant cavity, as shown in Fig. 3. After operation is over, there is no heat source from which heat transfers through the surface into the fluid. Since the system usually dissipates heat into ambient air and system temperature decreases slowly, it is reasonable to assume a uniform temperature and an equilibrium cooling process for the system. The curvature radius of the liquid-vapor interface can be positive or negative depending on whether the system is superheated or subcooled. When the system is subcooled, the vapor is condensed and the liquid-vapor interface recedes into the cavity with a concave curvature. Under equilibrium condition, the liquid-vapor interface may stay within the neck (Fig. 3 a) or within reservoir (Fig. 3 b), depending on the specific cavity shape and the contact angle(θ).

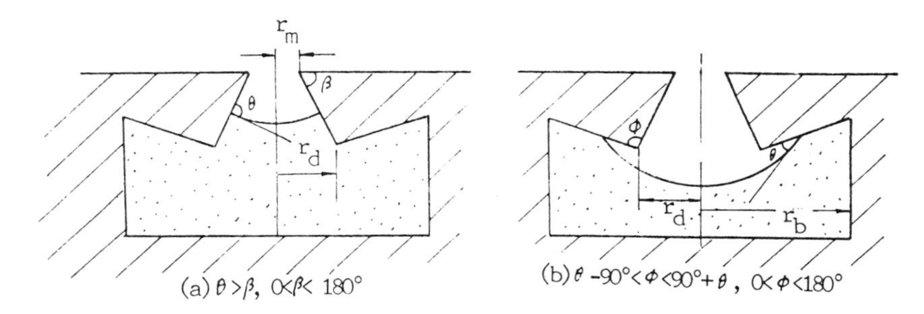

FIGURE 3. Curvature of a liquid-vapor interface with $\Delta T<0$.

With subcooling $\Delta T_{sb} = T_s - T_1$ replacing superheat ΔT, the radius R of the liquid-vapor interface is related to the subcooling of the liquid such as that given by Konev and Mitrovic [6].

$$R = \frac{2T_s \sigma}{\Delta h (T_s - T_1) \rho_v} \left(1 + \frac{T_s - T_1}{T_s} \omega \right) \tag{1}$$

and

$$\omega = \left(\frac{\Delta h \rho_1}{R^* T_s \Delta \rho} - 1 \right) \frac{\rho_v}{\Delta \rho}$$

Where T_s is the saturation temperature corresponding to the pressure of liquid, Δh is the enthalpy of evaporation, ρ_v and ρ_1 are the densities of vapor and liquid, respectively, $\Delta\rho = \rho_1 - \rho_v$ and R^* is the gas constant. Since ω is usually small, when $T_s - T_1 << T_s$, Eq. (1) reduces to $R = 2T_s\sigma/\Delta h \rho_v (T_s - T_1)$. For the liquid-vapor interface within the neck, $r_m/\sin(\theta-\beta) < R < r_d/\sin(\theta-\beta)$. The condition for a concave curvature is $\sin(\theta-\beta) > 0$, i.e. , $\theta > \beta$. For the liquid-vapor interface within the reservoir, $r_d/\cos(\phi-\theta) < R < r_b/\cos(\phi-\theta)$. The condition for a concave curvature is $\cos(\phi-\theta) > 0$, i.e. , $\theta - 90° < \phi < 90° + \theta$.

When the liquid-vapor interface radius is larger than that calculated from Eq.(1), the vapor will be condensed, and the interface will further recede into the cavity until the point where its radius is equal to or less than R in Eq.(1). When such radius do not exist, the cavity will be entirely flooded with liquid. Therefore, under certain conditions, the smallest radius the liquid-vapor interface could reach is crucial to the existence of vapor in the cavity. Within the neck of the cavity, the smallest possible radius is $r_m/\sin(\theta-\beta)$, while within the reservoir, the radius is $r_d/\cos(\Phi-\theta)$. Fig. 4 compares $\cos(\Phi-\theta)$ with $\sin(\theta-\beta)$ for different θ, Φ and β. It shows that for small θ, $\cos(\Phi-\theta)$ is much larger than $\sin(\theta-\beta)$. On the other hand, it is difficult for the cavity geometries of enhanced surfaces to meet the condition $\theta > \beta$ when θ is very small. Therefore, we choose $R = r_d/\cos(\Phi-\theta)$ as the flooding criterion for fluids with small θ, such as fluorocarbon liquid coolants.

Substituting $R = r_d/\cos(\Phi-\theta)$ into Eq.(1), we get

$$r_d = \frac{2\sigma T_s \cos(\Phi-\theta)}{\Delta h (T_s - T_l)\rho_v} \left(1 + \frac{T_s - T_l}{T_s}\,\omega\right) \tag{2}$$

for $\theta - 90° < \Theta < 90° + \theta$. When $(T_s - T_l)/T_s \ll 1$, Eq.(2) reduced to

$$r_d = 2\sigma T_s \cos(\Phi-\theta)/\Delta h (T_s - T_l)\rho_v \tag{2a}$$

For a two-dimension re-entrant cavity, the corresponding criterion is

$$r_{d2} = \frac{\sigma T_s \cos(\Phi-\theta)}{\Delta h (T_s - T_l)\rho_v} \left(1 + \frac{T_s - T_l}{T_s}\,\omega\right) \tag{3}$$

It is worth noticing that internal structures of re-entrant cavities are very important. Even though the re-entrant cavities with sufficiently small mouth radius may still be flooded when their internal structures do not satisfy the condition $\theta - 90° < \Phi < 90° + \theta$. Another aspect worth pointing out is that when θ is large, $\sin(\theta-\beta)$ may exceed $\cos(\Phi-\theta)$ (see Fig. 4.) and $r_d/\cos(\Phi-\theta)$ may be less than $r_m/\sin(\theta-\beta)$. Therefore re-entrant cavities may lose their superiorty of entrapping vapor. Even cavities, such as conical cavities and cylindrical cavities may retain vapor under larger subcooling than re-entrant cavities.

Fig. 5 and Fig. 6 represent the trapping conditions versus subcooling for R-113 and water, with the contact angles assumed to be 10° and 50° respectively. The hatched regions indicate the sizes of the potential vapor-trapping cavities.

Table 1 compares flooding criteria with experimental data of Marto and Lepere [3] and Bergles and Chyu [2]. The temperature overshoot(TOS) is defined as the temperature difference between the incipient boiling temperature and the corresponding temperature at the decreasing heat flux curve with the same heat flux. The floodong criteria confirm the experimenting conclusion that most surface cavities were flooded, leading to large temperature overshoot. Fluorocarbon liquids, such as R-113 are particularly prone to this behavior. Even with a slight subcooling of 0.5°C, most cavities of the High Flux tube were flooded and exhibited a large temperature overshoot. On the other hand, the temperature overshoot problem is much less severe when boiling in water. With a moderate subcooling of 1.5°C, most cavities of the High Flux tube could still trap vapor as demonstrated a very small temperature overshoot.

— cos (Φ - β)
---sin (θ - β)
θ =50°
θ =90°
θ =150°
θ =20°
θ =90°
θ =50°
θ =20°
θ =150°
1
0.5
0
20°
50°
90°
140°
Φ ; β

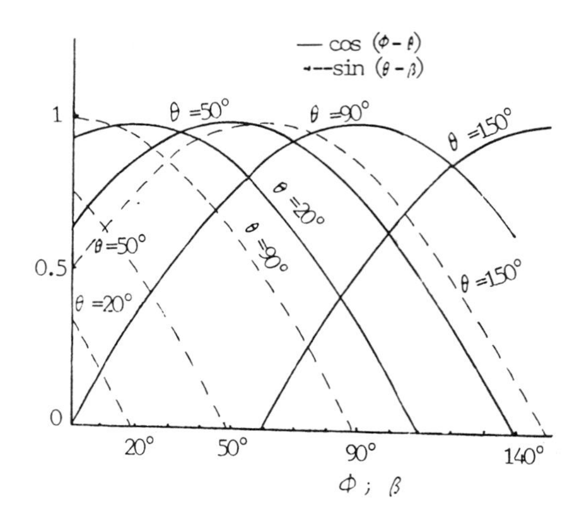

FIGURE 4. Comparision between $\cos(\phi-\beta)$ and $\sin(\theta-\beta)$ for different θ, ϕ, and β.

R-113
θ=10°
p=0.1MPa
Φ =70°
Φ=90°
Φ =95°
r_d
10^{-6}
m
10^{-7}
10^{-8}
0.5 1 2 5 10 20 40 80
T_s - T_l K

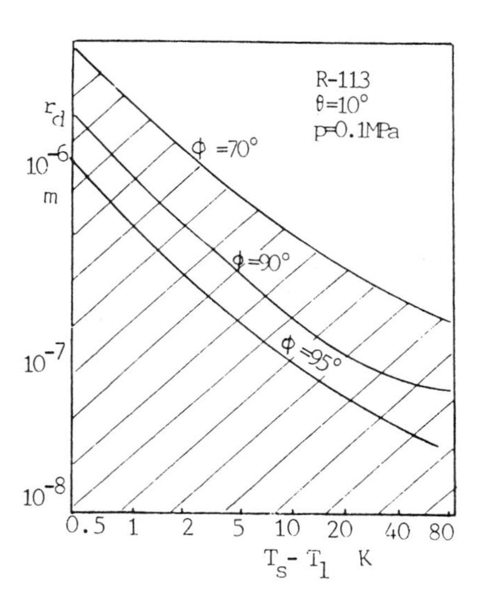

FIGURE 5. Region of potential vapor-trapping cavities versus subcooling for R-113.

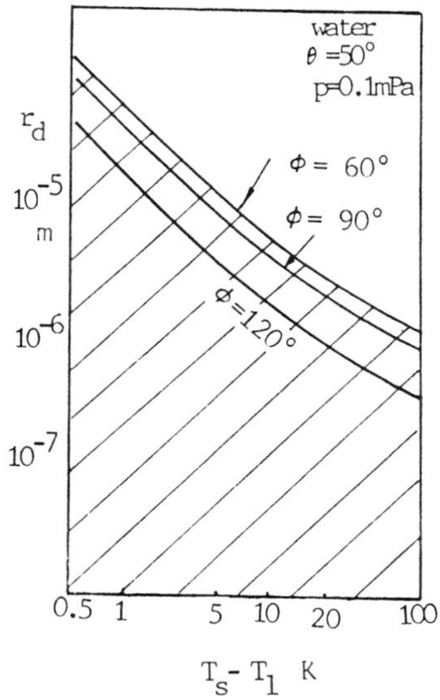

water
θ =50°
p=0.1mPa
Φ = 60°
Φ = 90°
Φ =120°
r_d
10^{-5}
m
10^{-6}
10^{-7}
0.5 1 5 10 20 100
T_s - T_l K

FIGURE 6. Region of potential vapor-trapping cavities versus subcooling for water.

Table 1 Experimental verification for flooding criteria

Surface Description (μm)	Fluid	Subcooling (°C)	TOS (°C)	Reference	Flooding Criteria (μm) (Φ= 90°)
Gewa-T, r_{d2}= 90	R-113	22.5	13.6	Marto[3]*	r_{d2}= 0.05***
Thermoexcel-E r_d= 50	R-113	22.5	12.0	" "	r_d= 0.01
High Flux, Average particle size d = 50 effective radius r_e = 7.05	R-113	22.5	8.4	" "	r_d= 0.10
idem	R-113	0.50	7.16	Bergles[2]	r_d= 3.32
idem	R-113	2.6	7.10	" "	r_d= 0.74
idem	Water	1.5	0.50	" "	r_d= 17 (θ = 50°)

Remarks: * Ambient temperature taken as 25°C .
** For R-113, θ taken as 10°.
*** r_e = 0.14d.
TOS Temperature Overshoot

BOILING INCIPIENCE WITHIN LIQUID FILLED CAVITIES

The wall inside a re-entrant cavity usually has the same surface condition as the wall outside the cavity. Since the wall surface has a large number of natural cavities with the lower limit of mouth diameters near zero, even though the re-entrant cavity is flooded, there still may have some vapor nuclei entrapped in very small cavities of the internal wall, which meet the requirements of vapor-trapping. The onset of boiling within liquid-filled cavities starts still by activating such pre-existing nuclei, not from heterogeneous nucleation [7]. The significant difference between the inside wall and the outside wall is that the liquid temperature field inside the cavity is much more uniform than that adjacent to the outside wall.

The liquid temperature distribution adjacent to the wall has significant effects on activation of cavities. Experiments conducted by Griffith and Wallis, using artificial conical cavities approximately 70 μm diameter, under condition of nonuniform superheat, commenced nucleating at a wall superheat of 11°C rather than the value of 1.6°C predicted by the well known "emergent superheat" equation, which has proved to be correct for a uniform temperature system [8].

Hsu [9] has provided a theoretical basis for the experimental observation mentioned above. The model is depicted in Fig. 7. The hypothesis presumes that the cavity meets the geometric requirements for the entrapment of vapor, and the nucleus for bubble growth is assumed to be located at the mouth of the cavity. When the liquid temperature exceeds the vapor temperature over the entire liquid-vapor interface, the nucleus begins to grow. The model uses the concept of "limiting thermal layer thickness", which is defined in a fashion similar to the fictitious but useful laminar sub-layer. The size range of active cavities is determined by setting $\tau=\infty$, i.e., $\theta_1/\theta_w = x/\delta$:

$$r_c^* = (\delta/4)\left[(1-\xi_{sat}) \pm ((1-\xi_{sat})^2 - (6.4 A'/\delta\theta_w))^{\frac{1}{2}}\right] \quad (4)$$

where $\theta_w = T_w - T_\infty$, $\theta_{sat} = T_s - T_\infty$, $A' = (2\sigma T_s/\rho_v \Delta h)$, $\xi_{sat} = \theta_{sat}/\theta_w$, r_{cmax} corresponds to the positive sign in Eq.(4) and r_{cmin} to the negative sign. For certain size range of active cavities, solving θ_w from Eq.(4), we get

$$\theta_w = 1.6 A'/r_c^* (2(1-\xi_{sat}) - 4 r_c^*/\delta) \quad (5)$$

It can be seen from Eq.(5) that θ_w has a strong dependence on δ, i.e., on the liquid temperature field adjacent to the wall. Larger δ indicating a more uniform temperature field corresponds to a smaller θ_w. Other authors such as Han and Griffith [10] and Bergles and Rohsenow [11] have slso derived the minimum boiling superheat relations based on the similar liquid temperature field concept with different boundary layer thickness expressions. The theoretical analyses above agree with experimental data well.

Within a re-entrant cavity filled with liquid, heat transfers from the internal wall into the liquid by natural convection prior to boiling incipience. Since the mouth diameter of the re-entrant cavity is small compared to the reservoir dimensions, the hotter fluid has difficulty escaping out of the cavity, while the colder bulk fluid has difficulty being sucked into the reservoir. The liquid temperature inside the cavity is apparently much higher, especially in the corners of the cavity, than that outside the cavity.

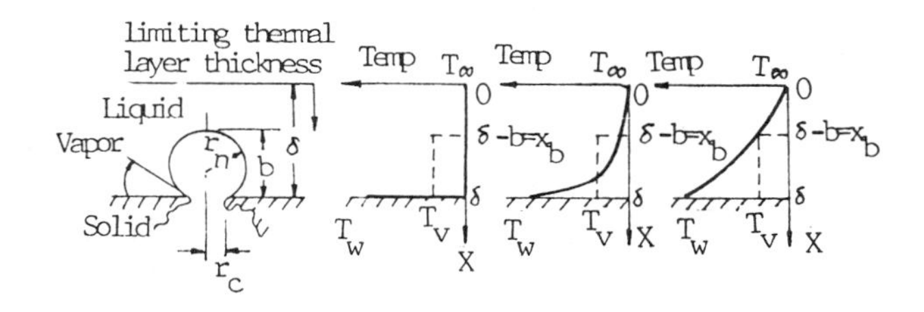

FIGURE 7. Model for nucleation hypothesis(Hsu[9]).

With mouth diameter being zero, i.e. , an enclosure with uniform wall temperature, the liquid temperature inside the cavity is uniform and equal to the wall temperature; with the mouth diameter being sufficiently large, the liquid temperature field will approach that adjacent to the surface outside the cavity. Since enhanced surfaces have numerous re-entrant cavities and the liquid temperature fields within the cavities are much more uniform, then require a much lower wall superheat for bubbles to form at the walls inside the cavities. Furthermore, because of the narrow space inside the cavities, the bubbles soon coalesce together and squeeze out of the cavity mouth, thereby insuring boiling at the lower wall superheat. This has been confirmed by many experiments. Incipient boiling superheats for High Flux, Thermoexcel-E and Gewa-T used by Marto and Lepere [3], For instance, are about 9°C, 12°C and 17°C respectively, while that of a corresponding smooth surface is about 21°C.

The structure of a cavity determines the liquid natural convection within the cavity, hence, determines the liquid temperature and the incipient boiling superheat. It is not the scope of this paper to study the natural convection within the cavity in detail. The parameter C=a/A, derived by Xin and Cao [12], is sufficient to describe qualitatively the temperature field within the cavity, assuming the same wall superheat, fluid and surface orientation. Here a is the opening area of the cavity, A is the heated surface area within the cavity. A smaller C corresponds to a more uniform liquid temperature distribution within the cavity. The validity of the parameter was also indirectly confirmed by numerical and experimental investigations of natural convection in open cavities [13].

Let's consider two typical structures, a High Flux surface and a dendritic surface [1], as shown in Fig. 8. The former exhibits a large boiling hysteresis, while the latter was reported to be almost boiling-hysteresis free even though its cavity mouth diameters are a little larger than those of the former. Because of the point contact of particles with the base wall due to sintering or brazing for the High Flux surface, heat transfers almost only from the base wall into the fluid. Moreover, the dendritic surface has a thicker coating than the High Flux surface. Therefore the heated area within the cavities of the dendritic surface is much larger than that of the High Flux surface, i.e., $C_d < C_h$. As a result, the dendritic surface may initiate boiling at a smaller superheat and even eliminate significant temperature overshoot.

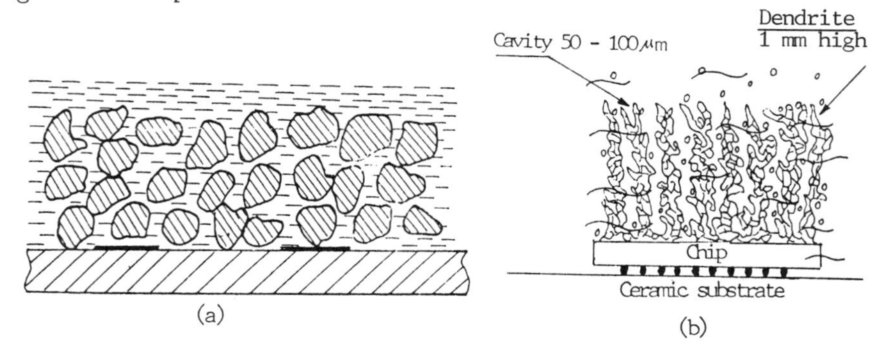

FIGURE 8. Comparision of two kinds of surface structure (a) High Flux, (b) Dendrite.

Another kind of porous surface which eliminates significant temperature overshoot might be explained based on the same concept above. Nakayama et al. [4] studied a porous stud as shown in Fig. 9. The stud is made by bonding several copper plates together, having a diameter of 1 cm. Each copper plate has grooves of 0.55 mm deep and 0.25 mm wide with the pitch of 0.55 mm on both sides of the plate. The result is a three-dimensional network of passages in the stud, with the possibility of the largest surface area allowable within the stud. Even though the passages are open to the environment with rectangular cross sections of 0.25×0.25 mm, and liquid flooding is inevitable, the parameter C can still be very small due to very large internal heated surface area, and the liquid temperature distribution within the complex internal passages may be uniform enough to eliminate significant temperature overshoot.

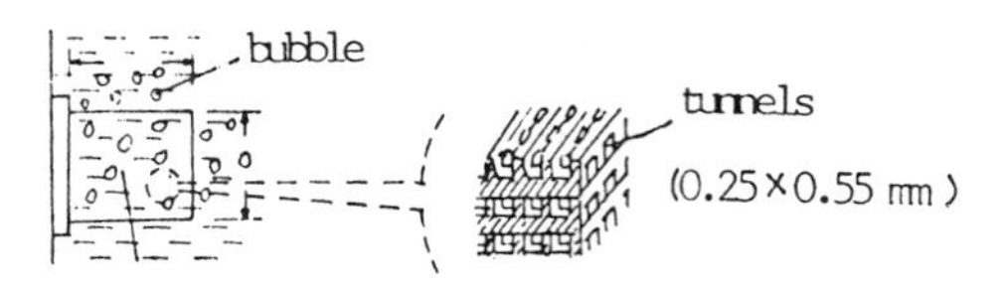

cylindrical heat sink stud

sketch of the multilayered porous structure

heat dissipation component

FIGURE 9. Porous stud attached to the face of a heat dissipating component.

CONCLUSIONS

Most commercial enhanced surfaces are subject to flood at a moderate subcooling with fluorocarbon fluids. Both contact angles and internal structure of re-entrant cavities have strong effects on initiation of boiling. Boiling incipience within a liquid-filled cavity is determined by the liquid temperature field inside the cavity, which can be described qualitatively with the structure parameter $C = a/A$. A smaller C corresponds to a more uniform temperature field within the cavity and demands a smaller superheat to initiate boiling. With appropriate structures, flooded enhanced surfaces may still eliminate significant temperature overshoot. But we must bear in mind that enhanced surfaces with smaller C may have a lower critical heat flux (CHF). Further research will concentrate on manufacturing the enhanced surfaces which can eliminate boiling hysteresis and extend the burnout point.

REFERENCES

1 Oktay, S., Departure from Natural Convection (DNC) in Low Temperature Boiling Heat Transfer Encountered in Cooling Microelectronic LSI Devices , Proceedings of the 7th International Heat Transfer Conference, Hemisphere Publishing Corp., Vol. 4. pp. 113-118, 1982.

2 Bergles, A.E. and Chyu, M.C., Characteristics of Nucleate Pool Boiling from Porous Metallic Coatings, J. Heat Transfer, Vol. 104, pp. 279-285, 1982.

3 Marto, P.J. and Lepere, V.J., Pool Boiling Heat Transfer from Enhanced Surfaces to Dielectric Fluids, J. Heat Transfer, Vol. 104, pp.292-299, 1982.

4 Nakayama, W., Thermal Management of Electronic Equipment: A Review of Technology and Research Topics, Appl. Mech. Rev., Vol. 39, pp.1847-1868, 1986.

5 Chu, R.C., Heat Transfer in Electronic Systems, Proceedings of the Eighth Int. Heat Transfer Conf., Vol. 1, pp.293-305, 1986.

6 Konev, S.V. and Mitrovic, J., An Explanation for the Augmentation of Heat Transfer During Boiling in Capillary Structures, Int. J. Heat Mass

Transfer, Vol. 29, pp.91-94, 1986.

7 Thormahlen, L., Superheating of Liquids at the Onset of Boiling, Proceedings of the Eight Int. Heat Transfer Conf., Vol. 4, pp. 2001-2006.

8 Cole, R., Boiling Nucleation, Advances in Heat Transfer, Vol. 10, Academic press, pp.127-157, 1974.

9 Van Stralen, S. D. and Cole, R., Boiling Phenomena, Vol. 1, Hemisphere Publishing Corporation 1979.

10 Han, C.Y., and Griffith, P., Int. J. Heat Transfer, Vol. 8, pp.887, 1965.

11 Bergles, A.E. and Rohsenow, M., J. Heat Transfer, Vol. 86, pp.365, 1964.

12 Xin, M. D. and Cao, Y. D., Analysis and Experiment of Boiling Heat Transfer on T-shaped Finned Surfaces, Chemical Engineering Communications, 50(1-6): 185-201, 1987.

13 Humphrey, J. A. C. and To, W.M., Numerical Simulation of Buoyant, Turbulent Flow-II, Free and Mixed Convection in a Heated Cavity, Int. J. Heat Mass Transfer, Vol. 29, pp. 593-610, 1986.

Heat Transfer and Pressure Drop for High Density Staggered Pin Fin Arrays with Liquid Coolants

T. R. CRAIG, F. P. INCROPERA, and S. RAMADHYANI
Heat Transfer Laboratory
School of Mechanical Engineering
Purdue University
West Lafayette, Indiana 47907, USA

ABSTRACT

An experimental study has been performed to assess the effect of pin fin configuration on heat transfer augmentation for a 50.8 mm × 50.8 mm heat source flush mounted to one wall of a rectangular channel. The pin fin geometry could be used as a cap for a liquid-cooled multi-chip array or for the enhancement of cold plate performance. The pin fins substantially reduce thermal resistance, with the largest reduction corresponding to large pin diameters and/or small pin pitches. Thermal resistances as small as 0.5 $cm^2 \cdot K/W$ and 0.15 $cm^2 \cdot K/W$ were obtained in FC-77 and water, respectively, for a pin diameter of 4.76 mm and a dimensionless pitch of 1.185. However, from the standpoint of heat transfer per unit pump power, improved performance is associated with decreasing pin diameter, as well as decreasing pin pitch.

INTRODUCTION

Considerable attention is currently being directed to the problem of cooling VLSI chips. Projections based on recent advances in microelectronic technology indicate the future possibility of chip heat fluxes as high as 200 W/cm^2. Such chips are typically interconnected in dense arrays, and conventional air cooling techniques are precluded by the space constraints. Since existing indirect liquid cooling techniques [1-3] are already approaching their limits, it has been suggested that future heat dissipation requirements will necessitate direct liquid (immersion) cooling of uncased chips [4-7]. Due to their superior electrical and chemical properties, fluorocarbons have been identified as the most suitable coolants.

Liquid immersion cooling can be effected by free convection, forced convection, or nucleate boiling. Potential problems with nucleate boiling include the high degree of superheat necessary to initiate nucleation with typical dielectric liquids and the low values of the critical heat flux associated with these liquids. Single-phase convection cooling avoids these problems, and chip thermal control is more easily achieved by control of the fluid inlet temperature and velocity. In this study, the potential for forced convection heat transfer enhancement through the use of short, closely packed, pin fins is examined. The study encompasses two different pin diameters, two pin lengths, and various pitches, with the pins arranged in a staggered array in each case. Water and an inert dielectric liquid (3M FC-77) are used as coolants, and the extent of heat transfer enhancement is assessed by comparing the augmented heat source results with those obtained for an unaugmented, flush-mounted heat source. Figure 1 is a schematic of the pin fin array and the base surface. Finned surfaces of this type could be used as caps on either individual chips or clusters of chips, as well as for the enhancement of cold plate performance.

Documentation on the use of extended surfaces with liquid coolants is sparse. Recent studies on the use of pin fins with liquid coolants have been reported by Kelecy et al. [8] and Ramadhyani and Incropera [9]. The former study employed small square heat sources, 12.7 mm on a side,

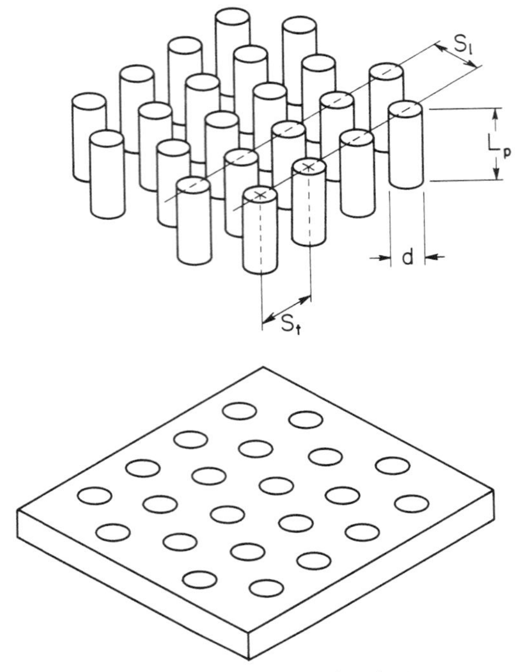

FIGURE 1. Characteristic parameters of a pin fin array.

equipped with in-line arrays of pin fins. The heat sources were mounted to one wall of a flow channel, while a small clearance was maintained between the fin tips and the opposite wall to promote heat transfer from the tips. Significant heat transfer enhancement was achieved, the thermal resistance between the heat source and the coolant being reduced by factors of six to ten in water and 15 to 25 in FC-77. Ramadhyani and Incropera employed 12.7 mm square heat sources equipped with in-line arrays of composite fins. These composite fins consisted of circular pins with a set of square plate fins attached at intervals along the pin length. The additional heat transfer enhancement achieved through the use of composite fins was small (approximately 20% higher than with pin fins), while the increased channel blockage increased the pressure drop by approximately 25%. The authors are unaware of any other reports on heat transfer enhancement with liquid coolants using pin fins.

The use of pin fins for heat transfer enhancement in air has been the subject of several investigations, and some of the complexities associated with the situation have been revealed by these studies. When a pin is attached to a flat surface, the flow near the pin-base junction acquires a complex three-dimensional character. As shown by Goldstein and Karni [10], a system of horseshoe vortices is formed in this region, resulting in local variations of the convection coefficient along the pin. The vortex system also significantly enhances the convection coefficient on the base surface. Local variations of the heat transfer coefficient on the base surface have been mapped out by Saboya and Sparrow [11,12] and by Lau et al. [13] using the naphthalene sublimation technique. The measured variations are consistent with the existence of the horseshoe vortex system.

In addition to the effects of the horseshoe vortex system, thermal interactions between the base and the pins promote nonuniformity in the local convection coefficients. For instance, fluid heated by the base surface depresses convection coefficients near the base of the pins as it flows past the pins. Such thermal interactions have not been studied in detail.

The effect of a clearance between the pin tips and the wall of the flow channel has been studied by Steuber and Metzger [14] for a large number of pin fin array configurations. Short pins (L/d < 1) were considered over a wide range of turbulent flow Reynolds numbers. In all cases, improved thermal performance was associated with zero tip clearance. However, when considered in terms of both heat transfer and pressure loss, tip clearance provided superior performance for several arrays. Sparrow et al. [15] obtained numerical solutions for fully developed laminar flow over a shrouded array of plate fins. Their results indicated that heat transfer coefficients were highly non-uniform over the fin and base surface. With zero tip clearance, the heat transfer coefficient reached a maximum at the center of the fin. With a tip clearance, the heat transfer coefficient increased monotonically from the base to the tip. Kadle and Sparrow [16,17] investigated turbulent flow and heat transfer in a shrouded array of plate fins both experimentally and numerically for zero tip clearance and experimentally with tip clearance. With zero tip clearance, the heat transfer coefficient was a maximum at the center of the fin. Although local variations were not determined in the tip clearance case, the heat transfer coefficient was found to decrease substantially with increasing tip clearance.

The row-by-row variation of convection coefficients in an array of pin fins was investigated by Sparrow and Ramsey [18] and Sparrow et al. [19]. Convection coefficients increased in the streamwise direction until the fourth row and remained constant thereafter, indicating a fully developed condition. Experiments by Simoneau and Van Fossen [20], in which heat transfer and turbulence intensity were measured for pin arrays, corroborate the findings of Sparrow and co-workers. Turbulence intensity and heat transfer coefficient were observed to increase with row number up to the fourth row.

APPARATUS AND EXPERIMENTAL PROCEDURES

This investigation examined the effects of pin diameter, pin height, and pin spacing on heat transfer augmentation and pressure drop in a rectangular flow channel for a staggered array of short pin fins. Eight configurations were studied, each of which consisted of a staggered array of cylindrical pin fins evenly spaced on a 50.8 mm square heated base surface which was flush-mounted to the bottom wall of a rectangular channel. In each of the pin arrays, the longitudinal and transverse pitches were equal, and the pin tip clearance with the upper wall of the channel was zero. In addition to the finned heat sources, a flat, unaugmented, flush-mounted source was tested to provide a baseline for comparison.

The fins and the base surface were integrally machined from solid blocks of oxygen-free copper by electro-discharge machining to eliminate thermal contact resistance between the pins and the base surface. In each case, the base was 50.8 mm on a side and 7.9 mm thick, the thickness being chosen to provide almost isothermal conditions on the base surface. The copper block was mounted within a Lexan module, and four 25.4 mm square electric resistance heaters were attached to the back of the block and held in place by nylon pressure screws. The heaters consisted of chromium alloy resistance elements sputtered on beryllium oxide substrates. The interface between each heater and the copper base was thinly coated with thermally conducting paste to minimize contact resistance. With the heaters in place, the Lexan module cavity was filled with a fine silica powder (Sylox 2) for thermal insulation and closed with a Lexan backplate. Heat losses through the Lexan module are estimated to be under 2% in the augmented surface tests and under 5% in the tests with the unaugmented source.

The heater/module assembly was mounted in the flow channel of Fig. 2. The channel formed a rectangular duct of height 11.9 mm and width 50.8 mm. Depending on the height of the pin fins, the channel height could be adjusted, by inserting additional Lexan pieces, to maintain a zero tip clearance in each case. For baseline tests involving an unfinned heat source, the channel height was maintained at 11.9 mm. To ensure fully developed turbulent flow conditions at the location of the heat source, the Lexan module was flush-mounted to the channel at a distance of 611 mm from the flow straightener.

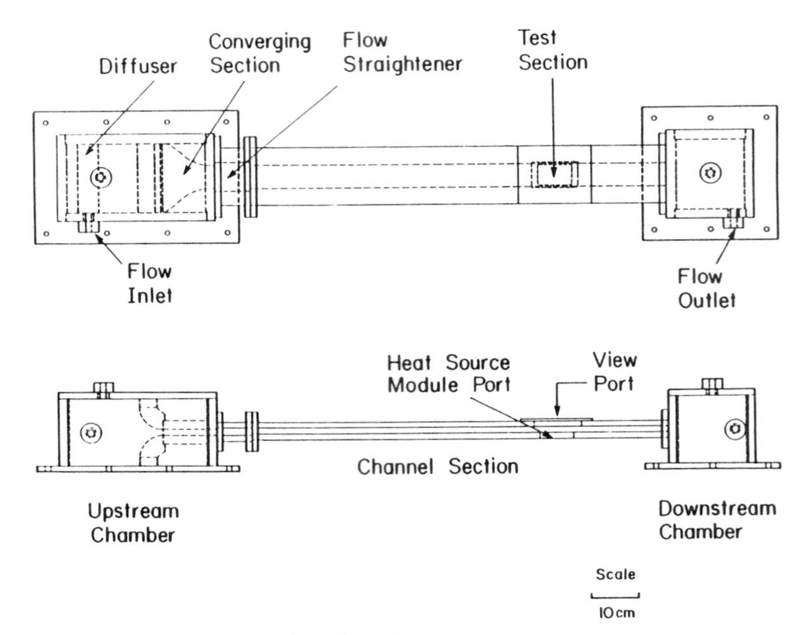

FIGURE 2 . Schematic of the flow channel .

Temperature measurements were made at the base surface, as well as at the pin tips, of the heater module. Six calibrated copper-constantan thermocouples were soldered to the surface of each of the copper base blocks. Thermocouples were also attached to the tips of selected pins through fine holes drilled along the axis of the pins.

A rheostat and current shunt were used to control and monitor the power dissipated in each heater. The flow channel was installed in a loop which included a storage tank, pump, turbine flowmeter, and heat exchanger. The heat exchanger was used to control the temperature, T_o, at the inlet of the channel. Before initiating the experiment, the channel and flow loop were filled with freshly degassed coolant (water or FC-77). When steady-state conditions were reached, a computer automatically read and stored the data for subsequent reduction. Pressure drop measurements were made with a manometer connected to two pressure taps in the upper wall of the channel.

EXPERIMENTAL RESULTS: UNAUGMENTED HEATER

Results obtained with a bare unaugmented heat source provide a suitable baseline for assessing benefits associated with use of the pin fins.

Data Reduction

The average Nusselt number for the unaugmented surface is defined as

$$\overline{Nu_L} = \frac{\bar{h}\, L_h}{k_o} = \frac{q\, L_h}{A\, \Delta T_{lm}\, k_o} \tag{1}$$

where q is the total heater power. The log mean temperature difference is defined as

regime. For water the resulting correlation is

$$R_T''\left(\frac{K\ cm^2}{W}\right) = 2236\ Re_D^{-0.73} \qquad (Re_D \geq 3000) \tag{8}$$

where the average and maximum deviations are 1.35% and 3.97%, respectively. For FC-77,

$$R_T''\left(\frac{K\ cm^2}{W}\right) = 15,216\ Re_D^{-0.74} \qquad (Re_D \geq 3000) \tag{9}$$

with average and maximum deviations of 1.46% and 4.05%. Due to its poorer heat transfer properties, the fluorocarbon yields substantially larger thermal resistances.

Average Nusselt numbers for water and FC-77 are plotted in Fig. 4 against a Reynolds number based upon heat source length. Using an error minimization technique to collapse the data, the following correlation was obtained:

$$\overline{Nu}_L / Pr^{0.37} (\mu_o/\mu_h)^{0.16} = 0.08 Re_L^{0.737} \tag{10}$$

The average deviation is 1.35% for $Re_L \geq 8000$. A Reynolds number of 8000 based on the heat source length is equivalent to a channel Reynolds number of approximately 3000. The results for water and FC-77 do not collapse as well at the lower flow rates. This behavior is attributed to buoyancy effects, which have a greater influence at small flow rates. In water, the ratio of the Grashof number to the Reynolds number squared is approximately 0.4 at a Reynolds number of 2000. The larger expansion coefficient and smaller viscosity of FC-77 make mixed convection effects even more pronounced in the low Reynolds number range.

EXPERIMENTAL RESULTS: PIN FIN ARRAYS

Data Reduction

In the study of pin fin configurations, flow through the array is characterized by two different velocity and length scales. To compare thermal resistance results for the pin arrays with those for the unaugmented source, the channel Reynolds number is used to characterize the flow. However, when considering pin fin data in terms of a Nusselt number, the Reynolds number is defined in terms of the maximum fluid velocity within the array and the pin diameter:

$$Re_{d,max} = U_{max} d/\nu \tag{11}$$

Fluid properties are evaluated at the film temperature, T_f, defined as

$$T_f = \frac{\overline{T}_h + \dfrac{(T_o + T_e)}{2}}{2} \tag{12}$$

Heat transfer results are reduced in terms of the thermal resistance for a unit base area and an effective Nusselt number. The Nusselt number is defined as

$$\overline{Nu}_{eff} = \overline{h}_{eff} d/k_f \tag{13}$$

where $\overline{h}_{eff}$ is the average effective heat transfer coefficient.

Although the local convection coefficient can vary over the surface of a staggered pin array, it is reasonable to assume separate, but spatially uniform, convection coefficients for the base ($\overline{h}_b$) and the pin ($\overline{h}_p$) surfaces. This approach is indicated schematically in Fig. 5. Assuming that all of the energy dissipated by the heaters, q, is transferred into the fluid by either the pins, q_p, or the base, q_b, it follows that

$$q = q_p + q_b \tag{14}$$

The heat dissipated by the base surface is expressed as

$$q_b = \overline{h}_b A_b \Delta T_{lm} \tag{15}$$

where $\overline{h}_b$ is the average convection coefficient associated with the exposed base.

$$\Delta T_{lm} = \left[(\overline{T}_h - T_o) - (\overline{T}_h - T_e)\right] / \ln \left[(\overline{T}_h - T_o) / \overline{T}_h - T_e\right] \quad (2)$$

where the exit temperature is calculated from the energy balance:

$$T_e = \frac{q}{\dot{m}\, c_p} + T_o \quad (3)$$

The Nusselt number is plotted against a Reynolds number based on the heat source length:

$$Re_L = U\, L_h / \nu \quad (4)$$

where the kinematic viscosity is evaluated at the upstream temperature and the mean fluid velocity U is defined as

$$U = Q / A_c \quad (5)$$

To facilitate comparisons of thermal performance for the bare and augmented heat sources, heat transfer results are also presented in terms of the thermal resistance

$$R''_T = \Delta T_{lm} \cdot A / q \quad (6)$$

Thermal resistances are plotted against a Reynolds number defined as

$$Re_D = U\, D / \nu \quad (7)$$

D being the hydraulic diameter of the channel.

Unaugmented Heat Source Results

Bare source thermal resistances for both water and FC-77 are shown in Fig. 3. Results are based on a unit base area to facilitate comparisons with other heat transfer surfaces. Experiments were performed in the Reynolds number range $1000 \leq Re_D < 10{,}000$, and the base surface was maintained isothermal to within 0.5°C. A typical result for water corresponds to a thermal resistance of 4.46 K cm^2/W at a Reynolds number of 5000. Hence, with a 10°C temperature difference between the surface and the fluid, a heat flux of 2.24 W/cm^2 could be maintained. Fully turbulent conditions are observed for channel Reynolds numbers above 3000. Below this value the flow is transitional, and the resistance has a weaker Reynolds number dependence. The solid lines indicate least square fits of the data within the fully turbulent

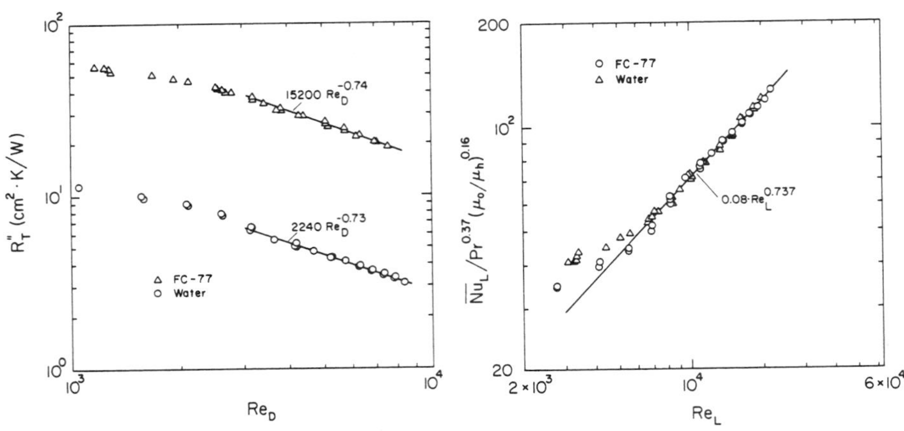

FIGURE 3. Thermal resistance (K cm^2/W) as a function of Reynolds number for bare (unaugmented) heat source.

FIGURE 4. Correlation of heat transfer data for the bare (unaugmented) heat source.

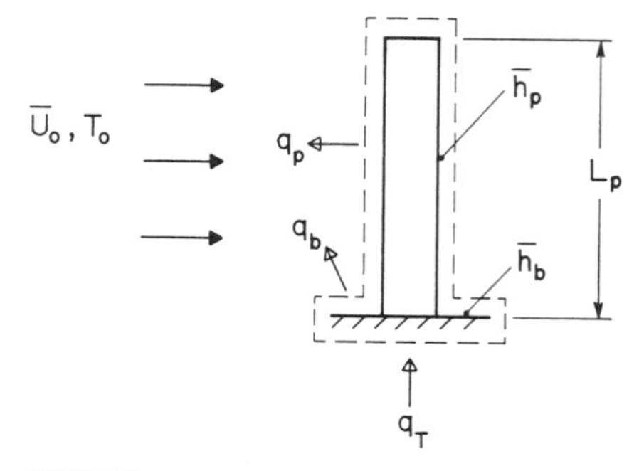

FIGURE 5. Schematic of heat transfer from a single pin.

Heat transfer from the pins is governed by the expression for one-dimensional conduction in a fin of uniform cross-section and an adiabatic tip [21].

$$q_p = M \tanh mL \tag{16}$$

where

$$M = (\bar{h}_p k \pi^2 d^3 / 4)^{1/2} \Delta T_{lm} \tag{17}$$

and

$$m = (4\bar{h}_p / kd)^{1/2} \tag{18}$$

Use of the foregoing equations to determine one of the convection coefficients, $\bar{h}_b$ or $\bar{h}_p$, dictates that the other coefficient be known. Kelecy et al. [8] performed experiments for which the base heat transfer coefficient was isolated by the use of thermally inactive Lexan pins. They found that the average coefficients associated with the base and the pins were approximately equal. Assuming applicability of this result to the staggered pin arrays of this study, an effective convection coefficient is defined as $\bar{h}_{eff} = \bar{h}_p = \bar{h}_b$ and equations (14)-(18) may be used to determine the effective coefficient. Heat transfer data are also reduced in terms of equation 6, and thermal resistance results are presented as a function of the channel Reynolds number.

The eight staggered pin fin configurations tested in this study are listed by module number in Table 1. Data were taken for both water and FC-77 over a channel Reynolds number range of $800 \leq Re_D \leq 8000$. In these experiments, the heater power was adjusted to give an approximate 10°C temperature difference between the surface and upstream fluid temperatures. Inlet fluid temperatures varied from approximately 14°C to 15°C for both water and FC-77. In addition to heat transfer experiments, pressure drop measurements were made for each module.

Pin Array Thermal Resistance Measurements

Thermal resistances for each of the four packing densities are shown in Figures 6 to 9, respectively. Each figure pertains to a particular packing density and provides thermal resistance data for the long and short pins in both water and FC-77. The solid lines indicate least squares best-fit correlations of the form

$$R''_T = B\, Re^a_{d,max} \tag{19}$$

where the best-fit coefficients B and a are summarized in Table 2. The table also lists the average and maximum deviations of the data from the correlations. Results for FC-77 are from 4 to 5 times larger than those for water, due to the relatively poor thermal conductivity of the FC-77.

Table 1 . Experimental pin fin configurations

Pin Fin Configurations *				
Module	Pin Diameter	Pin Length	Pin Pitch	Number of Pins
1	4.76	12.00	1.185	81
2	4.76	12.00	2.133	25
3	2.05	12.00	2.25	121
4	2.05	12.00	1.65	225
5	4.76	5.00	1.185	81
6	4.76	5.00	2.133	25
7	2.05	5.00	2.25	121
8	2.05	5.00	1.65	225

* All dimensions are in millimeters. The pin pitch is dimensionless.

For FC-77, thermal resistances for the shorter pins are consistently larger than those for the longer pins, indicating that the longer pins are preferred. The trend also characterizes the water data for the large diameter pins (Figures 6 and 7). However, for the smaller pin diameter, the thermal resistance of the long pins exceeds that of the short pins at large Reynolds numbers, with the tighter pin packing density (Figure 9) exhibiting this behavior at a lower Reynolds number than the sparser packing density (Figure 8). In fact, if the results are extrapolated to larger Reynolds number, it appears that the thermal resistance of the short pin would fall below that of the long pin for all test configurations in both water and FC-77, except perhaps for the tighter pin packing density corresponding to the larger pin diameter (Figure 6).

It is apparent that conditions exist for which the overall or effective heat transfer coefficient is larger for the short pins than for the long pins at equal inter-pin fluid velocities. For low flow rates in water or operation in FC-77, convection coefficients are relatively small, resulting in high fin efficiencies for the long, as well as the short, pins. Hence, all of the surface area remains active with increasing pin length, and irrespective of what changes may be occurring in local convection coefficients, the larger surface area dissipates more heat. With increasing flow rate, however, the fin efficiency decreases, the pin tip is less active, and reductions in thermal resistance with increases in pin length are less pronounced. This effect is most pronounced in water, due to its much larger convection coefficient. By itself, however, it cannot explain the cross-overs exhibited by the long and short pin data of Figures 8 and 9. If, however, there is an

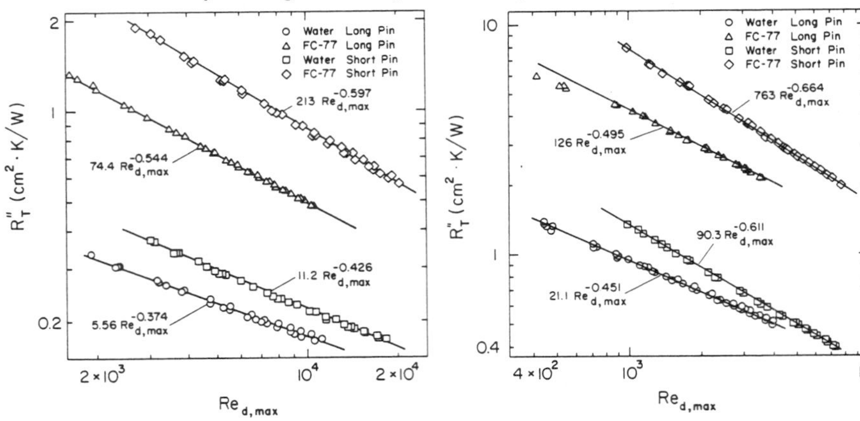

FIGURE 6. Thermal resistance (K cm^2/W) for modules 1 and 5 (d = 4.76 mm, S/d = 1.185).

FIGURE 7. Thermal resistance (K cm^2/W) for modules 2 and 6 (d = 4.76 mm, S/d = 2.133).

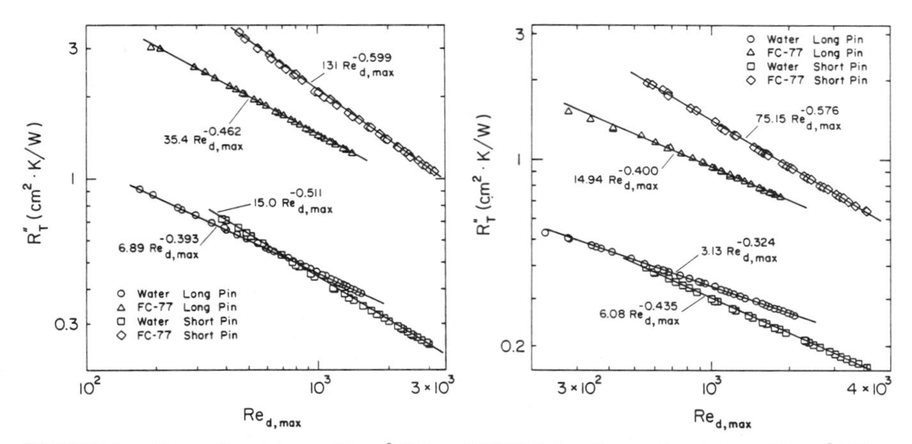

FIGURE 8. Thermal resistance (K cm^2/W) for modules 3 and 7 (d = 2.05 mm, S/d = 2.25).

FIGURE 9. Thermal resistance (K cm^2/W) for modules 4 and 8 (d = 2.05 mm, S/d = 1.65).

increase in local convection coefficients on the base and/or pin surfaces due to a decrease in pin length, this decrease could cause a reduction in thermal resistance. The reduction would occur inspite of the attendant reduction in the total surface area if a significant portion of the pin length were inactive. Although the absence of local convection coefficient data precludes precise confirmation, it is speculated that the observed cross-overs are due to an enhancement of local convection coefficients with decreasing pin length. Due to the larger convection coefficients associated with the tighter pin packing density and the lower pin efficiency of the smaller pin diameter, the cross-over between results for the long and short pins is first observed between Modules 4 and 8 in water.

In water, the pin fins reduce the thermal resistance by approximately an order of magnitude for Module 2 and by approximately a factor of 25 for Module 1. While the unaugmented heat source has a thermal resistance of 4.46 cm^2·K/W at a channel Reynolds number of 5000, Module 1 has a thermal resistance of 0.19 cm^2·K/W, which corresponds to a heat flux of 52.6 W/cm^2 for a 10 °C driving potential between the base surface and the upstream fluid

Table 2. Thermal resistance correlated as a function of $Re_{d,max}$

Thermal resistance correlations: $R''_T = B\, Re^a_{d,max}$					
Module	fluid	a	B	avg. dev.	max. dev.
1	water	-0.374	5.56	1.4%	3.3%
	FC-77	-0.544	74.4	.52%	2.3%
2	water	-0.451	21.1	1.2%	3.7%
	FC-77	-0.495	125.7	1.9%	5.2%
3	water	-0.393	6.89	0.4%	1.2%
	FC-77	-0.462	35.4	0.4%	3.0%
4	water	-0.324	3.13	0.4%	2.6%
	FC-77	-0.400	14.9	0.9%	3.0%
5	water	-0.426	11.15	0.9%	2.6%
	FC-77	-0.597	213.0	1.3%	2.8%
6	water	-0.611	90.3	0.66%	1.4%
	FC-77	-0.664	762.6	0.65%	2.3%
7	water	-0.511	14.99	0.5%	1.5%
	FC-77	-0.599	131.3	0.9%	2.9%
8	water	-0.435	6.08	0.4%	1.0%
	FC-77	-0.576	75.15	0.7%	1.7%

temperature. Compared to the results for water, heat transfer augmentation is more effective in FC-77 due to the increased pin efficiency associated with the smaller heat transfer coefficient of the fluorocarbon. The greatest gains in relative performance are realized for the larger packing densities.

Pin Array Nusselt Number Results

The effective Nusselt number for each module was determined from equations 13 to 18, where convection coefficients associated with the pin and base surfaces were assumed to be spatially uniform and equal, $\bar{h}_b \approx \bar{h}_p$. Results for both the long and short pins, in water and in FC-77, are presented for each of the four packing densities in Figures 10 to 13. The Nusselt number data are correlated in terms of $\overline{Nu}_{eff}/Pr^n$, where n was chosen to be the generally accepted value of 0.4. Except for Module 5, the short pin data for water and FC-77 were collapsed by this Prandtl number exponent. However, a Prandtl number exponent of approximately 0.54 was needed to collapse the long pin results of modules 1,3, and 4. This large Prandtl number exponent may be an unrealistic consequence of significant variations in the convection coefficient along the length of the pin.

Axial variations in the apparent convection coefficient along the pin surface could be caused by thermal boundary layer development on the base surface and by the presence of horseshoe vortex systems at the base and tip of each pin. Assuming that the heat transfer coefficient increases from base to tip, since the pin efficiency is large in FC-77 the larger convection coefficients near the pin tip would contribute proportionately more to the total pin heat loss in FC-77 than in water. For a Prandtl number exponent in the typical range $0.33 \leq n \leq 0.4$, this effect would cause the data reduction procedure to yield a larger value of $\overline{Nu}_{eff}$ in FC-77 than in water. The effect would be more pronounced with the smaller diameter pins due to their lower efficiency. The fact that the short pin Nusselt numbers exceed those of the long pins for only the small pin diameter (Figures 11 and 12) is consistent with the expectation that the effects of nonuniformities in the axial variation of the convection coefficient are most pronounced for the small pin diameter and large pin length. The fact that the effect is most pronounced in water may be due to inactivity of the pin tip for the longer pin and hence the inability to realize the effect of enhanced convection due to the vortex system at the tip.

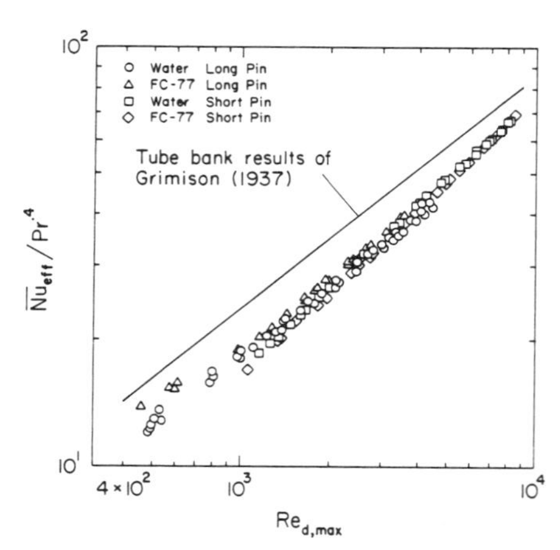

FIGURE 10. Comparison of Nusselt number results for modules 2 and 6 with the standard tube bank correlation (d = 4.76 mm, S/d = 2.133).

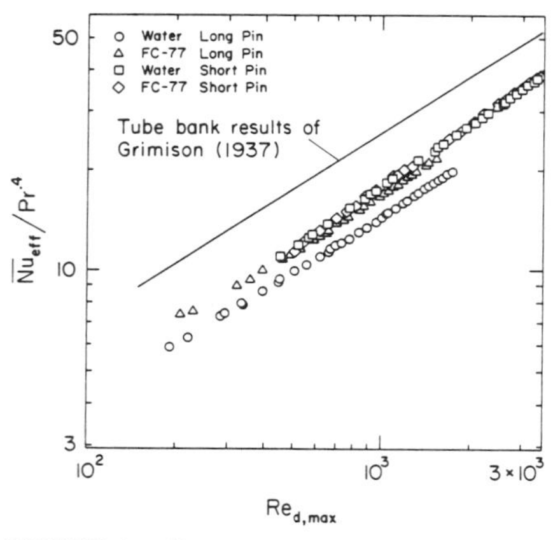

FIGURE 11. Comparison of Nusselt number results for modules 3 and 7 with the standard tube bank correlation (d = 2.05 mm, S/d = 2.25).

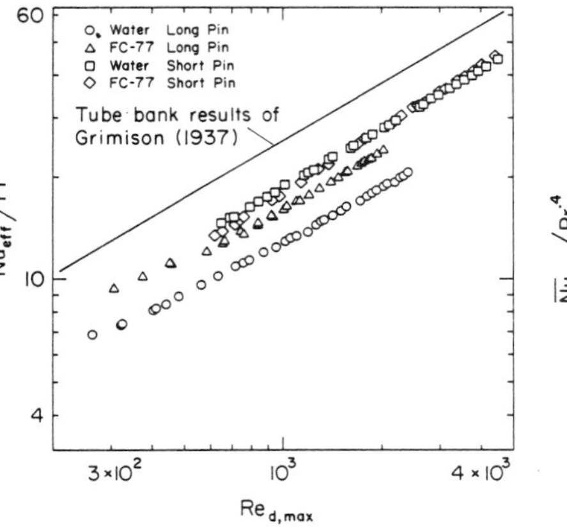

IGURE 12. Comparison of Nusselt number results for modules 4 and 8 with the standard tube bank correlation (d = 2.05 mm, S/d = 1.65).

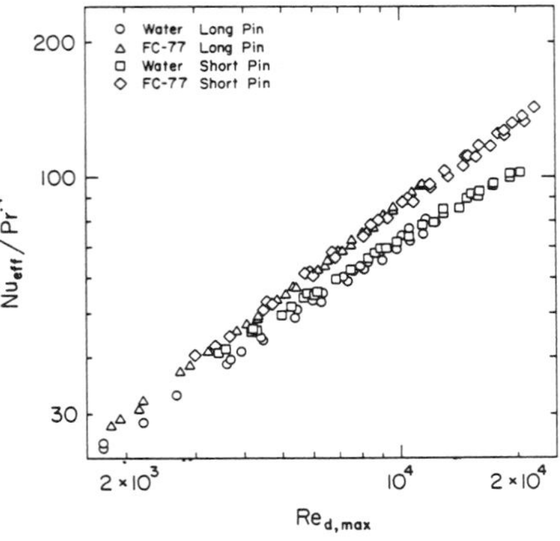

FIGURE 13. Nusselt number results for modules 1 and 5 (d = 4.76 mm, S/d = 1.185).

n Figures 10 to 12, the solid line indicates tube bank results correlated by Grimison [22]. The ightest packing density, Figure 13, does not fall within the range of existing tube bank eometries. Nusselt number data for the short pins fall approximately 35%, 31%, and 28% elow the tube bank results for Modules 6, 7, and 8, respectively. Thermal boundary layer evelopment and thermal wake formation on the base, as well as hydrodynamic boundary ayers on the channel walls, may combine to reduce the convection coefficient by more than the mount by which it is enhanced due to the wall/pin vortex system. Nusselt numbers approach ube bank results more closely with increasing Reynolds number. This trend may be due to hinning of the thermal and hydrodynamic boundary layers and the enhancement of the vall/pin vortex system with increasing flow velocity.

'in Array Pressure Drop Results

'ressure drop data were acquired for each of the eight pin configurations in water, using a nicromanometer filled with FC-77 (for low pressure drops) and U-tube manometers filled with 'C-77 and mercury for moderate and large pressure drops, respectively. Data are presented in erms of a friction factor defined as

$$= \frac{\Delta p}{N_L(0.5\rho_o U_{max}^2)} \quad (20)$$

nd representative results are presented in Figures 14 and 15. The results show that, for an quivalent value of $Re_{d,max}$ and hence U_{max}, friction factors for the short pins are consistently arger than those for the long pins. This trend is consistent with the expectation that, as the hannel hydraulic diameter decreases, the effect of the end walls on the pressure drop increases.

'he data of this study are also compared to existing tube bank results [23]. At moderate to arge Reynolds numbers, the tube bank friction factors are comparable to or less than results or the long pin arrays. The largest difference corresponds to the large diameter pins (Figure 4) and is attributed to the significant departure from tube bank conditions for the small L_p/d atio. Although low Reynolds number data for the small diameter pins fall below the tube ank predictions, the trend is inconsistent with expectations and is attributed to experimental ncertainty. The tube bank results should provide a lower limit to the data of this study.

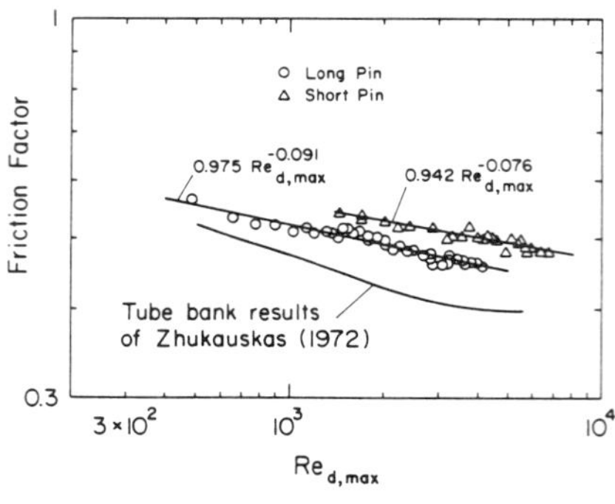

FIGURE 14. Comparison of tube bank friction factors with those of modules 2 and 6 (d = 4.76 mm, S/d = 2.133).

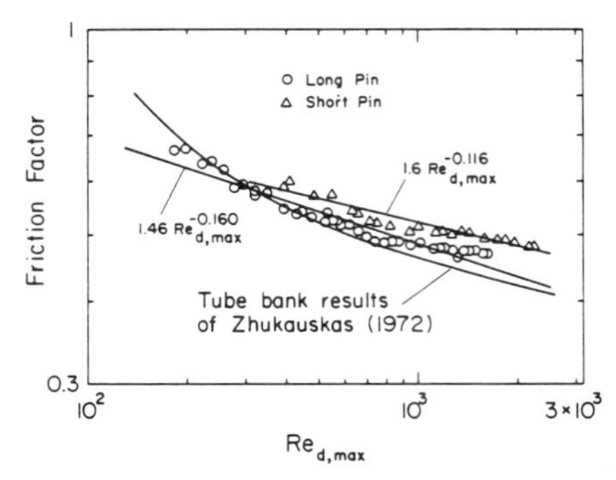

FIGURE 15. Comparison of tube bank friction factors with those of modules 3 and 7 (d = 2.05 mm, S/d = 2.25).

Pin Array Performance Evaluation

Since operating costs are directly proportional to pressure loss, it is desirable to minimize the pressure drop while maintaining the necessary heat transfer characteristics. In this section, performances of the eight pin fin configurations are compared in both water and FC-77. In particular, thermal performance, as measured by the reciprocal of the thermal resistance, $(R''_T)^{-1}$, is plotted as a function of pumping power.

The pump power, P_p, for an incompressible fluid is defined by

$$P_p = Q\Delta p \tag{21}$$

where Δp is the corresponding pressure drop across the pin array. Since heat transfer and pressure drop data were obtained in different experiments, the pressure drop corresponding to a particular heat transfer data point was obtained from a best-fit correlation to the friction factor data:

$$f = BRe^{a}_{d,max} \tag{22}$$

where B and a were determined by a least squares fit. The pressure drop across the array can then be expressed as

$$\Delta p = N_L(0.5\rho_o U^2_{max})BRe^{a}_{d,max} \tag{23}$$

This equation is substituted into equation (21) to obtain the following expression for the pumping power

$$P_p = QN_L(0.5\rho_o U^2_{max})BRe^{a}_{d,max} \tag{24}$$

Results are shown for each of the four packing densities in Figures 16-19. The results indicate that, under conditions for which the convection coefficient is small, that is, for operation at low flow rates or in FC-77, the long pins perform better than the short pins due to the increased surface area. However, for larger convection coefficients, the efficiency of the long pin is reduced, thereby negating the effect of added surface area near the tip. Even though friction factors are slightly higher, the performance of the short pins can surpass that of the long pins due to their larger convection coefficients.

Comparing results for modules having the same pin length and diameter, Figures 16,17 and Figures 18,19, it is apparent that performance gains can be realized by increasing the pin packing density. In each case, the thermal resistance for the larger pin packing density is significantly smaller at equal pumping power. In addition, the results for equal pin length and packing density, Figures 17 and 18, indicate that performance improves with a reduction in pin

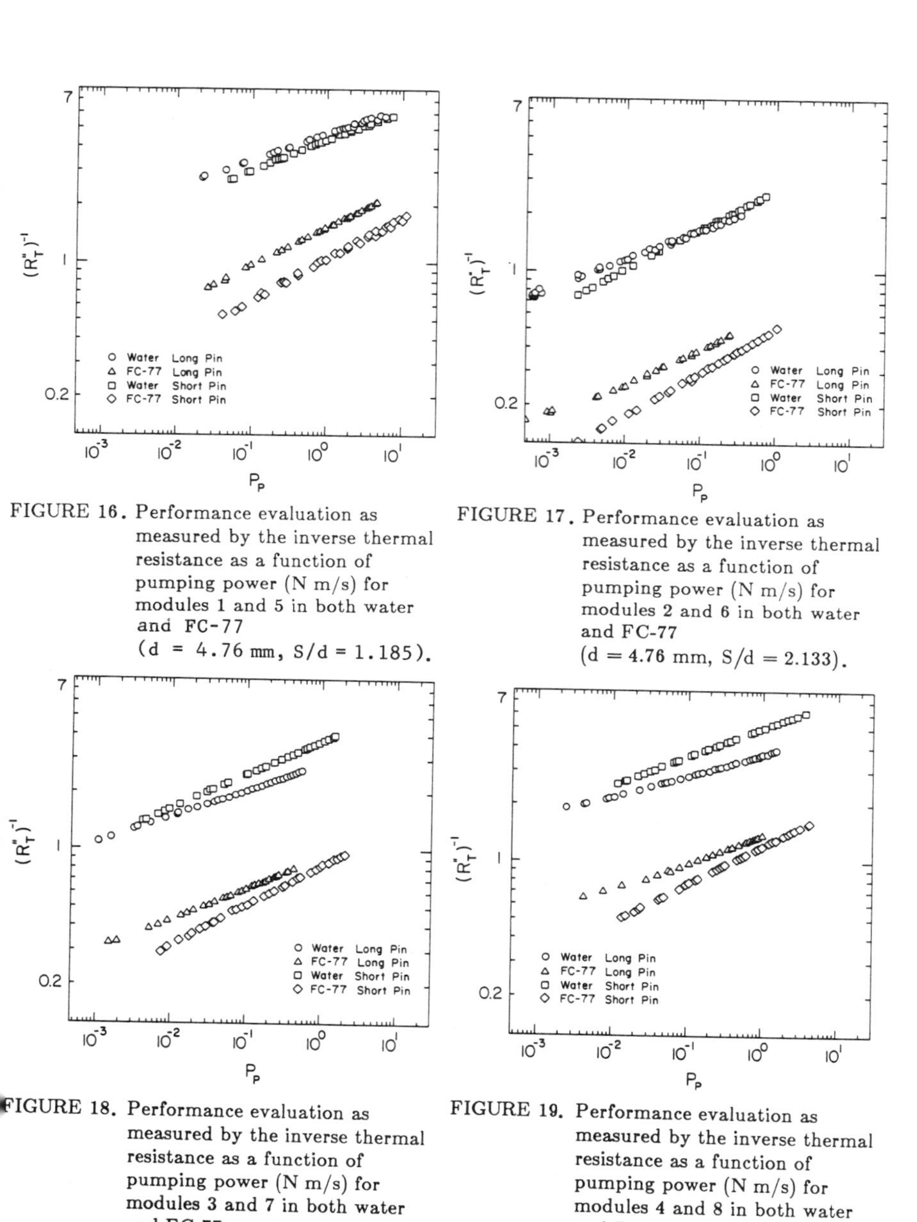

FIGURE 16. Performance evaluation as measured by the inverse thermal resistance as a function of pumping power (N m/s) for modules 1 and 5 in both water and FC-77 (d = 4.76 mm, S/d = 1.185).

FIGURE 17. Performance evaluation as measured by the inverse thermal resistance as a function of pumping power (N m/s) for modules 2 and 6 in both water and FC-77 (d = 4.76 mm, S/d = 2.133).

FIGURE 18. Performance evaluation as measured by the inverse thermal resistance as a function of pumping power (N m/s) for modules 3 and 7 in both water and FC-77 (d = 2.05 mm, S/d = 2.25).

FIGURE 19. Performance evaluation as measured by the inverse thermal resistance as a function of pumping power (N m/s) for modules 4 and 8 in both water and FC-77 (d = 2.05 mm, S/d = 1.65).

diameter. For Modules 2 and 3 in FC-77, performance improves by nearly 40% with a 57% reduction in pin diameter.

CONCLUSIONS

An experimental study has been performed to determine the effects of pin fin configuration on heat transfer augmentation for a 50.8 mm×50.8 mm heat source flush-mounted to one wall of a rectangular channel. Four different pin arrays provided a range of parameter values corresponding to pin diameters of 2.05 mm and 4.76 mm and dimensionless pin pitches ranging from 1.185 to 2.25. For each array, pin lengths of 5 mm and 12 mm were considered, and Reynolds numbers based on channel hydraulic diameter were varied from approximately 1000 to 10,000. Experiments were performed for both water and FC-77, and the key conclusions are as follows.

1. Without pin fins, the thermal resistance of the unaugmented heat source ranged from approximately 20 to 60 $cm^2 \cdot K/W$ in FC-77 and from approximately 3 to 10 $cm^2 \cdot K/W$ in water. Average convection coefficient data for FC-77 and water could be collapsed in terms of the parameter $\overline{Nu}_L/Pr^{0.37}$.
2. Use of the pin arrays provides a substantial reduction in the thermal resistance, with the largest reduction corresponding to large pin diameters and/or small pin pitches. The best performance was associated with a pin diameter and pitch of 4.76 mm and 1.185, respectively, for which thermal resistances ranged from approximately 0.5 to 1.3 $cm^2 \cdot K/W$ in FC-77 and 0.15 to 0.35 $cm^2 \cdot K/W$ in water. These results represent approximately forty- and twenty-fold reductions, respectively, in the corresponding ranges for the unaugmented heat source.
3. When comparisons are made on the basis of equivalent interpin velocities, the thermal resistance is consistently smaller for the longer pins in FC-77 and for the longer pins of large diameter in water. However, for large interpin velocities in water, the thermal resistance increases with pin length for the smaller pin diameter.
4. When reduced in terms of $\overline{Nu}_{eff}/Pr^n$, data for the short pins are larger (for the small pin diameter) or comparable (for the large pin diameter) to results for the long pin. For $S/d \geq 1.65$, data for water and FC-77 are well correlated by a Prandtl number exponent of $n = 0.4$.
5. Although results for $\overline{Nu}_{eff}/Pr^{0.4}$ are overpredicted by standard tube bank correlations, friction factors are comparable to results obtained from tube bank correlations.
6. From the standpoint of thermal performance per unit pump power, improved performance is associated with decreasing pin diameter and increasing pin packing density. Also, for conditions corresponding to small convection coefficients, as for operation in FC-77 or at low Reynolds numbers in water, improved performance is provided by the longer pins.

ACKNOWLEDGEMENT

Support of this work by a grant from the IBM Corporation is gratefully acknowledged.

NOMENCLATURE

A	surface area of the unfinned heat source
A_b	exposed base area of the finned heat source
A_c	channel cross-sectional area
A_{min}	channel cross-sectional area minus the frontal area of the pins
c_p	specific heat at constant pressure of the liquid
D	hydraulic diameter of the channel, $4A_c/2(W + L_c)$
d	pin diameter
f	friction factor
$\overline{h}$	average heat transfer coefficient over the unfinned surface

k_o	thermal conductivity of the fluid at the upstream temperature
L_p	pin length
L_c	channel height
L_h	length of the heat source in the streamwise direction
m	mass flow rate of coolant
$\overline{Nu}_L$	average Nusselt number on unfinned heat source
N_L	number of pin rows
$\overline{Nu}_{eff}$	effective Nusselt number on pin and base surfaces
	pressure
P_p	pump power
Q	volumetric flow rate of coolant
	heater power
q_b	base heat transfer rate
q_p	pin heat transfer rate
R''_T	thermal resistance
Re_D	Reynolds number based on channel hydraulic diameter
$Re_{d,max}$	Reynolds number based on pin diameter and maximum velocity
Re_L	Reynolds number based on heat source length
S_l, S_t	longitudinal and transverse pitches of the pin array
T_e	fluid exit temperature
T_f	film temperature
$\bar{T}_h$	average surface temperature of the base
T_o	fluid temperature upstream of the heat source
V	average fluid velocity in the channel
V_{max}	maximum fluid velocity in the pin array, Q/A_{min}
W	channel width
ΔT_{lm}	log mean temperature difference
μ_o, μ_h	dynamic viscosity of the coolant at the upstream and heater surface temperatures
	kinematic viscosity of the coolant
	coolant mass density

REFERENCES

1. Chu, R.C., Hwang, U.P., and Simons, R.E., Conduction Cooling for an LSI Package: A One-Dimensional Approach, *IBM J. Res. Develop.*, vol. 26, pp. 45-54, 1982.
2. Oktay, S., and Kammerer, H.C., A Conduction Cooled Module for High-Performance LSI Devices, *IBM J. Res. Develop.*, vol. 26, pp. 55-66, 1982.
3. Mizuno, T., Okano, M., Matsuo, Y., and Watari, T., Cooling Technology for the NEC SX Supercomputer, *Proc. Int. Symp. Cooling Tech. for Elec. Equip.*, Hawaii, pp. 110-125, 1987.
4. Chu, R.C., Bergles, A.E., and Seely, J.H., Survey of Heat Transfer Techniques Applied to Electronic Packages, IBM TR 00.2869, May 1977.

5. Preston, S.B., and Shillabeer, R.N., Direct Liquid Cooling of Microelectronics, *Proc. INTER/NEPCON, P, IX,* pp. 10-31, 1970.

6. Simons, R.E., and Moran, K.P., Immersion Cooling Systems for High Density Electronic Packages, *Proc. NEPCON*, Chicago, pp. 396-409, 1977.

7. Baker, E., Liquid Immersion Cooling of Small Electronic Devices, *Microelectronics and Reliability*, vol. 12, pp. 163-173, 1973.

8. Kelecy, F.J., Ramadhyani, S., and Incropera, F.P., Effect of Shrouded Pin Fins on Forced Convection Cooling of Discrete Heat Sources by Direct Liquid Immersion, *Proc. ASME-JSME Thermal Engineering Joint Conference*, Hawaii, vol. 3, pp. 387-394, 1987.

9. Ramadhyani, S., and Incropera, F.P., Forced Convection Cooling of Discrete Heat Sources With and Without Surface Enhancement, *Proc. Int. Symp. Cooling Tech. for Elec. Equip.*, Hawaii, pp. 249-264, 1987.

10. Goldstein, R.J., and Karni, J., The Effect of a Wall Boundary Layer on Local Mass Transfer from a Cylinder in Crossflow, *ASME J. Heat Transfer*, vol. 106, pp. 260-267, 1984.

11. Saboya, F.E.M., and Sparrow, E.M., Transfer Characteristics of Two Row Plate Fin and Tube Heat Exchanger Configurations, *Int. J. Heat and Mass Transfer*, vol. 19, pp. 41-49, 1976.

12. Saboya, F.E.M., and Sparrow, E.M., Local and Average Transfer Coefficients for One-Row Plate Fin and Tube Heat Exchanger Configurations, *ASME J. Heat Transfer*, vol. 96, pp. 265-272, 1974.

13. Lau, S.C., Kim, Y.S., and Han, J.C., Effects of Fin Configuration and Entrance Length on Local Endwall Heat/Mass Transfer in a Pin Fin Channel, ASME Paper 85-WA/HT-62.

14. Steuber, G.D., and Metzger, D.E., Heat Transfer and Pressure Loss Performance for Families of Partial Length Pin Fin Arrays in High Aspect Ratio Rectangular Ducts, *Heat Transfer-1986, Proc. Eighth International Heat Transfer Conf.*, vol. 6, pp. 2915-2920, 1986.

15. Sparrow, E.M., Baliga, B.R., and Patankar, S.V., Forced Convection Heat Transfer from a Shrouded Fin Array With and Without Tip Clearance, *ASME J. Heat Transfer*, vol. 100, pp. 572-579, 1978.

16. Kadle, D.S., and Sparrow, E.M., Numerical and Experimental Study of Turbulent Heat Transfer and Fluid Flow in Longitudinal Final Arrays, *ASME J. Heat Transfer*, vol. 108, pp. 16-23, 1986.

17. Sparrow, E.M., and Kadle, D.S., Effect of Tip-to-Shroud Clearance on Turbulent Heat Transfer from a Shrouded, Longitudinal Fin Array, *ASME J. Heat Transfer*, vol. 108, pp. 519-524, 1986.

18. Sparrow, E.M., and Ramsey, J.W., Heat Transfer and Pressure Drop for a Staggered Wall-Attached Array of Cylinders With Tip Clearance, *Int. J. Heat and Mass Transfer*, vol. 21, pp. 1369-1377, 1978.

19. Sparrow, E.M., Ramsey, J.W., and Altemani, C.A.C., Experiments on In-Line Pin Fin Arrays and Performance Comparisons With Staggered Arrays, *ASME J. Heat Transfer*, vol. 102, pp. 44-50, 1980.

20. Simoneau, R.J., and Van Fossen, G.J. Effect of Location in an Array on Heat Transfer to a Short Cylinder in Crossflow, *ASME J. Heat Transfer*, vol. 106, pp. 42-48, 1984.

21. Incropera, F.P., and DeWitt, D.P., *Fundamentals of Heat and Mass Transfer*, Wiley, New York, 1981.

22. Grimison, E.D., Correlation and Utilization of New Data on Flow of Gases over Tube Banks, *Trans. ASME*, vol. 59, pp. 583-594, 1937.

23. Zhukauskas, A., Heat Transfer from Tubes in Cross Flow, *Advances in Heat Transfer*, vol. 8, pp. 93-160, 1972.

Cold Plates for IBM Thermal Conduction Module Electronic Modules

U. P. HWANG and K. P. MORAN
International Business Machines Corporation
P.O. Box 390
Poughkeepsie, New York 12602, USA

ABSTRACT

This paper presents analytical and experimental works on the cold plate designs used in IBM's 308X, 3090, and 3090-E computer systems. The water-cooled cold plate is mechanically attached to the top surface of the thermal conduction module (TCM) by screws. Within the cold plate are coolant passages to guide and control the water flow through the cold plate. Heat dissipation from the TCM module is transferred by conduction through interface between TCM and cold plate, and cold plate fin base and fins, and finally by convection into the water. Thermal contact resistance was studied as a function of contact surface flatness, surface treatment, screw torque, and screw placement. The convection heat transfer and pressure drop characteristics were experimentally investigated. The system application was in the flow transition regime with a short entrance region. The maximun coolability of the TCM modules with the cold plates is calculated.

INTRODUCTION

With the development of very large scale integration (VLSI) technologies, the demand for faster circuits and increased capacity has given rise to both increased power dissipation per circuit and an increased number of circuits per packaging volume. This high density packaging has increased power densities at the chip, module, and system level. Many thermal techniques are required in order to maintain temperature limits within the electronic package throughout the test and assembly process [1,2].

In order to keep VLSI device junction temperature within both functionality and reliability limits, a thermal conduction module (TCM) [3,4] is used to provide heat transfer from the chip through module to the system heat sinks in IBM 308X,3090, and 3090-E systems. The TCM, shown in Figure 1, is a hermetically sealed module containing a 90 or 127 mm sq multilayer ceramic substrate designed to hold 100 to 132 semiconductor chips. The spring-loaded pistons contacting each chip allow for possible variations in chip height,

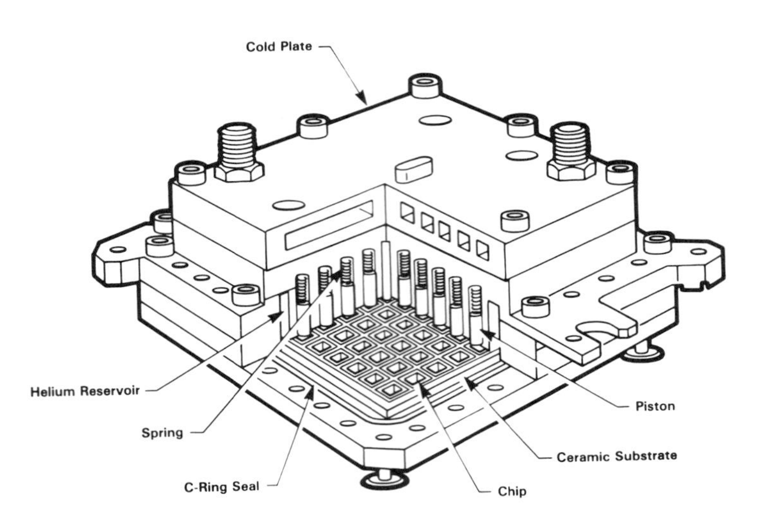

Figure 1. IBM Thermal Conduction Module (TCM) with water-cooled separable cold plate.

tilt, and location resulting from manufacturing tolerances. Heat is transferred from the chips across the helium gap via the pistons to the housing and eventually transferred to the cold plate. The water-cooled cold plate is mechanically attached to the top surface of the TCM module by screws. Within the cold plate are coolant passages to guide and control the water flow through the cold plate.

COLD PLATE DESIGNS

The basic structure of the cold plate can be realized as a thin plate with multiple straight fins. A housing covers the multiple fins so that water can be pumped in, and be circulated through fin spacing. The flow configurations are shown in Figure 2.

The cold plate material used is beryllium copper (BeCu), alloy CDA 175. BeCu gives high thermal performance, long life in the cold plate application, and easy manufacturability. BeCu has high corrosion resistance, is easy to machine, and can be satisfactorily joined by silver brazing. In addition, CDA 175 being greater than 97 percent copper may be electrolytic or electroless nickel plated and followed by tin plate. The latter plating is required in order to provide surface treatment to enhance the contact thermal conduction.

Table 1 lists key dimensions for cold plates used in IBM 308X, 3090, and 3090-E systems. Also listed are the application flow rates and heat transfer parameters that will be discussed in further detail.

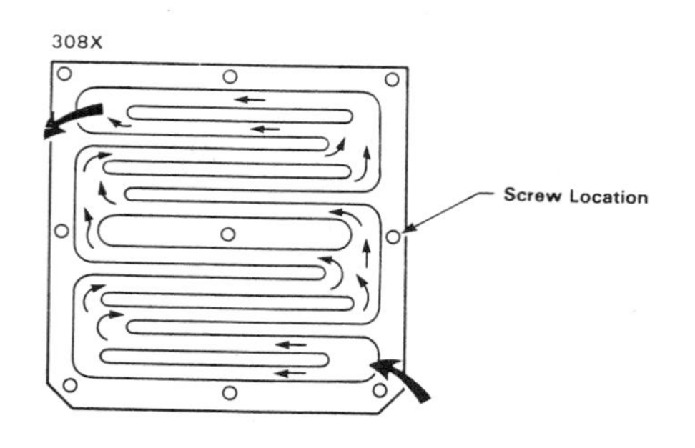

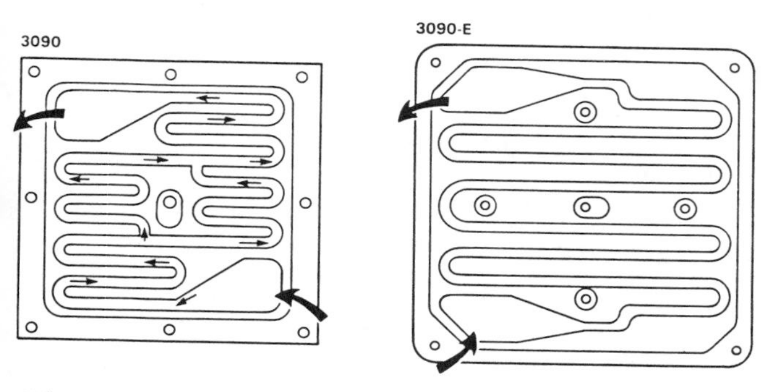

Figure 2. TCM Cold Plate Internal Flow Configuration.

Table 1. Cold Plate Dimensions.

		3080	3090	3090-E
Cold Plate Size	mm	107x104.5	106x106	115x115
Fin Height	mm	6.5	14.0	14.0
Fin Thickness	mm	3.0	4.3	6.3
Flow Width	mm	5.5	3.6	3.6
Flow Rate	l/min	2.3	3.4	3.4
Velocity	m/s	0.53	1.14	1.14
Hydraulic Diameter	mm	6.0	5.7	5.7
Fin Length	mm	76.0	76.0	88.0
Re		3653	7505	7505
Nu		51.5	92.5	92.5
Heat Transfer Coefficient W/m sq K		5431	10268	10268
Fin Efficiency	%	67.2	40.5	45.8

COLD PLATE THERMAL PERFORMANCE - ANALYTICAL CONSIDERATION

Heat is transferred from the TCM module to the cold plate through the interface between TCM and cold plate, and through the cold plate top surface to the water. The cold plate thermal performance can be determined by combining the interface thermal resistance and the cold plate thermal resistance, including conduction through fin base and fins, plus the convection from fins to water, including the water temperature rise.

The cold plate is mounted mechanically by screws on the TCM hat made of aluminum alloy 6061-T6. In general, this contact system consists of large contact surfaces, greater than 100 mm square. The average contact pressure based on the apparent cold plate contact area is estimated to be about 200 Newton/sq cm. The temperature of the contact surfaces was between 24-40°C, and the temperature drop across the interface was not to exceed 3°C. The interface ambient air pressure was about one atmosphere. Under these conditions, the radiation heat transfer across the interface is negligible. This thermal contact system can be classified as an interface formed by contacting dissimilar, non-conforming, large area surfaces in the presence of an air gap. There is elastic and plastic deformation of the bulk material and the contact surface asperities. There is no thermal contact resistance correlation which can be applied directly to this case. However, Yovanovich's correlation [5, 6] can be used to estimate the lower limit on the interface thermal resistance by considering the case of uniformly loaded, conforming, rough surfaces. The correlation is

$$\frac{1}{R} = 1.13\, K_s \left(\frac{P}{H}\right)^{0.94} \frac{m}{\delta} + \frac{K_f}{\eta\delta} \qquad (1)$$

where R = interface contact resistance; K_s = harmonic mean solid conductivity = $2\, K_1 K_2/(K_1 + K_2)$; K_1, K_2 = thermal conductivities of two contact solids; K_f = interstitial fluid conductivity; P = apparent contact pressure; H = hardness of the softer solid; δ = rms surface roughness = $\sqrt{\delta_1^2 + \delta_1^2}$; m = effective surface slope = $\sqrt{m_1^2 + m_2^2}$; and η ranges between 3 and 4 for most applications [5,6]. The first term of the equation correlated the solid contact conductance and the second term the interstitial fluid gap conductance.

For this application, typical values of the thermal and physical properties are listed below.

		BeCu	Al 6061-T6
Conductivity	W/m C	259.5	171.3
Hardness	N/m sq	19.6x10E8	9.32x10E8
Surface Roughness	m	1.06x10E-6	.56x10E-6

P = 200 N/cm sq
K_s = 206.4 W/m C
H = 9.32 x 10E8 N/m sq
K_f = 0.02675 W/m C
δ = 1.19 x 10E-6 m
m = .07 [5, 6]
η = 3.2

With the above values, the contact resistance can be calculated as

R = 0.194 C cm sq/W

This is a useful, approximate correlated value to provide a lower bound for the interface thermal resistance.

As previously noted, the cold plate thermal resistance is a series of resistances including the conduction through the fin base and fins, and the convection from fins to water, including the water temperature rise. In order to calculate the cold plate thermal resistance, the information needed is the convective heat transfer coefficient. Considering the system operating flow rates as shown in Table 1, the Reynolds numbers range between 3600 and 7500. Therefore, the convection heat transfer is in the transition zone. In addition, since the flow length to diameter ratios are less than 17, convection will be strongly enhanced by entrance effects. Although there were many heat transfer correlations to address internal flows for laminar, transitional, and turbulent conditions with entrance effect, there is no correlation that could be applied directly to this case. Although a numerical model [7] was used to to evaluate cold plate thermal resistance, experimental work was needed to provide a general correlation equation.

EXPERIMENTS

Experimental Set-Up

Experiments were conducted to measure the interface and cold plate thermal resistances. The experimental set-up is shown schematically in Figure 3. The closed loop flow system consists of a cold plate, flow meter, chiller (constant temperature bath), and flow control valve. To simulate the heat dissipation from a TCM, four 9.5 mm diameter cartridge heaters were installed in a 25.4 mm thick aluminum block. A power supply is used to power the four parallel connected heaters. The cold plates were plated with 2.54-5.08 μm

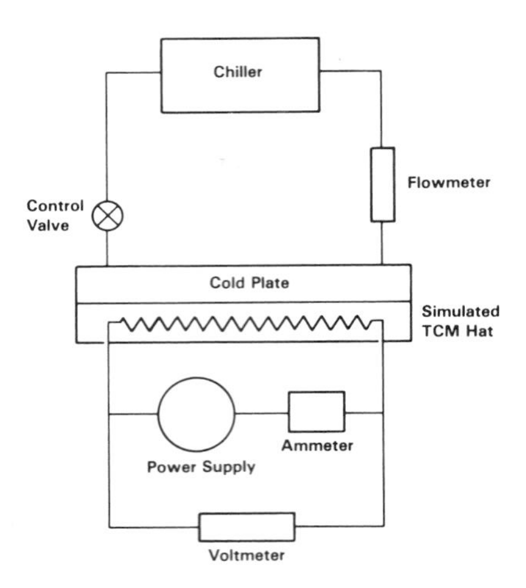

Figure 3. Electrical/Flow Schematic.

electroless nickel followed by 5.08-12.7 μm electroplated tin. The softness of tin provides for the easy deformation of surface asperities. The positions for locating screws used attached the cold plate to the heater block are shown in Figure 2. Nine M4 screws are used. Tests were performed at several torque levels. The clamping force from which clamping pressures can be obtained was computed using the equation T =0.2xdxF, where T is the screw torque, d is the screw major diameter, and F is the force. Based on this equation, a torque of 8.56 cm-Newton applied to the nine screws attaching the cold plate to the heater block resulted in an average clamping pressure of approximately 200 N/sq cm.

Since the contour of the mating surfaces affected the thermal interface resistance, profile measurements (tally-surf) and flatness contours (Flatness Analyzer by Tropel, Inc.) were taken of all cold plate and heater block test specimens. In general, the mating surfaces of all cold plates and heater blocks were concave. As a means of correlating the data, the maximum interface gap prior to tightening the screws attaching the cold plate to the heater block was computed from the profile (flatness measurements) as illustrated in Figure 4. A total of 20 temperatures were measured in the cold plate/heater block assembly with 30 gauge copper constantan thermocouple wire connected to an IBM PC data acquisition system. Nine thermocouples were installed in both the cold plate and heater block, close to the mating surfaces. One thermocouple for energy balancing was installed in both the inlet and outlet water lines of the cold plate.

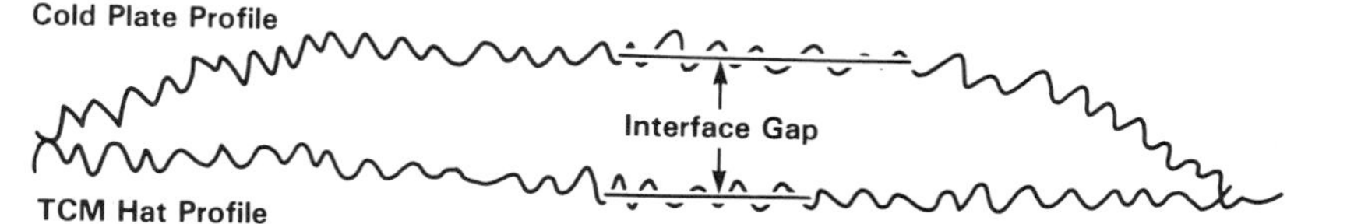

Figure 4. Characterization of Interface Gap.

Test Results

Figure 5 shows the results of the interface thermal resistance at 8.56 cm-Newton screw torque for 308X, 3090, 3090-E. The data scattered around in the plot of interface resistance vs. interface gap. The average interface thermal resistances are 0.009, 0.004, and 0.0024°C/W, respectively, for 308X, 3090, and 3090-E. Although the screw placements are identical for both 308X and 3090, the interface thermal resistance of the 308X cold plate is higher than that of the 3090. The reason may be due to the very thick solid bar located at the center of 308X cold plate yielding a stiff structure so that the mating surfaces can not be conformed by the center screw. The placement of five screws near the cold plate central portion for 3090-E did improve the interface thermal resistance, as expected.

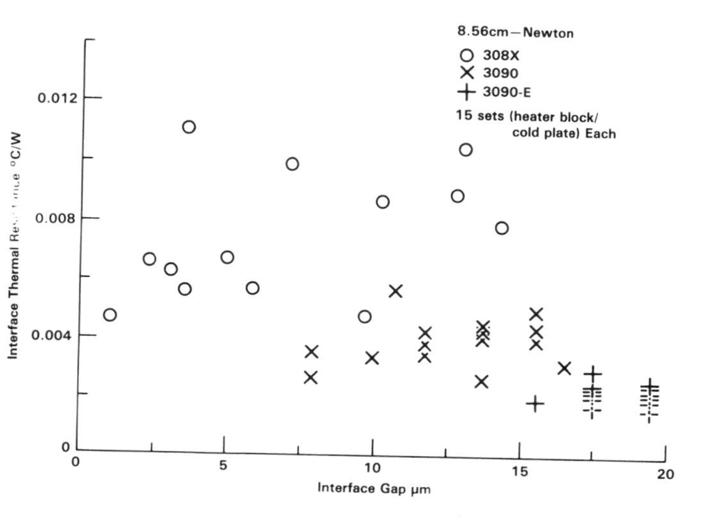

Figure 5. Interface Thermal Resistance vs. Interface Gap.

Figure 6 shows the results of the interface thermal resistance at various screw torque levels. The lower limit calculated by Yovanovich's equation is also shown for comparison. It can be seen that the 3090-E design is indeed approaching the calculated lower bound.

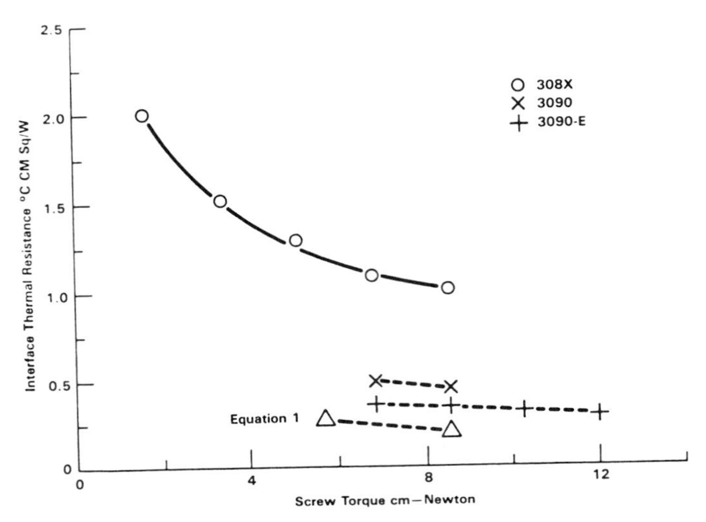

Figure 6. Interface Thermal Resistance vs. Screw Torque.

Figure 7 shows the results of the cold plate thermal resistances at various water flow rates for 308X, 3090, and 3090-E. The cold plate thermal resistance is the sum of the conduction resistance through fin base, the conduction/convection resistance from fins to water, and water temperature rise as illustrated in Figure 8. The conduction resistance through the fin base can be estimated by the conduction equation. The water temperature rise can be calculated by 1/mCp, where m = mass flow rate, and Cp = specific heat.

The conduction/convection resistance from fin to water can be expressed as

$$R_{fin} = \frac{1}{h(A_{base} + \eta_f A_{fin}} \tag{2}$$

and

$$\eta_f \quad \frac{\tanh \sqrt{2h/kt}\ (\ell + t/2)}{\sqrt{2h/kt}\ (\ell + t/2)} \tag{3}$$

where R fin =conduction/convection fin resistance, h = heat transfer coefficient, A base = area of fin base, A fin = area of fin,
η_f = fin efficiency, k = thermal conductivity of fin, t = fin thickness, ℓ = fin height.

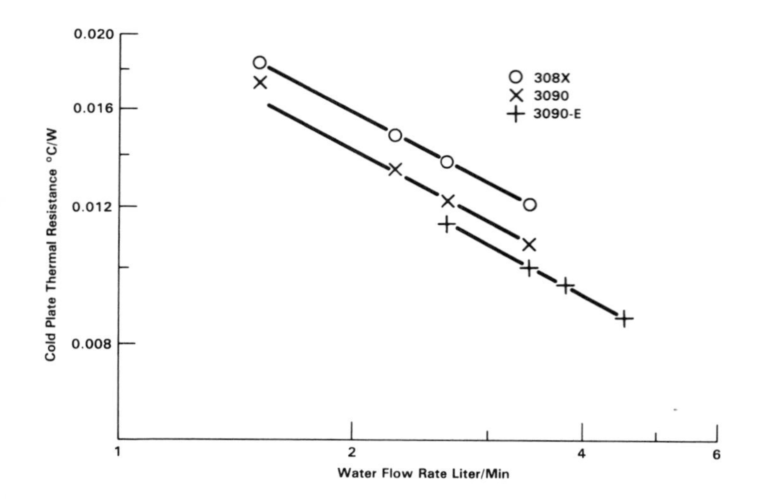

Figure 7. Cold Plate Thermal Resistance.

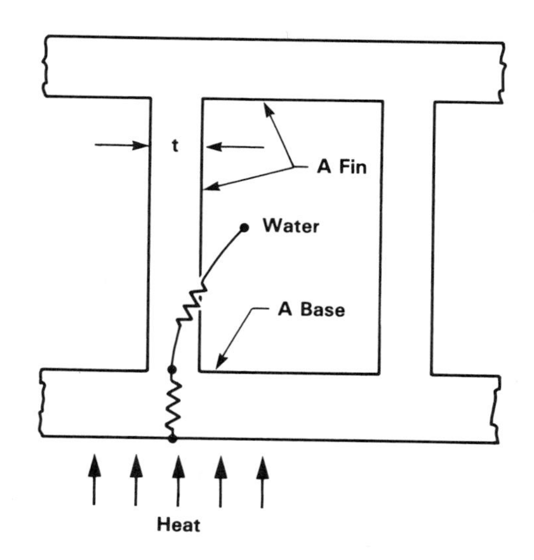

Figure 8. Cold Plate Heat Flow.

The heat transfer coefficient can be calculated from Figure 7 using the above relationship. The results are plotted in Figure 9 showing Nusselt number vs. Reynolds number.

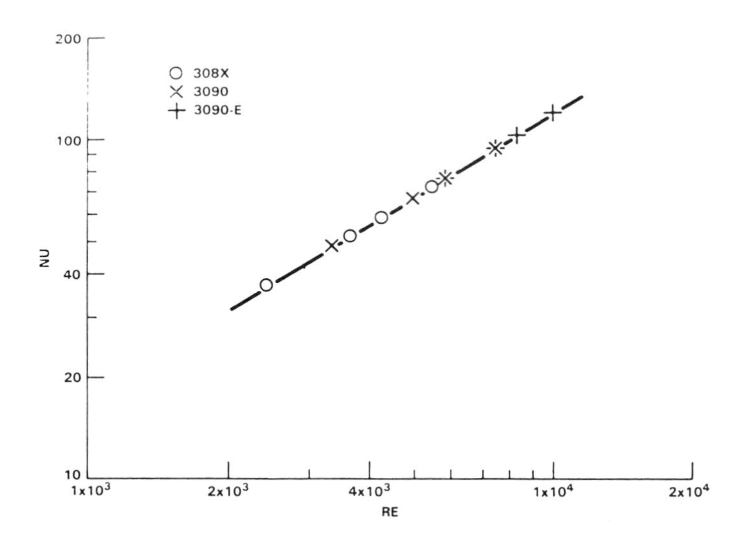

Figure 9. Heat Transfer Correlation.

The heat transfer coefficient is correlated by the following relationship:

$$Nu = 0.0316\, R_e^{\;0.8}\, P_r^{\;0.4} \qquad (4)$$

for this case of flow within the transition regime (Re=3600-7500) and for short entrance lengths length/diameter< 17).

Figure 10 shows the results of the water flow pressure drop characteristics of the cold plates. The pressure drop includes flow through fittings (4.6 mm I.D.), flow channel (fin spacing), and a series of 180 degree turns from channel to channel. The pressure drop is correlated as

$$\Delta P = 870 \times (\text{flow rate})^2 \qquad (5)$$

The results indicated that the major portion of the pressure drop is due to the head loss (fitting, and 180 degree turns). The head loss data are often expressed in the form of dimensionless loss coefficients, Kl, where

$$h_L = Kl\,\frac{V^2}{2g_c} \qquad (6)$$

where h_L = heat loss, V = velocity. In this case, the loss coefficients based on velocity at fitting can be computed to be 1.75. It is noted that the values of K is generally published in the range of 0.2 - 7.0 depends on geometry.

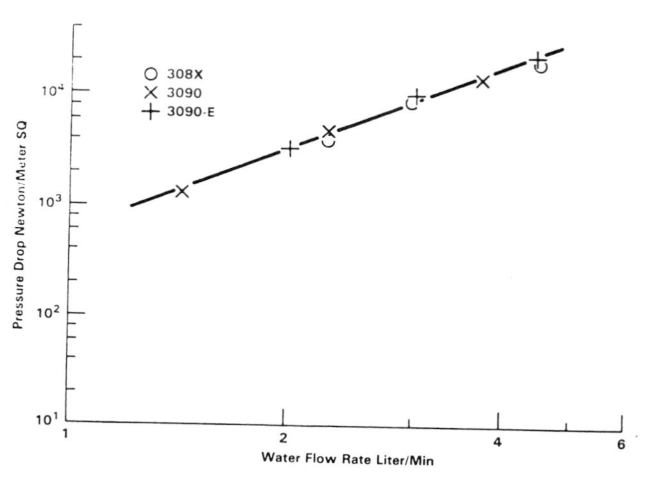

Figure 10. Pressure Drop Correlation.

COOLING LIMITS

The overall cold plate thermal performance is equal to the external thermal resistance which is the sum of interface and cold plate thermal resistances. Therefore, the external thermal resistances calculated from Figures 5 and 7 are 0.024, 0.0144, and 0.0122°C/W, respectively, for 308X, 3090, and 3090-E. With the fouling factors (0.004, 0.004, and 0.0031°C/W estimated), the end-of-life external thermal resistances of the cold plates are 0.028, 0.0184, and 0.0153°C/W for 308X, 3090, and 3090-E.

The maximum coolable chip power can be calculated from the following:

$$T_j = \Delta T_{j-c} + P_c R_{int} + P_m R_{ext} + T_w \tag{7}$$

$$\delta^2_{T_j} = \left(\frac{\partial T_j}{\partial P_c}\delta_{P_c}\right)^2 + \left(\frac{\partial T_j}{\partial P_m}\delta_{P_m}\right)^2 + \left(\frac{\partial T_j}{\partial R_{int}}\delta_{R_{int}}\right)^2 + \left(\frac{\partial T_j}{\partial R_{ext}}\delta_{T_{ext}}\right)^2 + \left(\frac{\partial T_j}{\partial T_w}\delta_{T_w}\right)^2 \tag{8}$$

$$= (\bar{R}_{int}\delta_{P_c})^2 + (\bar{R}_{ext}\delta_{P_m})^2 + (\bar{P}_c\delta_{R_{int}})^2 + (\bar{P}_m\delta_{R_{est}})^2 + \delta^2_{T_w},$$

where Tj = junction temperature of the chip, Δ Tjc = the temperature drop from the junction to the chip, 3°C typically, Pc = chip power, R int = internal thermal resistance from chip to module hat. Pm = module power, Rext = cold plate external thermal resistance, Tw = water temperature, δ = the standard deviation of each parameter, and the overscore = the mean value for each parameter.

To maintain maximum junction temperatures of 85°C, the statistical thermal parameters as shown in Table 2 will yield the maximum coolable power of 5.3, 8.1, and 9.0 Watts for 308X, 3090, and 3090-E.

Table 2. Thermal Parameters.

		3080		
R int	C/W	9.5 +/- 1.0	5.9 +/- .8	5.05 +/- .8
R ext	C/W	.028 +/- .005	.0184 +/-.002	.0153 +/-.002
Pc	Watts	? +/- 30%	? +/- 30%	? +/- 30%
Pm	Watts	300 +/- 10%	520 +/- 10%	770 +/- 10%
Tw	C	24 +/- 1.1	22.9 +/- 1.1	22.9 +/- 1.1

SUMMARY

This paper presents analytical and experimental works on the cold plate designs used in IBM's 308X, 3090, and 3090-E computer systems.

The cold plate made of beryllium copper CDA 175 is mounted by screws on the TCM hat made of aluminum alloy 6061-T6. This thermal contact system can be classified as an interface formed by contacting dissimilar, non-conforming, large area surfaces in the presence of an air gap. There is elastic and plastic deformation of the bulk material and the contact surface asperities. Yovonavich's correlation (1) provides a useful lower bound for the contact resistance. However, the major heat path of this contact system is the heat conduction through the air gap. To minimize this thermal resistance, the cold plate contact surface is plated with 2.54-5.08 μm electroless nickel followed by 5.08-12.7 μm electroplated tin. The softness of tin provides for the easy deformation of surface asperities. The large area contact surface also presents a flatness concern because of the difficulty of surface machining. A machined surface usually is concave, resulting in a large air gap at the center of the contact surfaces. This high contact thermal resistance is reduced by placing screws closer to the central portion of the contact surface.

The cold plate internal configuration can be described as a straight fin heat sink with water flowing through the fin gaps. The convection heat transfer is in the transition regime (Reynolds no. = 3600-7500) with a flow entrance-region of length to diameter ratio <17. The heat transfer is correlated by equation (4).

The water flow pressure drop through fittings, flow channel (fin spacing), and 180 degree turns from channel to channel is correlated by equation (5). Results indicated that the major portion of the pressure drop is due to the head loss (fittings, and 180 degree turns).

With these cold plate designs, maximum chip temperatures were maintained at 82°C while chip power increased from 5.3 to 8.1 to 9.0 watts, respectively, for IBM TCM modules in 308X, 3090, and 3090-E systems.

REFERENCES

1) Moran, K. P. and Simons, R. E., "Immersion Cooling Systems for High Density Electronic Packages," in Nat. Elec. Packaging and Production Conf. (NEPCON) Proc., February 1977, pp. 396-409.

2) Hwang, U. P. and Moran, K. P., "Boiling Heat Transfer of Silicon Integrated-Circuit Chips Mounted on a Substrate," ASME HTD-Vol. 20, pp. 53-60, November 1981.

3) Chu, R. C., Hwang, U. P., and Simons, R. E., "Conduction Cooling for an LSI Package: A One-Dimensional Approach," IBM J. Research and Development, Vol. 26, No. 1, pp. 45-54, January 1982.

4) Moran, K. P., et al., "Thermal Design of the IBM 3081 Computer," NEPCON West, 1982.

5) Yovanovich, M. M., "Thermal Contact Resistance in Microelectronics," in National Electronic Package and Production Conf. (NEPCON) Proc., pp. 177-188, 1978.

6) Antonetti, V. W. and Yovanovich, M. M., "Thermal Contact Resistance in Microelectronic Equipment," Int. J. for Hybrid Microelectronics, Vol. 7, No. 3, pp. 44-50, September 1984.

7) Agonafer, D., Chu, R. C., and Hwang, U. P., "Numerical Modeling of the Internal and External Thermal Resistance of a Thermal Conduction Module," Proc. of the Int. Symposium on Cooling Technology for Electronic Equipment, pp. 126-137, March 1987.

hhancement of Heat Transfer for Vapor ondensation in Evaporative Systems Cooling

GOGONIN and O. A. KABOV
titute of Thermophysics
erian Branch of the USSR
ademy of Sciences
vosibirsk, USSR

ABSTRACT

experimental results on heat transfer for film condensation of stationary e vapour on horizontal tubes with transverse fins of different dimensions reported. The height of liquid layer retained consistently between the fins measured. The ratio between interfin spacing and liquid capillary constant found to affect significantly the heat trasfer. For vapour condensation on tubes under the conditions of liquid flooding the heat transfer decrease is imum for the values of this ratio ranging from 0.4 to 1.0.

INTRODUCTION

our condensers are the integral components of liquid cooling systems for ctronic equipment, in which heat is removed in the process of boiling or porating of a liquid [1,2]. Refrigerants are frequently used as working ids. The decrease of specific quantity of metal in cooling systems is sible due to the enhancement of heat transfer processes occurring in them. surface finning is the most widely used method to enhance heat transfer for film condensation of vapour [3]. Of late a number of theoretical works [4,5], erimental investigations [6,7] and surveys [8] have been published.

aim of the present experimental study is:
o investigate the effect of geometrical finning characteristics, heat flux liquid properties on the heat transfer intensity at condensation of pure urated stationary vapour on horizontal tubes with transverse fins;
o study the effect of condensate flow rate on heat transfer for vapour densation over a vertical row of horizontal tubes.

EXPERIMENTAL TECHNIQUE

rigerants R11, R12 and R21 were used as working fluids in the experiments. The uration temperature fluctuated in the range from 20 to 100 °C. The finned es were mounted in a cylindrical condenser with an inner diameter of 400 mm length of 585 mm. The condenser design permitted mounting of from 1 to 10 es in it. Test tubes were arranged in a vertical row with spacing between axes being equal to 31 mm. The dependence of heat flux density on saturated our-wall temperature difference was studied for each type of tube. The t flux was determined on the basis of the enthalpy variation of cooling er. The average temperature of the tube wall was measured by thermocouples. efficiency factor of fins [3] was not less than 0.94; therefore, the fin surface was considered to be isothermal. Thermocouples with a thickness 0.15 mm were placed into two sections along the tube length in grooves c along the tube. Five thermocouples were imbedded in each section eve 45mm along the tube perimeter starting from the upper tube. The grooves we closed by a copper plate which followed precisely the fin profile. The er in defining the heat transfer coefficient averaged over the surface did r exceed 10% provided the values of the vapour-wall temperatu difference $\Delta T>2$ °C. A detailed description of the test rig utilized in experiments is presented in Reference [9].

In the experiments the fins were of rectangular and trapezoidal profiles (Figure 1). The heat transfer measurements were performed on fifteen fin tubes, the geometrical dimensions of which are presented in Table 1.

TABLE 1 Test tube dimensions.

Series	Material	D, mm	D_o, mm	h, mm	b, mm	a, mm	s, mm	φ, g
	Brass	19.78	17.72	1.03	0.50	0.25	0.75	0
	Copper	19.90	17.88	1.01	0.52	0.41	0.93	0
	Copper	19.86	17.68	1.09	0.50	1.00	1.50	0
1	Copper	19.75	17.73	1.01	0.46	2.14	2.60	0
	Copper	19.87	17.77	1.05	0.47	3.04	5.51	0
	Copper	19.89	17.73	1.08	0.56	6.94	7.50	0
	Copper	19.90	17.82	1.04	0.30	0.95	1.25	0
	Copper	19.80	17.80	1.00	0.40	0.96	1.35	0
2	Copper	19.89	17.77	1.06	0.75	1.00	1.75	0
	Copper	19.87	17.77	1.05	1.18	1.02	2.20	0
	Copper	19.90	17.82	1.04	2.24	0.97	3.21	0
	Copper	21.50	16.50	2.50	0.56	0.37	2.00	12.1
	Copper	20.30	17.10	1.60	0.54	0.51	1.92	15.2
3	Copper	20.60	17.78	1.41	0.47	0.46	2.00	20.8
	Copper	16.74	14.28	1.23	0.44	0.42	1.32	10.6

Three series of experiments were carried out. In the first series inter spacing was the only variable geometrical dimension. The ratio between inter spacing and capillary constant of the liquid varied in the range $0.24<\tilde{a}<11$ In the second series of experiments the width of the fin face was the o variable dimension. The parameter $\tilde{b}$ varied in the range from 0.28 to 3.8. the third series heat transfer on the tubes with the trapezoidal fins where fin height changed greatly ($0.9<\tilde{h}<3.3$) was investigated. A smooth tube mounted simultaneously with the finned tubes in a condenser to verify experimental technique. At the values of the film Reynolds number Re<20 measured Nusselt numbers were 5-15% greater than those calculated by Nusselt theoretical dependence [10].

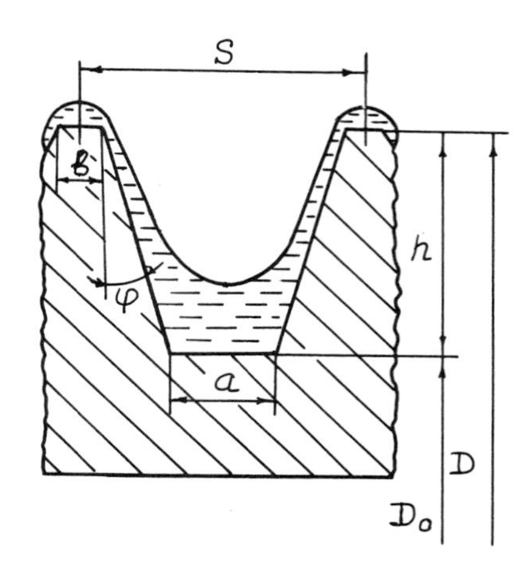

FIGURE 1. Finned surface profile.

3. LIQUID FLOW OVER HORIZONTAL FINNED CYLINDERS

The liquid flow for vapour condensation on a finned tube in the bank is shown schematically in Fig.2. One of the peculiarities of vapour condensation is the capillary retention of liquid between the fins. This process was studied both for R12 condensation and simulating the condensation process. The set-up allowed deluging the horizontal finned cylinder with a preset liquid flow rate. These experiments were carried out at atmospheric pressure under isothermal

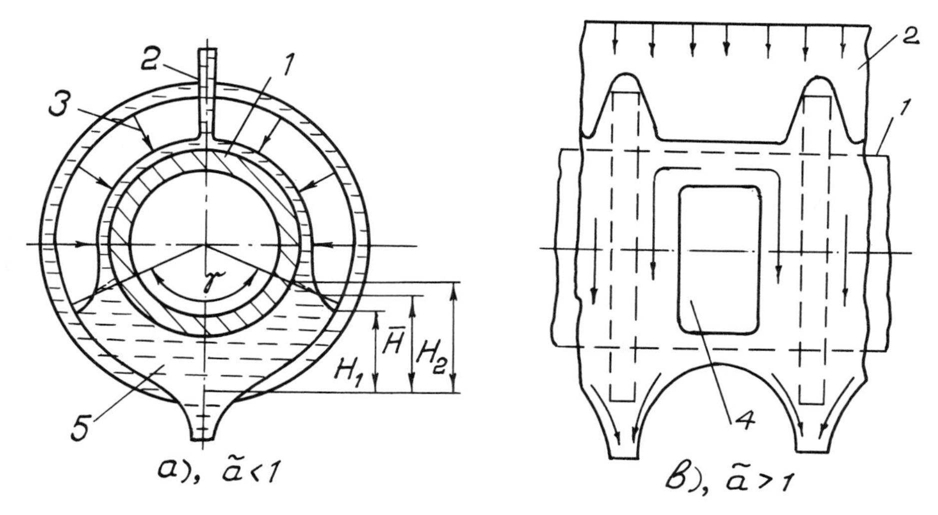

Figure 2. Liquid flow pattern for vapour condensation on a finned tube in a bank, 1 - finned tube, 2 - flowing condensate, 3 - direction of liquid flowing off the fin side surface, 4 - domain with thin condensate film, 5 - liquid layer retained costantly between fins.

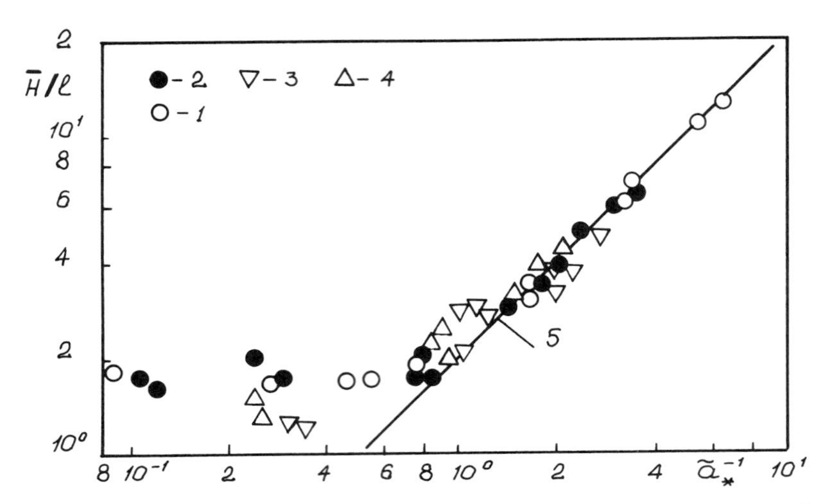

Figure 3. Capillary liquid retention height vs. interfin spacing, 1 - wate T =25°C, 2 - R12, T''=20-70°C - rectangular fins, 3 - water, T=25°C, 4 - 17% (mass) ethanol aqueous solution, T=25°C - trapezoidal fins, 5 - calculati according to (1).

conditions. Distilled water and ethanol aqueous solutions were used as worki fluids. The height of complete flooding H_1 and maximum flooding H_2 of the f side surface with liquid were measured by means of a cathetometer (see Fig. 2 The fin height varied over the range h=1.0-8.0 mm, the external diameter tubes changed within the range D=9.6-60 mm, and the inclination angle of t fin side surface varied over the range φ=7-50°.

The data obtained are generalized in Figure 3. Experimental results are comp red with the calculation using the Bressler and Wyatt approximate correlati [11] for complete flooding the channels with the liquid on a vertical pla

$$H_1=2\sigma/(\rho'-\rho'')ga_*,\quad a_*=\frac{a\cos\varphi+h\sin\varphi}{1-\sin\varphi}\,. \qquad (1)$$

Experimental data are in satisfactory agreement with the calculation for v lues of the parameter $\tilde{a}_*$>1.1. For rectangular fins $a_*\equiv a$ and Eq.(1) coincides with the relation for the height of capillary liquid rise in an inf nite vertical slot of width equal to a fin-to-fin spacing $H_0=2\sigma/(\rho'-\rho'')$ [12]. In this case the angle of the capillary liquid retention on the tube c be determined from the expression [12]

$$\gamma=2\arccos\left[\left(1-\frac{4\sigma}{(\rho'-\rho'')gaD}\right)/\frac{D}{(D-h)}\right],\qquad \tilde{a}<1.1.$$

Results of our experiments [12] are in good agreement with the data obtained reference [13].

In the case of deluging tubes with liquid for parameter values $\tilde{a}$<0.4 a Re>25 the height of the liquid layer retained constantly between the fins creases. At $\tilde{a}$<0.2 flooding of rectangular grooves along the entire perime of the cylinder took place at Re>15.

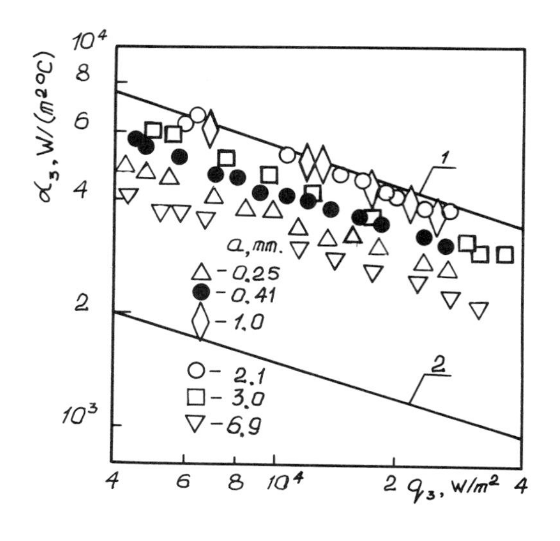

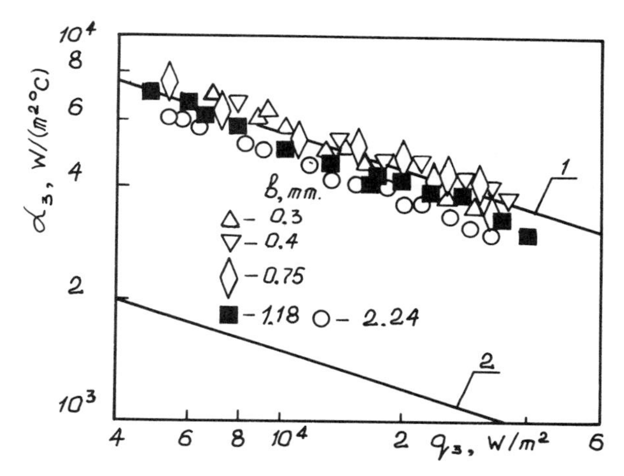

igure 4. Heat transfer for condensation of R12 on tubes with different inter-in spacing and different width of fin face, 1 - calculation according to 10] for a vertical plate with the height equal to that of a fin h=1 mm, 2 - alculation according to [10] for a smooth tube $D=D_0$.

n the experiments on tube banks the tubes were deluged by a stabili-ed continuous film. Streamlines were visualized with the help of small air ubbles.Three characteristic regimes can be singled out when liquid flows n the valley [14]. The film profile in the valley was of semi-circular hape for the parameter values $\tilde{a}<1$. For the parameter values $1<\tilde{a}<4$ and Re<200 fter flowing for some time in the valley the liquid was divided into two streams, ue to the effect of the capillary forces,and flowed down along the fins (see ig.2b).The domain with a thin liquid film was formed in the centre of the alley. The position of the film break point in the channel was a function of the parameters $\tilde{a}$ and Re. For tubes with the parameters $\tilde{a}>4$ and Re>40 portion of liquid flowed down across the centre of the valley. Analog patterns of the liquid flow were shown in reference [15] where liquid flo down into a rectangular vertcal channel with varying width.

4. VAPOUR CONDENSATION ON SINGLE HORIZONTAL TUBES

In the experiments with heat transfer on single tubes the heat flux varied the range q_3=3.9 10^3-7.8 10^4 W/m², the temperature difference changed wit ΔT=1.0-44.0 °C and the film Reynolds number varied from Re=5 to Re=240. Fi presents a portion of experimental data obtained in the first and second s of the experiments. A considerable enhancement of heat transfer for condensat on the finned surface is observed. At the same heat flux the heat trans coefficient averaged over the surface was higher by a factor of 3.7 than t on a smooth tube of the diameter equal to that of the fin root. Data satisfactorily described by the relations $\alpha \sim \Delta T^{-1/4}$ and $\alpha \sim q^{-1/3}$ obtained Nusselt for condensation on a smooth surface.

Three surfaces, i.e., the side surfaces of fins and horizontal surfaces of fa and valleys are involved in the process of heat transfer for vapour condensat on a finned tube. Heat fluxes on each of them may be quite different. fact that the side surface of fins occupies 60-90% of the total external t surface is an important peculiarity of the tubes utilized in engineering. As a result of the liquid film curvature on the fin the gradient of capill pressure appears which can exceed greatly the effect of gravitational fo [16]. The condensate flows off the side surfaces of the fin predominantly un the effect of capillary forces. The pattern of the liquid flow on the side s face of the fin is shown in Fig.2a. As a result, the heat transfer coeffici on the side surface of the fin can be much greater than that on horizontal sections of a tube. This led to the suggestion of a new technique processing experimental data based on the following assumptions [12]:

- predominantly heat removal is from the side surface of the fins;
- heat transfer on horizontal sections of the tube may be defined on the ba of the relation for a horizontal smooth tube at the same value of the f Reynolds number;
- liquid flows off the side surface of the fin into the valley towards the t radius. The fin height is the characteristic linear dimension of t surface.

The above assumptions permit calculation of the heat amount removed from fa and valleys and to determine the heat flux q_h and the heat transfer coeffici α_h on the side surface of the fin. Using these values, the Nusselt number Reynolds number of the film were calculated by

$$Nu_h^* = \frac{\alpha_h}{\lambda}\left(\frac{\nu^2}{g(1-\rho''/\rho')}\right)^{1/3}, \quad Re = \frac{q_h h}{\mu r(1+3/8K)\cos\varphi} \qquad ($$

Fig.5 shows the effect of the film Reynolds number on the enhancement of h transfer at the side surface of the fin. Experimentally determined Nuss numbers are compared with the calculation by theoretical Nusselt dependence a vertical plate in the case of gravitational film flow [10],

$$Nu_o^* = 0.925 Re^{-1/3}. \qquad ($$

The lines which average experimental data are sloped similar to the line obta by Nusselt $Nu_h^* \sim Re_h^{-1/3}$. Experimental data for tubes with different values parameter $\tilde{a}$ are not generalized by common dependence in these coordinates. of the reasons is the capillary retention of liquid in the interfin valleys as a result a portion of the tube surface is virtually excluded from h exchange.

ig.6 shows the effect of fin dimensions on heat transfer. For the data reduction it was assumed that in the domain of the capillary retention of liquid the vapour condensation is absent. The thickness of the condensate film in flooded domain is approximately ten times higher than that in the unflooded domain. At the parameter value $0.3<\tilde{a}<2$ the relative Nusselt number increases with the increase of interfin spacing $Nu^*_\gamma/Nu^*_o \sim \tilde{a}^{1/3}$. In our opinion it is attributed to the decrease of the liquid film thickness in the interfin valley and the enhancement of heat transfer in the valley for $\tilde{a}>1$. For the parameter values $\tilde{a}>3$ the heat transfer enhancement on the finned tubes ceases to increase.

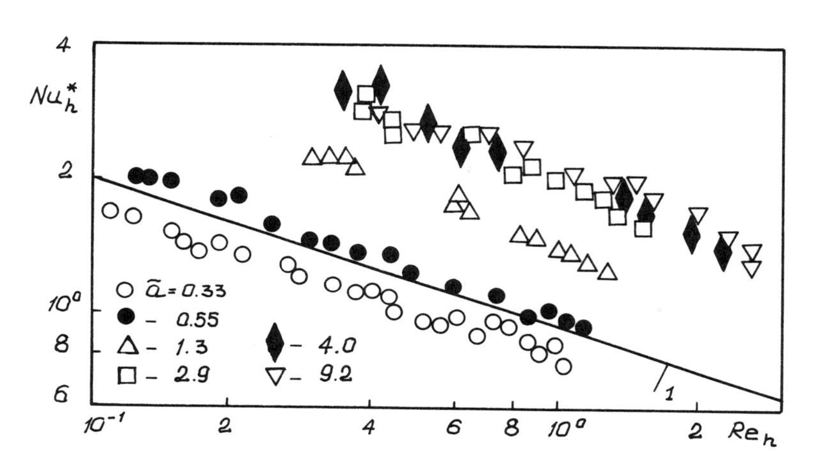

igure 5. Heat transfer for condensation on horizontal finned tubes, 12, $T''=40°C$, $\tilde{h}=1.4$, $\tilde{b}=0.7$, 1 - calculation by equation (3).

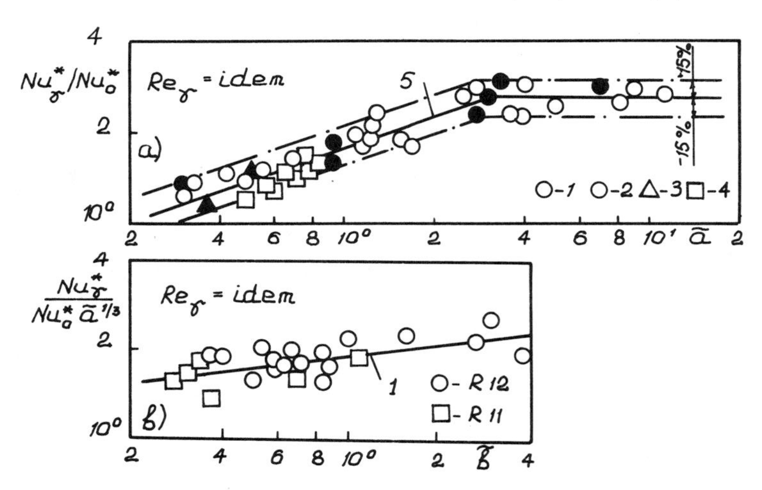

igure 6. Dependence of heat transfer on fin dimensions, a) interfin spacing effect, $0.4<\tilde{b}<1.0$, $0.9<\tilde{h}<3.3$, $T''=20-100°C$, 1 - R11, 2 - R12(rectangular fin profile), 3 - R21, 4 - R12(trapezoidal profile), 5 - equation $Nu^*_\gamma=1.65Re_\gamma^{-1/3}\tilde{a}^{1/3}$, b) effect of the fin face width, $0.3<\tilde{a}<1.7$, $T''=20-90°C$, - relation $Nu^*_\gamma=1.76Re^{-1/3}\tilde{a}^{1/3}\tilde{b}^{1/8}$.

The change of dimensionless fin height in the range $0.9<\tilde{h}<3.3$ does not result in a pronounced variation of the total heat transfer mechanism. Experiments on tubes with different thickness of the fins indicate that the width of the fin face slightly affects heat transfer ($Nu_{\gamma}^{*}\sim b^{1/8}$).

The application of the technique of processing experimental results which has been suggested by the authors and the introduction of parameter $\tilde{a}$ correlated all experimental data obtained in [12, 14, 17-22]. In these works experiments were carried out on thirty tubes with different dimensions of finning and for vapour condensation of five refrigerants (see Fig.7).

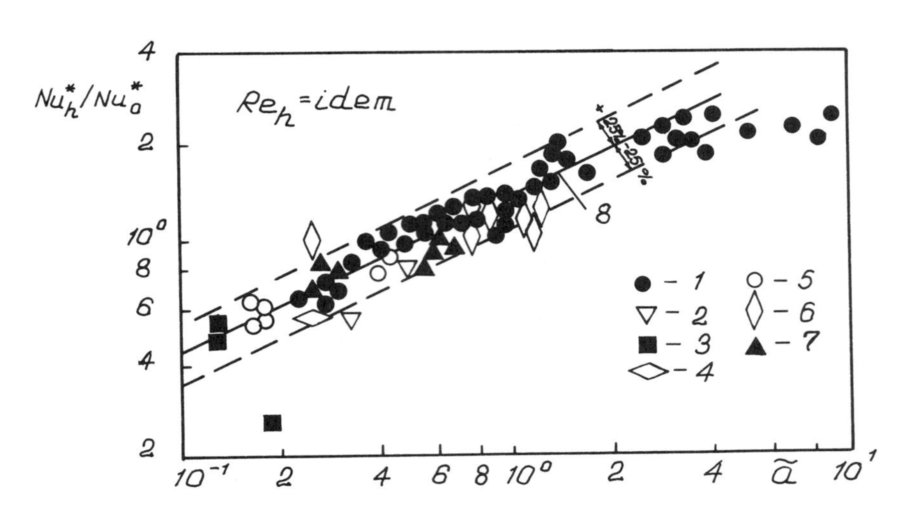

Figure 7. Vapour condensation on single finned tubes, 1 - R11, R12, R21, T''=20-100°C, 2 - R21, [17]; 3 - R113, [18]; 4 - R12, [19]; 5 - R11, [20]; 6 - R12, R22, [21]; 7 - R12, R22, [22]; 8 - calculation by equation (4).

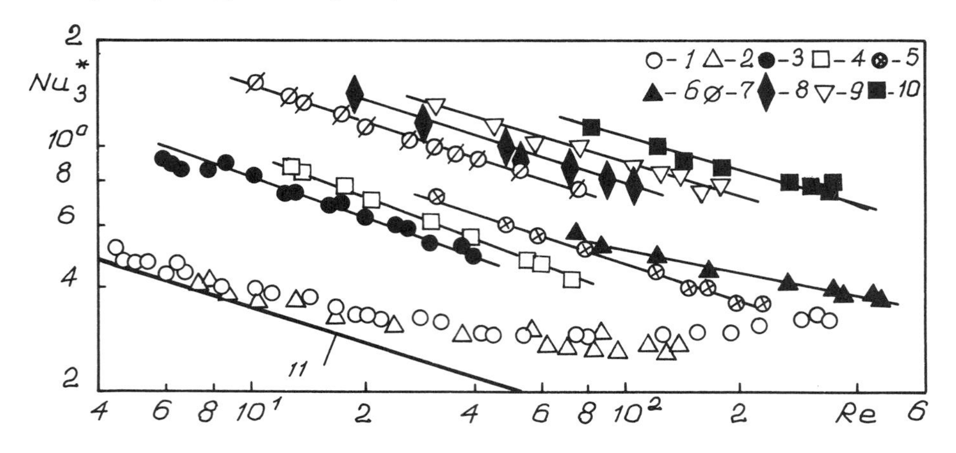

Figure 8. Heat transfer for condensation on tube banks, T''=40°C, 1 - R12; 2 - R11-bank of smooth tubes, D=16 mm; 3-6 - R12, $\tilde{a}$=9.2; 7-10 - R12, $\tilde{a}$=0.55; 11 - calculation by [10].

o calculate heat transfer we recommend the empirical relation

$$Nu_h^* = 1{,}25 Re_h^{-1/3} \tilde{a}^{1/2}, \quad (4)$$

hich is valid for the parameter values $0.1<\tilde{a}<3$, $0.13<\tilde{b}<4$, $0.9<\tilde{h}<3.3$

. EFFECT OF CONDENSATE FLOW RATE ON HEAT TRANSFER

he peculiarity of the liquid flow moving along a finned tube in the bank is hat the film thickness on a large portion of the side fin surface doesn't epend on the overhead flow rate and is a function of heat flux on a given ube. As a result the intensity of heat transfer is defined by three Reynolds umbers of the film: on the fin face, in the interfin valley Re_b, and on the side in surface Re_h. In Fig.8 the experimental data for two finned tubes are resented. There are quite different spacings between the fins on these ubes. The data are processed in the coordinates usually used to generalize ata on smooth tubes. The Nusselt number is derived on the basis of heat ransfer coefficient which is averaged for external surface of the finned tube. he results of calibration experiments on the banks of smooth tubes are given n the same figure. Experimental data on the banks of finned tubes in contrast o those on the banks of smooth ones are not generalized in these coordinates. he heat transfer is defined not only by condensate flow rate but also the mount of liquid condensed on the given tube.

he condensate film thickness on the greatest part of the fin side surface oesn't depend on the overhead flow rate which permits use of the Nusselt and eynolds numbers calculated by (2) for experimental data processing. The analy-is of the obtained data has shown that at the parameter value $0{,}4<\tilde{a}<1.0$ for ny tube of the bank the heat transfer is well described by the relation $Nu_h^* \sim Re_h^{-1/3}$.

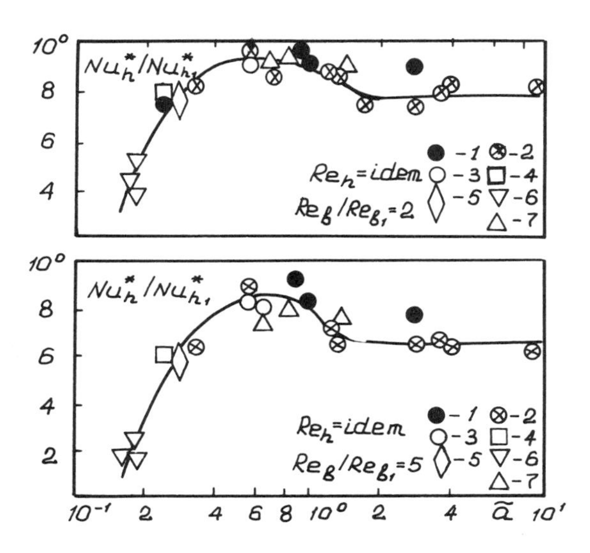

igure 9. Overhead flow rate effect on heat transfer versus dimensionless nterfin spacing, 1 - R11, 2 - R12-rectangular fins, 3 - R12-trapezoidal fins, - R12, [19], 5 - R12, [22], 6 - R11, [20], 7 - R12, n-butane, acetone, [23].

Fig.9 shows the processing of data obtained in [19,20,22-24] concerning th effect of parameter $\tilde{a}$ on the variation of heat transfer over the ban depth. Index 1 denotes the parameters for the tube without being supplied with condensate. The parameter Re_b/Re_{b_1} is the ratio of the film Reynolds number i the valley on a tube in the bank to that in the valley on a single tube. I characterizes the relative change of the liquid film thickness in the valle over the bank depth. In the range of the parameter value $0.4<\tilde{a}<1$ experimenta data are compared under condition Re_h=idem. In the regions $\tilde{a}<0.4$ and $\tilde{a}>1$ comparison is made under condition $Re_h=1.0$.

The obtained relation has a maximum at the parameter value $0.4<\tilde{a}<1.0$. Thi specific region is preferable when choosing an interfin spacing in condense consisting of several rows of tubes. Sharp decrease of the heat transfe intensity over the bank depth at the parameter value $\tilde{a}<0.3$ is due to floodir interfin valleys with a flowing liquid. The deterioration of heat transfer ove the bank depth at the parameter value $\tilde{a}>1$ may be explained by that of hea transfer in the interfin valley due to the reduction of the domain with a thi liquid film.

Fig.10 shows potential possibilities of heat transfer enhancement at vapou condensation on the tubes with the fins close to isothermal ones. The hea transfer coefficient on a finned tube (α_2) calculated on the basis of heat flu (q_2) carried to the surface of a smooth tube with the diameter $D=D_0$ is compare with the heat transfer coefficient on a smooth tube at the same Reynolds numbe of the film. The heat transfer on the bank of finned tubes may be ten time higher than that on the bank of smooth tubes when the choice of the surfac geometrical dimensions is correct. For condensation on single tubes th coefficient of heat transfer α_2 increased up to 14 times.

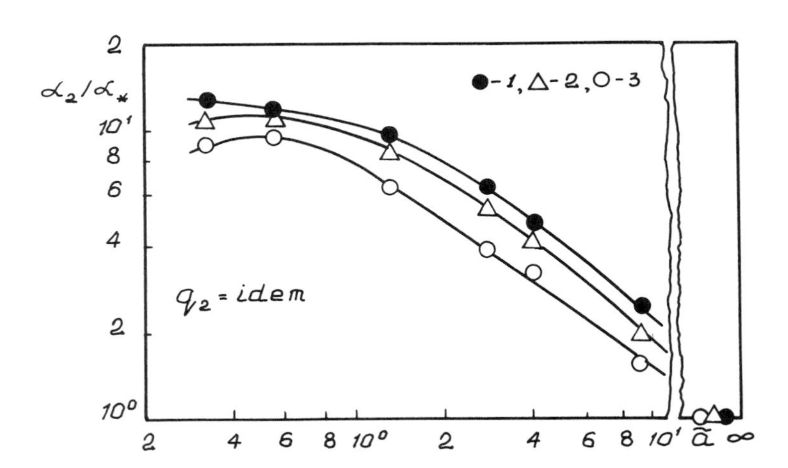

Figure 10. Comparison of heat transfer enhancement for condensation of vapou on banks of smooth and finned tubes, R12, $T''=40°C$, $q_2=3\ 10^4$ W/m². Singl tubes:1 - Re=26, tube bank: 2 - Re=75, 3 - Re=200.

6. NOMENCLATURE

ρ', ρ'' - density of liquid and vapour, kg/m³
σ - surface tension of liquid, N/m
μ, ν - dynamic and kinematic viscosity of liquid, (Ns)/m², m²/s

λ - thermal conductivity of liquid, W/m K
c - heat capacity of liquid, J/(kg K)
r - heat of phase transition, J/kg
g - free fall acceleration, m/s²
T'' - saturation temperature. °C
ΔT - temperature difference between saturated vapour and tube wall, K
q_3, α_3 - heat flux and heat transfer coefficient averaged with respect to tube surface, W/m², W/(m² K)
q_h, α_h - heat flux and heat transfer coefficient on the side surface of the fin, W/m², W/(m² K)
α_* - heat transfer coefficient on a smooth tube with $D=D_0$, W/(m² K)
Γ_0 - amount of liquid flowing onto 1 m long tube per second, kg/(ms)
$\gamma=1-\gamma/2\pi$ - fin side surface portion unflooded with the liquid retained between fins by capillary forces;
$l=\sqrt{\sigma/(\rho'-\rho'')g}$ - capillary constant of liquid, m
$\tilde{D}=D/l$, $\tilde{a}=a/l$, $\tilde{b}=b/l$, $\tilde{h}=h/l$ - dimensionless finning parameters
$Re=q_2\pi D_0/2\mu r(1+3/8K)+\Gamma_0/2\mu$ - total film Reynolds number on a tube
$Re_b=(\Gamma_0(s-b)+q_2\pi D_0 s/r)/2a\mu$ - film Reynolds number in an interfin valley
$K=r/c\Delta T$ - Kutateladze's criterion
$Nu^*_\gamma=Nu^*_h/\gamma$, $Re_\gamma=Re_h/\gamma$ - Nusselt and film Reynolds numbers on fin side surface calculated taking account of capillary liquid retention in interfin valley
$Nu^*_3=\alpha_3(\nu^2/(g(-\rho''/\rho')))^{1/3}/\lambda$ - Nusselt number on a finned tube calculated by heat transfer coefficient averaged with respect to surface

7. REFERENCES

1. Bergles, A.E., Liquid Cooling for Electronic Equipment, Presented at Int. Symp. on Cooling Technology for Electronic Equipment, Honolulu, March 1987.

2. Chu, R.C., Heat Transfer in Electronic Systems, Proc. 8th Int. Heat Transfer Conf., San Francisco, vol. 1, pp. 293-305, August 1986.

3. Hand-Book on Heat Exchangers: In 2v., v. 1 / Ed. B.S.Petukhov, V.K.Shirokov. - M.: Energoatomizdat, 1987, 560 p.

4. Hirasawa, S., Hijikata, K., Mori, Y., Nakayama, W., Effect of Surface Tension on Condensate Motion in Laminar Film Condensation (Study of Liquid Film in a Small Trough), Int. J. Heat Mass Transfer, vol. 23, No. 11, pp. 1471-1478, 1980.

5. Rifert, V.G., Barabash, P.A., Vizel, Ya.F., Trokoz, Ya.E. Surface Tension Effect on Hydrodynamics and Heat Transfer at Vapour Condensation on Shaped Surfaces, Promyshlennaya teplotekhnika, vol. 7, No. 2, pp. 20-25, 1985.

6. Webb, R.L., Rudy, T.M., Kedzierski, M.A., Prediction of the Condensation Coefficient on Horizontal Integral - Fin Tubes, Journal of Heat Transfer, vol. 107, No. 2, pp. 103-112, 1985.

7. Yau, K.K., Cooper, J.R., Rose, J.W., Effect of Fin Spacing on the Performance of Horizontal Integral - Fin Condenser Tubes, Journal of Heat Transfer, vol. 107, No. 2, pp. 113-120, 1985.

8. Marto P.J., Recent Progress in Enhancing Film Condensation Heat Transfer on Horizontal Tubes, Heat Transfer Engineering, vol. 7, No. 3, pp. 53-63, 1986.

9. Gavrilov, V.A., Gogonin, I.I., Kabov, O.A., Malkov, V.A., Solonenko, O.P., Padyukov, I.V., Tarasov, B.V., Computer-Aided Experimental Bench at Phase Transitions, Teplofizika i Gidrodinanika v Protswssakh Kipeniya i Kondensatsii, Collection, Novosibirsk, pp. 132-140, 1985.

10. Nusselt, W., Die Oberflächenkondensation das Wasserdampfes, Z. der VDI, vol. 60, No. 27-28,pp. 541-546, 569-575, 1916.

11. Bressler, R.G. and Wyatt, P.W., Surface Wetting Through Capillary Grooves, Transactions of the ASME, Ser. C, Journal of Heat Transfer, vol. 92, No. 2, pp. 132-139, 1970.

12. Gogonin, I.I., Kabov, O.A., Capillary Liquid Retention Effect on Heat Transfer at Condensation on Finned Tubes, Izv. SO AN SSSR, No. 8, Ser. Techn. Nauk, vol.2, pp. 3-9, 1983.

13. Rudy, T.M. and Webb, R.L., An Analytical Model to Predict Condensate Retention on Horizontal Integral-Fin Tubes, Journal of Heat Transfer, vol. 107, No. 2, pp. 94-103, 1985.

14. Gogonin, I.I., Kabov, .A., Padyukov, I.V., Film Condensation of Stationary Vapour on a Finned Surface, Inzh. Techn. Zhurnal, vol. 49, No. 5, pp. 709-717, 1985.

15. Sander-Beuermann, W., Schrooder, J.J., Investigation of Isotermal Falling-Film-Flow on Vertically Profile Surfaces, Proc. 6th Int. Symposium Fresh Water from the Sea, Las Palmas, vol. 1, pp. 173-182, 1978.

16. Gregorig, R., Hautkondensation an Feingewellten Oberflächen bei Berücksichtigung der Oberflächenspannungen, Zeitschrift für Angewandte Mathematik und Physik, vol. 5, No. 1, pp. 36-49, 1954.

17. Gogonin, I.I. and Dorokhov, A.R., Heat Transfer at R12 Condensation on Horizontal Tubes, Kholodilnaya Tekhnika, No.11, pp. 31-34, 1970.

18. Zozulya, N.V., Borovkov, V.P., Karkhu, V.A., Heat Enhancement at R-113 Condensation on Horizontal Tubes, Kholidilnaya Tekhnika, No. 9, pp. 25-28,1969.

19. Ivanov, L.P., Butyrskaya, S.T., Mamchenco, V.O., Heat Transfer at Condensation of R-12 Moving Vapour on Banks of Smooth and Finned Tubes,Kholodilnaya Tekhnika, No. 9, pp. 24-27, 1971.

20. Smirnov, G.F., Lukanov, I.I., Heat Transfer Investigation at R-11 Condensation in a Bank of Finned Tubes, Kholodilnaya Tekhnika, No. 5, pp. 31-33, 1971.

21. Khiznyakov, S.V., Heat Transfer at R-12 and R-22 Condensation on Smooth and Finned Tubes, Kholodilnaya Tekhnika, No. 1, pp. 31-34, 1971.

22. Chaikovsky, V.F., Bakhtiozin,P.A., Lukanov, I.I., Puchnov, V.V., Heat and Mass Transfer Investigation at Condensation of R-12 and R-22 Mixtures on Horizontal Finned Tubes, Kholodilnaya Tekhnika, No. 2, pp. 24-28, 1973.

23. Katz, D.L., Geist, J.M., Condensation on Six Pinned Tubes in a Vertical Row, Trans.ASME, vol. 70, No. 8, pp. 907-914, 1948.

24. Gogonin, I.I., Kabov, O.A., Condensate Flow Rate Effect on Heat Transfer at Condensation on Banks of Finned Tubes, Inzh.-Fiz. Zhurnal, vol. 51, No. 1, pp. 16-22, 1986.

An Immersion Cooling Unit: Condensation in a Remotely Subcooled Liquid

R. LETAN
Ben-Gurion University
Beer-Sheva, Israel

ABSTRACT

Immersion cooling units for electronic equipment were operated in the natural convection mode both in the zone of boiling on the heat dissipating surfaces as well as in the zone of condensation. The condensation was performed by employing a cold plate, or a submerged condenser. The performance of such units was limited by the condensive capacity of the units which degraded as vapor and noncondensibles enveloped the cooling surfaces.

In the presently described unit the operation is conducted in a forced flow mode in both zones in boiling and in condensation. The working liquid, subcooled, is introduced at the bottom of the heat dissipating zone flowing up along the surfaces and boiling. The so-generated vapor with the remaining liquid mix in the space above with a stream of liquid subcooled externally to a lower temperature. The condensation of the bubbles is achieved by direct contact with the liquid. The liquid exits at the top of the unit for subcooling and recirculation. In such mode of operation the condensive capacity of the unit is controlled by the temperature and flow rate of the remotely subcooled liquid.

A case study is herein conducted employing R113 as the dielectric working fluid. The parameters of significance are outlined but not investigated yet. The effect of subcooling on power density of the unit is illustrated by comparison of two cases. At a subcooling of 20C and 40C a power density of 12 and 29 MW/cubic meter were obtained respectively. These power densities exceed by an order of magnitude the performance of the immersion units operated in the natural convection mode.

INTRODUCTION

The present work relates to a study of a system and to a case study of the performance of the system. The system to be described is a device for immersion cooling of electronic components and is aimed to perform at high power densities, namely to provide compactness for its operation. The described system consists of : * zone of subcooled forced flow boiling on the heat dissipating surfaces; * a zone of subcooled flow condensation of bubbles by direct contact; * an external heat exchanger with a pump for liquid subcooling and recirculation to the two zones in the immersion unit.

A sytem operated in this mode is anticipated to be more effective than the conventional immersion units operated in the mode of natural convection employing a cold plate or submerged condenser for cooling. The effectiveness of such systems is expressed by its compactness measured in power density, namely as the ratio of the rate of heat absorbed and the volume of the unit.

The specific applications of the immersion cooling devices may dictate the design features of the unit. In some applications the evaporative zone is of main concern. In other cases the condensive zone volume may be focused on.

In applications like cooling of large scale computers the attention would be directed to uniform and high fluxes over the chip surfaces. In such cases the compactness of the condensive zone is of second-order importance. In air-borne equipment on the other hand the overall compactness of the unit would be focused on.

The modes of operation in the primary unit are subcooled flow boiling and subcooled flow condensation with the subcooling conducted externally. The variables to be considered in the subcooled flow boiling zone relate to the critical heat flux and to the effects of velocity, degree of subcooling, and the surface character on the heat fluxes. These parameters relate to the surfaces as experienced in unbounded fluids.

In practical devices which may consist of printed boards or other devices closely spaced, the dynamics of the two-phase flow exerts its effects too. The size of the bubbles, their velocity, the volumetric fraction of the vapor in the flow space are significant factors in the flow boiling regime.

The detachment and sweeping away of the bubbles, and maintaining the liquid temperature in the subcooled range depend on the flow velocity, the degree of subcooling, and the spacing of the heat-dissipating surfaces.

The parameters of primary significance in the zone of direct-contact condensation relate to the degree of subcooling, the mass flow rate of the cooling liquid, the mode of flow, cocurrent or countercurrent, vertical flow or tangential swirl, the size of bubbles to be condensed, the volumetric flow of the vapor, and the fraction of noncondensibles. The outlet temperature of the liquid with the condensate is also a design parameter which relates to the operational volume of the condensation zone, and to the inlet temperature into the boiling zone.

The present work focuses on the application of the system to cooling of electronic devices; therefore,the literature review and the case study conducted strictly relate to dielectric working fluids. In other kinds of heat-generating equipment polar liquids like water are employed.

RELATED STUDIES ON BOILING

Studies with air, dielectric fluid, water, water-alcohol mixtures on simulated microelectronic chips were conducted in modes of natural convection, forced convection, jet impingement, spray cooling, pool boiling, and flow boiling.

The usual range of air cooling heat fluxes is below 10^4 W/m^2. Immersion cooling in pool boiling of R-113 reaches up to $2x10^5$ W/m^2, and burnout of water-alcohol mixtures was recorded at $4x10^8$ W/m^2 (1).

Bergles and Rohsenow (2) suggested an additive superposition for the fluxes achievable in forced convection boiling for particular velocity and subcooling. Illustrative data for comparison of heat fluxes in pool boiling of water and in forced flow boiling showed that in pool boiling
at ΔT_{sat} = 17 C the heat flux is q"= $1.5x10^6$ W/m^2,
where in flow boiling at velocities of
u = 1.4 m/s the heat flux is q"= $1.5x10^6$ W/m^2
u = 2.6 m/s the heat flux is q"= $3x10^6$ W/m^2
u = 4.5 m/s the heat flux is q"= $6x10^6$ W/m^2.
These figures show that at a velocity of 4.5 m/s the heat flux is increased fourfold in comparison to pool boiling.

The critical heat flux in subcooled flow boiling appears to be much above the critical heat flux in pool boiling. The magnitude of those fluxes appear to be affected by the dynamics of the test system (3). Thus, the collapse of generated bubbles in the subcooled bulk, the sweeping away of bubbles from the surface, and the overall performance of the two-phase bulk may be interactive with the critical flux at the surface in subcooled forced flow boiling. Correlating equations have been suggested mostly for water at specific operating conditions and geometries (4).

In evaporative cooling of electronic components dielectric fluids are used. Those experimented with were freons and fluorinerts. Scott (5) compared the critical fluxes in pool boiling of water with fluxes in the fluorinated compounds, all at the saturation temperature:

Fluid:	Water	FC-78	FC-75	FC-43	R-113
T_{sat}, C:	100	50	102	174	48
$q"_{max}$, $\frac{W}{cm^2}$	135	35	42	54	16

Marto and Lepere (6) experimented with enhanced surfaces in pool boiling of FC-72, and R-113. The maximum heat fluxes achieved at $\Delta T(sat)$ = 10 C, were about q" = $2x10^5$ W/m^2.

Hino and Ueda (7) tested R-113 in subcooled boiling reaching also $q" = 2\times10^5$ W/m^2. These researchers used subcoolings of 10-30 C.

Oktay (8) described the Liquid Encapsulated Module - LEM, using the FC-86 fluid. The cooling was done by water flowing in a cold-plate. In these experiments a "dendritic" chip , 4.5x4.5 mm, dissipated 18 W at a temperature of 140 C in the FC-86 fluid which has a boiling point of 56.1 C.

Hwang and Moran (9) tested chip power in the range of 1-12 W at ΔT = 20-80 C. In pool boiling of FC-86 the heat fluxes were increased from 2 W/cm^2 to 30 W/cm^2 at ΔT = 30 C.

In a recent work, Maddox and Mudawwar (10) reported results obtained in single-phase and two-phase forced convection. Liquid FC-72 was circulated in a vertical rectangular channel using a single heat source flush mounted to one wall. The liquid was circulated at velocities up to 4.0 m/s and with subcooling up to 44 C. The results illustrated that increased velocity and higher degree of subcooling increased the critical heat fluxes. At a velocity of 2.25 m/s and subcooling of 44.2 C the value of CHF was 93.5 W/cm^2, showing an increase by a factor of 3.2 over the case of near-saturated boiling.

In another recent work Cho and Wu (11) described experiments in spray cooling with nucleate boiling over the heat-dissipating surface up to burnout. The liquid sprayed was R-113. The sprayed liquid velocities were 13.7 m/s, 10.1 m/s, and 5.9 m/s. The measured burnout heat fluxes were 115.2 W/cm^2, 93.8 W/cm^2, and 58.8 W/cm^2, respectively.

The experiments in flow boiling and subcooling performed with freons and fluorinerts by the numerous researchers indicate that in subcooled flow boiling heat fluxes of the order of 5 times the magnitudes reported in saturated pool boiling are achievable.

RELATED STUDIES ON CONDENSATION

Usually the immersion cooling devices employed submerged condensers or cold-plates at the top of the unit. Those devices are effective in their performance due to natural convection, and the "pumping" power of the bubbles. In the space of the unit a substantial fraction of the bulk fluid is occupied by the vapor bubbles. The higher the rate of heat dissipation, the larger is the fraction of vapor in the bulk fluid. Beyond the capacity of the unit the bubbles coalesce into slugs and accumulate at the top of the unit. The operational limits of submerged condensers were extensively investigated by Markowitz and Bergles (12). A condensive upper limit which corresponded to vapor space condensation was identified. Thus in the immersion units the condensation capacity of the unit controls its maximum power or effectiveness.

Another way to conduct condensation in an immersion unit is by direct contact condensation, namely by contacting the bubbles with a stream of a subcooled liquid. In this case the space above the evaporative zone is used for contacting the bubbles with an externally cooled liquid injected into that zone. Another way would also be to use highly subcooled liquid for the subcooled flow boiling, and allow the two-phase mixture to flow into the condensing zone to complete there the collapse of the vapor bubbles. The most pronounced advantage of such operation is the heat transfer across a renewable fluid interface with no accumulation of noncondensibles, and no degradation of performance.

The design of such units involves the knowledge and understanding of bubble dynamics. Numerous studies on collapse of bubbles were conducted in miscible and immiscible liquids. Most of the basic information gained in immiscible liquids appeared useful in miscible liquids, with respect to the flow field around the bubble, and the formation of condensate inside the bubble.

Theoretical and experimental studies on bubbles condensing in miscible and immiscible liquids were conducted by Chao and co-workers (13,14), by Sideman and co-workers (15,16), and by Letan and co-workers (17-20).

Florschuetz and Chao (13) considered a bubble collapsing in a miscible liquid (steam in water). The bubble was assumed to be stationary and the collapse spherically symmetrical. Wittke and Chao (14) studied a traversing bubble collapsing in its liquid. The bubble was postulated to move at a constant velocity, and to remain spherical at all times (the initial radius ranged from 0.15 to 0.30). The outer flow was assumed irrotational, and the noncondensibles uniformly distributed due to rapid diffusion of the vapor.

Isenberg and Sideman (15) assigned a thermal boundary layer to the immiscible bubble, and the viscous effects outside the bubble were semi-empirically accounted for by a "velocity factor". The condensate film inside the bubble was assumed to be negligible.

Lerner et al. (17) accounted both for the condensate film and the viscous effects outside the bubble. Visualization studies were conducted (18) to gain an insight into the mechanism and the physical phenomena governing the bubble collapse.

Of particular interest were the shadowgraphs and color-entrainment photographs illustrating a viscous boundary layer over the top of the bubble and a wake at the rear, at the initial stage of collapse. As the collapse progressed the bubble settled into its wake, and an envelope of vortices surrounded the bubble until the termination of the collapse.

As bubbles came closer together on their flow path their interaction became obvious by their higher velocity, and by variation of the collapse rate curve (19). Within the range studied experimentally the time of collapse remained constant. Studies were also conducted with bubbles collapsing in their own subcooled liquid, R113 (20).

Collapse rates were measured at temperature differences up to 10 C. The recorded rates were slower than those predicted by models in the literature. It seemed that the thermal resistance around the vapor might be higher than that predicted, due to the slowly moving deposited condensate film, and the viscous effects outside the condensate film. Within the range studied the time of collapse correlated inversely with the temperature difference. Such correlation was obtained in the range of 3-10 C (20) and was applied by extrapolation to higher temperature differences for estimates. The figures are to be presented and utilized in the case study to follow.

DESCRIPTION OF THE SYSTEM

The immersion unit system is illustrated in Fig.1. It consists of a two-zone unit of any shape. The lower part of the unit contains heat-dissipating elements like electronic components. The space above the heat-dissipating surfaces represents the condensive zone of the unit.

For operation of the unit the subcooled liquid is introduced at the bottom of the unit below the heat-dissipating elements. The liquid is forced to flow along the heat-dissipating surfaces. Boiling of the liquid takes place on these surfaces, and the so-generated bubbles of vapor are swept away from the surfaces by the forced flow along those elements. The vapor so generated and entrained into the subcooled stream is partly condensed in its flow. As the volumetric flow rate of the vapor increases in the evaporative zone so does the velocity of the two-phase mixture. That, in turn, enhances the detachment of the generated bubbles from the hot wall, and enhances their sweeping away into the subcooled liquid.

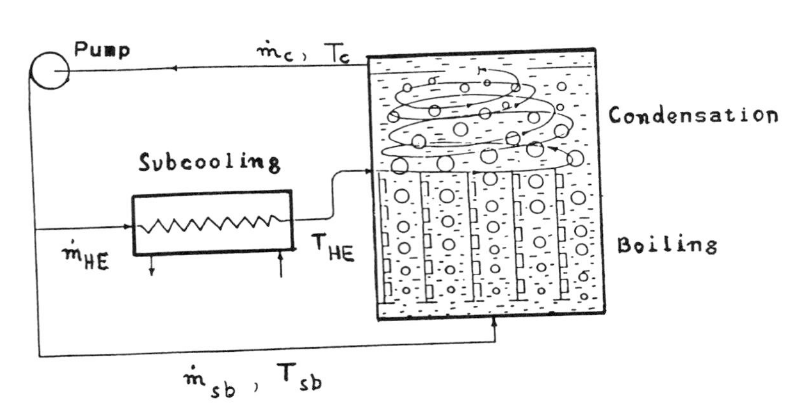

FIGURE 1. Immersion unit with remote subcooling.

The two-phase mixture of vapor bubbles in the subcooled liquid flows into the upper space above the heat-dissipating elements and into the condensive zone. There the mixture mixes into an incoming stream of a lower temperature and swirls up till the completion of collapse of all the vapor in the mixture. The incoming liquid is introduced into the condensive zone tangentially to induce the swirling motion and uniform mixing with the two-phase mixture which rises up vertically from the evaporative zone.

At the top of the condensive zone the subcooled liquid, which contains also the condensate is pumped out. The liquid stream out of the pump is divided into two streams: the heat exchanger and the evaporative zone of the immersion unit.

In the heat exchanger the liquid is cooled down giving up the entire heat rate dissipated by the elements in the immersion unit. This liquid is cooled down to the lowest temperature in the system, and flows from the heat exchanger into the condensive zone. It is injected closely above the evaporative zone. This way the largest volume of the rising up vapor is contacted with the coldest liquid to reduce rapidly the total volume of vapor.

The other stream from the pump directed to the bottom of the immersion unit remains at the temperature of the subcooled liquid withdrawn from the top of the condensive zone. This temperature is optional. It can be reduced by recirculating the whole stream or part of it through the heat exchanger.

The single pump and the single heat exchanger are utilized in the system for the forced convective boiling and for the direct contact condensation.

The external heat exchanger may be operated in a more versatile manner than a condenser submerged in a strictly limited bulk of liquid. The remote subcooling provides flexible and accurate control over the flow rate and temperature of the subcooled liquid. This can be used to extend the limits of the evaporative zone and the condensive capacity of the immersion unit: (1) The heat dissipating surfaces can be densly packaged because the subcooled forced flow boiling facilitates higher heat fluxes. (2) The direct contact condensation is not limited by accumulation of noncondensibles or built-up of vapor. The contacting with subcooled liquid at its lowest temperature in cocurrent flow yields high rates of collapse, smaller fractions of vapor, and is operable in small-volume units.

Fig. 2 schematically illustrates the cocurrent operation in the condensive zone of the unit, with and without noncondensibles in the collapsing vapor. In pure vapor the saturation temperature is constant and preserved till the final collapse of the bubbles. Then the condensate is cooled down to the temperature of the exit. On the other hand in the presence of noncondensibles the collapse of vapor progresses at a continuously decreasing temperature. Some vapor is left over with the noncondensibles at the final stage of collapse. The fraction of vapor left uncondensed decreases with decreasing temperature difference.

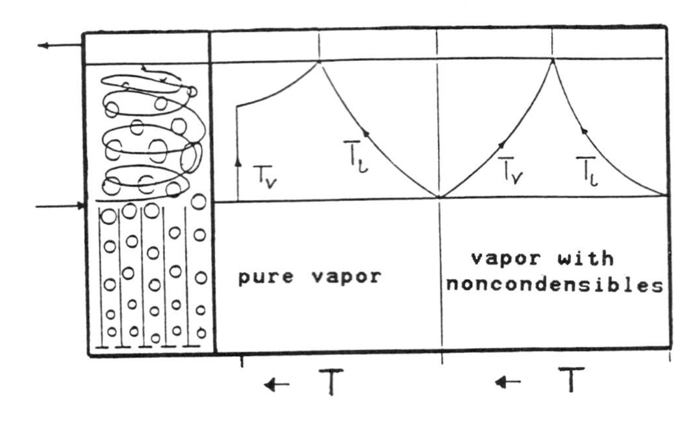

FIGURE 2. Cocurrent flow of vapor in subcooled liquid.

Fig. 3 shows this effect on the collapse rate of a bubble at two temperature differences. The larger is the temperature difference the faster is the collapse, and the smaller is the final radius of the left over bubble. In the cocurrent flow of the cooling liquid and the two-phase mixture the rapid decrease in bubble volume is attained near the interface between the evaporative and condensive zone. That gain in volume impacts on the condensation space volume.

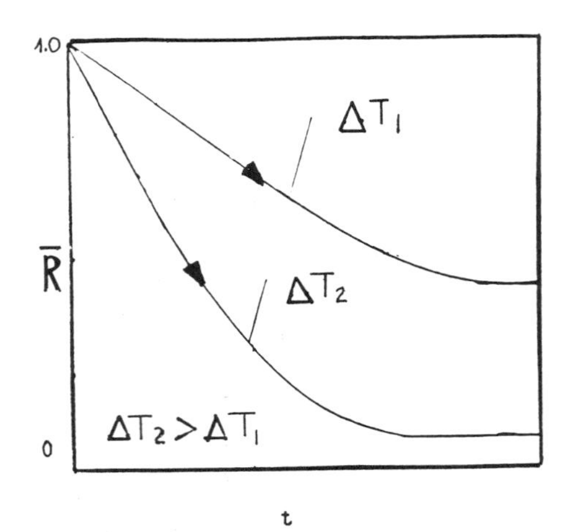

FIGURE 3. Bubble collapse at two subcoolings.

ANALYSIS OF OPERATION

The operational mode is schematically illustrated in Fig.1, and the symbols engaged in the analysis correspond to the notation of this figure. The quantitative analysis focuses on an elementary section in the evaporative zone, extended into the condensive zone. The elementary section consists of a channel of two heat-generating, vertically mounted plates. The analysis of a single "channel" can be extended to any modular design of an immersion unit. One plate or both plates may dissipate heat into the flow between them.
If the overall heat-dissipating surface in the channel is denoted A_b, and the average heat flux on those surfaces is denoted q", then the rate of heat generated is

$$q = q''. A_b \tag{1}$$

The mass flow rates in and out of the immersion unit,

$$\dot{m}_c = \dot{m}_{sb} + \dot{m}_{HE} \tag{2}$$

If none of the $\dot{m}_{sb}$ stream is recirculated through the heat exchanger, then,

$$T_{sb} = T_c \tag{3}$$

The recirculation of the stream $\dot{m}_{HE}$ in the heat exchanger lowers its temperature to

$$T_{HE} < T_c \tag{4}$$

If the heat dissipated at the surfaces of the channel is entirely used up by the liquid for vapor generation at a mass flow rate $\dot{m}_v$, then,

$$\dot{m}_v = q''. A_b/ h_{fg} \tag{5}$$

If, however, part of that generated vapor condenses in the subcooled liquid in the channel, and the temperature of the two-phase mixture rises on its way to the condensive zone, then,

$$\dot{m}_{vc} = (q''.A_b - \dot{m}_{sb}.c.(T_{bc} - T_{sb}))/h_{fg} \tag{6}$$

is the mass flow rate of vapor into the condensive zone.
T_{bc}, is the temperature of the two-phase mixture flowing into the condensive zone.

If all the vapor generated on the heat-dissipating surfaces condenses in the subcooled liquid in the channel then only liquid flows into the "condensive" zone which now performs as a direct-contact liquid-liquid heat exchanger,

$$\dot{m}_{HE}.c.T_{HE} + \dot{m}_{sb}.c.T_{bc} = \dot{m}_c.c.T_c \tag{7}$$

In all the three cases the exit temperature T_c, from the immersion unit is the same, being determined by the heat load of the heat-generating elements in the unit.

However, with regard to performance and design of the immersion unit the three cases are different. In the last case the intensive condensation in the channel may enhance higher heat fluxes and, therefore, be used for more dense packaging. The absence of vapor in the condensive zone makes the volume of this space much smaller. Thus the entire unit becomes much smaller.

By energy balance on the immersion unit,

$$\dot{m}_{HE} \cdot c \cdot (T_c - T_{HE}) = q'' \cdot A_b \tag{8}$$

Thus for a specified heat load in the immersion unit, and a defined operating temperature in condensation, the flow rate circulated through the heat exchanger and its outlet temperature are mutually dependent,

$$f(\dot{m}_{HE}, T_{HE}) = F(q, T_c) \tag{9}$$

If $\dot{m}_{HE}$, is selected then T_{HE} is determined. The lower is the temperature in the condensation zone the higher is the rate of collapse and the smaller is the vapor volume residing in the unit. The lower temperature T_c affects also the partial collapse in the evaporative channel as well as the heat fluxes on the surfaces. It also makes the needed flow rate smaller, and further makes the condensive zone smaller.
On the other hand the larger is the flow rate swirling into the condensation zone the more effectively it mixes into the upcoming two-phase mixture.

To analyze the volumetric flow in the evaporative zone, as well as into the condensive zone the rate of vapor generation has to be analyzed. Such analysis would yield the local velocities along the evaporative channel and the volumetric fraction of vapor inbetween the plates.

An energy balance on an infinitesimal element of height, dx, if set has to be integrated along the channel and several equations have to be solved simultaneously. Such effort presently is not justified, because some of the functions related to the physical variables are roughly approximated at this stage. Among those are the collapse rate curves, or distribution of bubble sizes detached from the hot wall. Therefore, presently a stepwise balance and solution have been adopted. The height of the evaporative zone is divided into N segments of height Δx. The analysis is conducted by segments from the leading edge to the trailing edge.

In the analysis it is assumed that the heat flux does not vary with height, q" = const. The vapor generated is a small mass fraction of the liquid introduced into the channel. The bubbles generated at the wall are detached along the channel at the same constant size, R_o. The detached bubbles are entrained into the main stream of the subcooled liquid, and collapse on their way up the channel. The collapse rate functions have to be known or approximated, $\bar{R}(t)$.

Spacing between the plates of the channel is L. Width of the plate is W. Height of the plate is H.

It is assumed that the vapor generated in any segment would begin to collapse in the segment above it. The volumetric flow rate of the vapor generated in the first segment is

$$\dot{V}_1 = q''.W.\Delta x.v/h_{fg} \quad (10)$$

$$\bar{R}_{01} = 1 \quad (10.1)$$

$$u_1 = u_l + \dot{V}_1/(W.L) \quad (10.2)$$

The vapor and the subcooled liquid flow into the next segment. The bubbles generated in segment, n = 1, now collapse within the time interval in this segment,

$$t_{1-2} = \Delta x/u_1 \quad (11.1)$$

The bubbles collapse to a radius,

$$\bar{R}_{12} = \bar{R}(t_{1-2}) \quad (11.2)$$

The vapor in the segment, n=2, is

$$\dot{V}_2 = (q''.W.\Delta x.v/h_{fg}).(1 + (\bar{R}_{12})^3) \quad (11.3)$$

$$u_2 = u_l + \dot{V}_2/(W.L) \quad (11.4)$$

From segment, n-1, to n, where $0 < n \leq N$, the velocity is

$$u_{n-1} = u_l + \dot{V}_{n-1}/(W.L) \quad (12.1)$$

The time for collapse in the segment is

$$t_{(n-1)-(n)} = \Delta x/u_{n-1} \quad (12.2)$$

There is a distribution of bubble sizes generated in the lower segments, and each bubble progresses in its collapse along the collapse curve within the time interval presented in eq. (12.2)

The radius of a bubble generated in segment m reaches in segment n the dimensionless value of

$$\bar{R}_i = \bar{R}(t_{m-n}) \quad (12.3)$$

where,

$$t_{m-n} = \sum_m^n t_i \quad (12.4)$$

For bubbles generated in segment n,

$$\bar{R}(t_{n-n}) = 1 \quad (12.5)$$

The volumetric flow rate of vapor in segment n, is, therefore,

$$\dot{V}_n = (q''.W.\Delta x.v/h_{fg}) \cdot \sum_1^n (\bar{R}_i)^3 \quad (12.6)$$

The volumetric fraction of vapor in the stream with no-slip condition is in segment n,

$$\phi_n = \dot{V}_n / (\dot{V}_n + u_l . W . L) \qquad (13)$$

The mass flow rate of vapor in each segment is

$$\dot{m}_v|_n = \dot{V}_n / v \qquad (14)$$

To illustrate the order of magnitude of the quantities involved in a unit of this kind a case study is to be presented.

CASE STUDY

Specifying the system as electronic components mounted on boards (plates) the following parameters are defined:
1. Temperature of the heat dissipating surfaces is 80 C.
2. R113 is selected as the working fluid, and is used for subcooled flow boiling.
3. Subcooling of 40 C is studied.
4. Liquid velocity between the plates is 2 m/s.
5. The heat load in the channel between the plates is 8000 W.
6. The channel geometry is L = 0.001 m, W = 0.1 m, H = 0.1 m.

The heat load, q, may be considered either dissipated from both sides of the channel, or from one plate. Thus, $q'' = 40\ W/cm^2$ or $q'' = 80\ W/cm^2$. In subcooled flow boiling such fluxes are attainable (10,11).

The following approximations are made in the study:
*The bubbles detached from the heat-dissipating surfaces are 2 mm in diameter.
*The collapse curve at 40 C subcooling is approximated by extrapolation of data at 10 C obtained with single bubbles.
*The heat flux along the plates is assumed constant.
*The velocity in the channel is high with no slip between the phases.

Properties of R113:

T_s, C	c, kJ/kg C	v, m^3/kg	ρ_l, kg/m^3	h_{fg}, kJ/kg	k, W/mC
47.6	0.89	0.134	1500	145	0.065

For the case study the height of the channel is divided to 5 segments, each 2cm high.

By eq. (1): $q''.(10\times10) = 8000W$; $\quad q'' = 80\ W/cm^2$

$\dot{m}_{sb} = u_l . \rho_l . (L.W) = 2.1500.(0.01.0.1) = 3\ kg/s$

By eq. (5): $\dot{m}_v = 8000/145\times10^3 = 0.055\ kg/s$

If $\dot{m}_v = 0$, then $\dot{m}_{sb}.c.\Delta T = 8000W$ and $\Delta T = 3\ C$

Jakob number is 49 at subcooling of 40 C. By the criterion established by Chao and co-workers (13,14) the collapse is heat transfer controlled ($B < 0.05$). By extrapolation (20) the time of collapse at 40 C subcooling was estimated as $t_f = 0.045$s.

Calculations:

by eq. (12.6) $\dot{V} = 1.5\times10^{-3}(\sum\bar{R}^3)$ m^3/s

by eq. (12.1) $u = 1.5\sum\bar{R}^3 + 2.0$ m/s

by eq (12.2) $t = 0.02/(2 + 1.5\sum\bar{R}^3)$, s

by eq. (13) $\phi = 1.5(\sum R^{-3})/(1.5\sum R^{-3} + 2.0)$

by eq. (14) $\dot{m}_v = 1.12\times10^{-2}(\sum\bar{R}^3)$, kg/s

Stepping from segment to segment up the channel:

	0-1	1-2	2-3	3-4	4-5
u, m/s		3.5	4.0	4.5	4.77
$\dot{V}$, m^3/s	0.0015	0.002	0.0025	0.0028	0.003
$\bar{R}_{0-1}$	1.0	0.7	0.62	0.51	0.45
$\bar{R}_{1-2}$		1.0	0.75	0.63	0.55
$\bar{R}_{2-3}$			1.0	0.78	0.65
$\bar{R}_{3-4}$				1.0	0.78
$\bar{R}_{4-5}$					1.0
$\sum\bar{R}^3$	1.0	1.34	1.66	1.85	2.0
t, s		0.0057	0.0050	0.0044	0.0042
ϕ		0.50	0.55	0.58	0.60
$\dot{m}_v$	0.0112	0.015	0.0186	0.021	0.0224

The two-phase mixture flows from the evaporative zone into the condensive zone. At the subcooling of 40 C, the inlet temperature to the immersion unit is 8 C. That is also the outlet temperature from the condensive zone. By eq. (8), a relation between $\dot{m}_{HE}$ and T_{HE} is obtained. Selecting the flow rate as 1.5 kg/s yields $T_{HE} = 2$ C. Thus the liquid flow rate to the condenser is $\dot{V}_1^{HE} = (3.0 + 1.5)/1500 = 3\times10^{-3}$ m^3/s . The vapor volumetric flow $\dot{V}_v$ is the one in (4-5). It collapses in the condenser; therefore, the average is 0.0015 m^3/s. The total volumetric flow in the condenser is

$\dot{V}_{tot} = 0.003 + 0.0015 = 0.0045\ m^3/s$

Volume of condenser = $\dot{V}_{tot} \cdot t_f = 0.0045 \times 0.045 = 2.02 \times 10^{-4}\ m^3$.

Volume of evaporative channel = $0.1 \times 0.1 \times 0.01 = 0.0001\ m^3$.

Total volume of the immersion unit = $3.02 \times 10^{-4}\ m^3$.

The power density of the unit = $8000/3.02 \times 10^{-4} = 26.5 \times 10^6\ W/m^3$.

The power density of this immersion unit, 26.5 MW/m^3 is very high in comparison to figures reported in the literature for conventional immersion units operated in the natural convection mode with submerged condenser. The figures reported for systems operated with cold plates or submerged condensers were in the range below 700 kW/m^3.

The herein described unit can be also operated with the heat exchanger submerged in the upper space. The pump could be used to circulate the dielectric liquid for subcooled flow boiling over the heat-generating elements, and forced flow in the condensation zone. However the the accumulation of non-condensibles on the heat transfer surfaces would still be there.

The case study presented in this work is only an illustration of the immersion unit and its operational mode. A parametric study of the geometric and operating conditions could show all the possibilities inherent in the characteristics of such an immersion unit.

NOMENCLATURE

A	surface area of heat-dissipating elements
c	specific heat capacity
H	height of evaporative zone
h_{fg}	latent heat
k	thermal conductivity
L	spacing between plates
$\dot{m}$	mass flow rate
q	heat transfer rate
q''	heat flux
R	radius of bubble
R_o	initial radius of bubble
$\bar{R}$	dimensionless radius of bubble, R/R_o
T	temperature
t	time
t_f	time of collapse
u	velocity
$\dot{V}$	volumetric flow rate
v	specific volume
W	width of plate
Δx	height of segment
ρ	density
ϕ	volumetric fraction of vapor

Subscripts

bc	at interface between boiling and condensive zones
c	condensive zone
HE	heat exchanger
l	liquid
s	saturation
sb	in the evaporative zone, subcooled
v	vapor
vc	vapor in condensive zone

REFERENCES

1. Bergles, A. E., Park, K.-A., Kim, C.-J. and Ma, C. F., Cooling Enhancement for Chip/Module Testing, Heat Transfer Laboratory Report HTL-43, Iowa State Univ., 1986. Cited in NSF Proc. on Research Needs in Electronic Cooling, p. 40, ed. F. P. Incropera, Dec. 1986.

2. Bergles, A.E. and Rohsenow, W.M., The Determination of Forced-Convection Surface-Boiling Heat Transfer, J. Heat Transfer, vol.86, pp.365-372, 1964.

3. Chen, J. C., Correlation for Boiling Heat Transfer to Saturated Liquids in Convective Flow, Ind.Eng.Chem.Proc. Des.Develop., vol.5, pp.322-329, 1966.

4. Gambill, W. R., Generalized Prediction of Burnout Heat Flux for Flowing Subcooled, Wetting Liquids, 5th Natl. Heat Transfer Conf., Houston, AIChE Reprint 17, 1962.

5. Scott, A. W., Cooling Electronic Equipment, p.163, Wiley 1974.

6. Marto, P. J. and Lepere, V. J., Pool Boiling Heat Transfer from Enhanced Surface to Dielectric Fluids, J. Heat Transfer, vol.104, pp. 292-299, 1982.

7 Hino, R. and Ueda, T., Studies on Heat Transfer and Flow Characteristics in Subcooled Flow Boiling. Part 1. Boiling Characteristics, Int. J. Multiphase Flow, vol.11, no.3, pp. 269-281, 1985.

8. Oktay, S., Departure from Natural Concection (DNC) in Low Temperature Boiling Heat Transfer Encountered in Cooling Microelectronic LSI Devices, Proc. 7th Int. Heat Transfer Conf., vol.4, pp. 113-118, 1982.

9. Hwang, U. P. and Moran, K. P., Boiling Heat Transfer of Silicon Integrated-Circuit Chips Mounted on a Substrate, ASME HTD-Vol. 20, pp. 53-60, 1981.

10. Maddox, D. E. and Mudawwar, I., Single and Two-Phase Convective Heat Transfer from Smooth and Enhanced Micro-Electronic Heat Sources in a Rectangular Channel, ASME Proc. Natl. Heat Transfer Conf., vol.1, HTD-Vol. 96, pp. 533-541, 1988.

11. Cho, C. S. and Wu, K., Comparison of Burnout Characteristics in Jet Impingement Cooling and Spray Cooling, ASME Proc. Natl. Heat Transf. Conf., vol.1, HTD-Vol. 96, pp. 561-567, 1988.

12. Markowitz, A. and Bergles, A. E., Operational Limits of a Submerged Condenser, Progress in Heat and Mass Transfer, vol.6, pp. 701-716, Pergamon Press, 1972.

13. Florschuetz, L. W. and Chao, B. T., On the Mechanics of Vapor Bubble Collapse, J. Heat Transfer, vol. 87, pp. 209-220, 1965.

14. Wittke, D. D. and Chao, B. T., Collapse of Vapor Bubbles with Translatory Motion, J. Heat Transfer, vol.89, pp. 17-24, 1967.

15. Isenberg, J. and Sideman, S., Direct Contact Heat Transfer with Change of Phase: Bubble Condensation in Immiscible Liquids, Int. J. Heat Mass Transfer, vol.13, pp. 997-1011, 1970.

16. Moalem, D., Sideman, S., Orell, A., and Hetsroni, G., Direct Contact Heat Transfer with Change of Phase: Condensation of a Bubble Train, Int. J. Heat Mass Transfer vol.16, pp. 2305-2319, 1973.

17. Lerner, Y., Kalman, H., and Letan, R., Condensation of an Accelerating-Decelerating Bubble: Experimental and Phenomenological Analysis, J. Heat Transfer, vol.109, pp. 509-517, 1987.

18. Kalman, H., Ullmann, A, and Letan, R., Visualization Studies of a Freon-113 Bubble Condensing in Water, J. Heat Transfer, vol.109, pp.543-545, 1987.

19. Lerner, Y., and Letan, R., Dynamics of Condensing Bubbles: Effects of Injection Frequency, ASME Paper 85-HT-47, 1985.

20. Kalman, H., and Letan, R., Condensation Characteristics in an Immersion Cooling Unit for Electronic Components, Israel Annual Conf. Mech. Eng., June 1987.

Enhancement of Immersion Cooling of Microelectronic Devices by Foreign Gas Jet Impingement

C. F. MA
Beijing Polytechnic University
Beijing, PRC

A. E. BERGLES
Rensselaer Polytechnic Institute
Troy, New York, USA

ABSTRACT

Experiments were performed to determine the heat transfer enhancement due to foreign gas impingement upon a vertical simulated microelectronic chip submerged in Freon 113. Heat transfer characteristics at the stagnation point were studied. The results showed that the heat transfer is greatly enhanced by foreign gas impingement, especially at lower heat fluxes, as mass transfer plays a very important role. Distributions of heat transfer coefficients were obtained in both horizontal and vertical directions. Nucleate boiling with air jet impingement was studied experimentally. A possible model for this two-phase two-component heat transfer process was presented which correlates all the non-boiling data.

INTRODUCTION

The rapid growth of microelectronics technology, especially the implementation of Very Large Scale Integrated (VLSI) components in new electronic devices, has led to very high power dissipation at the chip level. It is relatively common for today's state-of-the-art chips to dissipate heat fluxes on the order of $10^5 W/m^2$[1]. The high heat dissipation provides a great challenge for heat transfer engineers. Various advanced methods, including both air and liquid cooling, have been proposed and developed to meet the power dissipation requirements.

Research on heat transfer enhancement technology has been also stimulated by the development of new methods for microelectronic cooling. As the high heat flux dissipated from VLSL microelectronic packaging appears to make forced convection air cooling very difficult, the computer industry is considering changing over from forced-air to liquid cooling. A new cooling method was presented by the present authors to enhance the liquid cooling of microchips[2]. They placed a gas jet tube very close to a simulated microchip submerged in high volatility liquid (R-113 or ethanol). It was found that the heat transfer from the vertical heater to the liquid pool could be considerably enhanced by means of foreign gas jet impingement. In the case of high jet velocity, a heat flux as high as $3.3x10^4 W/m^2$ to

R-113 was recorded with the wall temperature lower than that of the coolant. This phenomenon was explained by evaporative cooling of the liquid film beneath the gas bubble on the wall based on the photographic study by Gan and Ma[3]. This active enhancement technique has already attracted interest among heat transfer engineers. Similar experiments were conducted by Yang with R-113, water or ethanol as coolant[4]. The results of their experimental and analytical research confirmed the conclusions of reference[2]. More recently, this enhancement method was investigated with a low volatility liquid as working fluid[5]. Compared with free convection data the heat transfer coefficient could be increased 12 times, although it was not observed that the heat flow was directed from wall to liquid with wall temperature lower than that of liquid. This method is of interest for microelectronic cooling as well as other electrical or industrial equipment.

The present work is concerned with air impingement perpendicular to a vertical simulated microchip submerged in R-113. Local heat transfer characteristics, including the heat transfer coefficient at the stagnation point and the heat transfer coefficient distributions, were studied experimentally both in free convection and pool boiling. Very high heat transfer coefficients were measured. This was a function of not only the air jet velocity but also the heat flux, and tended toward very high values because the temperature difference, T_w-T_l, approached zero. Negative temperature differences were recorded even in the relatively high heat flux ($3.5x10^4 W/m^2$) case. A possible model was set up, based on which a correlation method was presented. All the non-boiling data were well correlated. In this case mass transfer plays an important role. For the subcooled boiling case, hysteresis, including inverse temperature overshoot, was observed at low jet velocity and nucleation seemed to be suppressed at high jet velocity.

EXPERIMENTAL APPARATUS AND PROCEDURE

Apparatus. The experimental system is shown in Figs. 1 and 2. The working fluid was R-113. Dry air was supplied from a high pressure tank. The flow rate and temperature were respectively measured by a flow-meter and a 36 gage copper-constantan thermocouple located in the jet tube as depicted in Fig. 1. The jet temperature was always adjusted to equal the pool temperature. All the experiments were conducted at atmospheric pressure.

The sides and bottom of the test chamber shown in Fig. 1 were constructed of aluminum alloy. A visual port was provided for observation of the test section. In addition, the bottom section of the cover was transparent. A flexible polyethylene tube joined the two sections of the cover. The jet tube was attached to the upper aluminum section and remained stationary. The chamber could be adjusted by means of a milling machine table so that the position of the jet tube could be altered with respect to the test section in three dimensions. The placement was acomplished within 0.025mm. A 36 gage copper-constantan thermocouple located close to the test section was utilized to measure the temperature of R-113 in the pool. After impinging on the heated surface the air escaped through the reflux condenser within which the R-113 vapor was condensed.

The test section required considerable development effort. The final arrangement shown in Fig. 2 involved a strip of 10μm thick constantan foil with heated section of 5mm x 5mm exposed to the coolant. This active section was soldered to copper bus blocks which were in turn connected to power leads. The active section of the foil was cemented to a bakelite

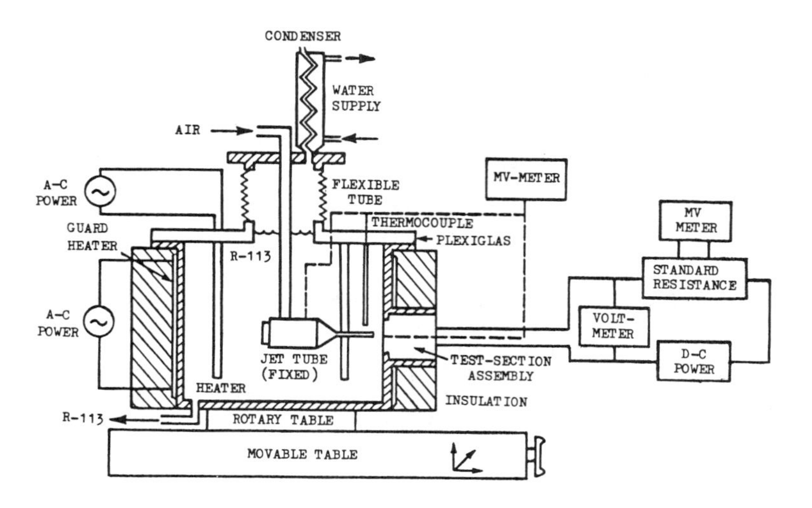

FIGURE 1. Test facility for enhanced heat transfer from a vertical simulated microchip to a liquid pool by foreign gas impingement.

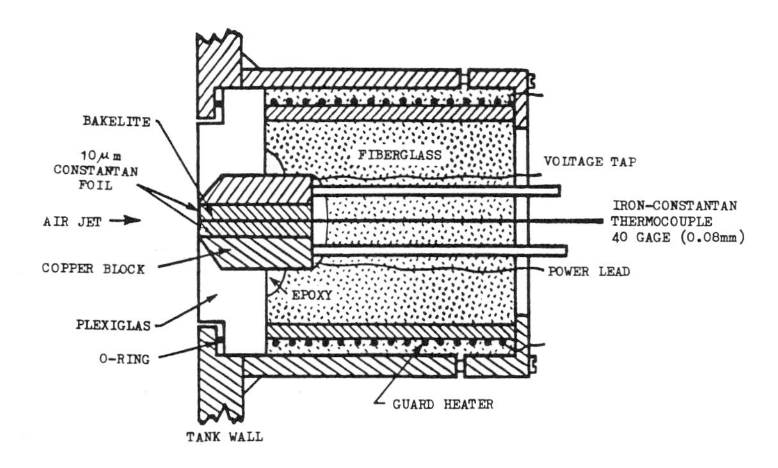

FIGURE 2. Test section details.

block inserted between the copper blocks. The temperature of the center of the inner surface of the heater was measured by a 40 gage iron-constantan thermocouple which was electrically insulated from the heater yet in close thermal contact. The diameter of the thermocouple junction was about 0.16mm. The test section assembly was cemented in a plexiglas disc of 45mm maximum diameter. The disc was inserted in the aluminum housing and retained by a plastic cylinder with a guard heater on the outer surface. The whole test section was heavily insulated with fiberglass. Direct current power to the test section was supplied by a 50 ampere power supply. The voltage drop across the test section was measured at the copper bus blocks.

Procedure. The area of the heat transfer surface was carefully measured by a microscope for each test section. The heated surface was left in the original highly polished condition and simply cleaned with acetone.

It was found by preliminary experiments that the guard heater had no important effect on the reading of the test section temperature in impingement and boiling cases. It was also determined in the preliminary experiments that the conduction heat losses out the backside of the test section as well as along the foil and thermocouple wires were less than 1% in these cases. Because of the sensitivity of the heat transfer behavior to the location of the jet on the heated surface, considerable attention was given to the centering. A special procedure was developed for this purpose whereby the jet was moved until the minimum temperature was recorded. As the heat transfer rate distribution is not symmetrical vertically for gas jet impingement on heaters submerged in liquid, an R-113 jet was used instead of an air jet for centering in a preliminary experiment as described in a previous paper[6].

The reading of the test section temperature was corrected for the very small temperature drop across the foil to get the surface temperature at the center of the simulated microchip. The heat flux was calculated from the measured values of the voltage drop, current and the area of one side of the heated surface. Due to the extremely small thickness of the foil the heat conduction parallel to the surface can be neglected. Hence the local heat transfer coefficient at the center of the heater can be calculated by the following formula:

$$h_t = q''_t/(T_w - T_l) \qquad (1)$$

EXPERIMENTAL RESULTS AND DISCUSSION

Stagnation Point Heat Transfer. Heat transfer characteristics at the stagnation point with and without boiling were investigated. The effects of jet velocity, heat flux, wall temperature and nozzle-to-plate spacing will be discussed in this section. Meanwhile, the data for unheated test sections will be reported to confirm and supplement the data obtained with heated surfaces.

(1) Negative Temperature Difference.

Fig. 3 presents the air jet impingement heat transfer data with q" vs. T_w-T_l in semilogarithmic coordinates. The most noticeable characteristic is that the "Negative Temperature Difference" is recorded even in cases of the heat flux as high as $3\times10^4 W/m^2$. That is why the semilogarithmic coordinate is chosen to present the data. The term "Negative Temperature Difference" accounts for the fact that heat flux can be directed from the wall to coolant at higher temperature. It should be probably emphasized that this phenomenon is not in contradiction with the second law of thermodynamics which can be expressed as follows: Heat will not flow of itself from a low-temperature object to a high-temperature object. In this case the heat flows from "Cold Wall" to "Hot Liquid" resulting from the work done by external force on the gas bubbles. This phenomenon has also been verified by some other investigators[4]. In order to verify the experimental result of negative temperature difference, wall temperature variations at the stagnation point were recorded at different jet velocities for Z/d=1 with an unheated

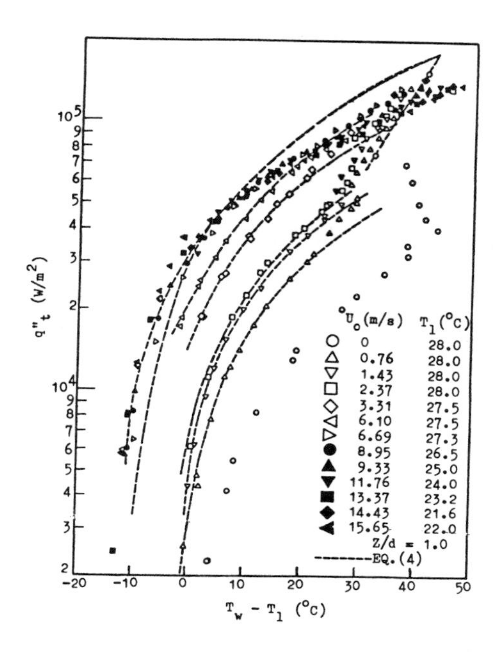

FIGURE 3. Heat transfer from vertical wall to liquid pool with foreign gas impingement.

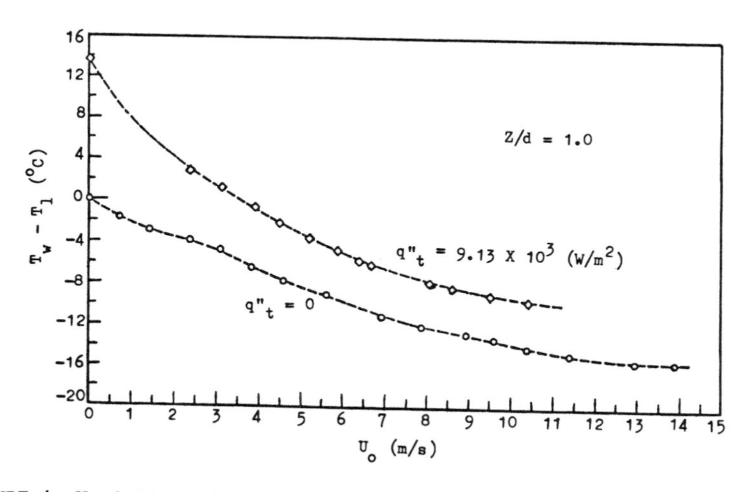

FIGURE 4. Variation of test section temperature with jet velocity.

test section. These data are plotted in Fig. 4 and show that the wall temperature decreases with increasing jet velocity and may be as much as 15°C below the pool temperature. A temperature curve with constant surface heat flux ($q''=9.13\times10^3 W/m^2$) is also presented in the same figure indicating the same trend. This fact shows the importance of mass transfer for the heat transmission as heat cannot be transferred from wall to liquid by convection while $T_w \leq T_1$. The reasons for this behaviour will be discussed in detail later.

(2) Heat Transfer Coefficient Variation with Wall Temperature and Heat Flux.

It was found that the heat transfer coefficient measured with air impingement was a function not only of the jet velocity but also of wall temperature or heat flux. Very strong effect of wall temperature on heat transfer coefficient in a non-boiling case is illustrated in Fig. 5. In fact, for higher jet velocities the heat transfer coefficients tend to infinity because the temperature difference (T_w-T_1) tends to zero and the heat transfer coefficients at low temperature difference may be as much as one order of magnitude greater than those for pool boiling. The influence of heat flux on heat transfer characteristics is demonstrated in Fig. 6. A similar downward trend of heat transfer coefficient with increased heat flux before the incipience of nucleate boiling is also observed. This fact suggests that the nature of the

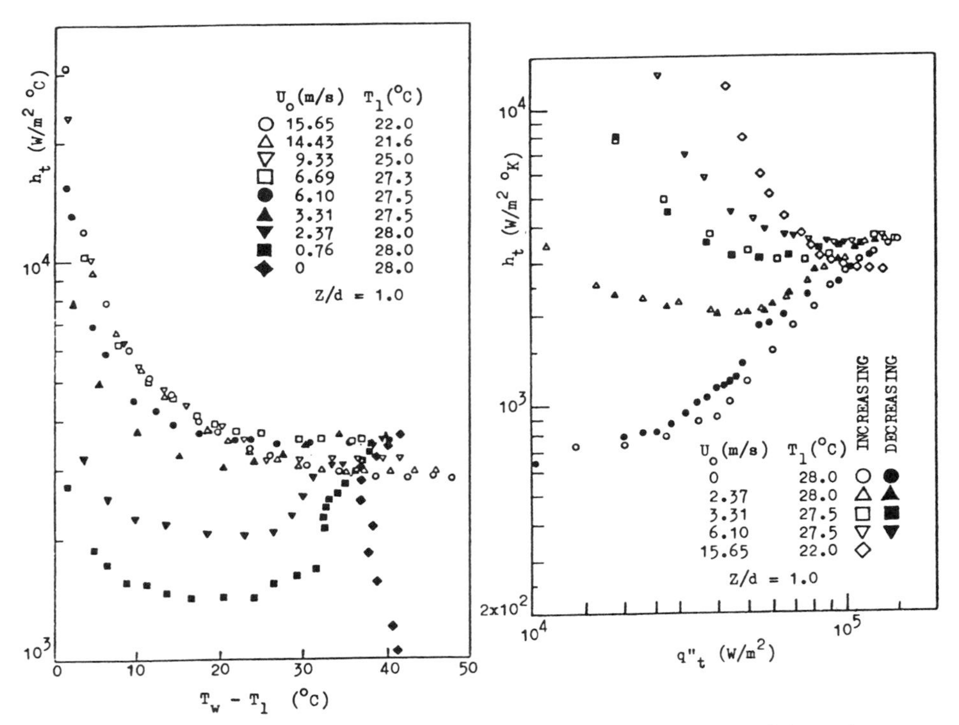

FIGURE 5. Variation of heat transfer coefficients with wall temperature.

FIGURE 6. Variation of heat transfer coefficients with heat flux.

heat transfer pattern in the present investigation must differ from both single phase convection and nucleate boiling. It is known that the heat transfer coefficient is essentially independent of heat flux for single-phase convection and proportional to q''^n ($n \sim 0.7$) for nucleate boiling. This particular feature of the heat transfer phenomenon in this study will be explained in the next section in terms of the mass transfer mechanism.

(3) Nucleate Boiling with Air Impingement.

At large temperature difference between wall and liquid, both the nucleation and the noboiling mechanism (including convection and evaporation) must be considered in the heat transfer process. As discussed in the previous section heat transfer coefficients decrease with increasing of heat flux (or wall temperature) before incipient boiling. Beyond the incipient boiling points the heat transfer coefficients may increase with increasing of heat flux if the jet velocity is not high. As shown in Figs. 5 and 6, with increasing heat flux (or wall temperature) the heat transfer coefficients decrease first then increase after they reach minimum values, which are dependent on jet velocity, for velocities lower than 3.3m/s. For high jet velocity, nucleation seems to be totally suppressed by gas jet impingement and the heat transfer coefficient tends toward a constant as illustrated in the figures. In order to examine the influence of nucleation some data are represented in Fig. 7 with q" vs $T_w - T_s$ instead of q" $T_w - T_l$. Subcooled pool boiling curves ($u_o = 0$) are also presented for comparison. Hysteresis is clearly observed for the boiling curves with zero or low jet velocities. It is noted that sudden jumps of wall temperature were recorded not only with increasing heat flux but also with decreasing heat flux. As shown in the figure, the wall temperature might suddenly increase about 8°C for subcooled pool boiling without jet impingement. This type of temperature over-shoot has been reported in reference [7] by Marto et al. However, they only observed very slow increase of wall temperature (0.37°C in 60 minutes) with decreasing heater power in pool boiling of liquid nitrogen. It may be attributed to a sudden decrease in the number of active nucleate sites when decreasing the heat flux. The departure from natural convection to nucleate boiling has been of concern to heat transfer engineers involved with microelectronic cooling, as maximum semiconductor junction temperature may be exceeded because of excessive temperature over-shoot. Effort has thus been devoted to develop anti-hysteresis methods. The gas jet impingement is a reliable active method to eliminate the temperature over-shoot at lower heat flux. As shown in Fig. 7, temperature over-shoot in either direction may be greatly reduced by impinging gas jet with low velocity of 0.76m/s and may be totally eliminated with higher velocity. In the cases of high heat fluxes nucleation seems to be suppressed with gas jets of high velocity and the heat transfer rate seems no longer to be influenced by the jet velocity.

(4) Effect of Jet Velocity.

The effect of jet velocity on heat transfer rate for given constant wall temperatures is demonstrated in Fig. 8. As shown in the figure, heat flux q" is proportional to u_o^n for constant wall temperature when the jet velocity, u_o, is lower than 7-8m/s. The exponent n is an empirical constant and depends on the temperature difference ranging from 1.2 to 0. When the jet velocity exceeds 7-8m/s, heat flux no longer varies with the velocity, as shown in Fig. 8.

(5) Effect of Nozzle-to-Plate Spacing.

Several Experiments were performed with different nozzle-to-plate spacings at constant jet velocities and constant heat fluxes. These results are presented in Fig. 9. It was found that the separation between nozzle and plate can have a great influence on the heat transfer coefficient. For high

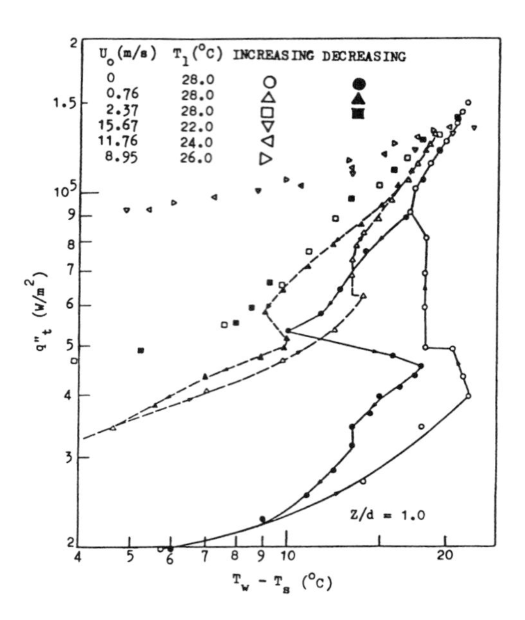

FIGURE 7. Pool boiling heat transfer with foreign gas impingement.

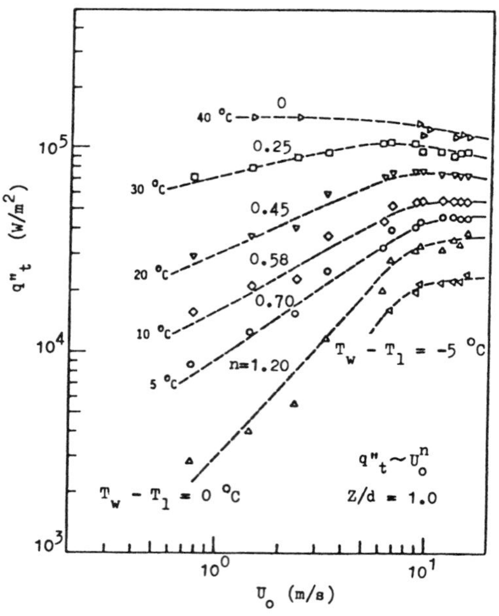

FIGURE 8. Effect of jet velocity on heat transfer behavior.

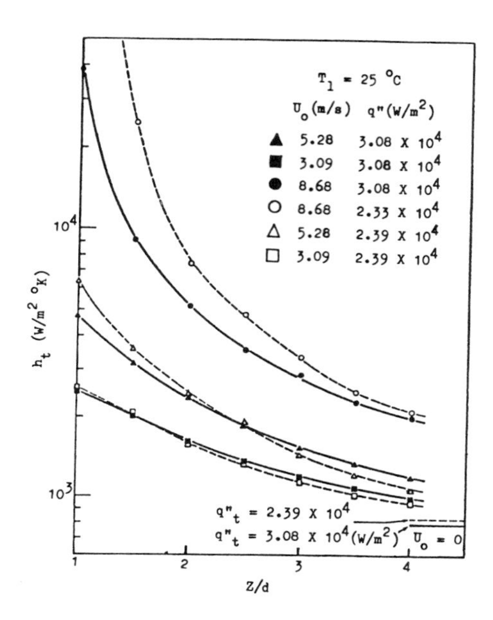

FIGURE 9. Effect of nozzle-to-plate spacing on heat transfer behavior.

jet velocity the heat transfer coefficient is very sensitive to the separation. For instance, the heat transfer coefficient with Z/d=1.0 may be as 20 times that with Z/d=4.0 for u_o=8.68m/s, as shown in the figure. In this case the heat transfer coefficient is also substantially affected by heat flux. With decreasing of the jet velocity, the influence of the nozzle-to-plate spacing on heat transfer becomes weaker. The effect of heat flux is also of less importance for lower jet velocity.

Heat Transfer Coefficient Distribution. The designers of industrial jet cooling or heating equipment always want to obtain not only high magnitudes of heat transfer rates but also uniform distributions of them. The present experimental system was designed in such a manner as to be used for measurement of the local heat transfer coefficient distribution. Detailed information on both vertical and horizontal distributions of heat transfer coefficients will be presented here.

(1) Horizontal Distribution of Heat Transfer Coefficients.

Horizontal distributions of heat transfer coefficients for two jet velocities with different heat fluxes are presented in Fig. 10. Because of the symmetry of the distribution, only half of the profiles are shown in the figure. The variation of heat transfer coefficients has a characteristic bell shape and is affected by the heat flux. As shown in Fig. 10, the curves become more steep when decreasing the heat flux. It is noted for the curve of $q''=7.62 \times 10^4 W/m^2$ in Fig. 10(b) that there exists a discontinuity at Z/d=2.16 which must be caused by incipience of nucleate boiling. It may be important for engineering application that heat transfer enhancement is still obvious even beyond Z/d=2.5 although no air bubbles were observed in that area. The enhancement of heat transfer is attributed to the liquid motion

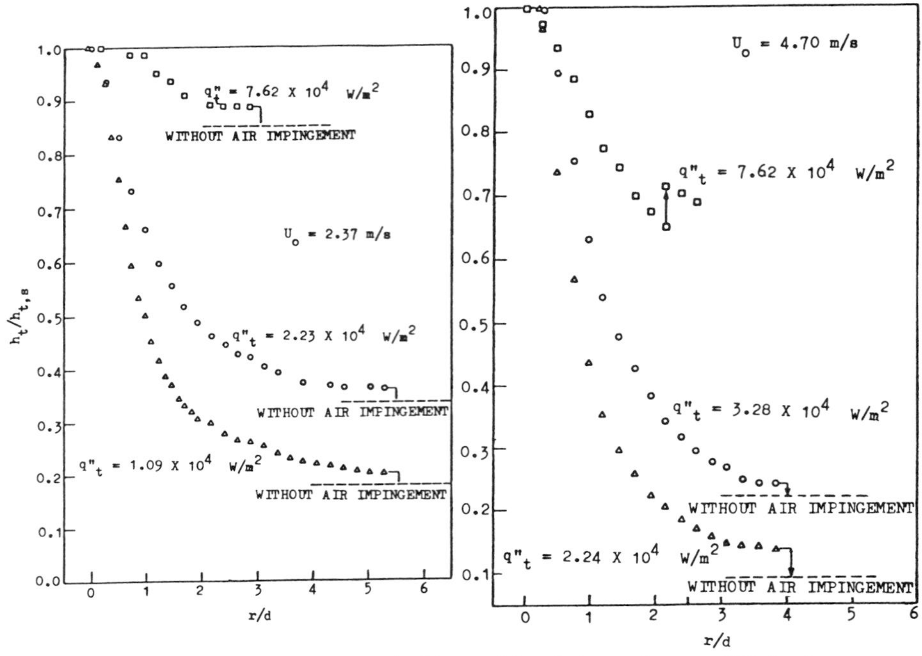

FIGURE 10. Horizontal distribution of heat transfer coefficients.

induced by air bubble wake flow.

(2) Vertical Distribution of Heat Transfer Coefficients.

The heat transfer coefficient distribution is not symmetrical in the vertical direction as shown in Fig. 11 because of the effect of buoyancy. The shapes of the distribution curves are strongly affected by jet velocity. The maximum heat transfer coefficients appear not at the stagnation point but above it. The position of the apex is dependent on the jet velocity, ranging from H/d=0.65 to 1.20 for velocity u_o=1.42-8.95m/s. The magnitude of the heat transfer rate peak may be over twice as large as that at the stagnation point. For high jet velocity the distribution curves are very steep around the peak point. But for lower jet velocities, a relatively uniform distribution may be obtained. These results suggest that in order to obtain uniform distribution, as well as high magnitude of heat transfer rates, the gas jet must be held not at the center of the heater but slightly below it.

Enhancement Mechanism and Data Correlation. Based on the photographical study, a possible physical model was presented in References [2,3]. It was postulated that the enhancement would be caused by

(1) Gas-Liquid Exchange.

Gas bubble formation, growth and departure caused rapid exchange of liquid in the region between the wall and the injector as shown by the high-speed motion pictures[2,3]. This mechanism of enhancement is similar to that of nucleate boiling.

(2) Induced Liquid Flow.

Bubble induces secondary flow in its wake region.

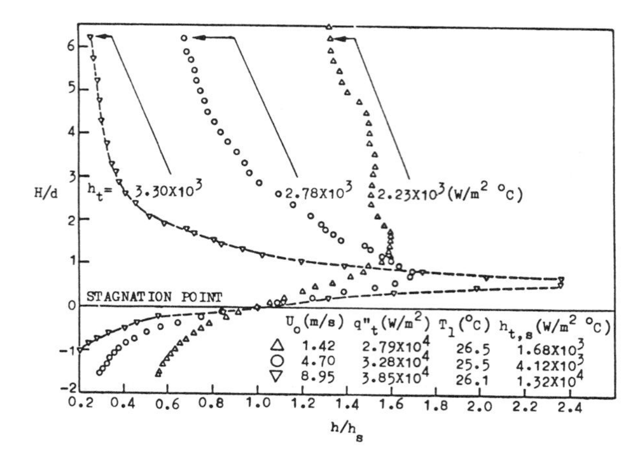

FIGURE 11. Vertical distribution of heat transfer coefficients.

(3) Evaporative Cooling.

At the base of the impacting and expanding gas bubble a liquid film would form and persist. The thermal resistance of the liquid film may be neglected as the film thickness is so small. The evaporative cooling of the film must play a very important role in the heat transfer process. The heat transfer coefficient of this simultaneous heat and mass transfer process may be expressed by [2]

$$q''_t=h_c(T_w-T_l)+h_d H_{fg}\frac{P_w\ M}{R_o\ T_w} \tag{2}$$

For non-heated cases (q"=0), the ratio of heat and mass transfer coefficients may be obtained from measured temperatures T_w and T_1 presented in Fig. 4:

$$r=\frac{h_c}{hd}=\frac{-H_{fg}P_w\ M}{R_o T_w(T_w-T_l)} \tag{3}$$

Combining Eqs. (3) and (2) we have

$$q''_t=h_c\ \frac{H_{fg}P_w\ M}{rR_oT_w}+(T_w-T_l) \tag{4}$$

With Eq. (4) all the non-boiling data are well correlated as shown in Fig. 3. This fact shows that the simplified model for the present two-phase two-component system is reasonable and may be adopted as the first approximation. However, further research, both experimental and analytical, is required for the clarification of the enhancement mechanism.

CONCLUSIONS

(1) Heat transfer from vertical heated surface to liquid may be greatly enhanced by foreign gas impingement, especially at low temperature differences. Negative or zero temperature difference (T_w-T_l) was recorded in the case of heat flux as high as $3.5\times10^4 W/m^2$.

(2) Stagnation point heat transfer behavior with and without boiling was experimentally studied. Vertical and horizontal distributions of the heat transfer coefficients were obtained. The results suggested that in order to get a high heat transfer rate as well as a uniform temperature distribution, the gas jet should be held very close to the wall and below the center of the heater.

(3) Nucleate boiling with foreign gas impingement was studied experimentally. As a reliable anti-hysteresis measure, gas jet impingement may effectively eliminate temperature over-shoot at low heat flux levels. At high heat flux, nucleate boiling was suppressed by an impinging gas jet of high velocity.

(4) Based on the simplified model presented in Reference 2, all the non-boiling data were well correlated with Eq. (4).

ACKNOWLEDGMENTS

This study was carried out in the Heat Transfer Laboratory of Iowa State University under the sponsorship of IBM Data Systems Division.

NOMENCLATURE

d	jet diameter
H_{fg}	heat of vaporization
h_c	convection heat transfer coefficient
h_d	mass transfer coefficient
M	molecular weight
p	pressure
q''	heat flux
R_o	general gas constant
T	temperature
u_o	exit velocity of jet
Z	distance between exit of jet nozzle and heated surface

SUBSCRIPTS

l	liquid
s	saturation or stagnation point
t	total
w	wall

REFERENCES

(1) Bar-Cohen, A., Kraus, A.D. and Davidson, S.F., "Thermal Frontiers in The Design and Packaging of Microelectronic Equipment", Mechanical Engineering, Vol.105, pp.53-59, 1983.

(2) Ma, C.F., Gan, Y.P., Tang, F.J. and Bergles, A.E., "A New Method of Heat Transfer Augmentation by Means of Foreign Gas Jet Impingement in Liquid Bath", Heat Transfer Science and Technology, Edited by B.X. Wang, Hemisphere Publishing Corporation, pp.789-797, 1987.

(3) Gan, Y.P. and Ma, C.F., "Photographical Study of A New Method of Heat Transfer Augmentation with Foreign Gas Jet Impingement in Liquid Pool" (in Chinese), Journal of Beijing Polytechnic University, Vol.12, No.2, 1986.

(4) Yang, N.S., "Enhancement of Heat Transfer from A Small Submerged Heater to Liquid with Foreign Gas Jet Impingement" (in Chinese), Ph.D. Dissertation, 1987. Dalian Institute of Technology, China.

(5) Tien, Y.C., Ma, C.F., Lee, S.C., La, D.H. and Bergles, A.E., "Enhanced Heat Transfer from Simulated Microchips to Surrounding Liquid of Low Volatility with Gas Jet Impingement", in Advances in Phase Change Heat Transfer, Edited by Xin Mingdao, International Academic Publishers, pp.419-424, 1988.

(6) Ma, C.F. and Bergles, A.E., "Boiling Jet Impingement Cooling of Simulated Microelectronic Chips", ASME Publication, HTD Vol.28, pp.5-11, 1983.

(7) Marto, P.J. et al., "Nucleate Pool Boiling of Nitrogen with Different Surface Conditions", Journal of Heat Transfer, Vol.90, pp.437-444, 1968.

(8) Oktay, S., "Departure from Natural Convection (DNC) in Low Temperature Boiling Heat Transfer Encountered in Cooling Mico-electronic LSI Devices", in Heat Transfer — 1982, Vol.4, Hemisphere, Washington, pp.113-118, 1982.

Closed Two-Phase Thermosyphon for Cooling of High-Power Semiconductors

M. LALLEMAND and V. DUYME
INSA, Lab. d'Energétique et Automatique
20 Av. Einstein
69621 Villeurbanne, France

U. ECKES and P. MEREL
ALSTHOM-TRV
11 Av. de Bel Air
69627 Villeurbanne, France

ABSTRACT

This work presents experimental results on closed two-phase thermosyphons for cooling of high-power semiconductors. In order to enhance the evaporation heat transfer, different kinds of materials (2) and structures of the surface (5) have been tested with R113 as working fluid. The lowest junction temperature of the semiconductor is obtained for an evaporator with a copper extended surface with tapped holes. The evaporator superheat increases with increasing the semiconductor electric power. The effect of air flow rate at the condenser and of its inclination are investigated.

INTRODUCTION

The application of phase change cooling of electronic components in railway traction equipment has been developed for about ten years. Advances in efficient cooling systems for high power semiconductors have been reviewed by Bergles (1).

With the aim of obtaining the smallest possible volume, a freon cooling immersion design was developed first. It used a hermetic tank, in which the components were submerged in the dielectric freon liquid, and up to twelve semiconductors were placed in each container. However, every time a device fails it is necessary to empty the container, open it, change the component, clean, evacuate the air and fill the container again. All these operations must take place in a special room, a "grey room", in order to avoid pollution.

In order to avoid maintenance problems and to keep a similar thermal performance, gravity-assisted heat pipes or two-phase thermosyphon systems, in which the components are not submerged, have been investigated and developed. They can be divided into two main types.

In the first type, which is already in use (2), the semiconductors are pressed in a sandwich pattern, between two small hollow heat sinks partially filled with liquid.

The vapor evaporated in these heat sinks (3 or 4 generally) is collected in a vapor tube and carried to a common condenser. After condensation of the vapor, the liquid returns by a liquid tube which guarantees a continual supply to the heat sinks. The advantage of this device is that the condenser only dissipates the mean losses from all the power semiconductors in a specific operating cycle.

In the second type, the semiconductors are mounted in the same way as in the previous case but each heat sink has its own condenser and the ascending vapor and falling liquid form a countercurrent two-phase flow in the same tube (3, 4). In this case, all the coolers are independent from one another and are cooled by forced air convection over the condenser fins.

EXPERIMENTAL APPARATUS AND PROCEDURE

A schematic diagram of the test apparatus is shown in Fig.1. The system is composed of three sub-systems : the test evaporator, the adiabatic section and the condenser. This system is basically a two-phase closed thermosyphon. The evaporator section can be dismantled and is composed of an aluminium block on which an evaporation plate is mounted. The semiconductor device (Ø 75 mm) and the jointed evaporator plate are pressed against the evaporator box with a pressure of 1 MPa.

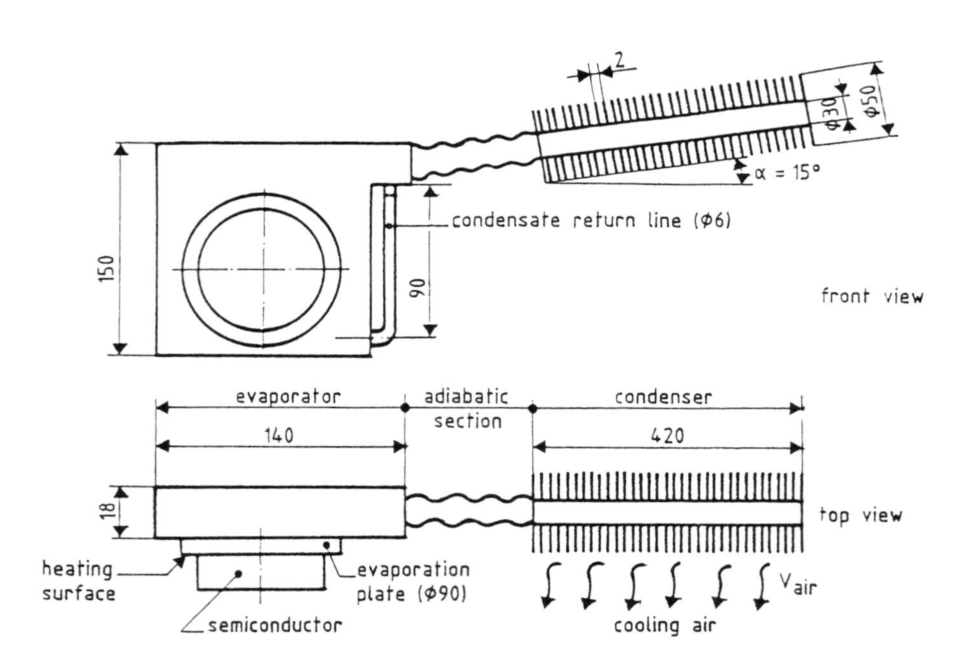

FIGURE 1 . Schematic of experimental set-up.

In order to investigate various materials and exchange area geometries, the evaporation plates can be removed. The thermal power dissipated by the semiconductor is carried via the evaporator wall to the evaporating fluid. The adiabatic section is a stainless steel corrugated tube of 0.1 m length. The 0.4 m long condenser is an external finned pipe with fins of 0.2 mm, cooled by forced convection. The evaporated liquid rises to the top of the pipe where it condenses. Since the condenser is positioned at about 15° from the horizontal, the condensate returns by gravity to the liquid pool along the pipe wall. Chromel-alumel (K-type) thermocouples are used to measure temperature on the equipment. In the condenser, two such sensors are attached to its outside surface and one is inserted into the inside of the tube to measure the vapor temperature. The heat flux in the evaporator is measured by a thermocouple on the wall of the evaporation plate and by one in the liquid inside the evaporator.

Before operation, the thermosyphon is vacuum dried. The thermosyphon is charged with a known amount of fluid and after the system reaches steady state, the temperature and the semiconductor power input Q are recorded.

The working fluid used was R113. This was chosen because although a number of coolants could be considered, from the viewpoint of boiling and condensing, R113 has a superior combination of insulation, incombustibility, operating temperature and vapor pressure properties. The allowable upper working temperature is determined by the semiconductor maximum allowable junction temperature (125°C).

In order to enhance the evaporation heat transfer into the evaporating medium between the wall and the fluid, enhanced heat exchange areas were used. Among the many possibilities for improving heat transfer, various kinds of evaporation plates were used in order to decrease the wall superheat required to initiate nucleate boiling. Investigations were carried out with plates made of different kinds of material and also with extended and rough surfaces. The plates with extended surfaces had grooves or a variable number of smooth holes or tapped holes, as shown in Fig. 2.

RESULTS AND DISCUSSION

Enhancement of boiling heat transfer

Evaporation plates with various kinds of smooth surfaces were tested and two materials were used namely : aluminium and copper, which have good compatibility with R113. The performance of these plates was then compared for similar cooling conditions. Figure 3 shows the temperature difference $\Delta T_{sat} = T_w - T_{sat}$ versus the thermal power supplied by the semiconductor (where T_w is the wall temperature at the center of the evaporation plate and T_{sat} the saturation temperature of the fluid). The best results were achieved with the copper evaporator which is because copper is a better thermal conductor than aluminium and as a result, the applied heat is distributed more evenly in the whole evaporator,

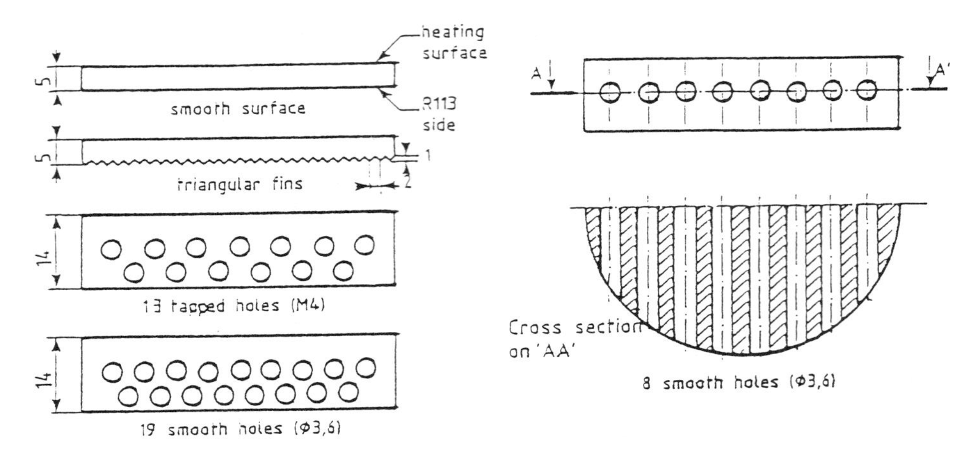

FIGURE 2 . The different kinds of evaporation plates.

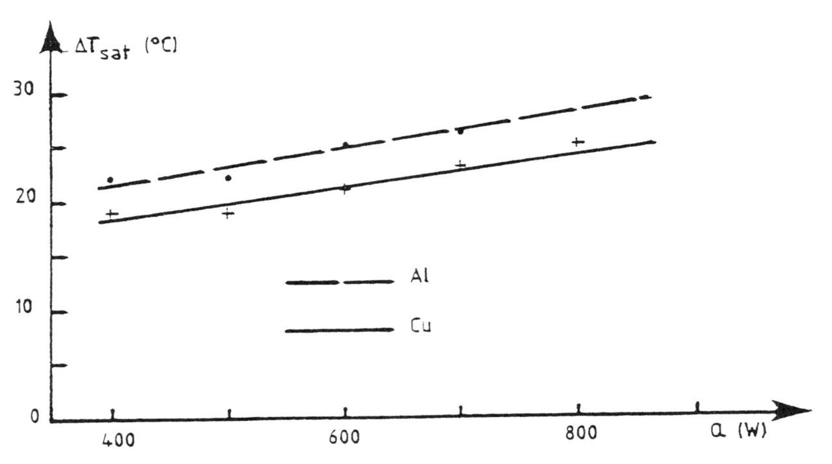

FIGURE 3 . ΔT_{sat} vs. semiconductor electric power for two materials.

and boiling occurs over a greater surface. For similar surface roughness, the wall superheat also decreases because, with copper, the junction temperature will be lower. The superheat increases with increasing semiconductor power as shown in Figure 4 which also shows the values of ΔT_{sat} obtained with different extended copper surfaces. For all these surfaces, ΔT_{sat} is lower than the ΔT_{sat} obtained for smooth surfaces. Part of this decrease is caused by an increase of the exchange area. But for the plates with small fins or tapped holes, this also comes from increased nucleation sites. In the best case, ΔT_{sat} is 40% lower for a prescribed surface heat flux of 800 W.

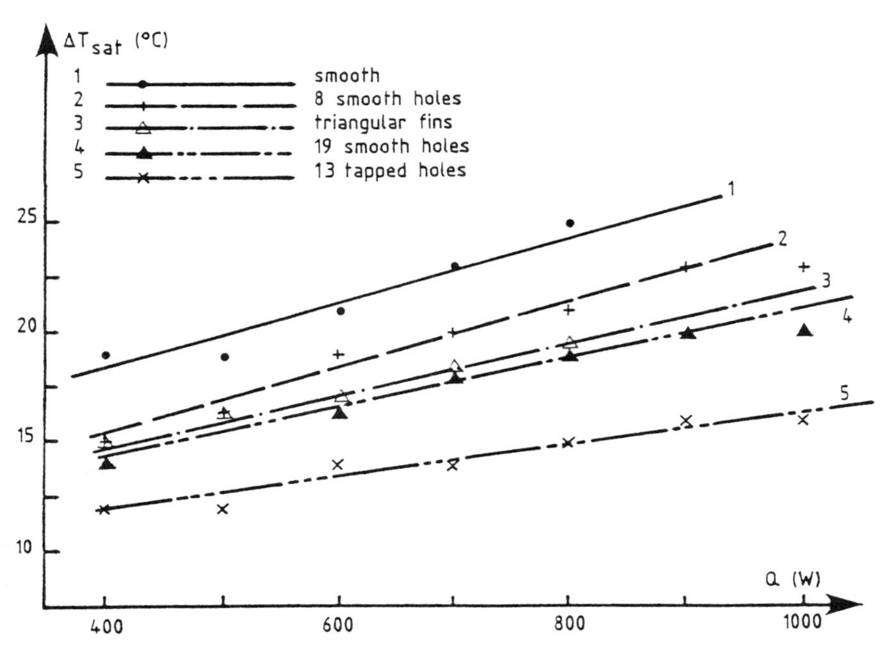

FIGURE 4 . Effect of extended surfaces on ΔT_{sat}.

In an attempt to enhance the heat transfer into the liquid, two roughend surface were tested. The first was a pitted copper surface and the second a porous metallic coating. Figure 5 shows that the heat transfer for the smooth surface was inferior to both roughend surfaces. This is due to the increased number of nucleation sites offered by the pitted or porous roughend surfaces.

Conditions influencing the performance of condensation

Condensation heat transfer is a function of condenser cooling and tilt angle. Figure 6 shows the influence of air flow rate through the fins on saturation temperature. T_{sat} decreases rapidly as air velocity rises, because heat transfer at the condenser is limited by the heat exchange with the air flow. As a result, when the air-side heat transfer coefficient increases, the overall heat transfer coefficient between the air and the R113 also increases and the saturation temperature decreases. At the evaporator, it is found experimentally that the temperature difference ΔT_{sat} remains constant for such a pressure variation range (0.5 - 1 bar). Thus, as T_{sat} decreases, the semiconductor temperature also decreases. Figure 7 shows the influence of condenser tilt angle. The saturation temperature rises when the condenser tends to the vertical position, as shown by Chen (5). Indeed, according to Nusselt's theory of film condensation, at the vertical position, there is an annular falling liquid film. At small tilt angle, the film tends to become stratified and in the upper part of the tube, the liquid film becomes thinner, forming rivulets, which arrive at the liquid pool with high kinetic energy.

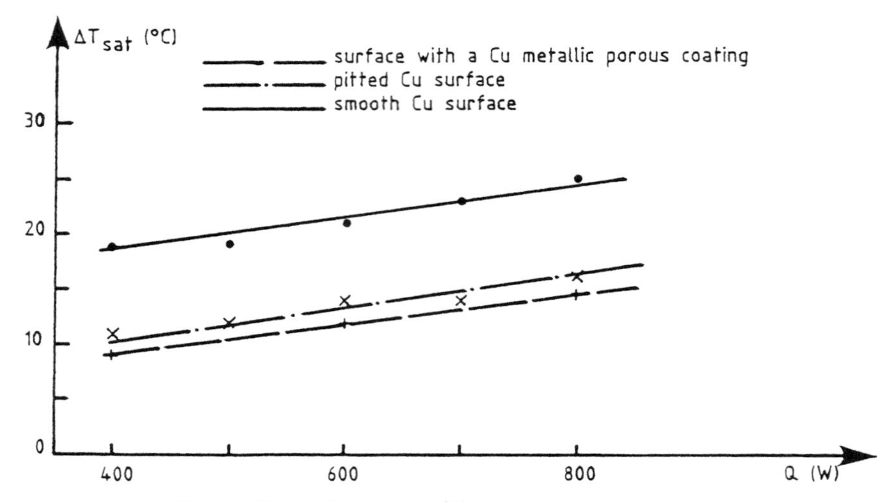

FIGURE 5 . Effect of roughness on ΔT_{sat}.

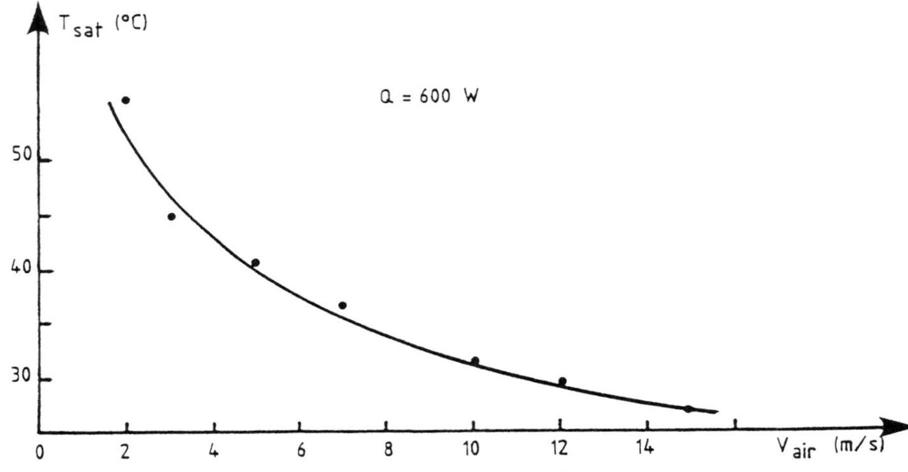

FIGURE 6 . Effect of air-flow rate on saturation temperature.

This, in turn, causes an increase of convection and the bubbles can be detached more easily from the wall and the rate of evaporation is increased.

Practical design

A prototype has been designed for industrial applications which is composed of a single copper block, with two rows of ten tapped holes (M5). It has been tested in the same conditions as the plate evaporator device. The variation of

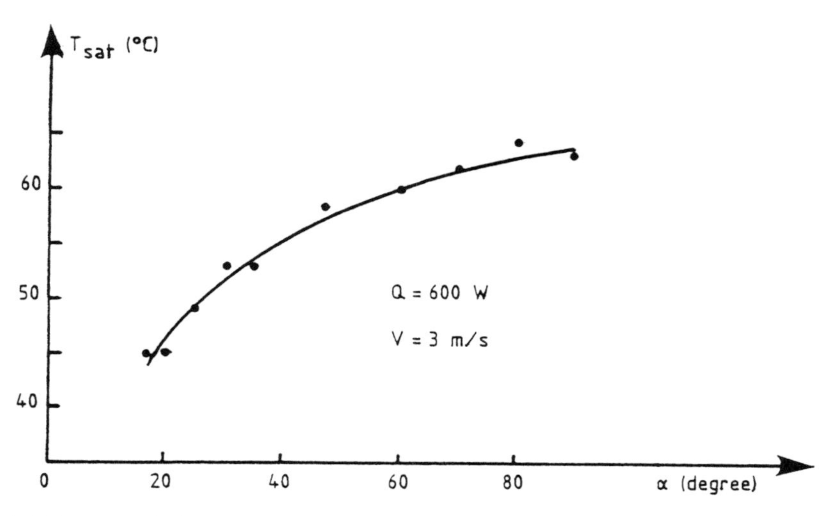

FIGURE 7 . Effect of tilt angle on saturation temperature.

ΔT_{sat} with electric power dissipated in the semiconductor is shown in Fig. 8. The difference between the saturation and wall temperatures is very low and as a result, the evaporator thermal resistance is also low. This can be due to a reduction of the thermal resistance between the semiconductor and the fluid. Indeed, the thermal resistance between the evaporator plate and the box is lower than for the previous case, because the evaporator is made of one block. In Fig. 9, the total thermal resistance R_{th} is plotted against the thermal power supplied to the fluid, Q_{th}. It is found that R_{th} varies with heat flux and also as a result of the air flow-rate variation, since it has been shown that T_{sat} decreases with increasing V_{air}. The thermal resistance values of the evaporator are similar to the ones found by Murase (5) and are independant of condenser cooling conditions. This device has been shown to easily maintain the junction temperature of the semi-conductor below the prescribed temperature of 125°C.

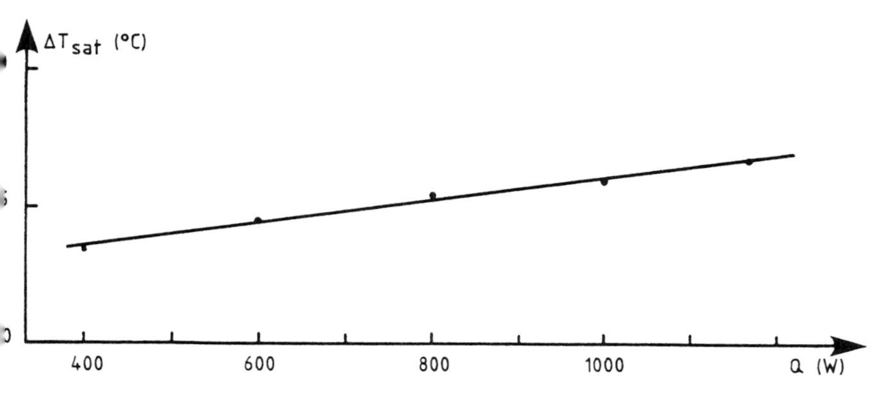

IGURE 8 . ΔT_{sat} vs. semiconductor electric power of the prototype.

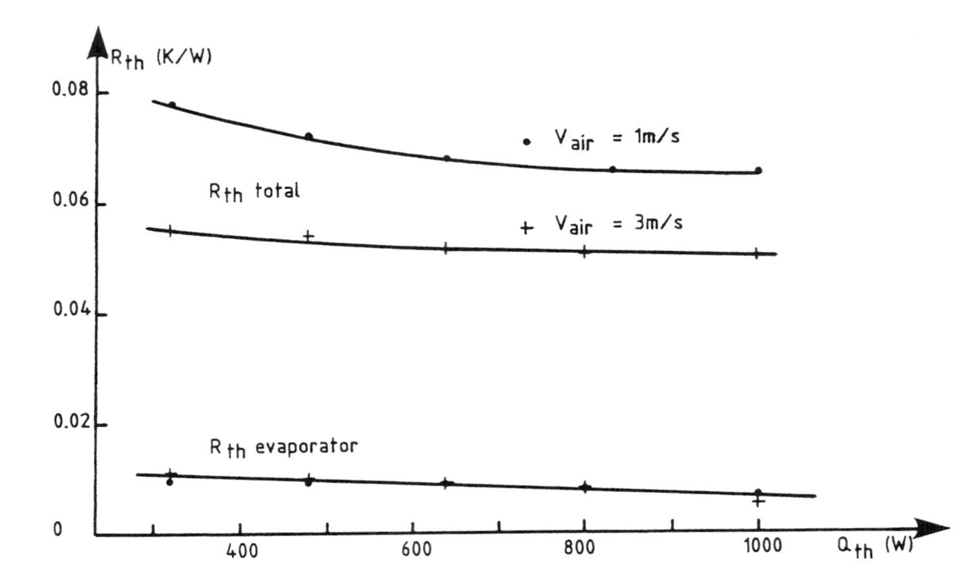

FIGURE 9 . Thermal resistance of the prototype vs. thermal power supplied to fluid.

CONCLUSION

Where a higher thermal performance is needed, phase change cooling can be used to advantage to dissipate waste heat from electronic equipment. A two-phase closed thermosyphon has been developed for cooling of semi-conductors such as GTO thyristors.

It is characterized by low internal thermal resistance and is capable of removing heat with a compact condenser. This device is also widely applicable to various types of power converters because of its modular geometry. This cooling system is particularly suitable when the thermal component losses are equal and when the operating cycle is variable. The insertion of such devices in subsystems with various electrical circuit configurations should be considerably simplified.

Investigations have been carried out with R113 as working fluid because this coolant is able to satisfy a number of criteria : suitable thermal characteristics, chemical stability, electrical insulation, toxicity, low price, etc. The results also have to be confirmed with other fluids, such as those carried out by Park and Bergles (6) who have tested fluorinert liquids. This heat pipe heat sink design complements the alternative immersion freon cooling in containers and modular freon systems (2) used for several years by Alsthom.

NOMENCLATURE

Q	electric power supplied to semiconductor	W
Q_{th}	thermal power supplied to fluid	W
R_{th}	thermal resistance	K/W
T_{sat}	saturation temperature	K
T_w	wall temperature	K

REFERENCES

1. BERGLES, A. E., Liquid cooling for electronic equipment, presented at International Symposium on Cooling Technology for Electronic Equipment, Honolulu, HI, March, 1987.

2. ECKES, U., MEREL, P., Modular 2-phase cooling system for power semiconductors, Proc. Second European Conference on Power Electronics and Applications, Grenoble, France, vol. 1, 1987.

3. GERAK, A., HORVATH, L., JELINEK, F., STULC, P., ZBORIL, V., Examples of heat pipe application in chemical, electrical and other industries, Sixth International Heat Pipe Conference, Grenoble, France, pp. 522-530, May 25-29, 1987.

4. MURASE, T., TANAKA, S., ISHIDA, S., Natural convection type long heat pipe heat sink "Powerkicker-N" for the cooling of GTO thyristor, Sixth International Heat Pipe Conference, Grenoble, France, pp. 537-542, May 25-29, 1987.

5. CHEN, M. M., Heat transfer performance of two-phase closed thermosyphons with different lengths, Sixth International Heat Pipe Conference, Grenoble, France, pp. 493-499, May 25-29, 1987.

6. PARK, K. A., BERGLES, A. E., Boiling heat transfer characteristics of simulated microelectronic chips with fluorinert liquids, Iowa State University Heat Transfer Laboratory Report, HTL-40, ISU-ERI-Ames-87022, August, 1986.

Nucleate Boiling Heat Transfer on Horizontal Cylinders with Re-entrant Grooved Surfaces

Y. HAMANO and Y. CHUJO
Department of Mechanical Engineering
Anan College of Technology
Minobayashi, Anan, Tokushima 774 Japan

ABSTRACT

This paper describes the nucleate boiling characteristics of horizontal cylindrical tubes having re-entrant cavities on the surface. The heated tubes consist of three types: tubes A have wound/rectangular grooved surface, tubes B have wound surface and tubes C have flat-braided lead-wound surface. Experiments have been conducted in a pool of saturated R-113, R-11 or water at atmospheric pressure. The experimental results indicate that the nucleate boiling characteristics of tubes A are improved as the ratio of the diameter of wire/the width of rectangular groove approaches 1, the heat transfer coefficient of tubes B is excellent at high heat fluxes and for the tube C of nominal cross section SQ=8 mm^2, the departure from natural convection(DNC) was not observed.

1. INTRODUCTION

It is generally known that the heat transfer coefficient of nucleate boiling is influenced considerably by the surface conditions and the constructions. In recent years porous surfaces for nucleate boiling have been used in the field of refrigeration and electronics cooling. These surfaces are classified roughly into the sintered metallic particles and the machined or formed structures. Various studies have been reported in papers by Bergles and Chyu[1], Nishikawa et al.[2], Marto and Lepere[3] and Nakayama et al.[4]. Ma et al.[5] have experimented with porous grooved surfaces covered by a metal screen using water and methanol. Oktay[6] discussed the nature of the transition point from natural convection to nucleate boiling in cooling microelectronic components where the problem is the excessive temperature overshoot on relatively low power chips. Nakayama et al.[7] reported on heat sink studs having enhanced boiling surfaces such as microfins and porous structure effect for the cooling of microelectronic components. Park and Bergles[8] simulated microelectronic chips with thin foil strips and showed that microholes of 0.71 mm diameter provided the best enhancement at a high heat flux. However sintered metallic particle surfaces have the disadvantage of being somewhat costly.

This paper gives a description of the nucleate boiling characteristics on grooved surfaces with re-entrant cavities made by a simple procedure easily carried out in the laboratory. The simple grooved shapes consist of the following three types: (A) wound/rectangular grooves, (B) copper wound grooves and (C) flat-braided lead wound grooves. Nucleate boiling heat transfer experiments have been conducted in pools of saturated R-113, R-11 and water at atmospheric pressure.

2. EXPERIMENTAL APPARATUS

Figure 1 is a schematic drawing of the test apparatus. The boiler vessel mainly consists of a square stainless steel container. The test apparatus is similar to the one used by Nishikawa et al.[2]. Boiling occurs on the simple grooved surface heated on the inside by a cylindrical silicon carbide heater with 8 mm diameter. The vapor is condensed and returned to the boiler by gravity from a pyrex glass condenser using cooling water in a tank. An acrylic thermostat bath(590 mm x 360 mm and 300 mm high) is filled up with water at saturated temperature as the working fluid.
The C-A thermocouples of diameter 1.6 mm are used in 3 sets to measure the wall temperature Tw at positions as shown in Figure 2. The temperature differences of the tube in the direction of the length and the circumference, for example, in the nominal cross section SQ=8 mm^2 of tube C, were 0.02 and 0.03 K at $q=1.28x10^2$ W/m^2, 1.5 and 0.4 K at $q=5.76x10^4$ W/m^2, respectively. The temperature Tw of the heating surface was calculated as that of the outer surface of copper cylinder from the value of the thermocouple in center of heating tubes. The heating surfaces were used

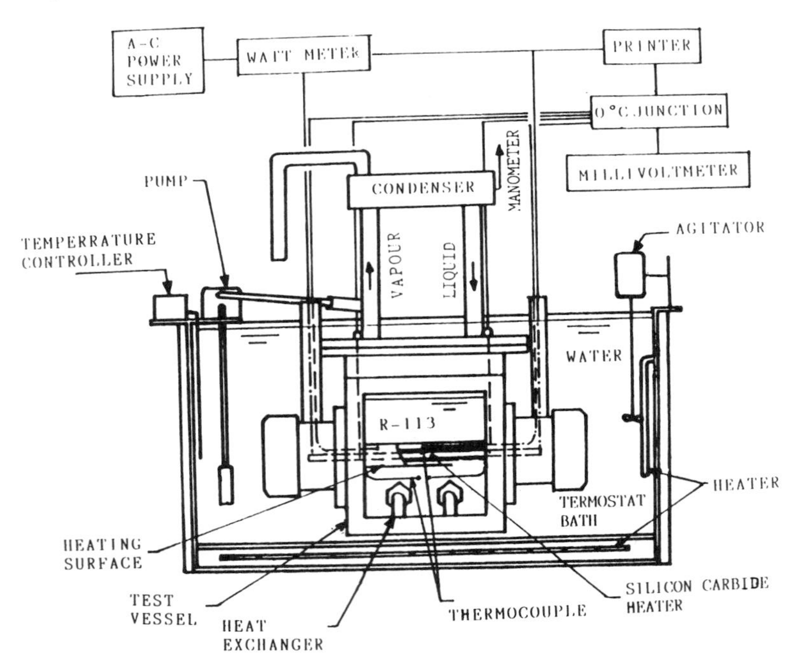

FIGURE 1. Experimental apparatus.

the cylindrical surface of diameter di=18 mm, length L=136 mm in case of tubes B and C. The maximum uncertainty in the wall superheat is estimated to be 0.06 K at (Tw-Ts)=0.2 K, 0.7 K at (Tw-Ts)=10 K.

2.1 The Types, Form and Dimensions of the Simple Grooved Tube.

The simple grooved tubes of the test tubes consist of the following three types. (A) wound/rectangular grooves: The tubes have a spiral shaped cut with rectangular cross sections and an annealed metallic wire is wound in contact with the groove-bottom surface and set in the center of the width of the rectangular grooves using a magnifying glass and is wound by suspending a weight of 0.7 kg. All the diameters of the rectangular groove-bottom are kept at a constant 18.5 mm. (B) copper wound grooves: The annealed copper wire is wound closely around the copper cylinder. (C) flat-braided lead wound grooves: The flat-braided tin-gilt covered lead wire is wound closely around the copper cylinder. In the following text these grooved tubes(grooved surfaces) are outlined A, B and C, respectively.

<u>The grooved tube A.</u> The form and dimensions of these grooved tubes are shown in Figure 2. The diameter of each rectangular groove-bottom is 18.5 mm and the width is 1.0 mm. The temperature of the cylinder on the rectangular groove-bottom is chosen as heating surface temperature Tw. The depth, the pitch and the lead of groove are variously changed for studying the effect on the nucleate boiling characteristics. In the case of boiling water, copper is used as material of the grooved tube A and the diameter of winding is changed from 0.6 to 0.9 mm. In the production of the rectangular grooves attention was paid to the right angle. The outer surface of the grooved tube was polished well using #1000 emery paper.

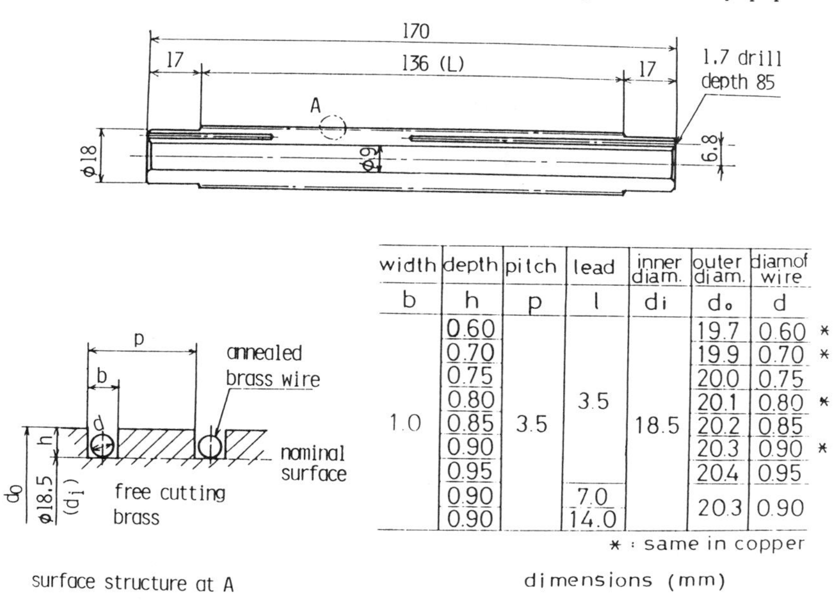

width	depth	pitch	lead	inner diam.	outer diam.	diam of wire	
b	h	p	l	di	do	d	
	0.60				19.7	0.60	*
	0.70				19.9	0.70	*
	0.75				20.0	0.75	
	0.80		3.5		20.1	0.80	*
1.0	0.85	3.5		18.5	20.2	0.85	
	0.90				20.3	0.90	*
	0.95				20.4	0.95	
	0.90		7.0		20.3	0.90	
	0.90		14.0				

* : same in copper

dimensions (mm)

FIGURE 2. Form and dimensions of grooved tube A.

The grooved tube B and the grooved tube C. The detail of copper cylinder and dimensions of the grooved tubes B, C are shown in Figures 3 and 4, respectively. The diameter of winding of the former is changed from 0.4 to 1.0 mm in steps of 0.2 mm. The latter wires consist of 1.25, 3.5 and 8.0 mm^2 with the nominal cross section(SQ). The diameter of the copper cylinder (polished in the manner described above) was 18.0 mm. The effective length of the heating surface of these tubes is 136 mm, as shown in Figure 2.

3. EXPERIMENTAL PROCEDURES.

The simple grooved tubes were set horizontally in the test vessel after the tubes were washed by a supersonic syringe in R-113. The height of the test liquid in the boiling vessel was 75 mm from the center level of the

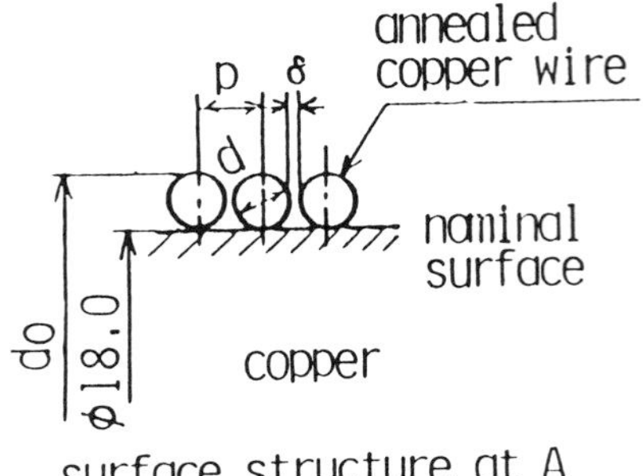

inner diam.	wire diam.	outer diam.	clearance
di	d	do	δ
18.0	0.4	18.8	0.012
	0.6	19.2	0.008
	0.8	19.6	0.002
	1.0	20.0	0.014

dimensions mm

FIGURE 3. Surface structure and dimensions of grooved tube B.

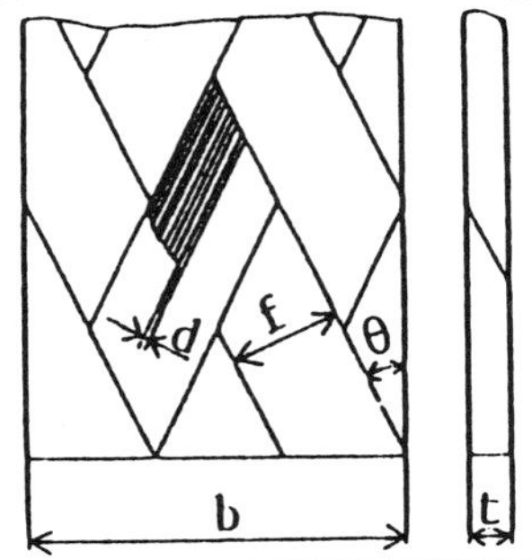

nominal cross section	width	thick-ness	width of bundle	number of bundle	number of wire	density of inter-section	porosity
SQ mm^2	b mm	t mm	f mm	N	n	c $1/cm^2$	ε
1.25	3.9	0.60	0.9	16	7	109	0.46
3.5	8.0	0.76	1.2	24	13	47	0.63
8.0	13.2	1.30	1.7	32	22	25	0.54

d=0.12 mm, θ=0.44 rad., 4 layers

FIGURE 4. Enlarged drawing of surface and dimensions on grooved tube C.

simple grooved tube. In order to eliminate the incondensable gas in reentrant cavities, preboiling was conducted at a heat flux $q=4 \times 10^4$ W/m^2 for 20 minutes. All the boiling data were measured, the heat flux increasing from 2×10^3 to 1×10^5 W/m^2. But for tubes C, the range of heat fluxes was from 100 to 7×10^4 W/m^2. Throughout the experiments, the temperature of the test liquids was controlled within 0.2 K compared with the saturation temperature at the atmospheric pressure. The boiling phenomena on the heating surfaces were observed with the naked eye and photograph by use of a camera. In the case of grooved tube A, the relation between the number of activated groove for nucleation at the groove parts and the heat flux was measured. In the case of grooved tube C, the temperature of test liquid R-113 was cooled to 30 or 40 K(ΔT_{sub}) from the saturation temperature and recovered at the saturation temperature, and the experiments were commenced.

4. EXPERIMENTAL RESULTS AND DISCUSSION

4.1 The Boiling Characteristics in the case of Grooved Tube A and B.

The boiling phenomena in grooved tube A. The generation of bubbles occurs in the clearance between the side wall of the rectangular groove and the winding and these bubbles are inevitably generated in the smaller clearance. The bubble populations increase as the diameter of winding of the rectangular groove approaches 1.0 mm wide. However, these populations occur only partially, if diameter has not reached to 0.9 mm. If the diameter of winding is over 0.9 mm, the bubbles occur uniformly from the grooved surfaces and are observed at many sites on the same circumferential groove, because the clearance is smaller. However, in the case of the diameter d=0.8 and 0.85 mm at the heat fluxes $q=(0.6-1.2) \times 10^4$ W/m^2, the activated sites for nucleation move up and down along the groove quite rapidly. In the interval, the number of activated grooves for nucleation does not increase and the generation of small bubbles from the outer surface of the tubes was observed. For at the grooved tubes below 0.8 mm diameter of winding, the boiling occurred vigorously at the same time from the number of grooves at the heat fluxes beyond about 1×10^4 W/m^2. At the low heat fluxes below 4×10^3 W/m^2, the bubble diameter at

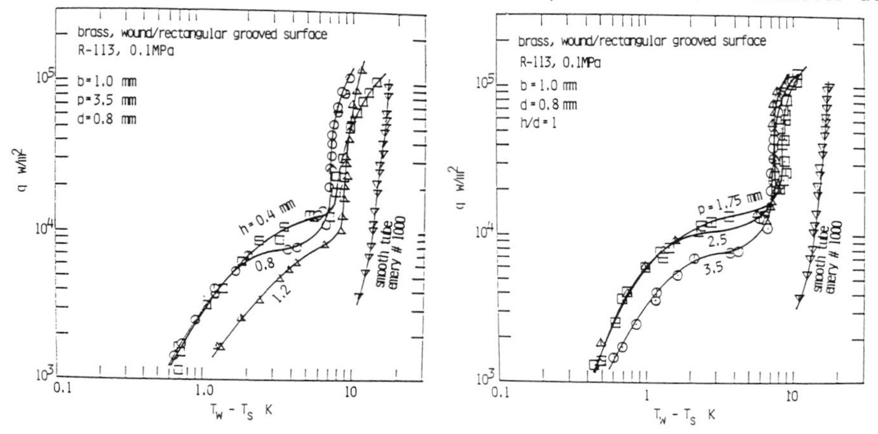

FIGURE 5. Effect of depth on groove tube A.

FIGURE 6. Effect of pitch on groove tube A.

departure becomes smaller and almost uniform, as the ratio d/b approaches unity.

Effect of the depth. Figure 5 shows the boiling characteristic curves when the depth of the rectangular grooves is varied from 0.4 to 1.2 mm for R-113. As shown in Figure 5, the heat transfer coefficient is relatively superior at low heat fluxes below $7x10^3$ W/m^2 as the depth becomes smaller. The boiling performance, however, is good at high heat fluxes above $1.5x10^4$ W/m^2 in the case of h=0.8 mm.

Effect of the pitch. Figure 6 shows that the heat transfer coefficient become a little greater as the pitch becomes smaller. But the effect of pitch is negligible at the heat fluxes beyond $2x10^4$ W/m^2.

Effect of the lead. The boiling characteristic varies almost not at all, when the lead is changed with 3.5, 7.0, 14.0 mm.

Effect on the diameter of winding. Figure 7(a) and 7(b) show the effect on the boiling curves of the wire diameter wound at grooved tube A for R-113 and R-11 respectively. The heat flux becomes larger with the same value of superheat and the shape of the boiling characteristic curve changes greatly with the ratio d/b. The heat transfer coefficient on the brass grooved tube A with the wire diameter of 0.95 mm is greatly superior to $3x10^4$ W/m^2 in comparison with the smooth tube. However, its effect is gradually inferior at the high heat fluxs. Figure 8 shows the relation between the number of activated grooves for nucleation in the lowest groove parts and the heat flux. The slope of the boiling characteristic curve is changed sharply at the range of the heat fluxes(0.6-1.2)$x10^4$ W/m^2 in case of d=0.8 mm for R-113 and R-11. This tendencyis observed to a slight degree at high heat fluxes in the case of d=0.9, 0.95 mm and is remarkable for R-11.Figure 8 shows the circumstances of these phenomena very well.

Figure 9 shows the relation between the number of boiling grooves in the lowest groove partd on grooved tube A and the superheat for R-113. When the diameter of winding is 0.95 mm, the average clearance between the side wall of the rectangular groove and the winding at the groove-bottom becomes $25x10^{-6}$ m and the superheat is less than 1 K. The boiling occurs from all grooves in which the superheat for the generation of nucleation has increased as the diameter of winding has become smaller. Especially, in the case of d=0.6 mm, boiling occurred in 70 % of groove parts at a

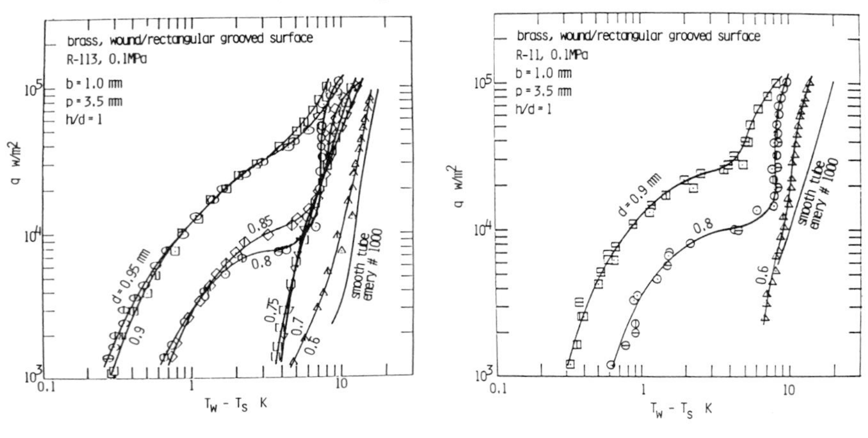

(a) R-113 (b) R-11

FIGURE 7. Effect of winding diameter for R-113 and R-11 on grooved tube A. (brass).

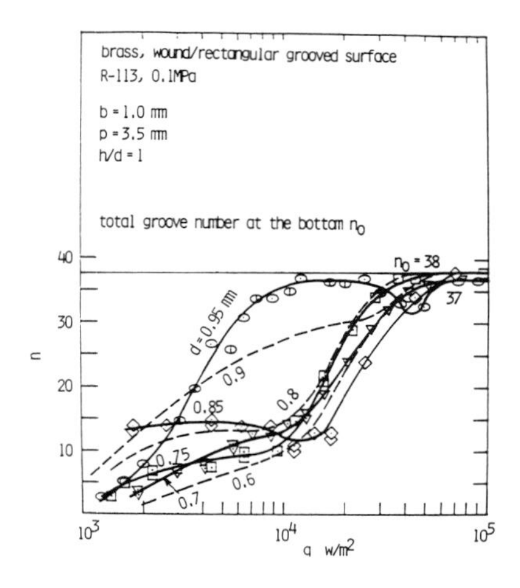

FIGURE 8. Relation between the number of boiling grooves in the lowest groove parts on grooved tube A and heat flux.

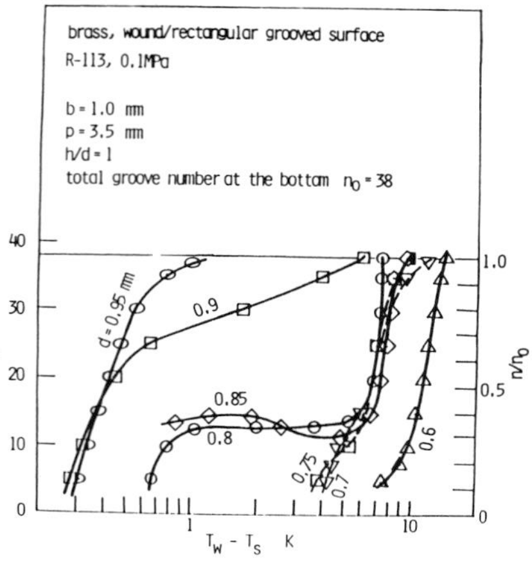

FIGURE 9. Relation between the number of boiling grooves in the lowest groove parts on grooved tube A and wall superheat.

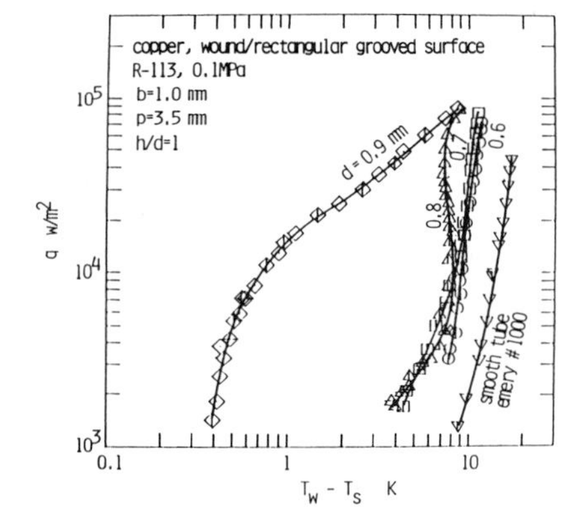

FIGURE 10. Effect of winding diameter for R-113 on grooved tube A (copper).

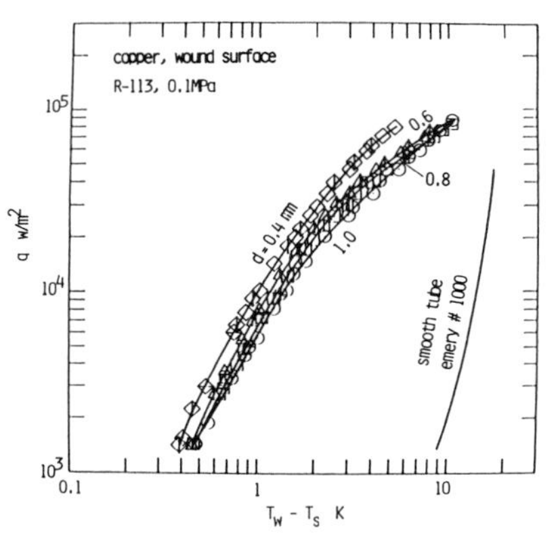

FIGURE 11. Effect of winding diameter on grooved tube B.

superheat beyond 10 K. It caused the heat capacity of R-113, which occupied the groove-space,to decrease as the diameter of winding became larger. The liquid of R-113 is superheated by only the small heat fluxes, as a small bubble nucleation trapped in re-entrant cavities grows by obtaining the latent heat of vaporization and the boiling occurs easily at the small superheat. To say it differently, as the diameter of winding is decreased, the heat capacity of R-113 in the groove parts becomes gradually larger. The temperature of the liquid does not rise at the same heat flux. Consequently the bubble does not grow and nucleable.

Effect of the material. Figure 10 shows the boiling curves in which the copper material used and the diameter of winding is varied from 0.6 to 0.9 mm in steps of 0.1mm. The heat transfer coefficient is inferior to that

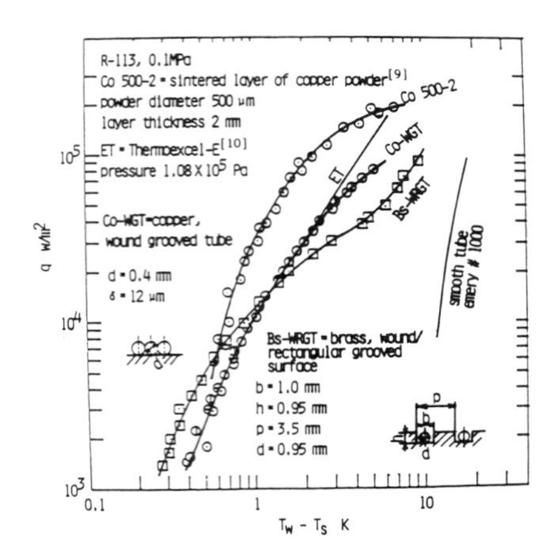

FIGURE 12. Comparison of grooved surfaces A and B to other porous surfaces for R-113.

of the brass wire in the diameter of 0.8 mm at low heat fluxs. It may be that this is because of the difference in the contact angles.
The boiling curve for grooved tube B. Figure 11 shows the effect of this construction on the boiling curves. The copper diameter of winding is varied from 0.4 mm to 1.0 mm in steps of 0. 2 mm for R-113. The heat transfer coefficient improves gradually in comparison with the smooth tube as the diameter of winding diminishes.
A comparison with other enhanced boiling surfaces(I). Figure 12 shows the boiling curves in comparison with Bs-WRGT in grooved tube A, Co-WGT in grooved tube B, Co 500-2, with 2 mm of layer thickness, 500×10^{-6}m of powder diameter[9], and Thermoexcel-E[10] shown by a symbol ET in Figure 12. The heat transfer coefficient of Bs-WRGT is especially excellent at low heat fluxs below 1.5×10^4 W/m^2. But beyond it, the boiling performance of Co-WGT is superior to Bs-WRGT, and becomes nearly equal to that of ET.

4.2 The Boiling Characteristics of Grooved Tube A for Water

The boiling phenomena. In the case of water, the sizes and forms of the bubble at the departure from the groove-parts are very different from these of R-113 and vary greatly. This may be because the surface tension of the water is about four times that of R-113.
The boiling curve. Figure 13 provides the boiling curve. The diameter of winding is varied from 0.6 to 0.9 mm in steps of 0.1 mm. These surfaces demonstrate a similar improvement in the heat transfer coefficient as the diameter of winding increases and the heat flux grows larger. But this tendency was very different in comparison with the characteristics of R-113, In particular,it was proved that the heat transfer coefficient at the middle heat fluxs was inferior to R-113. This can be understood from the relation between the number of boiling grooves and the heat flux as shown in figure 14. In water, that is, the number of boiling grooves is smaller in comparison with that of R-113 at heat fluxs below 3×10^4 W/m^2, and increases suddenly beyond that. Thus, the enhancement of heat transfer became excellent.

4.3 The Boiling Characteristics of Tube C for R-113

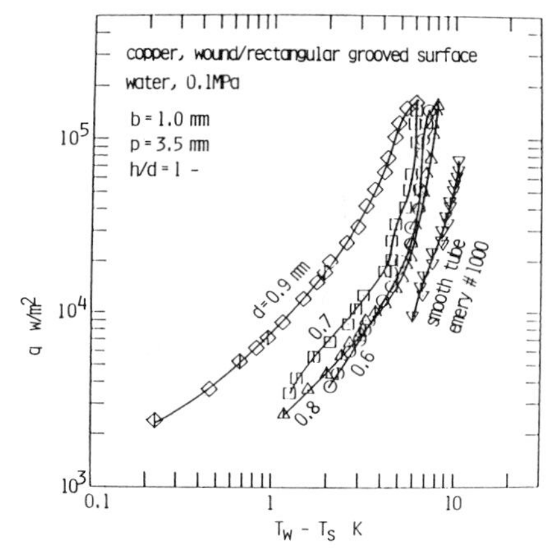

FIGURE 13. Effect of winding diameter on the tube A for water

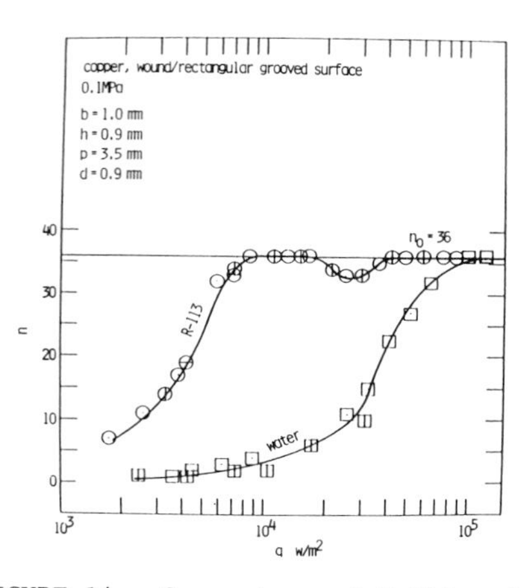

FIGURE 14. Comparison of R-113 and water, same relation in Fig. 8.

The boiling phenomena. In the surface of nominal cross section SQ=8 mm^2 which has the most excellent boiling performance, boiling was observed at several points with a bubble diameter of about 1 mm , at the heat flux $1.54x10^2$ W/m^2, and wall superheat Tw-Ts=0.32 K. Boiling occurred immediately upon being heated after the temperature of R-113 was cooled under 30 K from the saturation temperature(ΔTsub=30 K) and recovered then at the saturation temperature of R-113. In the heating surface of SQ=3.5 mm^2, the temperature overshoot from the natural convection was apparently observed, but the wall superheat was very small, less than 1.9 K.

The boiling curve. Figure 15 shows the boiling curve both for the copper smooth tube and the grooved tube C for R-113. For a comparison, the

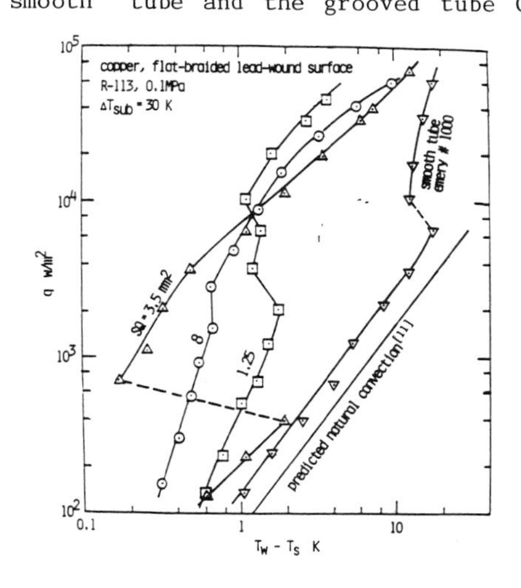

FIGURE 15. Effect of SQ on grooved surface C.

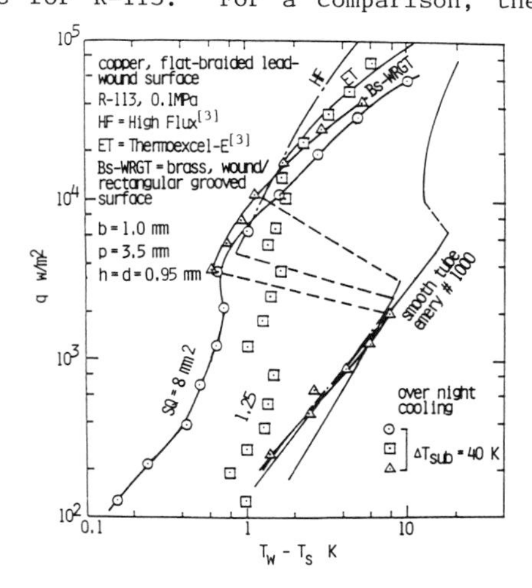

FIGURE 17. Comparison of grooved surface C and A with High Flux surface, Thermoexcel-E obtained by Marto and Lepere.

TABLE 1. The intersection opening of grooved tube C on the surface

nominal cross section	mean	maximum	minimum
SQ mm^2	x mm	xm mm	xs mm
1.25	0.21	0.37	0.13
3.5	0.55	0.89	0.37
8	1.07	1.44	0.53

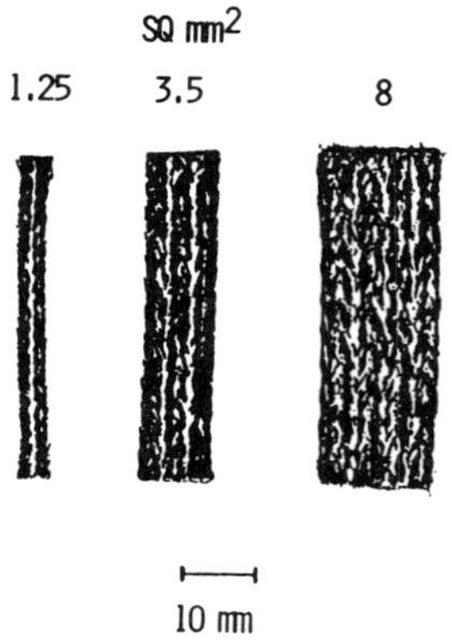

FIGURE 16. Contact condition of flat-braided lead wires.

natural convection correlation for a horizontal cylinder[11] lies slightly below the data measured. The point of departure from natural convection(DNC) was observed at a heating surface of SQ=3.5 mm^2, while the effect of the boiling heat transfer was most excellent in boiling regions below $8x10^3$ W/m^2. In the heating surface of SQ=1.25mm^2, the boiling curve became a line of discontinuity at low fluxes in relation to patch boiling. At fluxes beyond $1x10^4$ W/m^2,the intersection numbers of SQ=1.25 mm^2 become the largest at 109/cm^2 and the vigorous boiling consisted of small bubbles, so that the enhancement of heat transfer was best at the high heat fluxes, and the value of the maximum heat transfer coefficient became $1.2x10^4$ W/m^2K at $3.4x10^4$ W/m^2.
In SQ=8 mm^2, the regions of natural convection were non-existent and the heat transfer characteristic was excellent. Table 1 provides the results of the intersection opening measured by using a tool microscope on the grooved surface C. Figure 16 shows the dark and light by copying the flat-braided lead wires. The parts in contact with the plane are black, otherwise they are white. From these results, the intersection opening increases as the value of the nominal cross section becomes larger. In SQ=8 mm^2, the mean value of the intersection opening was 1.07 mm and the difference between the maximum and the minimum value of it increased to 0.91 mm. It is thought that the characteristics of the boiling heat transfer of the grooved tube C is closely related to the intersection opening and the density of the flat-braided lead wire. For highly wetting liquids such as fluorocarbons, a "doubly re-entrant" cavity has been suggested[12]. Therefore, a chip of "Dendrite" 1mm high has been experimented with by Oktay[6] for removing the temperature overshoot at the low heat flux with liquid of FC-86. In general,the heat transfer of flat-braided lead wire surfaces was carried out with almost as good a heat performance as with the "Dendritic" chip.
A comparison with other enhanced boiling surfaces(II). Figure 17 shows the comparison of boiling curve on some enhanced heating surfaces for R-113. The tube of grooved tube A with the diameter of winding 0.95 mm and the grooved tube of SQ=1.25, 8 mm^2 were cooled to Tsub=40 K in R-113 over night, and the saturated pool boiling experiment was conducted. However, High Flux tube and Thermoexcel-E was measured by Marto and Lepere[3] using the same procedure several years ago. The SQ=8 mm^2 boiled at $1.28x10^2$ W/m^2 immediately after heating, DNC was not observed in

comparison with High Flux, Thermoexcel-E and the Bs-WRGT at the low heat flux regions and the heat transfer coefficient were excellent at some high fluxes. Especially, in the case of SQ=8 mm^2, the superheat was improved five times on High Flux at the heat flux 200 W/m^2. In the case of SQ=1.25 mm^2, the wall superheat was less than 2 K at the fluxes below $1.4x10^4$ W/m^2, and the boiling performance was nearly equal to Thermoexcel-E at the flux regions beyond it.

5. CONCLUSION

The experimental investigation of pool boiling heat transfer on horizontal simple grooved tubes has led to the following conclusions.
(1) The effect of the depth, pitch, lead, and material of grooved tube A on the heat transfer coefficient is negligible. (2) The boiling characteristics of heat transfer by grooved tube A are improved as the ratio d/b approaches 1 and the curve of boiling characteristics greatly changes its shape by d/b. (3) In the region where the wire is in contact with the side of the rectangular groove, the vapor is trapped efficiently. Consequently the boiling is generated easily with a low degree of heating.(4) The boiling characteristics curves for water is widely different from type of R-113 and R-11. The promotion of boiling heat transfer is remarkable in the region of high heat flux. (5) The characteristics of heat transfer of the grooved tube B are excellent in the region of high heat fluxes in comparison with grooved tube A. (6) For the grooved tube C of the nominal cross section SQ=8 mm^2, the DNC point was not observed at this experiments and this tube has excellent heat transfer characteristics.

NOMENCLATURE

b	width of rectangular groove	mm
d	diameter of winding	mm
h	depth of rectangular groove	mm
L	effective length of heating surface	mm
n	number of boiling groove in the lowest groove parts on grooved tube A	
n_0	total groove number at the bottom	
p	pitch of rectangular groove	mm
q	heat flux	W/m^2
SQ	nominal cross section	mm^2
Tw	temperature of wall	K
Ts	saturation temperature of liquid	K
α	heat transfer coefficient	$W/(m^2K)$
ΔTsub	degree of subcooling	K
δ	clearance between windings	mm

REFERENCES

1. Bergles, A.E.,and Chyu, M.C., Characteristics of Nucleate Pool Boiling From Porous Metallic Coatings,ASME Journal of Heat Transfer, Vol.104, no.5,pp.279-285,1982.
2. Nishikawa, K.,Ito,T.,and Tanaka,K.,Augmented Heat Transfer by Nucleate Boiling at Prepared Surfaces,Proc. A.S.M.E.-J.S.M.E. Thermal Engineering Conference , Hawaii, Vol.1.pp.387-393,1983.
3. Marto, P.J.,and Lepere, V.J.,Pool Boiling Heat Transfer From Enhanced

Surfaces to Dielectric Fluid, ASME Journal of Heat transfer, Vol.104, no.5,pp.292-299, 1982.
4. Nakayama, W.,Daikoku,T.,and Nakajima, T.,Effects of Pore Diameters and System Pressure on Nucleate Boiling Heat Transfer from Porous Surface, ASME Journal of Heat Transfer, Vol.104, no.2,pp.286-291,1982.
5. Ma Tongze.,Liu Xin.,and Li Huigun., Effects of Geometrical Shapes and Parameters of Reentrant Grooves on Nucleate Pool Boiling Heat Transfer from Porous Surfaces, Proc.8th Int.Heat Transfer Conf., San Francisco, pp.2013-2018,1986.
6. Oktay, S.,Departure from Natural Convection(DNC)in Low-Temperature Boiling Heat Transfer Encountered in Cooling Micro-Electronic LSI Devices, Proc.7th Int. Heat Transfer Conf., Munchen, pp.113-118,1982.
7. Nakayama, W.,Nakajima, T.,and Hirasawa, S.,Heat Sink Studs Having Enhanced Boiling Surface for Cooling of Microelectronic Components, ASME Paper, 84-WA/HT-89, PP.1-8,1984.
8. Park, K.-A.,and Bergles, A.E.,Boiling Heat Transfer Characteristics of Simulated Microelctronic Chips with Detachable Heat Sinks,Proc.8th Int. Heat Transfer Conf., San Francisco, pp.2099-2104,1986.
9. Nishikawa, K.,Ito, T., and Tanaka, K.,Evaluation of High Performance Boiling Surface(First Report 1 Experiment on Surfaces Covered with Sintered Metal Powder),Proc.15th Annual Symposium Heat Transfer Society of Japan.,pp166-168,1978
10. Cat. No.A1-518C.,pp.5,Hitachi Cable,Ltd,1980.
11. McAdams, W.H.,Heat Transmission, 3rd., pp. 176, McGraw-Hill, New York, 1954.
12. Bergles, A.E., Collier, J.G., Delhaye, J.M., Heiwtt, G.F., and Mayinger, E., Pool Boiling, Two-Phase Flow and Heat Transfer in the Power and Process Industries, pp. 190-220, Hemisphere Publishing Corporation, New York,1981.

Boiling Heat Transfer Characteristics of Simulated Microelectronic Chips with Fluorinert Liquids

K.-A. PARK
Korea Standards Research Institute
Taejon, Korea

A. E. BERGLES
Rensselaer Polytechnic Institute
Troy, New York, USA

R. D. DANIELSON
3M Company
St. Paul, Minnesota, USA

ABSTRACT

The heat transfer characteristics of simulated microelectronic chips were investigated with direct immersion cooling in Fluorinert liquids (FC-72 and FC-87). The basic heater was a 4.5 mm x 4.5 mm electrically heated foil.

The basic plain heater and the heater with a detachable plain heat sink were tested in natural convection and boiling to provide a reference for the performance of the enhanced heat sinks. Small holes (microholes) or small fins (microfins) were machined in copper plates, which were tested as enhanced boiling heat sinks. Other heat sinks were made from the commercial Thermoexcel-E and High Flux surfaces. For all of these structured surface heat sinks, the boiling curves (heat flux versus wall superheat) for both Fluorinerts are in good agreement with each other. The best performance (lowest wall superheat) was obtained with the High Flux surface.

Boiling heat transfer characteristics for the flush heater and the heat sinks in FC-72 and FC-87 are compared with those for R-113. The boiling curves for the flush heater are in good agreement for all three liquids. Similar results were also obtained for plain, microhole, or microfin heat sinks. However, the boiling curves for the Thermoexcel-E and High Flux heat sinks for the Fluorinerts are slightly different from those for R-113.

A thermal chip was also tested. Again, the boiling curves for FC-72, FC-87, and R-113 are quite similar.

INTRODUCTION

The importance of electronic cooling, especially direct immersion cooling, and a discussion of pertinent literature are addressed in the survey paper by Incropera [1]. The Fluorinert liquids have proven themselves to be outstanding fluids for immersion cooling from the standpoints of dielectric strength and materials compatibility. The basic boiling heat transfer characteristics of the Fluorinert liquids have been established in a variety of tests with plain surfaces.

Some data for Fluorinert liquids with enhanced boiling surfaces have been reported. Dendritic "heat sinks" on thermal chips in FC-88 were tested by Oktay and Schmeckenbecher [2]. Also, Oktay [3] tested a "tunnel" heat sink (4.6 mm x 4.6 mm x 1 mm thick copper plate with 4 holes d = 0.8 mm) in FC-86. He observed considerable enhancement with the heat sinks on thermal chips. Surface modification of the chips (laser nucleation cavities) was investigated in FC-86 by Hwang and Moran [4]. It is difficult to compare the data for the thermal chips with data for regular surfaces because the heat fluxes reported for the thermal chips were not corrected for the considerable heat loss to the substrate.

Nakayama et al. [5] obtained data for several kinds of cylindrical structured heat sinks in FC-72. The heat input was provided by a cartridge heater.

Marto and Lepere [6] obtained data for plain, Gewa-T, Thermoexcel-E, and High Flux enhanced boiling surfaces in FC-72 and R-113. Heat transfer coefficients for FC-72 were lower than for R-113. The High Flux surface exhibited the best performance over the entire range of heat fluxes.

The primary objective of the present study was to determine boiling heat transfer characteristics of simulated microelectronic chips with or without heat sinks using Fluorinert liquids. These data are also compared with data previously obtained for R-113. The main reason for this comparison is to validate the use of the relatively inexpensive R-113 for extended tests that involve rather large liquid inventory and considerable evaporation loss. Additional tests with "thermal chips" were conducted to assess the boiling characteristics of ultrasmooth silicon surfaces.

EXPERIMENTAL APPARATUS AND PROCEDURE

Two working fluids, FC-72 and FC-87, were tested. These two dielectric fluids are members of a family of Fluorinerts used for direct immersion cooling of microelectronic components and have boiling points of 56°C and 28°C, respectively. Table 1 compares the properties of these fluids with those of R-113. The test sections were immersed in a pyrex tank 238 mm x 181 mm x 324 mm high, as shown in Fig. 1. This tank was heated by water flowing in an outer pyrex tank 308 mm x 308 mm x 304 mm high. Two condensers with chilled ethylene glycol/water (about 6°C for FC-72 and 0°C for FC-87) were used to condense the Fluorinert vapor. Also, silicone sealant (Dow Corning, Inc.) was used to seal the gap between the test tank and the plexiglas cover. The loss of fluid during these experiments was very small.

A circuit board (glass epoxy G-10) clad with copper on one side was the basis of the simulated chip test sections, as shown in Fig. 2. The copper cladding in the center was removed by a razor blade to make room for the foil heater (12.7 μm thick nichrome). M610 epoxy resin (Measurement Group, Inc.) was used to bond the foil to the circuit board. The foil and copper cladding were soldered to 25.4 μm brass foil for the electrical connections.

Five kinds of copper heat sinks were explored in this study: plain, microfin, microhole, Thermoexcel-E, and High Flux. The construction details of the heat sink test section are shown in Fig. 3. The dimensions, total surface areas, and ratios of total-to-projected areas are tabulated in Table 2. The test sections are the same as those used by Park and Bergles [7].

A single thermocouple (36 gauge copper-constantan) was placed at the back center of each test section. This thermocouple was connected to a multimeter through an ice point reference junction (MCG-T electronic ice point,

TABLE 1. Property Comparison of FC-72, FC-87, and R-113

a) at saturation temperature (760 mm Hg)

	FC-72	FC-87	R-113
Boiling point, °C	56	28	47.6
ρ_f, kg/m^3	1580	1630	1507
ρ_v, kg/m^3	12.5	11.9	7.5
h_{fg}, kJ/kg	88	88	144
σ, N/m	0.008	0.009	0.0147

b) at 25 °C

	FC-72	FC-87	R-113
k, W/mK	0.057	0.056	0.0747
μ, kg/m sec	6.72×10^{-4}	6.72×10^{-4}	6.64×10^{-4}
C_p, J/kg K	1050	1092	958
β, 1/K	0.0016	0.0016	0.00075
ρ_f, kg/m^3	1680	1630	1557

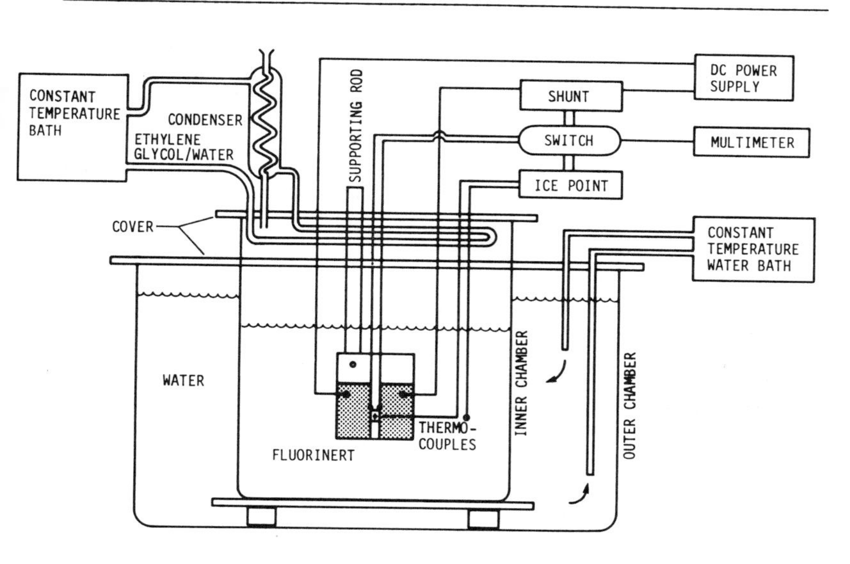

FIGURE 1. Experimental Apparatus.

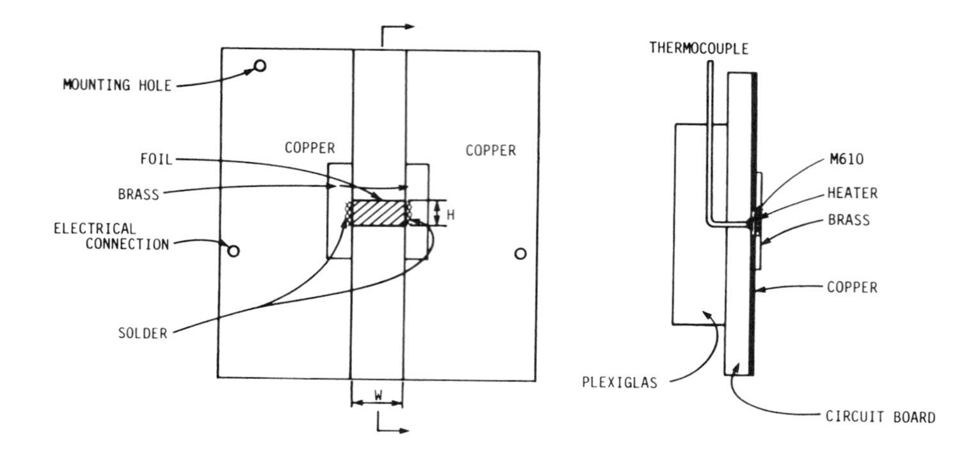

FIGURE 2. Test section details.

Omega Engineering, Inc.). The range of error of this reference junction is ± 0.1°C. A standard thermometer was placed in the test tank to measure pool temperature and check the thermocouple in the test section. Saturation temperatures were slightly lower than values given in the catalog, perhaps due to short chains of carbon and fluorine. The apparent saturation temperature and operating pressure are given in each data plot.

Visual observations were also emphasized during the FC-72 experiments. The black and white pictures are not reported here but are provided in a report [8].

The outer tank water was raised above the saturation temperature of the working fluid for degassing. High power (over 10^5 W/m^2) was also supplied to the test section and the process was carried out for about two hours. The water temperature was then reduced to about the saturation temperature of the working fluid. Power to the heater was shut off for 30 to 60 minutes, and then testing commenced. the experimental procedure and methods for data reduction were similar to those used previously [7].

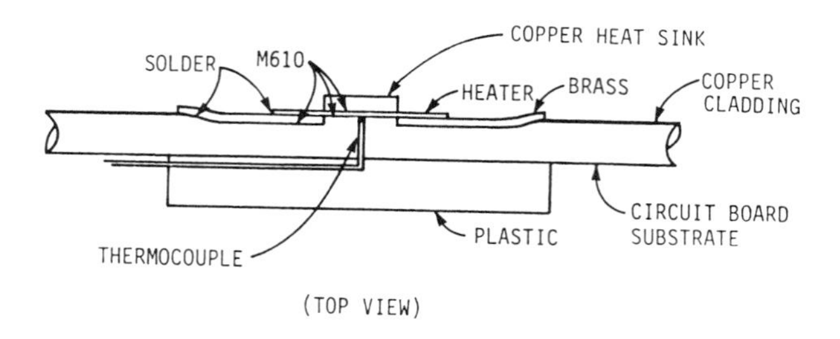

FIGURE 3. Construction details of heat sink test section.

TABLE 2. Dimensions of test sections

	Height mm	Width mm	Area Ratio	Comment
Flush Heater	4.4	4.4	----	----
Plain Heat Sink	4.4	4.5	1.90	----
Microfin Heat Sink	4.6	4.6	2.87	5 fins Fin gap = 0.49 mm Fin pitch = 0.99 mm
Microhole Heat Sink	4.5	4.3	3.8	Hole diameter = 0.71 mm 4 holes
Thermoexel-E Heat Sink	4.6	4.6	----	Pore diameter = 0.12 mm
High Flux Heat Sink	5.1	5.1	----	Porous layer = 0.7 mm Particle size 27 ~ 32 μm

The final tests were concluded with a silicon thermal chip 4.57 mm square and 0.38 mm thick. The surface of the chip appeared very smooth and shiny. The chip was mounted with an 11 x 11 solder pad matrix on the ceramic substrate (25.4 mm square and 1.6 mm thick), as shown in Fig. 4. The gap between the chip and substrate was 0.76 mm. Twelve temperature sensing diodes and seven 140 ohm resistors were distributed in the chip. The diodes were calibrated for temperature at a current of 100 μA.

The test rig for the thermal chip was similar to that used in the previous experiment. Data were obtained for saturated conditions after about five hours of degassing. The power to the chip was not corrected for substrate heat loss.

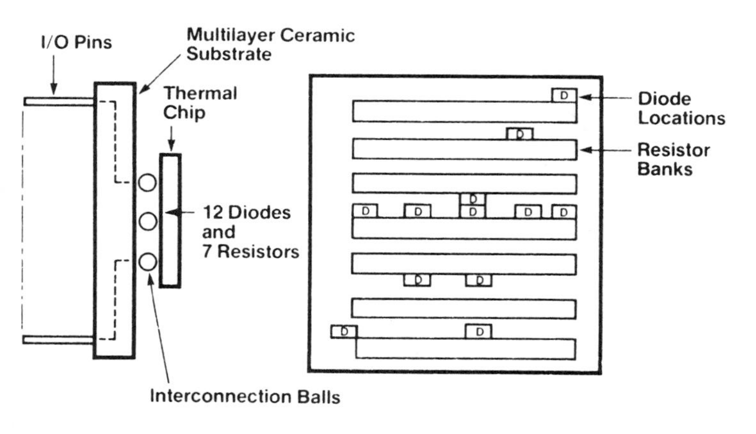

FIGURE 4. Typical structure of thermal chip (Hwang and Moran, 1981).

RESULTS AND DISCUSSION

The traditional boiling coordinates of heat flux versus wall superheat [9] were chosen. The base line in Fig. 5 is the Fujii and Fujii correlation of laminar boundary layer solutions for a vertical surface with constant heat flux:

$$Nu_c = \left(\frac{Pr}{4 + 9\sqrt{Pr} + 10Pr}\right)^{1/5} (Gr_c{}^* Pr)^{1/5} \tag{1}$$

The predictions for FC-72 and FC-87 are almost the same. Also, there are very small differences among the predictions for R-113 FC-72, and FC-87 as expected from the similarity in properties shown in Table 1. Since the data for natural convection for FC-72 and FC-87 are in good agreement with those for R-113, the size effects for natural convection for Fluorinerts seem to be similar to those for R-113 [7, 10].

The incipient boiling superheats and temperature excursions to established boiling flush heater in Fluorinerts are similar to those for R-113. Estab-

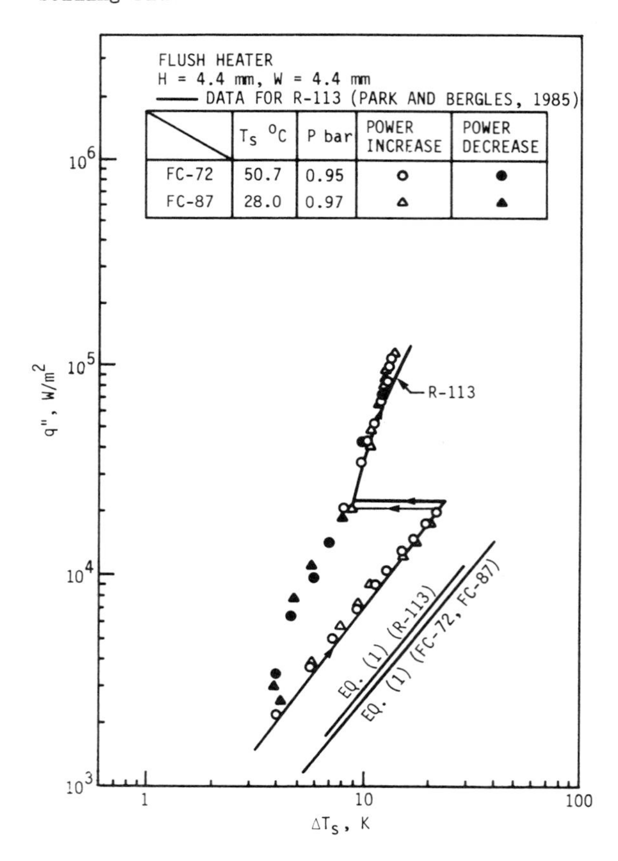

FIGURE 5. Comparison of boiling data for flush heater in FC-72, FC-87, and R-113.

lished boiling data for increasing and decreasing power for Fluorients are almost identical to those for R-113.

The visual observations for the flush heater in FC-72 indicate many boiling sites on the surface at the inception of boiling. At high heat flux, a large vapor mass was observed at the top of the heater, but still there was nucleate boiling at the bottom of the heater. The highest heat flux was chosen to be 80% of critical heat flux, predicted by the correlation of Zuber [11], to save the heater. Relatively few boiling sites were evident at low heat flux with decreasing power.

Data for the plain heat sink with Fluorinerts are similar to those for R-113, as shown in Fig. 6. Boiling took place at the top and side of the heat sink, but there were no boiling sites at the bottom of the heat sink at boiling inception. The visual observations also disclosed a large vapor mass at the top and nucleate boiling sites at the bottom of the plain heat sink, which was a similar situation to that observed with the flush heater. The number of boiling sites with decreasing power was almost identical to that for increasing power at similar wall superheat and heat flux.

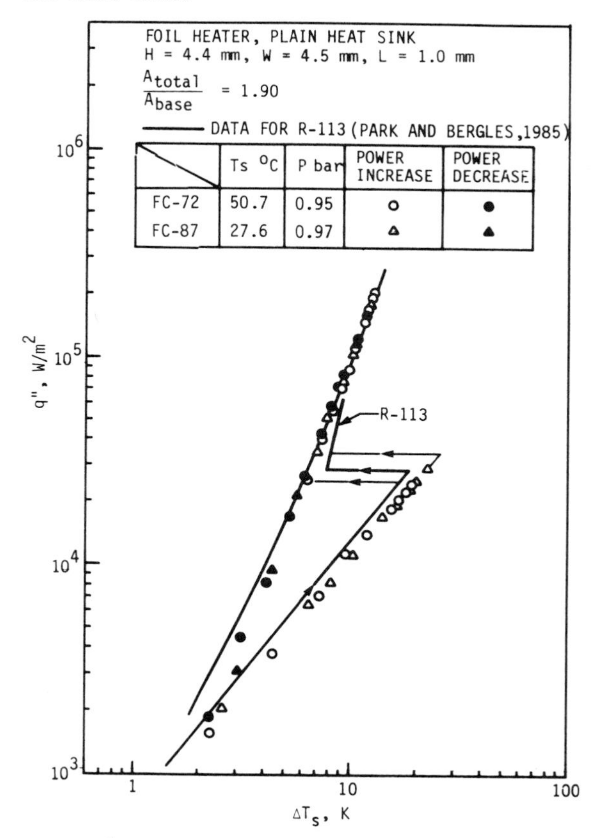

FIGURE 6. Comparison of boiling data for plain heat sink in FC-72, FC-87, and R-113.

Similar results were obtained for the microfin heat sink, as shown in Fig. 7. There is small shift of the boiling data relative to R-113. Also, a small two-step temperature excursion was observed. Small differences between the data for increasing and decreasing power at established boiling for the microfin heat sink were observed, apparently because a number of nucleation sites were inactive with increasing power but active with decreasing power. As shown in Fig. 8, data for the microhole heat sink in Fluorinerts shifted to higher superheat. The relatively low performance of the microhole heat sink in Fluorinerts compared with that for R-113 is likely due to the small differences of properties (especially surface tension, contact angle, and latent heat).

Substantial differences in performance were obtained for the Thermoexcel-E heat sink (Hitachi Ltd., Japan) and the High Flux heat sink (Linde Division of Union Carbide), as shown in Figs. 9 and 10, respectively. Data for natural convection and temperature excursions for the Thermoexcel-E and High Flux heat sink in Fluorinerts were similar to those for R-113. Boiling occurred initially at the right side of the top of the Thermo-

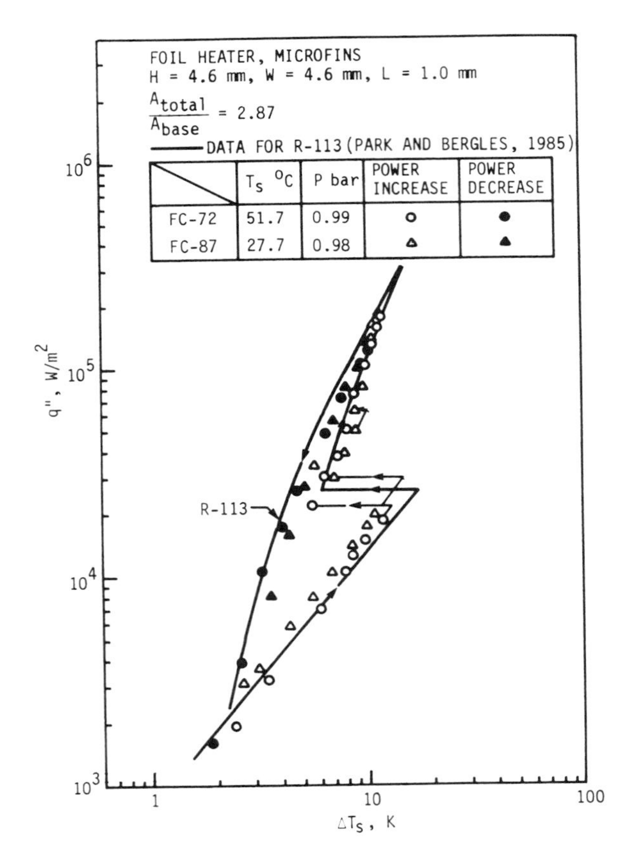

FIGURE 7. Comparison of boiling data for microfin heat sink in FC-72, FC-87, and R-113.

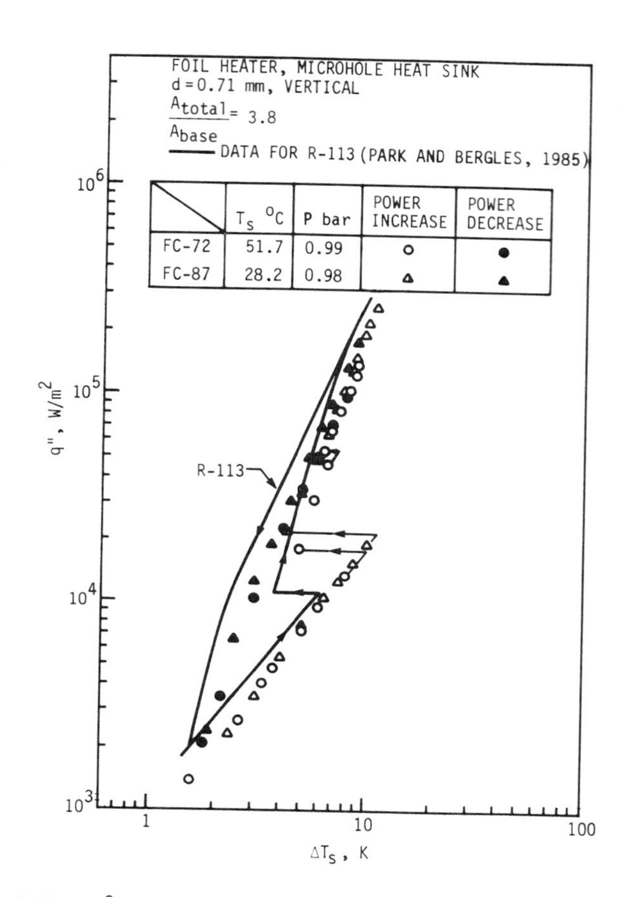

FIGURE 8. Comparison of boiling data for microhole heat sink in FC-72, FC-87, and R-113.

excel-E heat sink. Boiling slowly propagated to the bottom and left of the heater as power was gradually increased, which is a similar situation to that observed in R-113 [17]. There were many boiling sites at boiling inception and a large vapor mass at high heat flux.

The Thermoexcel-E and High Flux surfaces were prepared by the manufacturers to give the best performance in R-113. The present results suggest that the best surfaces for R-113 may not be the best for Fluorinerts.

Comparison of data (decreasing power) for the heaters tested in this study in FC-72 and FC-87 are shown in Figs. 11 and 12, respectively. It is evident that considerable enhancement was achieved by adding the heat sinks. The best performance was obtained with the High Flux heat sink. Data for the heat sinks seem to converge at high heat flux.

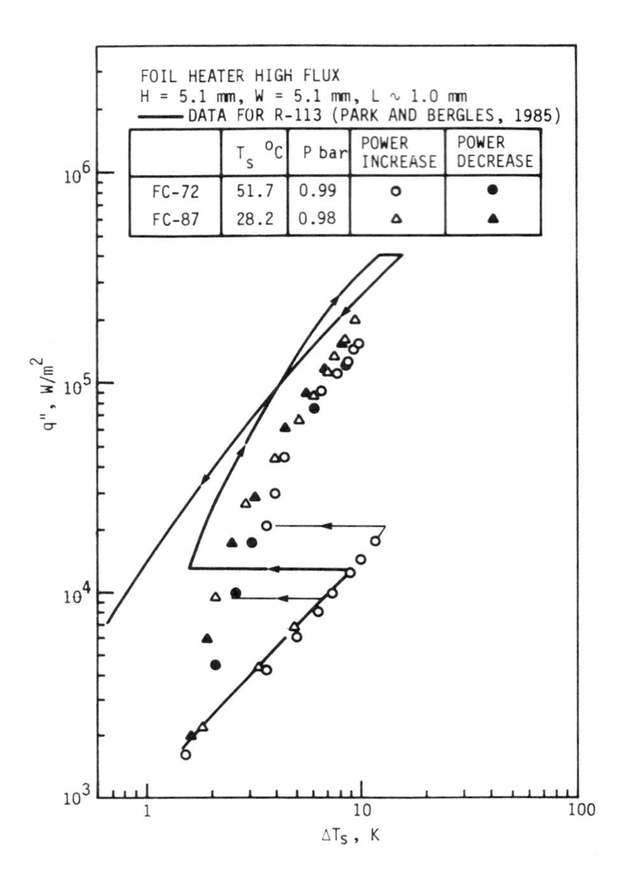

FIGURE 9. Comparison of boiling data for High Flux heat sink in FC-72, FC-87, and R-113.

RESULTS AND DISCUSSION FOR A THERMAL CHIP

Figure 13 presents the data for the thermal chip. Here, chip power is used rather than heat flux because the heat transfer area is uncertain. This uncertainty is due to the large heat conduction to the substrate through the solder bumps and to the convective heat transfer occurring in the gap itself.

The natural convection and nucleate boiling portions of the three sets of data are very similar as expected from the flush heater results depicted in Fig. 5. It is noted, however, that the data in Fig. 13 cannot be translated into boiling curves for comparison with Figs. 5 and 6 due to the uncertainty mentioned above. However, this uncertainty becomes smaller at higher heat flux because most of the energy goes out through the exposed surfaces of the chip.

Incipient boiling with the thermal chips is expected to be rather random depending on the behavior of the gap. There is a strong possibility that

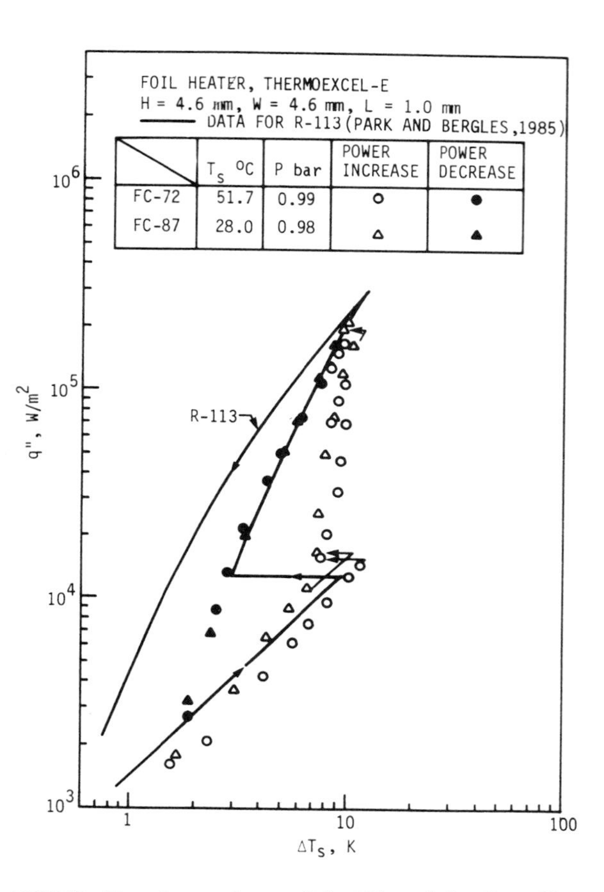

FIGURE 10. Comparison of boiling data for Thermoexcel-E sink in FC-72, FC-87, and R-113.

rogue nucleation sites exist within the solder bump matrix. In any event, the incipient superheats are similar for the three fluids. The high superheat expected from the silicon surface are not observed, apparently because the nucleation first occurs in the gap.

The chip power at CHF was about 5 W for both FC-72 and FC-87. This is expected due to the similarity in the properties that affect CHF. R-113 is expected to have a higher CHF power; however, limits on chip power precluded investigation of the CHF condition. At CHF, for both FC-87, wall temperatures jumped about 40°C due to the vapor blanketing. The operating point seemed reasonably stable as a greater fraction of the power was conducted to the substrate, but the power was rapidly reduced to avoid possible destruction of the chip. The regular nucleate boiling wall superheat was then reestablished.

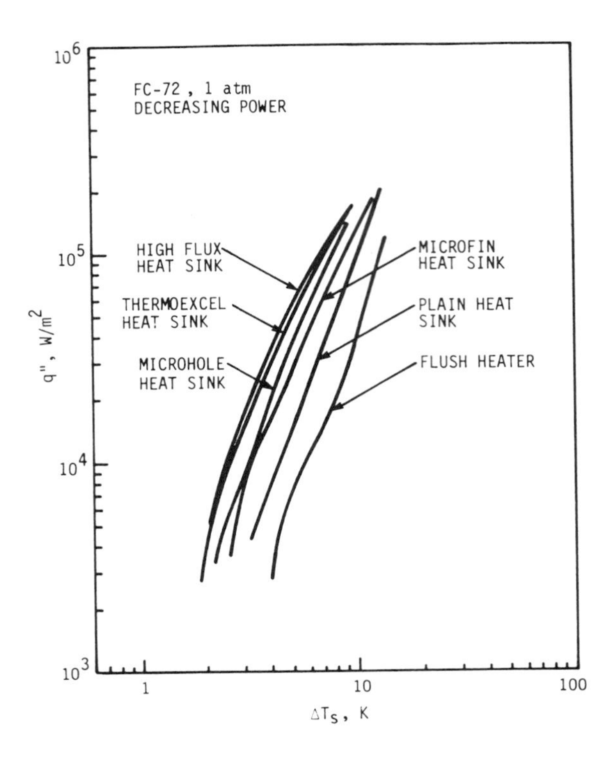

FIGURE 11. Comparison of data for flush heater, plain heat sink, microfin heat sink, Thermoexcel-E, and High Flux heat sink in FC-72.

CONCLUSIONS

The heat transfer characteristics for the flush heater for Fluorinert liquids (FC-72 and FC-87) are in very good agreement with those for R-113. The size effect for natural convection is also similar to that for R-113. A similar result was obtained for the plain heat sink. It is concluded that data for the plain surfaces in R-113 can be used to predict the heat transfer characteristics for Fluorinert liquids.

A shift of boiling data relative to R-113 for microfin or microhole heat sinks was very small and thus negligible. Boiling curves for the structured heat sink surfaces (Thermoexcel-E and High Flux surfaces) in Fluorinert liquids were quite different from those for R-113, probably due to small differences in properties. Boiling curves for Fluorinert liquids were shifted to the right of those for R-113; that is, the performance was lower.

Results similar to those obtained with simulated microelectronic chips were obtained for the thermal silicon chip using Fluorinerts and R-113. The incipient boiling superheat is surprisingly low, probably due to nucleation within the gap between the chip and the substrate.

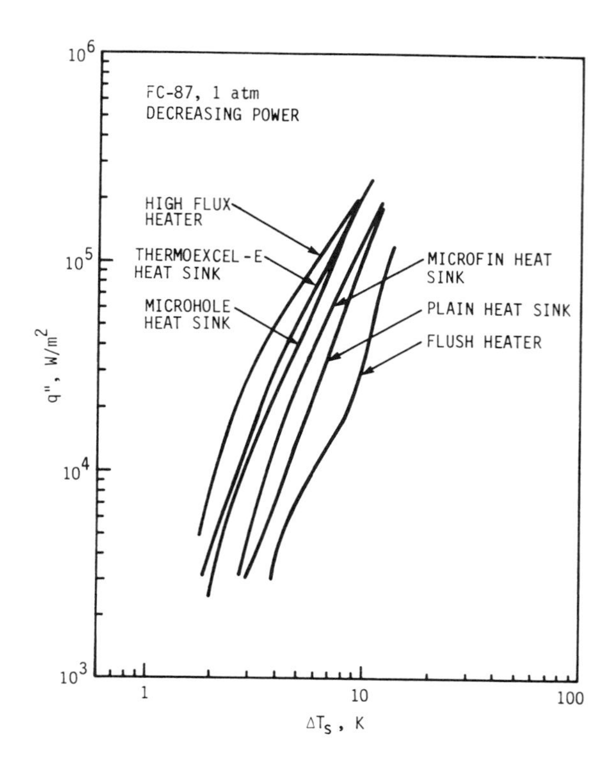

FIGURE 12. Comparison of data for flush heater, plain heat sink microfin heat sink, microhole heat sink, Thermoexcel-E, and High Flux heat sink in FC-87.

It is suggested that further work be done to determine the optimum structures of the detachable heat sinks used in immersion cooling with Fluorinerts. Boiling inception should be studied further with thermal chips to explore the effects of chip geometry, degassing procedure, and operating history.

ACKNOWLEDGMENTS

This study was carried out in the Heat Transfer Laboratory of Iowa State University with the support of 3M Commercial Chemicals Division and IBM Data Systems Division.

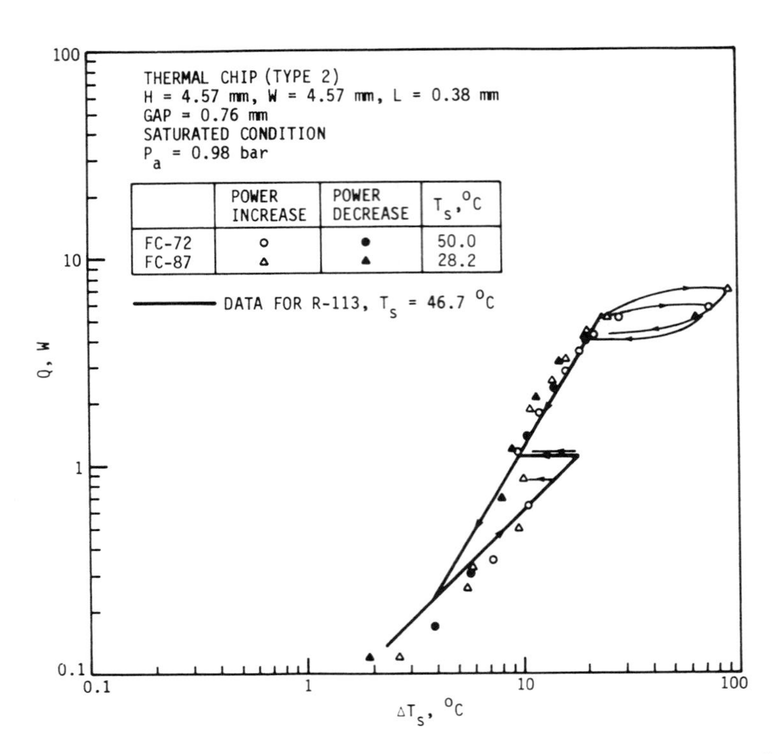

FIGURE 13. Comparison of boiling data for thermal chip in FC, FC-88, and R-113.

NOMENCLATURE

A_{total}	total surface area of the heat sink
A_{base}	base area of the heat sink
C_p	specific heat at constant pressure
c	H/2
d	diameter of microhole
e	fin gap
g	gravitational acceleration
Gr_c*	modified Grashof number ($g\beta q''c^4/k\nu^2$)
H	height of the test section
k	thermal conductivity
L	length of the heat sink
Nu_c	Nusselt number at the center of heater (hc/k)
P_a	atmospheric pressure
Pr	Prandtl number ($C_p\mu/k$)

Q	power
q''	heat flux
q''_{chf}	critical heat flux
Ra_c^*	local Rayleigh number ($Gr_c^* Pr$)
s	fin pitch
T_s	saturation temperature
T_w	wall temperature
ΔT_s	superheat, $T_w - T_s$
W	width of the test section
β	coefficient of expansion
μ	dynamic viscosity of liquid
ν	kinematic viscosity of liquid
ρ_f	density of liquid
ρ_v	density of vapor
σ	surface tension

All properties are evaluated at the film temperature, $(T_w + T_b)/2$, unless otherwise noted.

REFERENCES

1. Incropera, F. P., Liquid Immersion Cooling of Electronic Components, Keynote Lecture presented at the XXth ICHMT International Symposium, Dubrovnik, Yugoslavia, 1988.

2. Oktay, S., and Schmeckenbecher, A. F., Preparation and Performance of Dendritic Heat Sinks, Solid-State Science and Technology, vol. 121, no. 7, pp. 912-918, 1974.

3. Oktay, S., Departure from Natural Convection (DNC) in Low-Temperature Boiling Heat Transfer Encountered in Cooling Micro-Electronic LSI Devices, Heat Transfer 1982, Proceedings of the 7th International Heat Transfer Conference, vol. 4, Hemisphere, Washington, D.C., pp. 113-118, 1982.

4. Hwang, U. P., and Moran, K. P., Boiling Heat Transfer of Silicon Integrated Circuits Chip Mounted on a Substrate, in Heat Transfer in Electronic Equipment, HTD-vol. 20, pp. 53-59, 1981.

5. Nakayama, W., Nakajima, T., and Hirasawa, S., Heat Sink Studs Having Enhanced Boiling Surfaces for Cooling of Microelectronic Components, ASME Paper no. 84-WA/HT-89, 1984.

6. Marto, P. J., and Lepere, V. J., Pool Boiling Heat Transfer from Enhanced Surfaces to Dielectric Fluids, Journal of Heat Transfer, vol. 104, pp. 292-299, 1982.

7. Park, K.-A., and Bergles, A. E., 1985, Heat Transfer Characteristics of Simulated Microelectronic Chips under Normal and Enhanced Conditions, Iowa State University Heat Transfer Laboratory Report HTL-35, ISU-ERI-Ames-86211, Ames, Iowa, 1985.

8. Park, K. A. and Bergles, A. E., Boiling Heat Transfer Characteristics of Simulated Microelectronic Chips with Fluorinert Liquids, Iowa State University Heat Transfer Laboratory Report HTL-40, ISU-ERI-Ames-87022, Ames, Iowa, 1986.

9. Fujii, T., and Fujii, M., The Dependence of Local Nusselt Number on Prandtl Number in the Case of Free Convection along a Vertical Surface with Uniform Heat Flux, International Journal of Heat and Mass Transfer, vol. 19, pp. 121-122, 1976.

10. Park, K.-A., and Bergles, A. E., Natural Convection Heat Transfer Characteristics of Simulated Microelectronic Chips, Journal of Heat Transfer, vol. 109, pp. 90-96, 1987.

11. Zuber, N., On the Stability of Boiling Heat Transfer, Transactions of ASME, vol. 80, pp. 711-720, 1958.

On Performances of Nucleate Boiling Enhanced Surfaces for Cooling of High Power Electronic Devices

G. GLIELMINI, M. MISALE, and C. SCHENONE
Energy Engineering Department
University of Genoa,
Via all'Opera Pia 15/a
16145 (I) Genova, Italy

C. PASQUALI and M. ZAPPATERRA
Ansaldo S.p.A.
Divisione Ricerche
Corso Perrone 25
16152 (I) Genova, Italy

ABSTRACT

The cooling of power electronic components by nucleate boiling has been experimentally investigated. Heat sinks, obtained by soldering two layers of stainless steel wire mesh on plane copper nickel plated supports, were tested in a pool boiling of saturated R-113 at atmospheric pressure. The experimental results obtained show heat transfer enhancement of treated surfaces in comparison with plain surface. By varying the geometrical characteristics of porous layer, different boiling curves were observed whose trends were quite similar to those of commercial enhanced surfaces.

1. INTRODUCTION

In recent years, with the increasing power handling capacity of electronic equipment, the power dissipation of semiconductors has been on the rise. This trend has made it necessary to develop advanced thermal control techniques. One of the most effective means to remove heat from high power electronic components is provided by nucleate boiling, by direct immersion in a dielectric fluid [1,2]. In this area, even better results can be obtained by utilizing special treated surfaces able to enhance the nucleate pool boiling heat transfer.

The techniques used to obtain such surfaces are numerous and quite different from each other [3]. Mechanically worked surfaces, characterized by a regular structure of grooves and cavities, and surfaces coated with a porous matrix layer, whose structure is characterized by continuous internal cavities and a number of pores on the outer surface, are the two most common technological solutions; models and correlations have been suggested for them, and the enhancement and its mechanism for heat transfer have been discussed.

The bubble growth mechanism for a porous metallic matrix surface is different from that of normal cavity boiling. O'Neill et al. [4] describe a mechanism according to which vapor bubbles are formed in the interparticle spaces and bubbles grow on the surface of the thin layer of the liquid that wets the cavity walls surrounding the vapor nucleus. The liquid is continuously supplied mainly through channels and pores interconnected with those in which bubble growth occurs. The high performance of such surfaces, according to the model developed in [5], are mainly due to the contribution of two factors: a) the porous structure traps vapor nuclei characterized by high radius of the liquid-vapor interface, which considerably reduces the theoretical superheating required by nucleation; b) the porous structure generates a larger superficial area for the liquid film on which evaporation occurs than would be provided by a smooth surface.

That model was suggested by Bergles and Chyu [6] to describe the boiling mechanism on a commercial, porous coating surface (Linde High Flux); in particular, the authors point out the presence of preferential vapor escape channels inside the porous matrix, already stressed by Cohen [7] and Smirnov [8]. In the same study [6] it is established that the boiling curves for porous metallic matrix surfaces, together with a considerable heat exchange enhancement, present a phenomenon of hysteresis which is particularly pronounced with highly wetting liquids such as R-113. **Also** in the case of porous surfaces, the existence of different modes of bubble boiling has been observed by Afgan et al. [9]. At low heat fluxes, boiling takes place at the estuary of large pores, while at high fluxes boiling with a vapor film inside the porous layer occurs. The thermal crisis is observed at a much higher heat flux value than in the case of smooth surfaces.

Nakayama et al. [10,11] undertook a study with structured surfaces composed of interconnected internal cavities in the form of tunnels, connected with the liquid mass through pores. Heat transfer experimental results plus a visualization study, performed by means of a high speed movie camera, have pointed out the important contribution of bubble growth inside the tunnels in enhancing heat transfer. On the basis of these observations, the same authors have developed an analytical model of the dynamic cycle of bubble formation. The initial build-up phase of the pressure inside the tunnel is followed by the growth of bubbles connected with active pores. Finally, the full development of the bubble is followed by a short interval of pressure depression, in which the liquid flows in the tunnel through inactive pores. Developing the previous research, Nakayama et al. [12] have worked on surfaces having different diameter and population density of pores, noticing that boiling curves exhibit different trends as a function of the superficial geometrical parameters. Different hypotheses were made concerning bubble formation modes as the wall superheat was varied. High heat transfer performances were noted at low levels of heat flux where pores of the largest size play important roles.

Experimental studies on special modified structured surfaces have been made by a number of other authors, who have utilized new types of surfaces for heat exchange inside and outside tubes. Marto and Lepere [13] have compared the performances in pool boiling with dielectric fluids of copper tubes having different commercial surfaces (High Flux, Gewa-T and Thermoexcel-E). More recently, Ayub and Bergles [14] have studied heat exchange in pool boiling with distilled water and R-113 at atmospheric pressure, using numerous surfaces of the types Gewa-T and Gewa-K. Results were described by a heat transport model which refers to the analytical one by Nakayama et al. [10,11], assuming, however, that the entire bubble formation from a flat interface to the departure size occurs during a single phase. The heat exchange seems to be controlled by latent heat transport and agitated natural convection.

On the basis of the above observations, this experimental study has been developed, concerning surfaces obtained by fixing stainless steel wire meshes having different sizes on plane nickel plated copper supports so as to create conditions somewhat similar to those of porous coatings. Up to now, only a few studies have been performed on this type of surfaces [15-17] and no effective theory has yet been developed concerning the boiling mechanisms that takes place on them. In this study, some effects due to the geometry of superimposed mesh layers have been preliminarily investigated. Treated surfaces were tested in pool boiling of R-113 at atmospheric pressure.

2. EXPERIMENT

2.1 Experimental apparatus

The experimental apparatus was designed to study the behavior of wire mesh surfaces when they are assembled into one unit equipped with a vapor space condenser for the immersion cooling of high-power semiconductors. These systems were applied to the rectifier and speed controller of an electric train. An overall view of the experimental setup is shown in Fig. 1a. The semiconductor and heat sink assemblies are immersed in a

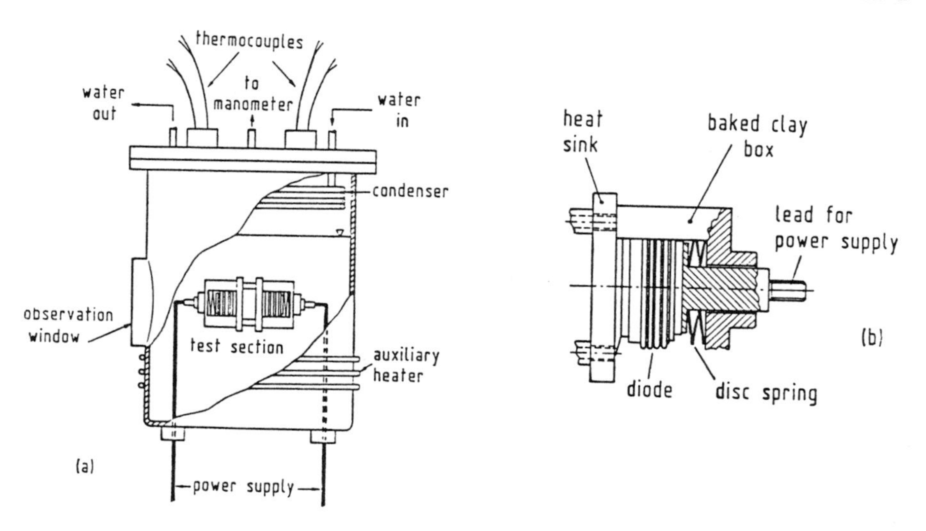

FIGURE 1. Schematic view of experimental apparatus (a); detail of the heat sink and diode assembly (b).

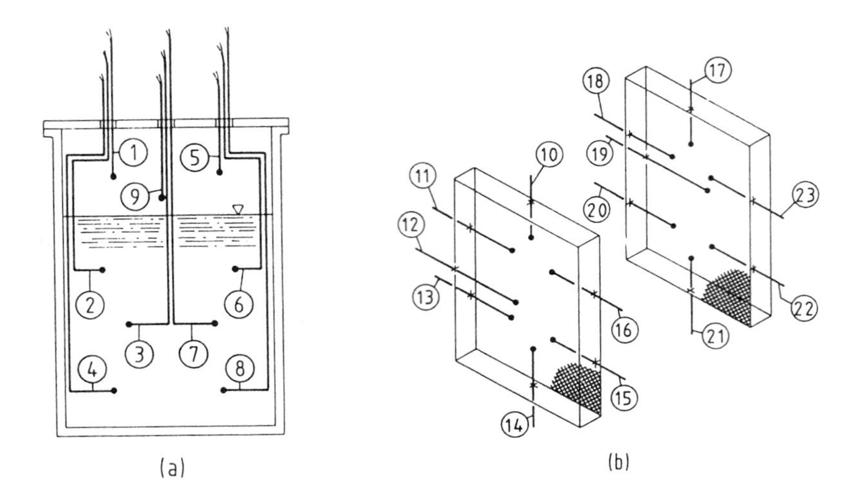

FIGURE 2. Positions of the thermocouples: in liquid and vapor bulk (a); in the heat sinks (b).

pool of R-113 and a watercooled condenser is placed over it, also inside the vessel. The latter consists of a stainless steel cylinder of 300 mm i.d. and 395 mm height. A stainless steel plate fitted with a rubber O-ring served as the cover. All around this vessel was wrapped an electric resistance that worked as auxiliary heater for maintaining the pool at saturation temperature and for degassing the liquid before every run. The vessel is thermally insulated at its lateral surface and on its bottom by mineral wool. During tests the liquid level is kept at about 50 mm above the test section so as to prevent the liquid from wetting the condenser. The vessel is equipped with two windows, arranged at a right angle to each other,to allow the view of the boiling process during the run.

In Fig. 1b a cross-sectional representation of the test section is shown. The configuration is symmetric: two vertically positioned heat sinks, facing each other, disperse the power dissipated in the diodes (type Ansaldo ARF-912). Disc springs maintain good contact between the heat sinks and diodes. Two baked clay boxes contain the electronic components in order to reduce the heat dispersions through their lateral walls. Each heat sink has a height of 80 mm, a width of 60 mm and is 10 mm thick.
Liquid, vapor and each heat sink have been provided with thermocouples of type J. Nine thermocouples were placed into liquid and vapor bulk and fourteen were located in the heat sinks. Figures 2a and 2b show the position of the thermocouples in liquid and vapor, and inside the heat sinks, respectively. The fixed fluid saturation conditions at various power dissipation levels are maintained by controlling temperature and pressure in the vessel and adjusting water flow rate through the condenser.

The power supplied to the semiconductors and auxiliary heater was controlled via powerstats. Test section power was evaluated by measuring direct current intensity and voltage by means of a precision digital voltmeter. The temperature and voltage data were logged by a data acquisition system complete with ice-point reference, a multichannel scanner, and a computer. A pressure transducer having a reading precision of ±0.01 bar was installed in the upper part of the vessel to make sure that the saturation conditions were the prefixed ones.

2.2 Test surfaces

The characteristics of the studied surfaces are shown in Table 1, where the first figure in each box refers to the lower wire mesh while the second one refers to the upper mesh.

Sample A has a smooth surface obtained by electrodeposition of a nickel layer about 15 μm thick on the copper heat sink. The nickel plating was applied to all heat sinks (except for sample C) to prevent deterioration of the surfaces due to ability of R-113 to chemically attack copper in case of prolonged contact.

A capillary porous structure consisting of two layers of stainless steel wire mesh was applied to the remaining five heat sinks. The mesh was fixed to the heat sinks by spot-welding or by brazing techniques. The latter procedure was carried out by the furnace melting of a thin nickel alloy sheet. The spot-welding was chosen to avoid the melted nickel alloy filling the lower grid apertures by capillarity. In particular, in sample F the two grids formed a 45° angle to each other; in sample C the plating nickel layer was obtained directly by the melting of the nickel alloy sheet.

In Fig. 3, microphotographs show the external appearance of the surfaces.

SAMPLE	WIRE MESH SIZE	WIRE FIXING	WIRE DIAMETER mm	INSIDE MESH APERTURE mm
A	Smooth	=	=	=
B	50-50	Brazing	0.14	0.38
C	50-50	Brazing	0.14	0.38
D	75-75	Brazing	0.14	0.20
E	50-130	Spot-welding	0.14-0.06	0.38-0.13
F	130-130	Spot-welding	0.06	0.13

TABLE 1. Characteristics of the tested surfaces

The alternation of full and empty spaces produces a superficial structure similar to the one that can be obtained by the deposition of solid particles on a plane matrix, for instance by sintering. The lower grid, fixed to the base surface, forms a series of intercommunicating channels that are similar to the ones typical of tunnel structure surfaces. Figures 3a and 3b refer to samples E and F, respectively.

In Figs. 4a and 4b, optical microphotographs of the same surface cross section are shown. Intercommunicating channels consisting of grid wires can be noted, as well as the large surface inside the structure which comes in contact with the liquid film. Figure 4b emphasizes in the cross section the rotation of the upper grid on the lower one.

2.3 Experimental procedure

Before each test, liquid was boiled for about one hour by means of the auxiliary heater, performing blowdowns at regular intervals in order to degas the liquid. By means of the auxiliary heater, the fluid was then brought to the pressure of 1.02 bar as monitored by the calibrated pressure transducer. After verifying the existence of uniform temperature conditions inside the vessel, the diodes were powered by varying by steps the intensity of the feeding electric current. Every time the dissipated power level was changed, measurements were taken as soon as steady state conditions had been reached and the uniformity of the internal temperature had been checked. During the data acquisition the auxiliary heater was off.

The temperature of the test surface was assumed equal to the mean value of the temperatures measured in the heat sinks (Fig. 2b), making the adjustment for the temperature drop over the distance between the thermocouple bead and the surface (about 3 mm). The heat flux transmitted to the fluid by the test surfaces was evaluated as a ratio of the power dissipated in the diodes to the lateral and frontal area of heat sinks, covered with porous layer.

3. RESULTS AND DISCUSSION

The experiments were performed on the test surfaces whose characteristics are reported in Table 1 for saturated pool boiling of R-113 at 1.02 bar.

Figure 5a shows a comparison between the results obtained with sample A, having plane surface and those obtained with porous wire mesh structures samples of types B and C. The two samples were both coated with a double 50 mesh grid layer, but differ from each other in that C was not initially nickel plated by electrodeposition before applying the two layers of mesh. The data presented in Fig. 5a indicate that the wire mesh surfaces are characterized by a pronounced phenomenon of hysteresis for which the increasing flux boiling curves are different from the decreasing flux curves. This phenomenon is quite similar to the one pointed out in [6] for porous surfaces and is due to the flooding of the wire mesh

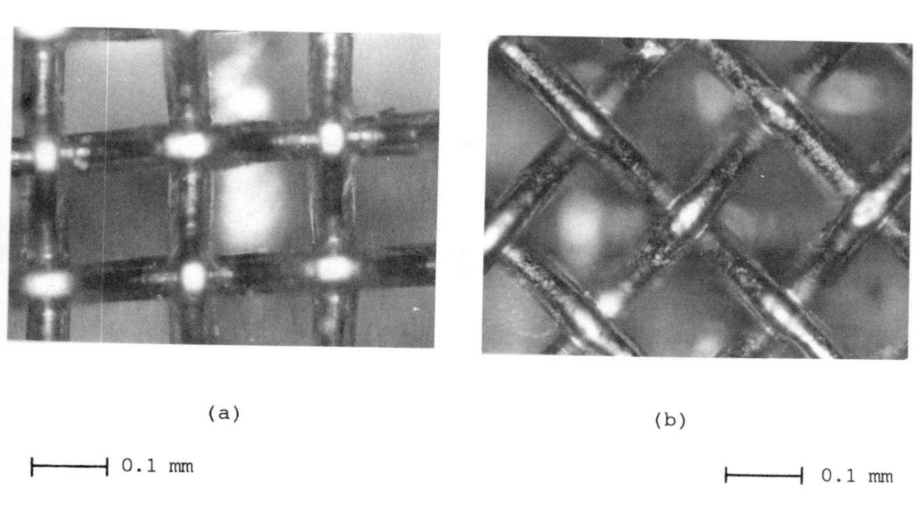

FIGURE 3. Microphotographs of the surface: sample E (a) and sample F (b).

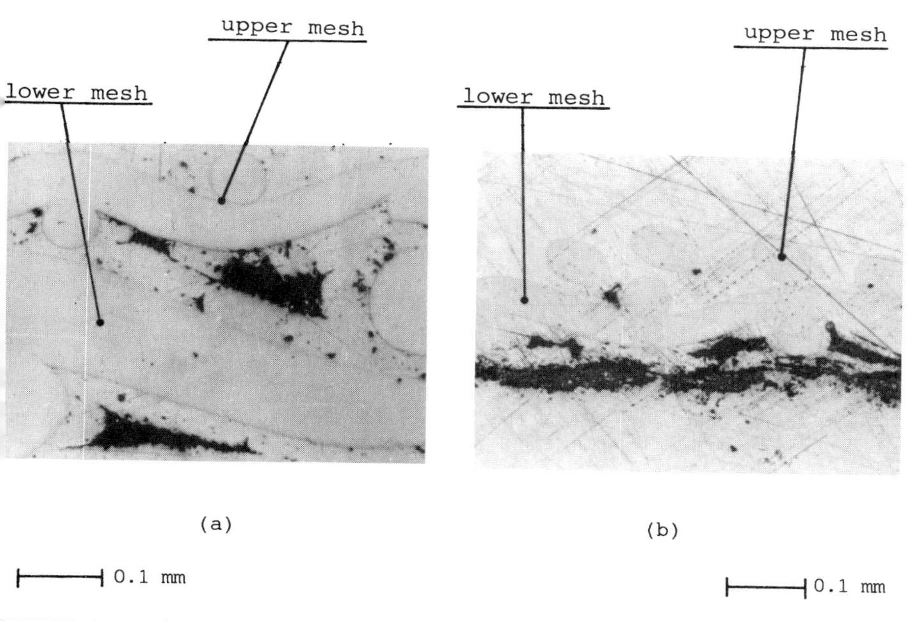

FIGURE 4. Microphotographs of the cross sections: sample E (a) and sample F (b).

structure by the liquid. Besides, the boiling curves show a considerable heat transfer enhancement, with ratios between the superheat of the smooth surface and the superheat of the treated ones of the order of 2.4 at a heat flux of 10^5 W/m^2 and up to 4.0 for a heat flux of $2 \cdot 10^4$ W/m^2. Enhancements are more pronounced at lower heat fluxes.

In Fig. 5b results are reported concerning the tests performed with sample D, again compared with data concerning A. As can be noticed, the boiling curve is similar to the curves obtained for samples B and C respectively, but there is less enhancement: 1.8 at 10^5 W/m^2 and 2.5 at $2 \cdot 10^4$ W/m^2. Sample D was made with two 75-mesh grids, the upper mesh being displaced by half mesh on the top of the lower one, so as to reduce considerably the inside mesh apertures of the latter grid.

Finally, Figs. 6a and 6b give the data obtained for samples E and F. Surface E was obtained by superimposing two grids having different meshes: a 50-mesh one in contact with the base surface and on top of it a finer (130 mesh) one. Thus, a porous matrix has been obtained having larger interstitial spaces in contact with the wall and smaller aperture superficial pores (of the order of 0.13 mm).
The behavior of surface E turned out to be similar to the behavior of samples B and C, with high enhancements at the lowest fluxes: 2.1 at 10^5 W/m^2 and 4.0 at $2 \cdot 10^4$ W/m^2. The best results at higher heat flux density

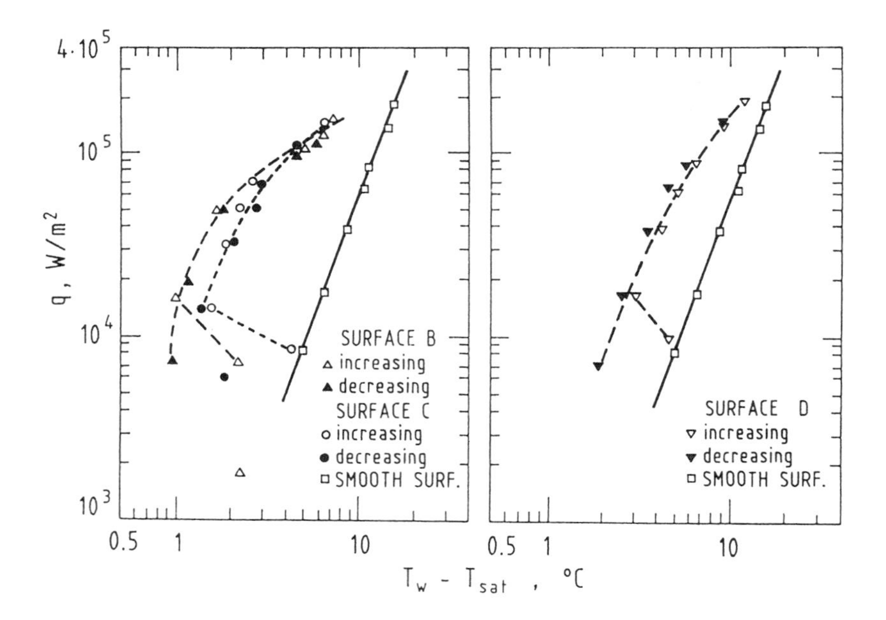

FIGURE 5. Boiling data for samples B and C (a) and for sample D (b).

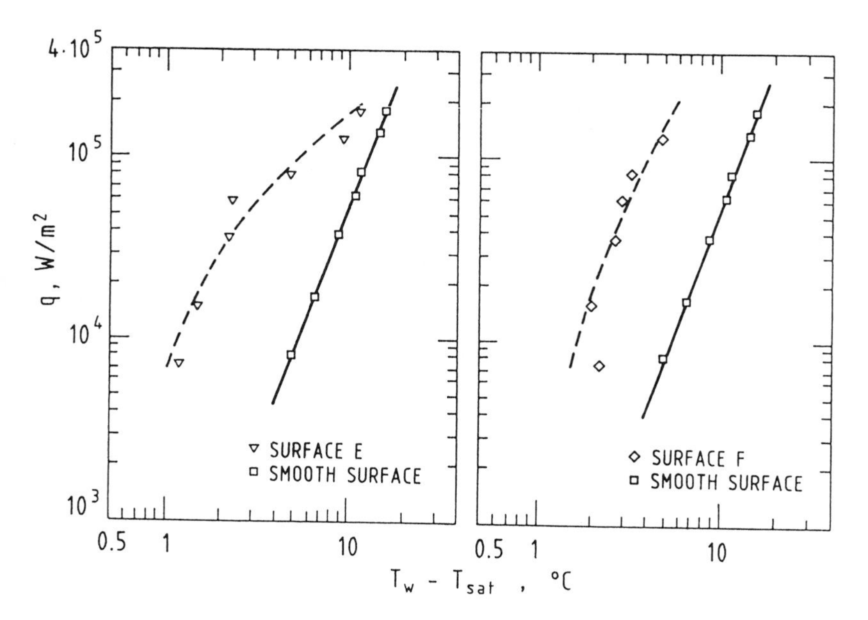

FIGURE 6. Boiling data for sample E (a) and for sample F (b).

were obtained for surface F, consisting of two fine-grid (130 mesh) layers, forming a 45° angle to each other. The enhancement obtained with this test section turned out to be 3.2 at 10^5 W/m^2 and 3.5 at $2 \cdot 10^4$ W/m^2, with the greatest enhancement at the first heat flux.

The comparison between the results obtained with wire mesh structures and those taken from the literature for enhanced commercial surfaces indicates improvements of the heat transfer coefficient of the same order of magnitude. In this connection , Fig. 7 compares the data concerning surfaces Thermoexcel-E and High Flux taken from [13] and the data concerning surface Gewa-T19B/F tested in [14], with those obtained for samples E and F. Even though the results were obtained for surfaces of different geometry (horizontal cylinders and plane vertical surfaces) and surface area the diagram in Fig. 7 shows a similar trends of all the boiling curves.

The results of this preliminary study on porous structure surfaces obtained with metallic grids seem to be quite promising, especially if the simplicity and low cost manufacturing are taken into account. The boiling curves obtained emphasize the influence of the wire mesh sizes and of the way in which they are superimposed to each other. A systematic study should be performed on this subject, in order to define the optimum geometry for a given fluid. The results of this work have pointed out

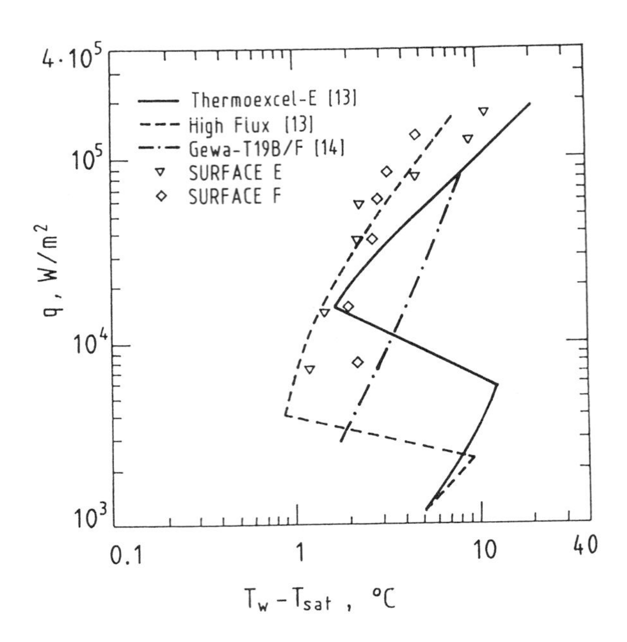

FIGURE 7. Comparison among boiling curves of some treated surfaces and boiling data of samples E and F.

that the highest enhancements at high heat fluxes correspond to grids having inside mesh apertures of the order of 0.13 mm, while at the lowest heat fluxes even larger mesh apertures (0.38 mm) turned out to be effective. These values seem also to be comparable with the experimental observations reported in [14] concerning the optimum aperture size for surfaces Gewa-T with R-113, and with the observations reported in [12], according to which the largest pores are more effective at low heat fluxes.

4. CONCLUSIONS

Structured surfaces made of two layers of stainless steel wire mesh were tested in a pool boiling of saturated R-113 at atmospheric pressure. The experimental results obtained showed a better heat transfer performance than the smooth plane surface.

By varying the layers from 50 to 130 mesh, and depending also on the mutual position of the two superimposed layers, enhancements of nucleate boiling heat transfer were obtained ranging from 1.8 to 3.2 at 10^5 W/m^2 and from 2.5 to 4.0 at $2 \cdot 10^4$ W/m^2. Besides, the tested surfaces showed a pronounced phenomenon of hysteresis, by which at the lowest heat fluxes a considerable temperature overshoot occurs, which is quite similar to

that reported in the literature for porous surfaces. These results indicate that the behavior of the wire mesh structured surfaces is close to the one of enhanced surfaces.
If the simplicity and low cost of manufacturing these surfaces are taken into account, the result appears to be rather promising. Further improvements seem to be obtainable through a systematic study in order to define the optimum geometry of porous coating.

NOMENCLATURE

q Heat flux density W/m^2
T_w Wall temperature °C
T_{sat} Saturation temperature °C

REFERENCES

1. Bar-Cohen, A., Thermal Design of Immersion Cooling Modules for Electronic Components, *Heat Transfer Engineering*, vol. 4, no.3-4, pp. 35-50, 1983.

2. Bergles, A. E., Liquid Cooling for Electronic Equipment, *Presented at Int. Sym. on Cooling Technology for Electronic Equipment*,Honolulu, Hawaii, March 1987.

3. Webb, R. L., The Evolution of Enhanced Surface Geometries for Nucleate Boiling, *Heat Transfer Engineering*, vol. 2, no. 3-4, pp. 46-69, 1981.

4. O'Neill, P.S., Gottzmann, C.P. and Terbot,J.W., Novel Heat Exchanger Increases Cascade Cycle Efficiency for Natural Gas Liquefaction, *Advances in Cryogenic Engineering*,vol.17, pp. 420-437, Plenum Press, 1972.

5. Gottzmann, C.P., Wulf, J.B. and O'Neill,P.S., Theory and Application of High Performance Boiling Surfaces to Components of Absorption Cycle Air Conditioners, *Proc. Conference on Natural Gas Research and Technology*, Chicago, Session V, Paper 3, February 1971.

6. Bergles, A. E. and Chyu, M.C., Characteristics of Nucleate Pool Boiling From Porous Metallic Coatings, *ASME J. Heat Transfer*, vol. 104, no. 2, pp. 279-285, 1982.

7. Cohen, P., Heat and Mass Transfer for Boiling in Porous Deposits with Chimneys, *AIChE Symposium Series*, vol. 70, no. 138, pp. 71-80, 1974.

8. Smirnov, G. F., Approximate Theory of Heat Transfer with Boiling on Surfaces Covered with Capillary-Porous Structures, *Teploenergetika*, vol. 24, no. 9, pp. 78-80, 1977.

9. Afgan, N.H., Jovic, L.A., Kovalev, S.A. and Lenykov, V.A., Boiling Heat Transfer from Surfaces with Porous Layers, *Int. J. Heat Mass Transfer*, vol. 28. no. 2, pp. 415-422, 1985.

10. Nakayama, W., Daikoku, T., Kuwahara, H. and Nakajima, T., Dynamic Model of Enhanced Boiling Heat Transfer on Porous Surfaces-Part I: Experimental Investigation, *ASME J. Heat Transfer*, vol.102, no. 3, pp. 445-450, 1980.

11. Nakayama, W., Daikoku,T., Kuwahara, H. and Nakajima,T., Dynamic Model of Enhanced Boiling Heat Transfer on Porous Surfaces-Part II: Analytical Modeling, *ASME J. Heat Transfer*, vol. 102,no. 3, pp. 451-456, 1980.

12. Nakayama, W., Daikoku, T. and Nakajima, T., Effects of Pore Diameters and System Pressure on Saturated Pool Nucleate Boiling Heat Transfer From Porous Surfaces,*ASME J. Heat Transfer*,vol. 104,no. 2, pp. 286-291, 1982.

13. Marto, P.J. and Lepere, V.J., Pool Boiling Heat Transfer From Enhanced Surfaces to Dielectric Fluids,*ASME J. Heat Transfer*,vol. 104, no. 2, pp. 292-299, 1982

14. Ayub, Z.H. and Bergles, A.E., Pool Boiling from Gewa Surfaces in Water and R-113,*Wärme und Stoffübertragung*,vol.21,no. 4,pp. 209-219, 1987.

15. Abhat, A. and Seban, R.A., Boiling and Evaporation From Heat Pipe Wicks With Water and Acetone, *ASME J. Heat Transfer*, vol.96, no. 3, pp. 331-337, 1974.

16. Rannenberg, M. and Beer, H., Heat Transfer by Evaporation in Capillary Porous Wire Mesh Structures, *Letters in Heat and Mass Transfer*, vol. 7, no. 6, pp. 425-436, 1980.

17. Xin Liu, Tongze Ma and Jipei Wu, Effects of Porous Layer Thickness of Sintered Screen Surfaces on Pool Nucleate Boiling Heat Transfer and Hysteresis Phenomena, *Proc. Int. Sym. on Heat Transfer*, vol.2, Tsinghua University Press, 1985.

Two-Phase Thermosyphon Devices for the Cooling of Large Thyristors

P. BORDIGNON and A. RAGNI
Power electronic Division
ANSALDO S.p.A.
V. le Sarca, 336, Milan, Italy

E. LATROFA and C. CASAROSA
Dipartimento di Energetica
Universita di Pisa
V. Diotisalvi, 2, Pisa, Italy

ABSTRACT

Due to the evolution in semiconductor technology, the development of two-phase closed thermosyphons as cooling devices in power electronics seems almost certain. There are several possible configurations of thermosyphon heat sinks. In this work only the separate condenser and the common condenser arrangements are considered. In particular, the general thermal and fluid dynamic criteria for the design of each component of these devices are first discussed. For the case of a separate condenser heat sink, a prototype using Freon 113 as working fluid is then described together with the tests for its characterization. As for the common condenser configuration, the performance of most of its components is also experimentally determined. The results show that for cooling thyristors 75-100 mm in diameter, thermal resistances lower than 10°C/kW may be achieved.

INTRODUCTION

The continuous increase of the power required for electric converters or inverters and the need of studying more and more compact and reliable conversion devices may find an answer in

- the use of power semiconductor components (thyristors , GTO, transistors) with higher and higher performance,
- the design of very effective cooling devices for heat transfer by means of natural or forced circulation of fluids or liquids in phase transition,
- the use of suitable technology for electric insulation by means of new materials - ceramic materials overall - and optical fibers for the electric triggering of the electronic components.

These are the main factors which take part in the design of the new power converters. These new technologies have revolutionized the constructive aspect of the different apparatus, improving, at the same time, the performance of the converters as regards required power, compacteness, reliability and costs.

In this paper we are only concerned with the aspects of the thermal control of the power apparatus; however, some reference will be made to the characteristics of the components to obtain a better insight into the problems and the specific constraints which must be satisfied.

GENERAL CONSIDERATIONS

The development of new semiconductors with higher and higher performance had a strong impulse in the last years; this can be considered the first reason for the innovation in other fundamental disciplines which are the basis of power electronics.
In Table 1 the limit performance of the actual power components are shown.

TABLE 1. Limit performance of the power electronic components

Component type	Parameter	Actual	Short/Mean Period Trend
	Wafer Diameter	100 mm	120-150 mm
Thyristor	Voltage (V_{DRM})	5.5 KV	8 KV
	Current (I_{RMS})	4 KA	? KA
	Triggering	Elect./Optic.	Optic./Elect.
	Wafer Diameter	75 mm	100 mm
GTO	Voltage (V_{DRM})	4.5 KV	6 KV
	Current (I_{TGR})	3 KA	3-4 KA
	Triggering	Electric	Unchanged ?
	Package	114X114 mm	
Modules (GTR)	Voltage	1 KV	
	Current	0.6 KA	

As it is shown in this table, the trend, which has been confirmed for several years, is that of developing components for higher and higher power levels in voltage and/or in current; this fact involved a necessary increase in the losses associated with component and an accompanying increase in the electric insulation required for its use.
Consequently, the whole exploitation of these new devices to obtain the maximum power requires the continuous change of technology in thermal control and in electric insulation, which are linked to evolution of the performance of the components.
For the maximum heat flux on the surface between the component and the dissipator the actual values are about $20\ 10^4\ W/m^2$ and the trend in near term period is towards $40\ 10^4\ W/m^2$.
Because the total temperature difference between the electronic component junction and the surroundings is determined, if the heat flux increases,the thermal resistance must decrease consequently.
For this reason it is necessary, to go down to 30 °C/KW , to use either water in forced convection or devices with intermediate fluid in natural or forced convection.

1.Cooling Configurations

Our experimental research concerns the use of the two-phase thermosyphon as heat sink for the cooling of the electronic component in different possible configurations.
Figure 1. shows the various apparatus considered in our experimental

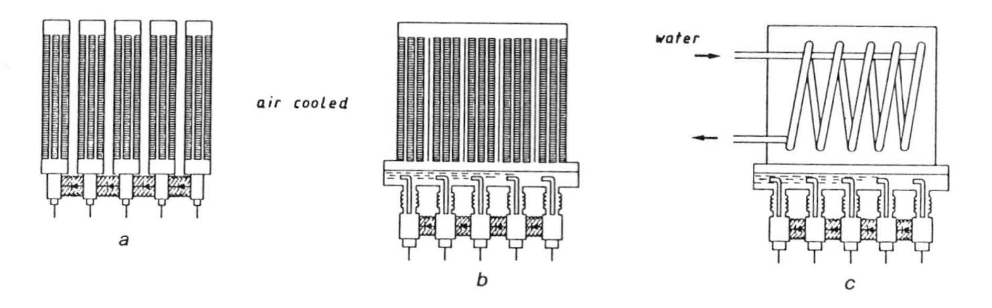

FIG. 1. Cooling Arrangements for Thermosyphon Heat Sink.

investigation.
Sketch **a** shows a packaging configuration using several components separated by two-phase thermosyphon heat sinks, each electrically insulated. The intermediate fluid, Freon 113, evaporates in proximity of the electronic component (evaporator with boiling fins) and condenses inside a final exchanger (condenser with internal condensing fins and external extended surface cooled by air in forced convection). The condensed liquid returns to the evaporator countercurrent with the vapor phase.
Sketch **b** shows a configuration having the same evaporators but a common condenser. This arrangement requires a separation between the evaporators and a separate return for condensed liquid. This solution is due to problems posed by flooding (see the experimental results and by the possible consequent dryout. However, the experimental results obtained by the previous configurations (sketch **a**) can be used to design this device once the technological problems posed by the electrical insulation have been solved.
The last configuration (sketch **c**) shows a condenser made with a water cooled coil. The condensation occurs on the external surface of the finned coil. As regards electric insulation and flooding, this configuration has the same problems as that shown in sketch **b** .

2.Technological and Manufacturing Features

The different devices presented in this paper have been designed and realized taking into account that one of the main problems is the use of industrial technologies that assure a high level of vacuum sealing to avoid infiltration of non condensable gases which might compromise the thermal performance of the condenser.
In order to achieve a high degree of sealing, the different parts of the heat sinks were carefully jointed by means of under-vacuum brazing, plasma, microplasma and friction welding (in such way aluminium-stainless steel and copper-stainless steel joints have been easily realized). These technological devices assure a sealing for the whole apparatus with leak rate of 10^{-8} - 10^{-9} standard cubic centimeters per second of Helium,at least.

In figure 2. one can see an electric insulated duct used for connecting the evaporators to the condenser in the common condenser arrangement. This connection is made by a bellows of stainless steel welded to a portion of a tube of Alumina which - from the point of view of electric insulation - isolates the evaporators from the condenser; another coaxial tube of Teflon is placed inside the above-mentioned tube and it supplies the return of the condensate, assuring, at the same time, the separation from vapor phase.

FIG. 2. Insulated Connection.

THERMAL AND FLUID DYNAMIC DESIGN

1. General Outlines

The heat sink based on the two-phase closed thermosyphon is a very simple cooling device and its basic components are

- the working fluid, utilized to transport heat by means of its evaporation and condensation,
- the container with various geometrical shapes.

However, in the thermal control of power electronic components, the

design of a two-phase thermosyphon, in spite of its simple hardware, may exhibit remarkable difficulties because high thermal powers must be dissipated with reduced dimensions of the whole apparatus, and high heat fluxes may occur on the case of the electronic component.
Moreover, the operating voltage differences of the thyristors are high, as well as the forces required for clamping the electronic component, its electrodes and the heat sink. Therefore, several problems of electrical insulation and mechanical strength are always present.
To solve these problems connected with electrical, mechanical and thermal aspects, several conflicting factors may inevitably arise, so that an optimum design is really a difficult task.
In this section we particularly examine the thermal aspects; but, obviously, in the design of the cooling devices, we also had to examine and to solve the electric problems and the strength problems.
For the design of heat sinks based on the two-phase thermosyphon, the first step is the selection of the working fluid and of the container material in relation to their compatibility. In reference [1] many criteria and useful information for this choice are reported. In our case Freon 113 was chosen as working fluid, while for the container material aluminium alloys, stainless steel and copper were selected.
Afterwards, one must consider the phenomena which can limit the thermal performance of a two-phase thermosyphon. The main phenomena to be considered are

- the dryout limit, which occurs when the amount of working fluid of the thermosyphon does not allow a continuous circulation of vapor and of condensate at a given heat flow [2],
- the critical heat flux limit, which occurs for large liquid filling and is similar to the burnout condition in pool boiling [2,3],
- the entrainment or flooding limit, which usually occurs in the adiabatic section or in the condenser because of the countercurrent two-phase flow [4,5].

The latter leads to different thermosyphon behaviours according to the operating conditions of the device. When the heat flow is imposed at the evaporator, as for the thyristor cooling, the thermosyphon behaviour turns unstable and may exhibit large oscillations [3]; on the other hand, if convective boundary conditions are imposed, the heat flow may be blocked at a threshold value (i.e., thermal blocking limit) and stable operation occurs [6].
At present, some theoretical models are avalaible and they seem able to predict the two-phase thermosyphon behaviour and the dryout limit as well as the flooding limit [7,8]. These models of fluid dynamic and heat transfer processes of a closed two-phase thermosyphon are constructed by a simple lumped parameter description of the system. The modeled geometry is so far cylindrical with constant cross section; an attempt of modeling more complex geometries was recently made [9].
In any case, it is probably premature to use the aforesaid theoretical models and the more usual approach of thermal resistances is, therefore, to be preferred [1]. However, it should not be neglected that a two-phase thermosyphon has a non-linear thermal characteristic and, therefore, the ratio between temperature difference and heat flow depends on the terms of the ratio itself; in addition, in order to avoid errors and misleading inferences, the heat flow should be one-dimensional.
For easy processing of experimental data, the evaporator resistance R_e as well as the condenser resistance R_c are often utilized. The first takes the conduction in the wall of the evaporator and the inside boiling of the working fluid into account; the second considers the working fluid condensation, the conduction in the wall of the condenser

and the heat exchange from the condenser to the surroundings, usually by natural or forced convection with air or water flow. If the possible difference of the working fluid temperature, linked to the pressure drop between evaporator and condenser, is considered, then the thermal analysis and the design of the cooling device may be carried out separately for the evaporator, the condenser and the adiabatic section by means of a thermal resistance network.

2. Analysis and Design of the Evaporator

The part at the bottom of the thermosyphon is the evaporator, which is the same for the different cooling configurations. For this reason, the general problems, posed by the design of this part of the devices, can be examined.

In the thermal control of power electronic components it is required to have a large safety margin to avoid damaging very expensive components. Therefore, in order to avoid the danger of local dryout, one is warned against low fillings of working fluid, and so the thermosyphons with liquid pool inside the evaporator are preferred.

Since for this condition in the evaporator there is a boiling in a bounded geometry with very complex hydrodynamics, it is difficult to predict the boiling heat flux limit. However, supposing that the current literature criteria for the burnout flux of pool boiling are conservative, the following equation can be used [10]

$$q''_{crit} = k_c \, \rho_v \, \Delta h_v \left[\frac{g \, \sigma (\rho_l - \rho_v)}{\rho_v^2} \right]^{.25} \quad (1)$$

where k_c is a coefficient depending on the fluid properties as well as on the boiling body geometry, and its value may range between 0.1 and 0.19 . Some references neglect the geometry effect and only give k_c for several fluids: the value of 0.10 for Freon 113 is suggested in [11]. For some other authors [10], k_c is a function of the Bond number, i.e. the dimensionless quantity

$$Bo = D \sqrt{\frac{g \, (\rho_l - \rho_v)}{\sigma}} \quad (2)$$

where D is a body characteristic dimension, and for cylinders it is

$$k_c = 0.118 \qquad \text{for} \qquad Bo > 2.34 \quad (3)$$

$$k_c = \frac{0.146}{Bo^{0.25}} \qquad \text{for} \qquad 0.24 < Bo < 2.34 \quad (4)$$

If for Freon 113 in the range of the saturation temperature 20-70 °C k_c is assumed to lie between 0.1 and 0.118, the results of Table 2 are obtained.

On the other hand the electric characteristics of the thyristors and of the usual electronic components allow us to estimate the heat fluxes that in the double-side cooling arrangement must be dissipated through the component case. The comparison with the burnout fluxes shows whether a boiling flat surface at the evaporator is a suitable solution.

TABLE 2. Critical heat flux

T_{sat} (°C)	q''_{crit} (kW/m^2)	
	k_c = 0.100	k_c = 0.118
20	112.	132
30	129	153
40	146	173
50	169	200
70	197	232

For thyristors 75 and 100 mm in diameter the heat fluxes are about 250 kW/m^2 and it is necessary to find suitable means to avoid a possible burnout. Among others, thick massive fins can be used for their convenient technological aspects [12,13]. The basic concept of this device (known as Vapotron) is to exceed the burnout limit: the heat flux is distributed over a surface more extended than the case base, and the fins avoid the film boiling, maintaining the surface of their interspaces in transition boiling.
To design boiling fins several references, based on the one-dimensional approach, are available [14,15,16]. Among others, some recent analytical results can be useful [17]. In any case, extending the hypothesis of one-dimensional conduction to the wall from which the fins protrude seems very doubtful and not at all rigorous. To reject this formulation it is necessary to solve a non-linear differential problem using a numerical method.
In the case of rectangular fins, a two-dimensional approach of the evaporator wall can be adequate. The boiling heat transfer coefficient can be assumed as a function of the local wall superheat [14], therefore causing the non-linearity of the differential problem. We turned our attention to the aforesaid differential problem using a finite-difference method with a boiling heat transfer coefficient suggested in [18]. The numerical results show that the required heat fluxes do not cause the wall burnout. Moreover, these results are probably conservative, since the hydrodynamic field in the bounded geometry may increase the pool boiling heat transfer coefficient. Here we describe some examples of rectangular fins, relative to a light alloy wall, with a thermal conductivity of 160 W/m K, and a wall thickness of 3 mm, cooled by Freon 113.
In Table 3 the following quantities are shown:
- the fin array dimensions, i.e., the interspace width between fins, a, the fin thickness, b, the fin length, L,
- the input heat flux, q_{th}, on the side of the thyristor case,
- the maximum heat flux, q_{bmax}, computed in the center of the fin interspace, for two working fluid temperatures,
- the average temperature, T_A, computed on the side in contact with the electronic component.

Table 3 shows that closely spaced fins are the most advantageous. Unfortunately, the dimension of the fins interspace is not arbitrary. Indeed, when this dimension is decreased, boiling in a narrow space occurs. Here not only the bubbles growth is geometrically limited, but also, when the bubble are growing on the opposite surfaces, a hydrodynamic interference is fairly probable. Consequently, the assumption that the boiling heat transfer depends only on the local conditions is no longer valid. Moreover, the boiling heat transfer coefficient has an undesirable decrease.

TABLE 3. Results of the numerical analysis.

Fin Dimension			Heat flux	Fluid Saturation Temperature			
				T_{sat} = 70 °C		T_{sat} = 40 °C	
a (mm)	b (mm)	L (mm)	q_{th} (KW/m^2)	q_{bmax} (KW/m^2)	T_A (°C)	q_{bmax} (KW/m^2)	T_A (°C)
2	2	10	200	166	83.4	164	54.4
2	1	3	200	154	83.2	152	53.7
1	1	3	250	172	84.4	168	55.9
1	0.5	2	250	161	84.3	159	54.7
0.5	0.5	2	250	123	83.7	120	54.1

Therefore, criteria for the minimum sizing of the fins interspace are indispensable. Fortunately for Freon 113 the experimental minimum of the fins interspace is available as 1.5 mm [19]. On this basis the inner fins of the evaporator were sized taking into account some other manufacturing constraints (e.g., the extrusion technology ones). So the dimensions of the rectangular fin array are: interspace, a = 2 mm; fin thickness, b = 1.5 mm; fin length, L = 4 mm .

3. Analysis and Design of the Condenser

In the design of the heat sink devices for all cooling configurations cooled by air or water flow, an important aim is to obtain a condenser having both a low thermal resistance and reduced weight and dimensions. For this purpose, the geometries and the flow arrangements of compact heat exchangers can be very useful [20].

The condenser thermal resistance has to be evaluated by means of the following equation [21]:

$$R_c = \frac{1}{C_f \, \varepsilon_c} \tag{5}$$

where C_f is the cooling fluid heat capacity rate and ε_c is the heat exchanger effectiveness. The latter is given by

$$\varepsilon_c = 1 - \exp\left(- U_c S_c / C_f\right) \tag{6}$$

where S_c is the condensation surface area and U_c the overall heat transfer coefficient between internal working fluid and external cooling fluid.

If the cooling fluid heat capacity rate is very high, for example, using liquid water, the condenser thermal resistance can be approximated as

$$R_c \cong \frac{1}{U_c S_c} \tag{7}$$

Obviously, the overall heat transfer coefficient contains the heat transfer coefficient of the working fluid condensation; this is a non-linear quantity, which may be calculated according to reference [12]. However, in the condensation of many working fluids, like the Freons, the heat transfer coefficients usually have small values (order of magnitude 1000 W/m^2 K); therefore, the thermal resistance due to the condensation may be too high unless suitable means are employed. Among these the simplest regard the condensation surface. Typically, we can use finned surfaces to increase the condensation surface area, or fluted surfaces that through the liquid surface tension are able to drain the condensate, to decrease the liquid film thickness and, consequently, to increase the heat transfer coefficient [12,13,22,23].
According to the possible cooling configurations of the thermosyphon heatsink previously shown, two condenser arrangements were employed: the separate condenser arrangement for separately working thermosyphons, and the common arrangement with the condenser shared by several electric insulated evaporators.
For the air cooling devices in the separate arrangement as well as in the common type, a compact heat exchanger of aluminium alloy was selected as thermosyphon condenser; it is a cross flow exchanger with unmixed fluids and it exhibits finned surfaces on the cooling air side as well as on the side of the condensing fluid.
For the water cooling devices in the common condenser arrangement, a coil heat exchanger was utilized with Freon 113 condensing on the external finned surface of the coil copper tube.
The thermal design of both these condensers (i.e., the air cross flow heat exchangers and the cooling water coil) was obviously carried out on the basis of the aforesaid design criteria.

4. Analysis of Flooding Limit

When a two-phase closed thermosyphon has a short adiabatic section or none, the thermal resistance due to the pressure drop between evaporator and condenser does not affect the device design. Consequently, only the check of the heat transfer limit due to the two-phase fluid dynamics (i.e., the flooding limit) is important.
Several criteria are available in the current literature [4,5], and the heat flow per cross section unit area at the flooding limit, or rather the axial limit heat flux, q_F, can be calculated by means of the following equations.
(a) Wallis correlation

$$q_{FW} = C_W^2 \sqrt{D_a}\, \Delta h_v \frac{\left[g \rho_v (\rho_l - \rho_v)\right]^{0.5}}{\left[1 + m_W (\rho_v/\rho_l)^{0.25}\right]^2} \tag{8}$$

where D_a is the adiabatic section diameter, C_W and m_W are 0.725 and 1, respectively.
(b) Kutateladze, Tien correlation

$$q_{FK} = 3.2 \tanh^2\left(0.5\, Bo^{0.25}\right) \Delta h_v \frac{\left[g\,\sigma\,(\rho_l - \rho_v)\right]^{0.25}}{\left[\rho_v^{-0.25} + \rho_l^{-0.25}\right]^2} \tag{9}$$

where the Bond number is calculated with the diameter D_a.

(c) Bezrodnyi correlation

$$q_{FB} = C_B\, K_p^n \sqrt{\rho_v}\, \Delta h_v \left[g\,\sigma\,(\rho_l - \rho_v)\right]^{0.25} \tag{10}$$

where

$$K_p = \frac{p}{\sigma}\left[\frac{\sigma}{g(\rho_l - \rho_v)}\right]^{0.5} \tag{11}$$

and the coefficients C_B and n for the three flooding regimes of Bezrodnyi [5], are shown in Table 4

TABLE 4. Coefficients of the Bezrodnyi correlation

Flooding Regimes	($K_p \le 10^4$) I	II	III	($K_p > 10^4$) I	II	III
C_B	5.72	8.20	10.30	0.94	1.35	1.70
n	-0.17	-0.17	-0.17	0.00	0.00	0.00

(d) Liu, Mc Carthy, Tien Correlation

$$q_{FL} = C_L\ \Delta h_v \frac{\left[g\,\sigma\,(\rho_l - \rho_v)\right]^{0.25}}{\left[\rho_v^{-0.5} + m_L\,\rho_l^{-0.5}\right]^2} \tag{12}$$

where C_L and m_L are 0.8 and 0.63, respectively.

All these criteria have to be carefully applied to the thermosyphon heat sink according to the cooling configuration.

As regards our selected condenser geometries, two cases had to be considered.

In the separate condenser arrangement, since every evaporator was welded to the bottom of its condenser, the adiabatic section was not present in the thermosyphon. Therefore, the flooding limit was only checked on the cross flow heat exchanger, i.e., on the internal ducts in which the working fluid condensation occurs.

On the contrary, in the common condenser arrangement all the evaporators were jointed to the condenser by suitable adiabatic connections with electric insulators. Therefore, a further flooding check was carried out particularly for these connections. Obviously, in the coil heat exchanger, none of the flooding checks was necessary for the condensation zone.

The results obtained for Freon 113 , at different saturation

THERMAL PERFORMANCE OF THE DEVICES

The two-phase thermosyphon heat sinks which have been examined are protypes working in the separate condenser configuration as well as in the common condenser arrangement.
In the first case, a stack of thyristors was cooled by air in forced convection by means of two-phase thermosyphons of the type shown in figure 3; this heat sink was employed as cooling device for thyristors 3" (ø 75 mm) and 4" (ø 100 mm) in diameter. For the thermosyphon heat sink, both the internal position and the side position are possible in the thyristor stack: the experimental performance in both positions

FIG. 3. Thermosyphon Heat Sink for Separate Configuration.

temperatures and for the selected condenser geometries, are reported in Tables 5 and 6; the first concerns the cross flow exchanger analysis, the second the insulated connection of the common condenser configuration.
The tables of the results clearly show that the flooding limit prediction is very uncertain, reaffirming that suitable experiments are necessary, in order to check the agreement betwen the design and the actual operating conditions of the device.

TABLE 5. Flooding limits for air cooling- Freon 113 condenser:cross flow heat exchanger type, in single and common arrangement

	Heat flow per cross section unit area (kW/m^2)					
			Bezrodnyi Regimes			
T_{sat} (°C)	q_{FW}	q_{FK}	$q_{FB(I)}$	$q_{FB(II)}$	$q_{FB(III)}$	q_{FL}
-10	351	432	1277	1830	2299	502
0	424	520	1500	2150	2745	618
10	498	610	1656	2374	2903	740
20	581	709	1757	2519	3165	880
30	657	800	1888	2706	3400	1013
40	727	884	1995	2862	3592	1139
50	817	991	2163	3102	3896	1307
70	921	1110	2238	3209	4029	1502

TABLE 6. Flooding limits for the common condenser arrangement: Freon 113 flowing into the electric insulated connection, 25 mm in inside diameter

	Maximum Heat Flow for Connection (W)					
			Bezrodnyi Regimes			
T_{sat} (°C)	Q_{FW}	Q_{FK}	$Q_{FB(I)}$	$Q_{FB(II)}$	$Q_{FB(III)}$	Q_{FL}
-10	568	472	627	898	1128	246
0	686	567	736	1055	1347	303
10	806	663	813	1165	1425	363
20	940	768	862	1236	1554	432
30	1063	863	927	1328	1669	497
40	1176	951	979	1405	1763	559
50	1322	1062	1062	1523	1912	642
70	1490	1180	1099	1575	1978	737

However, because through each connection a 2000 W heat flow is usual for cooling of thyristors 100 mm in diameter, it is evident from Table 6 that suitable devices are to be employed to exceed the flooding limits. A coaxial pipe inside the adiabatic section, i.e., in the connection duct, is a very simple device in which the condensed liquid can return into the evaporator without limiting interactions with the countercurrent vapor flow [24].
The device of the separate return was accordingly selected for all the thermosyphon heatsink in the common condenser configuration.

and for thyristors of the two aforesaid diameters is available in [21]; here only the performance of a heat sink in internal position is reported. In figure 4 the overall thermal resistance and the air flow pressure drop are shown as a function of the cooling air average velocity or flow rate.

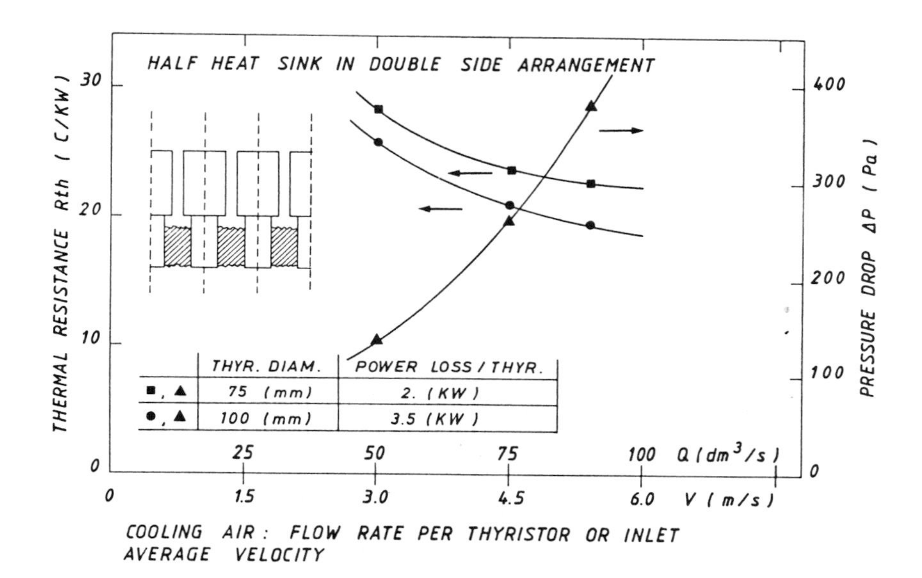

FIG. 4. Performance of the Heat Sink Cooled by Air Flow in Separate Condenser Arrangement.

In Table 7, an interesting comparison between this thermosyphon heat sink and other similar devices is reported. It is easy to conclude that proposed heat sink is charcterized by a low weight, a low air pressure drop across the condenser, a low value of thermal resistance and, moreover, some interesting technological features. However, further improvements, specially as regards the condenser, may be possible.
As for the common condenser cooling arrangement, the performance of the separate components of the thermosyphon heatsink has been experimentaly analyzed.
The evaporator of the thermosyphon was identical to that of the previous heat sink, i.e., an aluminium container with internal boiling fins (see fig.5), and its performance is shown in fig.6. Besides the characterization of the design steady state performance, the boiling critical heat flow was obtained by heating only one side of the evaporator and the results are shown in Table 8.
As the heat flow is not uniformly distributed on the evaporator surface, the heat fluxes corresponding to the critical boiling conditions are not exactly determined. A reasonable estimate, based on the flow area, yields approximately 30 10^4 W/m^2, a value which, when compared

TABLE 7. Comparison between some thermosyphon heat sinks for the cooling of Thyristors, ø 75-100 mm in diameter

Heat Sink Type	Size 10^{-3} (m^3)	Weight (kg)	Air Flow Rate 10^{-3} (m^3/s)	Pressure Drop (Pa)	Thermal Resistance (°C/kW)
Thermosyphon	5.0	4.5	125	180	14 (ø 100)
ANSALDO	5.0	4.5	100	135	19 (ø 75)
Heat Pipe	5.0	/	125	/	17 (ø 100)
Japanese	4.2	/	105	/	20 (ø 80)
Manufacture	4.0	13.0	105	90	19 (ø 75)
Some Usual	3.0	5.0	100	300	35 (ø 75)
Aluminium	2.5	4.0	100	200	35 (ø 75)
Heat sinks	2.5	4.0	100	200	32 (ø 100)
	8.5	/	100	10	42 (ø 75)

FIG. 5. Single Evaporator

with the data of Table 2, confirms the good performance of the evaporator internal fins.

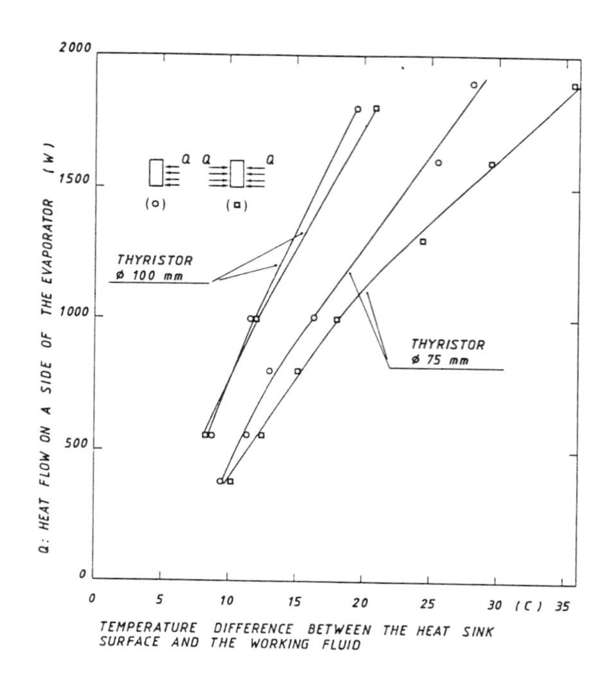

FIG. 6. Evaporator Performance.

TABLE 8. Boiling critical heat flow

Thyristor Diameter (mm)	Working Fluid Temperature (°C)	Critical Heat Flow (W)
75	51.1	2100
100	57.9	2800

As regards the common condenser heat sink, cooled by air flow, its performance is obviously the same as that of the separate condenser heat sink, for identical dimensions and conditions; but, since in the common condenser arrangement the condenser dimensions are not limited by the electric layout, it is evident that considerable improvements are achievable by increasing the heat transfer surface.
As shown in fig.1c, the last experimental set up refers to a condenser cooled by water in forced circulation inside a coil having an external finned surface over which the vapor condenses. In this case, a stack of six thyristors was immersed in Freon 113 in pool boiling, and the coil was placed above the pool in the containing box.
The condensation heat exchange was evaluated through a series of tests: in particular, in fig.7 the thermal resistance is given as a function

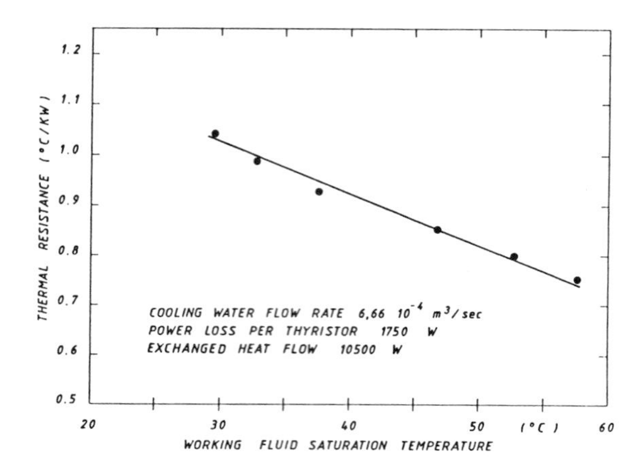

FIG. 7. Condenser Cooled by Water Flow in Common Arrangement

of the fluid saturation temperature at the evaporator. This resistance is obtained, according to equation (5), as the ratio between the temperature drop (i.e., the saturation temperature minus the inlet water temperature) and the total thermal power dissipated through the stack. If both the evaporator performance and the condenser performance are considered, the calculation yields that the thermal resistance achievable for each thyristor in the stack is less than 10 °C/kW.

CONCLUSIONS

From the results of the present investigation it is possible to maintain that two-phase closed thermosyphons can profitably be used for the cooling of large thyristors. Indeed, heat sinks employing these devices

- may be realized in different configurations, so that the constraints imposed by the electric layout may easily be overcome;
- may be realized employing light alloys, provided the technological precautions are taken to garantee the necessary safety of the vacuum sealing;
- have an overall thermal resistance which may be lower than 10°C/kW for the largest electronic components presently available.

NOMENCLATURE

a	interspace width between fins
b	fin thickness
Bo	Bond number
C	fluid heat capacity rate, parameter

D	characteristic dimension,diameter
g	gravity acceleration
k_c, K	coefficients
L	fin length
m	parameter
p	pressure
q_F	heat flow per cross section unit area at flooding limit
q''_{crit}	burnout in pool boiling
q_{th}	heat flux through the electronic component case
q_{bmax}	maximum heat flux at the center of the fin interspace
R	thermal resistance
T	temperature

Greek Symbols

Δhv	enthalpy of vaporization
ε	heat exchanger effectiveness
ρ	density
σ	surface tension

Subscripts

a	adiabatic
A	average
B	referred to Bezrodnyi
f	cooling fluid
K	referred to Kutateladze
L	referred to Liu
l	liquid
p	pressure
sat	saturation
v	vapor
W	referred to Wallis
I,II,III	different regimes

REFERENCES

1. Dunn,P. D., Reay, D. A. , *Heat Pipes*, 3d ed., Pergamon Press, Oxford, 1982 .

2. Bezrodnyi, M. K. , The Upper Limit of Maximum Heat Transfer Capacity of Evaporative Thermosyphons, *Teploenergetika*, 25, pp. 63-66, 1978 .

3. Fukano, T., Kadoguchi, K., Tien, C. L.,Oscillation Phenomena and Operating Limits of the Closed Two-Phase Thermosyphon, *Proc. 8th Int. Heat Transfer Conf., S.Francisco*, vol. 5, pp. 2325-2330 , 1986 .

4. Nguyen-Chi, H., Groll, M., Entrainment Flooding Limit in a Closed Two-Phase Thermosyphon, in *Advances in Heat Pipe Technology*, ed. D.A. Reay, pp.147-162, Pergamon Press, Oxford, 1981 .

5. Bezrodnyi, M. K. , Volkov, S. S., Study of Hydrodynamic Characteristics of a Two-Phase Flow in Closed Thermosyphons, in *Advances in Heat Pipe Technology*, ed. D.A. Reay, pp. 115-123, Pergamon Press, Oxford, 1981 .

6. Casarosa, C., Fantozzi, F., Latrofa, E., Bloccaggio Termico di un Termosifone Bifase Operante con Condizioni di Scambio Imposte, *Atti 41° Cong. Naz. ATI, Napoli*, Sez. VII, pp. 119-131, 1986 .

7. Dobran, F., Steady-State Characteristics and Stability Tresholds of a Closed Two-Phase Thermosyphon, *Int.J. of Heat & Mass Transfer*, vol.28 , pp. 949-957, 1985 .

8. Reed, J. G., Tien, C. L., Modeling of the Two-Phase Closed Thermosyphon, *J. of Heat Transfer*, vol. 109, pp 722-730, 1987 .

9. Casarosa, C., Dobran, F., Experimental Investigation and Analytical Modeling of a Closed Two-Phase Thermosyphon with Imposed Convection Boundary Conditions, to appear in *Int. J. of Heat & Mass Transfer.*

10. Rohsenow, W. M., Boiling , in *Handbook of Heat Transfer Fundamentals*, ed. W. M. Rohsenow, J. P. Hartnett, E. N. Ganic`, 2nd ed., pp. 12.1-12.94, McGraw-Hill, New York, 1985 .

11. Shah, M. M., Vaporization, in *Heat Transfer and Fluid Flow Data Book*, ed. D. A. Kaminski, sect. 507.1-507.6, General Electric Comp., Schenectady, N.Y., revis. 1987 .

12. Collier, J. G., *Convective Boiling and Condensation*, 2nd ed., pp. 310-393, McGraw-Hill, Oxford, 1981 .

13. Bergles, A. E., Techniques to Augment Heat Transfer, in *Handbook of Heat Transfer Applications*, ed. W. M. Rohsenow, J. P. Hartnett, E. N. Ganic`, 2nd ed., pp. 3.1-3.80 , McGraw-Hill, New York, 1985 .

14. Hsu, Y. Y., Analysis of Boiling on a Fin NASA Tech. Notes D-4797,Sept. 1968 .

15. Cash, D. R., Klein, G. J., Westwater, J. W., Approximate Optimum Fin Design for Boiling Heat Transfer, *J. of Heat Transfer*, vol.93, pp. 19-24, 1971 .

16. Takeyama, T., Endo, T., Owada, K., Analysis of Boiling Heat Transfer on a Fin, *Heat transfer Japan. Research*, vol.3, pp. 10-22, 1974 .

17. Son, A. K., Trinh, S., An Exact Solution for the Rate of Heat Transfer from a Rectangular Fin Governed by a Power Law-Type Temperature Dipendence, *J. of Heat Transfer*, vol.108, pp.457-459, 1986 .

18. Nishikawa, K., Fujita, Y., Ohta, H., Hidaka, S., Effect of the Surface Roughness on the Wide Range of Pressure, *Proc. 6th Int. Heat transfer Conf.,München*, vol.5, pp.61-66, Sept. 1982 .

19. Westwater, J. W., Development of Extended Surface for Use in Boiling Liquids, *AIChE Symp. Ser.*, vol.69, no.131, pp. 1-9, 1973 .

20. Bordignon, P., Latrofa, E., Casarosa, C., Martorano, L., Two-Phase Cooling of Large Diameter Power Thyristors, *Proc. European Conf. on Power Electronic and Applications, Brussels*, vol.2, pp. 5.105-5.109, Oct. 1985 .

21 Bordignon, P., Ragni, A., Latrofa, E., Casarosa, C., Two-Phase Thermosyphon Device for the Large Thyristors Cooling, *Proc. 19th Ann.IEE Power Electronics Specialists Conf., Kyoto*, vol.2, pp.1346-1352, April 1988 .

22. Burmeister, L. C., Vertical Fin Efficiency with Film Condensation, *J. of Heat Transfer*, vol.104, pp. 319-393, 1982 .

23. Patankar, S. V., Sparrow, E. M., Condensation on an Extended Surface, *J. of Heat Transfer*, vol. 101, pp. 434-440, 1979.

24. Casarosa, C., Fantozzi, F., Latrofa, E., Martorano, L., Pressure Drop and Flooding Limits fo a Two-Phase Closed Thermosyphon, 8th Int. Heat Transfer Conf., S. Francisco, Aug. 1986 .

Critical Heat Fluxes at Jet-Cooled Flat Surfaces

R. K. ŠKÉMA and A. A. ŠLANČIAUSKAS
Institute of Physical and Technical
Problems of Energetics
Academy of Sciences of the Lithuanian SSR
Kaunas, USSR

ABSTRACT

We present experimental results on the critical heat flux densities for boiling interaction of single circular jets or arrays of circular jets impinging on flat surfaces. The high range of heat flux densities ($q = 10^7$ W/m^2) with strongly subcooled water was studied. Jet velocities were u_o from 1 to 35 m/s at atmospheric pressure. The ratio of diameters D/d was from 0.5 to 8 for single jets, and relative pitches s/d were varied from 3 to 11 for jet arrays. Relative nozzle-surface distances h/d were varied from 2 to 4. A general relation to predict q_{CHF}, as a function of jet velocity and subcooling, was found for boiling at the stagnation point. The value of q_{CHF} was also related to D/d for single jets and to s/d for jet arrays.

NTRODUCTION

The constant improvements in modern electronics, which are accompanied by their power ncrease, maintain the interest of researchers on the problems of creating efficient power inks for high heat flux densities. Some assemblies of special electronic devices operate at q = = 10^7 W/m^2 and more, and traditional ways of liquid cooling can no longer cope with such oads.

The efficiency of cooling in such devices can be considerably increased by applying impinging ets of liquid. For other constant conditions, a hot surface-impinging jet interaction features onsiderably higher heat flux densities than other cooling modes. Among other advantages f jet cooling, note efficient operation on inaccessible surfaces, simple cooling rate control, s well as size and volume reduction of the devices. Cooling by water jets may also operate vith boiling, which means an additional heat transfer augmentation.

But to develop jet cooling technologies and to design the devices one must properly understand the processes involved, because they include ultimate heat flux densities. The main liming factor in jet cooling is the heat transfer crisis, which means, in practice, a sudden disintegration of the heated surface. The available literature on the heat transfer crisis at a jet-surface ontact is scarce and refers mainly to free jets [1-3]. But data on such processes are highly nportant, because the critical heat flux (CHF) value which causes the heat transfer crisis epresents the application limit of submerged jet cooling.

XPERIMENTAL APPARATUS AND PROCEDURE

he experiments were performed in a closed loop of water, Figure 1 whith test section constucted as a sealed vessel of 400 x 200 x 200 mm having observation windows in the side alls.

Different heated test surfaces could be mounted on the bottom of the test section for tests of different applications. Heated surfaces 2 were made of glass-textile-reinforced plastic and covered with copper sheets 30 μcm thick.

The heated surface was made of a rectangular copper sheet from 6 to 20 mm wide and from 9 to 20 mm long, continuously widened at the ends where conductors were connected. It constituted a strip of an imaginary disc, so that the length of the strip coincided with the diameter of the disc D. The widening profile was chosen to ensure a maximum and homogeneous heat emission from the rectangular heated surface. The ratio of the length of the rectangular heated part and of the nozzle outlet diameter D/d could be varied from 0.5 to 8. The heated surfaces were produced photochemically by the technique applied in plate printing so that all of them had identical geometrical parameters.

Heated surface was fixed on textile-reinforced plastic plate so that jet impinging was concentric, and was pressed by electric insulation contacts, which conducted electric current to the heated surface.

Chemically treated water was injected from a pipe through nozzles either as a single circular jet, or as array of jets. Single axisymmetric jets were injected through convergent nozzles with outlet diameters d=3.0, 9.0 and 18 mm. Arrays of circular jets were formed by divergent nozzles with removable bottom pieces, 80 mm diam., which had the necessary numbers of conic nozzles. Staggered and in-line arrays of two, four and more circular jets could be obtained with different pitches s. Relative pitch of the jet arrays s/d was varied from 3.0 to 1.1.

In order to achieve maximum critical heat flux densities, relative nozzle-surface distances h/d were varied from 2 to 4.

The traversing device could control the nozzle-surface distance within 0.1 mm.

Centrifugal pump supplied water at a controlled flow rate and pressure to the nozzle outlet. Its constant temperature was maintained by heat exchanger, mounted in storage tank. Water quality was controlled by filters. Discharge velocity was controlled by a turbine flow-rate meter, and average temperature was measured by thermocouples.

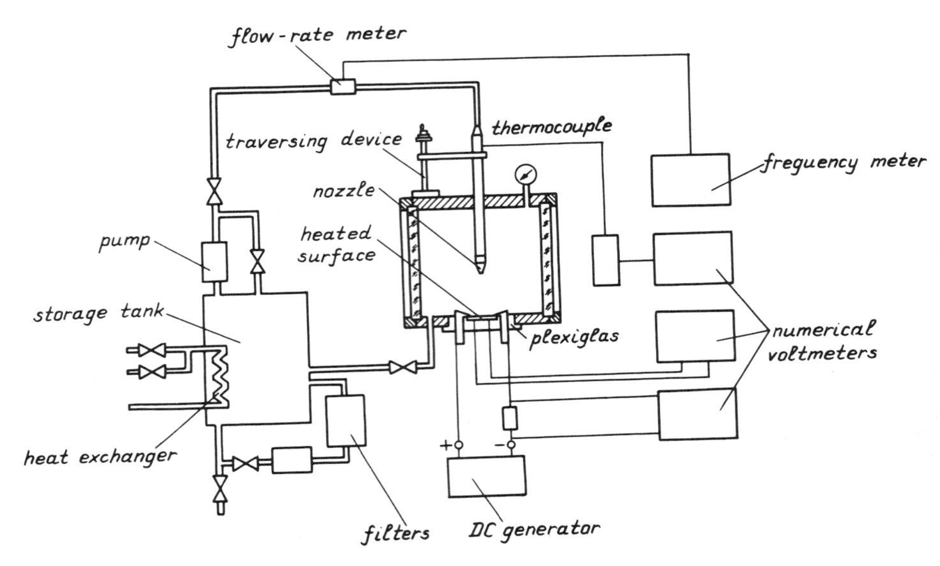

FIGURE 1. A schematic representation of the test rig.

A constant 200 mm thick layer of liquid was maintained on the heated surface throughout the experiment. Absolute pressure in the test section never exceeded $2 \cdot 10^5$ Pa.

The heat flux on the heated surface was created by a direct current of constant voltage. It was maintained by an electronic voltage control device. Relative heat flux in the heated area was determined from the electric power and voltage drop on the heated surface.

The temperature of the heated surface with boiling at the stagnation point was determined from the measured electric resistance of the surface. Accurate resistance - temperature relations were found thanks to the calibration of the tests surfaces in a constant temperature device.

The critical heat flux density was found from the maximum power applied preceding the failure of the heated test surface.

EXPERIMENTAL RESULTS AND DISCUSSION

Our study of the heat exchange of fully developed boiling at the contact of an impinging jet and a heated surface revealed a highly intensive process with the heat transfer coefficients up to $\alpha = 10^6$ W/(m^2.K). At higher heat loads the separate vapour bubbles begin to aggregate into a continuous film on the heated surface. This process develops to a crisis, when the heat transfer rate is abruptly decreased. This phenomenon is known as boiling crisis, and it constitutes the limiting admissible heat load before the failure of the heated surface.

Our tests in wide ranges of operation and geometric parameters (u_o from 1 to 35 m/s, t_f from 15 to 20 °C, D/d from 0.5 to 8, h/d from 2 to 4) suggest that the subcooled boiling crisis in the system considered is a case of subcooled and low-evaporation boiling and is noted by high critical heat flux densities, which reached $6 \cdot 10^7$ W/m^2 in some of the tests.

In our conditions q_{CHF} is a function of jet velocity, u_o, subcooling rate, ΔT_{sub}, and diameter ratio, D/d.

The values of q_{CHF} given in Figure 2 refer to the above parameters of pool boiling at the same rate of subcooling. The experimental results are in a satisfactory agreement with those for a free jet [1]. These results yielded description of q_{CHF} at the stagnation point($D/d \leqslant 0.5$) of a submerged jet:

$$q_{CHF\ sub.forced} = q_{CHF\ sub.pool} \left(1 + 0.92\, u_0^{0.44}\right) \qquad (1)$$

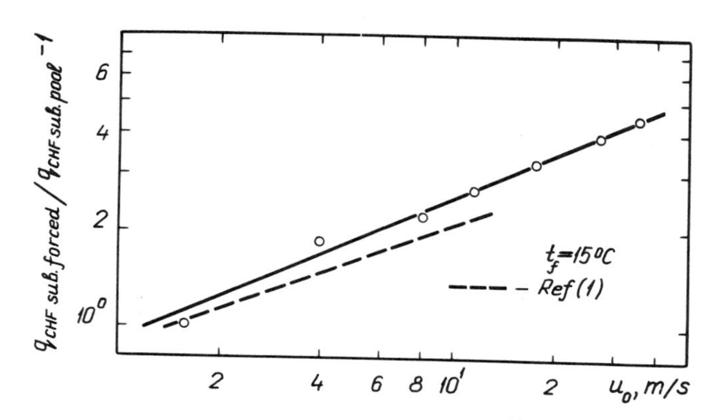

FIGURE 2. The effect of the jet velocity on q_{CHF}. d = 18 mm, D/d =0 .5, h/d = 2.

where $q_{CHF\ sub.forced}$, $q_{CHF\ sub.pool}$ are in W/m^2 and u_o in m/s.

Relationship between the critical heat flux and the subcooling is described by the correlation [1]

$$q_{CHF\ sub.pool} = q_{CHF\ sub\ =\ 0,\ pool} \left[1 + 0.112 \left(\frac{\rho_l}{\rho_v}\right)^{0.8} \cdot \left(\frac{c_{pl}\ \Delta T_{sub}}{L}\right)^{1.13} \right] \tag{2}$$

and the critical heat flux of saturated pool boiling can be found from the Kutateladze criterion [5]

$$q_{CHF\ sub\ =\ 0,\ pool} = 0.16\ L\ \rho_v \left[\frac{\sigma \cdot g(\rho_l - \rho_v)}{\rho_v^2} \right]^{1/4} \tag{3}$$

System pressure was calculated from the Bernoulli equation

$$P = P_\infty + \rho_l \cdot u_o^2 / 2 \tag{4}$$

where P_∞ is pressure in the test section.

An important factor for a heat sink of very high heat flux rates is the surface area which can be covered by a single jet. Our investigation of q_{CHF} in the range of D/d from 0.5 to 8 suggested a complicated relation between q_{CHF} and D/d Figure 3.

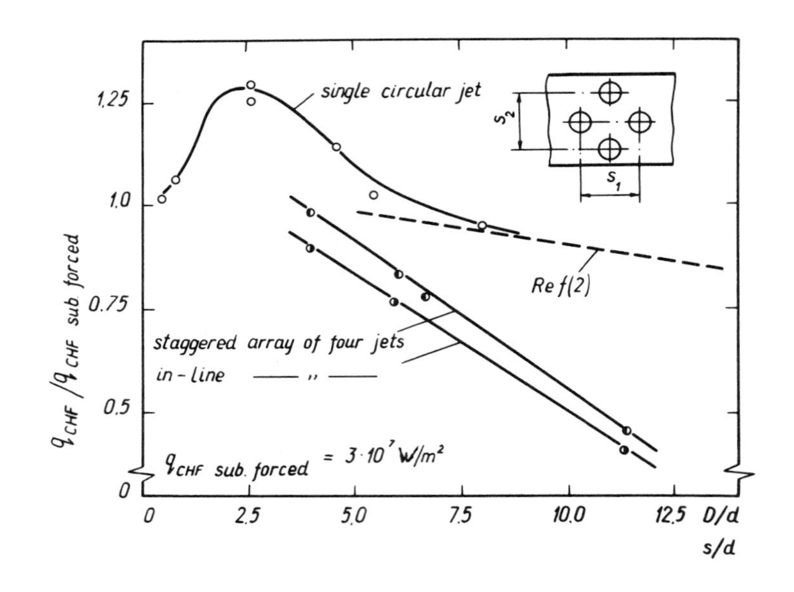

FIGURE 3. The relation of q_{CHF} to D/d for a single jet and to s/d for arrays of jets. $u_o = 8.0$ m/s, h/d = 2

With an increase of D/d from the stagnation point to 2 we observe an increase of q_{CHF} up to maximum at D/d = 2, which is followed by a monotonic decrease. This behaviour of q_{CHF} may be ascribed to a maximum velocity at the wall at D/d = 2 [4].

When large surface areas must be cooled, single-row and multi-row jet arrays are preferable. With this in mind, we studied the behaviour of q_{CHF} for submerged circular jets on flat surfaces in a row of four jets as a model of a multirow array. The study covered s/d from 3 to 11, h/d from 2 to 4, and u_o from 1 to 8 m/s. The discharge temperature, t_f, ranged from 15 to 17 °C.

The accumulated experimental results indicate that for an array of four circular jets the determining parameter of q_{CHF} is the relative pitch,as shown Figure 3.

With small relative pitches s/d from 3.0 to 4.0, the values of q_{CHF} are close to those at the stagnation points of single jets. With larger s/d, the decrease of q_{CHF} is sharper than for a single jet. Thus, an increase of s/d from 4.0 to 11 causes a 60 % decrease of q_{CHF}. The physical meaning of this is that longitudinal velocity decreases on the surface with the distance, and jet interaction decreases accordingly. Besides, heating rates are higher along the contact lines of separate jets, when their spacings are larger.

A comparison of q_{CHF} in different jet arrays ($s = s_1 = s_2$) indicated larger values of q_{CHF} in staggered arrays, Figure 3. This difference is not significant. For in-line arrangements studied, a maximum decrease of 10 % in q_{CHF} was observed. The decrease of q_{CHF} in in-line arrays may be ascribed to the fact that contact lines of separate jets run across the heated surfaces.

CONCLUSIONS

(1) This heat transfer study for subcooled bubble boiling at the interaction of submerged axisymmetric single jets or jet arrays revealed a very intensive process with the heat transfer coefficients as high as $\alpha = 10^6$ W/(m²K) and $q_{CHF} = 6 \cdot 10^7$ W/m².

(2) A relation predicting q_{CHF} (1) was found for boiling at the stagnation point (D/d $\leq$ 0.5) in terms of discharge velocity and subcooling rate.

(3) A relation between q_{CHF} and D/d was found. An increase of D/d from the stagnation point to 2 is accompanied by an increase of q_{CHF} up to a maximum, which is followed by a monotonic decrease with a further increase of D/d.

(4) The critical heat flux density at the interaction of four jet array is lower, than at the stagnation point of a single jet, and decreases sharply with an increase of s/d. A function curve of the heat flux decrease to be used with Figure 1 is given.

NOMENCLATURE

c_{pl}	specific heat capacity of liquid at constant pressure	J/(K kg)
d	jet nozzle diameter	m
D	heated surface diameter	m
g	local gravitational acceleration	m/s^2
h	jet nozzle - heated surface distance	m
L	latent heat	kJ/kg
P	system pressure	N/m^2
q	heat flux density	W/m^2
s	pitch of jet array	m
t_f	fluid temperature	C
u_o	impinging jet velocity	m/s
α	heat transfer coefficient	$W/(m^2 K)$
ρ_l	density of liquid	kg/m^3

ρ_v	density of vapor	kg/m^3
σ	surface tension	W/m

Subscripts

CHF	critical heat flux
forced	impinging jet boiling
pool	pool boiling
sub	subcooling

REFERENCES

(1) Miyasaka, I., Inada, S. and Owase, Y,. Critical heat flux and subcooled nucleate boiling in transient region between a two-dimensional water jet and a heated surface, J. of Chem. Eng. of Japan, vol. 13, N 1, pp. 29-35, 1980.

(2) Monde, M. and Okuma, Y., Critical heat flux in saturated forced convective boiling on a heated disk with an impinging jet - CHF and L - regime, Int. J. Heat Transfer, vol. 28, N 3, pp. 547-552, 1985.

(3) Monde, M. ,Nagae, O. and Ishibashi, Y., Critical heat flux in saturated forced convective boiling on a heated disk with impinging jet: In a high pressure region, Trans. JSME Ser.B, vol. 52, N 476, pp. 1799-1804, 1986.

(4) Dyban, E.P. and Mazur, A.I., Convection in Jet Cooling, Kiev: Naukova Dumka, 303 p., 1982.

(5) Kutateladze, S.S., Fundamentals of Heat Exchange, Moscow: Atomizdat, 416 p., 1979.

Thermophysical Aspects of Techniques Related to Enhancing Thermal Stability of Electronic Equipment Using Porous Coatings of Cooled Surfaces

M. A. STYRIKOVICH and S. P. MALYSHENKO
Institute of High Temperatures (IVTAN)
USSR Academy of Sciences
Moscow, USSR

ABSTRACT

Basic results of the investigations conducted at IVTAN in recent years into thermal processes in boiling on surfaces with low-conductivity porous coatings are reviewed. Thermophysical aspects of using porous coatings to enhance thermal stability of electronic equipment are analyzed.

I. INTRODUCTION

Thermal stabilization of radioelectronic equipment elements is determined by the relation of heat output, which is a function of temperature and other parameters e.g. $W = f(T,I \ldots)$ and the heat transfer rate $Q(T)$ from their surface. For elements cooled by boiling fluid the function $Q(T)$ is determined by the boiling curve. Thermal equilibrium of an element with uniform temperature will be stable, provided the following conditions are fulfiled:

$$W(T,I \ldots) = Q(T) \quad (1)$$

$$\frac{d}{dT}(Q - W) > 0. \quad (2)$$

It is clear that the intensification of boiling which brings about greater values of $Q(T)$ and $\frac{dQ}{dT}$, enhances thermal stability of the device. Porous coatings have been extensively used recently to enhance heat transfer in boiling. Depending on problems to be solved, the characteristics of these coatings should be optimized differently. Thus, it is necessary to carry out a detailed investigation of thermal processes in boiling with porous coatings of various types. Given in the following is a brief review of the principal results of such investigation carried out at the Institute of High Temperatures in recent years [1-10].

2. EXPERIMENTAL TECHNIQUE

Both integral characteristics of heat transfer (relationships between q and θ, nonstationary and transition processes, hystereses of boiling and those of transitions I and II, etc.), and the elementary processes (the dynamics of growth and departure of vapor bubbles, two-phase flow within porous coatings) were studied on surfaces with low-conductivity porous coatings of various types and geometry. With the known characteristics of coatings, this makes it possible to determine a relative role of different mechanisms of heat transfer in the coating, to generalize the experimental data and to adequately compare them with theoretical models of processes. The experiments were conducted in a broad range of heat loads q, superheating of the heated surface θ & pressures P in boiling of water, ethanol-water solutions & liquid helium under conditions of natural convection and the boundary conditions of the second kind (q = const).

Porous coatings were applied on the vapor-generating surfaces by way of sintering, plasma spraying and electrochemical precipitation. In the case of the first two types of coatings made of nickel-chromium and Al_2O_3 all basic parameters were experimentally measured, namely, heat conductivity, porosity, distribution of pores in size, pressure loss of gas filtration (permeability) in dry and wet conditions, the geometry of the pore lattice, particle size, the angles of wetting with a heat-transfer agent (water, ethanol-water solutions) & the height of capillary rise. The porous coatings obtained through electrochemical precipitation were used in the experiments with liquid helium, their characteristics were studied in lesser detail. The most important properties of the coatings involved in experiments are given in Table 1. More detailed information about them and detailed description of the experimental technique are outlined in [1-9]

TABLE 1. Basic Characteristics of Porous Coatings (Z = 6)

Material	Application technique	Mean particle size, D, μm	Porosity, ε, %	Mean pore dia, μm	Permeability $K \cdot 10^{11}$, m^2	Thermal conductivity, W/m2
Al_2O_3	Spraying	80	20-30	2.6	$(1.4-2.1)10^{-3}$	1.5-2.6
			50	-	-	0.7-1
NiCr	Spraying	80	36	15	$8 \cdot 10^{-2}$	2.8-3.1
			50	-		1.5-1.7
NiCr	Sintering	140-130	40-50	40	0.6-0.8	1.8-2.4
		270-250	40-50	65	1.2-1.5	1.8-2.5
		400-380	40-50	100	3.3-4.1	1.4-2.2
Zn	Electrochemical precipitation		20	10	10^{-3}	-

3. BOILING CURVE SHAPE, BOILING REGIMES

The generalized boiling curve according to data [1-9]′ is shown in Fig. 1. It is clear that the presence of low-conductivity porous coatings on a vapor-generating surface appreciably changes the dependence of heat flux q on the temperature drop $\theta = T_w - T_s$, where T_w is the temperature of heated surface under coating & T_s is saturation temperature. These distinctions can be briefly formulated as follows.

- in case of porous surface coatings boiling starts at considerably lower superheating of heated surface $\theta_{o.b.}$ and heat loads $q_{o.b.}$ than on smooth surfaces, and over the initial section of the curve of boiling the heat transfer coefficient $\alpha = q/\theta$ tends to increase appreciably (from 5 to 10 times),

- the bubble mechanism of vapor departure from the outer boundary of the coating is observed in two regimes of boiling, namely, when $q < q^* < q_c$, the zones of vapor generation within porous coatings are not connected to each other over the vapor-filled space, the heat transfer law $q \sim \theta^n$ corresponds to $n > 1$, heat transfer coefficients $\alpha = q/\theta$ exceed 4 to 10 times the corresponding values of α in case of boiling on smooth surfaces (regime 1); when $q^* < q < q_c$, the zones of vapor generation within the coating are connected to each other over the vapor-filled space, the heated surface is separated from the zone of evaporation by thin film of vapor stabilized in the coating body; the heat transfer law corr sponds to $n = 1$, in the areas of high superheating of heated surfaces corresponding to film boiling on smooth surfaces, the bubble mechanism of vapor departure from the outer boundary of the coating is preserved and the value of α exceeds 10 to 15 times the corresponding values of $\alpha(\theta)$ for a smooth surface (regime II) up to $q \sim q_c$;

- in case of characteristic heat loads $q = q^*$ there occurs a dissipative phase transition, first observed in Refs. [1-3], from one bubble boiling regime to the other which is accompanied by a drastic increase of the time of relaxation to a new stationary condition at $q \to q^*$, as well as by typical hysteresis and transition phenomena;

- in the general case, the boiling crisis, associated with the formation of a vapor film on the outer boundary of the coating, is observed at values of θ_{c1} and q_{c1} different than on the smooth surfaces; the values of θ_{c1} and q_{c1} depend on the characteristics of the coating; and, in case of optimal coatings one can increase q_{c1} 1.5 to 3 times compared to q^0_{c1} for smooth surfaces;

- the film boiling regime arises as a result of the crisis transition from bubble boiling regime II to the film one, the heat transfer coefficient in the film regime can be greater or smaller than in case of film boiling on smooth surfaces, depending on the coating characteristics and the working fluid properties;

- the values of the second critical heat load q_{c2} corresponding to the transition from the film boiling regime to the bubble one, given the porous coatings, appreciably exceed q^0_{c2} for smooth surface, in so doing, depending on the coating characteristics as q

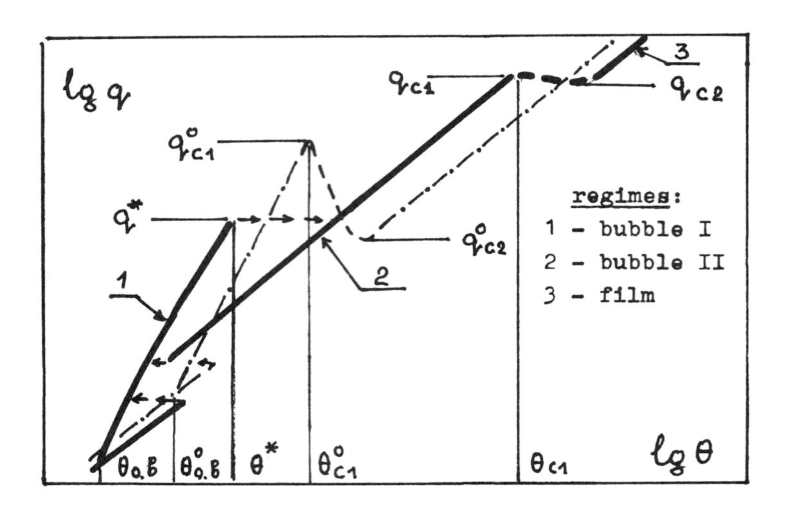

FIGURE 1. The shape of boiling curves on surfaces with low-conductivity porous coatings (▬) and on smooth surfaces (—·—).

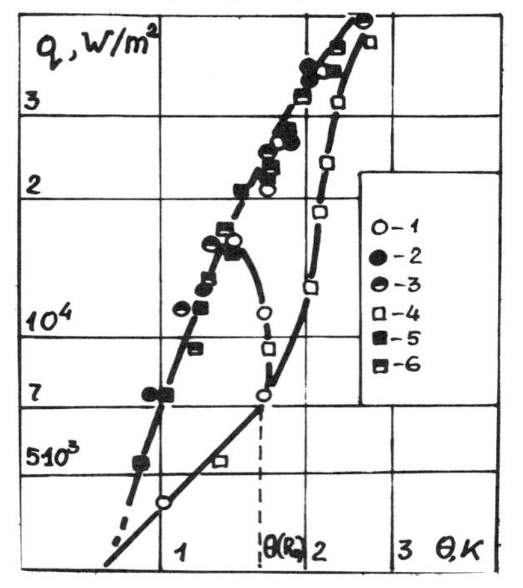

Figure 2. Initial section of water boiling curve on surface with sintered porous coating (h = 0.8 mm. D = 130-140 μm) at P = = 0.I MPa: 1-3 -- deactivation of centers by way of preliminary vacuum processing; 4-6 --- deactivation of centers by way of preliminary vacuum processing and prolonged boiling with cooling; 1,4 --- increase from q = 10^2 W/m², 2,5 --- decrease q after coming to regime of developed bubble boiling, 3,6 -- increase q from q ~ 10^2 W/m², 2,5 --- decrease q after coming to regime of developed bubble boiling, 3,6 --- subsequent increase q.

decreases, the film boiling regime gives place to bubble regime I or II;

The parameters of the points of boiling curves in the $q-\theta$ diagram corresponding to the location of boiling regimes are largely determined by the characteristics of the coating and the properties of the boiling fluid; the character of transitions during the replacement of boiling regimes depends on the relation between the size of finite clusters of the vapor-filled pores L_f and the coating thickness h.

We now briefly dwell on the basic peculiarities of thermal processes at separate sections of the boiling curves.

4. INITIAL REGION

The onset and the end of boiling on the surfaces with porous coatings are experimentally registered by the emergence of the first or disappearance of the last bubbles at the outer boundary of the coating. The values of $\theta_{o.b.}$, $q_{o.b.}$, $\theta_{e.b.}$ and $q_{e.b.}$ correspond to these points. These values, in general, are different, because the quantity and geometry of the vapor-filled pores in the body of the coating are different at the points of the onset and end of boiling.

In case of smooth surfaces the point of the onset of boiling is determined by a characteristic size - the throat radius of active cavities on the heated surface r_a. With porous coatings there exist two characteristic dimensions: r_a and the minimal radius of vapor-filled pores when the vapor "breaks through" the coating - r_p. It is by the lesser of them that the initial superheating is determined corresponding to the escape of the first bubbles of vapor to the outer surface of the coating:

$$\theta_{o.b} = \frac{2\sigma T_s}{\Delta H \cdot \rho'' \cdot r} \cos\gamma \qquad (3)$$

where σ is surface tension; γ angle of wetting; ΔH evaporation heat and ρ'' is vapor density.

Because in case of conventionally used porous coatings both r_a and r_p are substantially greater than r_a on smooth surfaces, $\theta_{o.b.}$, with the coatings is markedly smaller than on smooth surfaces. At the same time, given porous coatings, the processes of deactivation of active cavities are made appreciably easier, because their throat curvature radii are determined by the size of the particles of the coating's first layer, i.e., are sufficiently great. In this case, thermocapillary convection in the fluid films on the coating skeleton walls, caused, e.g., by the surface-inactive nonvolatile impurities in the fluid [10], is a fairly effective mechanism ensuring the filling of active cavities with liquid. This may give rise to a pronounced hysteresis of coming to a boil, in a number of cases even more extensive than on a smooth surface [5]. The type of the hysteresis loop at the initial section is determined by the relation of r_a and r_p as well as the relation between the number of possible escapes of vapor N_p and active cavities N_a per unit of the coating visible surface. The alteration of these relations, when using different procedures of deactivation of the centers of vapor formation, results in the variation of the hysteresis loop, as shown in Fig. 2 [5].

The minimal radius of vapor-filled pores r_p can be calculated if there is known the distribution of pores according to sizes f(r) and the geometry of the lattice of pores using the percolation theory technique. Presuming that vapor in the porous space spreads throughout the pores of the maximum radius, we obtain the condition for vapor to break through the coating,

$$\int_{r_p}^{r_{max}} f(r)\cdot dr = \beta \tag{4}$$

where β is the percolation threshold on connections in the pore lattice which can be estimated as

$$\beta = \frac{A}{(A-1)\,z} \tag{5}$$

where A is dimensionality of the pore lattice & z is its coordination number /11/. For three-dimensional lattices $\beta = \frac{3}{2}\cdot\frac{1}{z}$. Upon determining the values of r_p from Eqs.(4) and (5) one can calculate $\theta_{o.b.}$ from Eq. (3). The results of this calculation and their comparison with the experimental data are given in Fig. 3. The latter shows a satisfactory fit of the calculations according to Eqs. (3-5) to the experimental data.

Should films of fluid exist on the walls of the coating skeleton, the alteration of the geometry of the phase interface within the body of the coating corresponds to the point of the end of boiling. The cylinder meniscus in the throat of radius r_p changes to spherical [5,10]. This means that in this case $\theta_{e.b.}$ is half that of $\theta_{o.b.}$ [5,10]. This situation was repeatedly observed in experiments [1-5] for various types of coatings. In the absence of a fluid film on the channel walls of vapor escape $\theta_{o.b.} = \theta_{e.b.}$. This was also observed in the experiments for certain types of coatings [5].

5. THE TRANSITION FROM BUBBLE REGIME I TO BUBBLE REGIME II

In boiling on surfaces with low-conductivity porous coatings the non-equilibrium phase transition in the dissipative system caused by the emergence of the connectivity or link of the vapor generation zones as to the vapor-filled pores in the entire system was first observed in references [1,2]. At the point of transition for q= const. the basic parameters of the system undergo sudden changes, namely, superheating of the surface under the coating θ, departure bubble diameters & their location on the outer boundary of the coating & the index of n-th degree in the heat transfer law $q \sim \theta^n$ [1-4]. As the transition point is approached, the time of relaxation to the new steady-state condition (Fig. 4) is drastically increases, and the transition per se is characterized by the presence of the system "memory" and the

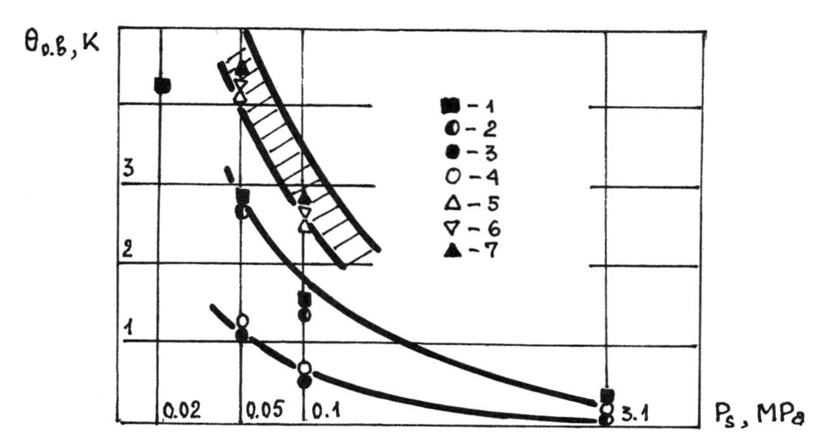

FIGURE 3. $\theta_{o.b}$ as a function of pressure for surfaces with sintered (1-4) and sprayed (5-7) porous coatings from ZnCr: 1-D=130-140 μm, h = 1.5 mm, 2-D=130-140 μm, h=0.8mm, 3-D=380-400μm, h-1.8 mm, 4-D=380-400 μm, h=0.8 mm, 5-porosity ε =0.32-0.36, h=0.2 mm, 6- ε =0.32-0.36, h=0.5 mm, 7- ε = 0.32-0.36, h = 1 mm. Solid lines - calculation according to (3)-(5).

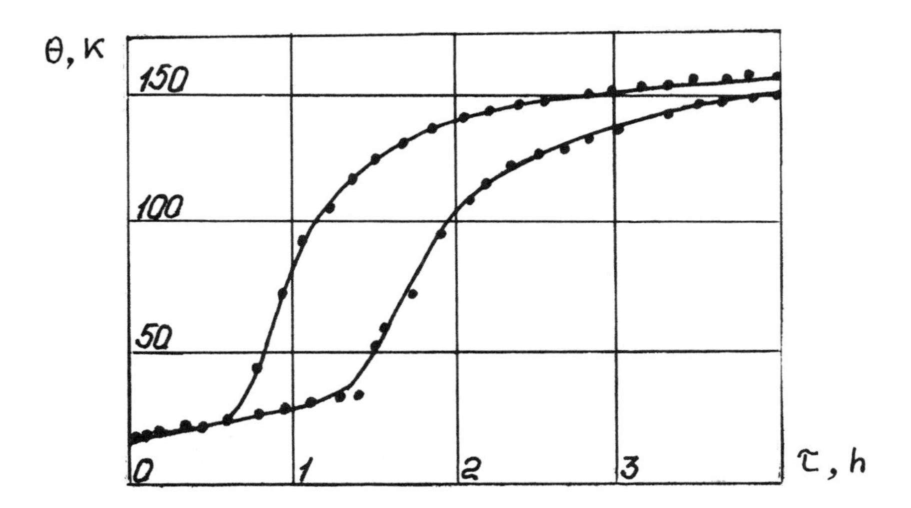

Figure 4. Relaxation θ to new steady state at $q=q^*$ (sprayed porous coating from ZnCr, ε = 0.5-0.6, h=1.4 mm, water, P = 0.I MPa).

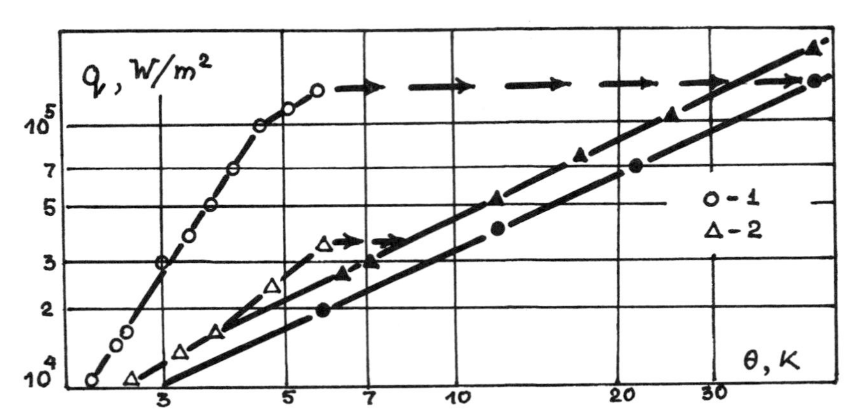

FIGURE 5. Hysteresis of heat transfer in transition from regime I to regime II. Water boiling at P = 0.1 MPa on surfaces with sprayed porous coatings h = I mm: 1 - ε = 0.5-0.6, 2 - ε = 0.32-0.36. Light marks - regime I, dark marks - regime II.

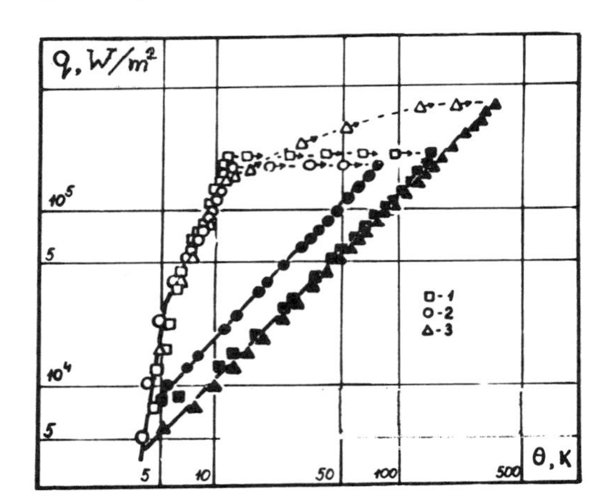

FIGURE 6. Hysteresis of heat transfer and "memory" effect in the system during transition I-II (water boiling at P = 0.1 MPa on surface with sprayed porous coating = 0.5-0.6, h=1.4 mm): I - transition at $q^* = 2.3 \times 10^5$ W/m^2 at upon attaining steady state in regime II ($\Delta\tau > 3$ hours); 2 - same with insufficient time of expectation ($\Delta\tau$ = 1 h); 3 - transition with $q > 2q^*$ and attainment of steady state II ($\Delta\tau \approx 0.5$ h). Light marks -- regime I, dark marks - regime II.

hysteresis of heat transfer (Figs. 5 & 6). These phenomena are studied in detail in [2,4].

The percolation transition from regime I to regime II comes into being in view of the attainment of void fraction φ in the porous structure, in its narrow layer close to the heated surface, which corresponds to the percolation threshold as regards the system of liquid-filled pores at the point of the onset of boiling. (By the void fraction φ here we mean the share of the pore space filled with vapor). This situation is possible in the case of a noticeable heterogeneity of void fraction in the body of the coating, i.e., for porous coatings with a relatively small permeability as to vapor and low heat conductivity. Increasing permeability in vapor, e.g., by way of creating special channels in the coating for the vapor escape brings about a change in the character of transition or even its disappearance (cf. Fig. 7). A decline in the heterogeneity of void fraction in the coating can be ensured by other factors, e.g., thermocapillary flows of fluid in the body of the coating caused by the gradient of surface tension coinciding in direction with the gradient of temperature in the body of the coating. Fig. 8 represents the results of experiments [5] with water ($\nabla\sigma \uparrow\downarrow \nabla T$) and ethanol-water solution ($\nabla\sigma \uparrow\uparrow \nabla T$) using one and the same sample. As was expected, in the second case the phase transition I-II at q = const. is absent, i.e., the regimes are substituted gradually within a broad range of heat loads.

The experiments [1-10] have been run with low-conductivity coatings. In porous coatings with high thermal conductivity of the skeleton the emergence of the heterogeneity of void fraction is hindered as to the coating thickness. Therefore, when boiling on surfaces coated with such structures, the transition I-II is normally not observed. It should be noted that the very presence of the transition I-II is not associated with the limited wettability of the structure skeleton with a heat transfer agent, because the experiments [6,7] with liquid helium absolutely wetting the structure, showed the presence of the transition in this case as well. However, the character of the heat transfer hysteresis, depending on the wetting characteristics according to the results [1-7] can be appreciably changed. In particular, in case of liquid helium the heat transfer hysteresis is actually absent during the transition I-II [6,7] while for water it is fairly great [1-5].

The radius of curviture of the interface in the body of the porous coating at the point of transition I-II can be estimated as the one corresponding to the percolation threshold in the lattice of the fluid-filled pores subsequent to the breaking through of the coating by vapor. The coordination number of this randomized lattice Z' is related to that of the initial lattice Z by the equation

$$Z' = Z(1-\beta) = Z - \frac{A}{A-1} \qquad (5')$$

where β is the percolation threshold in the initial lattice, A is its dimensionality. The minimal radius of vapor-filled pores r^* at the point of transition I-II can be determined from the relationship

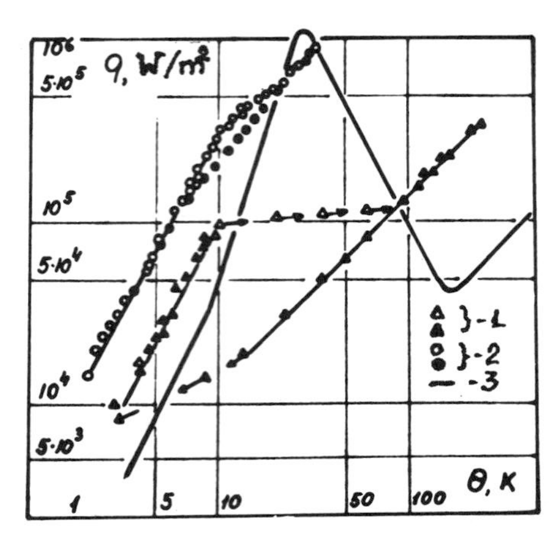

FIGURE 7. Alteration of the character of transition I-II when creating channels for vapor escape (water boiling at P = 0.1 MPa, h = 1.4 mm, ε = 0.5-0.6): 1 - sprayed coating from Al_2O_3, 2 - sprayed coating from Al_2O_3 with channels for vapor escape, 3 - boiling on smooth surface.

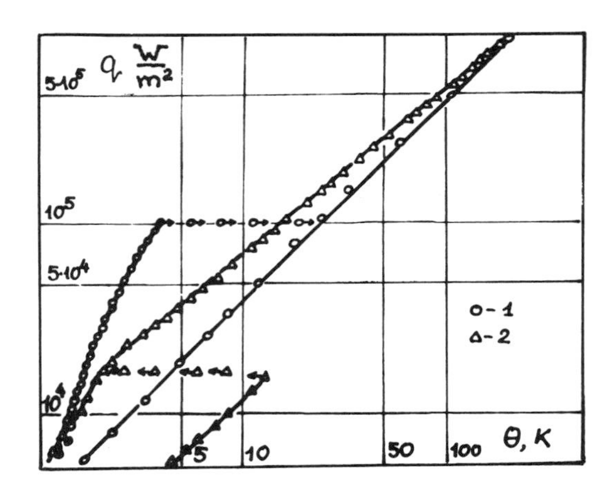

FIGURE 8. Boiling curves in bubble regimes I and II on surface with sintered porous coating (D = 130-140 μm, h = 0.8 mm) at P = 0.1 MPa: 1 - water, 2 - ethanol-water solution.

$$\int_{r^*}^{r_{max}} f(r) \cdot dr = \frac{\beta}{1-\beta} = \left[Z \frac{A-1}{A} - 1 \right]^{-1} \qquad (6)$$

whereupon using Eq. (3) one can find the value of θ_1^* corresponding to this r^*. Complete superheating of the surface to be heated at the point of the beginning of the transition θ^* is the sum of θ_1^* and θ_2^* - an average temperature drop from the heated surface under the coating to the evaporation zone. It is natural to assume that the average distance between the evaporation zone and the heating surface δ is determined by the position of the capillary throats in the first layer of particles comprising the coating. In this case, knowing the average size of particles D, the geometry of their packing and heat conductivity of skeleton λ one can estimate θ_2^* as $\theta_2^* = q^* \cdot \delta / \lambda$, where $\delta \approx D$ for the structures formed by tightly packed particles. The calculation of θ^* according to the above-identified diagram using the experimentally determined $f(r)$, r_{max}, Z, q_n^* & λ yields the values practically coinciding with the experimental ones [9]. One of the characteristic dimensions defining the processes of transport in the presence of the interface stabilized in the coating body is the dimension of the finite cluster of vapor-filled pores L_f. It can be calculated a model of branching pores for Bethe lattices [11,12] depending on the degree of proximity to the transition point. The amplitude of "blurring" the interface in the coating body is also specified by this dimension. It is clear that the character of the transition I-II largely depends on the relationship between the coating thickness h and L_f. In particular, at $h < L_f$ the transition occurs at the point q^*, θ^* on (q, θ)-diagram, and at $h \ll L_f$ is absent [9]. As L_f approaches the transition point, it sharply grows and the capillary pressure drop in this distance under the effect of which the phases are redistributed in the pore space $\Delta P_{cap} = 2\sigma \left(\frac{I}{r^*} - \frac{I}{r_p} \right) \cos \gamma$ remains finite. Therefore, close to the points of transition I-II the velocity of the interface motion in the pore space drastically decreases to bring about an anomalous growth of relaxation time, to new steady states as q changes [3]. Thus, the experimental boiling curves corresponding to different rates of q changes in the experiments are appreciably different (cf., Fig. 9), in particular, higher growth rates of q correspond to higher heat transfer coefficients $\alpha = \frac{q}{\theta}$. This circumstance may turn out rather useful when using porous coatings to thermally stabilize the elements of radioelectronic devices subjected to short term thermal disturbances.

6. BUBBLE REGIME II, HEAT TRANSFER CRISIS AND FILM REGIME

As was noted above, when boiling in bubble regime II, the superheating θ with a preset q is determined by the thermal resistance of a dried-up layer of the coating. According to the experimental data [1-5,9] the thickness of this layer depends on the coating characteristics and the working fluid properties, but in many cases it accounts for half the coating thickness for relatively thin coatings. This is quite natural because when the gas displaces the liquid in porous bodies at the initial stage of the

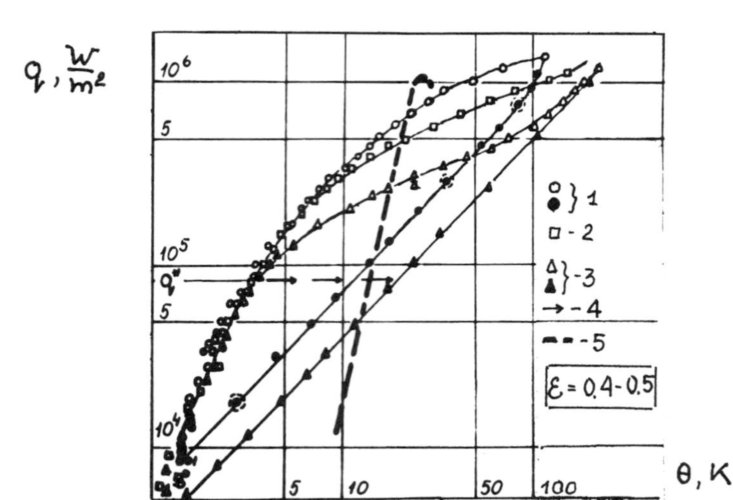

FIGURE 9. The effect of the rate of variation of heat load on the shape of water boiling curve on surfaces with sintered porous coating (P = 0.1 MPa, D = 130-140 μm, h = 0.8 mm) with increasing power (light marks) and **decrease** (dark marks) at expectation times : 1 - $\Delta\tau$ = 30 s; 2 - $\Delta\tau$ = 5 min; 3 - $\Delta\tau$ = 30 min; 4 - equilibrium transition from regime I to regime II, 5 - boiling on smooth surface.

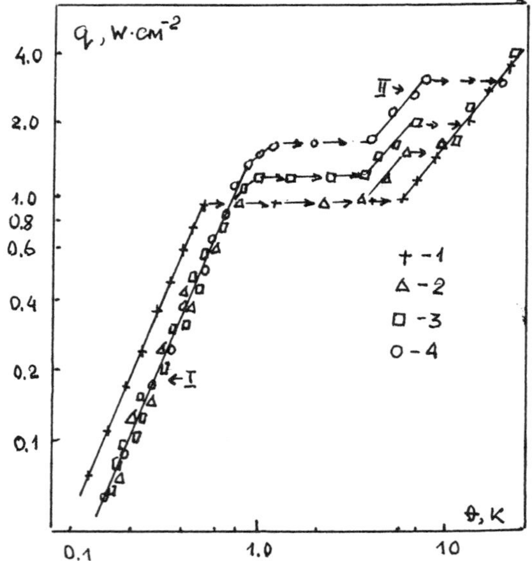

FIGURE 10. Helium boiling curves on surfaces with porous coatings of different thickness: 1 - smooth surface (h = 0), 2 - h = 10μm, 3 - h = 25 μm, 4 - h = 50 μm.

penetration of the gas phase into the body, the amplitude of "blurring" the interface according to the data [12] is close to the average depth of the gas phase penetration. Thus, with a middle position of the interface at a distance $h/2$ from the heating surface in case of thin coatings separate clusters of vapor-filled pores come out to the outer boundary of the coating and the middle position of the interface is stabilized at distances of the order of $h/2$.

Thermal resistance of the dried-up layer of the coating is determined by the heat conductivity of the structure skeleton and the heat transfer by the gas phase convection across the pore space. This convection is essential for low-boiling fluide with small viscosity and is particularly pronounced in the boiling of helium. In this case, as h grows, complete effective thermal resistance of the porous coating may not increase, but decrease in view of the growth of the heat transfer through convection. This result was observed experimentally [6,7] (Fig. 10).

The heat transfer crisis caused by complete drying of the coating and the formation of a vapor film on its outer boundary, as was noted above, develops at q_{cl}, greater than q^{0}_{cl} for smooth surfaces. The values of q_{cl}, are specified by the coating characteristics. In the region of small h the critical heat flux q_{cl} increases with the growth of h up to $\xi h \sim \frac{1}{2} L$, and declines with the further growth of the coating thickness (here ξ is tortuosity).

The character of the development of the crisis changes appreciably; it becomes "softer" & the time of its development drastically increases once $q = q_{cl}$. The rate of temperature growth on the heated surface under the coating T_w at $q = q_{cl}$, according to the results [7,8], declines, as compared to the smooth surface more than 10 times. This increase in the heat transfer stability under crisis heat loads may be fairly useful for many elements of radio-electronic equipment.

The film boiling regime on surfaces with porous coatings was studied in experiments [6-8] involving boiling helium for sufficiently thin coatings. According to these data the thin coatings actually do not affect the heat transfer coefficient $\alpha(\theta)$ in the film regime, but superheatings θ, under which this regime can be realized, drastically grow. It is naturally expected that in case of coatings with a thickness appreciably exceeding that of the vapor film at film boiling on smooth surfaces, the heat transfer coefficient α for the film boiling on surfaces with porous coating will be smaller than on surfaces without the coating, yet, this problem has not been sufficiently studied to date.

7. THE EFFECT OF POROUS COATING CHARACTERISTICS ON HEAT TRANSFER IN BOILING

The results of measurements [1-9] show that heat transfer in boiling on surfaces with porous coatings is defined not only by integral parameters of the coatings (thermal conductivity, permeability, porosity, distribution of pores as to size, thickness, etc.), but also the characteristics of elementary physical volume of the porous structure (diameters of particles forming the coating, type and coordination number of pore lattice, etc.).

These characteristics are interconnected, yet, the character of their relations is not universal and depends on the type of coating, the technology of application and cannot be unambiguously ascertained theoretically. Insignificant, practically uncontrolled deviations in the technology of coating application can bring about marked differences in heat transfer coefficients in boiling, particularly in regime I [9]. Under these conditions in order to generalize the experimental data it is rational to reduce them to a dimensionless form using the parameters of characteristic points on the boiling curves with subsequent approximation by the power law.

Characteristic points for regime I are the point of the onset (or the end) of boiling and the point of the initiation of transition I-II. Using the conventional heat flux as the sum of "convective" $q_{conv}(\theta)$ and "vaporizing" $q_v(\theta)$ components, the power law of heat transfer in dimensionless form can be written as follows:

$$\hat{q} = \hat{\theta}^{\,n} \qquad (7)$$

where $\hat{q} = \frac{q - q_{conv}(\theta)}{q^* - q_{conv}(\theta^*)}$ and $\hat{\theta} = \frac{\theta - \theta_{o.b.}}{\theta^* - \theta_{o.b.}}$ are the dimensionless heat flux and superheating of the surface below the coating, n is the index of a power determined by using the experimental data. When approximating the experimental data by the relationship (7) the component $q_{conv}(\theta)$ was determined by extrapolating the convective branch of the boiling curves to the region of considerable superheating. Using the relationship (7) one can approximate the entire bulk of experimental data in the range of $I \geq \hat{q} \geq 0.3$ with an error less than 20% with the value of $n = 1.35$ [9] (Cf. Fig. 11).

Note that the values of $\theta_{o.b.}$ and θ^* can be determined independently practically within experimental accuracy from Eqs. (4) and (6) if there is known the distribution of pores as to size, q^* - from the independent experiments on the filtration of gas through the wetted samples of coatings, and q_{conv} - from the data on thermal conductivity of coatings. This makes it possible to build the diagram for calculating the heat transfer in regime I using a limited amount of experimental data on the properties of porous structures forming the coating and to optimize the coating characteristics [13]. It is quite obvious that with all other identical parameters the thickness of coating corresponding to the maximum values of α when boiling in regime I is equal to $h_\alpha \simeq L_f/\xi$, where ξ is tortuosity. At the same time, $h_c \simeq \frac{3}{2} L_f \frac{I}{\xi}$ i.e., $h_c \sim \frac{3}{2} h_\alpha$ corresponds to the maximum values of α when boiling in regime II and to the maximum q_{cl}. Thus, generally speaking, depending on the problem to be solved, the optimal characteristics of coatings are changed. Taking into account the fact that in case of spherical coatings the calculations in the model of branching pores yield $L_f \simeq (4\text{-}6)\,D$, and the values of tortuosity $\xi \sim \sqrt{2}$, the thickness of isotropic spherical low-conductivity coatings ensuring an appreciable increase of both α and q_{cl} should be $h \simeq (3\text{-}5)\,D$.

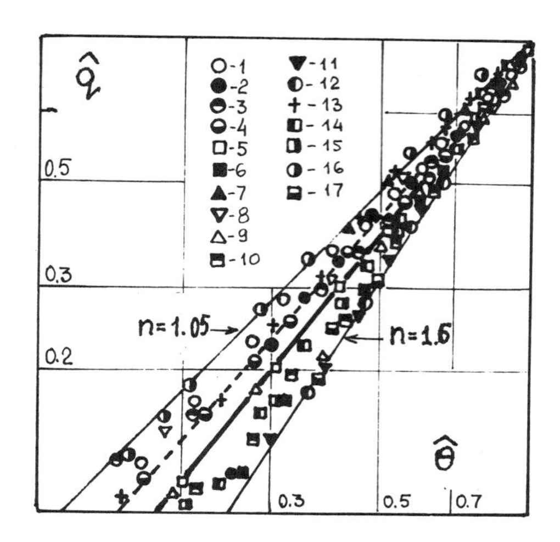

FIGURE 11. The generalization of experimental data on boiling in regime I on surfaces with isotropic porous coatings: 1-12 -- IVTAN data for water and ethanol-water solutions on surfaces with sprayed and sintered coatings of different thickness and porosity at P 0.05 and 0.1 MPa, 13 - data [20] for R-113, 14 - data [21] for R-113, 15 - data [21] for water, 16 - data [22] for water, 17 - data [23] for FC-72.

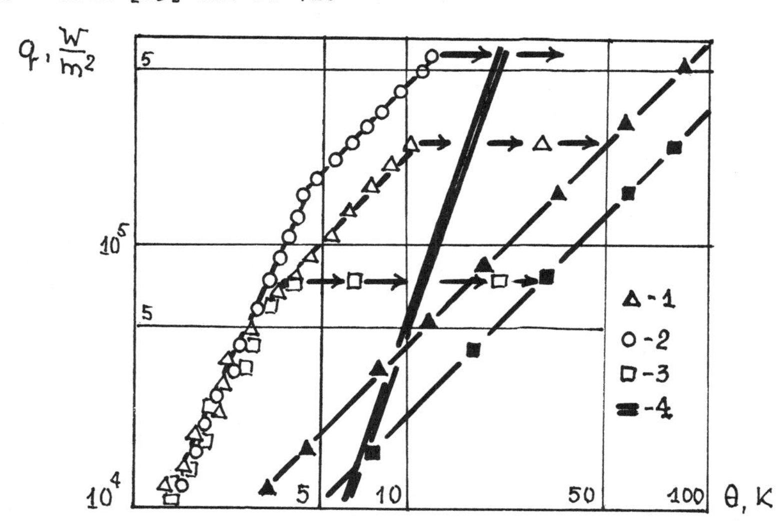

FIGURE 12. Heat transfer in water boiling at P= 0.1 MPa on surfaces with sintered porous coatings (h = 1.5 mm) 1 - coating with channels for vapor escape, 2 - two-layer anisotropic structure, 3 - isotropic coating D = 130-140 μm, 4 - smooth surface.

The values of α and q_{cl} can be further increased by using anisotropic coatings with lower hydraulic resistance to the vapor escape in the direction normal to the heating surface. This may be, e.g., two-layer structures from particles with different D or structures with vapor-escape channels (cf. Fig. 12).

The results obtained in references [1-10] show that the problem of optimization of the characteristics of low conductivity porous coatings despite its multiparameter nature, can be fairly effectively solved using the above-identified equations for $\theta_{o.b.}$, θ^* & L_f and model (or semi-empirical) equations for two-phase filtration obtained, e.g., in Refs [11, 12, 14-16].

8. PECULIARITIES OF USING POROUS COATINGS TO ENHANCE THERMAL STABILITY OF ELECTRONIC EQUIPMENT

The results of works [1-10,13] enable one to draw certain conclusions about the specifics of using porous coatings of cooled surfaces to enhance thermal stability of the elements in radioelectronic equipment.

The elements of equipment used in radioelectronic engineering vary in many basic parameters, namely, heat output and its dependencies on temperature and current, limiting temperatures at which their normal functioning is still possible, thermal conductivities of the material, etc. This explains the great variety of possible heat-transfer agents and porous coatings of cooled surfaces that effectively ensure thermal stabilization.

The general methods of analyzing thermal stability of objects cooled by a boiling heat transfer agent are fairly well developed [17]. In each particular case they can help formulate the requirements for boiling curves on cooled surfaces. At the same time, there are general peculiarities of using porous coatings for thermal stabilization of certain classes of devices.

In case of elements with a relatively low thermal conductivity and narrow range of working temperatures it is most important to ensure maximum values of $dq/d\theta$ in the bubble boiling regime, maximum values of q^*, q_{cl} and relatively low values of $\theta_{o.b.}$, θ^* and θ_{cl}, as well as a narrow temperature range of transition I-II or its absence. These requirements can be provided by way of using porous coatings with high thermal
skeleton, thickness of $h \simeq L_f$ and anisotropic structures having 2 maximums on the curve of pore distribution as to size.

For a number of applications it is very important to use porous coatings with the minimal hysteresis of incipient boiling. In this case, it is necessary to rule out the principal mechanisms bringing about the hysteresis of incipient boiling. First and foremost, one should ensure the absence of thermocapillary convection of the heat transfer agent in the porous structure directed along the temperature gradient [5,10]. The heat transfer agent should be free of surface-inactive nonvolatile and surface-active volatile impurities.

If for technical conditions of the element operation it is necessary to decrease or rule out the incipient boiling hysteresis, then, the use of freons and alcohols, for example, as coolants is undesirable because the above-mentioned impurities are usually present therein, which may lead to great incipient boiling hysteresis. The use of various types of anisotropic coatings also brings about a decrease in incipient boiling hysteresis.

In order to enhance thermal stability of elements with high heat conductivity and fairly broad ranges of working temperatures it is expedient that porous coatings be used which lead not only to an increase of α in the bubble boiling regime and an increase of q_{c1}, but also to an increase of q_{c2} and displacement of θ_{c1} and θ_{c2} to the region of high superheatings of cooled surfaces. Low-conductivity porous coatings can serve as such coatings.

In a one-dimensional case (e.g., thermal stabilization of a conductor with current) the temperature distribution (i.e., θ) along the element length is described by the equation of heat conduction

$$C(T)\cdot\rho\cdot\frac{\partial\theta}{\partial\tau} = \frac{\partial}{\partial x}\lambda(\theta)\cdot\frac{\partial\theta}{\partial x} - Q(\theta) + W(\theta, I, \ldots) \qquad (8)$$

where C, ρ and λ are heat capacity, density and heat conductivity of the conductor; W (θ ,I) is heat output (generation curve), $Q(\theta) = q(\theta)\cdot\frac{P}{\Omega}$ is heat transfer capacity, q(θ) is heat emission function from the cooled surface (boiling curve) & Ω and P are the cross section area and the cooled perimeter of the element, respectively

The conditions of Eqs (1) and (2) are met in steady-state for the elements with the temperature homogeneous lengthwise. Given the heterogeneity of temperature relative to the conductor length, if one end of the element is stable with a temperature θ', and the other - with temperature θ'', under the boundary conditions at the element ends $\partial\theta/\partial x = 0$, the velocity of temperature front movement along the element v is determined by the relationship

$$v = - S(\theta', \theta'', I, \ldots) / \int_{\theta'}^{\theta''} \lambda(\theta)\cdot C(\theta)\cdot\rho\cdot\frac{\partial\theta}{\partial x}\, d\theta \qquad (9)$$

where $S(\theta', \theta'', I, \ldots) = \int_{\theta'}^{\theta''} \lambda(\theta)\cdot[Q(\theta) - W(\theta, I, \ldots)]\, d\theta$

Assuming that $\theta'' > \theta'$ and the state with temperature θ'' corresponds to $x \to \infty$, and with θ' - $x \to -\infty$, the distribution of a hot zone along the element corresponds to the values $S < 0$, and its reduction - to $S > 0$. At $S = 0$, stationary profile of temperatures exists along the element. If λ = const., the theorem of equal areas [18] is fulfiled in this state:

$$S' = \int_{\theta'}^{\theta''} [Q(\theta) - W(\theta, I, \ldots)]\, d\theta = 0 \qquad (10)$$

It follows from relationships [9,10] that in order to enhance the thermal stability of elements cooled by boiling heat transfer agent and having a sufficiently broad range of working temperatures ($\theta'' > \theta^{o}_{c2}$), it is necessary to ensure an increase of heat extraction within a broad range of temperatures $[\theta', \theta'']$ so that the integral of the boiling curve $\int_{\theta'}^{\theta''} q(\theta)\cdot d\theta$ would take the maximum values. This problem is solved most effectively by way of applying low conductivity porous coatings with a thickness of $h \simeq \frac{3}{2} \frac{Lf}{\xi}$; in this case q_{c2}, θ_{c2} and the heat transfer coefficient in the region θ appreciably grow, which correspond to film boiling on smooth surfaces.

It is experimentally shown in papers [7,8,19] that the use of low-conductivity porous coatings makes it possible to substantially enhance the cryostatic stability of composite superconductors cooled by boiling helium. Thus, it has been possible to increase the minimal current of the normal zone distribution 1.5 to 1.8 times, decrease the velocity of the normal zone distribution 10-15 times and increase the minimal energy of critical thermal disturbance by one order (cf. Fig. 13). The theory of stability of superconductors with porous coatings cooled by boiling helium is evolved in detail in Ref. [7].

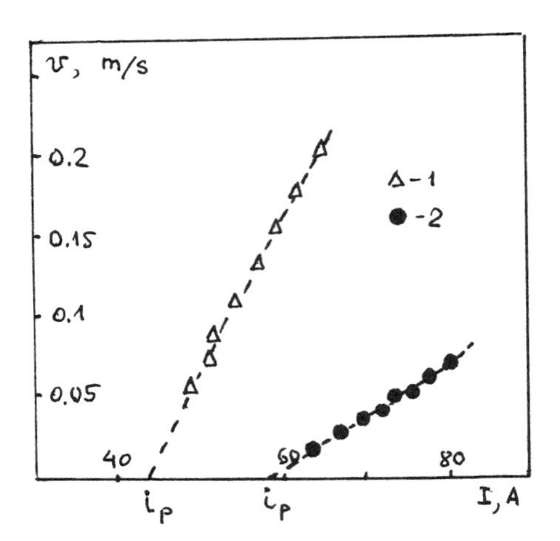

FIGURE 13. The effect of porous coatings on the rate of normal zone propagation in a combination superconductor, cooled by liquid helium (B = 4 T): 1 - conductor without coating, 2 - conductor with coating h = 50 μm.

9. CONCLUSIONS

The results of works [1-10,13,19] show that the use of low-conductivity porous coatings enables one to appreciably enhance the thermal stability of articles cooled by a boiling heat transfer agent. The coating characteristics depending on the problems being solved can be optimized on the basis of a semi-empirical percolation model developed at IVTAN using a limited set of characteristic parameters ($\theta_{o.b.}$, θ^* and L_f) computed on the basis of lattice models and the experimental data on hydraulic resistance of the wetted structures. The possibility of modifying the boiling curve by using porous coatings of various types makes their utilization a sufficiently universal step for solving a broad range of problems bearing on thermal stabilization of the elements of electronic equipment.

NOMENKLATURE

W	total heat output rate of element	W
Q	total heat transfer rate of element	W
T	temperature	K
q	heat flux density	W/m^2
$\theta = T - T_s(P)$	superheating	K
$T_s(P)$	temperature of vapor-liquid equilibrium	K
P	pressure	MPa
D	mean particle diameter	m
ε	porosity	-
k	premeability	m^2
Z	coordination number of pore lattice in porous coatings	-
h	thickness of coating	m
λ	heat conductivity	W/m^2
$\alpha = \frac{q}{\theta}$	heat transfer coefficient	$W/(m^2K)$
σ	surface tension	W/m
c	specific heat capacity	J/(K kg)
R	radius	m
ΔH	specific latent heat of evaporation	J/kg
ρ	density	kg/m^3

γ	angle of wetting	rad.
L_f	dimension of the finite cluster of vapor filled pores	-
A	dimensionality of pore lattice (A=3)	-
β	percolation threshold (critical number of percolation)	-
τ	time	s, h
ξ	tortuosity	-

REFERENCES

1. Styrikovich, M. A., Malyshenko, S. P., Andrianov, A. B., and Konovalov, S. D., Peculiarities of Boiling on Surfaces with Low-Conductivity Porous Coatings, Dokl.akad.nauk SSSR, vol.241, no. 2, p. 345, 1978.

2. Andrianov, A. B., Malyshenko, S. P., Sirenko, Ye. I., and Styrikovich, M. A., Hysteresis and Transition Phenomena in Boiling on Surfaces with Porous Coatings, Dokl.akad.nauk SSSR, vol.256, no. 3, p. 591, 1981.

3. Andrianov, A.B., Malyshenko, S. P., Styrikovich, M. A., and Talayev, I. V., Peculiarities of Transition Processes and the Shape of a Boiling Curve on Surfaces With Porous Coatings, Dokl.akad.nauk SSSR, vol.273, no. 4, p. 866, 1983.

4. Styrikovich, M. A., Malyshenko, S. P., Andrianov, A.B., and Talayev, I. V., Investigation of Boiling on Porous Surfaces, Heat Transfer - Soviet Research, vol.19, no. 1, Jan-Febr., pp. 23-29, 1987.

5. Malyshenko, S. P., and Andrianov, A. B., On the Initial Section of the Boiling curve on Surfaces with Porous Coatings and the Hysteresis of Boiling, Teplofiz. vys. temp., vol.25, no. 3, p. 563, 1987.

6. Andrianov, V. V., Bayev, V. P., Malyshenko, S. P., Muchnik, R. G., Peculiarities of Boiling Helium on the Surface with Porous Coatings, Dokl.akad.nauk SSSR, vol.297, no. 2, p. 354, 1987.

7. Andrianov, V. V., Bayev, V. P., Malyshenko, S. P., Muchnik, R. G., and Parizh, M. B., Enhancing Cryostatic Stabilization of Composite Superconductors Using Porous Coatings, IVTAN preprint, no. 4-227, 1987.

8. Andrianov, V. V., Bayev, V. P., Malyshenko, S. P., Muchnik, R. G., Enhancing the Efficiency of Stabilization of Combination Superconductors Using Porous Coatings, Dokl.akad. nauk SSSR, vol.293, no. 4, p. 856, 1987.

9. Andrianov, A. B., and Malyshenko, S. P., The Effect of Porous Coatings Characteristics on Heat Transfer in Boiling, Izv.akad.nauk SSSR, Energy and Transport, 1988.

10. Styrikovich, M. A., Leontjev, A. I., and Malyshenko, S. P., On the Mechanism of the Transfer of Non-volatile Admixtures in Boiling on Surfaces Covered with Porous Structure, Teplofiz.vys.temp., vol.15, no. 5, p. 998, 1976.

11. Kheifets, L. I., and Neimark, A. V., Mnogofaznye protsessy v poristykh sredakh (Multiphase Processes in Porous Media), Khimiya, Moscow, 1982.

12. Chizmadzev, Yu. A., Markin, V. S., Tarasevich, M. R., and Chirkov, Yu. G., Makrokinetika protsessov v poristykh sredakh (Macrokinetics of Processes in Porous Media), Nauka, Moscow, 1971.

13. Andrianov, A. B., Peculiarities of Thermal Processes in Boiling on Surfaces with Porous Coatings, Author's abstract, IVTAN, Moscow, 1987.

14. Kadet, V. V., and Selyakov, V. I., Percolation Model of a Two-Phase Filtration, Mekhanika zhidkosti i gaza, no. 1, p. 88, 1987.

15. Barenblatt, G. I., Entov, V. M., and Ryzhik, V. M., Dvizhenie zhidkostei i gazov v prirodnykh plastakh (The Motion of Liquids and Gas in Natural Beds), Nedra, Moscow, 1984.

16. Larson, R. G., Scriven, L. E., and Davis, H.T., Percolation Theory of Two-Phase Flow in Porous Media. Chemical Eng. Sci., vol.36, pp. 57-73, 1981.

17. Petukhov, B. S., Genin, L. G., and Kovalyov, S. A., Teploobmen v Yadernykh energeticheskikh ustanovkakh (Heat Transfer in Nuclear Energy Installations), Energoatomizdat, Moscow, 1986.

18. Maddock, B. J., James, G. B., and Norris, W. T., Superconductive Composites: Heat Transfer and Steady-State Stabilization, Cryogenics, vol.9, pp. 261-273, 1969.

19. Andrianov, V. V., Bayev, V. P., Malyshenko, S. P., and Muchnik, R. G., The Effect of Porous Coating on Characteristics of Partially Stabilized Composite Superconductors Cooled by Boiling Helium, Cryoprague-86, pp. 87-89, Prague, 1986.

20. Tekhver, Ya. Kh., and Sui, Kh. N., The Effect of Porous Coating Parameters on Hysteresis Phenomena in Boiling. Izv. akad. nauk Estonian SSR, Physics and Mathematics, vol.34, no. 4, pp. 413-418, 1985.

21. Bergles, A. E., and Chyu, M. C., Characteristics of Nucleate Pool Boiling from Porous Metallic Coatings. J. Heat Transfer, vol.104, pp. 279-285, 1982.

22. Abhat, A., and Seban, R., Boiling and Evaporation from Heat Pipe Wicks with Water and Acetone, J. Heat Transfer, vol.96, pp. 331-337, 1974

23. Marto, P. J., and Lepere, V. J., Pool Boiling Heat Transfer from Enhanced Surfaces to Dielectric Fluids, J. Heat Transfer, vol.104, pp. 292-299, 1982.

CONDUCTION ASPECTS

Fundamentals of Thermal Constriction (Spreading) Resistance for Electronic Cooling

M. MICHAEL YOVANOVICH
Microelectronics Heat Transfer Laboratory
University of Waterloo
Waterloo, Ontario, Canada N2L 3G1

ABSTRACT

A novel analytical-numerical surface element method is proposed for computing the centroidal and the average temperature rise of single and multiple contact areas subjected to uniform or distributed heat flux. Certain geometric parameters for an arbitrary surface element are developed using analytical solutions for right triangles. The accuracy and efficiency of the method is demonstrated by the computation of the centroidal temperature rise of single arbitrary areas. Geometric parameters are also developed for two interacting contact areas of arbitrary shape. These parameters are used to determine the constriction resistance of a set of identical contact spots. It is demonstrated that a continuous square contact area can be replaced by a finite set of identical contact spots.

INTRODUCTION

In a wide range of fields, from the microelectronics to the nuclear industry, the so-called thermal contact resistance needs to be determined to evaluate the overall thermal performance of systems. This contact or constriction resistance is defined as the average temperature rise of the contact area of the thermal source divided by the total heat flow rate from the source.
Specifically one may encounter single or multiple thermal contacts of simple or arbitrary geometries on the surface of a half-space. The flux distribution over the contacts may be uniform in one extreme or give a uniform contact temperature in the other extreme. This paper outlines how all of the above mentioned problems can be resolved conveniently and economically by a common approach.

Single Contact: Arbitrary Shape

Consider a single planar contact of arbitrary shape located on the surface of a semi-infinite half space as shown in Fig. 1. The governing differential equation is

$$\nabla^2 T = 0 \tag{1}$$

subject to the boundary conditions

$$-\lambda \frac{\partial T}{\partial z} = q \,, \quad \text{uniform flux within contact region} \tag{2}$$

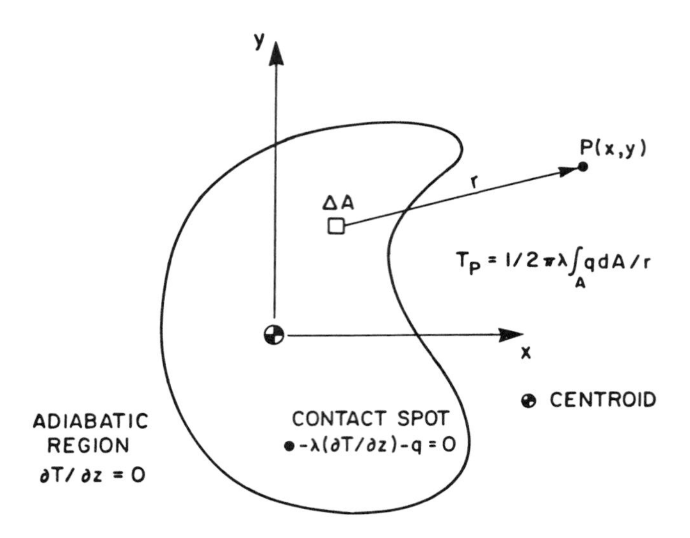

FIGURE 1. Schematic of Arbitrary Contact Area.

$$\frac{\partial T}{\partial z} = 0 \, , \qquad \text{insulated outside contact region} \tag{3}$$

$$\text{and} \quad T \to 0 \quad \text{as} \quad \sqrt{x^2 + y^2 + z^2} \to \infty \tag{4}$$

Superposition of the solution for a point source subject to (1), (2), (3) and (4) gives the temperature rise at any point P on the plane $z = 0$ as

$$T = \int_A \left\{ \frac{q}{2\pi\lambda r} \right\} dA \tag{5}$$

where r denotes the distance measured from P shown in Fig. 1.

For arbitrarily shaped contact areas, closed form solutions of Eq. (5) generally cannot be obtained. Nevertheless, if P lies outside the contact area, the temperature rise at P may be calculated accurately by a conventional numerical integration scheme. However this is not possible if P is a point within the contact area, since at this point $r = 0$ and the integrand is singular. This difficulty may be avoided by considering separately the inner and outer discretized triangular areas shown in Fig. 2. The inner triangular areas have a common apex at P and the temperature rise there due to each of these is given exactly by [1]

$$T_{IN} = \frac{q\delta}{2\pi\lambda} \ln \left\{ \frac{\tan\left[\frac{\pi}{4} + \frac{1}{2}\tan^{-1}\left(\frac{\overline{AC}}{\delta}\right)\right]}{\tan\left[\frac{\pi}{4} + \frac{1}{2}\tan^{-1}\left(\frac{\overline{BC}}{\delta}\right)\right]} \right\} \tag{6}$$

where the distances $\overline{AC}, \overline{BC}$ and δ are shown in Fig. 3.

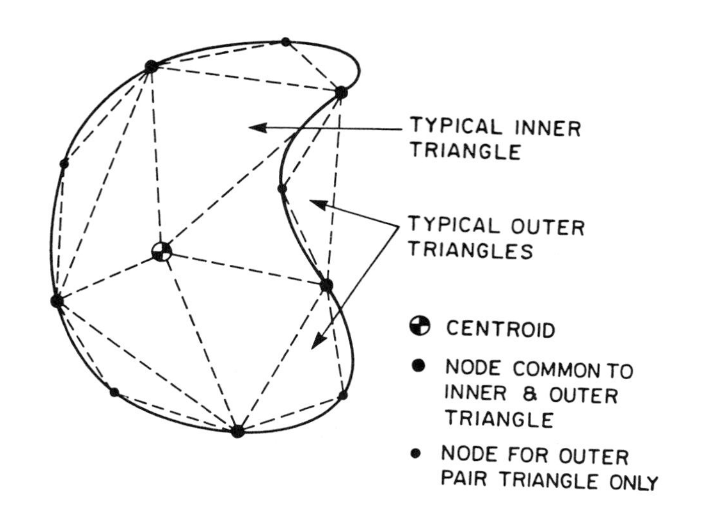

FIGURE 2. Discretization of Arbitrary Contact Area.

The outer triangles, each of which has one common side with an inner triangle, approximate the portion of the contact area between its perimeter and the inner triangles. For each outer triangle the temperature rise at P is given exactly by the infinite series (see Appendix).

$$T_{OUT} = \frac{q}{2\pi\lambda}\left\{\frac{A}{R} + \frac{2I_0 - 3I_{RR}}{2R^3} + \ldots\right\} \tag{7}$$

where A, R, I_0 and I_{RR} denote, respectively, the area, the distance from P to the centroid of the outer triangle, the polar second moment of area and the second moment of area with respect to the line of length R, all as shown in Fig. 4.

Summing the contributions from each triangle gives the required temperature rise at P. Thus

$$T_P = \sum T_{IN}^i + \sum T_{OUT}^i \tag{8}$$

the superscript denoting the contribution from the ith inner and outer regions.
The convenience, efficiency and accuracy of this technique for computations of the temperature rises and thermal constriction resistances will now be examined for the family of areas within the hyperellipses

$$\left(\frac{x}{a}\right)^n + \left(\frac{y}{b}\right)^n = 1$$

and for a semi-circle.
The boundary points by which the contact area was discretized into inner and outer triangles were positioned such that each inner triangle had the same angle at the centroid.

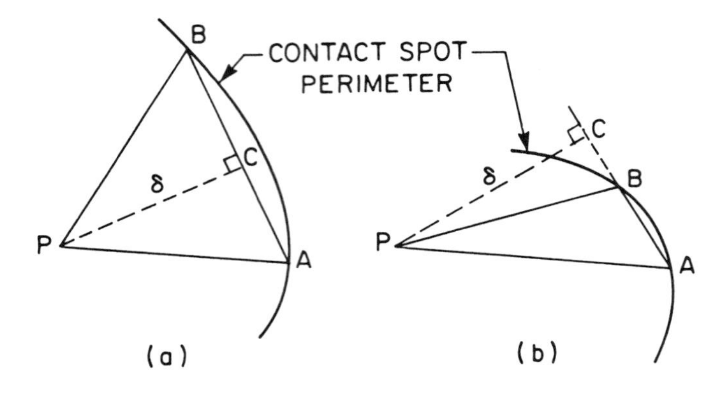

FIGURE 3. Inner Triangle Geometric Parameters.

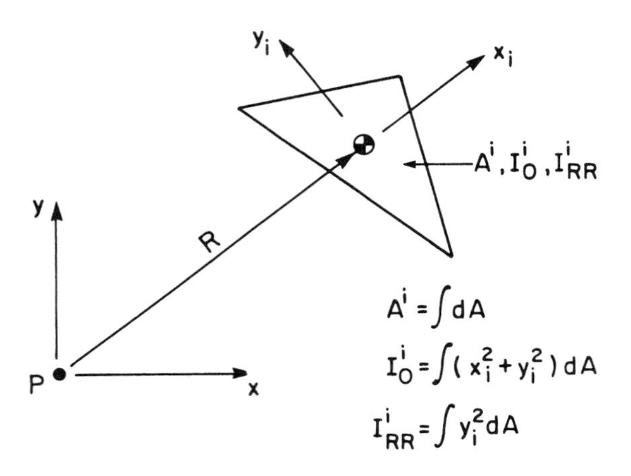

FIGURE 4. Outer Triangle Geometric Parameters.

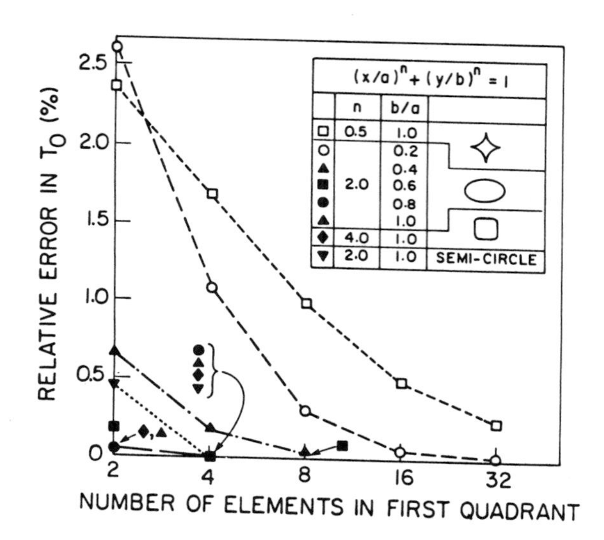

FIGURE 5. Effect of Discretization on Relative Error of Centroidal Temperature.

Symmetry permitted consideration of just the first quadrant for each shape. Fig. 5 shows the relative error of the centroidal temperature (or centroidal constriction resistance) versus the number of discretized inner-outer triangular elements.

For all shapes considered, the error decreases rapidly as the number of elements increases giving a more accurate approximation of the true contact area. With only four elements the maximum error was generally much less than 0.2%, except for the hyperellipse with $n = 1/2$ and the elongated ellipse, (shapes for which the contact area is poorly approximated with less than 8 elements).

In summary, with this method it is very simple to compute precisely the centroidal temperature rise for virtually any arbitrarily shaped contact area subjected to uniform flux.

Multiple Contacts: Arbitrary Shapes

Thermal contact resistance problems frequently involve multiple contacts located close to one another on the surface of a half-space. The temperature rise at the centroid of any one of the contacts and/or the total constriction resistance of the set are often needed.

Centroidal Temperatures

For convenience, the centroidal temperature rise T_i of the ith contact is written as the sum of the contribution from the flux on the ith contact, plus the contributions from all other contacts,

$$T_i = T_{ii} + \sum_{j=1, j \neq i}^{N} T_{ij} \tag{9}$$

The first of these, T_{ii}, can be calculated by the technique given in the previous section. Each T_{ij} is given by

$$T_{ij} = \frac{q_j}{2\pi\lambda}\left\{\frac{A^j}{R_{ij}} + \frac{2I_0^j - 3I_{RR}^j}{R_{ij}^3} + \dots\right\} \tag{10}$$

where R_{ij} is the distance between the centroids of the ith and jth contacts, while A^j, I_0^j and I_{RR}^j denote, respectively, the area, second polar moment of area and second linear moment of areas for the jth contact.

Constriction (Spreading) Resistance

In order to compute the total thermal constriction resistance, it is necessary to deal with average temperatures. As before the self and mutual effects are separated giving

$$\overline{T}_i = \overline{T}_{ii} + \sum_{j=1, j\neq i}^{N} \overline{T}_{ij} \tag{11}$$

where the overbar denotes an average value.
For thermal constriction resistance problems, individual contact spots would usually be modelled as circles or ellipses. For these shapes the $\overline{T}_{ii}$ have been determined [2] and no further computation is necessary. For future computations, it will be convenient to have $\overline{T}_{ii}$ expressed in the form

$$\overline{T}_{ii} = \frac{q_i}{\lambda} g_{ii} \tag{12}$$

As shown in Appendix I, $\overline{T}_{ij}$ is given exactly by

$$\overline{T}_{ij} = \frac{q_i}{\lambda} g_{ij} \tag{13}$$

where

$$\begin{aligned} g_{ij} &= \frac{1}{2\pi}\left\{\frac{A^j}{R_{ij}} + \frac{2I_0^j - 3I_{RR}^j}{2R_{ij}^3} + \left(\frac{A^j}{A^i}\right)\left(\frac{2I_0^i - 3I_{RR}^i}{2R_{ij}^3}\right)\right. \\ &+ \left.\left(\frac{12I_0^i - 15I_{RR}^i}{2R_{ij}^5}\right)\left(\frac{2I_0^j - 3I_{RR}^j}{2A^i}\right) + \dots\right\} \end{aligned} \tag{14}$$

with $A^j, A^i, I_0^j, I_0^i, I_{RR}^j, I_{RR}^i$ and R_{ij} defined as previously and shown in Fig. 6. Note that g_{ij} depends only on geometry.
For the full set of contacts, Eqs. (11), (12) and (13) lead to a system of linear algebraic equations of the form

$$\frac{1}{\lambda}[G_{ij}]\{q_j\} = \left\{\overline{T}_i\right\} \tag{15}$$

where $[G_{ij}]$ represents a matrix composed of the geometry dependent g_{ii} and g_{ij} terms defined above.
Note that Eq. (15), which was derived for the case of uniform flux on each contact spot, may also be used for isothermal or mixed boundary conditions if q_j is simply interpreted as the average flux on the jth spot.
For any of the above boundary conditions, the total constriction resistance of the system of contact spots is given by

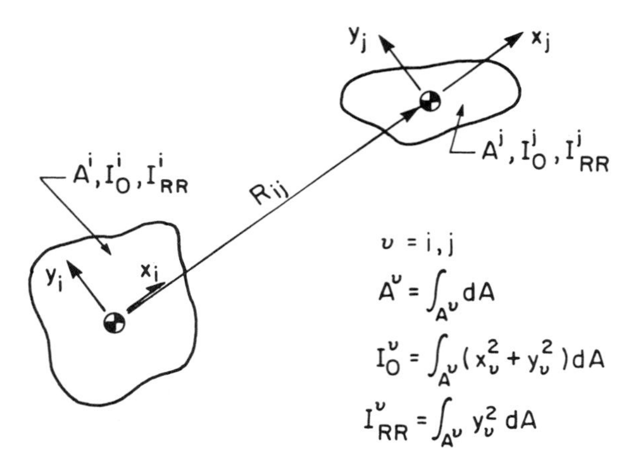

FIGURE 6. Geometric Parameters for Two Arbitrary Contact Areas.

$$\overline{R} = \frac{\overline{T}}{Q} = \frac{\frac{\sum \overline{T}_i A^i}{\sum A^i}}{\sum q_i A^i} \tag{16}$$

the summation taken over all spots.

Constriction Resistance of an Arbitrarily Shaped Contact

Consider a singly connected contact area such as shown in Fig. 7(a), and the system of discrete circular contacts shown in Fig. 7(b) which cover the same apparent area as the single contact. For a single arbitrary shape with arbitrary boundary conditions, the constriction resistance would be difficult to compute by conventional techniques. However, it is reasonable to anticipate that the difference in constriction resistance between the single contact and the discrete system will tend to zero as the area covered by the latter system increases towards that of the single contact. For the system of discrete contacts, calculation of the constriction resistance involves only a straightforward systematic computation by the technique described in the previous section.

For an example of this method, consider a square contact area as shown in Fig. 8. The apparent area has been discretized by circular contacts and the total area of the circular contacts is increased by uniformly increasing the number or density of the contacts. Table 1 contains the ratio of the resistance of the discrete system, $\overline{R}^*_{CALC}$, calculated by (15) and (16), to the theoretical resistance, $\overline{R}^*_{THEO} = 0.4732$ [2] as the number of contacts, N, and their percentage of the total apparent area, $A\%$, increases. In addition, a simple attempt to extrapolate the calculated $\overline{R}^*_{CALC}$ was made by fitting the latest 3 values of $\overline{R}^*_{CALC}$ to the model

$$C_0 + C_1 N^{-1} + C_2 N^{-2} = \overline{R}^*_{CALC}(N) \tag{17}$$

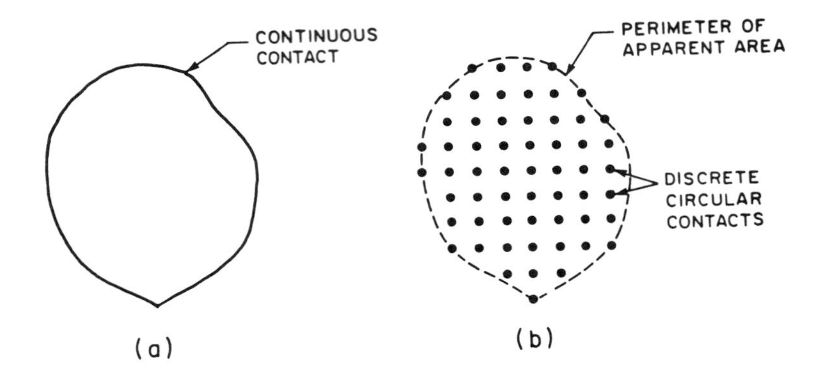

FIGURE 7. Approximation of Continuous Contact By Discrete Contact Spots.

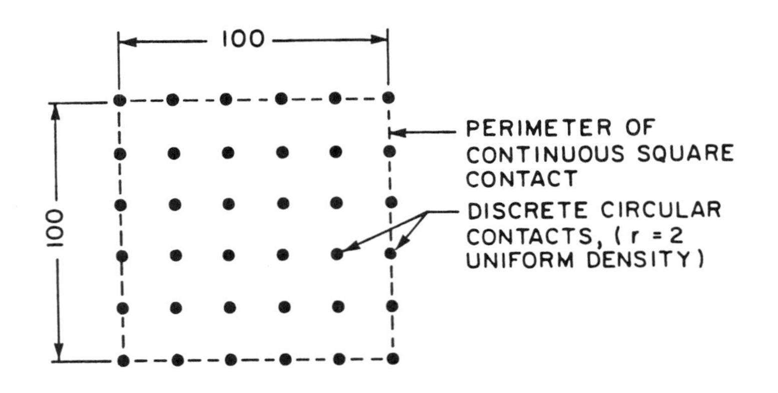

FIGURE 8. Discretization of Square Contact Area.

After determining the coefficients, an evaluation of Eq. (17) at $N = 625$ (where the contact density is such that the circles touch each other tangentially) produces the value $\overline{R}^*_{PRED}$ which is also reported in Table 1 as a ratio of $\overline{R}^*_{THEO}$.
Note from these results that with a discrete system covering approximately 1/4 of the total apparent area, the resistance of the system is only 2.45% above that of the continuous shape, and by using the simple extrapolation discussed, the resistance of the continuous shape is predicted to within 0.9%.

TABLE 1.

N	Area, %	$\overline{R}^*_{CALC}/\overline{R}^*_{THEO}$	$\overline{R}^*_{PRED}/\overline{R}^*_{THEO}$
9	1.1	3.730	
25	2.9	1.823	
49	5.7	1.344	0.903
81	9.4	1.166	0.951
121	14.0	1.086	0.971
169	19.6	1.046	0.982
225	26.1	1.025	0.991
289	33.6	1.013	0.996
361	41.9	1.006	0.999
441	51.2	1.003	1.000

CONCLUSIONS

An original analytical-numerical method has been developed for accurate and efficient computation of centroidal and average contact area temperature rise or resistance of arbitrary single or multiple areas. The method was used to determine the resistance of numerous geometries. It was also shown that the method can be used to determine the resistance of a single arbitrary area by means of a set of identical contact spots.

NOMENCLATURE

A	=	area
G_{ij}	=	matrix of geometric coefficients
g_{ij}	=	geometric coefficient
I_O	=	polar second moment of area
I_{RR}	=	radial second moment of area
N	=	number of contact spots
P	=	field point location
Q	=	total heat flow
q	=	heat flow per unit area
$\overline{R}^*$	=	non-dimensional constriction resistance
R_{ij}	=	distance between centroids of ith and jth contacts
r	=	radial coordinate
T	=	temperature rise
T_O	=	centroidal temperature rise
$\overline{T}$	=	average temperature rise
x, y, z	=	Cartesian coordinates

Greek Symbols

θ	=	polar coordinate
λ	=	thermal conductivity
π	=	pi
ρ	=	polar coordinate
ρ_0	=	distance from P to centroid of jth contact
∇^2	=	Laplacian operator

Subscripts and Superscripts

i, j	=	reference to ith and jth contacts

REFERENCES

1. Yovanovich, M.M., Thermal Constriction of Contacts on a Half-Space: Integral Formulation. Prog. in Astronautics and Aeronautics, Vol. 49, Edited by Allie M. Smith, AIAA, New York, 1976, pp. 397-418.

2. Yovanovich, M.M., Burde, S.S., and Thompson, J.C., Thermal Constriction Resistance of Arbitrary Planar Contacts with Constant Flux. Prog. in Astronautics and Aeronautics, Vol. 56, Edited by Allie M. Smith, AIAA, New York, 1977, pp. 127-139.

APPENDIX I

For the two planar contacts shown in Fig. 9, the temperature rise at any point P on the ith contact due to a uniform flux q_j on the jth contact is

$$T_p = \left(\frac{q_j}{2\pi\lambda}\right) \int_{A^j} \frac{dA^j}{\rho} \tag{18}$$

By the cosine law

$$\rho^2 = \rho_0^2 + r_j^2 - 2r_j\rho_0 \cos\theta_j \tag{19}$$

Hence

$$T_p = \left(\frac{q_j}{2\pi\lambda\rho_0}\right) \int_{A^j} \left(1 + \frac{r_j^2}{\rho_0^2} - \frac{2r_j}{\rho_0}\cos\theta_j\right)^{-1/2} dA^j \tag{20}$$

Since $r < \rho_0$, the binomial theorem may be used to expand the integrand, giving

$$T_p = \left(\frac{q_j}{2\pi\lambda\rho_0}\right) \int_{A^j} \left(1 - \frac{r_j^2}{2\rho_0^2} - \frac{r_j}{\rho_0}\cos\theta_j + \frac{3r_j^2\cos^2\theta_j}{2\rho_0^2} + \ldots\right) dA^j \tag{21}$$

Eq. (21) becomes

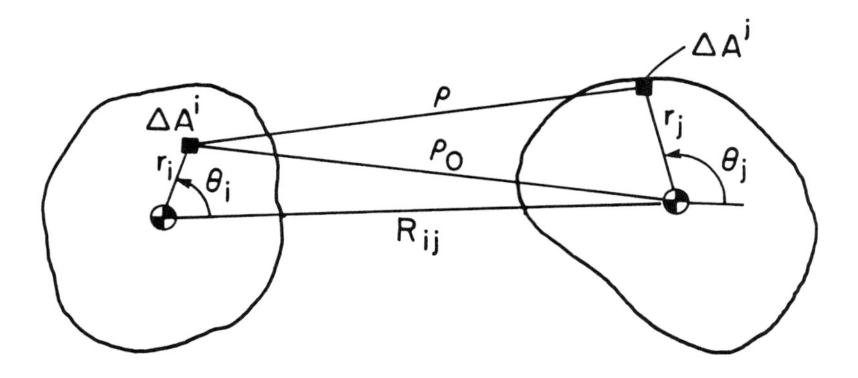

FIGURE 9. Local Polar Coordinates on ith and jth Contact Areas.

$$T_p = \left(\frac{q_j}{2\pi\lambda\rho_0}\right)\left\{\int_{A^j} dA^j + \left(\frac{1}{\rho_0}\right)\int_{A^j} r_j\cos\theta_j dA_j \right.$$
$$\left. + \int_{A^j}\left(\frac{2r_j^2 - 3r_j^2\sin^2\theta_j}{2\rho_0^2}\right) dA^j + \ldots\right\} \tag{22}$$

Since

$$\int_{A^j} dA^j = A^j \qquad \int_{A^j} r_j\cos\theta_j dA^j = 0 \tag{23}$$

$$\int_{A^j} r_j^2 dA^j = I_0^j \qquad \int_{A^j} r_j^2\sin^2\theta dA^j = I_{RR}^j \tag{24}$$

the temperature rise at P may conveniently be expressed in terms of easily computable geometric terms,

$$T_p = \left(\frac{q_j}{2\pi\lambda}\right)\left[\frac{A^j}{\rho_0} + \frac{2I_0^j - 3I_{RR}^j}{2\rho_0^3} + \ldots\right] \tag{25}$$

The average temperature rise on the ith contact is

$$\overline{T}_{ij} = \frac{1}{A^i}\int_{A^i} T_p dA^i \tag{26}$$

Thus

$$\overline{T}_{ij} = \left(\frac{1}{A^i}\right)\left(\frac{q_j}{2\pi\lambda}\right)\int_{A^i}\left[\frac{A_j}{\rho_0} + \frac{2I_0^j - 3I_{RR}^j}{2\rho_0^3} + \ldots\right] dA^i \tag{27}$$

The A^j and I_0^j are constant. In this derivation it is assumed that I_{RR}^j can be treated as a constant evaluated about the axis joining the centroids of the ith and jth contacts. Then for the terms of the infinite series as shown in (27), evaluation of $\overline{T}_{ij}$ requires the integrals,

$$\int_{A^i} \frac{dA^i}{\rho_0} \quad \text{and} \quad \int_{A^i} \frac{dA^i}{\rho_0^3}$$

These integrals are evaluated by making the substitution

$$\rho_0^2 = R_{ij}^2 + r_i^2 - 2r_i R_{ij} \cos\theta_i \tag{28}$$

expanding by the binomial theorem, and integrating as for (23) and (24), giving

$$\int_{A^i} \frac{dA^i}{\rho_0} = \frac{A^i}{R_{ij}} + \frac{2I_0^i - 3I_{RR}^i}{2R_{ij}^3} \qquad \int_{A^i} \frac{dA^i}{\rho_0^3} = \frac{A^i}{R_{ij}^3} + \frac{12I_0^i - 15I_{RR}^i}{2R_{ij}^5} \tag{29}$$

Substitution of (29) into (27) gives the average temperature in terms of readily computable quantities,

$$\begin{aligned} \overline{T}_{ij} &= \left(\frac{q_j}{2\pi\lambda}\right)\left\{\frac{A^j}{R_{ij}} + \frac{2I_0^j - 3I_{RR}^j}{2R_{ij}^3} + \left(\frac{A^j}{A^i}\right)\left(\frac{2I_0^i - 3I_{RR}^i}{2R_{ij}^3}\right)\right. \\ &+ \left.\left(\frac{2I_0^i - 15I_{RR}^i}{2R_{ij}^5}\right)\left(\frac{2I_0^j - 3I_{RR}^j}{2A^i}\right) + \ldots\right\} \end{aligned} \tag{30}$$

Thermal Contact Resistance in Electronic Equipment

VINCENT W. ANTONETTI*
International Business Machines Corporation
Poughkeepsie, New York, USA

ABSTRACT

This work describes the state-of-the-art of thermal contact resistance as related to electronic equipment. The focus is on three areas: progress made in the development of predictive theory, recent experimental studies, and examples of thermal interface management in current electronic equipment.

INTRODUCTION

The purpose of this work is to review recent progress made in understanding and solving the thermal contact resistance (TCR) problems encountered in electronic packages. Because recent advances in controlling the interfacial resistances in electronic packages are based, in large measure, on the body of knowledge developed by researchers in the last 30 years, it is appropriate to begin by citing some of the important past reviews that have appeared in the open literature.

One of the earliest contact heat transfer reviews was done by Williams [1] who covered the period from 1950 to 1966. More recently, Madhusudana and Fletcher [2] published a general review covering 1970 to 1980, and in 1984, Antonetti and Yovanovich [3] reported on TCR as specifically related to microelectronic equipment. It is fair to state that all the research to date has lead to an understanding of the basic mechanisms of contact heat transfer, and the collection of some worthwhile experimental data. Research in related fields such as electrical contacts, tribology, and solid mechanics has also lead to significant contributions in the areas of surface description and deformation. Several partially successful TCR correlations have been reported, notably in the area of flat, rough, conforming surfaces; but a general correlation, one covering waviness, directional effects, loading hysteresis, bolted joints, surface films, etc., has not been developed. This is not surprising in view of the large number of parameters involved.

In the sections that follow, three areas are reviewed. The first covers progress made in the development of predictive theory, the second covers recent experimental studies, and the last covers examples of thermal interface management in current electronic equipment.

---

*Currently at Manhattan College, Riverdale, New York, USA.

PROGRESS IN PREDICTIVE THEORY

As a rule, when two surfaces are pressed together the contact is imperfect and the real heat transfer area of the joint is only a small fraction of the apparent area. Heat transfer is by conduction across both the microscopic contacting solid spots and any interstitial fluid that may be present in the gap. Most often, the heat flow rate across the joint is assumed to consist of two independent parallel paths, and the total joint conductance is expressed as

$$h_j = h_s + h_g \tag{1}$$

where h_s is the conductance across the contacting solid spots and h_g is the conductance through the interstitial fluid in the gap.

Heat Transfer Across the Contacting Solid Spots

A majority of the available theory for heat transfer across the contacting solid spots at an interface is for flat, rough, conforming surfaces, and without going into detail that is what will be reviewed here. The usual approach is to divide the problem into three portions: metrological, mechanical, and thermal. In the metrological portion, the surface texture of the mating parts is determined by analyzing profilometer traces of the contacting surfaces. In the mechanical portion, the microhardness of the contacting surfaces is ascertained by measurement or from reference tables. Knowing the surface texture and microhardness allows the mean contact spot radius to be calculated. Then, an estimate of the real area of contact and more importantly the number, N, of discrete microcontacts over the projected, or apparent, contact area is obtained from a force balance at the interface. Finally, in the thermal portion of the analysis, the interfacial joint is modeled as N flux tubes arranged in parallel. The joint resistance across the solid microcontacts is determined by solving Laplace's equation for a discrete circular contact, located on the end of a concentric circular flux tube of semi-infinite length, and then adding the N resistances in parallel. (The conductance is of course the inverse of the resistance.) The relatively complex results of the preceding theoretical approach have been put in the form of simplified correlations by a number of investigators.

The conductance across the solid contacts of abutting surfaces is, therefore, frequently calculated using an available correlation, the most often employed being that of Yovanovich [4]

$$h_s = 1.25\, k_s \left(\frac{m}{\sigma}\right)\left(\frac{P}{H}\right)^{0.95} \tag{2}$$

where P is the apparent contact pressure, H is the microhardness of the softer metal, and when the abutting surfaces are given the subscripts 1 and 2, $\sigma = \sqrt{\sigma_1^2 + \sigma_2^2}$ is the RMS surface roughness, $m = \sqrt{m_1^2 + m_2^2}$ is the absolute average asperity slope, and $k_s = 2k_1k_2/(k_1 + k_2)$ is the harmonic mean thermal conductivity of the contacting materials.

It should be noted that most worked engineering surfaces exhibit a microhardness variation. Experiments have shown that the bulk material microhardness is appropriate for the interior of a specimen, but due to work hardening the microhardness is often much higher at the contacting surface of the part - and it is the surface microhardness which must be used in equation (2). The background needed to incorporate this important refinement into the analysis can be found in [5].

In addition, some recent limited data for a very smooth aluminum test section against a silicon chip [6] suggests that equation (2) is reasonably accurate over a much broader range of conditions, including those typically found in electronic equipment applications, than previously assumed. This is discussed in more detail in a later section of this paper.

Heat Transfer Across the Interstitial Fluid in the Gap

The simplest thermal model for the gap considers heat transfer as occurring across a fluid contained between two isothermal plates separated by an effective gap thickness $\delta = Y + M$. The gap conductance is then determined from

$$h_g = \frac{k_g}{Y + M} \tag{3}$$

where k_g is the thermal conductivity of the interstitial fluid, and Y, is the separation distance between the mean planes of the contacting surfaces which can be approximated from [7]

$$Y = 1.53\,\sigma \left(\frac{P}{H} \right)^{-0.097} \tag{4}$$

The gas parameter, M, is zero for liquids and greases, but for a gas, M accounts for the so-called "temperature-jump" effect, and is a complex function of the thermophysical properties of the interstitial gas in combination with the materials at the interface. The value of M is usually significant when the interstitial gas is at low pressure, and also for very smooth contacting surfaces in conjunction with a gas such as helium. For example, a typical value for very smooth aluminum ($\sigma \simeq 0.25\ \mu m$) and helium is $M \simeq 3.5\ \mu m$; whereas for smooth aluminum and air $M \simeq 0.3\ \mu m$ [6,8].

The gas parameter, M, can be evaluated from

$$M = \alpha\beta\Lambda \tag{5}$$

where the overall thermal accommodation coefficient, α, the parameter, β , and the mean free path, Λ, are determined from

$$\alpha = \frac{(2 - \alpha_1)}{\alpha_1} + \frac{(2 - \alpha_2)}{\alpha_2} \tag{6}$$

where α_1 and α_2 are the accommodation coefficients for gas-surface 1 and gas-surface 2 combinations, respectively, and the parameter β is defined as

$$\beta = \frac{2\gamma}{(\gamma + 1)} \, \frac{1}{Pr} \tag{7}$$

where γ is the specific heat ratio and Pr is the Prandtl Number of the gas.

$$\Lambda = \Lambda_0 \left(\frac{T}{T_0} \right) \left(\frac{P_0}{P} \right) \tag{8}$$

where the zero subscript denotes the reference state.

The determination of the accommodation coefficient has been a problem for real engineering surface-gas combinations. (By real engineering surfaces is meant those covered with layers of adsorbed gases.) Recently, however, Song and Yovanovich [9] developed the following thermal accommodation coefficient correlation which they recommend for real engineering surfaces.

$$\alpha = \exp\left[C_0\left(\frac{T_s - T_0}{T_0}\right)\right]\left(\frac{1.4M_g}{C_1 + 1.4M_g}\right) + \left\{1 - \exp\left[C_0\left(\frac{T_s - T_0}{T_0}\right)\right]\right\}\left[\frac{2.4\mu}{(1+\mu)^2}\right] \quad (9)$$

where M_g is the molecular weight of the gas, $\mu = M_g/M_s$, (M_s is the molecular weight of the solid), $C_0 = -0.57$, and $C_1 = 6.8$.

It is important to note that equations (2) through (4) were developed for ideal, nominally flat, rough, conforming surfaces. In practice, of course, many contact situations are far from this ideal condition, and involve out-of-flat surfaces, non-uniform interface pressure distribution, and so forth. Most of these situations cannot be handled with the available theory. Moreover, even under ideal conditions, the input values required to perform a theoretical calculation, or to use an available correlation may sometimes be difficult to obtain. As a consequence, more often than not an empirical approach is employed.

RECENT EXPERIMENTAL STUDIES

A review of TCR measurement techniques as applied to microelectronic equipment was done by Antonetti and Eid [8]. They classified TCR measurement approaches in the following manner.

1) Cut-Bar Technique: Used to measure TCR across laboratory-simulated mid-size interfaces (1 to 10 cm^2), where the heat flow across the interface is essentially one-dimensional.

2) Multipoint Temperature Averaging: Suitable for measuring the TCR across relatively large-scale interfaces (on the order of 100 cm^2), where a close mock-up of the actual equipment is important.

3) Small-Scale Techniques: Intended primarily for, but not limited to, very low contact pressure TCR measurements frequently occurring in conjunction with small contact area chip level applications. In this category, the authors also described the use of an integrated heat source, a simulated thermal chip, and an actual thermal chip.

It is a fact that most published experimental results are not directly applicable to the electronics industry. This is because the reported research concentrates on contact situations which are invariably too large, on idealized surface finishes, and on applied pressures that are much higher than those encountered in electronic packages. In [6], however, TCR measurements were made using the apparatus shown in Figure 1 for a 2x2 mm aluminum test section in contact with a silicon chip in air, argon, and vacuum environments. The contact pressure was varied from 27 to 500 kPa. The solid component of the interface resistance,

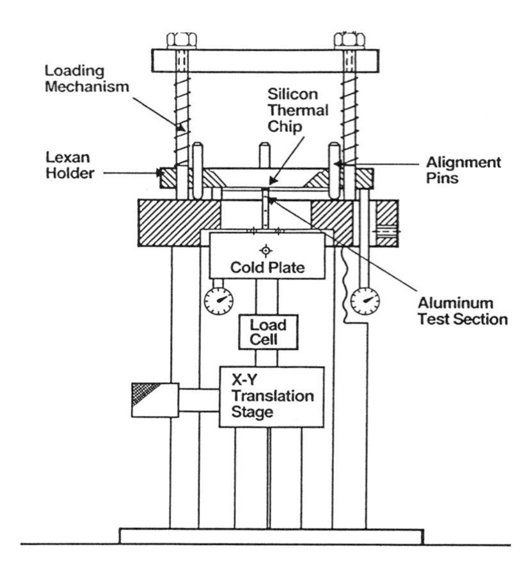

Figure 1. Schematic of Small-Scale Thermal Contact
Test Apparatus [6].

determined in vacuum tests, was found to be in close agreement with equation (2). The gap resistance in air and argon was in reasonable agreement with the predicted value from equation (3). The TCR data from this work are noteworthy for several reasons. First, the test surfaces were typical of those found in an actual application, i.e., the aluminum was ground and the silicon was polished, rather than the idealized bead-blasted surfaces normally reported. Second, the test specimen sizes, and hence the apparent contact areas were much smaller than any previously reported in the open literature. Third, the applied contact pressure ranged to a much lower level than any previously reported. And most remarkable was that the test data agreed with correlations (equations 2 and 3) which were not formulated for the imposed test conditions.

Another small-scale TCR experiment was reported by Peterson and Fletcher [10]. They conducted an experimental study to determine the TCR at the bonded joint between a silicon chip and an aluminum substrate. Seven conductive epoxies with thermal conductivities ranging from 0.27 W/mK to 1.93 W/mK were evaluated. The results of the investigation indicated that the TCR occurring at chip/bond and bond/aluminum interfaces was significant and increased dramatically as the thermal conductivity of the die-bond material decreased. Using the results of their experiments the authors developed an empirical expression for the TCR as a function of the thickness and thermal conductivity of the bonding material, and any void fraction present in the bonded joint.

Schwinkendorf and Moss [11] reported on the TCR of aluminum and alumina discs bonded with silicone resins filled with particles of either aluminum oxide, magnesium oxide, or boron nitride. The particle-loaded resins displayed a TCR which the authors attributed to incomplete wetting of the disc surfaces. In another paper concerning bonded joints, Hultmark et al. [12] described the use of a silicone rubber adhesive to attach the ceramic cap of a multichip module to an aluminum heat sink. This design resulted in a cost savings over the previously used soldering technique, with only a minimal increase in TCR.

Several approaches can be used to enhance contact conductance. Feldman, Hong, and Marjon [13] determined the thermal conductivity of a variety of thermal greases. In [12], a key feature in the thermal design was a paste with a thermal conductivity of 1.25 W/mK, which was used between the chip and module cap. In another development, metallic coatings were reported to be a very effective interface enhancement technique in [7].

Figure 2 depicts the elastic "soft-touch" contacts studied by Nakayama, et al, [14], which might be used between a chip and heat sink. Physical models were developed with constants determined by experiment. Among the schemes studied, a brush contactor made of fine copper wires and a design consisting of liquid gallium encapsulated in a film membrane, under contact pressures from 0.1 to 1.0 kg/cm^2, yielded conductances of 0.3 to 1.2 W/cm^2K. It was also noted that the TCR portion was a significant part of the total thermal resistance.

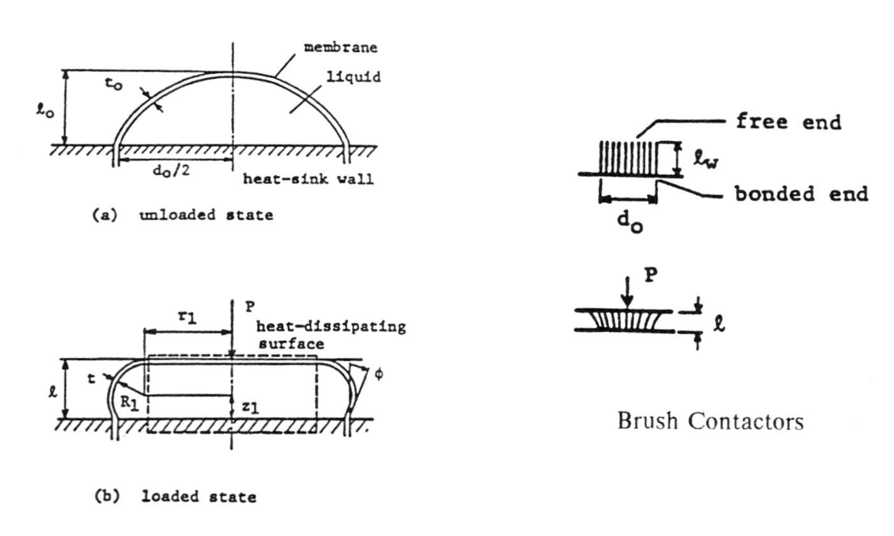

Figure 2. Elastic "Soft-Touch" Contacts [14].

Paal and Pease [15] studied chips attached to a substrate containing micro-capillary channels filled with a silicone oil (Figure 3). The paper describes the fabrication method used to make the channels. An average thermal contact conductance of 21.7 W/cm^2K was measured.

In avionic military systems, an important packaging approach utilizes electronic modules that are plugged into an air-cooled plenum. A continuing problem

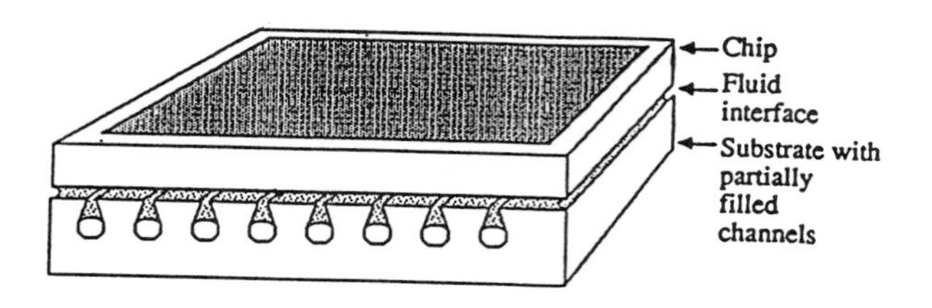

Figure 3. Microcapillary Attachment of Chip to Substrate [15].

in these designs is the interface resistance between the removable module and the rail of the plenum. In [16] Scott summarizes work on a new wedgeclamp designed to minimize TCR.

Finally, by noting that the thermal capacitance of a test interface is extremely small when compared with the capacitance of most test rigs, Antonetti and Eid [17] examined the merit of monitoring the instantaneous contact resistance rather than the specimen temperatures, in order to reduce the time required to determine the steady-state contact resistance. As shown in Figure 4, when the instantaneous heat flow was determined from the temperature gradients in the test specimens and then corrected analytically for heat loss and heat accumulation, the contact resistance for a relatively large specimen was determined in less than 10 minutes.

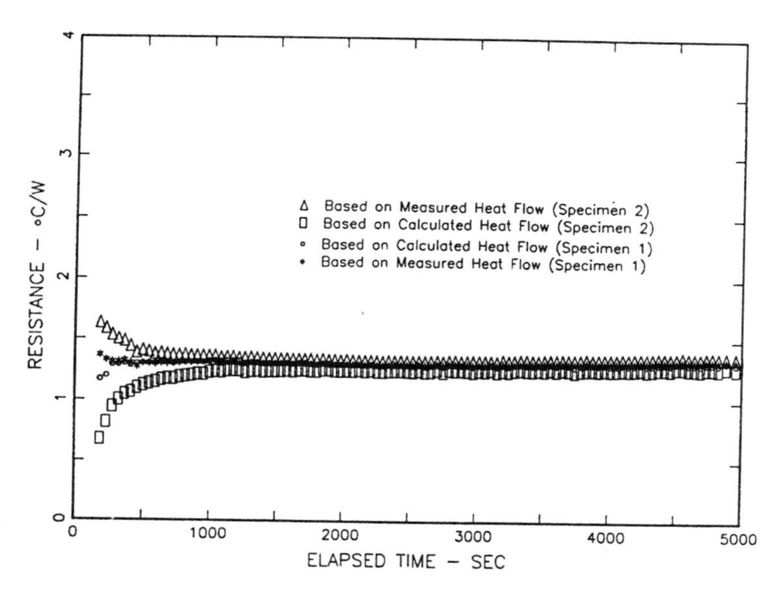

Figure 4. Apparent Thermal Contact Resistance vs. Elapsed Time [10].

THERMAL INTERFACE MANAGEMENT

In commercial equipment, thermal interface management is primarily an art dependent in large measure on innovation. Several examples, such as those shown in Figure 5, have been reported. Chu, Hwang, and Simons [18] describe the IBM thermal conduction module which employs an aluminum piston with a spherical tip pressed very lightly against a chip by means of a spring. Helium is used to fill the interface gaps. A thermal grease is used to fill the void between chip and cap in the air-cooled module described in [12,19]. Wilson [20] used a heat sink comprised of a water-carrying copper diaphragm abutting and cooling the back of a substrate. A metal stud pressed onto the face of a chip carrier with thermal grease at the interface is described by Watari and Murano [21].

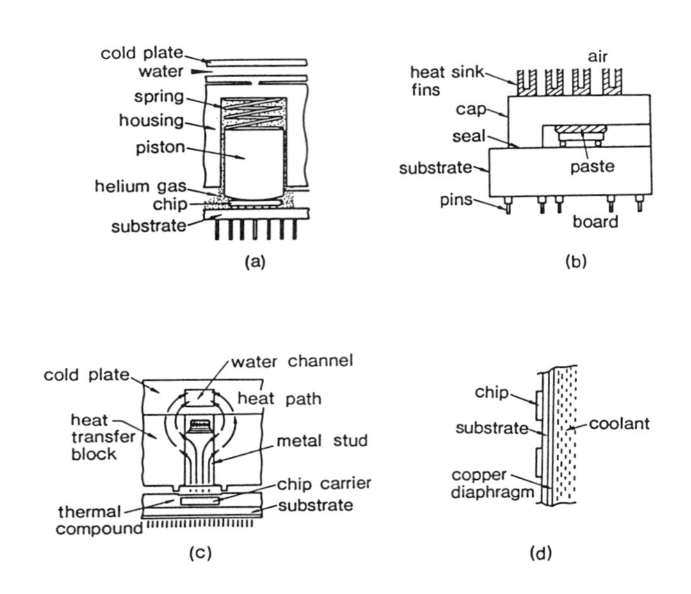

Figure 5. Examples of Thermal Interface Management [18-21].

Table 1 tabulates some of the manufacturing and servicing factors that Nakayama [22] recommends be considered in interface management. He notes that the manufacturing process may result in bowing, waviness, microroughness, surface misalignment, tilting, and void formation (as in die bonding). Not mentioned, but equally important, are scratches, and low levels of dust or other types of impurity at the contacting surfaces. In addition, during the service life of the product, the interstitial gap may vary due to thermal coefficient of expansion mismatch, or due to shock or vibration. Finally, the deterioration of any interstitial material (such as a grease) could adversely affect the thermal performance of a joint.

Phase	Parts	Factors
pre-assembling	parent	bow, wave, microroughness
	interstice	selection of material (gas, liquid, adhesive, solder)
assembling	parent	non-alignment, tilting, elastic-plastic deformation
	interstice	void formation
service	parent	variation of interstitial gap, cleavage formation, due to TEC mismatching of components
	interstice	deterioration of interstitial material

Table 1. Manufacturing and Servicing Factors Affecting Thermal Interface Management [22].

NOMENCLATURE

C_0, C_1	Constants in equation (9)
H	Microhardness
h_g, h_s, h_j	Thermal contact conductance of gas, solid, joint
k_g	Gap or gas thermal conductivity
k_s	Harmonic mean thermal conductivity of solids
m	Combined average absolute asperity slope
M	Gas parameter
M_g	Molecular weight of gas
M_s	Molecular weight of solid
P	Pressure
T	Temperature
Y	Separation distance of contact surface mean planes
α	Thermal accommodation coefficient
β	Defined by equation (7)
γ	Ratio of specific heats
δ	Effective thickness of gap = Y + M
μ	Gas molecular weight over solid molecular weight
σ	Combined RMS roughness
Λ	Mean free path

REFERENCES

1. A. Williams, "Heat Transfer Across Metallic Joints," Mechanical and Chemical Engineering Transactions, pp. 247-253, November 1968.

2. C.V. Madhusudana and L.S. Fletcher, "Contact Heat Transfer - The Last Decade," AIAA Journal, Vol.24, No.3, pp. 510-523, March 1986.

3. V.W. Antonetti and M.M. Yovanovich, "Thermal Contact Resistance in Microelectronic Equipment," International Journal for Hybrid Microelectronics, Vol. 7, No. 3, pp.44-50, September 1984.

4. M.M. Yovanovich, "New Contact and Gap Correlations for Conforming Rough Surfaces," AIAA-81-1164, presented at AIAA 16th Thermophysics Conference, Palo Alto, CA., June 1981.

5. S. Song and M.M. Yovanovich, "Explicit Relative Contact Pressure Expression: Dependence Upon Surface Roughness Parameters and Vickers Microhardness Coefficients," AIAA-87-0152, presented at AIAA 25th Aerospace Sciences Meeting, Reno, Nevada, January 1987.

6. J.C. Eid and V.W. Antonetti, "Small Scale Thermal Contact Resistance of Aluminum Against Silicon," Proceedings Eighth International Heat Transfer Conference, Vol. 2, pp. 659-664, San Francisco, CA, August 1986.

7. V.W. Antonetti and M.M. Yovanovich, "Using Metallic Coatings to Enhance Thermal Contact Conductance of Electronic Packages," ASME HTD-Vol. 28, pp. 71-77, November 1983.

8. V.W. Antonetti and J.C. Eid, "Thermal Contact Resistance Measurement Techniques for Microelectronic Packages," ASME HTD-Vol. 89, pp. 31-36, December 1987.

9. S. Song and M.M. Yovanovich, "Correlation of Thermal Accommodation Coefficient for "Engineering" Surfaces," Proceedings 1987 National Heat Transfer Conference, pp. 107-116, Pittsburgh PA, August 1987.

10. G.P. Peterson and L.S. Fletcher, "Thermal Contact Resistance of Silicon Chip Bonding Materials," Proceedings of International Symposium on Cooling Technology for Electronic Equipment," pp. 438-443, Honolulu, HI, March 1987.

11. W.E. Schwinkendorf and M. Moss, "Thermal Conductivity and Interface Resistance of Particle-Filled Resin," ASME Paper 84-HT-88.

12. E. Hultmark, J.L. Horvath, A. Trestman-Matts, and C. Parks, "The Use of Silicone Adhesives in Microelectronic Packaging," Proceedings of Sixth Annual International Packaging Society (IEPS) Conference, pp. 340-348, November 1986.

13. Feldman, K.J., Hong, Y.M., and Marjon, P.L., "Tests on Thermal Joint Compounds to 200°*C*," AIAA-80-1466, presented at AIAA 15th Thermophysics Conference, Snowmass, CO, July 1980.

14. W. Nakayama, N. Ashiwake, T. Daikoku, and M. Sato, "Heat Transfer Conductance and Contact Pressure of Elastic Contactors, ASME Paper 84-HT-87.

15. A. Paal and R.F. Pease, "Extending Microcapillary Attachment to Rough Surfaces," Proceedings of IEEE International Electronic Manufacturing Technology Symposium, pp. 169-172, San Francisco, CA, September 1986.

16. G.W. Scott, "A Multisegment Wedgeclamp for a Low Thermal Resistance Interface," Proceedings of International Symposium on Cooling Technology for Electronic Equipment," pp. 200-216, Honolulu, HI, March 1987.

17. V.W. Antonetti and J.C. Eid, "A Technique for Making Rapid Thermal Contact Resistance Measurements," Proceedings of International Sympo-

sium on Cooling Technology for Electronic Equipment," pp. 449-460, Honolulu, HI, March 1987.

18. R.C. Chu, U.P. Hwang, and R.E. Simons, "Conduction Cooling for an LSI Package: A One-Dimensional Approach," IBM Journal of Research and Development, Vol. 26, pp. 45-54, January 1982.

19. R.G. Biskeborn, J.L. Horvath, and E.B. Hultmark, "Integral Cap Heat Sink Assembly for the IBM 4381 Processor," Proceedings 1984 International Packaging Society (IEPS) Conference, pp. 468-474, November 1984.

20. E.A. Wilson, "Accommodating LSI in a High Performance Computer, Electronic Packaging and Production, pp. 142-152.

21. T. Watari and H. Murano, "Packaging Technology for the NEC SX Supercomputer," Proceedings IEEE Electronic Component Conference, pp. 192-198, 1985.

22. W. Nakayama, "Thermal Management of Electronic Equipment: A Review of Technology and Research Topics," Applied Mechanics Review, Vol. 39, No. 12, pp. 1847-1868, December 1986.

Thermal Contact Conductance—A Comparison of Methods

MÁRCIA BARBOSA HENRIQUES MANTELLI
and HANS-ULRICH PILCHOWSKI
INPE, Instituto de Pesquisas Espaciais
Av. dos Astronautas, 1758 C.P. 515
São José dos Campos, SP, Brasil

ABSTRACT

For the first Brazilian satellite thermal contact resistance calculation, it was necessary to determine, among several physical and mathematical models, the most suitable for the space conditions. First, three methods were selected and compared with literature experimental data, with the objective of stablishing the most appropriate one. It was concluded that the model developed by Mikic and Rohsenow [10] was the most suitable because it considers the heat flux macro and microconstriction influences.

1. INTRODUCTION

1.1. Thermal Contact Resistance Conception

The surfaces obtained in actual machining process present deformities, so that when put in physical contact, they touch each other at a restricted number of points. The surface roughness determines the true contact points and the waviness their geometrical distribution (see Figure 1); then a heat flux passing through this interface is first conducted to the contact point concentration zones and afterwards through these points. Due to this flux constriction, a temperature difference at the contact interface can be observed.

The thermal resistance conception is obtained from the analogy with Ohm's law for an electric circuit. The heat flux is analogous to an electrical flux and the temperature difference to electrical potential, so that:

$$\text{thermal resistance (R)} = \frac{\text{thermal potential difference } (\Delta T)}{\text{heat flux } (Q/A)}$$

The inverse of the thermal contact resistance is known as thermal contact conductance (h).

1.2. Objective of Work

The objective of this work is to compare the previously developed methods and to establish the mathematical model more suitable to use in the first Brazilian satellite.

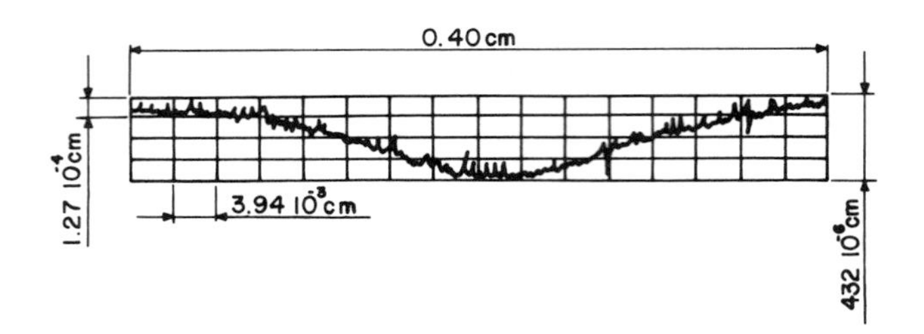

FIGURE 1. Typical surface linear profile showing waviness and roughness.

2. THERMAL CONTACT RESISTANCE WORKS

Too many parameters affect the thermal contact resistance, so it is very difficult to obtain a theoretical method that predicts, with precision, the thermal contact resistance for an arbitrary kind of surface contact; it explains the fact that so many thermal contact resistance works have been done around the world. Only the most important parameters are listed here:

(1) contact points number;

(2) contact points shape;

(3) contact points size;

(4) contact points physical arrangement;

(5) surface roughness;

(6) surface waviness;

(7) contact pressure;

(8) interstitial fluid thermal conductivity;

(9) material hardness;

(10) elasticity modulus;

(11) mean interface temperature;

(12) mean interstitial fluid pressure;

(13) interfilling junction material;

(14) directional effect;

(15) coupling history with respect to the number of compressions and decompressions;

(16) surface oxide films.

2.1. Chronological Evolution

During nineteen hundred forties, when the necessity of the study of the heat transfer through a contact interface emerged, the first efforts in thermal contact resistance were developed. As this phenomenon was not known until that time, it was first investigated experimentally. The first experimental works had as objective the investigation of specific contacts concerning the projects in which the authors were involved, this happened with Brunot and Buckland [1], and Barzelay et al. [2], [3]. They developed extensive works in experimental measurement of contacts found in aeronautical projects. With these results, some researchers tried to stablish mathematical models to estimate the thermal contact resistance. Fried and Costello [4] were among the first to study the thermal contact resistance problems for space applications. They made the hypothesis that the heat was transferred through basic mechanicms, considered as parallel resistances: conduction through physical contact points, conduction through intersticial fluids, and radiation. This model did not consider the flux lines deformation.

Fenech and Rohsenow [5] created the first important model to simulate this mechanism; they considered that all the contact points were: uniformly spread in surface contact, round, of uniform size, and cylindrical. Then, studying just one contact point was enough to evaluate the interface thermal contact resistance. Figure 2 shows the physical model that they adopted; D is the diameter of the unit cell, a_m and δ are the surface roughness mean radius and height, respectively.

Clausing [6], [7], [8] observed that, until then, the developed models presumed that the contact points were uniformly distributed in the coupling interface. This kind of distribution is found in interfaces between flat roughness surfaces, which are very difficult to obtain in the fabrication processes. The actual machined surfaces present waviness and roughness so that, when coupled, there are contact point concentration zones. Then the heat flux, to pass through the contact, suffers two kinds of constrictions: a microcontriction, expected in Fenech and Rohsenow model, and a macroconstriction. Clausing considered that the surface thermal contact resistance is formed by three resistances in series: the macroconstriction resistance, the microconstriction resistance, and surface oxide film resistance. In macrocontact area, this author conceived the surface waviness as spherical CALOTTES (see Figure 3) and used Hertz's theory [9] to calculate the contact area between pressured spherical surfaces, considering elastic deformation. He also assumed that the microcontacts are uniformly spread along the contact areas, and utilized the Hoess theory [6], [7], [8] to determine the microconstriction resitance. The oxide film resistance was not considered, because the tested surfaces were clean, i.e. without this film.

Mikic and Rohsenow [10] developed a mathematical model for a physical model similar to that of Clausing. They concluded that the macro and microconstrictions were similar physical phenomena that could be described by the same mathematical formulation, using typical dimensions for each case. The macrocontact area was calculated using Hertz's theory. These scientists verified that due to the roughness this area is a little greater than the estimated one and developed an equation to determine the area calculation more precisely. This theory can be used to calculate the thermal contact resistance together with its physical surface characteristics.

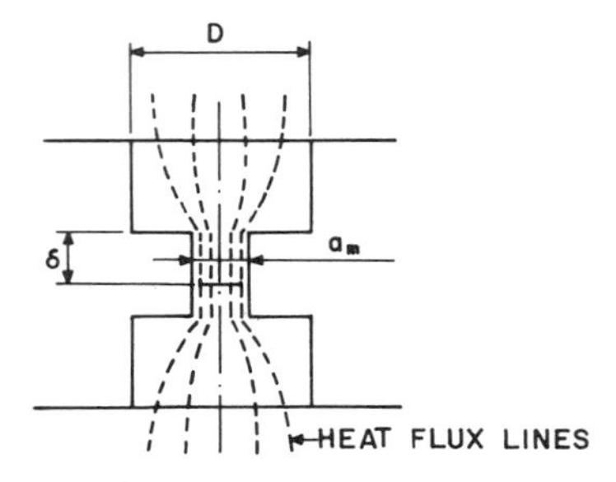

FIGURE 2. Fenech and Rohsenow physical model.

Yovanovich and Rohsenow [11], using Mikic and Rohsenow's equations, studied the contact point size variation along the interfaces and the contact point distribution monuniformity, for flat and rough surfaces, They analyzed the surface deformation and verified that there are regions with plastic deformation and others where the deformation is elastic.

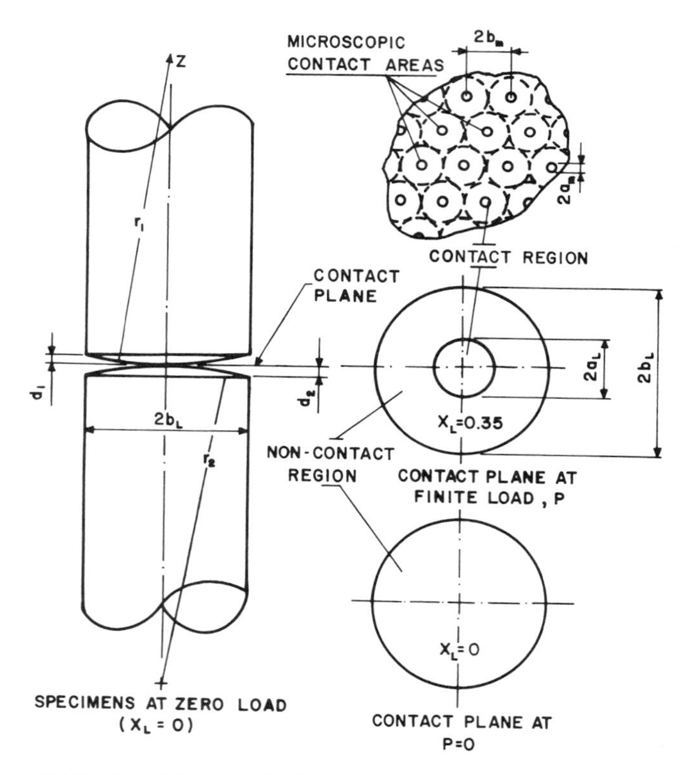

FIGURE 3. Macro and microcontacts,
Source: Clausing [8] p. 3.

At that time, some scientists, observing that the surface parameters like asperity height and microscopic areas have a Gaussian distribution for most of the surfaces, began to use statistical theory to get the necessary parameters for the application of the developed equations. Whitehouse and Archard [12] verified that many engineering surfaces had random characteristics, and treated their parameters statistically. Thomas and Sayles [13] used the statistic theory to analyze the waviness effect on thermal contact resistance. Jones et al. [14] developed a method to calculate the contact parameter from the coupled surface topography. Al-Astrabadi et al. [15] also utilized statistics to study the surface finishing.

2.2. Mathematical models

Some authors stablished mathematical models for their physical models, but only few works arrived to formulations able to be used for the theoretical calculation of the thermal contact resistance. The models developed by Fenech and Rohsenow [5], Clausing and Chao [16], and Mikic and Rohsenow [10] can be utilized to predict this resistance. The Fenech and Rohsenow model, the first to be developed, has some practical difficulties, like the surface parameters determination, that make its use unfeasible [17]. In this work Clausing and Mikic and Rohsenow models are used, described briefly below.

Clausing's model. Figure 3 shows the contact zone between two identical solid cylinders of length L and radius b_L. The macroscopic contact area dimensions are given by the bodies' elastic deformation. The waviness is represented by spherical CALLOTES of radius r_1 and r_2, placed in the top of the cylinders, in the contact zone. The distance between the CALLOTES base and its top (d) corresponds to the waviness height. X_L is the macrocontact area constriction rate ($X_L = a_L/b_L$), where a_L is the mean macrocontact area radius, a_m corresponds to the actual contact point mean radius and b_m its respective cylinder base radius.

For this physical model, Clausing considered that:

(1) L is big in comparison to b_L;

(2) there is perfect contact through the macrocontact area, in other words, $R_m << R_L$;

(3) the heat is transferred only through the macrocontact areas;

(4) the opposite cylinders base areas have uniform temperature;

(5) the surface materials are homogeneous, isotropic and their physical properties are constant with the temperature.

Mikic and Rohsenow model. Mikic and Rohsenow [10], in their physical model, considered the surface contact points round, with mean radius a_m uniformly spread in the macrocontact areas. The macrocontact areas are divided into circles of radius b_m with the elementary cylinder's base centered in the contact points as shown in Figure 4. On the other hand, the apparent contact area is divided into circles of radius b_L, centered in the macrocontact areas.

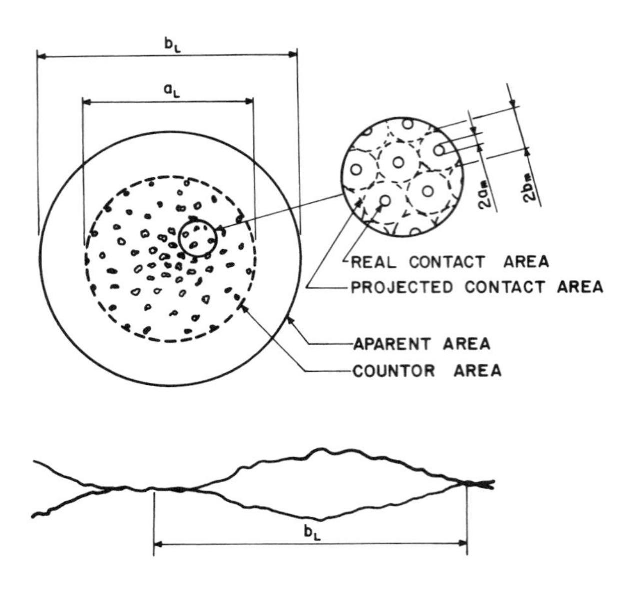

FIGURE 4. Spherical contact physical model.
Source: [10], p. 29

2.3. Thomas and Probert Experimental Correlation

For a better understanding of the phenomenon, experimental investigations were performed by some researchers. Thomas and Probert [18], grouped their data and obtained mean curves in an attempt to predict the thermal contact resistance from the mechanical properties and surfaces data. These scientists made the hypothesis that this resistance was affected by the surface hardness, material thermal conductivity and surface roughness, whose units have the following basic dimensions: mass, length, time, temperature and heat. There are three independent equations that relate these parameters and, by the π theorem, can be grouped in $5-3=2$ dimensionless sets. The aluminum mean curve constructed by Thomas and Probert was made using 240 experimental data. It is important to keep in mind that the unique surface parameter considered was the roughness. The correlation was performed using experimental data from different sources. Analyzing these works, it was verified that all the contacts such as found in enginnering, with waviness. Thus the noninclusion of the waviness parameter does not mean that the correlation is valid only for flat surfaces, but that it has mean waviness values inserted in its coefficients. It is difficult to estimate these values, because they result from the combined effects of the wavelength and the waviness maximum height; nevertheless one can expect the correlation to produce good results for mean waviness surfaces. The correlation curves are compared with the Clausing and Mikic and Rohsenow theoretical curves in the next section.

2.4. Comparison Between the Three Methods.

The aim of this section is to analyze the methods described above. This is made through the study of curve behavior versus the surface parameters variation, to determine the most convenient method for the thermal resistance calculation of the first Brazilian satellite contacts. In order to get this comparison, the methods were implemented in a computer. The program entries are the mechanical and surface properties of the coupled metals and the program output is the three thermal conductance curves as a function of the pressure, all in the same graph. The contacts are made of 2024 aluminum surfaces, and typical surface finishing variations of this material are used for this analysis. To organize this study the curves are grouped according to Tables 1 and 2, where:

(1) BL- wavelength;

(2) DT- maximum waviness height;

(3) RMS - root mean square;

(4) TGTT - tan(θ)

The Clausing [6], [7], [8] and Mikic and Rohsenow [10] theories are based on similar physical models. The main difference between them lies in the fact that Mikic and Rohsenow considered the influence of the microcontacts due to the roughness presence. It is expected that the curves are close for smooth (Figures 5.1a, 2.a, 3.a) and distant for rough surfaces (Figures 5.1.c, 2.c, 3.c), where the microconstriction importance grows. For flat surfaces (with small waviness), the distance between these curves is large (Figures 5.2.a, b, c); this effect is specially observed for flat and rough surfaces (Figures 5.2.c). It is interesting to note that the two theoretical curves have almost ever similar curvatures.

The influence of the maximum height variation in the waviness is bigger than the influence of the lenght variation, as it can be observed in Figures 5.5a, b and 6.a,b.

Table 1. Numerical values used in the sets from 1 to 6, as shown in Table 2.

		SMALL	MEDIUM	LARGE
ROUGHNESS PARAMETERS	ROUGHNESS RMS$\times 10^{-6}$ METERS	1.1003	2.1336	7.6200
	ASPERITY ANGLE PEAK TANG. TGTT	0.1200	0.1760	0.2670
WAVINESS PARAMETERS	WAVINESS LENG. BL - $\times 10^{-2}$ m	0.6096	1.2954	1.9812
	MAXIMUM HEIGHT DT - $\times 10^{-6}$ m	3.0480	21.6410	40.2340

Table 2. Curve sets used in the three method comparison.

SET	ROUGHNESS	WAVE LENGTH	MAXIMUM WAVE HEIGHT	ASPERITY ANGLE PEAK
1	VARIABLE	MEDIUM	MEDIUM	SMALL
2	VARIABLE	SMALL	SMALL	SMALL
3	VARIABLE	LARGE	LARGE	SMALL
4	MEDIUM	MEDIUM	MEDIUM	VARIABLE
5	MEDIUM	MEDIUM	VARIABLE	SMALL
6	MEDIUM	VARIABLE	MEDIUM	SMALL

Analyzing the experimental correlation behavior, it is verified that its curves are always more distant from the theoretical ones than the theoretical curves are from each other. This correlation estimates smaller conductance values than the other ones, for the same pressure, with the exception of Figure 5.1.a and 3.a. The Figure 5.3.a, which

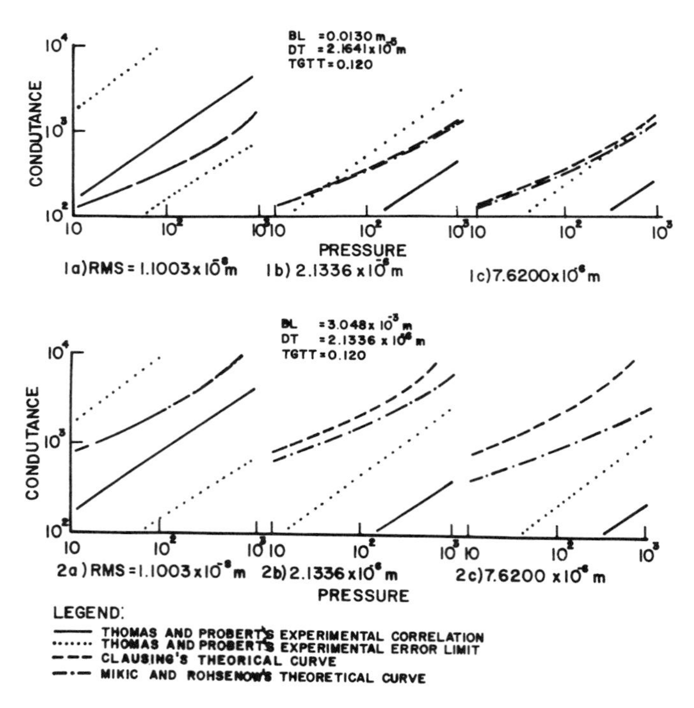

FIGURE 5. Three method comparison curves (first part).

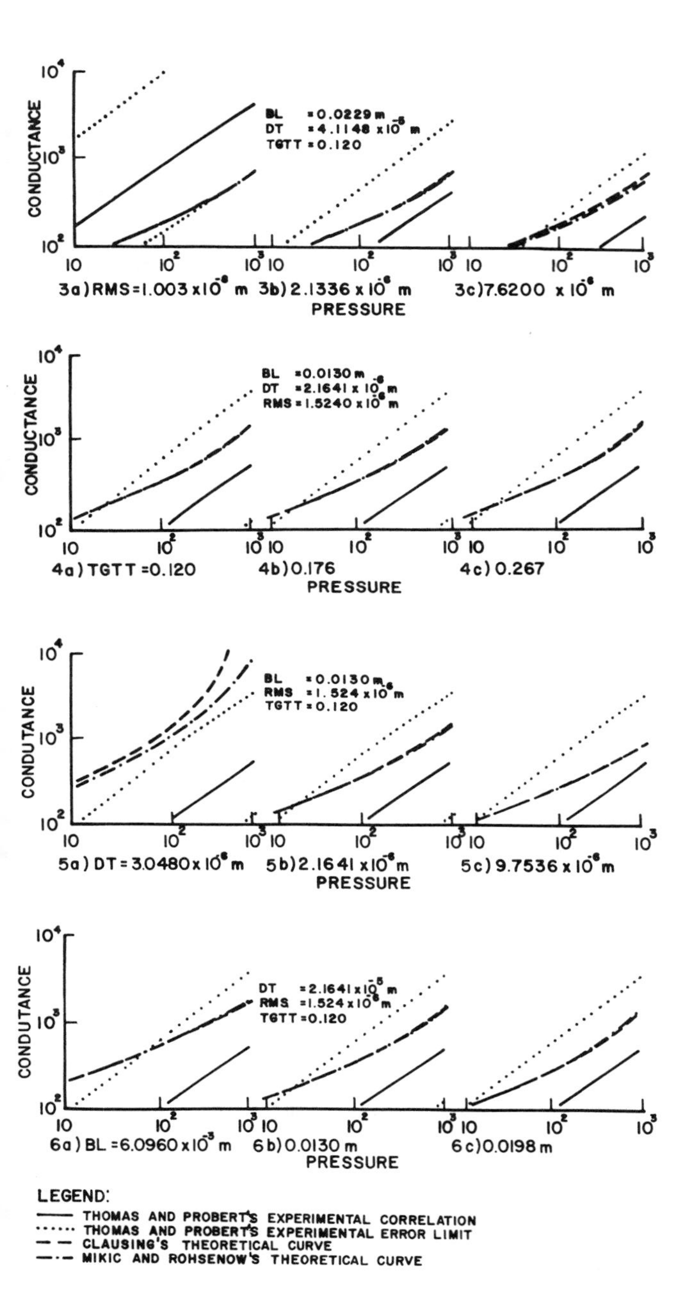

FIGURE 5. Three method comparison curves (second part).

represents contacts between surfaces with large waviness and medium roughness, presents the closest three method curves of all studied; it means that the contact surfaces considered when the correlation was developed probably had large waviness.

The asperity peak angle variation actually found, from 150° to 160°, causes a very small variation in the Mikic and Rohsenow theoretical curve, as it can be seen in Figure 5.4d. In this way, one can use $tg\theta = 0.120$, as suggested by these researchers, since otherwise it would be difficult to get this surface parameter from the contact bodies profile.

Note that for medium wavelength, medium roughness and small maximum waviness height surfaces (see Figure 5.6a), the theoretical curves are distant each from other, with a similar effect as for very rough surfaces. It can be explained by the fact that when the rate b_L/d_t grows, the waviness characterists changes, and the surface became very rough.

From this analysis one can arrive to the following conclusions: -

(1) for the studied contacts, the influence of the macroconstriction on thermal contact resistance is greater than that of microconstriction; this is observed from the fact that the Mikic and Rohsenow and Clausing curves have the same curvature and are close each other. It is also important to observe that Mikic and Roshenow in their theory, gave the same mathematical treatment to micro and macroconstriction;

(2) as the Mikic and Rohsenow theory includes both the micro and macroconstriction, it is expected that it reproduces the actual resistance with more precision than the Clausing theory, which considers only the macroconstriction. The same is expected, since the correlation just considers the roughness;

(3) the correlation can, for small waviness surfaces, be used as a lower limit for the contact conductance, when it is in favor of the safety factor;

(4) for small rough surface, the Clausing theory can be used, because both theoretical curves are very close:

(5) the asperity peak angle measurement can be omitted, and $tg\theta = 0.120$ can be assumed as a good mean value for practical use.

Table 3. Coupling surface parameters and physical properties.

COUPL	SURF	ROUG. RMS $\sigma \mu_m$	MAX. HEIG. $d \mu_m$	WAVI. LENG. BL- m	ASPER. ANGLE TANG.	MEAN THER CONDUCT. K W/m°C	HARD KN/m² x10⁶	ELAST KN/m² x10⁷
1	1A	1.257	3.810	0.014	0.120	120.230	1.500	6.895
	1B	1.397	2.540	0.014	0.120	120.230	1.500	6.895
2	2A	0.223	6.350	0.019	0.120	120.230	1.500	6.895
	2B	0.223	2.540	0.019	0.120	120.230	1.500	6.895

2.5. The Methods Comparison Applied to Literature Experimental Data

Fried and Kelley [19], using a thermal contact resistance measurement experimental apparatus, projected and built by them, studied, among other, two couplings similar to the contacts found in space applications, formed by 2024 T4 aluminum surfaces , described in Table 3. These researchers had the objective of studying the influence of the surface parameters, mechanical properties, and interface pressure. The Brazilian satellite contacts considered are made of 2024 T351 surfaces, with mechanical properties very similar to the 2024 T4 ones. Then the couplings showed in Table 3 are similar to the satellite ones, so that from the Fried and Kelley experimental results and the theoretical curves comparative study it is possible to indicate the most suitable method for the thermal contact resistance calculation of the Brazilian satellite.

Figure 6 shows the graphics used for comparison between the experimental data and theoretical curves for each coupling. Analyzing this figures, it is seen that most of experimental points are closer to the theoretical Clausing and Mikic and Rohsenow curves than to the experimental correlations. It is even noticeable that the correlation describes in a better way the less rough coupling (number 2) (Figure 6), according to the last section observations. Since the experimental points are more distant from the correlation curves than from the theoretical ones, the nonutilization of the experimental correlation is justified. The question that now arises is to verify which of the two theoretical methods is the most suitable for the calculation of thermal contact resistance of these satellite couplings.

There is no doubt that the experimental points are nearer to Mikic and Rohsenow than to Clausing curves for the coupling number 1. But this is not so clear for the coupling number 2, where at the right sight it could be seen that the Clausing curve is the best one. But for all the experimental points, it is easily observed that, on average, the Mikic and Rohsenow's is the most acceptable theoretical curve.

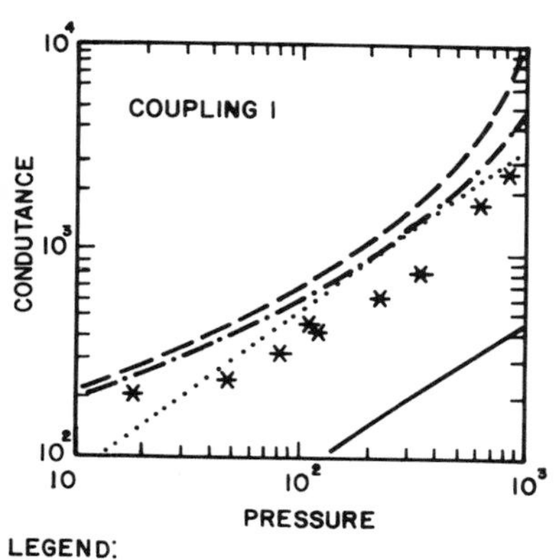

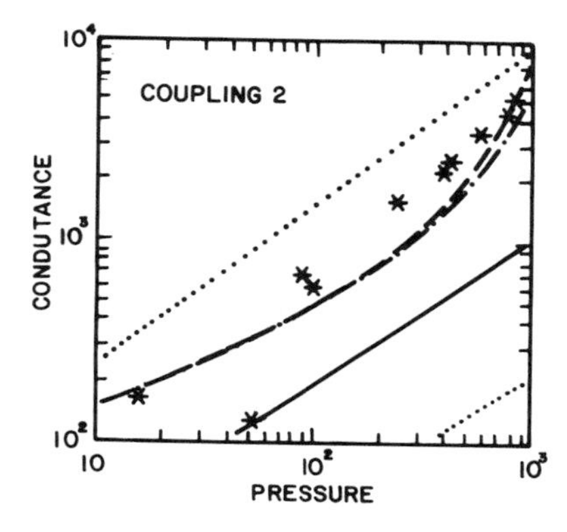

LEGEND:

—— THOMAS AND PROBERT EXPERIMENTAL CORRELATION
······ THOMAS AND PROBERT CORRELATION ERROR LIMIT
— — CLAUSING THEORETICAL CURVE
—·— MIKIC AND ROHSENOW THORETICAL CURVE
* FRIED AND KELLEY EXPERIMENTAL POINTS

FIGURE 6. Fried and Kelley experimental data and theoretical curves.

Therefore, it can be concluded that for the Fried and Kelley couplings, the Mikic and Rohsenow's proposed model is the most suitable for the experimental points, and as the satellite couplings are similar to these ones, it is recommended to use the Mikic and Rohsenow's theoretical curve for the calculation of the satellite thermal contact resistances.

3. CONCLUSION

The main conclusion is that the model developed by Mikic and Rohsenow [10] was the most acceptable because it considers the heat flux macro and microconstriction influence. It is important to note that this method has given good results in spite of the fact that it was found some difficulties to find up to date literature, where new and most appropriate experimental correlations were obtained. It is suggested to repeat this procedure using new literature methods.

From the comparative analysis, it could be concluded mainly that:

(1) the heat flux macroconstriction influence is greater than the microconstriction;

(2) the asperity peak angle determination is not necessary, and the literature suggested value can be used without problems.

As this work analyzed only one kind of aluminum coupling, it would be necessary to repeat this same procedure to qualify this method for a more general use.

NOMENCLATURE

a	microcontact area mean radius
A	apparent contact area
b	elementary cylinder mean radius
d	maximum waviness radius
D	unit cylinder diameter
h	thermal contact conductance
L	unit cylinder height
P	contact pressure
Q	heat flux
r	spherical CALLOTE radius, simulating the ondulations
R	thermal resistance
T	temperature
X	heat flux positive direction
x_L	macrocontact constriction rate
z	perpendicular direction to the contact surfaces;
δ	roughness mean height
θ	asperity angle peak
σ	rms roughness

Inferior index

L	macrocontacts
m	microcontacts
1	coupling surface number 1
2	coupling surface number 2

REFERENCES

1. Brunot, A.W., Buckland, F.F., Thermal Contact Resistance of Laminated and Machined Joints, *Transactions of ASME*, pp.253-256, 1948.

2. Barzelay, M.E., Tong, K.N., Holloway, G.F., Thermal Conductance of Contacts in Aircraft Joints, Washington, D.C., National Advisory Comittes for Aeronautics, 1954, (NACA Technical Note 3167).

3. Barzelay, M.E., Tong, K.N., Holloway, G.F., Effect of Pressure on Thermal Contact Resistance of Contact Joints, Washington, D.C., National Advisory Committee for Aeronautics, 1954, (NACA Technical Note 3167).

4. Fried, F., Costello, F.A., Interface Contact Resistance Problems in Space Vehicles, *ARS Journal*, vol.32, nº 2, pp.237-243, 1962.

5. Fenech, H., Rohsenow, W.M., Prediction of Thermal Conductance of Metallic Surfaces in Contact, *Journal of Heat Transfer*, vol. 85, nº 4 pp.15-24, 1963.

6. Clausing, A.M., An Experimental and Theoretical Investigation of the Thermal Contact Resistance, Urbana, I1., University of Illinois, 1966, (M.E. Technical Report nº 242-2).

7. Clausing, A.M. Heat Transfer at the Interface of Dissimilar Metals - the Influence of Thermal Strain, *International Journal of Heat and Mass Transfer*, vol.9, nº 8, pp. 791-801, 1966.

8. Clausing, A.M., Theoretical and Experimental Study of Thermal Contact Resistance in Vaccum Environment, Urbana, Il., University of Illinois, 1966.

9. Timoshenko, S.P., Goodyear, J.N., *Theory of Elasticity*, 3d. ed., Mc-Graw Hill Kogakusha, Tokyo, Japan, 1970.

10. Mikic, B.B., Rohsenow, W.M, Thermal Contact Resistance, Cambridge, M.T., MIT, 1966, (MIT Technical Report nº 4542-41).

11. Yovanovich, M.M., Rohsenow, W.M., Influence of Surface Roughness and Waviness upon Thermal Contact Resistance, Cambridge, M.T., MIT, 107 (MIT Technical Report nº 6361-48), 1967.

12. Whitehouse, D.J., Archard, J.F., The Properties of Random Surface of Significance in their Contact, *Proc. Royal Society of London*, 316-A, pp.97-121, 1970.

13. Thomas, T.R., Sayles, R.S., Random Process Analysis of the Effect of Waviness on Thermal Contact Resistance, *AIAA/ASME Thermophysics and Heat Transfer Conference, Boston*, MA., July 15-17 (AIAA Paper nº 74.691).

14. Jones, A.M., O'Callaghan, P.W., Probert, S.D., Prediction of Contact Parameters from the Topographies of Contacting Surfaces, *Wear*, vol.31, pp.89-107, 1975.

15. Al.Astrabadi, F.R., O'Callaghan, P.W., Probert, S.D., Effects of Surface Finish on Thermal Contact Resistance Between Different Materials, *AIAA Thermophysics Conference, Orlando, Fl.*, vol.14, June 4-6, 1979.

16. Clausing, A.M., Chao, B.T., Thermal Contact Resistance in a Vacuum Environment, *Journal of Heat Transfer*, vol.85, nº 4, pp.15-24, 1963.

17. Chen, M.S.H., Mantelli, M.B.H., Oliveira Filho, O.B., Estabelecimento de um Procedimento para Determinação de Parâmetros Necessários ao Cálculo da Resistência Térmica de Contactos no Interior de Satélites, São José dos Campos, INPE, Out 1985, (INPE-3690-RPE 1492).

18. Thomas, T.R., Probert, S.D., Correlations for Thermal Contact Conductance in Vacuum, *Int. Journal of Heat and Mass Transfer*, vol.13, pp.789-807, 1970.

19. Fried, F., Kelley, M.J., Thermal Conductance of Metallic Contacts in a Vacuum, *AIAA Thermophysics Specialists Conference*, Monterey, CA., Sept. 13-15, 1965. (AIAA Paper 65-661).

A Shape Factor Approach to Heat Conduction Problems of Composite Members in Electronic Equipment

W. NAKAYAMA, K. FUJIOKA, and Y. TAJIMA
Mechanical Engineering Research Laboratory
Hitachi, Ltd.
502 Kandatsu, Tsuchiura, Ibaraki, Japan

ABSTRACT

Printed wiring boards and ceramic substrates have densely packed conductor networks in a matrix of electrical insulation. The knowledge about heat conduction in such composite members is required in increasing detail as the requirement on the management of thermal stress in fine structures is becoming stringent. The analysis of heat conduction, however, encounters great difficulties due to the immense number of nodal points arising from high density conductor layouts. Proposed in this paper is a method of analysis where the need to deal with a large number of nodal points is reduced on the basis of geometric regularities of conductor arrangement.

The method is illustrated on an elementary model of a printed wiring board where conductor strips are bonded to the faces of a bakelite plate with a hole drilled at the center. The system is composed of two basic cell types, and in the first stage of analysis the sets of elementary solutions of temperature are found for those two basic cells. Also found are the correlations of heat exchange among conductor strips and insulator elements, called "shape factors". Then, a particular problem where heat flows from the heater in the center hole is considered, and the analysis is performed on the system represented by the shape factors. Comparison of the analytical results with the temperatures measured on several spots of the plate indicates the need to improve the physical modeling; nevertheless, the proposed approach is deemed promising.

INTRODUCTION

With the progress in circuit integration in electronic equipment the components at all structural levels of the equipment become densely populated by fine conductor networks. On the chip the feature sizes of transistors and conductor networks have been decreased to cram a great number of logic gates and memory cells in a given area. The substrate and the printed wiring board carrying those VLSI chips have to be equipped with fine conductor networks in many layers to provide signal transmission channels among the chips.

Conductor networks have to be imbedded in an electrical insulator; hence, one finds in those components laminates of fine metal strips and insulators. The insulator has a much lower thermal conductivity than the conductor, and often has a significantly different coefficient of thermal expansion from that of the conductor. Consequently, the heat flow in such composite members is regulated to a great extent by the arrangements of conductors, and a large thermal stress is likely to be caused in the conductor, the insulator, and their interfaces. It is becoming an increasingly challenging task for the manufacturer and the equipment designer to predict the temperature field in the component during manufacturing and equipment operation. The difficulty is amplified with the advances in the miniaturization of circuits, as more details of the temperature distribution need to be known to guarantee high product yield and reliability of delicate structures [1].

The length scale of temperature distribution around a particular point in the composite member is three to five orders of magnitude smaller than the overall size of the component, the latter being a representative length scale in the definition of the boundary condition for a heat transfer problem. Such a wide disparity of length scales between the thermal environment for the system and the objects of ultimate interest presents a serious challenge to one's attempt to perform numerical integration of heat conduction equation to find a temperature distribution of sufficient spatial resolution. If one discretizes the composite member into elements of microscopic scales in a straightforward manner, the discretization would result in an astronomical number of nodal points which even the today's most advanced supercomputers cannot effectively deal with.

One way to circumvent this difficulty is to first perform an analysis on the basis of "equivalent" thermal conductivity of the composite member to find a temperature distribution of coarse spatial resolution and then focus on a particular sub-area to find a detailed temperature field around the object of interest utilizing the macro system solution as a boundary condition. Pinto and Mikic [2] propose such a methodology. This so-called zooming technique requires the estimation of the equivalent thermal conductivity of composite materials, and for that subject there have been a number of reports published to date. The papers of Cheng and Vachon [3,4] contain the list of works on thermal conductivity of composite materials reported prior to 1970 and their own contributions to the archives. For composite materials involving randomness or high complexity in substructures there is an approach to locate an upper bound and lower bound of thermal conductivity, as employed by Schulgasser [4] for fiber reinforced materials. Those approaches would provide the basis of analyses for overall thermal systems on the substrate and the board; examples of such analyses are reported by Ellison [6] and Pinto and Mikic [2,7]. Considerable work is needed, however, to improve the modeling of microscopic thermal structures in electronic components, because those previous models of composite members are not adequate for the microsystems found in electronic components. In contrast to the situation assumed in most of the previous works where one of the constituents exists in a discontinuous phase, both conductors and insulator matrices in electronic composite materials are essentially continuous, and the equivalent thermal conductivity is itself often an inadeqate notion.

The present paper proposes an approach that does not utilize the notion of equivalent thermal conductivity; instead, it is based on certain geometric characteristics of the microstructures in electronic components.

THE CONCEPT OF A SHAPE FACTOR APPROACH

The general concept of the present approach will be explained using the illustrations of Figures 1 and 2. It is assumed that the object is a printed wiring board or a ceramic substrate bearing different conductor patterns in separate zones. In Figure 1 the board has three zones, A, B, and C, distinguished by different microscopic geometrical structures. The board is first split into the zones. Each zone has a conductor network formed by repetition of a particular pattern of sub-network, so that it is split into many but nearly identical subzones. Each subzone is further split into elementary cells of a microscopic scale. The number of cells could be large; however, the cell structures can be classified into a moderate number of types. Such geometrical regularity of the micro and medium scale structures, a characteristic generally found in electronic components, offers a great adavantage in solving heat conduction problems.

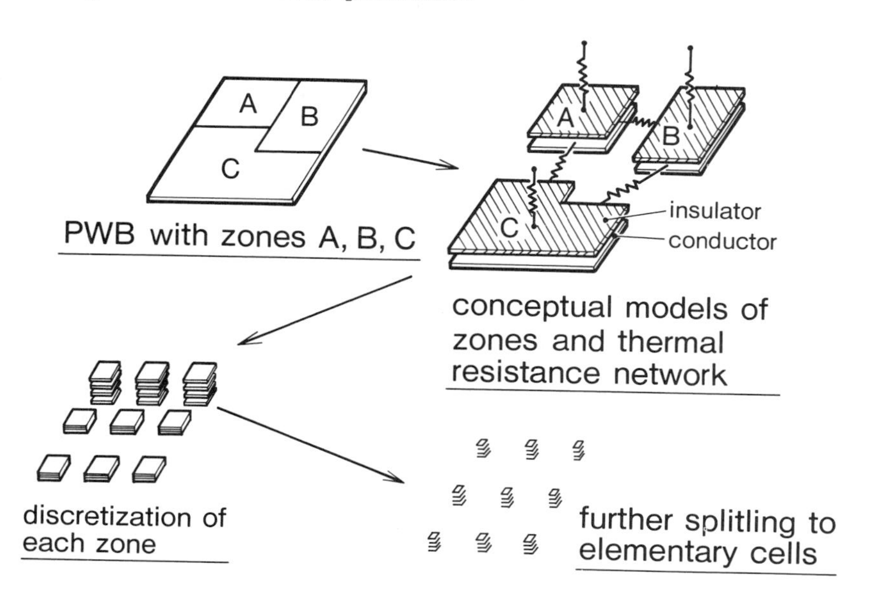

FIGURE 1. Discretization process.

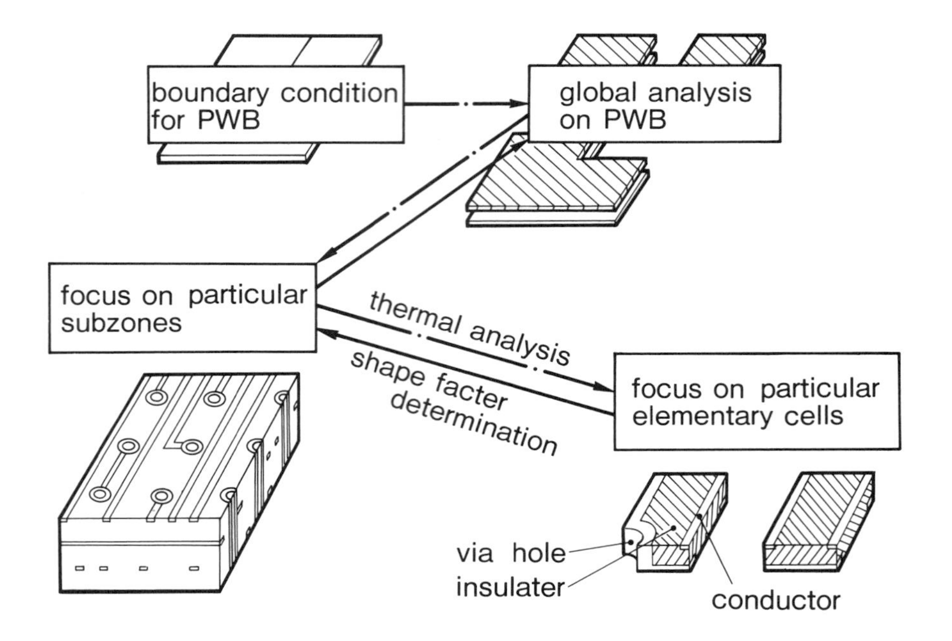

FIGURE 2. Steps of thermal analysis.

As illustrated in Figure 2, the attention is first directed to elementary cells. Each type of elementary cell is distinguished by the arrangement of conductor segments on an element of insulator. (The term "via hole" in the lower right of Fig. 2 stands for the drilled hole with electroplated metal on its wall. The function of the via hole, or simply the "via", is to provide electrical connections among different signal processing layers. The hole is often filled by solder or a metal pin, and of the heat transmission from the on-board components to the PWB a large fraction comes through the metals in the holes. In the absence of a better terminology the "via hole" will be used hereafter to designate the metal-filled hole.) The analysis is performed assuming a set of fundamental boundary conditions, which will be explained in the next section. The purpose of the analysis at this stage is to find the correlations of heat exchange among the conductor segments and the insulator element. The rule established by this analysis is represented by a parameter which is primarily a function of the shape of the cell and the geometrical arrangement of conductor segments, thus called "shape factor". The task to determine the shape factors is not overwhelming, because the number of cell types is not large. Once the shape factors are determined for typical cells, a similar procedure is applied to the subzone. Here, again, the number of subzone types is not large, and the rules to describe heat exchange among the same types of subzone, and among the subzones of different types, are termed "subzone shape factor". The entire thermal system is then built on the subzone shape factors, with a moderate number of nodal points representing subzones.

Once the boundary condition is specified for a macrosystem (the board), the analysis is first performed on the system, then on a particular subzone of interest, then on a particular cell, to find a temperature distribution having a fine spatial resolution. In this reverse procedure of analysis it is not necessary to solve heat conduction equations for the subzones and cells. The solutions obtained by the steps to determine the shape factors will be superposed to produce required temperature distributions in the subzones and cells.

THE PRINCIPLES OF THE SHAPE FACTOR APPROACH ILLUSTRATED ON A MODEL OF PWB

A Simulated Thermal System of PWB

A simple model of the printed wiring board is employed here to explain the principles of the analytical approach. The model is a bakelite plate, 10 cm x 10 cm and 8 mm thick, with rows of copper strips on its both faces as shown in Figure 3. Each copper strip is 2 mm wide and 0.5 mm thick, and the placement pitch of the strips is 7 mm. The copper strips on one side are set at right angles to the strips on the other side. At the center of the plate a hole of 10 mm diameter is bored, where a cylindrical heater is to be inserted. The dimensions

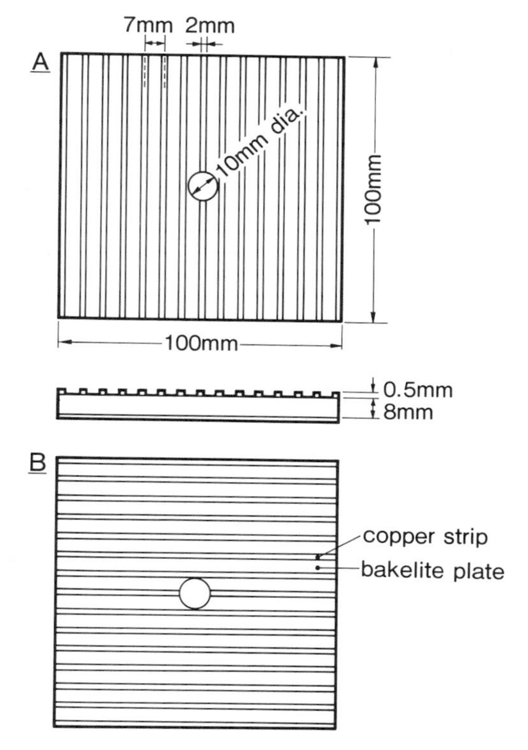

FIGURE 3. A model of PWB subzone.

of the structural parts are approximately ten times those on actual PWBs. However, the model incorporates the essential features observed on actual PWBs: the lamination of conductor strips separated by a layer of electrical insulator, with the strips in one layer set orthogonal to those in the adjacent layer, and the heat flow from the pin in the through-hole. The enlargement of geometric dimensions is designed to make the measurement of the temperature distribution easier than on actual PWBs.

It is assumed that the model is suspended in room air, setting the conductor-laden faces vertical. Heat supplied from the heater spreads in the plate by heat conduction while being carried away from the surfaces by natural convection of air. Natural convection heat transfer envelops the plate with a rather large thermal resistance; moreover, the variation of heat transfer coefficient over the plate is small. Hence, the pattern of heat flow in this composite system is determined primarily by the geometrical arrangement of heat conductive members. In fact, the thermograms of the surface obtained during heat transfer experiments reflect the arrangement of conductor strips; moreover, they show the symmetries of temperature field, with the lines of symmetry running vertically and horizontally through the center of the plate. Based on this observation only a quarter of this simulation model is considered in the analysis. The heat transfer coefficient on the conductor-laden faces is assumed to be uniform, while on the edges of the plate the adiabatic condition is assumed.

Coordinate System, Scales, and Discretization of the Zones

Figure 4 shows a quarter of the PWB model. The coordinate system (x, y, z) is defined with the origin at the center of a terminal face of the heater. All lengths are scaled by the pitch (p_*) of the conductor strip placement; in non-dimensional terms, the thickness of the plate is γ, the radius of the via hole is R, the width and thickness of the conductor strip are 2w and t, respectively, and the distance between the

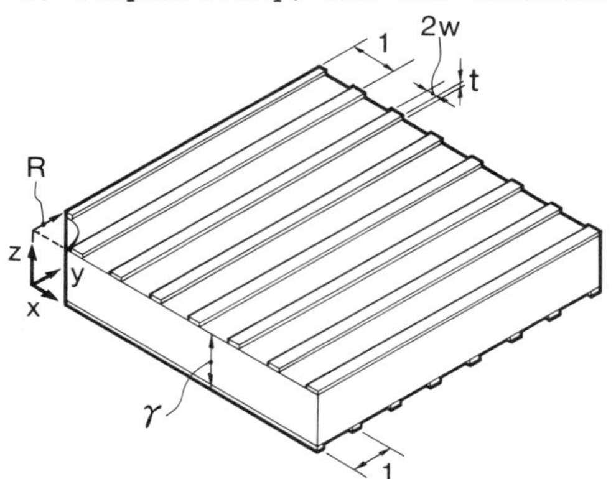

FIGURE 4. Coordinate system attached to the quarter of the PWB model.

centerlines of the neighboring strips is unity. The other scales of non-dimensionalization are

the temperature scale $\Delta T_* \equiv Q/\lambda_M p_*$,

the time scale $\tau_* \equiv \rho_M c_M p_*^2/\lambda_M$,

where Q is the rate of heat input to the heater, λ_M the thermal conductivity, ρ_M the density, and c_M the specific heat, with the suffix M denoting that these physical properties belong to the metal.

In what follows, equations are written in terms of the dimensionless quantities: the temperature of the conductor strip ϕ, that of the electrical insulator T, and

the thermal conductivity ratio $\sigma_\lambda \equiv \lambda_I / \lambda_M$,

the thermal diffusivity ratio $\sigma_\alpha \equiv (\lambda_I / \rho_I c_I)/(\lambda_M/\rho_M c_M)$,

the Biot number $B \equiv hp_*/\lambda_I$,

where λ_I is the thermal conductivity, ρ_I the density, and c_I the specific heat of the electrical insulator, respectively, and h is the heat transfer coefficient on the surface.

The entire zone is now discretized by the centerlines of the conductor strips, namely, by slicing the plate along the centerlines of the strips on one face, then, along those on the other side. Elementary cells, each having the size 1 x 1 x γ, are produced by this discretization process. All the elementary cells except the one at the via hole have an identical structure shown in Figure 5(a), the conductor strips lining the four edges of the insulator block. The cell of Fig. 5(a) will be called hereafter "common cell" or, simply, "cell". Another type of the elementary cell, named "via hole cell", is shown in Figure 5(b), where a quarter of the heater, approximated here by a rectangular body, is present at one of the corners.

Representation of Temperature in Elementary Cells

The cell of Fig. 5(a) has ten facial zones, numbered from 2 to 11. The faces numbered from 2 to 5 bear conductor strips, those of 6 and 7 are exposed to the room air, and those numbered from 8 to 11 are the interfaces with the adjacent cells, or the adiabatic terminal faces of the plate. On the via hole cell of Fig. 5(b) the faces of the via hole is given the number 1. The following assumptions are made regarding the thermal system of the elementary cell.

(1) The temperature is uniform on each conductor segment. This is a reasonable assumption for most electronic components, where the thermal conductivity of the insulator is three orders of magnitude smaller than that of the conductor.
(2) The thermophysical properties of the insulator are independent of temperature. This is also deemed reasonable

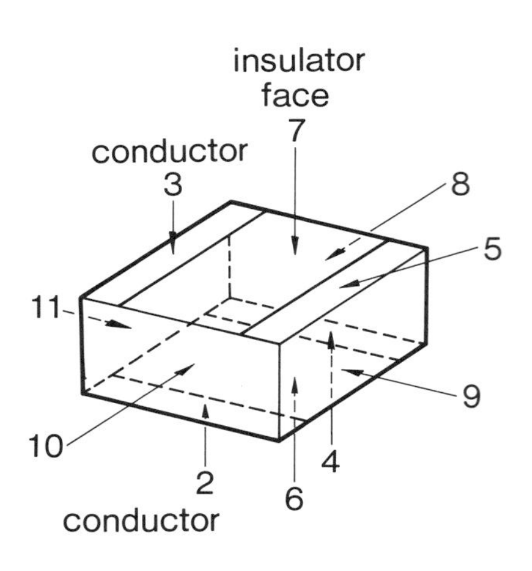

(a) (common) cell

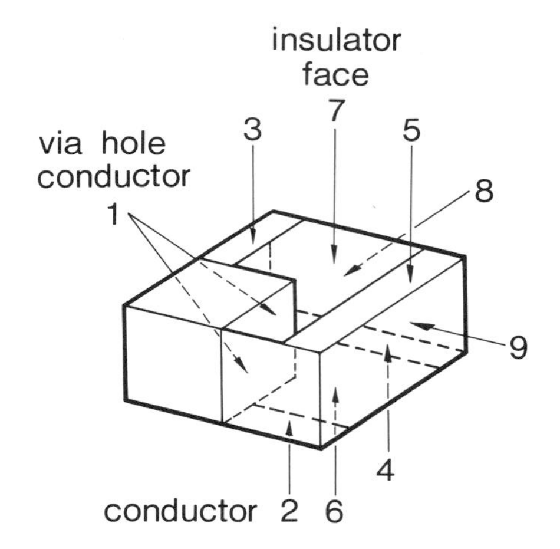

(b) via hole cell

FIGURE 5. Elementary cells : numbering of the faces on the cells

in view of actual situations encountered in electronic components.

(3) Thermal contact resistance at the interface between the conductor and the insulator is negligible.

(4) The details of temperature distribution around the via hole will not be felt in the cells which share the cross sections with the via hole cell. In those adjacent cells, and those cells further removed from the via hole, the topography of isotherms is dictated by the conductor strips on the cell edges, namely, it is insensitive to the details of temperature non-uniformity outside the cell.

Assumption (1) is equivalent to the statement that the conductor temperature varies with a length scale greater than the cell size; hence, it is determined only by the macrosystem analysis. In the representation of the temperature field of the cell the temperatures of the conductor segments are regarded as given boundary temperatures. In order to distinguish this nature of the conductor temperature in the micro-scale representation the temperature of the conductor segment is denoted here by Θ.

Within the cell the insulator temperature, T, has a distribution whose length scale is less than or comparable to the cell size. The distribution of T in the cell is determined under the mixed boundary conditions: isothermal boundaries on the faces S_1 to S_5, convective heat transfer on the faces S_6 and S_7, and conductive heat transfer or adiabatic condition on the faces S_8 to S_{11}, where the suffixes to the S correspond to the numbers defined in Fig. 5. Assumption (4) simplifies the treatment of the conductive heat transfer on the faces S_k (k = 8-11). It is assumed that each of those faces is exposed to an environment, whose temperature is uniform over the face, with the heat transfer coefficient $\lambda_I / p_* / 2$. The "equivalent" environment temperature is the insulator temperature of the adjacent cell, which is determined by the macrosystem analysis. In the microscale representation those environment temperatures are regarded as given quantities, and denoted by the symbols Θ_k (k = 8-11). The "equivalent" heat transfer coefficient represents the conductance between the face and the parallel mid-plane of the adjacent cell. In the non-dimensional formulation the "equivalent" Biot number will be used, which amounts to 2 according to its definition. For convenience of the formulation the Biot number is given a suffix k, where k is any of the face numbers from 8 to 11 defined in Fig. 5.

Based on the above conventions the boundary condition for the insulator temperature in the cell is written as follows:

$$T = \Theta_k \qquad \text{on } S_k , \quad 1 \leqq k \leqq 5, \tag{1}$$

$$\frac{\partial T}{\partial \rho} = B_k (T - \Theta_k) \qquad \text{on } S_k , \quad 6 \leqq k \leqq 11, \tag{2}$$

where ρ is the coordinate extending normal to the face, positive when directed inward.
(Note that the count of k starts from 2 for the cell of Fig. 5(a). This convention will also be applied hereafter.)

Assumption (2) and the linearity of the boundary condition allows the representation of insulator temperature by superposition of the elementary solutions θ_k (k = 1-11).

$$T = \sum_{k=1}^{5} \Theta_k \theta_k + \sum_{k=6}^{11} \beta_k \theta_k \tag{3}$$

where β_ks are the coefficients to be determined later.
The elementary solutions are defined under the following boundary conditions:

For $1 \leqq k \leqq 5$,

$$\theta_k = \begin{cases} 1 & \text{on } S_k , \\ 0 & \text{on } S_n , \quad n \neq k \leqq 5, \end{cases} \tag{4}$$
$$\frac{\partial \theta_k}{\partial \rho} = 0 \qquad \text{on } S_n , \quad 6 \leqq n \leqq 11.$$

For $6 \leqq k \leqq 11$,

$$\theta_k = 0 \qquad \text{on } S_n , \quad 1 \leqq n \leqq 5,$$
$$\frac{\partial \theta_k}{\partial \rho} = 0 \qquad \text{on } S_n , \quad 6 \leqq n \neq k \leqq 11, \tag{5}$$
$$\frac{\partial \theta_k}{\partial \rho} = B_k (\theta_k - 1) \qquad \text{on } S_k .$$

Numerical integration of the heat conduction equation on a system of discretized elements of the cell with the above boundary conditions will produce a group of elementary solutions.

It is apparent that equation (1) is satisfied because only one of the θ_ks , $1 \leq k \leq 5$, is unity on the face S_k and other solutions are zero there. The elementary solutions $\theta_k (1 \leq k \leq 5)$ depend only on the geometry of the cell. They are independent of particular boundary condition given to a macro-system, and also of the real dimension of the cell. Once they are determined for typical cells in the system, they are stored in the data file to be used repeatedly to analytically construct temperature fields in differnt macro thermal systems. The number of cell types is usually not large due to geometrical regularity of conductor arrangements on electronic components; hence, the required space for the storage of elementary solutions in the data file is not large.

The elementary solutions $\theta_k (6 \leq k \leq 11)$ serve to account for the heat flow on the faces S_6 to S_{11}. The rate of heat flow is governed by equation (2), where a pointwise representation of surface temperature will be introduced to facilitate the determination of the coefficients β_k. Namely, the T on the right-hand side of (2) is represented by the value of T at the center of the face, which is to be found from equation (3). A notation is now introduced for the value of the elementary solution at the center of the face: $\theta_{n.k}^*$ is the value of θ_n at the center of the face S_k. With this convention equation (3) is substituted in (2). The following equation results, which yields the values of $\beta_k (6 \leq k \leq 11)$ for a given set of "environment" temperatures:

$$[\ \mathfrak{D}\ ]\{\ \beta_k\ \} = \{\ \hat{\gamma}_k\ \} \tag{6}$$

$$[\ \mathfrak{D}\ ] = \begin{bmatrix} 1 & \theta_{7,6}^* & \cdots & \theta_{11,6}^* \\ \theta_{6,7}^* & 1 & \ddots & \theta_{11,7}^* \\ \cdot & & \cdot & \cdot \\ \cdot & & \cdot & \cdot \\ \cdot & & \cdot & \cdot \\ \theta_{6,11}^* & \cdot & \cdots & 1 \end{bmatrix} \tag{7}$$

$$\hat{\gamma}_k = \Theta_k - \sum_{n=1}^{5} \Theta_n \theta_{n.k}^* \tag{8}$$

Among the elementary solutions $\theta_k (6 \leq k \leq 11)$ those of $8 \leq k \leq 11$ have as wide an applicability as the solutions of $1 \leq k \leq 5$, because the Biot number (= 2) used in the boundary condition for $S_k (8 \leq k \leq 11)$ is independent of the material property and the dimension of the cell. Only the elementary solutions θ_6 and θ_7 depend on a particular thermal boundary condition given to the macrosystem, and, therefore, form a less versatile database than the rest of the elementary solutions.

Shape Factors

The shape factor, $F_{n.k}$, is defined on the conductor segment $S_k (1 \leq k \leq 5)$. It relates the unit temperature potential between the faces S_k and $S_n (1 \leq n \leq 5)$, or between the face S_k and the "environment" for the face $S_n (6 \leq n \leq 11)$, to the temperature gradient on S_k:

$$F_{n,k} = -\left(\overline{\frac{\partial \theta_n}{\partial \rho}}\right)_{S_k} \tag{9}$$

where the overbar on the right-hand side of equation (9) means that averaging is made over the face S_k.

Since the shape factors are based on the elementary solutions, they form a versatile database for cells of equivalent geometry. Except for those based on θ_6 and θ_7, the shape factors can be repeatedly used in analyses for various macroscale thermal boundary conditions. The shape factors based on θ_6 and θ_7 can be used for the cases where the Biot number is the same.

The differentiation of equation (3) with respect to the coordinate normal to the surface S_k yields the following expression for the temperature gradient on the surface S_k, which will be used in the macrosystem analysis:

$$G_k = -\left(\frac{\partial T}{\partial \rho}\right)_{S_k} = \sum_{n=1}^{5} \Theta_n F_{n,k} + \sum_{n=6}^{11} \beta_n F_{n,k} \tag{10}$$

Macrosystem Analysis

The numbering system for the macrosystem analysis is explained in Figure 6, where a cell and coductor segments are shown.

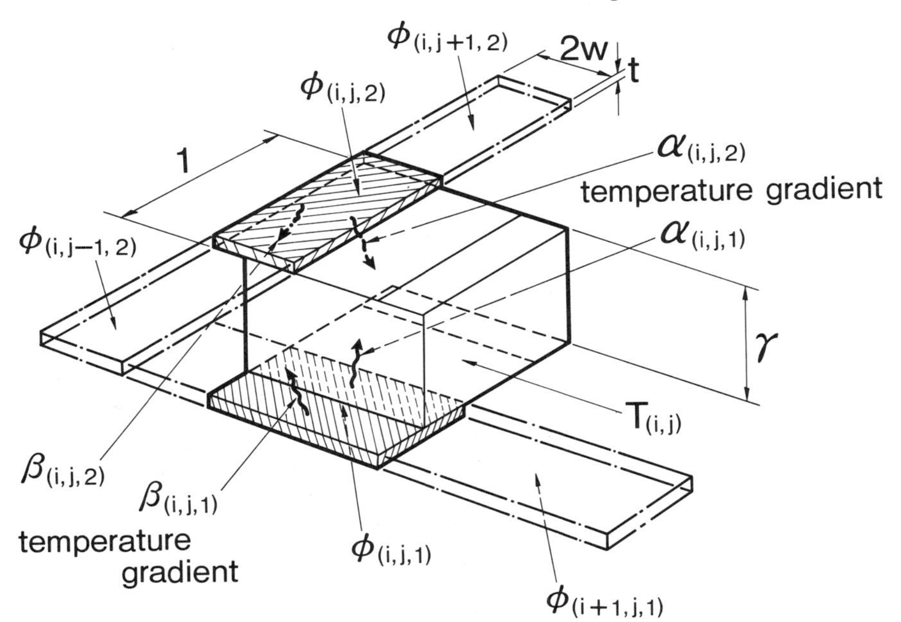

FIGURE 6. Element in the macrosystem analysis.

The index i pertains to the x-direction, and j to the y-direction. The conductor elements (i,j,1), (i,j,2), (i,j+1,1), and (i+1,j,2) are in direct contact with the insulator element (i,j), by half of the interfaces. The function G_k of (10) is given a notation suitable for the macrosystem analysis, namely, $\alpha_{i,j,k}$, or $\beta_{i,j,k}$. The $\alpha_{i,j,1}$ is G_2, and $\alpha_{i,j,2}$ is G_3 for the cell (i,j); $\beta_{i,j,1}$ is G_4 for the cell (i,j-1), and $\beta_{i,j,2}$ is G_5 for the cell(i-1,j).
The heat balance equations are written as follows:

For a conductor element,

$$V_M \frac{\phi_{i,j,k}^{(n+1)} - \phi_{i,j,k}}{\Delta\tau} = S_M(\Delta\phi_+ - \Delta\phi_-) - A_I\sigma_\lambda(\alpha_{i,j,k} + \beta_{i,j,k}) - A_M\sigma_\lambda B(\phi_{i,j,k} - T_\infty) + V_M\dot{q} \qquad (11)$$

where $\Delta\tau$ is non-dimensional time step, k = 1 or 2, V_M = 2wt x 1, S_M = 2wt, A_I = wt, A_M = 2w x 1, $\dot{q}$ is volumetric heat generation of the element, and

$$\Delta\phi_+ = \phi_{i+1,j,1} - \phi_{i,j,1} \quad \text{or} \quad \phi_{i,j+1,2} - \phi_{i,j,2}$$

$$\Delta\phi_- = \phi_{i,j,1} - \phi_{i-1,j,1} \quad \text{or} \quad \phi_{i,j,2} - \phi_{i,j-1,2}$$

For an insulator element,

$$\frac{V_I}{\sigma_\alpha} \cdot \frac{T_{i,j}^{(n+1)} - T_{i,j}}{\Delta\tau} = S_I(T_{i+1,j} + T_{i-1,j} + T_{i,j+1} + T_{i,j-1} - 4T_{i,j}) + A_I(\alpha_{i,j,1} + \alpha_{i,j,2} + \beta_{i,j+1,1} + \beta_{i+1,j,2}) - 2A_S B(T_{i,j} - T_\infty) \qquad (12)$$

where V_I = 1 x 1 x γ, S_I = 1 x γ, A_I = wt, and A_S = 1 x 1 - 2w x 1.

Equations (11) and (12) are written for initial value problems; $\phi_{i,j,k}^{(n+1)}$ and $T_{i,j}^{(n+1)}$ are the values at an instant $(n+1)\Delta\tau$, and those without the superscript are the values at a previous time step. Since the analysis for the elementary cells is performed on the assumption of a steady temperature field within the cell, the limit for the meaningful resolution of computed transient behavior is set by the diffusion time scale for the cell. This means that an analysis performed with the time step $\Delta\tau$ smaller than τ_*/σ_α does not necessarily guarantee a realistic prediction of transient temperature distribution over the macrosystem. A large time step, however, causes a problem of numerical instability. The stability can be gained by use of one of the implicit methods [8]. The transformation of equations (11) and (12) into implicit forms is not reproduced here to save space.

RESULTS AND DISCUSSION

The shape factors, $F_{n,k}$ and $\theta_{n,k}^*$, for the elementary cells of Fig. 5 are shown in Table 1. The non-dimensional dimensions of the conductor segments are $w = 0.1429$, $R = 0.7855$, and the thickness of the plate is $\gamma = 1.143$. The numerical analysis was performed on a 40 x 40 x 20 mesh system.

The macrosystem analysis is performed for the board of Fig. 3 using these shape factors and $\theta_{n,k}^*$. In the following the computed results are compared with the experimental data for steady temperature distributions. The insulator surface temperature is computed from equation (3), and its value at the center of the element surface is compared with the measured temperature. (The $T_{i,j}$ computed from (12) represents the temperature "level" of a lump of insulator, and is too crude to be compared with the measured data.)

Figure 7 shows the traces of infrared thermograms of faces A and B, where the power input to the heater is 1.9 W and the room temperature is 24.5 °C, and the joule heating is absent in the conductor strips ($\dot{q} = 0$). The computed isotherms are also shown in a quarter of the face A. The surface heat transfer coefficient is set at 5 W/(m² K) in the computation, and the thermophysical properties of the metal and the insulator are set at $\lambda_M = 398$ W/(m K), $\rho_M = 8880$ kg/m³, $c_M = 386$ J/(kg K), $\lambda_I = 0.23$ W/(m K), $\rho_I = 1360$ kg/m³, and $c_I = 1465$ J/(kg K). In the experiment the temperatures at some spots on the board were also measured by thermocouples (0.2 mm dia. copper-constantan) glued on the surface, with the estimated accuracy of ±0.1 °C. Those measured temperatures are regarded here as representing the temperatures of the elementary cells, or those of the conductor segments where the thermocouples are attached. The point-to-point comparisons of the computed and measured temperatures are shown in Table 2. The predictions

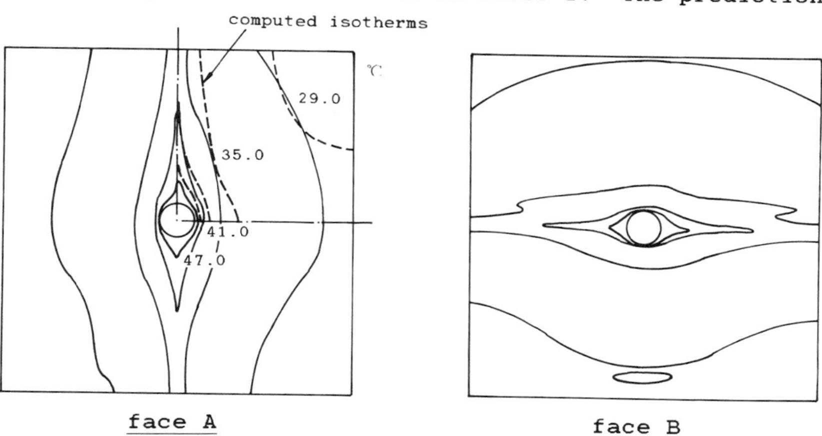

FIGURE 7. Thermograms of the faces and computed isotherms (heater power=1.9 W, room temperature=24.5 °C).

TABLE 1. Shape factors $F_{n,k}$ and $\theta_{n,k}^*$ for the elementary cells of Fig.5

γ=1.143, w=0.1429, t=0.05715, R=0.7855, B=0.30435
n=number of face element where θ=1, or convective boundary condition is given
k=number of face element where Fs or θ*s are defined

Common cell $F_{n,k}$

n \ k	2	3	4	5
2	3.058	-0.8932	-1.269	-0.8933
3	-0.8932	3.058	-0.8932	-1.269
4	-1.269	-0.8932	3.058	-0.8932
5	-0.8932	-1.269	-0.8932	3.058
6	-0.9118	-0.1739	-0.9119	-0.1740
7	-0.1739	-0.9118	-0.1740	-0.9119
8	-0.9572	-1.688	-2.952	-1.688
9	-1.688	-0.9572	-1.688	-2.952
10	-2.952	-1.688	-0.9572	-1.688
11	-1.688	-2.952	-1.688	-0.9572

Common cell $\theta_{n,k}^*$

n \ k	6	7	8	9	10	11
2	0.3591	0.1409	0.1820	0.2455	0.3269	0.2455
3	0.1409	0.3591	0.2455	0.1820	0.2455	0.3269
4	0.3591	0.1409	0.3269	0.2455	0.1820	0.2455
5	0.1409	0.3591	0.2455	0.3269	0.2455	0.1820
6	0.1319	0.02751	0.04727	0.05029	0.04725	0.05028
7	0.02751	0.1319	0.05029	0.04727	0.05028	0.04725
8	0.2063	0.2216	0.5647	0.2579	0.1777	0.2579
9	0.2216	0.2063	0.2579	0.5647	0.2579	0.1777
10	0.2063	0.2216	0.1777	0.2579	0.5647	0.2579
11	0.2216	0.2063	0.2579	0.1777	0.2579	0.5647

Via hole cell $F_{n.k}$

n \ k	1	2	3	4	5
1	1.624	-8.921	-8.921	-6.875	-6.875
2	-0.1756	8.941	-8.35E-6	-3.25E-3	-1.17E-3
3	-0.1756	-8.35E-6	8.941	-1.17E-3	-3.25E-3
4	-0.6370	-1.53E-2	-5.51E-3	6.887	-7.51E-3
5	-0.6370	-5.51E-3	-1.53E-2	-7.51E-3	6.887
6	-5.13E-2	-0.2791	-3.90E-5	-0.2261	-3.05E-4
7	-5.13E-2	-3.90E-5	-0.2791	-3.05E-4	-0.2261
8	-0.7006	-6.36E-3	-1.468	-1.422	-0.4406
9	-0.7006	-1.468	-6.36E-3	-0.4406	-1.422

The faces No.10 and 11 assumed adiabatic.

Via hole cell $\theta_{n.k}^*$

n \ k	6	7	8	9
1	0.2673	0.2673	0.9176	0.9176
2	5,522E-4	1.121E-5	4.461E-4	7.319E-4
3	1.121E-5	5.522E-4	7.319E-4	4.461E-4
4	0.7317	4.845E-4	0.04243	0.03896
5	4.845E-4	0.7317	0.03896	0.04243
6	0.03200	1.274E-5	9.969E-4	1.183E-3
7	1.274E-5	0.03200	1.183E-3	9.969E-4
8	0.02039	0.03730	0.3465	0.1868
9	0.03730	0.02039	0.1868	0.3465

TABLE 2. Comparison of analytical predictions with measured temperatures

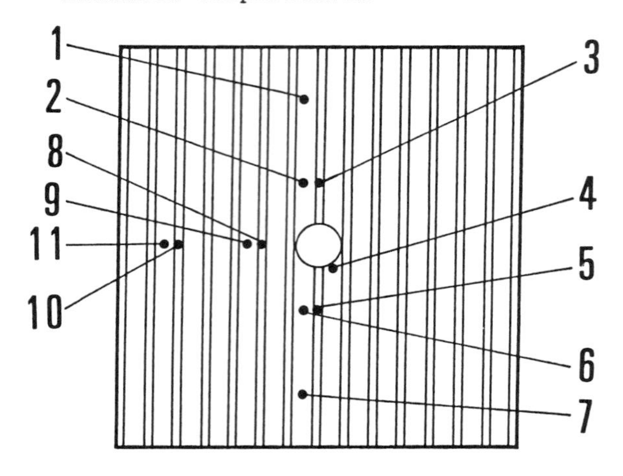

heater power=1.9 W, room temperature=24.5°C

location no.	T measured(°C)	T computed(°C)
1	36.2	37.0
2	39.0	39.9
3	46.9	53.9
4	42.3	44.1
5	44.9	53.9
6	37.3	39.9
7	34.8	37.0
8	33.4	32.6
9	33.1	34.3
10	30.5	29.9
11	30.7	31.9

of temperature by the macroscale analysis are in agreement with the measured data within ±20 %. Some uncertainties are involved in the estimation of heat transfer coefficient, thermophysical properties of the metal and the insulator, and power input to the heater. Also, a certain thermal resistance seems to exist between the heater and the center strip, although an attempt to secure their contact was made by soldering the strip to the via hole ring. Also another uncertainty exists in the approximation of the via hole by a rectangle in the analysis. Those uncertainties may contribute to rather poor accuracies of the prediction for some spots, particularly on 3 and 5. However, the analysis produces a temperature distribution which agrees generally with the measured distribution.

Figure 8 shows the temperature distribution in the via hole cell, which is obtained by substituting the results of the macroscale analysis, Θ_k and $\hat{\beta}_k$, into (3). Note that on the two side faces, which contains the axis of the heater, the adiabatic condition is assumed because of the observed symmetry of temperature field on both sides of each cross section. Even with this enlarged model it requires a certain technique to measure the temperature distribution of sufficient spatial resolution to warrant the comparison with the prediction. Until now such measurements have not yet been performed.

CONCLUDING REMARKS

This paper proposes a methodology of heat conduction analysis which bridges a gap of length scales between microscopic electronic elements and a macroscopic thermal system composed of a great number of microelements. It is pointed out that in most electronic components, such as VLSI chips, ceramic wiring substrates, and printed wiring boards, there exist certain regularities in the geometry of conductor network and the placement of heat generating components. The proposed

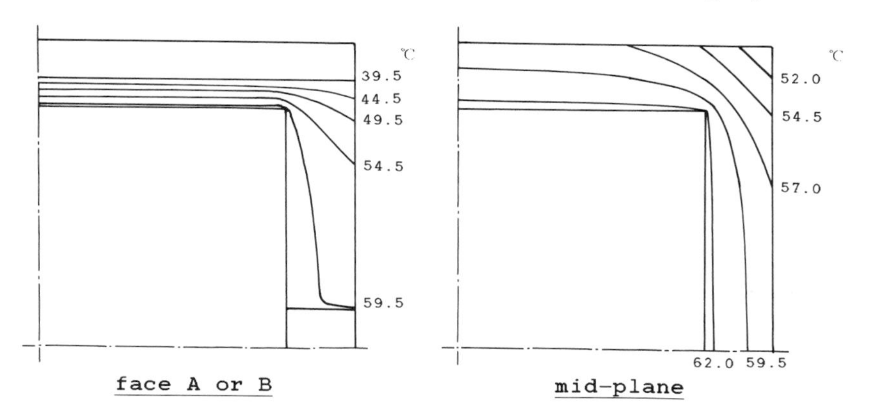

FIGURE 8. Computed temperature distributions in the via hole cell (heater power =1.9 W, room temperature=24.5 °C).

analytical approach exploits such geometrical regularity of microelement arrangement to reduce drastically the number of nodal points to be dealt in the analysis.

The method is applied to an enlarged model of printed wiring board. The temperature distribution measured on this model has proved the soundness of the proposed analytical method. More study is required to fully establish the methodology. One such study is the development of a diagnostic measure to estimate the accuracy of temperature distribution of micro-scales. More detailed measurements of the temperature distribution in an enlarged model will provide information with which the accuracy of the predicted microscale temperature distribution is estimated. Another important task is the establishment of a mathematical basis to support the shape factor approach which until now has been built on intuitive reasoning.

Finally, the potential power of the proposed approach is illustrated by drawing an example from the case of a VLSI chip. When a zone of 1 cm x 1 cm x 5 μm is discretized into meshes of 0.8 x 0.8 x 0.8 μm^3 cubes, the number of nodes becomes close to 10^9, which is far beyond the manageable limit. If the shape factor net is composed by elementary cells of 0.8 x 0.8 x 0.8 μm^3, the required number of nodes can be reduced by six orders of magnitude.

Nomenclature

(Those without the indication of units are non-dimensionalized quantities.)

A_I = half of the interfacial area between the conductor segment and the insulator (m^2)
A_M = surface area of conductor element exposed to the environment (m^2)
A_S = surface area of insulator element exposed to the environment (m^2)
B = Biot number, hp_*/λ_I
c = specific heat capacity (J/(kg K))
$F_{n,k}$ = shape factor defined by (9)
G_k = temperature gradient on the surface S_k, defined by (10)
h = heat transfer coefficient (W/(m^2 K))
p_* = placement pitch of conductor strips (m)
Q = power input to the heater (W)
$\dot{q}$ = volumetric heat dissipation density in a conductor element
R = radius of the via hole
S = surface, or area of the cross section
T = temperature of insulator (- or °C)
T_∞ = temperature of the environment
ΔT_* = temperature scale (°C)
t = thickness of conductor strip
V = volume
w = half of the width of conductor strip

Greek Symbols

α = temperature gradient on conductor segment
β = temperature gradient on conductor segment
$\hat{\beta}$ = coefficient in (3)
Θ = temperature of the medium next to a surface element of insulator cell
θ = elementary solution
$\theta_{n,k}^*$ = value of elementary solution at the center of the surface element S_k
γ = thickness of the board
λ = thermal conductivity (W/(m K))
ρ = density (kg/m^3), or coordinate normal to the surface of cell
σ_α = thermal diffusivity ratio
σ_λ = thermal conductivity ratio
τ_* = time scale (s)
$\Delta\tau$ = time step
ϕ = temperature of conductor element

Suffixes

I = insulator
M = conductor
k = number of surface element
n = number of surface element

Coordinates

x, y, z = Cartesian coordinates

References

1. Nakayama, W., Thermal Management of Electronic Equipment; A Review of Technology and Research Topics, Applied Mechanics Reviews, Vol.39, No.12, pp.1847-1868, 1986.

2. Pinto, E. J. and Mikic, B. B., Methodology for Evaluation of Temperature and Stress Fields in Substrates and Integral Circuit Chips, Heat Transfer In Electronic Equipment - 1986, ASME HTD-Vol.57, pp.209-217, 1986.

3. Cheng, S. C. and Vachon, R. L., The Prediction of the Thermal Conductivity of Two and Three Phase Solid Heterogeneous Mixtures, Int. J. Heat Mass Transfer, Vol.12, pp.249-264, 1969.

4. Cheng, S. C. and Vachon, R. L., A Technique for Predicting the Thermal Conductivity of Suspensions, Emulsions and Porous Materials, Int. J. Heat Mass Transfer, Vol.13, pp.537-546, 1970.

5. Schulgasser, K., On the Conductivity of Fiber Reinforced Materials, J. Mathematical Physics, Vol.17, No.3, pp.382-387, 1976.

6. Ellison, G. N., A Review of a Thermal Analysis Computer Program for Microelectronic Devices, ISHM Technical Monograph Series 6984-003, Thermal Management Concepts In Microelectronic Packaging, Furkay, S. S., Kilburn, R. F., and Monti, Jr., G., ed. The International Society For Hybrid Microelectronics, pp.295-312, 1984.

7. Pinto, E. J. and Mikic, B. B., Temperature Prediction on Substrates and Integrated Circuit Chips, Heat Transfer In Electronic Equipment - 1986, ASME HTD-Vol.57, pp.199-208, 1986.

8. Forsythe, G. E. and Wasow, W. R., Finite-Difference Methods For Partial Differential Equations, John Wiley & Sons, Inc., 1960, pp.101-107.

Conjugate Heat Transfer in Forced Convection Cooling of Chip Arrays

G. S. BAROZZI and E. NOBILE
Istituto di Fisica Tecnica
Facoltà di Ingegneria
Università di Trieste
Via A. Valerio, 10
34100 Trieste, Italy

ABSTRACT

Consideration is given to conjugate heat transfer for two-dimensional developing laminar flow over heat generating blocks protruding in a parallel plate channel. The blocks are intended to simulate IC components surface mounted on a printed circuit board. A contol volume based discretization procedure is used to solve the fully elliptic Navier-Stokes and energy equations. The effects of thermal coupling between the solid and the fluid are discussed.

1. INTRODUCTION

The trend of increasing circuit density and power dissipated per chip poses stringent heat removal problems in semiconductor devices. Reliability considerations actually require the temperature in the junction region not to exceed 125 °C [1], 85°C being now a common standard, and the temperature gradients throughout the device are to be limited so as to reduce thermally induced mechanical stresses. Increasingly more detailed and accurate predictions of the thermal behavior of electronic components are therefore demanded [2].

Thermal design of electronic equipment relies upon the solution of the conduction problem in the components. This is traditionally accomplished by thermal analyzers, which basically solve thermally equivalent resistor networks [3,4]. The technique tends now to be substituted or complemented by more refined 2D or 3D finite element discretizations. Very interesting applications of this kind have been presented [5-8]. In simple geometries, the effects of conduction can also be accounted for analytically, as suggested in [9,10] for the case of printed circuit boards (PCBs). In all of the above cases, however, convection coefficients must be supplied as input data at the solid-to-fluid interfaces.

In many electronic applications it is common practice to assemble integrated circuits (ICs) components into densely packed arrays, mounted on closely spaced PCBs. Forced air convection is the most widely used cooling technique in those cases. Heat transfer correlations for air cooled heat generating block arrays have been developed for various packaging configurations on a theoretical [11-13] or experimental [14-17] basis. It has been pointed out [18] that the interaction of conduction in the solid and convection in the cooling fluid may induce important modifications in the temperature distributions. Also, the application of standard heat transfer correlations, where either the wall temperature or the heat flux are assigned, may lead to unreliable predictions, provided that the thermal boundary conditions at the solid-to-fluid interfaces are not prescribed a priori. Thermal effects of this kind are denoted as conjugate in the heat transfer literature. After Zinnes [19], conjugate heat transfer with wall embedded heat sources have been investigated in [20,21]. The analysis of the flow field is more complex when the electronic components are surface mounted on the PCB and protrude into the air stream. Zebib and Wo [22] investigated 2D forced convection cooling of a single rectangular block in a channel formed by two contiguous PCBs. In the analysis, full account was taken of large inhomogeneities in the thermal conductivity of the materials, typical of electronic equipment. The convection coefficient was assumed consistent with heat transfer from the bottom surface of the PCB, while keeping insulated the upper boundary of the channel. Davalath and Bayazitoglu [23], considered conjugate heat transfer effects in a group of three homogeneous rectangular blocks. They stressed the importance of various geometrical factors, and supplied heat transfer data and correlations for that case. Heat generation was assumed uniform throughout the blocks. Relatively low values were adopted for the thermal conductivity of blocks and PCBs. The effect of heat losses from the PCB's bottom was evaluated by assuming that an equal quantity of heat was transferred to the air stream through the upper channel boundary. The occurrence of thermal periodicity in the cross-stream direction was skillfully accommodated by assuming that no interaction existed between the upper and lower thermal boundary layers. A 2D control volume formulation of the conservation equations was employed in both the analyses [22,23]. Ballister et al.[24] presented a numerical solution for a single cubical block in a air stream as a 3D application of spectral methods. No quantitative data were given.

This paper is intended to complement previous numerical experiments with finite volumes on conjugate heat transfer in shrouded chip arrays. As shown in Fig. 1a, a group of surface mounted IC components is assumed to be cooled by a steady and laminar forced air stream. As in [23], a group of three blocks is considered. In the present case, however, any single chip - chip carrier assembly is schematized as in Fig. 1b. The structure is not aimed at approximating any existing commercial device. It has rather been selected to assess the capabilities of the numerical scheme in the approach of a problem in thermofluids, presenting a few typical features of thermal control in microelectronics, i.e., geometrical complexity and small size of the components, wide variety of materials, and very high conductivity ratios between the solid and the fluid as well as within the solid. Conjugate heat transfer effects are investigated over the full range of laminar flow, and the effects of conduction along and through the PCBs are discussed.

2. STATEMENT OF THE PROBLEM

Periodicity of the flow and the temperature fields is assumed in the cross-stream direction. Consideration is therefore restricted to the single channel formed by two contiguous PCBs shown in Fig. 1.
The cooling fluid is air entering the gap at a uniform velocity and flowing steadily and laminarly through it. Heat is dissipated uniformly inside the junction region - layer 1 in Fig. 1b - at a constant rate per unit depth, Q_g.
From the generative area, S_gA, heat diffuses by conduction in the non-generating silicon layers 2 and 3, the eutectic solder sheet 4, the substrate 5, and the PCB 6. Values of the non-dimensional thickness, S, and the thermal conductivity ratio of the layers, K, are listed in Table 1. Lengths are referred to the channel width, h, and the thermal conductivty of air, k_f=0.026 W/mK, is assumed as a reference value for K's. Two values of K are considered for the PCB. The lower one is representative of epoxy materials, while the higher value is intended to

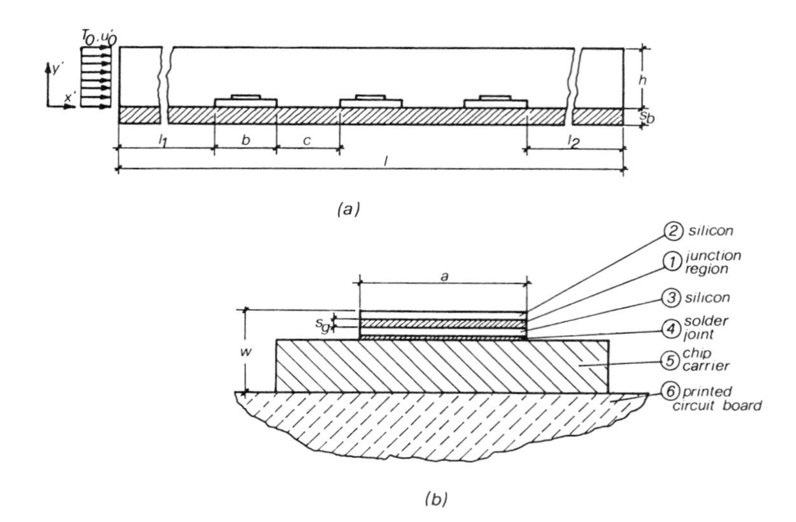

FIGURE. 1. Schematic of the problem: a) general view and coordinate system;
b) detail of the chip and the chip-carrier assembly.

account for the presence of thin metal sheets in a multilayered PCB. Heat is convected to the cooling fluid at any solid to fluid interface.
The fluid properties being held constant, the flow field is uncoupled by energy considerations and can be solved independently from them. The mathematical model of the fluid-dynamic problem is supplied by the two-dimensional continuity and momentum equations for incompressible elliptic flow. Using the non-dimensional quantities defined in the nomenclature, these are written as follows:

$$\frac{\partial u}{\partial x} + \frac{\partial v}{\partial y} = 0 \tag{1}$$

$$\frac{\partial (uu)}{\partial x} + \frac{\partial (vu)}{\partial y} = \frac{1}{Re}\left(\frac{\partial^2 u}{\partial x^2} + \frac{\partial^2 u}{\partial y^2}\right) - \frac{\partial p}{\partial x} \tag{2a}$$

$$\frac{\partial (uv)}{\partial x} + \frac{\partial (vv)}{\partial y} = \frac{1}{Re}\left(\frac{\partial^2 v}{\partial x^2} + \frac{\partial^2 v}{\partial y^2}\right) - \frac{\partial p}{\partial y} \tag{2b}$$

TABLE 1 - Non-dimensional thickness and thermal conductivity ratio of the materials (reference value: k_f=0.026 W/mK).

Layer Nr.	1	2	3	4	5	6
Thickness $S = s/h$	0.007	0.007	0.007	0.002	0.1	0.25
Therm.Cond. $K = k/k_f$	4000	4000	4000	10000	1000	10 4000

The solution of Eqs.(1) and (2) is subject to boundary conditions:

$u = v = 0$	at any solid boundary;
$u = 1 \; ; \; v = 0$	for $\{x=0 \; ; \; 0 \le y \le 1\}$;
$\partial u/\partial x = v = 0$	for $\{x=L \; ; \; 0 \le y \le 1\}$.

A unique form is used to express the conservation principle for energy over the fluid and the solid regions:

$$K(x,y) \left(\frac{\partial (uT)}{\partial x} + \frac{\partial (vT)}{\partial y} \right) = \frac{K(x,y)}{Pe} \left(\frac{\partial^2 T}{\partial x^2} + \frac{\partial^2 T}{\partial y^2} \right) - \frac{F(x,y)}{S_b \, Pe} \qquad (3)$$

Here $K(x,y)$, is the thermal conductivity ratio for the region under consideration ($K=1$ in the fluid region). $F(x,y)$ is unity over the generating layer 1 and zero everywhere else. Note that use has been made of the reference heat flux density, $q_r=Q_g/A$, when definining the non-dimensional temperature T.
Equation (3) holds over the entire computational domain $\{0 \le x \le L; \; -S_b \le y \le 1\}$, provided that the fluid velocity is set equal to zero in the solid.
Boundary conditions for the energy equation are

$T = 0$,	for $\{x=0 \; ; \; 0 \le y \le 1\}$;
$\partial T/\partial x = 0$	for $\{x=L \; ; \; 0 \le y \le 1\}$;
$\partial T/\partial x = 0$	for $\{x=0$ and $x=L \; ; \; -S_b \le y \le 0\}$;
$T(x,y) = T(x,y\pm(1+S_b))$	for $\{ 0 \le x \le L \; ; \; -S_b \le y \le 1\}$.

The last statement expresses the perodicity of the temperature field in the y-direction, and applies as a cyclic condition at the PCB's lower surface and at the upper boundary of the channel.

3. NUMERICAL METHOD

The solution domain is subdivided into small rectangular elements, the "control volumes". The SIMPLEC numerical procedure [25] is used to approach the conservation equations (1) to (3). These are discretized by an averaging procedure over any control volume surrounding a nodal point. A staggered grid structure [26] is adopted where pressure and temperature, as well as any other scalar variable, are defined in the center of the control volume - the nodal point -, while the velocity components are located at the centers of the control volume faces - the centers of the staggered grid. The arrangement is shown in Fig. 2.

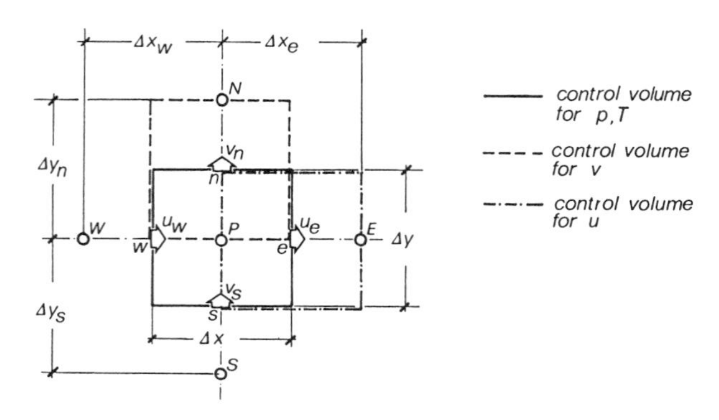

FIGURE. 2. Staggered control volume arrangement.

Following Patankar [27], any of the conservation equations (1) to (3) can be expressed in the generalized form

$$\frac{\partial(u\Phi)}{\partial x} + \frac{\partial(v\Phi)}{\partial y} = \frac{\partial}{\partial x}\left(\Gamma \frac{\partial \Phi}{\partial x}\right) + \frac{\partial}{\partial y}\left(\Gamma \frac{\partial \Phi}{\partial y}\right) + S^{\phi} \quad (4)$$

where Φ is a general variable, and Γ is a diffusion coefficient. The source term, S^{ϕ}, is linearized as follows:

$$S^{\phi} = S_c^{\phi} + \Phi_P \, S_b^{\phi} \, ,$$

and Eq. (4) discretized is

$$\Phi_P \, (a_P^{\phi} - S_b^{\phi}) = a_E^{\phi} \, \Phi_E + a_W^{\phi} \, \Phi_W + a_N^{\phi} \, \Phi_N + a_S^{\phi} \, \Phi_S + S_c^{\phi} \quad (5)$$

This form has been adopted for the energy equation (3).
The E-factor formulation suggested by Raithby [25] has been preferred for the solution of the momentum equations (2), and the discretized equation (4) accordingly rewritten

$$\Phi_p \, (a_p^{\phi} - S_b^{\phi}) \left(1 + \frac{1}{E}\right) = \sum\nolimits_{nb} a_{nb}^{\phi} \, \Phi_{nb} + S_c^{\phi} + \Phi_p^{o} \, (a_p^{\phi} - S_b^{\phi})/E \quad (6)$$

Here, as in (4), the summation extends to the surrounding -nb- control volumes and Φ_p^{o} is the value of Φ_p at the previous iteration cycle. The scheme represented by Eq. (6) is known to be definitely more stable than the one given by Eq. (5). In Eqs. (5) and (6), the coefficients are allowed to be functions of Φ, so as to account for non-linearities and possible inter-equation linkages. In the iterative process, intermediate solutions of the set of discretized equations are performed with "fixed" coefficients at each cycle. Convection-diffusion fluxes are approximated by the Power-law discretization scheme [27,28]. In previous numerical experiences [29,30], the procedure had proven to give a reasonable compromise between simpler, e.g., first order upwind, and more complex schemes, such as QUICK and its derivatives [31,32], in terms of accuracy and computational effort. In the course of the present work, values of E ranged from 5 to 20, depending on the value of Re.
By using the power-law scheme, the coefficients in Eqs. (5) and (6) are expressed as follows, $\langle\langle A, B \rangle\rangle$ denoting the maximum between A and B:

$$\begin{aligned}
a_E^{\phi} &= (\Gamma/\Delta x_e) \, \langle\langle 0, \{1 - \mathrm{mod}\,(u_e \Delta x_e/\Gamma)/10^5\} \rangle\rangle + \langle\langle -u_e, \; 0 \rangle\rangle \\
a_W^{\phi} &= (\Gamma/\Delta x_w) \, \langle\langle 0, \{1 - \mathrm{mod}\,(u_w \Delta x_w/\Gamma)/10^5\} \rangle\rangle + \langle\langle u_w, \; 0 \rangle\rangle \\
a_N^{\phi} &= (\Gamma/\Delta y_n) \, \langle\langle 0, \{1 - \mathrm{mod}\,(v_n \Delta y_n/\Gamma)/10^5\} \rangle\rangle + \langle\langle -v_n, \; 0 \rangle\rangle \qquad (7) \\
a_S^{\phi} &= (\Gamma/\Delta y_s) \, \langle\langle 0, \{1 - \mathrm{mod}\,(v_s \Delta y_s/\Gamma)/10^5\} \rangle\rangle + \langle\langle v_s, \; 0 \rangle\rangle \\
a_P^{\phi} &= \sum\nolimits_{nb} a_{nb}^{\phi}
\end{aligned}$$

Equation (6) gives a system of linear equations for the solution of the velocity components u and v at the interior nodes at each cycle. The solution is started with a guessed pressure field. The corresponding velocity field does not satisfy the continuity equation (1). Velocity and pressure are thus modified at the end of any iteration cycle in order to satisfy continuity. Use is made of the parameters

$$p = p^* + p' \; ; \quad u = u^* + u' \; ; \quad v = v^* + v' \tag{8}$$

Starred values represent the solution obtained with guessed pressure values p^*, and u', v' and p' are the corrections needed to satisfy continuity. Equations (8) are substituted in the momentum and the continuity equations. A linear system for the pressure correction p' is obtained after discretization in the form

$$a_P \, p_P' = a_E \, p_E' + a_W \, p_W' + a_N \, p_N' + a_S \, p_S' + b \tag{9}$$

The coefficients in Eq. (9) are written following the SIMPLEC algorithm scheme [25] as

$$a_E = \frac{(\Delta y)^2}{\left(a_e^u - \sum_{nb} a_{nb}^u\right)} \; ; \qquad a_W = \frac{(\Delta y)^2}{\left(a_w^u - \sum_{nb} a_{nb}^u\right)}$$

$$a_N = \frac{(\Delta x)^2}{\left(a_n^v - \sum_{nb} a_{nb}^v\right)} \; ; \qquad a_S = \frac{(\Delta x)^2}{\left(a_s^v - \sum_{nb} a_{nb}^v\right)} \tag{10}$$

$$a_P = \sum_{nb} a_{nb} \; ; \quad b = -(u_e^* \, \Delta y) + (u_w^* \, \Delta y) - (v_n^* \, \Delta x) + (v_s^* \, \Delta x)$$

The solution scheme for the flow field is as follows: 1. the velocity components are obtained from Eqs.(6), for guessed values of pressure, p^*; 2. the pressure correction equation (9) is solved; 3. new values are obtained for p, u, and v, from Eqs.(8); and 4. these values are used as entry values for the next cycle. The procedure is repeated until convergence is achieved. The Euclidean norm of the mass residual - the source term in Eq.(9) - referred to the total mass flow rate, is used here to control the convergence of the overall procedure. A criterion of 1×10^{-6} is used.
Since the solution of the energy equation does not affect the flow field, the set of equations (5), is solved for T only once, after the velocity distribution has been obtained. To account for the variation of the diffusion coefficient -the thermal conductivity in the energy equation- use has been made of the harmonic mean rule [33].

The solution of the momentum equations has been performed with the so-called Alternate Line Zebra Relaxation - ALZR - scheme. One application of ALZR consists of sequential relaxations of odd lines, even lines, odd columns and even columns. The technique has been specifically adapted for the staggered grids encountered in control volume formulations. The relaxation scheme is particularly suited for equations presenting strong anisotropies in the coefficients [34], and it has been proven to be advantageous for parallel processing on multi-processor machines. The pressure correction equation is a Poisson-like equation with Neumann's type boundary conditions. It is solved by a fast Additive Correction Multigrid -ACM- scheme [25,35]. The ALZR scheme is used here as a smoother for the multigrid cycling.
Even if linear, the solution of the energy equation has posed two specific problems: i- the boundary conditions at the lower PCB boundary and at the upper channel boundary are of the cyclic type. Namely, the wall temperature distributions at $y = -S_b$ and $y = 1$ are set equal, but the values are not given a priori. This approach is more general than the one used in [23], since it mantains its validity for any value of the Péclet number; ii - variations of the diffusion coefficients - the thermal conductivities of the materials - as large as 1×10^4 are encountered in this problem. Following Patankar et al. [36], the cyclic boundary conditions have been dealt with by using the Cyclic Tridiagonal Matrix Algorythm - CTDMA -, instead of the traditional Tridiagonal Matrix

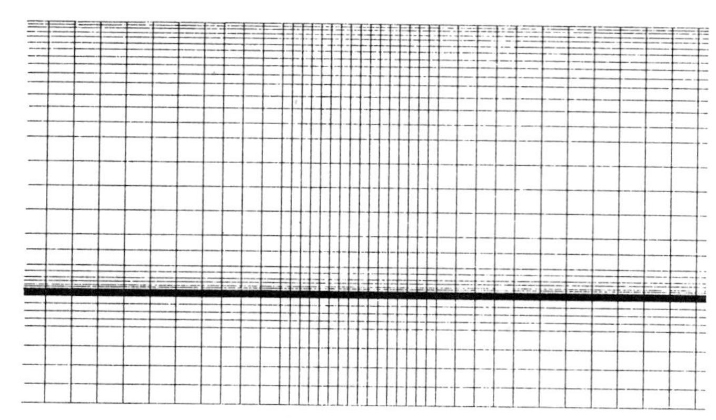

FIGURE. 3. View of the computational grid in the region of chip-carrier assembly.

Algorythm - TDMA -, in the iterative solution of the system of linear equations in the y direction. The second point gives rise to very strong anisotropies in the coefficients, which in turn render standard iterative solvers - SOR, SLOR - very inefficient in smoothing out the lowest frequency modes of the error [37]. That drawback had been encountered in early attempts of tackling the problem considered here. It seems also to be at the origin of the very low rate of convergence noticed in [22], where about 3000 iterations were needed to achieve a relatively low level of convergence. The problem has been solved through the Block Correction Procedure - BCP - devised in [38], specifically modified to deal with cyclic boundary conditions. In the solution of the energy equation, BCP is invoked after each application of ALZR.

The above procedures have been incorporated in the computer code CVFOR - Control Volume Forced Convection -, which had already undergone extensive testing and validation. Comparison with available solutions for elliptic entry flow in a parallel plane duct is of particular concern here. This has been done in [29], where agreement with the McDonald et al.[39] and Wang and Longwell [40] solutions was found to be very satisfactory. Comparison has also been done with Davalath and Bayazitoglu [23] graphical results for rectangular blocks, Re = 750 and 1500. Temperature contour plots have demonstrated a good agreement. Minor discrepancies should be imputed to the use of different meshes and different discretization schemes.
A non-uniform mesh is adopted, with higher concentration of nodes in the regions where the property gradients are steeper or the thinnest conductive layers are located. A detail of the grid in the region neighboring a chip - chip carrier assembly is shown in Fig. 3. The computational domain is subdivided into 136 control volumes along the x-direction and 48 control volumes in the y-direction.

Starting with a zero velocity distribution, the flow solution took 250-300 iterations, depending on the value of Re. In order to reduce the computational costs, restart capabilities of the code were fully employed. The temperature field was obtained with 200-300 iterations, depending on the Péclet number, to reduce the Euclidean norm residual to $1x10^{-8}$ of its initial value. To this level of convergence, the temperature at the nodal points remained constant up to seven digits.
Computations have been carried out on a CRAY X-MP/48 using CFT 1.15 BF3 FORTRAN 77 Compiler in single processor mode. The flow field solution required 60 to 80 CPU seconds, and 8 to 12 CPU seconds were needed to obtain the temperature field.

4. RESULTS AND COMMENTS

Computations have been performed for Re = 100, 250, 500, 750, 1000, and 1500, assuming for air Pr = 0.71. The following input data were used in all the cases: L = 17.5, L_1 = 3., L_2 = 9.5, A = 0.5 , B = 1., C = 1., S_b = 0.25, W = 0.123. Other thermal and geometrical properties are as given in Table 1.
A relatively high value has been choosen for L_2, to accommodate the outlet boundary conditions.

The velocity field for Re = 1500, is shown in Fig. 4 as a sample case. Over most of the domain, the velocity vector remains practically horizontal, indicating that the constriction effect is felt only in a relatively narrow fluid region neighboring the obstacles. This was expected, since the height of the blocks is small, less than 1/8 of the channel size. The influence of the obstacles increases slightly for lower Re, but always remains moderate. The most interesting features of the flow field are seen in Fig. 5, where the region between the second and the third block is enlarged, for Re = 250, and 1500. The recirculation bubble downstream of the chip-carrier progressively lengthens for increasing Re, and, for Re ≥ 1000, completely fills the gap between the obstacles. A very small recirculation region is also created downstream of the chip, at Re=1500. As in [22,23], no separation has been detected at the upstream corners of the chips and the substrates.

A parabolic velocity profile had been considered as the entry condition in [22,23]. This case has been tested here for the two cases, Re = 250, and 1500. When comparing the results for a uniform and a parabolic entry profile, only minor differences are observed in the flow field visualizations. They, however, definitely affect the pressure drop and heat transfer performances. Values of Δp for the two cases are listed in Table 2. For the sake of comparison, data for parallel plate ducts are also given from the literature [41]. The effect on the maximum chip-junction temperature can be seen in Table 3.

TABLE 2. Total pressure drop for parabolic and uniform entry, as compared to reference values, Δp_r , for parallel plate ducts [41]

Inlet Condition	Parabolic		Uniform	
Re	250	1500	250	1500
Δp	1.655	0.290	2.356	0.807
Δp_r	1.596	0.266	2.143	0.709

TABLE 3. Maximum junction temperature at various Re values (data in brakets refer to the parabolic entry condition)

Re	100	250	500	750	1000	1500
T_{max} (K_b = 10)	.051	.037 (.038)	.029	.025	.023	.019 (.021)
T_{max} (K_b=4000)	.023	.014 (.016)	.011	.0094	.0083	.0072 (.0087)

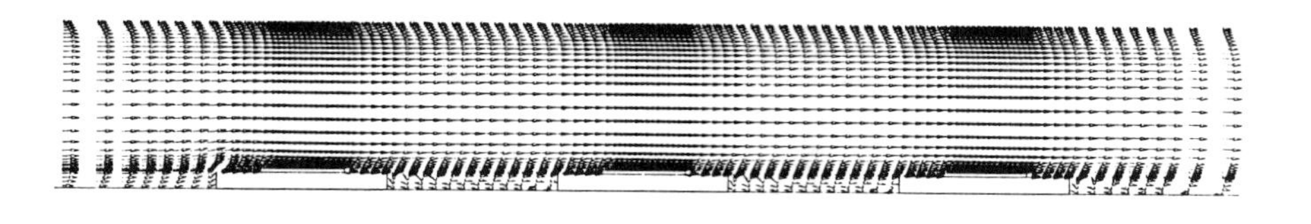

FIGURE. 4. Velocity field for Re = 1500.

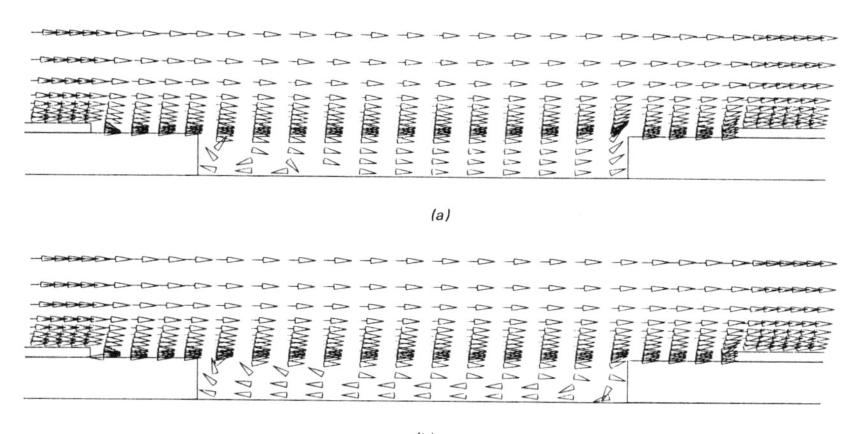

FIGURE 5 Velocity field ion the region between two chip-chip carrier assemblies, a) Re = 250 ; b) = 1500.

Temperature contour plots are presented in Figs. 6 and 7. These figures respectively refer to the case of low and high PCB conductivity. The higher concentration of isotherms observed in the proximity of the solid walls demonstrates that the efficiency of convection cooling improves for increasing Re. The effect of conduction along the PCB is also quite evident from the temperature plots. In fact, isotherms of equal value detach from the PCB surface well upstream for K_b = 4000 (e.g., compare contour level 0.01 in Figs. 6b and 7b). This indicates that the fin effect of the PCB increases with K_b. On the other hand, the conjugate heat transfer effect becomes stronger as the Péclet number is reduced. This is consistent with existing conjugate solutions for thermal entry in ducts. The upper and the lower thermal boundary layers show very little interference for K_b =10. This gives qualitative validation to the Davalath and Bayzitoglu [23] results. For K_b = 4000, however, the interaction becomes significant at the lower Re value, and the use of cyclic boundary conditions is necessary.

Temperature maxima occur inside the generating region of the chip. The influence of the board conductance is once again evident. While in fact the highest T-value is always found in the third block, for K_b = 10, the intermediate assembly becomes the more thermally stressed for K_b = 4000, case Re = 100 being the only exception. Enlarged views of temperature distributions in the hottest chip assemblies are presented in Fig. 8, for two cases. Data indicate that, due to the high conductivities of the materials, the chip, the solder and the substrate are nearly isothermal.

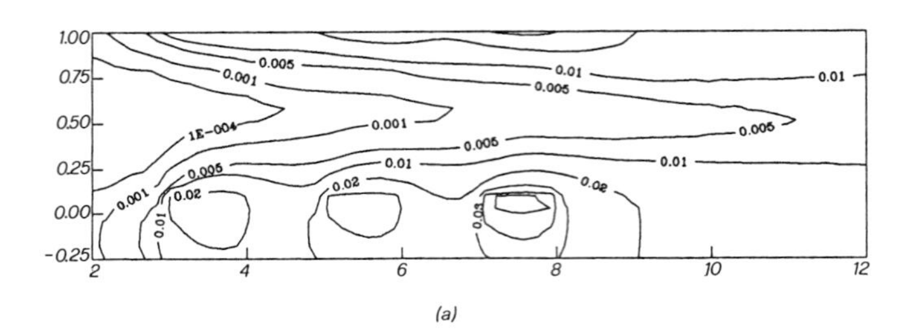

(a)

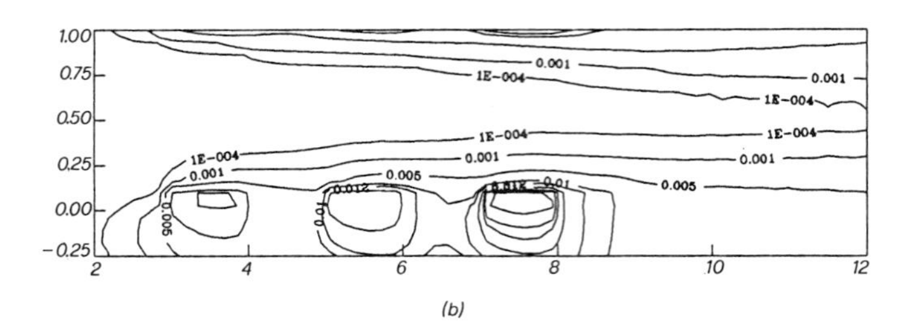

(b)

FIGURE 6 Temperature contour plots, K_b = 10, a) Re = 250 ; b) Re = 1500.

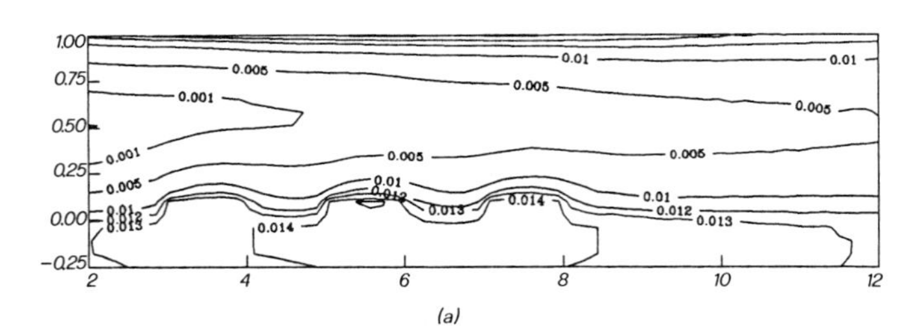

(a)

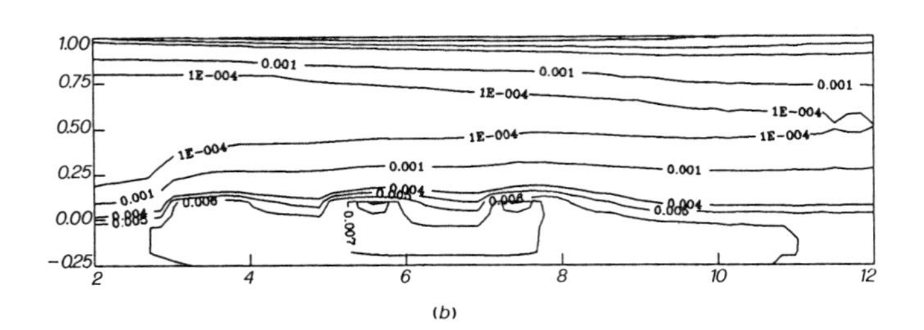

(b)

FIGURE 7 Temperature contour plots, K_b = 4000, a) Re = 250 ; b) Re = 1500.

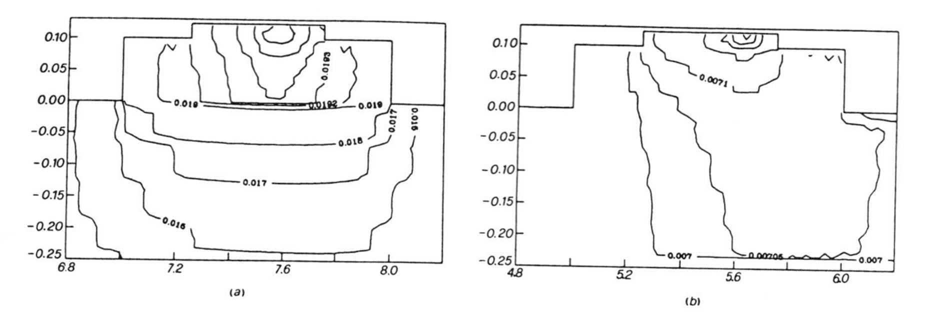

FIGURE 8 Temperature distribution in chip-chip carrier assemblies, Re = 1500. a) K_b = 10 (3rd block); b) K_b = 4000 (2nd block).

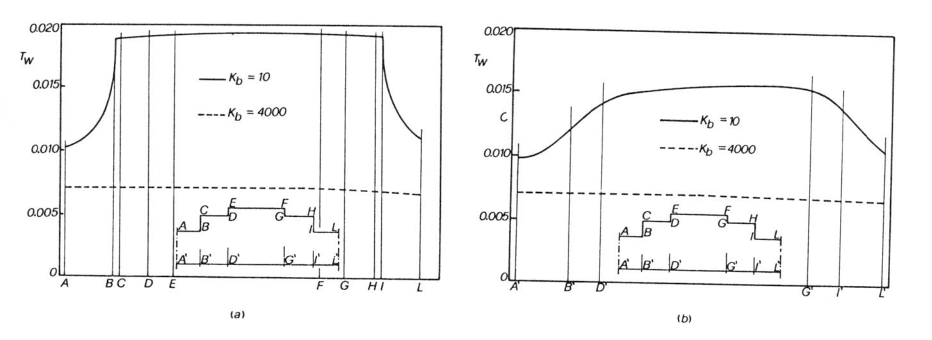

FIGURE 9 Distributions of wall temperature, T_w, a) at the chip surface;
b) at lower PCB surface (3rd block).

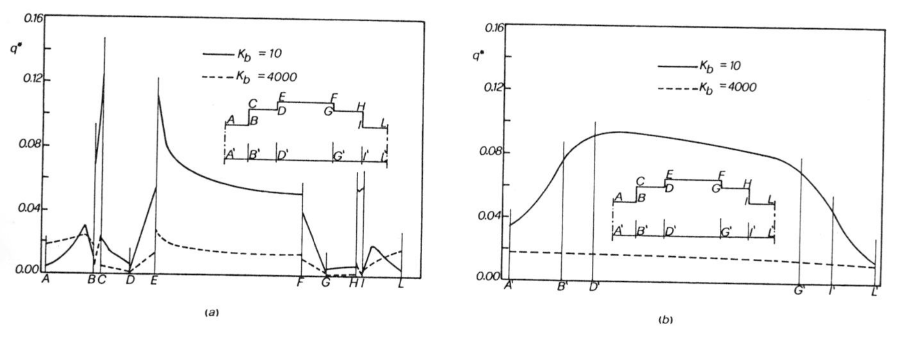

FIGURE 10 Distributions of normalized heat flux density, q*, a) at the chip surface; b) at lower PCB surface (3rd block).

Anyway, it may be worth to point out that the maximum value for T is found at about 0.66 A from leading edge of the chip. If a uniform value of the convection coefficient had been used, the maximum would be positioned at 0.5 A. Assuming a thermal boundary layer is developing over the chip surface, the hottest point would be closer to the downstream extremity. Maxima for T are presented in Table 3 for all the cases considered. The data confirm the influence of the flow rate, the PCB thermal conductivity, and the entry flow condition, as discussed above.

Heat transfer data in the form of Nusselt number distributions are of little practical significance in conjugate heat transfer problems, since both the heat flux density and the wall temperature are not specified. Plots of T_w and q^*, the normalized heat flux, at the chip and board surfaces are presented in Figs. 9 and 10, respectively. These are representative of typical heat transfer results for this problem.
For $K_b = 10$, the PCB acts as a thermal barrier causing the increase of the chip assembly temperature level shown in the figure. At $K_b = 4000$, the heat spreader effect of the PCB is quite strong, and the wall temperature becomes practically uniform, both at the upper and lower interfaces. The heat flux density distributes coherently. Plots in Fig. 10 indicate that at low values of K_b, most of the heat generated at the junction is either directely lost at the upper chip surface, or convected to the fluid through the lower PCB surface. For reducing K_b, the heat loss from the PCB surface is reduced on the whole, but concentrates below the chip-carriers. Fig. 10 also indicates some important thermal effects of the local flow configuration on heat transfer. The more effective regions seem to be the leading edge of the chip, and the vertical surfaces of the substrate. Further inspections with local grid refinements would be required to confirm those observations. The heat flux distribution is quite uniform for $K_b = 4000$, due to the strong fin effect of the PCB.

5. CONCLUDING REMARKS

A control-volume-based formulation of the conservation equations is presented, to deal with the numerical modelling of conjugate heat transfer problems encountered in the thermal control of surface mounted IC components.
The study complements previous investigations [22,23] on forced air convection in parallel plate channels formed by contiguous PCBs.
The procedure has been demonstrated to be computationally very efficient, even in the case of high anisotropies in the diffusion coefficients, and allows cyclic thermal boundary conditions to be easily accommodated.
Results are given for a sample geometrical configuration and composition of the materials. They indicate that the effect of the coupling of convection with conduction in the solid can strongly affect the thermal behavior of IC components, making the temperature and heat flux distributions very difficult to predict by conventional methods. The overall accuracy of the prediction is, however, found to be definitely influenced by a proper choice of boundary conditions for the flow field. Here the effect of the entry velocity profile has been pointed out.
Finally, consideration has been given to the effect of the PCB thermal conductivity for two extreme values.
The analysis has been restricted to 2D laminar forced convection. Further investigation should consider the effects of mixed convection, as well as the extension to transitional and fully turbulent regimes.

6. NOMENCLATURE

A	dimensionless chip length, a/h	
a	chip length	m
B	dimensionless chip-carrier length, b/h	
b	chip-carrier length	m
C	dimensionless distance between carriers, c/h	
c	distance between carriers	m
c_f	specific heat of the fluid	J/kgK
E	relaxation parameter in discretized Eq.(6)	
h	height of the channel	m
K	dimensionless thermal conductivity, k/k_f	
k	thermal conductivity of the material	W/mK
k_f	thermal conductivity of the fluid	W/mK
L	dimensionless length of the channel	
l	length of the channel	m
Pe	Peclet number, Re.Pr	
Pr	Prandtl number, $\mu.c_f/k_f$	
p	dimensionless pressure, $p'/\rho.u'_0{}^2$	
p'	pressure	Pa
Q_g	heat power dissipation per unit length	W/m
q_r	reference heat flux density, Q_g/a	W/m^2
q*	normalized heat flux, q/q_r	
q	heat flux density	W/m^2
Re	Reynolds number, $\rho.u'_0.h/\mu$	
s	thickness	m
S	dimensionless thickness, s/h	
T	dimensionless temperature $(T'-T'_0).k_f/(q_r.h)$	
T'	temperature	K
u	dimensionless axial fluid velocity, u'/u'_0	
u'	axial fluid velocity	m/s
v	dimensionless transverse fluid velocity v'/u'_0	
v'	transverse fluid velocity	m/s
x	dimensionless axial coordinate, x'/h	
x'	axial coordinate	m
y	dimensionless transverse coordinate, y'/h	
y'	transverse coordinate	m
W	dimensionless height of chip-carrier assembly, w/h	
w	height of chip-carrier assembly	m

Greek

Γ	generic diffusion coefficient, Eq. (4)	
μ	dynamic viscosity coefficient	Pa.s
ρ	fluid density	kg/m^3
Φ	generic transported property, Eq. (4)	

Subscripts

b	board
g	generating
o	initial value at x'= 0
W	wall

7. REFERENCES

1. A.D. Kraus, and A. Bar-Cohen - *Thermal Analysis and Control of Electronic Equipment*, Hemisphere Publishing Corporation, Washington, U.S.A, 1983.
2. S. Ramadhyani, E. Egan, B. Mikic, and J.I. Tustaniwskyj, Computer Modelling, Session 5, in *Procs. of the Workshop Research Needs in Electronic Cooling*, ed. F.P. Incropera,pp. 52-60, Andover, Mass., U.S.A., 1986.

3. D.S. Steinberg, *Cooling Techniques for Electronic Equipment, J.Wiley and Sons*, New York, U.S.A.,1980.
4. G.N. Ellison, *Thermal Computations for Electronic Equipment*, Van Nostrand Reinhold,New York, 1984.
5. N. Abuaf, and V. Kadambi, Effects of Voids on the Thermal Resistance of Power Chips, *Heat Transfer in Electronic Equipment*, ASME-HTD, vol.48, pp. 69-75, 1985.
6. V. Kadambi, and N. Abuaf, Numerical Thermal Analysis of Power Chip Packages, *Heat Transfer in Electronic Equipment*, ASME-HTD, vol.48, pp.77-84, 1985.
7. M.Bonnifait, and H. Charlier, Analyse thermique tridimensionelle par éléments finis. Application aux composants électroniques, *Revue General de Thermique*, no.280, pp. 455-463, 1985.
8. M.Bonnifait, and M. Cadre, Thermal Simulations for Electronic Components Using Finite Elements and Nodal Networks, *Heat Transfer in Electronic Equipment*, ASME-HTD, vol.57, pp.183-188, 1986.
9. L.M. Simeza, and M.M. Yovanovich, Application of BIEM to Thermal Analysis of Multiple Sources on PCBs, *Heat Transfer in Electronic Equipment*, ASME-HTD, vol.57, pp.161-166, 1986.
10. K.J. Negus and M.M. Yovanovich, Thermal Analysis and Optimization of Convectively Cooled Microelectronic Circuit Boards, *Heat Transfer in Electronic Equipment*, ASME-HTD, vol.57, pp.167-175, 1986.
11. M.E. Braaten, and S.V. Patankar, Analysis of Laminar Mixed Convection in Shrouded Arrays of Heated Rectangular Blocks, *Fundamentals of Natural Convection/Electronic Equipment Cooling*, ASME-HTD, vol.32, pp.77-84, 1984.
12. S. Habchi, and S. Acharia, Laminar Mixed Convection in Partially Blocked Vertical Channel, *Heat Transfer in Electronic Equipment*, ASME-HTD, vol.57, pp.189-197, 1986.
13. K.J. Kennedy, and A. Zebib, Combined Free and Forced Convection Between Horizontal Parallel Planes: Some Case Studies, *Int. J. Heat Mass Transfer*, vol.26, pp.471-474, 1983.
14. E.M. Sparrow, J.E. Niethammer, and A. Chaboki, Heat Transfer and Pressure Drop Characteristics of Arrays of Rectangular Modules Encountered in Electronic Equipment, *Int. J. Heat Mass Transfer*, vol.25, pp.961-973, 1982.
15. N. Ashiwake, W. Nakayama, T. Daikoku, and F. Kobayashi, Forced Convective Heat Transfer from LSI Packages in an Air-Cooled Wiring Card Array, *Heat Transfer in Electronic Equipment*, ASME-HTD, vol.28, pp.35-42, 1983.
16. G.L. Lehmann, and R.A. Wirtz, The Effect of Variations in Stream-wise Spacing and Length on Convection from Surface Mounted Rectangular Components, *Heat Transfer in Electronic Equipment*, ASME-HTD, vol.48,pp.39-47, 1985.
17. R.J. Moffat, D.E. Arvizu, and A. Ortega, Cooling Electronic Components: Forced Convection Experiments with an Air-cooled Array, *Heat Transfer in Electronic Equipment*, ASME-HTD, vol.48, pp.17-27, 1985.
18. R. Hannemann, F. Incropera, R. Simons, Single Phase Liquid Cooling, Session 2, in *Procs. of the Workshop Research Needs in Electronic Cooling*, ed.F.P. Incropera, pp. 6-25, Andover, Mass., U.S.A., 1986.
19. A.E. Zinnes, The Coupling of Conduction with Laminar Convection from a Vertical Flat Plate with Arbitrary Surface Heating, *Trans. ASME, Ser. C, Jl. Of Heat Transfer*, vol.92, pp. 507-516, 1970.
20. S. Ramadhyani, D.F. Moffat, and F.P. Incropera, Conjugate Heat Transfer from Small Isothermal Heat Sources Embedded in a Large Substrate, *Int. J. Heat Mass Transfer*, vol.28, pp.1945-1952, 1985.
21. D.F. Moffat, S. Ramadhyani, and F.P. Incropera, Conjugate Heat Transfer from Wall Embedded Sources in Turbulent Channel Flow, *Heat Transfer in Electronic Equipment*, ASME-HTD, vol.57, pp. 177-182, 1986.
22. A. Zebib, and Y.K. Wo, A Two-Dimensional Conjugate Heat Transfer Model for Forced Air Cooling of an Electronic Device, Procs.

International Electronic Packaging Conference, Orlando, Florida, pp.135-142, 1985.

23. J. Davalath, and Y. Bayazitoglu, Forced Convection Cooling Across Rectangular Blocks, *Trans. ASME, Ser. C, Jl. of Heat Transfer*, vol. 109,pp.321-328, 1987.
24. E.T. Bullister, G.E. Karniadakis, B.B. Milic, and A.T. Patera,A Spectral Method Applied to the Cooling of Electronic Components, *Heat Transfer in Electronic Equipment*, ASME-HTD, vol.57, pp.153-160, 1986.
25. J.P. Van Doormaal, and G.D. Raithby, Enhancements of the SIMPLE Method for Predicting Incompressible Fluid Flows, *Num. Heat Transfer*, vol.7, pp. 147-163, 1984.
26. F.H. Harlow, and J.E. Welch, Numerical Calculation of Time-Dependent Viscous Incompressible Flow of Fluid with Free Surface, *Phys. Fluids*, vol. 8, 2182-2189, 1965.
27. S.V. Patankar, *Numerical Heat Transfer and Fluid Flow*, McGraw-Hill, New York, 1980.
28. S.V. Patankar, A Calculation Procedure for Two-Dimensional Elliptic Situations, *Num. Heat Transfer*, vol. 4, pp.409-425, 1981.
29. G.S. Barozzi, and E. Nobile, Low Reynolds Number Heat Transfer and Fluid Flow in the Inlet region of Parallel Plates, *Procs. 6th UIT Conf.*, pp.141-152, Bari, Italy, 1988.
30. E. Nobile, F. Scotti, and C. Vecile, Analisi numerica di problemi di convezione naturale in cavità rettangolari: confronto fra volumi di controllo e elementi finiti, *Procs. 6th UIT Conf.*, pp.13-24, Bari, Italy, 1988.
31. B.P. Leonard, A Stable and Accurate Convective Modelling Procedure Based on Quadratic Upstream Interpolation, *Computer Meth. Appl. Mech. Eng.*, vol. 19, pp. 59-98, 1979.
32. A. Pollard, and A. L.-W. Siu, The Calculation of Some Laminar Flows Using Various Discretisation Schemes, *Computer Meth. Appl. Mech. Eng.*, vol. 35, pp. 293-313, 1982.
33. S.V. Patankar, A Numerical Method for Conduction in Composite Materials, Flow in Irregular Geometries and Conjugate Heat Transfer, *6th Int. Heat Transfer Conference*, vol. 3, pp. 297-302, Toronto, Canada, 1978.
34. P. Sonneveld, P. Wesseling, and P.M. de Zeeuw, Multigrid and Conjugate Gradient Methods as Convergence Acceleration Techniques, in *Multigrid Methods for Integral and Differential Equations*, edrs. D.J. Paddon, and H. Holstein , Clarendon Press, Oxford, 1985.
35. A. Brandt, Multi-Level Adaptive Solutions to Boundary-Value Problems, *Math. Comput.*, vol.31,1977. pp.333-390.
36. S.V. Patankar, C.H. Liu, and E.M. Sparrow, Fully Developed Flow and Heat Transfer in Ducts Having Streamwise-Periodic Variations of Cross-Sectional Area, *Trans. ASME, Ser. C, Jl. of Heat Transfer*, vol. 99, pp.180-186, 1977.
37. B.R. Hutchinson, and G.D. Raithby, A Multigrid Method Based on the Additive Correction Strategy, *Num. Heat Transfer*, vol.9, 511-537, 1986.
38. A. Settari, and K. Aziz, A Generalization of the Additive Correction Methods for the Iterative Solution of Matrix Equations, *SIAM J. Numer. Anal.*, vol. 10, pp. 506-521, 1973.
39. J.W. McDonald, V.E. Denny, and A.F. Mills, Numerical Solutions of the Navier-Stokes Equations in Inlet Regions, *Trans. ASME, J. App. Mech.*, vol.85, pp.873-878, 1972.
40. J.L. Wang, and P.A. Longwell, Laminar Flow in the Inlet Section of Parallel Plates, *AIChE J.*, vol.10, pp.323-329, 1964.
41. R.K. Shah, and A.L. London, *Laminar Flow Forced Convection in Ducts*, Academic Press, New York, 1978.

Transient Heating in Microelectronic Components

G. DE MEY, D. BHATTACHARYA, M. DRISCART, and L. ROTTIERS
Ghent State University
Sint-Pietersnieuwstraat 41
9000 Ghent, Belgium

ABSTRACT

Transient thermal calculations have been carried out in order to get a deeper insight in the typical time constants involved in thermal problems. These studies have been done for a transistor in a silicon chip and a screen printed resistor in a hybrid circuit. Very short time constants have been observed. For the transistor the influence of the velocity of sound has also been investigated. For the hybrid circuits a comparison between several ceramic substrate materials has been made.

1. INTRODUCTION

Thermal studies in microelectronics are generally limited to the steady-state analyses. The idea behind this is the theorem of thermal slowliness. The statement that the temperature cannot follow the rapid variations of the electric signals can be found in several textbooks. The wrong conclusion is made, however, that it is sufficient to consider just the mean dissipated power in order to calculate the temperature rise. This is certainly a misinterpretation of the thermal slowliness principle. If power is dissipated in a device, the temperature at a certain distance cannot follow the high frequency signals applied across the device. This is due to the finite diffusion time. However, in microelectronic thermal design one is interested in the peak temperatures, which will generally occur on the devices themselves. In this respect the temperature is able to follow the rapid variations of the generated heat.

In the recent literature several papers have been devoted to transient thermal analysis. Some of them consider the thermal capacity of the packaging materials, which usually leads to larger time constants [1][2][3][4]. Smaller thermal time constants have been calculated for power transistors [5], diodes [6], thin film fuses [7] and hybrid circuits [8]. Thermal capacity effects have also been used to delay attainment of the peak temperature, which may be useful for circuits with a limited operation time [9]. A more general survey can be found in some textbooks [10][11]. A good overview of possible mathematical techniques can be found in the book of Dean [12].

In the present contribution the transient thermal behavior of a transistor in a silicon integrated circuit will be calculated. Very short time constants are observed. Finally, some particular applications in hybrid circuits will be given.

One may wonder why a thermal transient can influence the electric behaviour of a circuit. As will be shown in the following, a device operating under pulsed conditions can exhibit much higher temperature peaks than the same circuit under DC operating conditions, even when the mean dissipated powers are exactly the same in both cases. Electrical characteristics usually depend upon temperature in a non linear way, it is clear that sharp thermal transients will influence a circuit in quite a different way than in the case where only a uniform constant temperature rise exists.

2. BASIC PROPERTIES

In a homogeneous medium the temperature distribution $T(x,y,z,t) = T(r,t)$ is given by the diffusion equation:

$$- \lambda \nabla^2 T + C_V \frac{\partial T}{\partial t} = p \tag{1}$$

where λ : thermal conductivity

C_V : volumetric specific heat and p : power density.

A lot of physical insight can be gained by considering the Green's function of equation (1), given by

$$G(\bar{r},t) = \frac{1}{8\, C_V (\pi \frac{\lambda}{C_V} t)^{3/2}} \exp\left(- \frac{\bar{r}^2}{4 \frac{\lambda}{C_V} t}\right) \tag{2}$$

where $\bar{r}^2 = x^2 + y^2 + z^2$. The function (2) is the solution of equation (1) if a Dirac impulse function $\delta(t)\ \delta(r)$ is put in the right hand member of (1). The function (2) can also be seen as the product of three Gaussian distributions (in x, y and z direction) with a mean square deviation proportional to $\sqrt{t}$. This means that the solution (2) changes rapidly at the beginning, giving the possibility for fast thermal transients.

The Green's function (2) also tells us how thermal characteristics are influenced by scaling, i.e., if all dimensions of a device or component are reduced by a given factor. Scaling effects have been studied extensively for the electrical characteristics of MOS transistors [13]. The same ideas were also applied to hybrid circuits [14]. A good initial insight is given by looking at the exponential factor in equation (2). Roughly an identical thermal behaviour will be found if x^2/t remains unchanged. Reducing all dimensions by a factor α means that the scale should be reduced by α^2. If a 3 μ technology is replaced by a 1 μ technology, thermal transients can be expected to be 9 times faster. If a screen printed resistor with typical dimensions of 1 mm has a thermal time constant of 1 ms, a 1 μ transistor will have a time constant of 1 ns. Note that this last value becomes comparable to the switching times in electric signals. This preliminary study is, of course, very crude because other parameters, such as thermal conductivity, cannot be neglected, but it gives a good indication of what can be expected if small dimensions are involved.

3. INTEGRATED CIRCUITS

For integrated circuits the power dissipations in the individual transistors consist of two parts, which have to be treated in different ways. First of all, one has to handle the mean power, the sum over all transistors being equal to the power delivered by the externally attached sources. The second part is the time dependent one, for which the sum is zero.

Regarding the D.C. power, the temperature of the silicon chip can be considered as isothermal. The temperature is then determined by the thermal resistance of the packaging materials, leads, connections on the outer surface, etc. This aspect of the problem will not be considered here. A more detailed study, including the nonhomogeneous power dissipation in a MOS transistor, has been done by Schütz et al. [15].

The transient power gives a totally different behavior for the temperature distribution. Due to the high frequencies in the electric signals, it is expected that the heated zones will diffuse in relatively small zones around each individual heat source. Hence, for the solution of equation (1), one may approximate the substrate by an infinite medium. The transistor is then modelled by a rectangular box with a homogeneous power density p. If a power pulse with duration t_m is applied, the temperature is then calculated by

$$T(\bar{r},t) = \frac{p}{c_v} \int_0^{\min(t,t_m)} dt' \int_0^a dx' \int_0^b dy' \int_0^c dz' \frac{\exp\left[-\dfrac{|\bar{r}-\bar{r}'|^2}{4\dfrac{\lambda}{c_v}(t-t')}\right]}{8\left(\pi\dfrac{\lambda}{c_v}(t-t')\right)^{3/2}} \qquad (3)$$

The integration of this equation with respect to x', y' and z' was done analytically. The time integration was carried out numerically. Some results are shown in fig.1 when a power pulse of t_m = 100 nsec is applied. The temperature rise is calculated at different points for a device with dimensions a = b = c = 1 μ. One observes that the temperature at the center point rises very fast in the first 10 nsec. This indicates that the temperature is able to follow rapid variations. For other points the temperature rise is slower. If the distance between two neighboring transistors is about 5 μ, one might say that the thermal coupling is weak in the nsec-range. Note also the large temperature gradients in the device : they will almost completely disappear in the steady state condition [15].

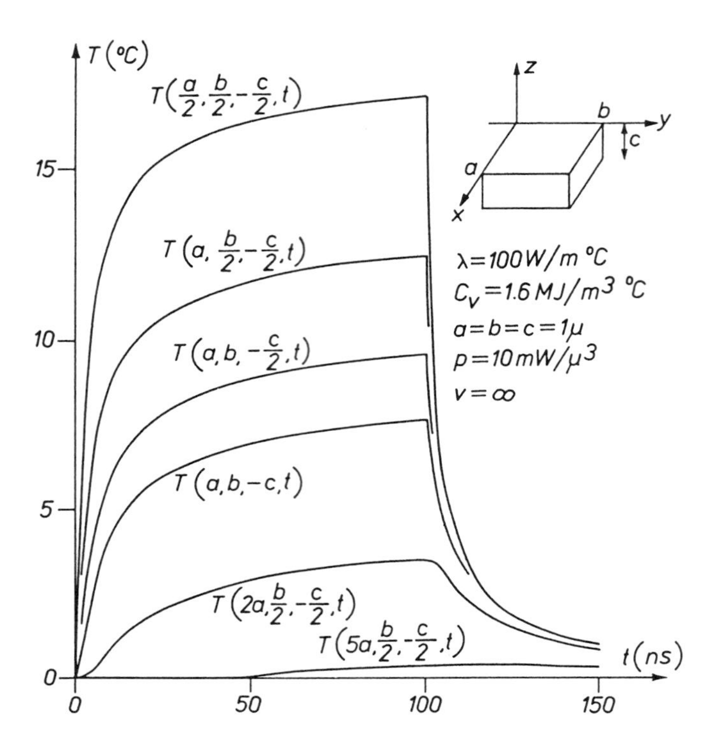

FIGURE 1. Temperature versus time for a transistor modeled by a box with dimensions (a,b,c).

Thermal diffusion is fundamentally a transport of mechanical energy in a solid. Hence, the maximum propagation velocity will be the velocity of sound, a fact that is not taken into account in (1) or (2). Assuming v = 5000 m/sec, a distance of 5 μ can be travelled in 1 nsec, so that this effect may not be neglected a priori. The diffusion equation should then be replaced by [16] :

$$\nabla^2 T - \frac{c_v}{\lambda}\frac{\partial T}{\partial t} - \frac{1}{v^2}\frac{\partial^2 T}{\partial t^2} = -\frac{p}{\lambda} \tag{5}$$

The Green's function of equation (5) is given by

$$G(\bar{r},t) = \frac{v}{r}\exp[-\frac{c_v v^2 t}{2\lambda}] \cdot \frac{c_v}{4\pi\lambda}$$

$$[\delta(vt-r) + \frac{c_v\, vr}{2\lambda\sqrt{r^2-v^2t^2}} J_1 [\frac{c_v v}{2\lambda}\sqrt{r^2-v^2t^2}]\, u(vt-r)] \tag{6}$$

where u is the unit step function and $r = |\bar{r}|$.
The expression for the temperature, equation (3), still holds provided the Green's function (2) is replaced by equation (6). Some results are shown on fig. 2. At first sight no remarkable difference with fig. 1 can be found. Fig. 3 gives an enlarged view for smaller times up to 1 nsec. The influence of finite propagation velocity turns out to be very small. If smaller dimensions are involved a = b = c = 0.5 μ, e.g., the diffusion time constant will be 4 times smaller whereas the transit time due to the propagnation velocity v will be reduced by 2. The influence of the parameter v will then be more pronounced.

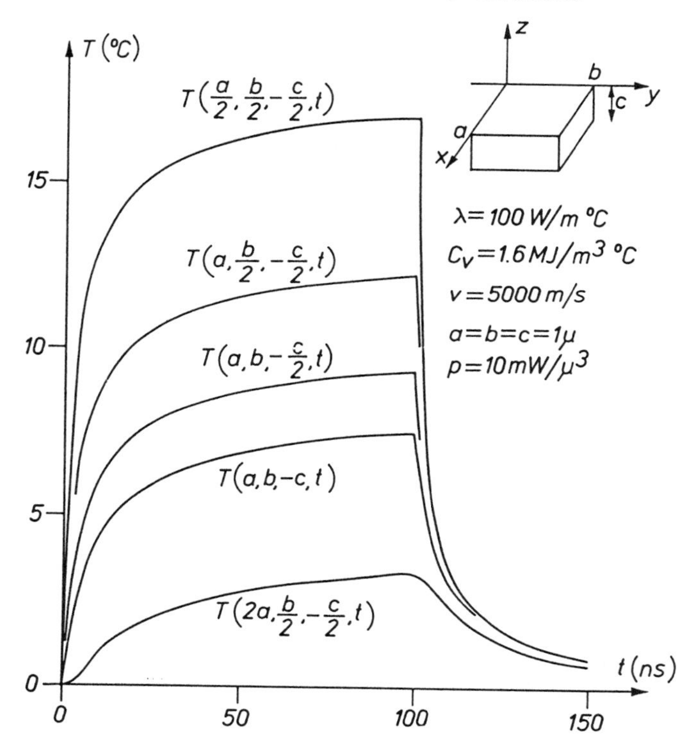

FIGURE 2. Temperature versus time for a transistor. In this case the finite propagation velocity v is taken into account. Note that the differences with figure 1 are almost negligible.

We assumed so far the transistor being placed in an infinite homogeneous medium. Actually, the silicon surface will be covered by an insulator SiO_2 or Si_3N_4. A worst-case analysis can be done by taking $\partial T/\partial z = 0$ at the top surface, $z = 0$. This can be analysed in a similar way by using image power sources in the upper half space $z > 0$. The formula (3) still holds.

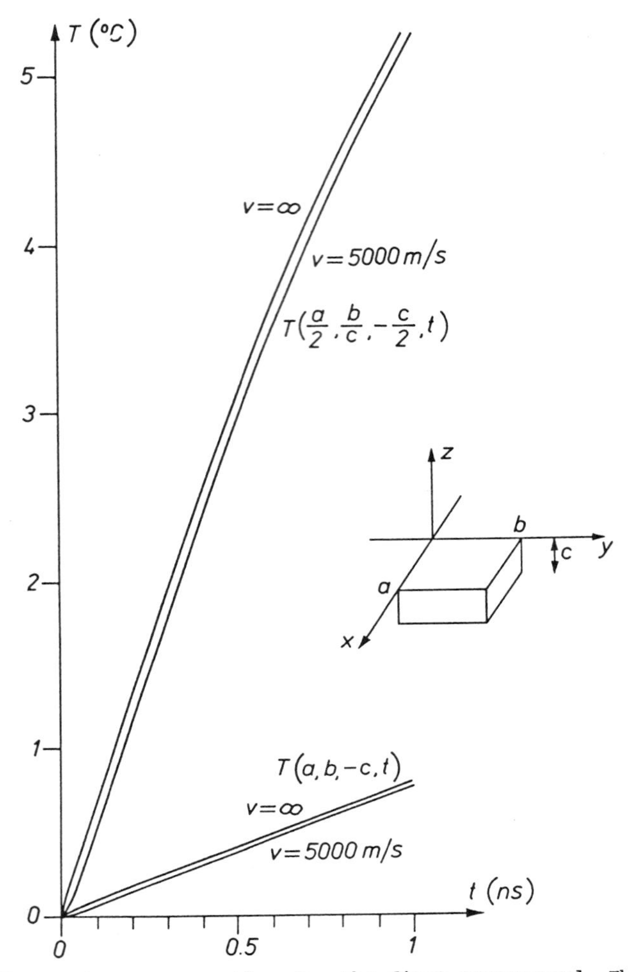

FIGURE 3. Temperature versus time for the first nanosecond. The difference between the curves obtained with a finite propagation velocity v and without (v=∞) remains small.

4. HYBRID CIRCUITS

In order to study thermal transient phenomena in hybrid circuits, we consider a ceramic substrate with a small screen printed resistor on the front side (fig. 4). Three ceramic substrate materials will be compared : Al_2O_3 (λ = 20), AlN (λ = 180) and BeO (λ = 250). Inside the substrate the diffusion equation still holds. At the front and the rear surfaces, convection has to be taken into account

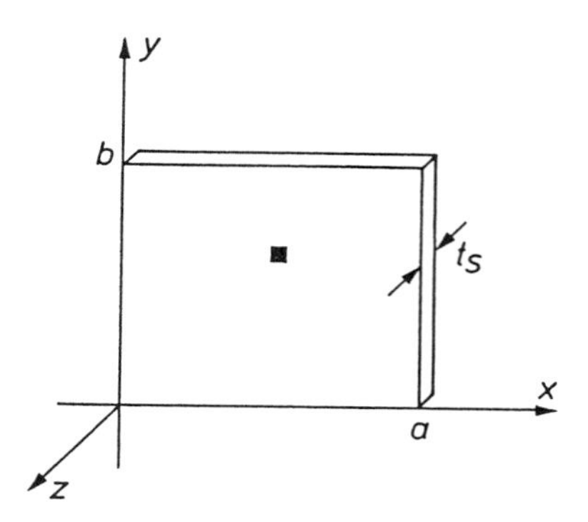

FIGURE 4. Ceramic substrate with a heat dissipating sreen printed resistor. Both front (z=0) and rear side ($z=-t_s$) are subjected to air convection.

$$\lambda \frac{\partial T}{\partial z} + \alpha T = P \qquad z = 0 \tag{7}$$

$$- \lambda \frac{\partial T}{\partial z} + \alpha T = 0 \qquad z = - t_s \tag{8}$$

where t_s is the substrate thickness and α the convection coefficient. The power density p (in W/m²) is zero outside the heat dissipating resistor.

Along the boundaries x = 0, x = a, y = 0 and y = b, the adiabetic condition is used, i.e., there is no heat loss along these sides, which is acceptable taking the small thickness t_s into account. Consequently, a two dimensional analysis seems quite appropriate at first sight. The temperature can then be calculated by a Fourier series expansion :

$$T = \sum_m \sum_n C_{mn} \cos \frac{m\pi x}{a} \cos \frac{n\pi y}{b} \left(1 - e^{-t/\tau_{mn}}\right) \tag{9}$$

It has been found that for short transients, the thermal field is limited to a small zone around the resistor. However, the transient phenomenon depends on the thickness t_s more than the larger dimensions a and b. It is then necessary to use a 3-D approach which can be evaluated by a series expansions similar to (9). Further details about the calculations have been published elsewhere [8].

Figure 5 and figure 6 show the temperature rise vs. time if a power step of 10 W is dissipated through a resistor placed in the middle of a 50 x 50 mm² substrate. The temperature was recorded in the center of the resistor. Note that for short transients the 2-D (fig. 5) and the 3-D (fig. 6) analyses give quite different results, whereas in steady-state conditions no appreciable difference could be detected.

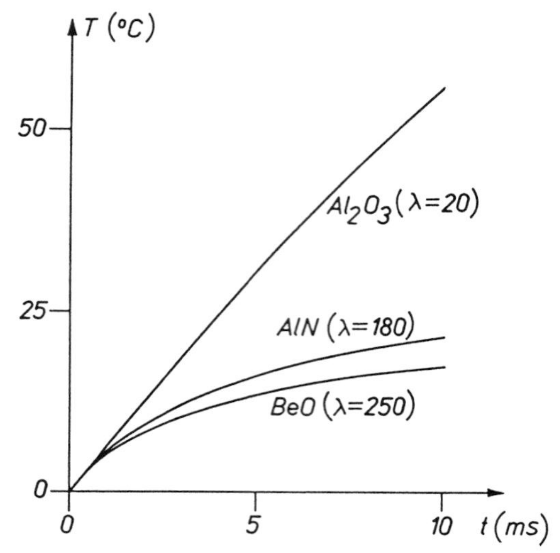

FIGURE 5. Temperature versus time in the middle of the resistor for three different substrate materials. These results are obtained with a two dimensional model.

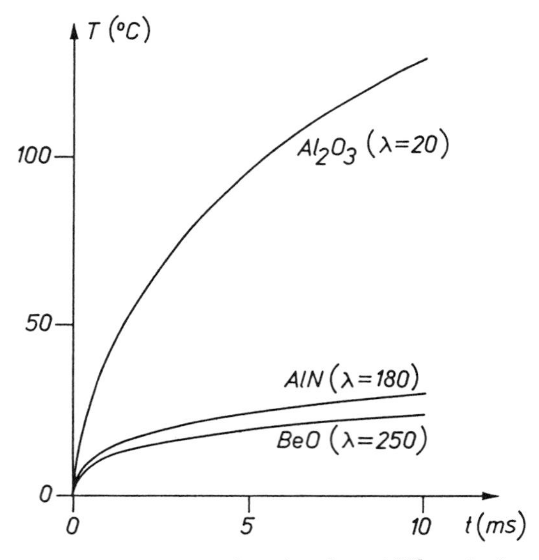

FIGURE 6. Temperature versus time in the middle of the resistor for three different substrate materials. These results are obtained with a three dimensional model.

For a 2-D approach, one assumes a vanishing temperature gradient in the z-direction whereas the 3-D approach includes the diffusion gradient $\partial T/\partial z$. As a consequence, higher temperatures will be found at the front side. The comparison between the three ceramic substrate materials is also remarkable. The highly conducting materials (AlN and BeO) seem to be very promising. Due to the high λ-values, heat is removed faster in the x and y directions. A greater part of the substrate can be used for cooling through convection. After 10 ms, Al_2O_3 gives T = 130° and only 30° was found for AlN. A steady state analysis was less favorable for Aln and BeO substrates [17], because the convection is then more important. Another important conclusion is that thermal transients within the msec range are possible.

As for the transistor analysis, it has also been observed that the temperature varies slowly at points further away from the heat source. Consequently, thermal coupling between neighbouring resistors or other devices will be small if thermal transients are involved.

5. CONCLUSIONS

In this paper, transient thermal problems in microelectronics have been studied theoretically. An important conclusion is that if one measures the temperature in the center of a heat dissipating device, very short time constants will be observed. For a 1-μ transistor even a nsec range was calculated. As a consequence, temperatures are able to follow the rapid variations of the electric signals. The statement that the temperature is only determined by the mean power is certainly wrong or a misinterpretation of the thermal inertia principle.

Thermal transient phenomena will also influence the reliability effects in microelectronics. If 10 Watt pulses are applied with a duty ratio 1/10 (average power is then 1 W) much higher peak temperatures can be obtained than the steady-state temperature due to a 1 W D.C. supply. Consequently the lifetime of the device will be shorter than expected.

Another result from the calculations is the slow temperature rise at points in the vicinity of a heated component. The thermal matching can be negligible as far as transients are concerned.

The study of a transistor showed us that the finite sound velocity has a negligible influence on the thermal behavior. Only for very small devices showed this effect can be taken into account. For the hybrids it is interesting to note that the high conducting substrates such as AlN and BeO are very promising with regard to thermal transient phenomena.

REFERENCES

1) V. Kadambi and N. Abuaf, An analysis of the thermal response of power chip packages, IEEE Transactions on Electron Devices, vol. ED-32, p. 1024-1033, 1985.

2) L. Mahalingam, J.A. Andrews and J.E. Drye, Thermal studies on pin grid array packages for high density LSI and VLSI logic circuits, IEEE Transactions on Components, Hybrids and Manufacturing Technology, Vol. CHMT-6, p. 246-256, 1983.

3) J.C. Mollendrof, The Applicability of approximate and exact transient heat transfer analyses to heating processes used to solder multilayer circuit boards, IEEE Transactions on Parts, Hybrids and Packaging, 1975, vol. PHP-11, p. 96-104.

4) M.L. Buller, Thermal transients in electronic packaging, IEEE Transactions on Components, Hybrids and Manufacturing Technology, 1980, vol. CHMT-3, p. 588-594.

5) G.K. Baxter, Transient temperature response of a power transistor, IEEE Transactions on Parts, Hybrids and Packaging, 1974, vol. PHP-10, p. 132-137.

6) D. De Cogan and S.A. John, Transmission line modelling and failure modes in power punch through diodes, Journal of Physics D, 1985, vol. 18, p. 497-505.

7) D. De Cogan and M. Henini, TLM modelling of thin film fuses on silica and alumina, Proceedings 3rd Conference on Electrical Fuses and their Applications, Eindhoven 11-13/5/87, p. 12-17.

8) M. Driscart, G. De Mey and L. Rottiers, Simulation of the transient thermal behaviour in hybrid circuits, Proceedings 6th European Hybrid Microelectronics Conference, Bournemouth 3-5/6/87, p. 212-217.

9) Y. Eukuoka, E. Matsumoto and M. Ishizuka, New package cooling technology using low melting point alloys, Proceedings 6th European Hybrid Microelectronics Conference, Bournemouth 3-5/6/87, p. 281-292.

10) H. Domingos, Transient temperature rise in microelectronic components, ISHM Technical Monograph series 6984-003, 1984, p. 67-82.

11) F.N. Sinmadurai, Thermal aspects of microelectronics packaging and interconnection, In Handbook of Microelectronics packaging and interconnection technologies, 1985, Electrochemical Publications, p. 222-262.

12) D.J. Dean,Thermal design of electronic circuit boards and packages, Electrochemical Publications, Ayr, 1985.

13) C. Mead and L. Conway, Introduction to VLSI systems, Adison Wesley, Reading, 1980.

14) G. De Mey and E. Boone, Scaling hybrid circuits, Hybrid Circuit Technology, 1987, vol. 12, p. 13-15.

15) A. Schütz, S. Selberherr and H. Pötzl, Temperature distribution and power dissipation in MOSFET's, Solid State Electronics, 1984, vol. 27, p. 394-395.

16) P. Morse and H. Feshbach, Methods of theoretical physics, Mc.Graw Hill, New York, 1953, p. 865-869.

17) G. De Mey, L. Rottiers, M. Driscart and E. Boone, Thermal studies on hybrid circuits, Proceedings Hybrid Microtech Conference, London, 1988, p. 25-26.

Conditions for Virtual Junction Temperature Calculation of Power Semiconductor Devices

ZVONKO BENČIĆ
School of Electrical Engineering
University of Zagreb
Unska 3, 41000 Zagreb, Yugoslavia

BORIS GRGURIĆ
Jugoturbina Institute
M. Švarča 155, 47000 Karlovac, Yugoslavia

ABSTRACT

A calculation of virtual junction temperature by the superposition principle is discussed. First, defining conditions for the calculation's starting parameters, i.e. forward characteristic and transient thermal impedance for constant power, were identified. This was followed by an investigation of the conditions under which those starting parameters could be applied in calculating virtual junction temperature.

1. INTRODUCTION

The current engineering practice in selecting power semiconductor devices regarding current is based exclusively on catalogue data [1,2]. Virtual junction temperature can be calculated after seeing that three ratings for a power semiconductor device have not been exceeded: (1) maximum allowable RMS value of on-state surrent, (2) maximum allowable repetitive on-state current and (3) critical rate of rise of on-state current. This calculation of the virtual junction temperature is used to check that the virtual junction temperature does not exceed the maximum allowable junction temperature at any instant of the loading duty. Exceeding the maximum allowable junction temperature could cause an unacceptable deterioration in electrical characteristics of the power semiconductor device. This calculation is made starting from the following characteristics of the power semiconductor device:

- Forward voltage-current characteristic;
- Transient thermal impedance for constant power.

The subject of this paper is an investigation of the conditions under which forward voltage-current characteristic and transient thermal impedance for constant power are defined and applicable to the calculation of virtual junction temperature.

2. FORWARD VOLTAGE-CURRENT CHARACTERISTIC

Forward voltage-current characteristic is a relationship between instantaneous value of on-state current and corresponding instantaneous value of voltage drop measured at power semiconductor device terminals. Voltage drop in a power semiconductor device consists of voltage drop on silicon wafer and of voltage drop on ohmic parts of the power semiconductor device's structure.

The forward voltage-current characteristic is defined under the following conditions:

- All parts of the power semiconductor device (silicon wafer, compensation plates, case) must have a uniform temperature;
- Spatial distribution of the charge carriers in the silicon wafer must be in balance with the loading current.

The first condition unambiguously defines temperature conditions in a power semiconductor device while the second condition indicates that we are dealing with the static forward voltage-current characteristic.

The power semiconductor device temperature uniformity condition is established because it is imposible to define the forward voltage-current characteristic as a function of the temperature field $T_V(x,y,z)$ in a power semiconductor device. The question is whether the forward voltage-current characteristic measured at a uniform temperature of the power semiconductor device can be used in calculating power losses in a power semiconductor device. The question arises because under real loading conditions with a constant or time-variable loading current a temperature drop occurs between silicon wafer and reference point on the power semiconductor device case. The answer to this question is affirmative because all parts of a power semiconductor device's structure outside the silicon wafer have a positive thermal coefficient of electrical resistance and because silicon wafer temperature under real loading conditions is practically uniform.

The condition of a charge carrier distribution balance, i.e., voltage drop associated with the static forward voltage-current characteristic is established because it is impossible to define the forward voltage-current characteristic as a function of spatial charge carrier distribution in the silicon wafer. This condition is only fulfilled when the loading current changes occur slowly. It is generally accepted that power losses caused by square-shaped current pulses, sinusoidal-shaped current pulses and "cut" sinusoidal-shaped current pulses, the duration of which is longer that 0.3 ms can be calculated by means of the static forward voltage-current characteristic.

3. TRANSIENT THERMAL IMPEDANCE FOR CONSTANT POWER

Transient thermal impedance for constant power is defined as a quotient of:

a) Change in temperature difference between virtual junction temperature and the reference point temperature on a power semiconductor device's case or in the cooling medium at a certain point in time, which is calculated from the start of power semiconductor device loading (from the thermal steady state) with constant power losses; and

b) Stepped change of power losses which caused this change in temperature difference [3].

Transient thermal impedance for constant power may be used to calculate virtual junction temperature for a power semiconductor device if the following conditions are satisfied:

- Uniformity of silicon wafer temperature;
- Linearity and time invariabillity of the thermal circuit of a power semiconductor device;
- Time constancy of the distribution of power losses in the silicon wafer.

3.1. Temperature Uniformity of Silicon Wafer

Virtual junction temperature can be compared with maximum allowable junction temperature only when it is defined in the same way as the latter.

Maximum allowable junction temperature is defined as a uniform silicon wafer temperature, because all characteristics of power semiconductor devices are defined at a uniform silicon wafer temperature.

The uniformity of silicon wafer temperature implies that current density in silicon wafer should be uniform. So, the definition of transient thermal impedance for constant power is only justified after the condition that conduction area has spread over the whole silicon wafer has been met. After thyristor triggering conduction area spreads from the gate over the silicon wafer, as is generally known [4,5].

3.2. Linearity of Power Semiconductor Device's Thermal Circuit and Time Independence of Power Loss Distribution in Silicon Wafer

The condition that power semiconductor device's thermal circuit be linear is necessary when using the superposition principle in a virtual junction temperature calculation. We shall now investigate conditions under which the superposition principle in virtual junction temperature calculation is applicable. The superposition principle would give the following expression for virtual junction temperature after the square-shaped power loss pulse shown in Fig. 1

$$\Delta T_{JC}(t) = \Delta T_{JH}(t+x) - \Delta T_{JH}(t) \qquad (1)$$

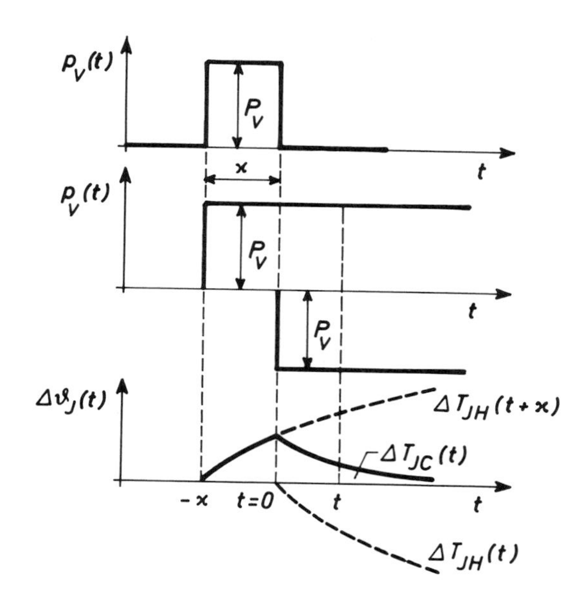

FIGURE 1. Explanation of virtual junction temperature calculation after the square-shaped power loss pulse using superposition principle.

or

$$\Delta T_{JH}(t)=\Delta T_{JH}(t+\varkappa)-\Delta T_{JC}(t) \tag{2}$$

The validity of equation (2) is in question. Its verification is based on the generalized equation for heat flow with a heat source present assuming that α and **k** are independent of temperature:

$$\nabla^2 T=\frac{1}{\alpha}\frac{\partial T}{\partial t}-\frac{1}{k}q(x,y,z,t) \tag{3}$$

After applying the operator ∇^2 to the left and to the right side of equation (2) it follows that

$$\text{div grad } \Delta T_{JH}(t)=\text{div grad }\left[\Delta T_{JH}(t\mp\varkappa)-\Delta T_{JC}(t)\right] \tag{4}$$

Cn taking equation (3) and the fact that during the cooling period power losses in a power semiconductor device equal zero, it follows that

$$\frac{1}{\alpha}\frac{\partial}{\partial t}\Delta T_{JH}(t)-\frac{1}{k}q(x,y,z,t)=$$

$$=\frac{1}{\alpha}\frac{\partial}{\partial t}\left[\Delta T_{JH}(t+\varkappa)-\Delta T_{JC}(t)\right]-\frac{1}{k}q(x,y,z,t+\varkappa) \tag{5}$$

By comparing equation (5) and equation (2), it can be seen that equation (2) is true if α and **k** are temperature independent and if spatial power loss distribution is not time-dependent.

Temperature independence of α and **k** means that the thermal circuit of the power semiconductor device is linear. The spatial variation of α and **k** does not affect linearity and superposition is possible.

Regarding spatial power loss distribution, we should emphasize spatial power loss distribution is time-dependent. The problem of spatial power loss distribution outside the silicon wafer should be solved by appropriately defining transient thermal impedance for constant power. The transient thermal impedance for constant power is measured/specified at a sufficiently low current to cause negligible power losses outside the silicon wafer. Such measured/specified transient thermal impedance for constant power is conservative, because power losses concentrated in the silicon wafer heat it more than any other spatial power loss distribution. Consequently, spatial power loss distribution in the silicon wafer should be constant with time. However, this is not so. In the silicon wafer there are sources and sinks of power loss. The solution of the time-dependent spatial power loss distribution problem in the silicon wafer implies that silicon wafer temperature should be uniform. If silicon wafer temperature is nearly uniform, the spatial power loss distribution in the silicon wafer is unimportant as different power loss distributions produce equal temperatures in the silicon wafer.

4. ANALYSIS OF VIRTUAL JUNCTION TEMPERATURE CALCULATION CONDITIONS

As mentioned, certain conditions have to be fulfilled when calculating virtual junction temperature of power semiconductor devices using transient thermal impedance for constant power. Conditions that should be investigated in more detail are silicon wafer temperature uniformity and the influence of temperature dependency of α and **k** on the maximum temperature of **the silicon wafer.**

In this paper temperature field calculation for a 1.8 kV/1 090 A-thyristor is made both for single-side and double-side cooling. It is made by means of the finite element method. Hewlett and Packard "Finite Element System - - Version 2.4" software has been used. It has been run on a computer model HP 9000.

Due to power semiconductor device's axial symmetry the calculation was worked out in cylindric coordinates z-r, Fig. 2, on the following assumptions:

- Power semiconductor device case is held at a constant temperature. This assumption is in accordance with the manufacturer's practice of declaring the device's characteristics and is introduced in order to enable comparisons of calculated results vs. catalog data;
- Heat conduction is accomplished through the cathode and anode cooling surfaces with side surfaces not contributing to heat conduction;
- Heat dissipation by radiation is neglected because of relatively small temperature differences between case surface and the environment;

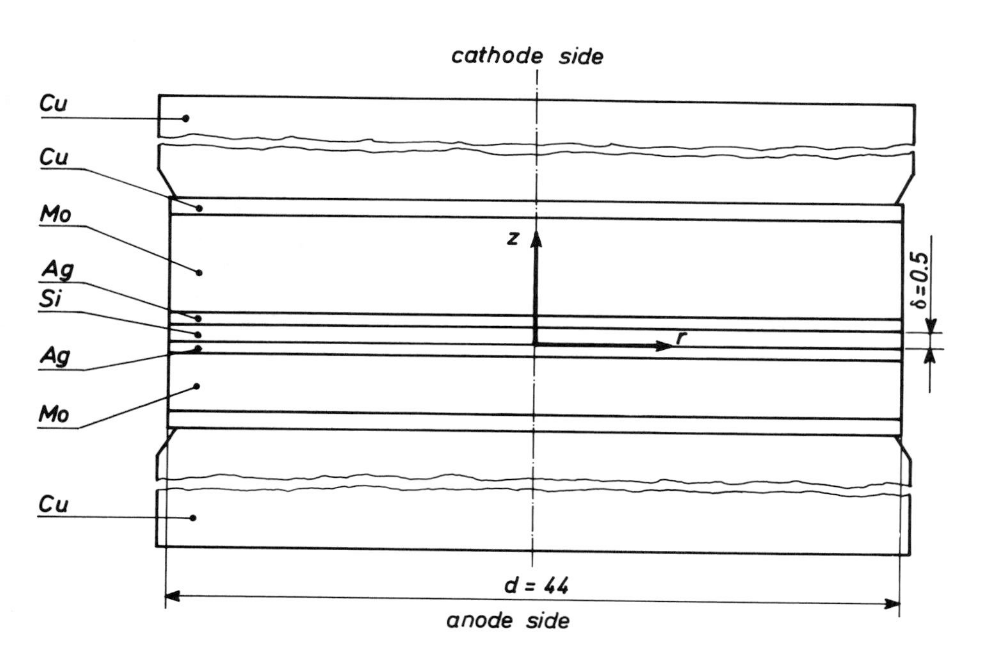

FIGURE 2. Structure of 1.8 kV/1 090 A-thyristor (in mm).

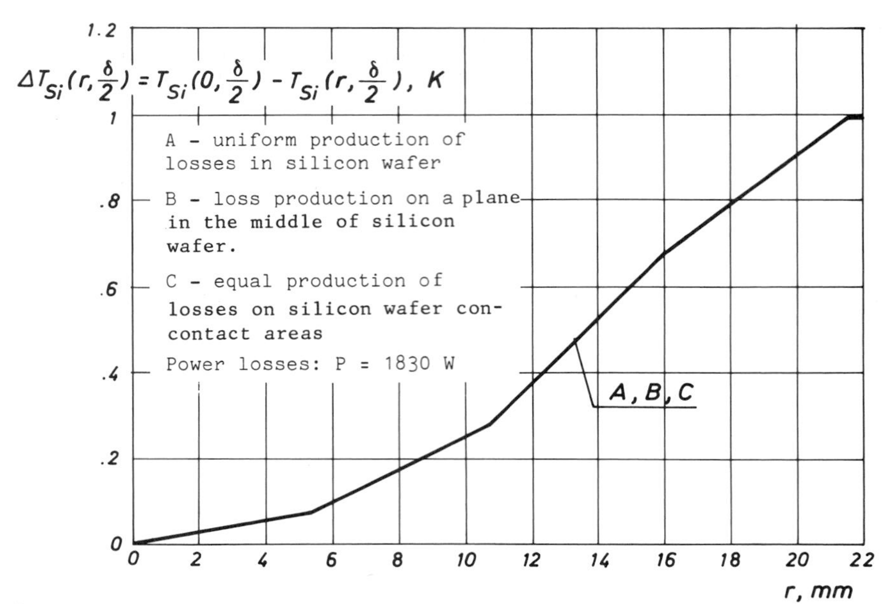

FIGURE 3. Temperature differences in the silicon wafer cross-section $z = \delta/2$ in the case of cathode-side cooling for different power loss distributions.

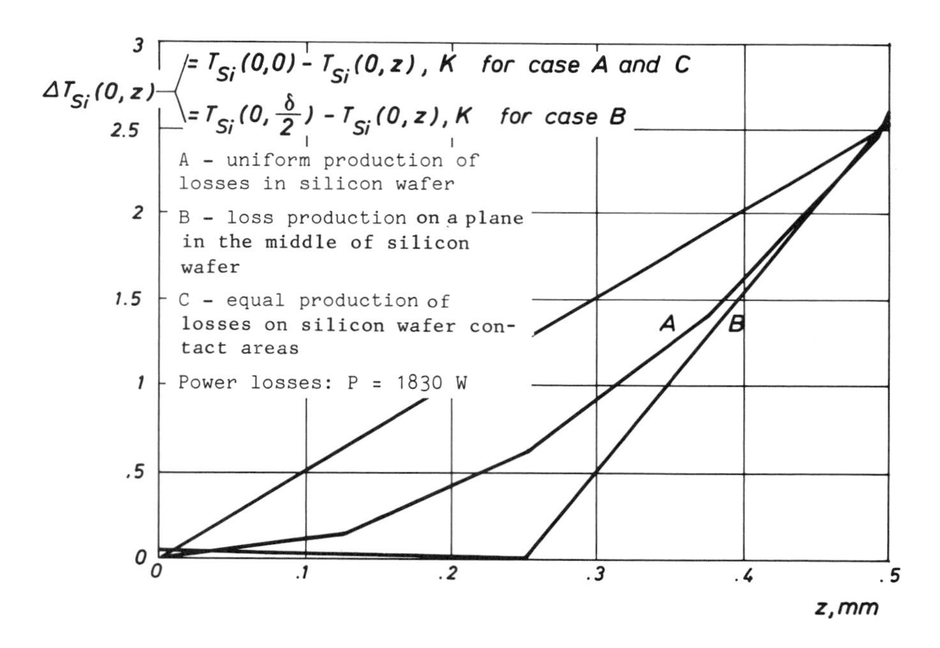

FIGURE 4. Temperature differences in the silicon wafer cross-section $r = 0$ in the case of cathode-side cooling for different power loss distributions.

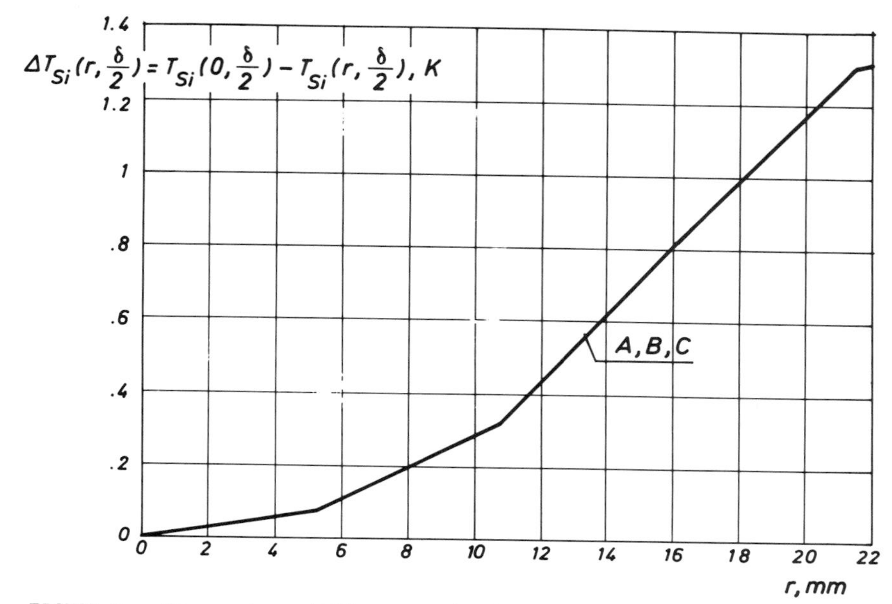

FIGURE 5. Temperature differences in the silicon wafer cross-section $z = \delta/2$ in the case of double-side cooling for different power loss distributions.

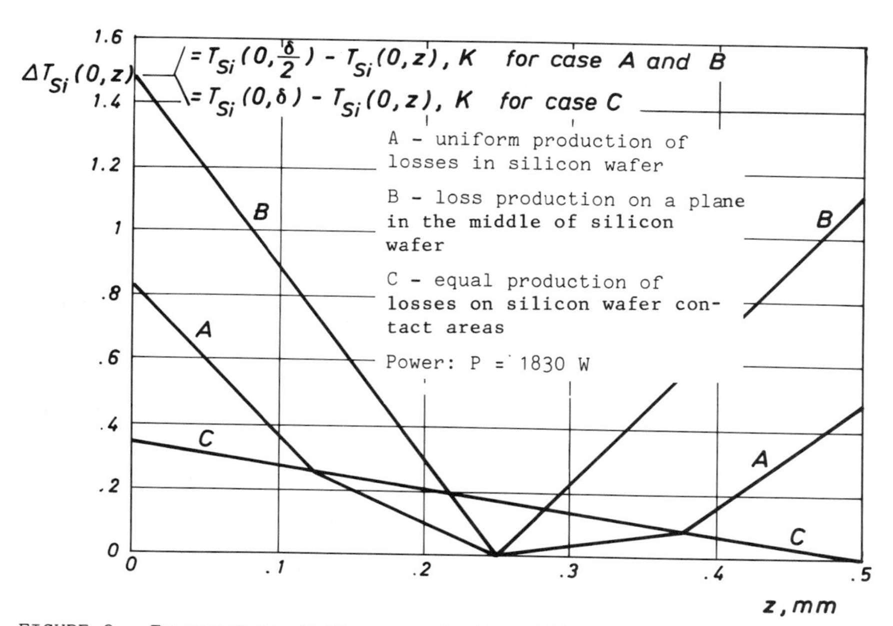

FIGURE 6. Temperature differences in the silicon wafer cross-section $r = 0$ in the case of double-side cooling for different power loss distributions.

- Thermal contact resistances on the adjacent surfaces inside the power semiconductor device's composite structure are disregarded. Under this assumption temperature differences in silicon wafer are larger;
- Power losses are generated in the silicon wafer only. This assumption fits in the definition of transient thermal impedance for constant power and assures conservatism;
- Current density (at maximum allowable constant current) is 1.2 A/mm^2 (total power losses P are 1 830 W);
- Case temperature T_C is 70°C.

This calculation was worked out for steady state. From the point of view of power loss sources three cases were investigated:

- Power losses produced uniformly over the whole volume of silicon wafer, case A;
- Power losses produced in the plane in the middle of silicon wafer, case B;
- Power losses produced equally on silicon wafer contact areas, case C.

Thermal conductivity values **k** used in this calculation are shown in Table 1 [8].

Table 1. Thermal Conductivity Values **k** at Different Temperatures for Constituent Parts of Power Semiconductor Devices

Temperature, °C	Thermal conductivity **k**, W/(m K)			
	Si	Ag	Cu	Mo
50	115	431	399	134
100	93	423	392	134
150	77	418	387	133

4.1. Analyzing Temperature Uniformity of Silicon Wafer

An answer to the question of whether this condition has been fulfilled should be sought in the results of temperature field calculation in silicon wafer. Analytical calculation for the thermal field in silicon wafer $T_{Si}(x,y,z)$ is very complex, even when dealing with constant power loading. In [6], the function $T_{Si}(x,y,z)$ of a 2.5 kV-thyristor is calculated, assuming that $T_{Si}(x,y,z) = T_{Si}(z)$. For example, in single-side cooling and current density value of 1.11 A/mm^2 the difference between minimum and maximum silicon wafer temperature is about 9 °C. In double-side cooling, the difference is less. In [7] the function $T_{Si}(x,y,z)$ of a 2.0 kV-diode is calculated, also assuming that $T_{Si}(x,y,z) = T_{Si}(z)$. For example, in double-side cooling and current density value of 2.18 A/mm^2 the difference between minimum and maximum silicon wafer temperature is about 5°C.

The results of the calculation for cases of single-side cooling are shown in Figs. 3, 4 and for those of double-side cooling are shown in Figs. 5, 6.

In cases of single-side cooling the maximum silicon wafer temperature difference at the chosen cross-section in the radial direction (r-coordinate) was 1.0 °C and in the axial direction (z-coordinate) 2.5 °C. The silicon wafer temperature distribution in the radial direction was equal in the

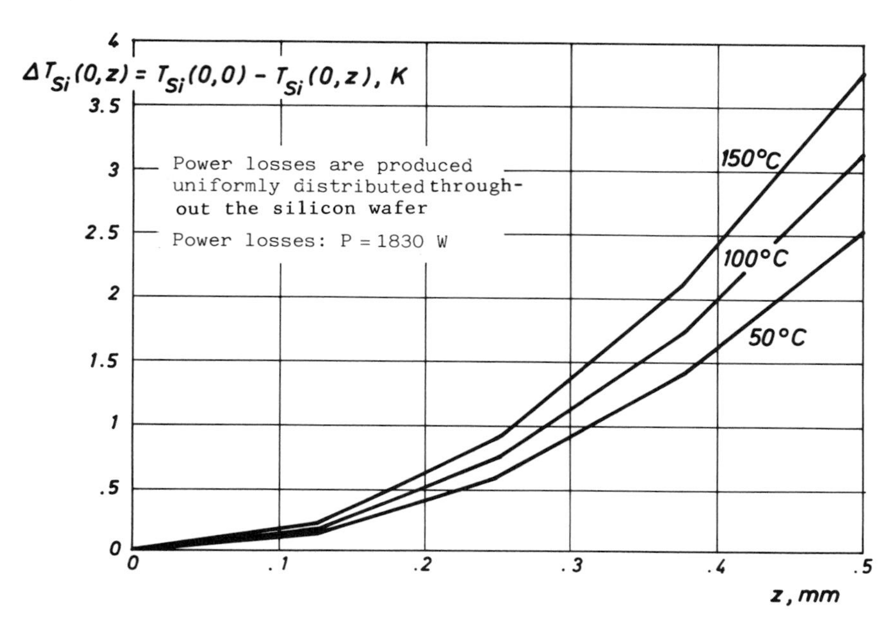

FIGURE 7. Temperature differences in the silicon wafer cross-section $r = 0$ in the case of cathode-side cooling for thermal conductivity of silicon at different temperatures.

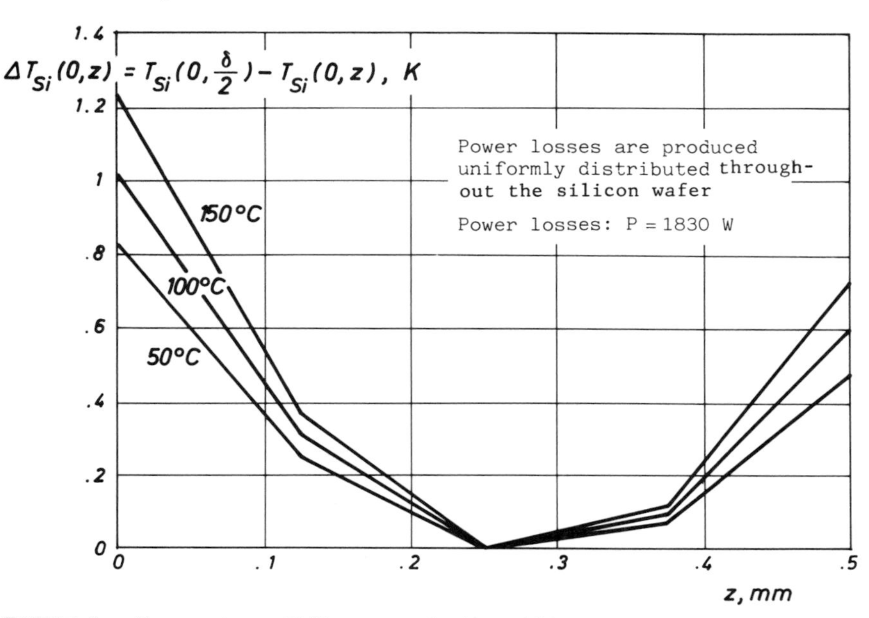

FIGURE 8. Temperature differences in the silicon wafer cross-section $r = 0$ in the case of double-side cooling for thermal conductivity of silicon at different temperatures.

three cases examined. For the entire silicon wafer volume the maximum temperature difference was 3.6 °C.

In cases of double-side cooling temperature uniformity was even better. The maximum silicon wafer temperature difference at the chosen cross-section in the radial direction was 1.3 °C whereas in the axial direction it ranged between 0.3 °C and 1.5 °C. The silicon wafer temperature distribution in the radial direction was equal in the three cases examined. In the axial direction the difference was the greatest assuming that the power losses are generated on a plane at the middle of the silicon wafer. For the entire silicon wafer volume the maximum temperature difference was 2.8 °C.

The uniformity of the silicon wafer temperature in the axial direction, that is, in the direction of heat flow, is much poorer as the silicon wafer thickness is larger. This can be accounted for by thermal resistance being proportional to the silicon wafer thickness.

The uniformity of the silicon wafer temperature in the radial direction resulted from device structure geometry. If the power semiconductor device structure geometry were equal to the cylinder geometry, the temperature differences in the radial direction would disappear. The latter is explainable by refering to the calculation's assumptions. A consequence of the homogeneous heat flow in the axial direction is a uniform temperature distribution in the radial direction.

4.2. Analysis of Linearity of Power Semiconductor Device's Thermal Curcuit

The effect of thermal conductivity **k** on the temperature field in the silicon wafer has been examined. A calculation was only made for uniform power loss distribution in the silicon wafer.

Computation results for single-side cooling are shown in Fig. 7 and for double-side cooling in Fig. 8. Temperature differences in the silicon wafer can be seen to rise with the fall in thermal conductivity. In single-side cooling maximum increase of temperature difference in the silicon wafer was 1.3 °C and in double-side cooling was 0.4 °C. This was due to the change in thermal conductivity of silicon in a temperature range between 50 °C and 150 °C. The change in thermal conductivities of other power semiconductor device's structural parts in the same temperature range resulted in an increase in maximum silicon wafer temperature of 1.2 °C in the case of single-side cooling and of 0.6°C in the case of double-side cooling.

Consequently, the influence of temperature depedence of thermal conductivity on the silicon wafer temperature is slight in conditions of nominal loading. Defining transient thermal impedance for constant power at maximum allowable junction temperature assures calculation's conservatism regarding thermal circuit non-linearity because thermal conductivity of all materials in the power semiconductor device's structure decreases with temperature increase.

5. CONCLUSION

Theoretically, it is feasible to calculate virtual junction temperature by using the superposition principle only under the following conditions:

- Silicon wafer temperature should be uniform;
- Thermal properties of semiconductor device's structure should be temperature-independent.

Uniform temperature is a fundamental condition. For this reason, temperature field in the silicon wafer of one 1.8 kV/1 090 A-thyristor for three different spatial power loss distributions in the silicon wafer was investigated. The maximum temperature difference in the silicon wafer was found to be 3.6 °C. It therefore follows that power semiconductor devices of equal current-classes but smaller voltage classes have a more uniform wafer temperature (power semiconductor device's of smaller voltage classes have silicon wafers of lesser thickness).

The condition of linearity for the thermal circuit can be met by conservatively defining transient thermal impedance for constant power. This last has to be defined at a sufficiently low current which causes a negligible power loss outside the silicon wafer, and this at maximum allowable junction temperature.

Other conditions are those governing forward voltage-current characteristic. From a practical point of view there are no other possibilities than to use the static forward voltage-current characteristic.

NOMENCLATURE

Symbol	Quantity	SI Unit
α	thermal diffusivity	m^2/s
d	diameter	m
k	thermal conductivity	W/(m K)
P	power	W
q	heat generation rate	W/m^3
T	temperature	°C
T_C	case temperature	°C
ΔT	temperature rise	°C
ΔT_J	virtual junction temperature rise	°C
T_{JC}	virtual junction temperature rise during cooling	°C
ΔT_{JH}	virtual junction temperature rise during heating	°C
ΔT_{Si}	temperature rise of the silicon wafer	°C

REFERENCES

1. Siemens, Silizium-Thyristoren-Netzthyristoren 30 A, Ausgabe 1983/84.
2. ASEA, YS 00-1 Y, 1981-82.
3. Gutzwiller F.W., and Sylvan T.P., Power semiconductor ratings under transient and intermittent loads, AIEE Trans, vol. 79, part 1, pp. 699-706, 1961.
4. Ruhl, H.J., Spreading velocity of the active area boundary in a thyristor, IEEE Trans. on Electron Devices, vol. ED-17, no. 9, pp. 672-679, 1970.

5. Somos, J., and Piccone, D.E., Plasma spread in high-power thyristors under dynamic and static conditions, IEEE Trans. on Electron Devices, vol. ED-17, no. 9, pp. 680-687, 1970.

6. Jaecklin, A.A., and Marek, A., Instantaneous temperature profiles inside semiconductor power devices, IEEE Trans. on Electron Devices, vol. ED-21, no. 1, pp. 50-53, part. 2, pp. 54-60, 1974.

7. Roulston, D.J., and Nakhla, M.R., Efficient modeling of thyristor static characterictics from device fabrication data, IEEE Trans. on Electron Devices, vol. ED-26, no. 2, pp. 143-147, 1979.

8. Goldsmith, A., Waterman, T.E., and Hirchorn, H.J., Handbook of thermophysical properties of solid materials, Revised Edition, vol. 1, Elements, The Macmillan Company, New York, 1961.

Thermal Problems in the Lasergyros

A. ALEXANDRE
Ecole Nationale de Mécanique et d'Aérotechnique
UA n°1098 CNRS
University of Poitiers
Poitiers, France

DHUICK ET RIGUET
Ingénieurs à la Société Française d'Equipements
de Navigation Aérienne
Châtellerault, France

ABSTRACT

A lasergyro is essentially made up of an optical block, including a small pipe with a laser beam running through the middle, the analysis of which permits the detection of angular variation of the optical block. To be perfect, the transmitted signal must be emitted from a block whose temperature at all points, on both left and right side, must be absolutely identical. In practice, this symmetry is never reached, so it is necessary to make corrections after the data is received from the thermal sensors which are located on the laser beam. The general design must take into account the thermal problems.
The proposed work shows how it is possible to develop a model of the complete gyrolaser (optical block and electronic card) with the help of nodal concept, using a thermal and electric network analyzer called ESACAP. We show how we can reduce thermal disrepancies by a design of the complete system based on the analysis of results and we show also that is necessary to use a thermal model to treat the signal delivered by the lasergyro.

1. INTRODUCTION

1.1. Lasergyros presentation

Lasergyros become the dominant solution for inertial navigation systems. The lasergyros detect rotation by measuring the frequency differences between two laser beams travelling in contrarotation around a closed path. The two beams are generated by one cathode and two anodes (Figure 1). The advantages of the lasergyros have now been extensively demonstrated and make them ideal for strapdown inertial navigation and guidance systems. In order to improve the performances of the lasergyros some particular studies have to be carried out. One of them is related to the thermal sensitivity and its modelling.

Schematically, a lasergyro is composed of an optical glass block (triangular form) with a dithering wheel in the center. The analysis of the detected signal is made by an electronic card fixed under the optical block which has several dissipating components. Around these pieces there is a metallic box with dry air inside to protect the system (see Figure 2).

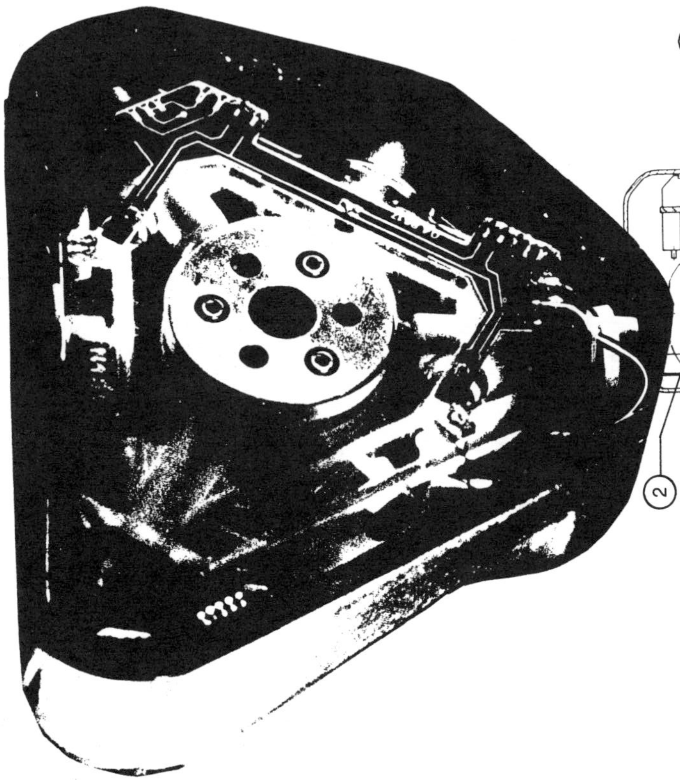

FIGURE 1 . General view of optical glass gyrolaser.

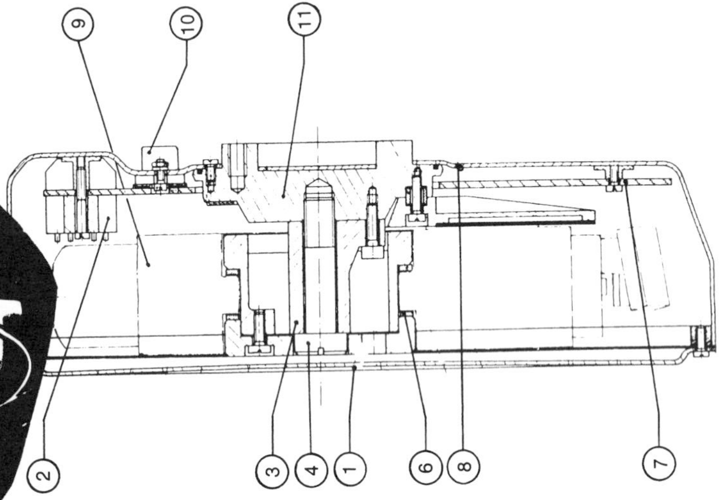

1 - Box
2 - Starting transformer
3 - Dithering wheel
4 - Screw
7 - Electronic card
8 - Box
9 - Optical block
10- Connector
11- Base

FIGURE 2 . Schematic view of gyrolaser.

1.2. Thermal problems to be solved

Lasergyros are very precise systems but the delivered signal depends on the temperature and, more particularly, on the differiential temperature between the two branches of the optical block. The thermal problems are caused by dissipating heat sources which do not have symmetrical positions and heat fluxes from the outside which do not have the same paths. The result is a temperature distribution that is not the same on the left and right branches.

The aim of this work is to study how to reduce discrepancies between these branches by a thermal analysis of the problem based on a modelling of the system. The thermal modelling permits the determination of the temperature behavior in the optical block when we manage the components on the electronic card and screens around it. We study also the interaction with the metallic box and, finally, we develop the technology which delivers the most symmetrical thermal profile in the two laser beam branches.

This study is considered to be proprietary and that is why all the graphics of this paper are shown with an arbitrary unit assigned to the initial thermal discrepancy between the two branches. The interest of the paper is not reduced by this fact because the most important results are based on relative values.

2. METHODOLOGY

2.1. The nodal concept

We deal with a very well known method which is usually used in our laboratory and which is able to very easily take into account all types of thermal transfers. For ten years the major fields of thermal problems have been solved by this approach (from the electronic component studies to the largest system analyses for space applications [1,2].
We need recall only that the discretization of the system generates n coupled differential equations that can be represented as

$$C_i \frac{dT_i}{dt} = \sum G_{ij} \ (T_j - T_i) + \sum Q_i$$

The conductance G_{ij} can be conduction, convection, radiation, or mass transport and it is generally a nonlinear term function of temperature, time, optional conditions, ...) and Q_i can be an arbitrary function.

This presentation (analog to electrical circuits) permits us to use directly the software of network analyzers. One of these is described now.

2.2. Simulation Tool : the Network Analyzer ESACAP

This is one of the powerful network analyzers that can treat very large networks in the field of electronic and thermal problems [3].
Networks larger than 1000 nodes (1000 coupled differential equations) and 5000 nonlinear couplings have been treated. The essential characteristics of this software are

- Possibility of expression with parameters defined by algebric expressions or tabulations.
- Powerful and easy description by the use of components like conductances (linear or nonlinear), radiative conductances (automatically taking into account the nonlinearity of these exchanges depending on T^4), sources and capacities (we describe only the components, without differential equation description).
- Powerful transient resolution by Gear method which can treat at the same time constants in the range of 1 to 10^9, in steady state, periodic and transient evolutions. In each case just one order is necessary so that the user needs to concentrate only on the description of the physical problem and not on the resolution method.
- Open code which can include FORTRAN subroutines if the logic can't be treated inside the software.

3. GYROLASER MODELLING

3.1. Optical Block Modelling

The decomposition of the optical block is realized according to geometry and dissipating heat sources. Figure 3 shows it (above view and detail of front view). Particular attention has been given to the region of the laser pipe.

There are around 400 nodes used for this sub-system. The couplings gererated include

- conduction in the glass,
- radiation in the cavities,
- heat sources along the path of laser beam, in cathodes and anodes,
- conduction towards the central dithering wheel and associated heat sources.

3.2. Electronic Card Modelling

The electronic card is composed of an epoxy plane covered by a copper circuit and a ground screen. Numerous passive components are fixed on it. They have been treated as they are incorporated to the card.

The active dissipating components have been treated separately and they exchange energy with the card by conduction through the connectors and by radiation because these components don't touch the card. A schematic nodal breakdown is shown on Figure 4. Indication of the value of heat rejection are also given in this picture.

When this card is fixed under the optical block (very close), we have treated the radiation interaction between the hot local component and the nine surfaces on the glass block which are just above and around it (Figure 5).

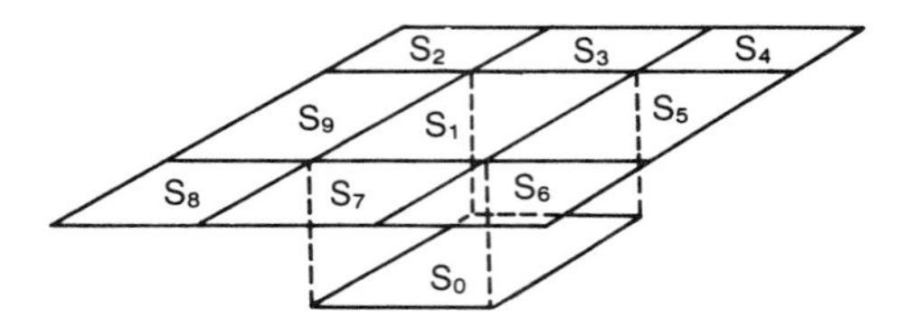

FIGURE 5 . Nodal arrangement for radiation interaction.

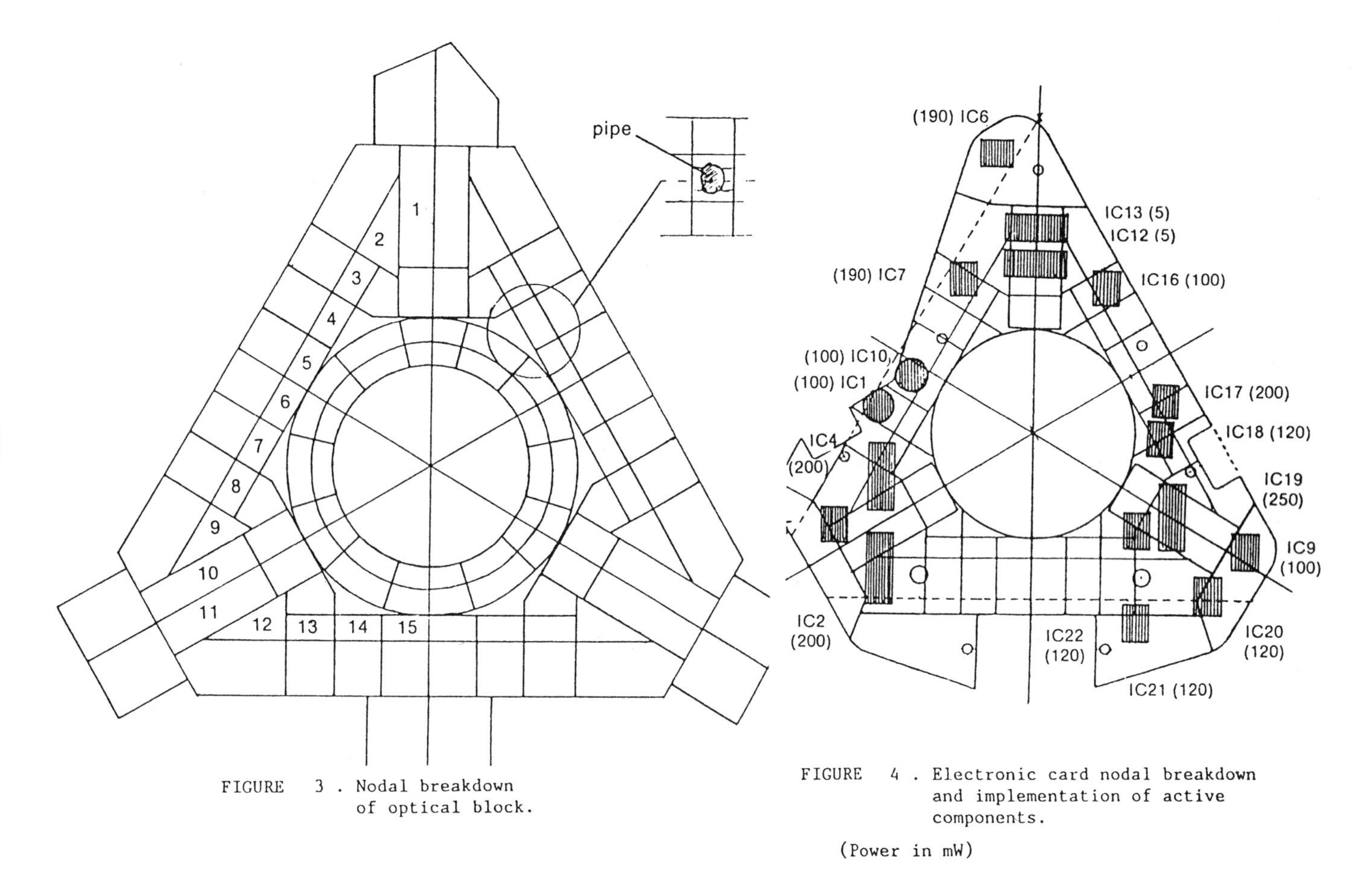

FIGURE 3 . Nodal breakdown of optical block.

FIGURE 4 . Electronic card nodal breakdown and implementation of active components.

(Power in mW)

These fine couplings are necessary in order to detect the very low temperature variations along the laser pipe.
With an enclosed air and very narrow paths, it is reasonable to treat the exchange through this air by conduction only.
A secondary electronic card is stacked above the optical block. It is composed only of passive components and, thus, the determination of couplings is easier.

3.3. Box

The box around the system can be modelled by five surfaces only (triangular prism) because the high conductivity gives a homogeneous plate of metal in each direction.
The total analysis is characterized by

620 nodes 2660 conductances	without protection screen	750 nodes 2900 conductances	with screen

4. SIMULATIONS

4.1. Optical Glass Block alone

This simulation permits us to check the couplings by the analysis of results for homogeneous thermal excitation. Secondly, we obtain the temperature profile along the laser pipe for the nominal heat dissipation. Figure 6 shows this profile and it is possible to see the large variations along the pipe and into the glass.

4.2. Complete Lasergyro

A first simulation has been realized with the optical block, the electronic card and the box. Figure 7 gives, with reduced units, the temperature differences between the left and right paths. In all the pictures we give the reduced temperature of the left path minus the right path. These temperature differences are too large and four corrective actions are described below.

4.3. Actions to Reduce the Discrepancies

4.3.1. Turn over of electronic card

A simple manner to reduce influences of components is to turn the card with components to face the box (and not to face the glass).
Figure 8 shows that we reduce temperature differences by a factor of about 3.

4.3.2. Screen of polished copper between glass and card

With the configuration of § 4.3.1, we put a screen of polished copper of 0.5 mm between the card and the glass. The effect is very important and the maximum difference is only 13 % of the initial value (Fig. 9). A sensitivity check on the plate thickness shows that we have interest in increasing the size of the plate because dissipating resistances localized far from the pipe influence this one when the thickness of copper increases. A better result than the result given by Figure 9 is obtained with a 0.1 mm plate (12 %).

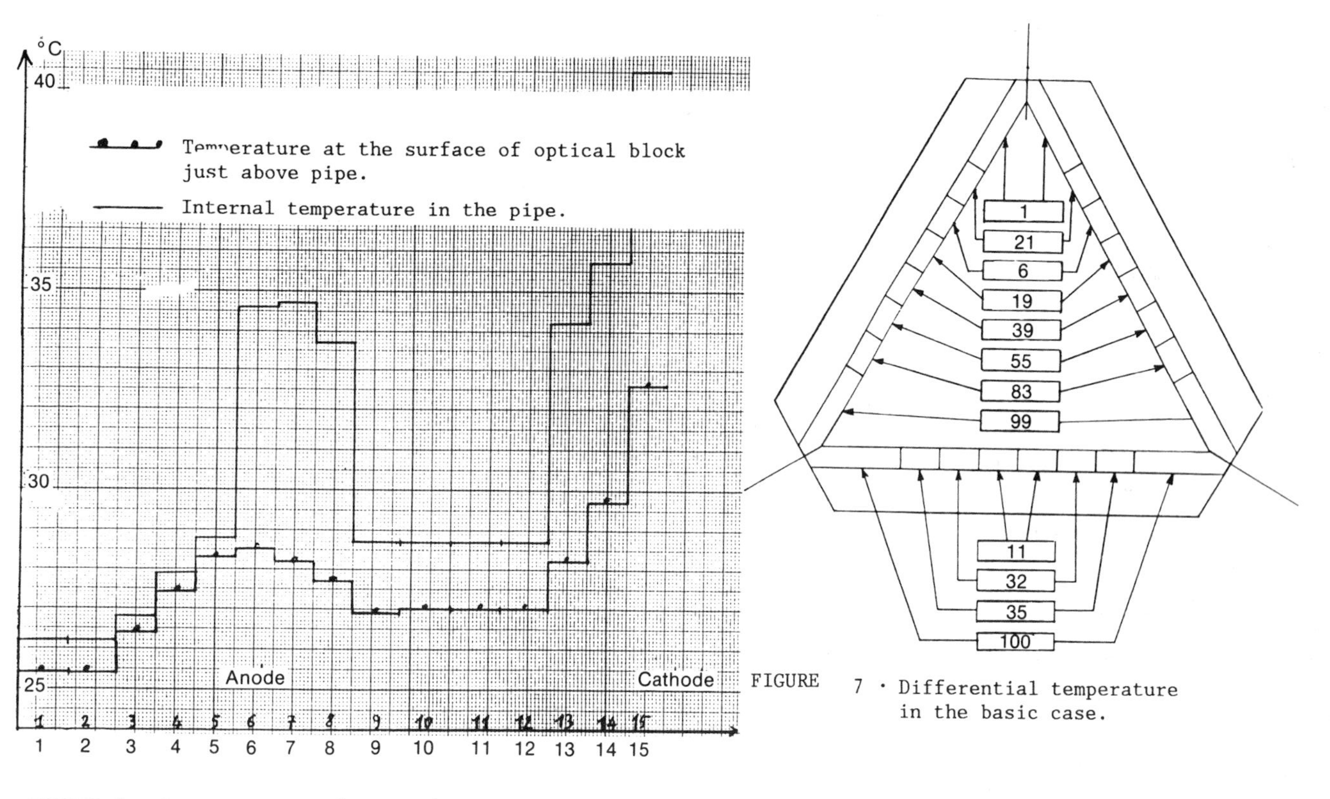

FIGURE 6 . Temperature evolution along the left beam pipe.

FIGURE 7 · Differential temperature in the basic case.

4.3.3. Screen of epoxy

We have here the same configuration as above but with a screen of epoxy (same material as the card). This gives better results with only 9 % of the initial discrepancy (Fig. 10). In conclusion, an epoxy screen is better that a copper screen. It is the result of heat transfer by conduction in a plate.

4.3.4. New implementation of electronic components

For technological reasons, a screen must be rejected and, so, we are obliged to solve the thermal problem by implementation of new components. Several design have been studied. Figures 11 and 12 show the last one with a new definition of the components. In this case, there is no screen; the components are in front of the box and this one is painted (base only) to increase the heat flux. We obtain 18 % of the initial descrepancy with a good homogeneity along the path (12 % to 18 %). So we have reduced the initial temperature differences by a factor of 5.

5. MODEL VALIDATION

The thermal model validation is not easy because we cannot measure the pipe temperature. When we put a thermal sensor in the glass the small heat flux dissipated by the sensor will modify the real local temperature. We must first study the influence of the thermal sensor.

5.1. Experimental Temperature Correction

In our thermal model we have to simulate the implementation of a thermal sensor (dissipating 5 mW introduced in a hole (1 mm diameter) above the pipe in the region 4 of Figure 3. This is shown in Figure 13.

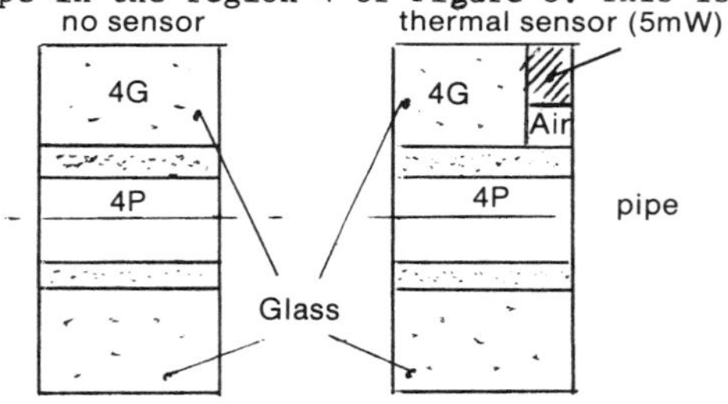

FIGURE 13 a . Schematic view through the glass.

Results given by the two approaches of Figure 13 are the following :

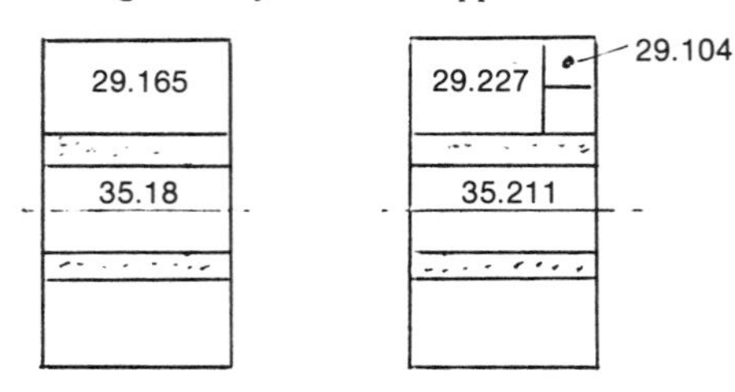

FIGURE 13b. Calculated temperatures in °C.

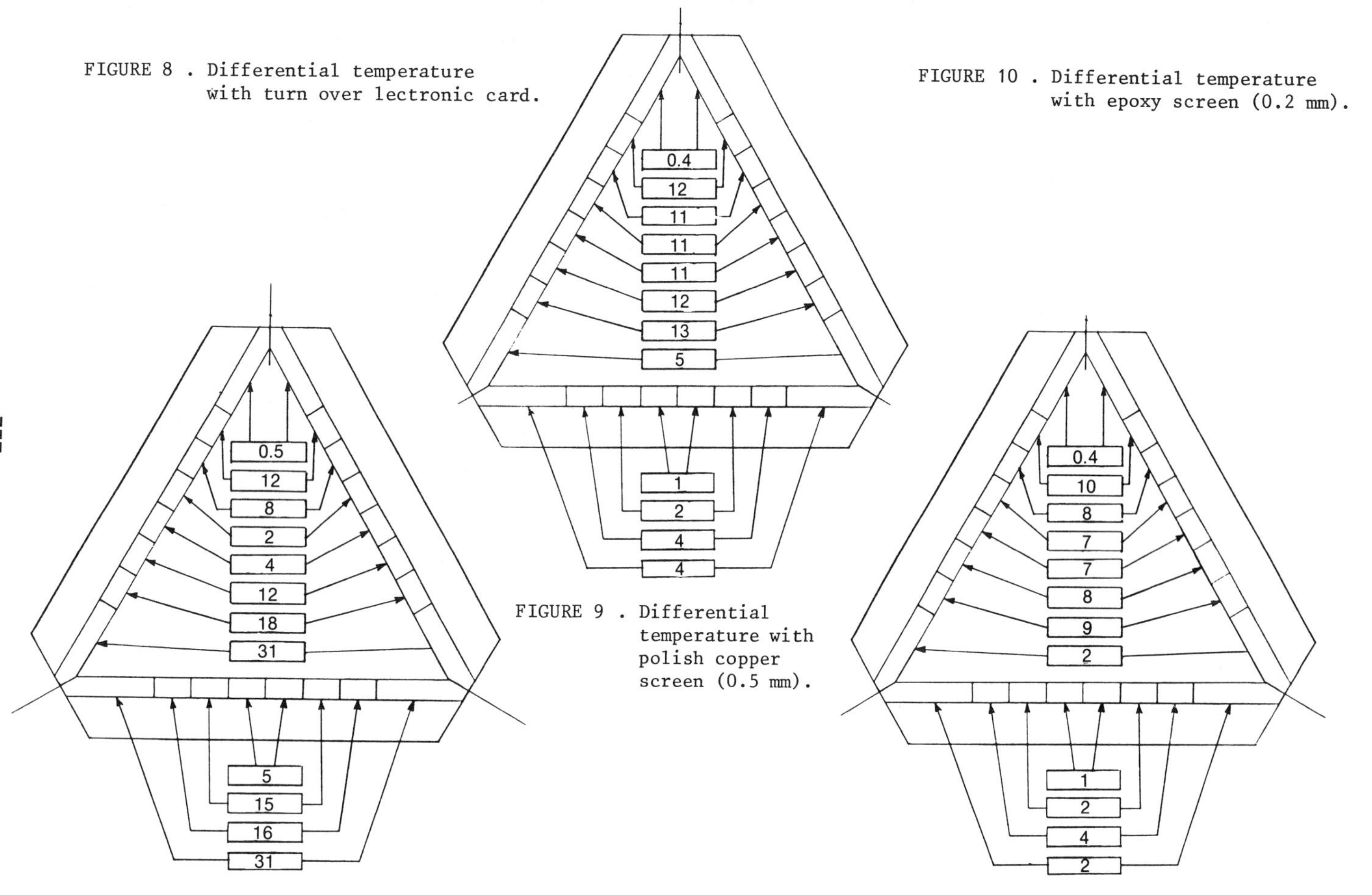

FIGURE 8 . Differential temperature with turn over lectronic card.

FIGURE 9 . Differential temperature with polish copper screen (0.5 mm).

FIGURE 10 . Differential temperature with epoxy screen (0.2 mm).

FIGURE 11 . Electronic card new implementation. (power in mW)

FIGURE 12 . Differential temperature with new components implementation.

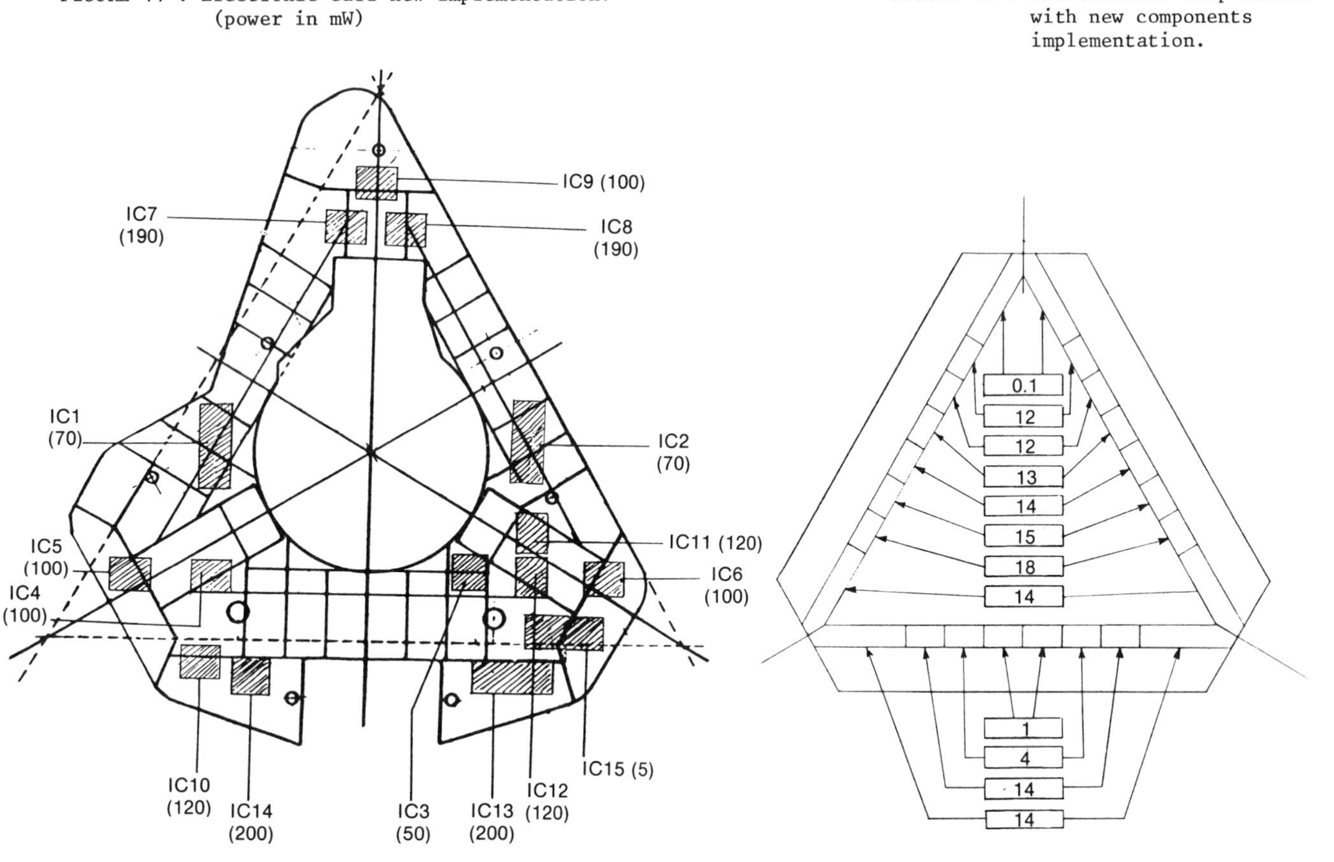

In conclusion, the sensor modifies slightly the local equilibrium (more if the sensor dissipate 10 mW) but the detected temperature is very far from the pipe temperature (6°C) and the electronics treating the lasergyro false zero must take account of this discrepancy.

5.2. Comparison of Experiment and Simulation

We have implemented several thermal sensors above the pipe in the glass in the same position as in § 5.1 and we have compared the transient behavior of the model and real lasergyro from startup. Four points are analyzed in Figures 14 to 17. The largest discrepancy (experimental model) is lower than 1°C during the entire transient evolution. It can be concluded that the developed model is in a good accordance with experiment and also that all conclusions about the differential temperatures in the left and right pipes predicted by the model are validated.
The knowledge of transient temperature in the left and right pipes as a function of the one detected by thermal sensors is very important because lasergyros are able to give information immediately after the power is turned on. Our model is able to furnish the necessary corrections to operate the real electronic system.

6. CONCLUSION

The thermal behavior of a lasergyro is very important to deliver a precise signal. We have shown in this article how we can develop a thermal model including all types of couplings in a complex system. We have shown that the behavior of the experiment is in a good accordance with the model. Additionally, have shown that it is possible to proceed on the lasergyro design with this thermal model and we are going towards a new approach where electronic design is associated at the starting point with thermal analysis to reduce temperature discrepancies in the two branches of the laser beam. This work contributes to the understanding of the thermal transient evolution in lasergyros just after power is applied. In this way, the basic needs of high performances when we switch on lasergyros can be treated.

NOMENCLATURE

T_i	temperature of the node i	°C
t	time	s
C_i	thermal capacity	J/°C
G_{ij}	conductance between nodes i and j	W/°C
Q_i	heat source generated on node i	W

REFERENCES

1. Saulnier, J.B., Alexandre, A., Modélisation et simulation en thermique, Congrès "La modélisation thermique par la méthode nodale: ses principes, ses succès et ses limites" ENSMA, Poitiers 1984.
2. Alexandre, A., Saulnier, J.B., Rapports de contrats ESA/ENSMA portant sur le modèle thermique de SPACELAB, ENSMA, Poitiers, 1981 to 1984.
3. Stangerup, P., ESACAP USER manual - Elektronikcentralen, Horsholm, Denmark,

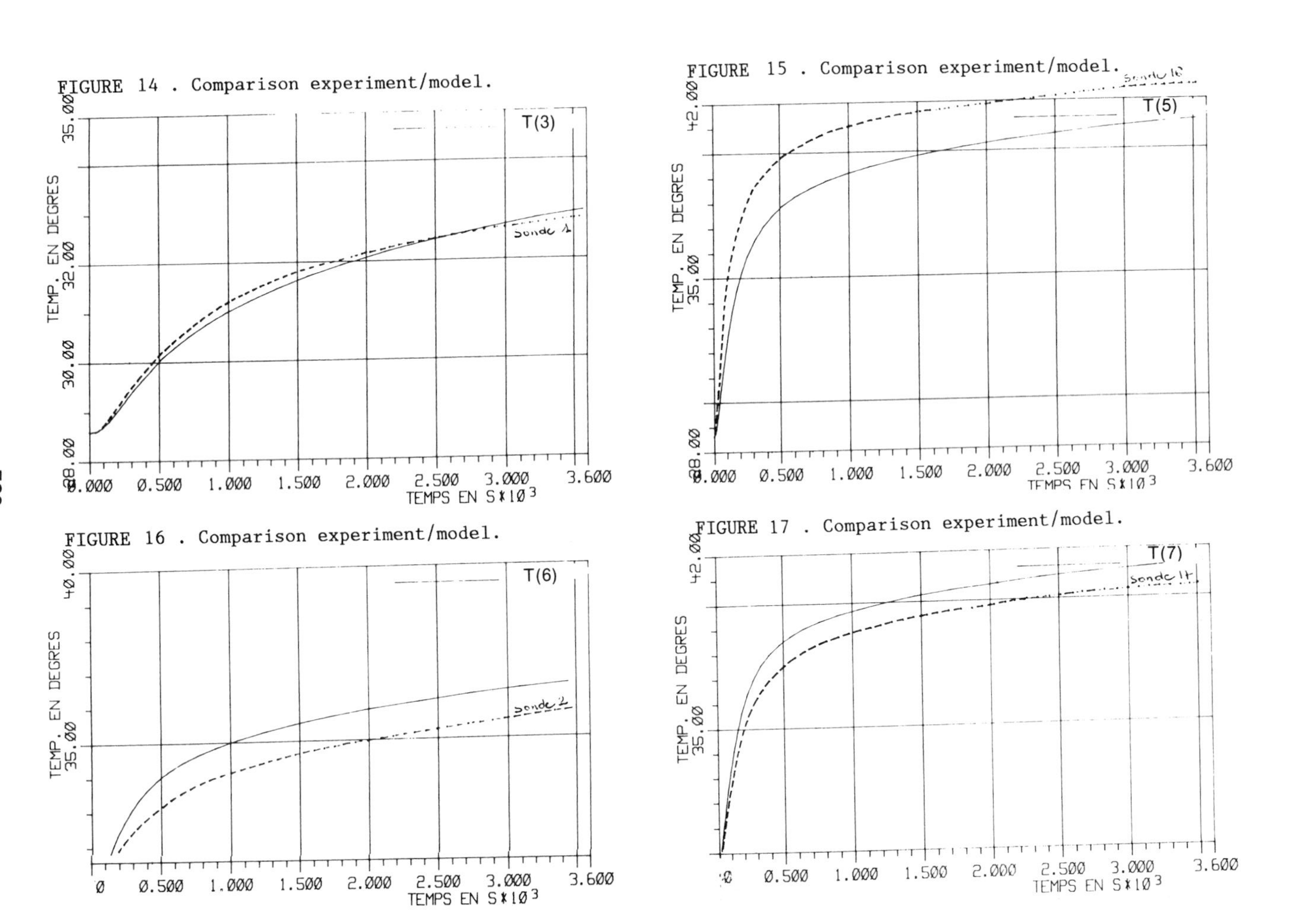

FIGURE 14 . Comparison experiment/model.

FIGURE 15 . Comparison experiment/model.

FIGURE 16 . Comparison experiment/model.

FIGURE 17 . Comparison experiment/model.

A Study on a Variable-Conductance Heat Pipe Using a Binary Mixture

KUNIO HIJIKATA, HIROMI HASEGAWA, and TAKAO NAGASAKI
Department of Mechanical Engineering Science
Tokyo Institute of Technology
Ohokayama 2-12-1, Meguro-ku, Tokyo 152, Japan

ABSTRACT

Heat transfer characteristics of a variable conductance heat pipe, which uses a binary mixture of R113 and R11, have been investigated. In a certain range of heat load, the more volatile substance, R11, is gathered in the upper part of the cooling section, and condensation does not occur in this part. The length of this non-condensing region decreases with increasing the heat load, where the temperature difference between the heated and cooled walls is kept constant. The same characteristics are also observed in the case of the mixture of R113 and air. However, the mixture of R113 and R11 has higher heat transfer rate and better performance for the constant temperature operation. A theoretical model is presented for the heat transfer in the cooling section of the heat pipe, and the calculated results well explain the experimental ones.

INTRODUCTION

The variable conductance heat pipe, which contains a noncondensable gas in the main working fluid, has been used for advanced thermal control, such as for temperature control of spacecraft equipment [1]. In such gas-loaded heat pipes the noncondensable gas is swept to the cold end of the pipe and makes some condensing surface area inactive. The working temperature of the heat pipe is kept nearly constant in a certain range of the heat load due to the variable condensing area. There have been many experimental and theoretical studies on gas-loaded heat pipes [2]-[5], and their characteristics have been clarified using a two-dimensional diffusion analysis [6]. Because the gas volume in the condenser, which determines the inactive condensing area, is controlled by the system pressure, the performance of these devices for constant temperature control is not so good. In order to improve their performance, another active or passive control system is required [1], but this makes the system complicated and impairs its reliability.

On the other hand, if a mixture of two condensable fluids, which have different boiling points, is used as the working fluid, much better performance for constant temperature control can be obtained under the condition that both components exist in the liquid and the vapor phases. If the heat flux through the heat pipe is low, the more volatile component is collected at the cold end of the pipe and acts like a noncondensable gas, so the condensing area changes according to the heat load. However, in contrast with gas-loaded heat pipes, both components are condensable, and the mixture of vapors acts like a single component vapor under high flux conditions. The operational characteristics of two-component heat pipes were studied experimentally and also predicted by a simplified analysis [7]. However, their constant temperature characteristics and the detailed mechanism have not yet been clarified. The purpose of this study is to investigate experimentally the characteristics of the two-component heat pipes with emphasis on their constant temperature operation and to develop an analytical model to predict their performance.

EXPERIMENTAL APPARATUS

Figure 1 shows a schematic diagram of the experimental apparatus. A simple vertical gravity-assisted heat pipe, or thermosyphon, was used. The inner diameter and total length of the pipe were 26mm and 1.62m, respectively. It consisted of three parts: heating section (brass tube), adiabatic section (glass tube) and cooling section (brass tube). The heating section (0.22m in length) was heated electrically with a uniform heat flux. The wall of the cooling section (0.43m in length) was cooled by water to maintain a uniform temperature of about 30°C. The heat load of the pipe was determined from the flow rate of the cooling water and its temperature increase by using mixing chambers and thermistors.

The working fluid was a mixture of R113 (T_b=47.6°C) and R11 (T_b= 23.8 °C). After degassing the working fluids, 200mℓ of R113 liquid was supplied to the heat pipe, which corresponded to 0.35m height in the pipe. Then a measured volume of degassed R11 liquid was charged. The wall temperature distribution along the pipe was measured for various heat loads and mixing ratios of the working fluids. The temperature distribution along the center line of the pipe was also measured by traversing a thermocouple. In addition to the experiment using binary mixtures, pure R113 and R113, with air as a noncondensable gas, were used as the working fluids for comparison.

EXPERIMENTAL RESULTS

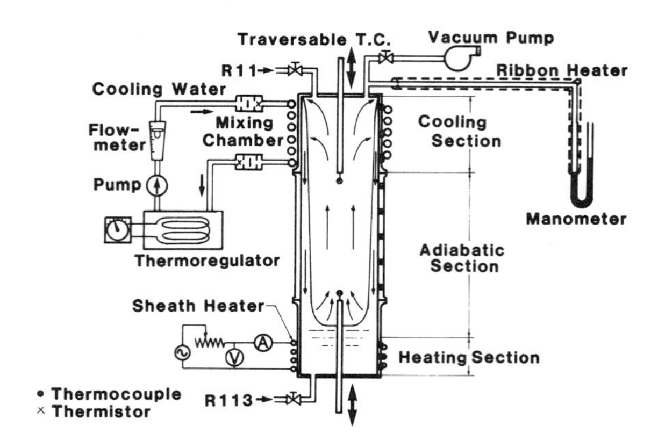

FIGURE 1. Experimental apparatus.

Temperature Distributions

To show the basic performance of the vertical heat pipe, the results for the pure R113 case are presented. Figure 2 shows the temperature variation along the wall and in the fluid along the centerline of the pipe, for two different heat loads, Q. Here z' is distance measured from the bottom of the pipe. For pure R113 the temperature at the centerline of the pipe is nearly constant throughout the adiabatic and cooling sections, because it is the temperature of the vapor corresponding to the saturation temperature for the system pressure.

On the other hand, the temperature distributions for the cases of R113 with air are shown in Fig.3 for two different heat load conditions. X_{air} denotes the initial mole fraction of air in the vapor phase without heat load, which is equal to 25% in this case for an initial total pressure of 57 kPa. For both heat load cases, a finite length region in which

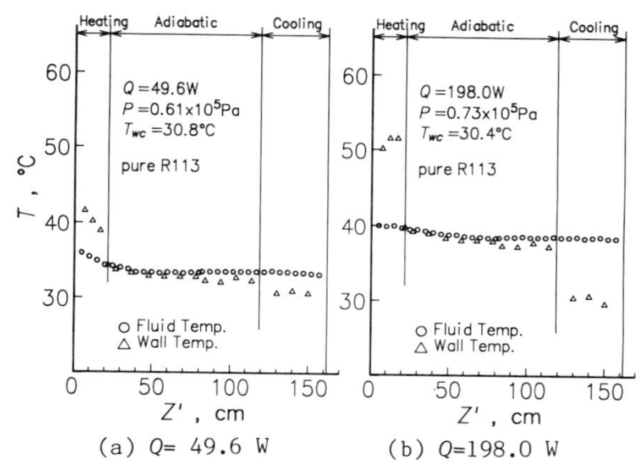

(a) Q= 49.6 W (b) Q=198.0 W

FIGURE 2. Temperature profiles (pure R113).

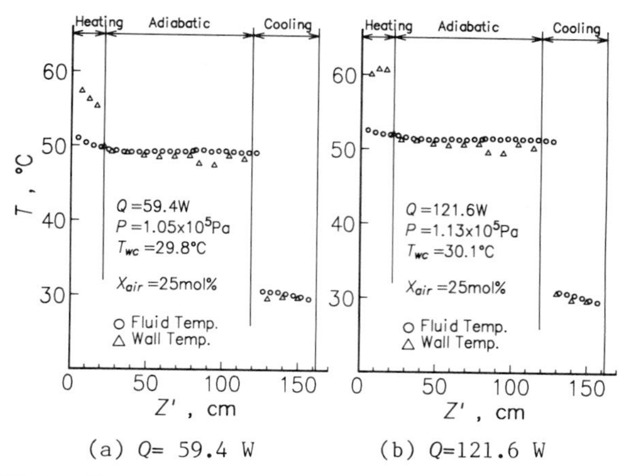

(a) Q= 59.4 W (b) Q=121.6 W

FIGURE 3. Temperature profiles (X_{air}=25%).

the fluid temperature equaled that of the wall remained near the cold end of the pipe. This cold fluid region was caused by the accumulation of air, and condensation did not occur in this region. The length of this region, namely the shut-off length, decreased with increasing heat load, as shown in Figs.3(a) and (b). The temperature difference between heating and cooling walls does not change very much with varying heat load. However, it is noted that the system pressure, P, and the total temperature difference of the pipe increase slightly with an increase of heat load.

The temperature distributions for the mixture of R113 and R11 are shown in Fig.4 for X_{R11}=24%, where X_{R11} is the mean mole fraction of R11 in the heat pipe. As shown in Figs.4(a) and (b), in the cases of low and intermediate heat loads, a non-condensing region, which can be distinguished by lower fluid temperature, is formed near the cold end, where the more volatile component, R11, accumulates. The shut-off length decreases with increasing heat loads in the same way as the R113-air mixture. However, in the case of high heat load shown in Fig.4(c), the non-condensing region disappears, and the whole cooling area contributes to the condensation. Since, as is stated later, the system pressure remains constant within the range of heat load where a non-condensing region exists, it is expected that the heat transfer characteristics of two-component heat pipe can be better than those of gas-loaded heat pipes over a wide range of heat load.

Heat Transfer Characteristics

The overall heat transfer characteristics for various combinations of working fluids are shown in Fig.5, which compares the heat load and the difference between average wall temperatures of the heating and cooling sections, ΔT. In the case of R113-air mixtures, denoted by triangular symbols, ΔT is larger than that of pure R113, shown by circular symbols, under the same heat load conditon. The heat transfer rate decreases with increasing initial content of air. However, ΔT does not change as much with heat load, compared with the pure R113 case. On the other hand, in the case of R113-R11 mixtures, shown by square symbols, ΔT is kept nearly constant in a certain range of heat load. This constancy of the

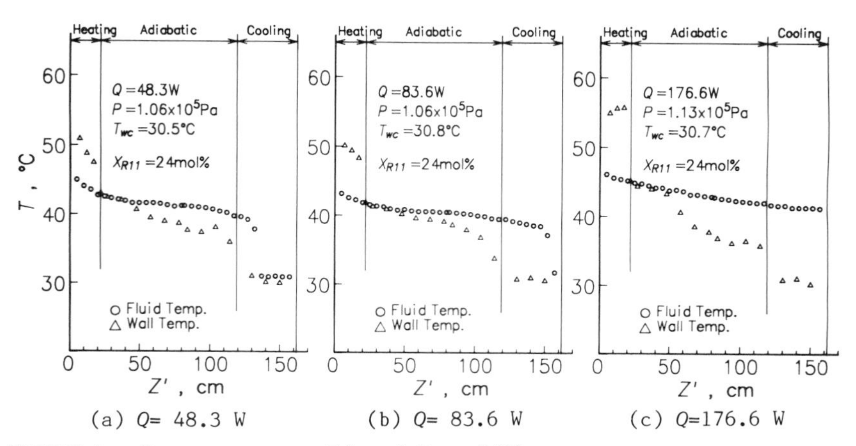

(a) Q= 48.3 W (b) Q= 83.6 W (c) Q=176.6 W

FIGURE 4. Temperature profiles (X_{R11}=24%).

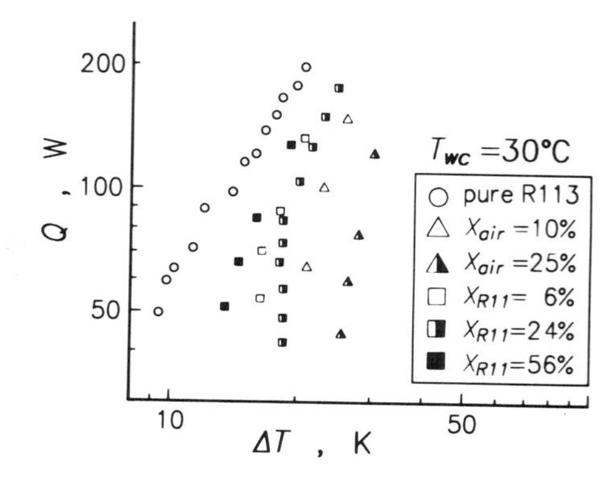

FIGURE 5. Overall heat transfer characteristics.

temperature difference under various heat load condition for the R113 and R11 mixture is superior to that of the R113-air mixture in this low power region. Further increases of the heat load bring changes of ΔT, and the heat transfer characteristics approach those of pure R113. The reduction of heat transfer rate caused by the addition of R11 is smaller than that of the R113-air mixture, which is also desirable. In the case of R113-air mixtures the heat transfer rate decreases monotonically with increasing concentration of air. However, for R113-R11 mixtures it decreases initially with increasing the concentration of R11. The heat transfer rate reaches a minimum value at X_{R11}=24% and increases again with an increase of the concentration of R11. Therefore, it is concluded that the reduction of heat transfer rate due to the addition of R11 is caused by a diffusional resistance in the vapor phase. Because the mole fraction of R11 in the vapor phase of cooling section is estimated to be about 50% when the liquid phase concentration of R11 is 24%, from the phase equilibrium diagram, the heat resistance due to diffusion of binary vapor mixture at cooling section is supposed to become a maximum in this case. Furthermore, the range of heat load in which ΔT remains constant becomes greater for this concentration of R11.

In order to clarify the mechanism controlling the constant temperature characteristics, the heat transfer characteristics of the heating section and cooling section are shown independently. Figure 6 compares the the wall heat flux and the temperature difference in the heating section, where ΔT_h is defined as the temperature difference between the average value at the heating wall and the value on the centerline at the exit of that section. As shown in Fig.6, there is no significant difference in the heat transfer characteristics of the heating section due to the variation of mixture. Such an insensitivity of the heat transfer rate to the concentration of binary mixture was also reported in a study on pool boiling of binary mixtures of volatile liquids, especially at low heat fluxes [7]. A conventional correlation for saturated pool nucleate boiling of R113 are also shown by a solid line [8], and the present data agree well with this correlation in the region of high temperature differences.

Since the heat transfer rate in the heating section is not influenced by the concentration, it is expected that the dependence of the overall heat transfer on the mixture percentage, as shown in Fig.5, is caused by the characteristics of cooling section. Figure 7 shows the relation between

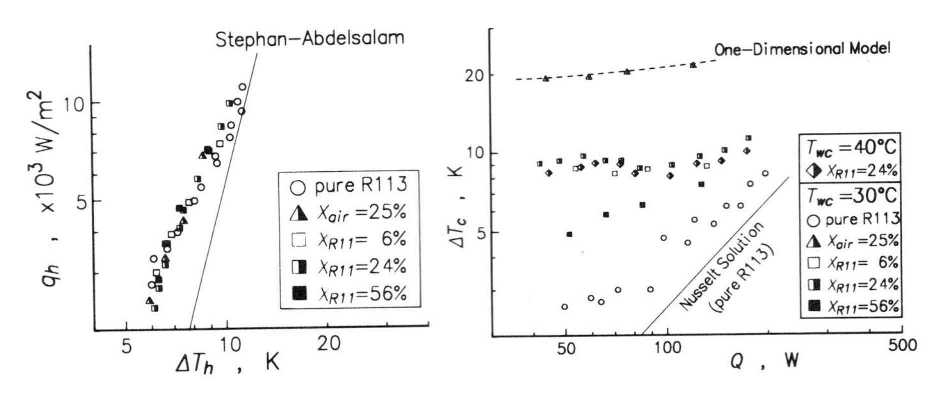

FIGURE 6. Heat transfer characteristics (heating section).

FIGURE 7. Heat transfer characteristics (cooling section).

the temperature difference and the heat load in the cooling section, in which ΔT_c is defined as the temperature difference between the average temperature of the cooling wall and the value at the centerline of the inlet of the cooling section. In the case of pure R113 and X_{R11}=56%, ΔT_c increases monotonically with increasing heat load. The solid line in this figure shows Nusselt's solution for filmwise condensation of R113. The measured condensation heat transfer for pure R113 is less than Nusselt's solution. One of the reasons of the disagreement might be the effect of corner edge of the upper end of the pipe. On the other hand, in the case of X_{R11}=6% and 24%, ΔT_c remains constant over a certain range of heat loads, which is caused by the change of condensing area under a constant temperature condition with varying heat loads as was shown in Fig.4. Also in the case of R113 and air, ΔT_c remains nearly constant. These characteristics of condensation heat transfer in the cooling section are responsible for the constant temperature characteristics of the heat pipe. In Fig.7, results for R113-R11 mixture when cooled wall temperature, T_{wc}, is equal to 40°C, are also shown. The temperature difference for T_{wc}=40°C is about 10% less than that for T_{wc}=30°C, which is considered to be caused by an increase of vapor density. It is about 20% larger than that for T_{wc}=30°C, due to the increase of system pressure.

The broken line in Fig.7 is a prediction by a simple model for an R113-air mixture, in which it is assumed that a saturated air/R113 mixture at the cooling water temperature plugs part of the condenser one-dimensionally, and condensation does not occur in that region. In this model the shut-off length changes due to the variation of system pressure. The agreement between the prediction and the experimental data is good, and the calculated shut-off length also agrees well with experimental results shown in Fig.3.

As mentioned previously, the main difference between using R11 and air as the additive is that R11 is condensable, even though volatile, but air is noncondensable. Therefore, R11 can be condensed without changing the system pressure, which means the shut-off length can change under constant pressure. However, air is noncondensable and the total mass of air must be conserved in the vapor phase. Thus, the system pressure increases with increasing heat loads. From this view point, the relations between system pressure and heat load are plotted for various mixtures in Fig.8. In the case of air, the system pressure increases gradually with

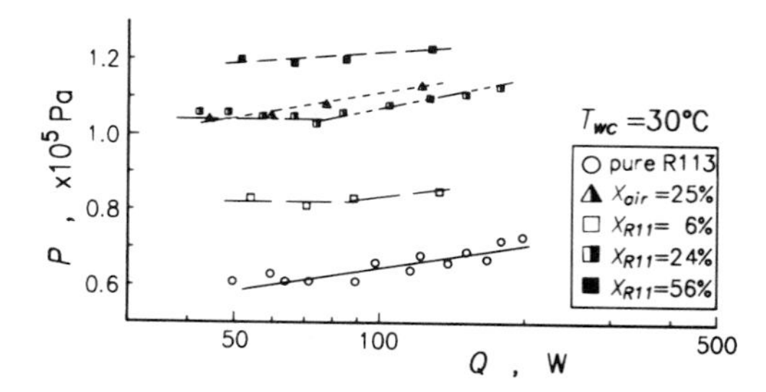

FIGURE 8. Effect of heat load on the system pressure.

heat load. However, in the case of R11, the system pressure is constant over a wide range of heat load. For example, it is constant when the heat load is lower than 80 W for X_{R11}=24%, in which a non-condensing region exists in the heat pipe. This confirms that variable conductance heat pipes using binary mixtures of condensable fluids have excellent performance for constant temperature operation because the shut-off length can vary without significant change of system pressure.

THEORETICAL ANALYSIS

Modeling of the Phenomena

In order to predict the operating characteristics of the volatile two-component heat pipe, an analytical model is presented. Based on the experimental results shown in Fig.4, the temperature profiles in the heat pipe change with increasing heat load as shown in Fig.9. That is, the temperature profiles for low and high heat loads are essentially similar to each other, and the only difference is the location of the interface of the active and inactive condensing regions. The interface moves toward the pipe end with increasing heat load, which results in the increase of active condensing area to compensate for the increased vapor flow. Such a movement of the similar temperature profile is possible only in the case of a binary mixture of vapor, because the volume of the inactive region can change freely under the constant system pressure by the condensation of the more volatile component. Therefore, the problem reduces to solving the axial profile which terminates in the inactive region and is uniquely determined for the given system pressure and wall temperature. Once the profile is determined, the length of the active region can be calculated for a given heat load to condense the vapor flow which enters the cooling section.

Figure 10 shows the analytical model of the condensing section of R113-R11 mixture, and following assumptions are made in the analysis;

1. The cooling wall temperature, T_{wc}, is constant. The pressure is also uniform throughout the system, and its value is given as a parameter.
2. The length of the cooling section is long enough, so an inactive region is formed, where the concentration of vapor phase is in an equilibrium state at the wall temperature, and no condensation occurs.
3. At the vapor-liquid interface, the temperature and concentration of the vapor and liquid phases correspond uniquely to each other via the

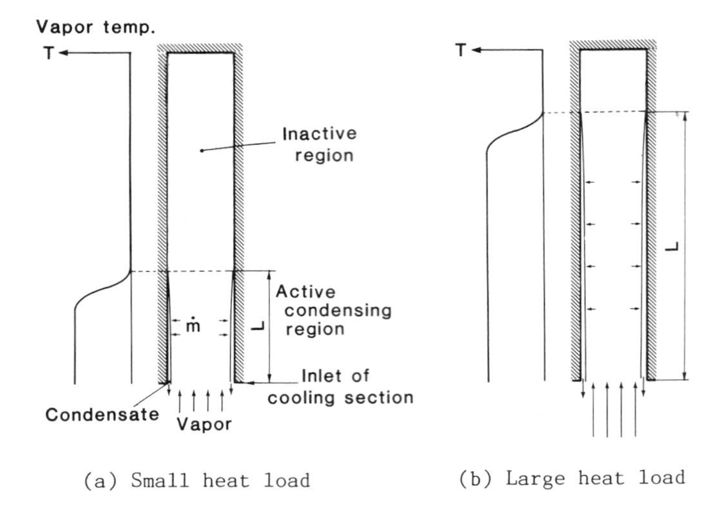

(a) Small heat load (b) Large heat load

FIGURE 9. Schematic of the change of temperature profile due to the variation of heat loads.

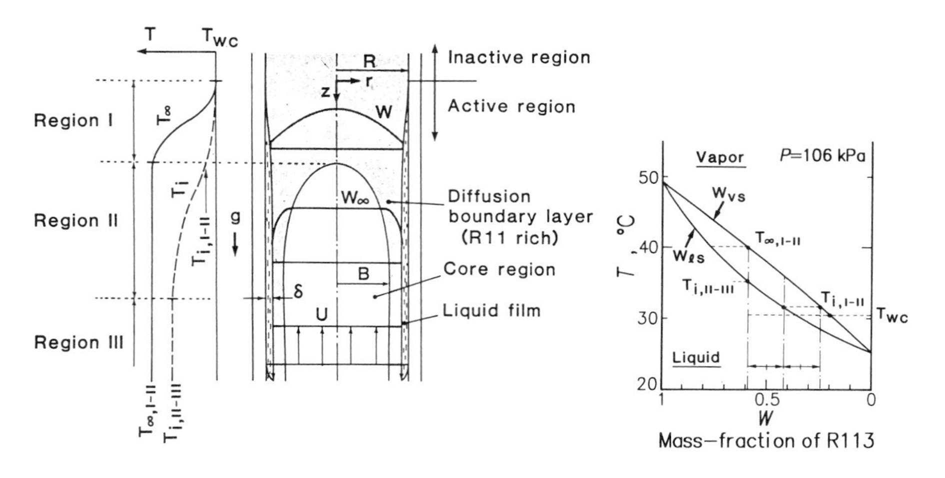

FIGURE 10. Analytical model of the cooling section.

FIGURE 11. Phase equilibrium diagram of a binary mixture of R113 and R11.

phase equilibrium diagram as shown in Fig.11.

4. The condensate liquid film is laminar, and the interfacial shear stress is neglected.
5. The thickness of the liquid film, δ, is small enough compared to the radius of the pipe, R, and can be neglected in the analysis of the vapor phase.
6. A boundary layer of radial diffusion is formed in the cross section of vapor phase in the region far from the inactive region, and the vapor concentration in the core region outside the boundary layer ($r<B$) is constant.
7. The velocity of vapor, U, is assumed uniform in the cross section as a first approximation, taking account of the suction effect at the surface of the liquid film.
8. Physical properties of the mixture are assumed to be constant and equal to those of pure R113.

Based on the above assumptions, fundamental equations are derived in the following. For the liquid film, the combination of mass and momentum conservation yields

$$\frac{\rho_\ell g}{\nu_\ell}\,\delta^2\frac{d\delta}{dz} = \dot{m} \quad , \tag{1}$$

in which $\dot{m}$ is mass flux of condensation at the liquid film surface and z is distance measured downward from the boundary of the active and inactive regions.

$$\dot{m} = \frac{\lambda_\ell}{h_{fg}}\,\frac{T_i-T_{wc}}{\delta} \tag{2}$$

Here, T_i is the temperature at the vapor-liquid interface. As for the vapor phase, the conservation equation of total mass and the transport equation of R113 is solved under the uniform axial velocity, U, in the cross section, as follows:

$$\rho\frac{dU}{dz} = -\frac{2}{R}\,\dot{m} \tag{3}$$

$$\frac{\partial}{\partial z}(rUW)+\frac{\partial}{\partial r}(rVW) = \frac{\partial}{\partial z}\left(rD\frac{\partial W}{\partial z}\right) + \frac{\partial}{\partial r}\left(rD\frac{\partial W}{\partial r}\right) \quad , \tag{4}$$

in which W and D denote the mass fraction of R113 in the vapor phase and diffusivity, respectively. Next, the vapor concentration profile of R113 is assumed as follows:

$$\begin{aligned} 0\leqq r\leqq B(z):&\quad W = W_\infty(z) \\ B(z)<r<R:&\quad W = W_i + (W_\infty-W_i)\left\{1-\left(\frac{r-B}{R-B}\right)^2\right\} \end{aligned} \tag{5}$$

Here B and W_∞ are the radius of the core region and concentration of R113 in it, respectively. W_i is the vapor concentration at the vapor-liquid interface. These variables are functions of the axial coordinate, z. Equation (4) is integrated once in the cross section by using the profile given in Eq.(5).

$$\frac{d}{dz}(UW_\infty)+\frac{d}{dz}\{f(B)U(W_i-W_\infty)\} = \frac{d}{dz}\left\{D\frac{d}{dz}\left(\frac{W_\infty+W_i}{2}\right)\right\} - \frac{2}{R}\left(VW-D\frac{\partial W}{\partial r}\right)\bigg|_{r=R} \tag{6}$$

$$f(B) \equiv \frac{(R-B)(3R+B)}{6R^2} \tag{7}$$

The second term of the right-hand side of Eq.(6) denotes the condensing flux of R113 at the vapor-liquid interface. The mass conservation across the interface is written as the following, when the diffusion in the liquid phase is neglected:

$$\rho(VW-D\frac{\partial W}{\partial r}) = W_{\ell i}\dot{m} \qquad \text{at } r=R \quad , \tag{8}$$

in which $W_{\ell i}$ is the liquid phase mass fraction of R113 at the interface. By using the profile given in Eq.(5), and the relation of $\rho V_{r=R} = \dot{m}$, Eq.(8) reduces to

$$\dot{m} = C(\frac{W_\infty - W_i}{W_{\ell i}-W_i}) \frac{2}{R-B} \rho D \tag{9}$$

Equation (9) represents the diffusional resistance in the vapor phase, and a constant factor, C, is introduced on the right-hand side of Eq.(9) because the simplified profile given by Eq.(5) cannot represent the true concentration gradient at the interface, which reflects the effect of suction or turbulence in the case of high Re number of vapor flow.

Eq.(6) is written by using Eq.(8), as follows:

$$\frac{d}{dz}(UW_\infty) + \frac{d}{dz}\{f(B)U(W_i - W_\infty)\} = \frac{d}{dz}\{D\frac{d}{dz}(\frac{W_\infty + W_i}{2})\} - \frac{2}{R}\frac{\dot{m}}{\rho} W_{\ell i} \tag{10}$$

In addition to the above equations, the phase equilibrium condition at the interface is used:

$$W_i = W_{vs}(T_i) \quad , \quad W_{\ell i} = W_{\ell s}(T_i) \tag{11}$$

Here, W_{vs} and $W_{\ell s}$ are equilibrium vapor and liquid mass fraction of R113, as given by Raoult's law for an ideal mixture.

When Eq.(10) is solved, the active condensing region is further divided into three parts as shown in Fig.10, namely, regions I, II and III. In region I, the boundary layer thickness in the vapor phase reaches the centerline of the pipe; therefore, B is equal to 0. In this region, W_∞ represents the concentration on the centerline, and its axial change is obtained from Eq.(10). On the other hand, in region II, a boundary layer is formed. Then the axial change of the core radius, B, is obtained from Eq.(10) under a constant value of W_∞ in region II. From a simple analysis of Eqs.(3),(9) and (10), it can be shown that W_∞ approaches a constant value with increasing z in region I,

$$z \to \infty : \quad W_\infty \to W_i + 2(W_{\ell i} - W_i) \tag{12}$$

In this asymptotic state, the concentration of R113 in the condensing vapor, that is, $W_{\ell i}$, is equal to the bulk concentration in the cross section; therefore, the vapor phase concentration is not changed by the condensation along the flow direction. The saturated vapor temperature corresponding to the asymptotic value of W_∞ is shown as $T_{\infty, I\text{-}II}$ in Figs.10 and 11. Thus, the transition from region I to II is defined to occur when the asymptotic value of W_∞ is reached in region I with increasing z, under the following condition:

$$W_\infty = W_i + 2(W_{\ell i} - W_i)\cdot 0.96 \tag{13}$$

The factor 0.96 is rather arbitrary; however, the calculated results were not affected significantly by the value. The length of region I is dominated mainly by axial diffusion of vapor, that is, the first term of right-hand side of Eq.(10). However, in succeeding regions II and III, the axial change of vapor concentration is small; thus, the axial diffusion term is neglected. Further, the value of the constant, C, in Eq.(9) is set equal to unity in region I based on its rigorous derivation; however in other regions, C is assumed to be 4 in this analysis to fit the calculated heat transfer rate to the experimental one. For the theoretical reasoning of this value, more complicated analysis should be made which precisely considers such effects as suction, turbulence and the waviness of the liquid film.

In region II, the temperature of vapor-liquid interface, T_i, increases with z due to the reduction of diffusional resistance in the vapor phase caused by the decrease of the boundary layer thickness, as shown in Fig.10. As the result, T_i reaches the critical value defined by the following equation:

$$W_{\ell s}(T_i) = W_\infty \tag{14}$$

When the condition of Eq.(14) is satisfied, the vapor in the core region condenses into the liquid film at the same composition as in the core, which means that the mixture of binary vapor can condense perfectly without diffusional resistance. Such a condition was also reported to occur in the case of natural-convective condensation of a binary mixture on a flat plate [10]. Therefore, once Eq.(14) is satisfied, the value of T_i is set constant at that value, namely $T_{i,II-III}$ in Fig.10 and 11, and this region of constant T_i is called region III.

After this, the unknown variables are δ, U, T_i, $\dot{m}$, and W_∞ (or B). T_i and $\dot{m}$ are solved to satisfy Eqs.(2) and (9) simultaneously, as required by the balance between heat resistance of the liquid film and diffusional resistance in the vapor phase. The other variables, δ, U and W_∞ (or B) are solved from the differential equations (1),(3) and (10). Since all the derivatives are first-order with respect to z, except for the first term of right-hand side of Eq.(10), a marching solution procedure is adopted with increasing z starting from the inactive region by assuming the initial axial gradient of W_∞ to be an arbitrarily small value, ε. However, since the behavior of solution in the region I is an asymptotic one, the variation of ε merely moves the virtual origin and doesn't affect the solution very much. The initial conditions are listed as the following:

$$z=0 : \quad \delta=0, \quad U=0, \quad W_\infty = W_{vs}(T_{wc}), \quad dW_\infty/dz=\varepsilon \ (\ll 1) \tag{15}$$

Analytical Results

As an example of calculated results, axial profiles are shown in Fig.12 for P=103 kPa, which corresponds to the experiment for X_{R11}=24%. As was mentioned previously, the centerline concentration, W_∞, approaches an asymptotic value with increasing z, as shown in Fig.12(a). After W_∞ reaches the constant value, $W_{\ell i}$ approaches W_∞ due to the increase of T_i, which is caused by the decrease of boundary layer thickness, $(R-B)$, as shown in Fig.12(b). Finally, after $W_{\ell i}$ reaches the constant value, which corresponds to a constant value of T_i, the liquid film thickness

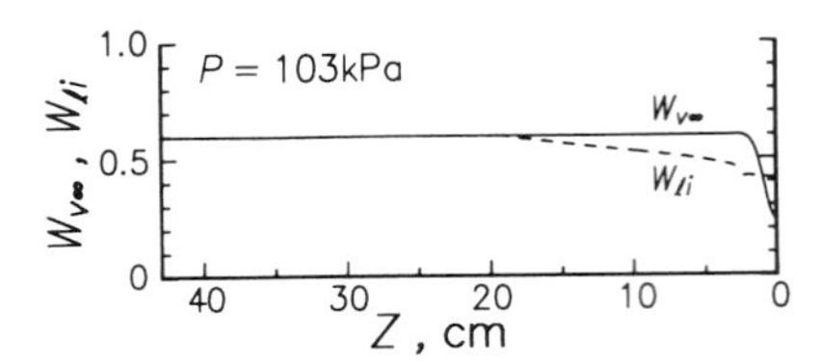

(a) Concentration of R113

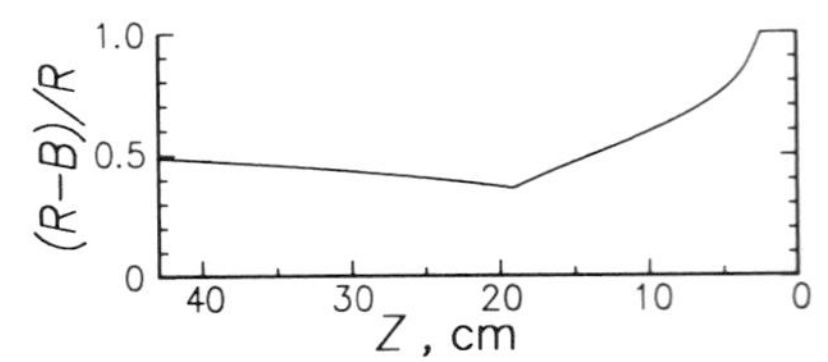

(b) Boundary layer thickness of diffusion in the vapor phase

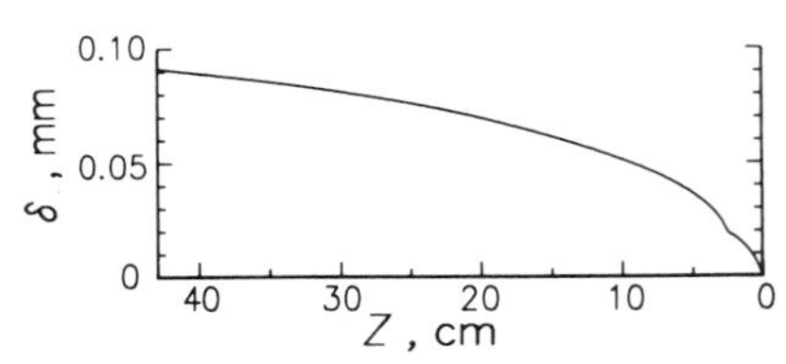

(c) Liquid film thickness

FIGURE 12. Calculated results of axial profiles.

increases with z to the power of 1/4 as shown in Fig.12(c). In this region of constant T_i, the boundary layer thickness in the vapor phase increases with z, in accordance with the increasing thermal resistance of the liquid film due to the increase of its thickness.

Axial concentration profiles along the centerline of the pipe are compared with experimental results in Fig.13 for two mean R11 concentration values, by assuming saturated conditions in the vapor. The abscissa of the figure is the distance from the inlet of the cooling section, and the calculated profiles are those for the same heat load and pressure as in the experiment. As shown in these figures, the present theory roughly predicts the characteristics of experimental results. It is noted that the concentration of vapor which enters the cooling section is uniquely determined by the system pressure, which means that system pressure is uniquely determined by the inlet vapor concentration and the wall temperature. The theory also predicts well the length of the active condensing region under various heat loads. However, the calculated axial temperature gradient near the inactive region is steeper than that of the experimental result. In order to make more precise prediction, a more rigorous analysis is necessary which deals with, for example, the two-dimensional fields of flow and mass transfer more correctly, the non-uniformity of physical properties, the resulting effect of a natural convection, etc. However, the essential characteristics of heat transfer in the cooling section of a two-component heat pipe, which control its constant temperature operation, can be explained by the present analysis.

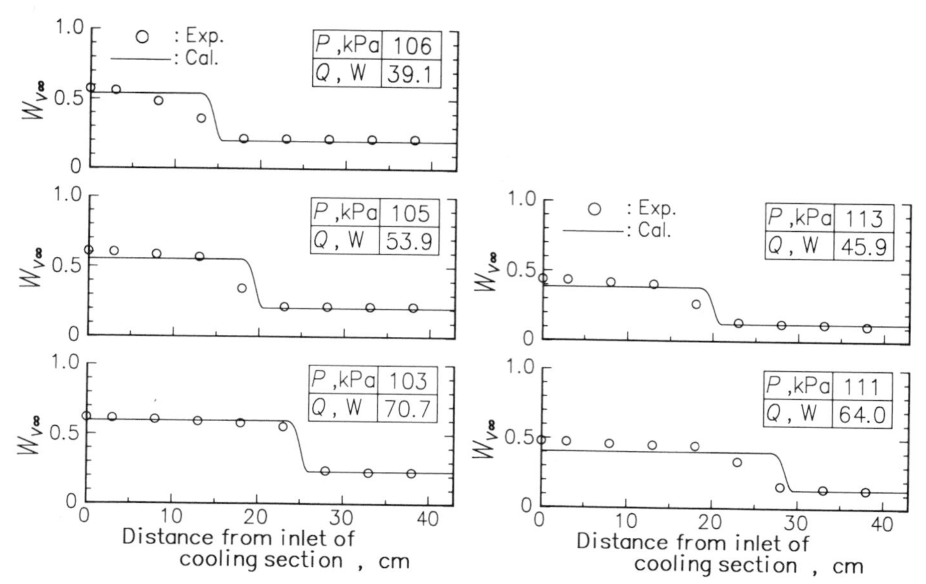

(a) Experimental data for X_{R11}=24 %   (b) Experimental data for X_{R11}=36 %

FIGURE 13. Comparison of axial concentration profiles with experimental results.

CONCLUSIONS

Characteristics of a variable conductance heat pipe which uses a binary mixture of R113 and R11 were investigated both experimentally and theoretically, and the following conclusions were obtained:

1. In the lower range of heat loads, the two-component heat pipe maintains a constant temperature difference while transmitting heat at a fluctuating rate. Its performance for constant temperature control is superior to that of a gas-loaded heat pipe.
2. In the higher range of heat loads, the heat transfer characteristics approach those of pure R113. The heat transfer rate is generally better than that of a gas-loaded heat pipe.
3. The heat transfer rate of the two-component heat pipe has a minimum value at a certain concentration of R11.
4. The constant temperature operation results from the formation of non-condensing region in the cooling section. The heat transfer characteristics of condensation of a binary mixture in the cooling section are explained well by the present analysis which employs an integral profile method.

ACKNOWLEDGEMENTS

This work is a Japan-side result carried out under the Japan-U.S. Cooperative Science Program.

NOMENCLATURE

B	radius of core region in vapor phase, m
D	diffusivity of R113 vapor, m^2/s
g	gravity acceleration, m/s^2
h_{fg}	latent heat of vaporization, J/kg
L	length of active condensing region, m
m	mass flux of condensation, kg/m^2s
P	pressure, Pa
Q	heat load, W
R	radius of pipe, m
r	radial coordinate, m
T	temperature, °C
T_b	saturation temperature at atmospheric pressure, °C
U	axial velocity, m/s
V	radial velocity, m/s
W	mass fraction of R113
X	mole fraction
z	distance measured downward from the interface of active and inactive condensing regions, m
z'	distance measured upward from bottom of the pipe, m
δ	liquid film thickness, m
λ	thermal conductivity, W/m°C
ν	kinematic viscosity, m^2/s

Subscripts

c	cooling section
h	heating section
i	vapor-liquid interface
ℓ	liquid
s	saturated value
v	vapor
w	wall

REFERENCES

1. Chi, S.W., *Heat Pipe Theory and Practice*, McGraw-Hill, New York, 1976.

2. Edwards, D.K., and Marcus, B.D., Heat and Mass Transfer in the Vicinity of the Vapor-Gas Front in a Gas-Loaded Heat Pipe, *Trans. ASME J. Heat Transfer*, vol.94, no.2, pp.155-162, 1972.

3. Rohani, A.R ., and Tien, C.L., Steady Two-Dimensional Heat and Mass Transfer in the Vapor-Gas Region of a Gas-Loaded Heat Pipe, *Trans. ASME J. Heat Transfer*, vol.95, no.3, pp.377-382, 1973.

4. Peterson, P.F., and Tien, C.L., Gas-Concentration Measurement and Analysis for Gas-Loaded Thermosyphons, *Proc. Int. Symp. Natural Circulation, ASME HTD*, Boston, pp.177-184, 1987.

5. Bobco, R.P ., Variable Conductance Heat Pipes: A First-Order Model, *J. Thermophysics and Heat Transfer*, vol.1, no.1, pp.35-42, 1987.

6. Hijikata, K., and Tien, C.L., Non-Condensable Gas Effect on Condensation in a Two-Phase Closed Thermosyphon, *Int. J. Heat Mass Transfer*, vol.27, no.8, pp.1319-1325, 1984.

7. Tien, C.L., and Rohani, A.R., Theory of Two-Component Heat Pipes, *Trans. ASME J. Heat Transfer*, vol.94, no.4, pp.479-484, 1972.

8. Stephan, K., and Abdelsalam, M., Heat-Transfer Correlations for Natural Convection Boiling, *Int. J. Heat Mass Transfer*, vol.23, no.1, pp.73-87, 1980.

9. Schlünder, E.U., Heat Transfer in Nucleate Boiling of Mixtures, *Proc. 18th Int. Heat Transfer Conf.*, vol.4, pp.2073-2079, 1986.

10. Hijikata, K., Mori, Y., Himeno, N., Inagawa, M., and Takahasi, K., Free Convective Filmwise Condensation of a Binary Mixture of Vapors, *Proc. 18th Int. Heat Transfer Conf.*, vol.4, pp.1621-1626, 1986.

Heat Pipes and Heat Pipe Heat Exchangers for Cooling of Electronic and Microelectronic Equipment

F. JELÍNEK and J. KABÍČEK
KOH-I-NOOR
Prague, Czechoslovakia

V. HLAVAČKA and F. POLÁŠEK
National Research Institute for Machine Design (SVÚSS)
Prague, Czechoslovakia

ABSTRACT

The paper describes a calculation of capillary and gravitational heat pipes as well as heat pipe heat exchangers from the point of view of their application in electronic and microelectronic equipment. Some examples of heat pipe application in electronics can give electronic experts a survey of Czechoslovak achievements in this field.

1. INTRODUCTION

The application of heat pipes in electronics and microelectronics is determined by their basic properties as follows:

- high thermal conductivity; that is to say, high heat transfer performance at low temperature difference;
- separation of heat sources from the cooling environment even over larger distances;
- more uniform temperature profiles of heat-transfer surfaces;
- possibilities of reversible transformation of heat fluxes as a rule;
  the transformation from high heat fluxes (in W m^{-2}) from heat sources to low heat fluxes with heat transfer into cooling environment is made use of;
- possibilities of stabilizing heat-transfer surface temperatures by means of variable conductance heat pipes as heat flow changes;
- possibilities of interruption of heat transfer with

pipes in the form of thermal diodes and thermal switches;
- low heat capacity of heat pipes, as their inner space is filled up with only a low quantity of the working fluid;
- low time constants, which means a rapid rise in the temperature of the entire heat pipe surface as the result of high velocity flow of vapor from the evaporator into the condenser section of the heat pipe;
- quiet operation and no demands on operationg personnel;
- possibilities of suitable design execution as far as lay-out is concerned.

The above characteristics explain why heat pipes began to be employed for cooling of equipment in electronics and microelectronics in the early 70´s, shortly after their development for space applications at Los Alamos in 1964.

While heat pipes find increasingly wider applications for waste heat recuperation in energy-conserving projects, views on the use of heat pipes in electronics differ, if such factors as prices of raw materials, advancement of technology, and the like are taken into account.

Equipment or complete systems with heat pipes installed are usually smaller in dimensions and lower in weight. However, heat pipes are generally more costly than other classical heat transfer elements such as air-cooled aluminium fin arrays.

In the Countries of Mutual Economic Assistance, Czechoslovakia ranks among pioneers both in research and in industrial production of heat pipes for various industrial applications, those in electronics and microelectronics included.

This contribution is aimed at providing experts dealing with cooling in electronics with a review of both research activities and industrial applications of heat pipes in Czechoslovakia. The paper backs up our views on the merits of heat pipes as applied in electronics.

2. CALCULATION OF HEAT PIPES

To cool electronic components and devices, either plain or, in most cases, finned (on the condenser section) heat pipes can be used. Heat transfer performance Q (expressed in terms of W) transferred by a heat pipe from a heat source (temperature t_Z) to cooling environments (temperature t_A) depends on temperature diference $t_Z - t_A$ and partial thermal resistances, as seen in Fig. 1.

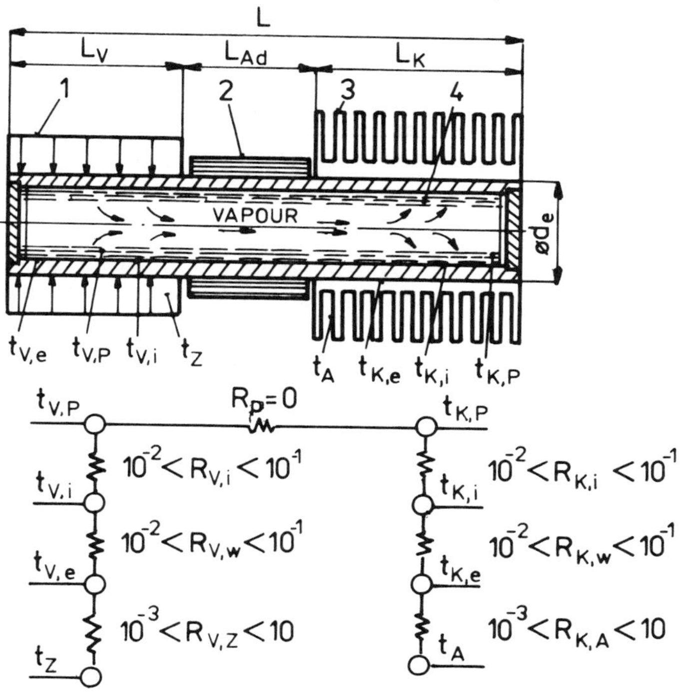

Fig. 1. Schematic diagram of a heat pipe and thermal resistances.

L; L_V; L_{Ad}; L_K - legths (total, evaporator, adiabatic part, condenser)

$R_{V,Z}$; $R_{V,w}$; $R_{V,i}$; R_P; $R_{K,i}$; $R_{K,w}$; $R_{K,A}$ - thermal resistances

- from a heat source to a heat pipe, of a evaporator wall, of evaporation, of vapor phase, of condensation, of condenser wall and from a heat pipe to cooling environment

1 - heat source; 2 - insulation; 3 - fins; 4 - wick

$$Q = \frac{t_Z - t_A}{R_{V,Z} + R_{V,w} + R_i + R_{K,w} + R_{K,A}} \quad (1)$$

where

$$R_i = R_{V,i} + R_P + R_{K,i} \quad (2)$$

is the inner heat pipe resistance.

Thermal resistance $R_{V,Z}$ determined by the way of connecting the heat pipe to the heat source and its calculation is dependent on the mode of heat transfer (conduction, convection, or combination of both). The thermal resistances of the heat pipe walls in the condenser and evaporator - $R_{K,w}$ and $R_{V,w}$ respectively - are determined by material and heat pipe dimensions.

To calculate thermal resistance $R_{K,A}$ for transfer of heat from the finned condenser section into the cooling surroundings (mostly air) by free or forced convection, well-known correlations reported in heat transfer handbooks can be employed [1] . The actual inner thermal resistance R_i of a heat pipe depends on thermal resistances in phase changes of evaporation ($R_{V,i}$) and condensation ($R_{K,i}$) of the working fluid and in vapor phase flow from the evaporator to condenser (R_P). Thermal resistance R_P is generally negligible when low-temperature heat pipes are used in electronics.

3. CAPILLARY HEAT PIPES

Thermal resistance for boiling and evaporation ($R_{V,i}$) of the working fluid in the evaporator of a capillary heat pipe (CHP) is recommended by Chi [2] to be calculated from the effective thermal conductivity $\lambda_{c,ef}$ and capillary system thickness δ_c as follows:

$$R_{V,i} = \frac{\delta_c}{\lambda_{c,ef} S_V} \quad (3)$$

where S_V is the heat-transfer surface on the phase boundary. Table 1 lists expressions for calculating $\lambda_{c,ef}$ of simple systems.

TABLE 1. Parameters of basic capillary structures [2]

Effective pore radius $r_{c,ef}$ m	Permeability K $[m^2]$	Effective thermal conductivity $\lambda_{c,ef}$ $[W\ m^{-1}\ K^{-1}]$	Capillary structure
$\frac{a + d_e}{2}$	$\frac{d_e^2\ \varepsilon^3}{122\ (1 - \varepsilon)^2}$	$\frac{\lambda_L \left[(\lambda_L + \lambda_c) - (1 - \varepsilon)(\lambda_L - \lambda_c)\right]}{\left[(\lambda_L + \lambda_c) + (1 - \varepsilon)(\lambda_L - \lambda_c)\right]}$	Screen; 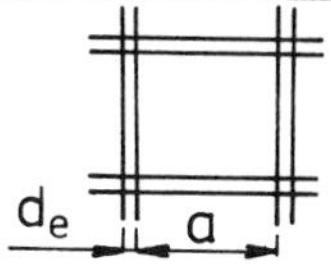 d_e, a
a	$\frac{2\ \varepsilon\ r_{H,L}^2}{f_L\ Re_L}$ $r_{H,L}$ and $f_L\ Re$ see [2]	$\frac{e\ \lambda_L\ \lambda_c\ h + a\ \lambda_L\ (0.185\ e\ \lambda_c + h\ \lambda_L)}{(a + e)\ (0.185\ e\ \lambda_c + h\ \lambda_L)}$ - EVAPORATOR $\varepsilon\lambda_L + (1 - \varepsilon)\ \lambda_c$ - CONDENSER	Grooves; 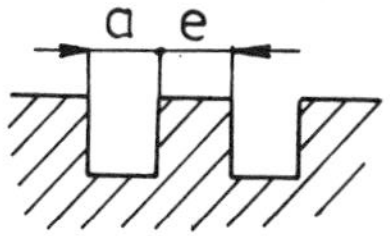 a, e $\varepsilon = \frac{a}{a + e}$
$0.41\ r_x$	$\frac{r_x^2\ \varepsilon^3}{37.5\ (1 - \varepsilon)^2}$	$\frac{\pi}{8}\left(\frac{r_y}{r_x}\right)^2 \lambda_c + \left[1 - \frac{\pi}{8}\left(\frac{r_y}{r_x}\right)^2\right]$ $\frac{\lambda_L\ \lambda}{\varepsilon'\ \lambda_c + \lambda_L\ (1 - \varepsilon)}$, $\varepsilon' = \frac{\varepsilon}{1 - \frac{\pi}{8}\left(\frac{r_y}{r_x}\right)^2}$	Sintered metal spheres; 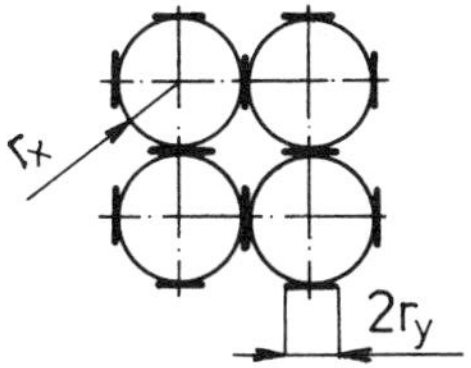 r_x, $2r_y$

This procedure is suitable for the process of surface evaporation from capillary system surfaces, without nucleate boiling, if the thickness δ_c is known. The thickness and, consequently, thermal resistance $R_{V,i}$ are difficult to determine in capillary systems under the conditions of nucleate boiling, unless suction ability of the capillary system is interfered with. In such a case, it is more appropriate to calculate the thermal resistance $R_{V,i}$ from the expression below using experimental values of heat transfer coefficient α_V.

$$R_{V,i} = \frac{1}{\alpha_V \, S_V} \tag{4}$$

Equations for the determination **of** maximum heat performance of CHPs under various boundary and working conditions, i.e., the dependence of CHP heat performance limits on vapor phase temperature, are given in the **Table** 2.

4. GRAVITATIONAL HEAT PIPES

A calculation procedure for the simplest type of heat pipes, the gravitational heat pipes (GHP) with smooth walls, has not been worked out completelly up to now in spite of the fact of their plentiful application in electronics. In the following part a simple and experimentally verified calculation of GHPs is given expecially from the point of view of their application in electronics.

Under the assumptions of Nusselt´s theory for a falling liquid film on a smooth vertical wall there are equations of the local film liquid thickness and the local heat flux in the distance x during film condensation of a saturated vapor

$$\delta_x = \left(\frac{4\lambda_L \; \nu_L \; x \; \Delta t_{K,i}}{l_v \; \varrho_L \; g}\right)^{1/4} \tag{5}$$

$$q_x = \lambda_L \frac{\Delta t_{K,i}}{\delta_x} = \left(\frac{l_v \; \varrho_L \; g \lambda_L^3}{4 \; x \; \nu_L}\right)^{1/4} \Delta t_{K,i} \tag{6}$$

where

$$\Delta t_{K,i} = t_P - t_{K,i} \tag{7}$$

TABLE 2. Heat pipe performance limits

1	Sonic limit	$Q_S = A_P \varrho_P l_v \left[\frac{R_G T_P \varkappa}{2 (\varkappa + 1)} \right]^{1/2}$		[2]
2	Flooding limit	$Q_{I,c} = A_P \left[\frac{\varrho_P \sigma l_v^2}{L^x} \right]^{1/2}$	$C^x = \sqrt{3.2}\ \mathrm{tgh}\ (0.5\ Bo^{1/4})$	[2]
		$Q_{I,g} = C^{x2} A_P l_v \frac{\left[g \sigma (\varrho_L - \varrho_P) \right]^{1/4}}{\left[\varrho_L^{-1/4} + \varrho_P^{-1/4} \right]^2}$	$Bo = d_p \left[\frac{g (\varrho_L - \varrho_P)}{\sigma} \right]^{1/2}$ L^x characters itics length [2]	[3]
3	Evaporation limit	$Q_{V,k} = \frac{2 \pi L_V \lambda_{c.ef} T_P}{l_v \varrho_P \ln \frac{d_i}{d_p}} \left(\frac{2 \sigma}{r_B} - \frac{2 \sigma \cos \phi}{r_{c,ef}} \right)$	r_B bubble radius $r_{c,ef}$ effective pore radius (see Table 1)	[2]
		$Q_{V,g} = \varphi l_v A_P \varrho_P^{1/2} \sqrt[4]{\sigma g (\varrho_L - \varrho_P)}$	$\varphi \langle 0.12 \div 0.16 \rangle$	[3]
4	Capillary limit	$Q_c = \frac{\frac{2 \sigma \cos \phi}{r_{c,ef}} - \Delta p_\perp \pm \varrho_L g L_{ef} \sin \beta}{\frac{\mu_L L_{ef}}{K A_K \varrho_L l_v}}$	$L_{ef} = \frac{L_V}{2} + L_{Ad} + \frac{L_K}{2}$ K permeability (see Table 1)	[2]

Table 3 demonstrates the development of a condensate layer of four working fluids most frequently employed in GHP. The difference between vapor temperature and inner wall is in all cases the same, i.e., $\Delta t_{K,i} = 5^{o}C$, which represents in water and ammonia a rather high value, while in substances like **ethanol or** Freons this is a rather adequate value.

TABLE 3. Development of a condensate layer ($\Delta t_{K,i} = 5^{o}C$)

Distance	[m]	0.1	0.2	0.6	1.0	1.6
Water						
$\delta_x x10^3$	m	0.070	0.083	0.109	0.123	0.139
$Re_{L,x}$	-	7.6	12.9	29.3	43.0	61.2
q_x	$W\ m^{-2}$	48700	40950	31120	27390	24350
Ammonia						
$\delta_x x10^3$	m	0.083	0.099	0.130	0.148	0.166
$Re_{L,x}$	-	16.9	28.7	64.9	95.8	135
q_x	$W\ m^{-2}$	26950	22660	17220	15150	13470
Ethanol						
$\delta_x x10^3$	m	0.066	0.078	0.103	0.117	0.132
$Re_{L,x}$	-	6.1	10.2	23.2	34.1	48.5
q_x	$W\ m^{-2}$	13130	11040	8390	7380	6565
Freon 12						
$\delta_x x10^3$	m	0.063	0.076	0.099	0.113	0.127
$Re_{L,x}$	-	38.2	64.3	146	215	306
q_x	$W\ m^{-2}$	6450	5325	4120	3630	3225

The mean temperature of the condensate has been chosen as $80^{o}C$ for water and ethanol, **but** as $40^{o}C$ for ammonia and Freon 12, due to a better description of the usual thermal regime of the pipes working with these substances in electronics. From Table 3 it will be understood that a purely laminar condensate film can be achieved only in the initial section; the remaining length of the condenser section develops a **pseudolaminar** flow. In Freon 12 the values can be $Re_{L,x} > 200$, which signifies transitional flow.

The condensate falls to the evaporator section mostly in the pseudolaminar regime+) and the liquid will evaporate along the whole evaporator part L_V and both the layer thickness and $Re_{L,x}$ will gradually be reduced. The Nusselt assumptions are maintained also in the evaporator section and the local film liquid thickness and the local heat flux will be defined as

$$\delta_x = \left[\frac{4\nu_L \ \lambda_L \ \Delta t_{V,i} \ (L_V - x)}{l_v \ \varrho_L \ g}\right]^{1/4} \tag{8}$$

$$q_x = \lambda_L \frac{\Delta t_{V,i}}{\delta_x} = \left[\frac{l_v \ \varrho_L g \ \lambda_L^3}{4 \ \nu_L \ (L_V - x)}\right]^{1/4} \Delta t_{V,i}^{3/4} \tag{9}$$

where

$$\Delta t_{V,i} = t_{V,i} - t_P \tag{10}$$

The mean heat fluxes in the evaporator and condenser are determined by performing the integration along L_K and L_V

$$\bar{q}_K = 0.943 \left(\frac{l_v \varrho_L g \lambda_L^3}{\nu_L \ L_K}\right)^{1/4} \Delta t_{K,i}^{3/4} = 0.943 \ K_L L_K^{-1/4} \ \Delta t_{K,i}^{3/4} \tag{11}$$

$$\bar{q}_V = 0.943 \left(\frac{l_v \varrho_L g \ \lambda_L^3}{\nu_L \ L_V}\right) \Delta t_{V,i}^{3/4} = 0.943 \ K_L L_V^{-1/4} \ \Delta t_{V,i}^{3/4} \tag{12}$$

where

$$K_L = \left(\frac{l_v \ \varrho_L \ g \ \lambda_L^3}{\nu_L}\right)^{1/4} \tag{13}$$

will now become the main factor influencing the heat transfer in the GHP. Its values for the usual working fluids in electronics and their dependance on temperature are shown in Table 4. The factor K_L is the principal criterion in selecting the GHP working fluids.

+) $20 < Re_L < 100$

TABLE 4. **Factor of thermophysical properties of GHP** $[W/(m^{7/4}K^{3/4})]$

Temperature (°C)	- 20	0	20	40	60	80	100
Freon 12	1880	1760	1640	1520	1400	1280	
Freon 22	2100	1860	1620	1390	1150	920	
Freon 113	1220	1250	1260	1250	1220		
Ammonia	8320	7690	7060	6440	5810	5180	
Methanol	2810	2900	3080	3310	3550	3750	3850
Ethanol	1830	1980	2220	2475	2690	2780	2735
Acetone	2760	2770	2810	2800	2770	2700	2550
Water				9790	10820	11670	12360

As the connecting section between the evaporator and condenser is considered adiabatic, the heat flow transferred through the pipe will be defined as

$$Q = \pi d_i L_K \bar{q}_K = \pi d_i L_V \bar{q}_V \tag{14}$$

Substituting for $\bar{q}_K$ and $\bar{q}_V$ and eliminating the vapor phase temperature t_P leads to the expression of the heat transfer performance of the GHP as

$$Q = K_G K_L \Delta t_i^{3/4} \tag{15}$$

where

$$\Delta t_i = t_{V,i} - t_{K,i} \tag{16}$$

and

$$K_G = 0.943 \pi d_i \left(\frac{L_K L_V}{L_K + L_V}\right)^{3/4} \tag{17}$$

is the geometrical factor of the GHP. The graphical presentation of equation (15) is in **Fig.** 2.

An important feature of the GHP is the inner thernal resistance R_i defined as

$$R_i = \frac{\Delta t_i}{Q} \tag{18}$$

Using equation (15) we get the following equations:

$$R_i = \frac{\Delta t_i^{1/4}}{K_G K_L} = \frac{Q^{1/3}}{(K_G K_L)^{4/3}} \tag{19}$$

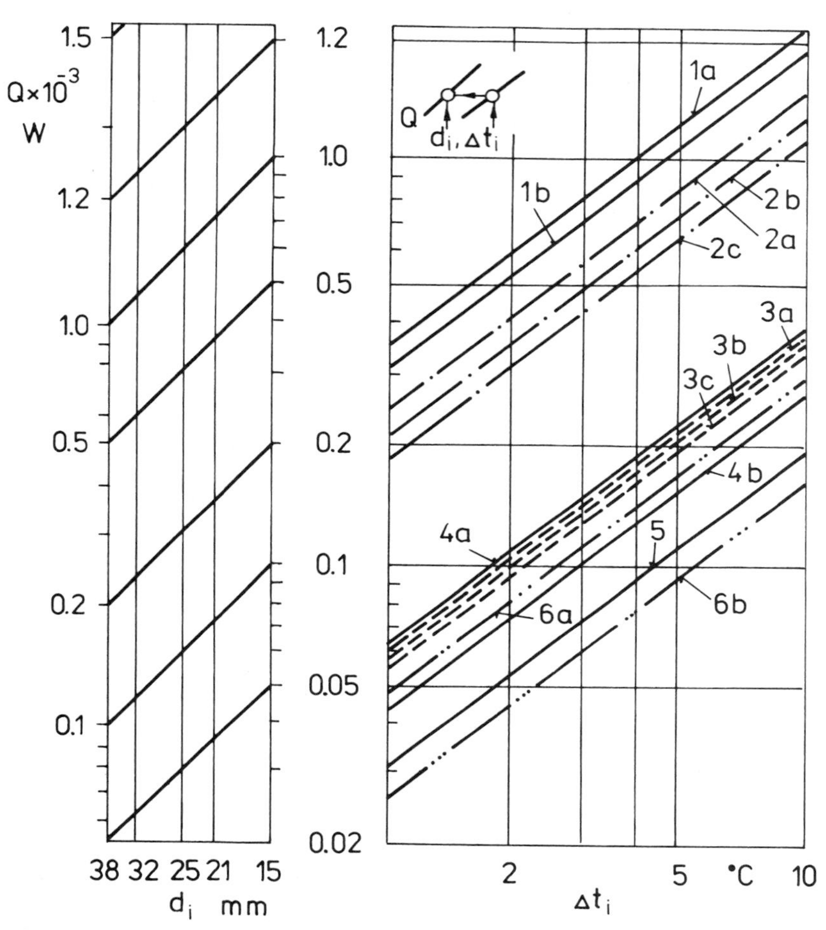

Fig. 2. Predicted heat performance of gravitational heat pipes.

1a - water (100 °C)	3c - ethanol (10 °C)
1b - water (60 °C)	4a - Freon 502 (-60 °C)
2a - ammonia (-40 °C)	4b - Freon 502 (20 °C)
2b - ammonia (-20 °C)	5 - Freon 113 (0 °C)
2c - ammonia (50 °C)	6a - Freon 12 (0 °C)
3a - ethanol (60 °C)	6b - Freon 12 (100 °C)
3b - ethanol (35 °C)	

The non-linearity in equation (15) and the dependence of the inner thermal resistance on the heat transfer **rate** may cause certain problems in the determination of heat pipe performance in the case of given outer conditions of heat transfer. A detailed assessment of the situation shows that at $\Delta t_i < 8$ to 10^oC it is quite feasible to employ instead of equation (15) the simpler relationship

$$Q = 0.72\ K_L K_G\ \Delta t_i \tag{20}$$

which is direct result of a simple linearization of equation (15) and of the approximate balance of the **measured** data (Fig. 3) with the best agreement at $\Delta t_i = 4$ to 6^oC. After such practical simplifica**tion** the inner thermal resistance of gravitational heat pipes can be defined as

$$R_i = \frac{1.4}{K_G K_L} \tag{21}$$

It is easy to see that this suggested procedure involves, within the given range of Δt_i, only minor errors for applications in electronics.

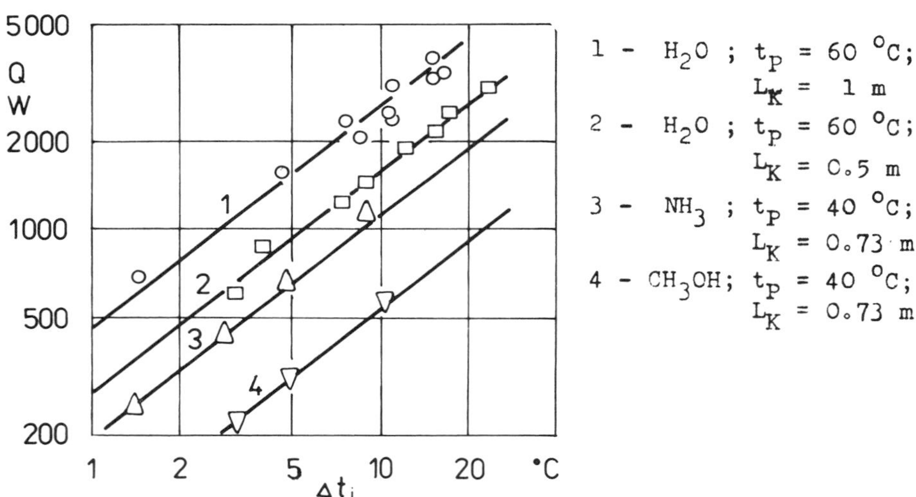

FIG. 3. Gravitational heat pipe performance as a function of temperature difference.

The relatively low values of inner thermal resistance are the reason why the heat pipe outwardly behaves as if it were made of highly conductive material. In water pipes usually $R_i < 0.01\ W^{-1}\ K$, the pipes containing less convinient thermophysical properties (e.g., Freons) have largely $R_i < 0.1\ W^{-1}K$. The inner conductivity λ_i of a heat pipe can be defined, e.g., as

$$\frac{\Delta t_i}{R_i} = \frac{\pi d_i^2}{4} \frac{\lambda_i}{0.5\ (L_K + L_V)} \Delta t_i \tag{22}$$

hence

$$\lambda_i = \frac{2\ (L_K + L_V)}{\pi d_i^2\ R_i} \tag{23}$$

The attained values of λ_i are generally within 10^3 to 10^4 $Wm^{-1}\ K^{-1}$.

Within the extensive research scope of gravitational heat pipe properties undertaken at SVÚSS, investigations were made into the heat performance characteristics of a large number of heat pipes of various dimensions and on application of most of suitable working fluid fillings. One of the main outcomes in the analysis of dependencies of the type of equation (15) is exemplified in Fig. 3. The tested heat pipes had the same length of the evaporator and condenser sections and the inner diameter d_i = 6 to 24 mm. In the first place it will be seen that the lines representing the mean value of measured figures have in all cases almost identical gradient satisfying the exponent 0.75 in equation (20).

Also the values of factor $K_G K_L$ resulting from the extrapolation of the curves to $\Delta t_i = 1^oC$ are described with adequate accuracy by equation (15). These facts indicate that the accepted simplified notion of the mechanism of inner transfer in GHP with $\Delta t_i < 10^oC$ is correct and will not lead to major errors in determining the heat flow transferred through these pipes.

It should also be pointed out that many tests have been made of the GHP operating in inclined positions. It has been found that the relationship (15) works very well up to 75° incline from the vertical position of the pipe.

The internal thermal resistance of a GHP for given heat pipe geometrical parameters and working fluid can be reduced only by enhancing heat transfer during evaporation, boiling, and condensation of the working fluid, i.e., by increasing the values of α_V and α_K. Equation (19) for the internal thermal resistance of GHP has been derived under assumptions that Nusselt´s theory is valid for the calculation of both α_V and α_K for falling films in the GHP. The values of α_V and α_K can be increased by using capillary-porous systems. Such systems, however, do not fulfill the function of capillary pump as capillary tubes do, but they cause nucleate boiling to begin at lower heat flux values by the action of a larger number of bubble nucleation centers and, in doing so, they enhance heat transfer. This area has been under study for several years already, especially from the standpoint of enhancing heat transfer during evaporation and boiling of coolants in pool boiling by means of sintered metal [4,5] , shaped grooves [6] , and special surfaces [7] . The experimental study of falling liquid films is, so far in early stage [8,9] . Therefore, the experimental program of the SVÚSS has been directed particularly at heat transfer enhancement during evaporation and boiling of Freons, methanol, ethanol and water both in pool boiling and in falling films using the following four selected capillary systems:

(a) capillary system consisting of mutually perpendicular rectangular grooves 0.4 x 0.4 x 0.4 mm (groove depth x groove width x tooth width);
(b) capillary system 2 mm in thickness made by sintering copper wires (length 2 ÷ 3 mm, dia. 60 μm). The porosity of the system was up to 80 %;
(c) capillary system 2 mm in thickness made by sintering copper file dust (appr. 0.5 mm in dimensions). The porosity of the system was up to 50 %;
(d) capillary system 2 mm in thickness made by sintering reduced copper powder of uniform particle size 0.063 mm. The porosity of the system was up to 60 %.

All of these capillary systems were prepared on a heating copper surface which simulated the operation of GHP with the heating surface in vertical position and all capillary systems showed an excellent heat transfer performance compared with the smooth tube. They were operating in nucleation regime for the entire range of heat fluxes. The sintered copper fibers had the best performance. Detailed information on the evaporation heat transfer enhancement of falling liquid films will be given at the Int. Heat Transfer Symposium, Varna 1989 [10] .

The performance limits of a GHP can be determined by means of the relations given in Table 2. There is the same equation for determination of the sonic limit of both the CHP and the GHP. The sonic limit has high values in the usual electronic applications in comparison with the evaporation and flooding limits especially at high heat fluxes in the evaporator ($q_V = {}^Q/S_V$, i.e., for small evaporator surface area) and at high heat fluxes in the vapor space ($q_P = {}^Q/A_P$, i.e., for low cross section of flow area). The evaporation limit of the GHP is defined as the heat flux at the transition from the nucleate to film boiling in the falling liquid films and it can be determined by Kutateladze´s equation (Table 2). An enhancement of evaporation limits can be also achieved by adding capillary structures on the inner smooth wall of the GHP.

5. HEAT PIPE HEAT EXCHANGERS (HPHE) FOR COOLING IN ELECTRONICS

Heat exchangers composed of heat pipes can be taken for a special case of already known heat-transfer systems provided with a heat-transfer intermediary (usually liquid) used for various purposes. The main example is three-component recovery circuits. The heat-transfer intermediary is passed through two heat exchangers, removing heat from one of them and dissipating it into the other one with the cooler environment. Compared to these systems, the HPHE (Fig.4) is characterized by a more compact arrangement (the hot and cool sections are close by or above each other) and the fact that heat transfer through the heat-transfer medium - heat pipe working fluid - occurs under boiling and condensation conditions. This feature represents an apparent enhancement of the process as compared to the heat transfer by convection in the flow of liquid.

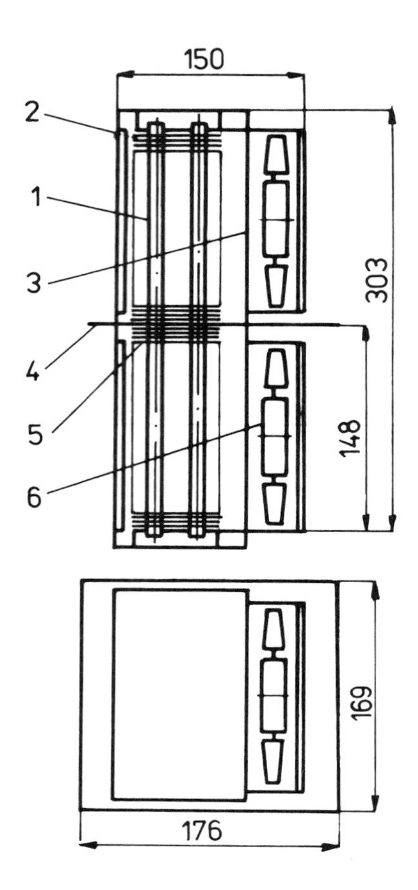

Fig. 4.

Schematic diagram of a heat pipe heat exchanger.
1 - heat pipes, 2 - frame, 3 - screen,
4 - partition, 5 - fins, 6 - fan

As opposed to classical, e.g., tubular heat exchangers, the internal heat-transfer surface of pipes in a heat-transfer intermediary circuit is transformed to the outer surface of one of exchangers. This tendency to increase overall heat transfer surface is supported by several circumstances:

(a) the possibility to use effective finned surfaces on the sides of both process media,
(b) higher values of the heat transfer coefficient for cross

flow around bundles of finned tubes,

(c) easier implementation of **the most effective countercurrent arrangement.**

These advantages predetermine HPHE particularly for heat transfer between gaseous environments, namely for ventilation and air conditioning equipment, various types of driers or for recuperation of heat from waste gases. In these areas of application, HPHE are frequently more appropriate than the other types of recuperative of regenerative equipment, when performance, size, process technology, and price are considered.

As far as thermal calculations are concerned, HPHE are accompanied with some peculiarities that are for the case of GHP explained further in compliance with the methodology elaborated by the SVÚSS in the framework of the R and D project dealing with heat pipes for large-scale applications in Czechoslovakia.

5.1 Calculation of Heat Pipe Heat Exchangers

Most of the experimental data obtained for GHP confirm the dependence between heat flow Q_1 transferred by the pipe and the difference in mean internal wall temperatures Δt_i in the evaporator and condenser sections to have the following form:

$$Q_1 = K_L K_G \Delta t_i^p \qquad (24)$$

the exponent p being 0.75 or so. This fact indicates the possibility of calculating internal thermal resistence R_{i1} from the expression for the condensation of vapor on a vertical wall. Such a procedure results in expressing the coefficient $K_L K_G$ as

$$K_L K_G = 0.56 \pi d_i \mu L_K^{3/4} \left(\frac{l_v \varrho_L \lambda_L^3 g}{\nu_L^2}\right)^{1/4} \qquad (25)$$

where

$$\mu = \frac{2 L_V}{L_V + L_K} \quad \frac{L_V + L_K}{L_V} - 1 \qquad (26)$$

is the approximate correlation, when the lengths of the evaporator and condenser sections are not the same. The internal thermal resistance of the pipe will be given by the following equation:

$$R_{i1} = \frac{\Delta t_i}{Q_1} = \frac{Q_1^{(1/p-1)}}{(K_L\, K_G)^{1/p}} \qquad (27)$$

For a heat pipe operating ideally, $R_{i1} = 0$, which means that $K_L K_G \rightarrow \infty$; this condition will be met, if $\lambda_L \rightarrow \infty$ or $\nu_L = 0$.

Overall heat transfer coefficients for a heat pipe operating ideally (i.e., if $R_{i1} = 0$) or for a heat exchanger assembled from such heat pipes can be defined, e.g., by the expression as follows:

$$k_o = \frac{1}{\frac{L_T}{L_K}\left(\frac{1}{\alpha_A} + \frac{S_I}{d_i}\,\frac{\delta_w}{\lambda_{K,w}}\right) + \frac{L_T}{L_V}\left(\frac{1}{\alpha_Z} + \frac{S_I}{\pi d_i}\,\frac{\delta_w}{\lambda_{V,w}}\right)} \qquad (28)$$

If $R_{i1} > 0$ is considered, the following formula can be derived for the overall heat transfer coefficient:

$$k = \frac{k_o}{1 + k_o\, S_I\, L_T\, R_{i1}} \qquad (29)$$

With the countercurrent arrangement of a heat exchanger, especially if the ratio of the heat capacities of both streams of heat transfer media approaches 1, the heat flow transferred by individual heat pipes will be almost the same.

With coefficient $K_L K_G$ for the mean temperature of the working fluid in heat pipe determined, the approximate heat performance of heat exchanger can be determined from the following equation:

$$Q = \frac{k_o\, (K_L K_G)^{1/p}\, m\, n\, L_T\, \theta_m\, S_I}{(K_L K_G)^{1/p} + k_o S_I L_T\, Q^{(1/p\, -\, 1)}} \qquad (30)$$

where the mean logarithmic temperature difference, θ_m, is connected with heat exchanger efficiency, η, according to

$$S_I\, k\, m\, n\, L_T \theta_m = W_A\, (t'_Z - t'_A)\, \eta \qquad (31)$$

if η is defined by

$$\eta = \frac{t''_A - t'_A}{t_Z - t_A} \qquad (32)$$

The iterative calculation of equation (30) can preferably start by substituting $Q = Q_o$, i.e., with heat transfer performance corresponding to a heat exchanger operating under ideal conditions.

A more precise calculating procedure relies on gradual establishment of the temperature profile in the heat exchanger which is calculated from one row of heat pipes to another one. The heat flow transferred by the j-th row of heat exchanger heat pipes will be given by the following equation:

$$Q_j = \frac{t_{Z,j} - t_{A,j}}{\frac{1}{k_j S_j} - \frac{1}{2}\left(\frac{1}{\dot{W}_{A,j}} + i\frac{1}{\dot{W}_{Z,j}}\right)} \qquad j = 1, \ldots n \qquad (33)$$

where the second term in the denominator represents a correction taking into account the fact that there are temperatures of the j-th points of the network in equation (33) in place of the mean temperatures of media on the j-th row. For co-current and countercurrent systems $i = +1$ and $i = -1$, respectively. Equation (33) is derived from the thermal balance of the j-th row:

$$Q_j = -i\,\dot{W}_{A,j}\,(t_{A,j} - t_{A,j+1}) =$$

$$= \frac{k_j S_j}{2}\,(t_{Z,j} + t_{Z,j+1} - t_{A,j} - t_{A,j+1}) =$$

$$= \dot{W}_{Z,j}\,(t_{Z,j} - t_{Z,j+1}) \qquad (34)$$

If we do not wish to compute the system of non-linear equations as a whole, the calculation can be made by rows with known temperatures of the two media for $k = 1$ with the first choice $Q_i = 1$ corresponding to the heat flow transferred by the first row at $R_{i1} = 0$.

6. EXAMPLES OF HEAT PIPE APPLICATIONS IN ELECTRONICS

6.1 Cooling of IC Boards

The increasing population density of boards with fast integrated circuits results in considerable demands on the cooling of the boards, because due to miniaturization of components there is not sufficient heat-transfer surface. That is why some companies dealing with electronics have made use of the favourable properties of heat pipes for cooling IC boards. For instance, the design of Hughes Aircraft [11] consists of making a thin capillary heat pipe in form of a plate, the surface of which is populated with intergrated circuits. As compared to aluminium and copper cold plates the decreases in the temperature of the IC surface are up to 26 K and 15 K, respectively [11] .

The design of the SVÚSS/VÚMS relies on the dissipation of heat from the board to the board edges by means of flat heat pipes (width 4 to 8 mm, height 2 mm) and in cooling the condenser sections by forced convection into the air.

To cool the heat sources of heat-stressed boards populated with electronic components, such as integrated circuits, gate banks, hybrid-type driver circuits, etc., the SVÚSS has designed and tested experimentally capillary heat pipes circular and rectangular in cross-section and different in diameter and length. Some results are:

(a) Using heat pipes 4 mm in diameter, e.g., the temperarature of a block provided with power electronic elements was maintained at 75°C even at full current loading, but when the block with electronic elements was connected to an air cooler by means of a metallic strip, the current loading had to be decreased more than 40 %.

(b) Flat heat pipes of maximum thickness 1.3 mm were tested for boards with integrated circuits and gate banks. When the temperature differences were compared in the operated lengths of the heat pipe and a copper strip of identical dimensions, the temperature difference on the heat pipe was 18 K lower than the one on the copper strip at full heat loading.

(c) With hybrid-type power driving circuits, the heat pipes permitted the shortest possible electrical connection to other circuits as possible.

6.2 Heat pipe cooling of power semiconductor elements

The heat pipes can be used to cool power semiconductor elements (PSE) in these ways:

a) cooling PSE electrodes through integrated capillary heat pipe coolers which make a compact module with the PSE,
b) cooling the PSE cover surfaces through contact gravitation or capillary heat pipe coolers (Fig. 5a)
c) cooling PSE or PSE systems by submersion in the liquid of the heat pipe evaporating part,
d) indirect cooling of PSE closed in boxes through a heat pipe exchangers cooling the air inside (Figs. 4 and 6).

For traction drives the current aluminium air coolers are heavy and thermally of low efficiency. As a result, the SVÚSS has started research of heat pipe coolers to be used in PSE wafers of diameter 30, 40 and 65 mm. In the first stage, contact face capillary heat pipe coolers were chosen, whose design is shown in Table 5. The coolers had either independent feeder bus-bars inserted between the heat pipe face and PSE cover, or the bus-bar formed the extended face of the heat pipe. The capillary system of the heat pipes was made by rectangular grooves 0.4 x 0.4 mm. The pipes were filled with water, methylalcohol, or their mixture. Two heat pipes can serve as a cooler of one semiconductor element, or there can be more heat pipes joined with wafer PSE in blocks.

Table 5 shows the basic geometrical parameters of the PSE heat pipe coolers as well as the thermal resistance for cooling air velocities of 6 $m.s^{-1}$ between fins.

While testing a tram pulse transducer consisting of three PSE of diameter 40 mm, two heat pipes 60 mm long and two 100 mm long, a maximum heat transfer performance of 2500 W was attained with a mean air velocity of 6 $m.s^{-1}$ between fins and an air inlet temperature of 20°C. During the tests, the temperature of the p-n junctions was lower than 140°C.

a) Gravitational heat pipes

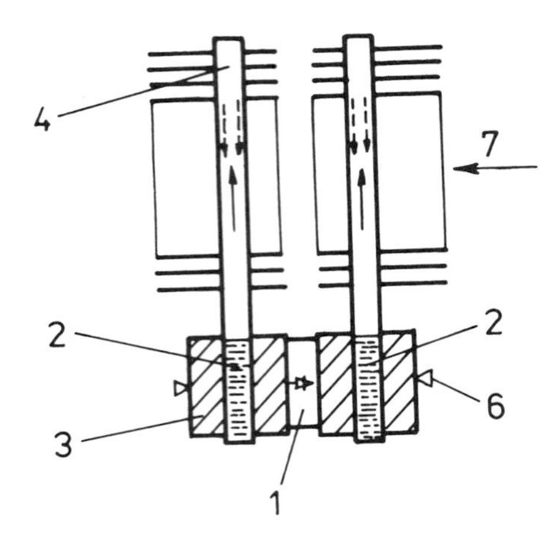

b) Heat pipe loops

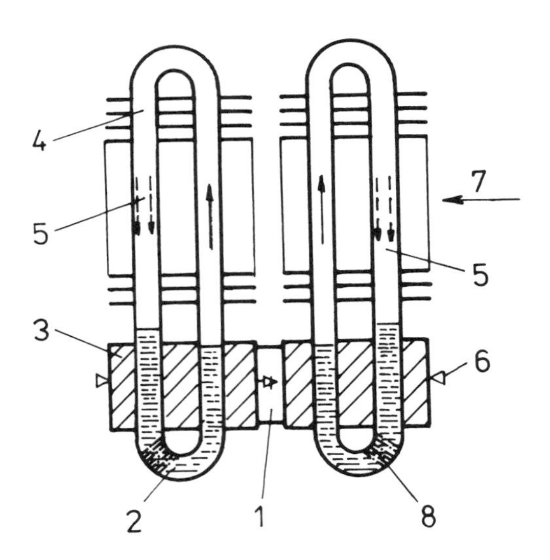

Fig. 5. Heat pipe cooler of power semiconductor elements.

1 - PSE, 2 - evaporator, 3 - block, 4 - condenser, 5 - liquid channel, 6 - tightening mechanism, 7 - cooling air, 8 - wick

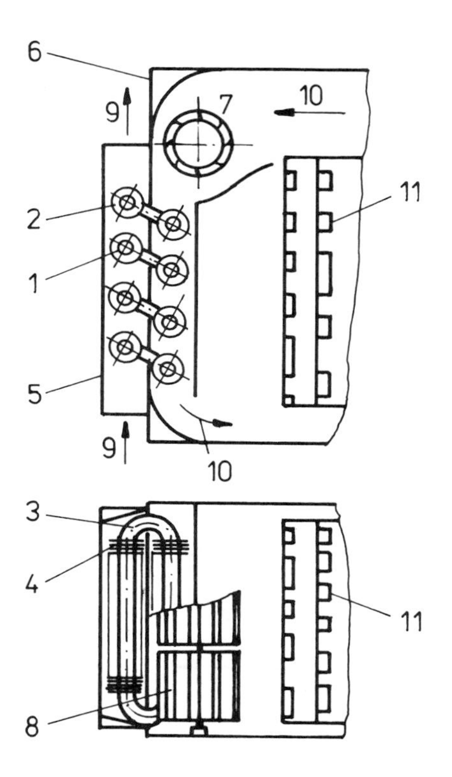

Fig. 6.

Schematic diagram of a side version of a heat pipe heat exchanger.

1 - looped type of heat pipes, 2 - radial fins, 3 - knee joint, 4 - condenser, 5 - air channel, 6 - frame, 7 - fan, 8 - fan, 9 - outer air, 10 - inner air, 11 - electronics elements

TABLE 5. Parameters of heat pipe coolers of power semiconductor elements

Type of the coolers					I	II	III	IV	V	VI
Dimension of two heat pipes	Tubes	Outer diameter	d_e	mm	30	40	40	40	65	65
		Length	L_t	mm	190	120	190	400	190	360
		Thickness	δ_t	mm	2.5				3.0	1.8
	Fins	Length	L_F	mm	100	100	100	100	100	100
		Width	h_F	mm	100	100	100	100	150	100
		Number	n	-	60	38	60	130	60	118
		Thickness	δ_F	mm	0.5	0.5	0.5	0.5	0.5	0.5
		Pitch	τ_F	mm	3	3	3	3	3	3
Minimal thermal resistance of the coolers at air velocity between fins 6 m s^{-1}			R_{min}	K W^{-1}	0.04	0.1	0.05	0.029	0.05	0.03

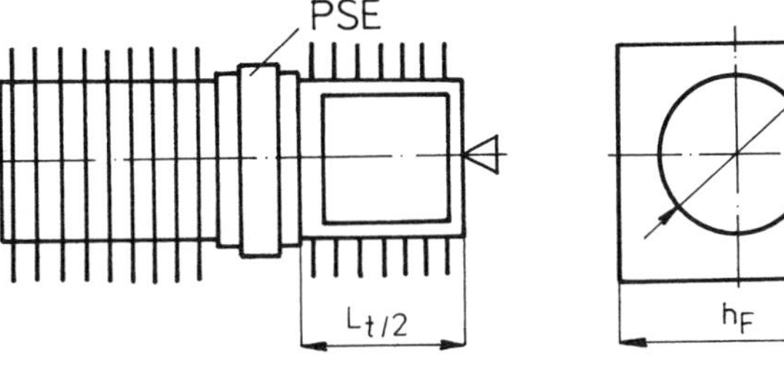

The next stage was designing and testing the functional samples of the gravitational variant of the cooler with independent channels for steam and liquid flow (see Fig. 5b). The merit of this method is the clamping of the PSE and the heat pipe over the massive body of the pipe´s evaporating part, which allows for clamping force as much as 40 000 N. The heat pipe´s condensing part is in the space above the evaporating part. While tested, the heat performance reached in a PSE of diameter 40 mm was 1600 W with mean air velocity between fins in the condensation part 6 $m.s^{-1}$ and inlet air temperature $20^{o}C$. The thermal resistance of the cooler was only 0.02 K W^{-1}.

Coolers of this type were subjected to tests also in a static DC converter of trolley. The tests proved the high functional reliability of the system under diverse operating conditions; compared to the standard design, the convertor was smaller and lighter and its operational characteristics improved. Special attention was given to the cooling efficiency of the heat pipes at low temperatures (down to $-40^{o}C$), with two different changes: water and methanol. At the lowest temperatures the efficiency of the water system deteriorated and could be compared to that of an inferior classical air heat-convection cooler. At slightly higher temperatures the heat pipes gradually warmed up along their whole length and their normal function was eventually resumed. The transients have been recorded in oscilographs. A critical situation with regard to the semiconductor components arises on transition from the frozen-up state to the normal cooling function of the heat pipes. The transients resulting from short-term thermal overloads of the cooler were treated as a separate problem. To cool potentialless modules, special coolers have been developed from copper heat pipes with sintered capillary system and water filling, whose evaporation parts are inserted into an aluminium block with three attached potentialless 431-type modules. This design of ČKD is shown in Fig. 7 [12].

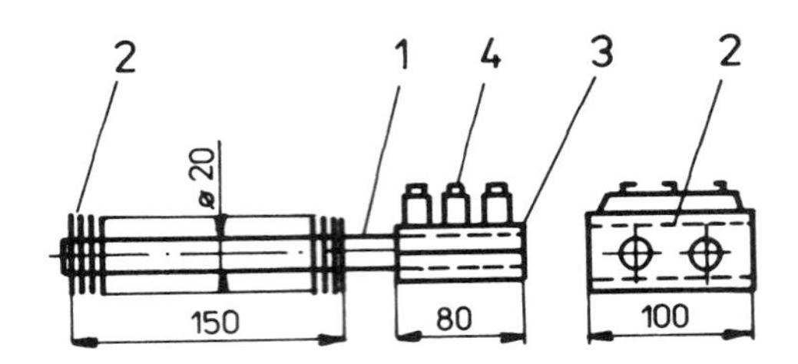

Fig. 7. Heat pipe cooler of potentialless modules.
1 - Cu/Water heat pipe, 2 - fins,
3 - Al block, 4 - potentialless modules

The manufactured PSE coolers have taken full credit for their thermal performance. Compared with the thermally equivalent aluminium air coolers used so far, they weigh about 50 percent less, take 60 percent smaller useful space and require a lower air flow. Technologically, however, they are more demanding than the standard air coolers, which must be taken into account in economic assessment.

The main advantage of heat pipe coolers lies in their being considered from the view point of the operator as air coolers, but which have high thermal parameters to be reached during the phase changes of the heat transmitting medium in the closed room of the cooler.

6.3 Cooling of sealed electronic boxes with heat pipe heat exchangers

To cool boxes containing electronic control elements in dusty and explosive surroundings presents considerable difficulties. For example electronic components, such as integrated circuit boards, may be exposed to corrosive action of the dusty or humid environments of metallurgical or chemical plants. Here, a bundle of heat pipes offers a practical solution, because the evaporating parts of the heat pipes are situated inside a sealed box, while the condensing parts work in a channel outside the box, cooled by the external cool but contaminated air. A heat exchanger consisting of a bundle of heat pipes can have arbitrary dimensions to suit the space conditions of the box and is usually located in its upper part, although a location near the side walls is not excluded (Fig. 6).

In view of dusty atmosphere and corrosive effects of the surroundings one can chose a heat pipe heat transfer surface which is conveniently extended. Table 6 shows some extended surfaces for small exchangers with heat performance from 100 to 500 W used jointly with fans of vane axial type ($\dot{V}_{max}$ = 0.05 m^3 s^{-1}, ΔP_{max} = 80 Pa). This table also gives the attained heat performances and pressure losses for face height of the evaporating and condensing part 124 x 124 mm and for face air velocities in both channels 3.25 m s^{-1}.

The compact exchangers with extended surface Al/Cu (No. 1 in **Table 6**) for heat transfer performance from 300 to 500 W are particularly suitable for clean working conditions, e.g., computing center and for cooling the warm air of sealed boxes of control and driving units of robots and manipulators.

Heat exchangers 200 to 500 W in heat performance utilizing vane axial fans having air flowrate up to 0.05 m^3 s^{-1} and maximum pressure drop 80 Pa are provided with tin-coated copper heat pipes (10/0.4 mm) filled up with water and copper fins 0.4 mm in thickness (finned area No. 4 in **Table 6**). For less aggressive environments, the tin-coated copper fins are replaced by aluminium ones 0.2 mm in thickness. The heat **exchangers** operate reliably in the boxes of the control system MARK-RZ (ČKD Polovodiče) and are fitted also into passanger and cargo ships.

7. CONCLUSION

The experience obtained so far from the operation of heat pipes provides a promising outlook for their wider application to cool electronic and microelectronic equipment. In Czechoslovakia heat pipe heat exchangers have been currently made use of to cool sealed control and regulation boxes of robots, passenger and cargo vessels, in metallurgy

TABLE 6. Comparison of predicted heat performance of heat pipe heat exchangers for cooling of closed boxes in electronics

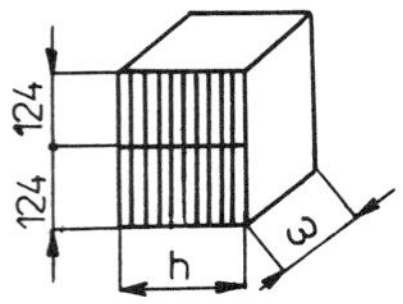

124 / 124 / h / ω

$\dot{V}_A = \dot{V}_Z = 0.5\ m^3/s$

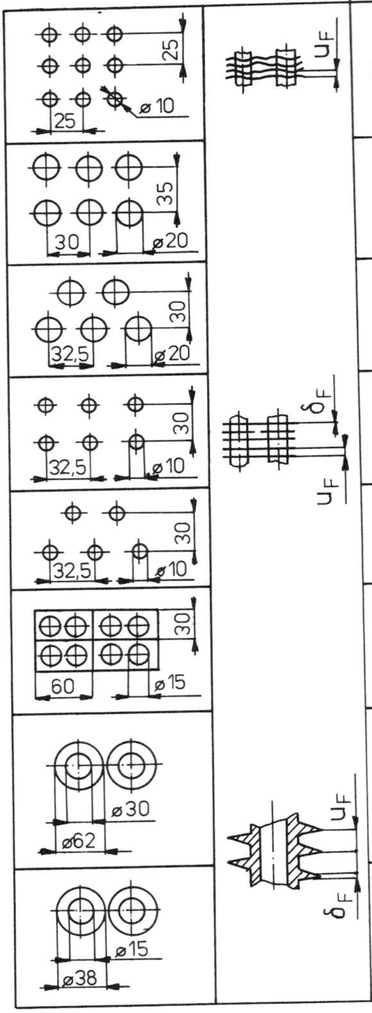

Tube arrangement	Fin type	u_F [mm]	δ_F [mm]	h [mm]	ω [mm]	Q [W]	ΔP_G [Pa]	tubes/fins
25, 25, ⌀10	u_F	2.0	0.2	125	75	323	56	Cu/Al
30, 35, ⌀20		3.5	0.5	120	70	299	113	Cu/Cu
32,5, 30, ⌀20		3.5	0.5	130	60	322	77	Cu/Cu
32,5, 30, ⌀10	δ_F, u_F	2.0	0.4	130	60	292	24	Cu/Cu
32,5, 30, ⌀10		2.0	0.4	130	60	259	35	Cu/Cu
60, 30, ⌀15		2.0	0.4	116	60	273	54	steel/Cu
⌀62, ⌀30	u_F, δ_F	3.0	0.65	124	62	142	19	Al/Al
				124	124	253	39	Al/Al
⌀38, ⌀15		3.0	0.65	114	76	209	37	Cu/Cu

chemical factories,etc.

8. NOMENCLATURE

Symbol	Quantity	Unit
A	cross section	m^2
a	width	m
Bo	Bond number	-
c^x	constant in Table 2	-
d	diameter	m
e	distance	m
f	Fanning friction factor	-
g	local gravitational acceleration	m/s^2
h	width, height	m
i	factor of heat exchanger arrangement	-
K_G	geometrical factor	$m^{7/4}$
K_L	factor of thermophysical properties	$W/(m^{7/4}\ K^{3/4})$
K	permeability	m^2
k	overall heat transfer coefficient	$W/(Km^2)$
L	**length**	m
L^x	characteristic length in Table 2	m
l_v	latent heat of vaporization	J/kg
m	number of tubes in a row	-
n	number of rows in an exchanger	-
P	pressure	N/m^2
Q	heat flow, heat transfer performance	W
q	heat flux	W/m^2
R	thermal resistance	K/W
r	radius	m
Re	Reynolds number	-
S	surface area	m^2
T	temperature	K
t	temperature	^oC
u	distance	m
$\dot{V}$	volume flow rate	m^3/s
$\dot{W}$	heat capacity	W/K
x	coordinate	m

R_G	gas constant	J/(kgK)
α	heat transfer coefficient	$W/(m^2K)$
β	angle of heat pipe inclination	o
δ	thickness	m
ε	void fraction	-
η	thermal efficiency	-
λ	thermal conductivity	W/(mK)
μ	parameter in Eq. (26)	-
μ	dynamic viscosity	kg/(sm)
ν	kinematic viscosity	m^2/s
ϱ	mass density	kg/m^3
σ	surface tension	W/m
Θ	difference	-
τ	pitch	m
φ	parameter in Table 2	-
ϕ	contact angle	o
$\varkappa$	ratio of specific heat capacities	-
ω	depth	m

SUBSCRIPTS

A	cold stream
Ad	adiabatic
B	bubble
c	capillary
ef	effective
e	external
F	fin
g	gravitational
G	gas
H	hydraulic
i	internal
I	interaction
K	condenser
L	liquid
max	maximum
min	minimum
n	mean
o	ideal

P	vapor
S	sonic
T	total
t	tube
V	evaporator
w	wall
x	direction
Z	heat source
1	related to one tube
I	related to 1 m of tube
′	inlet
′′	outlet
⊥	related to tube perimeter

9. REFERENCES

[1] Rohsenow, W.M., Hartnett, J.P. and Ganič, E.N., Handbook of Heat Transfer Fundamentals, McGraw-Hill, New York, 1985

[2] Chi, S.W., Heat Pipe Theory and Practice, Sourcebook, Hemisphere Publ. Corp., Washington, 1976

[3] Hlavačka, V., Polášek, F., Štulc, P. and Zbořil, V., Application of Heat Pipes in Electrical Engineering (in Czech), State Publ. House SNTL, Praha 1989

[4] Dunn, P.D., and Reay, D.A., Heat Pipes, Pergamon Press, London 1976

[5] Bergles, A.E., Chyu, M.C., Characteristics of Nucleate Pool Boiling from Porous Metallic Coatings, J. of Heat Transfer, Vol. 104, No. 2, pp. 279-285, 1982

[6] Marto, P.J., Hernandez, B., Nucleate Pool Boiling Characteristics of a GEWA-T Surface in Freon - 113, AICHE Symp. Ser. Vol. 70, No. 225, pp. 1-10, 1983

[7] Nakayama, W., Enhancement of Heat Transfer, Heat Transfer 1982, Proc. 7th Int. Heat Transfer Conference, Munich, Vol. 1, pp. 223-240, 1982

[8] Nakayama, W., Daikoku, T., Enhancement of Boiling and Evaporation on Structured Surfaces with Gravity Driven Film Flow of R-11, Heat Transfer 1982, Proc. 7th Int. Heat Transfer Conference, Munich, Vol. 4, pp. 409-414, 1982

[9] Fagerholm, N.E., Ghazanfari, A.R., Kivioja, E. and Järvinen, E., Boiling Heat Transfer Performance of Plain and Porous Tubes in Falling Flow of Refrigerant R 114, Wärme und Stoffübertragung, Vol. 21, No. 3, pp. 343-353, 1987

[10] Polášek, F., Heat Transfer Enhancement in Gravitational Heat Pipes, Int. Heat Transfer Symposium, Varna, 1989

[11] Nelson, L.A., Sekhon, K.S. and Fritz, J.E., Direct Heat Pipe Cooling of Semiconductor Devices, Proc. 3rd Int. Heat Pipe Conference, Palo Alto, 1978

[12] Polášek, F., Heat Pipe Research and Development in East European Countries, Proc. 6th Int. Heat Pipe Conference, Grenoble, 1987

Optimal Construction of Radioelectronic Devices of the Cassette Design According to Temperature Control

U. G. STOYAN, B. G. KOLODYAZHNY, A. S. TUGAYEV, and B. S. ELKIN
Kharkov State University
Kharkov, USSR

ABSTRACT

The main objective of the paper is to present the methods worked out for calculating temperature distribution of RED array of the cassette design functioning under natural or forced convection.

When choosing a mathematical model the Boussinesq approximation was used. An algorithm of solving a conjugated problem is suggested. A two-level optimization scheme and a hierarchical structure of mathematical models of multidimensional conjugate problems are used in optimal construction. The results of numerical calculations are provided.

1. INTRODUCTION

The problem of optimal construction of microelectronic devices (RED) of the cassette design according to dynamics of temperature control is investigated here. The RED array represents a bounded space region in which multilayer boards with discrete heat sources are located. The boards have the form of joined bodies of different configurations made of different materials. There is a movement of a cooling fluid (natural or forced convection) in clearances (channels) between elements of the device during the device operation.

The main optimization problem is to minimize the dimensions of the RED array by rationally locating elements of the device and heat sources on them. The condition must hold that the maximum values of the temperature in preassigned regions should not exceed specified values. Besides restrictions on the temperature distribution it is necessary to satisfy the conditions of the boards' mutual location and the heat sources on them, the conditions of the boards belonging to the region of the array and the heat sources belonging to the boards, and the conditions of the boards and the heat sources not interacting with forbidden regions.

This is a problem of nonlinear programming for systems with distributed parameters. While solving it, it is necessary to investigate complicated multidimensional nonstationary problems of mathematical physics - conjugate problems.

2. THE HIERARCHICAL STRUCTURE OF MATHEMATICAL MODELS

One of the possible ways of theory and practice of computation experiment development [1], which the authors regard as the most prospective one consists in creating and practically applying the hierarchical structure approach to mathematical models of complicated technical systems. The essence of it is as follows. The hierarchical tree of mathematical models is constructed instead of one mathematical model including all the multitude of physical fields, complete reological and geometrical-topological information about the system. It represents an internally noncontradicting system of mathematical physics boundary value problems closely connected with one another and jointly describing fully and adequately all the properties of the technical system being of interest for a designer. The place of each mathematical model on the hierarchical tree is determined by two interconnected factors, namely the complexity of the problem being posed and the cost of its numerical solution.

Advantages of the structural approach are most noticeably seen in problems of nonlinear optimization where at each step of the iteration process for finding the maximum of the objective function it is necessary to solve a complicated mathematical physics boundary value problem. It is much more expedient to use less complicated models situated on lower levels of hierarchical structure at the first steps of the iteration process instead of carrying out time consuming calculations according to a single, more nearly complete, but highly complicated model. This suggestion seems natural because demands for the accuracy of optimization calculations are not great in the first steps. With the increase of the degree of accuracy of determining optimized construction parameters it is necessary to involve more complicated mathematical models producing more accurate results [2,3]. It is possible and expedient to carry out even one-time calculations using the most complicated and complete mathematical models at the final stage. As it is easily seen from the above discussion the accuracy of calculations can be made adequate without substantial computer time expenditure. The time of optimization calculations with the use of hiearchical models is reduced approximately by a factor of 10.

It is worth noting that if necessity arises it is easy to organize multiple passes through all the hierarchies of models used in a specific calculation.

Below the advantages of the multilayer optimization use, taking as an example the problem of the cassette design RED array construction according to temperature control dynamics, are shown [3].

The first level is to optimize the location of the boards forming the channels while maintaining fixed positioning of the heat sources on each board.

The second level is to optimally lay out heat sources on each board by singling out the maximum heated points on them.

3. THE MATHEMATICAL MODEL OF THE FIRST LEVEL

The temperature distribution of the RED array on the first optimization level is described by a multidimensional conjugate nonstationary problem for the system of contours along which the cooling fluid passes. The middle cross section of the RED array is shown in Fig. 1.

In the process of constructing the mathematical model of the first level and developing the calculation method it was possible to get rid of some conventional restrictions [4]. Assuming negligible clearance between the heated zone and the RED case the nonstationary temperature regime of the multichannel system is considered. The distribution of the cooling fluid in the channels is determined in the process of solution of each time step. The location of the heat sources on the boards is arbitrary and the emitted power can vary with time. Differences of mass flow rates and of temperature distributions in different channels as well as the temperature distributions of different boards are taken into account. Taking into consideration the air flow velocity value and the fact that the channel's length does not exceed 0.2 m, it is natural to conclude that during the time of elementary portion of air passing along the channel the temperature of the boards will change by fractions of a degree only. That is why it is possible to neglect the inertia forces appearing in the cooling fluid because of the nonstationary character of the velocity field change. Having in mind the fact that the channel's length is much greater than its width, it is possible to regard the velocity field of the cooling fluid at the fixed moment of time in any section of the channel as the same, the cross section of the velocity field changing in time. It is evident that we have to neglect the fact that this condition does not hold at the entrance and at the exit of the channel [4]. It is also suggested that the direction of the cooling fluid movement is parallel to the channel axis and the pressure in any point of the fixed cross section of the channel is constant.

The temperature distribution in a solid body (the boards and the RED case) is described by the nonstationary heat conduction equation

$$c_T \rho_T \frac{\partial T}{\partial t} = \operatorname{div} (\lambda_T \mathbf{grad}\ T) + \mathfrak{P} (Q, t, T) \qquad (1)$$

The temperature distribution and the cooling fluid movement in the Boussinesq approximation [5] are described by the Navier-Stokes equation (2), the continuity equation (3) and

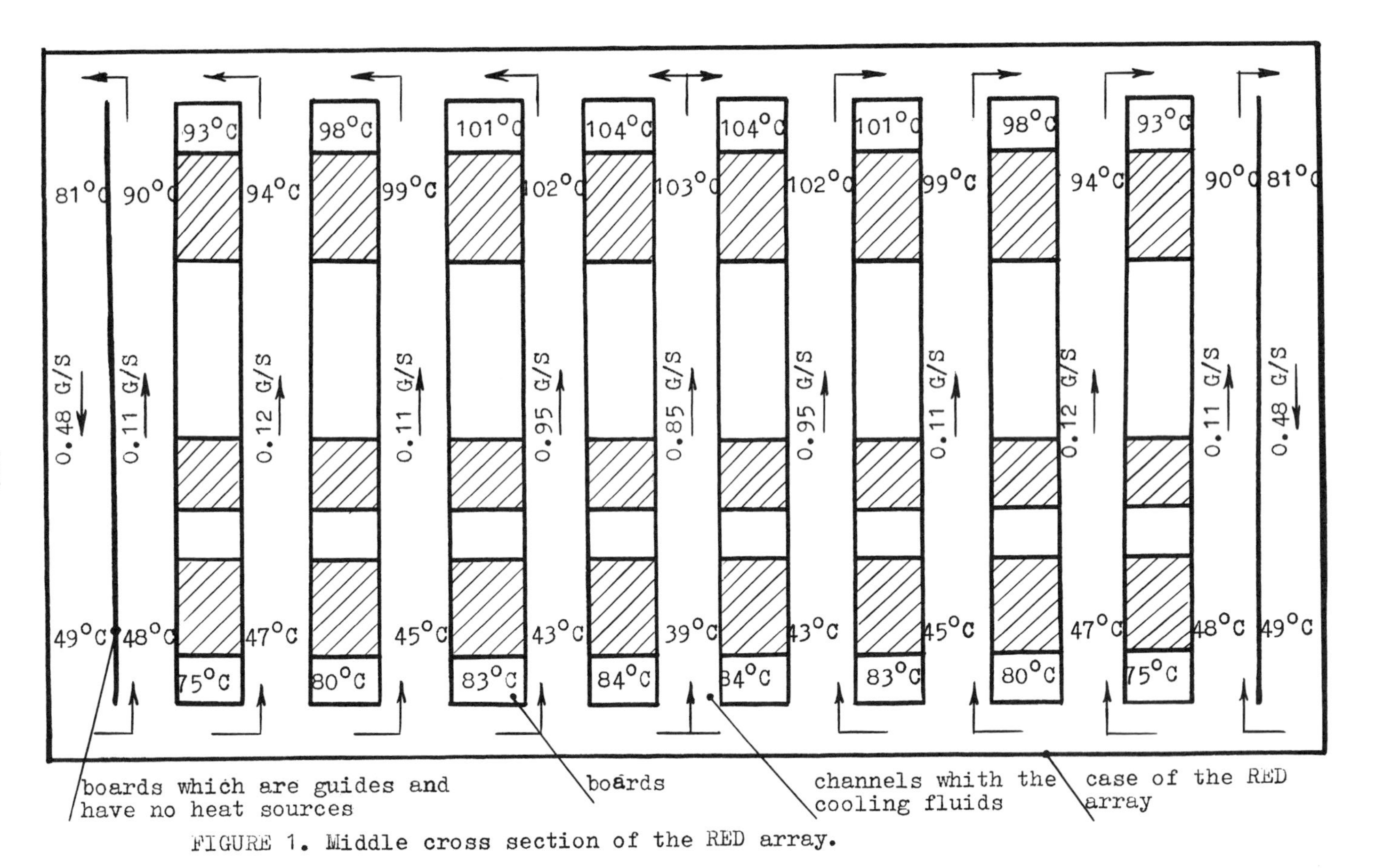

FIGURE 1. Middle cross section of the RED array.

the energy equation (4)

$$\frac{\partial p}{\partial z} = \mu \frac{\partial^2 V_z}{\partial x^2} + F_z \tag{2}$$

$$\text{div} (\rho_c \bar{V}) = 0 \tag{3}$$

$$c_c \rho_c V_z(x) \frac{\partial T_c}{\partial z} = \lambda_c \frac{\partial^2 T_c}{\partial x^2} \tag{4}$$

The conditions of conjugation (of the IV type) hold at the boundary Γ between the solid body and the cooling fluid

$$T\Big|_\Gamma = T_c\Big|_\Gamma \tag{5}$$

$$\lambda_T \frac{\partial T}{\partial n}\Big|_\Gamma = \lambda_c \frac{\partial T_c}{\partial n}\Big|_\Gamma \tag{6}$$

Convective heat exchange (the conditions of the III type) occurs at the boundary between the RED case and the surrounding space

$$\lambda_T \frac{\partial T}{\partial n} + \alpha T = \alpha T_s \tag{7}$$

To find the nonstationary temperature distribution of the system it is necessary to accomplish multiple solutions of three auxiliary problems:
- determining the cooling fluid mass flow rate in the channels under the assumption that the temperature distribution of the cooling fluid is known;
- determining the cooling fluid temperature distribution in the channels under the assumption that the cooling fluid flow rates are known and the temperature distribution of the boards and the RED case are known;
- determining the temperature distributions of the microelectronic devices and of the RED case under the assumption that the cooling fluid temperature distribution is known.

The problem of determining the cooling fluid flow rate in the channels of the RED array is solved for cases of natural and forced convection.

In case of a laminar flow (Fig. 1) which takes place under natural convection of the cooling fluid, the Navier-Stokes equation (2) and the continuity equation (3) is reduced to a system of N+2 linear algebraic equations (N is the number of boards which have heat sources) due to introduced assumptions [3]. The coefficients of proportionality according to which the cooling fluid flow rate in each of the channels is determined are unknowns. The obtained system is solved by the effective method for tridiagonal matrices [6].

In case of a turbulent flow which takes place under forced convection the situation is more complicated. In this case hydraulic resistances contained in the continuity equation (3) are proportional to the degree of cooling fluid velocity. A system of N+2 nonlinear algebraic equations is obtained. It is solved by the iterative method of the Newton type. At each iteration the process of solving the system of linear algebraic equations by the effective method for tridiagonal matrices is accomplished.

It is worth noting that the structure of the systems of equations and peculiarities of the effective method for tridiagonal matrices employed allows calculation of a large number of channels in reasonable time. While integrating the Naiver-Stokes equations (2) along closed contours the local hydraulic resistances arising from the mounting, layout of the devices and the entrance and exit of channels were taken into account.

The problem of determining the cooling fluid temperature distribution in the channels is solved under the assumption that the board temperature distributions and cooling fluid velocities in channels are known. The temperature distributions in a channel of width δ are described by the solution of the energy equation (4) with boundary conditions

$$\lambda_c \frac{\partial T_c}{\partial x}\bigg|_{x=\delta} = q_1 \quad \text{and} \quad -\lambda_c \frac{\partial T_c}{\partial x}\bigg|_{x=0} = q_o \qquad (8)$$

$$T_c\big|_{z=o} = T_c^o \qquad (9)$$

where q_o, q_1 are given heat fluxes from the solid body bordering on the channel to the cooling fluid (the temperature distribution of the boards and the RED case are considered to be known) and T_c^o is the cooling fluid temperature at the entrance into the channel. Equation (4) is of the parabolic type. The solution can easily be found by the effective method for tridiagonal matrices on every layer in z. There is a set of such equations in the problem under consideration, but temperatures at the channels entrances are not known. That is why defining the cooling fluid temperature distrtibution cannot be accomplished without joining the temperatures at the channels entrances and exit.

The set of the boundary value problems for the energy equations (4) and boundary conditions (8) and (9) for all the channels is solved on the basis of a specially designed scheme of computation which involves choosing a system of reference points [3]. The system of reference points is defined as the system of points with the following properties: if arbitrary values of temperature in these points are given and the energy equation is solved in the channels then it is possible to define the temperature distributions uniquely. The number of reference points is equal to the number of cooling fluid flow direction changes in the channels. The reference points in a variant of the RED array under consideration (Fig. 1) are the places of intersection of horizontal and vertical channels. The air temperature where flows from several channels converge is taken as the mixed-mean value. The desired temperature distribution in the RED array channels is defined as the linear combination of the temperature distribution calculated by solving homogeneous equations under conditions of assigning the combination of values 0 and 1 of the distribution to an arbitrary pair of reference points and the solution of the nonhomogeneous equation under the condition of assigning the value 0 to the reference points [3].

The proposed computation scheme is sufficiently effective from the point of view of setting up the algorithm especially for constructions when it is necessary to take into account numerous changes in direction of the cooling fluid movement in time.

The temperature distributions in the boards and the RED case are determined on the basis of the heat conduction equation in the solid body (1) with boundary conditions of the third type. To save the calculation time the hierarchical structure of mathematical models of thermal conduction in solid body is used. At the first optimization level the model describing heat flow in the boards and the RED case with a small degree of detalization is used and on the second optimization level separate parts of the system are singled out and a more detailed analysis of these parts with the complicated model is accomplished. In fact, on the first optimization level we use the model which is one or two steps lower in the hierarchy row of mathematical models than the one used on the second optimization level.

Since the thickness of the RED case and of the boards is small compared with their linear dimensions, the axial or spanwise temperature gradients are negligible. That is why it seems appropriate to consider the temperature distributions in the boards and in the RED case as one-dimensional with respect to space variables initially and to carry out the final computation according to a two-dimensional model with respect to space variables. This procedure substantially reduces the calculation time.

Determining the temperature distribution of the boards and the RED case is reduced to the solution of the system of linear algebraic equations at each time step by the effective method for tridiagonal matrices. The temperatures of the boards and of the RED case at the previous moment of time are taken as the initial values.

To define the nonstationary temperature distribution of the system as a whole the iteration process, including successive solution of the problems described earlier, is organized at each time step.

Let us consider in detail the method of solving the conjugate problem for generally locating the boards of the RED system. We choose the nodes of the network according to time. Transition from the n-th node to the (n+1)-th according to time is accomplished as follows. The temperature distributions of the boards and of the RED case are defined at the (n+1)-th moment of time under the assumption that the cooling fluid temperatures in the channels correspond to the n-th moment of time. To determine the temperature distribution and the mass flow rate of the cooling fluid in the channels the iterational process is organized at each time step. The heat flows from the boards and the RED case to the cooling fluid in the channels are determined from the proceeding temperature distributions of the boards and of the RED case at the (n+1)-th moment of time. The iteration process is accomplished as follows. The initial values of the proportionality coefficients, which define the flow distribution among the channels, are assigned. Mean values of the cooling fluid temperature in each channel are given. Having chosen the system of reference points we find the temperature in all the channels. Then using these temperatures we find the approximate values of the proportionality coefficients. The weighted coefficient is used in iterational procedure for stabilization of the process. The calculation process is iterated until the relative difference between two successive approximations becomes sufficiently small. In greater detail the algorithm is discussed in [3].

The convergence of the suggested iteration process has been checked both analytically and by numerical calculations. As is seen from the algorithm structure the converging iteration process can lead only to the solution of the initial problem [7].

4. THE MATHEMATICAL MODEL OF THE SECOND LEVEL

The mathematical model of the second level is intended for arranging the microelectronic devices in an optimal way by singling out the points of the maximum heating. The bounded region G_o which borders on N bounded regions G_i, each of them being divided into M_i subregions G_{ij} ($i=1,...,N$; $j = 1,...,M_i$) is considered. All the borders γ_{ij} dividing the regions are smooth (Fig.2). The temperature distribution T in the region G ($G = \bar{G}_o \cup \bar{G}_{11} \cup ... \cup \bar{G}_{1M} \cup ... \cup \bar{G}_{NM_N}$) ($\bar{G}$ is the closure of the region G) is described by the solution of the boundary value problem:

$$\operatorname{div} (\lambda_T \ \mathbf{grad}\ T) - qT - \mathfrak{P} = 0 \qquad (Q \in G) \tag{10}$$

$$\beta \frac{\partial T}{\partial n} + \alpha T = \alpha T_s \qquad (Q \in \partial G) \tag{11}$$

where β acquires meanings $\lambda(Q)$ or 0.

To investigate the mathematical model of the second level the following approach has been developed. First, the stationary temperature distribution of the microelectronic devices and the section of the board adjacent to it is found by the network method. Then the temperature distribution of multilayer system of devices with temperatures obtained by the same method taken as boundary conditions should be defined. The method of solution of this problem, which is three-dimensional with respect to space variables, using the modification of analog of the Schwartz alternation method has been devised [8,9]. This method consists of succesive solutions of the heat conduction equation (10) in each subregion G_{ij} by the method of finite integral transformations, i.e., by the approximation in the two-dimensional region. The method is most effective in the case when the solution of the heat conductiony equation in each subregion can be found by the method of dividing variables, e.g., for a system of parallelepipeds.

Finding the solutions $T(Q)$ of the boundary problem equations (10) and (11) amounts to finding the solutions $T_o(Q)$, $T_{ij}(Q)$ which are the restriction of the solution $T(Q)$ onto regions G_o and G_{ij} ($i=1,\dots,N;\ j=1,\dots,M_j$). The iteration process consists in constructing the sequence of functions $T_o^{(k)}, T_{ij}^{(k)}$ ($i=1,\dots,N;\ j=1,\dots,M_j;\ k=1,2,\dots$) which tends to T_o, T_{ij} in G_o, G_{ij} respectively. The conjugation of the solutions on the boundaries of subregions division is accomplished as follows. The vector-function $\varphi_{ij}^{(o)}=(\varphi_{11}^{(o)},\dots,\varphi_{ij}^{(o)},\dots,\varphi_{NM_N}^{(o)})$ is given, where $\varphi_{ij}^{(o)}$, is an arbitrary sufficiently smooth real-valued function given on γ_{ij} ($i=1,\dots,N;\ j=1,\dots,M_j$). The iteration process is constructed recurrently. Let the approximations $T_o^{(k)}, T_{ij}^{(k)}$ ($i=1,\dots,N;\ j=1,\dots,M_j$) be already constructed. After solving the boundary value problems in the regions G_o, G_{ij} according to a specially chosen scheme the vector

$$\tilde{\varphi}_{ij}^{(k+1)} = \left(-\lambda_{ij}\frac{\partial T_{ij}^{(k+1)}}{\partial n_{ij}} + \alpha_{ij}\, T_{ij}^{(k+1)}\right)\Bigg|_{\gamma_{ij}} \tag{12}$$

in calculated. Let us denote

$$\tilde{\varphi}^{(k+1)} = \left(\tilde{\varphi}_{11}^{(k+1)},\dots,\tilde{\varphi}_{ij}^{(k+1)},\dots,\tilde{\varphi}_{NM_N}^{(k+1)}\right) \tag{13}$$

and put

$$\varphi^{(k+1)} = (1-r)\tilde{\varphi}^{(k+1)} + r\tilde{\varphi}^{(k+1)} \tag{14}$$

where r is the parameter controlling the convergence of the iteration process.

Under the optimal choice of the parameter r the iteration process converges as an alternating-sign geometric progression [9]. Hence the error of the calculation is defined by the error of the choice of the mathematical model compared with real objects. Let us note that the chosen method allows us to find the highest temperatures of the system. The accomplished calculations showed that to reach necessary accuracy it is sufficient to make 3-5 iterations. The time of calculation for a five-layer system is approximately 5-10 times less than the time for calculations by the network or the variational methods, given the same accuracy requirement.

5. SOLUTION OF THE OPTIMIZATION PROBLEMS

The optimization problem of the first hierarchical level of complexity is solved by the method of meaningful variables. The synthesis of the RED array under the construction is accomplished as a result of solution of the following problem of nonlinear programming. First, determine

$$\min_{\bar{\delta}\in W} \varkappa(\bar{\delta}) = \min_{\bar{\delta}\in W} \left(\sum_{i=1}^{N+3} \delta_i \right) \tag{15}$$

where $\bar{\delta} = (\delta_1,\dots,\delta_{N+3})$; $\delta_i (i=1,2,\dots,N+3)$ are the dimensions of vertical channels. The region W is defined by the following system of inequalities:

$$0 < a \le \delta_i \le b, \quad (i=1,\dots,N+3) \tag{16}$$

$$\max_{\substack{(x,y)\in\Omega \\ t\in[0,+\infty[}} T (\delta_1,\dots,\delta_{N+3},x,y,t) \le T^{*} \tag{17}$$

It is worth noting that the objective function of the problem (15) is linear. The function T which is in the left hand side of equation (17) is nonlinear and, generally speaking, non-convex. Because of this the range of definition of the objective function is multiconnected and the problem (15) has multiple extrema and belongs to the class of problems of nonlinear mathematical programming of a special type. An essential feature of this class of problems is the fact that the function T is not represented analytically as the function of optimized parameters. Moreover, checking the restrictions of equation (17) for the RED array temperature distribution takes substantial time. Hence the search of the rational value of the objective function is accomplished by singling out the part of optimized parameters at every step - that is of the variables which are mainly responsible for the maximum value of the temperature distribution by the localization of its maximum value. Then, a repetitive process of changing parameters aimed at diminishing the maximum value of the distribution to the preassigned value T^* [3] is accomplished.

The optimization problem at the second level of complexity consists in the arranging the RED array elements by singling out the points with the maximum temperatures. From the point of view of this problem the microelectronic boards repesent multilayer systems of parallelepipeds (Fig. 2) and N heat sources with carriers S_i (i=1,...,N). The sources differ in intensity P_i (i=1,...,N). It is necessary to put one of the given heat sources into a given place in each multilayer subsystem so that the difference between the maximum and the minimum value of the temperature distribution would be the least. The function describing the emitted power is of the form

$$\mathfrak{P} = \begin{cases} \mathfrak{P}_i & (Q \in S_i,\ i=1,\dots,N) \\ 0 & (Q \notin S_i,\ i=1,\dots,N) \end{cases} \tag{18}$$

where $S_j \subseteq G_{ij}$ are the subregions of the multilayer system G_i in which the heat sources are located.

The problem of fixing amounts to the search of the minimum of the functional given on the multitude of permutations π of n elements

$$\min_{P\in\pi} \left(\max_{Q\in G} T(Q,P) - \min_{Q\in G} T(Q,P) \right) \tag{19}$$

where T(Q,P) is the solution of the boundary value problem equation (10) and (11), obtained when the sources are located as is written in the permutation P. This is a multiextremal problem belonging to the class of discrete optimization problems of mathematical programming.

As the space π is finite, the whole set of values of the functional could be used for the solution of the problem equation (19). But looking through n! of variants of the sources location is not possible even for contemporary high speed computers. That is why the combination of two effective methods of discrete optimization, that is of the vector of descent method and of the narrowing vicinities method is used to solve the optimization problem of the second level [10,11]. The similar structures of both methods made it possible to use their algorithms jointly while making the program of the second level optimization problem for computer calculation.

Figure 1 shows the distribution of the calculated temperatures and of the mass flow rate measured in units of the system SI for one of the variants of the constructed device on the first optimization level. Figure 3 shows the region of the projected solutions for the regular location of boards. Figure 4 shows the calculated temperature distributions and the optimal location of the heat sources on the second optimization level for one of the variants of the constructed device (Fig. 2).

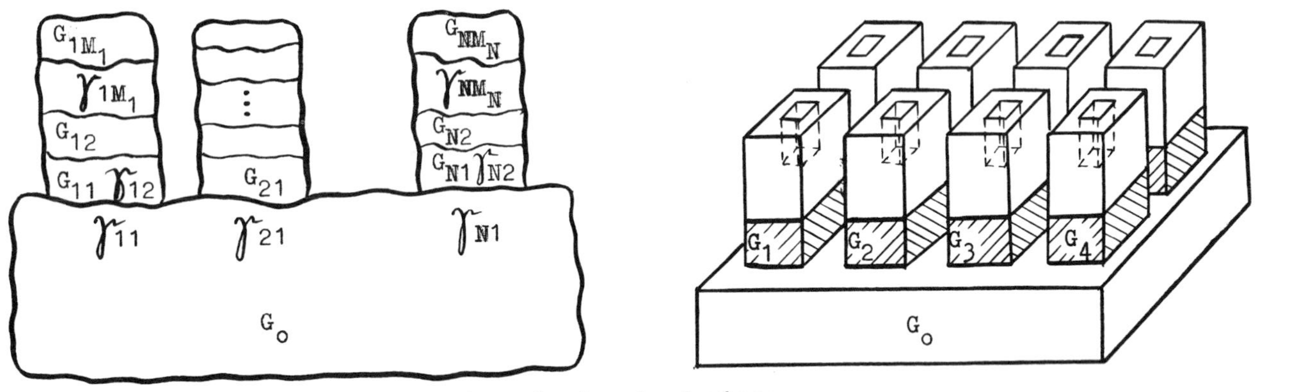

FIGURE 2. Multilayer models of microelectronic devices.

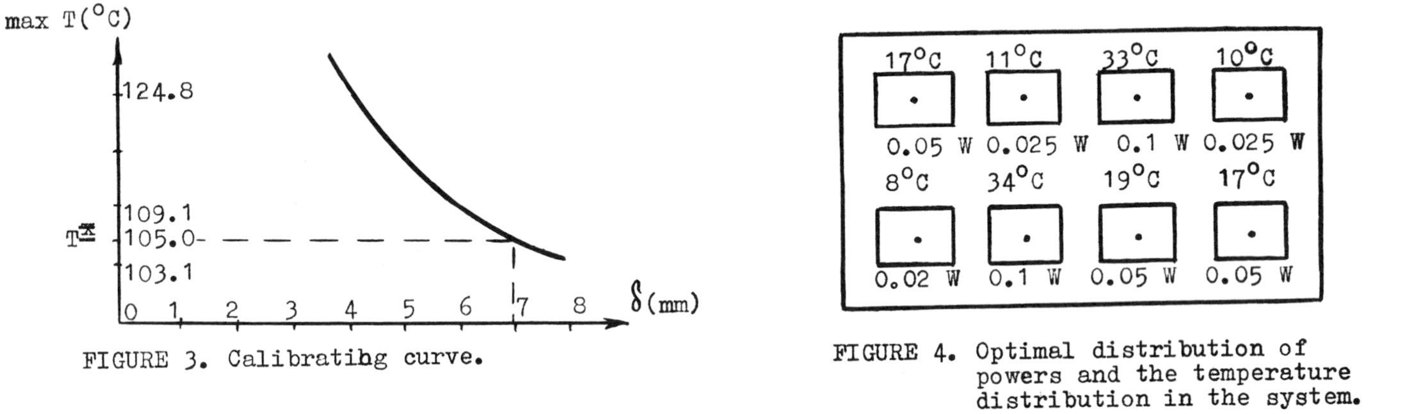

FIGURE 3. Calibratibg curve.

FIGURE 4. Optimal distribution of powers and the temperature distribution in the system.

CONCLUSION

A mathematical model and a method of solving a problem of optimal synthesis of the casette design RED array with discrete heat sources in laminar movement of the cooling fluid (Boussinesq approximation) are suggested with restrictions on the location of heat sources and maximum value of temperature distribution inside the system. An algorithm of solving the nonstationary conjugate problem has been worked out. A method of determining the temperature distributions in systems of heterogeneous regions has been developed.

All the methods and the algorithms of solution of both optimal problems and the mathematical physics problems mentioned above are carried out as programs for computer calculation using FORTRAN-IV as a programming languge. Both the experience of using these programs and analytical comparison of theoretical results with experimental data [3,4,11] made it possible to conclude that the problems designing of the RED package on the basis of hierarchical structures of the mathematical physics problems are solved sufficiently effectively with the necessary degree of accuracy.

NOMENCLATURE

a = minimal permitted dimension of channel m
b = maximal permitted dimension of channel m
c_c = cooler specific heat capacity J/(K kg)
c_T = specific heat capacity J/(K kg)
∂G = the border of the region G
$Q(x,y,z)$ = point of the region under consideration
q = the coefficient characterizing internal heat flows W/(K m^3)
F_z = gravitational force projection on the z-axis N/m^3
N = the number of boards which have heat sources
$\frac{\partial}{\partial n}$ = derivative with respect to the normal external to the region
P = the permutation of n elements
p = pressure N/m^3
T = temperature in solid body K
T^* = the maximal permitted temperature K
T_c = temperature in the channel K
T_s = the temperature of the surrounding space K
t = time s
V_z = z-component of the velocity vector $\bar{V}$ m/s

Greek symbols

α = the coefficient of heat transfer of the solid body into surrounding space W/(K m^2)
$\bar{\delta}$ = the dimensions of the vertical channels = $(\delta_1,\ldots,\delta_{N+3})$;

δ_i (i=1,2,...,N+3) m

λ_C = cooler heat conductivity W/(K m)

λ_T = heat conductivity W/(K m)

μ = dynamic viscosity kg/(s m)

π = the multitude of permutations of n elements

ρ_T = material density kg/m^3

Gothic symbol

$\mathfrak{P}$ = emited power density W/m^3

REFERENCES

1. Samarsky A.A., Computing Experiment in Technology Problems, *Vestnik of the USSR Academy of Sciences*, No 3, pp. 77-86, 1984.

2. Rvatchov V.L., Slesarenko A.P., *Algebro-logical and Projection Methods in Heat Exchange Problems*, Naukova Dumka, Kiev, 1978.

3. Stoyan U.G., Putyatin V.P., Elkin B.S., RED Block Optimization in Accordance With Heat Regime Dynamics and Arranging Characteristics, Preprint / the Ukrainian SSR Academy of Sciences, Institute of Engineering Industry Problems, Kharkov, 1983.

4. Dulnev G.N., *Heat and Mass Exchange in RED*, Vysshaya Shkola, Moscow, 1984.

5. Jaluria Y., *Natural Convection. Heat and Mass Transfer*, Pergamon Press, Oxford-New York-Toronto-Sydney-Paris-Frankfurt, 1980.

6. Samarsky A.A., *Introduction to the Theory of Difference Schemes*, Nauka, Moscow, 1971.

7. Menshikov V.V., Elkin B.S., Calculation Method of Stationary Temperature Fields in RED with Natural Convection, *Izvestiya of the Armenian SSR Academy of Sciences, Technical Sciences Section*, Vol.36, No 3, pp. 42-45, 1983.

8. Katsnelson V.E., Menshikov V.V., On an Analogue of Shwarts Alternating Method, *Theory of Functions, Functional Analysis and Their Application*, 17 issue, pp. 206-215,1973.

9. Elkin B.S., Stationary Temperature Field Calculation in Different Field Systems, *Engineering Industry Problems*, issue 21, pp. 59-63, 1984.

10. Stoyan U.G., Sokolovsky V.Z., *Some Multiextremal Problems Solution by Narrowing Neighbourhoods Method*, Naukova Dumka, Kiev, 1980.

11. Stoyan U.G., Putyatin V.P., Elkin B.S., Heat Sources

Fixing Optimization in a Region Having Different Heat Characterestics, Preprint / the Ukrainian
SSR Academy of Sciences Institute of Engineering Industry Problems, Kharkov, 1986.

Synthesis of Thermoelectric Thermostats for Radio Electronic Equipment and Their Automatic Design

M. U. SPOKOJNY, V. N. GALEV, N. V. KOLOMOETZ,
and M. A. LESENKINA
Research-production
Amalgamation, KVANT
Moscow, USSR

ABSTRACT

The paper examines thermal and mathematical models of thermoelectric thermostats as a basis for theoretical definition of static and dynamic errors of thermostatting.
Methods are developed for choosing construction parameters based on minimizing the thermostatting errors. A two-stage compensation scheme is proposed and analysed. The method of thermostat automated design is proposed and verified.

1. INTRODUCTIO

The design of new radioelectronic and computer equipment in modern industry encounters a tendency to use devices and systems with parameters strongly dependent on temperature. Besides, the increase of specific heat loads and growing demands for reliability require thermostabilization of certain elements and devices with simultaneons cooling of the equipment. In a wide range of applications, thermoelectric thermostatic devices are often the only acceptable and possible choice. A wide use of thermoelectric thermostats is restrained by the lack of satisfactory methods of their thermal design and methods of optimization of their basic features [1,3,4].
The thermoelectric thermostat design process is carried out by a trial-and-error method during experimental operational development of each thermostat.
The aim of the discussed studies is to develop methods of thermoelectric thermostat design, including construction optimization and their automatic design.

2.1 Theory

The procedure of synthesizing controlled temperature devices was developed in papers [5,6] under Prof.Dulnev's guidance. The method of thermoelectric thermostat (TET) synthesis is based on the consideration of thermal model for such devices.

The basic TET thermal model contains, Figure 1:
1) A temperature controlled object (TO) - a passive or heat-emitting physical body on the surface or in the volume of which it is necessary to maintain with certain accuracy a temperature designated.
2) A thermostat chamber (Ch)-is an enclosure for achieving uniform temperature in the thermostat and is also the thermostat bearing structure.
3) A thermomodule (TM) which is placed on the external surface of the thermostat chamber for heat absorption and heat release.

4) A sensor of a temperature control system (S) on the thermostat chamber surface.

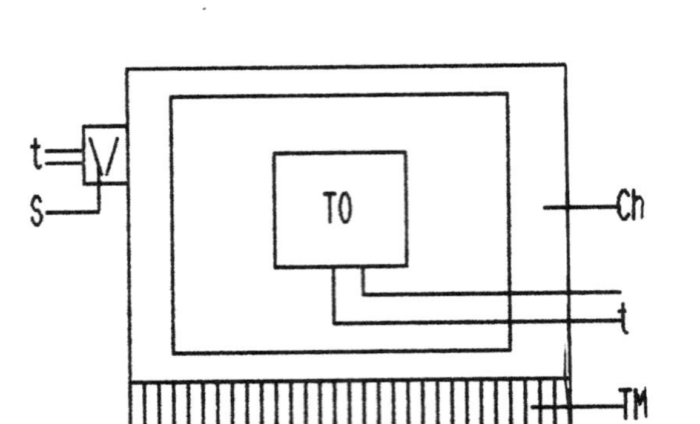

FIGURE 1. The basic thermal model of a thermoelectric thermostat.

For this system a mathematical model representing equations of thermal balance for each thermostat element can be expressed as

$$\frac{\lambda_{Ch} f_{Ch}}{L} \frac{d^2 T_{Ch}}{d\bar{Z}^2} - \sigma_{Ch,TO}(T_{Ch} - T_{TO}) - \sigma_{Ch,En}(T_{Ch} - T_{En}) = 0 \tag{1}$$

$$\sigma_{Ch,TO}\left(\int_0^1 T_{Ch}\, d\bar{Z} - T_{TO}\right) + Q_{TO} = \sigma_{En,TO}(T_{TO} - T_{En}) \tag{2}$$

$$\sigma_{Ch,S}\left(\frac{L}{h}\int_{\frac{L_S}{h}}^{\frac{(L_S+h)}{L}} T_{Ch}\, d\bar{Z} - T_S\right) = \sigma_{S,En}(T_S - T_{En}) \tag{3}$$

$$V_1[0.5R_1 I_1^2 - \alpha_1 T_{TM_1} I_1 + \sigma_{TM_1,TMh_1}(T_{TMh_1} - T_{TM_1})] = Q_0 \tag{4}$$

$$V_1[0.5R_1 I_1^2 + \alpha_{h_1} T_{TMh_1} I_1 - \sigma_{TM_1,TMh_1}(T_{TMh_1} - T_{TM_1})] = \sigma_{TMh_1,En}(T_{TMh_1} - T_{En}) \tag{5}$$

with boundary conditions

$$\left.\frac{dT_{Ch}}{d\bar{Z}}\right|_{\bar{Z}=1} = 0 \tag{6}$$

$$-\frac{\lambda_{Ch} f_{Ch_0}}{L} \left.\frac{dT_{Ch}}{d\bar{Z}}\right|_{\bar{Z}=0} = Q_0 \tag{7}$$

and

$$Q_0 = \sigma_{Ch_0,TM_1}\left(T_{TM_1} - T_{Ch}\big|_{\bar{Z}=0}\right) \tag{8}$$

Considering the mathematical model presentied in (1)-(8) we can define a static thermostatting error as

$$\Delta T_{TO}^{max} = (1 - \epsilon)(T_{En\,max} - T_{En\,min}) \qquad (9)$$

where ϵ - a thermostat compensation factor [5], defined by the expression

$$\epsilon = \frac{\left(1 + \frac{\sigma_{S,En}}{\sigma_{ch,S}}\right)\frac{h}{L}}{a_1 L_{S1} - a_2 \bar{L}_{S1} + \frac{h}{L} a_o} \qquad (10)$$

where

$$a_o = \frac{1}{1 + \frac{Bi_{ch,En}}{Bi_{ch,TO}}}; \qquad \varkappa_o = \sqrt{Bi_{ch,En} + Bi_{ch,TO}};$$

$$a_1 = \frac{\varkappa_o \exp(-\varkappa_o)}{\exp(\varkappa_o) - \exp(-\varkappa_o)}\left(1 - a_o + \frac{\sigma_{En,TO}}{\sigma_{ch,TO}}\right); \qquad a_2 = a_1 \frac{\exp(x_o)}{\exp(-x_o)}; \qquad (11)$$

$$L_{S1} = \frac{1}{x_o}\{\exp[x_o \frac{L_S + h}{L})] - \exp(x_o \frac{L_S}{L})\}; \bar{L}_{S1} = \frac{1}{x_o}\{\exp[-x_o \frac{L_S + h}{L})] - \exp(-x_o \frac{L_S}{L})\};$$

$$Bi_{ch,En} = \frac{\sigma_{ch,En} L}{\lambda_{ch} f_{ch}}; \qquad Bi_{ch,TO} = \frac{\sigma_{ch,TO} L}{\lambda_{ch} f_{ch}}$$

For the temperature control of an extensive object eq. (9) is ascertained by consideration of a temperature pattern of the object:

$$\frac{1}{L_{TO}^2} \lambda_{TO} f_{TO} \frac{d^2 T_{TO}}{dZ^2} + Q_{TO} = \sigma_{ch,TO} (T_{TO} - \hat{T}_{ch}) \qquad (12)$$

$$-\frac{1}{L_{TO}} \lambda_{TO} f_{TO} \frac{dT_{TO}}{dZ}\Big|_{Z=0} = \sigma_{ch_o,TO_o}(T_{ch}\big|_{Z=0} - T_{TO}\big|_{Z=0}) - \sigma_{En,TO}(T_{TO}|_{Z=0} - T_{En})$$

$$\frac{1}{L_{TO}} \lambda_{TO} \cdot f_{TO^o} \frac{dT_{TO}}{dZ}\Big|_{Z=1} = \sigma_{ch^o,TO^o}(T_{ch}\big|_{Z=L} - T_{TO}\big|_{Z=1}) \qquad (13)$$

For such a temperature controlled object the value of a static error can be deduced from eqs. (12) and (13) and has the form

$$\Delta \vartheta_{TO} = \hat{\epsilon}\, \epsilon\, \vartheta_S - \vartheta^{st} \qquad (14)$$

where

$$\hat{\epsilon} = \frac{sh\gamma_o \frac{L_{TO}}{L}}{\gamma_o \frac{L_{TO}}{L}} \gamma_1 + p; \quad \gamma_1 = \frac{\sigma_{Ch_o,TO_o} L_{TO}}{\lambda_{TO} f_{TO_o}}; \quad \gamma_o = \sqrt{\frac{\sigma_{ch,TO} L^2}{\lambda_{TO} f_{TO} L_{TO}}};$$

$$p_o = a_1 \mathcal{L}_L - a_2 \bar{\mathcal{L}}_L + a_o; \quad \mathcal{L}_L = \frac{1}{x_o}[\exp(x_o) - 1]; \bar{\mathcal{L}}_L = \frac{1}{x_o}[\exp(-x_o) - 1] \qquad (15)$$

2.2 Methods of choosing construction parameters

The analysis of equations (10) and (14) permits us to work out preliminary recommendations on the construction synthesis with regard to minimizing the static error. For this purpose, all parameters which influence the value of a static error should be divided into two groups: "controllable" and "uncontrollable" ones. The first group includes parameters that can be varied only in the design process - when choosing a construction: h/L; L_S/L; $\sigma_{En,TO}/\sigma_{Ch,TO}$; Bi. The second group includes "uncontrollable" parameters - $Bi_{Ch,TO}$.
These parameters which can be varied in the design process as well as during the adjustment of the assembled ejuipment refer to "controllable" parameters included in complex $\sigma_{S,En}/\sigma_{Ch,S}$.

The reasons for choosing uncontrollable construction parameters are
- the value of each parameter must lie in the range of allowed values P_{imin} P_{imax}, where P_i is one of parameters mentioned above. The range of allowed values is set for each development before designing and is defined by specifications, assumed manufacturing methods, etc;
- a set of chosen parameters must ensure the compensation of an error in temperature control, i.e.,to reduce to ϵ=1;
- the slope dependence of a compensation factor as each of these parameters must be the least possible. This demand is explained by the fact that for a small value of the slope it is possible to create a thermostat construction where the compensation factor (ϵ) is close to unity even if actual constructive parameters in the assembled thermostat deviate from their calculated value.

This difference between calculated and actual values is caused by the deviation from sizes, conditions of thermal contact, properties of materials used, etc. We have introduced a concept - the construction correctness - to define the slope dependence of the compensation factor on the construction parameters. The construction is considered correct (by analogy with a mathematical incorrectness theory) if the change of β_1% in any uncontrollable parameter (the value of thermal conductivity, geometric characteristics, and so on) causes the change in an output characteristic (the value of a static or dynamic error, starting operation time, etc.) of value β% which satisfies the inequality

$$\beta \leq \beta_1 \qquad (16)$$

Under the preset range of allowed values of each construction parameter, we assume the following condition for realization of a thermoelectric thermostat design with ϵ =1:

$$\begin{cases} \epsilon_{sup} > 1 \\ \epsilon_{inf} < 1 \end{cases} \qquad (17)$$

where ϵ_{sup} and ϵ_{inf} are the largest and the smallest values of the compensation factor, respectively.

The analysis of eq.(10) makes it possible to apply condition (17) to the discussed model considering construction parameters.

$$\epsilon_{sup} = f \begin{cases} \left(\dfrac{\sigma_{En,TO}}{\sigma_{ch,TO}};\ Bi_{ch,TO}\right)_{min} \\ \left(Bi_{ch,En};\ \dfrac{L_s}{L} + \dfrac{h}{L};\ \dfrac{\sigma_{s,En}}{\sigma_{ch,s}}\right)_{max} \end{cases} \quad \text{at} \quad \left(\dfrac{\sigma_{En,TO}}{\sigma_{ch,TO}}\right)_{min} < \dfrac{(Bi_{ch,En})_{max}}{6}$$

$$\begin{cases} \left(\dfrac{\sigma_{En,TO}}{\sigma_{ch,TO}}\right)_{min} \\ \left(Bi_{ch,TO};\ Bi_{ch,En};\ \dfrac{L_s}{L} + \dfrac{h}{L};\ \dfrac{\sigma_{s,En}}{\sigma_{ch,s}}\right)_{max} \end{cases} \quad \text{at} \quad \left(\dfrac{\sigma_{En,TO}}{\sigma_{ch,TO}}\right)_{min} < \dfrac{(Bi_{ch,En})_{max}}{6} \qquad (18)$$

$$\epsilon_{inf} = f \begin{cases} \begin{cases} \left(\dfrac{\sigma_{En,TO}}{\sigma_{ch,TO}};\ Bi_{ch,En};\ Bi_{ch,TO}\right)_{max} \\ \left(\dfrac{L_s}{L} + \dfrac{h}{L};\ \dfrac{\sigma_{s,En}}{\sigma_{ch,s}}\right)_{min} \end{cases} & \text{at} \begin{cases} \left(\dfrac{L_s}{L} + \dfrac{h}{L}\right)_{min} < 1 - \dfrac{1}{\sqrt{3}} \\ \left(\dfrac{\sigma_{En,TO}}{\sigma_{ch,TO}}\right)_{max} > \dfrac{(Bi_{ch,En})_{max}}{6} \end{cases} \\ \begin{cases} \left(\dfrac{\sigma_{En,TO}}{\sigma_{ch,TO}};\ Bi_{ch,En}\right)_{max} \\ \left(\dfrac{L_s}{L} + \dfrac{h}{L};\ \dfrac{\sigma_{s,En}}{\sigma_{ch,s}};\ Bi_{ch,TO}\right)_{min} \end{cases} & \text{at} \begin{cases} \left(\dfrac{L_s}{L} + \dfrac{h}{L}\right)_{min} < 1 - \dfrac{1}{\sqrt{3}} \\ \left(\dfrac{\sigma_{En,TO}}{\sigma_{ch,TO}}\right)_{max} < \dfrac{(Bi_{ch,En})_{max}}{6} \end{cases} \\ \begin{cases} \left(\dfrac{\sigma_{En,TO}}{\sigma_{ch,TO}};\ Bi_{ch,TO}\right)_{max} \\ \left(\dfrac{L_s}{L} + \dfrac{h}{L};\ \dfrac{\sigma_{s,En}}{\sigma_{ch,s}};\ Bi_{ch,En}\right)_{min} \end{cases} & \text{at} \begin{cases} \left(\dfrac{L_s}{L} + \dfrac{h}{L}\right)_{min} > 1 - \dfrac{1}{\sqrt{3}} \\ \left(\dfrac{\sigma_{En,TO}}{\sigma_{ch,TO}}\right)_{max} > \dfrac{(Bi_{ch,En})_{min}}{6} \end{cases} \\ \begin{cases} \left(\dfrac{\sigma_{En,TO}}{\sigma_{ch,TO}}\right)_{max} \\ \left(Bi_{ch,En};\ \dfrac{L_s}{L} + \dfrac{h}{L};\ \dfrac{\sigma_{s,En}}{\sigma_{ch,s}};\ Bi_{ch,TO}\right)_{min} \end{cases} & \text{at} \begin{cases} \left(\dfrac{L_s}{L} + \dfrac{h}{L}\right)_{min} > 1 - \dfrac{1}{\sqrt{3}} \\ \left(\dfrac{\sigma_{En,TO}}{\sigma_{ch,TO}}\right)_{max} > \dfrac{(Bi_{ch,En})_{max}}{6} \end{cases} \end{cases}$$

At the same time if $\epsilon_{sup} < 1$ or $\epsilon_{inf} > 1$ it is impossible to choose, in the preset range of allowed values, a thermostat design satisfying a compensation factor but only a set of construction parameters which provides the values of the compensation factor being most close to unity.

If condition (18) is satisfied the succession of choice of construction parameters will be the following:
We designate

$$\frac{L_S}{L} = 1 - \frac{h}{L} - \sqrt{\frac{\frac{\sigma_{S,En}}{\sigma_{ch,S}} - \frac{\sigma_{En,TO}}{\sigma_{ch,TO}} + \frac{Bi_{ch,En}}{6} + \frac{\sigma_{S,En}}{\sigma_{ch,S}} \frac{Bi_{ch,TO} + Bi_{ch,En}}{6}}{\frac{Bi_{ch,En}}{2} + \frac{(Bi_{ch,En} + B_{ch,TO})\,\sigma_{En,TO}}{2\,\sigma_{ch,TO}}}} \tag{19}$$

for all this, other construction parameters are assumed arbitrarily within the range of their allowed values. When it is impossible to choose L_s/L, we assume other construction parameters from equation (19) which satisfy relations

$$\frac{\sigma_{S,En}}{\sigma_{ch,S}} \geqslant \frac{\frac{\sigma_{En,TO}}{\sigma_{ch,TO}} - \frac{Bi_{ch,En}}{6}}{1 + \frac{Bi_{ch,En} + Bi_{ch,TO}}{6}} \tag{20}$$

$$\frac{\sigma_{S,En}}{\sigma_{ch,S}} \leqslant \frac{\left(\frac{Bi_{ch,En}}{2} + \frac{Bi_{ch,En} + Bi_{ch,TO}}{2} \cdot \frac{\sigma_{En,TO}}{\sigma_{ch,TO}}\right)\left(1 - \frac{h}{L}\right)^2 + \frac{\sigma_{En,TO}}{\sigma_{ch,TO}} - \frac{Bi_{ch,En}}{6}}{1 + \frac{1}{6}(Bi_{ch,En} + Bi_{ch,TO})} \tag{21}$$

Doing this, we assume h/L equal to $(h/L)_{min}$. To ascertain values of chosen parameters we must satisfy equation (16) for each of them:

$$\begin{cases} \dfrac{b_1 b_2}{\left(b_2 \dfrac{\sigma_{En,TO}}{\sigma_{ch,TO}} + b_3\right)^2} \leqslant \dfrac{1}{\dfrac{\sigma_{En,TO}}{\sigma_{ch,TO}}} \\ \dfrac{b_4 b_7 - b_6 b_5}{(b_6 Bi_{ch,En} + b_7)^2} \leqslant \dfrac{1}{Bi_{ch,En}} \\ \dfrac{b_4 b_{10} - b_8 b_9}{(b_9 Bi_{ch,TO} + b_{10})^2} \leqslant \dfrac{1}{Bi_{ch,TO}} \\ \dfrac{2 b_1 b_{11} \left(1 - \dfrac{L_S}{L} - \dfrac{h}{L}\right)^2}{\left[b_{11}\left(1 - \dfrac{L_S}{L} - \dfrac{h}{L}\right)^2 + b_{12}\right]^2} \leqslant \dfrac{1}{\dfrac{L_S}{L} + \dfrac{h}{L}} \\ \dfrac{\sigma_{S,En}}{\sigma_{ch,S}} \geqslant b_0 \end{cases} \tag{22}$$

where

$$b = \frac{1 + \frac{1}{6}(Bi_{ch,TO} + Bi_{ch,En})}{b_2 \frac{\sigma_{En,TO}}{\sigma_{ch,TO}} + b_3}\; ; \quad b_1 = \left(1 + \frac{\sigma_{S,En}}{\sigma_{ch,S}}\right)\left(1 + \frac{1}{6}(Bi_{ch,TO} + Bi_{ch,En})\right);$$

$$b_2 = 1 + \frac{1}{2}(Bi_{ch,En} + Bi_{ch,TO})\left(1 - \frac{L_S}{L} - \frac{h}{L}\right)^2; \quad b_3 = 1 + \frac{Bi_{ch,TO}}{6} + \frac{Bi_{ch,TO}}{2}\left(1 - \frac{L_S}{L} - \frac{h}{L}\right)^2;$$

$$b_4 = \frac{1}{6}\left(1 + \frac{\sigma_{S,En}}{\sigma_{ch,S}}\right); \qquad b_5 = \left(1 + \frac{\sigma_{S,En}}{\sigma_{ch,S}}\right)\left(1 + \frac{Bi_{ch,TO}}{6}\right); \tag{23}$$

$$b_6 = \frac{1}{2}\left(1 + \frac{\sigma_{En,TO}}{\sigma_{ch,TO}}\right)\left(1 - \frac{L_S}{L} - \frac{h}{L}\right)^2; \quad b_7 = 1 + \frac{Bi_{ch,TO}}{6} + \frac{\sigma_{En,TO}}{\sigma_{ch,TO}}\left[1 + \frac{1}{2} Bi_{ch,TO}\left(1 - \frac{L_S}{L} - \frac{h}{L}\right)^2\right];$$

$$b_8 = \left(1 + \frac{\sigma_{S,En}}{\sigma_{ch,S}}\right)\left(1 + \frac{Bi_{ch,En}}{6}\right); \qquad b_9 = \frac{1}{2}\left[\frac{1}{3}\frac{\sigma_{En,TO}}{\sigma_{ch,TO}}\left(1 - \frac{L_S}{L} - \frac{h}{L}\right)^2\right];$$

$$b_{10} = \left[1 + \frac{Bi_{ch,En}}{2}\left(1 - \frac{L_S}{L} - \frac{h}{L}\right)^2\right]\frac{\sigma_{En,TO}}{\sigma_{ch,TO}} + 1 + \frac{Bi_{ch,En}}{2}\left(1 - \frac{L_S}{L} - \frac{h}{L}\right)^2;$$

$$b_{11} = \frac{1}{2} Bi_{ch,En} + \frac{1}{2}(Bi_{ch,En} + Bi_{ch,TO})\frac{\sigma_{En,TO}}{\sigma_{ch,TO}}\; ; \quad b_{12} = 1 + \frac{\sigma_{En,TO}}{\sigma_{ch,TO}} + \frac{1}{6} Bi_{ch,TO}$$

The realization of these recommendations brings the value of the compensation factor of a designed thermostat as close to unity as possible, thus creating necessary conditions for its final compensation.

2.3 Methods of compensation for error in temperature control

The compensation of the static error can be realized due to an additional compensation thermomodule. The thermal model for such thermostat is shown in Figure 2 and the mathematical model has the following form:

$$\begin{cases} \frac{\lambda_{ch} f_{ch}}{L^2}\frac{d^2 T_{ch}}{d\bar{Z}^2} - \frac{\sigma_{ch,TO}}{L}(T_{ch} - T_{TO}) - \frac{\sigma_{ch,En}}{L}(T_{ch} - T_{En}) = 0 \\ \sigma_{ch,TO}\left(\int_0^1 T_{ch}\, d\bar{Z} - T_{TO}\right) + Q_{TO} = \sigma_{En,TO}(T_{TO} - T_{En}) \\ \sigma_{ch,S}\left(\frac{L}{h}\int_{L_S/L}^{(L_S+h)/L} T_{ch}\, d\bar{Z} - T_S\right) + Q_{ch} = \sigma_{S,En}(T_S - T_{En}) \\ V_1[0.5R_1 I_1^2 - \alpha_1 T_{TM_1} I_1 + \sigma_{TM_1,TMh_1}(T_{TMh_1} - T_{TM_1})] = Q_0 + Q_{TMh_2,TM_1} \\ V_1[0.5R_1 I_1^2 + \alpha_{h_1} T_{TMh_1} I_1 - \sigma_{TM_1,TMh_1}(T_{TMh_1} - T_{TM_1})] = \sigma_{TMh_1,En}(T_{TMh_1} - T_{En}) \\ V_2[0.5R_2 I_2^2 - \alpha_2 T_{TM_2} I_2 + \sigma_{TM_2,TMh_2}(T_{TMh_2} - T_{TM_2})] = \sigma_{S,TM_2}(T_{TM_2} - T_S) \\ V[0.5R_2 I_2^2 + \alpha_{h_2} T_{TMh_2} I_2 - \sigma_{TM_2,TMh_2}(T_{TMh_2} - T_{TM_2})] = Q_{TMh_2,TM_1} \end{cases} \tag{24}$$

The following boundary conditions apply on the chamber faces:

$$\begin{cases} \left.\dfrac{dT_{ch}}{d\bar{z}}\right|_{\bar{z}=1} = 0 \\ -\dfrac{\lambda_{ch} f_{ch_0}}{L} \left.\dfrac{dT_{ch}}{d\bar{z}}\right|_{\bar{z}=0} = Q_0 \\ Q_0 = \sigma_{ch_0,TM_1}(T_{TM_1} - T_{ch}|_{z=0}) \\ Q_{ch} = \sigma_{S,TM_2}(T_{TM_2} - T_S) \end{cases} \tag{25}$$

A numerical experiment carried out with more than 1000 thermostats showed t the described method for choosing construction parameters and realisation o two-stage compensation can transfer 80% of thermostats from a crude categor to a precise one.

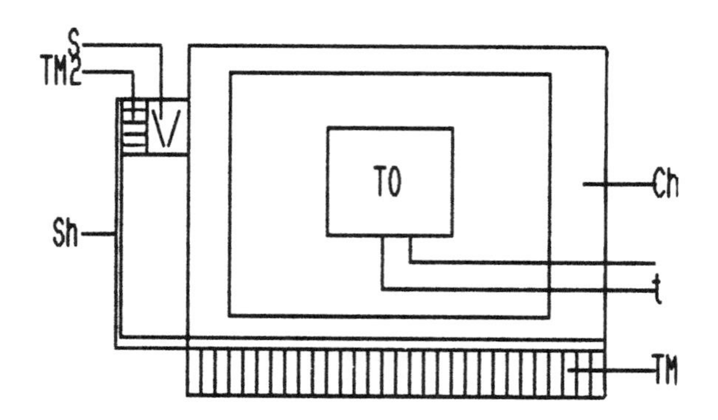

FIGURE 2. A thermal thermostat model with two-stage compensation.

2.4 Theory of time-dependent temperature control

In a number of thermostat applications it is important to maintain accurat temperature control in the dynamic operation modes. Examples of such modes are thermostat switching and a regime with sudden ambient temperature change.

The synthesis of a thermoelectric thermostat construction should be carried out with due regard for these processes. Therefore it is necessary
- to choose proper design parameters of the thermostat and the gain factor of a temperature controller which secures a preset time (time for reaching the opera-ting mode) to reach a required accuracy of temperature control;
- to ensure thermostat operation in the range of preset accuracy during sudden ambient temperature changes.

A mathematical model for analyzing processes of a thermostat coming to the ope-ration mode has the following form:

$$
\begin{cases}
c_{ch}\dfrac{d\vartheta_{ch}}{d\tau} = Q_o - \sigma_{ch,En}\vartheta_{ch} \\
Q_o = V_1\left[\dfrac{1}{2}\dfrac{U_o^2}{R} - \dfrac{\alpha_1 U_o \vartheta_{con}}{R} - \dfrac{\alpha_1 U_o T_{En}}{R} + \sigma_{TH_1,THh_1} M^2(\vartheta_{con} - \varphi_o \vartheta_{ch}) \right. : \\
c_{con}\dfrac{d\vartheta_{con}}{d\tau} = Q_1 - \sigma_{con,En}\vartheta_{con} \\
Q_1 = V_1\left[\dfrac{1}{2}\dfrac{U_o^2}{R} - \dfrac{\alpha_1 U_o \varphi_o \vartheta_{ch}}{R} + \dfrac{\alpha_1 U_o \varphi_o T_{En}}{R} - \sigma_{TH_1,THh_1} M^2(\vartheta_{con} - \varphi_o \vartheta_{ch}) \right. : \\
c_{TO}\dfrac{d\vartheta_{TO}}{d\tau} = \sigma_{ch,TO}(\vartheta_{ch} - \vartheta_{TO}) - \sigma_{En,TO}\vartheta_{TO} \\
c_S\dfrac{d\vartheta_S}{d\tau} = \sigma_{ch,S}(\varphi_1 \vartheta_{ch} - \vartheta_S) - \sigma_{S,En}\vartheta_S
\end{cases}
\tag{26}
$$

where φ_o and φ_1 are correction factors defined for sections of chamber Z=0 and chamber part $Z \in [Ls/L; (Ls/L+h/L)]$ correspondingly; index "p" applies to an introduced element of the thermostat, i.e., a radiator of thermomodule heat release junctions; and $M^2 = 1 + 1/2\, Z(\vartheta_{ch} + \vartheta_{con}) \approx 1 + Z\,(T_{En} - 10)$, where Z -is an object's thermoelectric efficiency.

The analysis of the above mathematical model leads to the definition of the object's temperature during the transient:

$$
\vartheta_{TO} = \frac{\sigma_{ch,TO}}{c_{TO}}\left[\frac{c_1}{r_1 + r_3} f_1(\tau) + \frac{c_2}{r_2 + r_3} f_2(\tau) + \frac{b_{i2}}{b_{i1} r_3} f_3(\tau)\right. \tag{27}
$$

where

$$
c_1 = \frac{\dfrac{1}{2}\dfrac{U_o^2}{R} - \dfrac{\alpha_1 U_o T_{En}}{R} + \dfrac{b_{i2}}{b_{i1}} r_2 \dfrac{c_{ch}}{V_1}}{\dfrac{c_{ch}}{V_1}(r_1 - r_2)}
$$

$$
c_2 = \frac{-\dfrac{1}{2}\dfrac{U_o^2}{R} + \dfrac{\alpha_1 U_o T_{En}}{R} - \dfrac{b_{i2}}{b_{i1}} r_1 \dfrac{c_{ch}}{V_1}}{\dfrac{c_{ch}}{V_1}(r_1 - r_2)} \tag{28}
$$

r_1 and r_2 are roots of equation $r^2 + b_{i0} r + b_{i1} = 0$

$$r_3 = \frac{\sigma_{ch,TO} + \sigma_{En,TO}}{C_{TO}}; \quad b_{i0} = \frac{V_1}{C_{ch}}[M^2\varphi_0\sigma_{TM_1,TMh_1} + \frac{\sigma_{ch,En}}{V_1}] + \frac{V_1}{C_{con}}[M^2\sigma_{TM_1,TMh_1} + \frac{\sigma_{con,En}}{V_1}]$$

$$b_{i1} = \frac{V_1^2}{C_{con}C_{ch}}[\frac{\alpha_1^2 U_0^2 \varphi_0}{R} + M^2\varphi_0\sigma_{TM_1,TMh_1}\frac{\sigma_{con,En}}{V_1} + M^2\sigma_{TM_1,TMh_1}\frac{\sigma_{ch,En}}{V_1} + \frac{\sigma_{con,En}\sigma_{ch,En}}{V_1}]$$

$$b_{i2} = \frac{V_1^2}{C_{con}C_{ch}}\{-\frac{1}{2}\frac{U_0^2}{R}[2M^2\sigma_{TM_1,TMh_1} + \frac{\sigma_{con,En}}{V_1} - \frac{\alpha_1 U_0}{R}] - \frac{\alpha_1 U_0 T_{En}}{R}[\varphi_0(\frac{\varphi_1 U_0}{R} -$$

$$M^2\sigma_{TM_1,TMh_1}) + M^2\sigma_{TM_1,TMh_1} + \frac{\sigma_{con,En}}{V_2}]\}; \quad f_1(\tau) = \exp(r_1\tau) - \exp(-r_3\tau);$$

$$f_2(\tau) = \exp(r_2\tau) - \exp(-r_3\tau); \qquad f_3(\tau) = 1 - \exp(-r_3\tau)$$

Equation (27) has allowed to add up to the method of thermoelectric thermostat synthesis regarding the time reduction for coming to operation mode with the following:

- the decrease of the object's heat capacity;
- the decrease of heat resistance betwen the chamber and the object;
- maximum approach of the thermomodule's sensor to the temperature controlled object aiming at reduction of heat resistance between them.

For the elaboration of recommendations on decrease of a dynamic error in transient processes (a jump of the ambient temperature), the mathematical model of equation (26) should be supplemented with the folloning equation for the controller:

$$Q = -K_{con}(T_S - T_{PS}) \tag{29}$$

The transient is described by the following expression:

$$\Theta_{TO} = \Theta_{TO}^{y}[1 - \frac{\Theta_{TO}^{y} + A_{TO}\sin\vartheta_1}{\Theta_{TO}^{y}}\exp(-\frac{\tau}{\varepsilon_{TO}})] + A_{TO}\exp(\chi\tau)\sin(\omega\tau + \gamma_1) \tag{30}$$

where

$$A_{TO} = \sqrt{E^2 + D^2}; \qquad E = \frac{(1 + \varepsilon_{TO}\chi)\sin\gamma_{ch} - \varepsilon_{TO}\omega\cos\vartheta_{ch}}{(1 + \varepsilon_{TO}\chi)^2 + (\varepsilon_{TO}\omega)^2}\frac{\mu A_{ch}}{\eta\varphi_1}$$

$$D = \frac{(1 + \varepsilon_{TO}\chi)\cos\gamma_{ch} + \varepsilon_{TO}\omega\sin\gamma_{ch}}{(1 + \varepsilon_{TO}\chi)^2 + (\varepsilon_{TO}\omega)^2}\frac{\mu A_{ch}}{\eta\varphi_1}; \quad \mu = \frac{\sigma_{ch,TO}}{\sigma_{ch,TO} + \sigma_{En,TO}};$$

$$\gamma_1 = \arcsin\frac{E}{\sqrt{E^2 + D^2}}; \qquad \gamma_{ch} = \gamma + \arcsin\frac{\varepsilon_S\omega}{\sqrt{(1 + \chi\varepsilon_S)^2 + (\varepsilon_S\omega)^2}};$$

$$\gamma = \arcsin \frac{-\omega}{\sqrt{\omega^2 + (\frac{N}{W} + \chi)^2}}; \quad \chi = -\frac{1}{2}(\frac{1}{\mathcal{E}_S} + \frac{1}{\mathcal{E}_{ch}});$$

$$A_{ch} = A\sqrt{(1 + \chi\mathcal{E}_S)^2 + (\mathcal{E}_S\omega)^2}; \quad A = \frac{\Theta_S^y \sqrt{\omega^2 + (\frac{N}{W} + \chi)^2}}{\omega}; \qquad (31)$$

$$\omega = -\frac{1}{4}(\frac{1}{\mathcal{E}_S} - \frac{1}{\mathcal{E}_{ch}})^2 + \frac{K_{con}\eta\varphi_1}{\mathcal{E}_{ch}\mathcal{E}_S\sigma_{ch,En}}; \quad W = \frac{\mathcal{E}_{ch}\mathcal{E}_S}{1 + \frac{K_{con}\eta\varphi_1}{\sigma_{ch,En}}}; \quad \eta = \frac{\sigma_{ch,s}}{\sigma_{ch,s} + \sigma_{S,En}};$$

$$\Theta_S^y = \frac{1 + \eta(\varphi_1 - 1)}{1 + \frac{K_{con}\eta\varphi_1}{\sigma_{ch,En}}}\Theta_{En}; \quad \mathcal{E}_{TO} = \frac{c_{TO}}{\sigma_{ch,TO} + \sigma_{En,TO}}; \quad \mathcal{E}_S = \frac{c_S}{\sigma_{ch,s} + \sigma_{S,En}}; \quad \mathcal{E}_{ch} = \frac{c_{ch}}{\sigma_{ch,En} + \sigma_{ch,TO}}$$

In case of transition from one steady state to another, no additional measures should be taken to increase the thermostatting accuracy according to the aperiodic law. Temperature will change abruptly only during an oscillating process.

The analysis of equation (30), comprising equations for the compensation thermomodule described above, imposes restrictions on the choice of the controller gain to assure aperiodic transient in the thermostat:

$$K_{con} \leq \frac{\mathcal{E}_{ch}}{4[\mathcal{E}_S \frac{\eta_{ch}\varphi_1}{\sigma_{ch,En}} - \frac{\xi\eta_{CTM}}{\sigma_{CTM,En}}\mathcal{E}_{ch}]} \qquad (32)$$

where ξ - is the compensation factor; index "CTM" is applied to the compensation thermomodule.

$$\eta_{CTM} = \frac{\sigma_{CTM,s}}{\sigma_{ch,s} + \sigma_{CTM,s} + \sigma_{S,En}}; \quad \eta_{ch} = \frac{\sigma_{ch,s}}{\sigma_{ch,s} + \sigma_{CTM,s} + \sigma_{S,En}}$$

3. Methods of thermostat automated design

A system of thermoelectric thermostat automated design is based on the suggested synthesis method [2].

The following parameters are used as initial data for the thermostat design:
- a range of an ambient temperature change $T_{En\,min}$, $T_{En\,max}$;
- rated controlled temperature T_{St};
- the accuracy of thermostatting Δ_T;
- an assumed time for going into operation τ_{as};
- data of the temperature controlled object (constructive features, connections

with an external equipment, a range of possible controlled temperatures, heat release capacity);
- assumed mass-dimension characteristics;
- assumed power consumption;
- conditions for reliability, strength, stability unification of different units.

The automatic design starts with the operation of the subsystem "Input". This subsystem realizes the input of formalized specifications, their common control and the correction of registered errors in source information, as shown in Figure 3.

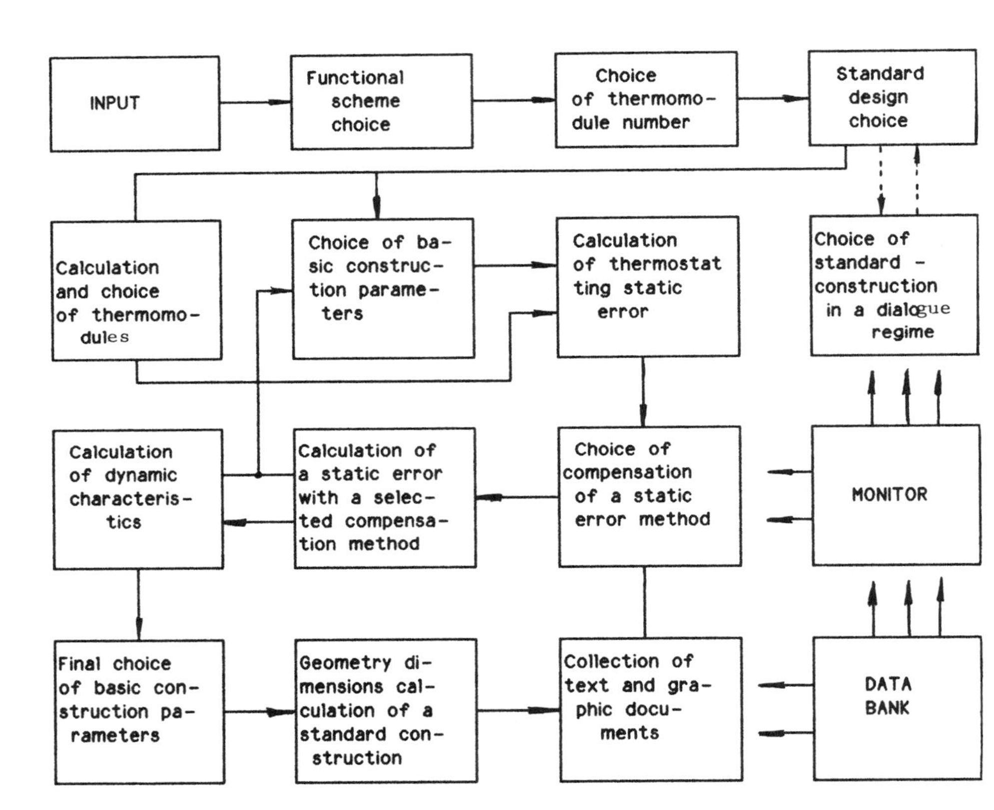

FIGURE 3. The thermostat structure scheme.

The subsystem "Functional scheme choice" is based on the analysis of formalized specifications and makes selection of the most advisable fuctional thermostat scheme in an interactive regime.

In subsystem "Choice of thermomodule quantity" at L_{T0}/L >0.08 - 0.15 we use expressions for the calculation of maximum temperature gradient along the thermostatting object. If calculated gradient exceeds the allowed value it is necessary to use two, three or more thermopiles in the construction.

When the choice of the standard thermostat construction is made, it becomes possible to make a preliminary choice of all construction thermostat parameters in accordance with the described method and, besides,to calculate thermomodules. The calculation of thermomodules includes the choice of standard thermoelectric modules, the determination of the operation current value and temperature difference between junctions, the determination of optimal geometric dimensions of the heat exchanger for heat release junctions in the thermomodule.

These functions are realized by subsystem "Calculation and choice of thermomodules" and "Choice of basic construction parameters".

Further operation of the system is dealing with providing with optimal static and dynamic characteristics of the developing thermostat due to an optimal choice method of a static error compensation and to the definition and final choice of the thermoelectric thermostat basic construction parameters. These tasks are solved by subsystems "Calculation of thermostatting static error", "Calculation of dynamic characteristics" and "Final choice of basic construction parameters". These operations are repeated several times to choose construction parameters which satisfy preset thermostatting accuracy. The final choice of thermoelectric thermostat construction parameters allows us to lay claims to the gain of a temperature controller and to calculate all geometric dimensions of the standard construction (subsystem "Geometry dimension calculation of a standard construction" and to get a full collection of design documents on the developing thermostat.

In accordance with the structure scheme a software for automatic design of a precision thermoelectric thermostat was developed and tested.

4. Conclusion

1) Thermal and mathematical models adequate to real constructions are developed and verified for a thermoelectric thermostat. The thermal fields are estimated in thermoelectric thermostats. The realization of this analysis makes it possible to propose a design calculation method for thermoelectric thermostats with an accuracy of temperature control in compliance with a prescribed specification.

2) Dependencies are achieved for defining a thermostatting static error in temperature control and a thermostat compensation factor. The condition of the temperature control static error compensation in thermoelectric thermostats is formulated. The achieved dependencies made it possible to determine the influence of thermostat construction parameters on the temperature control static error, thus providing the development of the method for choosing thermostat construction parameters at an early stage of design.

3) Calculating equations for the quantitive determination of the thermostatting static error in thermoelectric thermostats with a compensation thermomodule are achieved.

4) Several methods for the linear compensation of the thermostatting static er ror in thermoelectric thermostats are proposed and can be casily brought int practice. A system calculation algorithm of the linear compensation in thermo electric thermostats is developed and tested. It is showh that the linear com pensation can greatly increase the thermostatting accuracy of thermoelectri thermostats, thus transferring them from the category of medium accuracy to the category of precision thermostats.

5) A mathematical model is proved and it is a basis for a calculation method o dynamic characteristics in thermoelectric thermostats which makes it possib to choose final thermostat construction parameters ensuring optimal static an dynamic error of temperature control.

The value of temperature controller gain which is necessary for achieving preset temperature control accuracy is defined.

6) The conducted research, the developed algorithms and the calculation method for thermoelectric thermostats lead to the possibility of their automatic de sign by computer.

Nomenclature

T = are temperatures of elements
ϑ_i = is the overheat temperature of i-element relative to the ambient temper ture
θ_i = is an increment in temperature relatively steady temperature of i-element; θ_i^y= is a steady temperature value of the i-element.
ΔT_{TO}^{max} = is a static thermostatting error
T_{En}^{max} and T_{En}^{min} = maximum and minimum ambient temperatures, respectively
$\bar{T}_{ch}$=is the integral mean value of temperature in the chamber
super-and subscripts "o" refer to upper and lower face sections of the objec
T_{PS}= a preset thermostatting temperature on a set-point of a control system
$\sigma_{i,j}$ = heat conduction between i and j elements of a thermostat
λ_{ch} = a heat conductivity factor of a thermostat chamber
f_{ch} and f_{ch_o}= a cross-section of a thermostat chamber and its face sectio respectively
α_1 and α_{h_1}= thermo-emf factor of thermomodule heat absorption and heat releas junctions
C_i =is heat capacity of the thermostat's i-element
U_o =is thermomodule voltage
R = ohmic resistance of a thermobattery's thermopile
V = a number of thermoelement pairs in a thermomodule
I = thermomodule current
K_{con}= is a controller gain
L_s = a position coordinate of a sensor lower face
h,L = heights of sensor and thermostat chamber, respectively
$\bar{Z}$ = a coordinate; Z=Z/L - a reduced coordinate
τ = time

Subscript

TM1 and TM2=basic and compensation thermomodules
Sh=thermal shunt
t=terminals from the temperature controlled object and the sensor towards th external equipment
TO=temperature controlled object

Ch=thermostat chamber combined with thermostat's insulation
S=sensor of temperature control system
TM=face thermomodule
t=terminals from the temperature controlled object and from temperature sensor to external equipment
con=controller
En= environment
TM and TMh = for thermomodule heat absorption and heat release surface

REFERENCES

1. Winer, A.L., Thermoelektricheskiye Okhloditely, pp.1-176, Moscow, 1983.
2. Dendobrenko, B.I., Malika, A.C., Avtomatizatsiya Konstruirovanija REA; pp.1-384, Vysshaya Shkola, Moscow, 1980.
3. Jeno, J. and Cheroge, A., Termostatarea Prin Efect Peltier in Elektronica; Elektrotehn, Elektrons Outomat si Elektron, vol.23, no.1, pp.51-53, 1979.
4. Naer, V.A., Termoelektricheskiye Mikrotermostat, Pribory i Technika Eksperimenta, no.2, p.254, 1977.
5. Dulnev, G.N., Korenev, P.A., Spokojny, M.J., Automatizarovanniy Vybor Funktsionalnoy Schemy Termostata, Izvestiya Vuzov, Priborostroyeniye; vol.27, no.10, pp.90-95, 1984.
6. Dulnev, G.N., Korenev, P.A., Sharkov, A.B., Sintez Termostatiruyuschih Ustroystv, Inhenerno-Fizicheskiy Zhurnal, vol.51, no.3, pp.504-508, 1986.

Thermophysical Problems with Freon Evaporative Systems and Heat Pipes Employed in Cooling Electronic Equipment

G. V. REZNIKOV, G. F. SMIRNOV, and B. M. SHABANOV
Institute of Problems of Cybernetics
USSR Academy of Sciences
Moscow, USSR

ABSTRACT

Different freon conductive-evaporative cooling systems for computers are discussed. The first cooling loop of the systems incorporates heat pipes. Determined here are the tasks of optimizing such systems, which may be an indispensable part of an electronic computer.

1. INTRODUCTION

To improve the basic performance characteristics of electronic digital systems (speed, reliability and the like) calls for higher packaging density in the design arrangement and lower temperatures of electronic components, T_c. This results in greater heat fluxes, accounting for conversion from traditional air cooling apparatus to liquid or evaporating liquid systems.

At present there is a trend to reduce computer component temperatures down to 223 K, with the view to further decrease the temperature to 77 K. The reason for the above is that the major performance parameters of logic gates, such as operation delay and switching energy of the functions, are directly proportional to the operating temperature, T_c [1]. In the case of large scale electronic systems the total dissipated power is also going up. These factors make it mandatory to employ freon cooling systems built arround compression refrigeration units capable of removing thermal flows from the switching gates and the high dissipated power the computer.

Thus, we are facing the advent of a new field for applying vapor compression refrigerator units. To successfully solve the associated design and engineering problems, depends on how we will manage to resolve a number of thermophysical problems. The present paper is concerned with definition of these problems and some of the results obtained in an attempt to solve them.

2. SCHEMES COOLING SYSTEMS

In principle, a freon evaporative system intended for cooling a computer,which includes a vapor compression refrigerator unit, can be implemented through the following two basic designs:1-direct supply of refrigerant into the cooling channels,with the computer logic elements located on the channels(direct-flow system),FIGS.1a,1b,1c; 2-using intermediate heat-transfer circuits between the cooled components and the refrigerator unit(two-or three-loop circuits), FIGS .1d,1e,1f.

Figure 1 indicates direct supply evaporative and conductive -evaporative freon cooling systems:1a...1d-direct supply evaporative freon cooling systems;1e...1f-conductive-evaporative freon cooling systems;1-compressor;2-condenser;3-temperature control valve(thermo expansion valve);4-cooled unit (computer);5-super heater;6-pump ;7-regenerative heat exchanger; 8-liquid trap(reservoir); 9-heat pipe(HP).

To employ modular computer design, and ensure maintainability calls for the necessity of the following:a-joints(thermal or hydraulic) must be provided within the first loop of the liquid evaporative cooling system; b-thermal or hydraulic "circuits" designed to cool computer component parts must be arranged in such a way that the running of the entire refrigerating" circuit independent of whether separate units or modules are on or off.

The use of a direct-flow system offers high energy efficiency and reliable heat removal under elevated heat flows, involving, on the other hand, hydraulic joints, which produce an adverse effect upon the operating reliability of the cooling system.This drawback is eliminated when incorporating intermediate heat-removal loops into the system so as to ensure conductive heat removal from every unit or module in a computer.Intermediate heat-transfer loops having independent circulation are found to be more effective. These loops involve such devices as heat pipes(HP) or evaporative thermosiphons (ETS) that contain a working fluid undergoing an evaporating-condensing cycle.

To employ HPs or ETSs means to develop conductive-evaporative systems, being associated with unusual operating conditions such as:1)small heat-removal areas,2)small cross section of HPs and 3)the use of "heat" joints whose thermal resistances have some effect on thermal conditions. The use of direct-flow circuits calls for the solution to the following thermophysical problems: 1.To determine regularities in the local heat exchange and hydrodynamics within rectangular vapor -producing evaporator channels in refrigerator units having discrete heat supply, including the search for the boundaries of poorer heat exchange; 2.Determination and implementation of intelligent laws of refrigerant distribution

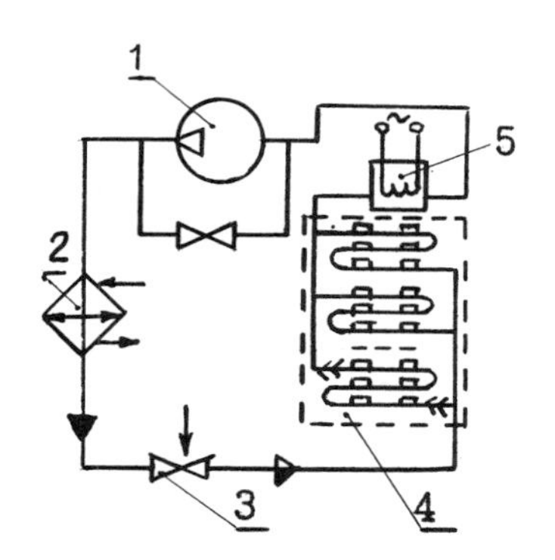

FIGURE 1a.

FIGURE 1b.

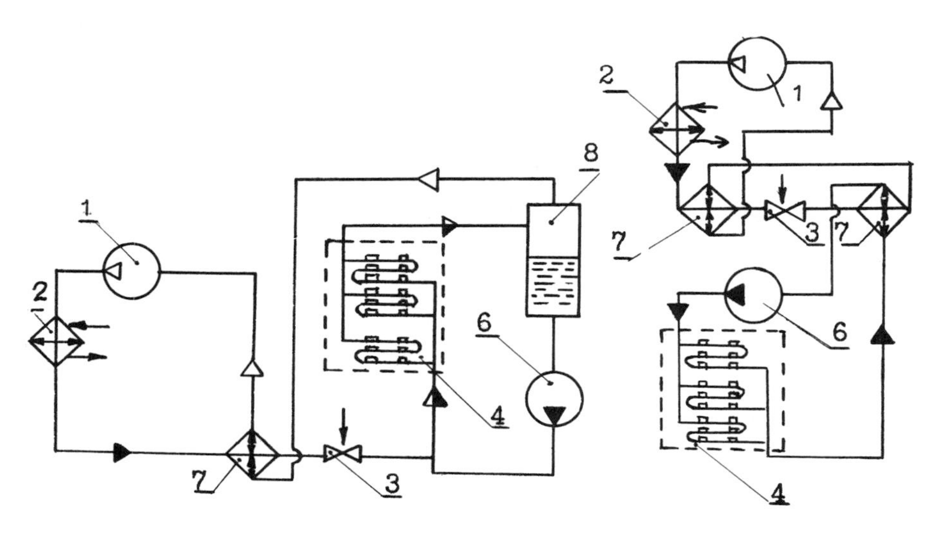

FIGURE 1c.

FIGURE 1d.

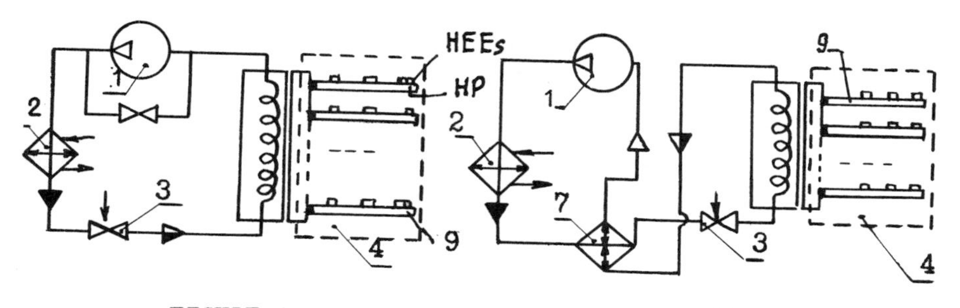

FIGURE 1e.

FIGURE 1f.

within the systems of parallel vapor-producing channels under nominal and unspecified modes of operation;
3. Determination of conditions for reliable operation of the cooling system and stabilization of thermal modes. The use of conductive-evaporative systems necessitates the following problems to be resolved: 4. Finding of regularites for heat and mass transfer within flat heat pipes having small condensation areas and minimal thermal resistances of ordinary and collector type; 5. Defining judicious shapes of the heat joints and their thermal resistances.

Results obtained in solving problems 1..5 enabled us to approach the task of making optimal design and engineering of the freon evaporative systems.

3. SOME ESTIMATIONS AND HEAT TRANSFER CORRELATION

A distinguishing feature of heat transfer conditions in evaporator channels of refrigerator units is that heat flows are being applied on a recurrent basis, thus accounting for strong interaction between the processes of heat transfer and heat exchange in the course of vaporization inside the channel. Relevant relationships were analyzed experimentally on full-scale cooling evaporator channels of rectangular cross section having equivalent diametrs from 3 to 6 mm, and lengths 2.3 and 4.6 m, as the local heat flow densities were being varied from 2 to 20 kW/m^2, mass velocities from 120 to 730 kg/m^2s, and saturation pressure of freon R-22 from 0.6 to 1.0 MPa.

In these experiments simulated were both real "discrete" conditions of heat supply and those of uniform heating of the wall, which mode is typical of vapor refrigerating machines. Representative plots are depicted in FIG.2. It can be seen that under even heating there occur heat transfer relationships which agree with the known generalizing functions. Under non-uniform heating, the heat transfer factors reduce, which can be attributed to the influence of the conduction process.

Rigorous quantitative analysis is impeded due to rather complicated and indefinite behavior of temperatures within a vapor-liquid flow under non-uniform heating further due to the lack of reliable data on the structure of a two-phase flow. That is why calculations were made for a certain extreme case of a "conventionally stratified" flow, wherein liquid is flowing as streamlet occupying a portion of the cross section equal to $1-\varphi$, while vapor is in the upper part of the channel. It this case, the heat transfer coefficient from the wall to the liquid α_1 is determined by pertinent relationships for liquid flow, while at wall-vapor interface as for vapor flow, α_2.

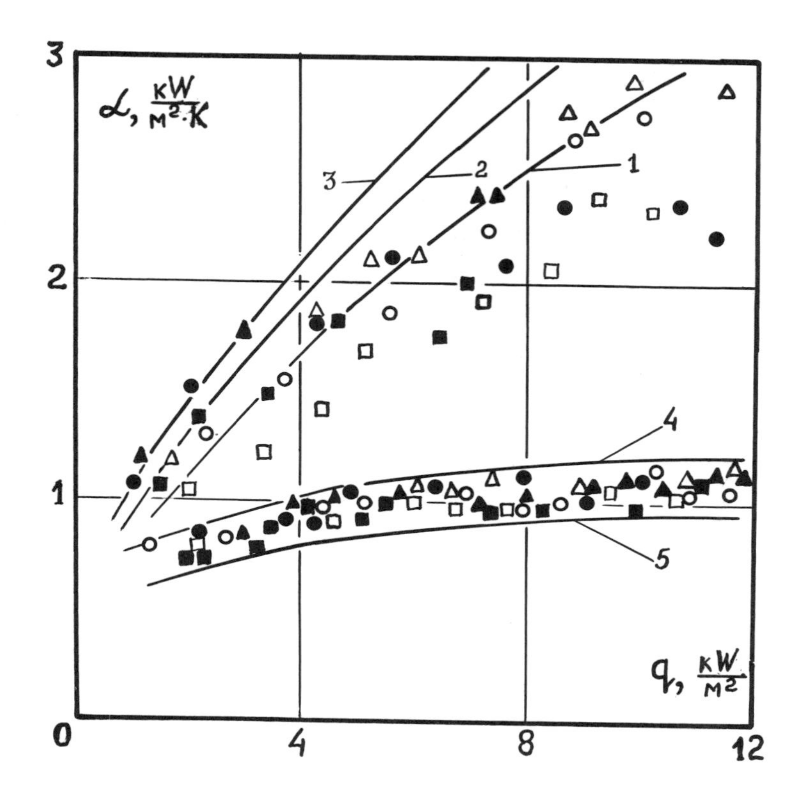

FIGURE 2. Convective heat transfer coefficient vs. heat flow density for continuous and discrete heat supply at t_o =20°C, d =4.7 mm: solid symbols for horizontaly oriented evaporator; open symbols for vertically oriented evaporator; 1-3 calculated after Bogdanov S.N., $\alpha = C \cdot q^{0,6} \cdot (\rho' w)^{0,2} \cdot d^{-0,2}$; 4, 5- heat transfer limits during the discrete heating[3]; C=1,47 for R-22, t=20°C; α, W/m²; q, W/m²; d, m; (ρ' w), kg/(m²s).

To a first approximation, when heat transfer coefficients are estimated to within the order of magnitude, these are taken as α - Const (α_1-Const, and α_2-Const). By neglecting temperature variations over wall thickness, we can reduce the three-dimensional heat-conduction problem under discussion to a two-dimensional one, the boundary conditions being as follows (FIG.3):

$$q = \text{const} \quad \text{for} \begin{cases} 0 < x < \frac{a}{2} \\ 0 < y < d \end{cases}$$

$$\frac{\partial t}{\partial x} = 0 \quad \text{for} \begin{cases} x = 0 \\ 0 < y < d+c \end{cases}$$

$$\frac{\partial t}{\partial y} = 0 \quad \text{for} \begin{cases} y = 0 \\ 0 < x < a+b \end{cases}$$

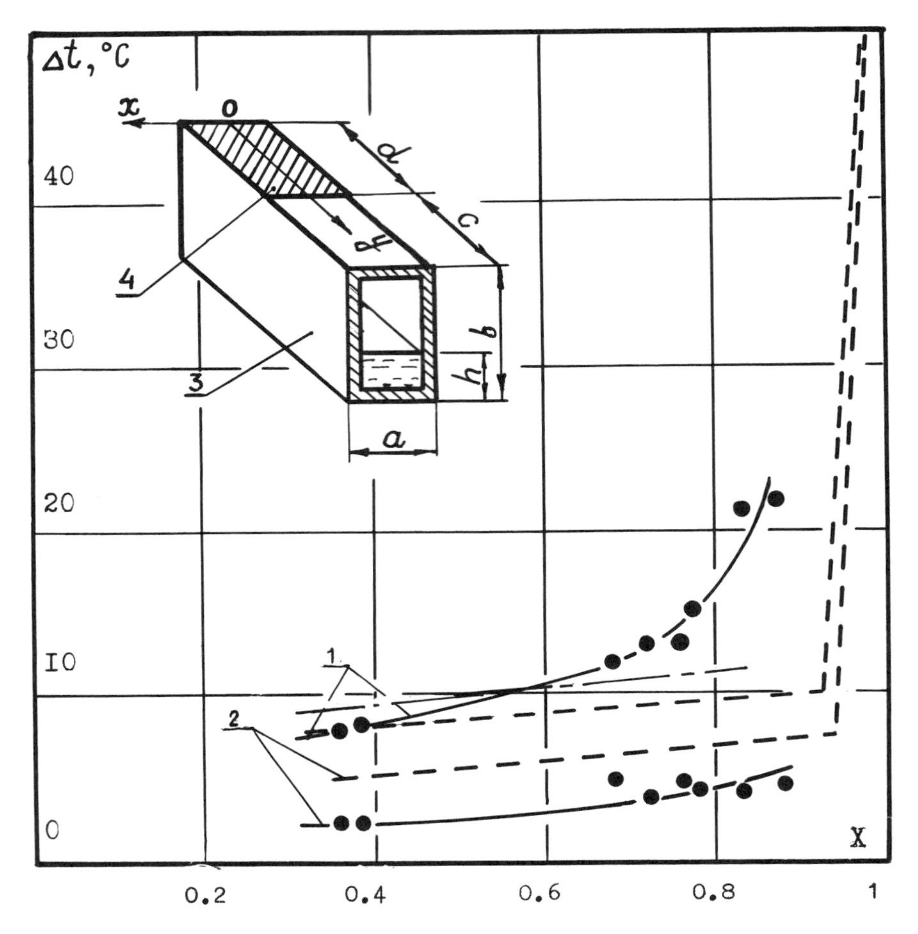

FIGURE 3. Comparison of experimental and calculated super-heat values for evaporator channel walls as a function of vapor content X : $q=10$ kW/m^2, $\rho' w = 400$ kg/m^{2}s, $t_o=20^\circ$C;

● ,—— experimental data, finite element method calculations;

\- - - - calculations for "stratified" flow of vapor-liquid;

—— - — calculations for α =Const along the channel perimeter;

1-max wall overheating ; 2- min wall overheating;
3-evaporator channel; 4-heater.
a=4.4 mm; b=8.4 mm; c=6.5 mm; d=6.3mm; h- liquid level.

Adiabatic conditions, $q = 0$, were specified at the external boundaries of the area under investigation, $\left\{ \begin{array}{l} 0 < x < (a+b) \\ 0 < y < (c+d) \end{array} \right.$, while boundary conditions of the third kind were given at the internal surface:

$$-\lambda\left(\frac{\partial t}{\partial n}\right)_{bdr} = \alpha_1(t_w - t_0) \quad \text{for} \quad x > \frac{a}{2} + b - h$$

$$-\lambda\left(\frac{\partial t}{\partial n}\right)_{bdr} = \alpha_2(t_w - t_o) \quad \text{for} \quad x > \frac{a}{2} + b - h$$

where, λ - thermal conductivity factor for the material of the evaporator wall; t_w - wall temperature.

Being thus defined the problem was being solved through the method of finite elements, and the results of the thus obtained solution are plotted in Fig. 3. Using experimental data on heat transfer coefficients as applied to respective values of q, and upon obtaining the results of numerical solution, we have juxtaposed them on the same drawing and a certain agreement is evident.

Of crucial importance for specifying recommendation in respect of the rational arrangement and design solution of the entire cooling system are the data on the boundaries where deteriorated heat-exchange conditions start to manifest themselves. These data are obtained experimentally. Experimental conditions and ranges of the variable factors were the same as those used in experiments on heat transfer. An additional factor here was the input mass flow vapor content, X_{in}.

The sought - for dependent variable X_b - boundary vapor content -- is understood as the vapor content found with known enthalpy or vapor content at the inlet with the aid of the energy equations , which was associated with the start of growth in temperature of wall beneath a heat-emitting element. In order to reliably detect this coordinate, and the respective value of X_b, we recorded temperatures under each of the heat-emitting element(HEE). It has been found that for certain values of $X=X_y < X_b$,temperature pulsation occured under the HEEs, the amplitude of these pulsations growing along the flow, and a marked growth in the wall temperature took place at $X=X_b$.

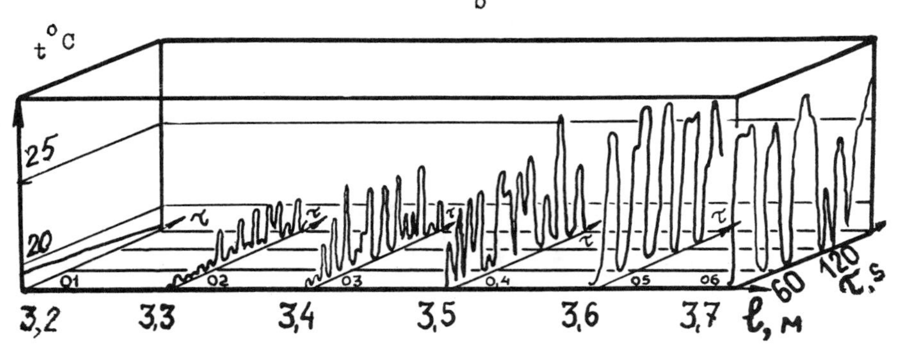

FIGURE 4. Temperature fluctuation along the evaporator chanal length; P_o=0,84 MPa, q=11,19kW/$_m$2, ρ'w=505 kg/$_m$2$_s$, τ - time axis,OI- heat release elements coordinates.

Figure 4 shows a typical picture of evolution of temperature pulsation, and the occurrence of deteriorated thermal conditions. By way of experiment we have obtained relationships between the boundaries of temperature pulsations X_y and of the deteriorated heat-exchange zone X_b, and the major factors depicted in FIG. 5. Empirical processing of experimental data has resulted in relationships for X_y and X_b as given below

$$x_i = 1 - C_i \, q/(\rho' w r) \,, \quad (x_i = x_y, \, x_b)$$

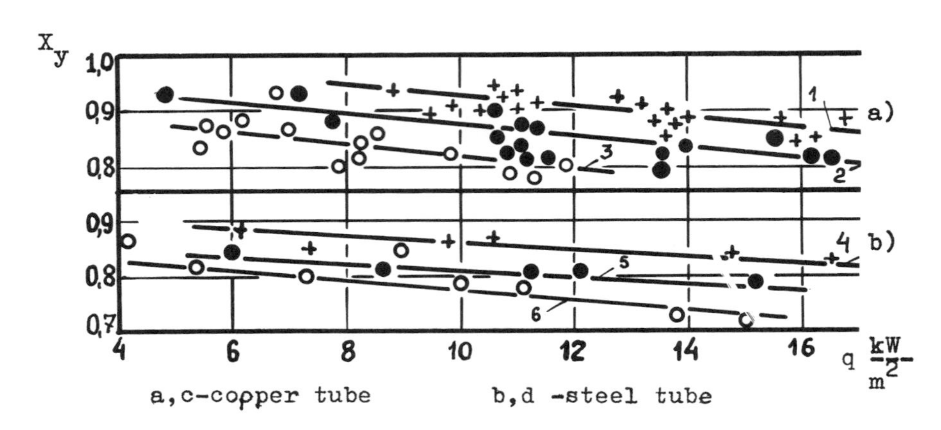

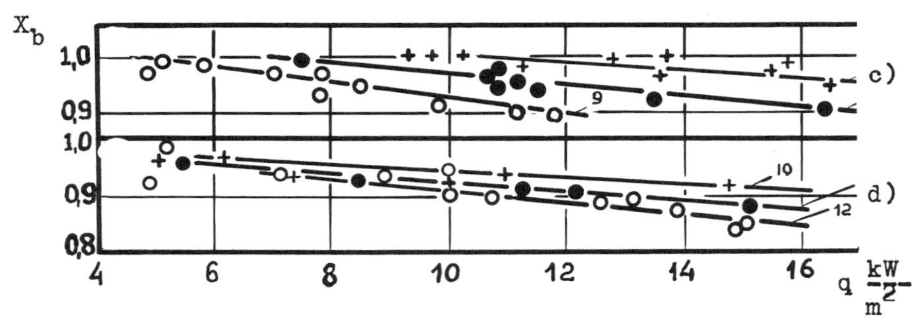

FIGURE 5. Worst case and border vapor content vs. heat flow at P_o = 0,76 - 0,85 MPa, o - $\rho' w$ = (230±10)kg/(m^2s), • - $\rho' w$ = (460 ± 10)kg/(m^2s), + - $\rho' w$ = (690 ± 10)kg/(m^2s).

The relationship thus found has enabled us to attack the problem of computing temperature conditions in the working channels of a direct-flow freon system intended for cooling a computer, wherein heat-emitting elements are positioned on the surface of the channels. Taken as a basis for solving these problems were mathematical models of heat exchange and hydrodynamics within a system of parallel vapor-producing channels [2].

The structure of these models are different from the channels integrated with common controls (TEV, pressure regulator and the like) to form a module, and for a group of modules connected in parallel. Certain distinctions can be also found in those cases where the system of parallel vapor-producing channels includes by-pass (adiabatic) lines for a vapor flow separated at the inlet to a group of channels, or where the channels are located in cold boards,with heat being brought thereto by heat pipes. But these distinctions are in no way fundamental. The following set of simultaneous equations forms the basis for all of these mathematical models:

1.Equation for two-phase hydraulics

$$\Delta p_o = A_i G_i^2 + B_i Q_i G_i + C_i Q_i^2 \ldots$$

where, A_i, B_i, C_i - coefficients dependent upon the conditions of entering the i-th channel, its geometry, thermophysical properties, vapor content, additional hydraulic resistances; ΔP_o - pressure drop within the system of n parallel vapor-producing channels; G_i, Q_i - refrigerant evaporated in the i-th channel.

The system of n simultaneous equations is complemented with the following relationships

$$G = \sum_{i=1}^{n} G_i$$

$$X_{2i} = X_{1i} + \frac{Q_i}{rG_i}$$

$$t_i = t_s + q_i/\alpha_i, \quad \alpha_i = \alpha'_i \text{ if } X_{2i} < X_b,$$

$$\alpha_i = \alpha''_i \text{ if } X_{2i} \geqslant X_b.$$

where, G -total flow rate of refrigerant; X_{1i} and X_{2i} - vapor content at the inlet to and outlet from the i-th vapor-producing channel (usually the following conditions are ensured at the inlet: X_{1i} - idem for $X_{1i} \rightarrow 0$); α'_i - heat transfer coefficient determined from the known generalizing relationships for heat transfer as applied to the flowing of a vapor-liquid stream through channels in a subcritical region ($X < X_b$); α_i'' - heat transfer coefficient in the region of deteriorated heat exchange conditions (to a first approximation it can be determined similarly as in the case of vapor flow).

On the basis of the above model the following problems were being solved: 1. Finding the values of additional hydraulic resistances at the inlet to the vapor-producing channels A_i which ensure reliable thermal conditions for all the channels when operating in the nominal mode. 2. Determination of variations in flow rates and temperature characteristics of the channels under typical unspecified modes of operation (shut-down of one or more channels).Typical results obtained in solving problem 2 are depicted in Figure 6.

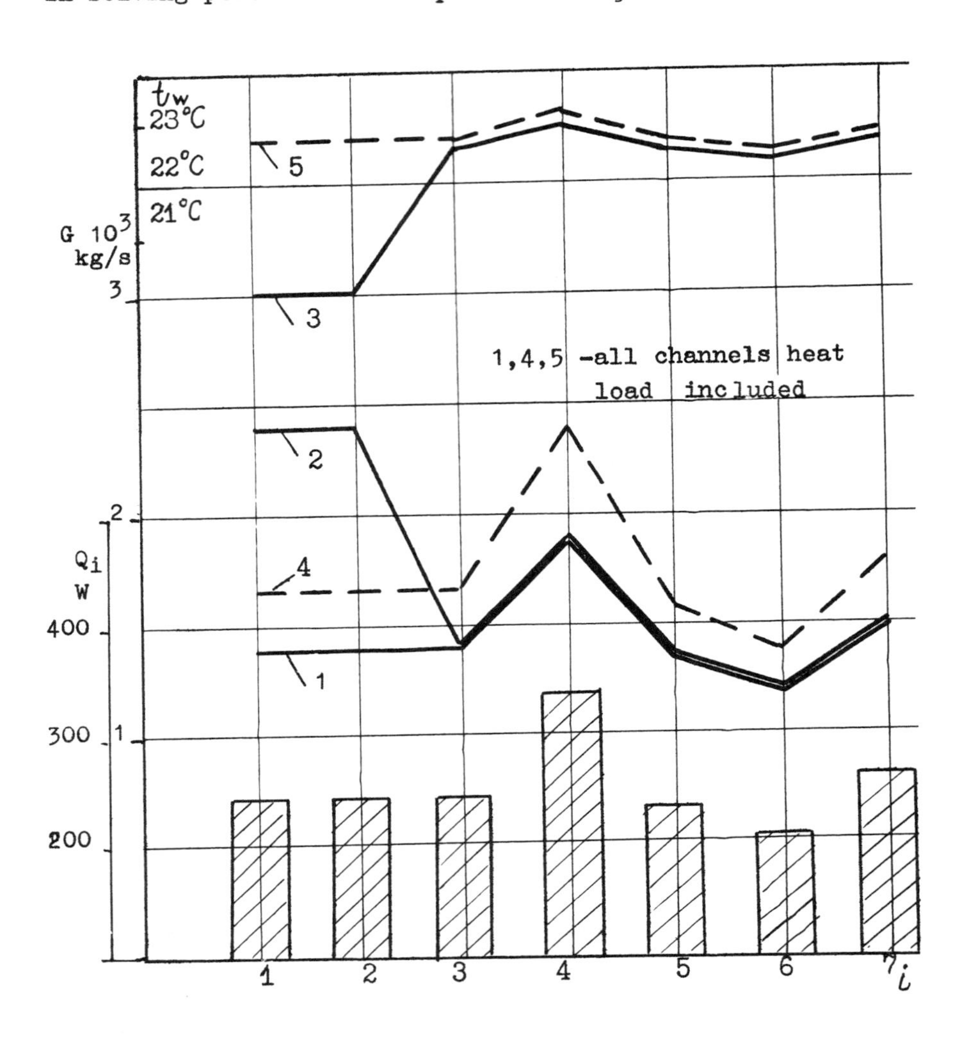

FIGURE 6. Coolant discharge distribution, wall temperature and a heat load bar graph for the first module incorporating 7 channels: nominal heat load $X_{out}=0,9$; 1,2 - coolant discharges at k = 1 and k = 1,2; 3 - channel wall temperature, partial heat load, $Q_1 = Q_2 = 0$; 4 - coolant discharge at k = 1,2; 5 - channel wall temperature (t_w(, k - coefficient store [2].

To find this form of a solution to the problems of flow distribution enables one to establish restrictions in terms of tolerated modes of operation of a refrigerator unit intended to cool computer assemblies through a direct-flow system, i.e., in essence, obtain an answer related to thermophysical problem 3 on the boundaries of reliable operation (stability) of the entire system designed to provide for the required thermal conditions of a computer.

Characteristics of heat pipes and evaporative thermo-siphons play an important role in analyzing freon systems that include such pipes or siphons. Characteristics of the HPs and ETSs are found through computations on the basis of the known generalizations amended with the aid of experimental data. Bearing in mind that the condensation area in a HP has a decisive role in heat transfer through the HPs under the conditions being discussed, we have made detailed analyses of the condensation process taking place on surfaces coated with porous structures.

Experimental investigations were made both on heat pipes and a model simulating the condensation area in a heat pipe. We have employed heat pipes having mesh-like and corrugated capillary structures, of flat and round shape, made of copper and stainless steel (refrigerant - water), producing heat flow densities from 1 to 10 W/cm^2, saturation temperatures from 303 to 373 K, varying geometrical parameters of the capillary structures.

Similar modes and types of capillary structures were simulated on a model made in the form of a water-cooled rectangular plate covered with various kinds of capillary structures under investigation. The results of experimental investigations performed on HPs and the model for identical structure are found to be in agreement (FIG.7), thus having provided evidence in favor of the following: 1. Heat transfer is found to be markedly lower in the case of condensation on a surface covered with capillary structures which are typical of the heat pipes, as compared to that on a smooth surface; 2. In the case of condensation on surfaces having mesh-like and corrugated structures, heat transfer coefficients are found to be dependent on the heat flux.

The above is at variance with the routine conceptions on the possibility of computing the HP thermal resistance on the assumption that heat transfer coefficients are invariable over the condensation area.

We have developed physical models of the heat exchange process taking place in the course of condensation under the above-mentioned conditions,proceeding from the ideas that when condensation occurs within capillary structures characteristic of HPs,these structures are completely submerged.The ratio between the capillary structure surface,wherein condensate is running down inside the structure,and the surface with external flow of the condensate in a thin sheet,varies as a function of the heat load,properties of refrigerant and capillary structure,and the saturation pressure.

It has been assumed that the flow within the internal space obeys Darcy's law, while the relationship for the laminar sheet flow hold true for the external space. It is only the external flow which had been taken into account for the corrugated structure. Heat transfer was based on the assumption that the key role is played by heat conduction taking place through both the wetted capillary structure and the sheet of condensate.

Figures 7 and 7d show a schematic representation of the adopted physical models. Predictions of these models have showen a certain degree of agreement with the experiments (Figures 7a and 7b). It follows from the analysis of the physical models and test data that to effectively improve heat transfer in the course of condensation on surfaces covered with porous structures, one has to clear the porous structure away from the cooled surface to the utmost degree. Such porous structure removal is restricted by the necessity of transporting the liquid.

To that end, use can be made of arterial capillary structure which are in contact with a bare condensation surface.The thus-designed capillary structures of heat pipes with short condensation areas have shown the possibility of reducing thermal resistance by a factor of 1.5 to 2.

Data on thermal resistances of the major heat transfer sections within a heat pipe, complemented by data on contact thermal resistances have enabled us to define and solve the problem of arriving at the optimum design of heat joints proceeding from analysis of the thermal resistance histograms. The more characteristic among the various variants considered are those depicted in Figure 8, accompanied by their respective histograms, manifesting appreciable possibilities of reducing total thermal resistances in choosing the optimal design solutions for the heat joints.

Figure 8 shows a cell of a electronic unit, wherein heat is being transferred with the aid of the heat pipe from the component parts to surface of a metal frame through a heat joint with one-sided conveyance of heat, and further from the metal frame a flat heat joint 3-4 to cold plate.

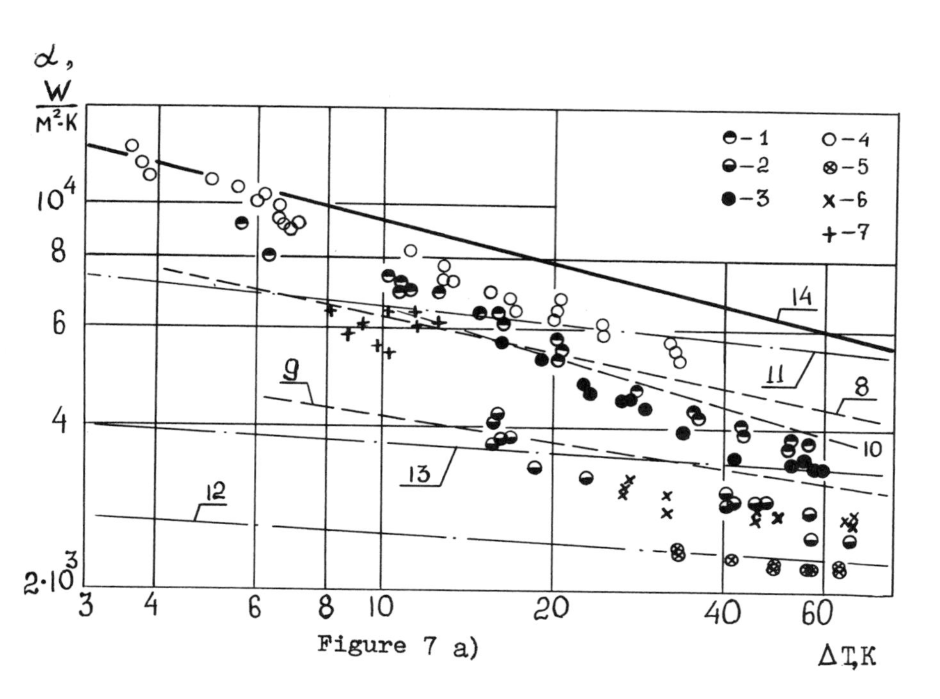
α,
W
M²·K
10⁴
8
6
4
2·10³
3
4
6
8
10
20
40
60
ΔT,K
1
2
3
4
5
6
7
14
11
8
10
9
13
12
Figure 7 a)

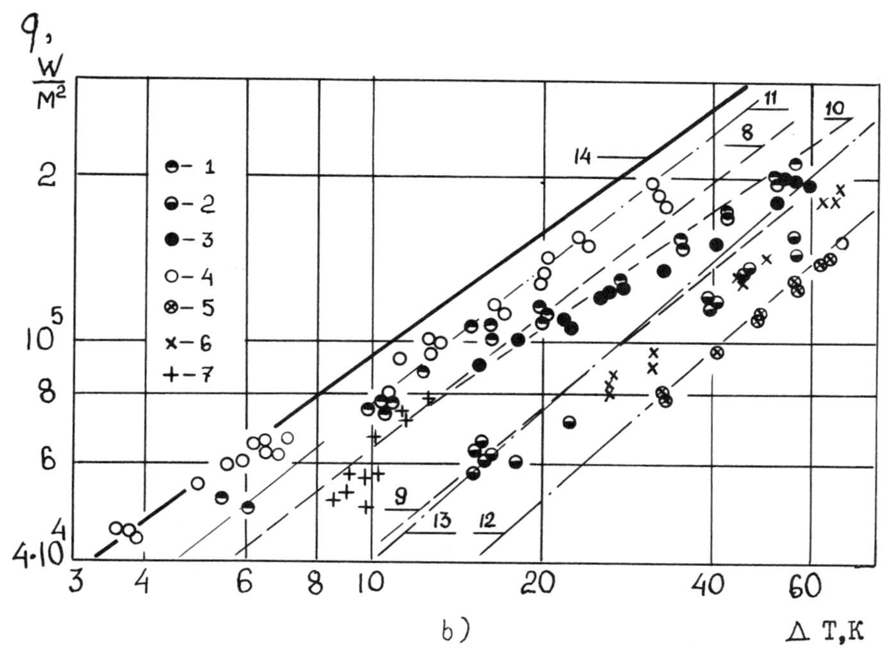
q,
W
M²
2
10⁵
8
6
4·10⁴
3
4
6
8
10
20
40
60
Δ T,K
1
2
3
4
5
6
7
14
11
8
10
9
13
12
b)

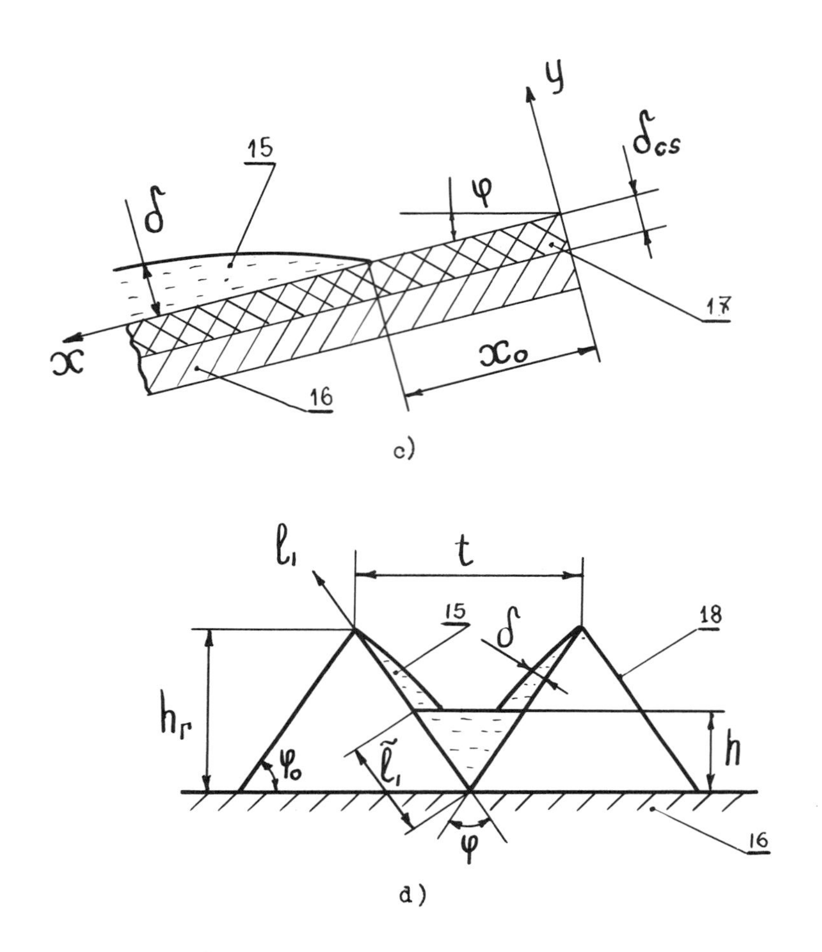

FIGURE 7. Experimental and calculated values of heat exchange with condensation: a,b- experimental data; c- condensation model for the wire capilary surface; d- condensation model for the corrugated capillary surface; 1,2-one layer and three layer steel wire netting having a mesh a=71.4μ m; 3-one layer steel netting with a=180μ m; 4,5-copper and steel corrugated capilary structure h_r=0.56 and 1.0mm; 6-steel corrugated structure; 7-heat release in heat pipes with copper corrugated structure; 8,9-calculated data for 1-and 3-layer steel netting with a=71.4μ m; 10-calculated data for 1-layer steel netting with a=180μ m; 11,12-calculated data for copper and steel corrugated structure; 13- calculations for 3-layer netting structure with a= 180μ m; 14- calculations for a smooth surface; 15- condensate film; 16- heat pipe wall; 17- capillary structure; 18- corrugated structure.

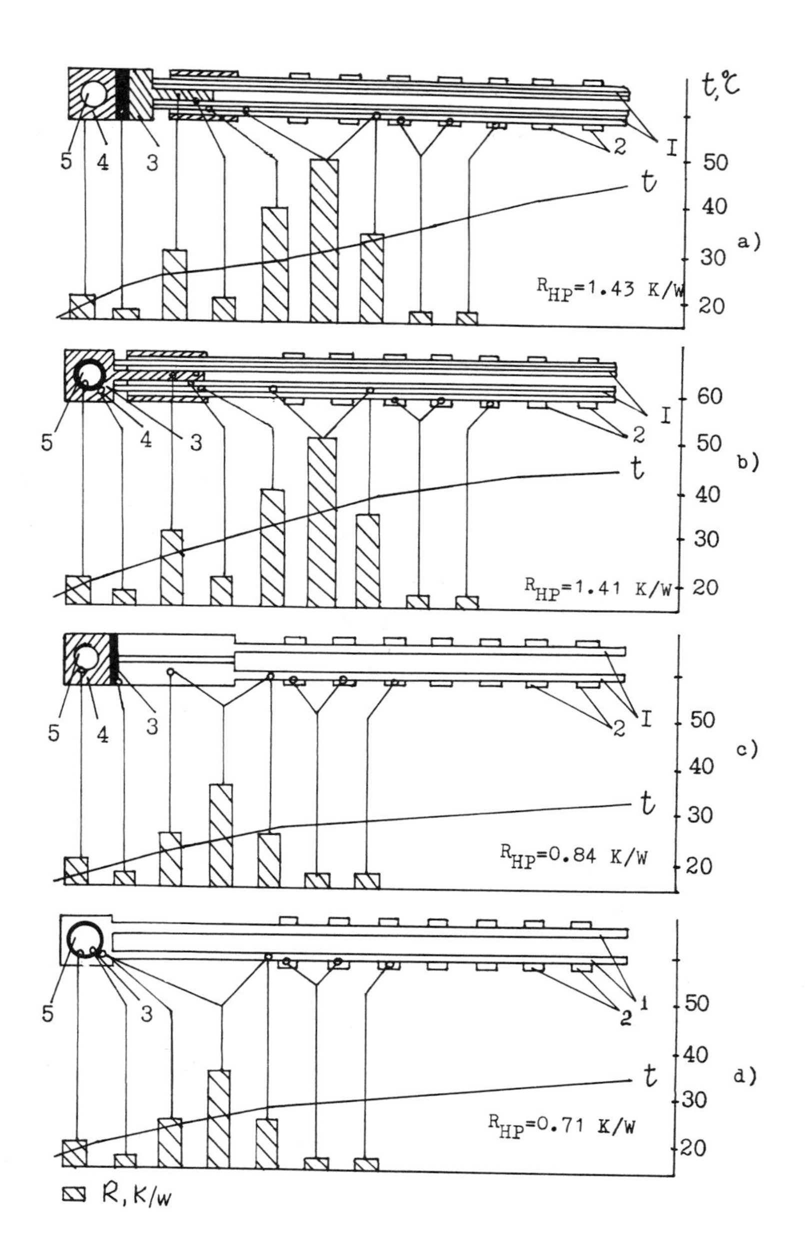

FIGURE 8. Total heat resistances components and temperature differences for heat sinks of various types: ▨ –R=0.1 K/W– heat resistance scale; R_{HP}– heat pipe resistance; 1-heat pipe; 2-heat emitting element; 3-fixing frame as a heat joint element; 4-cold plate; 5-evaporator channel of the refrigerator.

Figure 8b is a structural arrangement of a radio electronic unit, wherein a flat joint between a frame and a cold plate is replaced with a cylindrical heat joint between the frame and the wall of a cooled channel. It can be seen that substitution of another joint shape does not yield any substantial benefit in terms of the total thermal resistance. But to eliminate one of the heat joints due to the use of a collector-type HP (FIGURE 8c,8d) enables one to obtain drastic reduction in the total thermal resistance. In this case again, the shape of a heat joint makes no appreciable contribution to the total thermal resistance. Thus, a promising course of developing heat-transfer devices for computer-cooling systems is associated with the development of collector-type HPs.

4. OPTIMIZATION OF COOLING SYSTEMS

The entire package of solutions to the above-mentioned thermophysical problems of optimal thermophysical design of a cooling system has enabled us to attack the general problem of finding the optimum parameters of refrigerator units intended for cooling a computer. It was assumed that in the context of the above applications of the refrigerator units, the major requirements that shape the end function are as follows: a)-maximum speed of operation, b)-maximum reliability, c)-minimal energy and material expenditure. As the quantitative measure characterizing these requirements are indeed physical quantities of differing nature, we have formulated a comprehensive criterion of optimization as follows:

$$\Phi = q_{\tau}\frac{\tau}{\tau_o} + q_{\lambda}\frac{\lambda}{\lambda_o} + q_{n}\frac{N}{N_o} + q_{m}\frac{M}{M_o}$$

where τ, λ, N, M are characteristics values for a signal propagation time, failure rate, power consumption, mass of the cooling systems; index "O" corresponds to a certain initial alternative; q_i - weight proportions of respective partial criteria for optimization of the overall criterion.

Optimization problems were being solved on two levels: 1) on a unit level and 2) on a cooling system level. Taken into account should be the priority of the "weight" indices, q_{τ} and q_{λ}, as compared to q_n and q_m. For the purpose of specific estimates of degree we have assumed that $q_{\tau} = q_{\lambda} = 0.45$ and $q_n = q_m = 0.05$.

When making optimization on a unit level, those components of Φ which are related to variations in mass and energy expenditures,can be ignored. Given that,the functional can be presented as follows:

$$\Phi = q_{\tau}\frac{\tau_{\Sigma}}{\tau_{0\Sigma}} + q_{\lambda}\frac{\lambda_{\Sigma}}{\lambda_{0\Sigma}}$$

On the premise that delay time is proportional to the length of the interelement links,to the number of elements,and to the temperature,while it is being inversely proportional to the dissipated power,also assume that failure rate is proportional to the m-th power of temperature,where m=5..8,we have arrived at an equation that defines a relationship between Φ and the design parameters of the heat transfer devices (such as length,with, thickness and the like) and the temperature at the base of the functional components, T_c.

The optimum external and internal structural parameters of the heat sinks x_i are found from the solutions to a set of simultaneous equations of the type $\partial\Phi/\partial x_i = 0$ complemented by pertinent restrictions.As applied to heat sinks of the simplest configuration this set of equations takes on the following form:

$$A_0\frac{\partial}{\partial x_i}(a_0 l_k + b_0 b_k) + \left[B_0 + \frac{\partial\varphi(T)}{\partial T}\right]\frac{\partial T_c}{\partial x_i} = 0$$

where, A_0, B_0, a_0, b_0 are some complexes that take into account the process, design and functional factors producing certain effects on parameters τ and λ; $\varphi(T)$ is a temperature function that determines the dependence of λ from T; l_k and b_k are the length and width of the heat sink at the area of heat removal. In can be seen from the above equations that for the fixed geometry of a HP(l_k and b_k) the optimum parameters of the incasing devices of HP or ETS are determined by the condition $\partial T_c/\partial x_i = 0$,i.e., by the minimum of the total thermal resistance of a HP or ETS. To solve these equations for some simplified conditions yields simple relationships for optimal l_k and b_k, for example:

$$(l_k)_{opt} \sim \sqrt{R_{HP}} \quad \text{and} \quad b_k \sim \sqrt[3]{R_{HP}}.$$

Some of the approximate relationships are depicted in Figures 9a and 9b. Figure 9a shows the optimal HP length as a function of the HP thermal resistance. Figure 9b shows the optimal HP length as a function of the product of the operation delay of the logic element and the power dissipated.

The problem of optimizing a cooling system, i.e., determination of the optimum parameters of the cooling unit cycle, is considered for some specific applications of the cooling unit, where adequate information is available on the design of heat exchangers, geometry of evaporator channels, type and characteristic of the compressor, evaporator thermal load, temperature at the inlet to the condenser, and mimimum of the saturation temperature within the evaporator as limited by the dew point.

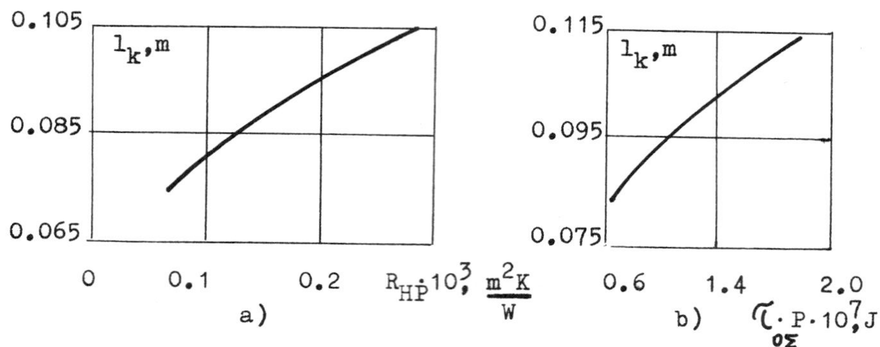

FIGURES 9a,9b. Optimal parameters selection of a heat pipe.

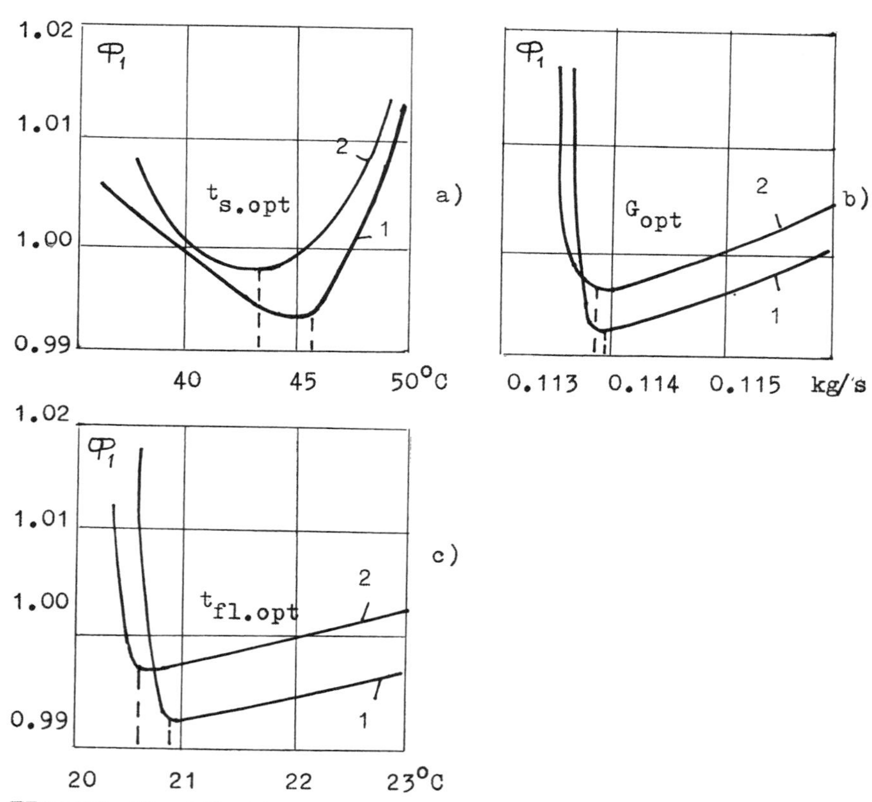

FIGURES 10a,10b,10c. Optimal parameters selection of refrigerator units: 1-without a regenerative heat exchanger; 2-with a regenerative heat exchanger; t_{fl}-temperature of subcooled freon in condenser; t_s-saturation temperature of freon in condenser.

Under these conditions, the initial mathematical model of the cooling system is a model for computing statistical characteristics for variable flow rate of the cooling water and condensation pressure, which are in fact the parameters being optimized. In this particular case, criterion Φ reduces to a simpler form

$$\Phi_1 = q_n \frac{N}{N_o} + q_g \frac{G_x}{G_{xo}},$$

where G_x and G_{xo} are flow rates of refrigerant for the optimum and initial alternatives. Priority of the "weight" share of q_g is reated to a strong influence of temperature field nonuniformity upon the reliability of the functional elements of the computer. Numerical solution of the optimization problem with respect to criterion Φ_1 has been found for a single variable saturation pressure P_s that is for saturation temperature t_s. Characteristic results of the optimization are given in FIGURES 10a, 10b, 10c.

5. CONCLUSIONS

Thus, solutions found to the above thermophysical problems have enabled us to lay down a procedure for searching for the optimum approaches to the layout and design of a computer cooling system built around compression vapor refrigerators. The thus-obtained techniques, and partial and general results will be hopefully useful to those concerned with thermal conditions of electronic equipment.

6. REFERENCE

1. Reznikov, G.V., Optimization of microelectronic computers operation delay and switching energy at temperature and heat limitations. Electronnaya Technica, ser.3, vol 1(113), pp.18-30, 1985.
2. Reznikov, G.V., Altman, E.I., Morkovkin, A.I., Smirnov, G.F, Selection of the circuits and basic parameters for refrigerators used in the computer cooling systems, Kholod. Tekhn. and technology, N 43, pp.21-27, 1986.
3. Reznikov, G.V., Calculation and construction of the computer cooling systems, Radio and Comm., Moscow, 1988.-227p.

Thermal Performance Analysis of a Power Module Heat Sink

MIROSLAV ŽIVANOVIĆ and ZORAN TATAROVIĆ
ILR, Lola Institute
Bulevar revolucije 84
11000 Belgrade, Yugoslavia

ABSTRACT

In this paper is presented the analysis of thermal performance of a power module heat sink, and based on it the algorithm for calculation of required air moving device characteristics. According to calculated characteristics, it is possible to choose the air spot cooling device (fan, blower, air mover with piezoelectric or electromagnetic transducer, etc.). In this way, the heat sink thermal performance will be improved and the correct functioning of power modules assured.

INTRODUCTION

Miniaturization and integration of electronic components have created a demand for high-efficiency heat sinks with reduced size. Heat sinks are designed to reduce the case-to-ambient thermal resistance of electronic components. They are important components since the junction-to-case thermal resistance and case-to-ambient resistance are in series, and it is obvious that the last-mentioned thermal resistance determines the semiconductor junction temperature rise. A semiconductor failure rate increases exponentially as the junction temperature rises.

Thus, power modules attached on printed circuit board require a high-efficiency heat sink. As the power dissipation and electronic component density is increased in comparison to a printed circuit board without power modules, the designer is faced with the problem of selecting the proper heat sink that will dissipate the required power within the space and air-flow constraints.

According to available space and higher electronic density packaging, it is necessary to apply a forced-air flow system instead of natural-air flow system. The forced-air flow system will improve the thermal performance of the power module heat sink, and in that way, enable its packaging in the available space. This system of air flow, so-called air spot cooling, demands a carefully selected air moving device since the space and weight are constrained.

As an air moving device for air spot cooling of electronics may be chosen a small fan or blower, an air mover with piezoelectric or electromagnetic transducer, a vortex tube, etc. To choose a suitable air mover for a particular application based only on available space and acceptable weight is not sufficient. It is also necessary to prepare a thermal performance analysis of the heat sink. Considering all factors, it is possible to make a selection among air moving devices that will improve thermal performance of particular heat sink.

THERMAL PERFORMANCE ANALYSIS

For the thermal performance analysis of a power module heat sink, the most influential thermophysical phenomena are accepted and the others with less influence are neglected. The thermal model has to be adequate for the accepted phenomena and may be expressed by mathematical equations.

Heat sinks are generally rated according to their thermal resistance. For a particular application is necessary to determine the thermal resistance that the heat sink must have in order to maintain a junction temperature that is not detrimental to the operation, performance, and reliability of the power module.

The basic relation for heat transfer may be stated as follows [1]:

$$P_d = \Delta T / \Sigma R_{th} \tag{1}$$

where P_d is the power dissipated by the power module, ΔT the temperature difference which causes the flow of heat, and ΣR_{th} sum of the thermal resistance of the heat flow path across which ΔT exists.

A simplified schematic of the heat flow from the power module to the ambient is shown in Fig.1. The above-mentioned relation (1) may be stated in the following forms:

$$P_d = (T_J - T_A) / (R_{thJC} + R_{thCS} + R_{thSA}) \tag{2}$$

$$P_d = (T_C - T_A) / (R_{thCS} + R_{thSA}) \tag{3}$$

$$P_d = (T_S - T_A) / R_{thSA} \tag{4}$$

where T_J is the junction temperature of the power module (maximum as stated by manufacturer), T_C the case temperature of the power module, T_S the temperature of the heat sink mounting surface in thermal contact with the power module, T_A the ambient temperature, R_{thJC} the thermal resistance from the power module junction to the mounting surface of its case, R_{thCS} the thermal resistance through the interface between the power module and the surface on which is mounted, and R_{thSA} the thermal resistance from the mounting surface

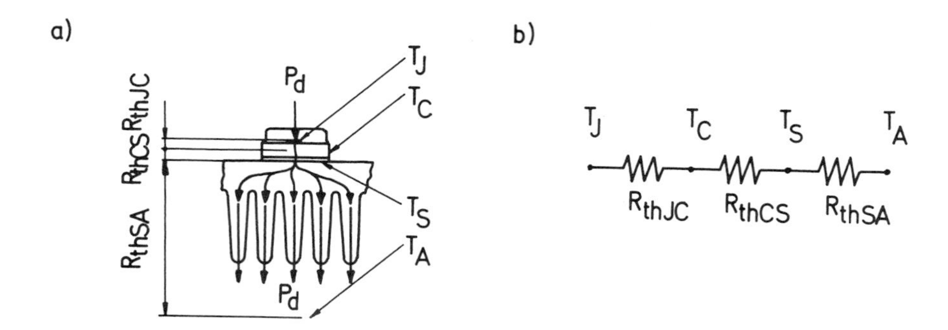

FIGURE 1. Heat flow from the power module to the ambient: a) schematic drawing which indicates the location of various heat flow paths, temperatures and thermal resistances; b) simplified representation of the system by a network of series connected resistances.

to ambient. The above relations are generally used to determine the required thermal resistance of the heat sink.

Heat transfer at the interface between a heat sink and a fluid (in this case, air) at different temperatures can be expressed by

$$P_d = \bar{h} A (T_S - T_A) \tag{5}$$

where $\bar{h}$ is the mean value of the heat transfer coefficient, A the surface area for convection, and $(T_S - T_A)$ the temperature difference between the surface of the heat sink and the ambient.

The surface area for convection can be expressed as follows:

$$A = P L \tag{6}$$

where P is the perimeter and L the length of a heat sink.

A heat transfer coefficient can be obtained from the Nusselt number:

$$Nu = \bar{h} L / k_f \tag{7}$$

where k_f is the thermal conductivity of the fluid.

First of all, it is necessary to determine the type of air flow over a heat sink surface - laminar or turbulent, which means the Reynolds number must be determined:

$$Re = u L / \nu_f \tag{8}$$

where u is the average air velocity and ν_f the kinematic viscosity of fluid.

In this analysis, the hydraulic, geometric and temperature constraints are as follows:

1. Average air velocity, u, must be less then 6 m/s.
2. Heat sink length, L, must be less than 0.225 m.
3. Heat sink surface temperature, T_S, must be less than 100°C.

According to the above constraints, the value of the Reynolds number must be less than the critical Reynolds number for an air flow on an isothermal plate, $Re_{crit} = 3x10^5$ [2]. For laminar boundary flow on an isothermal plate, the average Nusselt number is expressed as follows [2,3]:

$$Nu = 0.664\ Re^{0.5}\ Pr^{0.333} \quad (9)$$

where Pr is the Prandtl number. All fluid properties, such as k_f, ν_f, and Pr, are evaluated at the arithmetic average of the surface and ambient temperatures.

According to Eqs. (6-9), it is possible to express Eq. (5) as follows:

$$P_d = 0.664\ (u\ L\ /\ \nu_f)^{0.5}\ Pr^{0.333}\ k_f\ P\ (T_S - T_A) \quad (10)$$

The thermal resistance of the heat sink now, according to Eqs. (4) and (10), respectively, can be stated as:

$$R_{thSA} = (T_S - T_A)\ /\ P_d \quad (11)$$

and

$$R_{thSA} = [0.664\ (u\ L\ /\ \nu_f)^{0.5}\ Pr^{0.333}\ k_f\ P]^{-1} \quad (12)$$

According to the above thermal performance analysis of a heat sink and Eqs. (11) and (12), it is possible to present the air flow velocity as follows:

$$u = \frac{\nu_f}{0.441\ L\ Pr^{0.667}} \left[\frac{P_d}{k_f\ P\ (T_S - T_A)} \right]^2 \quad (13)$$

The functional dependence of the air flow velocity on thermal requirements, geometric characteristics and fluid properties is given by

$$u = f \left\{ \begin{array}{ll} [\ P_d,\ (T_S - T_A)\ ] & \text{- Thermal Requirements} \\ [\ L,\ P\ ] & \text{- Geometric Characteristics} \\ [\ k_f,\ \nu_f,\ Pr\ ] & \text{- Fluid Properties} \end{array} \right\} \quad (14)$$

On the basis of Eq. (14) and the design constraints, it is possible to make an algorithm to calculate the required air mover characteristics for improvement of heat sink thermal performance, which is given in Fig. 2.

EXPERIMENTAL CONFIRMATION

As a proof of the thermal performance analysis and adequate application of the algorithm when the air flow velocity value, u, is known, it is necessary

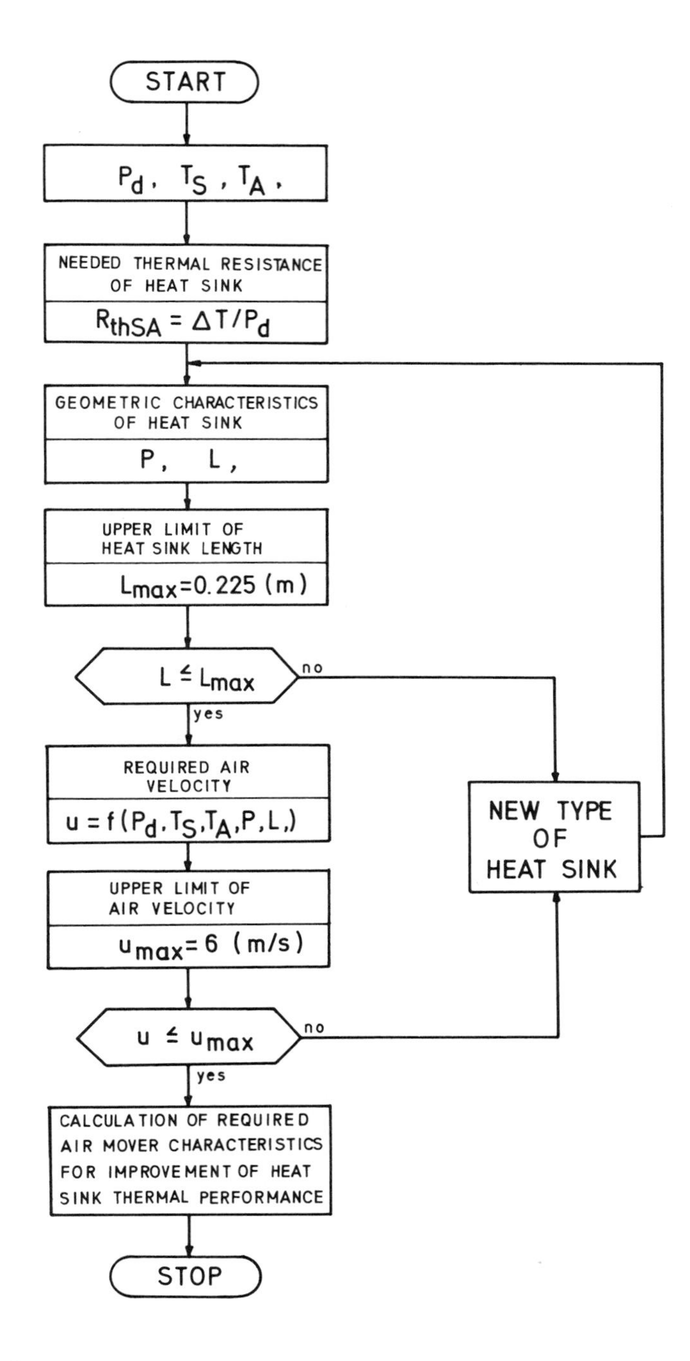

FIGURE 2. Simplified algorithm for calculation of required air mover characteristics.

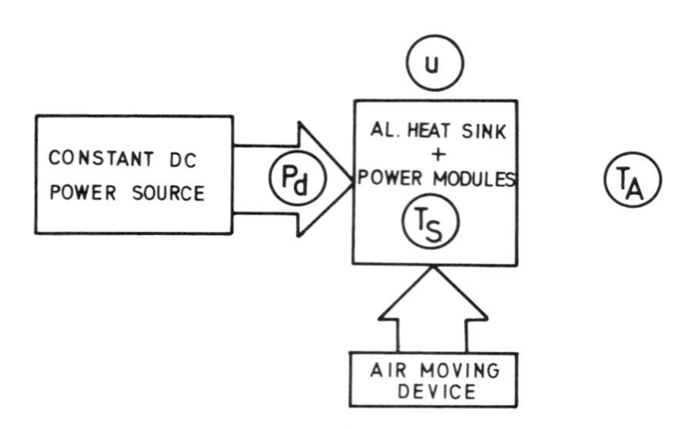

FIGURE 3. Functional scheme of experimental device with measuring points.

to check, according to Eq. (12), the calculated R_{thSA} value. For this purpose, it is necessary to form a test configuration consisting of power modules attached to a heat sink, an air moving device, and laboratory constant DC power sources, also necessary an instrumentation for measurement of power dissipation, P_d; temperature of the heat sink mounting surface, T_S; ambient temperature, T_A; and air velocity, u. A schematic of the test configuration with measuring points is shown in Fig. 3.

In the experimental test of the heat sink thermal performance, as power modules two Motorola MJ 10101 Darlington transistors with all needed equipment for their operation were used. Each of them was attached to the SEABA 012 3749/21000 heat sink which is manufactured by Nikola Tesla, Zagreb, Yugoslavia, with two M5 screws and 55 Ncm mounting screw torque. The interface between the power module and the heat sink is a 0.125 mm mica insulator with thermal compound.

As a air moving device a small axial fan, ISKRA type MIV 0201, was chosen. Its performance is: maximum airflow = 40×10^{-3} m^3/s, maximum static pressure = 55 Pa, and operating temperature from -10 to +70°C. The assembly of the heat sink with the power modules and the air moving device is shown in Fig. 4. The assembly which is used during the experimental test is a part of the Lola ATR DC motor speed controller.

The laboratory constant DC power sources which were used during the test were UNIS-ELKOS, Ljubljana, Yugoslavia, type RLU 01-30/10 and EI, Belgrade, Yugoslavia, type PS 3020-2. Power dissipation was calculated according to the measured values of collector-emitter voltage and collector current.

The temperature of the heat sink mounting surface was measured by a digital thermometer, Dalmacija, Dugi Rat, Yugoslavia, type DT I with surface probe TP 103, which consists of the thermocouple NiCr-Ni. The ambient temperature was measured by the same type of the thermometer but with immersion probe TP 101 for temperature measurement in gases.

The mean value of the air velocity at the outlet of the assembly was measured by a digital anemometer, Dalmacija, type DA 4000 with probe 275.

During the test, the ambient temperature was 25 ± 1.5°C, the power dissipation was changed from 10 to 60 W with steps of 10 W, and the air velocity was varied from 0.5 m/s to 3.5 m/s with steps of 0.5 m/s. The maximum value of the air velocity was limited by the fan. Experimental and theoretical data calculated from Eq. (12) are shown by the curves in Fig. 5. Also, the thermal resistance of the heat sink with natural convection was measured for the reference purpose and it is shown at the same Fig. 5. Each curve, a, b, and c, has direction arrows which show appropriate axes. The forced convection curves, a and b, refer to u and R_{thSA} axes, and the natural convection curve, c, refers to P_d and ΔT axes where ΔT is $(T_S - T_A)$.

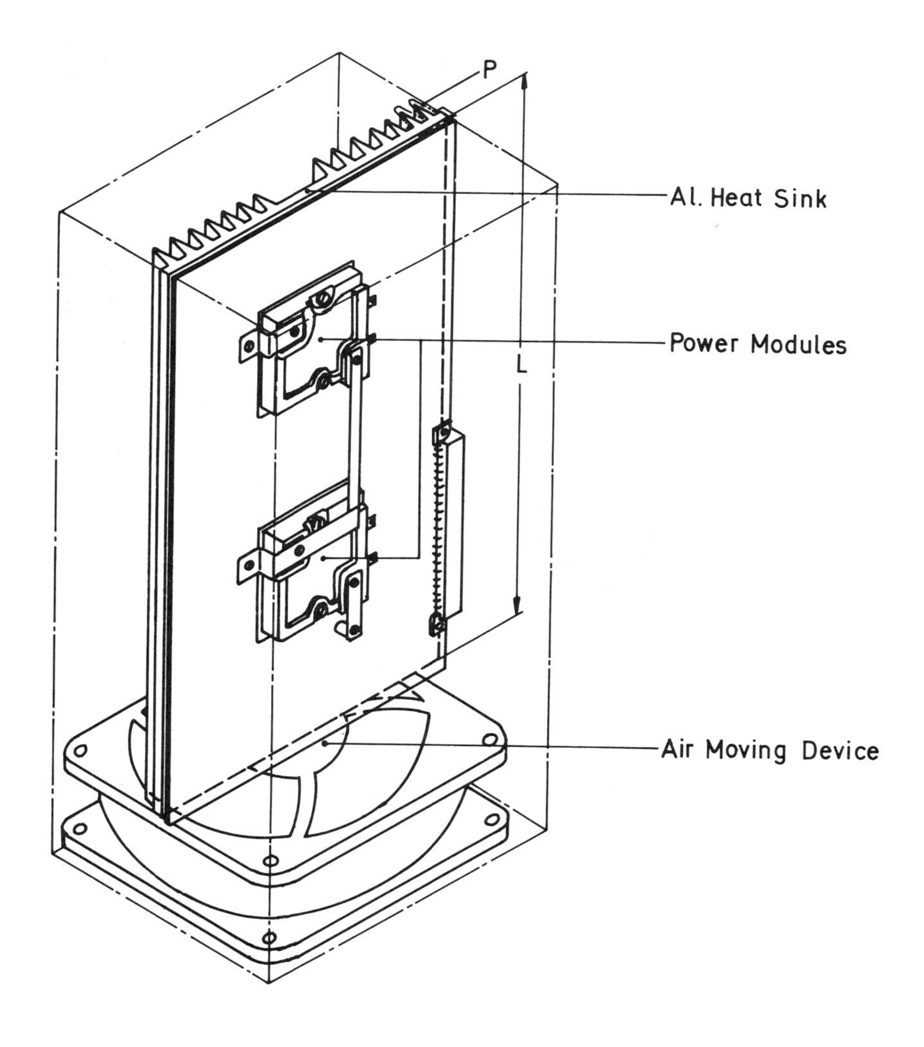

FIGURE 4. Schematic drawing of power modules, heat sink, and air mover assembly.

DISCUSSION

It is obvious from looking at the forced convection thermal performance of heat sink in Fig. 5 that the performance difference between the calculations and measurements is less than 10 %. The thermal performance analysis of power module heat sink is validated by experimental test.

According to Eq. (13) the air velocity required for a given thermal performance can be determined and it is possible to estimate the needed air flow rate of the air moving device. Since heat transfer is dependent on many variables in a real system and the thermal performance analysis of heat sink presented in this paper may not exactly match the actual application conditions, the thermal designer is advised to use it only as a first step in determining the thermal performance of the system.

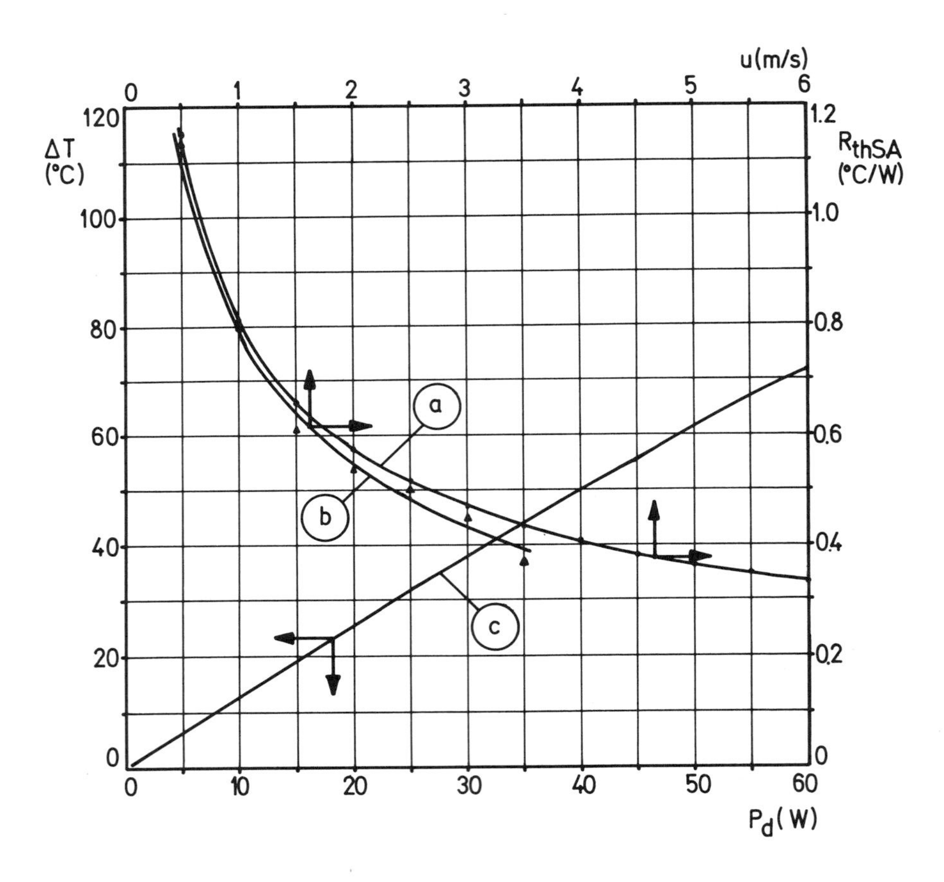

FIGURE 5. Forced and natural convection thermal performance of the heat sink: a-curve, forced convection, calculated data; b-curve, forced convection, measured data; c-curve, natural convection.

CONCLUSION

Most users of heat sinks are not thermal designers and must rely on outside information to accurately predict thermal performance of a heat sink with forced-air convection. This paper is useful for them because they may calculate the needed performances of an air moving device required to improve the thermal performance of a heat sink in order to maintain a junction temperature below the maximum.

NOMENCLATURE

Symbol	Quantity	SI Unit
A	area, cross section	m^2
a	thermal diffusivity	m^2/s
h	heat transfer coefficient	$W/(m^2K)$
k	thermal conductivity	$W/(m\ K)$
L	length	m
P_d	power dissipation	W
P	perimeter	m
R_{th}	thermal resistance	°C/W
T	temperature	°C
u	velocity	m/s
Greek Letters		
Δ	indicates a difference between variables	-
ν	kinematic viscosity	m^2/s
Superscripts		
$\bar{}$	indicates mean value condition	
Subscripts		
A	indicates ambient condition	
C	indicates case condition	
crit	indicates critical condition	
f	indicates fluid condition	
J	indicates junction condition	
S	indicates heat sink condition	

Symbol and Definition	Name
Dimensionless parameters	
$Nu = h\ L\ /\ k$	Nusselt number
$Pr = \nu\ /\ a$	Prandtl number

Symbol and Definition	Name
Dimensionless parameters (Continued)	
$Re = u L / \nu$	Reynolds number

REFERENCES

1. Ganić, E. N., Hartnett, J. P., and Rohsenow, W. M., Basic Concepts of Heat Transfer, in *Handbook of Heat Transfer Fundamentals,* eds. Rohsenow, W. M., Hartnett, J. P., and Ganić, E. N., ch. 1, pp. 1-1 to 1-13, McGraw-Hill, New York, 1985.

2. Kraus, A. D. and Bar-Cohen, A., *Thermal Analysis and Control of Electronic Equipment*, pp. 115-132, McGraw-Hill, New York, 1983.

3. Holman, J. P., *Heat Transfer, 6th ed.*, pp. 207-231, McGraw-Hill, New York, 1986.

Application of Laser Speckle Photography in the Study of Electronic Cooling

YAO-ZU SONG and ZENG-YUAN GUO
Department of Engineering Mechanics
Tsinghua University
Beijing 100084, PRC

ABSTRACT

In this paper, a novel measurement technique — laser speckle photography — that is applicable for experimental investigations of electronic cooling is presented in detail. The basic principle of the measurement of the temperature fields in fluids by laser speckle photography and its advantages over other methods are described. The experimental system of recording laser specklegrams and two methods of reconstruction of the specklegrams are proposed. For illustration, laser speckle photography has been used to study heat transfer with natural convection in channel that is a frequently encountered in electronic cooling. The experimental results are in good agreement with numerical solutions taking the property variations into account.

INTRODUCTION

The vertical channel formed by parallel plates or fins is a frequently encountered configuration in air natural convection cooling of electronic equipment, and typically occurs between convectively cooled rack-mounted printed circuit boards with heat dissipating components and/or modules on them. When an optical interferometric method is used for experimental investigations, the following two difficult cases often chosen. First, for most of theoretical analyses, the channel is assumed to be two dimensional. In order to verify these analytical results before they are applied to practical engineering situations, the test section must be constructed which is long enough to make the flow field in the channel two dimensional. As a result, a set of dense interferometric fringes usually occurs on the interferogram. The longer the channel is, the denser the fringes are. Because of this, it is difficult or impossible to count the fringe numbers and determine the temperature distribution. Even if the fringe numbers can be counted, the measurement accuracy of the temperature field decreases with the length of the test section because of the neglect of deflection of light beams through the long channel. Consequently, optical interferometric methods limit the length of the channel that can be studied. Second, some channels in actual electronic equipments are not long enough to make the flow field two-dimensional. As a result, the edge effect that results in a three dimensional characteristic of flow field has to be taken into account. Accordingly, experimentation is the main way for getting information about the heat transfer in the channel. Although some information about the flow field in the channel can be observed by using optical interferometric methods, quantitative results can not be obtained because of the three dimensional characteristic of the flow field. The following will demonstrate how the use of laser speckle photography can overcome these two difficulties.

EXPERIMENTAL APPARATUS FOR RECORDING A LASER SPECKLEGRAM

Laser speckle photography has been successfully applied to experimental investigations of the effects of variable properties on natural convection heat transfer with large surface-to-ambient temperature differences [1]. Furthermore, investigations show that laser speckle photography can also been applied to the experimental study of electronic cooling. The schematic of an experimental apparatus developed by us for recording a laser specklegram in study of electronic cooling is shown in Fig. 1. An expanded and collimated He-Ne laser beam passes through the test section that is imaged by means of a schlieren lens L_s onto a ground glass plate. This part of the experimental arrangement is similar to the optical schlieren system, except the edge is replaced by the ground glass plate. When the ground glass plate is illuminated by laser beams, the light beams scattered by the rough surface of the ground glass plate interfere with each other so as to form random patterns of interference fringes in space. They are called "Speckles". A speckle pattern on the defocused plane is recorded by an imaging lens L_i on a photographic plate with high resolution. This part of the experimental arrangement represents the subjective speckle recording system. Therefore, this experimental apparatus is referred to as a laser schlieren-subjective speckle photographic system. During the experiments, it is only required that a double exposure is taken on the same photographic plate for cases with and without natural convection. Because of the deflection of a light ray passing through the test section, the spatial speckle displacements that contain the information about flow and heat transfer inside the test section can be recorded on the photographic plate. It forms a specklegram.

The test section used in our experiments is a vertical two-dimensional channel formed by two parallel heated plates as shown in Fig. 2. Light ray propagates along the z-direction. According to

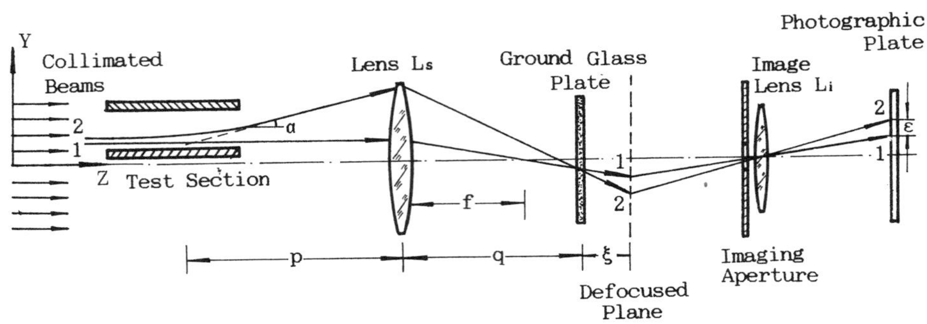

FIGURE 1. Top view of the laser schlieren-subjective speckle photographic system used in the study of electronic cooling.

the law of laser speckle movement in space and by the method of matrix optics[2], for the y-direction, the relation between the speckle displacement ε_y and the deflective angle α_y of light ray through the test section is given by

$$\varepsilon_y = M_i \cdot \xi \cdot \alpha_y / M_s \tag{1}$$

where the defocused distance ξ is the spacing between the ground glass plate and the defocused plane, M_s is a magnification of the schlieren lens L_s and M_i a magnification of the image lens L_i. According to the laws of ray optics, the relation between the light deflective angle α_y and the refractive index gradient $\partial n/\partial y$ in fluid is given by

$$\alpha_y = \int_0^L \frac{1}{n} \cdot \frac{\partial n}{\partial y} \cdot dz \approx \frac{1}{n} \int_0^L \frac{\partial n}{\partial y} \cdot dz \tag{2}$$

where L is the length of the test section, n is the refractive index of medium (for air $n \approx 1$). Thus,

$$\varepsilon_y = \frac{M_i}{M_s} \cdot \frac{\xi}{n} \int_0^L \frac{\partial n}{\partial y} \cdot dz \tag{3}$$

Eq.(3) reveals the relation between the speckle displacement ε_y on the specklegram and the refractive index $\partial n/\partial y$ of fluid in the test section for laser speckle photography.

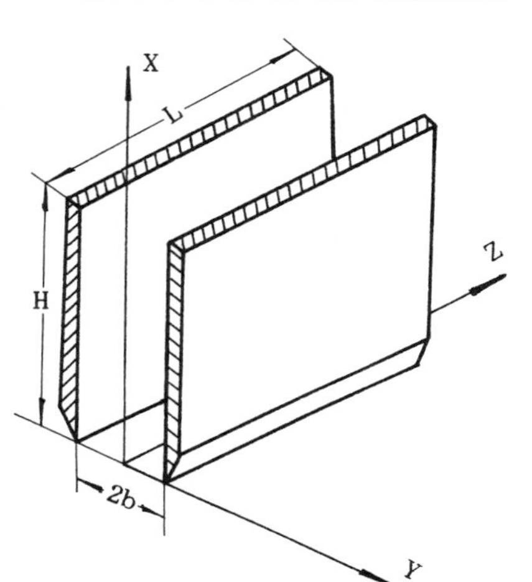

FIGURE 2. Schematic drawing of test section.

THE PRINCIPLE OF MEASURING GAS TEMPERATURE AND WALL HEAT FLUX

Suppose that the channel shown in Fig. 2 is long enough to neglect the edge effect of natural convection; then, the flow can be considered as two-dimensional and $\partial n/\partial y$ in Eq. (3) is not a function of the z-coordinate. Thus, for air, Eq. (3) can be represented as follows;

$$\varepsilon_y = \frac{M_i}{M_s} \cdot \frac{\partial n}{\partial y} \cdot \xi \cdot L \tag{4}$$

Using the Gladstone-Dale formula $n = \rho k + 1$ and the state equation of an ideal gas, the relation between the speckle displacement ε_y and temperature gradient $\partial T/\partial y$ is

$$\frac{\partial T}{\partial y} = -\frac{T^2}{\Omega} \cdot \frac{\varepsilon_y}{L} \cdot \frac{1}{\xi} \cdot \frac{M_s}{M_i} \tag{5}$$

where the constant $\Omega = MPK/R = 7.87 \times 10^{-2}$ [K] for air, K is the Gladstone-Dale constant, P = 1 atm, M is the molecular weight of air, R universal gas constant and T thermodynamic temperature.

In the same way, for the x-direction, the temperature gradient $\partial T/\partial x$ can also be deduced from the speckle displacement ε_x as follows:

$$\frac{\partial T}{\partial x} = -\frac{T^2}{\Omega}\cdot\frac{\varepsilon_x}{L}\cdot\frac{1}{\xi}\cdot\frac{M_s}{M_i} \tag{6}$$

For the isothermal boundary condition, two dimensional temperature profiles can be obtained for given wall temperature T_w and speckle displacements ε_y, by numerical integrations of equations (5) and (6). To take account of the effects of variable fluid properties on the heat transfer, local Nusselt number Nu_x and Rayleigh number Ra_x for the channel natural convection are defined, respectively, as

$$Nu_x = \frac{h\cdot b}{k_\infty} = \frac{-\left.\frac{\partial T}{\partial y}\right|_w\cdot b}{(T_w - T_\infty)}\cdot\frac{k_w}{k_\infty} \tag{7}$$

$$Ra_x = \frac{g\cdot b^4}{\nu_\infty{}^2\cdot x}\cdot\left(\frac{T_w}{T_\infty} - 1\right)\cdot Pr \tag{8}$$

For the constant heat flux cases, wall temperature T_w should first be determined by given temperature gradient $(\partial T/\partial y)_w$ at the wall surface as follows:

$$T_w = \left[\frac{L\cdot\Omega\cdot\xi}{(\varepsilon_y)_w}\cdot\frac{M_i}{M_s}\cdot\left.\frac{\partial T}{\partial y}\right|_w\right]^{1/2} \tag{9}$$

again two dimensional temperature profiles in channel can be obtained by numerical integrations of equations (5) and (6).

In many cases, the edge effects of natural convection in the channel must be considered, because the channels found in electronic equipment are often not long enough to assume two-dimensional flow and heat transfer. Under these circumstances, a quantitative evaluation of the temperature gradient at the wall surface $(\partial T/\partial y)_w$ is not possible because it changes in z-direction and Eq. (4) no longer holds. However, we can easily evaluate the heat transfer rate $Q_y(x)$ per unit of plate height by

$$Q_y(x) = -k_w\int_0^L \left(\left.\frac{\partial T}{\partial y}\right|_w\right)\cdot dz \tag{10}$$

because the speckle displacement itself characterizes the integration of the refractive index gradient. So long as the given T_w is constant in the z-direction, by means of Eq. (3), Gladstone-Dale formula and the ideal gas equation, the heat transfer rate $Q_y(x)$ of per unit plate height can be expressed as

$$Q_y(x) = \frac{k_w\cdot T_w{}^2\cdot(\varepsilon_y)_w\cdot M_s}{M_i\cdot\Omega\cdot\xi} \tag{11}$$

If the wall temperature in z-direction is slightly non-uniformed, it is reasonable to approximate it to some type of temperature distribution. For a linear distribution, the heat transfer rate $Q_y(x)$ is given by [3]:

$$Q_y(x) = \frac{k_w \cdot (\varepsilon_y)_w \cdot M_s}{M_i \cdot \Omega \cdot \xi} \left[\frac{a^2}{3} \cdot L^2 + abL + b^2 \right] \tag{12}$$

All these show that even if the edge effect of natural convection in channel should be taken into account and consequent the flow field in the channel is three-dimensional, the information about $Q_y(x)$ at the channel wall can still be obtained quantitatively by laser speckle photography. This reveals another advantage of laser speckle photography in the study of electronic cooling.

RECONSTRUCTION METHODS OF A SPECKLEGRAM

According to above discussions, it is clear that the direct measured variables for the speckle photography are the speckle displacements ε_x and ε_y in the x- and y-directions. by means of the reconstruction of the specklegram, the magnitude and direction of the speckle displacement at any point on the specklegram can be determined. There are two methods for reconstruction of specklegram, point-by-point analysis and a full view of flow field [4-5].

The optical system for point-by-point analysis of the specklegram is shown in Fig. 3. When re-illuminated with an un-expanded narrow laser beam, each portion of the specklegram reveals diffraction patterns consisting of a bright central spot and in the less intense diffraction halo, which is made up of a series of equally spaced Young's fringes. The fringe spacing d is given by

$$d = \lambda_0 \cdot S \,/\, \varepsilon \tag{13}$$

where λ_0 is the wavelength of the laser light used in reconstruction of the specklegram, S is a distance between the specklegram and a viewing screen and ε is the magnitude of the speckle displacement. The direction of the speckle displacement is normal to the direction of the Young's fringes. Fig. 4 shows the Young's patterns reconstructed from a specklegram of laminar natural convection between two heated vertical plates. The specklegram is more appropriate to computer image processing than the interferogram because of the regular nature of the Young's fringes. In our experiments, a video detector and a microcomputer for image processing have been used for measuring the

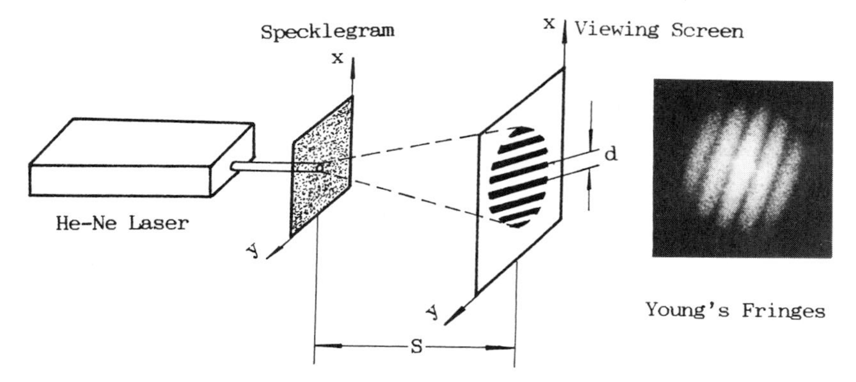

FIGURE 3. Optical arrangement for point-by-point analysis of a specklegram.

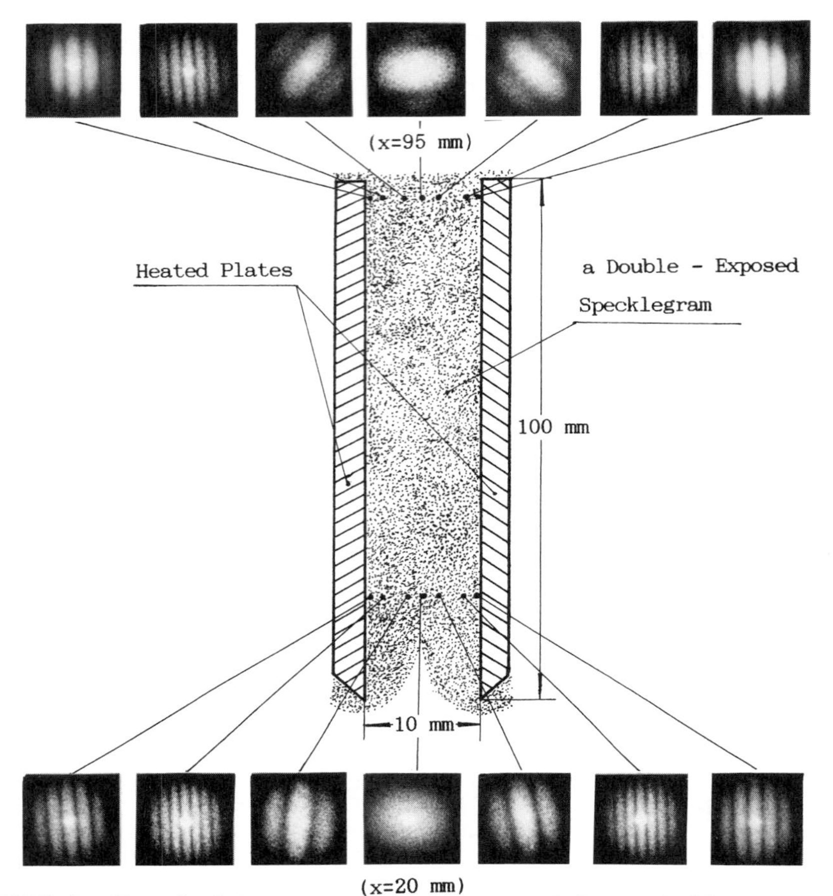

FIGURE 4. Young's fringe patterns reconstructed from a double exposure specklegram of laminar natural convection in a channel

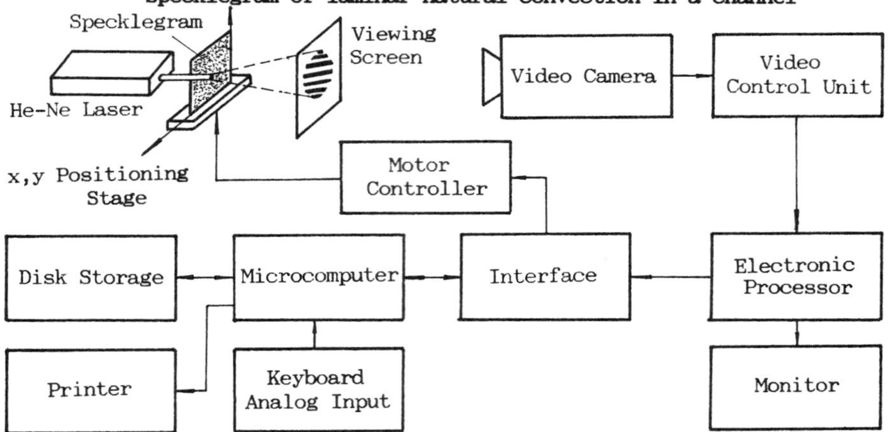

FIGURE 5. Block diagram of a video detector and image processing with a microcomputer for a specklegram.

direction and spacing of Young's fringes as shown in Fig.5. The measurement error of the image processing system is less than 5%.

An alternate reconstruction method is a full view of flow field from the specklegram by spatial filtering as shown in Fig.6. When the specklegram is illuminated with an expanded collimated beams, a Fourier transforming lens produces the spatial Fourier transform of the speckle pattern in the back focal plane of the lens. A small filtering aperture placed at the focal plane can be used to select diffraction orders and direction of speckle displacement. An imaging lens serves for imaging the specklegram, through the filtering aperture, onto the plane of observation. Fig.7 shows a full view of flow field for natural convection between two heated vertical plates by the spatial filtering. technique. An exact analysis of this spatial filtering reconstruction shows that the fringes are a set of curves of constant displacement components that are given by

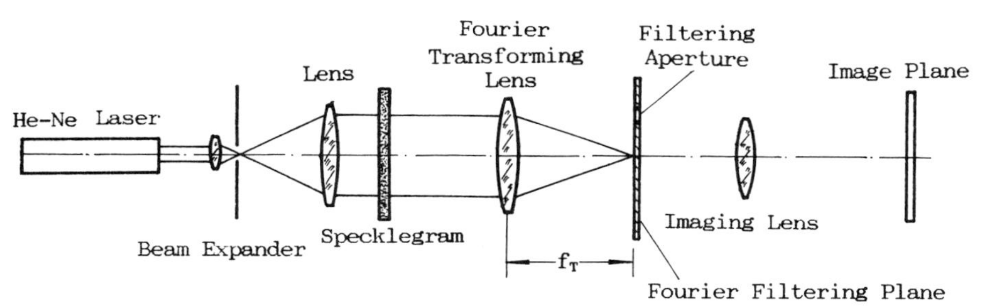

FIGURE 6. Schematic of optical arrangement for a full view of the flow field from a specklegram.

$$\varepsilon = \frac{N \cdot \lambda_0 \cdot f_T}{M_i \, \rho_a} \qquad N=1, 2, \ldots \qquad (14)$$

where N is the order of interference fringe, f_T a focal distance of Fourier transform lens, M_i is the magnification of image lens, λ_0 the laser wavelength and ρ_a a distance between optical axis and the filtering aperture on the Fourier filtering plane. The speckle displacement is obtained by fringe counting. The direction of the speckle displacement is in the direction of the radius from the center to the filtering aperture. The fringe pattern in Fig. 7 reflects a set of profiles of constant components of the density gradient of the fluid between two heated vertical plates. In most cases, the measurement accuracy for this type of reconstruction is lower than the accuracy for point-by-point analysis.

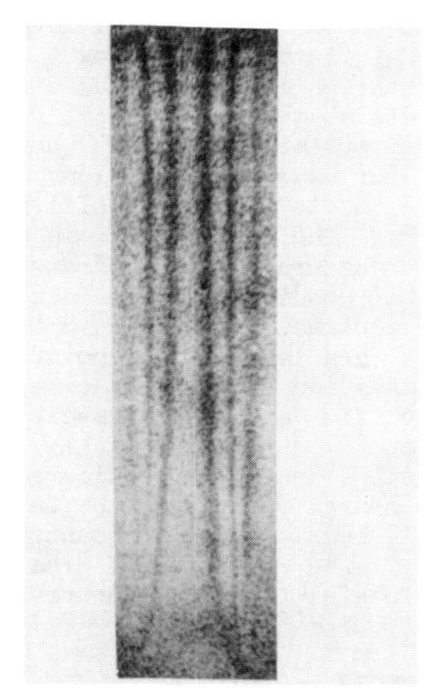

FIGURE 7. Image reconstructed from specklegram of natural convection in channel by spatial filtering.

EXPERIMENTAL DEMONSTRATIONS

For illustration, the laser speckle photography has been used to study heat transfer of laminar natural convection in a channel formed by two isothermal vertical plates with symmetric heating. In our experiments, the spatial resolution of the experimental optical system is about 0.2 mm. The minimum temperature gradients, i.e. the sensitivity, measured in the present experiments is about 1 K/mm. The uncertainties in the experimental results are about 5%. Some satisfactory experimental results have been obtained. The results have already been described in detail in reference [6]. We found that

(1) The effects of variable properties on the natural convection in a channel are much more profound than for a single plate, for example, the effects are reflected in the profiles of dimensional temperature gradient and dimensionless temperature, as well as the heat transfer expressions.

(2) The dimensionless correlating expression for heat transfer of developing laminar natural convection in a channel with symmetrical isothermal wall surfaces can be represented by

$$Nu_x = 0.76\,(\lg Ra_x) - 0.26 \qquad 250 > Ra_x > 1.0 \qquad (15)$$

$$Nu_x = 1.50\,(\lg Ra_x) - 1.96 \qquad 2000 > Ra_x > 150 \qquad (16)$$

(3) The existence of the thermal drag phenomenon in the channel with natural convection has been verified by laser speckle photography.

CONCLUSION

Based on our experimental investigations it may be shown that, in the study of electronic cooling, the laser speckle photography has the following three main advantages over the conventional optical interferometric methods. First, laser speckle photography is particularly suitable for cases where the deflections of the light beams is large, for example, in a long channel. The longer the channel, the larger the deflective angle of light beams through the channel and the more notable the effect of laser speckle is. Because of this, measurement accuracy can increase with the length of the channel, as long as the condition of correlation of laser speckles is maintained. Next, in contrast with the optical interferometric methods the temperature gradient at any point on the surface of the channel can be directly measured by using laser speckle photography. Therefore, the local heat flux and local Nusselt number can also be evaluated easily. Finally, and most interestingly, when the three dimensional flow characteristics must be considered, due to edge effects of channel or complex boundary conditions, the information of heat transfer rate $Q_y(x)$ can still be obtained quantitatively by using laser speckle photography. We have found that laser speckle photography is a very practical useful technique for research in the area of heat transfer in electronic equipment.

ACKNOWLEDGMENTS

The authors would like to acknowledge the financial support from Foundation of National Commission of Education of China.

NOMENCLATURE

Symbol	Quantity	SI Unit
a	thermal diffusivity	m^2/S
b	half wide of channel	m
d	Young's fringe spacing	m
g	local gravitational acceleration	m/s^2
H	channel height	m
h	heat transfer coefficient	W/(m K)
K	Gladstone-Dale Constant	m /kg
k	thermal conductivity	W/(m K)
L	channel length	m
M	molar mass	kg/mol
M_i	magnification of image lens	-
M_s	magnification of schlieren lens	-
n	refractive index of medium	-
P	atmospheric pressure	N/m
$Q_y(x)$	heat transfer rate per unit plate height	W/m
R	universal gas constant, R = 8.3144	J/(mol K)
T	temperature	K

Greek Letters

α	deflective angle of light beams	-
ε	speckle displacement	m
λ_0	laser wavelength	m
υ	kinematic viscosity	m /s
ξ	defocused distance	m
ρ	mass density	kg/m
ρ_a	distance [see Eq.(13)]	m
Ω	$= MPK/R = 7.87\times10^{-2}$ (for air) constant	K

Coordinates

x,y,z	cartesian coordinates

Subscripts

w	conditions at wall
∞	ambient conditions

Symbol and Definition	Name

Dimensionless parameters

$$Nu_x = \frac{h\cdot b}{k_\infty} = \frac{-\left.\frac{\partial T}{\partial y}\right|_w \cdot b}{(T_w - T_\infty)}\cdot\frac{k_w}{k_\infty}$$ local Nusselt number

$Pr = \upsilon / a$ Prandtl number

$$Ra_x = \frac{g \cdot b^4}{\nu_\infty^2 \cdot x} \cdot \left(\frac{T_w}{T_\infty} - 1 \right) \cdot Pr$$ Rayleigh number

REFERENCES

1. Guo, Z. Y. and Song, Y. Z., An Investigation in Heat Transfer of Natural Convection with Large Temperature Difference by Laser Speckle Photography, *Proc. the 1987 ASME Winter Annual Meeting*, Boston, MA, U.S.A., Dec. 13-18, 1987

2. Song, Y. Z., Zhou, X. G. and Guo, Z. Y., A Laser Schlieren Speckle Photography System Applicable to the Measurement of Large Changes of Refraction in Phase Object, *Proc. Int. Conference on Laser'1987*, Xiamen, Fujian, China, p.2-13, Nov. 15-19, 1987.

3. Song Y. Z., Laser Speckle Photography and Its Application in Thermophysics, Dissertation, Department of Engineering Mechanics, Tsinghua University, Beijing, China.

4. Farrel, P.V., and Hofeldt, D.L., Temperature Measurement in Gases Using Speckle Photography, *Applied Optics*, Vol. 23, No. 7, pp. 1055-1059, April 1984.

5. Wernekinck, U. and Merzkirch, W., Measurement of Natural Convection by Speckle Photography, *Proc. 8th Int. Heat Transfer Conf.*, San Francisco, U.S.A., Vol.2, pp.531-535, Aug. 17-23, 1986.

6. Guo, Z. Y., Song, Y. Z. and Zhao X. W., Experimental Investigations on Heat Transfer in a Channel by Laser Speckle Photography, to be presented at *First International Conference on Experimental Heat Transfer, Fluid Mechanics and Thermodynamics*, Dubrovnik, Yugoslavia, Sept. 4-9, 1988.

Thermal Performance of Leadless Chip Carriers Mounted on Cored PCB's

PETER JACKSON
Ferranti International Signal plc
Ferranti Computer Systems Limited
Bracknell, Berkshire, UK

ABSTRACT

Surface mounted components provide the mechanical engineer in the electronics industry with a number of advantages, such as smaller dimensions as compared to the dual-in-line equivalent. For the harsh environments experienced in military applications, ceramic leadless ship carriers are the most appropriate, but this style of component brings with it the problem of mismatch in expansion coefficients between the ceramic and the materials which are used in conventional multi-layer printed circuit boards.

Ferranti Computer Systems have produced a multi-layer printed circuit board construction which overcomes this problem. This paper presents the thermal performance of this type of printed circuit board construction when cooled by different cooling mechanisms, and describes analytical thermal models which can be used to predict component operating temperatures.

1. INTRODUCTION

Surface mounted components present the mechanical engineer in the electronics industry with a number of advantages. Among these advantages are:

i) A smaller volume occupied by the electronic components.
ii) The potential for reduced interconnection leading to greater reliability.

Surface mounted components come in a number of different package styles. However, if the mechanical engineer is designing a system to meet a harsh climatic environment, he is constrained to select components with the highest temperature rating and the greatest degree of hermeticity. Components that are able to meet these requirements are, in the main, only available in ceramic leadless chip carrier (LCC) form.

The use of ceramic LCC components, in conjunction with double sided or multilayer printed circuit techniques, introduces the problem of the mismatch in the expansion coefficients of the printed circuit board (PCB) material and the ceramic of the LCC's. If this problem is not addressed, excursions that can occur in temperature could cause the LCC's to fall off the PCB.

The Printed Circuit Group at Bracknell have been engaged in developing a multilayer PCB construction which can overcome this expansion coefficient mismatch. The construction (Metlam) employs copper clad Invar cores which causes the PCB to mimic the expansion coefficient of the ceramic.

Consequently, a multilayer PCB construction exists that can exploit LCC's. However, the space advantage provided by the use of LCC's will inevitably mean an increase in the power density on these PCB's, resulting in more severe thermal management problems. To this end, the Mechanical Design Group at Ferranti have been investigating the thermal performance of this cored PCB construction, and developing an analytical model that will allow the accurate prediction of component temperatures. This work will lead towards a maximisation of PCB reliability.

This paper is concerned with these thermal investigations. The work programme described is:

i) The design and manufacture of a cored PCB (Metlam) test board which lends itself to temperature measurement and analysis.
ii) Temperature measurements made under convection cooling and conduction cooling conditions.
iii) The preparation of and the logic behind the analytical thermal models for each cooling process.
iv) The comparison between the predicted temperatures resulting from the analytical models and the measured temperatures.

2. TEST PCB

The Metlam PCB used for these tests is shown in Figure 1. It is to the standard Double Eurocard dimensions of 233mm by 160mm.

FIGURE 1. Photograph of Metlam Printed Circuit Board.

The board construction is shown in Figure 2.

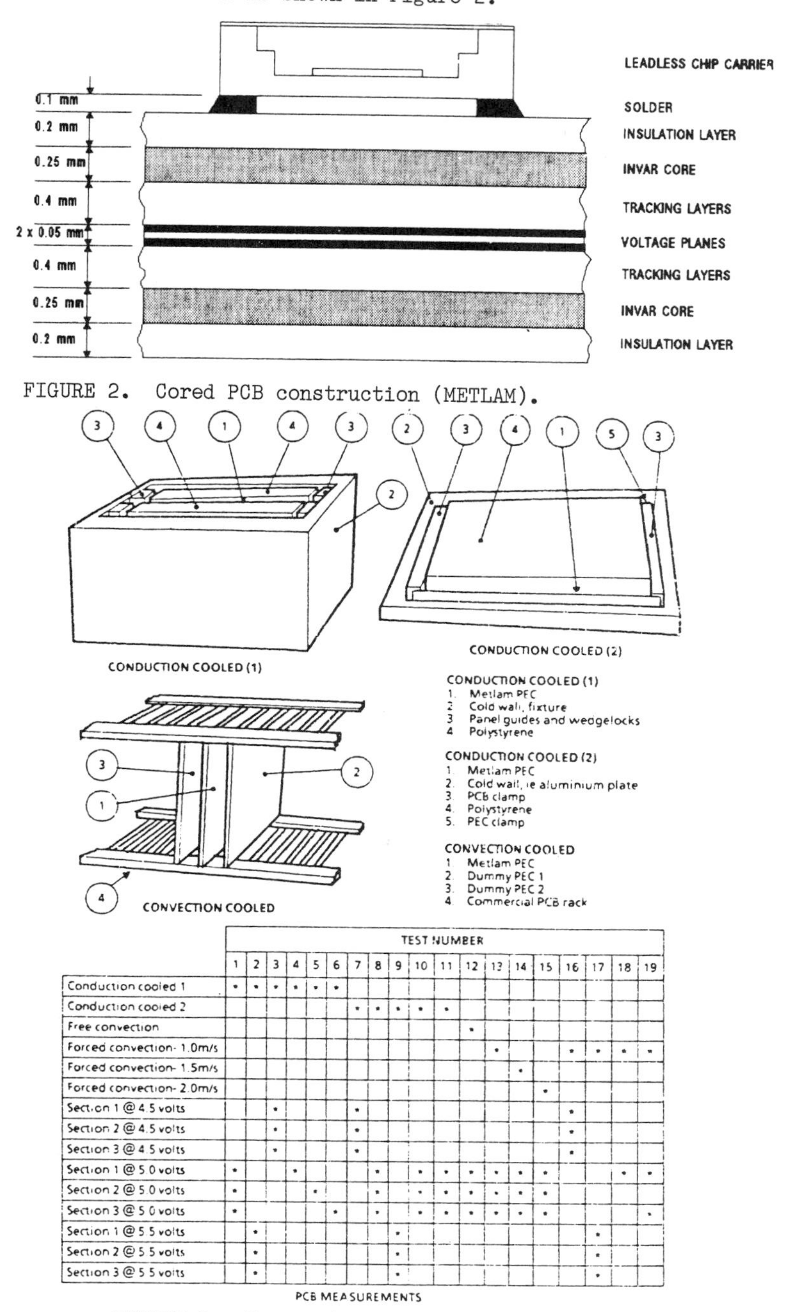

FIGURE 2. Cored PCB construction (METLAM).

	TEST NUMBER																		
	1	2	3	4	5	6	7	8	9	10	11	12	13	14	15	16	17	18	19
Conduction cooled 1	•	•	•	•	•	•													
Conduction cooled 2							•	•	•	•	•								
Free convection												•							
Forced convection- 1.0m/s													•			•	•	•	•
Forced convection- 1.5m/s														•					
Forced convection- 2.0m/s															•				
Section 1 @ 4.5 volts			•				•									•			
Section 2 @ 4.5 volts			•				•									•			
Section 3 @ 4.5 volts			•				•									•			
Section 1 @ 5.0 volts	•			•				•		•	•	•	•	•	•			•	•
Section 2 @ 5.0 volts	•				•			•		•	•	•	•	•	•				
Section 3 @ 5.0 volts	•					•		•		•	•	•	•	•	•				•
Section 1 @ 5.5 volts		•							•								•		
Section 2 @ 5.5 volts		•							•								•		
Section 3 @ 5.5 volts		•							•								•		

PCB MEASUREMENTS

FIGURE 3. Temperature measurements.

Three sizes of component were used and they were arranged in groups on the PCB. The largest components used contained 40 leads, the medium sized group 28 leads and the smallest group 20 leads.

3. TEMPERATURE MEASUREMENTS

Two cooling mechanisms and three cooling methods were investigated as shown in Figure 3. The cooling mechanisms were conduction and convection (forced and free). Two methods of conduction cooling were considered:

i) PCB clamped along two edges
ii) Rear face of PCB clamped to a cold wall

Although measurements were made with forced and free convection, the free convection case was not considered in the analyses.

Figure 3 also describes the full range of measurements that were made.

4. THERMOCOUPLE POSITIONS

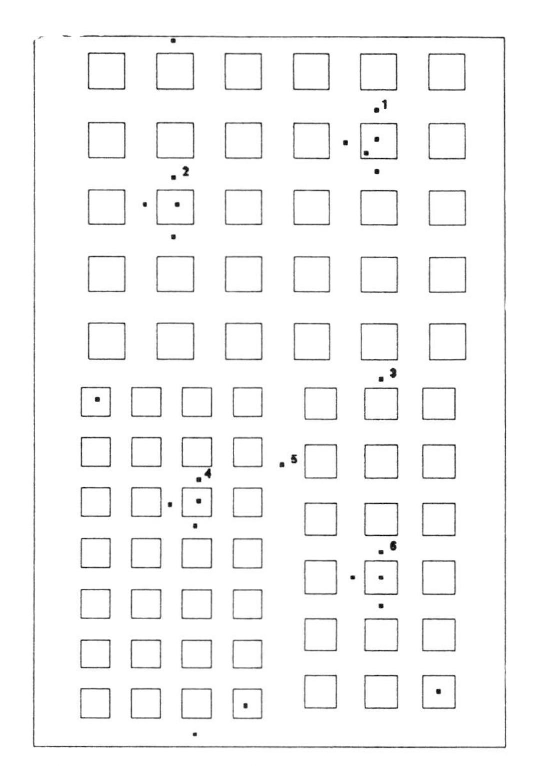

FIGURE 4. Thermocouple positions on PCB.

Figure 4 shows the thermocouple positions on the PCB. These positions were chosen with care, such that the resulting measurements could be used to their full advantage in determining the effectiveness of each of the available thermal paths.

The thermocouples used were type 'J' which is a combination of iron and constantan. They were fixed to the surface of the PCB by small epoxy 'blobs'.

5. ANALYTICAL THERMAL MODEL

The thermal resistance network technique was used to produce the analytical thermal model.

This technique demands that the thermal model is formed by defining each of the available thermal paths between the heat sources and the heat sink (which may be cold wall or cooling air) and assigning a value of thermal resistance to each one. From this, a resistance network is formed of series and parallel paths.

It is then necessary to solve this network to determine the overall thermal resistance between each heat source and the heat sink. With a knowledge of the heat dissipation of the heat source and the heat sink temperature, it is possible to determine the temperature of each heat source.

The network that results from applying this technique to a Metlam PCB contains about 500 thermal paths. It is not practical to solve this manually, and use is made of suitable software to speed the process. A number of computer programs are available specifically for this task. For this work, a programme called Newstead was used which was written for an MOD project at British Aerospace at Stevenage.

6. CONSTRUCTION OF THERMAL MODEL

From the description of the PCB construction it will be noted that the board is made up of a number of layers. Therefore, it is logical that the thermal model should reflect this fact.

Figure 2 shows the construction of the Metlam PCB. Only the Invar cores and the voltage planes are significant and therefore from a thermal point of view, the board construction can be considered as having 3 layers (two Invar core layers and one layer representing the voltage planes). The paths between the layers is formed by the epoxy/glassfibre construction of the PCB.

Each of these layers is divided into an identical grid system. The form of the grid system is defined by the component sizes and their layout on the PCB. It is important that the grid reflects the component layout; but to produce a convenient grid it may be necessary to adjust the positions of the components slightly as shown is Figure 5.

Once the grid is defined, each grid position is thermally connected with its neighbour in both the X and Y planes, and this is done for each layer. Each layer is then connected in the Z plane.

FIGURE 5. PCB grid relating to component layout.

If one then considers the leadless chip carrier heat sources and thermal connection with the Metlam PCB, the construction and the resulting resistance network is shown in Figure 6. It is then necessary to calculate the thermal resistance of each path.

If we first consider the conduction paths, the path from the chip to the LCC package is calculated from the relationship:

Thermal Resistance (R) = L/k x A

with the path being formed by a 0.12mm thickness of oxide loaded epoxy.

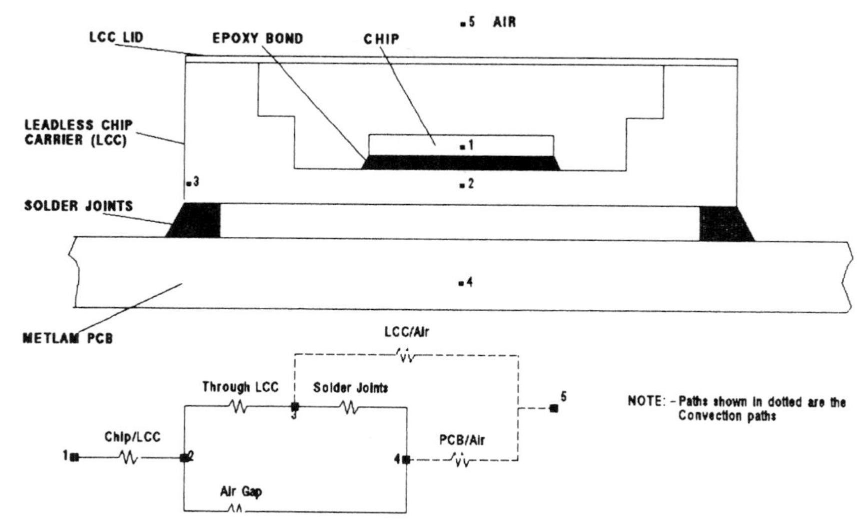

FIGURE 6. Leadless chip carrier thermal resistance network. Path 2 to 3 (Through X-Y plane of LCC) is modelled as shown in Fig. 7.

The LCC is considered as being partitioned into 4 sections and the thermal resistance determined for each section. The overall thermal resistance is determined by considering the four sections as parallel paths.

For the path through the solder joints, the dimensions of each pad on the LCC (and the mating pad on the PCB) were taken as 1mm x 0.5mm. The solder thickness, by measurement, is typically 0.12mm.

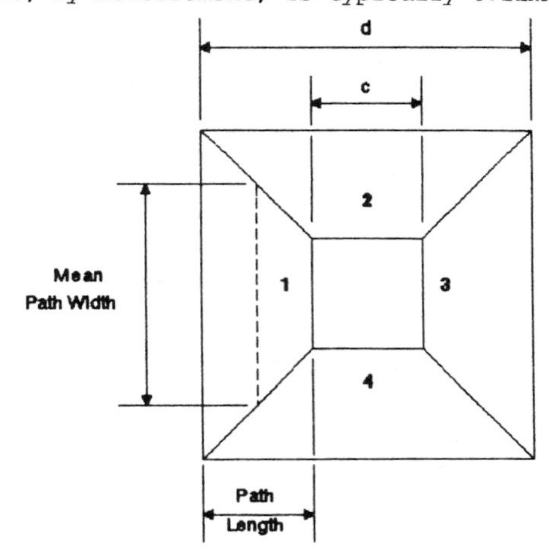

FIGURE 7. X-Y plane of LCC.

The air gap between the base of the LCC and the PCB was assumed to be the same as the solder thickness, i.e., 0.12mm. The area value used was the base area within the pads and the interface material was air (thermal conductivity 0.028 w/mk).

We then must consider the convection paths (when forced air is the cooling mechanism).

For forced convection where the flow is turbulent,

Nusselt No. (Nu) = C x Reynolds No. $(Re)^n$ x Prandtl No. $(Pr)^n$

where C and n are constants

Nu = h x De / k

where h = heat transfer coefficient

De = equivalent diameter of duct formed by PCBs

= 2 (a x b)/(a + b) where a and b are duct dimensions

k = thermal conductivity of air

therefore,

h = C x (Re) x (Pr) x k/De

convection thermal resistance (R)

= 1/(h x A)

where A = surface area exposed to the convective effect

For an LCC, the exposed surface area is made up of the four sides plus the lid. However, it would be incorrect to include the total lid area since the material from which it is made is thin, which will result in a marked temperature profile over the surface of the lid. Consequently, only 65% of the lid area is included. This effective area has been determined by measurement of the temperature profile over the surface of the lid.

In the case of the Metlam board, the rear surface is totally exposed, and therefore, the individual grid area is used when calculating this particular convective thermal resistance.

The front surface of the Metlam board is partially obscured by the components, and therefore the area used is the difference between the grid area and the component plan area.

Finally, we must consider the connection between the PCB and the cold wall when conduction is the cooling mechanism.

In these investigations, the Metlam board edges were clamped to cold walls by means of 'wedgelock' clamps. These clamps are normally fitted to the non-component side of the PCB and force the edge of the PCB against the cold wall. Experiments that have been carried out indicate that the resulting air gap between the PCB and the cold wall is 0.013mm and the thermal resistance for this path is based on this fact.

When the Metlam board is clamped on its edges to a rear mounted cold wall, the centre of the board is inclined to bow away from the cold face. The amount of bow is variable, but in the case of these tests, was found to be equivalent to 0.25mm over the surface. This figure was used when calculating this particular thermal resistance.

Having defined the resistance network reflecting the cooling process and calculated each of the thermal paths, it is possible to carry out the analysis and produce predicted temperatures.

7. RESULTS

THERMOCOUPLE POSITION (SEE FIG. 4)	FORCED CONVECTION		CONDUCTION - EDGE CLAMPS		CONDUCTION - REAR FACE	
	MEASURED	PREDICTED	MEASURED	PREDICTED	MEASURED	PREDICTED
1	51	50	60	57	43	40
2	55	56	69	73	46	45
3	62	56	75	80	50	43
4	57	54	61	64	41	39
5	60	56	72	74	47	43
6	63	58	70	72	48	46

NOTE:- Ambient Air Temperature For Convection Case - 32 degrees
Cold Wall Temperature for Conduction Cases - 30 degrees

FIGURE 8. Measured and predicted results.

For convenience, not all the measured and predicted temperatures have been included in the results. Selected positions on the surface of the Metlam PCB have been chosen, and the measured and predicted temperatures given for these positions only. The positions chosen are the numbered positions shown in Figure 4. The LCC surface temperatures have not been included for reasons that are given in the discussion section of this paper.

Figure 8 shows a table of the selected measured temperatures and the corresponding predicted values resulting from the analytical model described in Section 6.

8. DISCUSSION OF RESULTS

These investigations were carried out with two basic purposes in mind:

i) To compare the cooling performance of the cooling mechanisms and methods available for this configuration of PCB.
ii) To develop an analytical thermal model for this PCB configuration which accurately predicts the component temperatures for the various cooling methods.

8.1 Cooling Performance.

Before discussing the results that have been obtained, it is worth considering the benefits that surface mounted components provide and the consequences with respect to packaging concepts. The most obvious benefit, at least to a mechanical engineer, is the reduced size of the surface mounted package with respect to, say, the dual-in-line equivalent. Therefore, an electronic function can be performed in a smaller volume or a greater electronic function can be performed in the same volume. Whichever is the case, the component density will be greater.

A number of views have been expressed as to what this increase in component density will be. The most commonly expressed figure is an increase by a factor of 3. However, for the existing component technologies, this will also mean an increase in power dissipation density of the same value.

That is to say, that a double eurocard with what we would consider to be an acceptable but high power dissipation of 25 watts, could increase to 75 watts if surface mounted components were employed, and the full advantage was made of them. This possible step change in power dissipation cannot be ignored.

It will be noted from the results given in Figure 7 that the conduction cooling method with the rear of the PCB in contact with the cold wall ('Z' plane cooling) gives considerably lower temperatures than the other cooling methods. This is to be expected, since the path to the cold wall is small (i.e., the thickness of the PCB) compared to conduction cooling by clamping the PCB along its short edges ('X-Y' plane cooling) which was the other conduction cooling method investigated.

This rear face conduction cooling method has not been available with PCBs with conventional dual-in-line components, due to the solder spills on the rear face of the PCB. Therefore, the introduction of surface mounted components and a suitable printed circuit substrate has made available a cooling method that can accommodate much higher power densities.

It is a general rule that the temperature of a component is a function of its own power dissipation and the power dissipation of the surrounding components. This is certainly true for convection cooling and conduction cooling in the X-Y plane. However, in the case of Z plane cooling, it can be argued that providing the spacing between the components is equal to or greater than twice the thickness of the PCB substrate, then each component is unaffected by the surrounding components. Therefore, only the limitation of the individual component dissipation applies.

This argument is only absolutely correct for an homogeneous substrate which a Metlam PCB is not, but it is valid to say that the limitation provided by the overall PCB dissipation is greatly reduced. Consequently, the power dissipation density that can be accommodated by this cooling mechanism is significantly greater as the results of this investigation have confirmed.

It has already been said that Z plane cooling has been made viable by the fact that surface mounted components provide a 'clean' rear face to the PCB. This fact also opens the possibility of double sided PCB's which would further increase the component density and also the PCB power dissipation.

With respect to forced air cooling, the results have demonstrated that this cooling mechanism has a better performance than would have been expected. As with any cooling mechanism, surface area is an important factor in the equation. The smaller size of surface mounted components is a disadvantage in this respect. Also, when air flows within a passage formed by two PCB's in a shelf, there is a velocity distribution across the passage with the greatest velocity in the centre (where the air impedance is least) and reducing as the surface of the PCB is approached. As well as the surface area, the height of a surface mounted component (i.e., its distance from the surface of the PCB) is less than the dual-in-line equivalent. Consequently, the air velocity local to a surface mounted component will be less than for a dual-in-line component.

For a given PCB configuration, air velocity is the factor which has the greatest influence on the forced convection cooling effeciency. However,the component temperature decreases with increasing air velocity. This rate of decrease in component temperature reduces with increasing air velocity, and therefore, although an improvement in cooling performance can be obtained by providing more cooling air, one is dealing with the 'law of diminishing returns'. Also, acoustic noise is becoming a more critical factor in design, and for a given enclosure there is a direct correlation between air delivery and noise.

In the case of conduction cooling in the X-Y plane, one is bound to compare the performance of surface mounted components and Metlam, with conventional dual-in-line components and conduction plane PCB's. Before making this comparison, it is worth noting that the core material within Metlam (Invar) which effectively acts as the conduction plane, has a thermal conductivity one third that of copper, which is the normal material that we use for our conduction planes. Also, the total core thickness within the Metlam PCB that was used for the measurements was 0.5mm (2 x 0.25mm) compared to the normal 1mm thickness of a conventional conduction plane.

Based on these facts, the conduction cooling performance in the X-Y plane would be expected to yield temperature levels significantly higher than we would expect from conventional conduction cooled PCB's. Direct comparisons are difficult; but the temperature levels are lower than the above considerations would lead us to believe. The reason for this is the close proximity of an LCC to the Metlam substrate which offsets, to some extent, the core effects. This demonstrates the thermal advantage to be gained from leadless chip carriers as opposed to leaded chip carriers.

A greater total core thickness would result in a reduction in temperature levels in the case of conduction cooling in the X-Y plane. These investigations have indicated that an increase in the individual core thickness from 0.25mm to 0.5mm would result in a reduction in component temperature of about 10 degrees centigrade. If the ambient conditions are such that the maximum component junction temperatures are below 100 degrees, then from a thermal point of view this temperature reduction is of no real consequence, since it will have a negligible effect on reliability.

However, if the ambient temperature conditions were such that the operating junction temperature levels were higher than 100 degrees, then this temperature reduction resulting from the increased core thickness would result in a very significant improvement in reliability.

8.2 Analytical Thermal Model

The comparisons between measured and predicted temperatures for each of the three cooling methods show a good correlation. It is impossible to expect exact correlation, since apart from normal, experimental and analytical errors, the component manufacturers definition of component dissipation is open to variation. This is demonstrated by them giving maximum and typical values. For this reason, our aim is to achieve an accuracy of our predicted temperatures within 5 degrees centigrade of the measured values. It will be seen from the results that, in all but a few cases, this has been achieved.

It will be noted that the temperature positions used for comparison were all on the surface of the PCB. Measurements were made of the surface of the LCC's, but this temperature is only significant to the forced air cooling case, since for the conduction cooling cases this surface does not constitute a thermal path. Also, since the lid is made from a very thin material, there will be a significant temperature profile over the lid which makes this aspect of the thermal model of the LCC difficult to assess, at least with the measurements that have been made. It is possible to clarify this part of the model by using 'back emf' techniques of component temperature measurement. A detailed analytical model of an LCC has been produced which we have confidence in, but not until these measurements are made can that model be validated.

It is possible to produce an anlytical thermal model that produces accurate results for one particular case. If, then for example, the cooling air velocity were increased and the accuracy of the predicted temperatures went outside the accuracy limits set, then the thermal model describing the flow of heat within and away from the PCB would not be valid. To this end, measurements and predictions were obtained with variations in both air velocity (for the forced convection cooling case) and component power dissipation.

The thermal model that has resulted from these investigations, has been used to assess the effectiveness of thermal via holes which provide a thermal interconnection between the cores. The findings are that they provide no significant thermal advantage. That is to say that their use results in a 1 to 2 degrees centigrade reduction in temperature. The reason for this is that cross sectional area is a significant factor in conduction as well as the thermal conductivity of the material.

The core has a low thermal resistance in the X-Y plane compared to the epoxy/glassfibre between the cores. This low resistance of the core allows the easy spreading of heat in the X-Y plane, which increases the area of heat transfer between the cores and offsets, to a large extent, the low thermal conductivity of the epoxy/glassfibre. In short, the additional complication and reduction in tracking space resulting from the use of thermal vias is not justified by a thermal benefit.

9. NOMENCLATURE

SYMBOL	QUANTITY	S I UNIT
a and b	dimensions	m
A	area, cross section	m
De	equivalent diameter	m
h	heat transfer coefficient	w/m k
k	thermal conductivity	w/m k
L	length	m
n	exponent	
Nu	Nusselt number	
Pr	Prandtl number	
Re	Reynolds number	
T	temperature	k

10. REFERENCES

1. Finch, D.J., Goodacre, J.B., and Humphrey, A.T., Standard Equipment Pakaging MOD(N) Thermal Management, Marconi Research Laboratories, England 1981.

2. Steinberg, D.S., Cooling Techniques for Electronic Equipment. Wiley-Interscience Publication 1980.

3. Jackson, P., Dual-in-Line Component Surface Temperature Prediction on Convection Cooled PCB's, Internepcon Proceedings, Ferranti Computer Systems Limited (Bracknell) England. pp 29-46, 1982.

4. Cassidy, J., Metlam PEC Climatic and Thermal Test Record Sheets (No: 241) (Internal Document), Ferranti Computer Systems Limited (Bracknell) England 1982.

THERMAL ANALYSIS

The Synthesis of Thermostable Devices and Systems

G. N. DULNEV and A. V. SIGAILOV
Institute of Precision Mechanics and Optics
Leningrad, USSR

ABSTRACT
This paper presents a general approach to the problem of thermal analysis of complicated systems, mathematical models and calculation technique for the temperature fields in devices, methods of choice and optimization of the instrument design parameters. The above-mentioned aspects of this problem are considered in detail for PCB-based electronic boxes. A review of previous studies of solid state laser heads, optic-electronic devices and thermostatic devices is presented.

THE EFFECT OF THERMAL CONDITIONS ON THE DEVICE PERFORMANCE

The objects of the present investigation are microelectronic equipment, lasers, optic-electronic devices as well as complicated complexes integrating instruments and devices of different kinds (further on, "system" is used to designate such complexes). When a system is being designed it is necessary to solve the problem of normal thermal conditions, that is the temperatures of all the components of the system are to be in admissible ranges. A transient multi-dimensional temperature field results from the internal heat sources, external heat fluxes and the ambient medium temperature. The temperature effect in electronic devices results in components of the system failing due to the limited thermal stability of materials. Also, it results in the increase of the failure intensities and the decrease of the noise stability. Their electrical properties change as well.

In optical and optic-electronic devices the refractive indices of optical components do not coincide with their nominal values chosen for the most probable uniform temperature. Besides, the thermal expansion of materials produces changes in the radii and depths of lenses and mirrors, as well as in their spacing. The thermal stresses result in the photoelastic effect, that is the change of the index of refraction.

In lasers the temperature effect causes changes in the characteristics of the radiation generated. These changes are due to both the above-mentioned causes resulting in thermal aberrations and the dependence of the active medium spectral efficiency upon temperature. Sometimes, the temperature rise produces either the degradation of the laser rod or even its destruction.

The first calculations of temperature fields in electronic devices were made in the late 1950s. However, the models used were too idealized and the thermal interdependences of the components were not fully accounted for. Also, the influence of thermal processes upon the functional characteristics (optical, electrical, mechanical and others) of the device was not thoroughly analyzed. All this prevents one from using the models in question for the actual designs and the processes involved. It is generally recognized now that thermal and functional designs of the devices are inseparable and should not be done in sequence. It is often quite ineffective to try to determine the requirements for the thermal conditions of the equipment and to meet them after the device has been fully developed.

This paper presents a general approach to the problem of thermal analysis and synthesis of complicated systems, mathematical models and calculation techniques for the temperature fields in devices, methods of choice and optimization of the instrument design parameters, and software for a computer-aided design in terms of thermal conditions.

THE METHOD OF STEP-BY-STEP SIMULATION

The automated thermal design is based upon mathematical models and methods of thermal analysis at different stages of device design. To describe a general approach to the analyses of temperature fields in a complicated system the latter should be represented as a thermal model (Figure 1). A thermal model is referred to as an idealized object in which all significant processes and geometrical dimensions of the phenomenon investigated have been taken into account, whereas all insignificant parameters have been rejected. In Figure 1a placed in a shell (1) are systems 2-6, which are diagrammatic representations of various device implementations: nonuniform bodies (2), uniform components (4) inside the system (3), systems of shells (5), systems with regular arrangement of structural components (6). There may exist volumetric heat sources

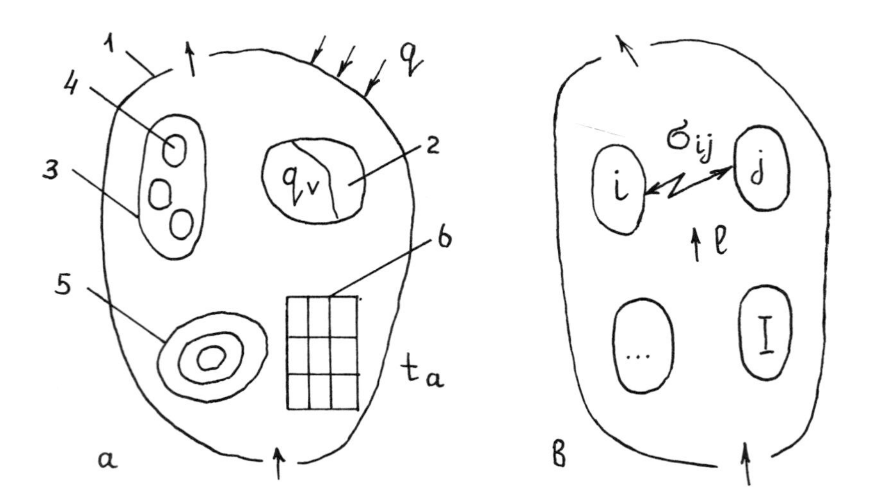

FIGURE 1. The model of a complicated system.

the fluids flowing out of other volumes of fluids. The number of unknown temperatures in the set of equations exceeds the number of equations. Therefore, to complete the system it is necessary to use some complementary equations to connect the derived temperatures and the boundary heat flux by a relationship. Several techniques of completion of a set of equations have been discussed in [3].

A simple technique is to assume that the temperature fields in bodies are uniform:

$$T_{vi} = T_{si}^{1} = \dots = T_{si}^{N} . \qquad (7)$$

In case the heat exchange can be described in terms of the mean surface temperature alone, $T_{si}^{1} = \dots = T_{si}^{N} = T_{si}$, which does not coincide with the mean volume temperature T_{vi}, the following relationship between T_{si} and T_{vi} is used:

$$T_{si} - \tilde{T}_i = \psi_i (T_{vi} - \tilde{T}_i) , \quad \tilde{T}_i = \sum_j \sigma_{ij} T_{sj} / \sum_j \sigma_{ij} , \qquad (8)$$

where ψ_i is the thermal nonuniformity coefficient of the body; $\tilde{T}_i$ is the conventional medium temperature. The nonuniformity coefficient can be estimated by [3]

$$\psi_i = (1 + 0.74)^{-1} , \quad H_i = \sum_j \sigma_{ij} K_i / \lambda_i V_i , \qquad (9)$$

where V_i is the volume; K_i is the configuration coefficient of the body i. The configuration coefficients for bodies of different shapes have been discussed in [1]. Given below are separate formulae describing differently shaped bodies: for a finite cylinder with the length h and radius R and also for a sphere with radius R.

$$K_c = ((\pi/h)^2 + (2.41/R)^2)^{-1} , \quad K_s = (R/\pi)^2 . \qquad (10)$$

For a parallelepiped of sides h_1, h_2, h_3

$$K_p = ((\pi/h_1)^2 + (\pi/h_2)^2 + (\pi/h_3)^2)^{-1} . \qquad (11)$$

To complete a set of equations for fluid flow one can use the assumptions $t_{v\ell} = t''_\ell$ or $t_{v\ell} = (t'_\ell + t''_\ell)/2$; one can also consider particular simulation studies. For example, one-dimensional flow of length h and perimeter u subject to a local heat transfer coefficient α , with the channel walls being heated uniformly, can be described by the relation

$$\psi_\ell = (t_{v\ell} - t'_\ell)/(t''_\ell - t'_\ell) = (1 - \exp(-\frac{\alpha u h}{cG}))^{-1} - \frac{cG}{\alpha u h} . \qquad (12)$$

In case the differences in mean temperatures of separate surface sections T_{si}^{1} , ... , T_{si}^{N} of the body i and to be taken into account, the following technique is used to complete the set of equations. First, an exact or approximate solution T_i (x, y, z) is constructed, which describes the temperature distribution in the body i with the heat transfer across the boundary surface sections to conventional media of temperature $\tilde{T}_i^n$. Subsequently, surface-and-volume averaging operations are carried out. This leads to the equations which connect the heat flux on the boundary surface sections P_i^n and the desired temperatures by a relationship. Generally, such equations take the form

$$P_i^n = \int_{S_i^n} \lambda_i \frac{\partial T_i}{\partial n} dS = f(T_{vi}, T_{si}^{1}, \dots, T_{si}^{N}) . \qquad (13)$$

with specific power q_{vi}, W/m^3, and external heat fluxes q_i, W/m^2. Let T_i denote the temperature of the i-th part of the system, with t_a denoting the temperature of the ambient medium. There may arise convective gas and liquid flows between individual parts of the system. Let t_l denote the temperature field of the l-th part of these flows. Figure 1b presents the same model in a more generalized way and shows thermal interactions between the system's components i and j and the fluid flux l.

The thermal interactions are described quantitatively by the heat conductances σ_{ij}^{nm} and σ_{il}^{n} [1]. The latter are equal to

$$\sigma_{ij}^{nm} = P_{ij}^{nm}/(T_{si}^{n} - T_{sj}^{m}), \quad \sigma_{il}^{n} = P_{il}^{n}/(T_{si}^{n} - t_{vl}), \tag{1}$$

where P_{ij}^{nm} and P_{il}^{n} are heat flux between the bodies i and j and the flow l; T_{si}^{n} and T_{sj}^{m} are mean surface temperatures of the parts n and m of the interacting bodies i and j; t_{vl} is the mean-volume temperature of the flow l.

In general, the thermal processes in individual components of the devices and the fluid flows are described by sets of partial differential equations with various boundary conditions. Thus, heat conduction equations for solid bodies and the equations for fluid flows are given by

$$c_i \rho_i \frac{\partial T_i}{\partial \tau} = \nabla(\lambda_i \nabla T_i) + q_{vi}, \quad i = 1, \ldots, I, \tag{2}$$

$$c_l \rho_l \left(\frac{\partial t_l}{\partial \tau} + \vec{v} \nabla t_l\right) = \nabla(\lambda_l \nabla t_l), \quad l = 1, \ldots, L, \tag{3}$$

where c_i, ρ_i, λ_i, c_l, ρ_l and λ_l are specific heat, density and thermal conductivity of the body i and the fluid l I, L are the number of the bodies and the fluid flows in a system.

For complete description of the system temperature fields in addition to the set of differential equations (2) and (3) one should formulate the boundary conditions between solid bodies and fluid flows as well as initial conditions.

To solve such a complicated mathematical model seems to be practically impossible even if a computer were used. Besides, the description of temperature fields in a complex system is likely to be inexpedient, not only because of the tremendous difficulties involved, but also because of uncertainty of input data (the power of heat sources, the thermal conductivities, the flow rate of the heat transfer agent). The numerical solution of the entire set of differential equations (2) and (3), if possible at all, produces a lot of unnecessary and quite expensive information. To overcome the difficulties mentioned it is necessary to develop approximate methods of the heat transfer analysis in complicated systems. This paper presents a new approach to the temperature field analysis in complicated systems which is suitable for thermal design procedures. This approach has been referred to as the method of step-by-step simulation [2].

According to this method the temperature field calculations are carried out by successive application of various mathematical models at different steps of the analysis. The number of subsystems simultaneously analyzed decreases with steps, while the detail of the temperature field description increases in going from one step to another. First, use is made of a model that describes the thermal conditions of the integrated system with the maximum of detail that is possible. This allows some mean temperatures of bodies or group of bodies to be determined. Also, some mean temperatures of liquid and gas flows can be estimated. For example, it is often necessary to evaluate the mean-volume T_{vi} and mean-surface T_{si} temperatures of the bodies under consideration. In this case the thermal model shown in Figure 1 is replaced by a rougher mathematical model whose description is given below.

From the mean temperatures of bodies and fluids at the step under consideration one can solve a number of problems of the device design, namely, to develop a general cooling system of the device and to determine its parameters, as well as to partly optimize the design. At the next steps of the analyses the integrated system is to be broken up into many individual parts, each of them being subjected to a more detailed thermal analysis. The information about mean temperatures and heat flux obtained at the previous steps of the analysis is used for the description of heat transfer between a particular component and the remainder of the system.

The following procedures are used to develop aggregate approximate models in the step-by-step simulation method:

- simplification of the body configurations without changing their integral characteristics (surface area, volume, perimeters, some characteristic sizes);
- transition from a subsystem with a complicated inner structure including components with various thermal properties to a quasihomogeneous body with effective thermal properties;
- substitution of complicated space distributions of internal and external thermal effects for more simple ones;
- reduction of the dimensions of the sets of equations describing thermal conditions of individual components and subsystems by means of averaging procedures.

The transition from the mathematical model (2), (3) to the model with lumped parameters has been taken as an example of the equation's dimension reduction. The lumped parameter model is meant to be for calculating mean temperatures of bodies and fluid flow. Let us apply the volume-averaging operator to equations (2) and (3):

$$I\left[f(x,y,z)\right] = \frac{1}{V}\int_V f(x,y,z)\,dV = f_V \qquad (4)$$

and use the definition of heat conductances (1). Some manipulation yields the following set of ordinary differential equations [3]:

$$C_i\frac{dT_{vi}}{d\tau} = \sum_{j=1}^{I}\sum_{m=1}^{N_j}\sum_{n=1}^{N_i}\sigma_{ij}^{nm}\left(T_{sj}^{m}-T_{si}^{n}\right)+\sum_{\ell=1}^{L}\sum_{n=1}^{N_i}\sigma_{i\ell}^{n}\left(t_{v\ell}-T_{si}^{n}\right)+P_i\,, \qquad (5)$$

$$C_\ell\frac{dt_{v\ell}}{d\tau} = \sum_{i=1}^{I}\sum_{n=1}^{N_i}\sigma_{i\ell}^{n}\left(T_{si}^{n}-t_{v\ell}\right)+c_\ell G_\ell\left(t'_\ell-t''_\ell\right), \qquad (6)$$

where C_i and C_ℓ - total heat capacities of the body i and the liquid volume ℓ; G_ℓ - mass flow rate; t'_ℓ and t''_ℓ - mean inlet and outlet temperatures of the fluid flow.

The mean temperatures of the bodies T_{vi}, T_{si}^{1}, ..., T_{si}^{N} and of the liquid flows $t_{v\ell}$ and t''_ℓ are unknown in equations (5) and (6). The inlet temperature t'_ℓ may be either assigned in advance if the flow enters the system from the outside, or calculated from the temperatures t''_K of

Thus, a closed system of equations is set up with equations (5) and (6) being included. Sometimes, a polynomial approximation of the temperature distribution of the body is used so that an equation of the form (13) is constructed.

Such a technique has been illustrated in [4]. This paper presents the relations which allow us to solve the problem of completion of the set of equations describing bodies with one-dimensional temperature fields, namely, rods and disks with heat transfer through their sides.

After the aggregate model has been constructed, one should turn to the analysis of temperature fields in the separate components the system is made up of. In so doing, the temperature distributions $T_i(x, y, z)$ and the boundary heat flux density $q_i(x, y, z)$ should be replaced by the values $\langle T_i^n \rangle$, $\langle q_i^n \rangle$ averaged at the boundary section Γ_i^n :

$$\langle f^n \rangle = \frac{1}{S_i^n} \int_{S_i^n} f(x,y,z)|_{\Gamma_i^n} dS, \qquad f = T,\ q. \tag{14}$$

Thus, the temperature distribution analysis of the i-th component of the system is made by using the approximately predetermined boundary conditions of the first, the second or the third form, as given below:

$$\lambda_i \frac{\partial T_i}{\partial n}\Big|_{\Gamma_i} = -\alpha_i^n \left(T_i|_{\Gamma_i^n} - \langle \tilde{T}_i^n \rangle \right) + \langle q_i \rangle, \tag{15}$$

with the mean values $\langle \tilde{T}_i^n \rangle$, $\langle q_i \rangle$ being estimated at the previous stage of the calculation.

The possibility to replace the spatial nonuniform heat actions at the body boundaries with the averaged ones is due to the local effect principle. According to this principle a local thermal disturbance can not extend to include distant regions of the temperature field [1, 2]. That is why it is possible to describe in less detail the heat transfer in the regions far from the ones under consideration. The quantitative estimate of the error caused by the replacement of spatial temperature distribution and the heat flux density with their surface-averaged values is suggested in [5].

Thus, the method of step-by-step modelling means both the enlargement of the initial model, that is the aggregation of the model components for the averaged characteristics to be estimated, and the decomposition, that is, breaking up of a complex system into many individual components or groups of components so that a more detailed analysis should be made possible. There are different methods of model aggregation and a lot of criteria for individual components to be integrated into subsystems. Both the model enlargement and its decomposion should be carried on in accordance with the hierarchy adopted for the design in question.

THE STEP-BY-STEP MODELLING OF THE ELECTRONIC EQUIPMENT THERMAL CONDITIONS

Taken as an example is the above-mentioned general approach applied to the thermal analysis of the electronic equipment whose typical designs are shown in Figure 2. According to the method suggested, peculiar stages of calculation usually correspond to the hierarchy level of the design arrangement:
the first level - ICs, detached semiconductor devices, electric cells;
the second level - PCB-based functional electronic blocks;

the third level – the integrated units of functional electronic blocks in a single chassis, that is, electronic boxes;
the fourth level – formed sheet-metal electronic assemblies (racks, cabinets, etc.);
the fifth level – the collection of electronic assemblies and detached electronic boxes in a stationary room or a payload compartment.

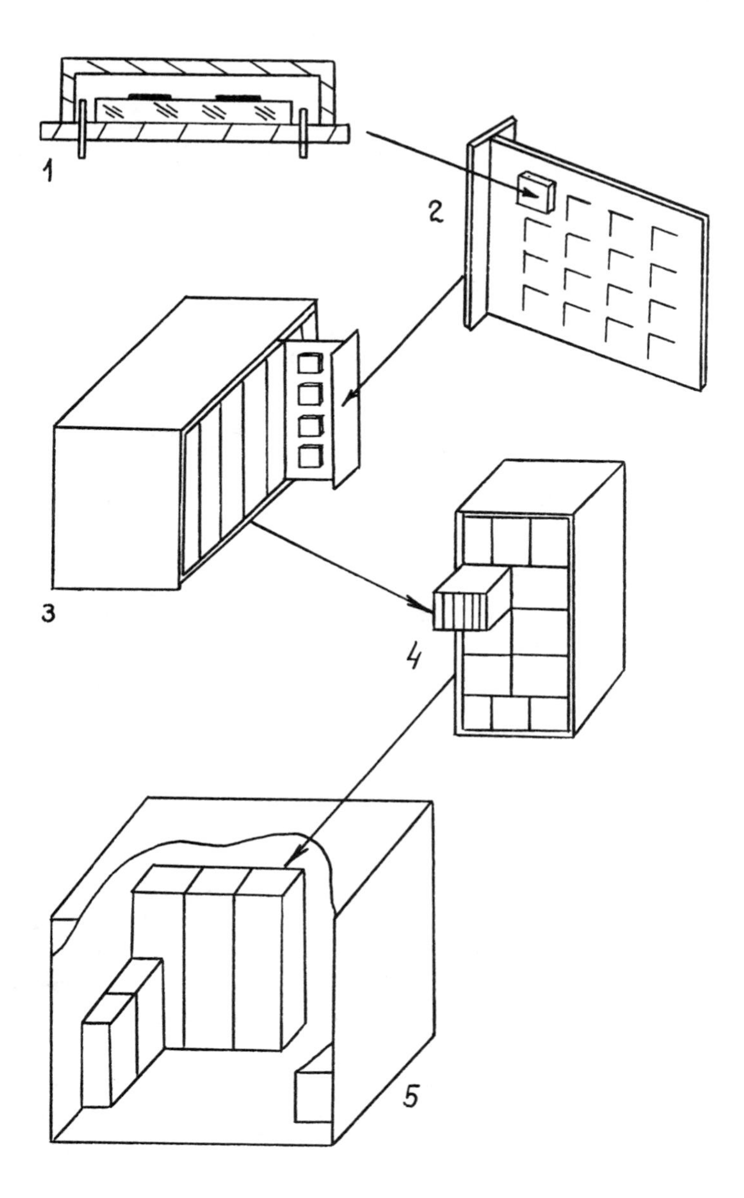

FIGURE 2. Design hierarchy of electronic equipment.

To make thermal calculations of the whole multi-level structure with due regard for the interdependence of modules of different hierarchy levels, it is necessary that the last (fifth) level simulate the first. At this step a number of factors are taken into account. Among them are the aggregate capacity of all the power supplies in devices, the exposure to the heat of the ambient medium and the performance of the general cooling system of the electronic equipment. The calculations are aimed at estimating mean-surface T_{si} and mean-volume T_{vi} temperatures of individual electronic boxes or groups of such boxes. The average temperature t_{vl} of the air bulk chosen as well as the temperatures t'_K and t''_K of the inlet and outlet flow through the analyzed sections of the complex-cooling system are to be estimated, too. To do it use is usually made of a lumped parameter model of equations (5) and (6). In case the mean temperature differences of surfaces of the electronic boxes T^1_{si}, ..., T^N_{si} are to be taken into account, quasihomogeneous bodies with effective thermal properties are substituted for the electronic boxes with regular inner structure [1], with the complementary equations of the form (8) or (13) being written for such bodies in the further calculation. The mean temperatures T_{si} and t_{vl} evaluated at the first stage are used then to estimate temperatures of apparent media $\tilde{T}^n_i$ for the n-th surface sections of the body i (boxes or racks). The multi-dimensional temperature distribution of that body is expected to be investigated in more detail at the next step of the analysis.

At the second step of the calculation the temperature distribution in individual electronic assemblies is analyzed, with the temperature of apparent media near the body surfaces and the inlet temperature of air and liquid flows being known. According to the peculiarities in the design one can utilize either a model with lumped parameters or models where a heterogeneous system is replaced by a set of quasihomogeneous components with effective thermal properties and heat sources distributed in the bulk [2]. The latter models are suitable for the analysis of the regularly structured system and allow calculation of the spatial temperature distribution in quasihomogeneous bodies.

The third stage covers the estimation of mean temperatures (or temperature distribution) of the PCBs in the electronic box. The temperature of apparent media in the neighbourhood of the wall sections of the box is known from the previous step of the analysis. Known also is the temperature of the fluid or air flows. The purpose of this step is to produce information indispensable for detailed analysis of temperature fields of individual PCBs with mounted components which is going to be done at the fourth step of the analysis.

The fourth step consists in obtaining mean temperatures of the electronic component packages (ICs, semiconductor devices), the temperature of the PCB spot under the electronic component, and the air temperature around the component. Given below are sample problems: the analysis of thermal conditions in electronic boxes with different cooling systems and the temperature field estimation for the PCBs with the components (the third and fourth steps of the calculation scheme suggested).

Finally, the fifth step embraces the estimation of multidimensional temperature distribution in ICs and semiconductor devices so that the temperatures of chips can be determined. In this case one should make a calculation of the three-dimensional temperature field in a heterogeneous configuration, i.e., a stack-mounted structure of multilayer parallelepipeds. The numerical methods of finite differences or finite elements are usually used for such calculation. For the quantitative assessment to be given, use is made of the model of a multilayer parallelepiped or a cylinder with local heat sources on the upper end face and heat transfer across the lower one. Such a simplified model is expected to be able to give an exact or approximate analytical solution. Depending on the model worked out when assigning the boundary thermal conditions one can use either package mean temperature or the PCB spot temperature under the component and the temperature of the air around the component package obtained at the fourth step.

THE THERMAL ANALYSIS OF ELECTRONIC BOXES

The basic cooling systems of PCB-based electronic boxes are represented in Figure 3. Different methods of air cooling are shown in diagrams a - d. Diagrams e and f show conductive heat sinks on PCBs which provide heat rejection to the electronic box housing (e) or to the liquid-cooled baseplate or substrate (f).

Let us consider the calculation techniques for the steady-state heat transfer in electronic boxes; these techniques have been used at the

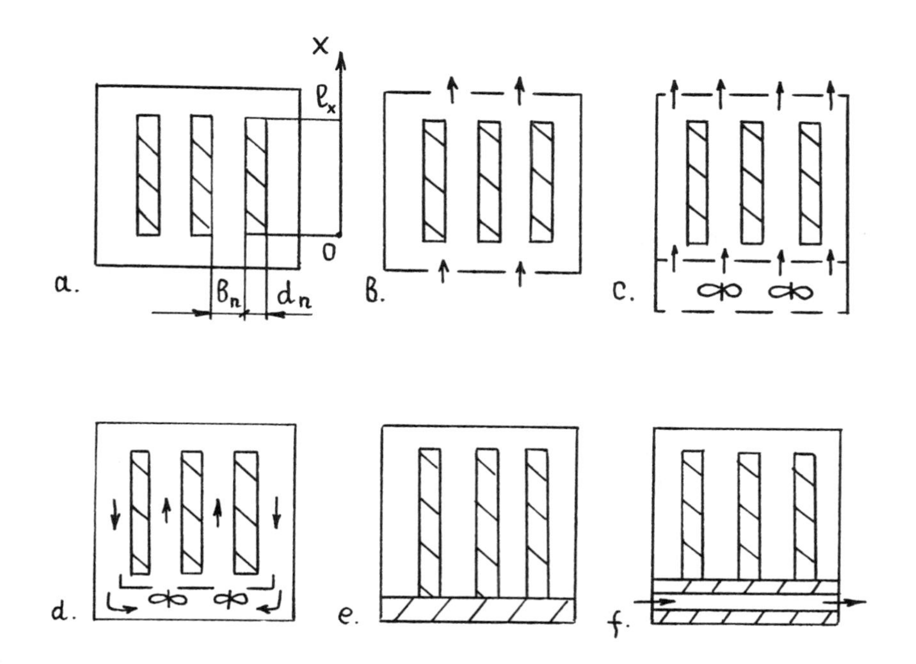

FIGURE 3. Basic schemes of electronic box cooling systems:
a - natural air cooling in a sealed housing; b - natural draft; c - induced draft; d - air inner stirring; e - conductive cooling; f - conductive heat rejection to the baseplate cooled by a liquid.

third step of the simulation scheme described above. First, air cooled electronic boxes are to be discussed. The simplest thermal model to estimate the mean temperature of all the PCBs is based upon the representation of an electronic box as a three-component system consisting of the heated zone (the collection of all the PCBs), the air and the box housing whose mean temperatures T_z, t_{air} and T_h appear in the equations. The mathematical model in question contains a set of algebraic equations of heat balance. In the case of natural convection it may be complemented by an equation which connects the mean temperature ($t_{air} - t_a$) and the air flow rate by a relationship. In [1, 2, 6] analytical relationships and function plotted curves for the mean temperatures (T_z, t_{air} and T_h) estimation are presented, with the cooling methods (a - d) being used.

A more detailed description of heat transfer in an electronic box presupposes the mean temperature estimation for individual PCBs $\bar{T}_n$ (n = 1), ..., N) and for the temperature of air in the channels between the PCBs $\bar{t}_n$. In this case the thermal model of a box is presented as a system of N parallel plates of sides l_x and l_y and of depth d_n. These plates are forming (N + 1) channels b_n wide and enclosed in a rectangular housing. Air cooled boxes (Figure 3, a - d) are believed to have negligible heat conduction between the plates and the housing, there being only convection and radiation heat transfer.

The heat transfer in an electronic box is described by a set of algebraic equations of heat balance for the PCBs, the air in the channels and the housing, written to solve for their mean temperatures $\bar{T}_n$, $\bar{t}_n$ and $\bar{T}_h$. The power P_n dissipated by a PCB is transferred by radiation to the adjacent ones (n + 1), (n - 1) and by convection to the enclosed air space (in channels n and (n+1)):

$$P_n = \sigma^r_{n,n-1}(\bar{T}_n - \bar{T}_{n-1}) + \sigma^r_{n,n+1}(\bar{T}_n - \bar{T}_{n+1}) + \sigma^c_{n,n}(\bar{T}_n - \bar{t}_n) + \sigma^c_{n,n+1}(\bar{T}_n - \bar{t}_{n+1}) + \sigma^r_{n,h}(\bar{T}_n - \bar{T}_h). \quad (16)$$

The thermal flux transferred to the air in the n-th channel from the PCBs heats the air:

$$\sigma^c_{n,n-1}(\bar{T}_{n-1} - \bar{t}_n) + \sigma^c_{n,n}(\bar{T}_n - \bar{t}_n) = 2cG_n(\bar{t}_n - t'_n). \quad (17)$$

The air temperature is presumed to change linearly with the length of the channel, i.e., $\bar{t}_n = (t'_n + t''_n)/2$ and $t''_n - t'_n = 2(\bar{t}_n - t'_n)$.

The total power of all the PCBs is equal to the sum of heat flows carried away by the air from boxes and transferred to the ambient medium from the surface of the box housing:

$$\sum_{n=1}^{N} P_n = 2c\sum_{n=1}^{N+1} G_n(\bar{t}_n - t'_n) + \sigma_{h,a}(\bar{T}_h - t_a). \quad (18)$$

For sealed electronic boxes the first term of the right side of equation (18) is omitted.

In equations (16) - (18) the following symbols are made use of: $\sigma^c_{n,n}$, $\sigma^c_{n,n+1}$ - convective heat conductances from the PCB to the air; $\sigma^r_{n,n\pm1}$ - radiation heat conductance between the PCBs; $\sigma^r_{n,h}$ - radiation heat conductance from the PCB to the housing; $\sigma_{h,a}$ - the heat conductance from the housing to the ambient medium; G_n - mass air flow rate in the n-th channel.

Electronic boxes with natural air cooling are represented by mathematical models which, in addition to the heat balance equations (16) – (18), contain algebraic equations derived from the principle of conservation of momentum for air flow and connecting the speed of moving air in the channels v_n and the temperature differences by a relationship. It should be noted that actual convection heat transfer is much more complicated than the simulative processes which were investigated on the basis of available data on the heat exchange intensities and hydraulic resistances in natural convection. That is why the adopted design relations for the heat transfer coefficient α and the coefficient of the hydraulic resistance ξ are very often the sources of the error of the method.

While calculating the average air speed in the channels of a sealed box (Figure 3a), one can use one of the two models in accordance with the Rayleigh number $Ra^* = (\beta g B^3 (\bar{T} - \bar{t}) Pr / \nu^2) B / \ell_x$. With Ra^* being small for narrow and long channels, the hydrodynamically stable laminar flow may be considered to appear in vertical channels. Then the loss of pressure due to friction in the channels may be expressed by [7]

$$\Delta P_n = \frac{96}{Re_n} \cdot \frac{\ell_x}{2B_n} \cdot \frac{\rho v_n^2}{2} = 12 \mu_n \ell_x v_n / B_n^2 = R_n v_n . \qquad (19)$$

The pressure balance equations are written for each pair of vertical channels:

$$R_{n+1} v_{n+1} - R_n v_n = \beta g \rho (\bar{t}_{n+1} - \bar{t}_n) \ell_x , \qquad (20)$$

where β , ρ and μ are the coefficient of volumetric expansion, the air density and the dynamic viscosity, respectively.

To determine (N + 1) velocities v_n, the system of N equations (20) must be complemented by the equation derived from the principle of mass conservation:

$$\sum_{n=1}^{N+1} \rho v_n B_n = 0 . \qquad (21)$$

It should be noted that correction factors are to be applied to equations (19) and (20) in case reliable data are available on hydraulic resistance increase due to the projections in the channels and due to the changes in inlet and outlet air flow direction.

With Ra^* being large for short and wide channels, the velocity v_n and the convective heat transfer coefficient α_n can be estimated individually for each channel by means of formulae for a single channel [1, 2].

In a perforated electronic box with natural draft (Figure 3B) the air flow rates are determined by the following equation:

$$\xi_n \frac{\rho v_n^2}{2} + (\xi' + \xi'') \frac{\rho v_{no}^2}{2} = \beta \rho g (\bar{t}_n - t_a) L_x \qquad (22)$$

where ξ , ξ' and ξ'' are the coefficients of hydraulic resistance in channels, when flowing into or out of the box; v_{no} is the air flow rate in the perforated holes; L_x is the height of an electronic box.

In electronic boxes with induced draft cooling the air flow rates G_n are usually derived from hydraulic analysis not connected with the thermal analysis, though sometimes the air heating is able to significantly affect the air distribution in channels.

Thus, the thermal analysis of an electronic box in terms of the model in question is reduced to the solution of a set of algebraic equations. The system of equations is nonlinear due to the functional relationship between thermal conductance and temperature and is solved by the method of iteration. During each iteration procedure the air velocities v_n; the thermal conductances, $\sigma^{r}_{n,n\pm1}$ and $\sigma^{c}_{n,n}$; and all temperatures, $\bar{T}_n$, $\bar{t}_n$ and $\bar{T}_h$, are calculated in succession.

Because of air heating in channels the temperatures of the PCBs change significantly with the height of the box (in the direction of x-axis). Therefore, it is not enough to know only mean temperatures $\bar{T}_n$ and $\bar{t}_n$ in order to pass to the next stage of the thermal analysis of the components on PCBs. Suppose that the air temperatures change linearly with the height,

$$t_n(x) = t'_n + 2(\bar{t}_n - t'_n)x/\ell_x , \quad t''_n = t_n(\ell_x), \tag{23}$$

and there is no heat conduction along the PCB int the x-direction. Let us assume that the density of thermal flux going from a PCB to air is constant all along the height of the PCB. Then, taking into account (23), one can formulate the following equation to estimate the PCB temperature $T_n(x)$ at the height x:

$$T_n(x) = \bar{T}_n + [(\bar{t}_n - t'_n)\sigma^{c}_{n,n} + (\bar{t}_{n+1} - t'_{n+1})\sigma^{c}_{n,n+1}] \cdot \tag{24}$$

$$\cdot (2x/\ell_x - 1)/(\sigma^{c}_{n,n} + \sigma^{c}_{n,n+1}).$$

To calculate the temperature distribution in air-cooled electronic boxes, one can use also complicated models with increasing detailing of the temperature field description. For example, [8] has suggested a model in which one-dimensional equations of heat conduction and energy are set up for every n-th PCB and for every n-th channel. These equations describe the temperature distribution $T_n(x)$ and $t_n(x)$ with the height of the box, the non-isothermality of the housing being also taken into account. An effective numerical calculation technique has been developed in [8] for this model to be realized.

[2] has investigated the generalized thermal model of an electronic box in which the heated zone is replaced by the quasihomogeneous anisotropic parallelepiped with effective thermal properties, the heat sources being distributed in the parallelepiped volume. The temperature distribution is such a model is described by the following set of equations:

$$\lambda_x \frac{\partial^2 T}{\partial x^2} + \lambda_y \frac{\partial^2 T}{\partial y^2} + \lambda_z \frac{\partial^2 T}{\partial z^2} + q_v - \alpha_v (T - t) = 0, \tag{25}$$

$$c\rho v_x \frac{\partial t}{\partial x} = \alpha_v (T - t), \tag{26}$$

where λ_x, λ_y and λ_z are effective thermal conductivity of the heated z one; α_v is the volumetric coefficient of the convective heat transfer (W/(m^3 K)) taking into account the air heat exchange inside the heated zone; v_x is the air velocity in the x-direction averaged over the cross section of the heated zone.

The boundary thermal conditions in which account is taken of the heat exchange with the housing are written for the surface of the heated zone. The heat conduction through the housing is calculated from a two-dimensional equation. The numerical calculation technique according to this model has been discussed in [9].

Now, let us proceed to the thermal analysis of electronic boxes with heat sinks (Figure 3e, f). In such structures the PCB temperature field is essentially nonuniform in the x-direction due to the heat transfer by conduction, the x-direction being at right angles to the cooled substrate. Therefore, in a mathematical model being constructed for the PCBs, one-dimensional heat conduction equations are written which take into account heat transfer across the PCB in the x-direction, and heat exchange between adjacent PCBs due to the convection and radiation through air gaps, as well as the heat release by the internal sources:

$$\lambda_{nx} d_n \frac{d^2 T_n}{dx^2} - \alpha_{n,n-1}(T_n - T_{n-1}) - \alpha_{n,n+1}(T_n - T_{n+1}) + P_n / \ell_x \ell_y, \qquad (27)$$

$$n = 1, \ldots, N,$$

where λ_{nx} and d_n are the effective thermal conductivity in the x-direction and the depth of the n-th PCB with heat sinks; $\alpha_{n,n\pm1}$ are the coefficients of heat transfer between adjacent PCBs.

The temperature fields of the housing $\overline{T}_h$ and the baseplate $\overline{T}_0$ are considered to be uniform. The heat flux transferred from the PCB end faces $x=0$ and from the housing to the baseplate, is transferred from the baseplate to the medium with the temperature t_{a1}. The thermal flux transferred from the PCBs to the housing flows into the medium with the temperature t_{a2}. Therefore, the heat balance equations for the baseplate and the housing acquire the form

$$\sum_{n=1}^{N} \sigma_{n,0}(T_n(0) - \overline{T}_0) + \sigma_{h,0}(\overline{T}_h - \overline{T}_0) = \sigma_{0,1}(\overline{T}_0 - t_{a1}),$$

$$\sum_{n=1}^{N} \sigma^r_{n,h}(\overline{T}_n - \overline{T}_h) + \sigma_{h,0}(\overline{T}_0 - T_h) = \sigma_{h,2}(\overline{T}_h - t_{a2}), \qquad (28)$$

where $\sigma_{n,0}$ are contact heat conductances between the end faces of the PCBs and the baseplate; $\sigma_{h,0}$, $\sigma_{0,1}$ and $\sigma_{h,2}$ are the "housing-baseplate", "baseplate-medium 1" and "housing-medium 2" heat conductances.

Boundary conditions for equation (27) when $x=0$ comprise the contact heat exchange with the baseplate, the heat transfer across the free end faces of the PCBs being neglected:

$$-\lambda_{nx} \frac{dT_n}{dx}\Big|_{x=0} + \frac{\sigma_{n,0}}{d_n \ell_y}(T_n(0) - \overline{T}_0) = 0\ ; \quad \frac{dT_n}{dx}\Big|_{x=\ell_x} = 0. \qquad (29)$$

The system of equations (27) - (29) can be solved numerically. If in equation (27) use is made of the mean temperatures $\overline{T}_{n-1}$, $\overline{T}_{n+1}$, instead of the local temperatures of the PCBs $T_{n-1}(x)$ and $T_{n+1}(x)$, the calculation can be reduced to the system of (3N + 2) algebraic equations solvable for temperatures $\overline{T}_h$, $\overline{T}_0$, $\overline{T}_n$, $T_n(0)$ and $T_n(\ell_x)$ where n = 1, ... , N. Subsequently, the temperature distribution $T_n(x)$ is to be determined by the exact solution of equation (27). Such a technique has been described in [10]. Another method of thermal analysis of electronic boxes with heat sinks consists of applying a quasihomogeneous body model (equation (25), $\alpha_V = 0$) in which heat exchange with the cooled baseplate is taken into account by assigning a corresponding boundary condition on the surface of the parallelepiped, $x=0$.

THE THERMAL ANALYSIS OF PCBs WITH MOUNTED COMPONENTS

The electronic box models discussed above enable us to estimate the mean temperatures $\overline{T}_n$ and $\overline{t}_n$ or temperature distributions $T_n(x)$ and $t_n(x)$ of the PCBs and air on assumption that the power P_n is distributed uniformly across the PCB. To further detail the thermal analysis of electronic equipment means to determine the package temperatures of individual components on PCBs taking into account the perculiarities in their mounting and actual power distribution among the electronic components. At this stage use is made of the model of a PCB on which J electronic components are mounted and dissipate P_j ($j = 1, \dots, J_1$ for the components on one side of the PCB and $j = J_1 + 1, \dots, J$ for the components on the other side). A thermal model of a PCB on it consists of a rectangular plate of sides ℓ_x and ℓ_y and depth δ with effective thermal conductivities λ_x and λ_y ; mounted on both sides of the plate are $J = J_1 + J_2$ electronic components with uniform thermal fields T_j and power P_j (Figure 4).

A heat balance equation is written for each of the electronic components. In accordance with this equation the power P_j is dissipated to the adjacent plate with mean temperature $T_{p\ell,k}$ by radiation and to the air with temperature $t_k(x_j)$ by convective heat transfer and enters the plate just on the place where the electronic component is mounted (heat flux $P_{p\ell,j}$):

$$P_j = \sigma_j^r (T_j - T_{p\ell,k}) + \sigma_j^c (T_j - t_k(x_j)) + P_{p\ell,j} , \qquad (30)$$

where k is the side of the plate index; $k = 1$ for the components if $j \leq J_1$ and $k = 2$ if $j > J_1$; σ_j^r and σ_j^c are radiant and convective heat conductances from the component package.

The heat flux $P_{p\ell,j}$, flowing from the component into the plate, is equal to

$$P_{p\ell,j} = (T_j - \overline{T}_{p\ell,j}) / R_j \qquad (31)$$

where $\overline{T}_{p\ell,j}$ is the mean temperature of the surface section of the plate ($\Delta x_j \, \Delta y_j$) under the j-th component; R_j is the thermal resistance "component-plate".

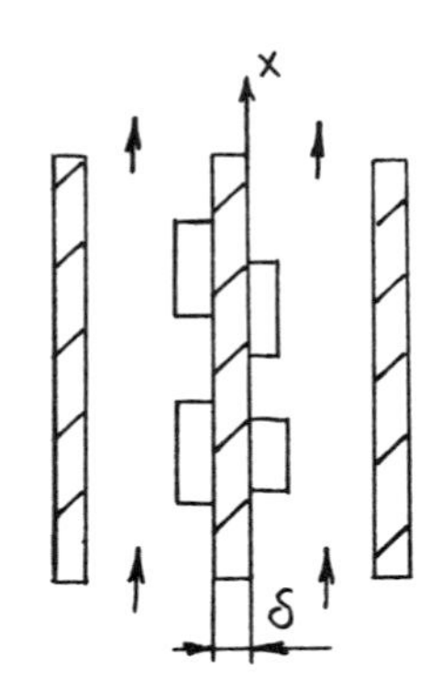

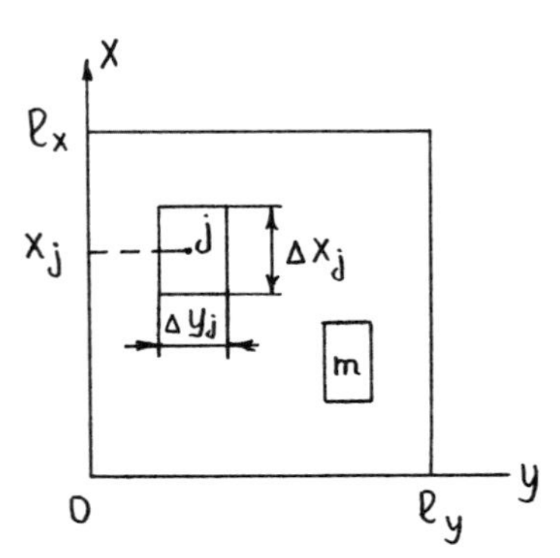

FIGURE 4. A model of PCB with mounted components.

The temperature field of the PCB is considered to be two-dimensional and described by the equation for the plate with local heat sources [2].

$$\frac{\partial}{\partial x}\left(\lambda_x \frac{\partial T_{pl}}{\partial x}\right) + \frac{\partial}{\partial y}\left(\lambda_y \frac{\partial T_{pl}}{\partial y}\right) - \frac{\tilde{\alpha}}{\delta}\left(T_{pl} - \tilde{T}\right) + \frac{q(x,y)}{\delta} = 0. \qquad (32)$$

In equation (32) the radiant and convective heat transfer across both sides of the plate is taken into account by introducing the apparent ambient temperature $\tilde{T}$ and the heat transfer coefficient to the apparent medium $\tilde{\alpha}$ which are described by the formulae

$$\tilde{\alpha} = \tilde{\alpha}_1 + \tilde{\alpha}_2, \qquad \tilde{\alpha}_k = \alpha_k^r + \alpha_k^c, \qquad k = 1,2$$

$$\tilde{T}(x) = \sum_{k=1,2} \left(\alpha_k^r T_{pl,k} + \alpha_k^c t_k(x)\right) / \tilde{\alpha},$$

where α^r_k, α^c_k are radiant and convective heat transfer coefficients from the side k of the plate. For that section of the plate surface $(\Delta x_j \Delta y_j)$ where the component is mounted, the heat transfer coefficient $\tilde{\alpha}_k$ from the corresponding side is considered to equal to zero.

The density distribution of the thermal flux is given by the function $q(x,y)$ in equation (32) and has the form

$$q(x,y) = \sum_{j=1}^{J} q_j(x,y), \qquad (33)$$

$$q_j(x,y) = \begin{cases} P_{pl,j}/(\Delta x_j \Delta y_j) & \text{in the projection zone of the component } j; \\ 0, & \text{outside the projection zone of the component } j; \end{cases}$$

According to the principle of superposition, the temperatures $T_{pl,j}$ of the surface sections under the electronic components can be written by the expression [2].

$$T_{pl,j} = \tilde{T}(x_j) + \sum_{m=1}^{J} F_{mj} P_{pl,j}, \qquad (34)$$

where $\tilde{T}(x_j)$ is the background temperature produced by the adjacent PCBs and the air in the channels; F_{mj} is the thermal resistance from the spot $\Delta x_m \Delta y_m$ of the source m to the spot $\Delta x_j \Delta y_j$.

The thermal resistances F_{mj} ($m, j = 1, \ldots, J$) are derived from equation (32). In so doing to estimate F_{mj}, $\tilde{T}$ and $P_{pl,j}$ ($j = 1, \ldots, J$; $j \neq m$) are assumed to be zero; from this the temperature $T_{pl,j}$ produced by the single source m is determined. Then, thermal resistances are derived from the formula $F_{mj} = T_{pl,j} / P_{pl,m}$.

In case the plate is homogeneous and $\tilde{\alpha}$ = const., one can use an approximate analytical solution given in [2] to calculate the temperature field. Otherwise, the estimation of the temperature field $T_{pl}(x,y)$ is to be done by numerical methods.

Thus, to calculate the temperatures of the conponent packages involves solving the set 3J of algebraic equations (30), (31) and (34) for the unknowns T_j, $T_{pl,j}$ and $P_{pl,j}$. Using determined mean temperatures of the packages T_j or temperatures $T_{pl,j}$ of the PCB and the air $t_k(x_j)$ around the component, one can proceed to the next step of the analysis, that is, the calculation of the temperature fields inside the components and the determination of the temperatures of the junctions.

The problems of thermal design are also solved in a step-by-step manner in accordance with the hierarchy levels of design. At every step of the design the device and cooling system parameters are specified, proceeding from the requirements to the representative temperatures of the modules of the level in question. For example, in developing the design and the cooling system of racks, the initial data are the requirements for the admissible mean temperatures of electronic boxes the heat-transfer medium (or agent) when flowing into the box. These admissible temperatures, in turn, are determined in analyzing the thermal conditions of electronic boxes and PCBs in terms of the requirements for the admissible temperatures of the electronis component packages. The admissible temperatures of the packages are determined by analyzing the temperature fields of the ICs.

Each stage of thermal design involves the solution to the following main problems:
- the choice of basic design and of cooling system scheme;
- the choice of the design and of cooling system parameters;
- the calculation of the temperature fields at the specified level and the check for meeting the requirements for the admissible temperatures;
- the optimization of the design and cooling system parameters.

At the initial stage of the design one should have a set of design procedures which requires only limited information on the power dissipated, the admissible temperatures and field conditions, as well as the size limitations for the basic designs and their principal parameters to be chosen. For example, the thermal design of electronic boxes leads to the following problems: the choice of the cooling system basic scheme and design principal parameters, there being no information on the power distribution among the PCBs and their arrangement in the box. The initial data for solution to this problem are the total power P dissipated in the electronic box, the maximum temperature of the medium t_a and the electronic block sizes ℓ_x and ℓ_y. In addition, some restrictions are imposed, for instance, on the allowable electronic box sizes, L_x, L_y, L_z, or limitations on the air flow rate in an induced draft, or requirements for the leak-proofness of the housing. [2] and [6] present probability relationships which connect the overheating of heated zone of the electronic boxes $(T_z - t_a)$ and heat flux per unit of the heated zone external surface P/S_z. The analysis of the electronic box models described above have resulted in more detailed relationships which connect the power P, the maximum of the mean overheating of the PCBs $\vartheta_{max} = \max(\bar{t} - t_a)$ and the basic design parameters.

Let the approach to the construction of such relations be considered. It is assumed that there are similar PCBs of width d and distance b from one another in an electronic box, all the PCBs dissipating equal amounts of energy $P_n = P/N$. Let the maximum of mean temperatures of the PCBs T_{max} be dependent upon M design parameters and operating conditions a_m: $T_{max} = f(a_1, \dots, a_M)$. Out of the entire set of parameters a_m, a number of principal ones a_m $(m = 1, \dots, M_0)$ is to be chosen, the ones whose values most substantially affect the thermal conditions of the electronic box. For all the rest (minor) parameters a_m $(m = M_0 + 1, \dots, M)$ the ranges of their probable changes are to be specified $/ a_m^{min}, a_m^{max} /$. With each chosen set of values of principal parameters, multiple calculations of T_{max} are made for L random sets of minor parameters. The set of random values obtained $\{\bar{T}^{max}_{l}\}_{l=1}^{L}$ is statistically processed, and the expected $E\{T^{max}\}$ and variance $\mathcal{D}\{T^{max}\}$ estimates are evaluated. These estimates determine the most probable thermal conditions with the specified principal parameters and possible deviation of temperature due to the differences in the minor parameter values.

The analysis showed that to generalize the results it is useful to introduce the parameter θ equal to the ratio of the central PCB overheat to the power per unit of area,

$$\theta = (\bar{T}_{max} - t_a) l_x l_y / P_n .$$

Let the dependence of θ upon principal parameters be presented for sealed and perforated electronic boxes with natural air cooling and for electronic boxes with induced draft (Figure 3a, b, c). B and l_x are chosen as principal parameters for the electronic boxes of type a; B and l_x and the perforation coefficient K (the ratio of the area of all the holes to the bottom and lid area of the housing) for the boxes of type b; B and G for the boxes of type c. The other parameters are changed in the following ranges: $l_y = (0.15 - 0.25)$ m, emissivity $\varepsilon = (0.7 - 0.9)$, the medium temperature $t_a = (10-60)$ °C, the power $P_n = (1-10)$ W (for natural cooling) and $P_n = (5-30)$ W (for induced draft). The results of the calculations are presented in Figure 5.

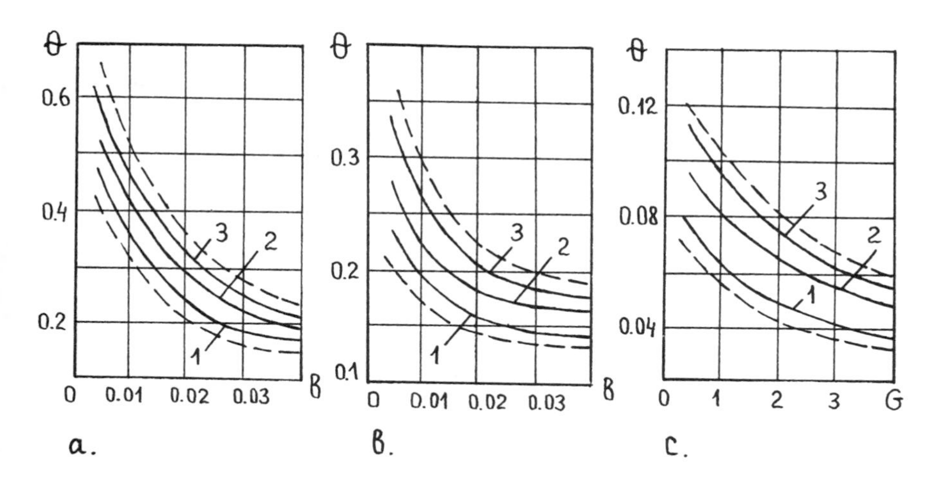

FIGURE 5. Relative temperature θ as a function of fundamental parameters of basic cooling schemes of electronic boxes:
a – box in a sealed housing; b – box with natural draft when a perforation coefficient is 0.2; c – box with induced draft; B – the distance between the PCBs, m; G – air flow rate, 10^{-4} kg/s; curves 1, 2, 3 in Figures 5a, b correspond to the height l_x = 0.1, 0.2, 0.3m; curves 1, 2 and 3 in Figure 5c correspond to the channel width B = 0.01 , 0.02, 0.03 m.

Full curves correspond to the expected estimate of θ, and dotted ones are plotted at the distance equal to the standard deviation from $E\{\theta\}$ of two marginal curves.

Similar dependences have been plotted for electronic boxes with conductive heat sinks in [10]. One can make use of such dependences to determine the applicability of one or another basic scheme of the electronic box cooling system and to choose its principal parameters.

The basic design having been chosen, the solution to the problems of module packaging and corrections of design parameters and operating conditions is to be made in the interactive mode with a computer by combining the nonformal exhaustive search and analysis of variants with formal optimization procedures. The software for this stage of design comprises packets of packaged programs which realize the calculation techniques of thermal conditions at different levels of device design. The stored information includes the basic data on the properties of materials, the parameters of standard design components, the characteristics of "thermal" hardware (radiators, fans, heat exchangers, heat pipes, etc.).

In solving some particular problems the application of formal optimization procedures may be effective. The following two problems may be presented as an example: the arrangement of the components in a device (microcircuits on a PCB, PCBs in an electronic box) and the optimization of the air flow rate distribution in systems with induced draft.

The problem on the arrangement of components in terms of the requirements for the thermal conditions is stated as follows. There are N electronic components dissipating various powers P_n ($n = 1, \ldots, N$). These components can be arranged in M fixed positions ($M \geq N$). It is required that such a variant of arrangement should be found that could allow the minimization of the efficiency function chosen φ (of the optimization criterion): $\varphi \to min$. The following functions, for instance, can be chosen as a criterion φ:

- the mean value of the temperature of the components

$$\varphi_1 = \overline{T} = \sum_{n=1}^{N} T_n / N \; ; \tag{35}$$

- the root-mean-square value of the deviation of temperatures from their mean value $\overline{T}$

$$\varphi_2 = \left(\sum_{n=1}^{N} (T_n - \overline{T})^2 / N \right)^{1/2} \; ; \tag{36}$$

- the maximum temperature of the components

$$\varphi_3 = \max_n T_n \tag{37}$$

The solution to the problem of arrangement by exhaustion of all the variants, with the temperature calculations being made for each of them, is not thought to be possible, as the number of arrangement variants is proportional to (N!). Therefore, a problem arises of developing the economically efficient arrangement algorithms. [11] has suggested such algorithms for the criteria φ_1 and φ_2. [11] has shown that for criterion φ_1 on the basis of the obtained values of the thermal resistance between the position j and i F_{ji} ($j, i = 1, \ldots, M$), one can calculate numbers of the positions i into which the electronic components with specified heat output P_n should be installed. In the case of the criterion φ_2 in [11] an algorithm has been suggested based upon the sequential improvement of arrangement by the pair rearrangement of components.

In arranging components in an electronic device it is necessary that account should be taken of the switching and thermal requirements simultaneously. Very often switching requirements dictate minimizing the total coupling length (electrical connections) between the electronic components:

$$\mathcal{D} = \sum_n \sum_K \ell_{nK} \, d_{i(n)j(K)} \to \min , \qquad (38)$$

where ℓ_{nK} is the number of connections between the components n and k; $d_{i(n)j(K)}$ is the coupling length between the positions i and j into which the components n and k are placed.

Various arrangement algorithms can be suggested which take account of both the thermal and switching requirements. For example, it is possible to introduce a criterion f, which is the combination of "thermal" and "switching" criteria $f = \alpha \varphi + \beta \mathcal{D}$, where α and β are certain weighting coefficients.

However, experience has shown that it is expedient to solve the arrangement problem in the following way. First, the arrangement is performed on account of switching requirements (the criterion $\mathcal{D}$ from (38)) using the algorithms, for instance, suggested in [12]. This results in various arrangements, with the optimal cooling length summed up as $\mathcal{D}_{opt}$. Then, the groups of components are assigned inside for which it is allowable to execute rearrangements on the basis of the thermal requirements. For these groups of components the arrangements are being improved with regard to the criteria φ_1, φ_2 and φ_3 and the analysis is made of the increase of the summed coupling length as compared with the optimal value ($\Delta\mathcal{D} = \mathcal{D} - \mathcal{D}_{opt}$). If the number of arrangement positions M exceeds the number of the component arrangements N, the effective technique for the improvement of the thermal conditions at an arrangement is to increase the distance between components of large power on account of "omission" of "empty" positions between them.

The determination of the optimal flow rate distribution in electronic equipment with induced draft is an example of another problem when it is useful to apply the optimization procedure to its solution. Let an electronic box with induced draft be considered (Figure 3). It is requires that the air mass fluxes G_n (n = 1, ... , N + 1) in the channels between the PCBs should be estimated for which the summed mass flux $G_\Sigma = \sum G_n$ is minimal, the temperatures of packages of all the microcircuits T_i^n being within the limits of their admissible values $T_{i,0}$.

The problem of optimization is formulated as follows. The effectiveness function is the sum of the modules of temperature deviations of the most heated components from their admissible values

$$\varphi = \sum_n \left| T_i^{(n)} - T_{i,0}^{(n)} \right| , \qquad (39)$$

where $T_i^{(n)}$ is the temperature of the component i in the channel n. In summing up in (39) for each channel n only one the most "critical" component is assigned for which the difference $\Delta T_i^{(n)} = T_i^{(n)} - T_{i,0}^{(n)}$ is maximum.

The mass fluxes in the channels G_n are estimated with the proviso that the effectiveness function is a minimum $\varphi = \varphi(G_1, \ldots, G_{N+1}) \to \min$. The calculation of the electronic component temperatures $T_i^{(n)}$ is made according to the technique described above, the effectiveness function φ being minimized by some gradient method.

THE DESIGNING OF THERMOSTABLE SOLID STATE LASER HEADS AND OPTIC-ELECTRONIC DEVICES

In optic-electronic equipment the limitations on temperature gradients are a consequence of the requirements on the allowable functional characteristic changes due to the temperature disturbances. That is why in optic-electronic devices it is necessary to jointly consider the thermophysical, thermomechanical and thermooptical processes. The relationship which connects the thermal conditions of an optic-electronic device and its functional characteristics leads to the fact that the special assignments of principal scheme, type and parameters of the device design is one of the most important ways to assure its thermostability. Let us investigate the peculiarities of the solution to the problem in the provision of optic-electronic device thermal conditions with solid state laser heads being taken as an example.

The solid state laser consists of the laser head and various optical components of the resonator-mirrors, prisms and lenses, as well as the components for controlling the radiation, namely, an electrooptical shutter, a nonlinear component, etc. The laser head includes a laser rod, an optical pumping source, a reflector and components for spectral filtration of the radiation. The laser heads may differ from one another in the number of pumping lamps, the type and shape of the reflector, the form of the laser rod, the types of the components for pumping light filtration, the applied technique for heat removal from the parts of the laser head. Examples of laser head designs are presented in Figure 6.

The characteristics of the laser radiation change under the influence of the laser head thermal conditions as a result of two main factors, one of which consists in thermal aberrations arising in the laser rod, the other is the temperature changes of the active medium spectroscopic parameters [12]. Thermal aberrations at nonuniform temperature fields are caused by

- the refractive index changes due to temperature;
- the influence of the photoelastic effect resulting in optical index disturbance due to the thermal stress;
- the laser rod deformation due to its nonuniform thermal expansion.

The temperature changes of the active medium spectroscopic parameters are mainly due to

- the temperature dependence of level lifetime and level population;
- the increase of spectral bandwidth and the corresponding decrease of stimulated transition cross section with the temperature rise.

In specifying the optimal laser head design it is useful to apply the methods of mathematical modelling. This makes it possible to investigate a number of different modifications of laser head designs in a short time, evaluating the joint influence of the processes progressing in it such as optical, thermophysical and thermomechanical ones upon the characteristics of laser radiation.

Let us consider the basic principles and stages of design. At present due to the uncertainty of the initial information and approximate character of the models used there are no calculation procedure enabling the absolute values of the oscillated radiation characteristics to be validly determined. Therefore, it is expedient to combine mathematical modelling with a small number of experiments. To do so, in calculation design a comparison is made of the oscillated radiation characteristics (output power, divergence, intensity distribution in the beam cross section, etc.) obtainable at this or that probable modification of the laser head design with similar radiation characteristics estimated from the experimental non-optimized breadboard model. The experimental model is to display some special features of the laser design and its performance that remain unchanged over the calculation: the type of active medium, the type of pumping lamp, and the duration and shape of pumping pulse. Such an approach to the design makes it possible to experimentally determine on the breadboard model a number of values indispensable for obtaining comparative numerical estimates. It also makes it possible to reduce the influence of the uncertainty in the initial data upon the result of the analysis.

There are two stages in the design. The first stage includes the choice of the basic laser head design. The choice consists in determining the number of pumping lamps, the reflector type and shape (monoblock, hollow, etc.), the type of the reflecting coatings (diffusively reflecting or mirror), the laser rod shape, the type of components for head removal. At the second stage of design the parametric optimization of the chosen basic design is performed. This results in determining the activator concentration, the sizes of the laser head parts and parameters of their mutual arrangement, as well as the parameters of thermal contact between the parts of the laser head. In choosing the basic design and parametric optimization account must be taken of the changes in the efficiency of the pumping system and the influence of thermal conditions upon the laser radiation characteristics. It is necessary to take into consideration the quite common tendency: the design modifications minimizing the influence of thermal conditions upon the radiation characteristics (for example, the increased number of pumping lamps, the sapphire pipe around the laser rod and so on) lead, as a rule, to the reduction of the pumping system efficiency.

Investigation into the laser head design quality is based upon the calculated results of distribution of heat sources and pumping light energy absorbed by the laser rod and the sequential calculation of the laser head thermal conditions. The heat source distribution is calculated by radiant heat transfer simulation by means of the Monte Carlo method using the proper system of computer programs [14].

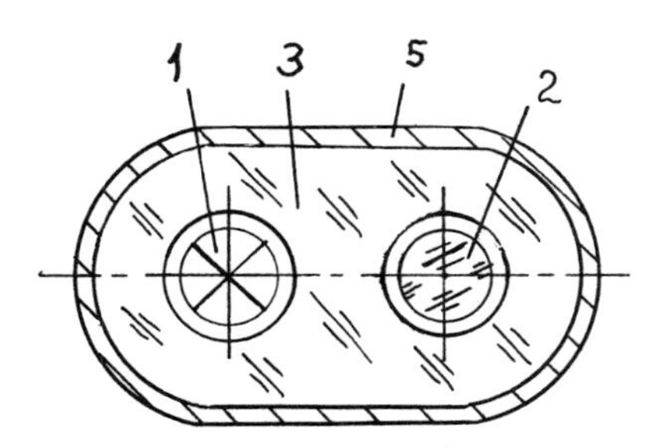

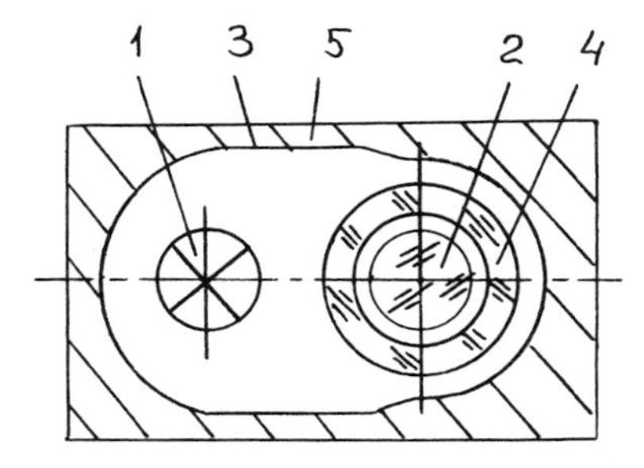

FIGURE 6. The schemes of laser heads:
1 - pumping lamp; 2 - laser rod; 3 - reflector; 4 - sapphire pipe; 5 - housing.

The mathematical model of laser head thermal conditions is a set of nonlinear multidimensional nonsteady equations (2) for the laser head parts interconnected through boundary conditions which represent the radiant, convective and contact heat transfer between the laser head parts [15]. To solve this set of equations use has been made of explicit and mixed finite-difference schemes. This or that scheme is to be used in accordance with the design peculiarities of a particular laser head [16]. Applying explicit and mixed schemes made it possible to realize the module principle in the program construction for the laser head modification being analyzed. According to this principle, these programs are constructed from ready-made subroutines.

Let the calculated results of the nonsteady-state temperature field of a one-lamp laser head (Figure 6a) be considered as an example of thermal conditions simulation carried out by means of the developed complex of programs. The three-dimensional temperature fields were analyzed when thermal calculations were being made for the laser rod, the reflector and the pumping lamp. Figure 10 presents the temperature distribution in the mid-section of the laser head.

As the main characteristics of the oscillated radiation is its output energy, the criteria, according to which a comparison is made of the quality of the laser head design modifications under consideration, normally contain the parameter

$$\eta = E(\tau)/E_0 ,$$

where $E(\tau)$ is the output energy time dependence calculated for the design variant being analyzed; E_0 is the experimental value of the output energy obtained by the breadboard model when operating in a single pulse mode.

The estimate for "relative" output energy η can be obtained on the basis of known approximate relations for the laser output energy [17] and temperature dependences of the active medium spectroscopic characteristics. Certain examples of relative analysis of different basic designs of laser head for η, running time and pulse recurrence frequency have

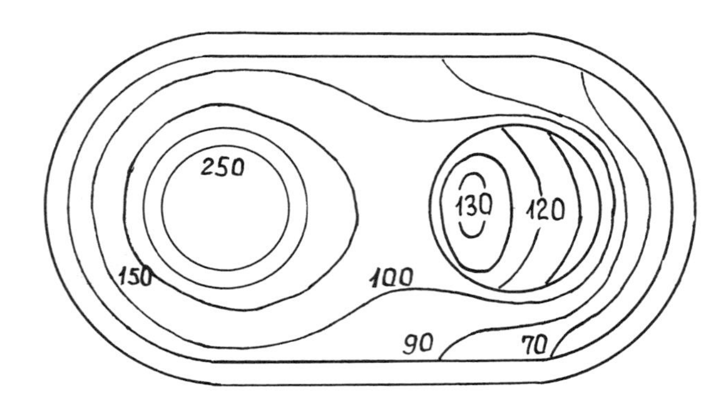

FIGURE 7. Calculation of the temperature field of a one-lamp laser head.

been given in [18, 19]. There often occurs a situation when an output energy value E_0 large enough has been obtained over the breadboard model, though the principal requirements to the radiation characteristics are connected with the beam divergence, the intensity distribution in the laser beam cross section, etc. It is useful here to choose a design producing minimal thermal aberrations with the constraints imposed on the minimal value of the relative output energy η .

The technique described above has been used for designing laser heads for a variety of solid-state lasers with various active media. In so doing not only the problems of parametric optimization of preset basic laser head designs were being solved, but also the complete design cycle was being performed with the choice of basic design and its succeeding optimization. In the former case it was usually possible to increase the output laser energy by 30–50% in comparison with the value obtained from the initial laser head variant that has been realized in an experimental breadboard model. In the latter case even more important results were achieved. To illustrate, the designing results of laser heads for the solid-state laser used in the laser mass-spectrometer of the international project "Phoebus" can be presented. Six modifications of one-lamp and two-lamp laser heads have been compared for the basic designs to be chosen. In designing a number of aspects were to be analyzed, such as the light-trapping efficiency of various laser heads, the decrease of the amplification factor due to laser rod heating, as well as thermal aberrations arising in the laser rod.

In addition to the analysis of thermal processes in laser heads and the synthesis of thermostable constructions of these devices, great attention has been lately paid to the analysis of heat transfer in optical and optic-electronic systems. For example, in 1980s the methods of joint analysis of thermal and optical processes in a system of devices designed for the observation of celestial bodies were developed. These methods were realized in the international projects "Vega" and "Phoebus" [20, 21]. The former project connected with the Halley comet study was implemented in 1986; the latter one connecting the exploration of the Mars satellite Phoebus is expected to be realized in 1988-1989.

In designing thermostable systems, a peculiar role is played by the thermostat facilities designed for producing stable thermal conditions in electronic, optical and other devices. The synthesis of thermostat facilities and the analysis of their thermal conditions are based upon the general approach discussed above in connection with electronic equipment and laser heads. The basic schemes of thermostat facilities for devices are presented in [22, 23]. Consideration is also taken of the methods of choice of such schemes in accordance with the request for a proposal. Mathematical models used for step-by-step modelling of thermostat thermal conditions are described in [24]. Application of lumped-parameter models enabling the mean temperatures of the thermostat parts to be calculated for the analysis of static and dynamic thermostat errors and the choice of the thermostat parameters are considered in detail in [25, 26]. One-dimensional models which make it possible to analyze spatial temperature distributions in parts of thermostats are suggested in [26]. [27] suggested the technique for the choice of heated thermostat parameters at which the thermostat error is minimum. The application of gas-controlled heat pipes is a very promising trend in creation of thermostat systems of electronic devices and components of laser systems. [28] is devoted to these problems.

CONCLUSION

Lately, the theory of heat transfer has found a new application connected with the analysis of temperature fields in various electronic devices. Peculiarities of such objects result in systematic approaches being indispensable for calculating temperature fields of a certain body, as the latter is thermally connected with a variety of other bodies. All this results in the joint analysis of a considerable number of equations describing heat transfer in a system of bodies. However, the heat transfer theory applications being discussed to the electronic devices cover only part of the problem. Now it becomes necessary to considerably widen the number of subjects by including into their scope optical and optic-electronic devices, lasers, mechanical devices, as well as instrument systems made up of the objects mentioned above. The final purpose in this case must be developing thermostable devices. This causes both heat transfer and optical, electrical and mechanical phenomena to be jointly analyzed. In conclusion, a new discipline must be developed. Let the initiative be taken by thermal physics engineers.

NOMENCLATURE

Symbol	Quantity	SI Unit
C	heat capacity	J/K
c	specific heat capacity	J/(K kg)
G	mass flow rate	kg/s
ℓ	length	m
P	power, heat flow rate	W
q	heat flux	W/m^2
q_v	volumetric heat flux	W/m^3
S	area	m^2
T	temperature of the solid body	K
t	temperature of the fluid	K
V	volume	m^3
v	velocity	m/s
x, y, z	Cartesian coordinates	
Greek Letters		
α	heat transfer coefficient	$W/(m^2\ K)$
β	volumetric expansion coefficient	K^{-1}
δ	thickness	m
μ	dynamic viscosity	kg/(s m
λ	thermal conductivity	W/(m K)
ν	kinematic viscosity	m^2/s
ρ	mass density	kg/m^3

Symbol	Quantity	SI UNIT
σ	thermal conductance	W/K
τ	time	s

REFERENCES

1. Dulnev, G.N., Tarnovsky, N.N., Thermal Conditions of Electronic Equipment, Energia, Leningrad, 1971.

2. Dulnev, G.N., Heat- and Mass Transfer in Radioelectronic Equipment, Vysshaya shcola, Moscow, 1984.

3. Dulnev, G.N., Sigalov, A.V., Step-by-step Modelling of a Thermal Regime of Complex Systems, J. Engng Phys., vol. 45, no. 4, pp.651-656, 1983.

4. Polshchikov, B.V., Ushakovskaya, E.D., Substitution Schemes of One-Dimensional Regions with Distributed Heat Sources, Izvestia Vuzov SSSR, Priborostroenije, vol. 24, no. 11, pp.92-96, 1981.

5. Dulnev, G.N., Sakhova, E.V., Sigalov, A.V., Local Effect Principle in the Step-by-Step Modelling Method, J. Engng Phys., vol. 45, no. 6, pp.1002-1008, 1983.

6. Rotkop, L.L., Spokoiny, Yu.E., Guaranteeing of Thermal Conditions in Designing of Radio-Electronic Equipment, Sov. Radio, Moscow, 1976.

7. Idelchik, I.E., Handbook on Hydraulic Resistances, Mashinostroenije, Moscow, 1975. Hemisphere, Washington, D.C., 1986.

8. Stoyan, Yu.G., Putyatin, V.P., Elkin, B.S., Optimization of Electronic Boxes in Terms of Thermal Dynamics and Arrangement Characteristics, Institut Problem Mashinostroenija AN SSSR, Kharkov, Paper no. 183, 1983.

9. Butko, E.F., Dulnev, G.N., Parfenov, V.G., Numerical Realization of General Thermal Model of Electronic Boxes, J. Engng Phys., vol. 40, no. 5, pp.876-882, 1981.

10. Dulnev, G.N., Sergeev, A.O., Sigalov, A.V., Analysis of the Thermal Regime and Design of Radio-Electronic Devices with Conductive Cooling, Izvestia Vuzov SSSR, Priborostroenije, vol. 30, no. 11, pp.82-87, 1987.

11. Dulnev, G.N., Sergeev, A.O., Location of Heat-Loaded Elements in a Radio-Electronic Device, J. Engng Phys., vol. 52, no. 3, pp.491-495, 1987.

12. Norenkov, I.P., Manichev, V.B., Systems of Computer-Aided Design of Electronic and Computing Equipment, Vysshaya shcola, Moscow, 1983.

13. Mezenov, A.V., Soms, L.N., Stepanov, A.I., Thermooptics of Solid State Lasers, Mashinostroenije, Leningrad, 1986.

14. Savintseva, L.A., Sharkov, A.V., Parfenov, V.G., Investigation of Power Distribution of Inner Heat Sources in the Elements of a Laser Head, Izvestija Vuzov SSSR, Priborostroenije, vol. 24, no. 8, pp.92-96, 1981.

15. Dulnev, G.N., Mikhailov, A.E., Parfenov, V.G., The Modelling of the Thermal Regimes of the Pump Systems, J. Engng Phys., vol. 53, no. 1, pp.107-113, 1987.

16. Egorov, V.I., Mikhailov, A.E., Parfenov, V.G., On a Choice of a Finite-Difference Scheme for Solution of a System of Heat Conduction Equations, J. Engng Phys., vol. 51, no. 2, pp.346-348, 1986.

17. Mak, A.A., Ananjev, I.A., Ermakov, B.A., Solid State Optical Quantum Oscillators, Uspekhi fizicheskikh nauk, vol. 92, no. 3, pp.373-426, 1967.

18. Dulnev, G.N., Mikhailov, E.E., Parfenov, V.G., Savintseva, L.A., Laser Pump System Variety Influence on Its Thermal Regime and Working Characteristics, Izvestia Vuzov SSSR, Priborostroenije, vol. 29, no. 6, pp.393-398, 1986.

19. Mikhailov, A.E., Parfenov, V.G., Savintseva, L.A., Thermal Regime and Radiation Energy of Solid-State Laser Constructions, Izvestia Vuzov SSSR, Priborostroenije, vol. 30, no. 4, pp.82-86, 1987.

20. Dulnev, G.N., Ushakovskaya, E.D., Thermal and Mathematical Modelling of Optic-Electronic Devices, J. Engng Phys., vol. 46, no. 4, pp.659-666, 1984.

21. Dulnev, G.N., Ushakovskaya, E.D., Tsukanova, G.I., Thermooptical Processes in Mirror-Lens Objectives, J. Engng Phys., vol. 53, no. 1, pp.101-106, 1987.

22. Dulnev, G.N., Korenev, P.A., Spokoiny, M.Yu., Automatic Selection of Thermostable Functional Scheme, Izvestia Vuzov SSSR, Priborostroenije, vol. 27, no. 10, pp.90-95, 1984.

23. Dulnev, G.N., Korenev, P.A., Sharkov, A.V., Synthesis of Thermostat Devices. Pt I. Base Model of a Thermostat, J. Engng Phys., vol. 51, no. 3, pp.504-508, 1986.

24. Dulnev, G.N., Korenev, P.A., Synthesis of Thermostat Devices. Pt 2. Mathematical Models, J. Engng Phys., vol. 51, no. 4, pp.660-667, 1986.

25. Ingberman, M.I., Fromberg, E.M., Grabov, L.P., Thermostatting in Communication Technique, Radio i Svjaz, 1979.

26. Yarishev, N.A., Andreeva, L.B., Thermal Calculation of Thermostats, Energoatomizdat, Leningrad, 1984.

27. Dulnev, G.N., Korenev, P.A., Sigalov, A.V., Solunin, A.N., Synthesis of Thermostat Devices. Pt 3. Minimization of a Thermostatting Error, J. Engng Phys., vol. 51, no. 5, pp.774-781, 1986.

28. Dulnev, G.N., Beljakov, A.P., Heat Pipes in Electronic Systems of Temperature Stabilization, Radio i Svjaz, Moscow, 1986.

Application of Computational Fluid Dynamics for Analyzing Practical Electronics Cooling Problems

C. PRAKASH
CHAM of North America, Inc.
1525-A Sparkman Drive
Huntsville, Alabama 35816, USA

ABSTRACT

Practical heat transfer problems related to the cooling of electronic equipment are complex. Hence, analysis of these problems using computational fluid dynamics (CFD) methods must proceed by one of the three approaches depending upon the level of detail to be predicted. These approaches may be called (i) the local analysis, (ii) the systems analysis and, (iii) the network analysis. Local analysis deals with the complete solution of the governing equations for a problem with well defined boundaries and boundary conditions. Typically, however, the domain in such problems represents only a segment of the total electronic cooling system. In contrast, systems analysis deals with the simulation of the entire systems but with local details modeled via empirical correlations. Network analysis is the simplest, and it deals with the representation of the system as network of currents and driving potentials. The purpose of this paper is to present examples in each of these three categories, and to emphasize, that from the standpoint of a electronics cooling specialist, it is highly desirable that a general purpose CFD code should allow one to perform all three kinds of analyses. It is believed that the PHOENICS code, which was used to solve all the examples presented herein, possesses such a capability.

1. INTRODUCTION

Heat transfer problems related to the cooling of electronic equipment are complex [1] because: (i) the geometries are complicated involving many heat producing electronic devices, finned heat sink passages, channels, various tortuous paths, etc., (ii) the problems are conjugate, requiring the simultaneous solution of heat transfer by conduction in the solid devices and by convection in the coolant fluid, (iii) mixed (i.e., forced plus natural) convection is involved, and (iv) flows can be laminar or turbulent, and often in the laminar-turbulent transition regime. The complexity of a practical system would be evident from Figure 1 which shows the inside of a typical computer.

Due to the enormity of practical electronics cooling problems, closed form analytical solutions are hard to obtain [1]. Hence, one must take recourse to the numerical solution of the governing equations using finite difference or finite element methods [2]. Indeed, since electronics cooling problems generally involve non-reactive single-phase flows for which the available turbulence models are relatively applicable, computational fluid dynamics (CFD) can be a useful tool for an electronics cooling specialist. This situation is in contrast to fields such as combustion where the role of CFD is limited primarily due to the lack of physical models describing phenomena such as turbulence, combustion, and two-phase flow.

In principle, CFD methods can be used to predict flow and heat transfer in systems as complicated as the one shown in Figure 1. However, if it is desired to predict the complete details of flow in every nook and corner, then one would need an extremely large number of grid nodes resulting in prohibitively high computational time and cost. Therefore, for all practical problems, one has to make a judicious compromise between the level of detail to be predicted and the affordable computer expense. In making such a compromise, one may take comfort from the realization that the best turbulence models available in literature are only tentative; hence, there is no point in doing an over-kill with a very fine grid for a problem for which basic turbulence models are not well established.

The above considerations suggest that one should maintain an engineer's prepective while using CFD methods to analyze electronic cooling problems. Three distinct approaches can be identified and these will be discussed below; success is to be attained only if one employs all three approaches in proper proportion.

2. LOCAL, SYSTEM AND NETWORK ANALYSES

2.1 Local Analysis

By local analysis we mean the detailed prediction of flow and heat transfer in a segment of the total system. Such a segment would be much smaller than the total system, but it would be a rather "clean" domain with well defined boundaries. Thus, within well affordable grid sizes, one should be able to obtain grid independent solutions to the governing equations. The equations are solved without any empiricism except that implied by the turbulence models.

FIGURE 1. A typical electronics system; the inside of a computer.

An example of local analysis would be to predict flow over heated blocks on a flat plate. For such a well defined problem, one can predict, with confidence, details like local hot spots, extent of flow separation and its effect on heat transfer, and the system pressure drop.

2.2 Systems Analysis

By systems analysis we refer to a simulation of the entire electronics cooling system. This entails, as for local analysis, the discretization of the domain by a family of grid nodes and the numerical solution of the governing equations. However, the computational grid is quite coarse, and it can only account for the presence of major components of the system. Thus, details at the local level of smaller components are not predicted. For example, a computational cell may contain a number of "tiny" heaters, and, hence, prediction of flow and heat transfer around such heaters is not possible.

Since local details are not captured by a systems analysis, it is necessary that the effect of smaller devices be modelled via source and sink terms in the governing transport equations. The necessary model can be empirical and established by experiments on individual components. From the hydrodynamic standpoint, these models must provide the resistance to the flow of the fluid, while for heat transfer purposes, one needs a heat transfer coefficient between the device and the surrounding fluid.

2.3 Network Analysis

Network analysis can be regarded as a further simplification of the system analysis where we completely dispense with the idea of numerically solving the governing differential equations. Instead, the major components of the system are represented via a system of inter-connected nodes, and one solves a network problem involving currents and driving potentials. For the hydrodynamic aspects of the problem, the driving potential is the pressure and the current is the fluid flow rate. For heat transfer, the potential is the temperature and the current is the heat flux. The current through any branch is equal to the potential difference across it divided by a resistance. Expression for these resistances have to be provided by a user, such information being generated experimentally. For fluid flow, the resistance is provided in terms of loss (or friction) coefficients [3]. For heat transfer, the resistance is related to the inverse of the overall heat transfer coefficient, i.e.,

$$R = \frac{1}{UA}$$

where R, U and A represent the thermal resistance, the overall heat transfer coefficient, and the heat flow area, respectively.

Unlike local and systems analysis, network analysis does not provide any detailed information about flow fields or isotherms; it merely establishes how the fluid and heat flow distribute themselves along major components of the device. Thus, local details regarding hot spots, for example, cannot be predicted by the network analysis.

Since network analysis is computationally the least expensive, it is very popular for analyzing electronics cooling systems. Indeed, text books dealing with electronics cooling are full of network diagrams [1].

2.4 The Combined Approach

A practicing electronics cooling specialist should use all three kinds of analysis discussed above since these approaches complement each other. Thus, for instance, a systems analysis on a coarse computational grid can be used to determine realistic boundary conditions for a detailed local analysis. In turn, local analyses can be used to establish modeling correlations, which are implemented into the systems and network analyses to empiricise the local details. As already stated, success is to be attained only if one uses a judicious combination of all three approaches, keeping an engineer's perspective on the problem.

It may be stated here that an engineer should not feel apologetic due to the lack of academic purity of a global systems type analysis; for, it should be remembered, that such an analysis is often performed to obtain a quick, qualitative, engineering feel for the problem without incurring exhorbitent computational costs. Of course, if one can afford it (which is rarely the case) one can, in principal, carry out a detailed local analysis for the entire system.

2.5 The Desirable Features Of A General Purpose CFD Code For Electronics Cooling Applications

The preceding discussion implies that a general purpose CFD code for electronics cooling applications should be such that it allows one to perform all three types of analyses (i.e., local, systems, and network). And, it should be possible to do so from just the user's (front) end of the code without any need to change the inside of the program. Such a code then saves the inconvenience of using different codes for different purposes.

A few general purpose CFD codes are commercially available in the market that can be used to perform local analysis. Likewise, some codes are available for performing network type of analysis for thermo-hydraulic problems [3]. However, the general purpose CFD code PHOENICS [4] is one that has the versatility that a user can perform all three types of analyses just from the user's end of the program. All the sample problems presented in this paper were solved using the PHOENICS code; more examples and greater computational details may be found in [5].

2.6 Comment About The Examples To Be Presented

In the following examples, computations were made using nominal grid sizes. Because, the intent was simply to illustrate what can be done, a detailed grid refinement study was not undertaken.

3. LOCAL ANALYSIS

As is to be expected, the academic community is much interested in detailed local analysis of rather well defined problems. Among others, one such active team is the heat transfer group at the University of Minnesota, working under the guidance of Dr. E.M. Sparrow and Dr. S.V. Patankar. This group, in a very timely fashion, moved from CFD application to compact heat exchanger problems to the closely similar electronics cooling problems; a large number of technical papers resulting from this effort may be found in journals such as the ASME Journal of Heat Transfer. The two examples of local analysis to be presented here are motivated by similar problems addressed by the Minnesota group.

3.1 Example 1: Buoyancy Induced Flow In A Strip Fin Passages

A generic electronics cooling problems is shown in Figure 2. It consists of a vertical plate on one side of which are mounted some heat generating electronic devices. On the other side of the plate we have parallel plate fins for the purpose of providing extended surface area for the dissipation of heat to the ambient fluid. In compact heat exchanger applications, flow through the inter-fin passages is forced; in contrast, in many electronic cooling situations, the flow through the fin passages in buoyancy induced (chimney effect).

It is a matter of common knowledge that heat transfer from the plate fins can be enhanced if they are cut into strips; for, the boundary layers must restart each time a strip is encountered, leading to high heat transfer coefficients. However, this enhancement in heat transfer comes at an expense because each time the boundary layer restarts, there is higher friction loss, resulting in increased pressure drop. In a purely buoyancy induced mode (as is often the case for electronics cooling problems), this means that the induced flow rate is reduced, which can off set the gains made by the enhancement of the heat transfer coefficient.

The efficiency of a strip fin passage in terms of heat transfer enhancement in the buoyancy induced flow node was investigated by Sparrow and Prakash [6]. An illustrative 2-D problem implemented in PHOENICS is shown in Figure 3(a). The flow is laminar and the usual Bousseinesq approximation is made for computing the buoyancy force. Sample results for part of the domain are shown in Figure 3(b). As is to be expected, the velocities are small near the finned region due to the no-slip condition. Once the fluid moves into the no-fin region, there is a homogenization of the flow field. Again, as expected, closely packed isotherms, implying high heat transfer coefficients, occur repeatedly at the tip of each strip fin. It was shown by Sparrow and Prakash [6] that for the same heat transfer area, a strip fin passage <u>can</u> dissipate more heat than a system of continuous plate fins, even in the buoyancy induced flow mode.

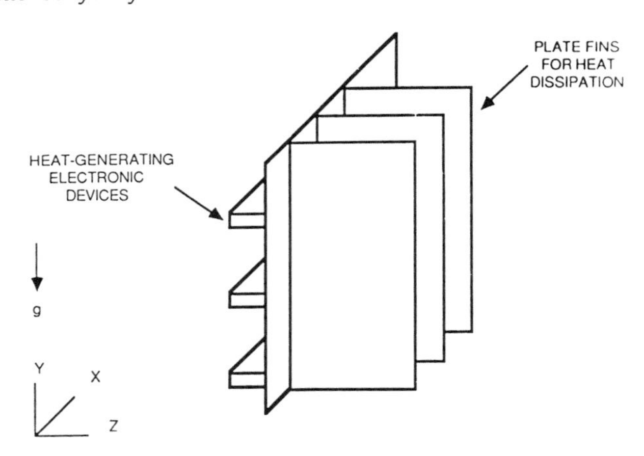

FIGURE 2. A generic electronics cooling problem.

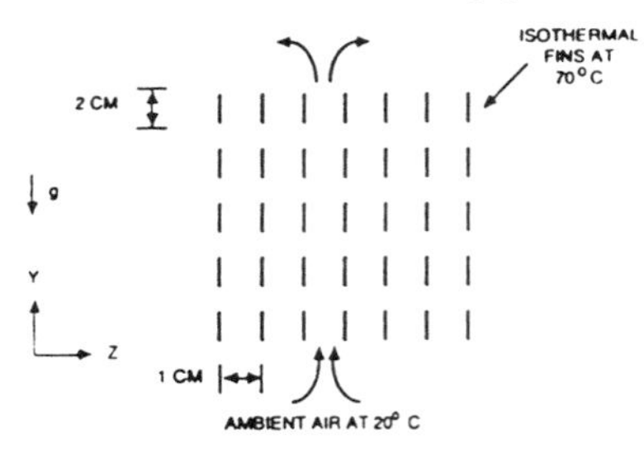

FIGURE 3 (a). Buoyancy induced flow in a strip fin passage.

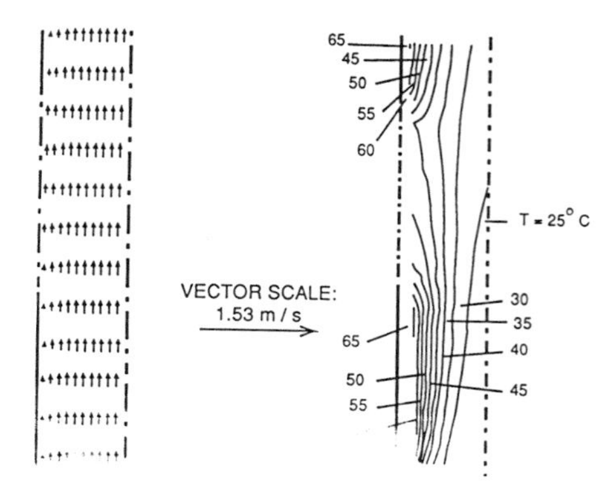

FIGURE 3 (b). Representative results over a part of the computational domain; velocity vectors (left), isotherms (right).

3.2 Example 2: Turbulent Flow Over Heated Blocks

Recalling Figure 2, we can now focus attention on the other side of the plate, i.e., on the side of heat generating block-like electronic components. A typical problem that can be investigated is shown in Figure 4(a). The shaded regions represent electronic components in which there is a volumetric heat generation rate Q. Only the periodically fully developed situation is analyzed in which the flow repeats itself over modules such as ABDC; i.e., conditions at the plane CD are identical to the conditions at the plane AB. Details of the formulation of such cyclic problems may be found in [7]. For the chosen situation, even the thermal aspects of the problem are cyclic and the temperature may be represented as

$$T(x,y) = \tilde{T}(x,y) + \Delta T \frac{x}{2L}$$

where $\tilde{T}$ is cyclic while ΔT is the temperature rise over a module of length 2L; ΔT is related to the flow rate and the heat generation rate Q.

The flow is assumed to be turbulent and the standard $k\sim\varepsilon$ model of turbulence is employed [8]. The problem is conjugate and conduction in the solid is analyzed simultaneously with convection in the fluid. For laminar flows, such conjugate problems can be analyzed easily by using the harmonic averaging of diffusive coefficients [2]. Such a facility is available in PHOENICS. For turbulent flow, however, the situation is different because at the solid-fluid interface one must account for the conductive resistance in the solid and the log-law-of-the-wall related heat transfer coefficient in the fluid. Fortunately, due to various facilities built into PHOENICS, all this can be handled from the user's segment of the program.

Some illustrative sample results are shown in Figures 4(b) and 4(c). These correspond to Reynolds number = 3.5×10^4, Prandtl number = 0.7, solid to fluid thermal conductivity ratio = 10, H/L = 1 and t/L = 0.25. The Reynolds number is based on the average velocity and the length L. As can be noted in Figure 4(b), the turbulent flow field is more uniform as compared to what one would expect for a laminar flow. The recirculation zone behind the block can be discerned. Isotherms of $\tilde{T}/\Delta T$ are shown in Figure 4(c). As is expected, greater temperature drop occurs within the solid block where heat is being generated. In the fluid, there is no heat generation, convection aids heat transfer, and the turbulent diffusivity is high, all of which leads to low temperature variation.

3.3 Closing Comments

The two examples presented above illustrate the kind of electronic cooling problems that can be handled in complete detail by CFD methods using affordable grid sizes. The only limitation, if any, is the lack of universally applicable turbulence models. Fortunately, however, due to small system dimensions, the flows in electronics cooling passages are often laminar. In any event, if the flow is turbulent, then an engineer must live with what best is available or, if he has the resources, use his CFD tools to enhance and develop new turbulence models.

4. SYSTEMS ANALYSIS

4.1 Desirable Features In a CFD Code Which Facilitate Systems Analysis

As already mentioned, systems analysis is much like the local analysis except that we now operate at a larger length scale and simulate the entire system rather than just a segment of it. Thus, as for local analysis, we discretize the computational domain by a system of grid nodes and numerically solve the governing partial differential equations. However, the physical dimensions of a computational cell can be greater than the size of the small devices in the system, and, hence, a mathematical model needs to be provided to account for the local effect of the devices on the global flow and temperature fields being predicted. Recalling Figure 1, it will be noted that the most important aspect of an electronic system is the large number of internal obstacles which provide obstruction and friction in the path of fluid and act as sources and sinks of heat.

In view of the above considerations, we can list the following two important features that a CFD code must possess in order to facilitate systems type analysis.

1. A user should be able to provide area and volume porosities for all the computational cells so that the presence of internal obstacles can be simulated. These porosities range from zero to one, with zero representing an entirely blocked face (or volume) and one indicating that the face (or volume) is completely available to flow. Consequently, in setting up the finite difference (or finite element) equations, all convective and diffusive fluxes must be multiplied by the free area (= geometrical area x porosity).

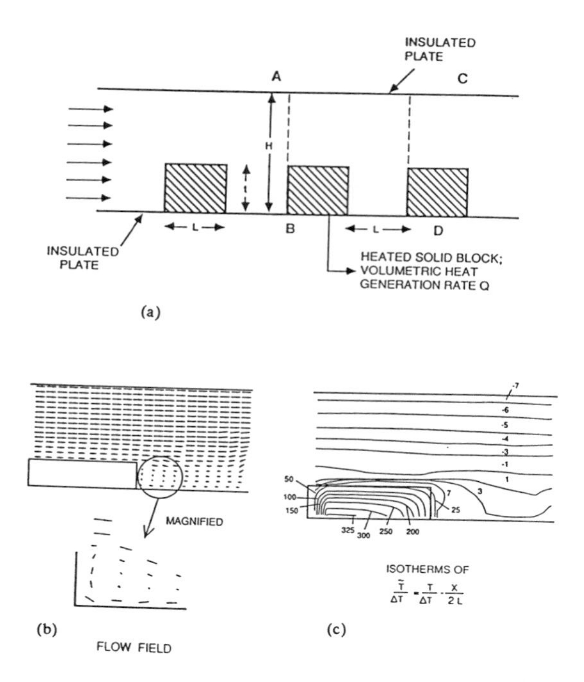

FIGURE 4. Cross-flow over heat generating solid blocks representing electronic devices; (a) schematics, (b) typical flow field, (c) typical isotherm plots.

2. A user should be able to provide any number of source and sink terms in the momentum and heat transfer equations to allow for the fluid-solid interaction at the boundaries as well as at the internal obstacles. The source terms can be on a per unit volume basis, or per unit cell face area basis. Activation of features such as laminar/turbulent wall functions should be permitted not just at the boundaries of the domain but also, if needed, in the inside of the calculation domain; this, again, is to allow the computation of solid-fluid interaction at internal obstacles.

The general purpose CFD code PHOENICS has both of the features listed above. These conveniences have been utlized not just for electronics cooling applications, but in practically every other class of problems that involve complex systems with a large number of internal obstacles; the most relevant example being flows through power plant steam generators and heat exchangers.

4.2 An Example of Systems Analysis: Buoyancy Induced Flow Through A CRT Monitor

Consider the CRT monitor shown in Figure 5(a). It is a cubical box containing a disk drive and a number of power boards stacked vertically and horizontally. Affixed on the power boards are a number of heat generating electronic devices. A typical cross-section of the monitor is shown in Figure 5(b). The flow is buoyancy induced; it enters the monitor through the mesh openings at the bottom and leaves the system through the openings at the top.

A systems analysis of the problem may proceed as follows:

1. First, the domain is discretized by a family of nodes and computational cells as shown in Figure 6(a). As is to be expected in a systems type analysis, the grid is quite coarse as compared to the physical dimensions of the small components in the system.

2. Next,we decide that the physical presence of only the major components shall be accounted for. These major components include the disk drive and the power boards. The presence of these components is simulated by setting the porosities of the cell faces coincident with their location as zero; doing so makes these cell faces impervious to flow, and, hence, the fluid has to turn and find a way around them.

3. The solid-fluid friction is provided via sink terms in the momentum equations. This is done not just at the enclosure walls of the CRT monitor, but also at the major internal obstacles including the disk drive and the power boards.

4. The physical presence of the "tiny" heaters is ignored. The heat dissipated by these heaters is provided as a source term in the enthalpy equation.

Some sample results at a vertical plane are shown in Figures 6(b) and 6(c). As is to be expected, the flow field is complex with recirculation zones around the major obstacles. Contrary to what may be expected, the velocities are larger in the vicinity of some power boards despite the solid-fluid friction there; this is a consequence of the fact that the heat producing devices are located on the power boards, and, hence, it is in that vicinity that the driving buoyancy forces are high. The isotherm pattern shown in Figure 6(c) can be used to identify the regions where high temperatures are to be expected.

4.3 Closing Remarks

Though it may raise some academic eyebrows, there are many the benefits of systems type analysis. The systems analysis provides:

1. A quick and inexpensive qualitative (and quantitative) description of the thermal field in the entire system. The results can often show features that are hard to guess otherwise.

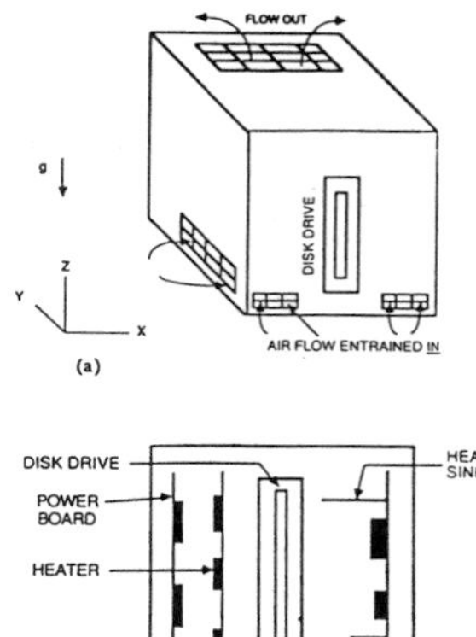

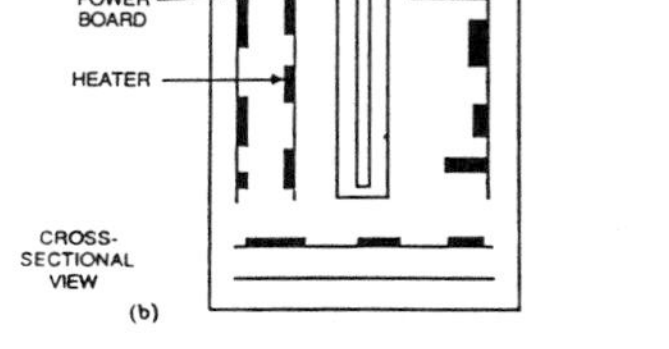

FIGURE 5. (a) CRT monitor (b) a typical vertical cross-section showing a number of power boards and heaters.

2. On examining the results of the systems analysis, one can identify the regions where interesting (and potentially harmful) things might be happening; then, a detailed local analysis of that region can be carried out.

3. Systems analysis can be used to infer proper boundary conditions for the local analysis.

As a final remark it may be mentioned that the term "grid refinement" must be interpreted carefully while performing a system analysis, for, once it has been decided to model certain local features, grid refinement below that level ceases to be meaningful.

5. NETWORK ANALYSIS

Network analysis is the simplest and most popular approach for analyzing electronics cooling problems. Thermal circuits, involving temperature as the driving potential and heat flux as the resulting current, are used to analyze heat transfer problems. Likewise, hydraulic circuits, involving pressure as the driving potential and flow rate as the resulting current, are used to predict flux distributions in channels. The hydraulic networks are essentially non-linear because the pressure difference is related to the square of the flow rate. An example of a flow network for a computer system is shown in Figure 7.

As already mentioned, commercial codes are available [3] that can be used to solve network problems. However, it will be highly desirable if a general purpose CFD code can perform network analysis in addition to the conventional local and systems analyses. How this can be achieved by the PHOENICS code will now be described.

5.1 <u>Adapting PHOENICS For Network Analysis</u>

All fluid flow codes determine the flow by solving the continuity and the momentum (the Navier Stokes) equations. The momentum equations contain the convection (inertia), the diffusion (viscous), the pressure gradient, and the body force terms. Now, if we turn off the convection and diffusion terms, and represent the body force terms as a suitably linearized function of the velocity, then, the momentum equations reduce to a current ~ potential difference relationship where pressure acts as the potential and velocity as the current. In addition, the problem can be arranged such that the continuity equation implies the net balance of current reaching a node. The anology is further completed if it is noted that codes such as PHOENICS use a staggered grid arrangement in which the velocity is driven by the difference between the two neighboring pressure nodes. The situation is described schematically in Figure 8.

It turns out that PHOENICS provides a user with switches that can be used to deactivate the convection and diffusion terms in various transport equation. It is this facility, in conjunction with the other realizations described above, that permit PHOENICS to be used for network type analysis. Note that since PHOENICS allows an iterative updating of the various source terms, both linear and non-linear (hydraulic) circuits can be analyzed. Further, for hydraulic problems, pumps can be simulated by equation very similar to those employed for resistances.

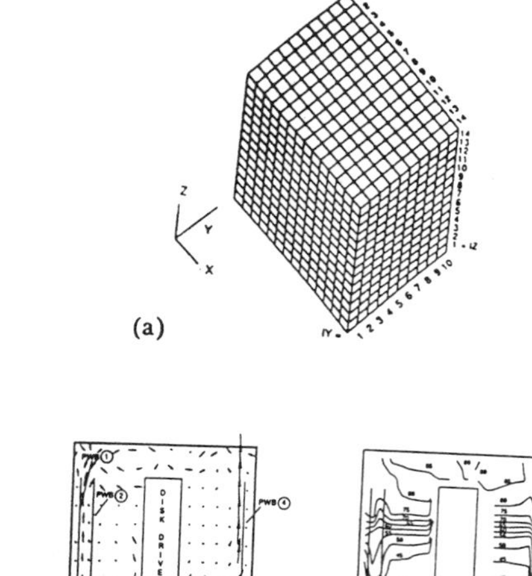

FIGURE 6. Buoyancy induced flow in a CRT monitor; (a) domain discretization, (b) representative flow field in a vertical plane, (c) representative temperature field in a vertical plane.

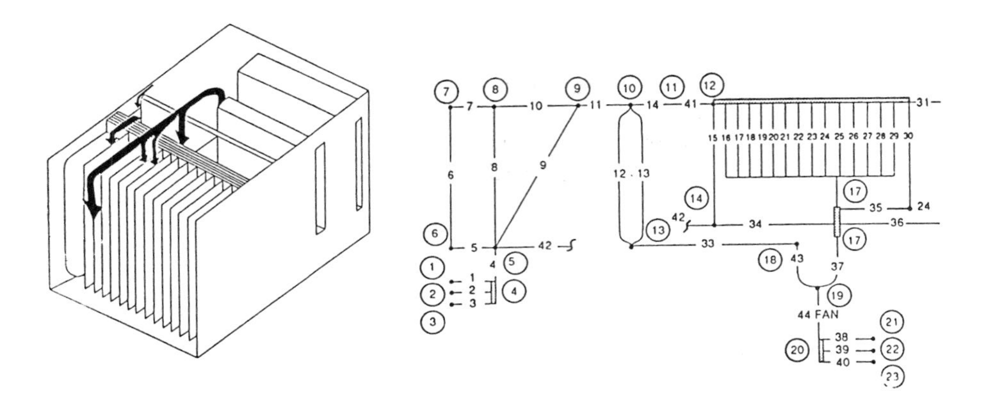

FIGURE 7. A typical computer system and the flow network.

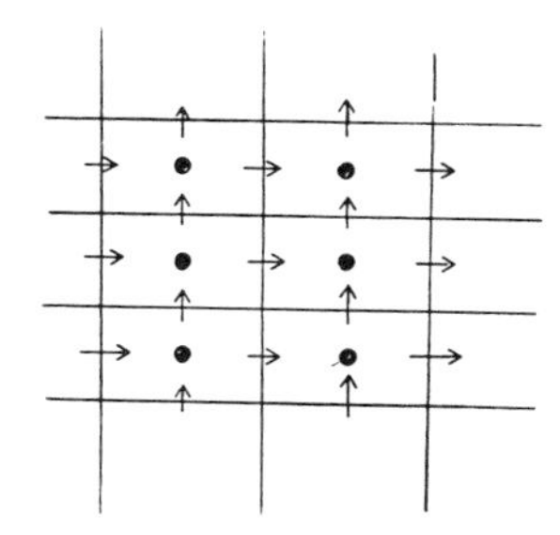

FIGURE 8. The staggered grid arrangement; (0) pressure or potential nodes; (↑) and (→) velocity or current locations.

In the following we present an example of a linear thermal circuit; example of hydraulic circuits may be found in [5].

5.2 An Example: Thermal Analysis Of A Chip Junction

The thermal network representing a chip junction is shown in Figure 9(a) and (b) and its PHOENICS analog is shown in Figure 9(c). The symbol Pij represents the pressure at node indentified by indices i and j. The index i refers to the grid location in the horizontal direction and the index j to the grid location in the vertical direction. The velocities U_{ij} and V_{ij} correspond to the heat flux in different branches. In this problem the ambient air temperature (T_A) and the heat generated by the chip are known, and the goal is to determine the temperature at the chip. The results of a sample calculation are shown in Figure 10.

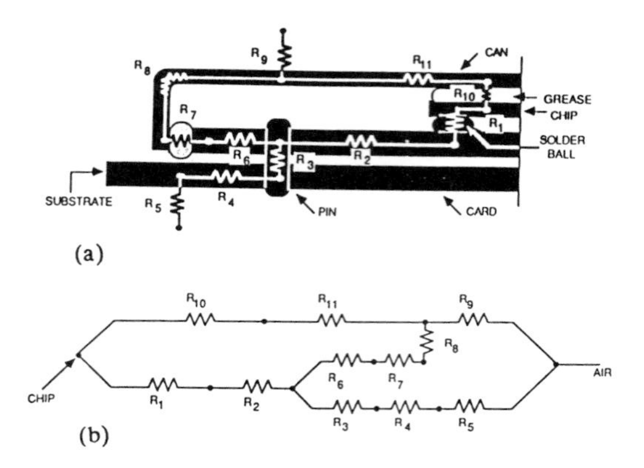

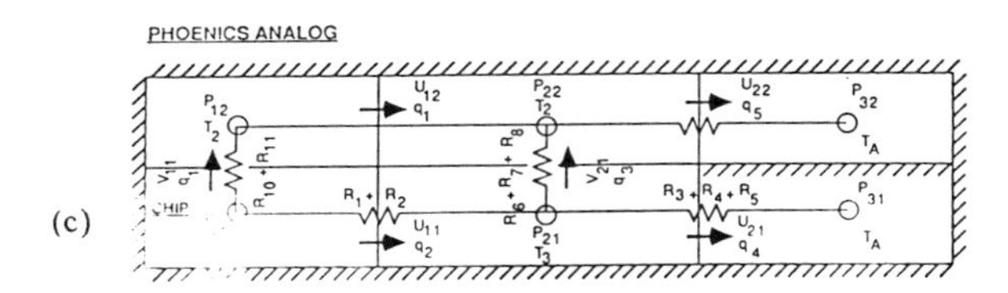

FIGURE 9. Network analog of an air-cooled thermal chip module; (a) module, (b) network representation, (c) PHOENICS implementation.

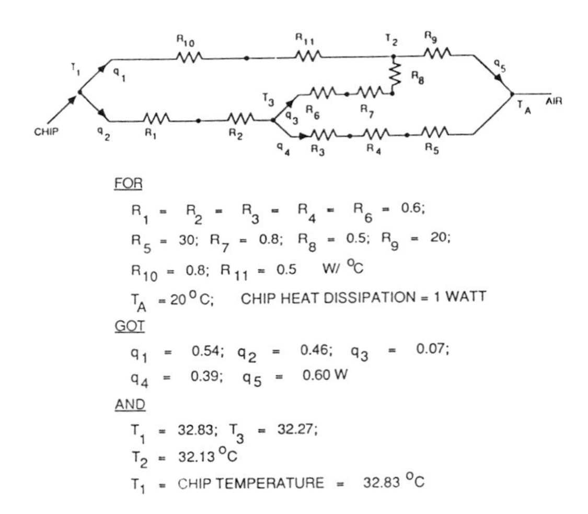

CHIP
AIR
FOR
R1 = R2 = R3 = R4 = R6 = 0.6;
R5 = 30; R7 = 0.8; R8 = 0.5; R9 = 20;
R10 = 0.8; R11 = 0.5 W/ °C
TA = 20 °C; CHIP HEAT DISSIPATION = 1 WATT
GOT
q1 = 0.54; q2 = 0.46; q3 = 0.07;
q4 = 0.39; q5 = 0.60 W
AND
T1 = 32.83; T3 = 32.27;
T2 = 32.13 °C
T1 = CHIP TEMPERATURE = 32.83 °C

FIGURE 10. Network analysis of a chip junction; results of a sample problem.

5.3 Closing Comments

Network analysis is an important procedure for analyzing practical electronics cooling problems. Though simple circuits can be analyzed by hand, computer codes are needed when the calculations have to be repeated frequently, the circuit is complex, or the problem is non-linear as in fluid-flow analysis. Thus a general purpose CFD code for electronics cooling should permit network analysis to be performed.

6. CONCLUDING REMARKS

This paper emphasizes that detailed fine grid analysis of real electronic cooling systems is very expensive (and practically impossible) using the best CFD methods and computers available today. Hence, an engineer must employ a judicous combination of large scale network and system analyses and the detailed local analysis. Each approach has its merits, and they all complement each other. Thus, CFD application for electronic cooling problems is similar to taking a picture........one should start off by taking a global picture and then zoom onto interesting local details. A good CFD code should, therefore, act like a good camera and allow a user to operate at different levels. It has been shown that the PHOENICS code provides all such necessary facilities.

7. NOMENCLATURE

A heat transfer area
R thermal resistance
T temperature
U overall heat transfer coefficient

8. REFERENCES

1. Kraus, A.D. and Bar-Cohen, A., Thermal Analysis and Control of Electronic Equipment, McGraw-Hill, New York, 1983.

2. Patankar, S.V., Numerical Heat Transfer and Fluid Flow, McGraw-Hill, New York, 1980.

3. Cullimore, B.A. and Lin, C.H., FLUINT: General Fluid System Analysis with SINDA' 85, NASA-CR, 1986.

4. Rosten, H.I. and Spalding, D.B., The PHOENICS Beginner's Guide, CHAM TR-100, 1987.

5. Prakash, C., PHOENICS Demonstration Calculations for Electronics Cooling Applications, CHAM NA Report, 1987.

6. Sparrow, E.M. and Prakash, C., Enhancement of Natural Convection Heat Transfer by a Staggered Array of Discrete Vertical Plates, ASME Journal of Heat Transfer, Vol. 102, NO. 2, pp. 215-220, 1980.

7. Patankar, S.V., Liu, C.H. and Sparrow, E.M., Fully Developed Flow and Heat Transfer in Ducts Having Streamwise - Periodic Variation of Cross-Sectional Area, ASME Journal of Heat Transfer, Vol. 99, pp. 180-186, 1977.

8. Launder, B.E. and Spalding, D.B., The Numerical Computation of Turbulent Flow, Comp. Methods Appl. Mechanical Engineering, Vol. 3, pp. 269-289, 1974.

Analysis of the Heat Transfer in Coupled Electronic Boards by a New Global Network Representation

J. BASTOS and J. B. SAULNIER
Laboratoire de Thermique, ENSMA
UA 1098 CNRS
University of Poitiers
Poitiers, France

ABSTRACT

This work deals with a new way of modelling the conduction / convection problems using just a network of conductances. This technique is called Global Nodal Method and it doesn't need the knowledge of the heat exchange coefficients. Up to know we are interested in the simulation of "channels" boards configurations in forced convection. Besides, we suppose that the dynamic field is known and the network just solves the energy equation.

1. INTRODUCTION

The convection analysis around electronic boards can sometimes, and as a first approach, be represented by the classical concept of heat exchange coefficient. Unfortunately in most cases, this coefficient is poorly known in this environment (discretized heat sources, diffusion in the boards, unknown local fluid temperature, interaction between boards) and a detailed analysis of the fluid flow is then necessary.
The classical treatment of this kind of problem consists of modelling the structure parts by a Thermal Network Analyzer and the fluid flow (generally represented by the boundary layer equations) by a finite-difference method [1, 2]. This technique provides us with a set of reference solutions ; however a systematic exploitation of this model seems difficult due to the C.P.U. time consumption.
A more original approach dealing with both the convection and the conduction by a single network of conductances will be presented in this paper. The convection is represented here by a network of appropriate fluid transport conductances and conduction conductances. This new method is called "The Global Nodal Method" and it is less time consuming than the first one (Nodal Method Plus Finite Difference).
Two examples in a forced convection configuration are shown to illustrate the results we have reached with this new technique:

1. Influence of an aluminium heat spreader on the temperatures of several stacked boards.
2. Modelling of boards in a realistic environment (including sockets, walls of boxes, ambient conditions).

2. THE GLOBAL NODAL METHOD

2.1. Principles of the Nodal Method

The Nodal Method is based on the concept of "Conductance" (connecting 2 isothermal volumes called nodes) which is defined by the relation

$$\emptyset = G_{ij} \, (T_j - T_i) \tag{1}$$

In equation (1) the heat flux $\emptyset$ is proportional to the temperature difference $(T_j - T_i)$ between the 2 nodes. Each node exchanges energy with its neighbors and the set of thermal balance equations obtained are of the following form:

$$C_i T_i = \Sigma \, G_{ij} (T_j - T_i) + Q_i \tag{2}$$ *

According to the phenomenological laws governing the different heat exchanges, the following expressions can be obtained for the conductances:

CONDUCTION => $G^{\ell}_{ij} = \dfrac{\lambda S_{ij}}{L_{ij}}$

RADIATION => $G^{r}_{ij} = \varepsilon_i \alpha_j SF_{ij} \, \sigma (T_j^2 + T_i^2)(T_j + T_i)$

CONVECTION => $G^{c}_{ij} = hS$

ENTHALPY TRANSPORT => $G^{f}_{ij} = \dot{m} \, C_p$

In our problem, as we mentioned before, the classical heat exchange coefficients (h) are unknown and the corresponding convective conductances are then replaced by a set of conduction and fluid transport conductances. Their value will be determined in the next section.

2.2. Modelling the Fluid Flow

According to the basic principle of the nodal method, the fluid flow region is divided into several volumes that are each considered as isothermal and of uniform velocity.
Each node of the grid is connected to the adjacent nodes by fluid transport conductances (in the fluid flow direction - x and perpendicular to this fluid flow - y) and conductive conductances (only perpendicular to the fluid flow - y). This network of conductances is represented in Figure 1 for a two-dimensional flow in a rectangular duct.
The fluid transport conductances in the x direction are calculated by integrating the velocity profile (u) over the isothermal volume:

$$\dot{m} = \rho \int_{y_o}^{y_o + \Delta y} u \, dy \tag{3}$$

$$GF = \rho \, C_p \int_{y_o}^{y_o + \Delta y} u \, dy \tag{4}$$

---

* If we consider the steady-state problem, the left hand of equation (2) is equal to zero.

Let us observe here that we suppose the physical properties of the fluid uniform. This means that the velocity and the thermal fields are decoupled and we can evaluate the velocity profile before the temperature field computation. This was done in a semi-analytical way for laminar flow [3] and with a finite-difference method for turbulent flow [4].
The transverse fluid transport conductances (associated with the transverse velocity v) are calculated by the mass balance applied to each node of the grid (Figure 2).
The conductive conductances ($GL = \lambda \ \Delta x/\Delta y$) are related to the heat diffusion between the wall and the fluid layer near this wall, and also between two successive fluid layers.
The thermal conductivity, for laminar flow, is an intrinsic function of the fluid. In this case, the conductive conductances are the same for all points of the grid. In turbulent flows, the thermal conductivity depends on the fluid and on the flow as well. The conductive conductances vary then from point to point and they are calculated with the aid of a turbulent model.
The network shown in Figure 1 is valuable for the entrance region of the channel. For fully developed flow, the dynamic field has assumed its definitive parabolic shape and the transverse velocities (v) are equal to zero. So, the transverse fluid transport conductances in this region vanishes.
With this technique, we are now able to represent the convective phenomenon without any knowledge a priori of the heat exchange coefficient. Let us also mention here that this parameter can be precisely evaluated as a particular result of our computation. More details about this procedure can be found in reference [7].

3. VALIDATION OF THE GLOBAL NODAL METHOD

We show in this paragraph a comparison, for 3 different configurations, between the results obtained with the global nodal method and the finite-difference plus nodal method [2]. We indicate that the finite-difference plus nodal method was tested for different cases of heat flux distributions in laminar and turbulent flows and the results obtained were in good agreement with those in the literature.
The geometry of the problem is presented in Figure 3. In order to verify our method, we shall first apply it to the well known convective model of uniform flux on the solid/fluid interfaces. Then we present the results concerning non-conventional cases with stepwise increasing and decreasing heat flux distributions.

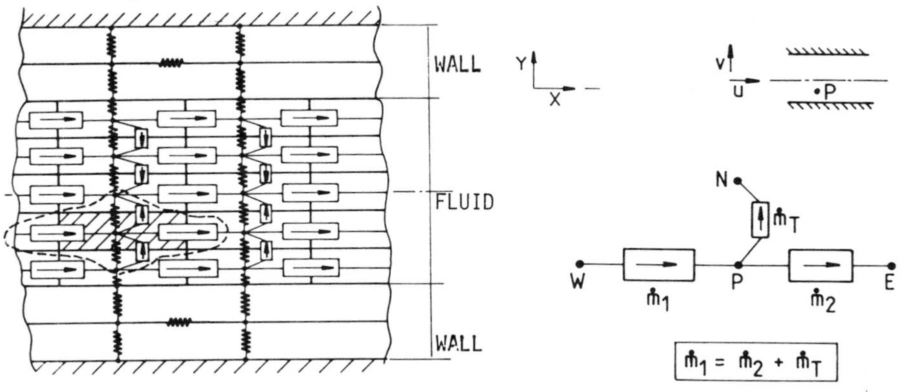

FIGURE 1. The network of conductances FIGURE 2 - Mass balance on node P.

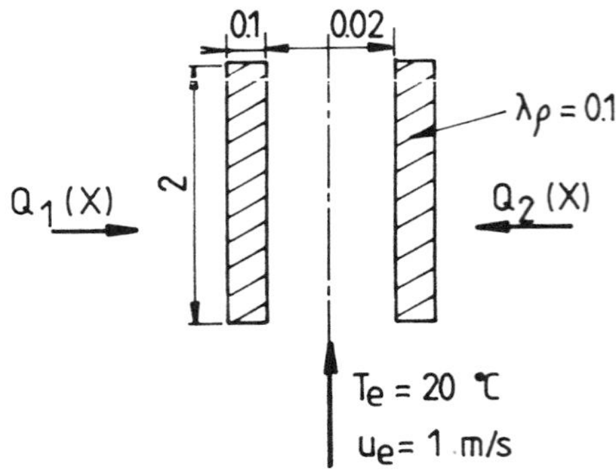

FIGURE 3. Channel with conducting walls.

For each configuration we present the interface, the bulk and the centerline temperatures.

3.1. Uniform Heat Flux

For this first configuration we impose a flux density of 200 W/m^2 over the entire length of the exposed surfaces. We consider as well very low conductive walls (λ_p = 0.1 W/m°C) to get the same uniform flux at the solid/fluid interfaces.

The results obtained (Figure 4) are identical to those of the finite-difference plus nodal method (in dotted lines).

3.2. Stepwise Flow Distributions

We show here two different cases of heat flux distributions over the surface walls:

- Stepwise increasing distribution : the heat flux increases downstream in the duct (Figure 5).
- Stepwise decreasing distribution : the heat fluxes are larger in the entrance region and decreases downstream (Figure 6).

The total dissipated flux is the same in both cases. Only its local distribution is different.

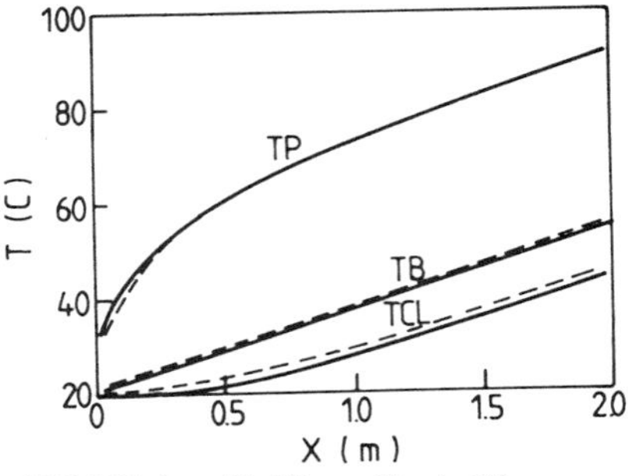

FIGURE 4. Uniform Heat Flux.

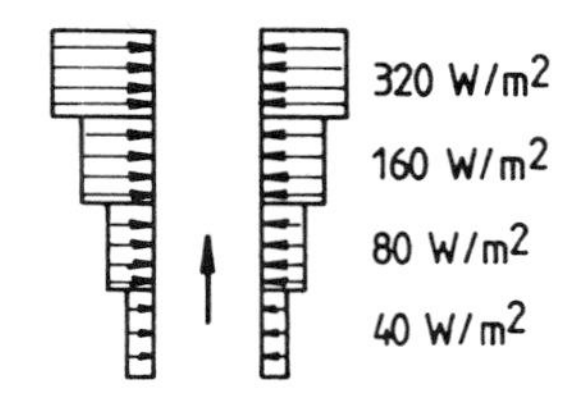

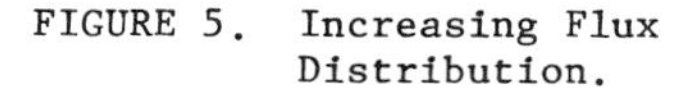

FIGURE 5. Increasing Flux Distribution.

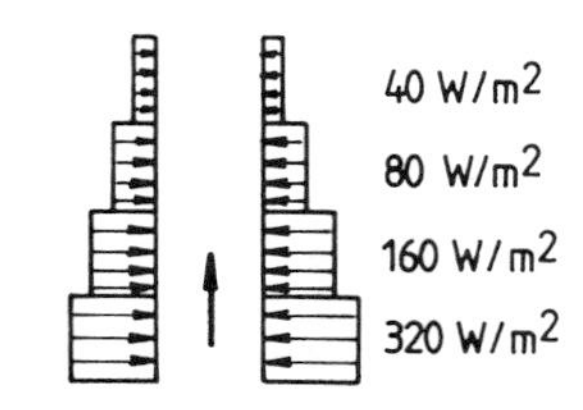

FIGURE 6. Decreasing Flux Distribution.

The wall conductivity, as for the first example, is very low (λ_p = 0.1 W/m C).
We can note for both configurations (Figures 7 and 8) an excellent evaluation of the interface solid/fluid temperatures. This agreement is also rather good for the bulk and the centerline temperatures.
Comparing these two configurations from a more physical point of view, we can say that the maximum interface solid/fluid temperature for the case of increasing flux is 100°C and for the case of decreasing heat flux it is never larger than 80°C. It is clear that, for the same total heat flux, the local temperature distribution is very sensitive to the way this flux is distributed on the wall.

4. MODELLING ELECTRONIC BOARDS BY THE GLOBAL NODAL METHOD

We have shown in paragraph 3 some good results obtained with the global nodal method for simple academic examples. Our objective now is to simulate with this same technique more realistic cases concerning heat exchange in electronic equipment.

4.1. Thermal Field of Several Stacked Boards. Influence of a Heat Spreader on Board Temperature

This study deals with the temperature sensitivity of an electronic board on an aluminium heat spreader.
The configuration treated is composed of three electronic boards limited by two aluminum walls (Figure 9).

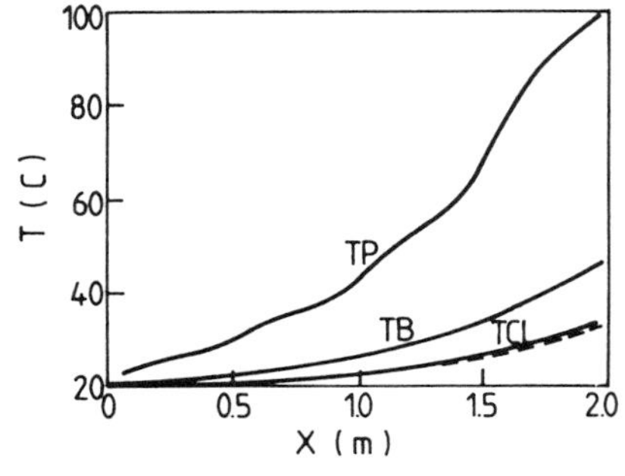

FIGURE 7. Increasing heat flux.

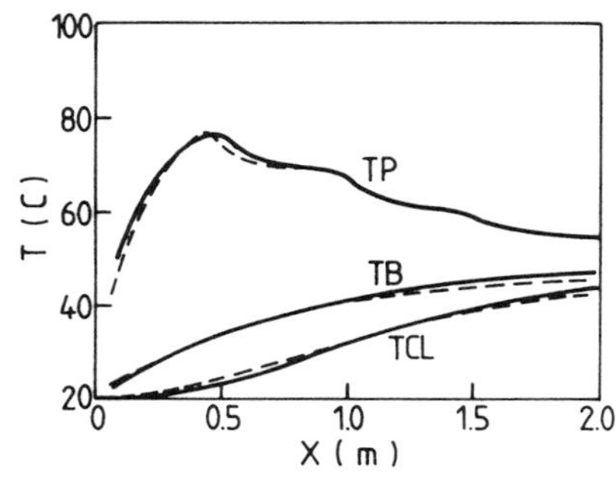

FIGURE 8. Decreasing heat flux.

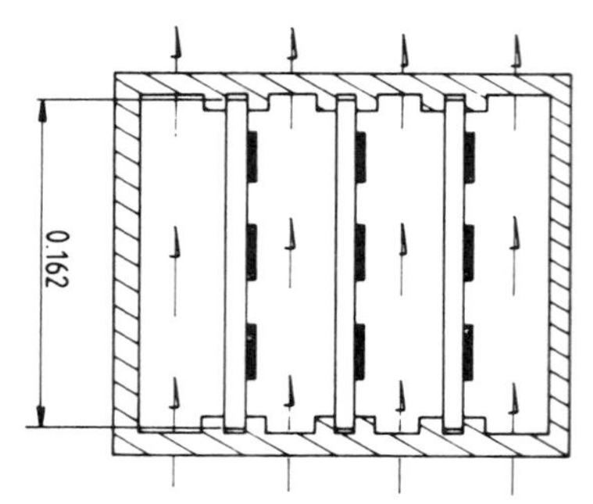

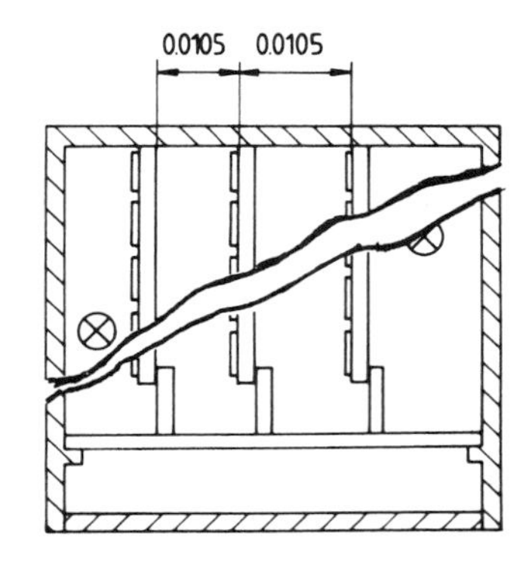

FIGURE 9. Stacked boards problem.

The following exchanges are involved in this problem:

- conduction in the structure parts,
- forced convection in "rectangular channel" configurations in the box,
- natural convection between the outside surface of the box and the ambient air.

All the components are supposed to dissipate the same heat flux (0.25 W) and their distribution on the boards is shown in Figure 10.
In order to reduce the model's size, we consider it as two- dimensional. Then the modelling will be restricted to a slice of the total board represented in Figure 10 by the region between the dotted lines.
Figure 11 shows a schematic representation of the model with the fluid flow entrance conditions and the thermal boundary conditions. We notice that for this first model, the components have no physical (we mean volumetric expansion) representation. They are considered here just as a heat flux applied to the board surfaces.
In Figures 12a and b we can see the evolution of the bulk and the board temperatures obtained with this model. The bulk temperatures of channels 2 and 3 are higher than channels 1 and 4. This is explained by the fact that channels 2 and 3 have two dissipative walls whereas channels 1 and 4 have just one (the other wall is non-dissipative). On the other hand, the surfaces temperature are the same for all boards (in Figure 12b, Tp_1 represents the heated surfaces and T_{p2} the opposite ones). This means that there is no interaction between the boards. In the same way, the enclosure has no influence on the board temperatures. We have here an entrance region configuration and there is no interaction between the boundary layers.

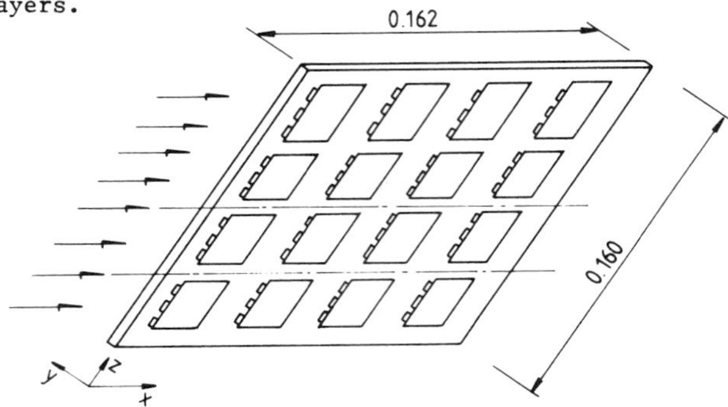

FIGURE 10. Distribution of the components on the board's surface.

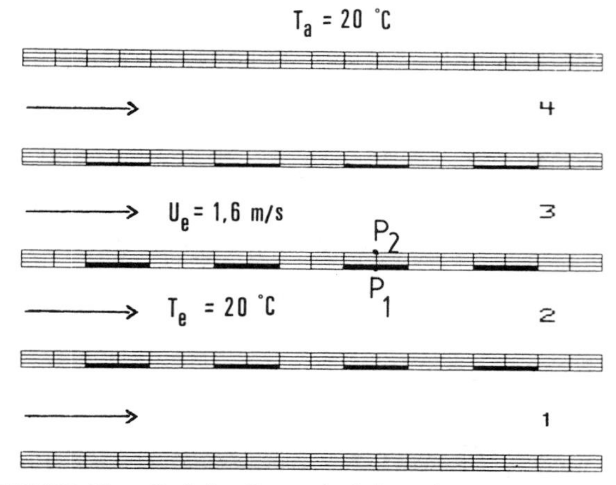

FIGURE 11. Model of stacked boards.

It is interesting to observe that in that case the thermal periodicity of the board temperatures gives us the possibility, with the same accuracy, to reduce the model's size to represent just one channel.

Refining the component's representation

Up to now we considered the chip as a heat source applied on the board's surface with no other physical interaction. Reducing the size of the model, by the modelling of just one channel, we are able now to give to the chip a more realistic representation (Figure 13). We have used here local networks to give to the integrated circuits a more detailed representation.

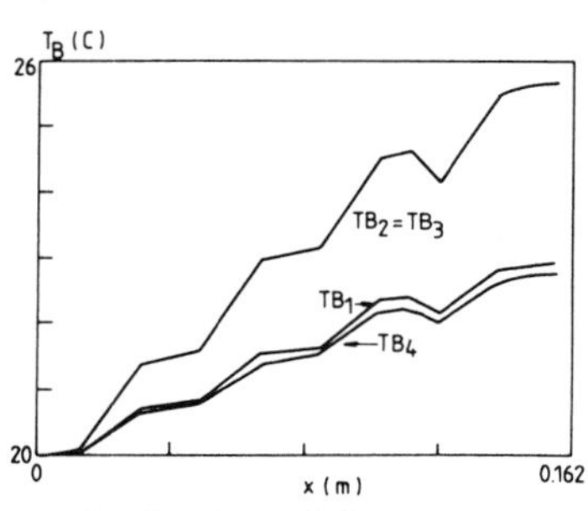

a - Evolution of T_B.

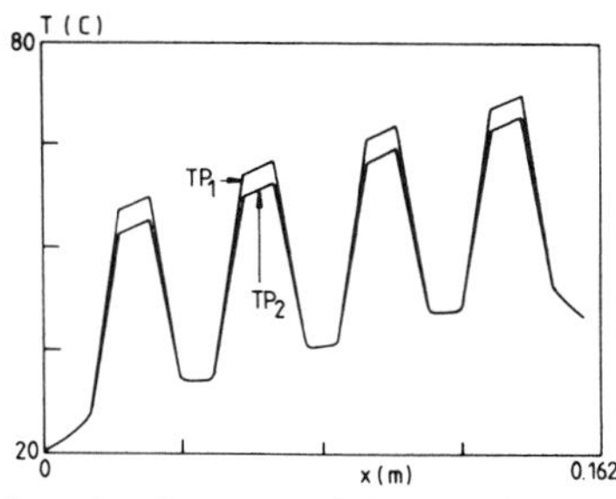

b - Evolution of T_{p1} and T_{p2}.

FIGURE 12. Thermal field for stacked boards.

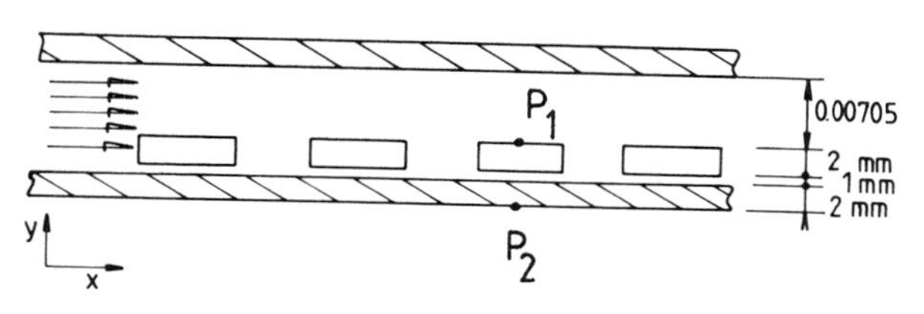

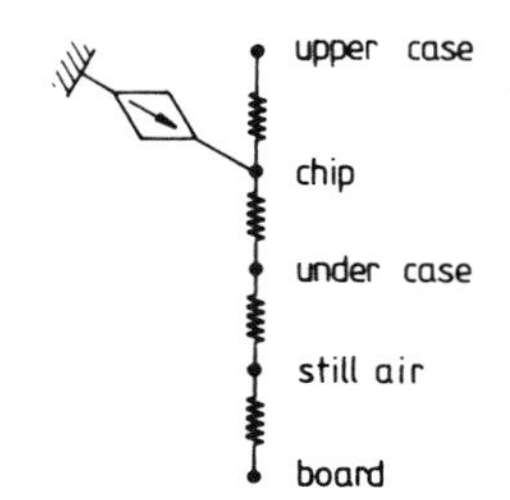

FIGURE 13. New board's representation.

The heat fluxes here are applied directly to the chip. The air enclosed between the chips and the boards and in the gap between two chips are considered as still air (with only conduction inside). The model's temperatures are presented in Figures 14a and b.
The temperature differences between the two board's surfaces (T_{p1} and T_{p2}) are much larger for this model than for the first one. The larger gradients found here are due to the still air layer between the board's surface and the components. This result seems in better agreement with some experimental measurements obtained in the laboratory.
We show as well (Figure 14b) the heat flux exchanged between each board surface and the cooling air. At the trailing edge of the channel, on surface 1, the heat flux becomes negative. This means that locally the air heats the board. This analysis is confirmed by the fluid temperature profile in this region (Figure 15).

Introduction of the aluminium heat spreader

The heat spreader increases the thermal conductivity of the electronic board. That way we expect to equalize the temperatures between the regions submitted or not submitted to the heat fluxes. We consider in this study the same model described by Figure 13 in which we added a 0.5 mm aluminium layer. Two configurations are tested:

- the heat spreader is placed at the opposite side of the component.
- it is placed on the same side of the component.

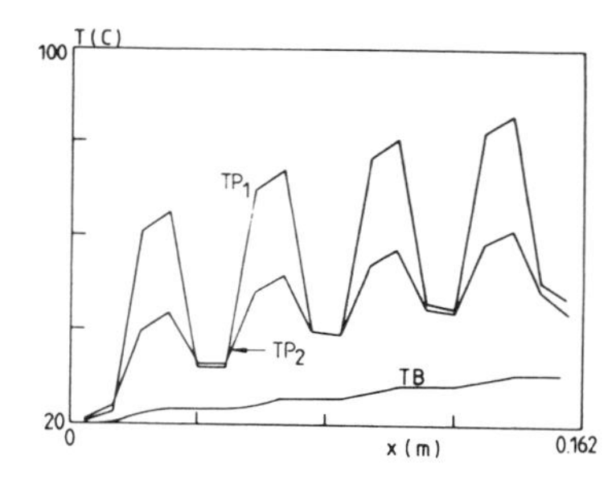

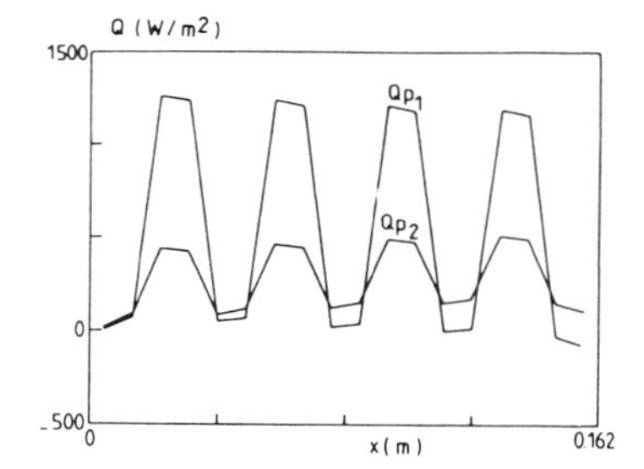

a - Temperature field T_{p1}, T_{p2}. b - Heat flux field Q_{p1}, Q_{p2}.

FIGURE 14. Thermal results with the new model.

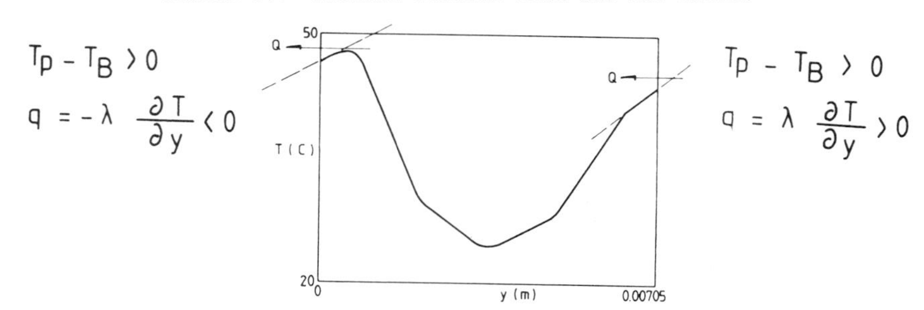

FIGURE 15. Thermal fluid flow profile in the outlet section (trailing edges).

Both cases are represented in Figure 16.

FIGURE 16. Two different configurations for the heat spreader.

In Figure 17 we present the evolution of the hottest board's surface (T_{p1}) for 3 different models : boards without a heat spreader (configuration 1) and with a heat spreader (configurations 2 and 3). For both cases 2 and 3 the maximum board surface temperature decreases:

- for case (2) the board temperature, at the outlet region, decreases from 8°C but there is no difference for the entrance region.
- for case 3 the board temperature decreases from 20°C for the channel outlet region and more than 5°C for the entrance region.

4.2. Modelling of boards in a realistic environment

In this example we model a complete electronic component box composed by boards, sockets, sliders, bus, etc. The box scheme is presented in Figure 18.

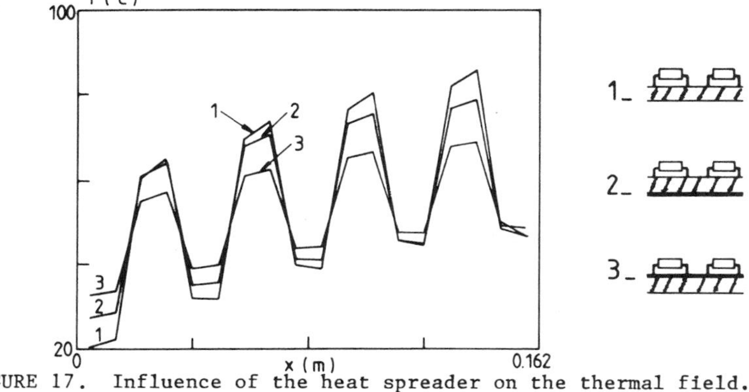

FIGURE 17. Influence of the heat spreader on the thermal field.

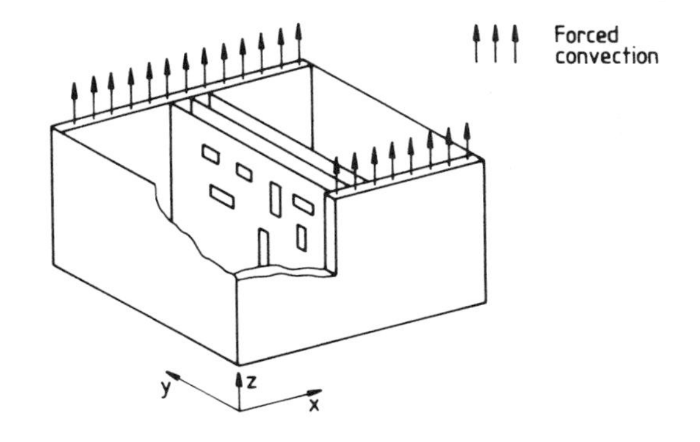

FIGURE 18. Box schematic.

The air inside the box is still, and the heat fluxes generated by the electronic components are dissipated by convection in the double-sided walls situated on two lateral surfaces of the box (turbulent forced convection) and by radiation and convection on the other external sides of the box (natural convection on plane plates).
The model's geometry is represented in Figure 19 and the following hypothesis are considered:

- periodicity of the thermal field.
- uniform heat flux distribution on the boards corresponding to 40% of the total heat flux generated by the components.
- uniform heat flux distribution in the air near the boards corresponding to 60% of the total heat flux.
- a plane of symmetry is installed in the center of the box parallel to the double-sided walls.
- the air in the box is still and the heat exchanges are purely conductive (according to the Van der Held hypothesis [5]).

The electric connection between the bus and the board is realized by the socket represented in Figure 20. This connector is composed of a bloc of epoxy resin and 144 brass-wires ($\emptyset$ = 0.5 mm, ℓ = 28 mm). In the model the conductances representing these sockets are evaluated thanks to the equivalent conductivity concept.

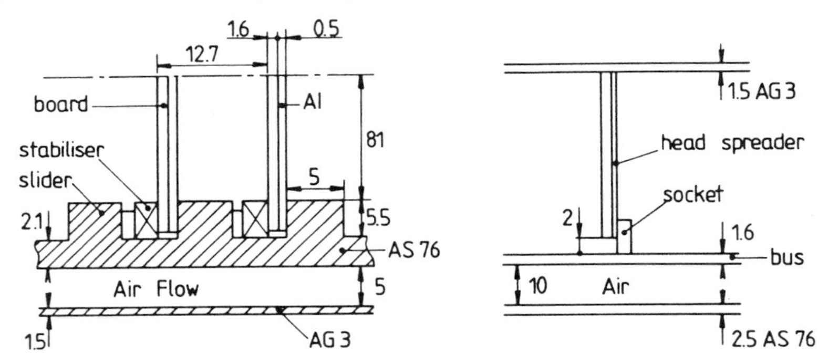

FIGURE 19. The electronic box scheme.

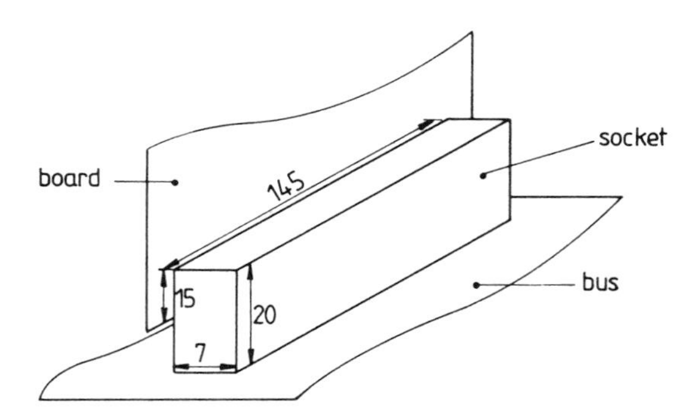

FIGURE 20. Connection between the bus and the boards.

A contact between the heat spreaders and the sliders develops a thermal resistance. Its evaluation depends on the surface roughness and the contact pressure (6).
The thermal conductivity of the materials involved in this model are presented in Table 1.

material	λ (W/m°C)
AS 7 G	159
AG3	125
Epoxy	0.4
Al	203
Air	0.03

TABLE 1. Material thermal conductivity.

The fluid flow in the double-sided walls is represented by a network of 600 nodes : 20 on the fluid flow direction and 30 between the two walls. The velocity profile is calculated previously by a finite-difference method : we have here a turbulent configuration, and a mixing-length model is used to estimate the turbulent diffusivities.
The results obtained for an ambient temperature of 20°C, a global dissipation of 6 W and an air velocity of 35 m/s are presented in Figures 21a and b.
We can see that the maximum board temperature is 43°C which represents a Δt of 23°C with regard to the ambient temperature. The horizontal gradients are larger than the vertical ones showing that the heat flux is accommodated mostly by convection in the double-sided walls. Actually, the bulk temperature of the air at the channel outlet is T_{BS} = 21.84°C. The heat flux transmitted from the structure to the flow of air is found to be equal to Q = 5.4 W. This value corresponds to 90% of the total heat flux dissipated inside the box by the board (Q = 6.0 W).
In Figure 21b we show the air temperature near the board. The temperature level is here higher than for the board. This result is a consequence of the heat flux distribution : 40% of the total heat flux generated by the components is imposed on the board and 60% in the air.

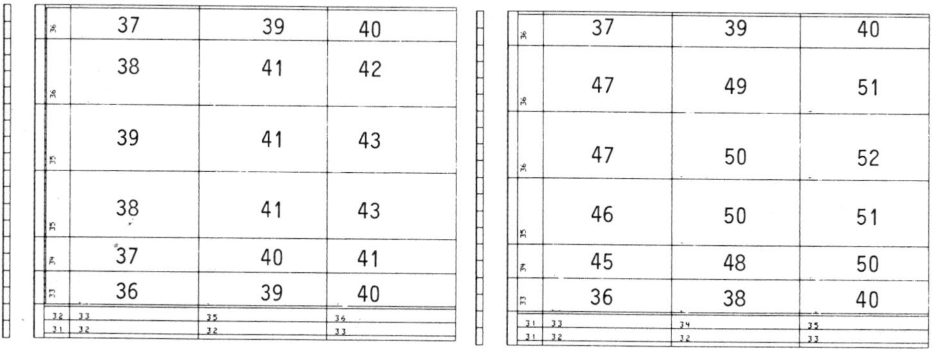

37	39	40
38	41	42
39	41	43
38	41	43
37	40	41
36	39	40

a - temperature field on the board.

37	39	40
47	49	51
47	50	52
46	50	51
45	48	50
36	38	40

b - temperature field for the air.

FIGURE 21. Sectional elevation of the box.

5. CONCLUSIONS

We have shown in this paper a new way of modelling the coupled conduction convection problem, using only the nodal method (global nodal method: GNM).
This non conventional approach has given us excellent results for some academic cases: we have achieved the validation of the GNM method by comparing the solutions obtained with those calculated by a more classical technique (finite-difference for the fluid and nodal method for the structure).
This new global nodal method allows the simulation of large models because the convection's representation is performed with fewer nodes than by a finite-difference method. The consequence is that we have been able to simulate the thermal behavior of a set of several boards at the same time.
The first example we have chosen concerned 3 boards located in a small box. Because of the air velocity between the boards, the thermal boundary layers were still thin compared to the distance between the boards and there is practically no thermal interaction between them (entrance region).
In the second example, we could potentially simulate large models, by taking into account some realistic details of the board's environment: sliders, connectors, bus, heat spreader, etc...
Besides in every case, we could easily resolve the lack of correlations for the heat exchange coefficient, which is, in fact, now a particular result of our simulations.

NOMENCLATURE

x	abscissa along the fluid flow
y	ordinate perpendicular to the fluid flow
T_i	temperature of node i
T_e	fluid entrance temperature
T_a	ambient temperature
T_p	surface temperature
T_B	bulk temperature
h	heat exchange coefficient
Q_i	heat source
G_{ij}	thermal conductance between nodes i and j
u	axial velocity
u_e	inlet velocity
v	transverse velocity
C_p	specific heat
$\dot{m}$	mass flow rate
$\dot{m}_T$	transverse mass flow rate

Greek symbols

λ	conductivity
ρ	density
ΔT	temperature difference

REFERENCES

1. J.C. HUCLIN, J.B. SAULNIER, New developments in the simultaneous computation of the conduction / convection partial differential equations. IMACS Congress, Bethlehem, Pennsylvania, 1984.

2. J. BASTOS, J.B. SAULNIER, Modelling the steady state coupling conduction / convection in a rectangular channel : simulation by a thermal network analyzer. Congrès IMACS, Paris, France (18/22 July 1988).
3. M.S. BHATTI, C.W. SAVERY, Heat transfer in the entrance region of a straight channel: laminar flow with uniform wall heat flux. ASME-AIChE Heat Transfer Conference, Saint Louis, August 9-11 1976.
4. A.F. EMERY, F.B. GESSNER, The numerical prediction of turbulent flow and heat transfer in the entrance region of a parallel plate duct. Journal of Heat Transfer, November 1976.
5. F. KREITH, Transmission de la chaleur et thermodynamique. Masson et Cie Editions 1987.
6. General Electric, Heat transfer and fluid flow data book, 1987.
7. J. BASTOS, Extension de la méthode nodale à la modélisation des transferts couplés conduction / convection dans un canal: Application à la thermique du bâtiment et aux cartes électroniques. Thèse de l'Université de Poitiers, France, 1988.

On the Development of Efficient Interactive CAD Techniques for Circuit Boards

N. OTTAVY and G. PIERRA
Ecole Nationale Supérieure de Mécanique et d'Aérotechnique
Laboratoire Informatique
20 rue Guillaume VII 86034 Poitiers, Cédex, France

ABSTRACT

The lifetime of an electronic circuit board depends closely on the temperature of its components. A difference of a few degrees from the nominal design temperature reduces the component lifetime considerably. This emphasizes the great importance of having a high-speed interactive function to take the thermal aspects of the board into account right from the design phase. Since existing codes are not fast enough, the French Telephone Research Agency (CNET) decided to work with the ENSMA Computer Science and Thermics Laboratories in developing a 2D finite element model of the problem for use in an interactive task. The purposes of the study are to compare the different finite elements types that can be used for the problem at hand, to evaluate the feasibility of a code offering a reasonable precision within the set time constraints, and to analyze the possibilities of stepping up the speed substantially during the interactive design phase, during both the original designing of the board and during later modifications. This paper is divided into three parts. In the first two, we show that the classical finite element method cannot satisfy all three essential objectives simultaneously (i.e., precision, interactive mode and component mobility). In the third part, we introduce a new method which consists of building up the finite element method mesh by means of superimposition of independant local meshes. We demonstrate the possibility of achieving all three objectives by means of this Mesh Superimposition Method (MSM).

GENERAL PRESENTATION

Reference Problem

What we want to do is to compute the temperature distribution on an electronic board carrying components on a rectangular epoxy support. To facilitate the modeling and its validation, the reference board comprises twelve heating elements, regularly distributed over a six-row/four-column area, with each element dissipating a different power. With this regular geometry, the analytical process was first carried out on a "board" with a single heating element in the center, and then on the whole card with twenty-four components.

The test board support is described by a 3D shape $[0,a] \times [0,b] \times [0,e]$. The values of the sides a , b and thickness e are shown by Figure 1. It is presumed to be made of epoxy, the conductivity of which is $k = 0.45$ W/mK. Each heating element is also a 3D shape $[0,a'] \times [0,a'] \times [0,e']$. The side of the square surface is $a' = 1$ cm and the thickness is here $e' = 17$ μm. The heating zone is in fact made of a very thin layer of resistive ink, and it can dissipate a heat flux up to 2 W (heat density 2.10^4 W/m^2). The experiments were conducted with forced convection corresponding to an air flow velocity $V = 4$ m/s.

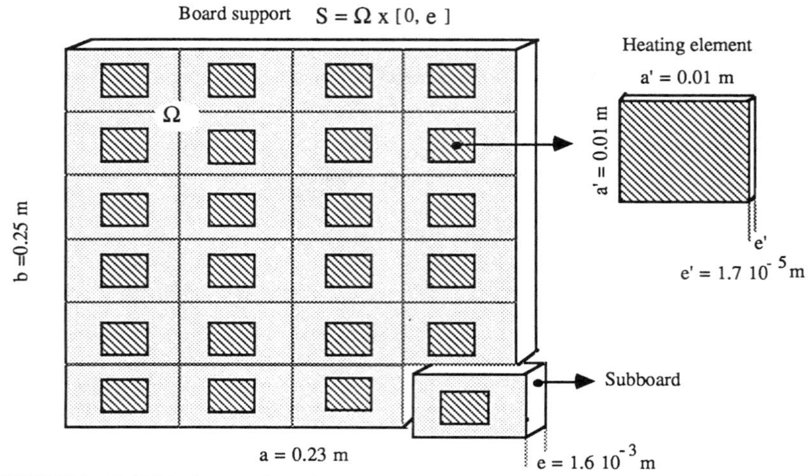

FIGURE 1. Full 24-element board.

The ambient temperature was equal to 20°C, and for these conditions we shall assume a convection heat transfer coefficient $\alpha = 30$ W/m^{2}K. In fact, this value takes account both of convection and of radiation (assumed to be linearized). The heat transfer on an edge of the support will be neglected, and the problem will be considered as a 2 D conduction/convection one. Let us mention that already existing methods of thermal model have been applied to this kind of problem (network analyzers : Kraus and Bar-Cohen [1], Saulnier and Bertin [2] ; finite elements : Patureau [3]). But our main objective is to reduce computation time and to facilitate (thanks to a high degree of interactivity) any modification in the integrated circuit implementation.

Modeling

Using the notation: T = temperature, Δ = Laplacian, W = flux dissipated by the components, $\Omega = [0,a] \times [0,b]$ and $\partial\Omega$=boundary of Ω, the two-dimensional equation classically used with the above hypotheses is written as

$$- ke\, \Delta T(x,y) + 2\alpha\, [T(x,y) - T_0] = W\,(x,y) \qquad ; \quad (x, y) \in \Omega$$
$$(dT/dn)(x,y) = 0 \qquad ; \quad (x, y) \in \partial\Omega$$

The above equation is the strong form of the classical Newman problem. In its weak form, the problem admits a unique solution in $H^1(\Omega)$ when W is in $L^2(\Omega)$ Ciarlet [4], Dold and Eckmann [5] and Lions [6].

Reference Solution

The solution of this problem is a generalization of a method presented in Carslaw and Jaeger [7] concerning a board with a uniform value of the dissipated power. It can easily be shown that the solution to the present problem admits a series expansion on the orthonormal base F_{ij} $(i,j \in N)$:

$$T(x,y) = T_0 + \Sigma\, T_{ij}\; F_{ij}\;(x,y) \qquad ; (x,y) \in \Omega$$

in which

$$F_{ij}(x,y) = (d_{ij}/\,(ab)^{\,0.5})\;\cos\,(i\pi x/a)\;\;\cos\,(j\pi y/b)$$

$$d_{ij} = \begin{cases} 1 & \text{if } i = 0 \text{ and } j = 0 \\ (2)^{0.5} & \text{if } i \times j = 0 \text{ and } i + j \neq 0 \\ 2 & \text{if } i \times j \neq 0 \end{cases}$$

$$T_{ij} = W_{ij}\;(k\;ec_{ij} + 2\alpha)\;c_{ij} = (i\pi/a)^2 + (j\pi/b)^2$$

$$W_{ij} = \int_\Omega (WF_{ij})\,(x,y)\;dx\;dy$$

As the component support is rectangular, the coefficients W_{ij} can easily be expressed as a finite sum of sines. This series has a two-fold advantage. On the one hand, it provides a simple way of computing the exact solution for the whole card, allowing us to measure precisely the performance of each finite element type without resorting to experimental data. On the other hand, it has revealed itself to be a very interesting alternative to the finite element method for computing the whole board when the physical characteristics (conductivity, conduction) are constant.

CLASSICAL FINITE ELEMENT METHOD

Reference Finite Elements

Considering the specific geometry of the support and the components, the most appropriate elements are straight subparametric elements. The requirement that temperature modifications due to the interactive movement of a component be computed rapidly, and therefore without changing the grid, calls for a grid that is rectangular in the x and y directions. Using the multi-index concept, we denote by Q_n (resp. $Q_{0,3}$) the space of the two-variable polynomials ($X = (x,y)$) with real coefficients a_β, in which $\beta = (\beta_1, \beta_2)$:

$$Q_n = \{\; \Sigma\;\; a_\beta\; X^\beta\,;\;\; \beta \leq n\;\}$$

$$Q_{0,3} = \{\; \Sigma\;\; a_\beta\; X^\beta\;\;;\;\; \beta \leq 3 \;\text{ and }\; |\beta| \leq 4 \;\;\text{ and } \beta \neq (2,2)\;\}\;\;.$$

We chose to use two Lagrange finite elements and two Hermite finite elements, every one with rectangular supports, Ciarlet [4], Cook [8], Dhatt and Touzot [9], Zienkiewicz [10]. The first (resp. the second) Lagrange element is the full bilinear (resp. full quadratric) element with interpolation space Q_1 (resp. Q_2) which we will hereafter denote by $L[Q_1]$ (resp. $L[Q_2]$). The first (resp. the second) Hermite element is the full cubic (resp. incomplete cubic) element of Bogner, Fox and Schmit with interpolation space Q_3 (resp. $Q_{0,3}$) which we will hereafter denote by $H[Q_3]$ (resp. $H[Q_{0,3}]$).

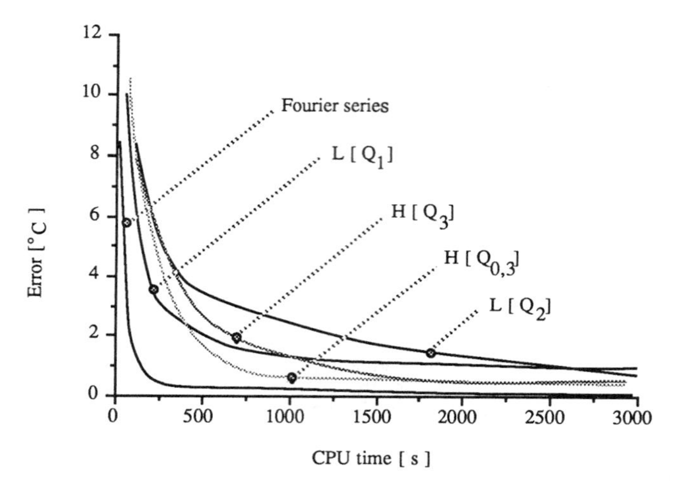

FIGURE 2. Results obtained for classical uniform meshes.

The first computations were made for a subcard with a single heating element having a flux of 1.16 W/m^2, and the results were compared with the exact solution. Three parameters δT, u_t and t were used for the comparison. δT is the maximum absolute error observed over 72 test points covering the critical area of the heating element, u_t is the number of unknowns used for the computations and t is the CPU time in seconds. The computer used is the Bull SPS7, without wired floating operator, with a manufacturer-specified power of 0.07 MIPS Whestone. The $\delta T(t)$ curve in Figure 2 visualizes the results obtained. We retained only ten representative tests for each method, for values of u_t distributed regularly between zero and 2500.

Although rarely used in industrial codes, the continuous-flux elements clearly offer the best time/accuracy ratio as the number of variables increases. The numerical problem studied is a "stiff" problem. The difficulties encountered are due to the large variations of temperature over small nonstationary areas (component mobility). This aspect of the problem leads to a high uniform density of unknowns. To meet the constraint $\delta T \leq 2^{\circ}C$, a relatively large number of variables and long computation times are needed even for the one-component subboard (Table 1).

TABLE 1. Maximum absolute error observed over 72 test points covering the critical area

		Maximum error Total number of unknowns used		$\delta T = 2^{\circ}C$ u_t
Methods	Fourier	$H[Q_{0,3}]$	$L[Q_1]$	$L[Q_2]$
u_t	0360	0510	1050	1350
CPU time	0098	0478	0603	2678

So a favorable finite element requires, for a single subcard, 500 unknowns and a CPU time of 500 seconds (or 2.5 seconds on a 1.5 MIPS Whestone work station). Though these values may seem reasonable, they become much less so when considering the complete board. By simply extrapolating these results, we find we need 8000 to 10000 unknowns and a computation time of the order of three hours (or nine minutes on a 1.5 MIPS work station), so the full card was not studied with a uniform mesh. It therefore seems impossible, using a uniform grid method, to achieve the desired interactivity, accuracy and component mobility, unless of course oversized hardware is used, Ottavy[11].

Fourier Series Method

The Fourier series method was initially expected to serve as a reference solution for the finite element computations ; but even without optimization, this method was found to perform much better than the finite element method. The desired precision for the subboard (Figure 2 and Table 1) was achieved with 360 unknowns in 98 seconds (or about four seconds on a 1.5 MIPS work station).

Classical Mesh Overlay

As we have just pointed out, the most appropriate finite elements for a uniform grid cannot solve the problem with the desired precision as part of an interactive task. Conventionally, when this difficulty is encountered, a specific mesh is constructed for the support to reduce the interpolation errors and the number of variables Ciarlet [4], Zienkiewicz and Phillips [12], Bernadou [13] and Sadek [14]. This possibility cannot be used for the present problem because we have to be able to move the components. This is why we were led to use a method of local overlapping of the elements encountering the singularity (Figure 3).This classical method amounts to a recursive breakdown of the elements containing the singularity. The approximation space V_h ($h = (h_1, h_2)$) associated with the two meshes is defined by $V_h = V_{h_2} + V_{h_1}$. V_{h_1} is constructed from the approximation space V_{h_1} of the coarse grid by suppressing the base functions whose support is included in the overlapped area, and V_{h_2} is constructed from the interpolation space V_{h_2} of the fine mesh, by suppressing the base functions on the edge of the overlapped area. Using this method requires, in particular, the computation of the apparent stiffness matrix A_h associated with the overlapped coarse elements (macro-element technique), the assembly of the coarse grid and the solution of the resulting linear system.

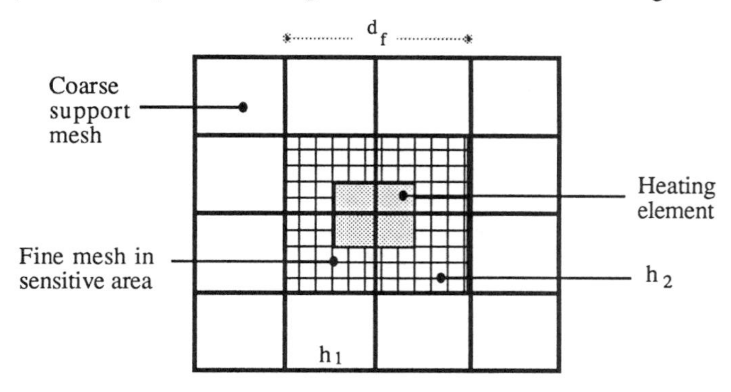

FIGURE 3. Local mesh overlay.

Results Obtained For Classical Mesh Overlay

For these tests, we assumed that the heating elements were all geometrically identical. The coarse grid was chosen such that each component was located at the center of a single mesh square of this grid. The overlapped area for each component thus consists of a finer breakdown of the mesh rectangle containing it. To optimize the computation time significantly, we used the idea of macro-elements, which makes many pretreatments possible such as preintegration and preinversion, to arrive at the elementary stiffness matrix for each mesh element quickly. Let us note that macroelements do away with internal variables and thereby considerably reduce the dimension of the linear system to be solved. Although very restrictive in practice, the above hypotheses make it possible in theory to measure the maximum performance that can be expected of the overlapping method in an interactive task environment.

The results obtained for the subcard, with the element $L[Q_1]$ (Table 2), are denoted u_c, u_f (resp. d_f) which are the number of variables respectively used on the coarse and fine grids (resp. the length of the mesh square side centered on the component and used as support for the fine grid).

The advantages the grid overlap method has over the uniform grid methods appear clearly in Table 2. For the precision $\delta T = 2^{\circ}C$, we use 164 variables instead of 1050 and the computation time is divided by six. In reality, though, considering the method used, the number of variables actually used for the computations is 64 and the time saved is greater than it appears because computing δT requires about one minute of CPU time, which gives a corrected factor of the order of ten for the CPU time required.

The results are even more favorable on the complete card (Table 3). As could be expected, the ratio of the number of unknowns used for the two methods is of the same order of magnitude as before; but considering the non-linearity of the time cost of the computation, the time gain is much greater since the corrected ratio is better than twenty.

TABLE 2. Results obtained for the subcard (three different precisions δT are studied for classical mesh overlay)

* Classical mesh overlay, # classical uniform mesh
Total number of unknowns used $u_t = u_f + u_c$

Methods	d_f	u_f	u_c	u_t	δT	CPU time
$L[Q_1]$ *	2.0	100	64	0164	4.5	105
$L[Q_1]$ *	2.2	100	64	0164	3.0	105
$L[Q_1]$ *	2.4	100	64	0164	1.5	105
$L[Q_1]$ #	-	-	-	1050	2.0	603
Fourier	-	-	-	0360	2.0	098

TABLE 3. Results obtained for the complete card (four different precisions δT are studied for classical mesh overlay)

* Classical mesh overlay, # classical uniform mesh
Total number of unknowns used $u_t = 12u_f + u_c$

Methods	d_f	u_f	u_c	u_t	δT	CPU time
$L[Q_1]^*$	2.0	100	560	1760	6.10	330
$L[Q_1]^*$	2.0	100	315	1515	7.70	255
$L[Q_1]^*$	2.2	100	315	1515	5.30	255
$L[Q_1]^*$	2.0	100	140	1340	12.0	180
$L[Q_1]^\#$	-	-		2400	13.0	2600
$H[Q_{0,3}]^\#$	-	-		1300	12.5	1900

The advantages the grid overlap method has over the uniform grid methods appear clearly in this case (Table 3). For the precision $\delta T = 12^\circ C$, we use 1340 variables instead of 2400 and the CPU time is divided by 14. In reality, though, considering the method used, the number of variables actually used for the main computations is 140 instead of 1340. Moreover,the time saved is greater than it appears because computing δT requires about one minute of CPU time, which gives a corrected factor of the order of twenty for the CPU time required.

Critical Analysis

In comparison with the uniform mesh methods, the local grid overlap method offers two essential advantages. Firstly, it reduces the number of useful variables by distributing them better. Secondly, the linear system it generates has a diagonal block structure that makes the system easier to solve and also makes it possible to retain all or part of the computations already made when a component is moved.

Despite these advantages, the actual possibilities this method offers in an interactive framework appear to be very limited. The overlap used for a component depends on the coarse mesh and on the position of the component, which means that, unless very restrictive hypotheses are adopted (Figure 4), it is impossible to insert any pretreatments such as preliminary mesh generation, preintegration, preinversion, etc. , which would accelerate the process in the interactive phase. Moreover, the fact of associating the overlapping with the elements of the coarse grid enclosing the singularity may significantly increase the number of variables depending on the size of the critical area and of the steps h_1 and h_2. Finally, computing the stiffness matrices means crossing all of the base functions of a coarse element on the edge with all of the base functions of V_{h_2} having the support contained in this element.

The above remarks are applicable to the classical methods for finite-element computation of singularities. These methods consist in transforming the geometrical mesh (h-methods) or of modifying the order of the interpolation polynomials (p-methods), or of a combination of the two (hp-methods), Babuska [15], Ciarlet [4], Zienkeiwicz and Phillips [12]. Using one of these methods to solve the problem of the random continuous motion of a singularity is an effective method in terms of precision, but catastrophic in terms of time. The random, continuous character of the motion prohibits any preprocessing and often means the problem has to be recomputed completely at each modification. To get around this difficulty, we can then use preprocessing techniques (macro-elements, etc.) and specific mesh techniques (overlap, multigrid, etc.); the price paid for such improvements by this technique are always the severe limitations on the movements.

Mesh Superimposition Method(MSM)

Principle

As we have just shown, using the most appropriate finite elements is not sufficient to solve with the desired precision the problem in an interactive task environment. This is why we were led to create a method of computation by finite elements that takes better account of the specific nature of the problem. This approach, which to our knowledge is original, is based on the use of geometrically independent meshes (Figure 4) which are superimposed. This is called the Mesh Superimposition Method (MSM).

Taken separately, the meshes used are typically finite element meshes and are treated as such. The essential difference from classical grids is that they are geometrically independent. Conceptually, the "fine grid" is related to the component and covers the critical area of strong temperature field variation, while the "coarse grid" is linked to the support. For the MSM, the approximation space $V_h(h = (h_1, h_2))$ is defined by the classical approximation space V_{h_1} (resp. V_{h_2}) defined on Ω_1 (resp. Ω_2) by the coarse (resp. fine) grid to which a sufficiently regular weighting function λ_- (resp. λ_+) is associated by $V_h = \lambda_- V_{h_1} + \lambda_+ V_{h_2}$. The functionals λ_- and λ_+ partition unity for Ω_1 ($\lambda_+ + \lambda_- = 1$) such that the support of λ_+ is included in Ω_2 (Figure 4). There are three advantages to this method. First, the components can be moved easily without modifying the grids. Secondly, the degrees of freedom are easy to distribute since the support and the components are standard elements that are perfectly described physically. And finally, a great many preintegrations and preinversions can be made, in particular on fine grids. The solution entails a simple assembly on the coarse grid using the links induced by the fine grids, which considerably reduces the volume of the problem in space and time.

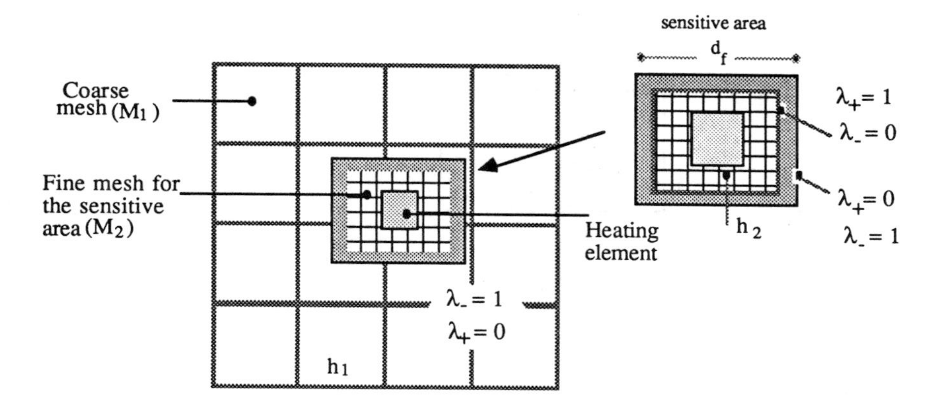

FIGURE 4. Mesh Superimposition Method (applied to the subboard).

General Scope Of MSM

Let us note that there may be many ways of using MSM depending on the field of application considered and the exact nature of the problem to be solved. As concerns the fields of our present application, the MSM was designed to allow a better integration of the finite elements into an interactive task environment. Depending on the goal we try to reach, there are three typical fields of application of the method.

The first of these fields relates to the classical problem of modifying a given mesh M_1 of pitch h_1 that is recognized to be coarse though in an area where the solution exhibits strong variations. To solve this problem, the MSM consists of superimposing a second mesh M_2, of pitch h_2, on mesh M_1. M_2 is a local, fine mesh of the area in question and is in principle completely independent of grid M_1. This approach both simplifies the meshing problem considerably and offers a good approximation of the solution, that is an error of only $0(h_2)^k$ on the M_2 mesh. Moreover, when using iterative methods Ottavy [16], the solution on the coarse grid can serve as a starting point in the search for a new solution using the superimposed grids.

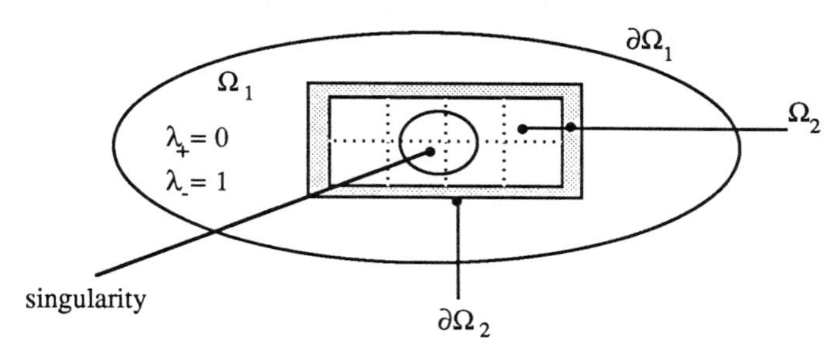

FIGURE 5. Displacement.

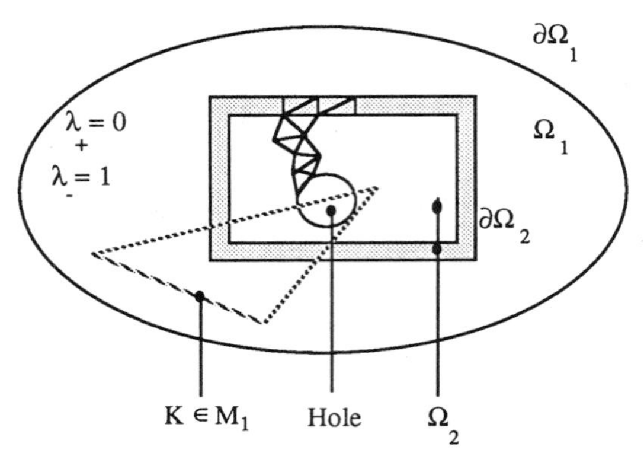

FIGURE 6. MSM mesh.

The second field of application relates to the problem of introducing, deleting or moving a "physical element" on a support Ω. The physical elements are, for example, sources of heat, obstacles to flow, mechanical anchoring points, etc. We then have situations that are similar to the previous one in form, since the physical elements are sources of singularities for the solution to the problem, which also means modifying the mean grid in the areas these elements occupy. This "displacement" problem is a typical one in computer-aided design. It has no satisfactory solution in the framework of classical grids which require a complete remodeling of the problem for each new movement. The MSM, on the other hand, consists in assigning a domain, a physical representation and a specific grid to each element in the problem that is allowed to move. The main support Ω is then regarded as a specific physical element on which other independant physical elements can be introduced, removed or moved. The MSM links these elements together "functionally". This domain of application differs profoundly from the first by the fact that the physical description of each element is attached to the local mesh of the element. This way, many preliminary computations (mesh, numbering, assembly, etc.) can be made for each element, thereby reducing considerably the volume of computations needed in solving the final problem. As before, a manipulation of a physical element in the problem does not mean recomputing the other elements.

The third field of application is related to the meshing problem itself. In this case, the MSM can considerably simplify the construction of the mesh. Inside the window (Figure 4), the supports K of the coarse grid ($K \in M_1$) are inactive and as such can, in particular, ignore the singularities inside the window, such as the hole in Figure 6.

Numerical Results

To compute the entire card by the MSM, we use the $L[Q_1]$ element and treat the problem by four different procedures for the sake of easier comparison with the previous results, and to get an idea of what the MSM can do.

TABLE 4. This table shows the computation time t as a function of the precision δT obtained for the four situations above. Computations are carried out on a VAX 8250 operating at 1.5 MIPS Whestone and two different precisions δT are studied for MSM.

CPU time					
	δT	PROC. 1	PROC. 2	PROC. 3	PROC. 4
MSM	10.79	128	53	11.46	4.15
MSM	05.08	283	84	16.57	8.11
Classical uniform mesh	13.07	182	-	-	-

Procedure 1. Problem fully computed without any preprocessing both by MSM and classical uniform mesh.

Procedure 2. Full computation using the preprocessing possibilities the MSM offers.

Procedure 3. Full computation with the movement of a single component, without using the preprocessing of procedure 2.

Procedure 4. Full computation with the movement of a single component, using the preprocessing possibilities of procedure 2.

The analysis of the above results shows that, for a given precision, great time savings can be made between procedures 1 and 2 (resp. 3 and 4), essentially because of the preprocessing possibilities the MSM offers. These results are in particular very favorable for procedure 4, the problem of random, continuous movement of a singularity. The result is that the MSM is appropriate for an interactive task in a CAD environment if we consider that the basic time t associated with a movement (10 s for a required precision of two or three degrees Celsius) can be further reduced by optimizing the code in the compilation phase (25%) and by optimizing the algorithms specific to the code (about 30%) which gives a mean component movement time of five seconds on a machine operating at 1.5 MIPS.

CONCLUSIONS

One conclusion to be drawn from this study is that the most appropriate finite element types for solving the reference problem are those that account for the underlying physics of the problem by ensuring flux continuity. The increase in the order of the interpolation polynomials is found to be a very effective operation in terms of precision and cost, but only when it serves to ensure the continuity of the first derivatives. In the other cases, this operation is effective in terms of precision but catastrophic in terms of cost.

The second conclusion is that a classical computation by finite elements cannot be an effective interactive electronic board design tool. Only specific methods like MSM can achieve this objective. This is because the process of designing an object consists essentially of a succession of modifications Gardan [17], Newman and Sproull [18]. Introducing an interactive tool using classical finite elements in a CAD environment would require a complete and rapid solution to the problem each time a modification is made, which is not feasible in practice considering the cost of the computation involved in the finite element method. To get around this difficulty, a large part of the computations already made must be preserved when there is a modification. This method goes along with this, and the numerical results from the MSM are extremely encouraging, at least for the reference problem. Strictly from the point of view of the finite element computation, in addition to the simplification for the user, who no longer has to create a complex grid, the MSM offers a computation time saving by a factor of at least ten. From the CAD point of view, the possibility of being able to consider the finite element modeling of an object (grid, stiffness matrix, etc.) as a specific attribute of the object makes the operations of adding, moving or eliminating easier. These operations can be carried out without disturbing the existing linear system greatly. This approach should make it possible to integrate the finite element method as an interactive function in a CAD system. As we have already emphasized, there are many ways of putting the MSM to use. So the road ahead is far from being entirely explored. In particular, the best strategy for using the MSM remains to be defined in each case.

NOMENCLATURE

d_f	length of the fine mesh square side
a, b	length sides of rectangular board
a'	length side of square heating element
e, e'	thickness
h, h_1, h_2	steps of meshes
k	thermal conductivity
M, M_1, M_2	meshes
n	natural number
T	board temperature
T_0	ambient temperature
u_f	number of unknowns (fine grid)
u_c	number of unknowns (coarse grid)
u_t	total number of unknowns used
W(x, y)	flux dissipated by the components
x, y	cartesian coordinates

Greek letters

α	heat transfer coefficent
β	multi-index
Δ	Laplacian operator
$\lambda, \lambda_+, \lambda_-$	MSM weighting functions
Ω	domain
$\partial\Omega$	Ω boundary

Others symbols

FEM	finite element method
$H^1(\Omega)$	Sobolev space
$L^2(\Omega)$	Sobolev space
MSM	mesh superimposition method
N	natural number set
$Q_n, Q_{0,n}$	polynomials spaces
V_h	FEM or MSM approximation space
V_{h_1}, V_{h_2}	FEM or MSM approximation spaces

REFERENCES

1. Kraus, A.D., and Bar-Cohen, A., *Thermal Analysis and control of electronic equipment.*, Hemisphere and McGraw-Hill, New York, 1983.

2. Saulnier, J.B., and Bertin, Y., *The detailed analysis of the conductivelconvective heat transfer in a single electronic board. XX th International Symposium "Heat Transfer in Electronic and Microelectronic Equipment" International Centre for Heat and Mass Transfer.* August 29 -September 1988, Dubrovnik, Yugoslavia.

3. Patureau, M.J., *Modelisation thermique d'une carte de composants électroniques : un outil d'aide et de conception. Journées d'etudes sur les Aspects Thermiques dans les Matériels de Télécommunication* , CNET, Perros Guirec, France, December 1980.

4. Ciarlet, P.G., *The finite element method for elliptic problems*, North-Holland Publishing Company, New York, 1978.

5. Dold, A., and Eckmann, B., *Mathématical aspects of finite element method*, Lecture Notes in Math. Springer-Verlag, New York, 1977.

6. Lions, J.L. , *Problèmes aux limites non-homogènes et application*, Vol. 1, Dunod, Paris, 1968.

7. Carslaw, H.S., and Jaeger, J.C., *Conduction of heat in solids* , Oxford, 1959.

8. Cook, R. D., *Concept and applications of finite element analysis*, John Wiley & Sons, New York, 1981.

9. Dhatt, G., and Touzot, G., *Présentation de la méthode des éléments finis*, Maloine S.A., Paris, 1982.

10. Zienkiewicz, O.C., *The finite element method in engineering*, Mc Graw-Hill, New York, 1971.

11. Ottavy, N., *Etude des possibilités offertes par la méthode des éléments-finis en CAO, lors de la conception de cartes électroniques*, convention CNET n° 84 8B 00 790 9245 LB/SER/EVV 84 PE 0 464, Report 6, ENSMA, Poitiers, 1986.

12. Zienkiewicz, O.C., and Phillips, D.V. , *An automatic mesh generation schema for plane and curved surfaces by isoparametric coordinates*, Int. Jou. for Numerical Methods in Engineering, vol. 3, pp. 519-528, 1971.

13. Bernadou, J., *Modulef bibliothèque modulaire d'éléments finis*, Inria, Paris, 1985.

14. Sadek, E.A., *A schema for the automatic generation of triangular finite elements*, Int. Jou. for Numerical Methods in Engineering, vol. 15, pp. 1813-1822, 1980.

15. Babuska, I. , *A posteriori error analysis of finite element solutions of one dimensional problems*, SIAM, J. Numer. Anal., Vol 18, pp. 565-589, 1981.

16. Ottavy, N., *Strong convergence of projection-like methods in Hilbert spaces*, Journal of Optimization Theory and Applications, vol. 56 no. 3, pp. 433-461, March 1988.

17. Gardan, Y., *la CFAO, introduction technique et mise en œuvre*, Hermes, Paris, 1986.

18. Newman, W.M., and Sproull, R.F., *Principales of interactive computer graphics*, Mc Graw-Hill, New York, 1979.

A Comparison of Finite Element Conduction Heat Transfer Codes

KENNETH S. MANNING and DEBORAH A. KAMINSKI
Rensselaer Polytechnic Institute
Troy, New York, USA

ABSTRACT

Computer codes for the purpose of thermal design and analysis were judged for suitability and four were chosen for in depth study. Evaluation was made on the basis of both objective and subjective criteria by using each code in the solution of four heat transfer codes.

INTRODUCTION

Finite element computer codes are powerful and versatile tools for analyzing new designs for the cooling of electronic equipment. Designs can be modeled and variables adjusted more quickly and inexpensively than can be accomplished with a physical test. The purpose of this study was to examine many of the commercial codes available for solving heat transfer problems. From among these codes four were chosen for extensive evaluation. Objective criteria such as accuracy, speed, quality of documentation, pre-processor flexibility, and reputation were considered. Subjective judgements of friendliness and readability are also reported.

PROCEDURE

A literature search produced only one compehensive paper [1] that gave a survey of heat transfer codes available on the market. This survey was based on questionnaires that were submitted by the developers of each code. The results were compiled and tabulated, and a written description was supplied for each code, including size, cost to purchase, language used, etc. No attempt was made in [1] to actually judge the codes in any way. The table in [1] was used as the basis for choosing the four codes that would actually be evaluated in this project. Features that were judged to be most useful for thermal analysis of integrated circuits were chosen and weighted for ranking of the codes. Points were given to the codes for the features each possessed, ranging from zero points for not very useful or critical, to three points for highly desirable. The scores were totaled and judgement made to determine which codes would be studied further.

The desired features and the weighting assigned to each are shown in Table 1. Once the codes were ranked by overall score several other characteristics were used to eliminate codes, as shown in Table 2. A code could not be too small (less than about 20 000 lines) as such codes do not show the versatility re-

TABLE 1. Desired Features and Weights

---

3 points (most desirable)
- finite element code
- conduction
- forced convection
- temperature dependent flux boundary conditions
- forced convection boundary conditions
- fluid nodes
- steady state direct solution
- uncoupled thermal stress
- triangular elements
- restart capability

2 points
- linear steady state
- linear transient
- non-linear steady state
- capacitance
- free convection
- isotropic
- anisotropic
- multi-layered
- temperature dependent conductivity
- temperature dependent convection coefficient
- variable initial conditions
- steady state temperature boundary condition
- steady state flux boundary conditions
- free convection boundary conditions
- narrow gap radiation
- elemental heat generation
- nodal heat generation
- contact resistance
- field data transfer
- symmetry boundary conditions
- user defined output
- maxima, minima flags
- temperature limit flags

1 point
- general purpose code
- radiation
- inter-elemental convection-radiation
- coupled thermal stress
- automatic renumbering
- data plotting
- model plotting

TOTAL POINTS 87 (not all features are listed)

quired. In deference to the sponsors of this project the code must be available for use on the VAX system, and the code must be well supported in the United States. And finally, to avoid actually purchasing the codes a service was found that allowed each of the codes to be used on the same system. This service was the Scientific Information Service (SIS) supplied by the Control Data Corporation (CDC), and the system was the SUPRA CRAY X-MP. These final constraints

TABLE 2. Rank of All Codes

Rank	Code	Score	Size	Notes
1	NASTRAN	78	430	
2	TAU	77	18	a,b,e
3	BERSAF	76	100	c,e
4	MARC	73	80	
5	SAMCEF	72	40	c,e
5	SPAR	72	18	a,e
7	ANSYS	71	100	
7	TRUMP	71	10	a,b,e
9	ABAQUS	70	75	
10	WECAN	68	155	b,e
11	ANDES	66	9	a,b,e
12	ADINA-T	63	57	
12	HEATRAN	63	10	a,b,e
14	SINDA	63	30	d
15	MITAS	62	150	b,e
15	TEMP	62	75	b,e
17	ASASHEAT	60	35	e
18	PAFEC	57	--	e
18	TACO	57	12	a,b,e
20	HEATING-6	52	15	a,b,e
21	SAHARA	51	9	a,b,e
21	SSPTA	51	5	a,e
23	HEATING-5	49	--	e
24	SESAM	45	110	b,c
25	NTEMP	44	10	a,b,e
26	THTD	40	22	a,b,e
27	TAC-3D	37	150	b,e
28	TAC-2D	35	120	b,e
29	THACSIP	33	5	a,b,e
30	NNTB	31	2	a,e
30	AGTAP	31	1	a,b,e
32	CAVE-3	26	3	a,b,e
33	TANG	20	2	a,b,e
33	CAVE-1	20	3	a,b,e
33	CAVE-2	20	3	a,b,e
36	FLUX-2	11	1	a,b,e
37	CONFAC	10	1	a,b,e

notes:

a) small source code
b) not on VAX
c) non-USA code
d) not a finite element code
e) not on CDC SIS

eliminated all but four codes. The ranking and the final criteria are shown in Table 2. The four codes chosen for evaluation were ABAQUS, ANSYS, MARC, and NASTRAN [2-5]. The user's manuals were then purchased and the evaluation begun.

THE TESTS

Three test cases, in increasing complexity and comprehensiveness, were used to evaluate the codes.

Test 1

The purpose of the first test was to examine the ease of use of the codes, to evaluate the clarity and accessibility of the manuals, and to determine the accuracy and speed of the codes. The basic model was that of a solid copper bar undergoing various thermal and mechanical loadings. Four different loadings were used, all with available analytical solutions.

The codes were chosen in random order, the manuals read as needed, and the inputs to describe the loadings were written. It was a simple matter to compare the time required to complete the inputs of each of the four loadings, and the accuracy, speed, and system cost of each code.

Loading 1. The temperatures of the sides of the solid were held constant: $T_1=100^oC$ and $T_2=0^oC$, and there was no heat generation. The top and the bottom of the solid were adiabatic, as shown in Figure 1. This was then modeled by a five-by-nine node mesh. Analytically, the solid should be isothermal from top to bottom, and the temperature profile should be linear from end to end.

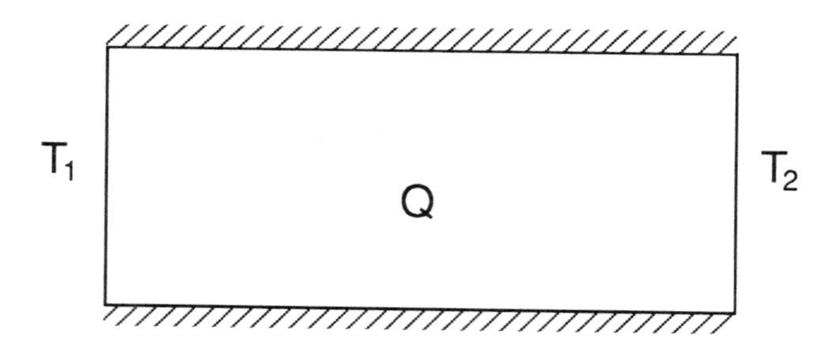

FIGURE 1. The physical model used for test 1, loadings 1 and 2.

Loading 2. The side temperatures were equal: $T_1=T_2=100^oC$, and there was heat generation, $Q=80\ 000\ w/m^2$. Again the sides were adiabatic. The calculational mesh remained the same as in loading 1. The analytical solution was again isothermal across the solid, but the longitudinal profile would be parabolic and symmetric about the center of the model.

Loading 3. This model was a one-dimensional bar with very high conductivity. The ends were completely constrained when the solid was 0°C. The ends were instantaneously changed to 100°C, and the codes were used to calculate the temperature distribution. The resulting mechanical stresses were then determined by the codes. This loading specifically tested the ease with which a code can be made to reach a temperature solution, and from that calculate an uncoupled stress solution.

Loading 4. A long tube was surrounded by an insulating material, as shown in Figure 2. The inside and outside surfaces had convection boundary conditions. The model was then reduced to a two-dimensional pie-shaped wedge.

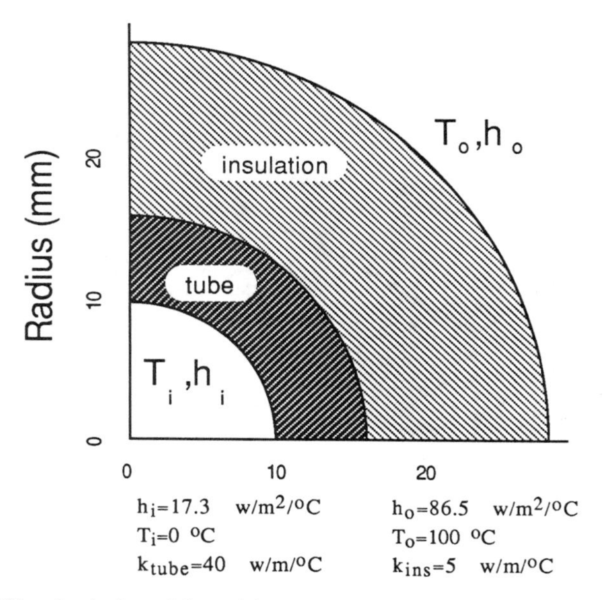

FIGURE 2. The physical model used for test 1, loading 4.

Test 2

Loading 1 used in test 1 was modified from 40 to 170 nodes, and then again to 340 nodes to assess the ability of each code to handle larger problems. Little effect was expected on the accuracy of the results, but significant changes were expected in the speed and cost as the nodes increased. The trends could easily be compared for the four codes.

Test 3

The third test was a three-dimensional model of a multi-layered circuit board with one face undergoing forced convection as though cooled by an airstream. The opposite face had areas of surface heat flux representing chips. All other surfaces were adiabatic. This model is shown in Figure 3. The problem was first executed to obtain a steady state solution. Then, in a separate run, the convective heat transfer coefficient was reduced to approximately 40% of the original value, representing decreased air flow. The problem was then run as a transient from an isothermal initial condition that approximated the steady state results of the first part. No analytical solution was available for this problem, but the codes were compared with each other as a determination of accuracy. Again, the relative times and costs of the runs were expected to be of great interest.

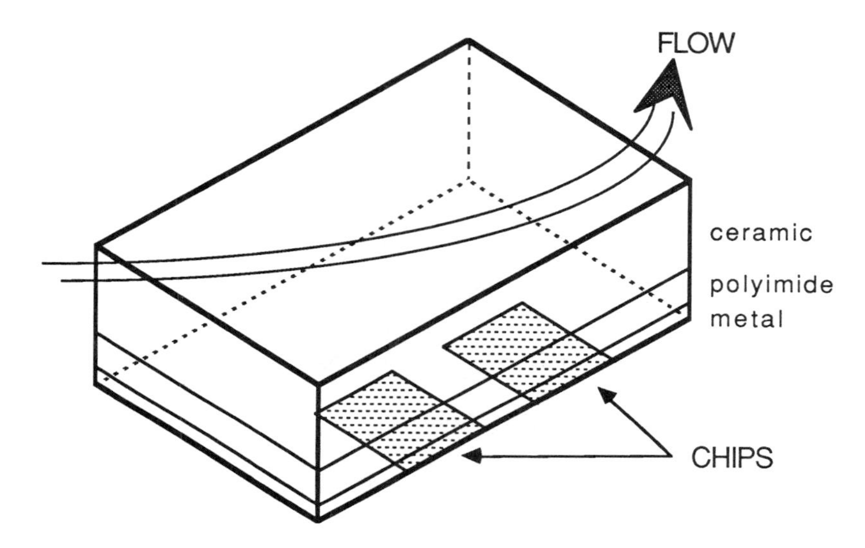

FIGURE 3. The physical model used for test 3.

RESULTS

Test 1

The most striking of the test 1 results was that the NASTRAN code was found to be inaccessible to the new user. After many hours studying the manual, not one line of input code could be written. The manuals were verbose, with no easily determined starting point past the introduction, and had no illuminating examples. From this point on, only ABAQUS, ANSYS, and MARC were included in this study.

The objective results show the accuracy, speed, and efficiency of the codes. All three codes were accurate within one tenth of one percent when compared to the analytical solution, as shown in Table 3. There were large differences in the speed and system cost of the codes. MARC was consistently faster and more economical than the others. MARC took about 0.3 CP seconds, ANSYS required

Table 3. Results of Test 1.

Loading 1

distance	theory	Temperature (°C) ANSYS	ABAQUS	MARC
0.0000	0.0000	0.0000	0.0000	0.0000
0.0025	14.287	14.287	14.29	14.286
0.0050	28.571	28.572	28.57	28.571
0.0075	42.858	42.859	42.86	42.858
0.0100	57.143	57.144	57.14	57.143
0.0125	71.429	71.429	71.43	71.429
0.0150	85.714	85.715	85.71	85.714
0.0175	100.000	100.000	100.000	100.000
CP seconds		0.762	2.426	0.303
SBU's		149	67	35

Loading 2

distance	theory	Temperature (°C) ANSYS	ABAQUS	MARC
0.0000	100.00	100.00	100.00	100.00
0.0025	137.40	137.41	137.4	137.41
0.0050	162.34	162.35	162.3	162.34
0.0075	174.81	174.82	174.8	174.81
all were symmetric for the other half of the model				
CP seconds		0.633	2.425	0.332
SBU's		143	67	41

Loading 3

distance	theory	ANSYS	Stress (Pa) ABAQUS	MARC
0.0	30 000	30 000	30 000	30 000
1.0	30 000	30 000	30 000	30 000
CP seconds		0.653	2.142	0.345
SBU's		144	62	68

Loading 4

distance	theory	Temperature (°C) ANSYS	ABAQUS	MARC
0.010	34.571	34.598	34.60	34.598
0.028	38.258	38.217	38.22	38.217
CP seconds		0.672	2.347	0.307
SBU's		142	66	40

0.7 seconds, and ABAQUS needed 2.4 seconds. System cost was determined from the system billing units (SBU's) required to solve the problems. This relative measure involves the compilers, tapes, files, printers, etc. for the entire problem. MARC required 35 to 68 SBU's, ABAQUS needed about 65, and ANSYS used about 145, as shown if figure 4.

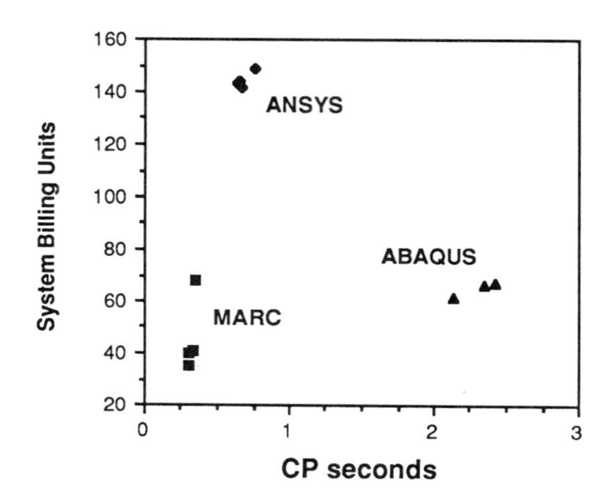

FIGURE 4. The results of the first test, all four loadings.

Test 2

Loading 1 from test 1 was then analyzed using increasingly finer meshes. The temperatures calculated differed insignificantly from those found in test 1. However, important changes were found when time and cost were compared for the codes.

As seen in Figure 5, MARC was the fastest and the least costly for the first two runs of 40 and 170 nodes, while ABAQUS was slower but less expensive than ANSYS was. However, at 340 nodes MARC became the most expensive of the three, but was still the fastest. The trends for ANSYS and ABAQUS were essentially linear. The exact conclusions that can be drawn from this might depend on the computer environment in which the codes are used, so none are made here.

Test 3

In the first part of test 3 each code calculated a steady state solution. The steady state problem was linear, so no iteration was required. Solutions from all three codes agreed to within a tenth of a degree at every node. The input decks required for the steady state runs differed greatly in size. The ANSYS

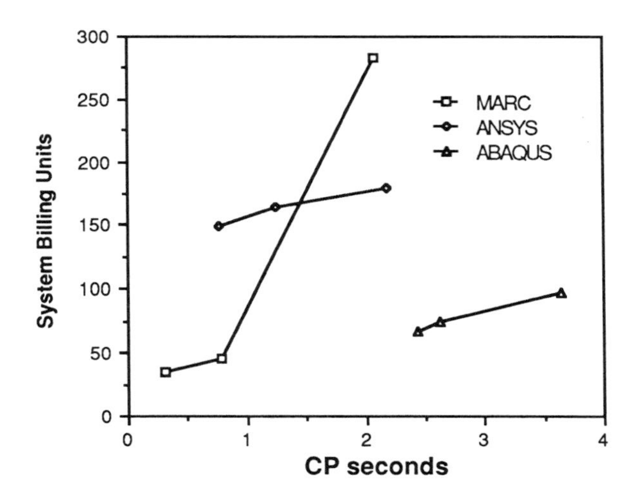

FIGURE 5. Time and Costs for Increasing Nodes: 40, 170, and 340 nodes left to right. deficiencies in MARC and ANSYS.

deck was only 89 lines long, while the ABAQUS deck was over three times longer at 291 lines. MARC input was 118 lines.

A special problem was encountered in setting up the model for ANSYS for test 3. ANSYS does not allow the specification of surface heat flux as a boundary condition. Extra nodes and elements had to be attached to the face where surface flux was desired, and these elements were assigned an equivalent volumetric heat generation rate. The thermal conductivity of the extra elements was set at a very low value, and the density and specific heat (needed for the transient runs) were set at very large values.

The original intent of test 3 was to use the steady state solution, including the geometry, as restart input for the transient, and in this way test both the transient and the restart set-up for each code. However, great difficulty was encountered in doing this with all three codes. It was then decided that each vendor should be called for advice, without letting them know the comparative nature of this study. In this way, the vendors could also be assessed for their ability to provide support.

The MARC technicians recommended that the geometry be regenerated, rather than read in, but this increased the length of both input and output, and for very large problems would be prohibitively wasteful. In fact, special system measures were required to receive the lengthy MARC output, and these were not always successful. Attempts to reduce the output, with the aid of the technicians, were unprofitable, and the full transient was never run.

The ANSYS technicians admitted to being puzzled by the problems encountered. They offered advice to check the validity of the data read in, but did not help otherwise. Again, no transient was completed. The technicians were also uncertain which restart to use (normal or alternate) until the alternate restart was described to them, at which time they agreed it should be used.

In the first of two calls to the ABAQUS technicians useful advice was obtained that allowed the transient to be attempted with only minor problems remaining. Upon a second call to solve these, a second technician offered advice that was in direct opposition to that given by the first. However, the second technician said he would take a look at the input and output files through the SIS service (the only technician to offer to do so), and he would call back. He returned the call over a week later, by which times the problems were corrected. It must be re-emphasized that ABAQUS was the only code that was running a transient without help from the vendors.

At this point it was decided to simplify the transient in order to keep the study moving. This time the initial temperature condition would be isothermal, and not the steady state solution from part 1. The geometry was regenerated for each. There was no connection to any results from the steady state runs, except that the initial temperature for the transient was the average temperature from the steady state. The transients were all set-up and begun successfully from this point.

ABAQUS and ANSYS could terminate the transient when the time limit was reached, the maximum time steps were taken, or upon steady state, by user's choice. ABAQUS was the only code where this was done with success. Halting ANSYS at steady state was complicated by the presence of the extra nodes needed to model the surface flux. These nodes were still changing in temperature after the main body nodes had reached steady state criteria, making the transient run too long. Surface flux could also have been modeled by using a convection boundary condition with a sink temperature higher than the anticipated surface temperature by several magnitudes, giving an essentially constant temperature difference, and thus surface flux. Either method unnecessarily complicated the model. MARC would allow termination upon time limit, step limit, or when all temperatures were either above or below a set value, again by user's choice. These termination difficulties were considered serious deficiencies in MARC and ANSYS.

All three codes offered time-step optimization options to reduce the number of time increments as the transient progressed. The ABAQUS procedure was to increase the time step as long as the temperature change at that time increment was less than a user set limit. This lead to absurdly long time intervals when the temperature trace flattened out near steady state. Ultimately, the ABAQUS transient as, well as the other two, were run with twenty time steps of five seconds each for a 100 second transient, and the final temperature fields of the three codes could be compared. Due to the size of the transient output that had to be downloaded by modem it was very important to limit the amount of output generated, especially any concerning the geometry as this had already been verified in the steady state runs. ANSYS allowed liberal paring of the output, but ABAQUS still delivered considerable undesirable output, while MARC output was unmanagable, as previously mentioned. The longest MARC transient downloaded was 20 seconds, and generated more output and used more system time than the entire ABAQUS 100 second transient.

In other words, all three codes required that the transient be revised to fit the code, rather than the other way around. Indeed, MARC and ANSYS were only run with some fairly precise knowledge of the final state of the transient, information received from ABAQUS. As can be seen in figure 6, ABAQUS was slower in CP time than the other two, but substantially less expensive.

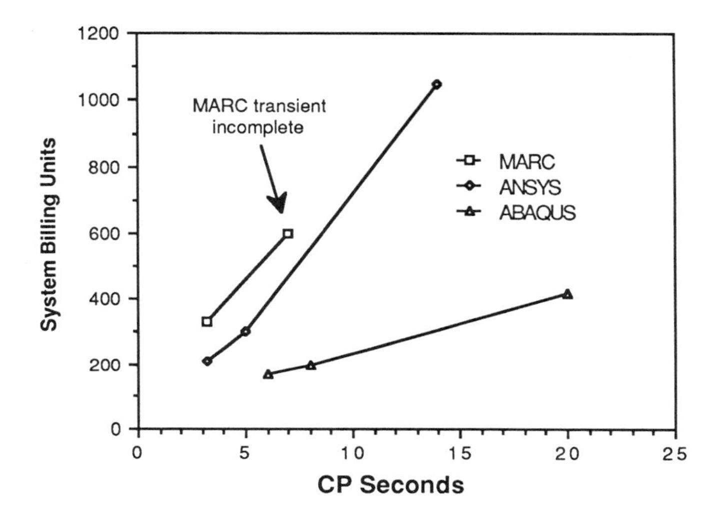

FIGURE 6. Time and Costs for Increasing Time Increments. Increments are steady state, 3 increments, and 20 increments left to right.

Subjective results

Each code has strengths and weaknesses. The manuals were read with the belief that, even though each code is supported by seminars and classes, not every user would attend these, nor would any user employ the codes evenly through time. In other words, engineers using the codes would come and go, and a code will be used extensively for a period, then be shelved for an extended period. It is advantageous to have a code that can be picked up and quickly used after some lay-off.

Physically, all manuals were standard 8.5 x 11 inch size. ABAQUS was a single large volume. ANSYS was two volumes, each about three-quarters the size of the ABAQUS volume. MARC was divided into four volumes, the total about that of ANSYS. The volumes were usefully separated into A) User Information, B) MARC Element Library, C) Program Input, and D) User Subroutines and Special Routines. The size and division of the manuals is important because they are referenced frequently, often several sections at a time. Thus, the manuals are on and off the shelf often, and for this reason the MARC manuals are considered the most easily accessed.

It is important that a manual contain a table of contents, an introduction, and an index, for easy reference. All three had these features, with some differences in the all-important index. Only ANSYS had both a subject and a command index, but there was only one index for the two volumes. MARC had an index for each volume, either by subject or by command, as appropriate for each volume, and a table of contents for each section (several per volume). ABAQUS had a command index and an element type index. All manuals supplied tabbed dividers between sections.

Only ANSYS offered a suggested starting place, other than just going to the first page and beginning from there. In order to run the codes it was not necessary to understand all the theory of finite elements, nor the theory of stress analysis, most of which is inapplicable to heat transfer. All three codes treated heat transfer as an add-on feature, often refering to temperature as displacement and heat flux as force, as the governing equations are similar in this way. To a heat transfer engineer with little structural or finite element background this would be confusing. ANSYS instructed the reader to skip all the theory, if preferred. However, MARC offered an extensive introduction by was of Volume 1, even including an interesting company history, but much of it was finite element theory. These codes will often be run by entry-level engineers, and the theory of finite elements (and of structural mechanics, for that matter) was too often distracting, even misleading.

Perhaps most useful to the actual writing of the input was the availability of examples. Only ANSYS offered more than one, most illustrating the generation of meshes. MARC had only one, and it contained an error. ABAQUS had none. It is known that example manuals are available, but these are the size and cost of the user's manuals themselves. If funds and space are available, the example manuals are recommended.

Pre-processors

MARC had the most useful command terminology. For instance, MARC assigned temperature boundary conditions with "FIXED TEMPERATURE", whereas ANSYS used "NT", and ABAQUS used a "BOUNDARY 11". MARC would then assign the temperature to several nodes with the command, for example, "1 TO 81 BY 10", while ANSYS and ABAQUS would need "1,81,10,". These kind of features determine how easily an input sequence could be debugged. All three codes allowed the use of comment cards throughout the decks.

All three codes used fixed or free format input, at the user's discretion. Free format was the most easily used, but care was required with the last number in any line. ABAQUS and MARC might reformat any number with less than 10 digits unless there was a terminating comma, but only MARC warned of this. ANSYS terminated the number at the first space following the number, which was the most trouble free approach.

The majority of the time spent on any problem modeled on a computer is in the design and input of the model itself, especially in the mesh generation. Therefore, the quality of the mesh generators is paramount. The manual must explain the generator clearly, the commands must have some logical order, and terminology should avoid any tedious and repitious input. ANSYS allowed the use of a repeat command that would arithmetically increment the card that preceeded it any number of times by a chosen amount. This command was instrumental in reducing the ANSYS input to the smallest of the three.

ANSYS and MARC each offered two mesh generators, but neither manual made it clear how to choose one over the other. Presumably, there are times when one is definitely an advantage to use, but it is human nature to try the first one encountered, and never try the second.

Both ANSYS and ABAQUS could handle the thermal-stress problem with only a few extra lines of code. MARC required two separate runs, with the temperature solution of the heat transfer run written to a file, and that file used as input to the stress run. The calculational mesh could not be carried over and had to be regenerated. This would be burdensome with a very large model.

For ABAQUS to move directly from a thermal solution to a stress solution in one step required the use of compatible thermal and stress elements. This was possible in the thermal-stress problem used here, but was not when using a 3-D eight node brick like those required in tests 2 and 3. Only a twenty node brick could be used, greatly increasing the size and complexity of the problem, and its input and output

All codes offered model plotters, but it was beyond the resources of this project to evaluate those. As the codes were run via a subscription service the plotters would have to be run interactively at great expense.

The times required to take each code all the way through all four loadings of test 1, from the initial reading of the manuals to completion of the last loading, were as follows:

ABAQUS	35 hours
ANSYS	37 hours
MARC	55 hours

Output structure

Once the problems ran successfully, the solutions were easily found at the end of the output. The output of each code could be customized somewhat to fascilitate recognition of the desired information. ANSYS and MARC allowed the printing of sorted output that made areas of temperature extremes easier to spot. As noted before, it was often very difficult, if not impossible, to eliminate certain large amounts of unwanted output from both MARC and ABAQUS. ANSYS was very easy to streamline.

All codes presented the temperatures in a form that could be processed by the user's own routines. ABAQUS could not sort the data into ascending or descending order to fascilitate the search for maxima or minima, and included the most unnecessary output. Only MARC offered an exit code that could be looked up in the manual to debug problems and/or check the success of a run. This proved very helpful. None of the output from any of the codes was in 80 column format, nor could it be made so, for normal PC viewing and printing.

CONCLUSIONS

Four finite element computer codes were selected and tested for application to heat transfer problems. The four codes tested were ABAQUS, ANSYS, MARC, and NASTRAN. Three tests were designed to determine the friendliness, speed, accuracy, and costs of running the codes.

The NASTRAN code was found to be inaccessible under the guidelines of this study. The manuals were verbose and poorly written. There was no guidance provided in the introduction as to how to begin a problem. No examples were supplied to illustrate the input structure, and no lines of input were written.

The accuracy of each of the remaining codes was excellent when analytical soulutions were available. When such solutions were not available the codes agreed amoung themselves to a high degree. Large differences were found when the codes were judged for speed as determined by the central processor time used during runs. For 2-D steady state runs MARC was the fastest, ABAQUS was the slowest. For 3-D steady state and transient runs ANSYS was the fastest. No conclusion could be drawn whether MARC or ABAQUS was the faster as the MARC transient was not successfully run to completion due to the prohibitive size of the output and the great system cost of running the transients.

ABAQUS was the only code that did not require that the transient problem be redesigned to achieve results. MARC had inadequate options for terminating transients by not allowing cessation upon steady state conditions. ANSYS did allow such termination, but this feature could not be used. Accessory nodes and elements were required to model surface flux, and these nodes did not respond as the rest of the model did as they had to be given non-realistic material constants.

ABAQUS was the only code of the four that had no major shortcomings. The ABAQUS input was, however, the longest and most cumbersome in every problem.

RECOMMENDATIONS

If purchase was imminent, ABAQUS would be the code of choice under the guidelines of this study. However, no examination was made of any codes available that were not mainframe codes. Some of these may be more suitable for some designers. All the manuals need rewriting with the practicing engineer in mind and not the finite element theorist. Further, the heat transfer capabilities of these codes should be taken more seriously, as temperature is not displacement and heat flux is not force to any heat trasnfer engineer. Any code designed with the modeling of electronic cooling in mind should fare very well against the present competition.

ACKNOWLEDGEMENT

The authors gratefully acknowledge the support of the Digital Equipment Corporation in pursuit of this project.

NOMENCLATURE

Symbol	Quantity	SI unit
T	temperature	°C
Q	volumetric heat flux	w/m^3
h	heat transfer coefficient	$w/m^2/^oC$
k	thermal conductivity	$w/m/^oC$
SBU	system billing units	
CP	central processor	

Subscripts

i	inside
o	outside

REFERENCES

1. Noor, Ahmed, Computer programs for Heat Tranfer Analysis, *Heat Transfer and Fluid Flow Design Book*, section 501.6, pp. 1-45, Genium Publishing Corporation, Schenectady, New York, 1983.

2. The ABAQUS computer code, Hibbitt, Karlson and Sorenson, Incorporated, 100 Medway Street, Providence, Rhode Island, version 4-5-179.

3. The ANSYS computer code, Swanson Analysis Systems, Incorporated, Houston, Pennsylvania, version 4.2 B.

4. The MARC computer code, MARC Analysis Corporation, Incorporated, 260 Sheridan Avenue, Suite 314, Palo Alto, California, version K.2-3.

5. The NASTRAN computer code, MacNeal-Schwendler Corporation, 815 colorado Boulevard, Los Angeles, California, version 65.

Thermal Analysis of Hybrid Integrated Circuits

MITRA LUGOVIĆ
CVTŠ
Ilica 256-b
41000 Zagreb, Yugoslavia

PETER BILJANOVIĆ
Faculty of Electrical Engineering
Unska 3
41000 Zagreb, Yugoslavia

ABSTRACT

A model of heat exchange in hybrid high-power integrated circuits is developed and presented. The model is based on solution of the homogeneous partial differential equations system by the method of separation of variables with the defined boundary and initial conditions. The stationary model of heat exchange is developed under the following simplifying assumptions: the only mechanism of heat exchange is conduction, thermal conductivity coeficients are temperature independent, all construction elements and heat sources are rectangular and homogeneous, and constant temperature at the bottom of the hybrid package. From the assumed model and theoretical solution is developed a high speed algorithm for temperature determination at any point on the ceramic substrate for any distribution of high - power devices. The influence of parameters of the assumed model on the temperature distribution is analyzed. The heat exchange model is verified on real hybrid integrated circuits. Computed results of the research are tested by a thermovision measurement system. Theoretical results of heat analysis are compared with the results of measurements and significant agreement is shown. Thermovisually measured pictures are digitalized and a posteriore error of the assumed model of heat exchange in hybrid high - power integrated circuits is estimated. The average relative error is less than 10%.

1. INTRODUCTION

Hybrid integrated circuits (HIC) used separately built in some other electronic industry final products are being more often used instead of discrete circuits. HIC are getting more complex with the development of materials and fabrication technology. At the same time, higher demands are put on greater reliability at all stages: designing, producing and utilizing the circuits [1,2].

Reliability, as one of the most important parameters of the quality of a product, is highly dependent on temperature. For the majority of failure mechanisms, according to Arrhenius, the degradation rate is approximately doubled for each 10 K of temperature increase.

Fast thermal analysis of HIC, giving the overall temperature distribution on the substrate surface, would enable the designers to change the place of temperature - dependent elements at the stage of mask design as well as the elements which dissipate high power, avoiding in that way possible thermal overloading [3].

Indirectly,through the thermal analysis of the circuits, the analysis of possible failure mechanisms caused by temperature increase can be made, resulting in production of high power HICs with lower failure rates, therefore increasing the reliability.

In this paper, the best compromise among the exactness of description relation to the model complexity and the solution of the system of equations and the calculation time is sought.

The research concept in this paper is based on a theoretical and experimental program, defining the thermal model or theoretical-experimental exchange of high power HICs, solving it theoretically and experimentally verifying the results obtained on the basis of the assumed models of thermal analysis.

2. THERMAL ANALYSIS MODEL DEFINITION

HIC are microelectronic circuits in which different techniques and technologies are combined. Passive components of the circuit are made by either thin film or thick film techniques. Active components are manufactured mostly in standard semiconductor technology and mounted like discrete microelectronic components [4]. The passive chip components are resistors and capacitors, the other chip-like passive components being not so frequently used. Of the active chip-like components, the most common are different types of transistors and monolithic circuits of all degrees of integration. Ceramic substrates serve as a foundation on which the circuit is built, providing electric isolation among circuit components and good thermal conductivity.

In complex physical processes it is very difficult to determine these relationships when the properties very considerably in space and time.

The complexity of the final circuit is such that thermal analysis model of high power HICs must be developed under the following simplifying assumptions [5,6]:

- The only mechanism of heat exchange in a microcircuit is conduction; radiation and convection are disregarded. The heat conduction takes place in all construction elements.The latter two heat paths are disregarded in the model since the power conducted in this way is estimated to be of order of several hundred milliwatts whereas the total power conducted to the heat-sink is of order of several watts. This relation is primarily a consequence of small cross-section area of the elements.
- All construction elements essential for heat-exchange analysis are rectangular and homogeneous.
- There is a constant heat flow through the ceramic surface.
- The metal heat sink to which the ceramic substrate is attached has a constant temperature equal to zero centigrade degrees.

It is necessary to make the specification of some parameters of the assumed model with as few as possible errors which result from simplifying the heat-exchange model.

The assumed high power HIC model (Figure 2.1) is described with ten parameters as follows: adhesive thickness (w_1), ceramic substrate thickness (w_2), ceramic substrate length (L_1), ceramic substrate width

(M_1), coordinates of the geometrical center of the heat source (x_i,y_i), heat source length (L_2), heat source width (M_2), power dissipation of the heat source (P_i), adhesive thermal conductivity (λ_1) and thermal conductivity of the ceramic substrate (λ_2).

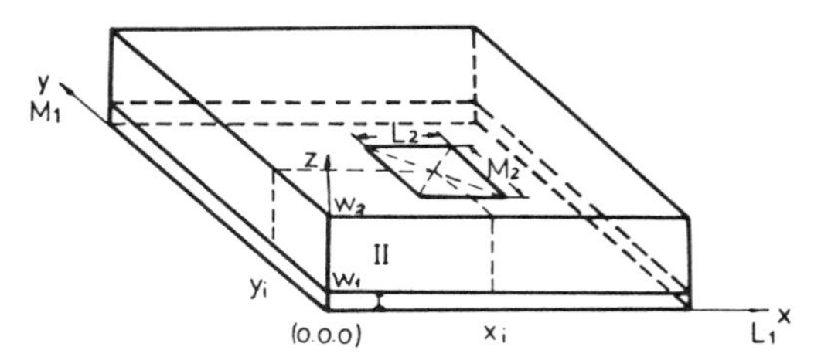

FIGURE 1. Thermal model of a HIC.

The conductivity phenomenon is described by the general differential conductions equation [7,8,9]. Finally, for steady-state conduction in the absence of inner heat sources the general equation of heat conduction takes the form of the Laplace equation:

$$\frac{\partial^2 T}{\partial x^2} + \frac{\partial^2 T}{\partial y^2} + \frac{\partial^2 T}{\partial z^2} = 0 \tag{1}$$

The overall thermal distribution in steady-state can be described by the solution of the homogeneous partial differential equations system

$$\begin{aligned} T_1\ (x,y,z) &= 0 \\ T_2\ (x,y,z) &= 0 \end{aligned} \tag{2}$$

with the defined boundary and initial conditions where $T_1(x,y,z)$ is the temperature distribution in the adhesive defined for the first region (I) and $T_2(x,y,z)$ is the temperature distribution in the ceramic defined for the second region (II).

The problem of the system of equations with the given boundary and initial conditions is solved by the development of the functions of the temperature field $T_1(x,y,z)$, i.e., $T_2(x,y,z)$ in a Fourier's series.

3. THEORETICAL SOLUTION

The system of equations (2) is solved by the method of separation of variables with the defined boundary and initial conditions [10]. First, it is necessary to determine all the solutions of the Laplace equations (2), i.e., all the solutions with the separated variables. Based on the principle of linear superposition, the solution which satisfies the boundary and initial conditions is obtained. The problem is solved by the development of the functions $T_1(x,y,z)$ and $T_2(x,y,z)$ in Fourier's series according to eigenfunctions X_n, Y_m and Z_{nm}.

By determination of the development coefficient and its substitution the final solution of the problem, the temperature distribution on the ceramic surface induced by several sources can be determined from the superposition principle. Based on the assumed model and theoretical solution, the algorithm for temperature determination at any point on the ceramic plate surface is developed for any placement of high power elements.

The program is written in FORTRAN V for CDC CYBER 170/820 computer. The flow chart is shown in Figure 2.

The program is tested on the example of high power HIC being described with the following input parameters: $L_1=45\cdot10^{-3}$m, $M_1=30\cdot10^{-3}$m, $L_2=7.5\cdot10^{-3}$m, $M_2=5.3\cdot10^{-3}$m, P=10 W, $w_1=0.15\cdot10^{-3}$m, $w_2=0.8\cdot10^{-3}$m, $\lambda_1=15\cdot10^{-2}$ W/mK and $\lambda_2=25$ W/mK.

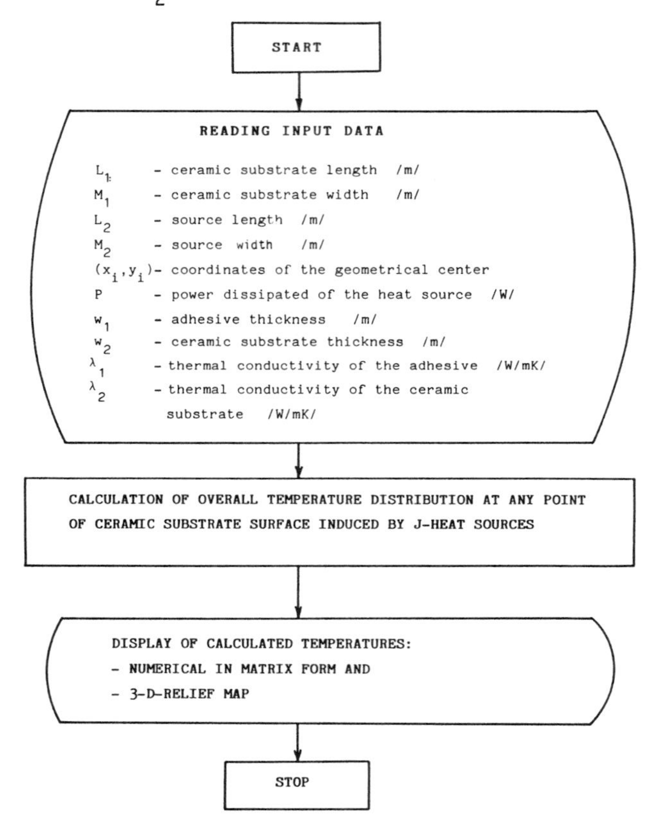

FIGURE 2. The flow chart.

The results of overall temperature distribution for one heat source, P=10 W, are given in Figure 3.

x										
1.0	1.18	1.29	1.48	1.64	1.67	1.58	1.35	1.06	.81	.67
5.5	1.83	2.13	2.69	3.26	3.49	3.18	2.51	1.79	1.24	.95
10.0	3.46	4.49	6.70	9.46	10.86	9.34	6.41	3.90	2.33	1.58
14.5	5.66	8.29	15.42	29.87	40.91	29.71	15.01	7.41	3.82	2.34
19.0	6.83	10.57	21.90	51.84	79.57	51.64	21.41	9.51	4.58	2.67
23.5	5.64	8.27	15.40	29.85	40.90	29.69	14.99	7.39	3.80	2.32
28.0	3.39	4.42	6.63	9.40	10.82	9.28	6.35	3.84	2.28	1.54
32.5	1.66	1.95	2.49	3.07	3.33	3.00	2.33	1.64	1.12	.84
37.0	.73	.80	.93	1.06	1.12	1.02	.85	.66	.50	.41
41.5	.35	.37	.40	.43	.46	.41	.35	.30	.24	.21
	1.00	4.00	7.00	10.00	13.00	16.00	19.00	22.00	25.00	28.00
										y

IZVOR BR: 1   7.50   5.30   19.00   13.00   10.00

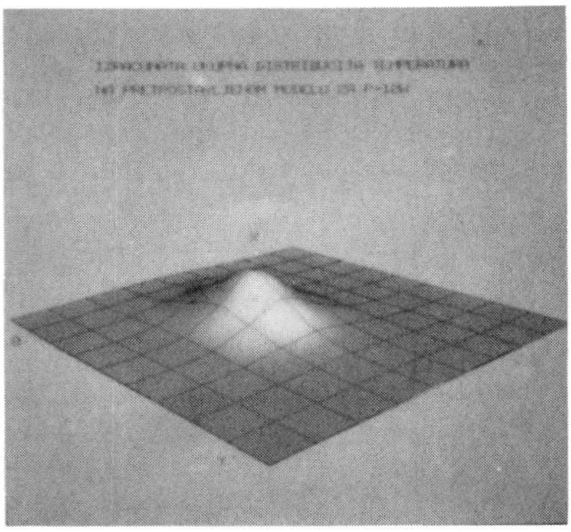

a)   *b)*

FIGURE 3. Overall temperature distribution for P = 10 W.
a) temperature map, b) 3-D-relief map

Overall temperature distribution on the same circuit induced by two heat sources, P = 3 W and P = 4 W shown in Figure 4.

x										
1.0	.56	.62	.74	.87	1.17	1.00	1.11	.86	.75	.67
5.5	.85	1.00	1.29	1.63	2.00	1.98	1.92	1.56	1.26	1.06
10.0	1.57	2.04	3.05	4.39	5.44	5.75	5.13	4.01	2.84	2.12
14.5	2.57	3.74	6.85	12.48	17.29	18.07	15.49	10.73	6.05	3.77
19.0	2.95	4.71	9.99	24.81	75.23	38.54	60.34	20.61	8.94	5.18
23.5	2.56	3.73	6.84	12.47	17.22	18.06	15.43	10.72	6.04	3.76
28.0	1.54	2.00	3.01	4.35	5.28	5.70	5.00	3.97	2.80	2.08
32.5	.76	.90	1.19	1.52	1.70	1.85	1.66	1.44	1.16	.96
37.0	.35	.39	.46	.55	.54	.63	.55	.54	.46	.41
41.5	.17	.18	.20	.23	.17	.25	.19	.23	.21	.19
	1.00	4.00	7.00	10.00	13.00	16.00	19.00	22.00	25.00	28.00
										y

1	2.00	2.00	19.00	13.00	4.00
2	2.00	2.00	19.00	19.00	3.00

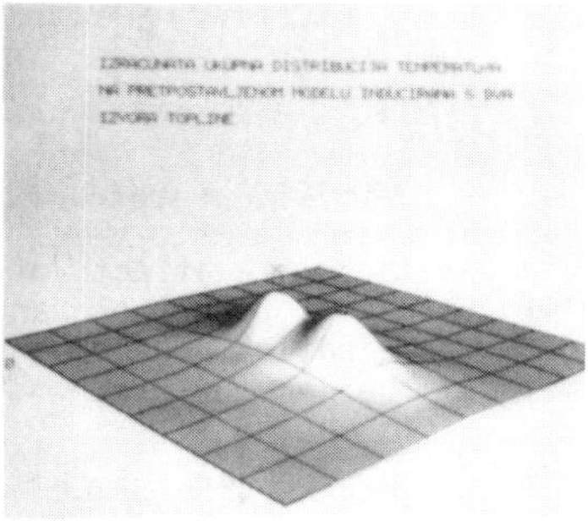

a)   *b)*

FIGURE 4. Overall temperature distribution for P = 3 W and P = 4 W.

a) temperature map, b) 3-D-relief map

4. ANALYSIS OF THE ASSUMED HIGH POWER HIC MODEL EFFECTS ON OVERALL TEMPERATURE DISTRIBUTION

An analysis of the effects of temperature parameters describing the assumed high power HIC model on the overall temperature distribution is performed. The research is done with the constantly defined parameters which describe the assumed model, and the effect of discrete changes of each parameter on distributed overall temperature level is analyzed.

To avoid the "individual" interpretation of the results and to base the analysis system on statistical principles, it is possible to make a Pareto diagram which illustrates the hierarchic order of some parameters effects expressed in percentages, Figure 5.

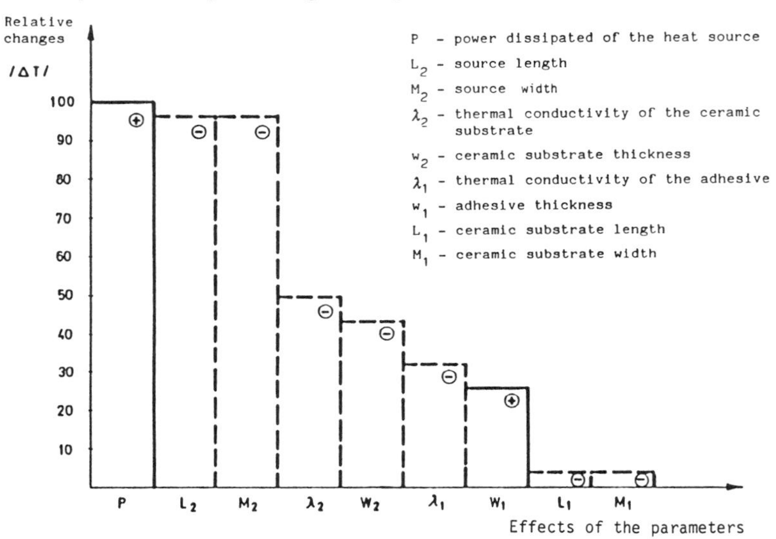

FIGURE 5. Pareto diagram of some parameters effects.

On abscissa there are affects of the parameters and on ordinate there are absolute values of relative changes of maximum temperature in geometrical center of dissipation element induced by a 100% increase of each parameter value.

5. COMPUTER THERMAL ANALYSIS OF THE REAL HIC

Thermal analysis verification will be carried out on the real HIC without a metal heat sink (Figure 6). Hybrid structure is made in the Laboratory of Hybrid Microelectronic Technology in RIZ-IETA, Zagreb.

Ceramic plate surface of 96% Al_2O_3 is used. HIC has only one dissipation element - thick film resistor which is described with the following input parameters: $L_1=25\cdot 10^{-3}$m, $M_1=25\cdot 10^{-3}$m, $L_2=2.5\cdot 10^{-3}$m, $M_2=5.0\cdot 10^{-3}$m, $w_1=0{,}83\cdot 10^{-3}$m, $w_2=0{,}63\cdot 10^{-3}$m, $\lambda_1=32.0\cdot 10^{-3}$W/mK, $\lambda_2=24.0$ W/mK. Temperature of the environment is 22°C.

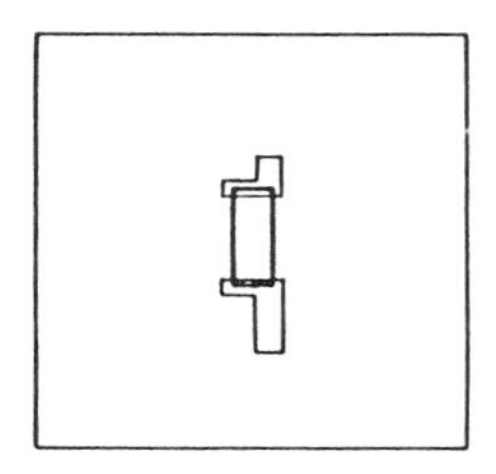

FIGURE 6. HIC model without metal heat sink.

Temperature distribution for the power source of: 3W, 2W and 1W is calculated, as shown in Figures 7-9.

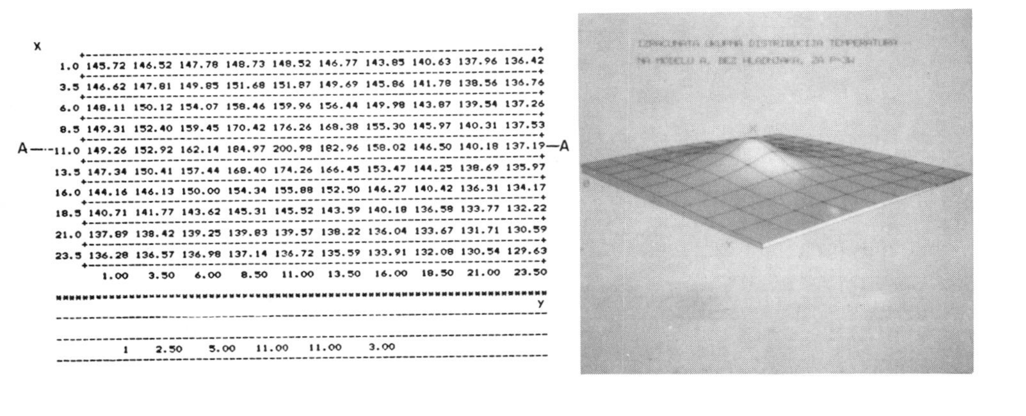

x \ y	1.00	3.50	6.00	8.50	11.00	13.50	16.00	18.50	21.00	23.50
1.0	145.72	146.52	147.78	148.73	148.52	146.77	143.85	140.63	137.96	136.42
3.5	146.62	147.81	149.85	151.68	151.87	149.69	145.86	141.78	138.56	136.76
6.0	148.11	150.12	154.07	158.46	159.96	156.44	149.98	143.87	139.54	137.26
8.5	149.31	152.40	159.45	170.42	176.26	168.38	155.30	145.97	140.31	137.53
A—11.0	149.26	152.92	162.14	184.97	200.98	182.96	158.02	146.50	140.18	137.19—A
13.5	147.34	150.41	157.44	168.40	174.26	166.45	153.47	144.25	138.69	135.97
16.0	144.16	146.13	150.00	154.34	155.88	152.50	146.27	140.42	136.31	134.17
18.5	140.71	141.77	143.62	145.31	145.52	143.59	140.18	136.58	133.77	132.22
21.0	137.89	138.42	139.25	139.83	139.57	138.22	136.04	133.67	131.71	130.59
23.5	136.28	136.57	136.98	137.14	136.72	135.59	133.91	132.08	130.54	129.63

1 2.50 5.00 11.00 11.00 3.00

a) *b)*

FIGURE 7. Calculated overall temperature distribution for P = 3W. a) temperature map, b) 3-D-relief map

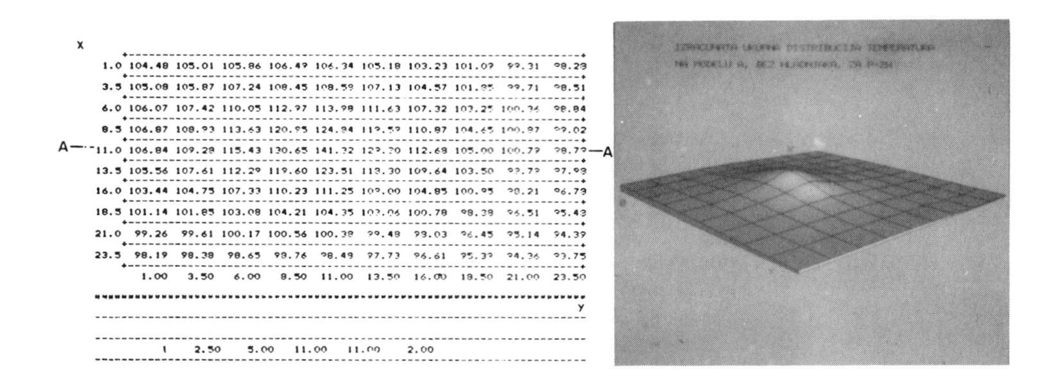

x	1.00	3.50	6.00	8.50	11.00	13.50	16.00	18.50	21.00	23.50
1.0	104.48	105.01	105.86	106.49	106.34	105.18	103.23	101.09	99.31	98.29
3.5	105.08	105.87	107.24	108.45	108.59	107.13	104.57	101.85	99.71	98.51
6.0	106.07	107.42	110.05	112.97	113.98	111.63	107.32	103.25	100.36	98.84
8.5	106.87	108.93	113.63	120.95	124.94	119.59	110.87	104.65	100.97	99.02
A---11.0	106.84	109.28	115.43	130.65	141.32	129.30	112.68	105.00	100.79	98.79---A
13.5	105.56	107.61	112.29	119.60	123.51	118.30	109.64	103.50	99.79	97.98
16.0	103.44	104.75	107.33	110.23	111.25	109.00	104.85	100.95	98.21	96.79
18.5	101.14	101.85	103.08	104.21	104.35	103.06	100.78	98.39	96.51	95.43
21.0	99.26	99.61	100.17	100.56	100.39	99.48	98.03	96.45	95.14	94.39
23.5	98.19	98.38	98.65	98.76	98.49	97.73	96.61	95.39	94.36	93.75

y

1 2.50 5.00 11.00 11.00 2.00

a) *b)*

FIGURE 8. Calculated overall temperature distribution for P = 2 W.
a) temperature map, b) 3-D-relief map

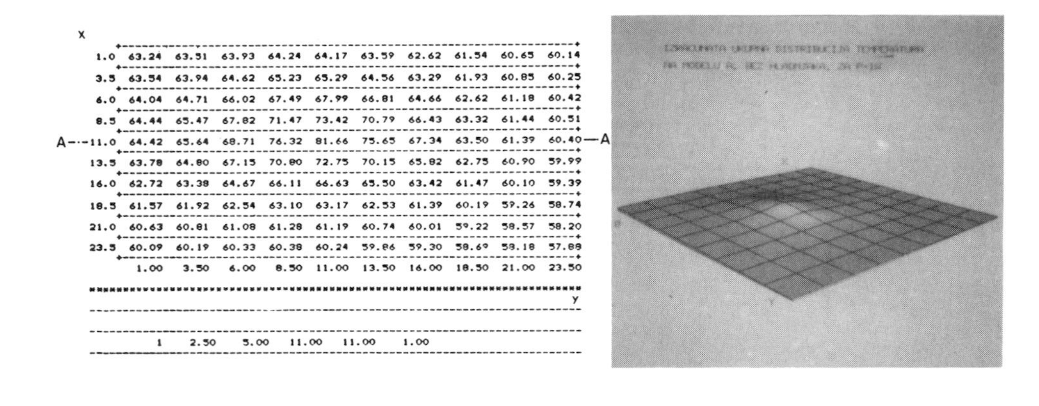

x	1.00	3.50	6.00	8.50	11.00	13.50	16.00	18.50	21.00	23.50
1.0	63.24	63.51	63.93	64.24	64.17	63.59	62.62	61.54	60.65	60.14
3.5	63.54	63.94	64.62	65.23	65.29	64.56	63.29	61.93	60.85	60.25
6.0	64.04	64.71	66.02	67.49	67.99	66.81	64.66	62.62	61.18	60.42
8.5	64.44	65.47	67.82	71.47	73.42	70.79	66.43	63.32	61.44	60.51
A---11.0	64.42	65.64	68.71	76.32	81.66	75.65	67.34	63.50	61.39	60.40---A
13.5	63.78	64.80	67.15	70.80	72.75	70.15	65.82	62.75	60.90	59.99
16.0	62.72	63.38	64.67	66.11	66.63	65.50	63.42	61.47	60.10	59.39
18.5	61.57	61.92	62.54	63.10	63.17	62.53	61.39	60.19	59.26	58.74
21.0	60.63	60.81	61.08	61.28	61.19	60.74	60.01	59.22	58.57	58.20
23.5	60.09	60.19	60.33	60.38	60.24	59.86	59.30	58.69	58.18	57.88

y

1 2.50 5.00 11.00 11.00 1.00

a) *b)*

FIGURE 9. Calculated overall temperature distribution for P = 2 W.
a) temperature map, b) 3-D-relief map

6. EXPERIMENTAL VERIFICATION OF THE THEORETICALLY OBTAINED RESEARCH RESULTS

Verification of the theoretically obtained research results is carried out by remote sensing temperature measurement on HIC surface with the thermovision AGA system.

The experiment is performad in the same order as in section 5. and on the same already described model for all three power source values. The results are shown in Figures 10-12.

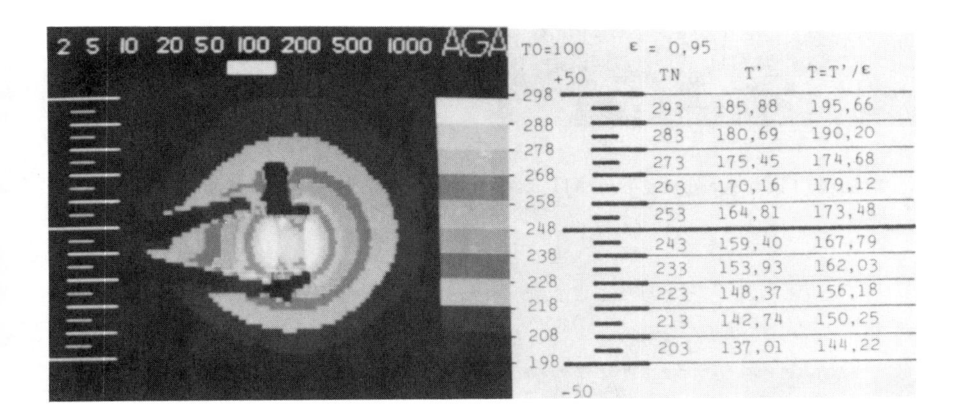

FIGURE 10. Measured overall temperature distribution for P = 3 W.

TO - *temperature range*,
TN - *temperature level*,
T' - *measured temperature*,
T - *real temperature*,
ε - *emissivity*.

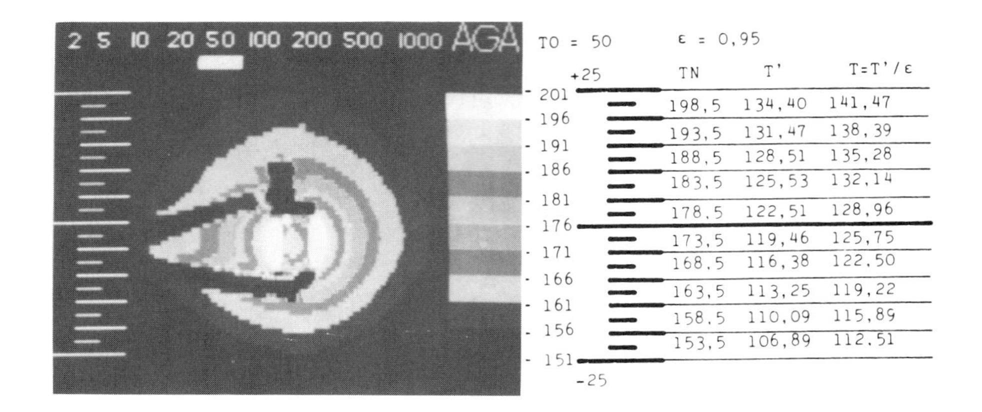

FIGURE 11. Measured overall temperature distribution for P = 2 W.

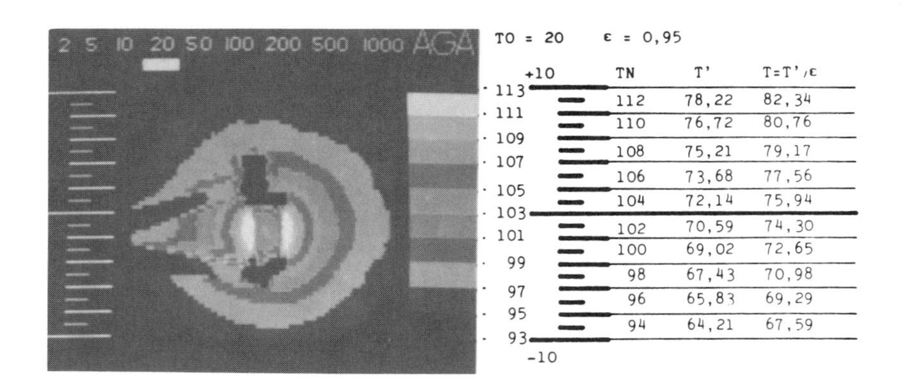

FIGURE 12. Measured overall temperature distribution for P = 1 W.

7. DIGITALIZATION OF THE PICTURES OBTAINED BY MEASUREMENT AND COMPARISON OF THEORETICAL AND EXPERIMENTAL RESEARCH RESULTS

To compare theoretically obtained results of the computer HIC thermal analysis with the measurement results, the pictures in Figures 10-12 in 10x10 ppints are digitalized.

The results are shown in Figures 13-15.

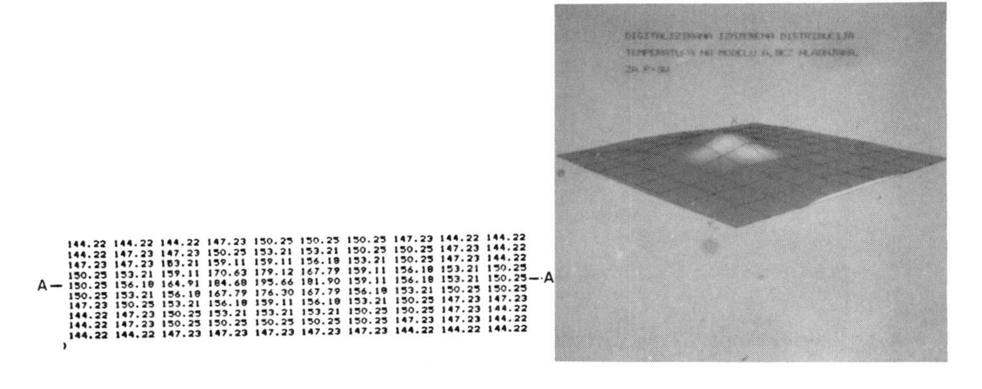

	144.22	144.22	144.22	147.23	150.25	150.25	150.25	147.23	144.22	144.22	
	144.22	147.23	147.23	150.25	153.21	153.21	150.25	150.25	147.23	144.22	
	147.23	147.23	183.21	159.11	159.11	156.18	153.21	150.25	147.23	144.22	
	150.25	153.21	159.11	170.63	179.12	167.79	159.11	156.18	153.21	150.25	
A—	150.25	156.18	164.91	184.68	195.66	181.90	159.11	156.18	153.21	150.25	—A
	150.25	153.21	156.18	167.79	176.30	167.79	156.18	153.21	150.25	150.25	
	147.23	150.25	153.21	156.18	159.11	156.18	153.21	150.25	147.23	147.23	
	144.22	147.23	150.25	153.21	153.21	153.21	150.25	150.25	147.23	144.22	
	144.22	147.23	150.25	150.25	150.25	150.25	150.25	147.23	147.23	144.22	
	144.22	144.22	147.23	147.23	147.23	147.23	147.23	144.22	144.22	144.22	

a) *b)*

FIGURE 13. Distribution of temperature obtained by digitalization of the picture 6.1. for P P = 3 W.
a) temperature map, b) 3-D-relief map

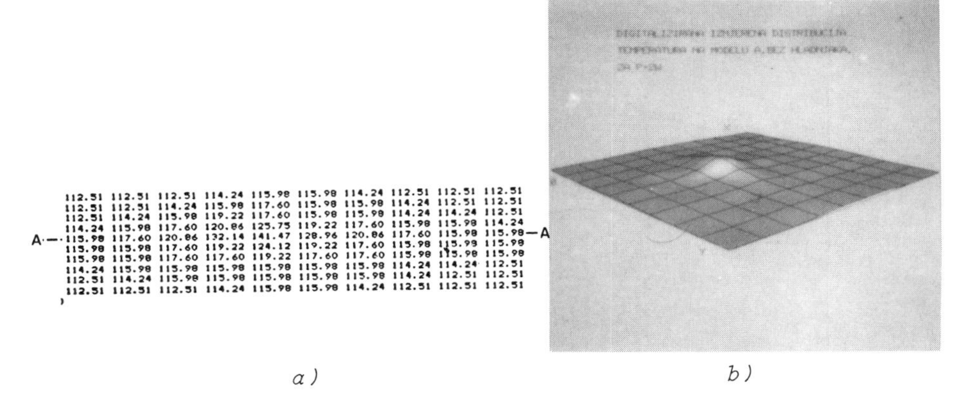

	112.51	112.51	112.51	114.24	115.98	115.98	114.24	112.51	112.51	112.51	
	112.51	112.51	114.24	115.98	117.60	115.98	115.98	114.24	112.51	112.51	
	112.51	114.24	115.98	119.22	117.60	115.98	115.98	114.24	114.24	112.51	
	114.24	115.98	117.60	120.86	125.75	119.22	117.60	115.98	115.98	114.24	
A—	115.98	117.60	120.86	132.14	141.47	128.96	120.86	117.60	115.98	115.98	—A
	115.98	115.98	117.60	119.22	124.12	119.22	117.60	115.98	115.99	115.98	
	115.98	115.98	117.60	117.60	119.22	117.60	117.60	115.98	115.98	115.98	
	114.24	115.98	115.98	115.98	115.98	115.98	115.98	114.24	114.24	112.51	
	112.51	114.24	115.98	115.98	115.98	115.98	115.98	114.24	112.51	112.51	
	112.51	112.51	112.51	114.24	115.98	115.98	114.24	112.51	112.51	112.51	

a) *b)*

FIGURE 14. Distribution of temperature obtained by digitalization of the picture 6.2 for P = 2 W.
a) temperature map, b) 3-D-relief map

	67.59	67.59	67.59	68.44	69.29	69.29	68.44	67.59	67.59	67.59	
	67.59	67.59	69.44	70.13	70.13	70.13	70.98	69.29	67.59	67.59	
	67.59	68.44	71.81	71.81	71.81	71.81	70.98	70.13	69.29	67.59	
	68.44	70.13	71.81	72.65	75.94	72.65	71.81	70.13	70.13	68.44	
A —	69.29	70.98	72.65	79.17	82.34	78.36	74.30	71.81	70.13	69.29	— A
	69.29	70.98	72.65	72.65	75.94	73.47	72.65	71.81	69.29	69.29	
	68.44	70.13	71.81	72.65	72.65	71.81	71.81	70.98	69.29	68.44	
	68.44	69.29	70.98	71.81	71.81	71.81	70.98	70.13	69.29	67.59	
	67.59	68.44	70.13	70.13	70.13	70.13	70.13	69.29	68.44	67.59	
	67.59	67.59	68.44	69.29	69.29	69.29	69.29	67.59	67.59	67.59	

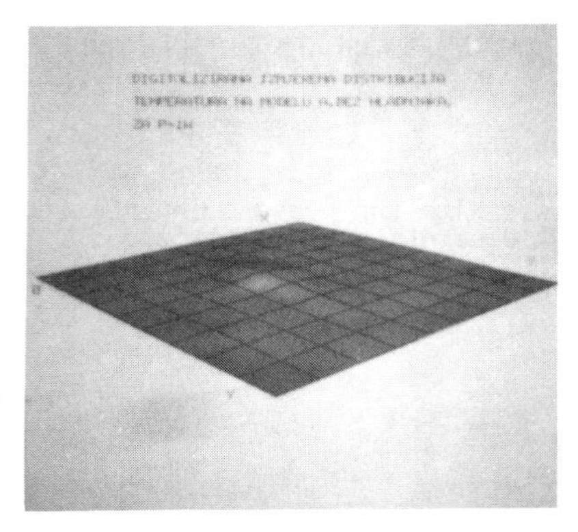

a) b)

FIGURE 15. Distribution of temperature obtained by digitalization of the picture 6.3 for P = 1 W.
a) temperature map, b) 3-D-relief map

7.2. *Comparison of theoretical and experimental research results*

If A-A cross-sections of theoretically obtained temperature distribution in Figures 7-9 are done and compared with A-A cross-sections of temperature distribution obtained by measurement in Figures 10-12, the temperature distribution for all three heat sources will be obtained, Figure 16.

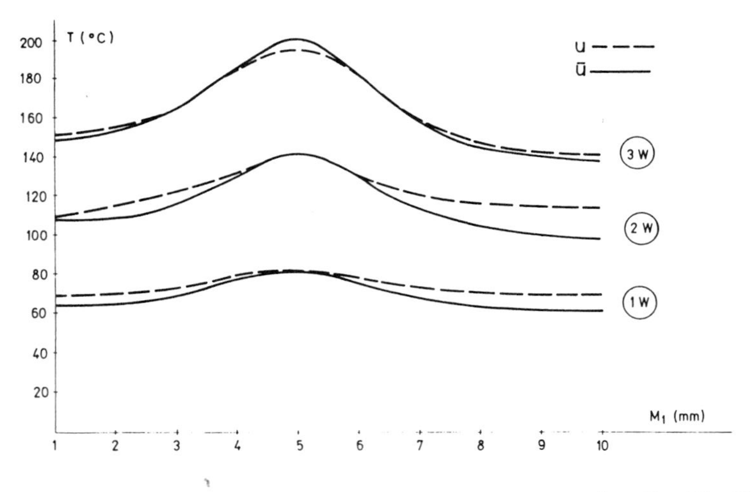

FIGURE 16. Comparison of theoretical obtained results $\bar{u}$ with the measurement results u for all three cases.

Comparing theoretically obtained results based on the assumed model of thermal analysis $\bar{u}$ with the measurement results u (Figure 16), it may be concluded:

- temperature distributions are qualitativelly being overlapped;
- they are symmetric in relation to maximum in geometrical center of the heat source;
- quantitative deviations are increasing with decreasing heat source power due to the simplifying assumptions in the development of the thermal analysis model.

7.3. *A posteriori error estimation of the heat exchange assumed model*

According to Figure 16 it is assumed that $\bar{u}$ solution is theoretically obtained on the basis of the assumed model of thermal analysis and u solution is experimentally obtained.

If $x_1, x_2, \dots, x_N$ are the points at which u values are measured then the norm

$$||u - \bar{u}|| = \sum_{i=1}^{N} [u(x_i) - \bar{u}(x_i)]^2 \qquad (3)$$

describes error estimation of the heat exchange assumed model according to [11].

The calculated relative error of the heat exchange assumed model for all three cases mentioned is given in Table 1.

Table 1. Relative error of the heat exchange assumed model for all three power sources

	P = 1W	P = 2W	P = 3W
Relative error p(%)	8.07	7.88	4.19

Relative error of the theoretically obtained overall temperature distribution on the ceramic plate surface under the simplifying assumptions is lower than 10% on average.

Therefore, it may be concluded that fast algorithm for determination of overall temperature distribution on the ceramic plate surface obtained on the basis of the assumed model and theoretical solution, with high reliability of 90%,describes overall temperature distribution of the analyzed model.

8. CONCLUSION

Verification methods, numerical results processing and a posteriori error estimation of theoretical model of thermal analysis are very important since they have proved the assumptions, on which the obtained thermal analysis high power HIC model is based and theoretical analysis is performed, to be correct.

Based on the results of the important effects analysis of certain parameters change which describe the assumed thermal analysis model, it is possible, at the stage of professional HIC designing, to make particular corrections of the circuit parameters lowering in that way thermal overloadings and decelerating degradation failures. The relationship between life cycle and operating temperature is experimentally proved.

Failure rate for most failure mechanisms according to Arrhenius is approximately doubled per 10K of temperature rise.

Analysis of the failure mechanisms is of vital importance for ensuring HIC reliability. The results of the analysis can be successfully applied to improve HIC designing and technology as well as to predict and increase reliability.

Fast determination of temperature distribution on the HIC ceramic plate surface for any placement of elements dissipating higher power will make it possible even at the mask designing stage by replacement of temperature sensitive elements and high power elements to predict and avoid possible heat overloadings, increasing indirectly HIC reliability.

NOMENCLATURE

Symbol	Quantity	SI Unit
w_1	adhesive thickness	m
w_2	ceramic substrate thickness	m
L_1	ceramic substrate length	m
M_1	ceramic substrate width	m
x_i, y_i	coordinates of the geometrical center	-
L_2	source length	m
M_2	source width	m
P	power	W
λ_1	thermal conductivity of the adhesive	W/mK
λ_2	thermal conductivity of the ceramic substrate	W/mK
T	temperature distribution	^oC,K
x,y,z	cartesian coordinates	-
T_1	temperature distribution in the adhesive	^oC,K
T_2	temperature distribution in the ceramic	^oC,K
X_n, Y_m, Z_{nm}	eigenfunctions	-
ε	emissivity	-

REFERENCES

1. Hamer, D.W., Biggers, J.V. Thick Film Hybrid Microcircuit Technology, Wiley - Interscience, 1972.
2. Harper, C.A., Handbook of Thick Film Hybrid Microelectronics, McGraw-Hill Book Company, 1974.
3. Širbegović, S., Mazalica, M., Krčmar, R., Temperature Verification of Hybrid Microelectronic Circuit Design, Third European Hybrid Microelectronics Conference 1981., Avignon, 51-56.
4. Biljanović, P., Mikroelektronika, Integrirani elektronički sklopovi (Microelectronics, Integrated Electronic Circuits), Školska knjiga, Zagreb, 1983.
5. David, R.F., Computerized Thermal Analysis of Hybrid Circuits, IEEE Transactions on Parts, Hybrids, and Packaging, 13, (3), 283-290,1977.
6. Maly, V., Piotrowski, A.P., Heat Exchange Optimization Technique for High-Power Hybrid IC'S, IEEE Transactions on Components, Hybrids, and Manufacturing Technology, 2, (2), 226-231, 1979.
7. Carslaw, H.S, Jaeger, J.C., Conduction of Heat in Solids, Oxford University Press, 1959.
8. Grigull, U., Sandner, H., Wärmeleitung, Springer - Verlag, Berlin, Heidelberg, New York, 1979.
9. Isachenko, V.P., Osipova, V.A., Sukomel, A.S., Heat Transfer, Mir Publishers, 1977.
10. Andrews, L.C., Elementary Partial Diferential Equations with Boundary Value Problems, Academic Press, Orlando, 1986.
11. Scheid, F., Numerical Analysis, Schaum's Outline Series, McGraw-Hill Book Company, 1968.

Thermal Analysis of Microelectronic Systems Using Finite Element Modeling

H. HARDISTY and J. ABBOUD
School of Mechanical Engineering
University of Bath
Claverton Down
Bath BA2 7AY, UK

ABSTRACT

The paper describes heat transfer research into forced convection air cooling of a regular array of electronic modules on a PCB. An experimental rig together with a data acquisition system has been constructed. A realistic FE model of a row of modules on a common substrate has been developed and used to correlate heat transfer data.

A systematic series of experiments has been carried out for the important case when the modules on the PCB are non-uniformly powered. Results from these tests show the effects of thermal wakes and conduction on the heating of both powered and unpowered modules. The FE model is used to analyse the test data.

1. INTRODUCTION

The trend in the manufacture of microelectronic systems continues to be towards greater packing density. In addition to increasing the number of circuits per chip, components are packed more closely together to reduce signal transit time. In some cases circuit power per device has also been raised to achieve faster operating speeds, Chu [1]. Although the power dissipated by the devices may not be great in absolute terms (a few Watts per chip), because of the miniature size of the components heat fluxes may be extremely high, Oktay et al. [2]. Because of the high packing density, the miniature nature of the components and complex three-dimensional nature of the heat flow paths, the thermal analysis and design of microelectronic systems present formidable problems [3].

In a large scale computer, it may be justifiable to use liquid cooling, or to design a special cooling module [3]. However, because of its cheapness and ease of application, air cooling is the most widely used technology. This paper describes research in which air in forced convection is used to cool a rectangular array of dual-in-line packages (DIPs) mounted on a printed circuit board (PCB). Although such a configuration is widely used throughout industry, relevant publications either of general design methods, or of general experimental data, remain sparse.

Empirical heat transfer correlations have been derived for the case of all modules on the PCB uniformly powered. Finch and Goodacre [4] investigated air cooling of PCBs mounted in cabinets, for fully developed flow. Wills [5] used a simple lumped parameter network model to correlate his experimental heat transfer data from air cooled PCBs. Although this paper provides useful guidance for industrial design, it contains little information either of experimental method, or of data on which the correlation

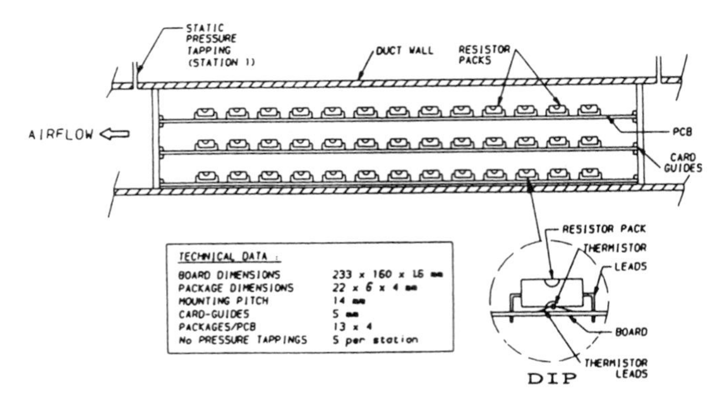

FIGURE 2. Test section of rig.

undersurface, Fig. 2.

The PCB itself, measuring 160 mm x 230 mm x 1.6 mm, was of composite structure, and was manufactured by gluing together two double-sided G10 fibre glass cards to form a 7 layer board.

For each steady state test it was necessary to set and measure the level of power dissipation, and to monitor the temperature of each of the 52 packages. The automatic data acquisition system consisted of three main components : a) A 'Mowlem' Autonomous Data Acquisition Unit (ADU), b) A specially constructed Interface Unit, and c) A Microcomputer.

The ADU is a proprietary device (7), based on a microprocessor which accesses 16 Kbyte of internal memory and runs under its own operating system. The host microcomputer communicates with the ADU and downloads a program to perform a specific test procedure.

3. THERMAL ANALYSIS

3.1 Energy Balance

To eliminate heat losses from the test section its walls were constructed of thick perspex which acted as an insulator. At the relatively low temperatures at which the packages were operated heat transfer from them by radiation to the side walls was generally relatively small. To reduce this to negligible proportions the inside surfaces of the test section were covered with a reflecting material. At any point x, the cumulative temperature rise of the airstream above the inlet ΔT_x was calculated from an energy balance. Assuming complete mixing of the air

$$\Delta T_x = \frac{Q_x}{mc_p} \tag{1}$$

Although for most tests ΔT_x is small in magnitude (2-4 °C over the entire PCB) energy balance calculations confirmed that, within experimental error, heat losses were negligibly small.

3.2 General Remarks on Calculating Heat Transfer Coefficients

The surface heat transfer coefficient is defined in the usual manner:

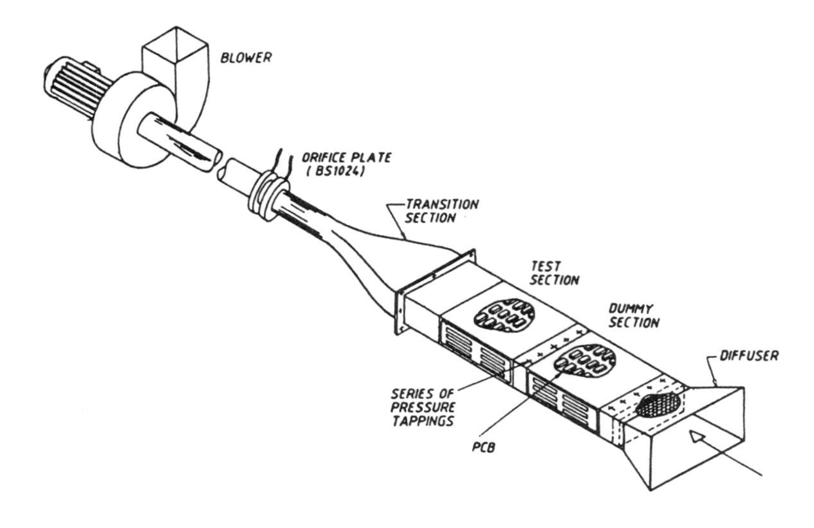

FIGURE 1. General construction of experimental rig.

was based. Hardisty et al. [6] used the finite element method (FEM) to develop a realistic thermal model of 13 DIPs on a PCB, which was then successfully used to correlate their experimental heat transfer results.

Even less data has been published for the industrially interesting case when the modules are not powered uniformly; this is discussed further in Section 3.5.

2. HEAT TRANSFER RESEARCH RIG

The general arrangement of the rig, showing the principal components, is shown in Fig. 1. Ambient air from the laboratory is drawn through the test section by a fan placed downstream of the working section; for a more detailed description of rig and instrumentation see Hardisty et al. [6]. Dummy boards (plastic replicas of the PCB) were fitted upstream of the test section, and on either side of the test board, Fig. 2. The upstream boards served to promote developed flow and to establish well defined inlet conditions. The air temperature was measured using an array of thermistors mounted at the inlet and outlet of the test section.

The PCB used for the present series was a multi-layer double Euro card, on which is mounted a rectangular array (13 x 4) of 16-pin DIPs, Fig. 2. To facilitate control and measurement of the power dissipated by the DIPs resistor packs were used (ceramic, wire wound Beckman 898-3-R680). The central layer of each resistor pack consists of eight 680 Ohm resistances connected in parallel giving an effective resistance of 85 Ohm/pack. The central layer in which heat is generated is sandwiched between two blocks of ceramic and is connected to the board by 2 rows of 8 soldered-in leads.

Research has demonstrated that the temperature of the ceramic body of the package is substantially uniform. Also, because a realistic FE model is used to evaluate the heat transfer coefficients, the actual position of the temperature sensor attached to the package is not of primary importance. In order to minimise flow disturbances the surface temperature of each DIP was measured by a miniature bead thermistor glued to its

$$Q = hA(T_S - T_A) \quad (2)$$

$$h = \frac{Q}{A(T_S - T_A)} \quad (3)$$

The air velocities used in most of these tests were quite low (approx. 2.5 m/s) so a check was necessary to ensure the absence of mixed free/forced convection effects. This check (see Hardisty et al. [6]) demonstrated these effects to be negligible. The surface area A was taken as the area of the package together with its associated rectangular area of board (the combination constituting a module) extending to lines midway between packages.

Difficulties were encountered when the above deceptively simple equations were used to evaluate h values from experimental data. First, even though the total power P dissipated by the chip, and a single surface temperature can be measured with accuracy, the complex distribution of T_S (and associated Q) over the surface of the package and substrate is unknown. Second, although a thermal boundary layer will develop from the first powered row, because of the interrupted nature of the flow, calculations of the thermal boundary layer are extremely difficult. The unknown extent of mixing in the thermal wake behind heated module entails that the local air temperature does not have a single unambiguous value. Thermal boundary layer analysis is especially difficult when the board is heated non-uniformly.

The first difficulty, associated with the complex 3-dimensional nature of the temperature distribution, was largely overcome by using the FEM to develop a realistic thermal model of the module. The experimental data for non-uniformly powered boards presented later, and the inferences drawn from them, is a contribution towards overcoming the second difficulty. Both these aspects are now examined in greater detail.

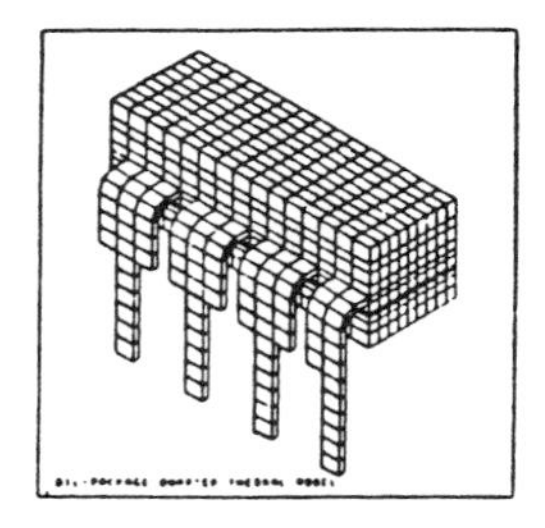

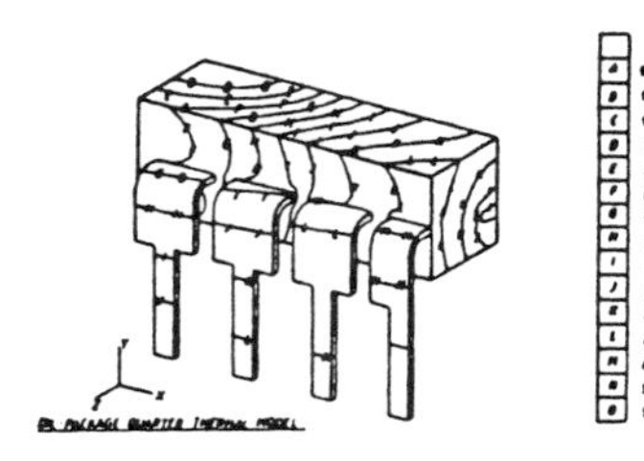

FIGURE 3. a) DIP FE model b) Temperature distribution in a DIP

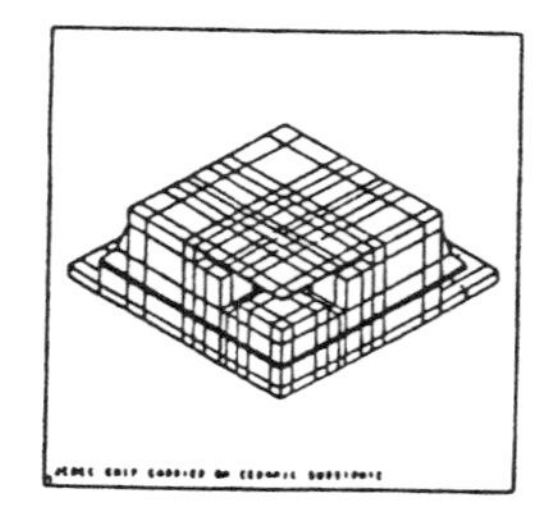

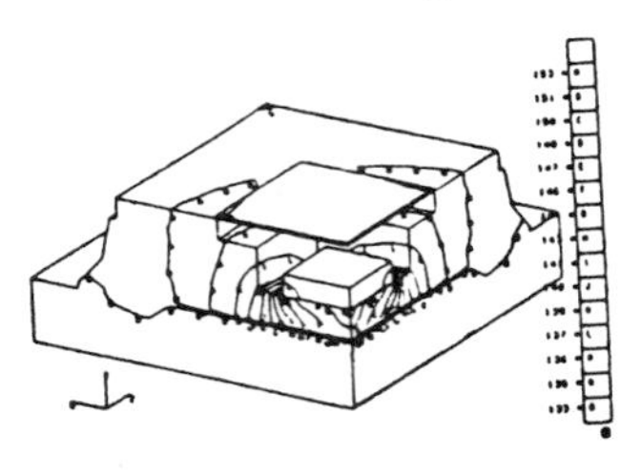

FIGURE 4. a) LCCC FE model b) Temperature distribution in LCCC

3.3 Theoretical (FE) Modelling

Because modules generally have a complex shape, analytical models of them usually involve such gross oversimplifications of geometry, as to cast doubt on the relevance of the solution. This criticism may also be directed at the simpler network and finite difference models. To construct more realistic thermal models it was decided to use the FEM. To facilitate construction of the rather complex mesh the widely used commercial ANSYS package [8] was employed for the investigations to be described here.

The FE technique has been applied to the construction of realistic thermal models of individual packages in previous publications. The complex 3-dimensional nature of the temperature distribution in a DIP can be seen in Fig. 3, from Hardisty and Abboud [9]. This study demonstrated that the internal thermal resistance of the package is not constant, but changes as the internal heat flow paths adjust themselves in response to changes in the magnitude of the external heat transfer coefficient. A corresponding thermal analysis of a surface mounted device (SMD) on a substrate was carried out in Abboud and Hardisty [10]. This revealed that for a leadless ceramic ship carrier (LCCC) with high internal thermal conductivity and low surface area, heat transfer is dominated by the large external thermal resistance. A typical temperature distribution in a LCCC taken from that latter analysis is shown in Fig. 4.

For the present investigation the FE model represented a 13 x 4 array of resistor packs mounted on a PCB. To reduce the size of the FE mesh only one line of thirteen resistor packs together with the associated PCB were modelled. Fig. 5a shows the individual elements used to construct one of the thirteen modules. Three-dimensional isoparametric thermal elements were used throughout, except for the leads which were modelled by 3D heat conducting bar elements. Energy is dissipated uniformly in the layer of resistor elements. Data from a computational fluid dynamics model have indicated that air flow through the 1mm gap beneath the package is laminar in character. Because the thermal resistance of such a slow moving air layer will be extremely high, it was modelled as a stationary layer.

The conductivities of various materials used in the FE model are presented in Table 1 below:

TABLE 1. Thermal Conductivity, k, W/m C

Material	Application	k
Ceramic	package	20.0
G10 Fibre glass	board	0.29
Copper	board/leads	385.0
Air	air gap	0.026
Resistors	resistor elements	50.0

The actual PCB was of 7 layer composite construction, copper and fibre glass layers interrupted by plated-through holes. Detailed calculations showed that the thermal conductivity normal to the plane of the board (k_n) was 0.31 W/m C, while parallel to the board (k_p) it was 30.2 W/m C. For a module, a 35mm x 15mm x 1.6mm section of board was modelled as a single layer of elements, each with calculated conductivities given above.

The complete system model was constructed by replicating the above module 13 times; this is shown in Fig. 5b.

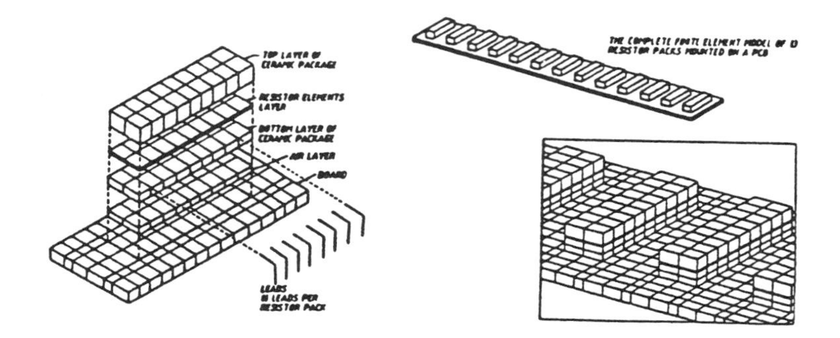

FIGURE 5. a) Construction of FE model b) 13 Packs on PCB

3.4 The Momentum and Thermal Boundary Layers

Simple boundary layer theory for flow over an isothermal flat plate cannot be applied to flow over an interrupted surface with non-uniform heating, but the general concepts provide valuable insight. The momentum (velocity) boundary layer begins to form at duct inlet and its thickness progressively increases with streamwise (x) distance until the flow becomes fully developed. It should be noted that preliminary flow visualisation studies have shown that a boundary layer progressively forms on the tops of a row of modules despite the spaces which separate them.

The convective rate equation (Eq.2) may be rewritten as

$$Q = \frac{(T_S - T_A)}{\frac{1}{hA}} = \frac{(T_S - T_A)}{R} \qquad (4)$$

where $R = 1/hA$ = convective thermal resistance.

The thermal resistance term ($1/hA$) depends essentially on the thickness of the momentum boundary layer, which in turn depends on the Reynolds no. (Re) of the flow. It should be noted that when the flow conditions, and consequently, h, are constant, then the surface-to-fluid temperature difference is directly proportional to power dissipation. When this latter condition has been achieved then the measured surface temperature of a module is independent of its position on the board.

Heat transferred from a powered module is conducted and convected downstream by the flow and forms a thermal boundary layer. The relative thickness of momentum and thermal boundary layers will be a function of the Prandtl no. (Pr) of the fluid. The surface temperature of each module increases with downstream distance. The heat transfer coefficient is found to be highest at inlet to the test section and to decrease with distance downstream, Hardisty et al. [6].

3.5 Calculation Method for Non-Uniform Power Dissipation

With all modules on the PCB powered to the same level heat transfer coefficients can be calculated in a relatively straightforward manner. For details of the method used, based on the FE model, see Section 4.2 below. However, when the PCB is powered unevenly, two additional effects require attention :

i) Heat is conducted through the PCB from the powered to unpowered rows.
ii) Unheated modules downstream are heated by the presence of high temperature thermal wakes convected from powered modules upstream.

One of the advantages of using a realistic FE model of 13 packages on a common PCB is that the inter-module conduction effects of (i) above are implicitly modelled without difficulty. Thermal wake effects require further analysis.

Wills (5) has suggested that the temperature T_n of any module in row n can be expressed by an equation of the form

$$T_n = T_A + R_{on}P_n + \Sigma Z_{ni}P_i \quad (5)$$

where

T_A = local air temperature °C
P_n = power dissipation of row n, W
R_{on} = convective thermal resistance of row n, all other powers zero, °C/W
Z_{ni} = temperature rise at n due to 1 Watt power at row i, °C/W
P_i = power dissipation at row i, W

In Eq. (5) boundary layer convection is broken down into two separate effects: first, the evaluation of the thermal resistance of a single powered module; second, the evaluation of the influence coefficients Z_{ni}, i.e., the effect of the thermal wakes. Based on the method of superposition Wills presents a method of deducing all values of R_{on} and Z_{ni} from standard heat transfer correlations for a uniformly powered board, but gives no experimental data to validate his method. Experimental results from research presented in this paper, see in particular Section 4.5, appear to suggest that the effect of uneven heating is to introduce asymmetries into the problem which prevents using the method of superposition in a straightforward manner. A similar approach, based on the principle of superposition, is proposed by Arvizu and Moffat [11].

It is the objective of this paper to present experimental data, which when correlated by the FE model, will allow solution of this difficult problem.

4. EXPERIMENTAL INVESTIGATION

4.1 Preliminary

The average thermistor reading from the four packages in a row was taken as the module temperature of that row. The 'Temperature Rise' shown on Figs. 6-10 was calculated by subtracting the inlet air temperature from the average module temperature for that row. In the various tests described below the principal experimental variable was the power dissipated in the chip; unless stated otherwise a module was either switched on to a power of 1 W/chip, or switched off. The air velocity was maintained as close as practical to a constant value of 2.5 m/s for all tests. The FE model used to correlate the test data was the model of 13 resistor packs on a PCB shown in Fig. 5.

4.2 Test Series 1 : All Modules Uniformly Powered

Typical row-to-row temperature distributions for tests in which the power dissipation per chip was maintained constant for all modules on the board are shown in Fig. 6. Three tests were performed at powers of 0.5, 1.0 and 1.5 W/chip. It can be seen that module temperatures are lowest at the leading edge of the board where the heat transfer coefficient h is highest and the thermal boundary layer is undeveloped. It can be seen that after about row 8 the value of h becomes almost constant. It will be noted that for a particular row temperatures increase almost directly in proportion to chip power. This result follows from Eq. 4, and the analysis of Section 3.4 above.

4.3 Test Series 2 : Modules Switched Off in Sequence, From Row 1

In the first test in this sequence of 13, all modules were uniformly powered at 1 W/chip giving a temperature distribution similar to that of Fig. 6. For subsequent tests, rows were switched off in sequence beginning with row 1. For the last test only row 13 was powered.

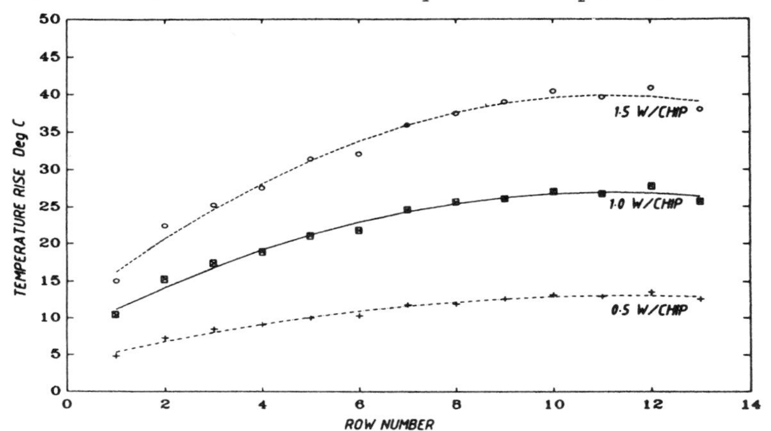

FIGURE 6. Test Series 1: All modules uniformly powered.

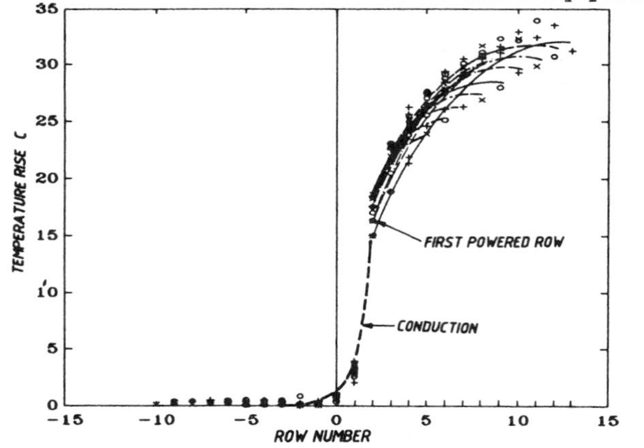

FIGURE 7. Test Series 2: Modules switched off in sequence from row 1.

Provided that the fluid conditions are the same for all modules, then the heat transfer coefficient of the first module of any group of powered modules, and the thermal boundary layer which develops from it, should be the same for all tests in this series. To test this hypothesis, temperature profiles for all 13 tests were plotted as shown in Fig. 7. Here the first powered row of any test is plotted in position 1 regardless of its actual row number. Negative row numbers refer to unpowered, upstream rows.

The results show that the hypothesis to be approximately true. It is considered that the drop in temperature which occurs at row 13 is a consequence of slight irregularities in flow conditions at outlet from the test section. The effects of conduction, extending upstream for about four rows, can be clearly seen.

4.4 Test Series 3: Varying Power on a Single Row; Remainder Uniform

In each of this sequence of tests all rows were uniformly powered at 1 W/chip except for one of the rows, the power of which was varied in 4 steps, 0, 0.5, 1.0 and 1.5 W/chip. This procedure was repeated for each of rows 3, 5, 8, 11, giving 16 tests in total. Fig. 8 shows typical results for row 8.

As has been previously mentioned in Section 3.1, the bulk mean temperature rise of the air calculated from Eq. 1 is quite small, approx. 3 °C for

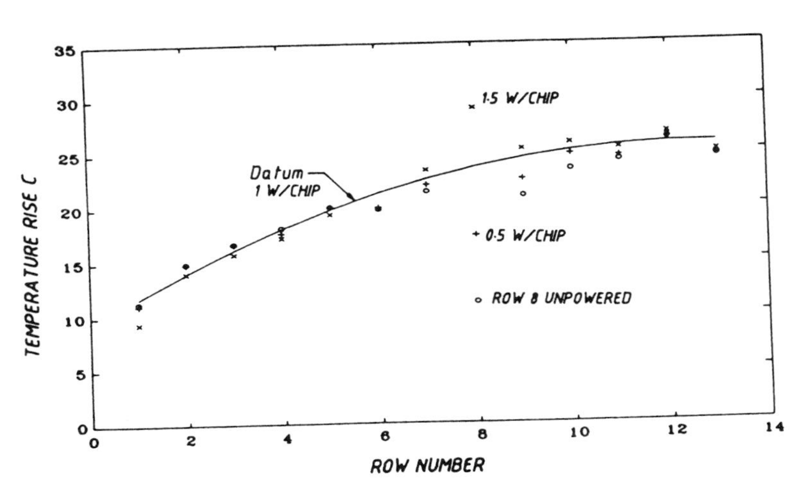

FIGURE 8. Test Series 3: Varying power on a single row, remainder uniform.

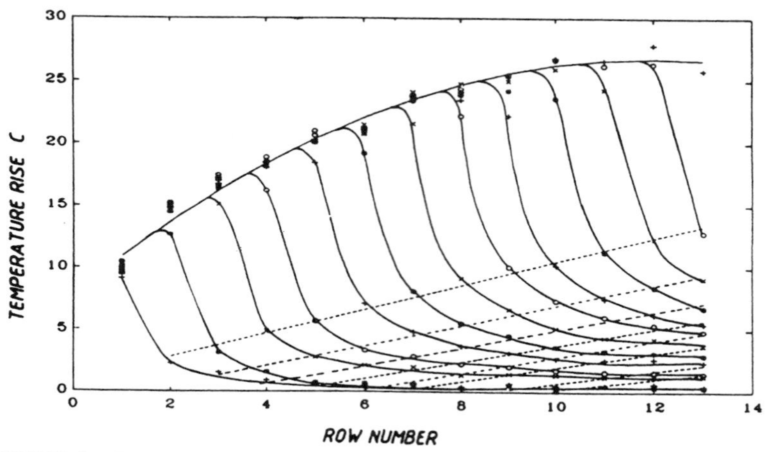

FIGURE 9. Test Series 4: Modules switched off in sequence, from row 13.

these tests. It can be seen from Fig. 8 that the temperature of an unpowered module lies considerably above this mean air temperature. There appear to be two explanations for this. First, heat is conducted into the unpowered module from the hotter modules on either side of it. Second, the unpowered module is heated by the high temperature thermal wakes (boundary layers) convected from the powered modules upstream. The temperature rise of a row above its unpowered temperature is roughly in direct proportion to the amount of power dissipated.

4.5 Test Series 4: Modules Switched Off in Sequence, From Row 13

The principal objective of this series of tests was to determine the effect of thermal wakes (boundary layers) on the unpowered modules downstream. The series began by unpowering row 13, additional rows were then switched off in sequence until for the last test only row 1 remained powered. The temperature profiles for all 13 tests are shown in Fig. 9; the temperature profile for all modules uniformly powered has been added as a comparative datum.

The following points should be noted

a) The temperature of row 13 remains high even when unpowered, a consequence of the thermal wakes shed by the 12 upstream rows. Roughly half of the temperature rise of row 13 when powered arises from the effect of upstream thermal wakes.

b) For powered rows (the datum curve) thermal boundary layer effects extend from row 1 to row 12.

c) For unpowered rows thermal wakes decay to zero after about 5 rows.

d) The effects of conduction is to lower the temperature of downstream modules.

4.6 Test Series 5: An Individual Powered Row

Fig. 10 shows the effect on module temperatures of powering row 3 only, to three different power levels. Similar tests were carried out for all 13

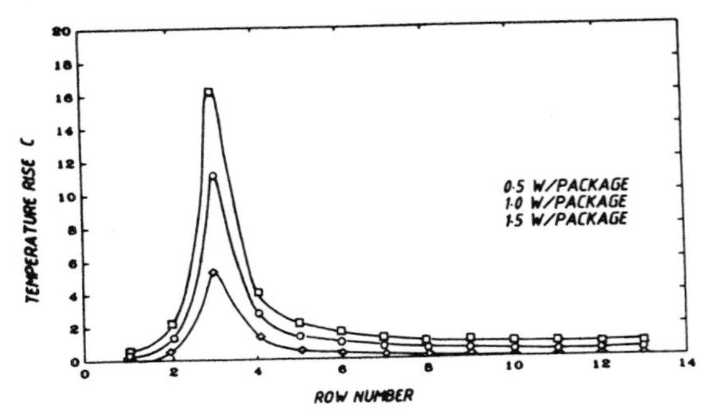

FIGURE 10. Test Series 5: Individual powered row, temperature profile for row 3.

rows, all yielding roughly similar results. It can be seen that once again module temperature is approximately directly proportional to the module power dissipation. The effect of conduction in raising the temperature of adjacent rows is apparent. The effect of thermal wakes is to increase downstream temperatures and to make the profiles asymmetrical.

5. COMPARISON OF EXPERIMENTAL DATA WITH PREDICTIONS FROM FE MODEL

5.1 Test Series 1: All Modules Uniformly Powered

The FE model was previously successfully used to correlate heat transfer coefficients for this case, Hardisty et al. [6]. Experience showed that the use of a realistic FE model was necessary to overcome a difficulty which arises when Eq. 3 is used to calculate h values from test data. FE studies have demonstrated that the temperature varies significantly over the surface of package and substrate, see Figs. 3 and 4. Direct substitution into Eq. 3 of the temperature registered by the thermistor will result in a calculation of an erroneous value of h. To overcome this difficulty a heat transfer correlation applicable to all 13 modules was incorporated in the FE program. For a computer run corresponding to a given experimental test, a temperature from the FE model at the correct geometrical location was matched to the thermistor temperature. The values of the empirical constants in the correlation were adjusted until a best fit of FE predictions to the experimentally determined temperatures of all 13 modules was obtained. Finally, the results from a large number of runs were correlated by means of a least squares fit.

The above method yielded the following correlation for Nu_x, the Nusselt Number at a distance x from the leading edge of the board:

$$Nu_x = 0.323\ Re^{0.65}\ Pr^{0.33} \qquad (6)$$

The computer predictions of temperature rise shown in Fig. 11 were obtained by using h values from the above correlation in the FE model, up to row 8. After row 8 the flow was considered to be fully developed and a constant value of h was used for all subsequent rows. It can be seen that the agreement between FE predictions and experiment is excellent.

5.2 Test Series 2: Modules Switched Off in Sequence, From Row 1

The test when rows 1-5 were switched off and row 6 was the first powered

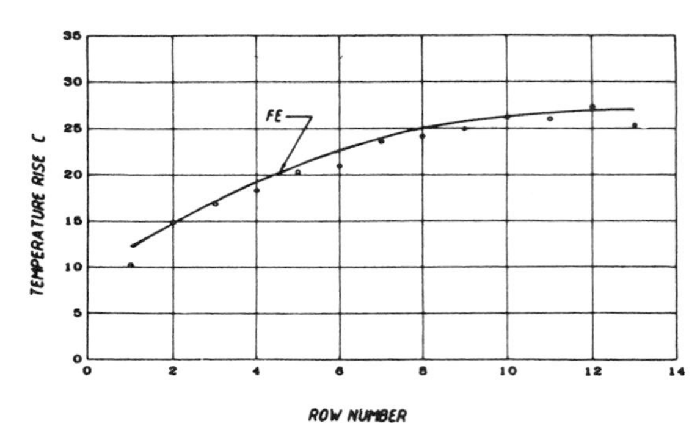

FIGURE 11. FE Predictions - Test Series 1.

row, was selected for comparison with predictions from the FE model. Heat transfer coefficients from Eq. 5 were used in the model. Initially the zero point for the distance x used in this correlation was taken as row 5. However this yielded a rather poor agreement with experiment. The computer predictions of temperature rise shown in Fig. 12 were obtained with the zero of the x distance located on row 4. This can be tentatively justified on the basis that upstream conduction from row 6 extends to row 4. With this proviso the agreement between experiment and predictions from the FE model is quite good.

Fig. 12 also shows the effect of eliminating conduction to the upstream rows in the FE model. This was achieved by assigning a very low conductivity value to the elements in the PCB at the interface between rows 5 and 6. It can be seen that the effect of conduction is to round out the effect of a discontinuity at row 6; the high temperatures are reduced and to compensate the low temperatures are increased by approximately the same amount.

5.3 Test Series 3: Varying Power on a Single Row; All Other Rows Uniformly Powered

The test from this series selected for computer prediction was that in which the power on row 5 was reduced to zero. When heat transfer coefficients from Eq. 6 were used without modification in the FE model, although the fit for most rows was reasonable, the temperature predicted for the unpowered row was some 3 degrees higher than the test temperature. The h

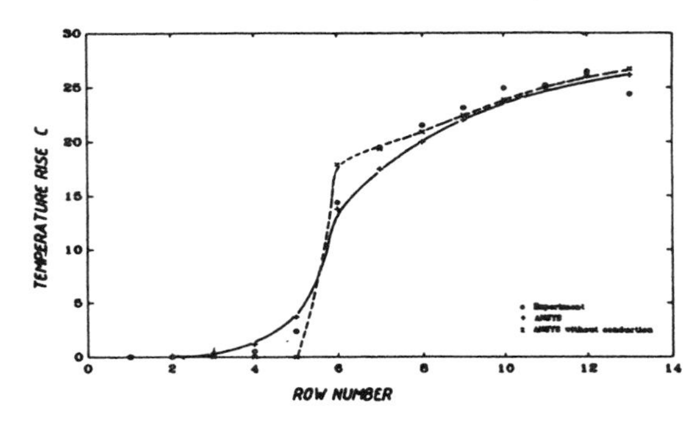

FIGURE 12. FE Predictions - Test Series 2.

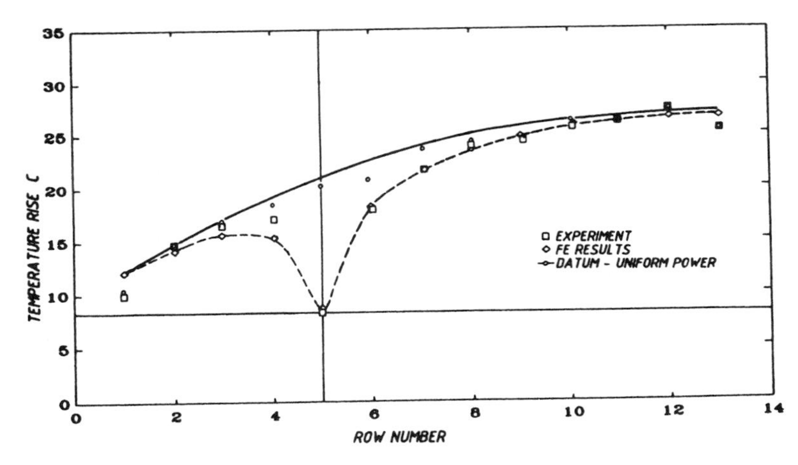

FIGURE 13. FE Predictions - Test Series 3.

value of row 5 was then increased until a good temperature fit at row 5 was obtained. It can be seen from Fig. 13 that with this adjustment of h at row 5, the agreement between prediction and experiment for all 13 rows is excellent.

5.4 FE Predictions for Test Series 4 and 5

In each of these series of tests the effects of both the thermal wakes on downstream unpowered rows, and of inter-module conduction, are present to a marked degree. To date a limited number of computer runs have been made in an attempt to correlate the test data. Various estimates of h values and of thermal wakes, have been tested in the FE model. Although agreement between computer predictions and experimental data is qualitatively reasonable, predictions are not yet sufficiently accurate to constitute a validation of any particular method. It should be noted that to achieve a good fit of the test data of Figs. 9 and 10, constitutes quite a severe test of a theoretical model. Improvements in the theory to attain this objective are the subject of current research.

6. CONCLUSIONS

1) A comprehensive series of tests have been carried out to measure the temperatures of a rectangular array of DIPs on a PCB cooled by forced convection. Because of the complex nature of the flow conditions such experimental data is quite scarce and the research results represent an extension of design knowledge in this field.
2) Because of the complex nature of the temperature and heat flux distribution in the multi-module system an accurate theoretical model is an essential prerequisite for accurate determination of h values. Research results have demonstrated that the FE methods well suited to the construction of such realistic thermal models.
3) By using the FE model the heat transfer coefficients measured with all modules on the PCB uniformly powered were successfully correlated.
4) Temperature distributions have been presented for a comprehensive series of tests with the modules on the PCB non-uniformly powered. These test results provide valuable data on thermal boundary layer effects and on inter-module conduction.

5) Good agreement between test results and predictions from the FE model has been achieved for conditions when the downstream modules are powered. Agreement between theory and experiment is less satisfactory for the cases when the downstream modules are unpowered, or when only a single row is powered. Computer results showed that the use of the principle of superposition to predict temperature profiles led to inconsistencies. Research is currently being undertaken to modify the thermal model of the thermal boundary layer to improve the theoretical predictions.

7. ACKNOWLEDGEMENTS

The research reported in this paper forms part of a programme of research into the thermal analysis and cooling of microelectronic systems currently supported by a grant from the UK Science and Engineering Research Council.

NOMENCLATURE

Symbol	Quantity	SI Unit
A	Surface area	m^2
C_p	Specific heat	J/kg K
h	Heat transfer coefficient	W/m^2 K
k	Thermal conductivity	W/m K
m	Mass flow rate of air	kg/s
P	Power dissipation	W
Pr	Prandtl number	$\frac{\mu C_p}{k}$
Q	Heat transfer rate	W
Re	Reynolds number	$\frac{VL\rho}{\mu}$
R	Convective thermal resistance, 1/hA	°C/W
T_A	Temperature of air	°C
T_S	Temperature of surface	°C
V	Air velocity	m/s
Z	Influence coefficient	°C/W
x	Streamwise distance along PCB	m
μ	Absolute viscosity	kg/m s
ρ	Density	kg/m^3

REFERENCES

1. Chu, R.C., Heat Transfer in Electronic Systems, Proc 8th Int. Heat Transfer Conf., Ed Tien C.L., Hemisphere 1986.

2. Oktay, S., Hanneman, R., and Bar-Cohen, A., High Heat from a Small Package, Mechanical Engineering, March 1986.

3. Bar-Cohen, A., Kraus, A.D., and Davidson, S.F., Thermal Frontiers in the Design and Packaging of Microelectronic Equipment, Mechanical Engineering, pp.53-59, June 1983.

4. Finch, D.J. and Goodacre, J.B., The Thermal Management of Printed Circuit Board Assemblies, Marconi Review, vol. XLV 227, Fourth Quarter, 1982.

5. Wills, M., Thermal Analysis of Air-Cooled PCBs, 4 Parts, Electronic Production, May-August 1983.

6. Hardisty, H., Abboud, J. and Vertman, S., Forced Convection Air Cooling of an Array of Electronic Modules, Using the FEM to Analyse Module Heat Transfer, Inst. Mech. Engrs, accepted for pub., Feb.88.

7. Mowlem Microsystems Ltd., Hemel Hempstead, Herts, HP2 7HP.

8. ANSYS, Swanson Analysis Systems Inc., Johnson Road, Huston, U.S.A. Support distributor : STRUCOM, London, U.K.

9. Hardisty, H. and Abboud, J.B., Thermal Analysis of a Dual-in-Line Package using the Finite Element Method, IEE proc., vol. 134, Pt.I, no. 1, Feb. 1987.

10. Abboud, J. and Hardisty,H., The Finite Element Method Applied to the Thermal Analysis of Microelectronic Packaging, Proc. INTERNEPCON, Electronic Packaging Conf., Brighton, Oct. 87, U.K.

11. Arvizu, D.B. and Moffat, R.J., The Use of Superposition in Calculating Cooling Requirements for Circuit Board Mounted Electronic Components, Electron Components Conf., vol. 32, pp 133-144, IEEE, 1982.

Thermal Analysis of Multilayer Printed Circuit Boards Using a Hybrid Integral-Thermal Network Method

L. M. SIMEZA and M. M. YOVANOVICH
Microelectronics Heat Transfer Laboratory
University of Waterloo
Waterloo, Ontario, Canada N2L 3G1

ABSTRACT

A two-dimensional method for analysis of multi-layer printed circuit boards is developed. The method is a combination of a one-dimensional integral method along the length of the layers, which are coupled by thermal networks. The temperatures of the surface sources and within the layers are evaluated. The effect of the copper layers in the board is analyzed. It is shown that in densely populated boards the position of the copper is not important. It is also shown that an effective thermal conductivity for a multi-layer printed circuit board cannot be easily found, except for densely populated boards where the effective thermal conductivity across the board is applicable. The possibility of using an orthotropic model for the board is also investigated.

INTRODUCTION

Analysis of heat conduction in multi-layer printed circuit boards is of prime importance in the thermal design of electronic systems. It has generally been possible to analyze heat transfer in electronics by using simple elementary heat transfer equations as given in several papers published on the determination of shape factors for this problem [1–4]. However, most of these are valid only for cylinders with holes of small radii. Smith and Lind [1] correlated electric analog results for a hollow square cylinder. Their correlation is valid for cylinders of small radius/apothem ratio and is useful up to a ratio of about 0.80. Other workers have developed approximate analytical solutions, notably, Balcerzak and Raynor [2], who based their solution on approximate mapping and point matching on the outer boundary. Laura and Susemihl [3] developed a similar solution using conformal mapping. Both methods give nearly identical results which are accurate only up to a radius/apothem ratio of about 0.8 for the case of a hollow square cylinder. Dugan [4] developed a much more accurate solution using the boundary residual technique. This solution, however, requires the solution of simultaneous equations whose number must be increased as the inner hole radius, although no simple way has been found of establishing and evaluating this apparent thermal conductivity. It can be determined experimentally, although the value obtained is usually the apparent thermal conductivity across the board, which does not represent the spreading resistance from the sources very well. A more realistic approach is to try and represent the board as having orthotropic properties with

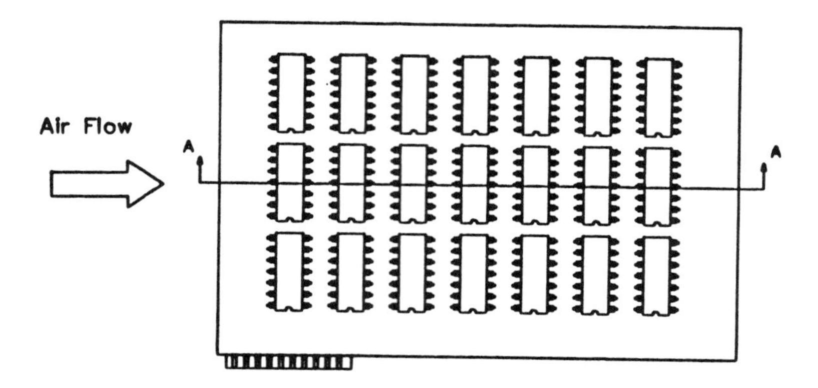

FIGURE 1. Typical Printed Circuit Board with Sources Mounted on the Surface.

different effective thermal conductivities along the length of the board and across the board. There is, however, a need to validate this approach when applied to multi-layer printed circuit boards.
In this paper heat conduction in a multi-layer printed circuit board is analyzed using a model which represents the layers in detail. The method is called the Hybrid One Dimensional Boundary Element - Thermal Network Method. It is a combination of one-dimensional integral representation along the length of the layer and the layers are then coupled by thermal resistive networks. From the results the importance of the copper layer in the multi-layer boards is examined. The results are then compared with those found by considering the board to be orthotropic with one effective thermal conductivity along the length of the board and another conductivity across the board. The conditions under which this approximation can be used are determined.

STATEMENT OF PROBLEM

Consider a multi-layer printed circuit board which is cooled by a flow of air along the length of the board. If the sources in a row are considered to be of the same strength a two-dimensional model, as shown in Figure 1, can be used to analyze the heat transfer in the board. The model can further be simplified to the form shown in Figure 2, where only one source is considered and it is flush with the surface of the board. It is required to determine the maximum temperature at the source when a certain level of convection is maintained over the board. The method of solution must take into account the effect of the different thermal conductivities in the various layers. The results will then be compared with those obtained when the board is approximated as orthotropic.
The board section used in the numerical evaluations in this paper was 100 mm long and

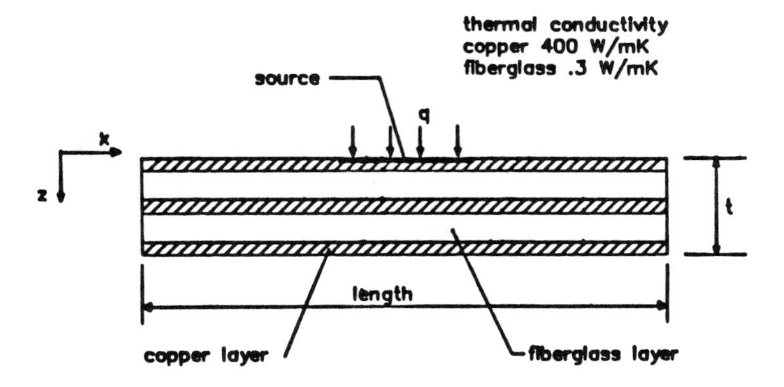

FIGURE 2. Two-dimensional Representation of a Multi-layer PCB.

2 mm thick. The copper layer was 0.068 mm thick. This is called a two ounce copper layer. Two fiberglass layers 0.966 mm each made up the rest of the board. A source 20 mm long and strength 2000 W/m^2 was placed at the centre of the board section. This case is called the sparsely populated board. A second board section of length 25 mm and thickness 2 mm was also analyzed. The source was 20 mm in length and 2000 W/m^2 strength and again placed at the center of the board. This was called the densely populated model. The variables compared in all these boards were the temperature profiles in the boards.

SOLUTION

Heat conduction in the board is two-dimensional in the z and x directions. However, since the layers are thin the multi-layer board in Figure 2 can be replaced by the model shown in Figure 3 where the layers are represented as one-dimensional in the x-direction and are connected by thermal resistances in the z-direction. These resistances account for the heat flow in the z-direction and are developed in detail later. The layers are discretized into elements and the resistances are connected to the nodes of the elements. The nodes are at the centroids of the elements.

The governing differential equation for each of the layers is Laplace's equation. However, due to the fact that the layers are considered one-dimensional in the x-direction, the boundary conditions in the other direction can be included as source terms. The equation that has to be solved is Poisson's equation:

$$\frac{d^2T}{dx^2} = -\rho \tag{1}$$

where ρ , the source term, is made up of the net heat transfer into the layer due to various heat transfer processes in the direction perpendicular to the x direction.

One Dimensional Boundary Element Method (BEM)

The temperature profile along the length of each layer is determined by one-dimensional BEM. The details of this method are discussed below.
The general form of the boundary integral equation for steady-state heat conduction with domain sources can be found in several references, (Brebbia [5-7]; Banerjee and Butterfield [8]). It is

$$C_i T_i = \int_{\Gamma} (G \frac{\partial T}{\partial n} - T \frac{\partial G}{\partial n}) d\Gamma - \int_{\Omega} G \nabla^2 T \, d\Omega \tag{2}$$

For one-dimensional problems this reduces to

$$C_i T_i = \sum_{j=1}^{2} (G \frac{\partial T}{\partial n} - T \frac{\partial G}{\partial n}) - \int_{S} G \nabla^2 T \, dS \tag{3}$$

There are only two boundary points in each layer which are at the ends of the domain. The domain integral over S is a line integral along the length in the x direction. Before application of the integral equation to a layer of the multi-layer board, the length of the layer is discretized into elements with nodes at their centroids. Two boundary nodes are placed on elements at the ends of the layers. The size of these boundary elements is equal to the thickness of the layer. The boundary nodes are required to determine the length of the layer. The discretized form of the integral equation applied to the whole board then is

$$C_i T_i = \sum_{j=1}^{N} (G \frac{\partial T}{\partial n} - T \frac{\partial G}{\partial n}) - \sum_{p=1}^{M} \int_{S_p} G \nabla^2 T \, dS \tag{4}$$

where i is the field point located at all the nodes; p and j are the source points which are also at the nodes; M is the number of interior points; and N is the total number of boundary points, one at each end of a layer. At the boundary points the constant C_i has the value 0.5, while at interior points it has the value 1.0. The fundamental solution, G, of Laplace's equation in a one-dimensional domain is $r/2$ per unit thickness of region, where r is the distance between the source and field points $|x_i - x_j|$. To account for the thickness of a layer the fundamental solution used is

$$\frac{r}{2 * thickness\ of\ layer}$$

When Equation (4) is applied to the layers of a multi-layer printed circuit board, a set of algebraic equations, which are coupled by the domain integral terms, is generated. The domain integral terms are functions of thermal resistances across the layers. These will now be derived.

Thermal Network

Consider a section of the board which has each layer represented as an element with a node at the middle of the layer as shown in Figure 3. The positive direction of the z coordinate

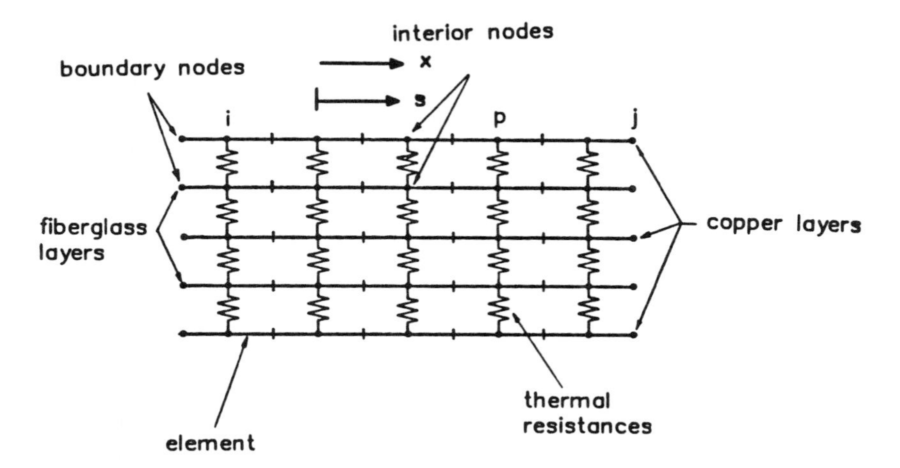

FIGURE 3. Hybrid One-dimensional BEM Thermal Network Representation of the Multi-layer Printed Circuit Board.

is designated (1) and called the bottom, while the opposite direction is designated (2) and is the top. Heat can get into a layer by 1) conduction from the top or bottom layers, 2) convection from a fluid on the top of the layer when the layer is at the top or from the bottom fluid when the layer is at the bottom, and 3) direct input into the layer from a specified flux source.

Using Fourier's conduction law and Newton's cooling law, the source term can be expressed as

$$\rho = \alpha_2 \frac{(T_t - T)}{R_t} + \beta_2 h_2 (T_{ft} - T) - \alpha_1 \frac{(T - T_b)}{R_b} - \beta_1 h_1 (T - T_{fb}) + \gamma \tag{5}$$

where T is the nodal temperature of the layer under consideration; T_b is the bottom layer temperature; T_t is the top layer temperature; T_{fb} is the bottom fluid temperature; T_{ft} is the top fluid temperature; α_1 and α_2 are constants which take on the value one if conduction into the layer from the adjacent layers at the bottom or top is present, and zero if conduction is absent; β_1 and β_2 are also constants which take on the value one if convection into the layer, from the bottom or top, respectively, is present, and are zero when it is absent; γ represents the net specified source strength into the layer, h_1 and h_2 are convective film coefficients on the bottom and on the top of the layer, respectively; and R_b and R_t are the thermal resistances between layers in the z-direction. They will be derived in detail later.

Therefore, the governing differential equation for a layer is

$$\nabla^2 T = -\rho = \alpha_1 \frac{(T - T_b)}{R_b} + \beta_1 h_1 (T - T_{fb}) - \alpha_2 \frac{(T_t - T)}{R_t} + \beta_2 h_2 (T_{ft} - T) - \gamma \tag{6}$$

The thermal resistance from a node to a corresponding node in the layer above is called R_t. The resistance to the corresponding node in the layer below is R_b. In the z-direction, which is the direction through the board thickness, the variation in temperature between the layer elements is included by determining these thermal resistances. The expressions of these resistances are slightly different for layers that are at the top and bottom of the board, which are called boundary layers and the interior layers. The resistances are given by the equations

$$R_b = \frac{a}{kA} + \frac{a_b}{k_b A} \tag{7}$$

$$R_t = \frac{a}{kA} + \frac{a_t}{k_t A} \tag{8}$$

where a is half the thickness of an interior layer; k is the thermal conductivity of the interior layer, a_b is half the thickness of the bottom layer and k_b is its thermal conductivity; a_t is half the thickness of the top layer and k_t is its thermal conductivity; and A is the normal area in the z-direction.

When the top element is on the boundary, a_t becomes the total thickness of the layer, and when the bottom element is on the boundary, a_b becomes the total thickness of the bottom layer. For the layers that are on the boundary the nodes are considered to be on the boundary and either R_b or R_t disappears and a becomes the total thickness of the current layer.

The top and bottom boundary conditions on the board are usually convective or flux specified. It is not necessary to make special arrangements to incorporate these boundary conditions since they are included in the source term of Equation (6). The boundary conditions are, therefore, incorporated into the governing differential equation in a manner similar to that used in the finite-difference method.

The boundary nodes at the end of the board belong to the boundary element model. These can be either temperature specified or a function of the derivative of temperature in the x-direction. They are included as the T or the $\partial T/\partial n$ in the complete boundary integral equation, Equation (11).

The multi-layer PCB can hence be solved by the Hybrid BE-Thermal Network model shown in Figure 3. The heat conduction in the x-direction is solved by the one-dimensional BEM while the heat conduction in the z-direction is accounted for by coupling the layers with a thermal network represented by the domain terms in the integral

$$\int_S G \nabla^2 T dS \tag{9}$$

which in discretized form is

$$\sum_{p=1}^{M} \int_{S_p} G \nabla^2 T dS \tag{10}$$

When $\nabla^2 T$ is replaced by ρ as given in Equation (6), the layers are coupled by terms which contain temperature. The resulting boundary integral equation is

$$C_i T_i = \sum_{j=1}^{N} (G\frac{\partial T}{\partial n} - T\frac{\partial G}{\partial n}) + \sum_{p=1}^{M} \int_{S_p} G \rho dS \tag{11}$$

and when applied to each node in the model shown in Figure 3, a total of (N+M) algebraic equations with unknown nodal values are generated. There are M interior nodes, where temperature is unknown, and N boundary nodes at which either the temperature or it derivative in the x-direction is unknown. Rearrangement into matrix form yields a matrix set of the form

$$[A][T] = [B] \tag{12}$$

where $[T]$ is the vector of the unknown values. The matrix equation can be solved by the various inversion techniques.

RESULTS AND DISCUSSION

Multi-layer PCBs are constructed with either a copper or dielectric layer on the surface. Normally they have copper layers coated with a negligible amount of dielectric on each surface. The copper layers should be placed where they can spread heat as much as possible. The effect of the position of the copper layers was investigated by analysis of boards with only one layer of copper. The temperature distributions when the copper layer is at the top, or in the middle and at the bottom of the board, were evaluated. Two equal size layers of dielectric make up the rest of the board. The boundary conditions were convective everywhere, except over the source and the ends where no heat was allowed to leave the board. The thermal conductivity of the copper was 400 W/mK , while that of the dielectric is 0.3 W/mK, (fiberglass). The problem was solved using the Hybrid BE-Thermal Network model with 20 interior nodes and two boundary nodes per layer. The convective heat transfer over the board was varied by adjusting the convective coefficient to values between 10 W/m^2K and 100 W/m^2K , which are typical values for natural convection and forced convection regimes in air. The effect of the copper position in the board is seen in Figures 4, 5 and 6, where the temperatures in the board are plotted for a convective heat transfer coefficient of 100 W/m^2K . These plots are for the cases where the copper is placed at the top, middle and bottom of the board, respectively.

The board temperatures over the source are higher in the latter two cases because the heat has to conduct through the dielectric, which has a low thermal conductivity, before it reaches the copper in which it can spread. It can be noted from the plots that the largest temperature difference between the top and bottom layers occurs under the source area. This difference becomes small a short distance away from the source. It is also interesting to note that the heat seeks the copper layer. This is evident in Figures 5 and 6 where the temperature of the copper layer is higher than that of the fiberglass everywhere except under the source. Most of the heat, therefore, seeks the copper layer and then conducts to the fiberglass and eventually to the ambient air. When the copper layer is in the middle or the bottom of the board the temperature rise is largest under the source and it decays quickly away from the source. The temperature outside the source can be represented by a fin solution. Therefore, only the region under the source requires a two-dimensional solution. Figure 7 shows the values of the average source temperature for the three positions of the copper layer, top, middle and bottom of the board, plotted against varying convective heat transfer coefficient. The heat transfer improved as the copper was moved to the top of the board. It is, therefore, concluded that in boards which are sparsely populated, where convection areas exist on both the top and bottom faces of the board,

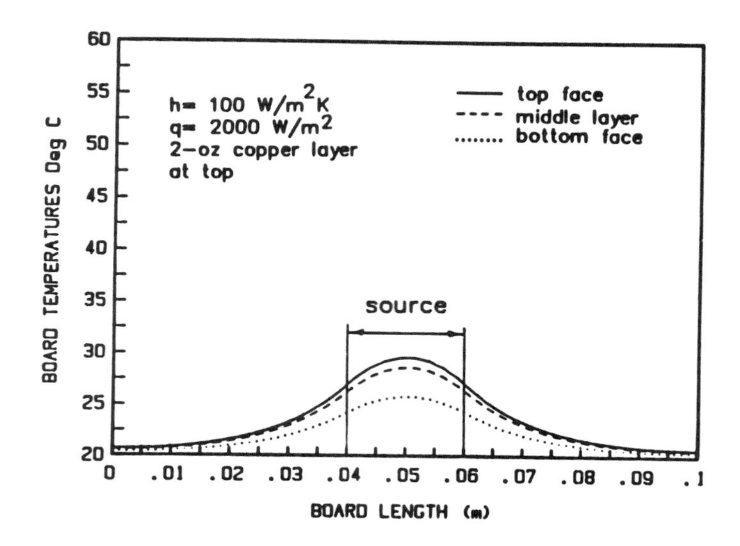

FIGURE 4. Plot of Temperature Profiles at Top, Middle and Bottom of the PCB for the Case Where the Copper Layer is at the Top.

the copper must be placed near the top of the board in order to use its heat spreading characteristic.

Usually printed circuit boards are fully populated with only a small area between the sources left to interact with the ambient fluid. When all the sources are operating, the heat flow is predominantly across the board from the top to the bottom.

Figure 8 shows that the temperatures in the three boards, when the board length is only 25 mm, are largely independent of the position of the copper layers. This is because the heat conducts predominantly across the board. Since the resistances of the copper and the dielectric are in series when heat is flowing directly across the board, the order of the layers is not important. Therefore, for this situation, which represents a fully populated board with all the sources operating, an effective thermal conductivity can be employed. Its value is equal to the effective thermal conductivity across the PCB in the z-direction.

Orthotropic Approximation

It is generally accepted that multi-layer regions can in many cases be approximated as being regions with orthotropic properties. This approximation was applied to multi-layer PCBs and the conditions when the approximation can be used identified. The effective thermal conductivity in the direction along the board can be expressed as (Yovanovich [8]; Bear [9])

$$k_{xe} = \sum_{i=1}^{N} \frac{k_i A_i}{A_T} \tag{13}$$

where k_{xe} is the effective thermal conductivity in the x-direction, k_i is the thermal con-

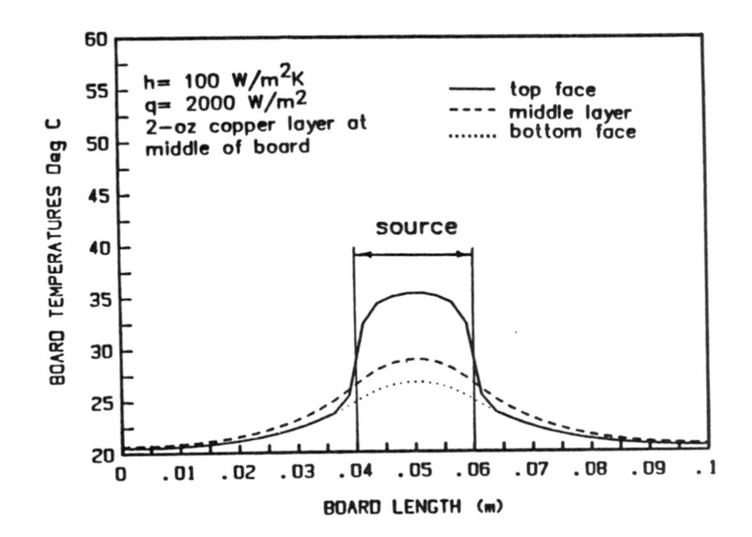

FIGURE 5. Plot of Temperature Profiles at Top, Middle and Bottom of the PCB for the Case Where the Copper Layer is at the Middle.

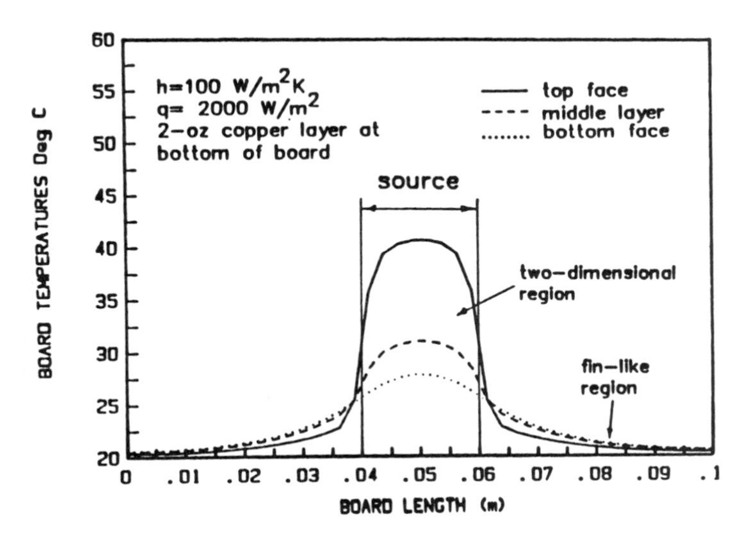

FIGURE 6. Plot of Temperature Profiles at Top, Middle and Bottom of the PCB for the Case Where the Copper Layer is at the Bottom.

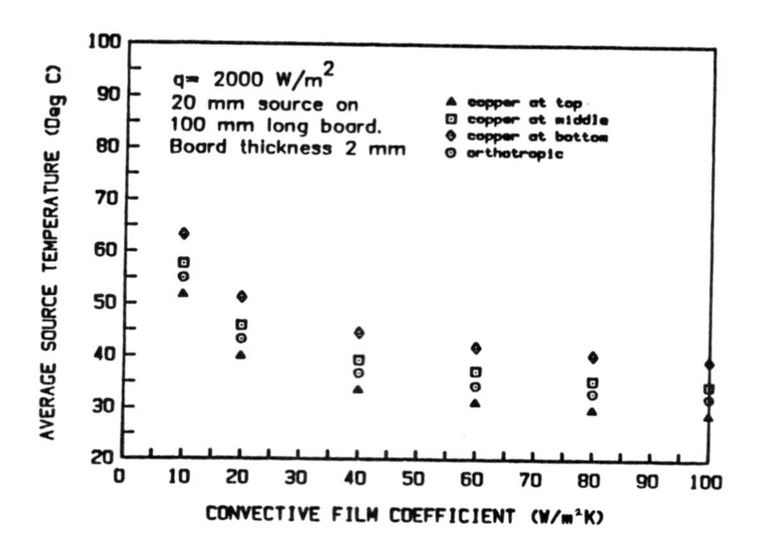

FIGURE 7. Comparison of Orthotropic Model Results With Those From the Detailed Hybrid One-dimensional BE-Thermal Network Method (single source).

ductivity of the ith layer; A_T is the total area normal to the x-direction; A_i is the area of the ith layer normal to the x-direction; and N is the total number of layers. The effective thermal conductivity across the board is

$$k_{ze} = \delta_T / \sum_{j=1}^{N} \frac{\delta_i}{k_i} \tag{14}$$

where δ_i is the thickness of the layer i; δ_T is the total thickness of the board; k_i is the thermal conductivity of the ith layer; and k_{ze} is the effective thermal conductivity in the z-direction. The two equations, Equation (13) and Equation (14), are easily evaluated if the properties of the layers are known. It is important to note that the order of the layers is not important in these equations.

The above effective thermal conductivities were used in the classical two-dimensional BEM (Brebbia 1980) code and temperatures at the boundary nodes on the surface of the multi-layer board determined. The results showing the value of the average source temperature, plotted against the convective film coefficient, are shown in Figure 7, where they are compared with those found by the Hybrid One-Dimensional BE-Thermal Network Method. In Figure 7 the results of the average source temperature from the orthotropic approximation are compared with those from the detailed hybrid method for the case of a single source on a large section of a PCB. It can be seen that the orthotropic model predicts a temperature that lies between the case of a PCB with copper at the top and that with copper in the middle. There is a significant difference in the temperatures predicted by the two models. Since the hybrid method is considered correct, it can be concluded that the orthotropic approximation gives inaccurate results when there is only one source on a large section of a PCB.

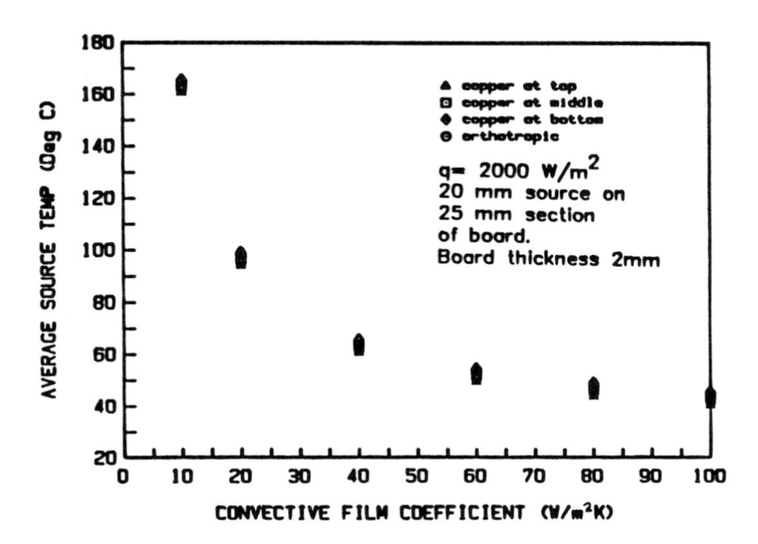

FIGURE 8. Comparison of Orthotropic Model Results With Those From the Detailed Hybrid One-dimensional BE-Thermal Network Method for the Case of Multiple Sources.

Figure 8 compares the results from the orthotropic model with those from the hybrid method for the case of multiple sources. The difference between the results of the two methods is now negligible. The orthotropic model can be used for this case. The main advantage of this approach is that it is not necessary to experimentally determine an effective thermal conductivity to be used in the analysis. In general, however, it is concluded that the orthotropic model cannot be confidently used to predict temperatures in multi-layer printed circuit boards.

Also included on the two graphs are the results obtained using the effective thermal conductivity along the board only and that across the board as the thermal conductivity for the whole board. From these plots it can be concluded that for closely packed sources the apparent thermal conductivity across the board can be used confidently. However, as the spacing becomes more sparse neither effective thermal conductivity is accurate. There is, therefore, a need for detailed layered models, such as the one used in this paper, to represent the layers in multi-layer printed circuit boards correctly.

SUMMARY

A detailed two-dimensional model which couples a one-dimensional integral method to a thermal network method, has been used to determine the temperature profiles in multi-layer printed circuit boards. Both sparsely populated and densely populated boards were analyzed. The effect of the position of the copper layer in the boards was found to be important only in sparsely populated boards. For this case the heat seeks the copper, which conducts it easily, spreading it over the length of the board, hence improving heat dissipation, and this results in lower source temperatures. In densely populated boards, the heat conducts predominantly across the board and hence the position of the copper is not important. The temperature profiles obtained when the board is considered to be orthotropic were compared with those of the detailed layered model and it has been shown that this approximation is valid only for densely populated boards.

ACKNOWLEDGEMENTS

The authors would like to acknowledge the financial support of NSERC under operating grant A7445.

NOMENCLATURE

A - area
$\vec{A}$ - a matrix
$\vec{B}$ - column vector
C - a constant
G - fundamental solution
h - convective heat transfer coefficient
k - thermal conductivity
M - number of internal elements
$\vec{n}$ - normal
N - number of boundary nodes
q - heat flux
R - thermal resistance
S - line
T - temperature
x - Cartesian coordinate
z - Cartesian coordinate

GREEK SYMBOLS

α - function
β - coordinate
Γ - boundary of region
δ - thickness of layer
ρ - source term

SUBSCRIPTS

b - bottom
e - effective
f - fluid
i - point i
j - point j
n - normal
p - point inside a domain
s - surface
t - top
1 - bottom direction
2 - top direction

REFERENCES

1. Smith, J.C., Lind, J.E. and Lermond, D.S., Shape factors for Conductive Heat Flow *A.I.Ch.E. Journal*, Vol. 4, No. 3, Sept. 1958. pp. 330-331.
2. Balcerzak, M.J. and Raynor, S., Steady State Heat Flow and Temperature Distribution in Prismatic Bars with Isothermal Boundary Conditions. *International Journal of Heat and Mass Transfer*, Vol. 3, 1961, pp. 113-125.
3. Laura, P.A. and Susemihl, E.A., Determination of Heat Flow Shape Factors for Hollow Regular Polygonal Prisms. *Nuclear Engineering and Design*, Vol. 25, 1973, pp. 409–412.
4. Dugan, J.P., Heat Flow in Prismatic Cylinders with Isothermal Boundary Conditions. ASME Paper No. 74WA/HT-36 presented at the Winter Meeting New York, N.Y., Nov. 17–22, 1974.
5. Brebbia, C.A. *The Boundary Element Method for Engineers.* Wiley, New York, 1978.
6. Brebbia, C.A., *The Boundary Element Method for Engineers.* 2nd *rev. ed.* Pentech Press, London, 1980.
7. Brebbia, C.A., *Boundary Element Techniques in Engineering.* Newnes-Butterworths, Boston, London, 1980.
8. Brebbia, C.A., Telles, J.C.F. and Wrobel, L.C., *Boundary Element Techniques; Theory and Applications in Engineering.* Springer Verlag, Berlin, 1984.
9. Banerjee, P.K. and Butterfield, R., *Boundary Element Methods in Engineering Science.* McGraw Hill, London, 1981.
10. Yovanovich, M.M., On the Temperature Distribution and Constriction Resistance in Layered Media. *Journal of Composite Materials*, Vol. 4, p. 567, 1970.
11. Bear, J., *Dynamics of Fluids in Porous Media*, American Elsevier, New York, 1972.

Index